Arithmetic Operations:

$ab+ac = a(b+c)$

$\frac{a}{b} + \frac{c}{d} = \frac{ad+bc}{bd}$

$\frac{a+b}{c} = \frac{a}{c} + \frac{b}{c}$

$\frac{\left(\frac{a}{b}\right)}{\left(\frac{c}{d}\right)} = \frac{ad}{bc}$

$a\left(\frac{b}{c}\right) = \frac{ab}{c}$

$\frac{a-b}{c-d} = \frac{b-a}{d-c}$

$\frac{ab+ac}{a} = b+c, \ a \neq 0$

$\frac{\left(\frac{a}{b}\right)}{c} = \frac{a}{bc}$

$\frac{a}{\left(\frac{b}{c}\right)} = \frac{ac}{b}$

Exponents and Radicals:

$a^0 = 1, \ a \neq 0$

$\frac{a^x}{a^y} = a^{x-y}$

$\left(\frac{a}{b}\right)^x = \frac{a^x}{b^x}$

$\sqrt[n]{a^m} = a^{m/n} = (\sqrt[n]{a})^m$

$a^{-x} = \frac{1}{a^x}$

$(a^x)^y = a^{xy}$

$\sqrt{a} = a^{1/2}$

$\sqrt[n]{ab} = \sqrt[n]{a}\sqrt[n]{b}$

$a^x a^y = a^{x+y}$

$(ab)^x = a^x b^x$

$\sqrt[n]{a} = a^{1/n}$

$\sqrt[n]{\left(\frac{a}{b}\right)} = \frac{\sqrt[n]{a}}{\sqrt[n]{b}}$

Algebraic Errors to Avoid:

$\frac{a}{x+b} \neq \frac{a}{x} + \frac{a}{b}$ (To see this error, let a = b = x = 1.)

$\sqrt{x^2+a^2} \neq x + a$ (To see this error, let x = 3 and a = 4.)

$a-b(x-1) \neq a-bx-b$ (Remember to distribute negative signs. The equation should be $a-b(x-1) = a-bx+b$.)

$\frac{\left(\frac{x}{a}\right)}{b} \neq \frac{bx}{a}$ (To divide fractions, invert and multiply. The equation should be

$$\frac{\frac{x}{a}}{b} = \frac{\frac{x}{a}}{\frac{b}{1}} = \left(\frac{x}{a}\right)\left(\frac{1}{b}\right) = \frac{x}{ab}.)$$

$\sqrt{-x^2+a^2} \neq -\sqrt{x^2-a^2}$ (We can't factor a negative sign outside of the square root.)

$\frac{\not{a}+bx}{\not{a}} \neq 1+bx$ (This is one of many examples of incorrect cancellation. The equation should be $\frac{a+bx}{a} = \frac{a}{a} + \frac{bx}{a} = 1 + \frac{bx}{a}$.)

$\frac{1}{x^{1/2}-x^{1/3}} \neq x^{-1/2}-x^{-1/3}$ (This error is a sophisticated version of the first error.)

$(x^2)^3 \neq x^5$ (The equation should be $(x^2)^3 = x^2x^2x^2 = x^6$.)

Conversion Table:

1 centimeter = 0.394 inches	1 joule = 0.738 foot-pounds	1 mile = 1.609 kilometers
1 meter = 39.370 inches	1 gram = 0.035 ounces	1 gallon = 3.785 liters
= 3.281 feet	1 kilogram = 2.205 pounds	1 pound = 4.448 newtons
1 kilometer = 0.621 miles	1 inch = 2.540 centimeters	1 foot-lb = 1.356 joules
1 liter = 0.264 gallons	1 foot = 30.480 centimeters	1 ounce = 28.350 grams
1 newton = 0.225 pounds	= 0.305 meters	1 pound = 0.454 kilograms

Trigonometry

FOURTH EDITION

Roland E. Larson **Robert P. Hostetler**

The Pennsylvania State University
The Behrend College

With the assistance of

David E. Heyd

The Pennsylvania State University
The Behrend College

HOUGHTON MIFFLIN COMPANY Boston New York

Sponsoring Editor: Christine B. Hoag
Senior Associate Editor: Maureen Brooks
Managing Editor: Catherine B. Cantin
Assistant Editor: Carolyn Johnson
Supervising Editor: Karen Carter
Associate Project Editor: Rachel D'Angelo Wimberly
Editorial Assistant: Caroline Lipscomb
Production Supervisor: Lisa Merrill
Art Supervisor: Gary Crespo
Marketing Manager: Charles Cavaliere
Marketing Associate: Ros Kane
Marketing Assistant: Kate Burden Thomas

Cover design by Harold Burch Design, NYC

Composition: Meridian Creative Group

Printed in the U.S.A.

Library of Congress Catalog Card Number: 96-076658

ISBN: 0-669-41737-8

23456789–DC–00 99 98 97

Preface

A firm foundation in algebra and trigonometry is necessary for success in college-level mathematics courses. *Trigonometry,* Fourth Edition, is designed to help students develop their proficiency in trigonometry, and so strengthen their understanding of the underlying concepts. Although the basic concepts of algebra are reviewed in the text, it is assumed that most students taking this course have completed two years of high school algebra.

The text takes every opportunity to show how algebra with trigonometry is a modern modeling language for real-life problems. Examples, exercises, and group activities—many using real data—provide a real-life context to help students grasp mathematical concepts. As appropriate, graphing technology is utilized throughout the text to enhance student understanding of mathematical concepts.

New to the Fourth Edition

All text elements in the previous edition were considered for revision, and many new examples, exercises, and applications were added to the text. Following are the major changes in the Fourth Edition.

Improved Coverage Chapter P, "Prerequisites," contains material that the student should have studied in earlier courses. All or part of this review material may be covered or omitted, offering greater flexibility in designing the course syllabus.

As a result of user requests, Chapter 6, "Topics in Analytic Geometry," has been reordered so that the topic of parametric equations is discussed before polar equations.

CD-ROM To accommodate a variety of teaching and learning styles, *Trigonometry,* Fourth Edition, is also available in a multimedia, CD-ROM format. *Interactive Trigonometry* offers students a variety of additional tutorial assistance, including examples and exercises with detailed solutions; pre-, post-, and self-tests with answers; and *TI-82* and *TI-83* graphing calculator emulators. (See pages xvi–xviii for more detailed information.)

Technology The new Fourth Edition acknowledges the increasing availability of graphing technology by offering the opportunity to use graphing utilities throughout, without requiring their use. This is achieved through a combination of features, including—at point of use—many opportunities for exploration using technology (see page 84); graphing utility instructions in the text, (see page 31); and clearly labeled exercises that require the use of a graphing utility (see page 365). In addition, *Interactive Trigonometry* offers the text in a CD-ROM format, as well as additional tutorial and technology enhanced features.

Data Analysis and Modeling Throughout the Fourth Edition, students are offered many more opportunities to collect and interpret data, to make conjectures, and to construct mathematical models. Students are encouraged to use mathematical models to make predictions or draw conclusions from real data (see page 392); invited to compare models (see page 403); and asked to use curve-fitting techniques to write models from data (see page 50). This edition encourages greater use of charts, tables, scatter plots, and graphs to summarize, analyze, and interpret data.

Applications To emphasize for students the connection between mathematical concepts and real-world situations, up-to-date, real-life applications are integrated throughout the text. Appearing as chapter introductions with related exercises (see pages 213 and 229), examples (see page 164), exercises (see page 169), Group Activities (see page 50), and Chapter Projects (see pages 210–211), these applications offer students frequent opportunities to use and review their problem-solving skills. The applications cover a wide range of disciplines including areas such as physics, chemistry, the social sciences, biology, and business.

Group Activities Each section ends with a Group Activity. These exercises reinforce students' understanding by exploring mathematical concepts in a variety of ways, including interpretation of mathematical concepts and results (see page 199); problem posing and error analysis (see page 132); and constructing mathematical models, tables, and graphs (see page 392). Designed to be completed in class or as homework assignments, the Group Activities give the students the opportunity to work cooperatively as they think, talk, and write about mathematics.

Connections In addition to highlighting the connections between algebra and areas outside mathematics through real-world applications, this text emphasizes the connections between algebra and other branches of mathematics, such as geometry (see page 203) and statistics (see Appendix A). Many examples and exercises throughout the text also reinforce the connections through graphical, numerical, and analytical representations of important algebraic concepts (see page 164).

There are many other new features of the Fourth Edition as well, including Exploration, Study Tips, Historical Notes, Focus on Concepts, and Chapter Projects. These and other features of the Fourth Edition are described in greater detail on the following pages.

Features of the Fourth Edition

Chapter Opener Each chapter opens with a look at a real-life application. Real data is presented using graphical, numerical, and algebraic techniques. In addition, a list of the section titles shows students how the topics fit into the overall development of algebra and trigonometry.

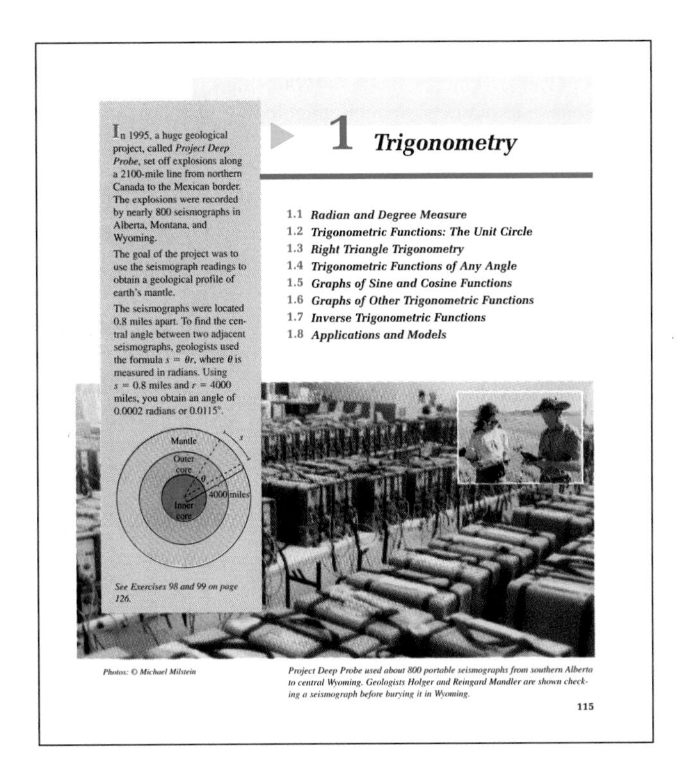

In 1995, a huge geological project, called *Project Deep Probe*, set off explosions along a 2100-mile line from northern Canada to the Mexican border. The explosions were recorded by nearly 800 seismographs in Alberta, Montana, and Wyoming.

The goal of the project was to use the seismograph readings to obtain a geological profile of earth's mantle.

The seismographs were located 0.8 miles apart. To find the central angle between two adjacent seismographs, geologists used the formula $s = \theta r$, where θ is measured in radians. Using $s = 0.8$ miles and $r = 4000$ miles, you obtain an angle of 0.0002 radians or 0.0115°.

See Exercises 98 and 99 on page 126.

1 Trigonometry

1.1 *Radian and Degree Measure*
1.2 *Trigonometric Functions: The Unit Circle*
1.3 *Right Triangle Trigonometry*
1.4 *Trigonometric Functions of Any Angle*
1.5 *Graphs of Sine and Cosine Functions*
1.6 *Graphs of Other Trigonometric Functions*
1.7 *Inverse Trigonometric Functions*
1.8 *Applications and Models*

Photos: © Michael Milstein

Project Deep Probe used about 800 portable seismographs from southern Alberta to central Wyoming. Geologists Holger and Reingard Mandler are shown checking a seismograph before burying it in Wyoming.

115

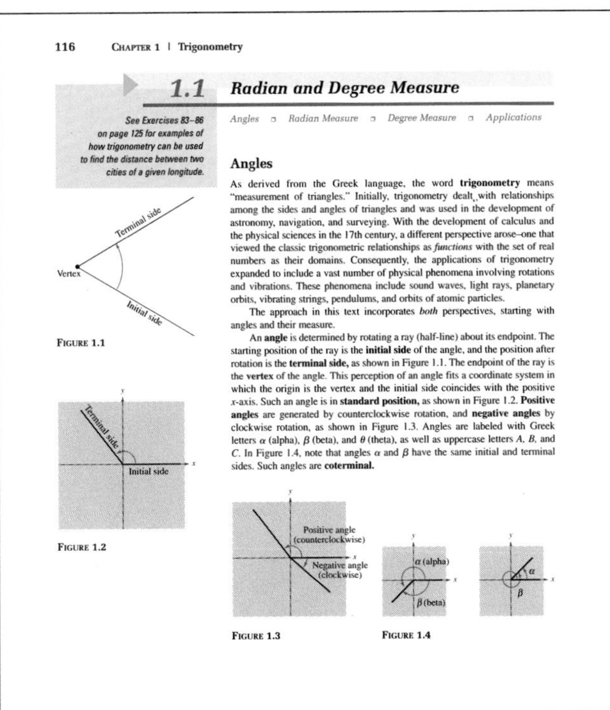

116 CHAPTER 1 | Trigonometry

1.1 Radian and Degree Measure

Angles ▫ *Radian Measure* ▫ *Degree Measure* ▫ *Applications*

See Exercises 83–86 on page 125 for examples of how trigonometry can be used to find the distance between two cities of a given longitude.

Angles

As derived from the Greek language, the word **trigonometry** means "measurement of triangles." Initially, trigonometry dealt with relationships among the sides and angles of triangles and was used in the development of astronomy, navigation, and surveying. With the development of calculus and the physical sciences in the 17th century, a different perspective arose–one that viewed the classic trigonometric relationships as *functions* with the set of real numbers as their domains. Consequently, the applications of trigonometry expanded to include a vast number of physical phenomena involving rotations and vibrations. These phenomena include sound waves, light rays, planetary orbits, vibrating strings, pendulums, and orbits of atomic particles.

The approach in this text incorporates *both* perspectives, starting with angles and their measure.

An **angle** is determined by rotating a ray (half-line) about its endpoint. The starting position of the ray is the **initial side** of the angle, and the position after rotation is the **terminal side,** as shown in Figure 1.1. The endpoint of the ray is the **vertex** of the angle. This perception of an angle fits a coordinate system in which the origin is the vertex and the initial side coincides with the positive x-axis. Such an angle is in **standard position,** as shown in Figure 1.2. **Positive angles** are generated by counterclockwise rotation, and **negative angles** by clockwise rotation, as shown in Figure 1.3. Angles are labeled with Greek letters α (alpha), β (beta), and θ (theta), as well as uppercase letters A, B, and C. In Figure 1.4, note that angles α and β have the same initial and terminal sides. Such angles are **coterminal.**

FIGURE 1.1

FIGURE 1.2

FIGURE 1.3

FIGURE 1.4

Section Outline Each section begins with a list of the major topics covered in the section. These topics are also the subsection titles and can be used for easy reference and review by students. In addition, an exercise application that uses a skill or illustrates a concept covered in the section is highlighted to emphasize the connection between mathematical concepts and real-life situations.

Graphics Visualization is a critical problem-solving skill. To encourage the development of this ability, the text has nearly 2300 figures in examples, exercises, and answers to exercises. Included are graphs of equations and functions, geometric figures, displays of statistical information, scatter plots, and numerous screen outputs from graphing technology. All graphs of equations and functions are computer- or calculator-generated for accuracy, and they are designed to resemble students' actual screen outputs as closely as possible. Graphics are also used to emphasize graphical interpretation, comparison, and estimation.

Theorems, Definitions, and Guidelines All of the important rules, formulas, theorems, guidelines, properties, definitions, and summaries are highlighted for emphasis. Each is also titled for easy reference.

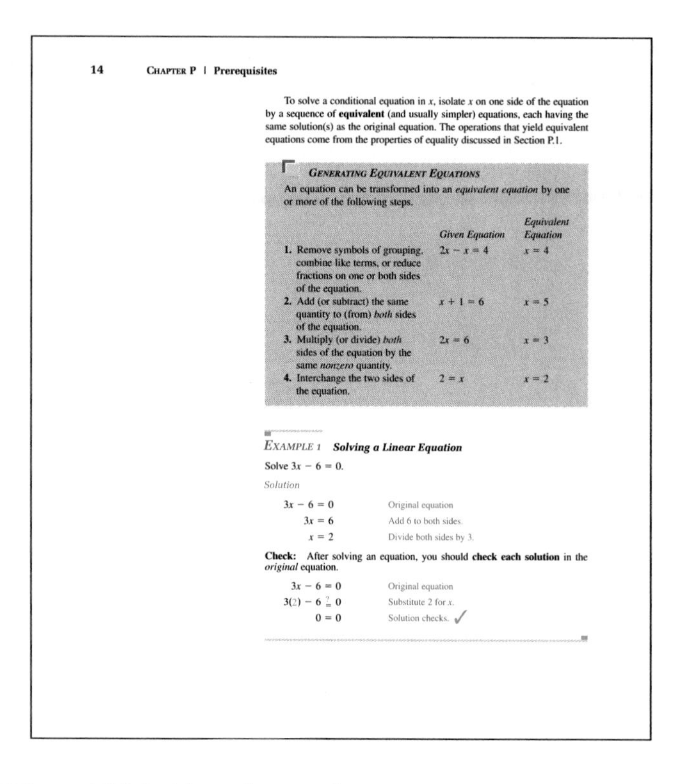
14 CHAPTER P | Prerequisites

To solve a conditional equation in x, isolate x on one side of the equation by a sequence of **equivalent** (and usually simpler) equations, each having the same solution(s) as the original equation. The operations that yield equivalent equations come from the properties of equality discussed in Section P.1.

GENERATING EQUIVALENT EQUATIONS

An equation can be transformed into an *equivalent equation* by one or more of the following steps.

	Given Equation	*Equivalent Equation*
1. Remove symbols of grouping, combine like terms, or reduce fractions on one or both sides of the equation.	$2x - x = 4$	$x = 4$
2. Add (or subtract) the same quantity to (from) *both* sides of the equation.	$x + 1 = 6$	$x = 5$
3. Multiply (or divide) *both* sides of the equation by the same *nonzero* quantity.	$2x = 6$	$x = 3$
4. Interchange the two sides of the equation.	$2 = x$	$x = 2$

EXAMPLE 1 ***Solving a Linear Equation***

Solve $3x - 6 = 0$.

Solution

$3x - 6 = 0$ Original equation

$3x = 6$ Add 6 to both sides.

$x = 2$ Divide both sides by 3.

Check: After solving an equation, you should **check each solution** in the *original* equation.

$3x - 6 = 0$ Original equation

$3(2) - 6 \stackrel{?}{=} 0$ Substitute 2 for x.

$0 = 0$ Solution checks. ✓

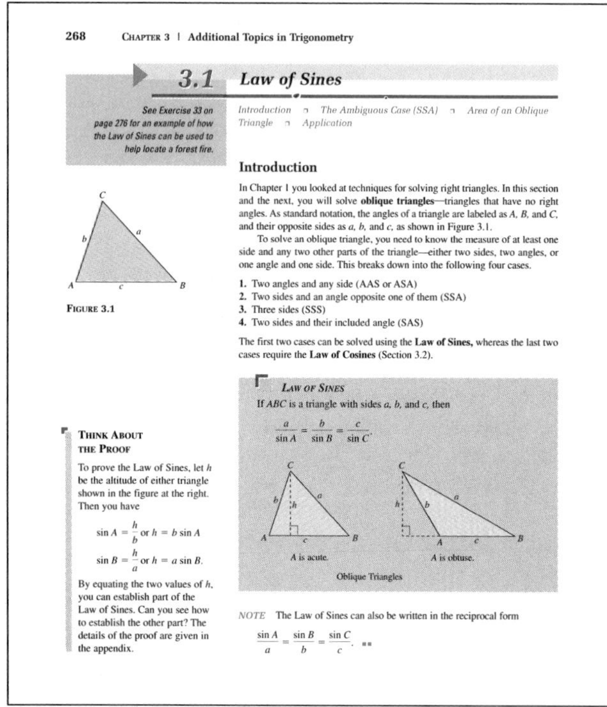
268 CHAPTER 3 | Additional Topics in Trigonometry

3.1 *Law of Sines*

Introduction ▫ *The Ambiguous Case (SSA)* ▫ *Area of an Oblique Triangle* ▫ *Application*

See Exercise 33 on page 276 for an example of how the Law of Sines can be used to help locate a forest fire.

Introduction

In Chapter 1 you looked at techniques for solving right triangles. In this section and the next, you will solve **oblique triangles**—triangles that have no right angles. As standard notation, the angles of a triangle are labeled as A, B, and C, and their opposite sides as a, b, and c, as shown in Figure 3.1.

To solve an oblique triangle, you need to know the measure of at least one side and any two other parts of the triangle—either two sides, two angles, or one angle and one side. This breaks down into the following four cases.

1. Two angles and any side (AAS or ASA)
2. Two sides and an angle opposite one of them (SSA)
3. Three sides (SSS)
4. Two sides and their included angle (SAS)

The first two cases can be solved using the **Law of Sines,** whereas the last two cases require the **Law of Cosines** (Section 3.2).

FIGURE 3.1

LAW OF SINES

If ABC is a triangle with sides a, b, and c, then

$$\frac{a}{\sin A} = \frac{b}{\sin B} = \frac{c}{\sin C}.$$

A is acute. A is obtuse.

Oblique Triangles

THINK ABOUT THE PROOF

To prove the Law of Sines, let h be the altitude of either triangle shown in the figure at the right. Then you have

$\sin A = \frac{h}{b}$ or $h = b \sin A$

$\sin B = \frac{h}{a}$ or $h = a \sin B$.

By equating the two values of h, you can establish part of the Law of Sines. Can you see how to establish the other part? The details of the proof are given in the appendix.

NOTE The Law of Sines can also be written in the reciprocal form

$$\frac{\sin A}{a} = \frac{\sin B}{b} = \frac{\sin C}{c}.$$

Think About the Proof Located in the margin adjacent to the corresponding theorem, each Think About the Proof feature offers strategies for proving the theorem. Detailed proofs for selected theorems are given in Appendix B.

Technology Instructions for using graphing utilities appear in the text at point of use. They offer convenient reference for students using graphing technology, and they can easily be omitted if desired. Additionally, problems in the Exercise Sets that require a graphing utility have been identified with the icon ▦.

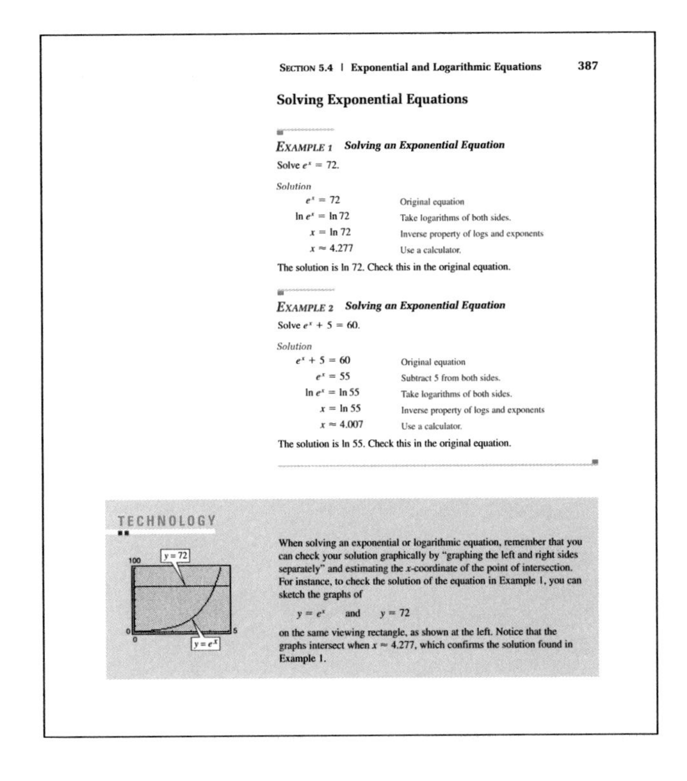

SECTION 5.4 | Exponential and Logarithmic Equations 387

Solving Exponential Equations

EXAMPLE 1 ***Solving an Exponential Equation***

Solve $e^x = 72$.

Solution

$e^x = 72$	Original equation
$\ln e^x = \ln 72$	Take logarithms of both sides.
$x = \ln 72$	Inverse property of logs and exponents
$x \approx 4.277$	Use a calculator.

The solution is ln 72. Check this in the original equation.

EXAMPLE 2 ***Solving an Exponential Equation***

Solve $e^x + 5 = 60$.

Solution

$e^x + 5 = 60$	Original equation
$e^x = 55$	Subtract 5 from both sides.
$\ln e^x = \ln 55$	Take logarithms of both sides.
$x = \ln 55$	Inverse property of logs and exponents
$x \approx 4.007$	Use a calculator.

The solution is ln 55. Check this in the original equation.

TECHNOLOGY

When solving an exponential or logarithmic equation, remember that you can check your solution graphically by "graphing the left and right sides separately" and estimating the x-coordinate of the point of intersection. For instance, to check the solution of the equation in Example 1, you can sketch the graphs of

$$y = e^x \quad \text{and} \quad y = 72$$

on the same viewing rectangle, as shown at the left. Notice that the graphs intersect when $x \approx 4.277$, which confirms the solution found in Example 1.

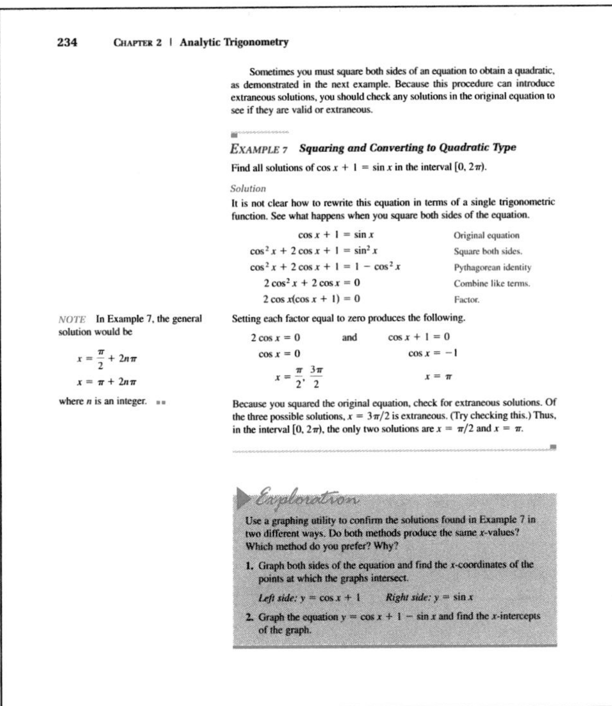

234 CHAPTER 2 | Analytic Trigonometry

Sometimes you must square both sides of an equation to obtain a quadratic, as demonstrated in the next example. Because this procedure can introduce extraneous solutions, you should check any solutions in the original equation to see if they are valid or extraneous.

EXAMPLE 7 ***Squaring and Converting to Quadratic Type***

Find all solutions of $\cos x + 1 = \sin x$ in the interval $[0, 2\pi)$.

Solution

It is not clear how to rewrite this equation in terms of a single trigonometric function. See what happens when you square both sides of the equation.

$\cos x + 1 = \sin x$	Original equation
$\cos^2 x + 2\cos x + 1 = \sin^2 x$	Square both sides.
$\cos^2 x + 2\cos x + 1 = 1 - \cos^2 x$	Pythagorean identity
$2\cos^2 x + 2\cos x = 0$	Combine like terms.
$2\cos x(\cos x + 1) = 0$	Factor.

Setting each factor equal to zero produces the following.

$2\cos x = 0$	and	$\cos x + 1 = 0$
$\cos x = 0$		$\cos x = -1$
$x = \frac{\pi}{2}, \frac{3\pi}{2}$		$x = \pi$

Because you squared the original equation, check for extraneous solutions. Of the three possible solutions, $x = 3\pi/2$ is extraneous. (Try checking this.) Thus, in the interval $[0, 2\pi)$, the only two solutions are $x = \pi/2$ and $x = \pi$.

NOTE In Example 7, the general solution would be

$$x = \frac{\pi}{2} + 2n\pi$$
$$x = \pi + 2n\pi$$

where n is an integer.

Exploration

Use a graphing utility to confirm the solutions found in Example 7 in two different ways. Do both methods produce the same x-values? Which method do you prefer? Why?

1. Graph both sides of the equation and find the x-coordinates of the points at which the graphs intersect.

   Left side: $y = \cos x + 1$ *Right side:* $y = \sin x$

2. Graph the equation $y = \cos x + 1 - \sin x$ and find the x-intercepts of the graph.

Exploration Throughout the text, the Exploration features encourage active participation by students, strengthening their intuition and critical thinking skills by exploring mathematical concepts and discovering mathematical relationships. Using a variety of approaches, including visualization, verification, use of graphing utilities, pattern recognition, and modeling, students develop conceptual understanding of theoretical topics.

Notes Notes anticipate students' needs by offering additional insights, pointing out common errors, and describing generalizations.

Historical Notes To help students understand that algebra has a past, historical notes featuring mathematicians and their work and mathematical artifacts are included in each chapter.

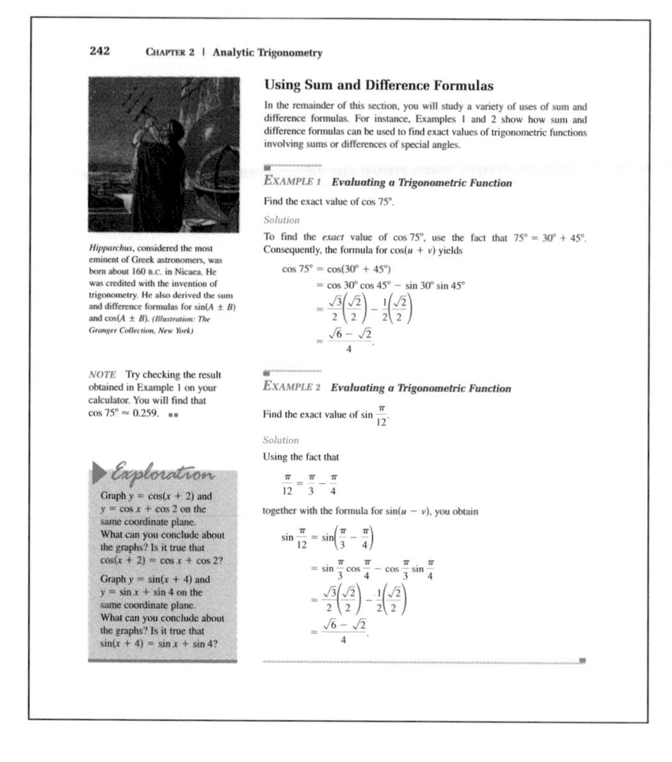

242 CHAPTER 2 | Analytic Trigonometry

Hipparchus, considered the most eminent of Greek astronomers, was born about 160 B.C. in Nicaea. He was credited with the invention of trigonometry. He also derived the sum and difference formulas for $\sin(A \pm B)$ and $\cos(A \pm B)$. (Illustration: The Granger Collection, New York)

Using Sum and Difference Formulas

In the remainder of this section, you will study a variety of uses of sum and difference formulas. For instance, Examples 1 and 2 show how sum and difference formulas can be used to find exact values of trigonometric functions involving sums or differences of special angles.

EXAMPLE 1 ***Evaluating a Trigonometric Function***

Find the exact value of $\cos 75°$.

Solution

To find the *exact* value of $\cos 75°$, use the fact that $75° = 30° + 45°$. Consequently, the formula for $\cos(u + v)$ yields

$$\begin{aligned}\cos 75° &= \cos(30° + 45°)\\ &= \cos 30° \cos 45° - \sin 30° \sin 45°\\ &= \frac{\sqrt{3}}{2}\left(\frac{\sqrt{2}}{2}\right) - \frac{1}{2}\left(\frac{\sqrt{2}}{2}\right)\\ &= \frac{\sqrt{6} - \sqrt{2}}{4}.\end{aligned}$$

NOTE Try checking the result obtained in Example 1 on your calculator. You will find that $\cos 75° \approx 0.259$.

EXAMPLE 2 ***Evaluating a Trigonometric Function***

Find the exact value of $\sin \frac{\pi}{12}$.

Solution

Using the fact that

$$\frac{\pi}{12} = \frac{\pi}{3} - \frac{\pi}{4}$$

together with the formula for $\sin(u - v)$, you obtain

$$\begin{aligned}\sin \frac{\pi}{12} &= \sin\left(\frac{\pi}{3} - \frac{\pi}{4}\right)\\ &= \sin \frac{\pi}{3} \cos \frac{\pi}{4} - \cos \frac{\pi}{3} \sin \frac{\pi}{4}\\ &= \frac{\sqrt{3}}{2}\left(\frac{\sqrt{2}}{2}\right) - \frac{1}{2}\left(\frac{\sqrt{2}}{2}\right)\\ &= \frac{\sqrt{6} - \sqrt{2}}{4}.\end{aligned}$$

Exploration

Graph $y = \cos(x + 2)$ and $y = \cos x + \cos 2$ on the same coordinate plane. What can you conclude about the graphs? Is it true that $\cos(x + 2) = \cos x + \cos 2$?

Graph $y = \sin(x + 4)$ and $y = \sin x + \sin 4$ on the same coordinate plane. What can you conclude about the graphs? Is it true that $\sin(x + 4) = \sin x + \sin 4$?

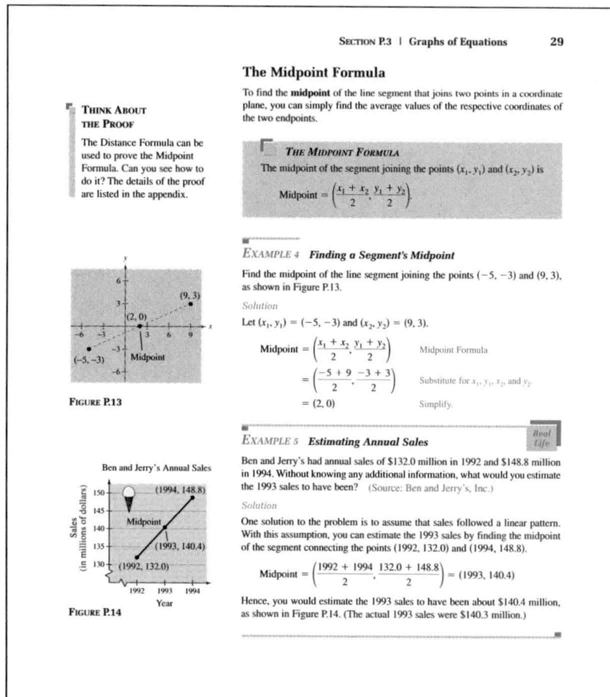

SECTION P.3 | Graphs of Equations 29

The Midpoint Formula

To find the **midpoint** of the line segment that joins two points in a coordinate plane, you can simply find the average values of the respective coordinates of the two endpoints.

THINK ABOUT THE PROOF

The Distance Formula can be used to prove the Midpoint Formula. Can you see how to do it? The details of the proof are listed in the appendix.

THE MIDPOINT FORMULA

The midpoint of the segment joining the points (x_1, y_1) and (x_2, y_2) is

$$\text{Midpoint} = \left(\frac{x_1 + x_2}{2}, \frac{y_1 + y_2}{2}\right).$$

EXAMPLE 4 ***Finding a Segment's Midpoint***

Find the midpoint of the line segment joining the points $(-5, -3)$ and $(9, 3)$, as shown in Figure P.13.

Solution

Let $(x_1, y_1) = (-5, -3)$ and $(x_2, y_2) = (9, 3)$.

$$\begin{aligned}\text{Midpoint} &= \left(\frac{x_1 + x_2}{2}, \frac{y_1 + y_2}{2}\right) && \text{Midpoint Formula}\\ &= \left(\frac{-5 + 9}{2}, \frac{-3 + 3}{2}\right) && \text{Substitute for } x_1, y_1, x_2, \text{ and } y_2.\\ &= (2, 0) && \text{Simplify.}\end{aligned}$$

FIGURE P.13

EXAMPLE 5 ***Estimating Annual Sales*** Real Life

Ben and Jerry's had annual sales of \$132.0 million in 1992 and \$148.8 million in 1994. Without knowing any additional information, what would you estimate the 1993 sales to have been? (Source: Ben and Jerry's, Inc.)

Solution

One solution to the problem is to assume that sales followed a linear pattern. With this assumption, you can estimate the 1993 sales by finding the midpoint of the segment connecting the points (1992, 132.0) and (1994, 148.8).

$$\text{Midpoint} = \left(\frac{1992 + 1994}{2}, \frac{132.0 + 148.8}{2}\right) = (1993, 140.4)$$

Hence, you would estimate the 1993 sales to have been about \$140.4 million, as shown in Figure P.14. (The actual 1993 sales were \$140.3 million.)

FIGURE P.14

Applications Real-life applications are integrated throughout the text in examples and exercises. These applications offer students constant review of problem-solving skills, and they emphasize the relevance of the mathematics. Many of the applications use recent, real data, and all are titled for easy reference. Photographs with captions in the introduction to the chapter and throughout the text also encourage students to see the link between mathematics and real life.

Study Tips Study Tips appear in the margin at point of use and offer students specific suggestions for studying algebra.

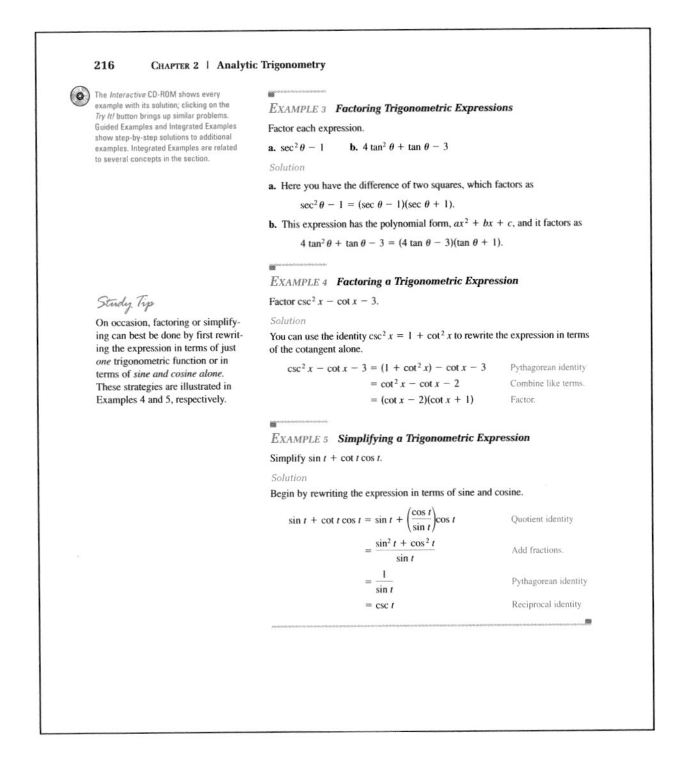

216 CHAPTER 2 | Analytic Trigonometry

The *Interactive* CD-ROM shows every example with its solution; clicking on the *Try It!* button brings up similar problems. Guided Examples and Integrated Examples show step-by-step solutions to additional examples. Integrated Examples are related to several concepts in the section.

EXAMPLE 3 ***Factoring Trigonometric Expressions***

Factor each expression.

a. $\sec^2\theta - 1$ **b.** $4\tan^2\theta + \tan\theta - 3$

Solution

a. Here you have the difference of two squares, which factors as

$$\sec^2\theta - 1 = (\sec\theta - 1)(\sec\theta + 1).$$

b. This expression has the polynomial form, $ax^2 + bx + c$, and it factors as

$$4\tan^2\theta + \tan\theta - 3 = (4\tan\theta - 3)(\tan\theta + 1).$$

Study Tip

On occasion, factoring or simplifying can best be done by first rewriting the expression in terms of just *one* trigonometric function or in terms of *sine and cosine alone.* These strategies are illustrated in Examples 4 and 5, respectively.

EXAMPLE 4 ***Factoring a Trigonometric Expression***

Factor $\csc^2 x - \cot x - 3$.

Solution

You can use the identity $\csc^2 x = 1 + \cot^2 x$ to rewrite the expression in terms of the cotangent alone.

$$\csc^2 x - \cot x - 3 = (1 + \cot^2 x) - \cot x - 3 \quad \text{Pythagorean identity}$$
$$= \cot^2 x - \cot x - 2 \quad \text{Combine like terms.}$$
$$= (\cot x - 2)(\cot x + 1) \quad \text{Factor.}$$

EXAMPLE 5 ***Simplifying a Trigonometric Expression***

Simplify $\sin t + \cot t \cos t$.

Solution

Begin by rewriting the expression in terms of sine and cosine.

$$\sin t + \cot t \cos t = \sin t + \left(\frac{\cos t}{\sin t}\right)\cos t \quad \text{Quotient identity}$$
$$= \frac{\sin^2 t + \cos^2 t}{\sin t} \quad \text{Add fractions.}$$
$$= \frac{1}{\sin t} \quad \text{Pythagorean identity}$$
$$= \csc t \quad \text{Reciprocal identity}$$

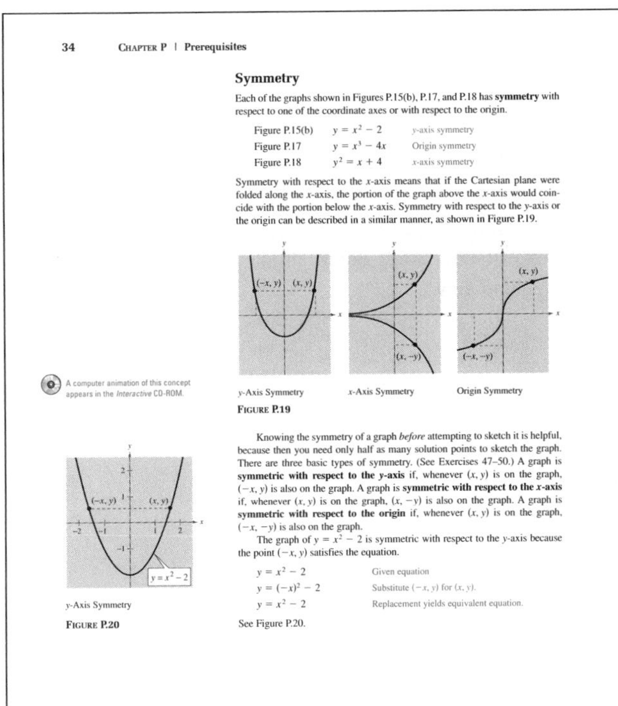

34 CHAPTER P | Prerequisites

Symmetry

Each of the graphs shown in Figures P.15(b), P.17, and P.18 has **symmetry** with respect to one of the coordinate axes or with respect to the origin.

Figure P.15(b)	$y = x^2 - 2$	y-axis symmetry
Figure P.17	$y = x^3 - 4x$	Origin symmetry
Figure P.18	$y^2 = x + 4$	x-axis symmetry

Symmetry with respect to the x-axis means that if the Cartesian plane were folded along the x-axis, the portion of the graph above the x-axis would coincide with the portion below the x-axis. Symmetry with respect to the y-axis or the origin can be described in a similar manner, as shown in Figure P.19.

y-Axis Symmetry x-Axis Symmetry Origin Symmetry

FIGURE P.19

A computer animation of this concept appears in the *Interactive* CD-ROM.

Knowing the symmetry of a graph *before* attempting to sketch it is helpful, because then you need only half as many solution points to sketch the graph. There are three basic types of symmetry. (See Exercises 47–50.) A graph is **symmetric with respect to the *y*-axis** if, whenever (x, y) is on the graph, $(-x, y)$ is also on the graph. A graph is **symmetric with respect to the *x*-axis** if, whenever (x, y) is on the graph, $(x, -y)$ is also on the graph. A graph is **symmetric with respect to the origin** if, whenever (x, y) is on the graph, $(-x, -y)$ is also on the graph.

The graph of $y = x^2 - 2$ is symmetric with respect to the y-axis because the point $(-x, y)$ satisfies the equation.

$$y = x^2 - 2 \quad \text{Given equation}$$
$$y = (-x)^2 - 2 \quad \text{Substitute } (-x, y) \text{ for } (x, y).$$
$$y = x^2 - 2 \quad \text{Replacement yields equivalent equation.}$$

See Figure P.20.

y-Axis Symmetry

FIGURE P.20

CD-ROM The icon refers to additional features of *Interactive Trigonometry* that enhance the text presentation, such as exercises, computer animations, examples, tests, and *TI-82* and *TI-83* graphing calculator emulators.

Examples Each of the over 300 text examples was carefully chosen to illustrate a particular mathematical concept, problem-solving approach, or computational technique, and to enhance students' understanding. The examples in the text cover a wide variety of problem types, including theoretical problems, real-life applications (many with real data), and problems requiring the use of graphing technology. Each example is titled for easy reference, and real-life applications are labeled. Many examples include side comments in color that clarify the steps of the solution.

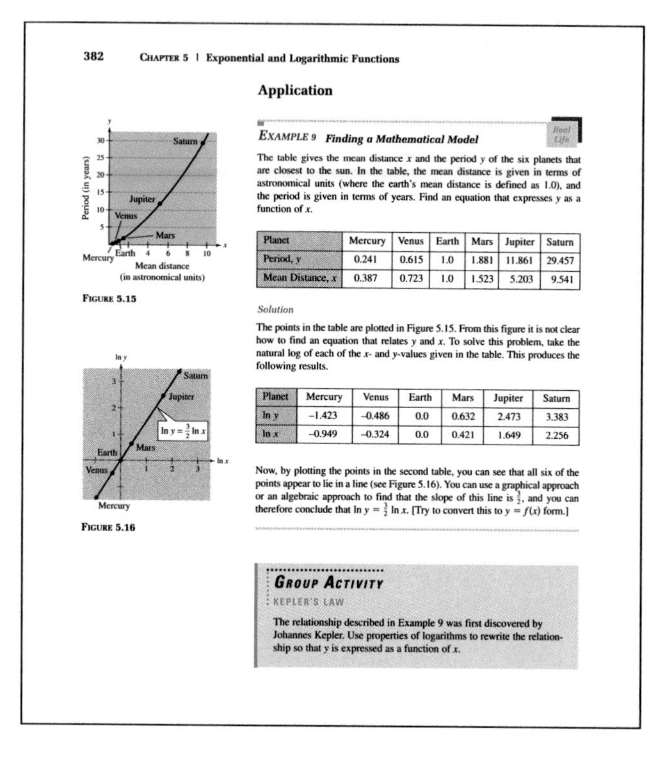

382 CHAPTER 5 | Exponential and Logarithmic Functions

Application

EXAMPLE 9 *Finding a Mathematical Model* Real Life

The table gives the mean distance x and the period y of the six planets that are closest to the sun. In the table, the mean distance is given in terms of astronomical units (where the earth's mean distance is defined as 1.0), and the period is given in terms of years. Find an equation that expresses y as a function of x.

Planet	Mercury	Venus	Earth	Mars	Jupiter	Saturn
Period, y	0.241	0.615	1.0	1.881	11.861	29.457
Mean Distance, x	0.387	0.723	1.0	1.523	5.203	9.541

FIGURE 5.15

Solution

The points in the table are plotted in Figure 5.15. From this figure it is not clear how to find an equation that relates y and x. To solve this problem, take the natural log of each of the x- and y-values given in the table. This produces the following results.

Planet	Mercury	Venus	Earth	Mars	Jupiter	Saturn
$\ln y$	−1.423	−0.486	0.0	0.632	2.473	3.383
$\ln x$	−0.949	−0.324	0.0	0.421	1.649	2.256

FIGURE 5.16

Now, by plotting the points in the second table, you can see that all six of the points appear to lie in a line (see Figure 5.16). You can use a graphical approach or an algebraic approach to find that the slope of this line is $\frac{3}{2}$, and you can therefore conclude that $\ln y = \frac{3}{2} \ln x$. [Try to convert this to $y = f(x)$ form.]

GROUP ACTIVITY

KEPLER'S LAW

The relationship described in Example 9 was first discovered by Johannes Kepler. Use properties of logarithms to rewrite the relationship so that y is expressed as a function of x.

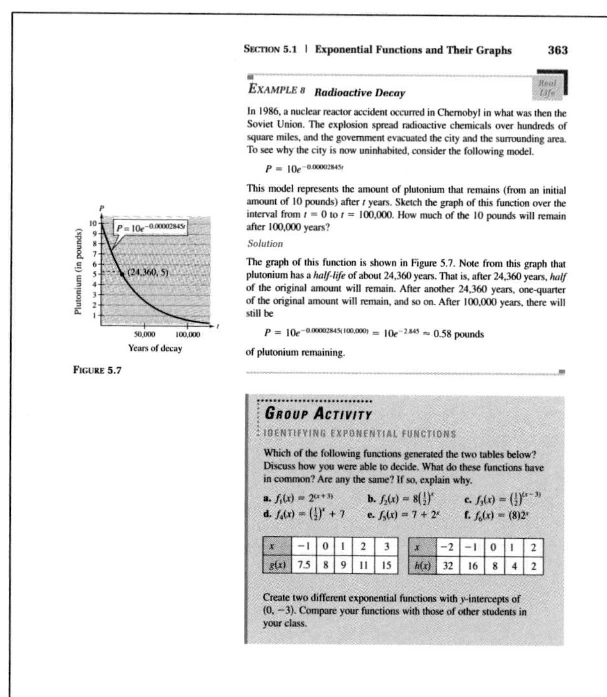

SECTION 5.1 | Exponential Functions and Their Graphs 363

EXAMPLE 8 *Radioactive Decay* Real Life

In 1986, a nuclear reactor accident occurred in Chernobyl in what was then the Soviet Union. The explosion spread radioactive chemicals over hundreds of square miles, and the government evacuated the city and the surrounding area. To see why the city is now uninhabited, consider the following model.

$$P = 10e^{-0.00002845t}$$

This model represents the amount of plutonium that remains (from an initial amount of 10 pounds) after t years. Sketch the graph of this function over the interval from $t = 0$ to $t = 100{,}000$. How much of the 10 pounds will remain after 100,000 years?

Solution

The graph of this function is shown in Figure 5.7. Note from this graph that plutonium has a *half-life* of about 24,360 years. That is, after 24,360 years, *half* of the original amount will remain. After another 24,360 years, one-quarter of the original amount will remain, and so on. After 100,000 years, there will still be

$$P = 10e^{-0.00002845(100{,}000)} = 10e^{-2.845} \approx 0.58 \text{ pounds}$$

of plutonium remaining.

FIGURE 5.7

GROUP ACTIVITY

IDENTIFYING EXPONENTIAL FUNCTIONS

Which of the following functions generated the two tables below? Discuss how you were able to decide. What do these functions have in common? Are any the same? If so, explain why.

a. $f_1(x) = 2^{(x+3)}$ b. $f_2(x) = 8\left(\frac{1}{2}\right)^x$ c. $f_3(x) = \left(\frac{1}{2}\right)^{(x-3)}$
d. $f_4(x) = \left(\frac{1}{2}\right)^x + 7$ e. $f_5(x) = 7 + 2^x$ f. $f_6(x) = (8)2^x$

x	−1	0	1	2	3
$g(x)$	7.5	8	9	11	15

x	−2	−1	0	1	2
$h(x)$	32	16	8	4	2

Create two different exponential functions with y-intercepts of $(0, -3)$. Compare your functions with those of other students in your class.

Group Activities The Group Activities that appear at the ends of sections reinforce students' understanding by studying mathematical concepts in a variety of ways, including talking and writing about mathematics, creating and solving problems, analyzing errors, and developing and using mathematical models. Designed to be completed as group projects in class or as homework assignments, the Group Activities give students opportunities to do interactive learning and to think, talk, and write about mathematics.

Warm Ups Each section (except Section P.1) contains a set of 10 warm-up exercises that students can use for review and practice of the previously learned skills that are necessary for mastery of the new skills and concepts presented in the section. All warm-up exercises are answered in the back of the text.

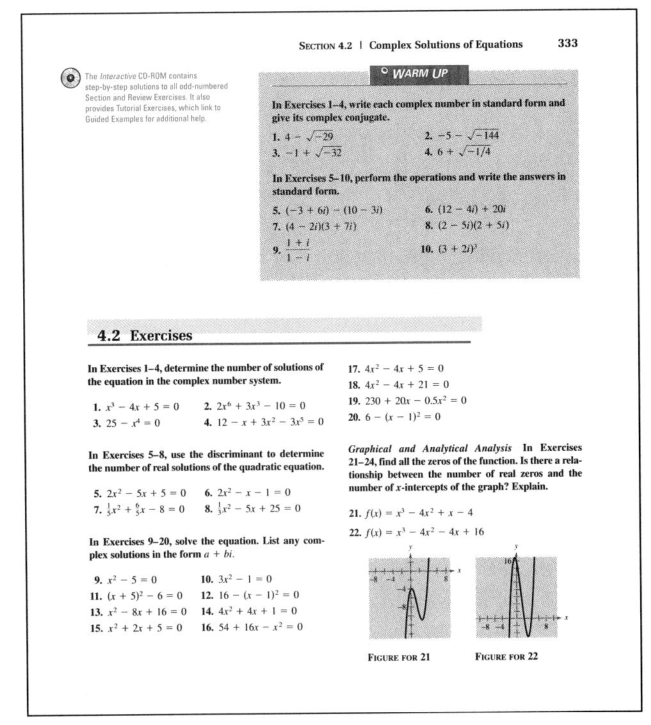

SECTION 4.2 | Complex Solutions of Equations 333

The *Interactive* CD-ROM contains step-by-step solutions to all odd-numbered Section and Review Exercises. It also provides Tutorial Exercises, which link to Guided Examples for additional help.

WARM UP

In Exercises 1–4, write each complex number in standard form and give its complex conjugate.

1. $4 - \sqrt{-29}$
2. $-5 - \sqrt{-144}$
3. $-1 + \sqrt{-32}$
4. $6 + \sqrt{-1/4}$

In Exercises 5–10, perform the operations and write the answers in standard form.

5. $(-3 + 6i) - (10 - 3i)$
6. $(12 - 4i) + 20i$
7. $(4 - 2i)(3 + 7i)$
8. $(2 - 5i)(2 + 5i)$
9. $\dfrac{1 + i}{1 - i}$
10. $(3 + 2i)^3$

4.2 Exercises

In Exercises 1–4, determine the number of solutions of the equation in the complex number system.

1. $x^3 - 4x + 5 = 0$
2. $2x^6 + 3x^3 - 10 = 0$
3. $25 - x^4 = 0$
4. $12 - x + 3x^2 - 3x^5 = 0$

In Exercises 5–8, use the discriminant to determine the number of real solutions of the quadratic equation.

5. $2x^2 - 5x + 5 = 0$
6. $2x^2 - x - 1 = 0$
7. $\frac{1}{5}x^2 + \frac{6}{5}x - 8 = 0$
8. $\frac{1}{3}x^2 - 5x + 25 = 0$

In Exercises 9–20, solve the equation. List any complex solutions in the form $a + bi$.

9. $x^2 - 5 = 0$
10. $3x^2 - 1 = 0$
11. $(x + 5)^2 - 6 = 0$
12. $16 - (x - 1)^2 = 0$
13. $x^2 - 8x + 16 = 0$
14. $4x^2 + 4x + 1 = 0$
15. $x^2 + 2x + 5 = 0$
16. $54 + 16x - x^2 = 0$
17. $4x^2 - 4x + 5 = 0$
18. $4x^2 - 4x + 21 = 0$
19. $230 + 20x - 0.5x^2 = 0$
20. $6 - (x - 1)^2 = 0$

Graphical and Analytical Analysis **In Exercises 21–24, find all the zeros of the function. Is there a relationship between the number of real zeros and the number of x-intercepts of the graph? Explain.**

21. $f(x) = x^3 - 4x^2 + x - 4$
22. $f(x) = x^3 - 4x^2 - 4x + 16$

FIGURE FOR 21 FIGURE FOR 22

Geometry Geometric formulas and concepts are reviewed throughout the text in examples, group activities, and exercises. For reference, common formulas are listed inside the back cover of this text.

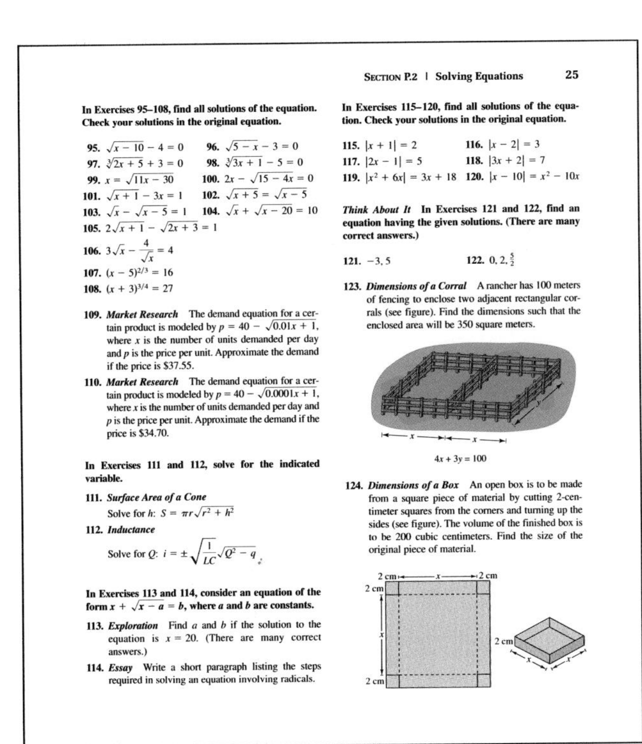

SECTION P.2 | Solving Equations 25

In Exercises 95–108, find all solutions of the equation. Check your solutions in the original equation.

95. $\sqrt{x - 10} - 4 = 0$
96. $\sqrt{5 - x} - 3 = 0$
97. $\sqrt[3]{2x + 5} + 3 = 0$
98. $\sqrt[3]{3x + 1} - 5 = 0$
99. $x = \sqrt{11x - 30}$
100. $2x - \sqrt{15 - 4x} = 0$
101. $\sqrt{x + 1} - 3x = 1$
102. $\sqrt{x + 5} = \sqrt{x - 5}$
103. $\sqrt{x} - \sqrt{x - 5} = 1$
104. $\sqrt{x} + \sqrt{x - 20} = 10$
105. $2\sqrt{x + 1} - \sqrt{2x + 3} = 1$
106. $3\sqrt{x} - \dfrac{4}{\sqrt{x}} = 4$
107. $(x - 5)^{2/3} = 16$
108. $(x + 3)^{3/4} = 27$

109. ***Market Research*** The demand equation for a certain product is modeled by $p = 40 - \sqrt{0.01x + 1}$, where x is the number of units demanded per day and p is the price per unit. Approximate the demand if the price is $37.55.

110. ***Market Research*** The demand equation for a certain product is modeled by $p = 40 - \sqrt{0.0001x + 1}$, where x is the number of units demanded per day and p is the price per unit. Approximate the demand if the price is $34.70.

In Exercises 111 and 112, solve for the indicated variable.

111. ***Surface Area of a Cone***
Solve for h: $S = \pi r\sqrt{r^2 + h^2}$

112. ***Inductance***
Solve for Q: $i = \pm\sqrt{\dfrac{1}{LC}}\sqrt{Q^2 - q}$

In Exercises 113 and 114, consider an equation of the form $x + \sqrt{x - a} = b$, where a and b are constants.

113. ***Exploration*** Find a and b if the solution to the equation is $x = 20$. (There are many correct answers.)

114. ***Essay*** Write a short paragraph listing the steps required in solving an equation involving radicals.

In Exercises 115–120, find all solutions of the equation. Check your solutions in the original equation.

115. $|x + 1| = 2$
116. $|x - 2| = 3$
117. $|2x - 1| = 5$
118. $|3x + 2| = 7$
119. $|x^2 + 6x| = 3x + 18$
120. $|x - 10| = x^2 - 10x$

Think About It **In Exercises 121 and 122, find an equation having the given solutions. (There are many correct answers.)**

121. $-3, 5$
122. $0, 2, \frac{5}{2}$

123. ***Dimensions of a Corral*** A rancher has 100 meters of fencing to enclose two adjacent rectangular corrals (see figure). Find the dimensions such that the enclosed area will be 350 square meters.

$4x + 3y = 100$

124. ***Dimensions of a Box*** An open box is to be made from a square piece of material by cutting 2-centimeter squares from the corners and turning up the sides (see figure). The volume of the finished box is to be 200 cubic centimeters. Find the size of the original piece of material.

Exercises The exercise sets were completely revised—and expanded by over 20%—for the Fourth Edition. The text now offers nearly 4000 exercises with a broad range of conceptual, computational, and applied problems to accommodate a variety of teaching and learning styles. Included in the section and review exercise sets are multi-part, writing, and more challenging problems with extensive graphics that encourage exploration and discovery, enhance students' skills in mathematical modeling, estimation, and data interpretation and analysis, and encourage the use of graphing technology for conceptual understanding. Applications are labeled for easy reference. The exercise sets are designed to build competence, skill, and understanding; each exercise set is graded in difficulty to allow students to gain confidence as they progress. Detailed solutions to all odd-numbered exercises are given in the *Study and Solutions Guide*; answers to all odd-numbered exercises appear in the back of the text.

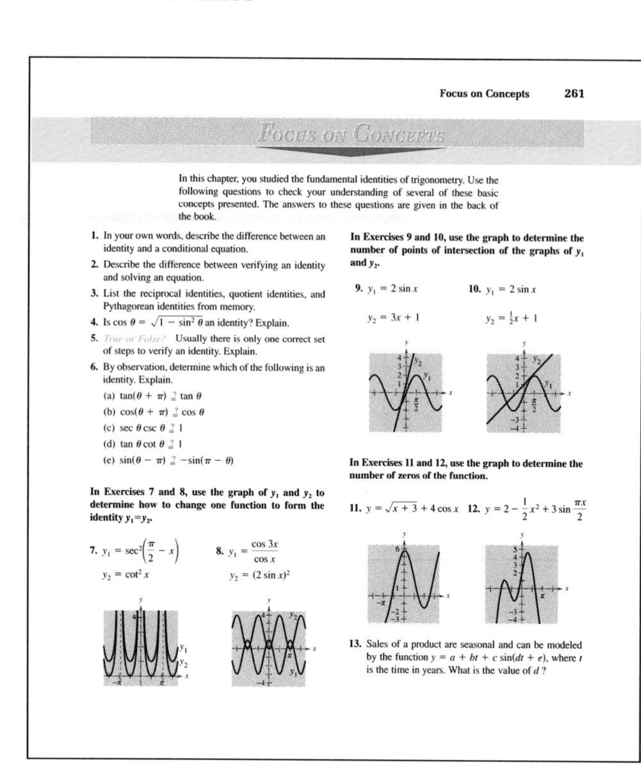

Focus on Concepts 261

FOCUS ON CONCEPTS

In this chapter, you studied the fundamental identities of trigonometry. Use the following questions to check your understanding of several of these basic concepts presented. The answers to these questions are given in the back of the book.

1. In your own words, describe the difference between an identity and a conditional equation.
2. Describe the difference between verifying an identity and solving an equation.
3. List the reciprocal identities, quotient identities, and Pythagorean identities from memory.
4. Is $\cos\theta = \sqrt{1-\sin^2\theta}$ an identity? Explain.
5. *True or False?* Usually there is only one correct set of steps to verify an identity. Explain.
6. By observation, determine which of the following is an identity. Explain.
   (a) $\tan(\theta + \pi) \stackrel{?}{=} \tan\theta$
   (b) $\cos(\theta + \pi) \stackrel{?}{=} \cos\theta$
   (c) $\sec\theta\csc\theta \stackrel{?}{=} 1$
   (d) $\tan\theta\cot\theta \stackrel{?}{=} 1$
   (e) $\sin(\theta - \pi) \stackrel{?}{=} -\sin(\pi - \theta)$

In Exercises 7 and 8, use the graph of y_1 and y_2 to determine how to change one function to form the identity $y_1 = y_2$.

7. $y_1 = \sec^2\left(\frac{\pi}{2} - x\right)$, $y_2 = \cot^2 x$
8. $y_1 = \frac{\cos 3x}{\cos x}$, $y_2 = (2\sin x)^2$

In Exercises 9 and 10, use the graph to determine the number of points of intersection of the graphs of y_1 and y_2.

9. $y_1 = 2\sin x$, $y_2 = 3x + 1$
10. $y_1 = 2\sin x$, $y_2 = \frac{1}{2}x + 1$

In Exercises 11 and 12, use the graph to determine the number of zeros of the function.

11. $y = \sqrt{x+3} + 4\cos x$
12. $y = 2 - \frac{1}{2}x^2 + 3\sin\frac{\pi x}{2}$

13. Sales of a product are seasonal and can be modeled by the function $y = a + bt + c\sin(dt + e)$, where t is the time in years. What is the value of d?

Focus on Concepts Each Focus on Concepts feature is a set of exercises that test students' understanding of the basic concepts covered in the chapter. Answers to all questions are given in the back of the text.

Chapter Projects Chapter Projects are extended applications that use real data, graphs, and modeling to enhance students' understanding of mathematical concepts. Designed as individual or group projects, they offer additional opportunities to think, discuss, and write about mathematics. Many projects give students the opportunity to collect, analyze, and interpret data.

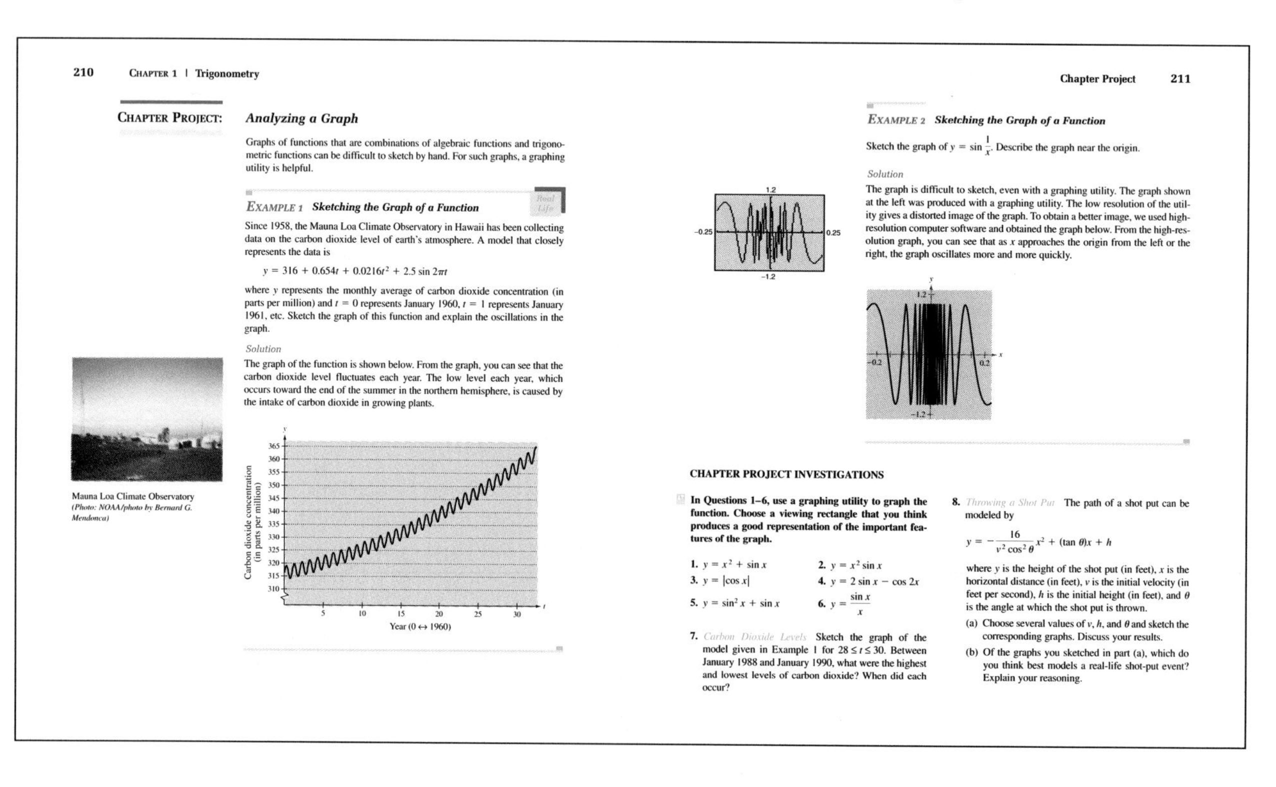

210 CHAPTER 1 | Trigonometry

CHAPTER PROJECT: *Analyzing a Graph*

Graphs of functions that are combinations of algebraic functions and trigonometric functions can be difficult to sketch by hand. For such graphs, a graphing utility is helpful.

EXAMPLE 1 ***Sketching the Graph of a Function*** (Real Life)

Since 1958, the Mauna Loa Climate Observatory in Hawaii has been collecting data on the carbon dioxide level of earth's atmosphere. A model that closely represents the data is

$$y = 316 + 0.654t + 0.0216t^2 + 2.5\sin 2\pi t$$

where y represents the monthly average of carbon dioxide concentration (in parts per million) and $t = 0$ represents January 1960, $t = 1$ represents January 1961, etc. Sketch the graph of this function and explain the oscillations in the graph.

Solution

The graph of the function is shown below. From the graph, you can see that the carbon dioxide level fluctuates each year. The low level each year, which occurs toward the end of the summer in the northern hemisphere, is caused by the intake of carbon dioxide in growing plants.

Carbon dioxide concentration (in parts per million); Year (0 ↔ 1960)

Mauna Loa Climate Observatory (*Photo: NOAA/photo by Bernard G. Mendonca*)

Chapter Project 211

EXAMPLE 2 ***Sketching the Graph of a Function***

Sketch the graph of $y = \sin\frac{1}{x}$. Describe the graph near the origin.

Solution

The graph is difficult to sketch, even with a graphing utility. The graph shown at the left was produced with a graphing utility. The low resolution of the utility gives a distorted image of the graph. To obtain a better image, we used high-resolution computer software and obtained the graph below. From the high-resolution graph, you can see that as x approaches the origin from the left or the right, the graph oscillates more and more quickly.

CHAPTER PROJECT INVESTIGATIONS

In Questions 1–6, use a graphing utility to graph the function. Choose a viewing rectangle that you think produces a good representation of the important features of the graph.

1. $y = x^2 + \sin x$
2. $y = x^2\sin x$
3. $y = |\cos x|$
4. $y = 2\sin x - \cos 2x$
5. $y = \sin^2 x + \sin x$
6. $y = \frac{\sin x}{x}$
7. *Carbon Dioxide Levels* Sketch the graph of the model given in Example 1 for $28 \le t \le 30$. Between January 1988 and January 1990, what were the highest and lowest levels of carbon dioxide? When did each occur?
8. *Throwing a Shot Put* The path of a shot put can be modeled by
$$y = -\frac{16}{v^2\cos^2\theta}x^2 + (\tan\theta)x + h$$
where y is the height of the shot put (in feet), x is the horizontal distance (in feet), v is the initial velocity (in feet per second), h is the initial height (in feet), and θ is the angle at which the shot put is thrown.
   (a) Choose several values of v, h, and θ and sketch the corresponding graphs. Discuss your results.
   (b) Of the graphs you sketched in part (a), which do you think best models a real-life shot-put event? Explain your reasoning.

Review Exercises The Review Exercises at the end of each chapter offer students an opportunity for additional practice. Answers to odd-numbered review exercises are given in the back of the text.

Review Exercises 411

Review Exercises

In Exercises 1–6, match the function with its graph. [The graphs are labeled (a) through (f).]

1. $f(x) = 4^x$
2. $f(x) = 4^{-x}$
3. $f(x) = -4^x$
4. $f(x) = 4^x + 1$
5. $f(x) = \log_4 x$
6. $f(x) = \log_4(x - 1)$

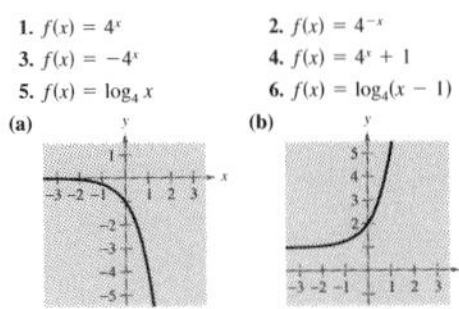

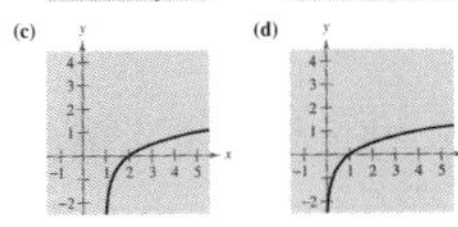

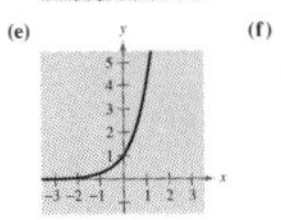

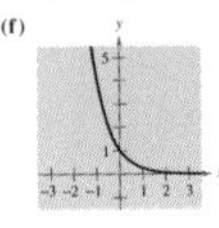

In Exercises 7–12, sketch the graph of the function.

7. $f(x) = 0.3^x$
8. $g(x) = 0.3^{-x}$
9. $h(x) = e^{-x/2}$
10. $h(x) = 2 - e^{-x/2}$
11. $f(x) = e^{x+2}$
12. $s(t) = 4e^{-2/t}, \quad t > 0$

In Exercises 13 and 14, use a graphing utility to graph the function. Identify any asymptotes.

13. $g(x) = 200e^{4/x}$
14. $f(x) = \dfrac{10}{1 + 2^{-0.05x}}$

In Exercises 15 and 16, complete the table to determine the balance A for P dollars invested at rate r for t years and compounded n times per year.

n	1	2	4	12	365	Continuous
A						

15. $P = \$3500,\ r = 10.5\%,\ t = 10$ years
16. $P = \$2000,\ r = 12\%,\ t = 30$ years

In Exercises 17 and 18, complete the table to determine the amount P that should be invested at rate r to produce a balance of \$200,000 in t years.

t	1	10	20	30	40	50
P						

17. $r = 8\%$, compounded continuously
18. $r = 10\%$, compounded monthly
19. ***Waiting Times*** The average time between incoming calls at a switchboard is 3 minutes. The probability of waiting less than t minutes until the next incoming call is approximated by the model
$F(t) = 1 - e^{-t/3}$.
If a call has just come in, find the probability that the next call will be within
(a) $\frac{1}{2}$ minute. (b) 2 minutes. (c) 5 minutes.
20. ***Depreciation*** After t years, the value of a car that cost \$14,000 is given by
$V(t) = 14{,}000(\frac{3}{4})^t$.
(a) Use a graphing utility to graph the function.
(b) Find the value of the car 2 years after it was purchased.
(c) According to the model, when does the car depreciate most rapidly? Is this realistic? Explain.

416 CHAPTER 5 | Exponential and Logarithmic Functions

Chapter Test

Take this test as you would take a test in class. After you are done, check your work against the answers given in the back of the book.

The *Interactive* CD-ROM provides answers to the Chapter Tests and Cumulative Tests. It also offers Chapter Pre-Tests (which test key skills and concepts covered in previous chapters) and Chapter Post-Tests, both of which have randomly generated exercises with diagnostic capabilities.

1. Sketch the graph of the function $f(x) = 2^{-x/3}$.
2. Determine the horizontal asymptotes of the function $f(x) = \dfrac{1000}{1 + 4e^{-0.2x}}$.
3. Determine the amount after 30 years if \$5000 is invested at $6\frac{1}{2}\%$ compounded (a) quarterly and (b) continuously.
4. Determine the principle that will yield \$200,000 when invested at 8% compounded daily for 20 years.
5. Write the logarithmic equation $\log_4 64 = 3$ in exponential form.
6. Write the exponential equation $5^{-2} = \frac{1}{25}$ in logarithmic form.
7. Sketch a graph of the function $g(x) = \log_3(x - 2)$.
8. Use the properties of logarithms to expand $\ln\left(\dfrac{6x^2}{\sqrt{x^2 + 1}}\right)$.
9. Use the properties of logarithms to condense $3 \ln z - [\ln(z + 1) + \ln(z - 1)]$.
10. Use the properties of logarithms to simplify $\log_6 \sqrt{360}$.

In Exercises 11–14, solve the equation. Round the solution to three decimal places.

11. $e^{x/2} = 450$
12. $\left(1 + \dfrac{0.06}{4}\right)^{4t} = 3$
13. $5 \ln(x + 4) = 22$
14. A truck that costs \$28,000 new has a depreciated value of \$20,000 after 1 year. Find the value of the truck when it is 3 years old by using the exponential model $y = Ce^{kt}$.

In Exercises 15–17, the population of a certain species t years after it is introduced into a new habitat is given by $p(t) = 1200/(1 + 3e^{-t/5})$.

15. Determine the population size that was introduced into the habitat.
16. Determine the population after 5 years.
17. After how many years will the population be 800?
18. By observation, identify the equation that corresponds to the graph shown in the figure. Explain your reasoning.
(a) $y = 6e^{-x^2/2}$ (b) $y = \dfrac{6}{1 + e^{-x/2}}$ (c) $y = 6(1 - e^{-x^2/2})$

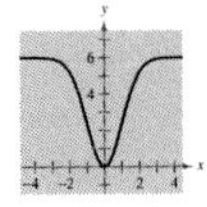

FIGURE FOR 18

Chapter Tests Each chapter that is not followed by a Cumulative Test ends with a Chapter Test, an effective tool for student self-assessment.

318 CUMULATIVE TEST FOR CHAPTERS 1–3

Cumulative Test for Chapters 1–3

Take this test as you would take a test in class. After you are done, check your work against the answers given in the back of the book.

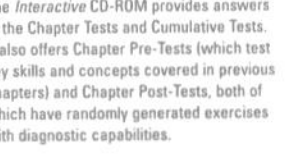

The *Interactive* CD-ROM provides answers to the Chapter Tests and Cumulative Tests. It also offers Chapter Pre-Tests (which test key skills and concepts covered in previous chapters) and Chapter Post-Tests, both of which have randomly generated exercises with diagnostic capabilities.

1. Consider the angle $\theta = -120°$.
(a) Sketch the angle in standard position.
(b) Determine a coterminal angle in the interval $[0°, 360°)$.
(c) Convert the angle to radian measure.
(d) Find the reference angle θ'.
(e) Find the exact values of the six trigonometric functions of θ.
2. Convert the angle of magnitude 2.35 radians to degrees. Round the answer to one decimal place.
3. Find $\cos\theta$ if $\tan\theta = -\frac{4}{3}$ and $\sin\theta < 0$.
4. Sketch the graphs of (a) $f(x) = 3 - 2\sin\pi x$ and (b) $g(x) = \dfrac{1}{2}\tan\left(x - \dfrac{\pi}{2}\right)$.
5. Find a, b, and c such that the graph of the function $h(x) = a\cos(bx + c)$ matches the graph in the figure.
6. Write an algebraic expression equivalent to $\sin(\arccos 2x)$.
7. Subtract and simplify: $\dfrac{\sin\theta - 1}{\cos\theta} - \dfrac{\cos\theta}{\sin\theta - 1}$.
8. Prove the identities.
(a) $\cot^2\alpha(\sec^2\alpha - 1) = 1$ (b) $\sin(x + y)\sin(x - y) = \sin^2 x - \sin^2 y$
(c) $\sin^2 x \cos^2 x = \frac{1}{8}(1 - \cos 4x)$
9. Find all solutions of the equations in the interval $[0, 2\pi)$.
(a) $2\cos^2\beta - \cos\beta = 0$ (b) $3\tan\theta - \cot\theta = 0$
10. Find the remaining angles and side of the triangle shown in the figure.
(a) $A = 30°,\ a = 9,\ b = 8$ (b) $A = 30°,\ b = 8,\ c = 10$
11. From a point 200 feet from a flagpole, the angles of elevation to the bottom and top of the flag are 16° 45′ and 18°, respectively. Approximate the height of the flag to the nearest foot.
12. An airplane is flying at an airspeed of 500 kilometers per hour and a bearing of N 30° E. The wind at the altitude of the plane has a velocity of 50 kilometers per hour and a bearing of N 60° E. What is the true direction of the plane, and what is its speed relative to the ground?
13. Find the projection of **u** onto **v** if **u** = ⟨−4, 3⟩ and **v** = ⟨−1, 5⟩.

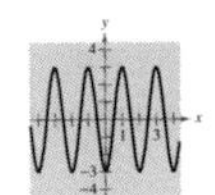

FIGURE FOR 5

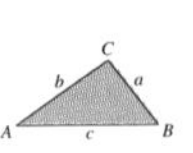

FIGURE FOR 10

Cumulative Tests The Cumulative Tests that follow Chapters 3 and 6 help students judge their mastery of previously covered material, as well as reinforce the knowledge they have been accumulating throughout the text—preparing them for other exams and for future courses.

Supplements

Trigonometry, Fourth Edition, by Larson and Hostetler, is accompanied by a comprehensive supplements package. Most items are keyed to the text.

Printed Resources

FOR THE STUDENT

Study and Solutions Guide by Dianna Zook, Indiana University/Purdue University—Fort Wayne

- Section summaries of key concepts
- Detailed, step-by-step solutions to all odd-numbered exercises
- Practice tests with solutions
- Study strategies

Graphing Technology Keystroke Guide

- Keystroke instructions for many graphing calculators from Texas Instruments, Sharp, Casio, and Hewlett-Packard, including *TI-83, TI-92, HP-38G,* and *Casio CFX-9800G.*
- BestGrapher instructions for both IBM and Macintosh
- Examples with step-by-step solutions
- Extensive graphics screen output
- Technology tips

FOR THE INSTRUCTOR

Instructor's Annotated Edition

- Includes the entire student edition of the text, with the student answers section
- Instructor's Answers section: Answers to all even-numbered exercises, and answers to all Explorations, Technology exercises, Chapter Project exercises, and Group Activities
- Annotations at point of use offer specific teaching strategies and suggestions for implementing Group Activities, point out common student errors, and give additional examples, exercises, class activities, and group activities.

Complete Solutions Guide

- Detailed, step-by-step solutions to all section, review, Focus on Concepts, and Chapter Project exercises

Test Item File and Instructor's Resource Guide

- Printed test bank with nearly 1600 test items (multiple-choice, open-ended, and writing) coded by level of difficulty
- Technology-required test items coded for easy reference
- Bank of chapter test forms with answer keys
- Two final exam test forms

- Notes to the instructor, including materials for alternative assessment and managing the multicultural and cooperative-learning classrooms

Problem Solving, Modeling, and Data Analysis Labs by Wendy Metzger, Palomar College

- Multipart, guided discovery activities and applications
- Keystroke instructions for Derive and *TI-82*
- Keyed to the text by topic
- Funded in part by NSF (National Science Foundation, Instrumentation and Laboratory Improvement) and California Community College Fund for Instructional Improvement

Media Resources

FOR THE STUDENT

Interactive Trigonometry (See pages xvi–xviii for a description, or visit the Houghton Mifflin home page at http://www.hmco.com for a preview.)

- Interactive, multimedia CD-ROM format
- IBM-PC for Windows

Tutor software

- Interactive tutorial software keyed to the text by section
- Diagnostic feedback
- Chapter self-tests
- Guided exercises with step-by-step solutions
- Glossary

Videotapes by Dana Mosely

- Comprehensive, text-specific coverage keyed to the text by section
- Real-life application vignettes introduced where appropriate
- Computer-generated animation
- For media/resource centers
- Additional explanation of concepts, sample problems, and applications
- Instructional graphing calculator videotape also available

FOR THE INSTRUCTOR

Computerized Testing (IBM, Macintosh, Windows)

- New on-line testing
- New grade-management capabilities
- Algorithmic test-generating software provides an unlimited number of tests.
- Nearly 1600 test items
- Also available as a printed test bank

Transparency Package

- 40 color transparencies color-coded by topic

Interactive Trigonometry

To accommodate a variety of teaching and learning styles, *Trigonometry* is also available in a multimedia, CD-ROM format. In this interactive format, the text offers the student additional tutorial assistance with

- Complete solutions to all odd-numbered text exercises.
- Chapter pre-tests, self-tests, and post-tests.
- *TI-82* and *TI-83* emulators.
- Guided examples with step-by-step solutions.
- Editable graphs.
- Animations of mathematical concepts.
- Warm-up, section, and tutorial exercises.
- Glossary of key terms.

These and other pedagogical features of the CD-ROM are illustrated by the screen dumps shown below.

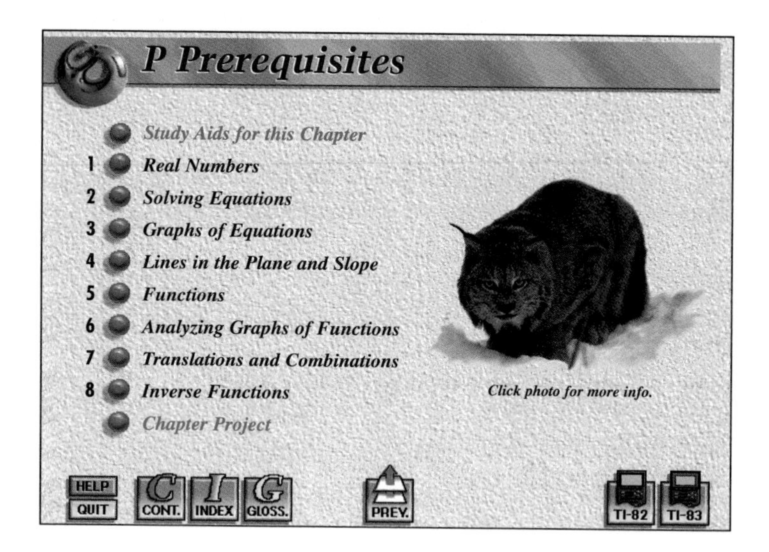

Chapter Topics Each chapter begins with an outline of the topics to be covered. Using the buttons at the bottom of the screen, the student can quickly move to the appropriate section.

Introductory Chapter Application Each chapter opens with a real-data application that illustrates the key concepts and techniques to be covered. Clicking on the photo, the student can access additional data and background information that frames the real-world context for a mathematical concept.

Chapter Project Each chapter is accompanied by a Chapter Project. This offers the student the opportunity to synthesize the algebraic techniques and concepts studied in the chapter. Many projects use real data and emphasize data analysis and mathematical modeling.

Study Aids Each section offers the student an array of additional study aids, including Chapter Pre-, Post-, and Self-Tests, Review Exercises, and Focus on Concepts. With diagnostics, complete solutions, or answers, these helpful features promote the focused practice needed to master mathematical concepts. Short, informative video segments are also included.

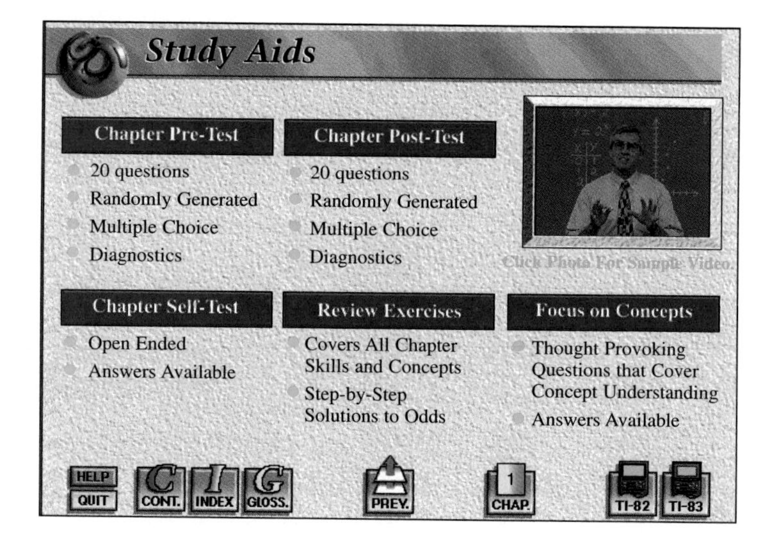

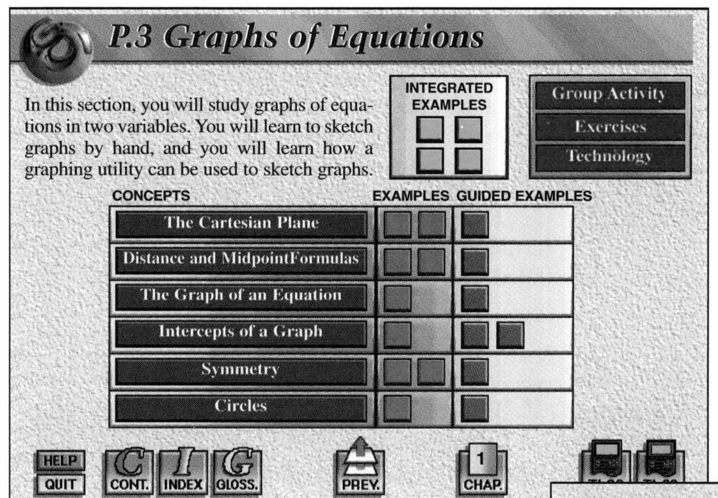

Examples *Interactive Trigonometry* illustrates mathematical concepts by featuring all of the Examples found in the text. Some selected Examples are enhanced by editable graphs. Guided Examples with step-by-step solutions that appear one line at a time offer additional opportunities for practice and skill development. Group Activities, Exercises, and Integrated Examples help synthesize the concepts in the section.

Graphs Some examples are accompanied by editable graphs for exploration and discovery. Using the keys below the graph, the student can change the function, the graphing window, and the x- and y-scales, and can trace and zoom.

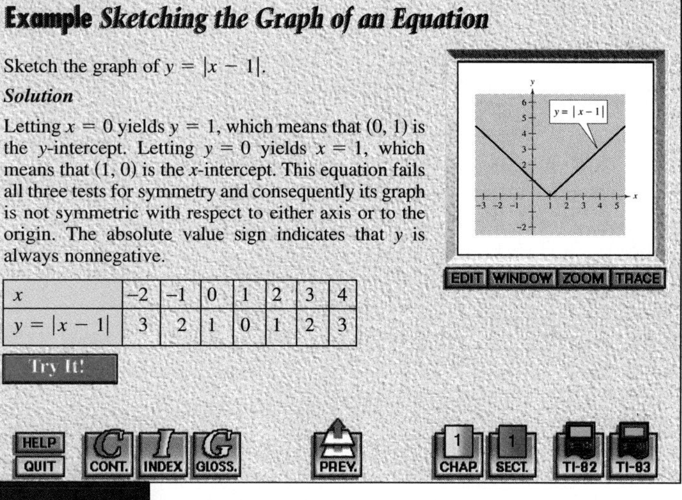

Try It! After studying the worked-out example, the student can use the Try It! button to access similar examples—with solutions following on separate screens—to test his or her mastery of mathematical concepts and techniques.

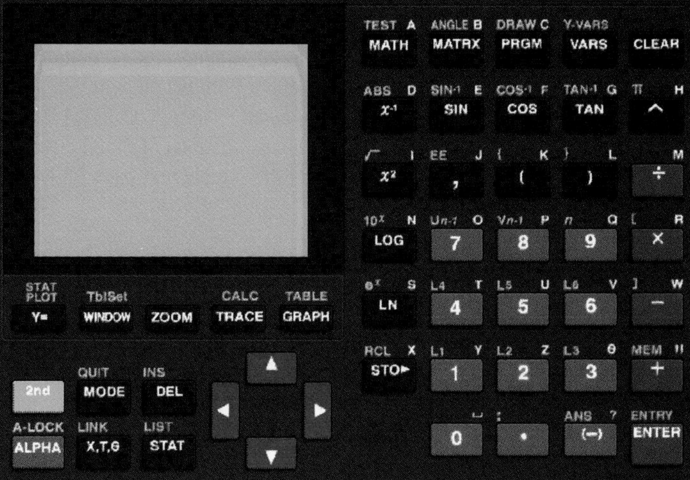

***TI-82* and *TI-83* Emulators** Accessible on every screen, the *TI-82* and *TI-83* emulators give instant access to graphing utilities as tools for computation and exploration. They are also available for working exercises in the text that require the use of a graphing utility, all of which are clearly marked by the icon . Instruction on using the emulators is also available at the click of a button.

These emulators were developed and copyrighted by Meridian Creative Group with the prior written permission of Texas Instruments.

In Exercises 1 and 2, complete a table of values. Use the solution points to sketch the graph of the equation.

1. $y = -\frac{1}{2}x + 2$
2. $y = x^2 - 3x$

In Exercises 3–12, sketch the graph *by hand*.

3. $y - 2x - 3 = 0$
4. $3x + 2y + 6 = 0$
5. $x - 5 = 0$
6. $y = 8 - |x|$
7. $y = \sqrt{5 - x}$
8. $y = \sqrt{x + 2}$
9. $y + 2x^2 = 0$
10. $y = x^2 - 4x$
11. $y = \sqrt{25 - x^2}$
12. $x^2 + y^2 = 10$

HELP QUIT CONT. INDEX GLOSS. PREV. CHAP. SECT.

Section Exercises Each section is accompanied by a comprehensive set of exercises promoting skills mastery and conceptual understanding. Solutions to all odd-numbered exercises are available for instant feedback.

Tutorial Exercises Every section has a set of exercises in a multiple-choice format that offer students additional practice. Examples and diagnostics enhance this guided practice.

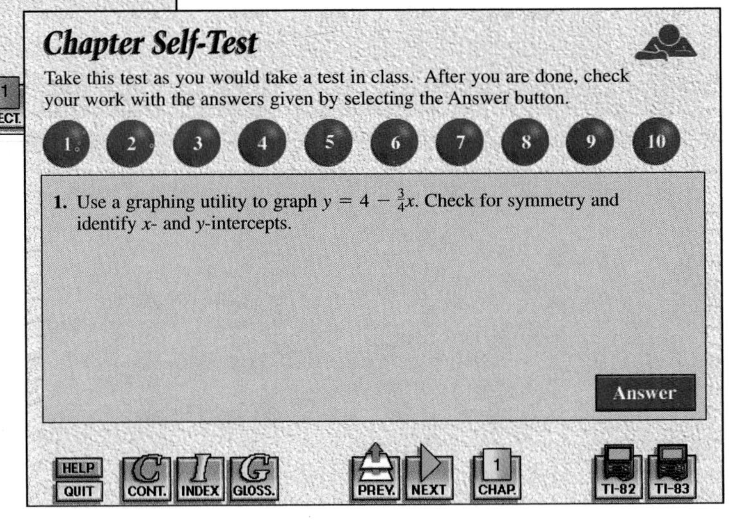

Tests Every chapter of the interactive text includes tests that are different from those in the textbook: Chapter Pre-Tests (testing key skills and concepts covered in previous chapters), and Chapter Post-Tests test mastery of the material covered in the textbook. The Chapter Self-Tests from the text are also included. Answers to all tests are included.

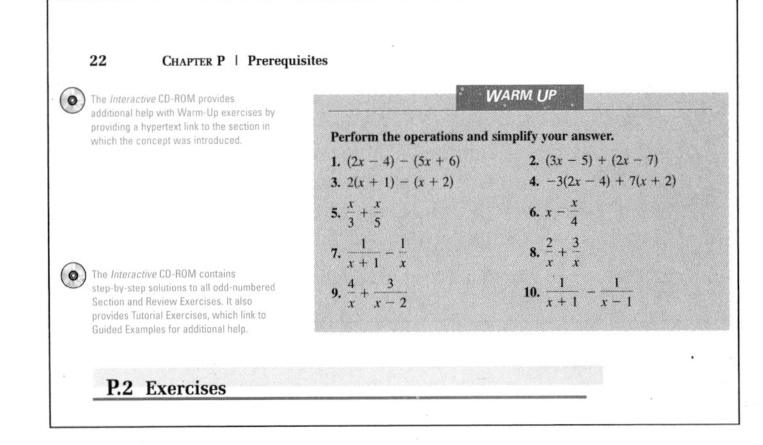

22 CHAPTER P | Prerequisites

The *Interactive* CD-ROM provides additional help with Warm-Up exercises by providing a hypertext link to the section in which the concept was introduced.

WARM UP

Perform the operations and simplify your answer.

1. $(2x - 4) - (5x + 6)$
2. $(3x - 5) + (2x - 7)$
3. $2(x + 1) - (x + 2)$
4. $-3(2x - 4) + 7(x + 2)$
5. $\frac{x}{3} + \frac{x}{5}$
6. $x - \frac{x}{4}$
7. $\frac{1}{x + 1} - \frac{1}{x}$
8. $\frac{2}{x} + \frac{3}{x}$
9. $\frac{4}{x} + \frac{3}{x - 2}$
10. $\frac{1}{x + 1} - \frac{1}{x - 1}$

The *Interactive* CD-ROM contains step-by-step solutions to all odd-numbered Section and Review Exercises. It also provides Tutorial Exercises, which link to Guided Examples for additional help.

P.2 Exercises

Interactive Trigonometry supports the mathematical presentation in the text *Trigonometry,* Fourth Edition, with a variety of tutorial, diagnostic, and demonstration features. Throughout both the student text and the Instructor's Annotated Edition, CD-ROM icons identify these additional functions of the interactive text, as illustrated by the sample text page (at left).

Acknowledgments

We would like to thank the many people who have helped us at various stages of this project to prepare the text and supplements package. Their encouragement, criticisms, and suggestions have been invaluable to us.

Fourth Edition Reviewers: Joby Milo Anthony, University of Central Florida; Sudhir Kumar Goel, Valdosta State University; Kathy B. Hamrick, Augusta College; Steven Z. Kahn, Anne Arundel Community College; Anne Landry, Dutchess Community College; Sue Little, North Harris Community College; Giles Maloof, Boise State University; Steven E. Martin, Richard Bland College; Carol Paxton, Glendale Community College (CA); Joan N. Powell, Auburn University; Michael P. Scanlon, Fairleigh Dickinson University; Patricia Shelton, North Carolina Agricultural and Technical State University; Laurence Small, Los Angeles Pierce College; Patricia B. Taylor, Thomas Nelson Community College; Gary Thomasson, Palm Beach Community College; Jeffrey X. Watt, Indiana University–Purdue University, Indianapolis; and Jacci Wozniak, Brevard Community College.

Fourth Edition Survey Respondents: Marwan A. Abu-Sawwa, Florida Community College at Jacksonville; Barbara C. Armenta, Pima Community College—East; Gladwin E. Bartel, Otero Junior College; Carole A. Bauer, Triton College; Joyce M. Becker, Luther College; Marybeth Beno, South Suburban College; Charles M. Biles, Humboldt State University; Ruthane Bopp, Lake Forest College; Tim Chappell, North Central Missouri College; Michael Davidson, Cabrillo College; Diane L. Doyle, Adirondack Community College; Donna S. Fatheree, University of Louisiana at Lafayette; John R. Formsma, Los Angeles City College; John S. Frohliger, Saint Norbert College; Gary Glaze, Eastern Washington University; Irwin S. Goldfine, Truman College; Elise M. Grabner, Slippery Rock University; Donnie Hallstone, Green River Community College; Lois E. Higbie, Brookdale Community College; Susan S. Hollar, Kalamazoo Valley Community College; Fran Hopf, Hillsborough Community College; Margaret D. Hovde, Grossmont College; John F. Keating, Massasoit Community College; John G. LaMaster, Indiana University–Purdue University at Fort Wayne; Giles Wilson Maloof, Boise State University, Kenneth Mangels, Concordia State University; Peggy I. Miller, University of Nebraska at Kearney; Gilbert F. Orr, University of Southern Colorado; Jim Paige, Wayne State College; Elise Price, Tarrant County Junior College; Doris Schraeder, McLennan Community College; Linda Schultz, McHenry County College; Fay Sewell, Montgomery County Community College; Patricia G. Shelton, North Carolina A & T State University; Hazel Shows, Hinds Community College; Joseph F. Stokes, Western Kentucky University; Diane Van Nostrand, University of Tulsa; and Raymond D. Wuco, San Joaquin Delta College.

Thanks to all of the people at Houghton Mifflin Company who worked with us in the development and production of the text, especially Chris Hoag, Sponsoring Editor; Cathy Cantin, Managing Editor; Maureen Brooks, Senior Associate Editor; Carolyn Johnson, Assistant Editor; Karen Carter, Supervising Editor; Rachel Wimberly, Associate Project Editor; Gary Crespo, Art Supervisor; Lisa Merrill, Production Supervisor; Ros Kane, Marketing Associate; and Carrie Lipscomb, Editorial Assistant.

We would also like to thank the staff at Larson Texts, Inc. who assisted with proofreading the manuscript, preparing and proofreading the art package, and checking and typesetting the supplements.

On a personal level, we are grateful to our wives, Deanna Gilbert Larson and Eloise Hostetler, for their love, patience, and support. Also, special thanks go to R. Scott O'Neil.

If you have suggestions for improving the text, please feel free to write to us. Over the past two decades, we have received many useful comments from both instructors and students, and we value these very much.

Roland E. Larson
Robert P. Hostetler

Contents

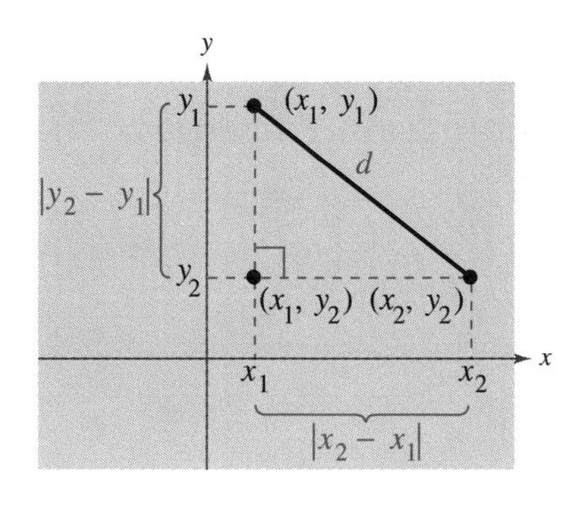

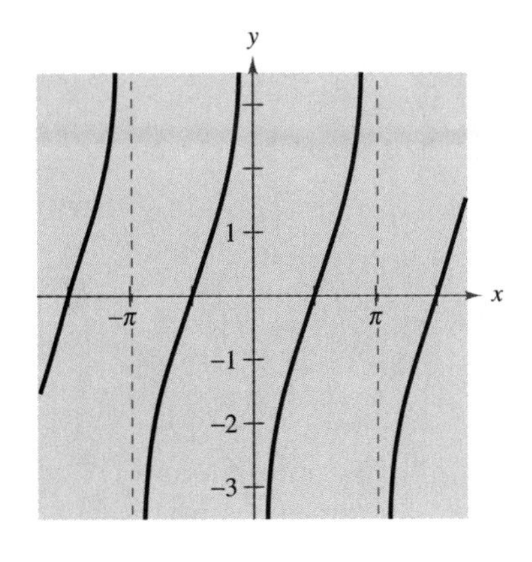
y
x
1
−1
−2
−3
−π
π

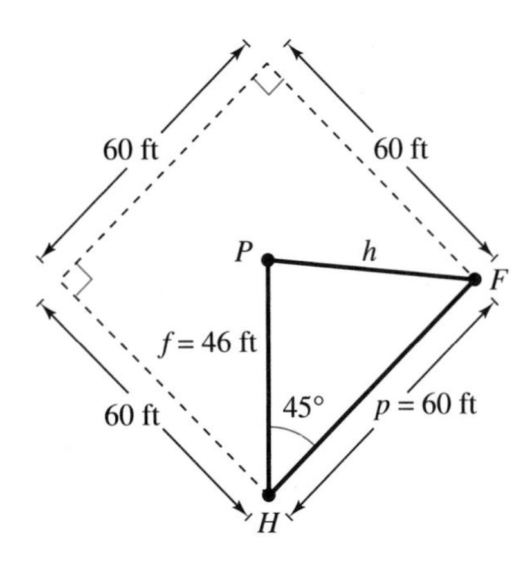
60 ft
60 ft
60 ft
P
h
F
f = 46 ft
45°
p = 60 ft
H

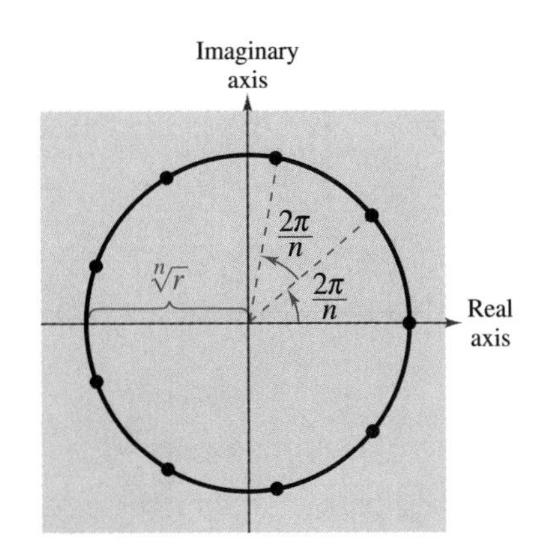
Imaginary axis
Real axis
$\frac{2\pi}{n}$
$\frac{2\pi}{n}$
$\sqrt[n]{r}$

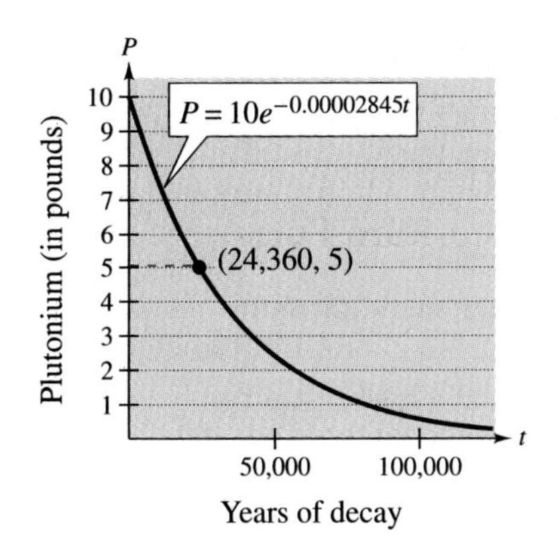

CHAPTER 5 *Exponential and Logarithmic Functions* *355*

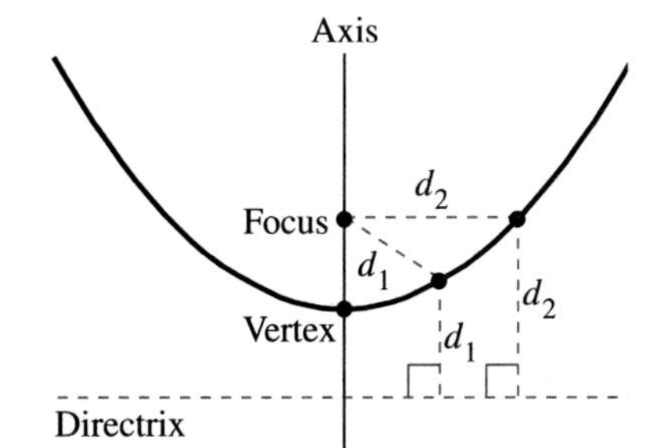

CHAPTER 6 *Topics in Analytic Geometry* *417*

P Prerequisites

Many wildlife populations follow a cyclical "predator-prey" pattern. One example is the populations of snowshoe hare and lynx in the Yukon Territory. The researchers shown in the photo kept track of the lynx and hare populations from 1988 through 1995. Lynx numbers in a 350-square-kilometer region of the Yukon are shown below.

1988 (10)	1989 (16)
1990 (50)	1991 (60)
1992 (28)	1993 (15)
1994 (9)	1995 (8)

The hare population was low in 1986, increased to a high in 1990, and then decreased to a low again in 1992.

The numbers of lynx and snowshoe hare are functions of the year. For instance, the graph below depicts the number of lynx as a function of the year.

The data was supplied by Mark O'Donoghue, as part of the Kluane Boreal Forest Ecosystem Project.

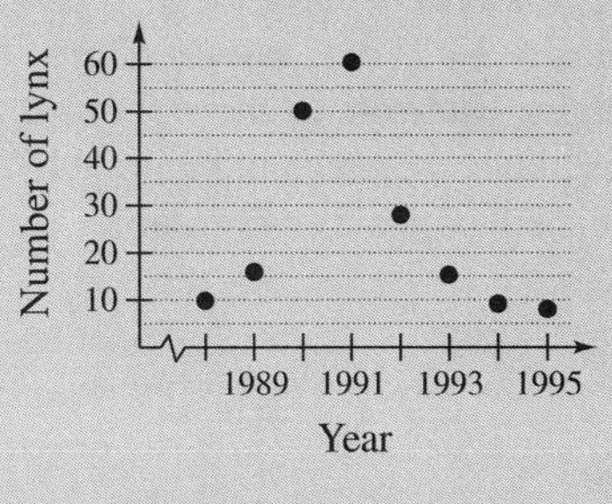

See Exercise 90 on page 70.

Photo: Alejandro Frid/Biological Photo Service

Husband and wife researchers Elizabeth Hofer and Peter Upton are measuring a lynx that has been trapped and sedated. The researchers work in the Yukon Territory, Canada.

P.1 Real Numbers

See Exercises 67–70 on page 11 for an example of how real numbers and absolute value are used to solve a budget variance problem.

Real Numbers ▫ *Ordering Real Numbers* ▫ *Absolute Value and Distance* ▫ *Algebraic Expressions* ▫ *Basic Rules of Algebra*

Real Numbers

Real numbers are used in everyday life to describe quantities such as age, miles per gallon, container size, and population. To represent real numbers, you can use symbols such as

$$9, 0, \frac{4}{3}, 0.666 \ldots, 28.21, \sqrt{2}, \pi, \text{ and } \sqrt[3]{-32}.$$

Here are some important subsets of the real numbers.

$\{1, 2, 3, 4, \ldots\}$	Set of natural numbers
$\{0, 1, 2, 3, 4, \ldots\}$	Set of whole numbers
$\{\ldots -3, -2, -1, 0, 1, 2, 3, \ldots\}$	Set of integers

A real number is **rational** if it can be written as the ratio p/q of two integers, where $q \neq 0$. For instance, the numbers

$$\frac{1}{3} = 0.3333 \ldots, \quad \frac{1}{8} = 0.125, \quad \text{and} \quad \frac{125}{111} = 1.126126 \ldots$$

are rational. The decimal representation of a rational number either *repeats* (as in $3.1454545 \ldots$) or *terminates* (as in $\frac{1}{2} = 0.5$). A real number that cannot be written as the ratio of two integers is called **irrational.** Irrational numbers have infinite *nonrepeating* decimal representations. For instance, the numbers

$$\sqrt{2} \approx 1.4142136 \quad \text{and} \quad \pi \approx 3.1415927$$

are irrational. (The symbol $\approx$ means "is approximately equal to.")

Real numbers are represented graphically by a **real number line.** The point 0 on the real number line is the **origin.** Numbers to the right of 0 are positive, and numbers to the left of 0 are negative, as shown in Figure P.1. The term **nonnegative** describes a number that is either positive or zero.

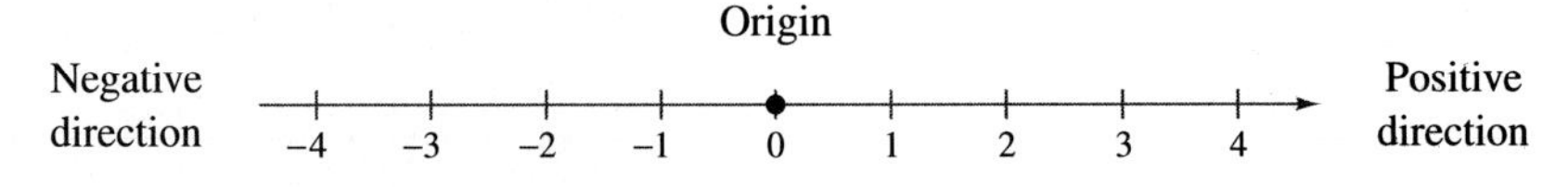

FIGURE P.1 The Real Number Line

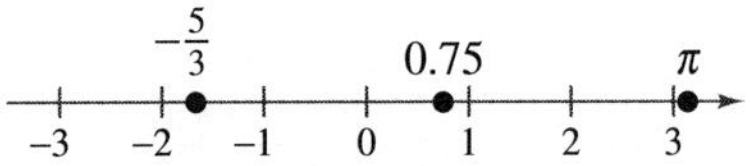

As illustrated in Figure P.2, there is a *one-to-one correspondence* between real numbers and points on the real number line.

$-\frac{5}{3}$ 0.75 π

−3 −2 −1 0 1 2 3

Every real number corresponds to exactly one point on the real number line.

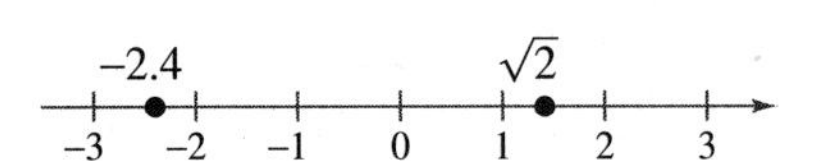

Every point on the real number line corresponds to exactly one real number.

FIGURE P.2 One-to-One Correspondence

Ordering Real Numbers

One important property of real numbers is that they are **ordered.**

> ***DEFINITION OF ORDER ON THE REAL NUMBER LINE***
>
> If a and b are real numbers, a is **less than** b if $b - a$ is positive. This order is denoted by the **inequality**
>
> $$a < b.$$
>
> This can also be described by saying that b is **greater than** a and writing $b > a$. The inequality $a \leq b$ means that a is **less than or equal to** b, and the inequality $b \geq a$ means that b is **greater than or equal to** a. The symbols <, >, ≤, and ≥ are **inequality symbols.**

FIGURE P.3 $a < b$ if and only if a lies to the left of b.

Geometrically, this definition implies that $a < b$ if and only if a lies to the *left* of b on the real number line, as shown in Figure P.3.

(a) $x \leq 2$

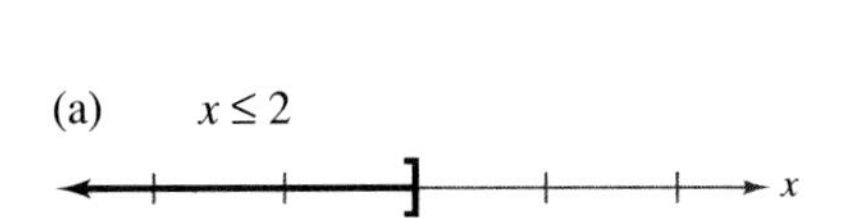

(b) $-2 \leq x < 3$

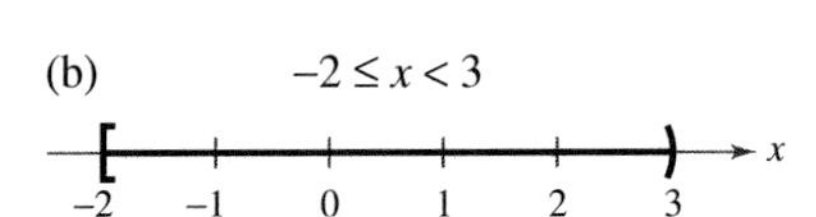

FIGURE P.4

EXAMPLE 1 *Interpreting Inequalities*

a. The inequality $x \leq 2$ denotes all real numbers less than or equal to 2, as shown in Figure P.4(a).

b. The inequality $-2 \leq x < 3$ means that $x \geq -2$ *and* $x < 3$. The "double inequality" denotes all real numbers between -2 and 3, including -2 but *not* including 3, as shown in Figure P.4(b).

Inequalities can be used to describe subsets of real numbers called **intervals.**

NOTE In the bounded intervals at the right, the real numbers a and b are the **endpoints** of each interval.

BOUNDED INTERVALS ON THE REAL NUMBER LINE

Notation	*Interval Type*	*Inequality*	*Graph*
$[a, b]$	Closed	$a \leq x \leq b$	a b
(a, b)	Open	$a < x < b$	a b
$[a, b)$	Half-open	$a \leq x < b$	a b
$(a, b]$	Half-open	$a < x \leq b$	a b

NOTE The symbols ∞, **positive infinity,** and $-\infty$, **negative infinity,** do not represent real numbers. They are simply convenient symbols used to describe the unboundedness of an interval such as $(1, \infty)$ or $(-\infty, 3]$.

Unbounded Intervals on the Real Number Line

Notation	*Interval Type*	*Inequality*	*Graph*
$[a, \infty)$	Half-open	$x \geq a$	
(a, ∞)	Open	$x > a$	
$(-\infty, b]$	Half-open	$x \leq b$	
$(-\infty, b)$	Open	$x < b$	
$(-\infty, \infty)$	Entire real line		

Example 2 *Using Inequalities to Represent Intervals*

Use inequality notation to describe each of the following.

a. c is at most 2.

b. All x in the interval $(-3, 5]$

Solution

a. The statement "c is at most 2" can be represented by $c \leq 2$.

b. "All x in the interval $(-3, 5]$" can be represented by $-3 < x \leq 5$.

Example 3 *Interpreting Intervals*

Give a verbal description of each interval.

a. $(-1, 0)$ **b.** $[2, \infty)$ **c.** $(-\infty, 0)$

Solution

a. This interval consists of all real numbers that are greater than -1 and less than 0.

b. This interval consists of all real numbers that are greater than or equal to 2.

c. This interval consists of all negative real numbers.

The **Law of Trichotomy** states that for any two real numbers a and b, *precisely* one of three relationships is possible:

$$a = b, \quad a < b, \quad \text{or} \quad a > b. \qquad \text{Law of Trichotomy}$$

Exploration

Absolute value expressions can be evaluated on a graphing utility. To evaluate $|-4|$ with a *TI-83*, use these keystrokes:

MATH (NUM) (1:abs ( ) (-) 4 ENTER.

To evaluate $|-4|$ with a *TI-82*, use these keystrokes:

ABS (-) 4 ENTER.

When evaluating an expression such as $|3 - 8|$, parentheses should surround the entire expression. Evaluate each expression below. What can you conclude?

a. $|6|$ **b.** $|-1|$

c. $|5 - 2|$ **d.** $|2 - 5|$

NOTE The absolute value of a real number is either positive or zero. Moreover, 0 is the only real number whose absolute value is 0. Thus, $|0| = 0$.

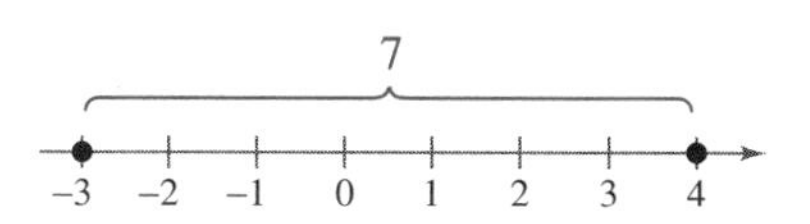

FIGURE P.5 The distance between −3 and 4 is 7.

Absolute Value and Distance

The **absolute value** of a real number is its *magnitude*.

DEFINITION OF ABSOLUTE VALUE

If a is a real number, then the **absolute value** of a is

$$|a| = \begin{cases} a, & \text{if } a \geq 0 \\ -a, & \text{if } a < 0. \end{cases}$$

Notice from this definition that the absolute value of a real number is never negative. For instance if $a = -5$, then $|-5| = -(-5) = 5$.

EXAMPLE 4 *Evaluating the Absolute Value of a Number*

Evaluate $\dfrac{|x|}{x}$ for (a) $x > 0$ and (b) $x < 0$.

Solution

a. If $x > 0$, then $|x| = x$ and $\dfrac{|x|}{x} = \dfrac{x}{x} = 1$.

b. If $x < 0$, then $|x| = -x$ and $\dfrac{|x|}{x} = \dfrac{-x}{x} = -1$.

PROPERTIES OF ABSOLUTE VALUES

1. $|a| \geq 0$ **2.** $|-a| = |a|$

3. $|ab| = |a||b|$ **4.** $\left|\dfrac{a}{b}\right| = \dfrac{|a|}{|b|}, \quad b \neq 0$

Absolute value can be used to define the distance between two numbers on the real number line. For instance, the distance between −3 and 4 is $|-3 - 4| = |-7| = 7$, as shown in Figure P.5.

DISTANCE BETWEEN TWO POINTS ON THE REAL LINE

Let a and b be real numbers. The **distance between a and b** is

$$d(a, b) = |b - a| = |a - b|.$$

Algebraic Expressions

One characteristic of algebra is the use of letters to represent numbers. The letters are **variables,** and combinations of letters and numbers are **algebraic expressions.** Here are a few examples of algebraic expressions.

$$5x, \qquad 2x - 3, \qquad \frac{4}{x^2 + 2}, \qquad 7x + y$$

DEFINITION OF AN ALGEBRAIC EXPRESSION

A collection of letters (**variables**) and real numbers **(constants)** combined using the operations of addition, subtraction, multiplication, division, and exponentiation is an **algebraic expression.**

The **terms** of an algebraic expression are those parts that are separated by *addition.* For example,

$$x^2 - 5x + 8 = x^2 + (-5x) + 8$$

has three terms: x^2 and $-5x$ are the **variable terms** and 8 is the **constant term.** The numerical factor of a variable term is the **coefficient** of the variable term. For instance, the coefficient of $-5x$ is -5, and the coefficient of x^2 is 1.

To **evaluate** an algebraic expression, substitute numerical values for each of the variables in the expression. Here are two examples.

Expression	*Value of Variable*	*Substitute*	*Value of Expression*
$-3x + 5$	$x = 3$	$-3(3) + 5$	$-9 + 5 = -4$
$3x^2 + 2x - 1$	$x = -1$	$3(-1)^2 + 2(-1) - 1$	$3 - 2 - 1 = 0$

Study Tip

When evaluating an algebraic expression, the Substitution Principle is used. It states, "If $a = b$, then a can be replaced by b in any expression involving a." In the first evaluation shown at the right, for instance, 3 is *substituted* for x in the expression $-3x + 5$.

Basic Rules of Algebra

There are four arithmetic operations with real numbers: **addition, multiplication, subtraction,** and **division,** denoted by the symbols $+$, $\times$ or $\cdot$, $-$, and $\div$. Of these, addition and multiplication are the two primary operations. Subtraction and division are the inverse operations of addition and multiplication, respectively.

Subtraction

$$a - b = a + (-b)$$

Division

If $b \neq 0$, then $a \div b = a\left(\frac{1}{b}\right) = \frac{a}{b}$.

In these definitions, $-b$ is the **additive inverse** (or opposite) of b, and $1/b$ is the **multiplicative inverse** (or reciprocal) of b. In the fractional form a/b, a is the **numerator** of the fraction and b is the **denominator.**

The French mathematician Nicolas Chuquet (ca. 1500) wrote Triparty en la science des nombres, *in which a form of exponent notation was used. Our expressions* $6x^3$ *and* $10x^2$ *were written as* $.6.^3$ *and* $.10.^2$*. Zero and negative exponents were also represented, so* x^0 *would be written as* $.1.^0$ *and* $3x^{-2}$ *as* $.3.^{2.m}$*. Chuquet wrote that* $.72.^1$ *divided by* $.8.^3$ *is* $.9.^{2.m}$*. That is,* $72x \div 8x^3 = 9x^{-2}$.

Be sure you see that the following **basic rules of algebra** are true for variables and algebraic expressions as well as for real numbers. Try to formulate a verbal description of each property. For instance, the first property states that *the order in which two real numbers are added does not affect their sum.*

BASIC RULES OF ALGEBRA

Let a, b, and c be real numbers, variables, or algebraic expressions.

Property		*Example*
Commutative Property of Addition:	$a + b = b + a$	$4x + x^2 = x^2 + 4x$
Commutative Property of Multiplication:	$ab = ba$	$(4 - x)x^2 = x^2(4 - x)$
Associative Property of Addition:	$(a + b) + c = a + (b + c)$	$(x + 5) + x^2 = x + (5 + x^2)$
Associative Property of Multiplication:	$(ab)c = a(bc)$	$(2x \cdot 3y)(8) = (2x)(3y \cdot 8)$
Distributive Properties:	$a(b + c) = ab + ac$	$3x(5 + 2x) = 3x \cdot 5 + 3x \cdot 2x$
	$(a + b)c = ac + bc$	$(y + 8)y = y \cdot y + 8 \cdot y$
Additive Identity Property:	$a + 0 = a$	$5y^2 + 0 = 5y^2$
Multiplicative Identity Property:	$a \cdot 1 = a$	$(4x^2)(1) = 4x^2$
Additive Inverse Property:	$a + (-a) = 0$	$5x^3 + (-5x^3) = 0$
Multiplicative Inverse Property:	$a \cdot \frac{1}{a} = 1, \quad a \neq 0$	$(x^2 + 4)\left(\frac{1}{x^2 + 4}\right) = 1$

NOTE Because subtraction is defined as "adding the opposite," the Distributive Properties are also true for subtraction. For instance, the "subtraction form" of $a(b + c) = ab + ac$ is

$$a(b - c) = ab - ac.$$

As well as formulating a verbal description for each of the following basic properties of negation, zero, and fractions, try to gain an *intuitive sense* for the validity of each.

PROPERTIES OF NEGATION

Let a and b be real numbers, variables, or algebraic expressions.

Property	*Example*
1. $(-1)a = -a$	$(-1)7 = -7$
2. $-(-a) = a$	$-(-6) = 6$
3. $(-a)b = -(ab) = a(-b)$	$(-5)3 = -(5 \cdot 3) = 5(-3)$
4. $(-a)(-b) = ab$	$(-2)(-x) = 2x$
5. $-(a + b) = (-a) + (-b)$	$-(x + 8) = (-x) + (-8)$
	$= -x - 8$

NOTE Be sure you see the difference between the *opposite of a number* and a *negative number.* If a is already negative, then its opposite, $-a$, is positive. For instance, if $a = -5$, then $-a = -(-5) = 5$.

NOTE The "or" in the Zero-Factor Property includes the possibility that either or both factors may be zero. This is an **inclusive or,** and it is the way the word "or" is generally used in mathematics.

PROPERTIES OF ZERO

Let a and b be real numbers, variables, or algebraic expressions.

1. $a + 0 = a$ and $a - 0 = a$
2. $a \cdot 0 = 0$
3. $\frac{0}{a} = 0, \quad a \neq 0$
4. $\frac{a}{0}$ is undefined.
5. Zero-Factor Property: If $ab = 0$, then $a = 0$ or $b = 0$.

NOTE In Property 1, the phrase "if and only if" implies two statements. One statement is: If $a/b = c/d$, then $ad = bc$. The other statement is: If $ad = bc$, where $b \neq 0$ and $d \neq 0$, then $a/b = c/d$.

PROPERTIES OF FRACTIONS

Let a, b, c, and d be real numbers, variables, or algebraic expressions such that $b \neq 0$ and $d \neq 0$.

1. Equivalent Fractions: $\frac{a}{b} = \frac{c}{d}$ if and only if $ad = bc$.
2. Rules of Signs: $-\frac{a}{b} = \frac{-a}{b} = \frac{a}{-b}$ and $\frac{-a}{-b} = \frac{a}{b}$
3. Generate Equivalent Fractions: $\frac{a}{b} = \frac{ac}{bc}, \quad c \neq 0$
4. Add or Subtract with Like Denominators: $\frac{a}{b} \pm \frac{c}{b} = \frac{a \pm c}{b}$
5. Add or Subtract with Unlike Denominators: $\frac{a}{b} \pm \frac{c}{d} = \frac{ad \pm bc}{bd}$
6. Multiply Fractions: $\frac{a}{b} \cdot \frac{c}{d} = \frac{ac}{bd}$
7. Divide Fractions: $\frac{a}{b} \div \frac{c}{d} = \frac{a}{b} \cdot \frac{d}{c} = \frac{ad}{bc}, \quad c \neq 0$

EXAMPLE 5 *Properties of Fractions*

a. $\frac{x}{5} = \frac{3 \cdot x}{3 \cdot 5} = \frac{3x}{15}$ — Generate equivalent fractions.

b. $\frac{x}{3} + \frac{2x}{5} = \frac{5 \cdot x + 3 \cdot 2x}{15}$ — Add fractions with unlike denominators.

c. $\frac{7}{x} \div \frac{3}{2} = \frac{7}{x} \cdot \frac{2}{3} = \frac{14}{3x}$ — Divide fractions.

PROPERTIES OF EQUALITY

Let a, b, and c be real numbers, variables, or algebraic expressions.

1. If $a = b$, then $a + c = b + c$. — Add c to both sides.
2. If $a = b$, then $ac = bc$. — Multiply both sides by c.
3. If $a + c = b + c$, then $a = b$. — Subtract c from both sides.
4. If $ac = bc$ and $c \neq 0$, then $a = b$. — Divide both sides by c.

If a, b, and c are integers such that $ab = c$, then a and b are **factors** or **divisors** of c. A **prime number** is an integer that has exactly two positive factors: itself and 1. For example, 2, 3, 5, 7, and 11 are prime numbers. The numbers 4, 6, 8, 9, and 10 are **composite** because they can be written as the product of two or more prime numbers. The number 1 is neither prime nor composite. The **Fundamental Theorem of Arithmetic** states that every positive integer greater than 1 can be written as the product of prime numbers in precisely one way (disregarding order). For instance, the *prime factorization* of 24 is $24 = 2 \cdot 2 \cdot 2 \cdot 3$.

When adding or subtracting fractions with unlike denominators, you have two options. You can use Property 5 of fractions as in Example 5(b), or you can rewrite the fractions with like denominators. Here is an example.

$$\frac{2}{15} - \frac{5}{9} + \frac{4}{5} = \frac{2(3)}{15(3)} - \frac{5(5)}{9(5)} + \frac{4(9)}{5(9)}$$ The LCD is 45.

$$= \frac{6 - 25 + 36}{45}$$

$$= \frac{17}{45}$$

GROUP ACTIVITY

DECIMAL APPROXIMATIONS OF IRRATIONAL NUMBERS

At the beginning of this section, it was pointed out that $\sqrt{2}$ is not a rational number. There are, however, rational numbers whose squares are very close to 2. For instance, if you square the rational number

$$\frac{140}{99}$$

you obtain 1.9998. Try finding other rational numbers whose squares are even closer to 2. Write a short paragraph explaining how you obtained the numbers.

P.1 Exercises

In Exercises 1–6, determine which numbers are (a) natural numbers, (b) integers, (c) rational numbers, and (d) irrational numbers.

1. $-9, -\frac{7}{2}, 5, \frac{2}{3}, \sqrt{2}, 0, 1$
2. $\sqrt{5}, -7, -\frac{7}{3}, 0, 3.12, \frac{5}{4}$
3. $2.01, 0.666\ldots, -13, 0.010110111\ldots$
4. $2.30300030003\ldots, 0.7575, -4.63, \sqrt{10}$
5. $-\pi, -\frac{1}{3}, \frac{6}{3}, \frac{1}{2}\sqrt{2}, -7.5$
6. $25, -17, -\frac{12}{5}, \sqrt{9}, 3.12, \frac{1}{2}\pi$

In Exercises 7–10, use a calculator to find the decimal form of the rational number. If it is a nonterminating decimal, write the repeating pattern.

7. $\frac{5}{8}$
8. $\frac{1}{3}$
9. $\frac{41}{333}$
10. $\frac{6}{11}$

In Exercises 11 and 12, approximate the numbers and place the correct symbol (< or >) between them.

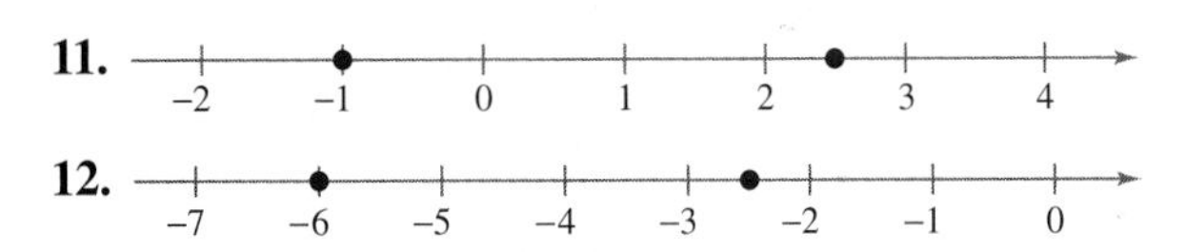

In Exercises 13–18, plot the two real numbers on the real number line. Then place the appropriate inequality sign (< or >) between them.

13. $\frac{3}{2}, 7$
14. $-3.5, 1$
15. $-4, -8$
16. $1, \frac{16}{3}$
17. $\frac{5}{6}, \frac{2}{3}$
18. $-\frac{8}{7}, -\frac{3}{7}$

In Exercises 19–28, verbally describe the subset of real numbers represented by the inequality. Then sketch the subset on the real number line. State whether the interval is bounded or unbounded.

19. $x \leq 5$
20. $x \geq -2$
21. $x < 0$
22. $x > 3$
23. $x \geq 4$
24. $x < 2$
25. $-2 < x < 2$
26. $0 \leq x \leq 5$
27. $-1 \leq x < 0$
28. $0 < x \leq 6$

In Exercises 29 and 30, use a calculator to order the numbers from smallest to largest.

29. $\frac{7071}{5000}, \frac{584}{413}, \sqrt{2}, \frac{47}{33}, \frac{127}{90}$
30. $\frac{26}{15}, \sqrt{3}, 1.7320, \frac{381}{220}, \sqrt{10} - \sqrt{2}$

In Exercises 31–36, use inequality notation to describe the set.

31. x is negative.
32. z is at least 10.
33. y is nonnegative.
34. y is no more than 25.
35. The person's age A is at least 30.
36. The annual rate of inflation r is expected to be at least 2.5%, but no more than 5%.

In Exercises 37–46, evaluate the expression.

37. $|-10|$
38. $|0|$
39. $|3 - \pi|$
40. $|4 - \pi|$
41. $\dfrac{-5}{|-5|}$
42. $-3 - |-3|$
43. $-3|-3|$
44. $|-1| - |-2|$
45. $-|16.25| + 20$
46. $2|33|$

In Exercises 47–52, place the correct symbol (<, >, or =) between the pair of real numbers.

47. $|-3| \quad \square \quad -|-3|$ **48.** $|-4| \quad \square \quad |4|$

49. $-5 \quad \square \quad -|5|$ **50.** $-|-6| \quad \square \quad |-6|$

51. $-|-2| \quad \square \quad -|2|$ **52.** $-(-2) \quad \square \quad -2$

In Exercises 53–60, find the distance between a and b.

53.

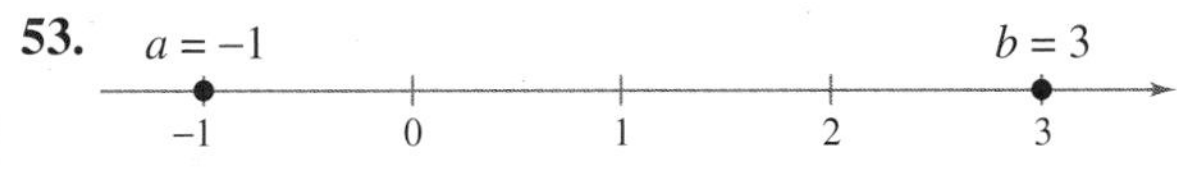

54.

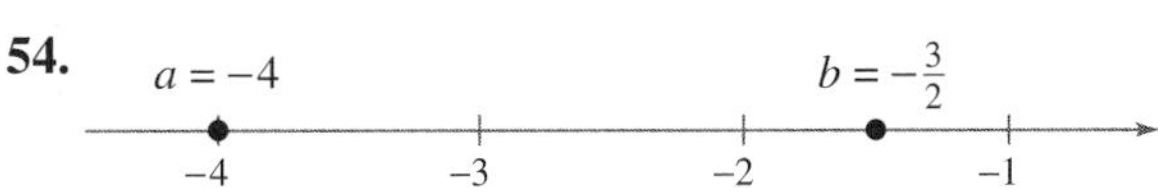

55. $a = -\frac{5}{2}$ $b = 0$

−3 −2 −1 0

56. $a = \frac{1}{4}$ $b = \frac{11}{4}$

0 1 2 3

57. $a = 126, b = 75$ **58.** $a = -126, b = -75$

59. $a = \frac{16}{5}, b = \frac{112}{75}$ **60.** $a = 9.34, b = -5.65$

In Exercises 61–66, use absolute value notation to describe the situation.

61. The distance between x and 5 is no more than 3.

62. The distance between x and -10 is at least 6.

63. While traveling, you pass milepost 7, then milepost 18. How far do you travel during that time period?

64. While traveling, you pass milepost 103, then milepost 86. How far do you travel during that time period?

65. y is at least six units from 0.

66. y is at most two units from a.

Budget Variance **In Exercises 67–70, the accounting department of a company is checking to see whether the actual expenses of a department differ from the budgeted expenses by more than \$500 or by more than 5%. Fill in the missing parts of the table, and determine whether the actual expense passes the "budget variance test."**

	Budgeted Expense, b	*Actual Expense, a*	$\lvert a - b \rvert$	$0.05b$
67. Wages	\$112,700	\$113,356		
68. Utilities	\$9400	\$9772		
69. Taxes	\$37,640	\$37,335		
70. Insurance	\$2575	\$2613		

Federal Deficit **In Exercises 71–74, use the bar graph, which shows the receipts of the federal government (in billions of dollars) for selected years from 1960 through 1993. In each exercise you are given the outlay of the federal government. Find the magnitude of the surplus or deficit for the year.** (Source: U.S. Treasury Department)

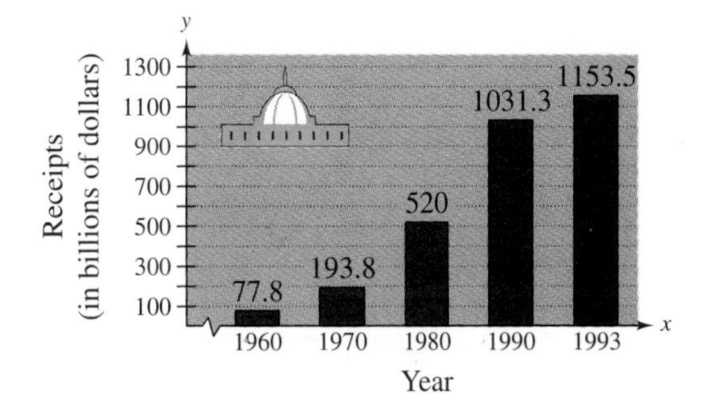

	Income, y	*Outlay, x*	$\lvert y - x \rvert$
71. 1960		\$92.2 billion	
72. 1980		\$590.9 billion	
73. 1990		\$1252.7 billion	
74. 1993		\$1408.2 billion	

75. ***Exploration*** Consider $|u + v|$ and $|u| + |v|$.

(a) Are the values of the expressions always equal? If not, under what conditions are they unequal?

(b) If the two expressions are not equal for certain values of u and v, is one of the expressions always greater than the other? Explain.

76. ***Think About It*** Is there a difference between saying that a real number is positive and saying that a real number is nonnegative? Explain.

In Exercises 77–80, identify the terms of the expression.

77. $7x + 4$

78. $3x^2 - 8x - 11$

79. $4x^3 + x - 5$

80. $3x^4 + 3x^3$

In Exercises 81–86, evaluate the expression for the values of x. (If not possible, state the reason.)

	Expression	*Values*	
81.	$4x - 6$	(a) $x = -1$	(b) $x = 0$
82.	$9 - 7x$	(a) $x = -3$	(b) $x = 3$
83.	$x^2 - 3x + 4$	(a) $x = -2$	(b) $x = 2$
84.	$-x^2 + 5x - 4$	(a) $x = -1$	(b) $x = 1$
85.	$\dfrac{x+1}{x-1}$	(a) $x = 1$	(b) $x = -1$
86.	$\dfrac{x}{x+2}$	(a) $x = 2$	(b) $x = -2$

In Exercises 87–96, identify the rule(s) of algebra illustrated by the equation.

87. $x + 9 = 9 + x$

88. $2\left(\frac{1}{2}\right) = 1$

89. $\dfrac{1}{h+6}(h + 6) = 1, \quad h \neq -6$

90. $(x + 3) - (x + 3) = 0$

91. $2(x + 3) = 2x + 6$

92. $(z - 2) + 0 = z - 2$

93. $1 \cdot (1 + x) = 1 + x$

94. $x + (y + 10) = (x + y) + 10$

95. $x(3y) = (x \cdot 3)y = (3x)y$

96. $\frac{1}{7}(7 \cdot 12) = \left(\frac{1}{7} \cdot 7\right)12 = 1 \cdot 12 = 12$

In Exercises 97–100, evaluate the expression. (If not possible, state the reason.)

97. $\dfrac{81 - (90 - 9)}{5}$

98. $10(23 - 30 + 7)$

99. $\dfrac{8 - 8}{-9 + (6 + 3)}$

100. $15 - \dfrac{3 - 3}{5}$

In Exercises 101–110, perform the operations. (Write fractional answers in reduced form.)

101. $(4 - 7)(-2)$

102. $\dfrac{27 - 35}{4}$

103. $\frac{3}{16} + \frac{5}{16}$

104. $\frac{6}{7} - \frac{4}{7}$

105. $\frac{5}{8} - \frac{5}{12} + \frac{1}{6}$

106. $\frac{10}{11} + \frac{6}{33} - \frac{13}{66}$

107. $\frac{4}{5} \cdot \frac{1}{2} \cdot \frac{3}{4}$

108. $\frac{11}{16} \div \frac{3}{4}$

109. $12 \div \frac{1}{4}$

110. $\left(\frac{3}{5} \div 3\right) - \left(6 \cdot \frac{4}{8}\right)$

In Exercises 111–114, use a calculator to evaluate the expression. (Round your answer to two decimal places.)

111. $-3 + \frac{3}{7}$

112. $3\left(-\frac{5}{12} + \frac{3}{8}\right)$

113. $\dfrac{11.46 - 5.37}{3.91}$

114. $\dfrac{\frac{1}{5}(-8 - 9)}{-\frac{1}{3}}$

115. Use a calculator to complete the table.

n	1	0.5	0.01	0.0001	0.000001
$5/n$					

116. ***Think About It*** Use the result of Exercise 115 to make a conjecture about the value of $5/n$ as n approaches 0.

117. Use a calculator to complete the table.

n	1	10	100	10,000	100,000
$5/n$					

118. ***Think About It*** Use the result of Exercise 117 to make a conjecture about the value of $5/n$ as n increases without bound.

P.2 Solving Equations

See Exercise 42 on page 23 for an example of how a linear equation can be used to model the number of married women in the civilian work force in the United States

Equations and Solutions of Equations ❐ *Linear Equations* ❐ *Quadratic Equations* ❐ *Polynomial Equations of Higher Degree* ❐ *Radical Equations* ❐ *Absolute Value Equations*

Equations and Solutions of Equations

An **equation** is a statement that two algebraic expressions are equal. For example, $3x - 5 = 7$, $x^2 - x - 6 = 0$, and $\sqrt{2x} = 4$ are equations. To **solve** an equation in x means to find all values of x for which the equation is true. Such values are **solutions.** For instance, $x = 4$ is a solution of the equation $3x - 5 = 7$, because $3(4) - 5 = 7$ is a true statement.

The solutions of an equation depend on the kinds of numbers being considered. For instance, in the set of rational numbers, $x^2 = 10$ has no solution because there is no rational number whose square is 10. However, in the set of real numbers the equation has the two solutions $\sqrt{10}$ and $-\sqrt{10}$.

An equation that is true for *every* real number in the domain of the variable is called an **identity.** For example, $x^2 - 9 = (x + 3)(x - 3)$ is an identity because it is a true statement for any real value of x, and $x/(3x^2) = 1/(3x)$, where $x \neq 0$, is an identity because it is true for any nonzero real value of x.

An equation that is true for just *some* (or even none) of the real numbers in the domain of the variable is called a **conditional equation.** For example, the equation $x^2 - 9 = 0$ is conditional because $x = 3$ and $x = -3$ are the only values in the domain that satisfy the equation. Learning to solve conditional equations is the primary focus of this section.

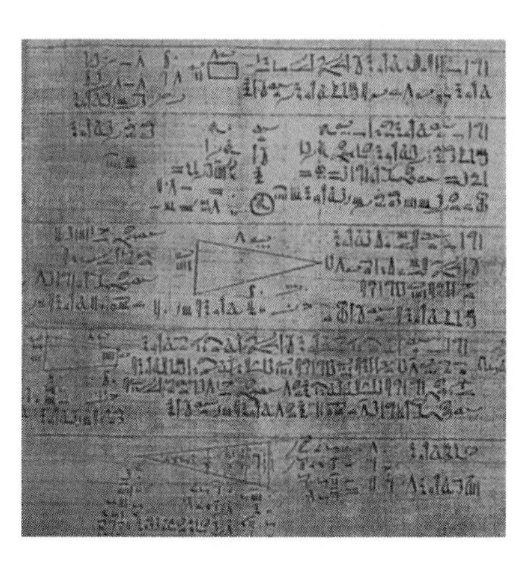

This ancient Egyptian papyrus discovered in 1858 contains one of the earliest examples of mathematical writing in existence. The papyrus itself dates back to around 1650 B.C., but it is actually a copy of writings from two centuries earlier. The algebraic equations on the papyrus were written in words. Diophantus, a Greek who lived around A.D. 250, is often called the Father of Algebra. He was the first to use abbreviated word forms in equations. *(Photo: © British Museum)*

Linear Equations

A **linear equation** in one variable x is an equation that can be written in the standard form

$$ax + b = 0$$

where a and b are real numbers, with $a \neq 0$. A linear equation has exactly one solution. To see this, consider the following steps. (Remember that $a \neq 0$.)

$ax + b = 0$	Original equation
$ax = -b$	Subtract b from both sides.
$x = -\dfrac{b}{a}$	Divide both sides by a.

To solve a conditional equation in x, isolate x on one side of the equation by a sequence of **equivalent** (and usually simpler) equations, each having the same solution(s) as the original equation. The operations that yield equivalent equations come from the properties of equality discussed in Section P.1.

Generating Equivalent Equations

An equation can be transformed into an *equivalent equation* by one or more of the following steps.

	Given Equation	*Equivalent Equation*
1. Remove symbols of grouping, combine like terms, or reduce fractions on one or both sides of the equation.	$2x - x = 4$	$x = 4$
2. Add (or subtract) the same quantity to (from) *both* sides of the equation.	$x + 1 = 6$	$x = 5$
3. Multiply (or divide) *both* sides of the equation by the same *nonzero* quantity.	$2x = 6$	$x = 3$
4. Interchange the two sides of the equation.	$2 = x$	$x = 2$

Example 1 **Solving a Linear Equation**

Solve $3x - 6 = 0$.

Solution

$3x - 6 = 0$ Original equation

$3x = 6$ Add 6 to both sides.

$x = 2$ Divide both sides by 3.

Check: After solving an equation, you should **check each solution** in the *original* equation.

$3x - 6 = 0$ Original equation

$3(2) - 6 \stackrel{?}{=} 0$ Substitute 2 for x.

$0 = 0$ Solution checks. ✓

To solve an equation involving fractional expressions, find the least common denominator of all terms and multiply every term by this LCD.

Study Tip

Students sometimes tell us that a solution looks easy when we work it out in class, but that they don't see where to begin when trying it alone. Keep in mind that no one—not even great mathematicians—can expect to look at every mathematical problem and immediately know where to begin. Many problems involve some trial and error before a solution is found. To make algebra work for you, you must put in a lot of time, you must expect to try solution methods that end up not working, and you must learn from both your successes and your failures.

EXAMPLE 2 *An Equation Involving Fractional Expressions*

Solve $\frac{x}{3} + \frac{3x}{4} = 2$.

Solution

$$\frac{x}{3} + \frac{3x}{4} = 2 \qquad \text{Original equation}$$

$$(12)\frac{x}{3} + (12)\frac{3x}{4} = (12)2 \qquad \text{Multiply by the LCD of 12.}$$

$$4x + 9x = 24 \qquad \text{Reduce and multiply.}$$

$$13x = 24 \qquad \text{Combine like terms.}$$

$$x = \frac{24}{13} \qquad \text{Divide both sides by 13.}$$

The solution is $\frac{24}{13}$. Check this in the original equation.

When multiplying or dividing an equation by a *variable* quantity, it is possible to introduce an **extraneous** solution.

NOTE An extraneous solution is one that does not satisfy the original equation.

EXAMPLE 3 *An Equation with an Extraneous Solution*

Solve $\frac{1}{x-2} = \frac{3}{x+2} - \frac{6x}{x^2-4}$.

Solution

The LCD is $x^2 - 4$ or $(x + 2)(x - 2)$. Multiply every term by this LCD.

$$\frac{1}{x-2}(x+2)(x-2) = \frac{3}{x+2}(x+2)(x-2) - \frac{6x}{x^2-4}(x+2)(x-2)$$

$$x + 2 = 3(x - 2) - 6x, \qquad x \neq \pm 2$$

$$x + 2 = 3x - 6 - 6x$$

$$4x = -8$$

$$x = -2$$

In the original equation, $x = -2$ yields a denominator of zero. Therefore, $x = -2$ is an extraneous solution, and the original equation has *no solution.*

Quadratic Equations

A **quadratic equation** in x is an equation that can be written in the standard form

$$ax^2 + bx + c = 0$$

where a, b, and c are real numbers, with $a \neq 0$. A quadratic equation in x is also known as a **second-degree polynomial equation in *x*.**

You should be familiar with the following four methods for solving quadratic equations.

Solving a Quadratic Equation

Method	*Example*
Factoring: If $ab = 0$, then $a = 0$ or $b = 0$.	$x^2 - x - 6 = 0$ $(x - 3)(x + 2) = 0$ $x - 3 = 0 \Rightarrow x = 3$ $x + 2 = 0 \Rightarrow x = -2$
Square Root Principle: If $u^2 = c$, where $c > 0$, then $u = \pm\sqrt{c}$.	$(x + 3)^2 = 16$ $x + 3 = \pm 4$ $x = -3 \pm 4$ $x = 1$ or $x = -7$
Completing the Square: If $x^2 + bx = c$, then $x^2 + bx + \left(\frac{b}{2}\right)^2 = c + \left(\frac{b}{2}\right)^2$ $\left(x + \frac{b}{2}\right)^2 = c + \frac{b^2}{4}$.	$x^2 + 6x = 5$ $x^2 + 6x + 3^2 = 5 + 3^2$ $(x + 3)^2 = 14$ $x + 3 = \pm\sqrt{14}$ $x = -3 \pm \sqrt{14}$
Quadratic Formula: If $ax^2 + bx + c = 0$, then $x = \frac{-b \pm \sqrt{b^2 - 4ac}}{2a}$.	$2x^2 + 3x - 1 = 0$ $x = \frac{-3 \pm \sqrt{3^2 - 4(2)(-1)}}{2(2)}$ $= \frac{-3 \pm \sqrt{17}}{4}$

NOTE The Quadratic Formula can be derived by completing the square with the general form

$$ax^2 + bx + c = 0.$$

EXAMPLE 4 *Solving Quadratic Equations by Factoring*

a. $2x^2 + 9x + 7 = 3$ — Original equation

$2x^2 + 9x + 4 = 0$ — Standard form

$(2x + 1)(x + 4) = 0$ — Factored form

$2x + 1 = 0 \Rightarrow x = -\frac{1}{2}$ — Set 1st factor equal to 0.

$x + 4 = 0 \Rightarrow x = -4$ — Set 2nd factor equal to 0.

The solutions are $-\frac{1}{2}$ and -4. Check these in the original equation.

b. $6x^2 - 3x = 0$ — Original equation

$3x(2x - 1) = 0$ — Factored form

$3x = 0 \Rightarrow x = 0$ — Set 1st factor equal to 0.

$2x - 1 = 0 \Rightarrow x = \frac{1}{2}$ — Set 2nd factor equal to 0.

The solutions are 0 and $\frac{1}{2}$. Check these in the original equation.

Be sure you see that the Zero-Factor Property works *only* for equations written in standard form (in which the right side of the equation is zero). Therefore, all terms must be collected on one side *before* factoring. For instance, in the equation $(x - 5)(x + 2) = 8$ it is *incorrect* to set each factor equal to 8. Can you solve this equation correctly?

EXAMPLE 5 *Extracting Square Roots*

a. $4x^2 = 12$ — Original equation

$x^2 = 3$ — Divide both sides by 4.

$x = \pm\sqrt{3}$ — Extract square roots.

The solutions are $\sqrt{3}$ and $-\sqrt{3}$. Check these in the original equation.

b. $(x - 3)^2 = 7$ — Original equation

$x - 3 = \pm\sqrt{7}$ — Extract square roots.

$x = 3 \pm \sqrt{7}$ — Add 3 to both sides.

The solutions are $3 \pm \sqrt{7}$. Check these in the original equation.

EXAMPLE 6 *The Quadratic Formula: Two Distinct Solutions*

Use the Quadratic Formula to solve $x^2 + 3x = 9$.

Solution

$$x^2 + 3x = 9 \qquad \text{Original equation}$$

$$x^2 + 3x - 9 = 0 \qquad \text{Standard form with } a = 1, b = 3, c = -9$$

$$x = \frac{-b \pm \sqrt{b^2 - 4ac}}{2a} \qquad \text{Quadratic Formula}$$

$$x = \frac{-3 \pm \sqrt{(3)^2 - 4(1)(-9)}}{2(1)} \qquad \text{Substitute.}$$

$$x = \frac{-3 \pm \sqrt{45}}{2} \qquad \text{Simplify.}$$

$$x = \frac{-3 \pm 3\sqrt{5}}{2} \qquad \text{Simplify.}$$

The equation has two solutions:

$$x = \frac{-3 + 3\sqrt{5}}{2} \quad \text{and} \quad x = \frac{-3 - 3\sqrt{5}}{2}.$$

Check these in the original equation.

EXAMPLE 7 *The Quadratic Formula: One Repeated Solution*

Use the Quadratic Formula to solve $8x^2 - 24x + 18 = 0$.

Solution

$$8x^2 - 24x + 18 = 0 \qquad \text{Original equation}$$

$$4x^2 - 12x + 9 = 0 \qquad \text{Standard form}$$

$$x = \frac{-b \pm \sqrt{b^2 - 4ac}}{2a} \qquad \text{Quadratic Formula}$$

$$x = \frac{12 \pm \sqrt{144 - 4(4)(9)}}{2(4)} \qquad \text{Substitute.}$$

$$x = \frac{12 \pm \sqrt{0}}{8} \qquad \text{Simplify.}$$

$$x = \frac{3}{2} \qquad \text{Repeated solution}$$

The solution is $\frac{3}{2}$. Check this in the original equation.

Polynomial Equations of Higher Degree

The methods used to solve quadratic equations can sometimes be extended to polynomials of higher degree.

Study Tip

A common mistake that is made in solving an equation such as that in Example 8 is dividing both sides of the equation by the variable factor x^2. This loses the solution $x = 0$. When using factoring to solve an equation, be sure to set each factor equal to zero. Don't divide both sides of an equation by a variable factor in an attempt to simplify the equation.

EXAMPLE 8 *Solving a Polynomial Equation by Factoring*

Solve $3x^4 = 48x^2$.

Solution

First write the polynomial equation in standard form with zero on one side, factor the other side, and then set each factor equal to zero.

$$3x^4 = 48x^2 \quad \text{Original equation}$$
$$3x^4 - 48x^2 = 0 \quad \text{Standard form}$$
$$3x^2(x^2 - 16) = 0 \quad \text{Factor.}$$
$$3x^2(x + 4)(x - 4) = 0 \quad \text{Factored form}$$
$$3x^2 = 0 \Rightarrow x = 0 \quad \text{Set 1st factor equal to 0.}$$
$$x + 4 = 0 \Rightarrow x = -4 \quad \text{Set 2nd factor equal to 0.}$$
$$x - 4 = 0 \Rightarrow x = 4 \quad \text{Set 3rd factor equal to 0.}$$

Check:

$$3x^4 = 48x^2 \quad \text{Original equation}$$
$$3(0)^4 = 48(0)^2 \quad \text{0 checks. ✓}$$
$$3(-4)^4 = 48(-4)^2 \quad \text{−4 checks. ✓}$$
$$3(4)^4 = 48(4)^2 \quad \text{4 checks. ✓}$$

After checking, you can conclude that the solutions are 0, −4, and 4.

EXAMPLE 9 *Solving a Polynomial Equation by Factoring*

Solve $x^3 - 3x^2 - 3x + 9 = 0$.

Solution

$$x^3 - 3x^2 - 3x + 9 = 0 \quad \text{Original equation}$$
$$x^2(x - 3) - 3(x - 3) = 0 \quad \text{Factor by grouping.}$$
$$(x - 3)(x^2 - 3) = 0 \quad \text{Distributive Property}$$
$$x - 3 = 0 \Rightarrow x = 3 \quad \text{Set 1st factor equal to 0.}$$
$$x^2 - 3 = 0 \Rightarrow x = \pm\sqrt{3} \quad \text{Set 2nd factor equal to 0.}$$

The solutions are 3, $\sqrt{3}$, and $-\sqrt{3}$. Check these in the original equation.

Radical Equations

The steps involved in solving the remaining equations in this section will often introduce *extraneous solutions.* Operations such as squaring both sides of an equation, raising both sides of an equation to a rational power, or multiplying both sides by a variable quantity all have this potential danger. Thus, when you use any of these operations, checking is crucial.

EXAMPLE 10 *Solving an Equation Involving a Rational Exponent*

Solve $4x^{3/2} - 8 = 0$.

Solution

$4x^{3/2} - 8 = 0$	Original equation
$4x^{3/2} = 8$	Add 8 to both sides.
$x^{3/2} = 2$	Isolate $x^{3/2}$.
$x = 2^{2/3}$	Raise both sides to $\frac{2}{3}$ power.
$x \approx 1.587$	Round to three decimal places.

The solution appears to be $2^{2/3}$. You can check this as follows.

Check:

$4x^{3/2} - 8 = 0$	Original equation
$4(2^{2/3})^{3/2} \stackrel{?}{=} 8$	Substitute $2^{2/3}$ for x.
$4(2) \stackrel{?}{=} 8$	Property of exponents
$8 = 8$	Solution checks. ✓

NOTE The essential technique used in Example 10 is to isolate the factor with the rational exponent, and raise both sides to the *reciprocal power*. In Example 11, this is equivalent to isolating the square root and squaring both sides.

EXAMPLE 11 *Solving an Equation Involving a Radical*

$\sqrt{2x + 7} - x = 2$	Original equation
$\sqrt{2x + 7} = x + 2$	Isolate the square root.
$2x + 7 = x^2 + 4x + 4$	Square both sides.
$0 = x^2 + 2x - 3$	Standard form
$0 = (x + 3)(x - 1)$	Factored form
$x + 3 = 0 \Rightarrow x = -3$	Set 1st factor equal to 0.
$x - 1 = 0 \Rightarrow x = 1$	Set 2nd factor equal to 0.

By checking these values, you can determine that the only solution is 1.

Absolute Value Equations

To solve an equation involving an absolute value, remember that the expression inside the absolute value signs can be positive or negative. This results in *two* separate equations, each of which must be solved.

EXAMPLE 12 *Solving an Equation Involving Absolute Value*

Solve $|x^2 - 3x| = -4x + 6$.

Solution

First Equation

$x^2 - 3x = -4x + 6$		Use positive expression.
$x^2 + x - 6 = 0$		Standard form
$(x + 3)(x - 2) = 0$		Factored form
$x + 3 = 0$	➡ $x = -3$	Set 1st factor equal to 0.
$x - 2 = 0$	➡ $x = 2$	Set 2nd factor equal to 0.

Second Equation

$-(x^2 - 3x) = -4x + 6$		Use negative expression.
$x^2 - 7x + 6 = 0$		Standard form
$(x - 1)(x - 6) = 0$		Factored form
$x - 1 = 0$	➡ $x = 1$	Set 1st factor equal to 0.
$x - 6 = 0$	➡ $x = 6$	Set 2nd factor equal to 0.

Of the possible solutions -3, 2, 1, and 6, a check will show that only -3 and 1 are actual solutions.

GROUP ACTIVITY

SOLVING EQUATIONS

Choose one of the equations below and write a step-by-step explanation of how to solve the equation, without using another equation in the explanation. Exchange explanations with another student— see if he or she can correctly solve the equation just by following your instructions.

a. $x - 2 + \dfrac{3x - 1}{8} = \dfrac{x + 4}{4}$ **b.** $t - \{7 - [t - (7 + t)]\} = 27$

The *Interactive* CD-ROM provides additional help with Warm-Up exercises by providing a hypertext link to the section in which the concept was introduced.

The *Interactive* CD-ROM contains step-by-step solutions to all odd-numbered Section and Review Exercises. It also provides Tutorial Exercises, which link to Guided Examples for additional help.

WARM UP

Perform the operations and simplify your answer.

1. $(2x - 4) - (5x + 6)$

2. $(3x - 5) + (2x - 7)$

3. $2(x + 1) - (x + 2)$

4. $-3(2x - 4) + 7(x + 2)$

5. $\frac{x}{3} + \frac{x}{5}$

6. $x - \frac{x}{4}$

7. $\frac{1}{x+1} - \frac{1}{x}$

8. $\frac{2}{x} + \frac{3}{x}$

9. $\frac{4}{x} + \frac{3}{x-2}$

10. $\frac{1}{x+1} - \frac{1}{x-1}$

P.2 Exercises

In Exercises 1–6, determine whether the values of x are solutions of the equation.

Equation	*Values*	
1. $5x - 3 = 3x + 5$	(a) $x = 0$	(b) $x = -5$
	(c) $x = 4$	(d) $x = 10$
2. $7 - 3x = 5x - 17$	(a) $x = -3$	(b) $x = 0$
	(c) $x = 8$	(d) $x = 3$
3. $3x^2 + 2x - 5 =$	(a) $x = -3$	(b) $x = 1$
$2x^2 - 2$	(c) $x = 4$	(d) $x = -5$
4. $5x^3 + 2x - 3 =$	(a) $x = 2$	(b) $x = -2$
$4x^3 + 2x - 11$	(c) $x = 0$	(d) $x = 10$
5. $\frac{5}{2x} - \frac{4}{x} = 3$	(a) $x = -\frac{1}{2}$	(b) $x = 4$
	(c) $x = 0$	(d) $x = \frac{1}{4}$
6. $\sqrt[3]{x - 8} = 3$	(a) $x = 2$	(b) $x = -5$
	(c) $x = 35$	(d) $x = 8$

In Exercises 7–12, determine whether the equation is an identity or a conditional equation.

7. $2(x - 1) = 2x - 2$

8. $3(x + 2) = 5x + 4$

9. $-6(x - 3) + 5 = -2x + 10$

10. $3(x + 2) - 5 = 3x + 1$

11. $x^2 - 8x + 5 = (x - 4)^2 - 11$

12. $3 + \frac{1}{x+1} = \frac{4x}{x+1}$

13. ***Think About It***

(a) What is meant by equivalent equations? Give an example of two equivalent equations.

(b) In your own words, describe the steps used to transform an equation into an equivalent equation.

14. Justify each step of the solution.

$$3(x - 4) + 10 = 7$$
$$3x - 12 + 10 = 7$$
$$3x - 2 = 7$$
$$3x - 2 + 2 = 7 + 2$$
$$3x = 9$$
$$\frac{3x}{3} = \frac{9}{3}$$
$$x = 3$$

In Exercises 15–40, solve the equation (if possible) and check your solution.

15. $2(x + 5) - 7 = 3(x - 2)$

16. $2(13t - 15) + 3(t - 19) = 0$

17. $\frac{5x}{4} + \frac{1}{2} = x - \frac{1}{2}$

18. $\frac{x}{5} - \frac{x}{2} = 3$

19. $0.25x + 0.75(10 - x) = 3$

20. $0.60x + 0.40(100 - x) = 50$

21. $x + 8 = 2(x - 2) - x$

22. $3(x + 3) = 5(1 - x) - 1$

23. $\frac{100 - 4u}{3} = \frac{5u + 6}{4} + 6$

24. $\frac{17 + y}{y} + \frac{32 + y}{y} = 100$

25. $\frac{5x - 4}{5x + 4} = \frac{2}{3}$

26. $\frac{10x + 3}{5x + 6} = \frac{1}{2}$

27. $10 - \frac{13}{x} = 4 + \frac{5}{x}$

28. $\frac{15}{x} - 4 = \frac{6}{x} + 3$

29. $\frac{1}{x - 3} + \frac{1}{x + 3} = \frac{10}{x^2 - 9}$

30. $\frac{1}{x - 2} + \frac{3}{x + 3} = \frac{4}{x^2 + x - 6}$

31. $\frac{x}{x + 4} + \frac{4}{x + 4} + 2 = 0$

32. $\frac{2}{(x - 4)(x - 2)} = \frac{1}{x - 4} + \frac{2}{x - 2}$

33. $\frac{7}{2x + 1} - \frac{8x}{2x - 1} = -4$

34. $\frac{4}{u - 1} + \frac{6}{3u + 1} = \frac{15}{3u + 1}$

35. $(x + 2)^2 + 5 = (x + 3)^2$

36. $(x + 1)^2 + 2(x - 2) = (x + 1)(x - 2)$

37. $(x + 2)^2 - x^2 = 4(x + 1)$

38. $(2x + 1)^2 = 4(x^2 + x + 1)$

39. $4 - 2(x - 2b) = ax + 3$

40. $5 + ax = 12 - bx$

41. *Exploration*

(a) Complete the table.

x	-1	0	1	2	3	4
$3.2x - 5.8$						

(b) Use the table in part (a) to determine the interval in which the solution to the equation $3.2x - 5.8 = 0$ is located. Explain your reasoning.

(c) Complete the table.

x	1.5	1.6	1.7	1.8	1.9	2
$3.2x - 5.8$						

(d) Use the table in part (c) to determine the interval in which the solution to the equation $3.2x - 5.8 = 0$ is located. Explain how this process can be used to approximate the solution to any desired degree of accuracy.

42. *Using a Model* The number of married women y in the civilian work force (in millions) in the United States from 1988 to 1992 can be approximated by the model

$$y = 0.43t + 30.86$$

where $t = 0$ represents 1990 (see figure). According to this model, during which year did this number reach 30 million? Explain how to answer the question graphically and algebraically. (Source: U.S. Bureau of Labor Statistics)

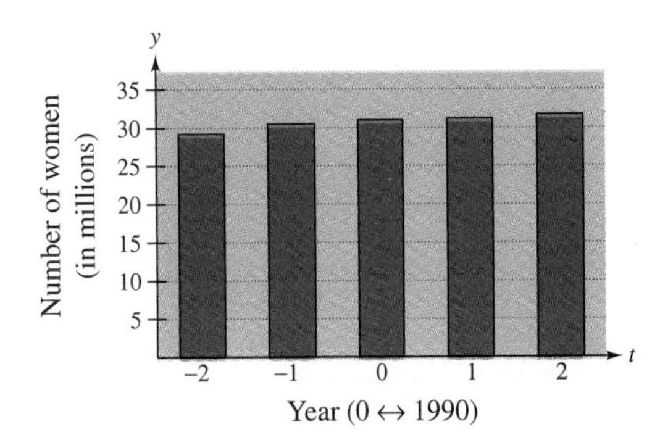

In Exercises 43–56, solve the equation by factoring.

43. $6x^2 + 3x = 0$

44. $9x^2 - 1 = 0$

45. $x^2 - 2x - 8 = 0$

46. $x^2 + 10x + 25 = 0$

47. $3 + 5x - 2x^2 = 0$

48. $16x^2 + 56x + 49 = 0$

49. $2x^2 = 19x + 33$

50. $(x + a)^2 - b^2 = 0$

51. $2x^4 - 18x^2 = 0$

52. $20x^3 - 125x = 0$

53. $x^3 - 2x^2 - 3x = 0$

54. $x^3 - 3x^2 - x + 3 = 0$

55. $2x^4 - 15x^3 + 18x^2 = 0$

56. $x^3 + 2x^2 - 3x - 6 = 0$

In Exercises 57–64, solve the equation by extracting square roots. List both the exact solution *and* the decimal solution rounded to two decimal places.

57. $x^2 = 16$

58. $x^2 = 144$

59. $3x^2 = 36$

60. $9x^2 = 25$

61. $(x - 12)^2 = 18$

62. $(x + 13)^2 = 21$

63. $(x + 2)^2 = 12$

64. $(x - 5)^2 = 20$

In Exercises 65–70, solve the quadratic equation by completing the square.

65. $x^2 - 2x = 0$

66. $x^2 + 4x = 0$

67. $x^2 + 6x + 2 = 0$

68. $x^2 + 8x + 14 = 0$

69. $8 + 4x - x^2 = 0$

70. $4x^2 - 4x - 99 = 0$

In Exercises 71–82, use the Quadratic Formula to solve the equation.

71. $2x^2 + x - 1 = 0$

72. $2x^2 - x - 1 = 0$

73. $x^2 + 8x - 4 = 0$

74. $4x^2 - 4x - 4 = 0$

75. $12x - 9x^2 = -3$

76. $16x^2 + 22 = 40x$

77. $3x + x^2 - 1 = 0$

78. $36x^2 + 24x - 7 = 0$

79. $28x - 49x^2 = 4$

80. $9x^2 + 24x + 16 = 0$

81. $8t = 5 + 2t^2$

82. $25h^2 + 80h + 61 = 0$

83. ***True or False?*** If $(2x - 3)(x + 5) = 8$, then $2x - 3 = 8$ or $x + 5 = 8$. Explain.

84. ***Exploration*** Solve the equation

$$3(x + 4)^2 + (x + 4) - 2 = 0$$

in two ways.

(a) Let $u = x + 4$, and solve the resulting equation for u. Then solve the u-solution for x.

(b) Expand and collect like terms in the equation, and solve the resulting equation for x.

(c) Which method is easier? Explain.

85. ***Exploration*** Solve the equations, given that a and b are not zero.

(a) $ax^2 + bx = 0$

(b) $ax^2 - ax = 0$

86. ***Dimensions of a Building*** The floor of a one-story building is 14 feet longer than it is wide. The building has 1632 square feet of floor space.

(a) Draw a rectangle that gives a visual representation of the floor space. Represent the width as w and show the length in terms of w.

(b) Write a quadratic equation in terms of w.

(c) Find the length and width of the building floor.

In Exercises 87–94, solve the equation of quadratic type. Check your solutions in the original equation.

87. $x^4 - 4x^2 + 3 = 0$

88. $4x^4 - 65x^2 + 16 = 0$

89. $\frac{1}{t^2} + \frac{8}{t} + 15 = 0$

90. $6\left(\frac{s}{s + 1}\right)^2 + 5\left(\frac{s}{s + 1}\right) - 6 = 0$

91. $2x + 9\sqrt{x} = 5$

92. $6x - 7\sqrt{x} - 3 = 0$

93. $3x^{1/3} + 2x^{2/3} = 5$

94. $9t^{2/3} + 24t^{1/3} + 16 = 0$

In Exercises 95–108, find all solutions of the equation. Check your solutions in the original equation.

95. $\sqrt{x - 10} - 4 = 0$ **96.** $\sqrt{5 - x} - 3 = 0$

97. $\sqrt[3]{2x + 5} + 3 = 0$ **98.** $\sqrt[3]{3x + 1} - 5 = 0$

99. $x = \sqrt{11x - 30}$ **100.** $2x - \sqrt{15 - 4x} = 0$

101. $\sqrt{x + 1} - 3x = 1$ **102.** $\sqrt{x + 5} = \sqrt{x - 5}$

103. $\sqrt{x} - \sqrt{x - 5} = 1$ **104.** $\sqrt{x} + \sqrt{x - 20} = 10$

105. $2\sqrt{x + 1} - \sqrt{2x + 3} = 1$

106. $3\sqrt{x} - \dfrac{4}{\sqrt{x}} = 4$

107. $(x - 5)^{2/3} = 16$

108. $(x + 3)^{3/4} = 27$

109. ***Market Research*** The demand equation for a certain product is modeled by $p = 40 - \sqrt{0.01x + 1}$, where x is the number of units demanded per day and p is the price per unit. Approximate the demand if the price is \$37.55.

110. ***Market Research*** The demand equation for a certain product is modeled by $p = 40 - \sqrt{0.0001x + 1}$, where x is the number of units demanded per day and p is the price per unit. Approximate the demand if the price is \$34.70.

In Exercises 111 and 112, solve for the indicated variable.

111. ***Surface Area of a Cone***

Solve for h: $S = \pi r\sqrt{r^2 + h^2}$

112. ***Inductance***

Solve for Q: $i = \pm\sqrt{\dfrac{1}{LC}}\sqrt{Q^2 - q}$

In Exercises 113 and 114, consider an equation of the form $x + \sqrt{x - a} = b$, where a and b are constants.

113. ***Exploration*** Find a and b if the solution to the equation is $x = 20$. (There are many correct answers.)

114. ***Essay*** Write a short paragraph listing the steps required in solving an equation involving radicals.

In Exercises 115–120, find all solutions of the equation. Check your solutions in the original equation.

115. $|x + 1| = 2$ **116.** $|x - 2| = 3$

117. $|2x - 1| = 5$ **118.** $|3x + 2| = 7$

119. $|x^2 + 6x| = 3x + 18$ **120.** $|x - 10| = x^2 - 10x$

Think About It **In Exercises 121 and 122, find an equation having the given solutions. (There are many correct answers.)**

121. $-3, 5$ **122.** $0, 2, \frac{5}{2}$

123. ***Dimensions of a Corral*** A rancher has 100 meters of fencing to enclose two adjacent rectangular corrals (see figure). Find the dimensions such that the enclosed area will be 350 square meters.

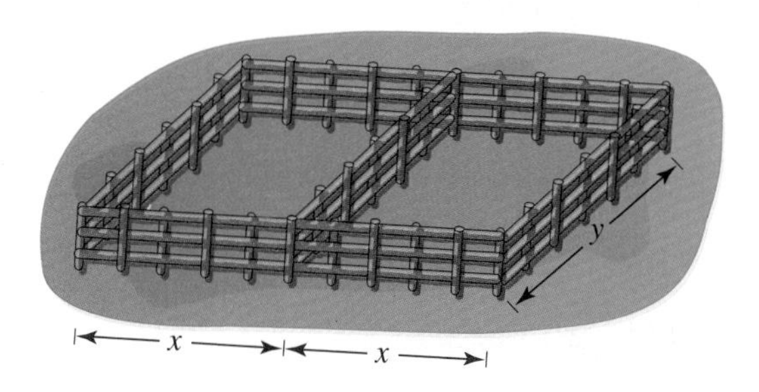

$4x + 3y = 100$

124. ***Dimensions of a Box*** An open box is to be made from a square piece of material by cutting 2-centimeter squares from the corners and turning up the sides (see figure). The volume of the finished box is to be 200 cubic centimeters. Find the size of the original piece of material.

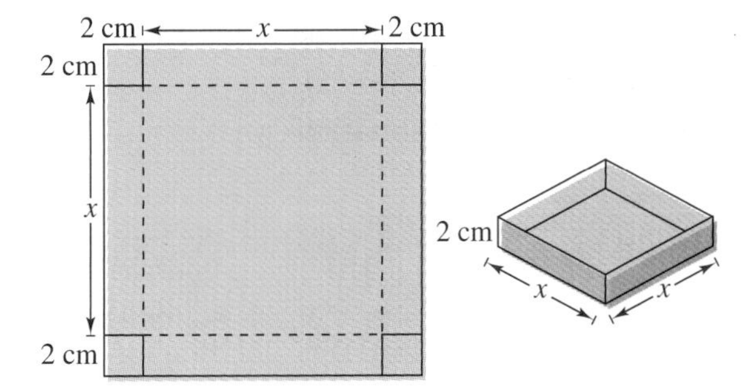

P.3 Graphs of Equations

See Example 2 on page 27 for an example of how to represent real-life data graphically.

The Cartesian Plane ▫ *The Distance Formula* ▫ *The Midpoint Formula* ▫ *The Graph of an Equation* ▫ *Intercepts of a Graph* ▫ *Symmetry* ▫ *Circles*

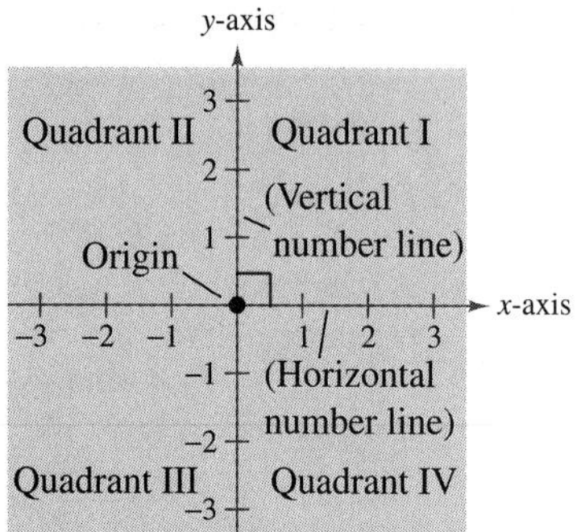

FIGURE P.6

The Cartesian Plane

Just as you can represent real numbers by points on a real number line, you can represent ordered pairs of real numbers by points in a plane called the **rectangular coordinate system,** or the **Cartesian plane,** after the French mathematician René Descartes (1596–1650).

The Cartesian plane is formed by using two real lines intersecting at right angles, as shown in Figure P.6. The horizontal real line is usually called the ***x*-axis,** and the vertical real line is usually called the ***y*-axis.** The point of intersection of these two axes is the **origin,** and the two axes divide the plane into four parts called **quadrants.**

Each point in the plane corresponds to an **ordered pair** (x, y) of real numbers x and y, called **coordinates** of the point. The ***x*-coordinate** represents the directed distance from the y-axis to the point, and the ***y*-coordinate** represents the directed distance from the x-axis to the point, as shown in Figure P.7.

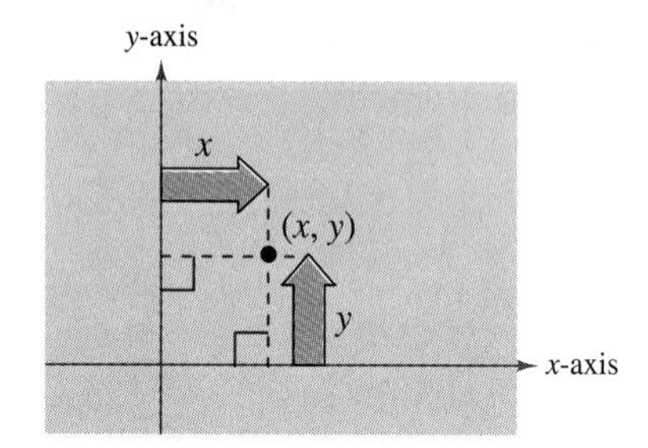

FIGURE P.7

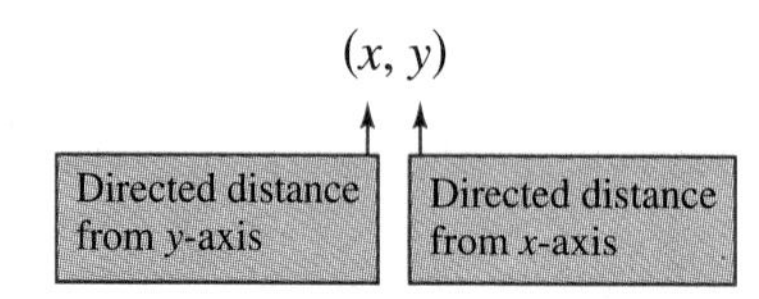

NOTE The notation (x, y) denotes both a point in the plane and an open interval on the real line. The context tells you which meaning is intended.

EXAMPLE 1 *Plotting Points in the Cartesian Plane*

Plot the points $(-1, 2)$, $(3, 4)$, $(0, 0)$, $(3, 0)$, and $(-2, -3)$.

Solution

To plot the point

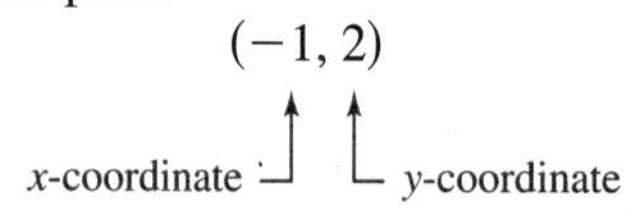

imagine a vertical line through -1 on the x-axis and a horizontal line through 2 on the y-axis. The intersection of these two lines is the point $(-1, 2)$. The other four points can be plotted in a similar way, and are shown in Figure P.8.

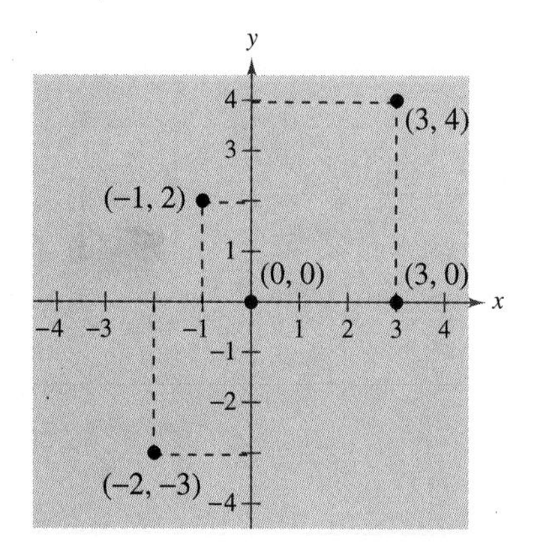

FIGURE P.8

NOTE In Example 2, you could have let $t = 1$ represent the year 1984. In that case, the horizontal axis would not have been broken, and the tick marks would have been labeled 1 through 10 (instead of 1984 through 1993).

The beauty of a rectangular coordinate system is that it allows you to see relationships between two variables. It would be difficult to overestimate the importance of Descartes's introduction of coordinates to the plane. Today, his ideas are in common use in virtually every scientific and business-related field.

EXAMPLE 2 *Sketching a Scatter Plot*

The amounts A (in millions of dollars) spent on fishing tackle in the United States in the years 1984 to 1993 are given in the table, where t represents the year. Sketch a scatter plot of the data. (Source: National Sporting Goods Association)

t	1984	1985	1986	1987	1988	1989	1990	1991	1992	1993
A	616	681	773	830	766	769	776	711	678	685

Solution

To sketch a *scatter plot* of the data given in the table, you simply represent each pair of values by an ordered pair (t, A) and plot the resulting points, as shown in Figure P.9. For instance, the first pair of values is represented by the ordered pair (1984, 616). Note the break in the t-axis, which indicates that the numbers between 0 and 1984 have been omitted.

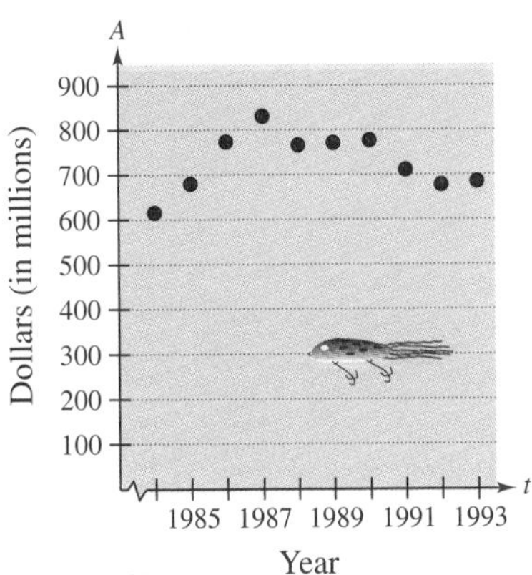

FIGURE P.9

TECHNOLOGY

The scatter plot in Example 2 is only one way to represent the data graphically. Two other techniques are shown at the right. The first is a *bar graph* and the second is a *line graph*. All three graphical representations were created with a computer. If you have access to a graphing utility, try using it to represent graphically the data given in Example 2.

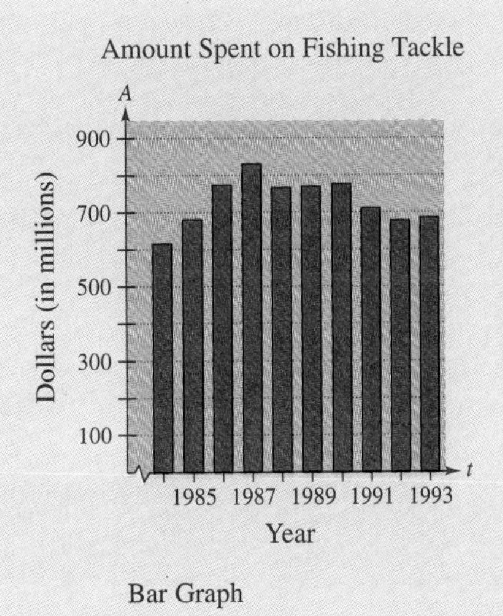

Bar Graph

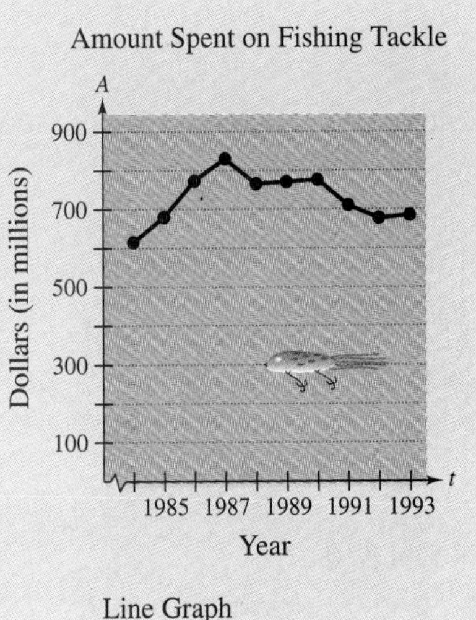

Line Graph

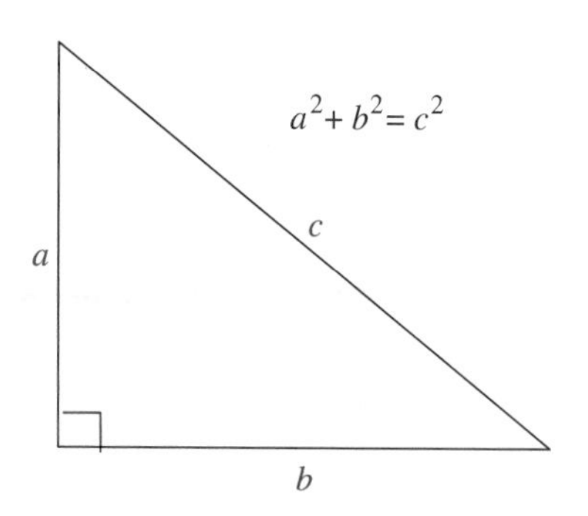

FIGURE P.10

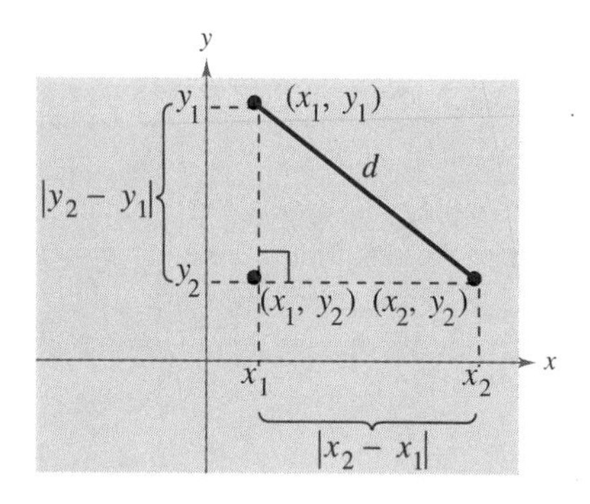

FIGURE P.11

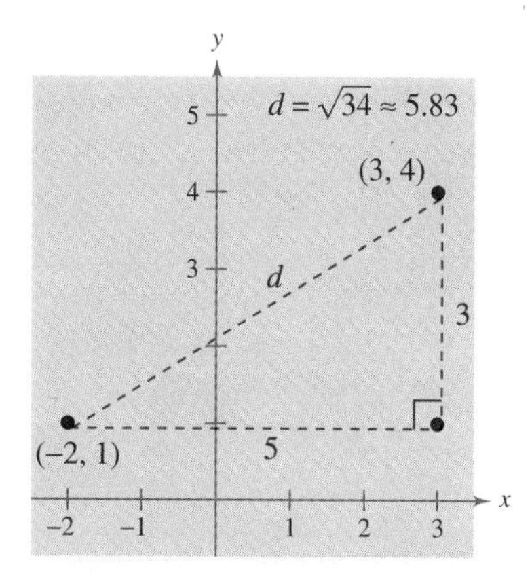

FIGURE P.12

The Distance Formula

Recall from the Pythagorean Theorem that, for a right triangle with hypotenuse of length c and sides of lengths a and b, you have

$$a^2 + b^2 = c^2 \qquad \text{Pythagorean Theorem}$$

as shown in Figure P.10. (The converse is also true. That is, if $a^2 + b^2 = c^2$, the triangle is a right triangle.)

Suppose you want to determine the distance d between two points (x_1, y_1) and (x_2, y_2) in the plane. With these two points, a right triangle can be formed, as shown in Figure P.11. The length of the vertical side of the triangle is $|y_2 - y_1|$, and the length of the horizontal side is $|x_2 - x_1|$. By the Pythagorean Theorem, you can write

$$d^2 = |x_2 - x_1|^2 + |y_2 - y_1|^2$$

$$d = \sqrt{|x_2 - x_1|^2 + |y_2 - y_1|^2}$$

$$d = \sqrt{(x_2 - x_1)^2 + (y_2 - y_1)^2}.$$

This result is the **Distance Formula.**

THE DISTANCE FORMULA

The distance d between the points (x_1, y_1) and (x_2, y_2) in the plane is

$$d = \sqrt{(x_2 - x_1)^2 + (y_2 - y_1)^2}.$$

EXAMPLE 3 Finding a Distance

Find the distance between the points $(-2, 1)$ and $(3, 4)$.

Solution

Let $(x_1, y_1) = (-2, 1)$ and $(x_2, y_2) = (3, 4)$. Then apply the Distance Formula as follows.

$$d = \sqrt{(x_2 - x_1)^2 + (y_2 - y_1)^2} \qquad \text{Distance Formula}$$

$$= \sqrt{[3 - (-2)]^2 + (4 - 1)^2} \qquad \text{Substitute for } x_1, y_1, x_2, \text{ and } y_2.$$

$$= \sqrt{(5)^2 + (3)^2} \qquad \text{Simplify.}$$

$$= \sqrt{34}$$

$$\approx 5.83 \qquad \text{Use a calculator.}$$

Note in Figure P.12 that a distance of 5.83 looks about right.

The Midpoint Formula

To find the **midpoint** of the line segment that joins two points in a coordinate plane, you can simply find the average values of the respective coordinates of the two endpoints.

THINK ABOUT THE PROOF

The Distance Formula can be used to prove the Midpoint Formula. Can you see how to do it? The details of the proof are listed in the appendix.

THE MIDPOINT FORMULA

The midpoint of the segment joining the points (x_1, y_1) and (x_2, y_2) is

$$\text{Midpoint} = \left(\frac{x_1 + x_2}{2}, \frac{y_1 + y_2}{2}\right).$$

EXAMPLE 4 *Finding a Segment's Midpoint*

Find the midpoint of the line segment joining the points $(-5, -3)$ and $(9, 3)$, as shown in Figure P.13.

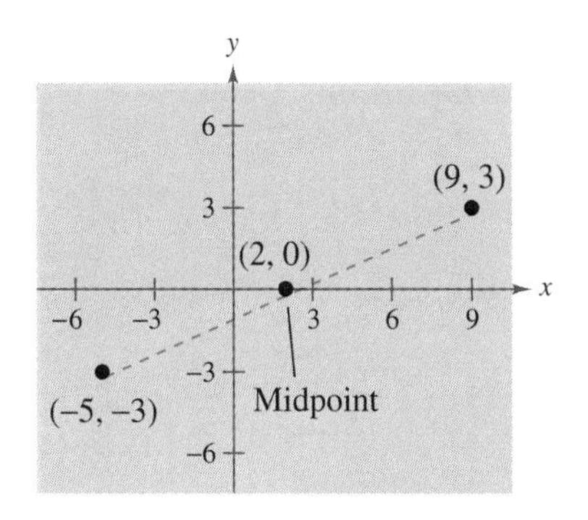

FIGURE P.13

Solution

Let $(x_1, y_1) = (-5, -3)$ and $(x_2, y_2) = (9, 3)$.

$$\begin{aligned}\text{Midpoint} &= \left(\frac{x_1 + x_2}{2}, \frac{y_1 + y_2}{2}\right) && \text{Midpoint Formula}\\ &= \left(\frac{-5 + 9}{2}, \frac{-3 + 3}{2}\right) && \text{Substitute for } x_1, y_1, x_2, \text{ and } y_2.\\ &= (2, 0) && \text{Simplify.}\end{aligned}$$

EXAMPLE 5 *Estimating Annual Sales*

Real Life

Ben and Jerry's had annual sales of \$132.0 million in 1992 and \$148.8 million in 1994. Without knowing any additional information, what would you estimate the 1993 sales to have been? (Source: Ben and Jerry's, Inc.)

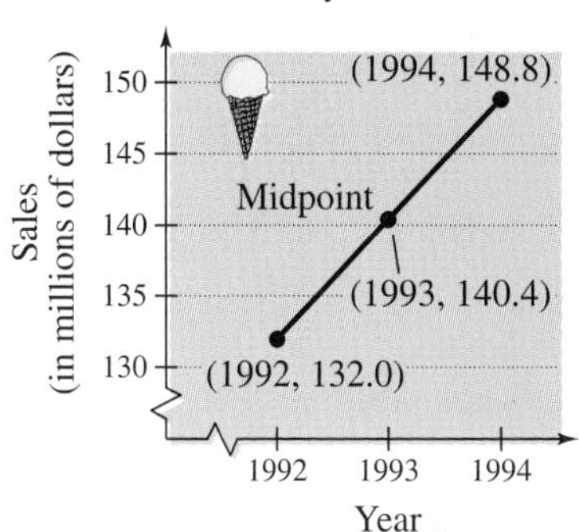

FIGURE P.14

Solution

One solution to the problem is to assume that sales followed a linear pattern. With this assumption, you can estimate the 1993 sales by finding the midpoint of the segment connecting the points (1992, 132.0) and (1994, 148.8).

$$\text{Midpoint} = \left(\frac{1992 + 1994}{2}, \frac{132.0 + 148.8}{2}\right) = (1993, 140.4)$$

Hence, you would estimate the 1993 sales to have been about \$140.4 million, as shown in Figure P.14. (The actual 1993 sales were \$140.3 million.)

The Graph of an Equation

In Example 2, you used a coordinate system to represent graphically the relationship between two quantities. There, the graphical picture consisted of a collection of points in a coordinate plane.

Frequently, a relationship between two quantities is expressed as an **equation** in two variables. For instance, $y = 7 - 3x$ is an equation in x and y. An ordered pair (a, b) is a **solution** or **solution point** of an equation in x and y if the equation is true when a is substituted for x and b is substituted for y. For instance, $(1, 4)$ is a solution of $y = 7 - 3x$ because $4 = 7 - 3(1)$ is a true statement.

In this section, you will review some basic procedures for sketching the graph of an equation in two variables. The **graph** of an equation is the set of all points that are solutions of the equation.

EXAMPLE 6 *Sketching the Graph of an Equation*

Sketch the graph of $y = x^2 - 2$.

A computer animation of this concept appears in the *Interactive* CD-ROM.

Solution

Begin by constructing a table of values.

x	-2	-1	0	1	2	3
$y = x^2 - 2$	2	-1	-2	-1	2	7

Next, plot the points given in the table, as shown in Figure P.15(a). Finally, connect the points with a smooth curve, as shown in Figure P.15(b).

NOTE The graph shown in Example 6 is a **parabola.** The graph of any second-degree equation of the form

$$y = ax^2 + bx + c, \qquad a \neq 0$$

has a similar shape.

(a)

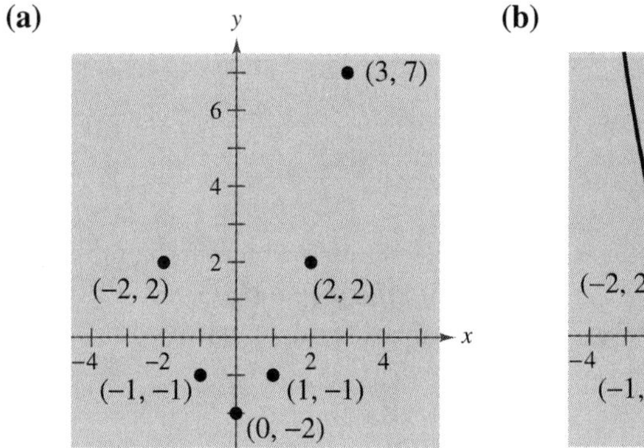

(b)

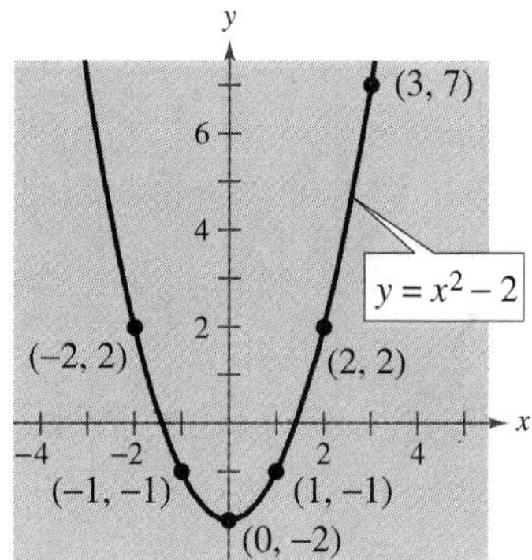

FIGURE P.15

TECHNOLOGY

Creating a Viewing Rectangle

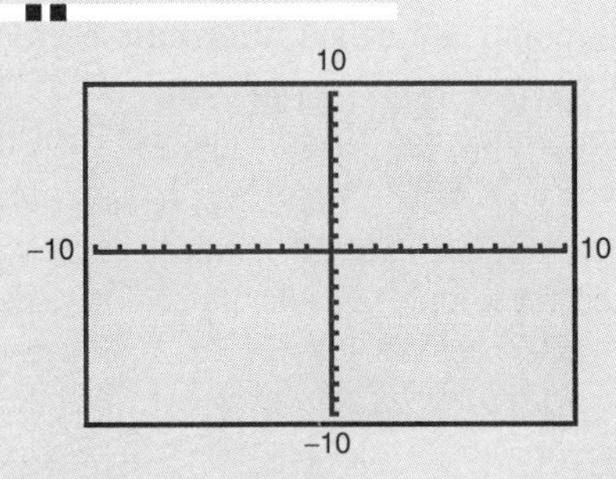

A **viewing rectangle** for a graph is a rectangular portion of the coordinate plane. A viewing rectangle is determined by six values: the minimum x-value, the maximum x-value, the x-scale, the minimum y-value, the maximum y-value, and the y-scale. When you enter these six values into a graphing utility, you are setting the **range** or **window.** Some graphing utilities have a standard viewing rectangle, as shown at the left.

By choosing different viewing rectangles for a graph, it is possible to obtain very different impressions of the graph's shape. For instance, below are four different viewing rectangles for the graph of

$$y = 0.1x^4 - x^3 + 2x^2.$$

Of these, the view shown in part (a) is the most complete.

(a)

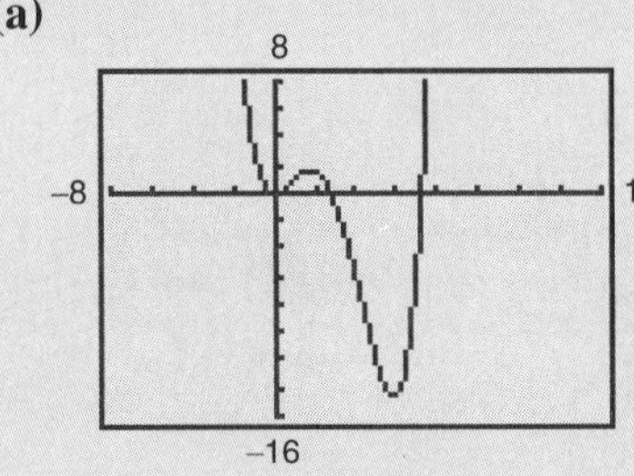

(b)

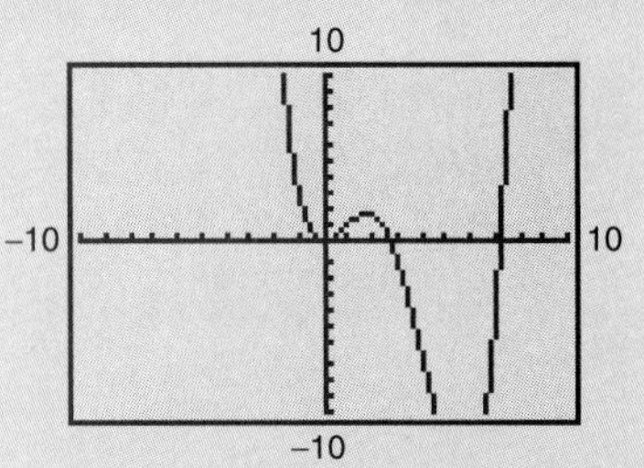

(c)

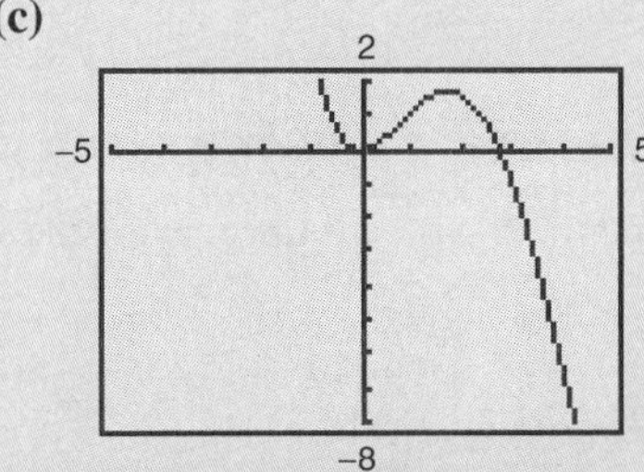

(d)

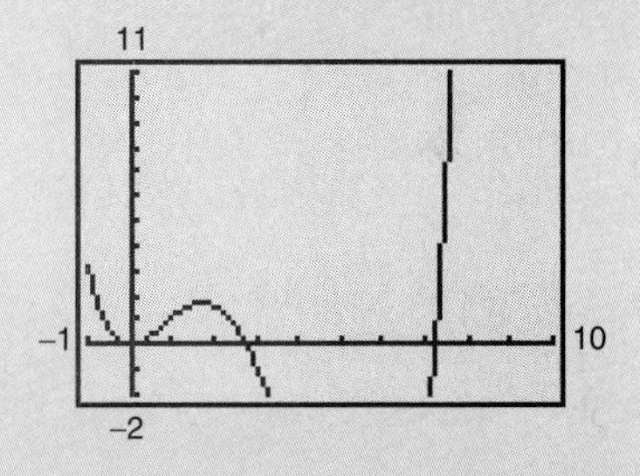

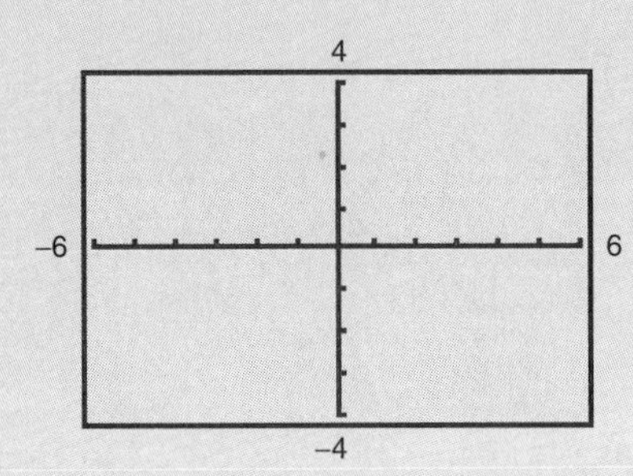

On most graphing utilities, the display screen is two-thirds as high as it is wide. On such screens, you can obtain a graph with a true geometric perspective by using a **square setting**—one in which

$$\frac{Y_{max} - Y_{min}}{X_{max} - X_{min}} = \frac{2}{3}.$$

One such setting is shown at the left. Notice that the x and y tick marks are equally spaced on a square setting, but not on a standard setting.

Intercepts of a Graph

It is often easy to determine the solution points that have zero as either the x-coordinate or the y-coordinate. These points are called **intercepts** because they are the points at which the graph intersects the x- or y-axis.

It is possible for a graph to have one intercept, several intercepts, or no intercepts, as shown in Figure P.16.

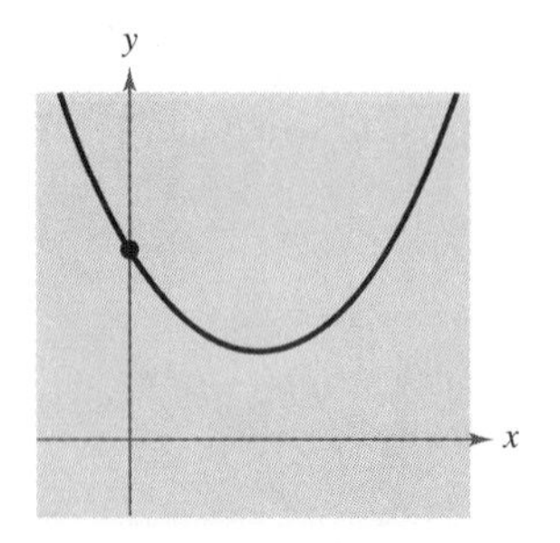

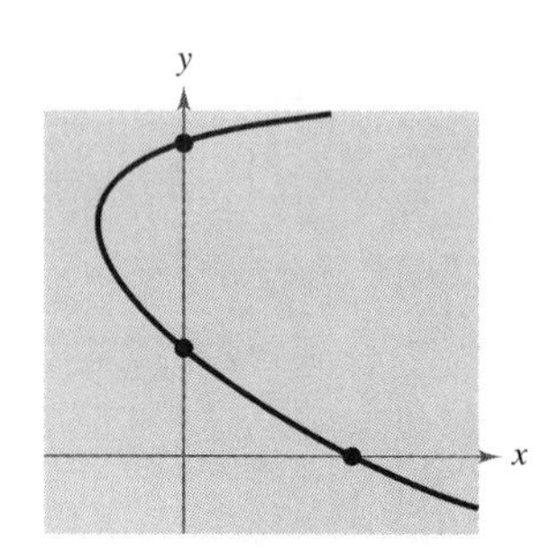

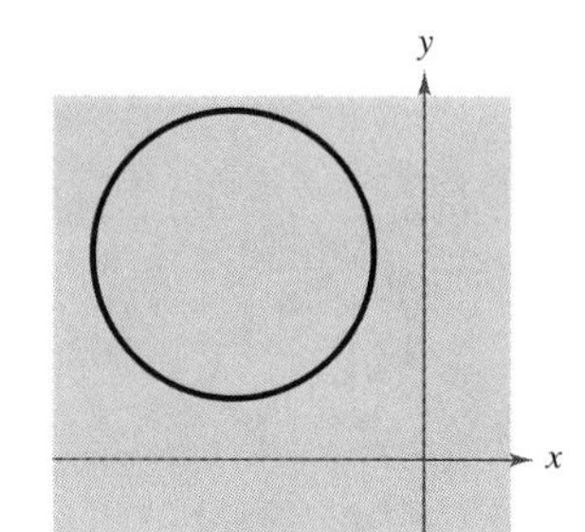

FIGURE P.16

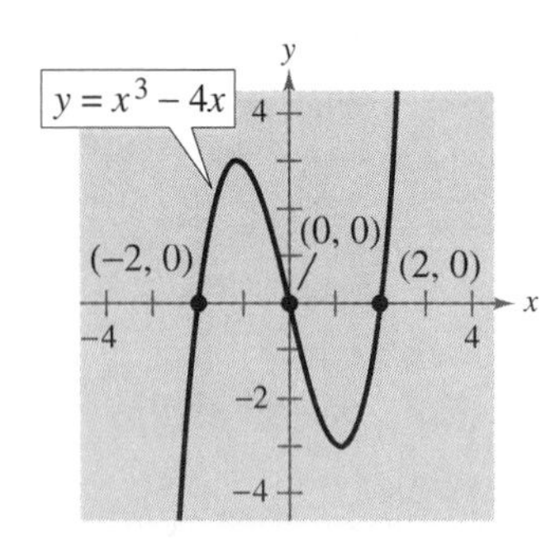

FIGURE P.17

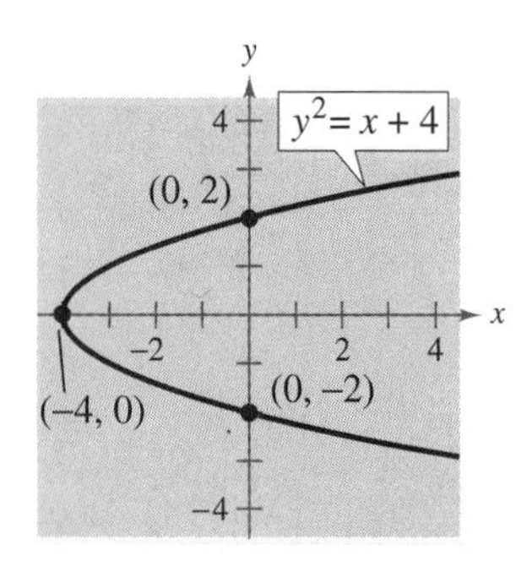

FIGURE P.18

FINDING INTERCEPTS

1. To find x-intercepts, let $y = 0$ and solve the equation for x.
2. To find y-intercepts, let $x = 0$ and solve the equation for y.

EXAMPLE 7 Finding x- and y-Intercepts

Find the x- and y-intercepts for the graph of each equation.

a. $y = x^3 - 4x$ **b.** $y^2 = x + 4$

Solution

a. Let $y = 0$. Then $0 = x(x^2 - 4)$ has solutions $x = 0$ and $x = \pm 2$.

x-intercepts: $(0, 0), (2, 0), (-2, 0)$

Let $x = 0$. Then $y = 0$.

y-intercept: $(0, 0)$ (See Figure P.17.)

b. Let $y = 0$. Then $-4 = x$.

x-intercept: $(-4, 0)$

Let $x = 0$. Then $y^2 = 4$ has solutions $y = \pm 2$.

y-intercepts: $(0, 2), (0, -2)$ (See Figure P.18.)

TECHNOLOGY

Zooming in to Find Intercepts

You can use the **zoom** feature of a graphing utility to approximate the x-intercept(s) of a graph. Suppose you want to approximate the x-intercept(s) of the graph of

$$y = 2x^3 - 3x + 2.$$

Begin by graphing the equation, as shown below in part (a). From the viewing rectangle shown, the graph appears to have only one x-intercept. This intercept lies between -2 and -1. By zooming in on the intercept, you can improve the approximation, as shown in part (b). To three decimal places, the solution is $x \approx -1.476$.

(a)

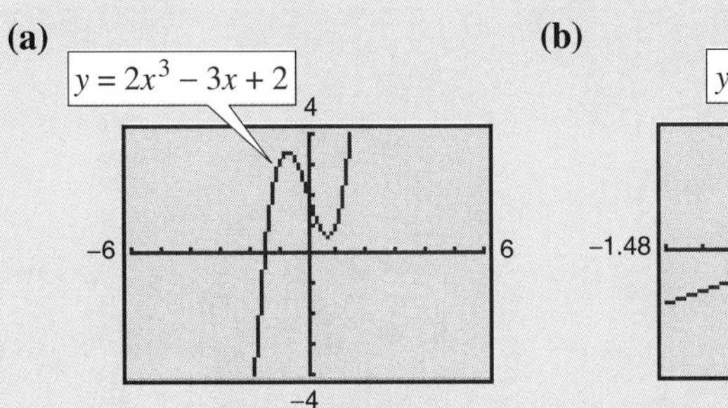

(b)

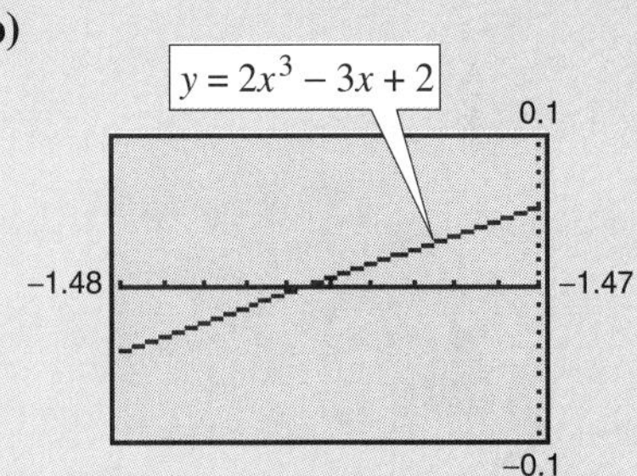

Here are some suggestions for using the zoom feature.

1. With each successive zoom-in, adjust the x-scale so that the viewing rectangle shows at least one tick mark on each side of the x-intercept.
2. The error in your approximation will be less than the distance between two scale marks.
3. The **trace** feature can usually be used to add one more decimal place of accuracy without changing the viewing rectangle.

Part (a) below shows the graph of $y = x^2 - 5x + 3$. Parts (b) and (c) show "zoom-in views" of the two intercepts. From these views, you can approximate the x-intercepts to be $x \approx 0.697$ and $x \approx 4.303$.

(a)

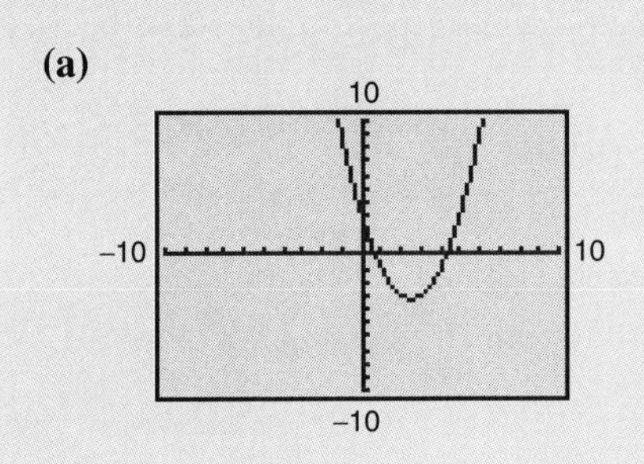

(b)

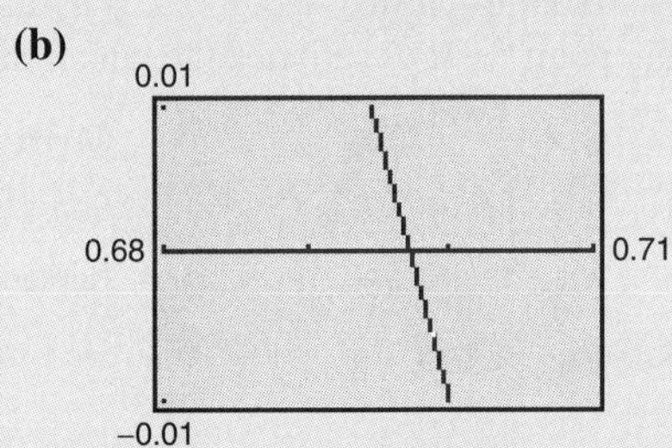

(c)

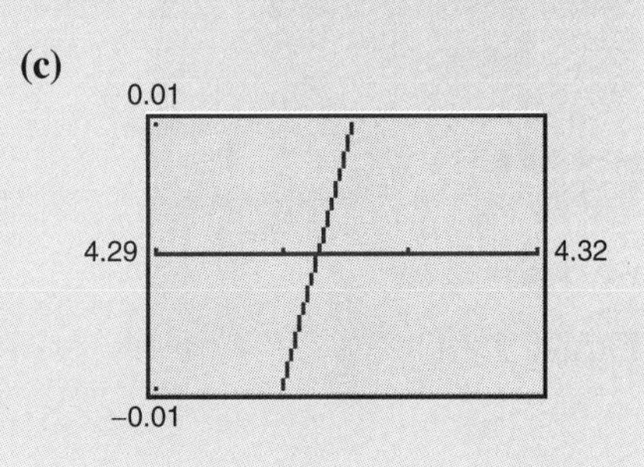

Symmetry

Each of the graphs shown in Figures P.15(b), P.17, and P.18 has **symmetry** with respect to one of the coordinate axes or with respect to the origin.

Figure P.15(b)	$y = x^2 - 2$	y-axis symmetry
Figure P.17	$y = x^3 - 4x$	Origin symmetry
Figure P.18	$y^2 = x + 4$	x-axis symmetry

Symmetry with respect to the x-axis means that if the Cartesian plane were folded along the x-axis, the portion of the graph above the x-axis would coincide with the portion below the x-axis. Symmetry with respect to the y-axis or the origin can be described in a similar manner, as shown in Figure P.19.

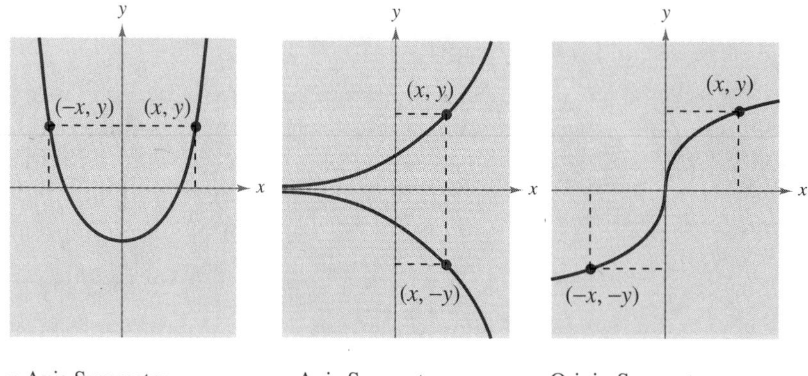

y-Axis Symmetry x-Axis Symmetry Origin Symmetry

FIGURE P.19

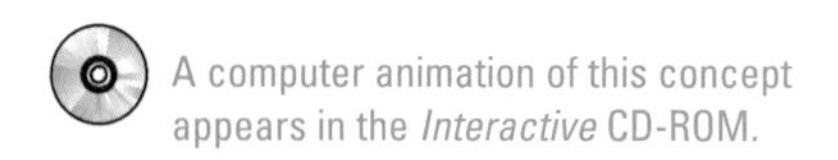

Knowing the symmetry of a graph *before* attempting to sketch it is helpful, because then you need only half as many solution points to sketch the graph. There are three basic types of symmetry. (See Exercises 47–50.) A graph is **symmetric with respect to the y-axis** if, whenever (x, y) is on the graph, $(-x, y)$ is also on the graph. A graph is **symmetric with respect to the x-axis** if, whenever (x, y) is on the graph, $(x, -y)$ is also on the graph. A graph is **symmetric with respect to the origin** if, whenever (x, y) is on the graph, $(-x, -y)$ is also on the graph.

The graph of $y = x^2 - 2$ is symmetric with respect to the y-axis because the point $(-x, y)$ satisfies the equation.

$y = x^2 - 2$	Given equation
$y = (-x)^2 - 2$	Substitute $(-x, y)$ for (x, y).
$y = x^2 - 2$	Replacement yields equivalent equation.

See Figure P.20.

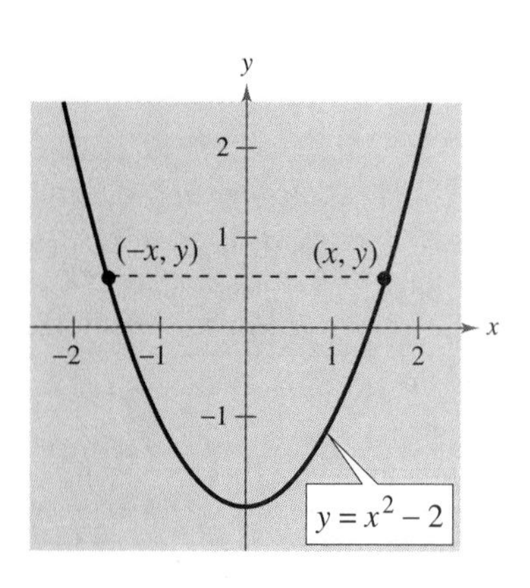

y-Axis Symmetry

FIGURE P.20

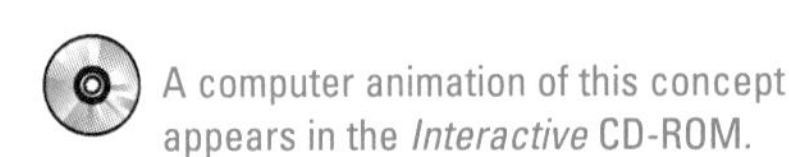

A computer animation of this concept appears in the *Interactive* CD-ROM.

> ***TESTS FOR SYMMETRY***
>
> 1. The graph of an equation is symmetric with respect to the *y-axis* if replacing x with $-x$ yields an equivalent equation.
> 2. The graph of an equation is symmetric with respect to the *x-axis* if replacing y with $-y$ yields an equivalent equation.
> 3. The graph of an equation is symmetric with respect to the *origin* if replacing x with $-x$ *and* y with $-y$ yields an equivalent equation.

EXAMPLE 8 *Using Intercepts and Symmetry as Sketching Aids*

Use intercepts and symmetry to sketch the graph of $x - y^2 = 1$.

Solution

Letting $x = 0$, you can see that $-y^2 = 1$ or $y^2 = -1$ has no real solutions. Hence, there are no y-intercepts. Letting $y = 0$, you obtain $x = 1$. Thus, the x-intercept is $(1, 0)$. Of the three tests for symmetry, the only one that is satisfied is the test for x-axis symmetry. Thus, the graph is symmetric with respect to the x-axis. Using symmetry, you need only to find the solution points above the x-axis and then reflect them to obtain the graph, as shown in Figure P.21.

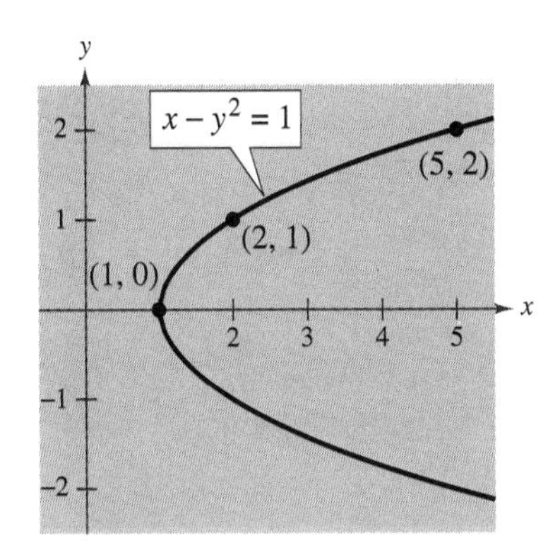

FIGURE P.21

y	0	1	2
$x = y^2 + 1$	1	2	5

EXAMPLE 9 *Sketching the Graph of an Equation*

Sketch the graph of $y = |x - 1|$.

Solution

Letting $x = 0$ yields $y = 1$, which means that $(0, 1)$ is the y-intercept. Letting $y = 0$ yields $x = 1$, which means that $(1, 0)$ is the x-intercept. This equation fails all three tests for symmetry and consequently its graph is not symmetric with respect to either axis or to the origin. The absolute value sign indicates that y is always nonnegative.

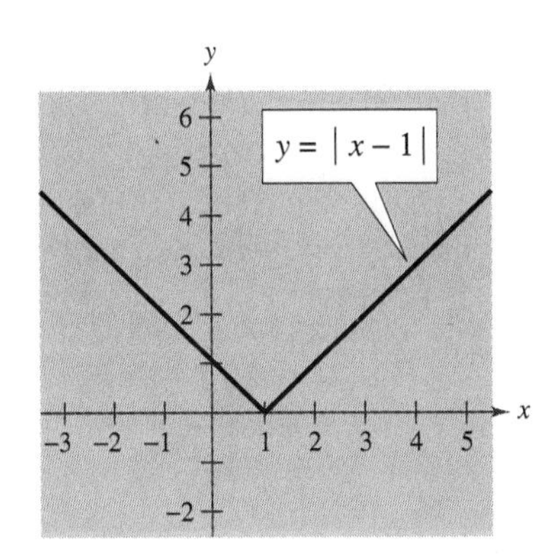

FIGURE P.22

x	-2	-1	0	1	2	3	4
$y = \|x - 1\|$	3	2	1	0	1	2	3

The graph is shown in Figure P.22.

Circles

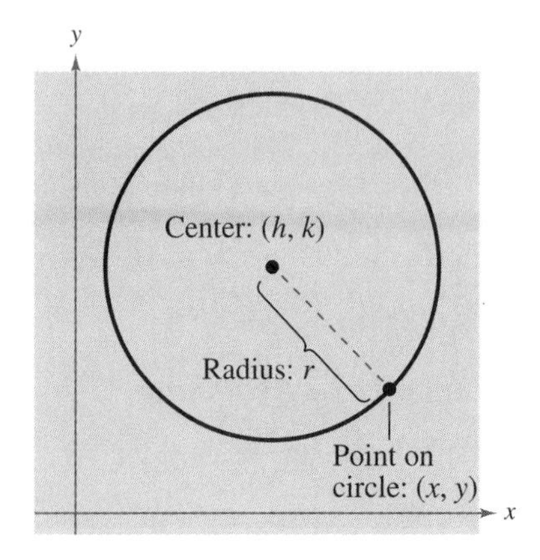

FIGURE P.23

Throughout this course, you will learn to recognize several types of graphs from their equations. For instance, you will learn to recognize that the graph of a second-degree equation of the form $y = ax^2 + bx + c$ is a parabola (see Example 6). Another easily recognized graph is that of a **circle.**

Consider the circle shown in Figure P.23. A point (x, y) is on the circle if and only if its distance from the center (h, k) is r. By the Distance Formula,

$$\sqrt{(x-h)^2 + (y-k)^2} = r.$$

By squaring both sides of this equation, you obtain the **standard form of the equation of a circle.**

STANDARD FORM OF THE EQUATION OF A CIRCLE

The point (x, y) lies on the circle of **radius** r and **center** (h, k) if and only if

$$(x-h)^2 + (y-k)^2 = r^2.$$

From this result, you can see that the standard form of the equation of a circle with its center at the origin, $(h, k) = (0, 0)$, is simply

$$x^2 + y^2 = r^2. \qquad \text{Circle with center at origin}$$

EXAMPLE 10 *Finding the Equation of a Circle*

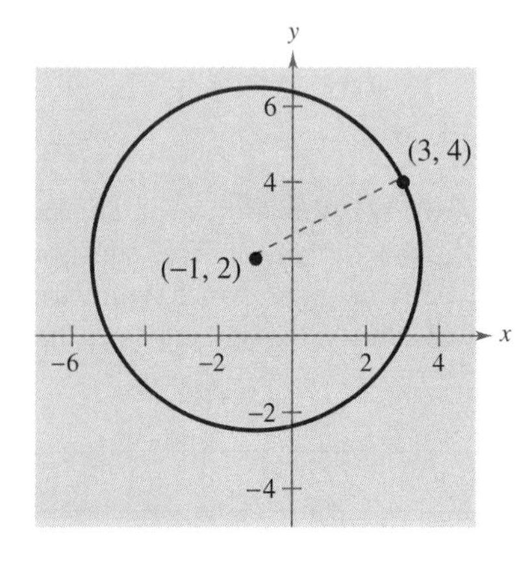

FIGURE P.24

The point $(3, 4)$ lies on a circle whose center is at $(-1, 2)$, as shown in Figure P.24. Find the standard form of the equation of this circle.

Solution

The radius of the circle is the distance between $(-1, 2)$ and $(3, 4)$.

$$r = \sqrt{[3-(-1)]^2 + (4-2)^2} \qquad \text{Distance Formula}$$
$$= \sqrt{16 + 4} \qquad \text{Simplify.}$$
$$= \sqrt{20} \qquad \text{Radius}$$

Using $(h, k) = (-1, 2)$ and $r = \sqrt{20}$, the equation of the circle is

$$(x-h)^2 + (y-k)^2 = r^2$$
$$[x-(-1)]^2 + (y-2)^2 = \left(\sqrt{20}\right)^2 \qquad \text{Substitute for } h, k, \text{ and } r.$$
$$(x+1)^2 + (y-2)^2 = 20. \qquad \text{Standard form}$$

GROUP ACTIVITY

CLASSIFYING GRAPHS OF EQUATIONS

The graphs below represent six common types of models. Match each equation with its corresponding graph.

(a)

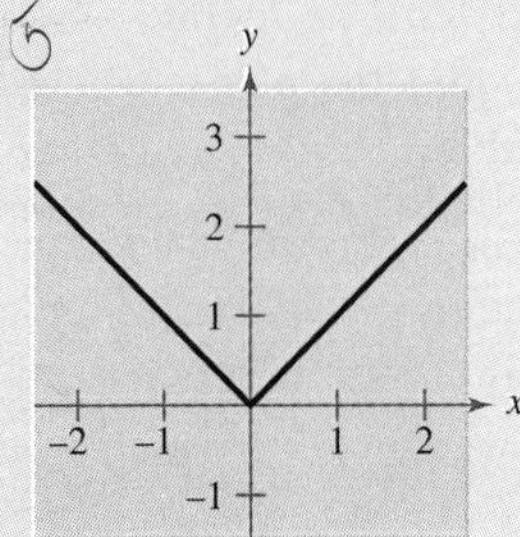

(b)

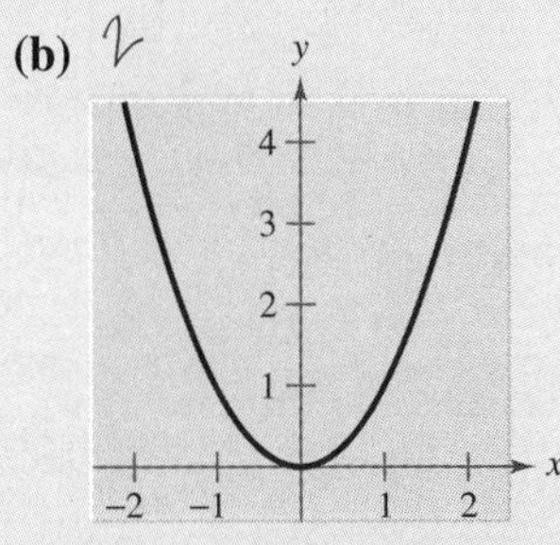

(c)

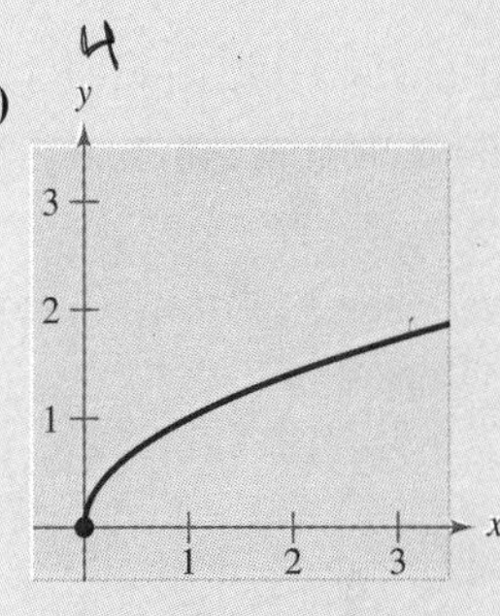

(d)

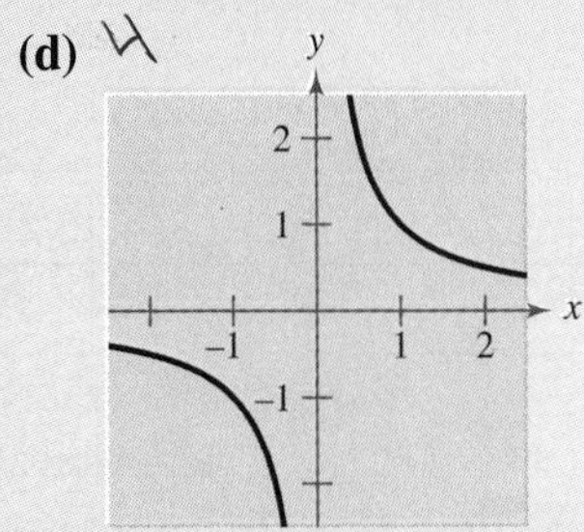

(e)

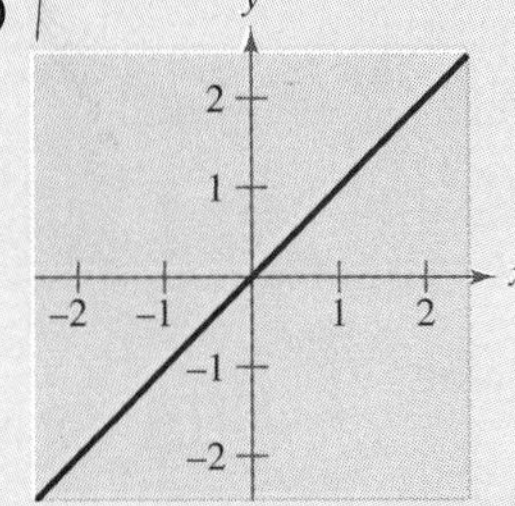

(f)

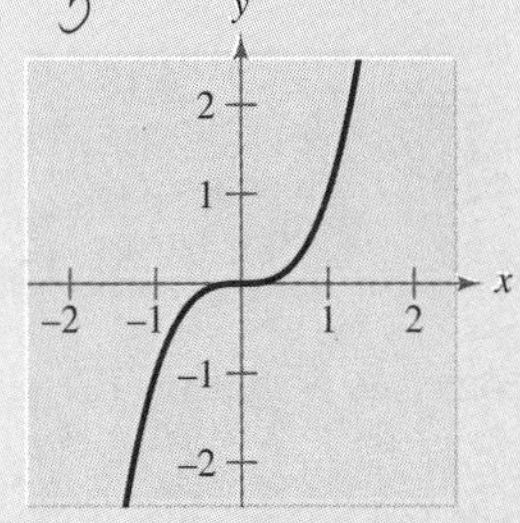

1. Linear model: $y = x$

2. Quadratic model: $y = x^2$

3. Cubic model: $y = x^3$

4. Square root model: $y = \sqrt{x}$

5. Absolute value model: $y = |x|$

6. Rational model: $y = \dfrac{1}{x}$

WARM UP

Simplify the expression.

1. $4(x - 3) - 5(6 - 2x)$

2. $-s(5s + 2) + 5(s^2 - 3s)$

3. $3y(-2y^2)^3$

4. $2t(t + 1)^2 - 4(t + 1)$

5. $\sqrt{150x^4}$

6. $\sqrt{4x + 12}$

Simplify the equation.

7. $-y = (-x)^3 + 4(-x)$

8. $(-x)^2 + (-y)^2 = 4$

9. $y = 4(-x)^2 + 8$

10. $(-y)^2 = 3(-x) + 4$

P.3 Exercises

In Exercises 1 and 2, sketch the polygon with the indicated vertices.

1. Triangle: $(-1, 1), (2, -1), (3, 4)$
2. Parallelogram: $(5, 2), (7, 0), (1, -2), (-1, 0)$

In Exercises 3 and 4, approximate the coordinates of the points.

3.

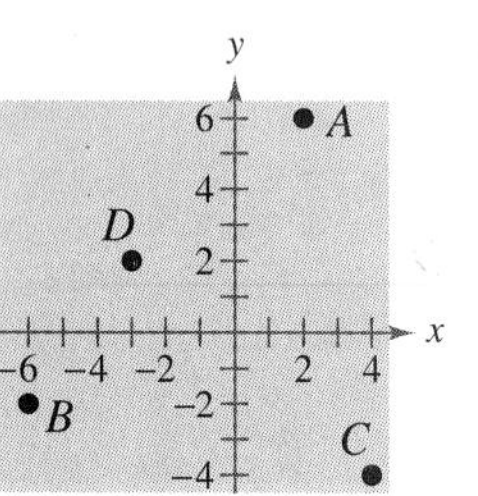

4.

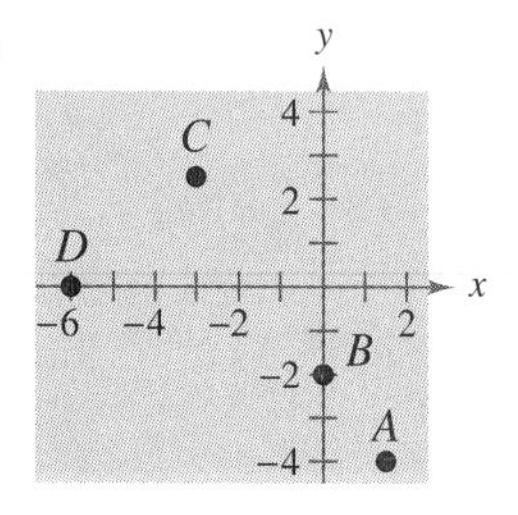

In Exercises 5–8, find the coordinates of the point.

5. The point is located three units to the left of the y-axis and four units above the x-axis.
6. The point is located eight units below the x-axis and four units to the right of the y-axis.
7. The point is located five units below the x-axis and the coordinates of the point are equal.
8. The point is on the x-axis and 12 units to the left of the y-axis.

In Exercises 9–12, determine the quadrant(s) in which (x, y) is located so that the condition(s) is (are) satisfied.

9. $x > 0$ and $y < 0$
10. $x > 2$ and $y = 3$
11. $(x, -y)$ is in the second quadrant.
12. $xy < 0$

13. ***Think About It*** What is the y-coordinate of any point on the x-axis? What is the x-coordinate of any point on the y-axis?

14. ***Think About It*** When plotting points on the rectangular coordinate system, is it true that the scales on the x and y axes must be the same? Explain.

In Exercises 15 and 16, plot the points whose coordinates are given in the table.

15. ***Normal Temperatures*** The normal temperature y (in degrees Fahrenheit) in Duluth, Minnesota, for each month x, where $x = 1$ represents January, is given in the table. (Source: NOAA)

x	1	2	3	4	5	6
y	6	12	23	38	50	59

x	7	8	9	10	11	12
y	65	63	54	44	28	14

16. ***Wal-Mart*** The number y of Wal-Mart stores for each year x from 1985 through 1994 is given in the table. (Source: Wal-Mart Annual Report for 1994)

x	1985	1986	1987	1988	1989
y	745	859	980	1114	1259

x	1990	1991	1992	1993	1994
y	1399	1568	1714	1850	1953

In Exercises 17–20, (a) find the length of each side of a right triangle, and (b) show that these lengths satisfy the Pythagorean Theorem.

17.

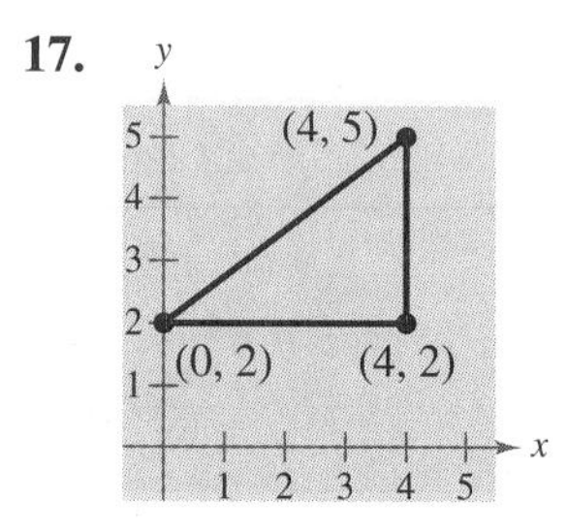

18.

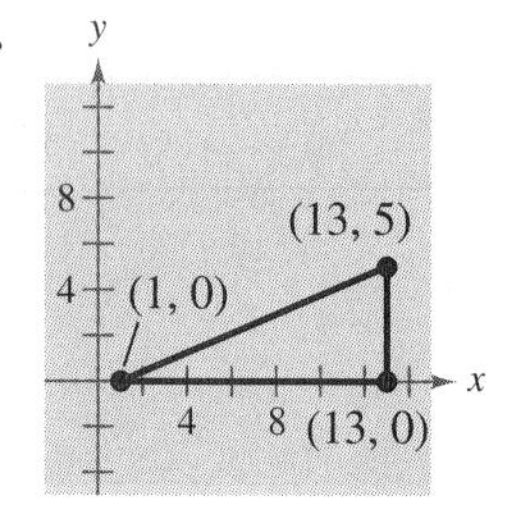

19.

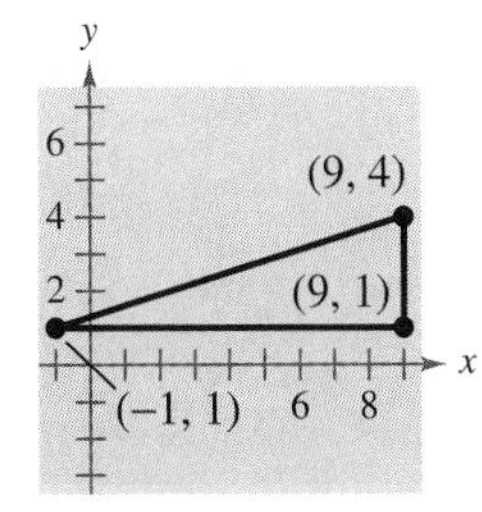

20.

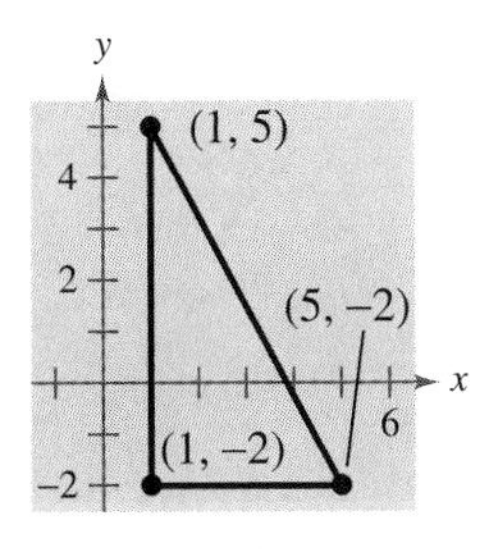

In Exercises 21–26, (a) plot the points, (b) find the distance between the points, and (c) find the midpoint of the line segment joining the points.

21. $(1, 1), (9, 7)$

22. $(1, 12), (6, 0)$

23. $(-1, 2), (5, 4)$

24. $\left(-\frac{1}{3}, -\frac{1}{3}\right), \left(-\frac{1}{6}, -\frac{1}{2}\right)$

25. $(6.2, 5.4), (-3.7, 1.8)$

26. $(-16.8, 12.3), (5.6, 4.9)$

In Exercises 27 and 28, use the Midpoint Formula to estimate the sales of a company in 1993, given the sales in 1991 and 1995. Assume that the sales followed a linear pattern.

27.

Year	1991	1995
Sales	\$520,000	\$740,000

28.

Year	1991	1995
Sales	\$4,200,000	\$5,650,000

29. *Make a Conjecture* Plot the points $(2, 1)$, $(-3, 5)$, and $(7, -3)$ on a rectangular coordinate system. Then change the sign of the y-coordinate of each point and plot the three new points on the same rectangular coordinate system. What conjecture can you make about the location of a point when the sign of the y-coordinate is changed?

30. *Football Pass* A quarterback throws a pass from the 15-yard line, 10 yards from the sideline (see figure). The pass is caught on the 40-yard line, 45 yards from the same sideline. How long is the pass?

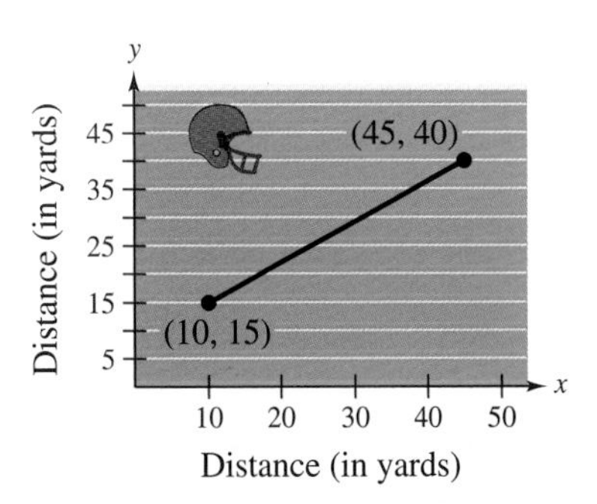

FIGURE FOR 30

In Exercises 31–34, determine whether the points lie on the graph of the equation.

	Equation	*Points*	
31.	$y = \sqrt{x + 4}$	(a) $(0, 2)$	(b) $(5, 3)$
32.	$y = x^2 - 3x + 2$	(a) $(2, 0)$	(b) $(-2, 8)$
33.	$x^2y - x^2 + 4y = 0$	(a) $\left(1, \frac{1}{5}\right)$	(b) $\left(2, \frac{1}{2}\right)$
34.	$y = \dfrac{1}{x^2 + 1}$	(a) $(0, 0)$	(b) $(3, 0.1)$

In Exercises 35 and 36, complete the table. Use the resulting solution points to sketch the graph of the equation. Use a graphing utility to verify the graph.

35. $y = \frac{3}{2}x - 1$

x	-2	0	$\frac{2}{3}$	1	2
y					

36. $y = x^2 - 2x$

x	-1	0	1	2	3
y					

In Exercises 37–40, use a graphing utility to graph the equation. Use a standard setting. Approximate any x- or y-intercepts of the graph.

37. $y = x - 5$

38. $y = 9 - x^2$

39. $y = x\sqrt{x + 6}$

40. $y = \dfrac{2x}{x - 1}$

In Exercises 41–46, check for symmetry with respect to both axes and to the origin.

41. $x^2 - y = 0$

42. $xy^2 + 10 = 0$

43. $y = \sqrt{9 - x^2}$

44. $xy = 4$

45. $y = \dfrac{x}{x^2 + 1}$

46. $y = x^4 - x^2 + 3$

In Exercises 47–50, use symmetry to sketch the complete graph of the equation.

47.

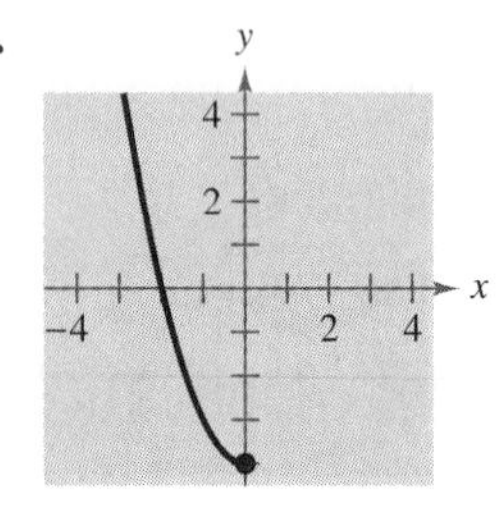

y-Axis Symmetry

48.

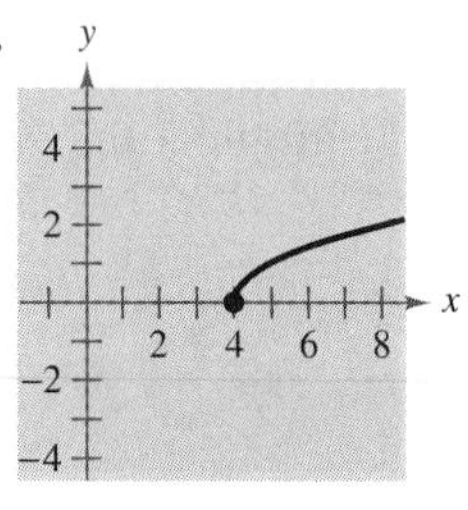

x-Axis Symmetry

49.

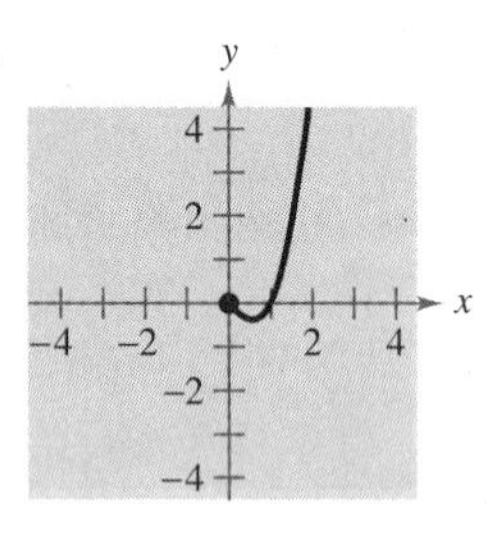

Origin Symmetry

50.

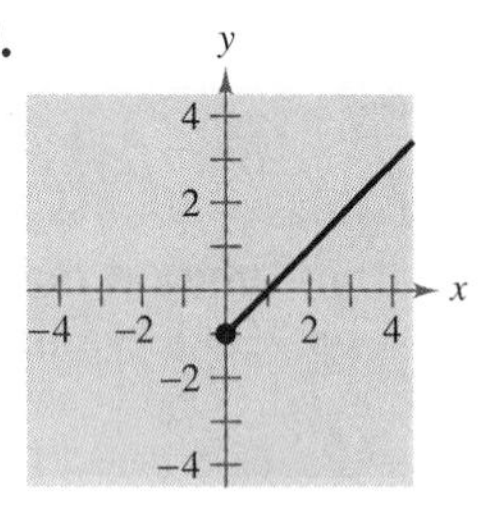

y-Axis Symmetry

In Exercises 51–56, match the equation with its graph. [The graphs are labeled (a), (b), (c), (d), (e), and (f).]

(a)

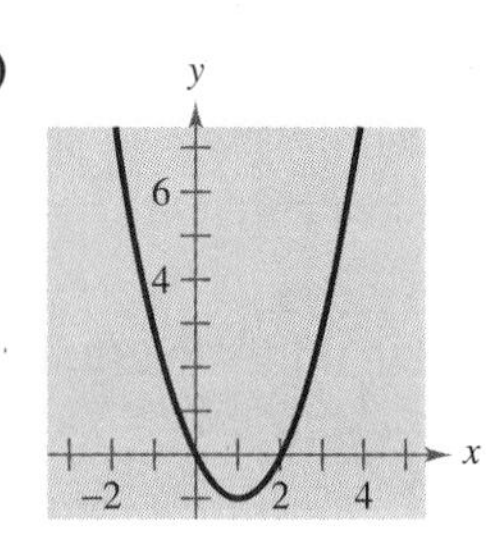

(b)

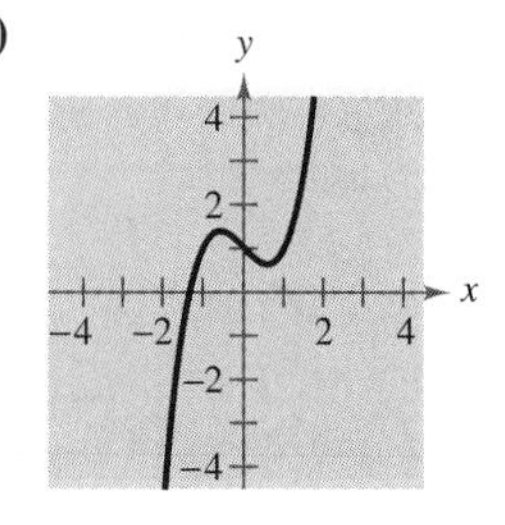

(c)

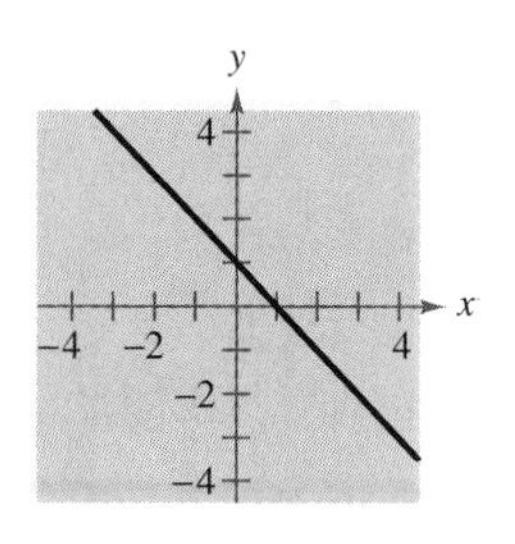

(d)

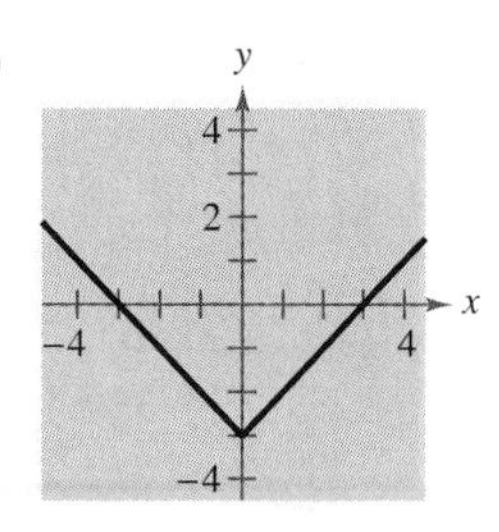

(e)

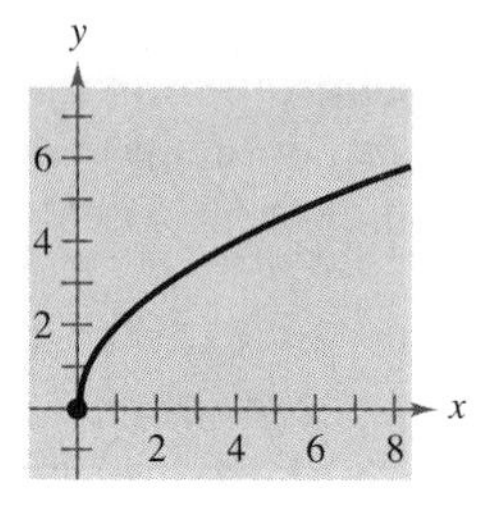

(f)

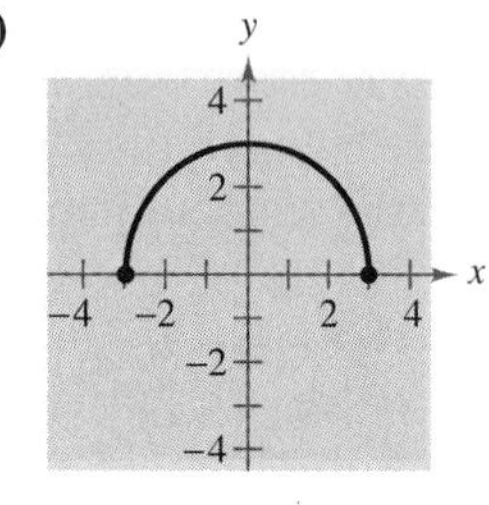

51. $y = 1 - x$

52. $y = x^2 - 2x$

53. $y = \sqrt{9 - x^2}$

54. $y = 2\sqrt{x}$

55. $y = x^3 - x + 1$

56. $y = |x| - 3$

In Exercises 57–68, sketch the graph of the equation. Test for symmetry.

57. $y = -3x + 2$

58. $y = 2x - 3$

59. $y = x^2 - 3x$

60. $y = -x^2 - 4x$

61. $y = x^3 + 2$

62. $y = x^3 - 1$

63. $y = \sqrt{x - 3}$

64. $y = \sqrt{1 - x}$

65. $y = |x - 2|$

66. $y = 4 - |x|$

67. $x = y^2 - 1$

68. $x = y^2 - 4$

In Exercises 69 and 70, use a graphing utility to sketch the graph of the equation. Begin by using a standard setting. Then graph the equation a second time using the specified setting. Which setting is better? Explain.

69. $y = \frac{5}{2}x + 5$

Xmin = 0
Xmax = 6
Xscl = 1
Ymin = 0
Ymax = 10
Yscl = 1

70. $y = -3x + 50$

Xmin = -1
Xmax = 4
Xscl = 1
Ymin = -5
Ymax = 60
Yscl = 5

In Exercises 71–74, find a setting on a graphing utility such that the graph of the equation agrees with the given graph.

71. $y = 4x^2 - 25$

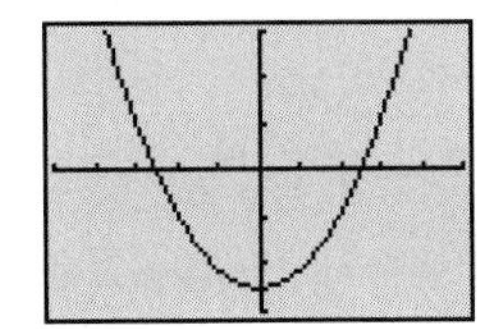

72. $y = x^3 - 3x^2 + 4$

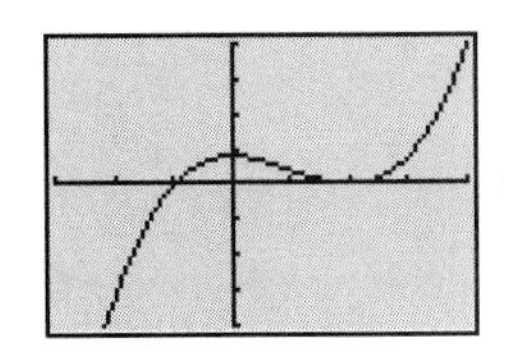

73. $y = |x| + |x - 10|$

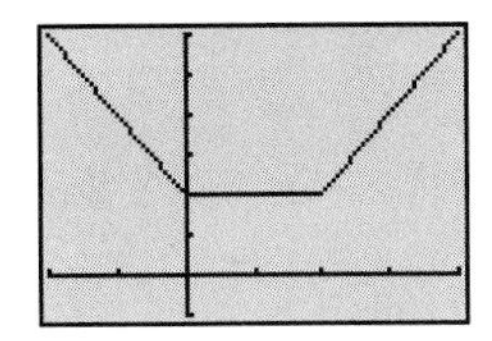

74. $y = 8\sqrt[3]{x - 6}$

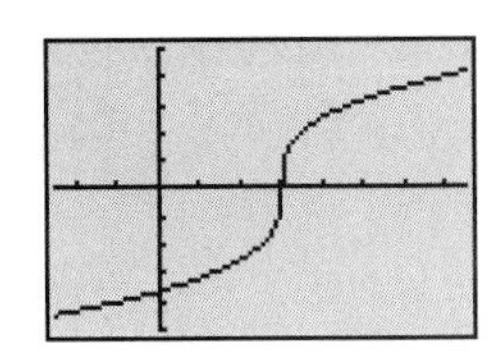

In Exercises 75–82, use a graphing utility to graph the equation. Use the graph to approximate any x-intercepts of the graph. Set $y = 0$ and solve the resulting equation. Compare the result with the x-intercepts of the graph.

75. $y = 12 - 4x$

76. $y = 3(x - 5) + 9$

77. $y = \dfrac{x + 2}{3} - \dfrac{x - 1}{5} - 1$

78. $y = \frac{3}{2}x + \frac{1}{4}(x - 2) - 10$

79. $y = 1 - (x - 2)^2$

80. $y = -4x^2 + 4x + 3$

81. $y = x^3 - 9x^2 + 18x$

82. $y = x^4 - 4x^2$

In Exercises 83–90, find the standard form of the equation of the specified circle.

83. Center: $(0, 0)$; radius: 3

84. Center: $(0, 0)$; radius: 5

85. Center: $(2, -1)$; radius: 4

86. Center: $\left(0, \frac{1}{3}\right)$; radius: $\frac{1}{3}$

87. Center: $(-1, 2)$; solution point: $(0, 0)$

88. Center: $(3, -2)$; solution point: $(-1, 1)$

89. Endpoints of a diameter: $(0, 0)$, $(6, 8)$

90. Endpoints of a diameter: $(-4, -1)$, $(4, 1)$

91. ***Depreciation*** A manufacturing plant purchases a new molding machine for \$225,000. The depreciated value y after t years is given by

$$y = 225{,}000 - 20{,}000t, \qquad 0 \le t \le 8.$$

Sketch the graph of the equation over the given interval for t.

92. ***Dimensions of a Rectangle*** A rectangle of length x and width w has a perimeter of 12 meters.

(a) Show that the width of the rectangle is $w = 6 - x$ and its area is $A = x(6 - x)$.

(b) Use a graphing utility to graph the equation for the area.

(c) From the graph of part (b), estimate the dimensions of the rectangle that yield a maximum area.

93. ***Life Expectancy*** The table gives the life expectancy of a child (at birth) for selected years from 1920 to 1990. (Source: Department of Health and Human Services)

Year	1920	1930	1940	1950
Life Expectancy	54.1	59.7	62.9	68.2

Year	1960	1970	1980	1990
Life Expectancy	69.7	70.8	73.7	75.4

A mathematical model for life expectancy during this period is

$$y = \frac{t + 66.93}{0.01t + 1}$$

where y represents life expectancy and t represents time in years, with $t = 0$ corresponding to 1950.

(a) Use a graphing utility to graph the data points and the model.

(b) Use the model to predict y for the years 1998 and 2000.

P.4 Lines in the Plane and Slope

See Exercise 31 on page 52 for an example of how a linear equation can be used to model the earnings per share for General Mills stock from 1987 through 1994.

Using Slope

The simplest mathematical model for relating two variables is the **linear equation** $y = mx + b$. The equation is called *linear* because its graph is a line. (In mathematics, the term *line* means *straight line*.) By letting $x = 0$, you can see that the line crosses the y-axis at $y = b$, as shown in Figure P.25. In other words, the y-intercept is $(0, b)$. The steepness or slope of the line is m.

$$y = mx + b$$

Slope (pointing to m) — y-intercept (pointing to b)

The **slope** of a nonvertical line is the number of units the line rises (or falls) vertically for each unit of horizontal change from left to right, as shown in Figure P.25.

Exploration

Use a graphing utility to compare the slopes of the lines $y = mx$ where $m = 0.5$, 1, 2, and 4. Which line rises most quickly? Now, let $m = -0.5, -1, -2$, and -4. Which line falls most quickly? Use a square setting to obtain a true geometric perspective.

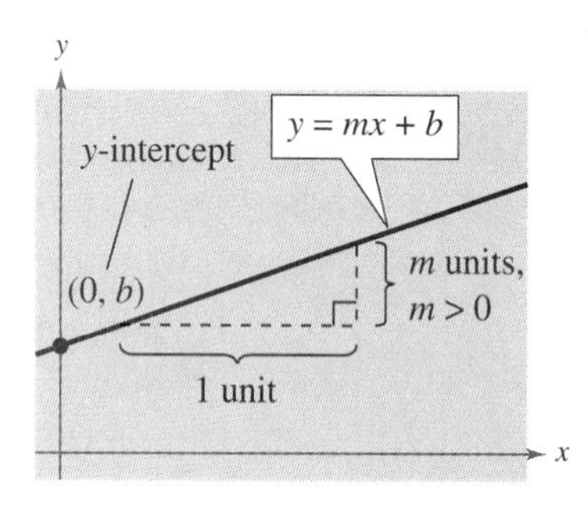

Positive slope, line rises.

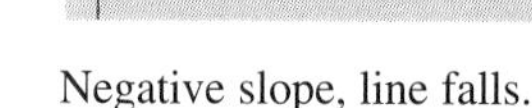

Negative slope, line falls.

FIGURE P.25

A linear equation that is written in the form $y = mx + b$ is said to be written in **slope-intercept form.**

THE SLOPE-INTERCEPT FORM OF THE EQUATION OF A LINE

The graph of the equation

$$y = mx + b$$

is a line whose slope is m and whose y-intercept is $(0, b)$.

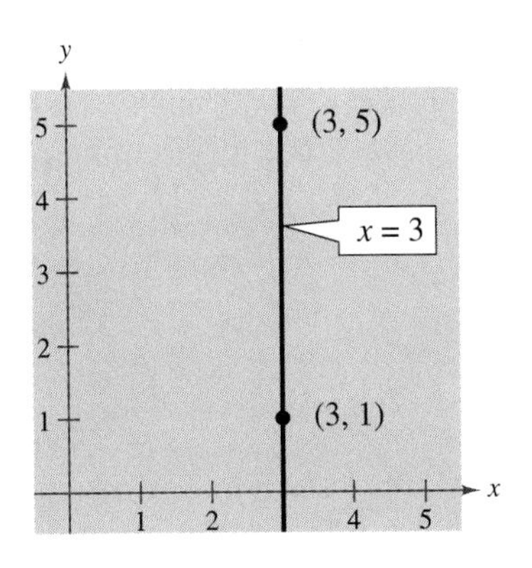

FIGURE P.26 Slope is undefined.

Once you have determined the slope and the y-intercept of a line, it is a relatively simple matter to sketch its graph. In the following example, note that none of the lines is vertical. A vertical line has an equation of the form

$$x = a. \qquad \text{Vertical line}$$

The equation of a vertical line cannot be written in the form $y = mx + b$ because the slope of a vertical line is undefined, as indicated in Figure P.26.

EXAMPLE 1 *Graphing a Linear Equation*

Sketch the graph of each linear equation.

a. $y = 2x + 1$ **b.** $y = 2$ **c.** $x + y = 2$

Solution

a. Because $b = 1$, the y-intercept is $(0, 1)$. Moreover, because the slope is $m = 2$, the line *rises* two units for each unit the line moves to the right, as shown in Figure P.27(a).

b. By writing this equation in the form $y = (0)x + 2$, you can see that the y-intercept is $(0, 2)$ and the slope is zero. A zero slope implies that the line is horizontal—that is, it doesn't rise *or* fall, as shown in Figure P.27(b).

c. By writing this equation in slope-intercept form

$$x + y = 2 \qquad \text{Original equation}$$
$$y = -x + 2 \qquad \text{Subtract } x \text{ from both sides.}$$
$$y = (-1)x + 2 \qquad \text{Slope-intercept form}$$

you can see that the y-intercept is $(0, 2)$. Moreover, because the slope is $m = -1$, this line *falls* one unit for each unit the line moves to the right, as shown in Figure P.27(c).

(a) When m is positive, the line rises.

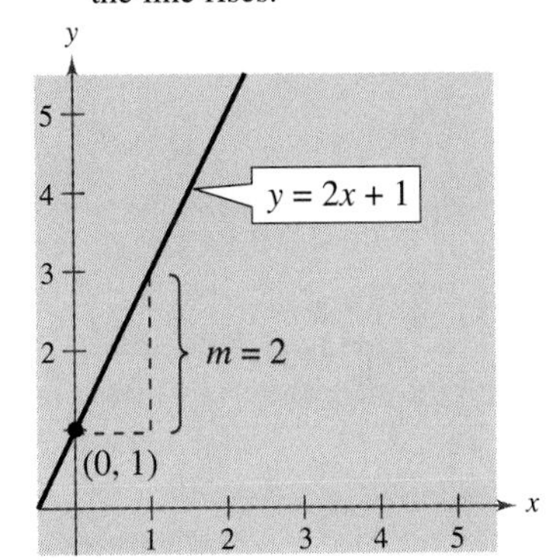

(b) When m is 0, the line is horizontal.

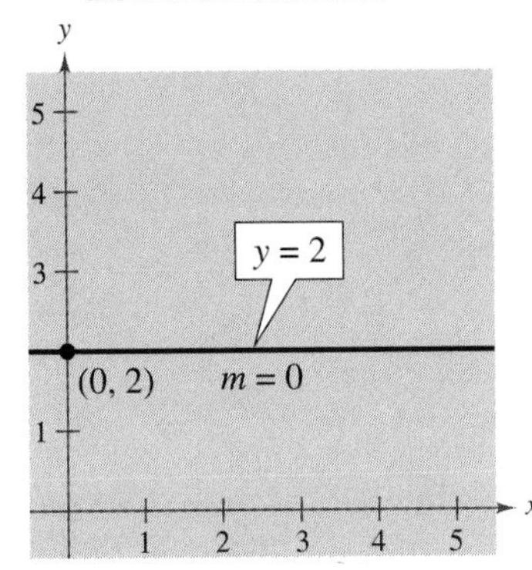

(c) When m is negative, the line falls.

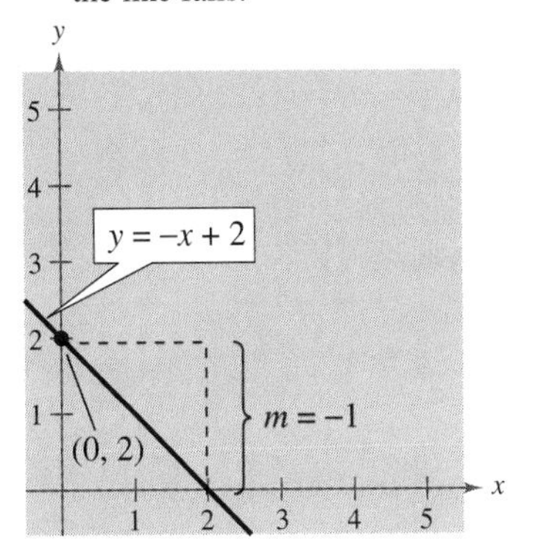

FIGURE P.27

In real-life problems, the slope of a line can be interpreted as either a *ratio* or a *rate*. If the x-axis and y-axis have the same unit of measure, then the slope has no units and is a **ratio.** If the x-axis and y-axis have different units of measure, then the slope is a **rate** or **rate of change.**

EXAMPLE 2 *Using Slope as a Ratio*

Real Life

The maximum recommended slope of a wheelchair ramp is $\frac{1}{12}$. A business is installing a wheelchair ramp that rises 22 inches over a horizontal length of 24 feet. Is the ramp steeper than recommended? (Source: *American Disabilities Act Handbook*)

Solution

The horizontal length of the ramp is 12(24) or 288 inches, as shown in Figure P.28. Thus, the slope of the ramp is

$$\text{Slope} = \frac{\text{vertical change}}{\text{horizontal change}} = \frac{22 \text{ in.}}{288 \text{ in.}} \approx 0.076 < 0.08\overline{3} = \frac{1}{12}.$$

Thus, the slope is not steeper than recommended.

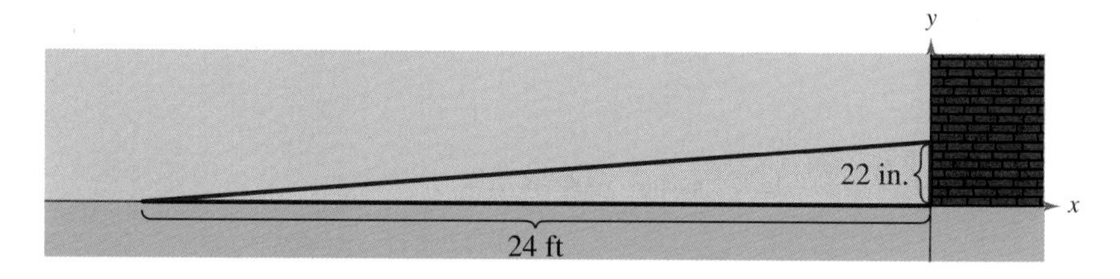

FIGURE P.28

EXAMPLE 3 *Using Slope as a Rate of Change*

A manufacturing company determines that the total cost in dollars of producing x units of a product is

$$C = 25x + 3500. \qquad \text{Cost equation}$$

Describe the practical significance of the y-intercept and slope of this line.

Solution

The y-intercept (0, 3500) tells you that the cost of producing zero units is \$3500. This is the **fixed cost** of production—it includes costs that must be paid regardless of the number of units produced. The slope of $m = 25$ tells you that the cost of producing each unit is \$25, as shown in Figure P.29. Economists call the cost per unit the **marginal cost.** If the production increases by one unit, then the "margin" or extra amount of cost is \$25.

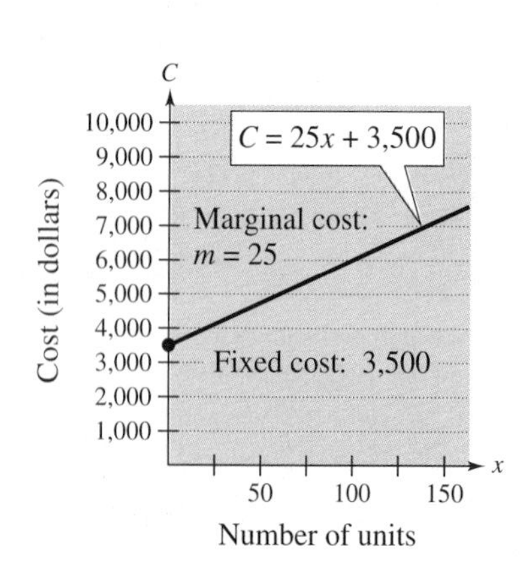

FIGURE P.29 Production cost

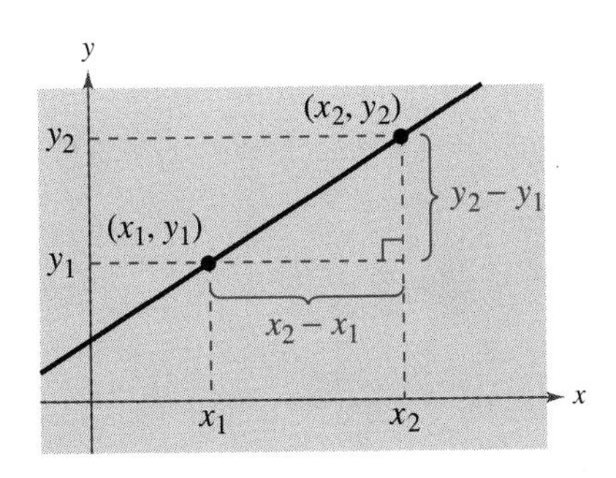

FIGURE P.30

Finding the Slope of a Line

Given an equation of a line, you can find its slope by writing the equation in slope-intercept form. If you are not given an equation, you can still find the slope of a line. For instance, suppose you want to find the slope of the line passing through the points (x_1, y_1) and (x_2, y_2), as shown in Figure P.30. As you move from left to right along this line, a change of $(y_2 - y_1)$ units in the vertical direction corresponds to a change of $(x_2 - x_1)$ units in the horizontal direction.

$$y_2 - y_1 = \text{the change in } y = \text{rise}$$

and

$$x_2 - x_1 = \text{the change in } x = \text{run}$$

The ratio of $(y_2 - y_1)$ to $(x_2 - x_1)$ represents the slope of the line that passes through the points (x_1, y_1) and (x_2, y_2).

$$\text{Slope} = \frac{\text{change in } y}{\text{change in } x} = \frac{y_2 - y_1}{x_2 - x_1}$$

NOTE The French verb meaning to mount, to climb, or to rise is *monter*. Because Descartes was largely responsible for the development of analytical geometry, his use of *m*—short for *monter*—to indicate the slope became the accepted term among European mathematicians.

THE SLOPE OF A LINE PASSING THROUGH TWO POINTS

The **slope** m of the nonvertical line through (x_1, y_1) and (x_2, y_2) is

$$m = \frac{y_2 - y_1}{x_2 - x_1}$$

where $x_1 \neq x_2$.

When this formula is used for slope, the *order of subtraction* is important. Given two points on a line, you are free to label either one of them as (x_1, y_1) and the other as (x_2, y_2). However, once you have done this, you must form the numerator and denominator using the same order of subtraction.

$$\underbrace{m = \frac{y_2 - y_1}{x_2 - x_1}}_{\text{Correct}} \qquad \underbrace{m = \frac{y_1 - y_2}{x_1 - x_2}}_{\text{Correct}} \qquad \underbrace{m = \frac{y_2 - y_1}{x_1 - x_2}}_{\text{Incorrect}}$$

For instance, the slope of the line passing through the points (3, 4) and (5, 7) can be calculated as

$$m = \frac{7 - 4}{5 - 3} = \frac{3}{2}$$

or

$$m = \frac{4 - 7}{3 - 5} = \frac{-3}{-2} = \frac{3}{2}.$$

The *Interactive* CD-ROM shows every example with its solution; clicking on the *Try It!* button brings up similar problems. Guided Examples and Integrated Examples show step-by-step solutions to additional examples. Integrated Examples are related to several concepts in the section.

EXAMPLE 4 Finding the Slope of a Line

Find the slope of the line passing through each pair of points. (See Figure P.31.)

a. $(-2, 0)$ and $(3, 1)$ **b.** $(-1, 2)$ and $(2, 2)$
c. $(0, 4)$ and $(1, -1)$ **d.** $(3, 4)$ and $(3, 1)$

Solution

a. Letting $(x_1, y_1) = (-2, 0)$ and $(x_2, y_2) = (3, 1)$, you obtain a slope of

$$m = \frac{y_2 - y_1}{x_2 - x_1} = \frac{1 - 0}{3 - (-2)} = \frac{1}{5}.$$

b. The slope of the line passing through $(-1, 2)$ and $(2, 2)$ is

$$m = \frac{2 - 2}{2 - (-1)} = \frac{0}{3} = 0.$$

c. The slope of the line passing through $(0, 4)$ and $(1, -1)$ is

$$m = \frac{-1 - 4}{1 - 0} = \frac{-5}{1} = -5.$$

d. The slope of the vertical line passing through $(3, 4)$ and $(3, 1)$ is not defined because division by zero is undefined.

(a)

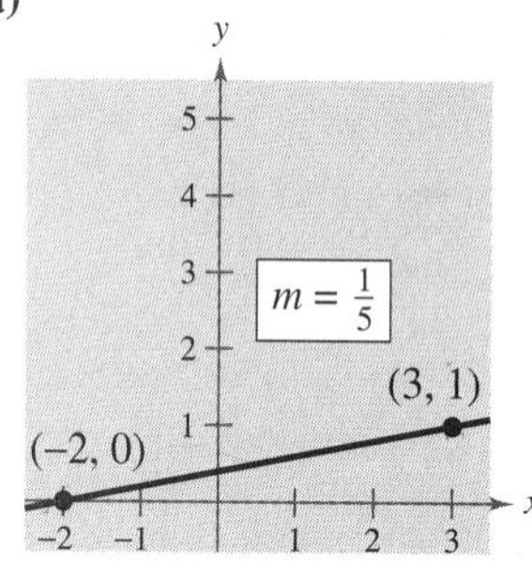

(b)

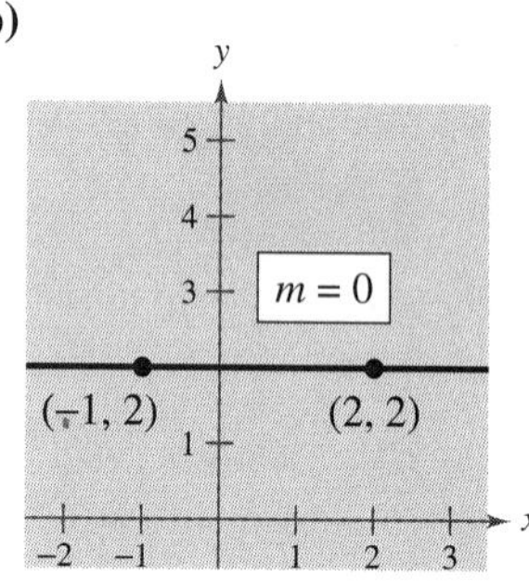

(c)

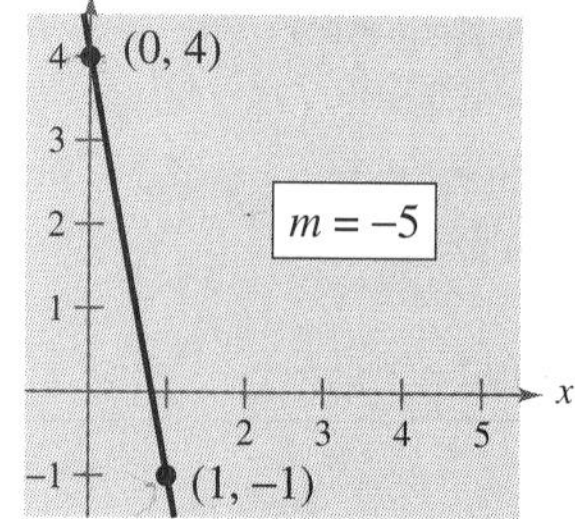

(d)

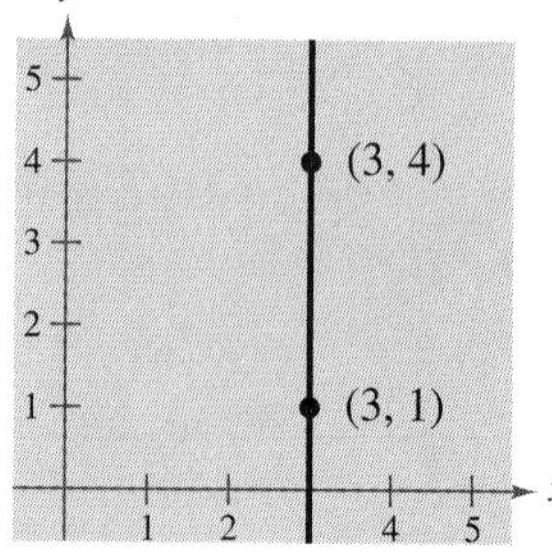

FIGURE P.31

NOTE In Figure P.31, note the following relationships between slope and the description of the line.

a. Positive slope; line rises from left to right

b. Zero slope; line is horizontal

c. Negative slope; line falls from left to right

d. Vertical line; undefined slope

Writing Linear Equations

If (x_1, y_1) is a point lying on a line of slope m and (x, y) is *any other* point on the line, then

$$\frac{y - y_1}{x - x_1} = m.$$

This equation, involving the variables x and y, can be rewritten in the form $y - y_1 = m(x - x_1)$, which is the **point-slope form** of the equation of a line.

POINT-SLOPE FORM OF THE EQUATION OF A LINE

The equation of the line with slope m passing through the point (x_1, y_1) is

$$y - y_1 = m(x - x_1).$$

The point-slope form is most useful for *finding* the equation of a line. You should remember this formula.

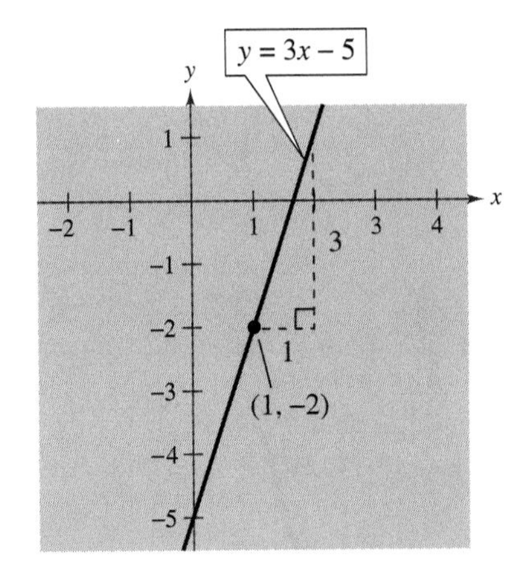

FIGURE P.32

EXAMPLE 5 *Using the Point-Slope Form*

Find the equation of the line that has a slope of 3 and passes through the point $(1, -2)$, as shown in Figure P.32.

Solution

Use the point-slope form with $m = 3$ and $(x_1, y_1) = (1, -2)$.

$y - y_1 = m(x - x_1)$	Point-slope form
$y - (-2) = 3(x - 1)$	Substitute for m, x_1, and y_1.
$y + 2 = 3x - 3$	Simplify.
$y = 3x - 5$	Slope-intercept form

The point-slope form can be used to find an equation of the line passing through points (x_1, y_1) and (x_2, y_2). To do this, first find the slope of the line

$$m = \frac{y_2 - y_1}{x_2 - x_1}, \qquad x_1 \neq x_2,$$

and then use the point-slope form to obtain the equation

$$y - y_1 = \frac{y_2 - y_1}{x_2 - x_1}(x - x_1). \qquad \text{Two-point form}$$

This is sometimes called the **two-point form** of the equation of a line.

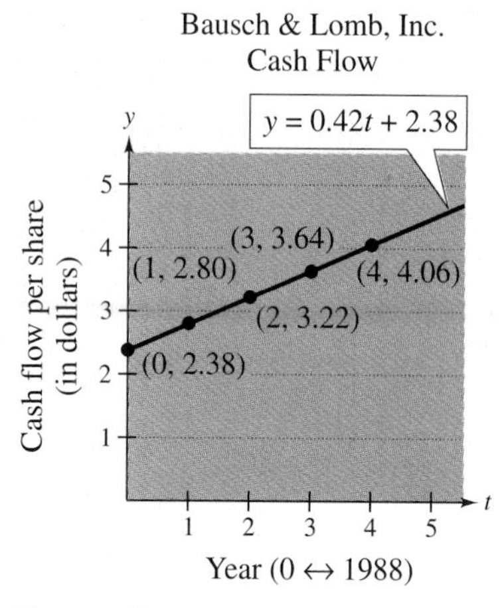

FIGURE P.33

EXAMPLE 6 *Predicting Cash Flow Per Share*

The cash flow per share for Bausch & Lomb, Inc. was \$2.38 in 1988 and \$2.80 in 1989. Using only this information, write a linear equation that gives the cash flow per share in terms of the year. (Source: Bausch & Lomb, Inc.)

Solution

Let $t = 0$ represent 1988. Then the two given values are represented by the ordered pairs (0, 2.38) and (1, 2.80). The slope of the line passing through these points is

$$m = \frac{2.80 - 2.38}{1 - 0} = 0.42.$$

Using the point-slope form, you can find the equation that relates the cash flow C and the year t to be

$$y = 0.42t + 2.38.$$

You can use this model to predict future cash flows. For instance, it predicts the cash flows in 1990, 1991, and 1992 to be \$3.22, \$3.64, and \$4.06, respectively, as shown in Figure P.33. (In this case, the predictions are quite good—the actual cash flows in 1990, 1991, and 1992 were \$3.38, \$3.65, and \$4.16, respectively.)

(a) Linear Extrapolation

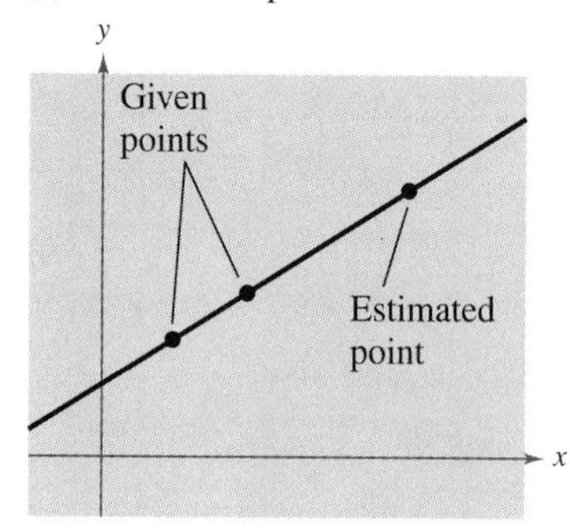

(b) Linear Interpolation

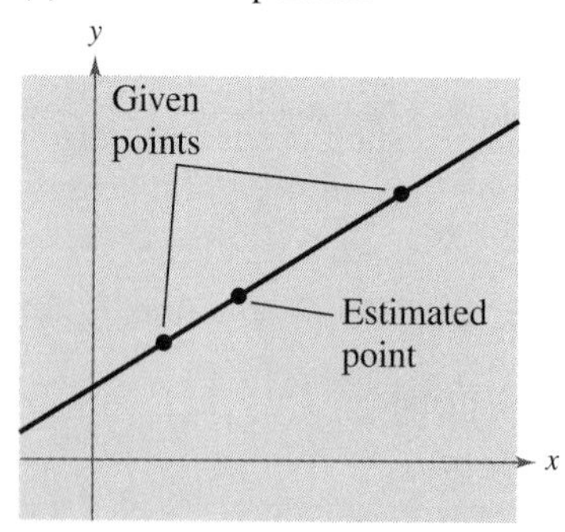

FIGURE P.34

The prediction method illustrated in Example 6 is called **linear extrapolation.** Note in Figure P.34(a) that an extrapolated point does not lie between the given points. When the estimated point lies between two given points, as shown in Figure P.34(b), the procedure is called **linear interpolation.**

Because the slope of a vertical line is not defined, its equation cannot be written in slope-intercept form. However, every line has an equation that can be written in the **general form**

$$Ax + By + C = 0 \qquad \text{General form}$$

where A and B are not both zero. For instance, the vertical line given by $x = a$ can be represented by the general form $x - a = 0$.

EQUATIONS OF LINES

1. General form: $Ax + By + C = 0$
2. Vertical line: $x = a$
3. Horizontal line: $y = b$
4. Slope-intercept form: $y = mx + b$
5. Point-slope form: $y - y_1 = m(x - x_1)$

Parallel and Perpendicular Lines

> **TECHNOLOGY**
>
> On a graphing utility, lines will not appear to have the correct slope unless you use a viewing rectangle that has a square setting. For instance, try graphing the lines in Example 7 using the standard setting $-10 \le x \le 10$ and $-10 \le y \le 10$. Then reset the viewing rectangle with the square setting $-9 \le x \le 9$ and $-6 \le y \le 6$. On which setting do the lines $y = \frac{2}{3}x - \frac{5}{3}$ and $y = -\frac{3}{2}x + 2$ appear perpendicular?

PARALLEL AND PERPENDICULAR LINES

1. Two distinct nonvertical lines are **parallel** if and only if their slopes are equal. That is, $m_1 = m_2$.
2. Two nonvertical lines are **perpendicular** if and only if their slopes are negative reciprocals of each other. That is, $m_1 = -1/m_2$.

EXAMPLE 7 Finding Parallel and Perpendicular Lines

Find an equation of the line that passes through the point $(2, -1)$ and is (a) parallel to and (b) perpendicular to the line $2x - 3y = 5$.

Solution

By writing the equation in slope-intercept form

$$2x - 3y = 5 \qquad \text{Original equation}$$
$$-3y = -2x + 5 \qquad \text{Subtract } 2x \text{ from both sides.}$$
$$y = \tfrac{2}{3}x - \tfrac{5}{3} \qquad \text{Slope-intercept form}$$

you can see that it has a slope of $m = \frac{2}{3}$, as shown in Figure P.35.

a. Any line parallel to the given line must also have a slope of $\frac{2}{3}$. Thus, the line through $(2, -1)$ that is parallel to the given line has the following equation.

$$y - (-1) = \tfrac{2}{3}(x - 2) \qquad \text{Point-slope form}$$
$$3(y + 1) = 2(x - 2) \qquad \text{Multiply both sides by 3.}$$
$$3y + 3 = 2x - 4 \qquad \text{Distributive Property}$$
$$2x - 3y - 7 = 0 \qquad \text{General form}$$
$$y = \tfrac{2}{3}x - \tfrac{7}{3} \qquad \text{Slope-intercept form}$$

b. Any line perpendicular to the given line must have a slope of $-1/(2/3)$ or $-\frac{3}{2}$. Thus, the line through $(2, -1)$ that is perpendicular to the given line has the following equation.

$$y - (-1) = -\tfrac{3}{2}(x - 2) \qquad \text{Point-slope form}$$
$$2(y + 1) = -3(x - 2) \qquad \text{Multiply both sides by 2.}$$
$$3x + 2y - 4 = 0 \qquad \text{General form}$$
$$y = -\tfrac{3}{2}x + 2 \qquad \text{Slope-intercept form}$$

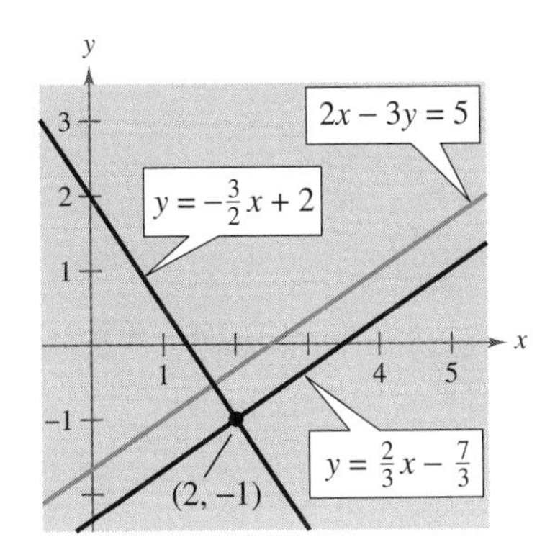

FIGURE P.35

NOTE Most business expenses can be deducted the same year they occur. One exception is the cost of property that has a useful life of more than 1 year. Such costs must be **depreciated** over the useful life of the property. If the *same amount* is depreciated each year, the procedure is called **linear depreciation.** The **book value** is the difference between the original value and the total amount of depreciation accumulated to date.

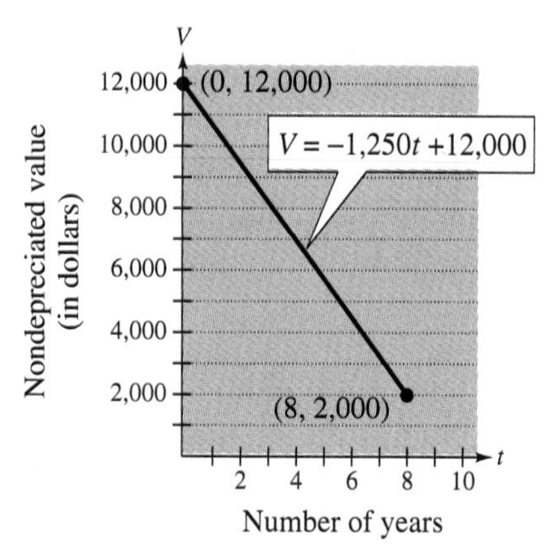

FIGURE P.36 Straight-Line Depreciation

Application

EXAMPLE 8 Depreciating Equipment

Your company has purchased a $12,000 machine that has a useful life of 8 years. The salvage value at the end of 8 years is $2000. Write a linear equation that describes the book value of the machine each year.

Solution

Let V represent the value of the machine at the end of year t. You can represent the initial value of the machine by the ordered pair (0, 12,000) and the salvage value of the machine by the ordered pair (8, 2000). The slope of the line is

$$m = \frac{2000 - 12,000}{8 - 0} = -\$1250$$

which represents the annual depreciation in *dollars per year.* Using the point-slope form, you can write the equation of the line as follows.

$$V - 12,000 = -1250(t - 0) \qquad \text{Point-slope form}$$

$$V = -1250t + 12,000 \qquad \text{Slope-intercept form}$$

The table shows the book value at the end of each year, and the graph of the equation is shown in Figure P.36.

t	0	1	2	3	4	5	6	7	8
V	12,000	10,750	9500	8250	7000	5750	4500	3250	2000

GROUP ACTIVITY

MODELING LINEAR DATA

x	y
3	83.3
4	83.8
5	84.9
6	85.9
7	87.4
8	88.6

x	y
9	90.4
10	92.1
11	93.1
12	92.1
13	93.1
14	94.2

The table at the left shows the total number y (in millions) of households in the United States that owned at least one TV set during each year x from 1983 through 1994, where $x = 3$ represents 1983. (Source: Nielsen Media Research)

Sketch a scatter plot of the data, and use a straight-edge to sketch the best-fitting line through the points. Find the equation of the line. Interpret the slope and y-intercept in the context of the data. Compare your model to those obtained by other students. Are all the different models valid? Explain. Use your model to estimate the number of TV households in 1997, and compare your estimate with those of other students.

WARM UP

Simplify the expression.

1. $\dfrac{4 - (-5)}{-3 - (-1)}$
2. $\dfrac{-5 - 8}{0 - (-3)}$
3. Find $-1/m$ for $m = 4/5$.
4. Find $-1/m$ for $m = -2$.

Solve for y in terms of x.

5. $2x - 3y = 5$
6. $4x + 2y = 0$
7. $y - (-4) = 3[x - (-1)]$
8. $y - 7 = \frac{2}{3}(x - 3)$
9. $y - (-1) = \dfrac{3 - (-1)}{2 - 4}(x - 4)$
10. $y - 5 = \dfrac{3 - 5}{0 - 2}(x - 2)$

P.4 Exercises

In Exercises 1 and 2, identify the line that has the specified slope.

1. (a) $m = \frac{2}{3}$ (b) m is undefined. (c) $m = -2$
2. (a) $m = 0$ (b) $m = -\frac{3}{4}$ (c) $m = 1$

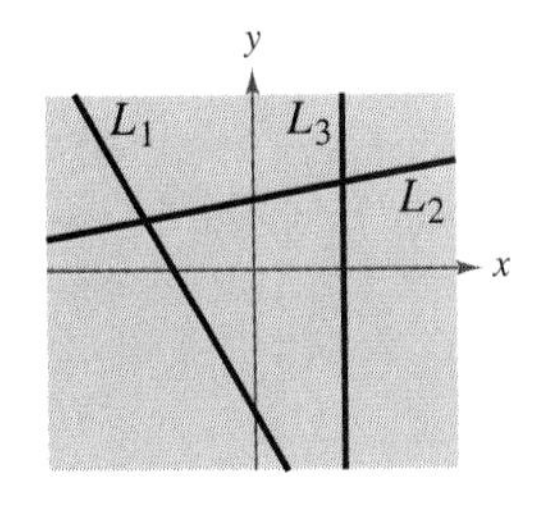

FIGURE FOR 1

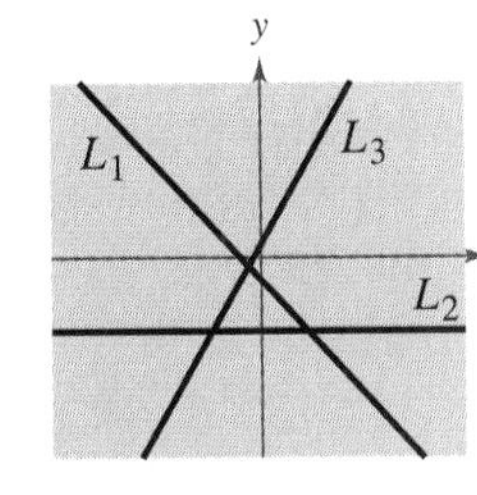

FIGURE FOR 2

In Exercises 3 and 4, sketch the graphs of the lines through the given point with the indicated slopes. Make the sketches on the same set of coordinate axes.

Point	*Slopes*			
3. (2, 3)	(a) 0	(b) 1	(c) 2	(d) -3
4. $(-4, 1)$	(a) 3	(b) -3	(c) $\frac{1}{2}$	(d) undefined

In Exercises 5–10, estimate the slope of the line.

5.
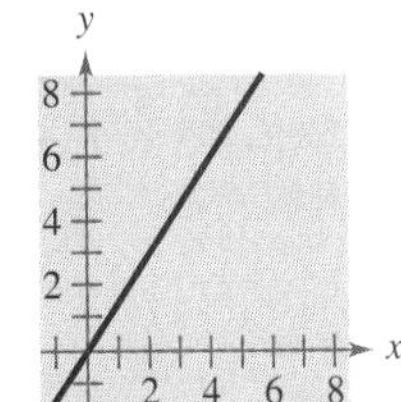

6.
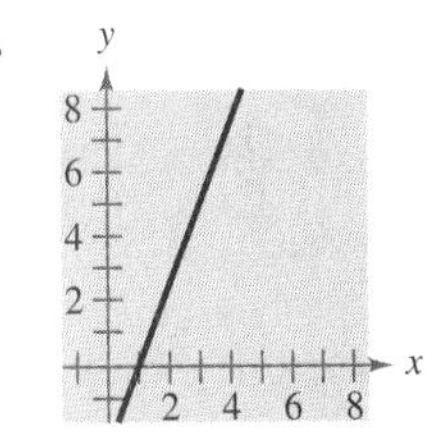

7.
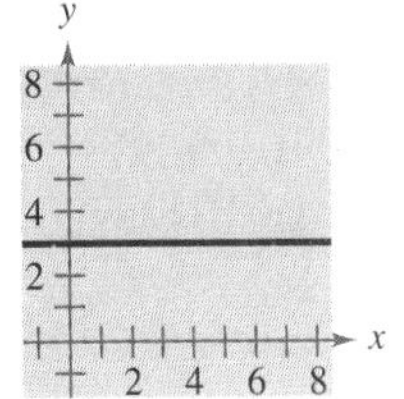

8.
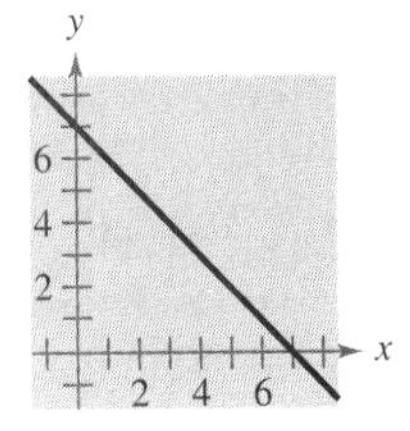

9.
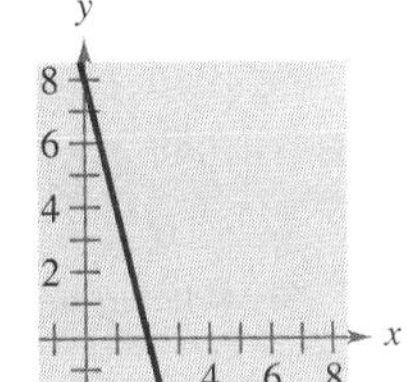

10.
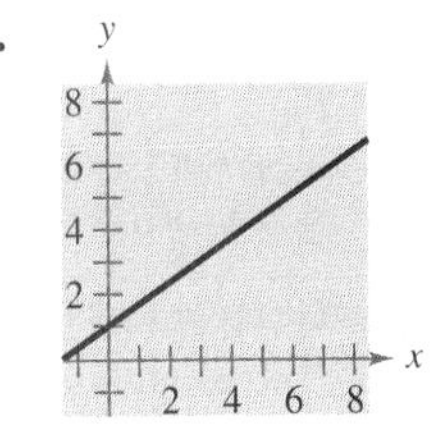

In Exercises 11–16, plot the points and find the slope of the line passing through the pair of points.

11. $(-3, -2), (1, 6)$ **12.** $(2, 4), (4, -4)$

13. $(-6, -1), (-6, 4)$ **14.** $(0, -10), (-4, 0)$

15. $(1, 2), (-2, -2)$ **16.** $\left(\frac{7}{8}, \frac{3}{4}\right), \left(\frac{5}{4}, -\frac{1}{4}\right)$

In Exercises 17–22, use the point on the line and the slope of the line to find three additional points through which the line passes. (The solution is not unique.)

	Point	*Slope*
17.	$(2, 1)$	$m = 0$
18.	$(-4, 1)$	m is undefined.
19.	$(5, -6)$	$m = 1$
20.	$(10, -6)$	$m = -1$
21.	$(-8, 1)$	m is undefined.
22.	$(-3, -1)$	$m = 0$

In Exercises 23–26, determine if the lines L_1 and L_2 passing through the pairs of points are parallel, perpendicular, or neither.

23. L_1: $(0, -1), (5, 9)$
L_2: $(0, 3), (4, 1)$

24. L_1: $(-2, -1), (1, 5)$
L_2: $(1, 3), (5, -5)$

25. L_1: $(3, 6), (-6, 0)$
L_2: $(0, -1), \left(5, \frac{7}{3}\right)$

26. L_1: $(4, 8), (-4, 2)$
L_2: $(3, -5), \left(-1, \frac{1}{3}\right)$

27. ***Essay*** Write a brief paragraph explaining whether or not any pair of points on a line can be used to calculate the slope of the line.

28. ***Think About It*** Is it possible for two lines with positive slopes to be perpendicular? Explain.

29. ***Rate of Change*** The following are the slopes of lines representing annual sales y in terms of time x in years. Use the slopes to interpret any change in annual sales for a 1-year increase in time.

(a) The line has a slope of $m = 135$.

(b) The line has a slope of $m = 0$.

(c) The line has a slope of $m = -40$.

30. ***Rate of Change*** The following are the slopes of lines representing daily revenues y in terms of time x in days. Use the slopes to interpret any change in daily revenues for a 1-day increase in time.

(a) The line has a slope of $m = 400$.

(b) The line has a slope of $m = 100$.

(c) The line has a slope of $m = 0$.

31. ***Earnings per Share*** The graph gives the earnings per share of common stock for General Mills for the years 1987 through 1994. Use the slope to determine the years when earnings (a) decreased most rapidly and (b) increased most rapidly. (Source: General Mills)

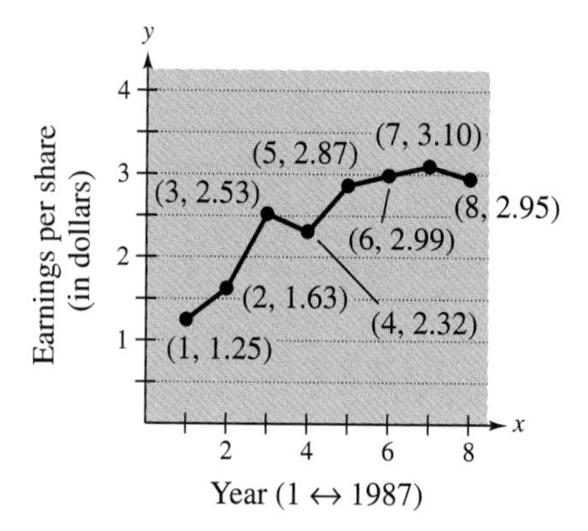

32. ***Dividends per Share*** The graph gives the declared dividend per share of common stock for the Procter and Gamble Company for the years 1987 through 1994. Use the slope to determine the year when dividends increased most rapidly. (Source: Procter and Gamble)

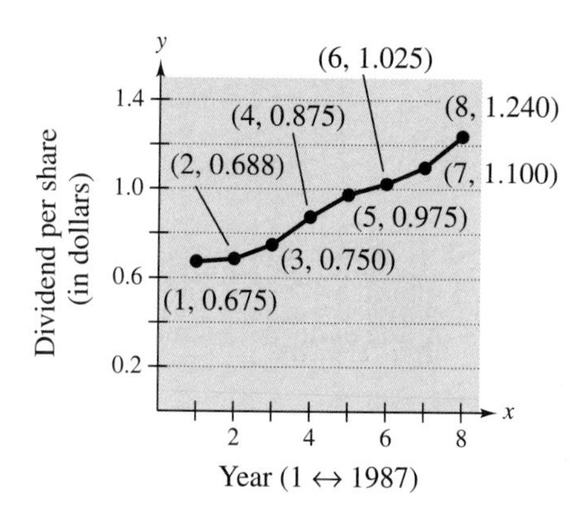

33. ***Mountain Driving*** When driving down a mountain road, you notice warning signs indicating that it is a "12% grade." This means that the slope of the road is $-\frac{12}{100}$. Approximate the amount of horizontal change in your position if you note from elevation markers that you have descended 2000 feet vertically.

34. ***Attic Height*** The "rise to run" in determining the steepness of the roof on a house is 3 to 4. Determine the maximum height in the attic of the house if the house is 32 feet wide (see figure).

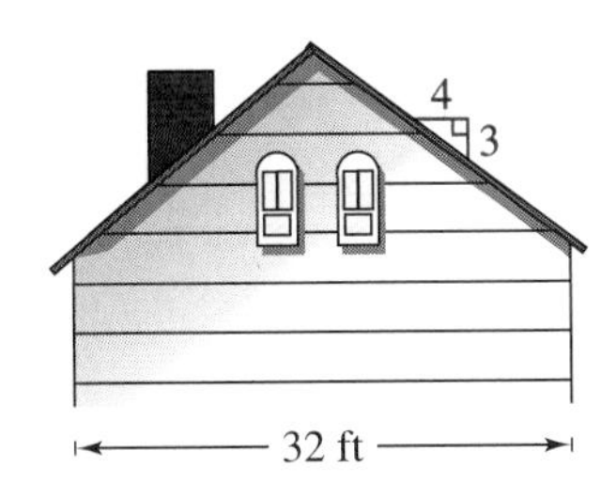

In Exercises 35–40, find the slope and *y*-intercept (if possible) of the equation of the line. Sketch a graph of the line.

35. $5x - y + 3 = 0$

36. $2x + 3y - 9 = 0$

37. $5x - 2 = 0$

38. $3y + 5 = 0$

39. $7x + 6y - 30 = 0$

40. $x - y - 10 = 0$

In Exercises 41–48, find an equation of the line passing through the points and sketch a graph of the line.

41. $(5, -1), (-5, 5)$

42. $(4, 3), (-4, -4)$

43. $\left(2, \frac{1}{2}\right), \left(\frac{1}{2}, \frac{5}{4}\right)$

44. $(-1, 4), (6, 4)$

45. $(-8, 1), (-8, 7)$

46. $(1, 1), \left(6, -\frac{2}{3}\right)$

47. $(1, 0.6), (-2, -0.6)$

48. $(-8, 0.6), (2, -2.4)$

In Exercises 49–58, find an equation of the line that passes through the given point and has the indicated slope. Sketch a graph of the line.

	Point	*Slope*
49.	$(0, -2)$	$m = 3$
50.	$(0, 10)$	$m = -1$
51.	$(-3, 6)$	$m = -2$
52.	$(0, 0)$	$m = 4$
53.	$(4, 0)$	$m = -\frac{1}{3}$
54.	$(-2, -5)$	$m = \frac{3}{4}$
55.	$(6, -1)$	m is undefined.
56.	$(-10, 4)$	$m = 0$
57.	$\left(4, \frac{5}{2}\right)$	$m = \frac{4}{3}$
58.	$\left(-\frac{1}{2}, \frac{3}{2}\right)$	$m = -3$

In Exercises 59–64, use the intercept form to find the equation of the line with the given intercepts. The intercept form of the equation of a line with intercepts $(a, 0)$ and $(0, b)$ is

$$\frac{x}{a} + \frac{y}{b} = 1, \qquad a \neq 0, b \neq 0.$$

59. x-intercept: $(2, 0)$
y-intercept: $(0, 3)$

60. x-intercept: $(-3, 0)$
y-intercept: $(0, 4)$

61. x-intercept: $\left(-\frac{1}{6}, 0\right)$
y-intercept: $\left(0, -\frac{2}{3}\right)$

62. x-intercept: $\left(\frac{2}{3}, 0\right)$
y-intercept: $(0, -2)$

63. Point on line: $(1, 2)$
x-intercept: $(a, 0)$
y-intercept: $(0, a)$, $a \neq 0$

64. Point on line: $(-3, 4)$
x-intercept: $(a, 0)$
y-intercept: $(0, a)$, $a \neq 0$

In Exercises 65–70, write equations of the lines through the given point (a) parallel to the given line and (b) perpendicular to the given line.

	Point	*Line*
65.	$(2, 1)$	$4x - 2y = 3$
66.	$(-3, 2)$	$x + y = 7$
67.	$(-6, 4)$	$3x + 4y = 7$
68.	$\left(\frac{7}{8}, \frac{3}{4}\right)$	$5x + 3y = 0$
69.	$(-1, 0)$	$y = -3$
70.	$(2, 5)$	$x = 4$

In Exercises 71–76, use a graphing utility to graph the pair of equations on the same viewing rectangle. Are the lines parallel, perpendicular, or neither? Use a square setting.

71. $L_1: y = \frac{1}{3}x - 2$
$L_2: y = \frac{1}{3}x + 3$

72. $L_1: y = 2x - 1$
$L_2: y = 2x + 1$

73. $L_1: y = \frac{1}{2}x - 3$
$L_2: y = -\frac{1}{2}x + 1$

74. $L_1: y = -\frac{4}{5}x - 5$
$L_2: y = \frac{5}{4}x + 1$

75. $L_1: y = \frac{2}{3}x - 3$
$L_2: y = -\frac{3}{2}x + 2$

76. $L_1: y = -1.8x + 3.1$
$L_2: y = 2.8x - 4.5$

Graphical Interpretation **In Exercises 77 and 78, use a graphing utility to graph the equation on each viewing rectangle. Which viewing rectangle is better? Explain your reasoning.**

77. $y = 0.5x - 3$

Xmin = -5	Xmin = -2
Xmax = 10	Xmax = 10
Xscl = 1	Xscl = 1
Ymin = -1	Ymin = -4
Ymax = 10	Ymax = 1
Yscl = 1	Yscl = 1

78. $y = -8x + 5$

Xmin = -5	Xmin = -5
Xmax = 5	Xmax = 10
Xscl = 1	Xscl = 1
Ymin = -10	Ymin = -80
Ymax = 10	Ymax = 80
Yscl = 1	Yscl = 20

Graphical Interpretation **In Exercises 79–82, use a graphing utility to graph the three equations on the same viewing rectangle. Adjust the viewing rectangle so the slope appears visually correct. Identify any relationships that exist among the lines.**

79. (a) $y = 2x$ (b) $y = -2x$ (c) $y = \frac{1}{2}x$

80. (a) $y = \frac{2}{3}x$ (b) $y = -\frac{3}{2}x$ (c) $y = \frac{2}{3}x + 2$

81. (a) $y = -\frac{1}{2}x$ (b) $y = -\frac{1}{2}x + 3$ (c) $y = 2x - 4$

82. (a) $y = x - 8$ (b) $y = x + 1$ (c) $y = -x + 3$

Rate of Change **In Exercises 83 and 84, you are given the dollar value of a product in 1996 *and* the rate at which the value of the item is expected to change during the next 5 years. Use this information to write a linear equation that gives the dollar value V of the product in terms of the year t. (Let $t = 6$ represent 1996.)**

	1996 Value	*Rate*
83.	\$2540	\$125 increase per year
84.	\$156	\$4.50 increase per year

Graphical Interpretation **In Exercises 85–88, match the description with its graph. Also determine the slope and how it is interpreted in the situation. [The graphs are labeled (a), (b), (c), and (d).]**

85. A person is paying \$20 per week to a friend to repay a \$200 loan.

86. An employee is paid \$8.50 per hour plus \$2 for each unit produced per hour.

87. A sales representative receives \$30 per day for food plus \$0.32 for each mile traveled.

88. A word processor that was purchased for \$750 depreciates \$100 per year.

(a)
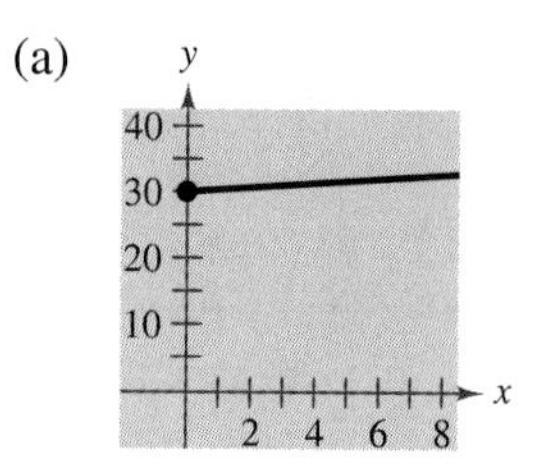

(b)
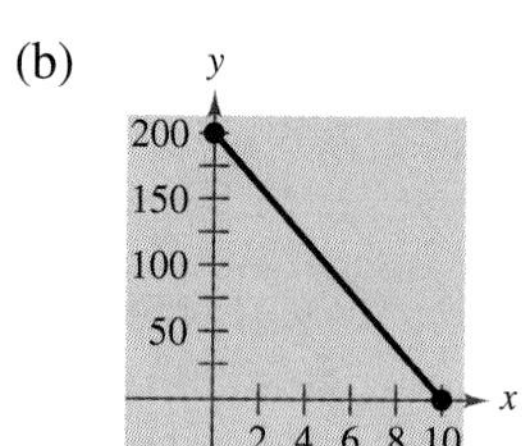

(c)
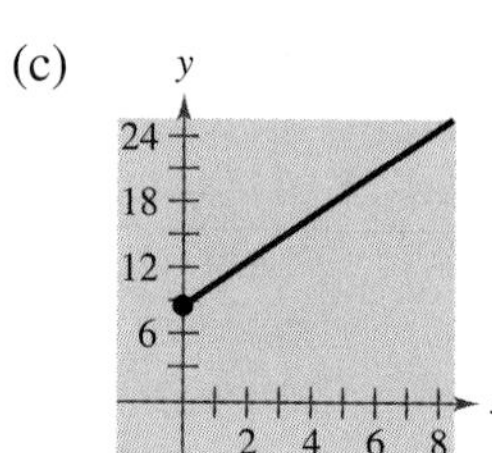

(d)
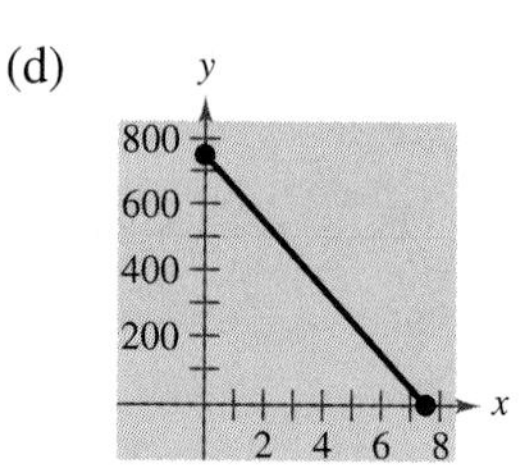

In Exercises 89 and 90, find a relationship between x and y so that (x, y) is equidistant from the two points.

89. $(4, -1), (-2, 3)$

90. $\left(3, \frac{5}{2}\right), (-7, 1)$

91. ***Temperature*** Find the equation of the line giving the relationship between the temperature in degrees Celsius C and degrees Fahrenheit F. Remember that water freezes at 0° Celsius (32° Fahrenheit) and boils at 100° Celsius (212° Fahrenheit).

92. ***Temperature*** Use the result of Exercise 91 to complete the table.

C		$-10°$	$10°$			$177°$
F	$0°$			$68°$	$90°$	

93. ***Annual Salary*** Your salary was \$28,500 in 1994 and \$32,900 in 1996. If your salary follows a linear growth pattern, what will your salary be in 1999?

94. ***College Enrollment*** A small college had 2546 students in 1994 and 2702 students in 1996. If the enrollment follows a linear growth pattern, how many students will the college have in 2000?

95. ***Straight-Line Depreciation*** A small business purchases a piece of equipment for \$875. After 5 years the equipment will be outdated and have no value. Write a linear equation giving the value V of the equipment during the 5 years it will be used.

96. ***Straight-Line Depreciation*** A small business purchases a piece of equipment for \$25,000. After 10 years the equipment will have to be replaced. Its value at that time is expected to be \$2000. Write a linear equation giving the value V of the equipment during the 10 years it will be used.

97. ***Discount*** A store is offering a 15% discount on all items. Write a linear equation giving the sale price S for an item with a list price L.

98. ***Hourly Wages*** A manufacturer pays its assembly line workers \$11.50 per hour. In addition, workers receive a piecework rate of \$0.75 per unit produced. Write a linear equation for the hourly wages W in terms of the number of units x produced per hour.

99. ***Contracting Purchase*** A contractor purchases a piece of equipment for \$36,500. The equipment requires an average expenditure of \$5.25 per hour for fuel and maintenance, and the operator is paid \$11.50 per hour.

(a) Write a linear equation giving the total cost C of operating this equipment for t hours. (Include the purchase cost of the equipment.)

(b) Assuming that customers are charged \$27 per hour of machine use, write an equation for the revenue R derived from t hours of use.

(c) Use the formula for profit ($P = R - C$) to write an equation for the profit derived from t hours of use.

(d) ***Break-Even Point*** Use the result of part (c) to find the number of hours this equipment must be used to yield a profit of 0 dollars.

100. ***Real Estate Purchase*** A real estate office handles an apartment complex with 50 units. When the rent per unit is \$580 per month, all 50 units are occupied. However, when the rent is \$625 per month, the average number of occupied units drops to 47. Assume that the relationship between the monthly rent p and the demand x is linear.

(a) Write the equation of the line giving the demand x in terms of the rent p.

(b) Use this equation to predict the number of units occupied if the rent is \$655.

(c) Predict the number of units occupied if the rent is \$595.

101. ***Perimeter*** The length and width of a rectangular garden are 15 meters and 10 meters, respectively. A walkway of width x surrounds the garden.

(a) Draw a figure that gives a visual representation of the problem.

(b) Write the equation for the perimeter y of the walkway in terms of x.

(c) Use a graphing utility to graph the equation for the perimeter.

(d) Determine the slope of the graph in part (c). For each additional 1-meter increase in the width of the walkway, determine the increase in its perimeter.

102. ***Sales Commission*** A salesperson receives a monthly salary of $2500 plus a commission of 7% of sales. Write a linear equation for the salesperson's monthly wage W in terms of monthly sales S.

103. ***Daily Cost*** A sales representative of a company using a personal car receives $120 per day for lodging and meals plus $0.26 per mile driven. Write a linear equation giving the daily cost C to the company in terms of x, the number of miles driven.

104. ***Simple Interest*** An inheritance of $12,000 is invested in two different mutual funds. One fund pays $5\frac{1}{2}\%$ simple interest and the other pays 8% simple interest.

(a) If x dollars is invested in the fund paying $5\frac{1}{2}\%$, how much is invested in the fund paying 8%?

(b) Write the annual interest y in terms of x.

(c) Use a graphing utility to graph the function in part (b) over the interval $0 \le x \le 12{,}000$.

(d) Explain why the slope of the line in part (c) is negative.

105. ***Baseball Salaries*** The average annual salaries of major league baseball players (in thousands of dollars) from 1984 to 1994 are shown in the scatter plot. Find the equation of the line that you think best fits this data. (Let y represent the average salary and let t represent the year, with $t = 0$ corresponding to 1984.) (Source: Major League Baseball Players Association)

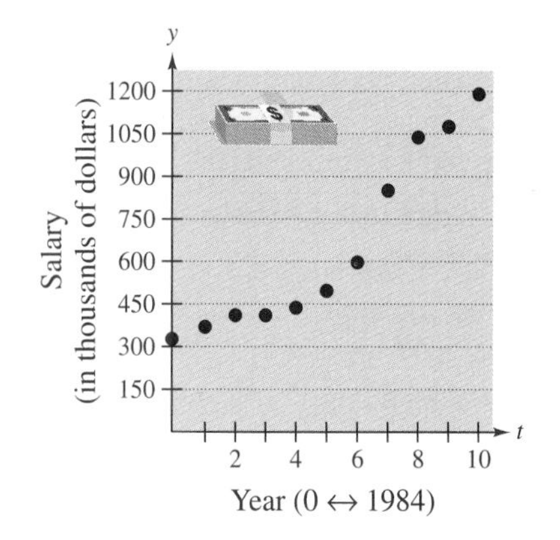

106. ***Data Analysis*** An instructor gives regular 20-point quizzes and 100-point exams in a mathematics course. Average scores for six students, given as ordered pairs (x, y) where x is the average quiz score and y is the average test score, are (18, 87), (10, 55), (19, 96), (16, 79), (13, 76), and (15, 82). [*Note*: The answers are not unique for parts (b)–(d).]

(a) Sketch a scatter plot of the data.

(b) Use a straight edge to sketch the "best-fitting" line through the points.

(c) Find an equation for the line sketched in part (b).

(d) Use the equation of part (c) to estimate the average test score for a person with an average quiz score of 17.

(e) If the instructor added 4 points to the average test score of everyone in the class, describe the changes in the positions of the plotted points and the change in the equation of the line.

Review **Solve Exercises 107–110 as a review of the skills and problem-solving techniques you learned in previous sections. Match the equation with its graph. [The graphs are labeled (a), (b), (c), and (d).]**

(a)
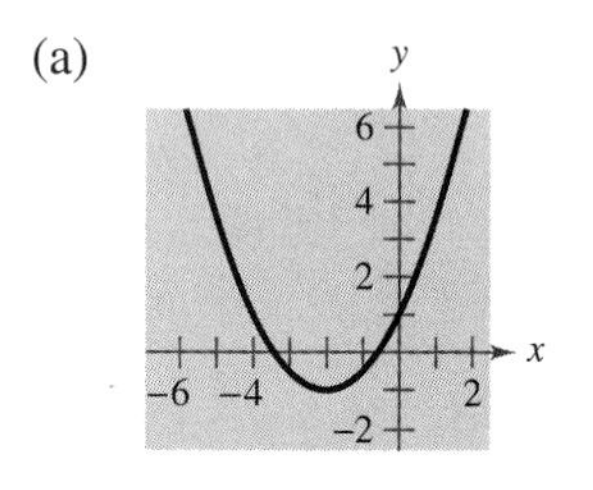

(b)
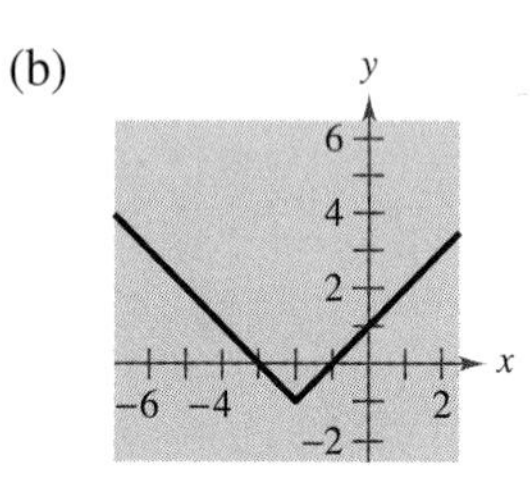

(c)
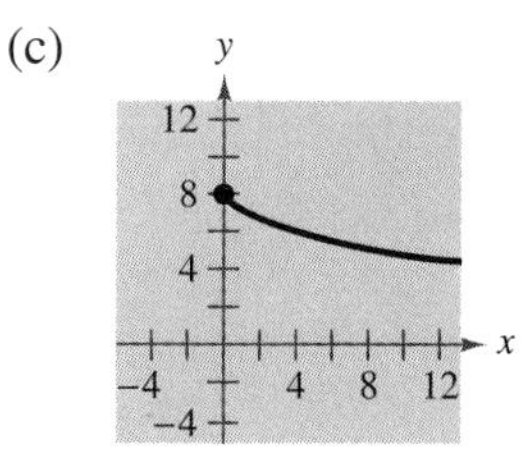

(d)
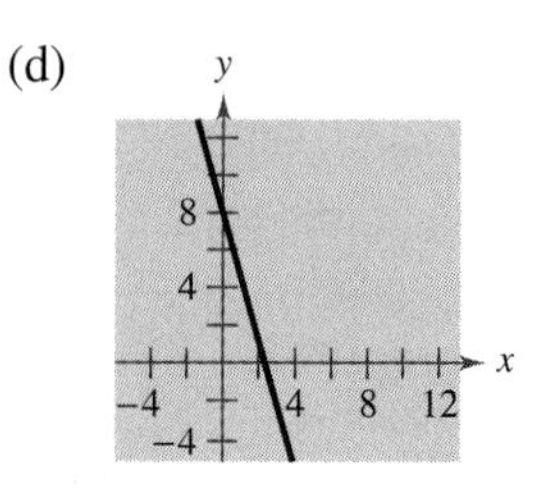

107. $y = 8 - 3x$

108. $y = 8 - \sqrt{x}$

109. $y = \frac{1}{2}x^2 + 2x + 1$

110. $y = |x + 2| - 1$

P.5 Functions

See Exercise 84 on page 69 for an example of how a piecewise-defined function can be used to model the price of mobile homes from 1974 through 1993.

Introduction to Functions ▫ *Function Notation* ▫ *The Domain of a Function* ▫ *Applications*

Introduction to Functions

Many everyday phenomena involve two quantities that are related to each other by some rule of correspondence. Here are some examples.

1. The simple interest I earned on \$1000 for 1 year is related to the annual interest rate r by the formula $I = 1000r$.
2. The distance d traveled on a bicycle in 2 hours is related to the speed s of the bicycle by the formula $d = 2s$.
3. The area A of a circle is related to its radius r by the formula $A = \pi r^2$.

Not all correspondences between two quantities have simple mathematical formulas. For instance, people commonly match up NFL starting quarterbacks with touchdown passes and hours of the day with temperature. In each of these cases, however, there is some rule of correspondence that matches each item from one set with exactly one item from a different set. Such a rule of correspondence is called a **function.**

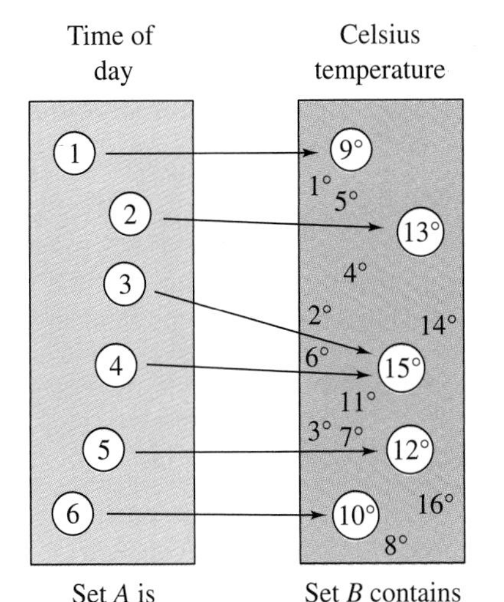

FIGURE P.37

DEFINITION OF A FUNCTION

A **function** f from a set A to a set B is a rule of correspondence that assigns to each element x in the set A exactly one element y in the set B. The set A is the **domain** (or set of inputs) of the function f, and the set B contains the **range** (or set of outputs).

To help understand this definition, look at the function illustrated in Figure P.37. This function can be represented by the following ordered pairs.

$$\{(1, 9°), (2, 13°), (3, 15°), (4, 15°), (5, 12°), (6, 10°)\}$$

In each ordered pair, the first coordinate is the input and the second coordinate is the output. In this example, note the following characteristics of a function.

1. Each element in A must be matched with an element of B.
2. Some elements in B may not be matched with any element in A.
3. Two or more elements of A may be matched with the same element of B.

The converse of the third statement is not true. That is, an element of A (the domain) cannot be matched with two different elements of B.

In the following example, you are asked to decide whether different correspondences are functions. To do this, you must decide whether each element in the domain A is matched with exactly one element in the range B. If any element in A is matched with two or more elements in B, the correspondence is not a function.

EXAMPLE 1 *Testing for Functions*

Let $A = \{a, b, c\}$ and $B = \{1, 2, 3, 4, 5\}$. Which of the following sets of ordered pairs or figures represent functions from set A to set B?

a. $\{(a, 2), (b, 3), (c, 4)\}$

b. $\{(a, 4), (b, 5)\}$

c.

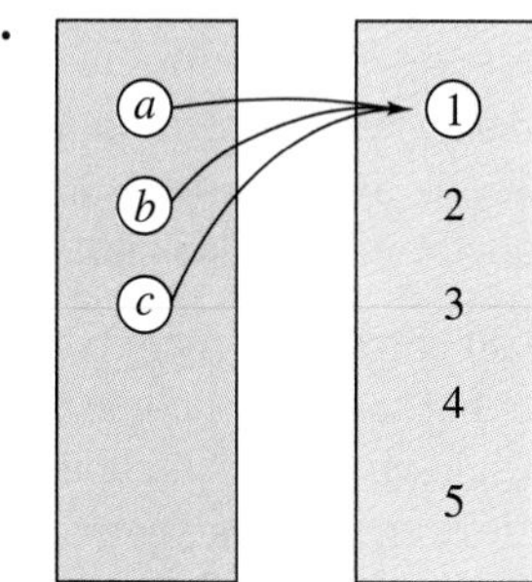

d.

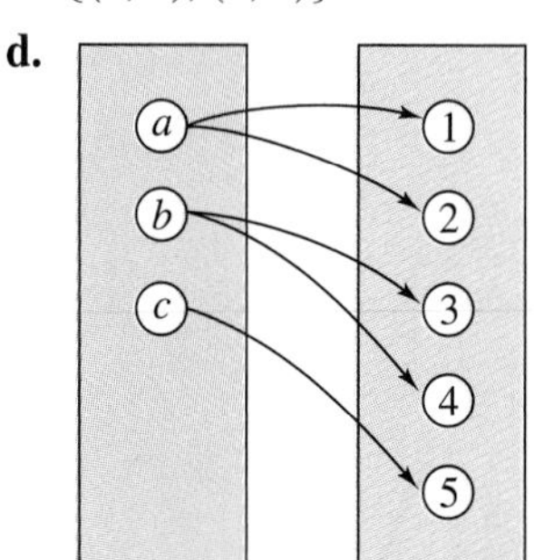

Solution

a. This collection of ordered pairs *does* represent a function from A to B. Each element of A is matched with exactly one element of B.

b. This collection of ordered pairs *does not* represent a function from A to B. Not every element of A is matched with an element of B.

c. This figure *does* represent a function from A to B. It does not matter that each element of A is matched with the same element of B.

d. This figure *does not* represent a function from A to B. The element a in A is matched with *two* elements, 1 and 2, of B. This is also true of the element b.

Representing functions by sets of ordered pairs is common in *discrete mathematics*. In algebra, however, it is more common to represent functions by equations or formulas involving two variables. For instance, the equation

$$y = x^2 \qquad y \text{ is a function of } x.$$

represents the variable y as a function of the variable x. In this equation, x is the **independent variable** and y is the **dependent variable.** The domain of the function is the set of all values taken on by the independent variable x, and the range of the function is the set of all values taken on by the dependent variable y.

Leonhard Euler (1707–1783), a Swiss mathematician, is considered to have been the most prolific and productive mathematician in history. One of his greatest influences on mathematics was his use of symbols, or notation. The function notation $y = f(x)$ was introduced by Euler.

EXAMPLE 2 ***Testing for Functions Represented by Equations***

Which of the equations represent(s) y as a function of x?

a. $x^2 + y = 1$ **b.** $-x + y^2 = 1$

Solution

To determine whether y is a function of x, try to solve for y in terms of x.

a. Solving for y yields the following.

$$x^2 + y = 1 \qquad \text{Original equation}$$

$$y = 1 - x^2 \qquad \text{Solve for } y.$$

To each value of x there corresponds exactly one value of y. Thus, y *is* a function of x.

b. Solving for y yields the following.

$$-x + y^2 = 1 \qquad \text{Original equation}$$

$$y^2 = 1 + x \qquad \text{Add } x \text{ to both sides.}$$

$$y = \pm\sqrt{1 + x} \qquad \text{Solve for } y.$$

The $\pm$ indicates that to a given value of x there correspond two values of y. Thus, y *is not* a function of x.

Function Notation

When an equation is used to represent a function, it is convenient to name the function so that it can be referenced easily. For example, you know that the equation $y = 1 - x^2$ describes y as a function of x. Suppose you give this function the name "f." Then you can use the following **function notation.**

Input	*Output*	*Equation*
x	$f(x)$	$f(x) = 1 - x^2$

The symbol $f(x)$ is read as **the value of f at x** or simply **f of x.** The symbol $f(x)$ corresponds to the y-value for a given x. Thus, you can write $y = f(x)$. Keep in mind that f is the *name* of the function, whereas $f(x)$ is the *value* of the function at x. For instance, the function given by

$$f(x) = 3 - 2x$$

has *function values* denoted by $f(-1), f(0), f(2)$, and so on. To find these values, substitute the specified input values into the given equation.

For $x = -1$, $\quad f(-1) = 3 - 2(-1) = 3 + 2 = 5.$

For $x = 0$, $\quad f(0) = 3 - 2(0) = 3 - 0 = 3.$

For $x = 2$, $\quad f(2) = 3 - 2(2) = 3 - 4 = -1.$

Although f is often used as a convenient function name and x is often used as the independent variable, you can use other letters. For instance,

$$f(x) = x^2 - 4x + 7, \quad f(t) = t^2 - 4t + 7, \quad \text{and} \quad g(s) = s^2 - 4s + 7$$

all define the same function. In fact, the role of the independent variable is that of a "placeholder." Consequently, the function could be described by

$$f(\quad) = (\quad)^2 - 4(\quad) + 7.$$

EXAMPLE 3 Evaluating a Function

Let $g(x) = -x^2 + 4x + 1$ and find the following.

a. $g(2)$ **b.** $g(t)$ **c.** $g(x + 2)$

Solution

a. Replacing x with 2 in $g(x) = -x^2 + 4x + 1$ yields the following.

$$g(2) = -(2)^2 + 4(2) + 1 = -4 + 8 + 1 = 5$$

b. Replacing x with t yields the following.

$$g(t) = -(t)^2 + 4(t) + 1 = -t^2 + 4t + 1$$

c. Replacing x with $x + 2$ yields the following.

$$\begin{aligned} g(x + 2) &= -(x + 2)^2 + 4(x + 2) + 1 \\ &= -(x^2 + 4x + 4) + 4x + 8 + 1 \\ &= -x^2 - 4x - 4 + 4x + 8 + 1 \\ &= -x^2 + 5 \end{aligned}$$

NOTE In Example 3, note that $g(x + 2)$ is not equal to $g(x) + g(2)$. In general, $g(u + v) \neq g(u) + g(v)$.

EXAMPLE 4 A Piecewise-Defined Function

Evaluate the function when $x = -1, 0$, and 1.

$$f(x) = \begin{cases} x^2 + 1, & x < 0 \\ x - 1, & x \geq 0 \end{cases}$$

Solution

Because $x = -1$ is less than 0, use $f(x) = x^2 + 1$ to obtain

$$f(-1) = (-1)^2 + 1 = 2.$$

For $x = 0$, use $f(x) = x - 1$ to obtain

$$f(0) = (0) - 1 = -1.$$

For $x = 1$, use $f(x) = x - 1$ to obtain

$$f(1) = (1) - 1 = 0.$$

NOTE A function defined by two or more equations over a specified domain is called a **piecewise-defined** function.

The Domain of a Function

Exploration

Use a graphing utility to graph $y = \sqrt{4 - x^2}$. What is the domain of this function? Then graph $y = \sqrt{x^2 - 4}$. What is the domain of this function? Do the domains of these two functions overlap? If so, for what values?

The domain of a function can be described explicitly or it can be *implied* by the expression used to define the function. The **implied domain** is the set of all real numbers for which the expression is defined. For instance, the function given by

$$f(x) = \frac{1}{x^2 - 4}$$

has an implied domain that consists of all real x other than $x = \pm 2$. These two values are excluded from the domain because division by zero is undefined. Another common type of implied domain is that used to avoid even roots of negative numbers. For example, the function given by

$$f(x) = \sqrt{x}$$

is defined only for $x \geq 0$. Hence, its implied domain is the interval $[0, \infty)$. In general, the domain of a function *excludes* values that would cause division by zero *or* result in the even root of a negative number.

The *Interactive* CD-ROM offers graphing utility emulators of the *TI-82* and *TI-83*, which can be used with the Examples, Explorations, Technology notes, and Exercises.

EXAMPLE 5 *Finding the Domain of a Function*

Find the domain of each function.

a. $f\colon \{(-3, 0), (-1, 4), (0, 2), (2, 2), (4, -1)\}$

b. $g(x) = \dfrac{1}{x + 5}$

c. Volume of a sphere: $V = \frac{4}{3}\pi r^3$

d. $h(x) = \sqrt{4 - x^2}$

Solution

a. The domain of f consists of all first coordinates in the set of ordered pairs.

$$\text{Domain} = \{-3, -1, 0, 2, 4\}$$

b. Excluding x-values that yield zero in the denominator, the domain of g is the set of all real numbers $x \neq -5$.

c. Because this function represents the volume of a sphere, the values of the radius r must be positive. Thus, the domain is the set of all real numbers r such that $r > 0$.

d. This function is defined only for x-values for which $4 - x^2 \geq 0$. By solving this inequality, you can conclude that $-2 \leq x \leq 2$. Thus, the domain is the interval $[-2, 2]$.

NOTE In Example 5(c), note that the domain of a function may be implied by the physical context. For instance, from the equation $V = \frac{4}{3}\pi r^3$, you would have no reason to restrict r to positive values, but the physical context implies that a sphere cannot have negative radius.

Applications

EXAMPLE 6 *The Dimensions of a Container*

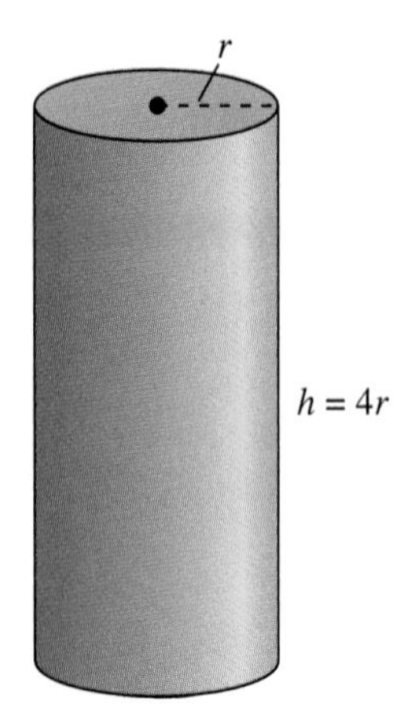

FIGURE P.38

You work in the marketing department of a soft-drink company and are experimenting with a new soft-drink can that is slightly narrower and taller than a standard can. For your experimental can, the ratio of the height to the radius is 4, as shown in Figure P.38.

a. Express the volume of the can as a function of the radius r.

b. Express the volume of the can as a function of the height h.

Solution

a. $V = \pi r^2 h = \pi r^2(4r) = 4\pi r^3$ $\quad$ V as a function of r

b. $V = \pi\left(\dfrac{h}{4}\right)^2 h = \dfrac{\pi h^3}{16}$ $\quad$ V as a function of h

EXAMPLE 7 *The Path of a Baseball*

A baseball is hit at a point 3 feet above ground at a velocity of 100 feet per second and an angle of 45°. The path of the baseball is given by the function

$$y = -0.0032x^2 + x + 3$$

where y and x are measured in feet, as shown in Figure P.39. Will the baseball clear a 10-foot fence located 300 feet from home plate?

Solution

When $x = 300$, the height of the baseball is given by

$$y = -0.0032(300)^2 + 300 + 3 = 15 \text{ feet.}$$

Thus, the ball will clear the fence.

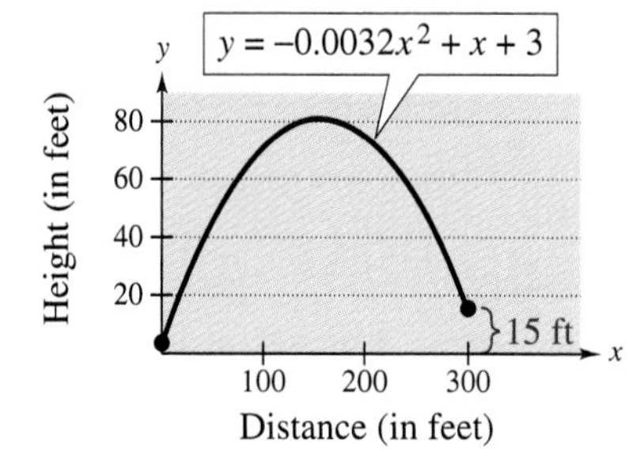

FIGURE P.39

NOTE In the equation in Example 7, the height of the baseball is a function of the distance from home plate.

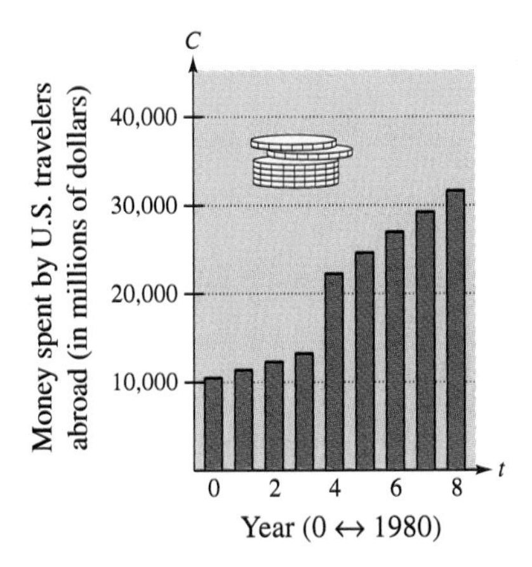

FIGURE P.40

EXAMPLE 8 *U.S. Travelers Abroad*

The money C (in millions of dollars) spent by U.S. travelers in other countries increased in a linear pattern from 1980 to 1983, as shown in Figure P.40. Then, in 1984, the money spent took a sharp jump and until 1988 increased in a *different* linear pattern. These two patterns can be approximated by the function

$$C = \begin{cases} 10{,}479 + 917.1t, & 0 \le t \le 3 \\ 12{,}808 + 2350.4t, & 4 \le t \le 8 \end{cases}$$

where $t = 0$ represents 1980. Use this function to approximate the total amount spent by U.S. travelers abroad between 1980 and 1988. (Source: U.S. Bureau of Economic Analysis)

Solution

From 1980 to 1983, use the formula $C = 10{,}479 + 917.1t$.

$$\underbrace{\$10{,}479}_{1980},\ \underbrace{\$11{,}396}_{1981},\ \underbrace{\$12{,}313}_{1982},\ \underbrace{\$13{,}230}_{1983}$$

From 1984 to 1988, use the formula $C = 12{,}808 + 2350.4t$.

$$\underbrace{\$22{,}210}_{1984},\ \underbrace{\$24{,}560}_{1985},\ \underbrace{\$26{,}910}_{1986},\ \underbrace{\$29{,}261}_{1987},\ \underbrace{\$31{,}611}_{1988}$$

The total of these nine amounts is \$181,970, which implies that the total amount spent was approximately \$181,970,000,000.

EXAMPLE 9 *From Calculus: Evaluating a Difference Quotient*

For $f(x) = x^2 - 4x + 7$, find $\dfrac{f(x+h) - f(x)}{h}$.

Solution

$$\begin{aligned}\frac{f(x+h) - f(x)}{h} &= \frac{[(x+h)^2 - 4(x+h) + 7] - (x^2 - 4x + 7)}{h} \\ &= \frac{x^2 + 2xh + h^2 - 4x - 4h + 7 - x^2 + 4x - 7}{h} \\ &= \frac{2xh + h^2 - 4h}{h} \\ &= \frac{h(2x + h - 4)}{h} \\ &= 2x + h - 4, \quad h \ne 0\end{aligned}$$

NOTE One of the basic definitions in calculus employs the ratio

$$\frac{f(x+h) - f(x)}{h}, \quad h \ne 0$$

called a **difference quotient,** as illustrated in Example 9.

SUMMARY OF FUNCTION TERMINOLOGY

Function: A **function** is a relationship between two variables such that to each value of the independent variable there corresponds exactly one value of the dependent variable.

Function Notation: $y = f(x)$

f is the **name** of the function.
y is the **dependent variable.**
x is the **independent variable.**
$f(x)$ is the **value of the function at x.**

Domain: The **domain** of a function is the set of all values (inputs) of the independent variable for which the function is defined. If x is in the domain of f, we say that f is **defined** at x. If x is not in the domain of f, we say that f is **undefined** at x.

Range: The **range** of a function is the set of all values (outputs) assumed by the dependent variable (that is, the set of all function values).

Implied Domain: If f is defined by an algebraic expression and the domain is not specified, the **implied domain** consists of all real numbers for which the expression is defined.

GROUP ACTIVITY

MODELING WITH PIECEWISE-DEFINED FUNCTIONS

The table at the right shows the monthly revenue (in thousands of dollars) for one year of a landscaping business, with $x = 1$ representing January.

x	y
1	5.2
2	5.6
3	6.6
4	8.3
5	11.5
6	15.8

x	y
7	12.8
8	10.1
9	8.6
10	6.9
11	4.5
12	2.7

A mathematical model that represents this data is:

$$f(x) = \begin{cases} -1.97x + 26.33 \\ 0.5|x^2 - 1.47x + 6.3|. \end{cases}$$

For what values of x is each part of the piecewise-defined function defined? How can you tell? Explain your reasoning.

Find $f(5)$ and $f(11)$, and interpret your results in the context of the problem. How do these model values compare with the actual data values?

WARM UP

Simplify the expression.

1. $2(-3)^3 + 4(-3) - 7$

2. $4(-1)^2 - 5(-1) + 4$

3. $(x + 1)^2 + 3(x + 1) - 4 - (x^2 + 3x - 4)$

4. $(x - 2)^2 - 4(x - 2) - (x^2 - 4)$

Solve for y in terms of x.

5. $2x + 5y - 7 = 0$

6. $y^2 = x^2$

Solve the inequality.

7. $9 - 2x \geq 0$

8. $3x + 2 \geq 0$

9. $9 - x^2 \geq 0$

10. $x^2 - 3x + 2 \geq 0$

P.5 Exercises

In Exercises 1–4, is the relationship a function?

1.

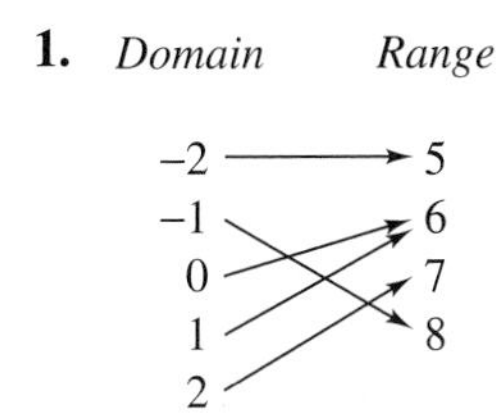

2.

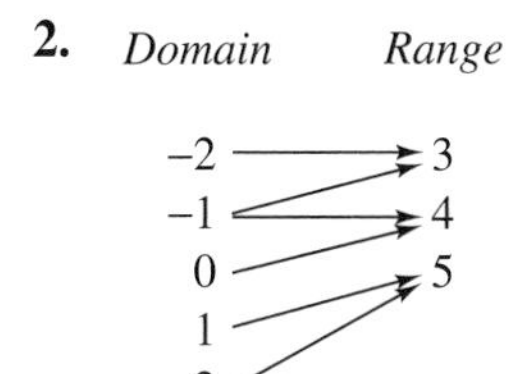

3. **4.**

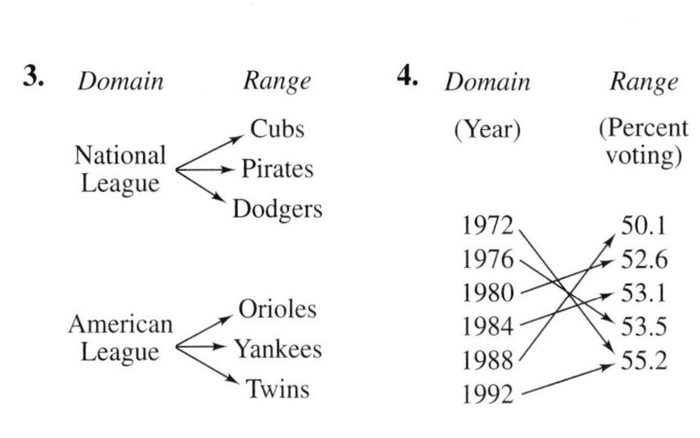

In Exercises 5–8, does the table describe a function? Explain your reasoning.

5.

Input Value	−2	−1	0	1	2
Output Value	−8	−1	0	1	8

6.

Input Value	0	1	2	1	0
Output Value	−4	−2	0	2	4

7.

Input Value	10	7	4	7	10
Output Value	3	6	9	12	15

8.

Input Value	0	3	9	12	15
Output Value	3	3	3	3	3

In Exercises 9 and 10, which sets of ordered pairs represent function(s) from A to B? Explain.

9. $A = \{0, 1, 2, 3\}$ and $B = \{-2, -1, 0, 1, 2\}$

(a) $\{(0, 1), (1, -2), (2, 0), (3, 2)\}$

(b) $\{(0, -1), (2, 2), (1, -2), (3, 0), (1, 1)\}$

(c) $\{(0, 0), (1, 0), (2, 0), (3, 0)\}$

(d) $\{(0, 2), (3, 0), (1, 1)\}$

10. $A = \{a, b, c\}$ and $B = \{0, 1, 2, 3\}$

(a) $\{(a, 1), (c, 2), (c, 3), (b, 3)\}$

(b) $\{(a, 1), (b, 2), (c, 3)\}$

(c) $\{(1, a), (0, a), (2, c), (3, b)\}$

(d) $\{(c, 0), (b, 0), (a, 3)\}$

Circulation of Newspapers **In Exercises 11 and 12, use the graph, which shows the circulation (in millions) of daily newspapers in the United States.** (Source: Editor & Publisher Company)

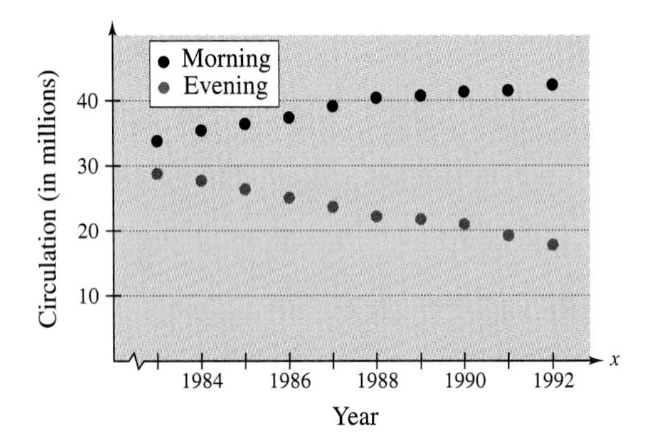

11. Is the circulation of morning newspapers a function of the year? Is the circulation of evening newspapers a function of the year? Explain.

12. Let $f(x)$ represent the circulation of evening newspapers in year x. Find $f(1988)$.

In Exercises 13–22, determine if the equation represents y as a function of x.

13. $x^2 + y^2 = 4$ **14.** $x = y^2$

15. $x^2 + y = 4$ **16.** $x + y^2 = 4$

17. $2x + 3y = 4$ **18.** $(x - 2)^2 + y^2 = 4$

19. $y^2 = x^2 - 1$ **20.** $y = \sqrt{x + 5}$

21. $y = |4 - x|$ **22.** $|y| = 4 - x$

In Exercises 23 and 24, fill in the blanks using the specified function and the given values of the independent variable.

23. $f(s) = \dfrac{1}{s + 1}$

(a) $f(4) = \dfrac{1}{(\quad) + 1}$

(b) $f(0) = \dfrac{1}{(\quad) + 1}$

(c) $f(4x) = \dfrac{1}{(\quad) + 1}$

(d) $f(x + c) = \dfrac{1}{(\quad) + 1}$

24. $g(x) = x^2 - 2x$

(a) $g(2) = (\quad)^2 - 2(\quad)$

(b) $g(-3) = (\quad)^2 - 2(\quad)$

(c) $g(t + 1) = (\quad)^2 - 2(\quad)$

(d) $g(x + c) = (\quad)^2 - 2(\quad)$

In Exercises 25–36, evaluate the function at the specified values of the independent variable and simplify.

25. $f(x) = 2x - 3$

(a) $f(1)$ (b) $f(-3)$ (c) $f(x - 1)$

26. $g(y) = 7 - 3y$

(a) $g(0)$ (b) $g\left(\frac{7}{3}\right)$ (c) $g(s + 2)$

27. $h(t) = t^2 - 2t$

(a) $h(2)$ (b) $h(1.5)$ (c) $h(x + 2)$

28. $V(r) = \frac{4}{3}\pi r^3$

(a) $V(3)$ (b) $V\left(\frac{3}{2}\right)$ (c) $V(2r)$

29. $f(y) = 3 - \sqrt{y}$

(a) $f(4)$ (b) $f(0.25)$ (c) $f(4x^2)$

30. $f(x) = \sqrt{x + 8} + 2$

(a) $f(-8)$ (b) $f(1)$ (c) $f(x - 8)$

31. $q(x) = \dfrac{1}{x^2 - 9}$

(a) $q(0)$ (b) $q(3)$ (c) $q(y + 3)$

32. $q(t) = \dfrac{2t^2 + 3}{t^2}$

(a) $q(2)$ (b) $q(0)$ (c) $q(-x)$

33. $f(x) = \dfrac{|x|}{x}$

(a) $f(2)$ (b) $f(-2)$ (c) $f(x - 1)$

34. $f(x) = |x| + 4$

(a) $f(2)$ (b) $f(-2)$ (c) $f(x^2)$

35. $f(x) = \begin{cases} 2x + 1, & x < 0 \\ 2x + 2, & x \ge 0 \end{cases}$

(a) $f(-1)$ (b) $f(0)$ (c) $f(2)$

36. $f(x) = \begin{cases} x^2 + 2, & x \le 1 \\ 2x^2 + 2, & x > 1 \end{cases}$

(a) $f(-2)$ (b) $f(1)$ (c) $f(2)$

In Exercises 37–42, complete the table.

37. $f(x) = x^2 - 3$

x	-2	-1	0	1	2
$f(x)$					

38. $g(x) = \sqrt{x - 3}$

x	3	4	5	6	7
$g(x)$					

39. $h(t) = \frac{1}{2}|t + 3|$

t	-5	-4	-3	-2	-1
$h(t)$					

40. $f(s) = \dfrac{|s - 2|}{s - 2}$

s	0	1	$\frac{3}{2}$	$\frac{5}{2}$	4
$f(s)$					

41. $f(x) = \begin{cases} -\frac{1}{2}x + 4, & x \le 0 \\ (x - 2)^2, & x > 0 \end{cases}$

x	-2	-1	0	1	2
$f(x)$					

42. $h(x) = \begin{cases} 9 - x^2, & x < 3 \\ x - 3, & x \ge 3 \end{cases}$

x	1	2	3	4	5
$h(x)$					

In Exercises 43–46, find all real values of x such that $f(x) = 0$.

43. $f(x) = 15 - 3x$

44. $f(x) = \dfrac{3x - 4}{5}$

45. $f(x) = x^2 - 9$

46. $f(x) = x^3 - x$

In Exercises 47–50, find the value(s) of x for which $f(x) = g(x)$.

47. $f(x) = x^2, \quad g(x) = x + 2$

48. $f(x) = x^2 + 2x + 1, \quad g(x) = 3x + 3$

49. $f(x) = \sqrt{3x} + 1, \quad g(x) = x + 1$

50. $f(x) = x^4 - 2x^2, \quad g(x) = 2x^2$

In Exercises 51–60, find the domain of the function.

51. $f(x) = 5x^2 + 2x - 1$

52. $g(x) = 1 - 2x^2$

53. $h(t) = \dfrac{4}{t}$

54. $s(y) = \dfrac{3y}{y + 5}$

55. $g(y) = \sqrt{y - 10}$

56. $f(t) = \sqrt[3]{t + 4}$

57. $f(x) = \sqrt[4]{1 - x^2}$

58. $h(x) = \dfrac{10}{x^2 - 2x}$

59. $g(x) = \dfrac{1}{x} - \dfrac{3}{x + 2}$

60. $f(s) = \dfrac{\sqrt{s - 1}}{s - 4}$

In Exercises 61–64, assume that the domain of f is the set $A = \{-2, -1, 0, 1, 2\}$. Determine the set of ordered pairs representing the function f.

61. $f(x) = x^2$

62. $f(x) = \dfrac{2x}{x^2 + 1}$

63. $f(x) = \sqrt{x + 2}$

64. $f(x) = |x + 1|$

65. ***Think About It*** In your own words, explain the meanings of *domain* and *range*.

66. ***Think About It*** Describe an advantage of function notation.

Exploration **In Exercises 67–70, select a function from $f(x) = cx$, $g(x) = cx^2$, $h(x) = c\sqrt{|x|}$, and $r(x) = c/x$ and determine the value of the constant c such that the function fits the data given in the table.**

67.

x	-4	-1	0	1	4
y	-32	-2	0	-2	-32

68.

x	-4	-1	0	1	4
y	-1	$-\frac{1}{4}$	0	$\frac{1}{4}$	1

69.

x	-4	-1	0	1	4
y	-8	-32	Undef.	32	8

70.

x	-4	-1	0	1	4
y	6	3	0	3	6

In Exercises 71–76, find the difference quotient and simplify your answer.

71. $f(x) = x^2 - x + 1, \quad \dfrac{f(2 + h) - f(2)}{h}, h \neq 0$

72. $f(x) = 5x - x^2, \quad \dfrac{f(5 + h) - f(5)}{h}, h \neq 0$

73. $f(x) = x^3, \quad \dfrac{f(x + c) - f(x)}{c}, c \neq 0$

74. $f(x) = 2x, \quad \dfrac{f(x + c) - f(x)}{c}, c \neq 0$

75. $g(x) = 3x - 1, \quad \dfrac{g(x) - g(3)}{x - 3}, x \neq 3$

76. $f(t) = \dfrac{1}{t}, \quad \dfrac{f(t) - f(1)}{t - 1}, t \neq 0$

77. ***Area of a Circle*** Express the area A of a circle as a function of its circumference C.

78. ***Area of a Triangle*** Express the area A of an equilateral triangle as a function of the length s of its sides.

79. ***Exploration*** An open box of maximum volume is to be made from a square piece of material, 24 centimeters on a side, by cutting equal squares from the corners, and turning up the sides (see figure).

(a) Complete six rows of a table. (The first two rows are shown.) Use the result to guess the maximum volume.

Height x	Width	Volume V
1	$24 - 2(1)$	$1[24 - 2(1)]^2 = 484$
2	$24 - 2(2)$	$2[24 - 2(2)]^2 = 800$

(b) Plot the points (x, V). Is V a function of x?

(c) If V is a function of x, write the function and determine its domain.

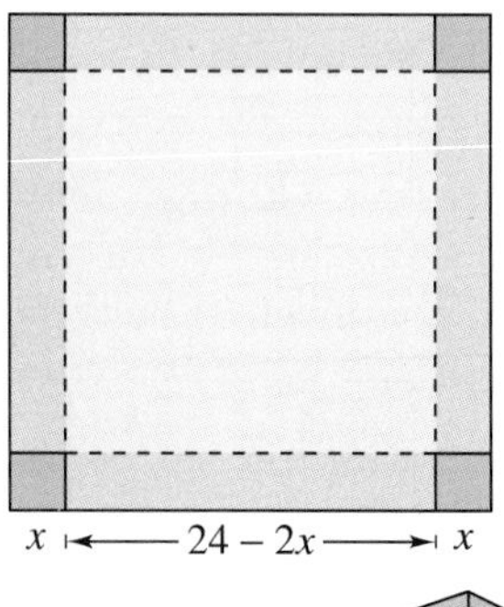

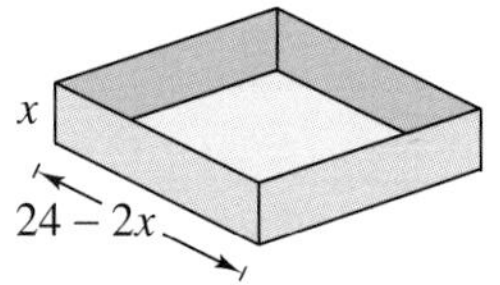

80. ***Exploration*** The cost per unit in the production of a certain radio model is \$60. The manufacturer charges \$90 per unit for orders of 100 or less. To encourage large orders, the manufacturer reduces the charge by \$0.15 per radio for each unit ordered in excess of 100 (for example, there would be a charge of \$87 per radio for an order size of 120).

(a) Complete six rows of the table. Use the result to estimate the maximum profit.

Units x	Price p	Profit P
102	$90 - 2(0.15)$	$xp - 102(60)$
104	$90 - 4(0.15)$	$xp - 104(60)$

(b) Plot the points (x, P). Is P a function of x?

(c) If P is a function of x, write the function and determine its domain.

81. ***Area of a Triangle*** A right triangle is formed in the first quadrant by the x- and y-axes and a line through the point $(2, 1)$ (see figure). Write the area of the triangle as a function of x, and determine the domain of the function.

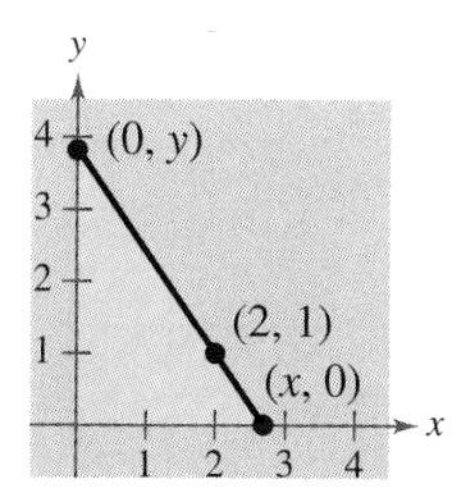

82. ***Area of a Rectangle*** A rectangle is bounded by the x-axis and the semicircle $y = \sqrt{36 - x^2}$ (see figure). Write the area of the rectangle as a function of x, and determine the domain of the function.

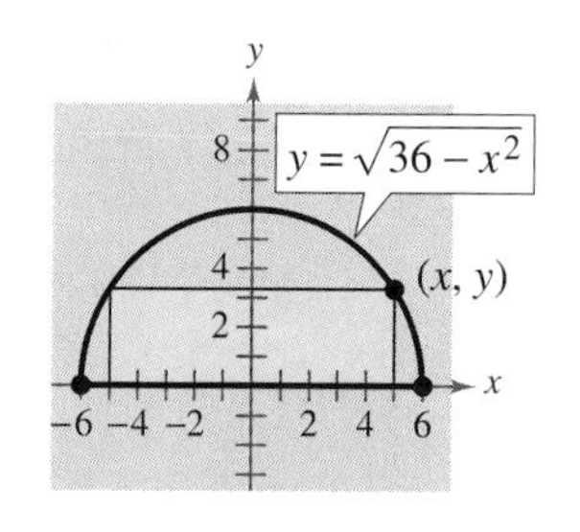

FIGURE FOR 82

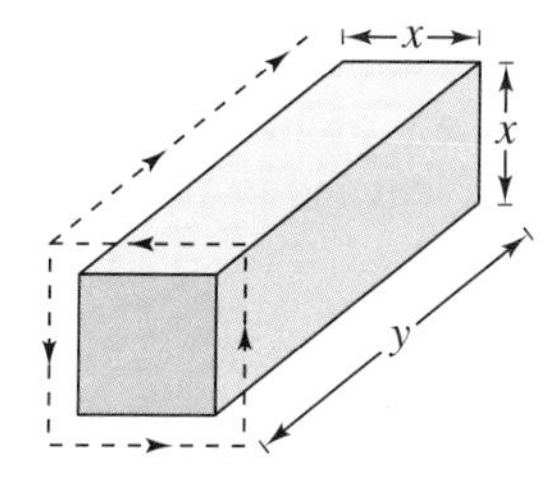

FIGURE FOR 83

83. ***Volume of a Package*** A rectangular package to be sent by a postal service can have a maximum combined length and girth (perimeter of a cross section) of 108 inches (see figure). Write the volume of the package as a function of x. What is the domain of the function?

84. ***Price of Mobile Homes*** The average price p (in thousands of dollars) of a new mobile home in the United States from 1974 to 1993 can be approximated by the model

$$p(t) = \begin{cases} 19.247 + 1.694t, & -6 \le t \le -1 \\ 19.305 + 0.427t + 0.033t^2, & 0 \le t \le 13 \end{cases}$$

where $t = 0$ represents 1980 (see figure). Use this model to find the average price of a mobile home in 1978, 1988, and 1993. (Source: U.S. Bureau of Census, Construction Reports)

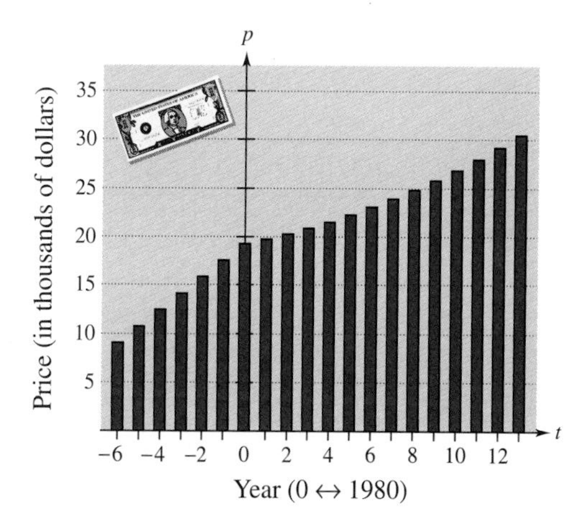

85. ***Cost, Revenue, and Profit*** A company produces a product for which the variable cost is \$12.30 per unit and the fixed costs are \$98,000. The product sells for \$17.98. Let x be the number of units produced and sold.

(a) Write the total cost C as a function of the number of units produced.

(b) Write the revenue R as a function of the number of units sold.

(c) Write the profit P as a function of the number of the units sold. [*Note*: $P = R - C$.]

86. ***Cost Analysis*** The inventor of a new game believes that the variable cost for producing the game is \$0.95 per unit and the fixed costs are \$6000. The inventor sells each game for \$1.69. Let x be the number of games sold.

(a) Write the total cost C as a function of the number of games sold.

(b) Write the average cost per unit $\overline{C} = C/x$ as a function of x.

87. ***Charter Bus Fares*** For groups of 80 or more people, a charter bus company determines the rate per person according to the formula

$$\text{Rate} = 8 - 0.05(n - 80), \qquad n \geq 80$$

where the rate is given in dollars and n is the number of people.

(a) Express the revenue R for the bus company as a function of n.

(b) Use the function from part (a) to complete the table. What can you conclude?

n	90	100	110	120	130	140	150
$R(n)$							

88. ***Fluid Force*** The force F (in tons) of water against the face of a dam is a function given by

$$F(y) = 149.76\sqrt{10}y^{5/2}$$

where y is the depth of the water in feet. Complete the table.

y	5	10	20	30	40
$F(y)$					

(a) What can you conclude from the table?

(b) Use the table to approximate the depth at which the force against the dam is 1,000,000 tons. How could you find a better estimate?

89. ***Height of a Balloon*** A balloon carrying a transmitter ascends vertically from a point 3000 feet from the receiving station.

(a) Draw a figure that gives a visual representation of the problem. Let h represent the height of the balloon and let d represent the distance between the balloon and the receiving station.

(b) Express the height of the balloon as a function of d. What is the domain of the function?

90. ***Chapter Opener*** Use the data on the opening page of this chapter. Let $f(t)$ represent the number of lynx in year t.

(a) Find $f(1992)$.

(b) Find

$$\frac{f(1994) - f(1991)}{1994 - 1991}$$

and interpret the result in the context of the problem.

(c) An approximate formula for the function is

$$N(t) = \frac{434t + 4387}{45t^2 - 55t + 100}$$

where N is the number of lynx and t is time in years, with $t = 0$ corresponding to 1990. Complete the table and compare the result with the data.

t	1988	1989	1990	1991
N				

t	1992	1993	1994	1995
N				

Review **Solve the equations in Exercises 91–94 as a review of the skills and problem-solving techniques you learned in previous sections.**

91. $\dfrac{t}{3} + \dfrac{t}{5} = 1$

92. $\dfrac{3}{t} + \dfrac{5}{t} = 1$

93. $\dfrac{3}{x(x+1)} - \dfrac{4}{x} = \dfrac{1}{x+1}$

94. $\dfrac{12}{x} - 3 = \dfrac{4}{x} + 9$

P.6 Analyzing Graphs of Functions

See Exercise 67 on page 81 for an example of how a step function can be used to model the cost of a telephone call.

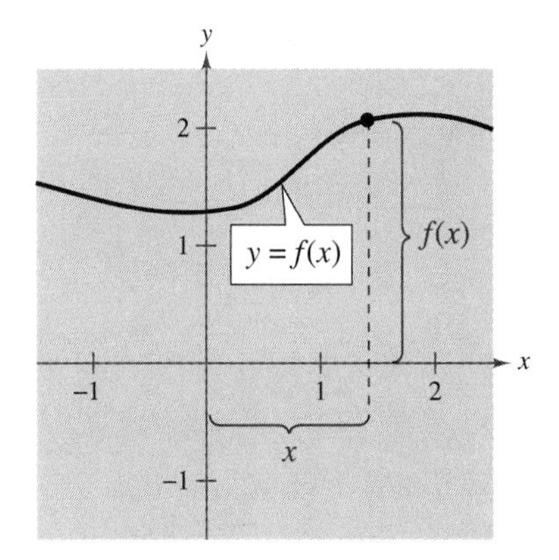

FIGURE P.41

The Graph of a Function

In Section P.5 you studied functions from an algebraic point of view. In this section, you will study functions from a geometric perspective. The **graph of a function f** is the collection of ordered pairs $(x, f(x))$ such that x is in the domain of f. As you study this section, remember that

$$x = \text{the directed distance from the } y\text{-axis}$$
$$f(x) = \text{the directed distance from the } x\text{-axis}$$

as shown in Figure P.41. If the graph of a function has an x-intercept at $(a, 0)$, then a is a **zero** of the function. In other words, the zeros of a function are the values of x for which $f(x) = 0$. For instance, the function given by $f(x) = x^2 - 4$ has two zeros: -2 and 2.

EXAMPLE 1 *Finding the Domain and Range of a Function*

Use the graph of the function f, shown in Figure P.42, to find (a) the domain of f, (b) the function values $f(-1)$ and $f(2)$, and (c) the range of f.

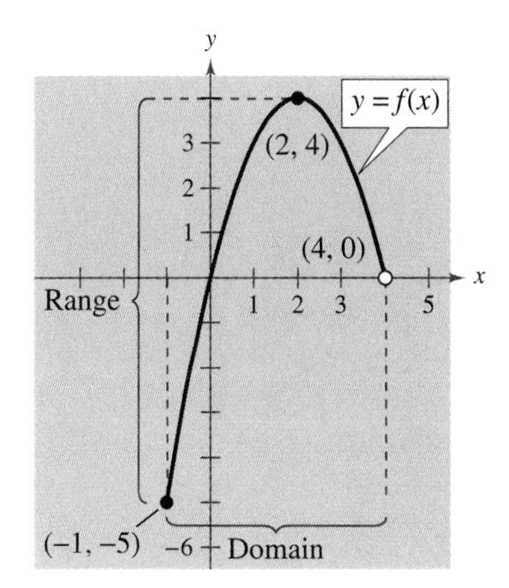

FIGURE P.42

Solution

a. A closed dot (on the left) indicates that $x = -1$ is in the domain of f, whereas the open dot (on the right) indicates $x = 4$ is not in the domain. Thus, the domain of f is all x in the interval $[-1, 4)$.

b. Because $(-1, -5)$ is a point on the graph of f, it follows that

$$f(-1) = -5.$$

Similarly, because $(2, 4)$ is a point on the graph of f, it follows that

$$f(2) = 4.$$

c. Because the graph does not extend below $f(-1) = -5$ or above $f(2) = 4$, the range of f is the interval $[-5, 4]$.

NOTE The use of dots (open or closed) at the extreme left and right points of a graph indicates that the graph does not extend beyond these points. If no such dots are shown, assume that the graph extends beyond these points.

By the definition of a function, at most one y-value corresponds to a given x-value. It follows, then, that a vertical line can intersect the graph of a function at most once. This observation provides a convenient visual test called the **Vertical Line Test** for functions.

> ***VERTICAL LINE TEST FOR FUNCTIONS***
>
> A set of points in a coordinate plane is the graph of y as a function of x if and only if no vertical line intersects the graph at more than one point.

EXAMPLE 2 Vertical Line Test for Functions

Use the Vertical Line Test to decide whether the graphs in Figure P.43 represent y as a function of x.

(a)

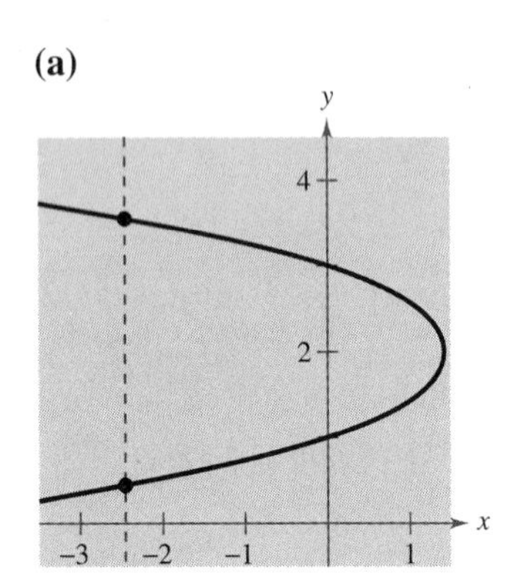

(b)

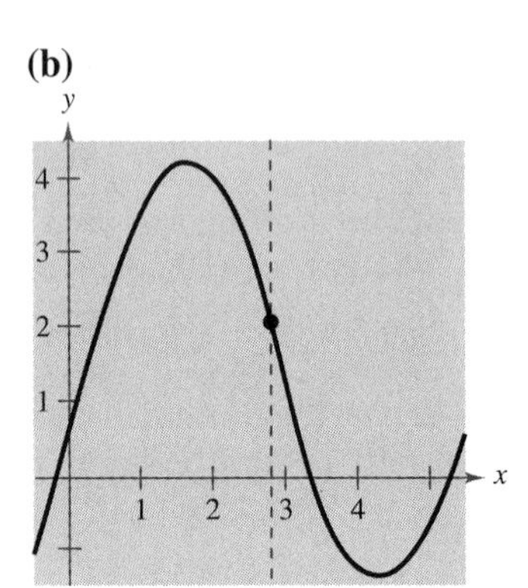

(c)

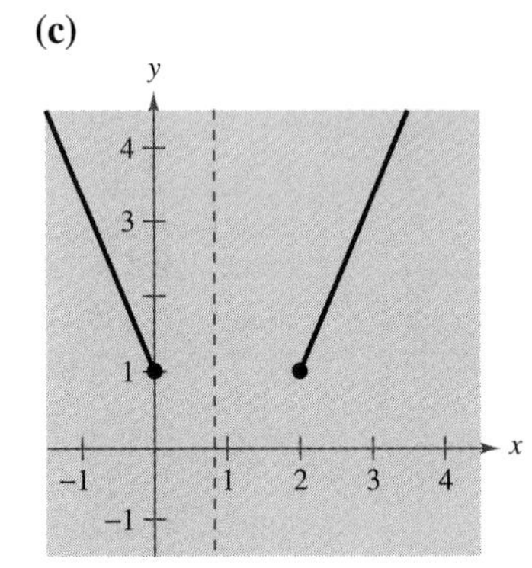

FIGURE P.43

Solution

a. This *is not* a graph of y as a function of x because you can find a vertical line that intersects the graph twice.

b. This *is* a graph of y as a function of x because every vertical line intersects the graph at most once.

c. This *is* a graph of y as a function of x. (Note that if a vertical line does not intersect the graph, it simply means that the function is undefined for that particular value of x.)

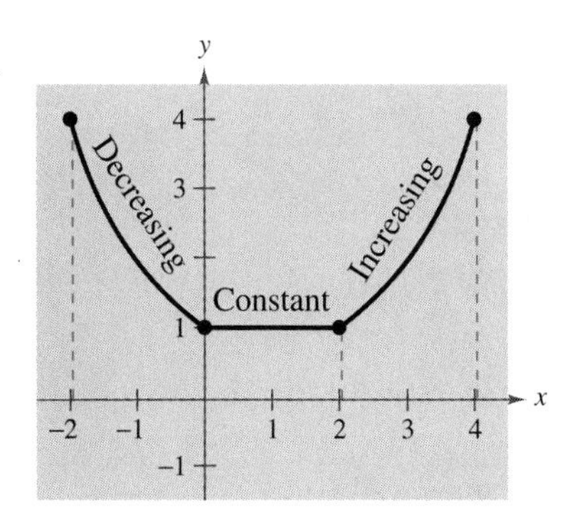

FIGURE P.44

Increasing and Decreasing Functions

The more you know about the graph of a function, the more you know about the function itself. Consider the graph shown in Figure P.44. As you move from *left to right,* this graph decreases, then is constant, and then increases.

> *INCREASING, DECREASING, AND CONSTANT FUNCTIONS*
>
> A function f is **increasing** on an interval if, for any x_1 and x_2 in the interval, $x_1 < x_2$ implies $f(x_1) < f(x_2)$.
>
> A function f is **decreasing** on an interval if, for any x_1 and x_2 in the interval, $x_1 < x_2$ implies $f(x_1) > f(x_2)$.
>
> A function f is **constant** on an interval if, for any x_1 and x_2 in the interval, $f(x_1) = f(x_2)$.

EXAMPLE 3 *Increasing and Decreasing Functions*

In Figure P.45, describe the increasing or decreasing behavior of the function.

Solution

a. This function is increasing over the entire real line.

b. This function is increasing on the interval $(-\infty, -1)$, decreasing on the interval $(-1, 1)$, and increasing on the interval $(1, \infty)$.

c. This function is increasing on the interval $(-\infty, 0)$, constant on the interval $(0, 2)$, and decreasing on the interval $(2, \infty)$.

(a)

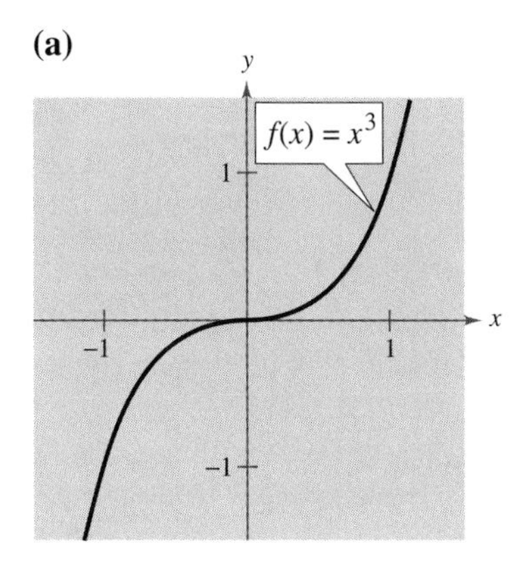

(b)

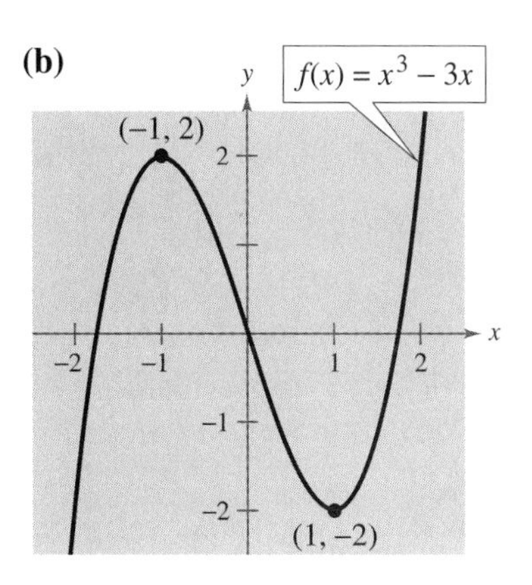

(c)

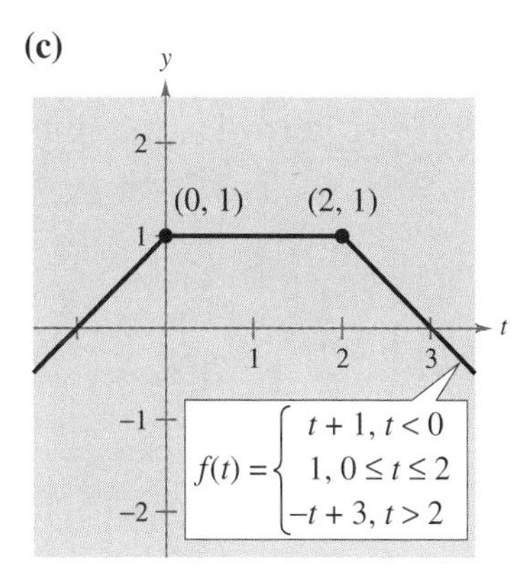

FIGURE P.45

When analyzing the graph of a function, you want to find the points at which the function changes its increasing, decreasing, or constant behavior. These points often identify *maximum* or *minimum* values of the function.

EXAMPLE 4 The Price of Diamonds

Real Life

During the 1980s, the average price of a 1-carat polished diamond decreased and then increased according to the model

$$C = -0.7t^3 + 16.25t^2 - 106t + 388, \qquad 2 \le t \le 10$$

where C is the average price in dollars (on the Antwerp Index) and t represents the calendar year, with $t = 2$ corresponding to 1982. According to this model, during which years was the price of diamonds decreasing? During which years was the price of diamonds increasing? Approximate the minimum price of a 1-carat diamond between 1982 and 1990. (Source: Diamond High Council)

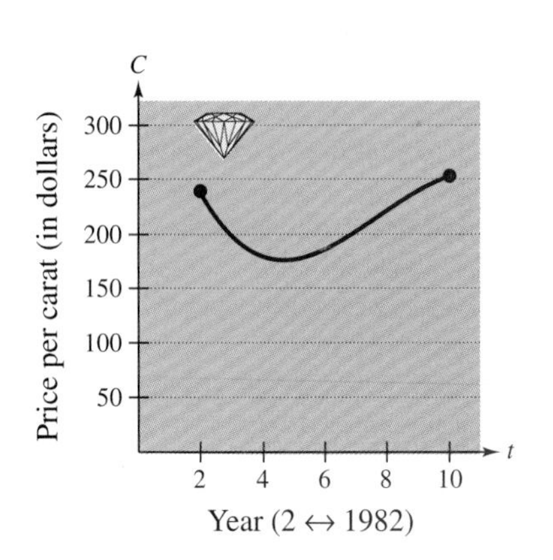

FIGURE P.46

Solution

To solve this problem, sketch an accurate graph of the function, as shown in Figure P.46. From the graph, you can see that the price of diamonds decreased from 1982 until late 1984. Then, from late 1984 to 1990, the price increased. The minimum price during the 8-year period was approximately \$175.

TECHNOLOGY

A graphing utility is useful for determining the minimum and maximum values of a function over a closed interval. For instance, graph

$$C = -0.7t^3 + 16.25t^2 - 106t + 388, \qquad 2 \le t \le 10$$

as shown below. By using the trace feature, you can determine that the minimum value occurs when $x \approx 4.7$.

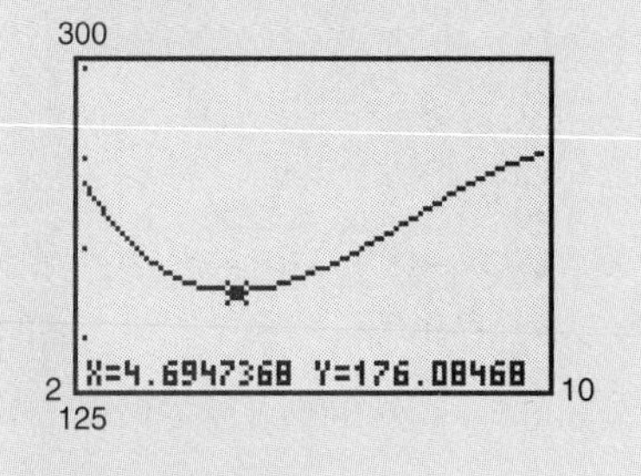

Step Functions

EXAMPLE 5 *The Greatest Integer Function*

The **greatest integer function** is denoted by $[\![x]\!]$ and is defined by

$$[\![x]\!] = \text{the greatest integer less than or equal to } x.$$

The graph of this function is shown in Figure P.47. Note that the graph of the greatest integer function jumps vertically one unit at each integer and is constant (a horizontal line segment) between each pair of consecutive integers. The greatest integer function is an example of a category of functions called **step functions.** Some values of the greatest integer function are as follows.

$$[\![-1]\!] = -1 \qquad [\![-0.5]\!] = -1$$
$$[\![0]\!] = 0 \qquad [\![0.5]\!] = 0$$
$$[\![1]\!] = 1 \qquad [\![1.5]\!] = 1$$

The range of the greatest integer function is the set of all integers.

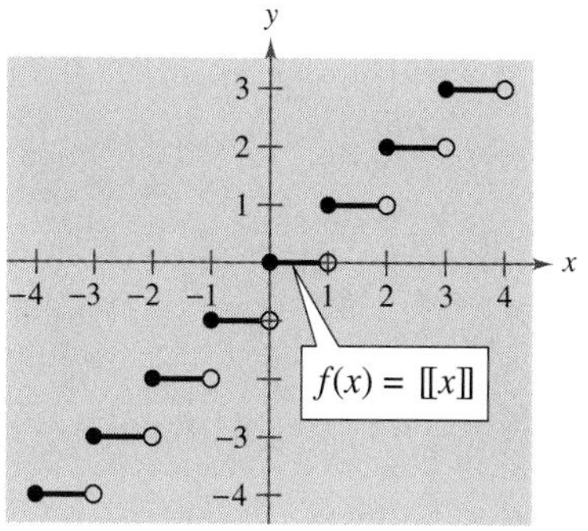

FIGURE P.47

NOTE If you use a graphing utility to sketch a step function, you should set the utility to *Dot* mode rather than *Connected* mode.

EXAMPLE 6 *The Price of a Telephone Call*

The cost of a telephone call between Los Angeles and San Francisco is \$0.50 for the first minute and \$0.36 for each additional minute (or portion of a minute). The greatest integer function can be used to create a model for the cost of this call, as follows.

$$C = 0.50 + 0.36[\![t]\!], \qquad t > 0$$

where C is the total cost of the call in dollars and t is the length of the call in minutes. Sketch the graph of this function.

Solution

For calls up to 1 minute, the cost is \$0.50. For calls between 1 and 2 minutes, the cost is \$0.86, and so on.

Length of Call	$0 < t < 1$	$1 \le t < 2$	$2 \le t < 3$	$3 \le t < 4$	$4 \le t < 5$
Cost of Call	\$0.50	\$0.86	\$1.22	\$1.58	\$1.94

Using these values, you can sketch the graph shown in Figure P.48.

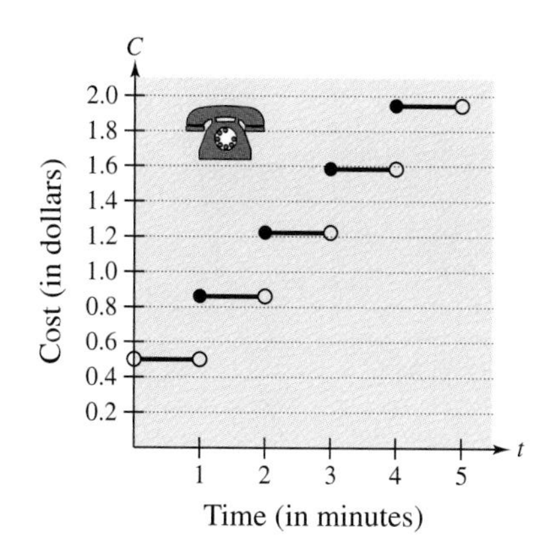

FIGURE P.48

Even and Odd Functions

In Section P.3, you studied different types of symmetry of a graph. In the terminology of functions, a function is said to be **even** if its graph is symmetric with respect to the y-axis and to be **odd** if its graph is symmetric with respect to the origin. The symmetry tests in Section P.3 yield the following tests for even and odd functions.

Tests for Even and Odd Functions

A function given by $y = f(x)$ is **even** if, for each x in the domain of f,

$$f(-x) = f(x).$$

A function given by $y = f(x)$ is **odd** if, for each x in the domain of f,

$$f(-x) = -f(x).$$

Example 7 *Even and Odd Functions*

a. The function $g(x) = x^3 - x$ is odd because

$$g(-x) = (-x)^3 - (-x) = -x^3 + x = -(x^3 - x) = -g(x).$$

b. The function $h(x) = x^2 + 1$ is even because

$$h(-x) = (-x)^2 + 1 = x^2 + 1 = h(x).$$

The graphs of the two functions are shown in Figure P.49.

(a) Symmetric to Origin

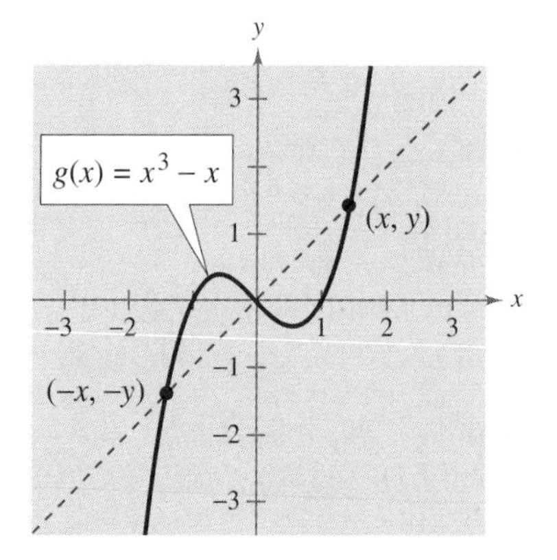

(b) Symmetric to y-Axis

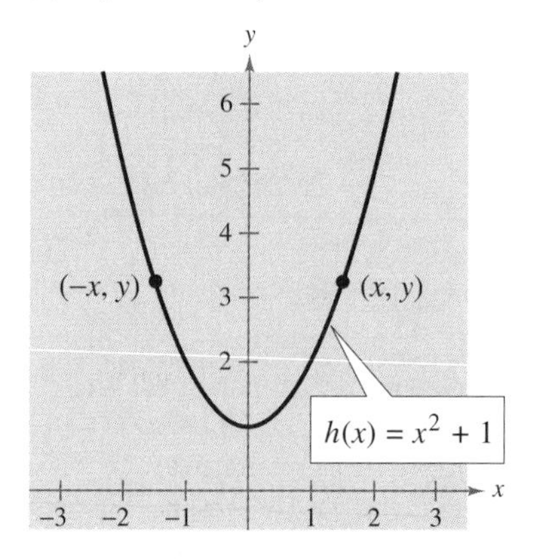

FIGURE P.49

Summary of Graphs of Common Functions

Figure P.50 shows the graphs of six common functions. You need to be familiar with these graphs.

(a) Constant Function

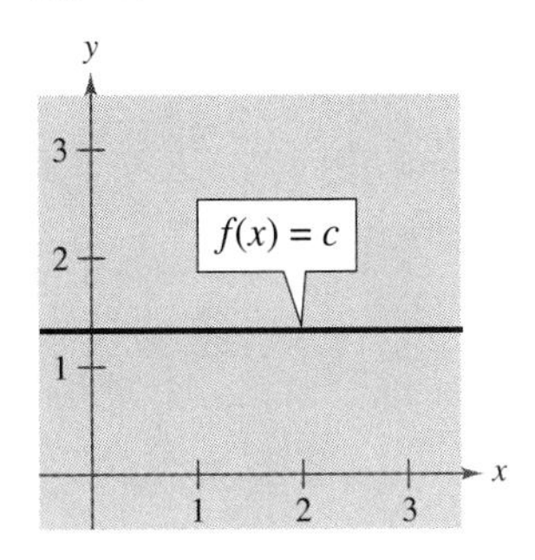

(b) Identity Function

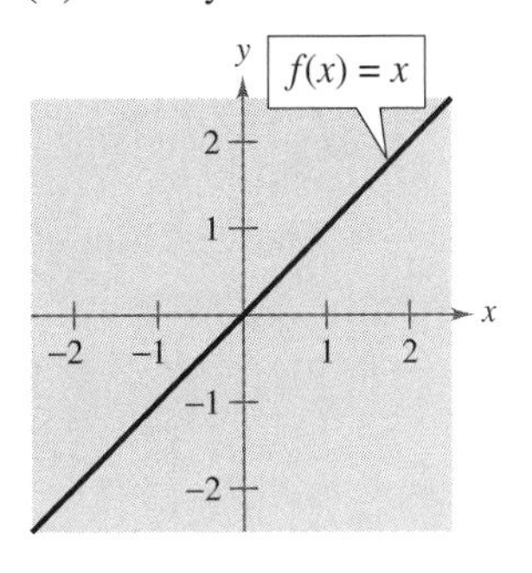

(c) Absolute Value

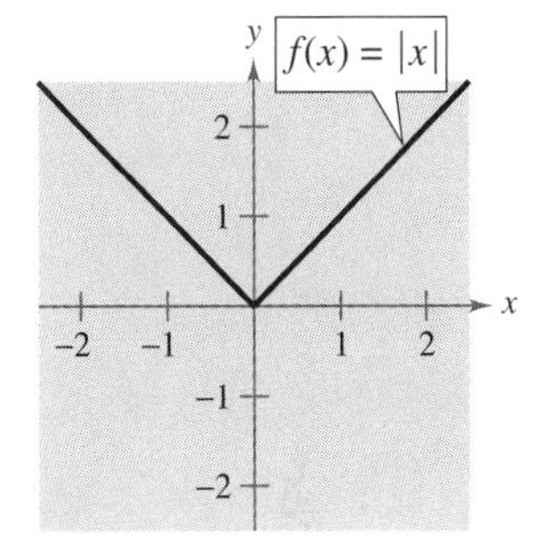

(d) Square Root

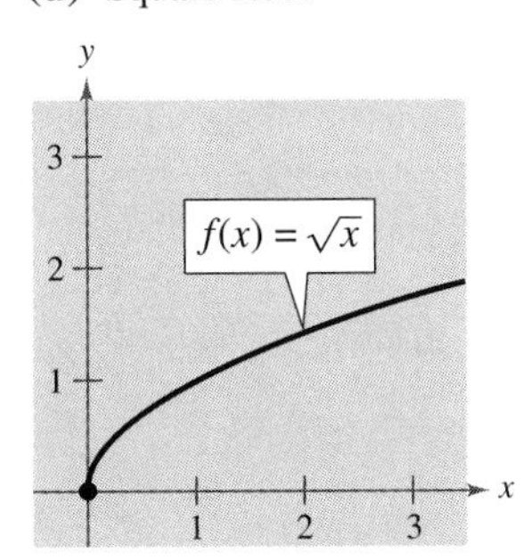

(e) Squaring Function

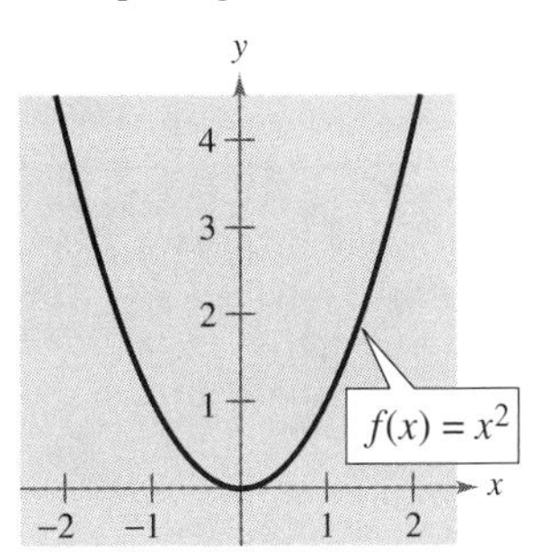

(f) Cubing Function

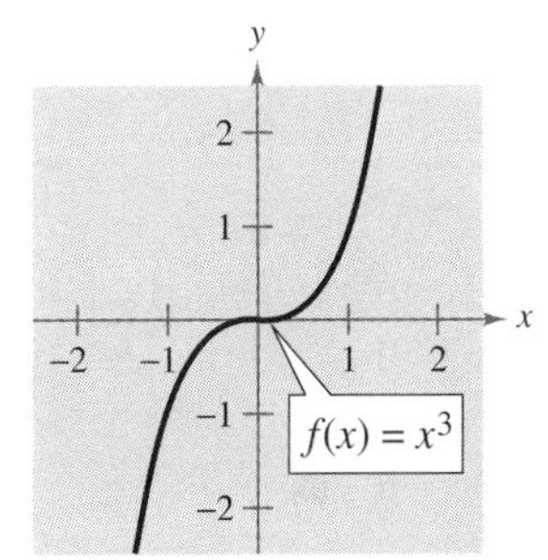

FIGURE P.50

GROUP ACTIVITY

IDENTIFYING FUNCTIONS

The table gives the circulation y (in millions) and the annual acquisitions expenditures x (in millions of dollars) for top public libraries in the United States in 1993. Discuss whether the data itself represents a function. Explain your reasoning. Sketch and discuss the scatter plot of the data. On your scatter plot, draw what you think is the best-fitting model. Is the model that best represents the data a function? Defend your position. (Source: Public Library Association)

x	y
3.0	15.9
5.7	10.1
5.3	9.3
1.2	1.5
2.8	6.3

x	y
5.0	7.8
1.3	3.1
5.7	13.2
4.2	6.5
2.0	5.4

WARM UP

1. Find $f(2)$ for $f(x) = -x^3$.
2. Find $f(6)$ for $f(x) = x^2 - 6x$.
3. Find $f(-x)$ for $f(x) = \frac{3}{x}$.
4. Find $f(-x)$ for $f(x) = x^2 + 3$.

Solve for x.

5. $x^3 - 16x = 0$
6. $2x^2 - 3x + 1 = 0$

Find the domain of the function.

7. $g(x) = \dfrac{4}{x - 4}$
8. $f(x) = \dfrac{2x}{x^2 - 9x + 20}$
9. $h(t) = \sqrt[4]{5 - 3t}$
10. $f(t) = t^3 + 3t - 5$

P.6 Exercises

In Exercises 1–6, find the domain and range of the function.

1. $f(x) = 1 - x^2$

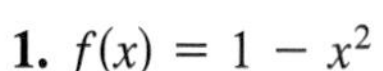

2. $f(x) = \sqrt{x - 1}$

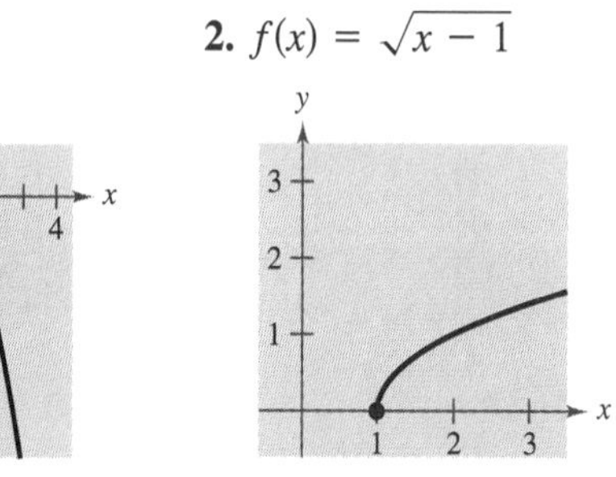

3. $f(x) = \sqrt{x^2 - 1}$

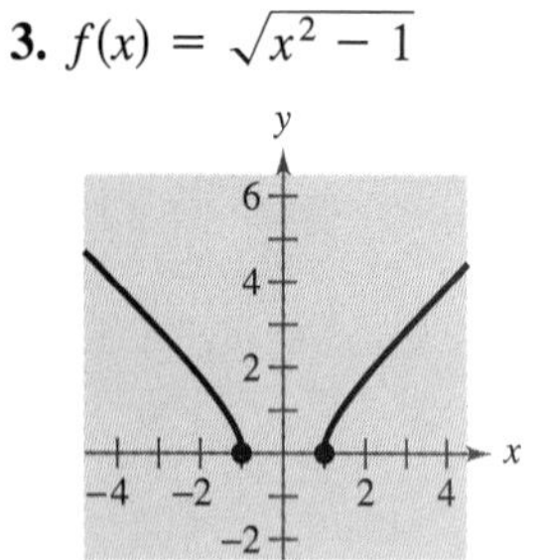

4. $f(x) = \frac{1}{2}|x - 2|$

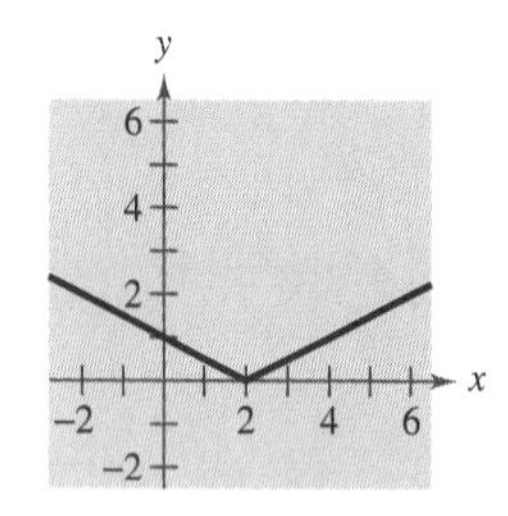

5. $h(x) = \sqrt{16 - x^2}$

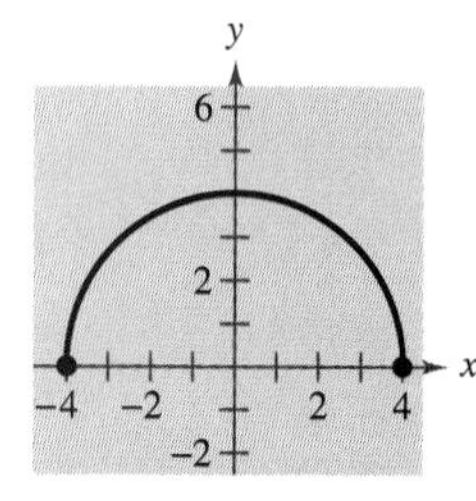

6. $g(x) = \dfrac{|x - 1|}{x - 1}$

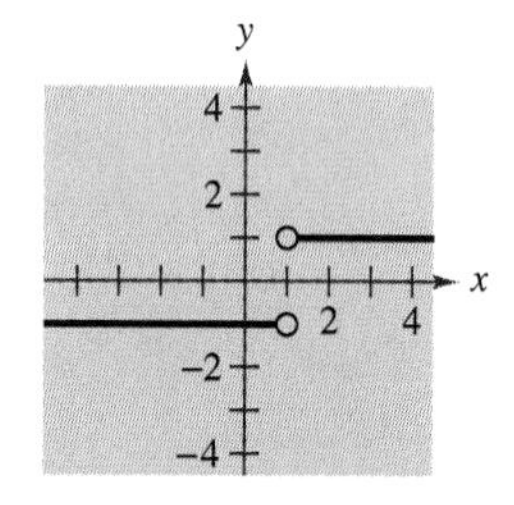

In Exercises 7–12, use the Vertical Line Test to determine whether y is a function of x.

7. $y = \frac{1}{2}x^2$

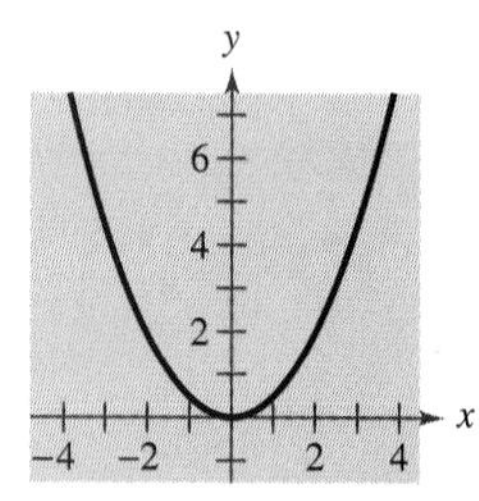

8. $y = \frac{1}{4}x^3$

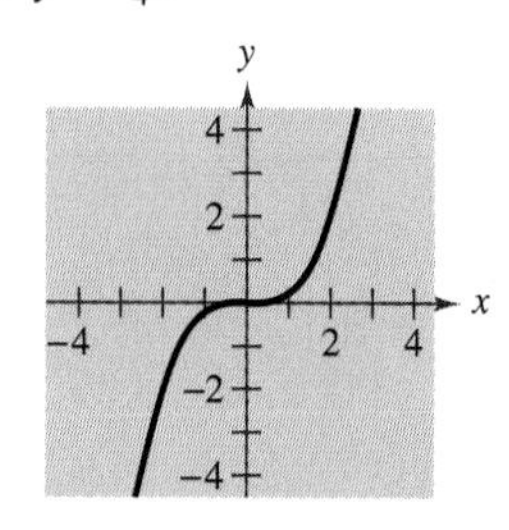

9. $x - y^2 = 1$

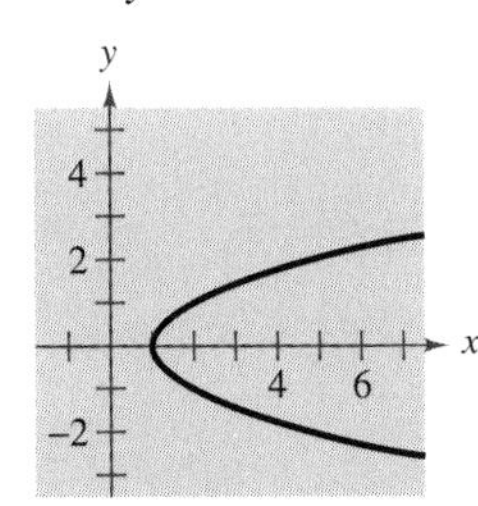

10. $x^2 + y^2 = 25$

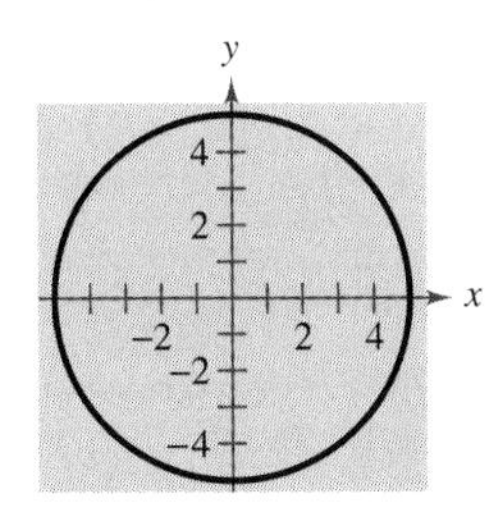

11. $x^2 = 2xy - 1$

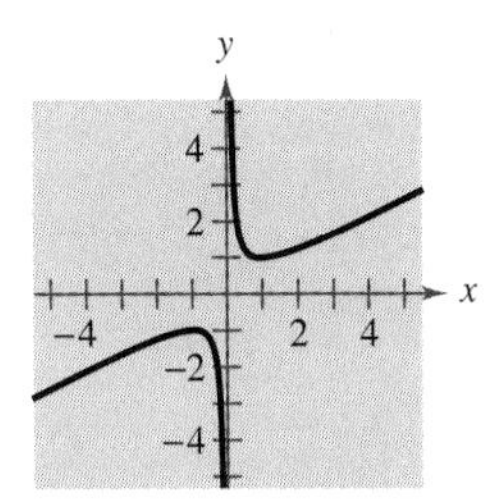

12. $x = |y + 2|$

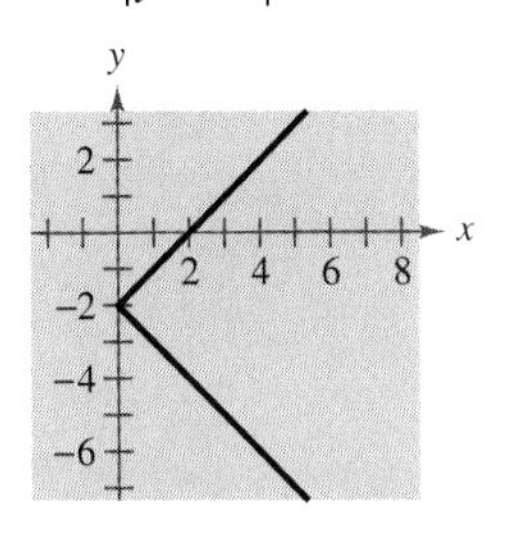

13. ***Think About It*** Does the graph in Exercise 9 represent x as a function of y? Explain.

14. ***Think About It*** Does the graph in Exercise 10 represent x as a function of y? Explain.

In Exercises 15–18, select the viewing rectangle that shows the most complete graph of the function.

15. $f(x) = -0.2x^2 + 3x + 32$

(a)	(b)	(c)
Xmin = -2	Xmin = -10	Xmin = 0
Xmax = 20	Xmax = 30	Xmax = 10
Xscl = 1	Xscl = 5	Xscl = 0.5
Ymin = -10	Ymin = -5	Ymin = 0
Ymax = 30	Ymax = 50	Ymax = 200
Yscl = 4	Yscl = 5	Yscl = 25

16. $f(x) = 6[x - (0.1x)^5]$

(a)	(b)	(c)
Xmin = -500	Xmin = -25	Xmin = -20
Xmax = 5000	Xmax = 25	Xmax = 20
Xscl = 50	Xscl = 5	Xscl = 5
Ymin = -500	Ymin = -25	Ymin = -100
Ymax = 500	Ymax = 25	Ymax = 100
Yscl = 50	Yscl = 5	Yscl = 20

17. $f(x) = 4x^3 - x^4$

(a)	(b)	(c)
Xmin = -2	Xmin = -50	Xmin = 0
Xmax = 6	Xmax = 50	Xmax = 2
Xscl = 1	Xscl = 5	Xscl = 0.2
Ymin = -10	Ymin = -50	Ymin = -2
Ymax = 30	Ymax = 50	Ymax = 2
Yscl = 4	Yscl = 5	Yscl = 0.5

18. $f(x) = 10x\sqrt{400 - x^2}$

(a)	(b)	(c)
Xmin = -5	Xmin = -20	Xmin = -25
Xmax = 50	Xmax = 20	Xmax = 25
Xscl = 5	Xscl = 2	Xscl = 5
Ymin = -5000	Ymin = -500	Ymin = -2000
Ymax = 5000	Ymax = 500	Ymax = 2000
Yscl = 500	Yscl = 50	Yscl = 200

In Exercises 19–22, (a) determine the intervals over which the function is increasing, decreasing, or constant, and (b) determine if the function is even, odd, or neither.

19. $f(x) = \frac{3}{2}x$

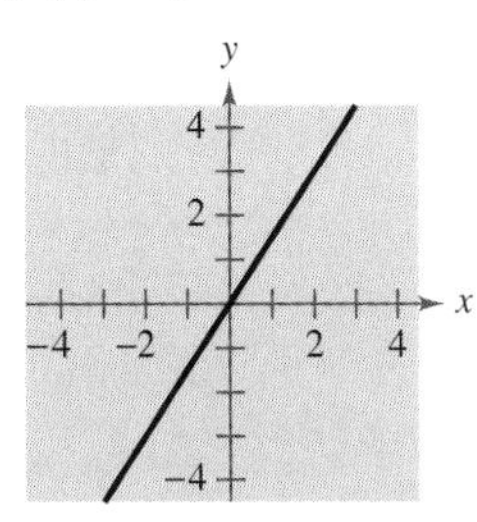

20. $f(x) = x^2 - 4x$

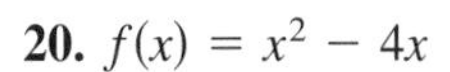

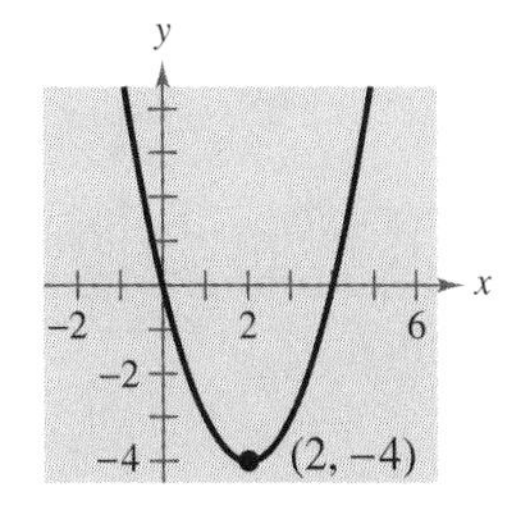

21. $f(x) = x^3 - 3x^2 + 2$

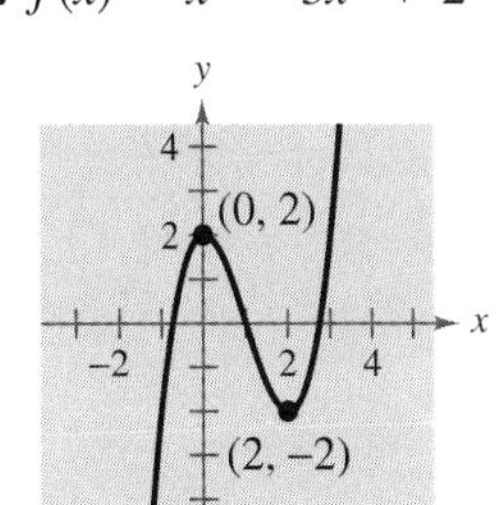

22. $f(x) = \sqrt{x^2 - 1}$

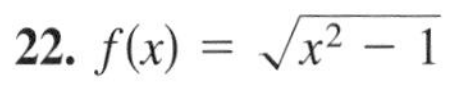

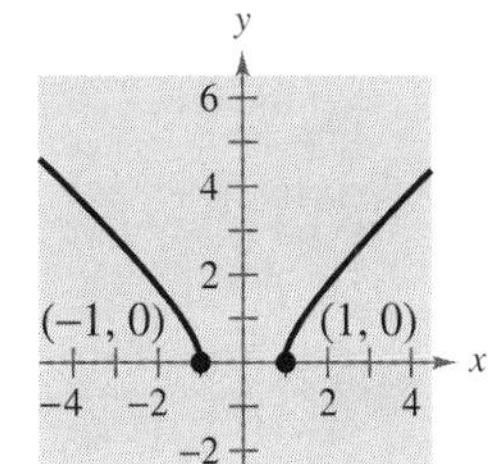

In Exercises 23–26, (a) use a graphing utility to graph the function, (b) determine the intervals over which the function is increasing, decreasing, or constant, and (c) determine if the function is even, odd, or neither.

23. $f(x) = 3x^4 - 6x^2$
24. $f(x) = x^{2/3}$
25. $f(x) = x\sqrt{x + 3}$
26. $f(x) = |x + 1| + |x - 1|$

In Exercises 27–32, determine whether the function is even, odd, or neither.

27. $f(x) = x^6 - 2x^2 + 3$
28. $h(x) = x^3 - 5$
29. $g(x) = x^3 - 5x$
30. $f(x) = x\sqrt{1 - x^2}$
31. $f(t) = t^2 + 2t - 3$
32. $g(s) = 4s^{2/3}$

Think About It **In Exercises 33 and 34, find the coordinates of a second point on the graph of a function f if the given point is on the graph and the function is (a) even and (b) odd.**

33. $\left(-\frac{3}{2}, 4\right)$
34. $(4, 9)$

In Exercises 35–46, sketch the graph of the function and determine whether the function is even, odd, or neither.

35. $f(x) = 3$
36. $g(x) = x$
37. $f(x) = 5 - 3x$
38. $h(x) = x^2 - 4$
39. $g(s) = \dfrac{s^2}{4}$
40. $f(t) = -t^4$
41. $f(x) = \sqrt{1 - x}$
42. $f(x) = x^{3/2}$
43. $g(t) = \sqrt[3]{t - 1}$
44. $f(x) = |x + 2|$
45. $f(x) = \begin{cases} x + 3, & x \le 0 \\ 3, & 0 < x \le 2 \\ 2x - 1, & x > 2 \end{cases}$
46. $f(x) = \begin{cases} 2x + 1, & x \le -1 \\ x^2 - 2, & x > -1 \end{cases}$

In Exercises 47–56, graph the function and determine the intervals for which $f(x) \ge 0$.

47. $f(x) = 4 - x$
48. $f(x) = 4x + 2$
49. $f(x) = x^2 - 9$
50. $f(x) = x^2 - 4x$
51. $f(x) = 1 - x^4$
52. $f(x) = \sqrt{x + 2}$
53. $f(x) = x^2 + 1$
54. $f(x) = -(1 + |x|)$
55. $f(x) = -5$
56. $f(x) = \frac{1}{2}(2 + |x|)$

In Exercises 57–60, graph the function.

57. $f(x) = \begin{cases} 2x + 3, & x < 0 \\ 3 - x, & x \ge 0 \end{cases}$
58. $f(x) = \begin{cases} \sqrt{4 + x}, & x < 0 \\ \sqrt{4 - x}, & x \ge 0 \end{cases}$
59. $f(x) = \begin{cases} x^2 + 5, & x \le 1 \\ -x^2 + 4x + 3, & x > 1 \end{cases}$
60. $f(x) = \begin{cases} 1 - (x - 1)^2, & x \le 2 \\ \sqrt{x - 2}, & x > 2 \end{cases}$

In Exercises 61 and 62, use a graphing utility to graph the function. State the domain and range of the function.

61. $f(x) = |x + 3|$
62. $h(t) = \sqrt{4 - t^2}$

In Exercises 63 and 64, use a graphing utility to graph the function. State the domain and range of the function. Describe the pattern of the graph.

63. $s(x) = 2\left(\frac{1}{4}x - [\![\frac{1}{4}x]\!]\right)$
64. $g(x) = 2\left(\frac{1}{4}x - [\![\frac{1}{4}x]\!]\right)^2$

65. ***Essay*** Use a graphing utility to graph each function. Write a paragraph describing any similarities and differences you observe among the graphs.
(a) $y = x$ (b) $y = x^2$
(c) $y = x^3$ (d) $y = x^4$
(e) $y = x^5$ (f) $y = x^6$

66. ***Conjecture*** Use the results of Exercise 65 to make a conjecture about the graphs of $y = x^7$ and $y = x^8$. Use a graphing utility to graph the functions and compare the results with the graph drawn by hand.

67. ***Comparing Models*** The cost of a telephone call between two cities is \$0.65 for the first minute and \$0.40 for each additional minute.

(a) It is required that a model be created for the cost C of a telephone call between the two cities lasting t minutes. Which of the following is the appropriate model? Explain.

$$C_1(t) = 0.65 + 0.4[\![t - 1]\!]$$
$$C_2(t) = 0.65 - 0.4[\![-(t - 1)]\!]$$

(b) Graph the appropriate model. Determine the cost of a call lasting 18 minutes and 45 seconds.

68. ***Cost of Overnight Delivery*** Suppose that the cost of sending an overnight package from New York to Atlanta is \$9.80 for under one pound and \$2.50 for each additional pound. Use the greatest integer function to create a model for the cost C of overnight delivery of a package weighing x pounds, $x > 0$. Sketch the graph of the function.

69. ***Maximum Profit*** The marketing department of a company estimates that the demand for a product is given by $p = 100 - 0.0001x$, where p is the price per unit and x is the number of units. The cost of producing x units is given by $C = 350{,}000 + 30x$, and the profit for producing and selling x units is given by

$$P = R - C = xp - C.$$

Use a graphing utility to graph the profit function and estimate the number of units that would produce a maximum profit.

70. ***Fluorescent Lamp*** The number of lumens (time rate of flow of light) L from a fluorescent lamp can be approximated by the model

$$L = -0.294x^2 + 97.744x - 664.875, \qquad 20 \le x \le 90$$

where x is the wattage of the lamp. Use a graphing utility to graph the function and estimate the wattage of a bulb necessary to obtain 2000 lumens.

In Exercises 71–74, write the height h of the rectangle as a function of x.

71.

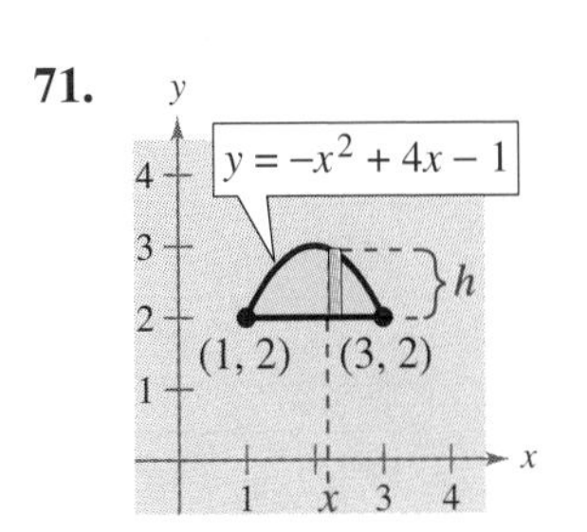

72.

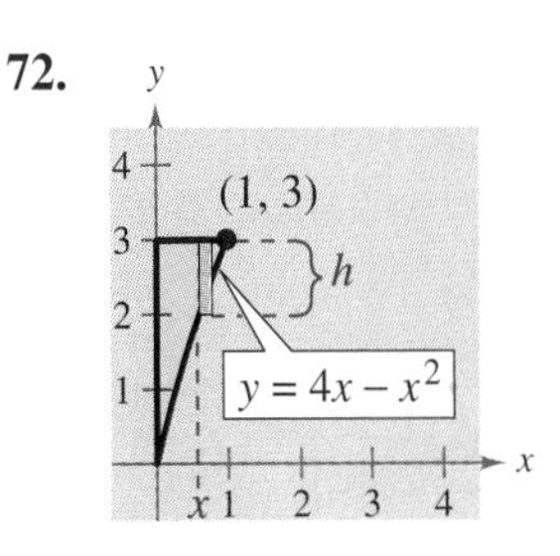

73.

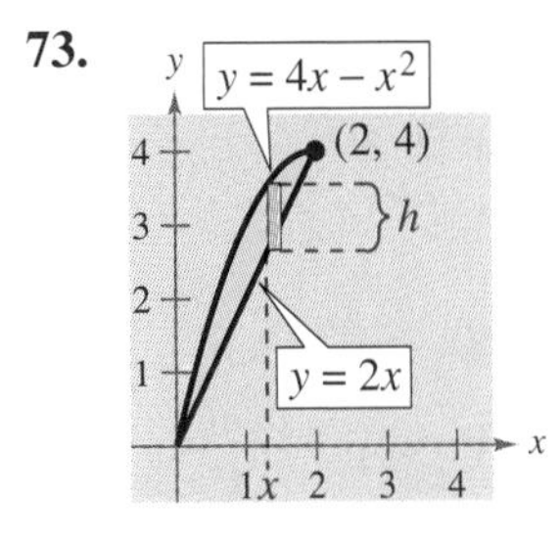

74.

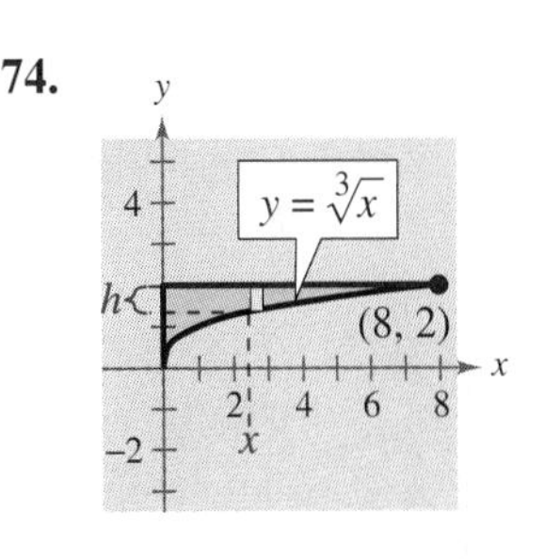

In Exercises 75–78, write the length L of the rectangle as a function of y.

75.

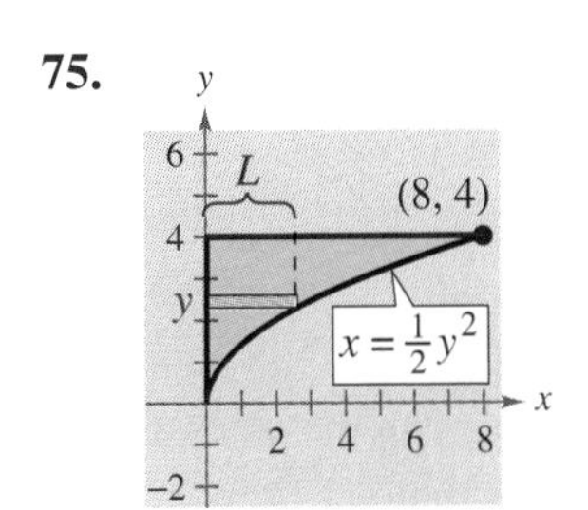

76.

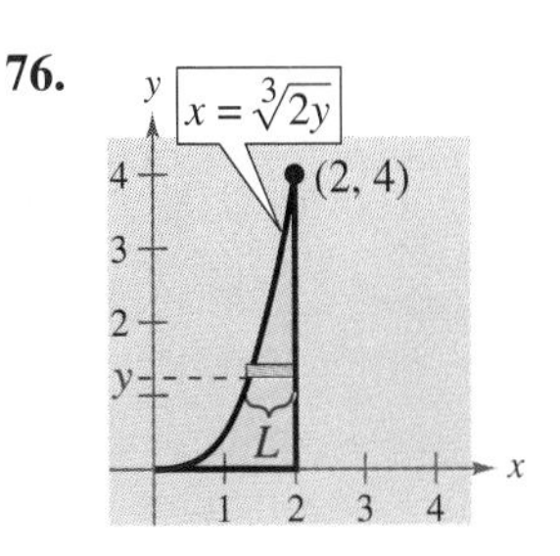

77.

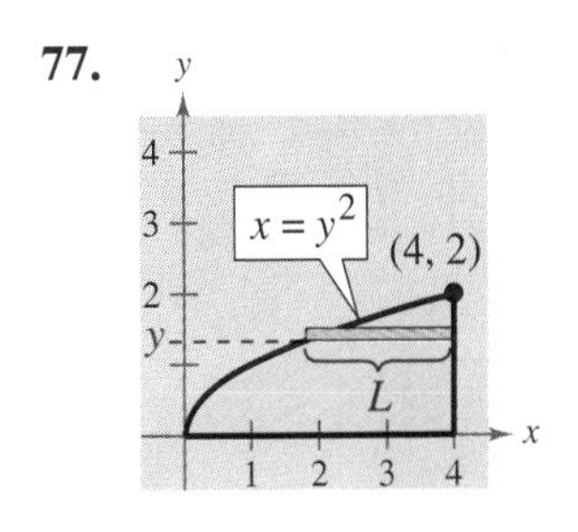

78.

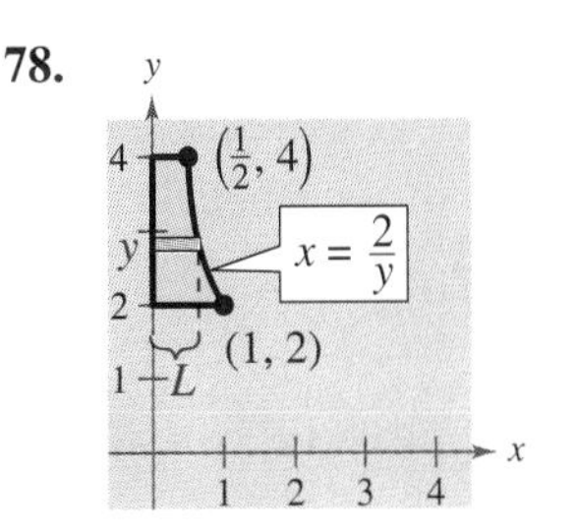

79. *Data Analysis* The table gives the amounts y (in billions of dollars) of the merchandise trade balance of the United States for the years 1986 through 1993. (Source: U.S. Bureau of the Census)

Year	1986	1987	1988	1989
y	-152.7	-152.1	-118.6	-109.6

Year	1990	1991	1992	1993
y	-101.7	-65.4	-84.5	-115.8

A model for this data is given by

$$y = -87.49 + 16.28t - 4.82t^2 - 1.20t^3$$

where $t = 0$ represents 1990.

(a) What is the domain of the model?

(b) Use a graphing utility to graph the data and the model on the same viewing rectangle.

(c) For which year does the model most accurately estimate the actual data? During which year is it least accurate?

(d) Why would economists be concerned if this model remained valid in the future?

80. *Geometry* Corners of equal size are cut from a square with sides of length 8 meters (see figure).

(a) Write the area A of the resulting figure as a function of x. Determine the domain of the function.

(b) Use a graphing utility to graph the area function over its domain. Use the graph to find the range of the function.

(c) Identify the resulting figure for the maximum value of x in the domain of the function. What is the length of each side of the figure?

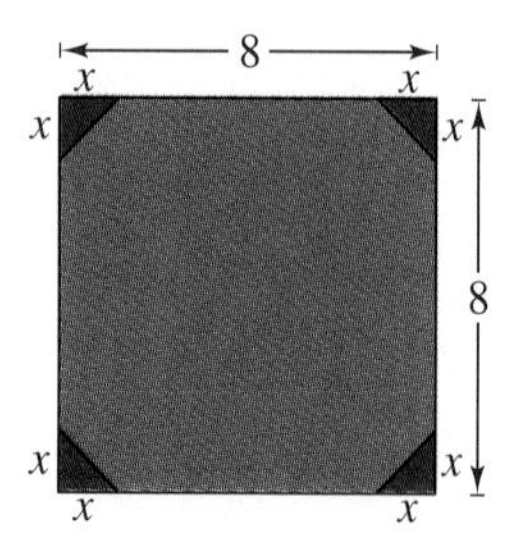

81. *Coordinate Axis Scale* It is necessary to graph the function $f(t)$, which models the specified data for the years 1980 through 1996, with $t = 0$ corresponding to 1980. State a possible scale for the vertical axis (e.g., hundreds, thousands, millions, etc.) of the graph and give a reason for your answer.

(a) $f(t)$ represents the average salary of college professors.

(b) $f(t)$ represents the U.S. population.

(c) $f(t)$ represents the percent of the civilian work force that is unemployed.

82. *Fluid Flow* The intake pipe of a 100-gallon tank has a flow rate of 10 gallons per minute, and two drainpipes have flow rates of 5 gallons per minute each. The figure shows the volume V of fluid in the tank as a function of time t. Determine the pipes in which the fluid is flowing in specific subintervals of the 1 hour of time shown on the graph. (There is more than one correct answer.)

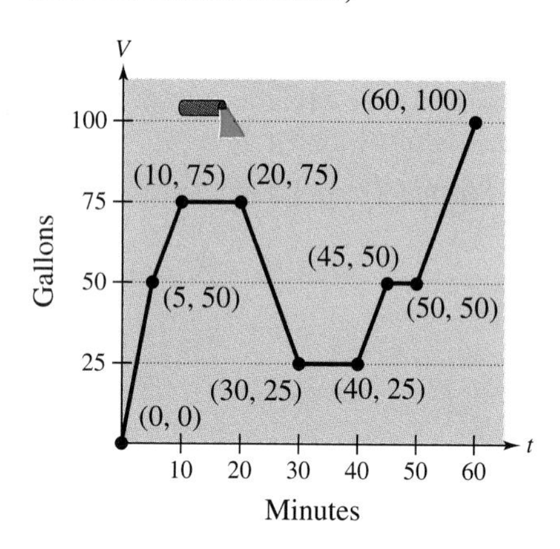

83. Prove that a function of the following form is odd.

$$y = a_{2n+1}x^{2n+1} + a_{2n-1}x^{2n-1} + \cdots + a_3x^3 + a_1x$$

84. Prove that a function of the following form is even.

$$y = a_{2n}x^{2n} + a_{2n-2}x^{2n-2} + \cdots + a_2x^2 + a_0$$

Review **Solve the equations in Exercises 85–88 as a review of the skills and problem-solving techniques you learned in previous sections.**

85. $x^2 - 10x = 0$

86. $100 - (x - 5)^2 = 0$

87. $x^3 + x = 0$

88. $16x^2 - 40x + 25 = 0$

P.7 Translations and Combinations

See Exercises 47 and 48 on pages 94 and 95 for examples of how combinations of functions can be used to analyze the costs of automobile upkeep from 1985 through 1991.

Shifting, Reflecting, and Stretching Graphs ❒ Arithmetic Combinations of Functions ❒ Composition of Functions ❒ Applications

A computer animation of this concept appears in the *Interactive* CD-ROM.

Shifting, Reflecting, and Stretching Graphs

Many functions have graphs that are simple transformations of the common graphs summarized on page 77. For example, you can obtain the graph of $h(x) = x^2 + 2$ by shifting the graph of $f(x) = x^2$ *up* two units, as shown in Figure P.51. In function notation, h and f are related as follows.

$$h(x) = x^2 + 2 = f(x) + 2 \qquad \text{Upward shift of 2}$$

Similarly, you can obtain the graph of $g(x) = (x - 2)^2$ by shifting the graph of $f(x) = x^2$ to the *right* two units, as shown in Figure P.52. In this case, the functions g and f have the following relationship.

$$g(x) = (x - 2)^2 = f(x - 2) \qquad \text{Right shift of 2}$$

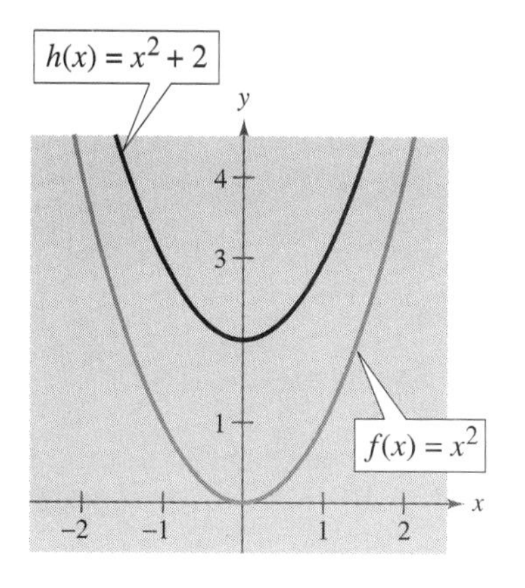

FIGURE P.51

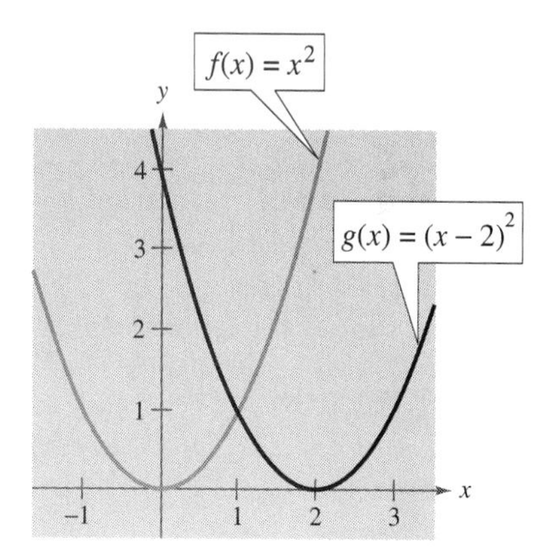

FIGURE P.52

VERTICAL AND HORIZONTAL SHIFTS

Let c be a positive real number. **Vertical and horizontal shifts** in the graph of $y = f(x)$ are represented as follows.

1. Vertical shift c units **upward:** $h(x) = f(x) + c$
2. Vertical shift c units **downward:** $h(x) = f(x) - c$
3. Horizontal shift c units to the **right:** $h(x) = f(x - c)$
4. Horizontal shift c units to the **left:** $h(x) = f(x + c)$

NOTE In items 3 and 4, be sure you see that $h(x) = f(x - c)$ corresponds to a *right* shift and $h(x) = f(x + c)$ corresponds to a *left* shift for $c > 0$.

Some graphs can be obtained from a combination of vertical and horizontal shifts. This is demonstrated in Example 1(b).

EXAMPLE 1 *Shifts in the Graph of a Function*

Use the graph of $f(x) = x^3$ to sketch the graph of each function.

a. $g(x) = x^3 + 1$

b. $h(x) = (x + 2)^3 + 1$

NOTE Vertical and horizontal shifts such as those shown in Example 1 generate a *family of functions*, each with the same shape but at different locations in the plane.

Solution

Relative to the graph of $f(x) = x^3$, the graph of $g(x) = x^3 + 1$ is an upward shift of one unit, and the graph of $h(x) = (x + 2)^3 + 1$ involves a left shift of two units *and* an upward shift of one unit. The graphs of both functions are compared with the graph of $f(x) = x^3$ in Figure P.53.

(a)

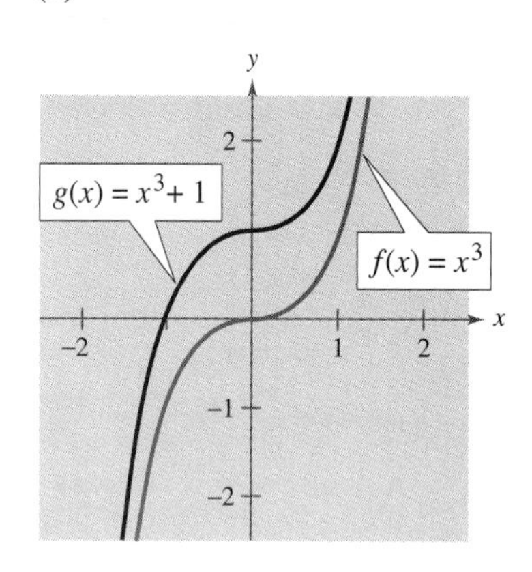

(b)

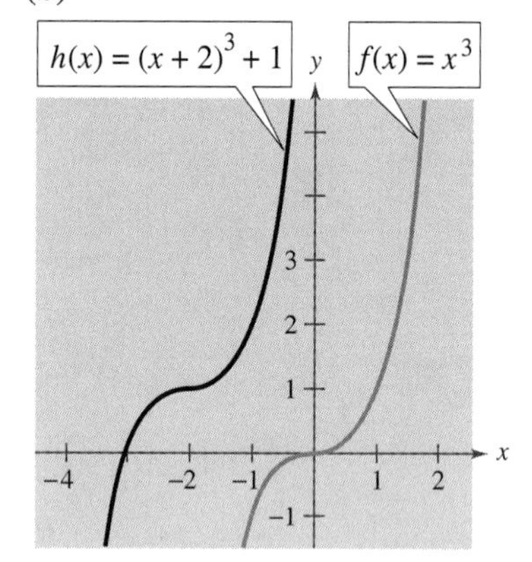

FIGURE P.53

Study Tip

In part (b) of Figure P.53, notice that the same result is obtained if the vertical shift precedes the horizontal shift *or* if the horizontal shift precedes the vertical shift.

Exploration

Graphing utilities are ideal tools for exploring translations of functions. Graph f, g, and h on the same screen. Before looking at the graphs, try to predict how the graphs of g and h relate to the graph of f.

a. $f(x) = x^2, \quad g(x) = (x - 4)^2, \quad h(x) = (x - 4)^2 + 3$

b. $f(x) = x^2, \quad g(x) = (x + 1)^2, \quad h(x) = (x + 1)^2 - 2$

c. $f(x) = x^2, \quad g(x) = (x + 4)^2, \quad h(x) = (x + 4)^2 + 2$

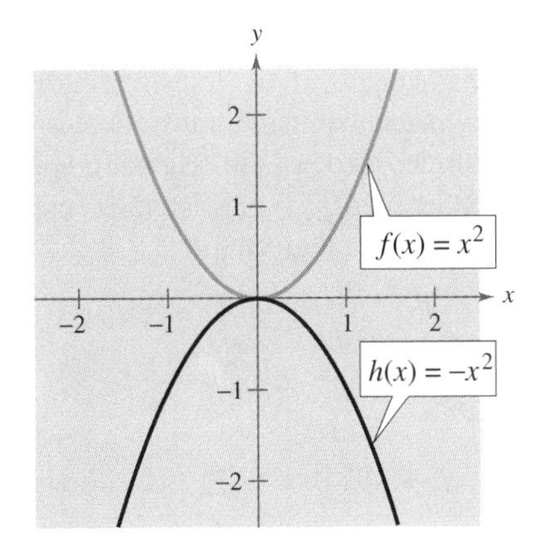

FIGURE P.54

A computer animation of this concept appears in the *Interactive* CD-ROM.

TECHNOLOGY

In the appendix, you will find programs for a variety of graphing calculator models that will give you practice working with reflections, horizontal shifts, and vertical shifts. These programs will sketch a graph of the function

$$y = R(x + H)^2 + V$$

where $R = \pm 1$, H is an integer between -6 and 6, and V is an integer between -3 and 3. Each time you run the program, different values of R, H, and V are possible. From the graph, you should be able to determine the values of R, H, and V. After you have determined the values, press the enter key to see the answer. (To look at the graph again, press the graph key.)

The second common type of transformation is a **reflection.** For instance, if you consider the x-axis to be a mirror, the graph of

$$h(x) = -x^2$$

is the mirror image (or reflection) of the graph of $f(x) = x^2$, as shown in Figure P.54.

REFLECTIONS IN THE COORDINATE AXES

Reflections in the coordinate axes of the graph of $y = f(x)$ are represented as follows.

1. **Reflection in the x-axis:** $h(x) = -f(x)$
2. **Reflection in the y-axis:** $h(x) = f(-x)$

EXAMPLE 2 *Reflections and Shifts*

Compare the graphs of each function with the graph of $f(x) = \sqrt{x}$.

a. $g(x) = -\sqrt{x}$

b. $h(x) = \sqrt{-x}$

Solution

a. The graph of g is a reflection of the graph of f in the x-axis because

$$g(x) = -\sqrt{x} = -f(x).$$

b. The graph of h is a reflection of the graph of f in the y-axis because

$$h(x) = \sqrt{-x} = f(-x).$$

The graphs of both functions are shown in Figure P.55.

(a)

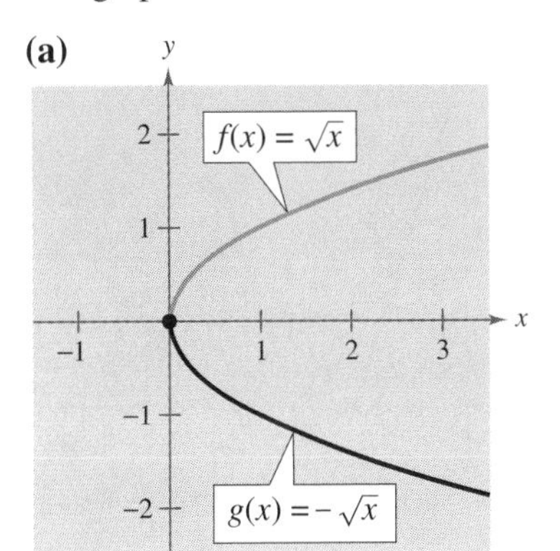

(b)

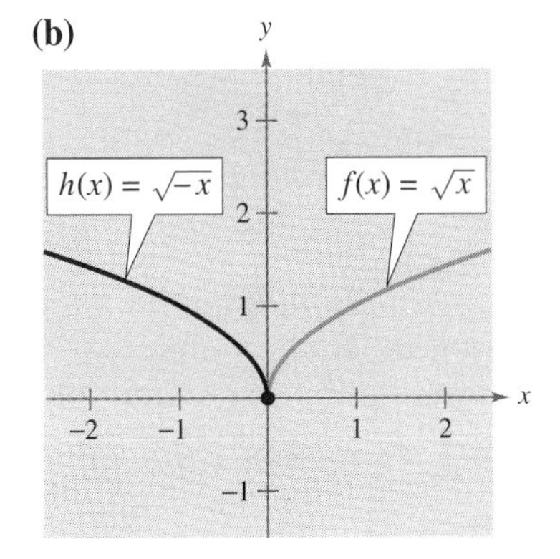

FIGURE P.55

Horizontal shifts, vertical shifts, and reflections are **rigid** transformations because the basic shape of the graph is unchanged. These transformations change only the *position* of the graph in the xy-plane. **Nonrigid** transformations are those that cause a *distortion*—a change in the shape of the original graph. For instance, a nonrigid transformation of the graph of $y = f(x)$ is represented by $g(x) = cf(x)$, where the transformation is a **vertical stretch** if $c > 1$ and a **vertical shrink** if $0 < c < 1$.

Exploration

Sketch the graph of $f(x) = 2x^2$. Compare this graph with the graph of $h(x) = x^2$. Describe the effect of multiplying x^2 by a number greater than 1. Then graph $g(x) = \frac{1}{2}x^2$. Compare this with the graph of $h(x) = x^2$. Describe the effect of multiplying x^2 by a number less than 1. Can you think of an easy way to remember this generalization?

EXAMPLE 3 Nonrigid Transformations

Compare the graph of each function with the graph of $f(x) = |x|$.

a. $h(x) = 3|x|$

b. $g(x) = \dfrac{1}{3}|x|$

Solution

a. Relative to the graph of $f(x) = |x|$, the graph of

$$h(x) = 3|x| = 3f(x)$$

is a vertical stretch (multiply each y-value by 3) of the graph of f.

b. Similarly, the function

$$g(x) = \frac{1}{3}|x| = \frac{1}{3}f(x)$$

indicates that the graph of g is a vertical shrink of the graph of f.

The graphs of both functions are shown in Figure P.56.

(a)

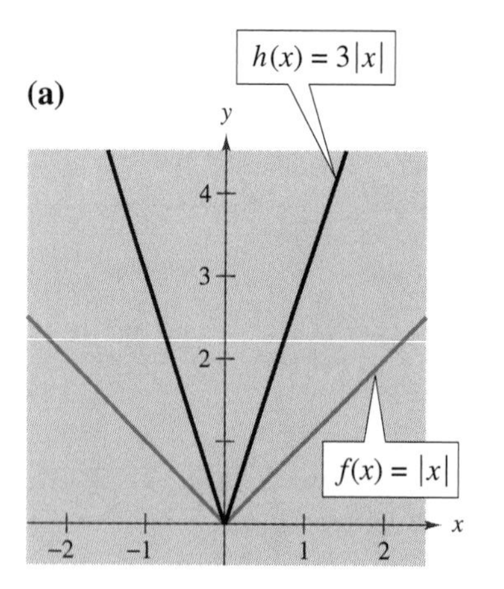

(b)

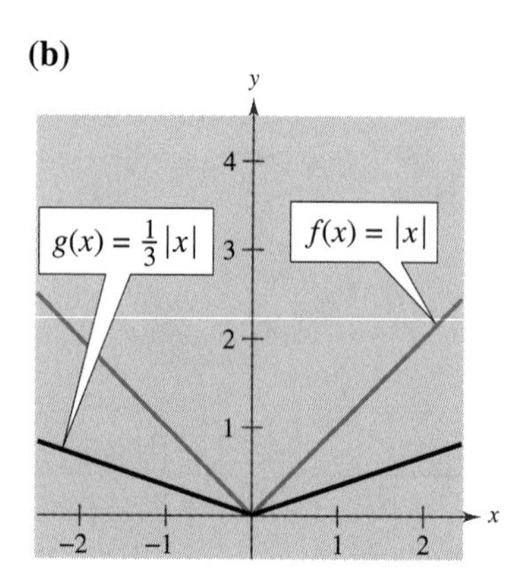

FIGURE P.56

Arithmetic Combinations of Functions

Just as two real numbers can be combined by the operations of addition, subtraction, multiplication, and division to form other real numbers, two *functions* can be combined to create new functions. For example, the functions

$$f(x) = 2x - 3 \quad \text{and} \quad g(x) = x^2 - 1$$

can be combined to form the sum, difference, product, and quotient of f and g as follows.

$$f(x) + g(x) = (2x - 3) + (x^2 - 1) = x^2 + 2x - 4 \qquad \text{Sum}$$
$$f(x) - g(x) = (2x - 3) - (x^2 - 1) = -x^2 + 2x - 2 \qquad \text{Difference}$$
$$f(x)g(x) = (2x - 3)(x^2 - 1) = 2x^3 - 3x^2 - 2x + 3 \qquad \text{Product}$$
$$\frac{f(x)}{g(x)} = \frac{2x - 3}{x^2 - 1}, \quad x \neq \pm 1 \qquad \text{Quotient}$$

The domain of an arithmetic combination of functions f and g consists of all real numbers that are common to the domains of f and g. In the case of the quotient $f(x)/g(x)$, there is the further restriction that $g(x) \neq 0$.

SUM, DIFFERENCE, PRODUCT, AND QUOTIENT OF FUNCTIONS

Let f and g be two functions with overlapping domains. Then, for all x common to both domains, the **sum, difference, product,** and **quotient** of f and g are defined as follows.

1. *Sum:* $(f + g)(x) = f(x) + g(x)$
2. *Difference:* $(f - g)(x) = f(x) - g(x)$
3. *Product:* $(fg)(x) = f(x) \cdot g(x)$
4. *Quotient:* $\left(\frac{f}{g}\right)(x) = \frac{f(x)}{g(x)}, \quad g(x) \neq 0$

EXAMPLE 4 *Finding the Sum of Two Functions*

Given $f(x) = 2x + 1$ and $g(x) = x^2 + 2x - 1$, find $(f + g)(x)$.

Solution

$$\begin{aligned}(f + g)(x) &= f(x) + g(x) \\ &= (2x + 1) + (x^2 + 2x - 1) \\ &= x^2 + 4x\end{aligned}$$

EXAMPLE 5 *Finding the Difference of Two Functions*

Given $f(x) = 2x + 1$ and $g(x) = x^2 + 2x - 1$, find $(f - g)(x)$. Then evaluate the difference when $x = 2$.

Solution

The difference of f and g is given by

$$\begin{aligned}(f - g)(x) &= f(x) - g(x)\\ &= (2x + 1) - (x^2 + 2x - 1)\\ &= -x^2 + 2.\end{aligned}$$

When $x = 2$, the value of this difference is

$$(f - g)(2) = -(2)^2 + 2 = -2.$$

In Examples 4 and 5, both f and g have domains that consist of all real numbers. Thus, the domains of $(f + g)$ and $(f - g)$ are also the set of all real numbers. Remember that any restrictions on the domains of f and g must be considered when forming the sum, difference, product, or quotient of f and g.

EXAMPLE 6 *Finding the Quotient of Two Functions*

Find the domains of $(f/g)(x)$ and $(g/f)(x)$ for the functions

$$f(x) = \sqrt{x} \qquad \text{and} \qquad g(x) = \sqrt{4 - x^2}.$$

Solution

The quotient of f and g is given by

$$\left(\frac{f}{g}\right)(x) = \frac{f(x)}{g(x)} = \frac{\sqrt{x}}{\sqrt{4 - x^2}}$$

and the quotient of g and f is given by

$$\left(\frac{g}{f}\right)(x) = \frac{g(x)}{f(x)} = \frac{\sqrt{4 - x^2}}{\sqrt{x}}.$$

The domain of f is $[0, \infty)$ and the domain of g is $[-2, 2]$. The intersection of these domains is $[0, 2]$. Thus, the domains of f/g and g/f are as follows.

$$\text{Domain of } \frac{f}{g}\text{: } [0, 2) \qquad \text{Domain of } \frac{g}{f}\text{: } (0, 2]$$

Can you see why these two domains differ slightly?

Composition of Functions

Another way of combining two functions is to form the **composition** of one with the other. For instance, if $f(x) = x^2$ and $g(x) = x + 1$, the composition of f with g is given by

$$f(g(x)) = f(x + 1) = (x + 1)^2.$$

This composition is denoted as $f \circ g$.

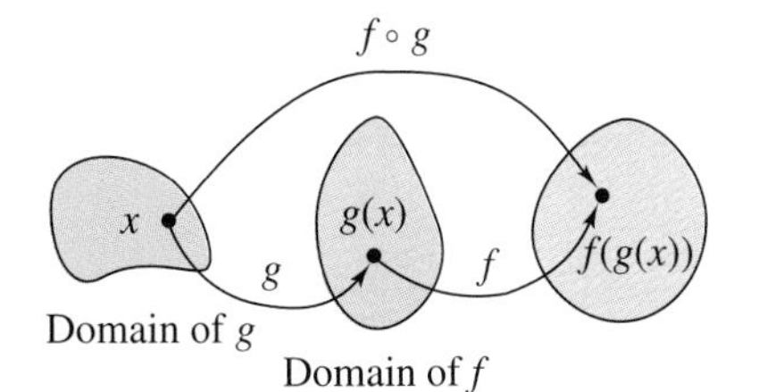

FIGURE P.57

> ***DEFINITION OF COMPOSITION OF TWO FUNCTIONS***
>
> The **composition** of the function f with the function g is given by
>
> $$(f \circ g)(x) = f(g(x)).$$
>
> The domain of $(f \circ g)$ is the set of all x in the domain of g such that $g(x)$ is in the domain of f. (See Figure P.57.)

EXAMPLE 7 *Composition of Functions*

Given $f(x) = x + 2$ and $g(x) = 4 - x^2$, find the following.

a. $(f \circ g)(x)$

b. $(g \circ f)(x)$

Solution

a. The composition of f with g is as follows.

$(f \circ g)(x) = f(g(x))$	Definition of $f \circ g$
$= f(4 - x^2)$	Definition of $g(x)$
$= (4 - x^2) + 2$	Definition of $f(x)$
$= -x^2 + 6$	Simplify.

b. The composition of g with f is as follows.

$(g \circ f)(x) = g(f(x))$	Definition of $g \circ f$
$= g(x + 2)$	Definition of $f(x)$
$= 4 - (x + 2)^2$	Definition of $g(x)$
$= 4 - (x^2 + 4x + 4)$	Expand.
$= -x^2 - 4x$	Simplify.

Note that, in this case, $(f \circ g)(x) \neq (g \circ f)(x)$.

NOTE The following tables of values help illustrate the composition of the functions f and g given in Example 7.

x	0	1	2	3
$g(x)$	4	3	0	−5

$g(x)$	4	3	0	−5
$f(g(x))$	6	5	2	−3

x	0	1	2	3
$f(g(x))$	6	5	2	−3

Note that the first two tables can be combined (or "composed") to produce the values given in the third table.

Applications

EXAMPLE 8 *Political Makeup of the U.S. Senate*

Consider three functions R, D, and I that represent the numbers of Republicans, Democrats, and Independents in the U.S. Senate from 1965 to 1995. Sketch the graphs of R, D, and I, and the sum of R, D, and I, in the same coordinate plane. The numbers of Republicans and Democrats in the Senate are shown below. (Source: Secretary of the Senate)

Year	Republicans	Democrats
1965	32	68
1967	36	64
1969	43	57
1971	44	54
1973	42	56
1975	37	60
1977	38	61
1979	41	58

Year	Republicans	Democrats
1981	53	46
1983	54	46
1985	53	47
1987	45	55
1989	45	55
1991	44	56
1993	43	57
1995	53	47

Solution

The graphs of R, D, and I are shown in Figure P.58. Note that the sum of R, D, and I is the constant function $R + D + I = 100$. This follows from the fact that the number of senators in the United States is 100 (two from each state).

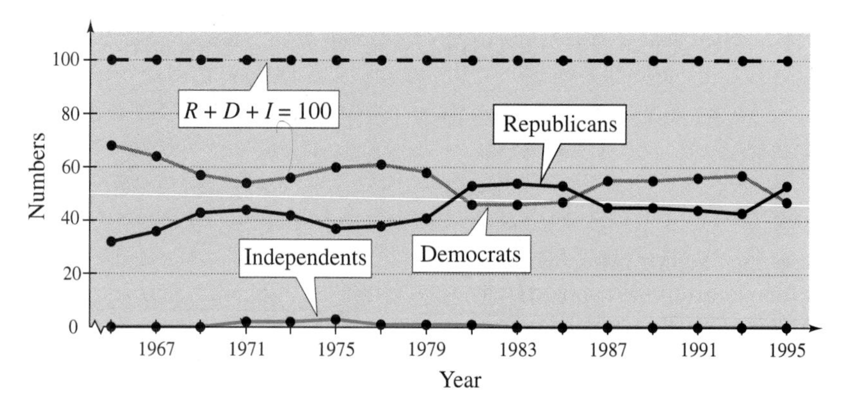

FIGURE P.58

EXAMPLE 9 Bacteria Count

Real Life

The number of bacteria in a refrigerated food is given by

$$N(T) = 20T^2 - 80T + 500, \qquad 2 \le T \le 14$$

where T is the temperature of the food. When the food is removed from refrigeration, the temperature is given by

$$T(t) = 4t + 2, \qquad 0 \le t \le 3$$

where t is the time in hours. Find (a) the composite $N(T(t))$ and interpret its meaning in context and (b) the time when the bacteria count reaches 2000.

Solution

a. $$\begin{aligned} N(T(t)) &= 20(4t + 2)^2 - 80(4t + 2) + 500 \\ &= 20(16t^2 + 16t + 4) - 320t - 160 + 500 \\ &= 320t^2 + 320t + 80 - 320t - 160 + 500 \\ &= 320t^2 + 420 \end{aligned}$$

The composite function represents the number of bacteria in the food as a function of time.

b. The bacteria count will reach 2000 when $320t^2 + 420 = 2000$. Solve this equation to find that the count will reach 2000 when $t \approx 2.2$ hours.

NOTE If you solve the equation in part (b) of Example 9, you will find that $t \approx \pm 2.2$. However, the negative value is rejected because it is not in the domain of the composite function.

GROUP ACTIVITY

ANALYZING COMBINATIONS OF FUNCTIONS

a. Use the graphs of f and $f + g$ to make a table showing the values of $g(x)$ when $x = 1, 2, 3, 4, 5$, and 6. Explain your reasoning.

b. Use the graphs of f and $f - h$ to make a table showing the values of $h(x)$ when $x = 1, 2, 3, 4, 5$, and 6. Explain your reasoning.

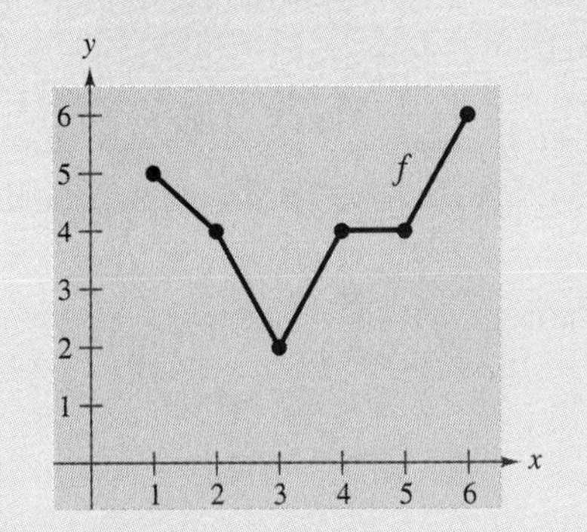

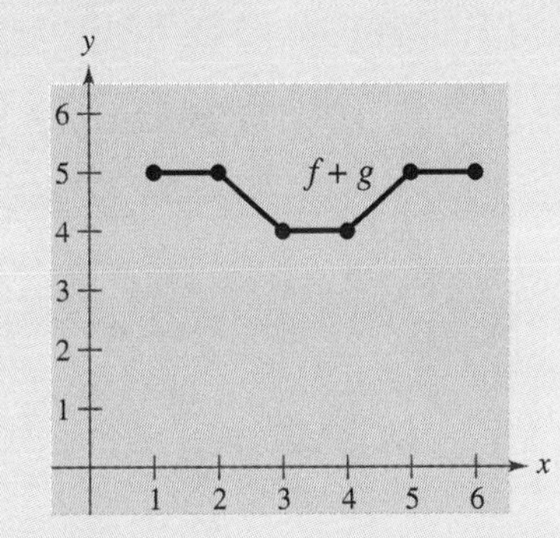

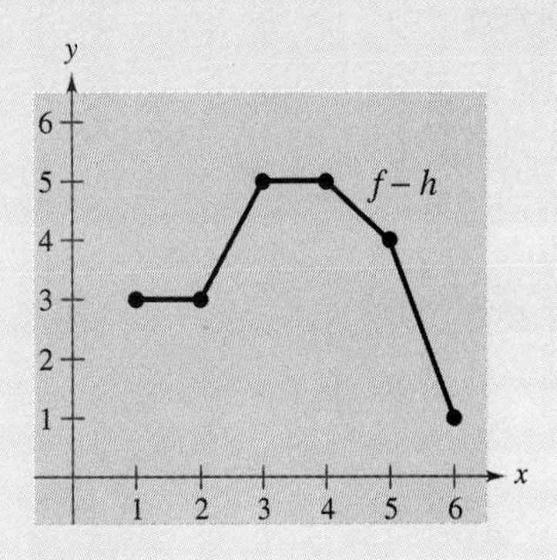

WARM UP

Perform the operations and simplify.

1. $\dfrac{1}{x} + \dfrac{1}{1 - x}$

2. $\dfrac{2}{x + 3} - \dfrac{2}{x - 3}$

3. $\dfrac{3}{x - 2} - \dfrac{2}{x(x - 2)}$

4. $\dfrac{x}{x - 5} + \dfrac{1}{3}$

5. $(x - 1)\left(\dfrac{1}{\sqrt{x^2 - 1}}\right)$

6. $\left(\dfrac{x}{x^2 - 4}\right)\left(\dfrac{x^2 - x - 2}{x^2}\right)$

7. $(x^2 - 4) \div \left(\dfrac{x + 2}{5}\right)$

8. $\left(\dfrac{x}{x^2 + 3x - 10}\right) \div \left(\dfrac{x^2 + 3x}{x^2 + 6x + 5}\right)$

9. $\dfrac{(1/x) + 5}{3 - (1/x)}$

10. $\dfrac{(x/4) - (4/x)}{x - 4}$

P.7 Exercises

1. Sketch (on the same set of coordinate axes) a graph of f for $c = -2, 0,$ and 2.

(a) $f(x) = x^3 + c$

(b) $f(x) = (x - c)^3$

2. Sketch (on the same set of coordinate axes) a graph of f for $c = -2, 0,$ and 2.

(a) $f(x) = x^2 + c$

(b) $f(x) = (x - c)^2$

3. Sketch (on the same set of coordinate axes) a graph of f for $c = -1, 1,$ and 3.

(a) $f(x) = |x| + c$

(b) $f(x) = |x - c|$

(c) $f(x) = |x + 4| + c$

4. Sketch (on the same set of coordinate axes) a graph of f for $c = -3, -1, 1,$ and 3.

(a) $f(x) = \sqrt{x} + c$

(b) $f(x) = \sqrt{x - c}$

(c) $f(x) = \sqrt{x - 3} + c$

5. Use the graph of f (see figure) to sketch the graphs.

(a) $y = f(x) + 2$ (b) $y = -f(x)$

(c) $y = f(x - 2)$ (d) $y = f(x + 3)$

(e) $y = f(2x)$ (f) $y = f(-x)$

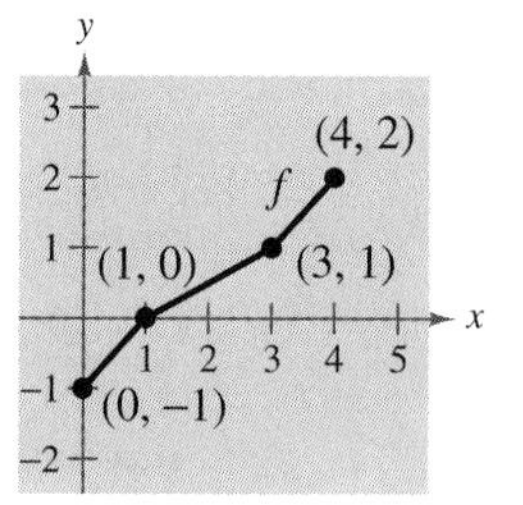

FIGURE FOR 5

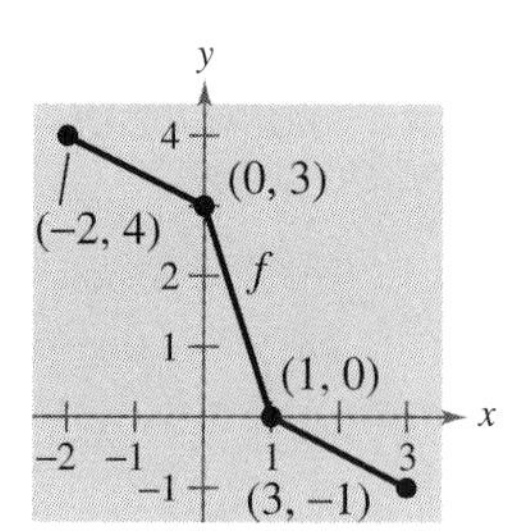

FIGURE FOR 6

6. Use the graph of f (see figure) to sketch the graphs.

(a) $y = f(x) - 1$ (b) $y = f(x + 1)$

(c) $y = f(x - 1)$ (d) $y = -f(x - 2)$

(e) $y = f(-x)$ (f) $y = \frac{1}{2}f(x)$

7. Use the graph of $f(x) = x^2$ to write formulas for the functions whose graphs are shown below.

(a)

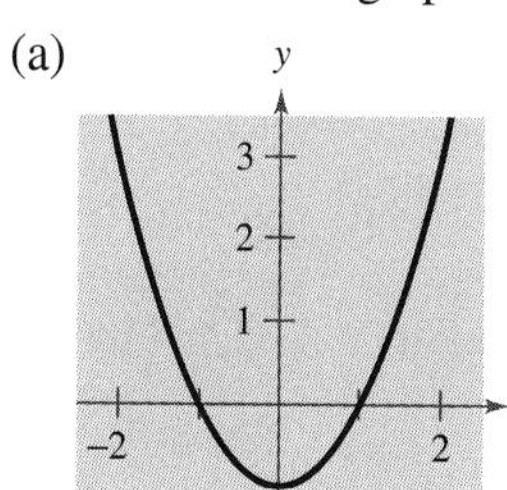

(b)

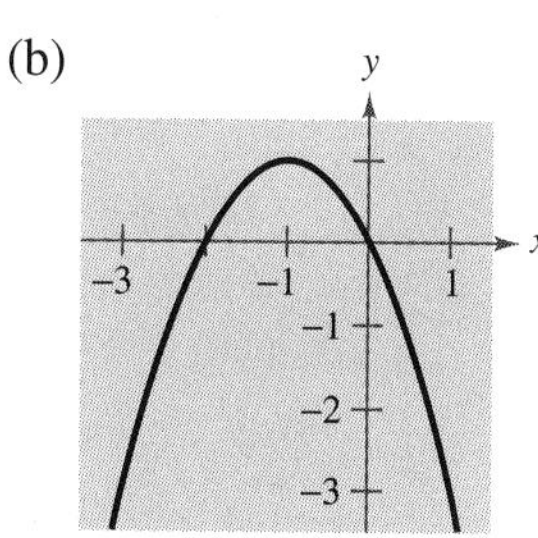

8. Use the graph of $f(x) = x^3$ to write formulas for the functions whose graphs are shown below.

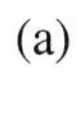

(a)

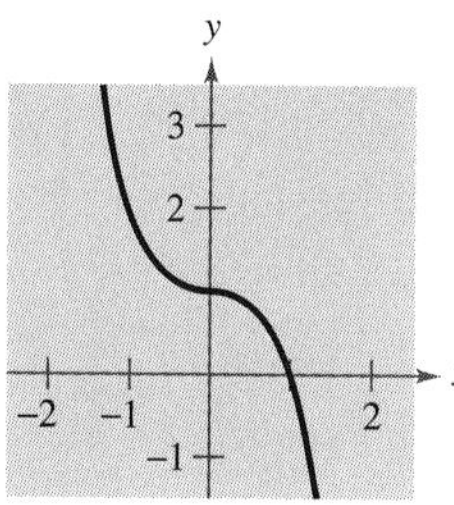

(b)

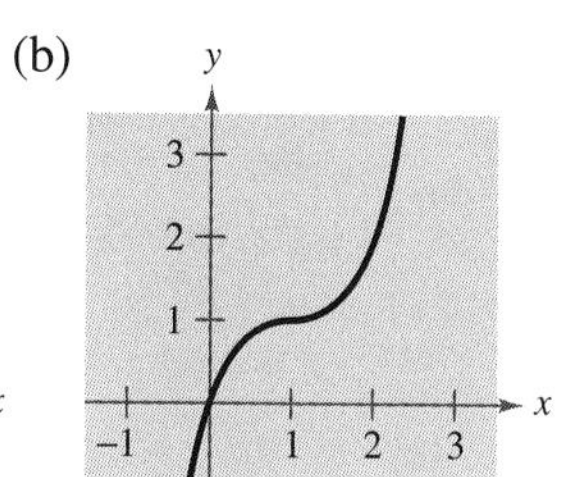

In Exercises 9–14, identify the common function and the transformation shown in the graph. Write the formula for the graphed function.

9.

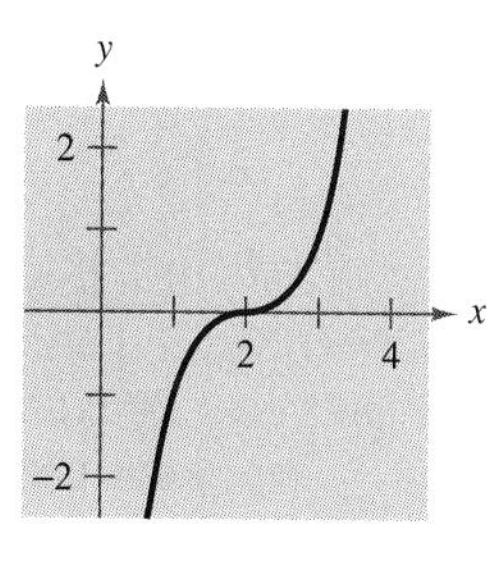

10.

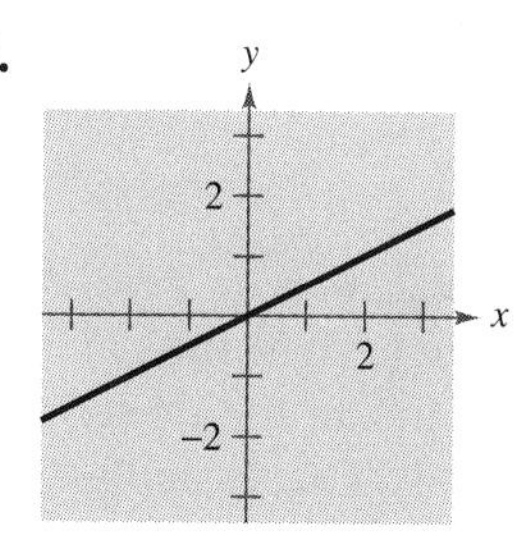

11.

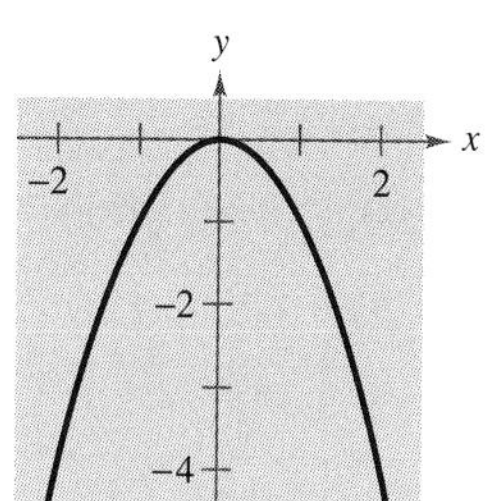

12.

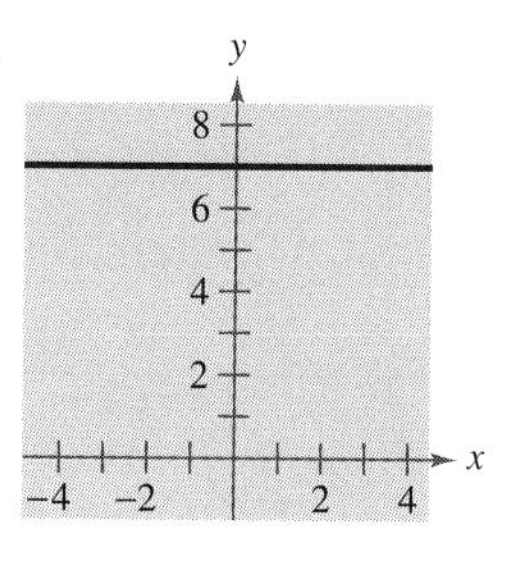

13.

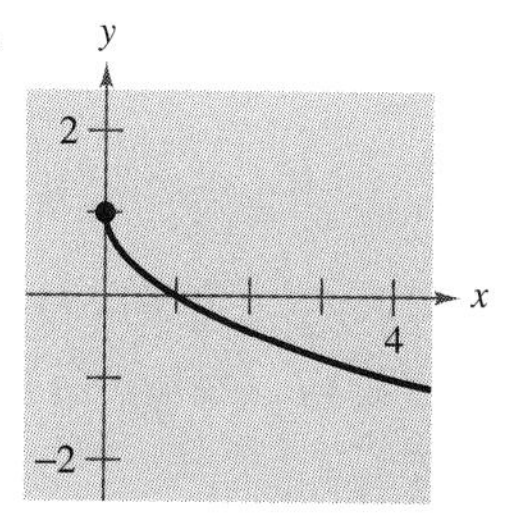

14.

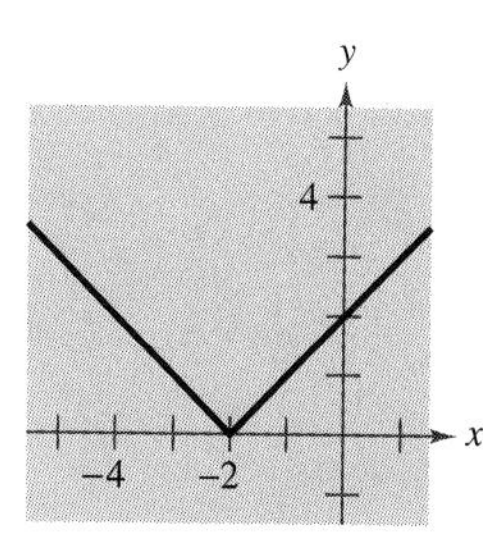

In Exercises 15–18, use the graphs of f and g to graph $h(x) = (f + g)(x)$.

15.

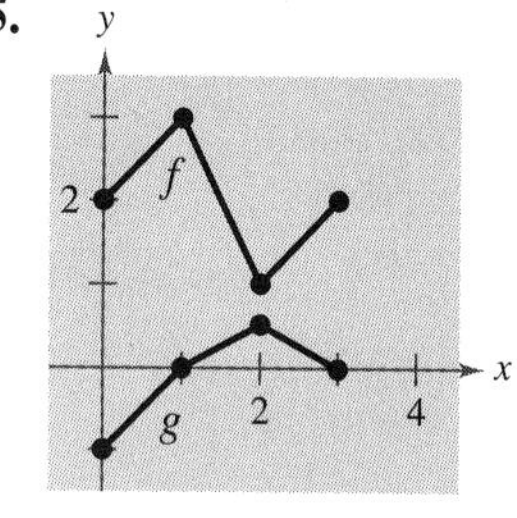

16.

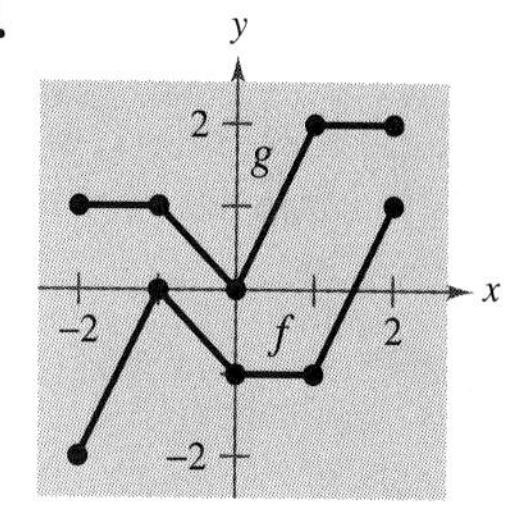

17.

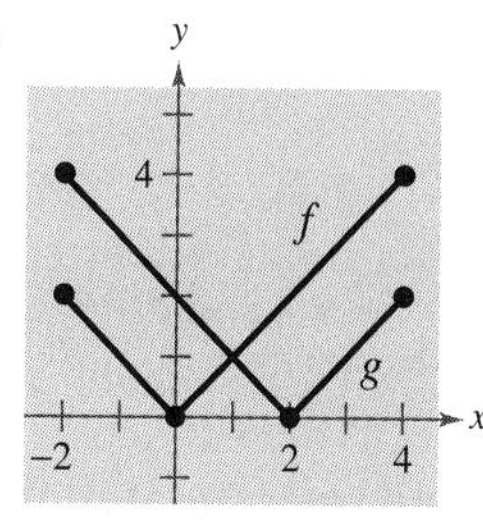

18.

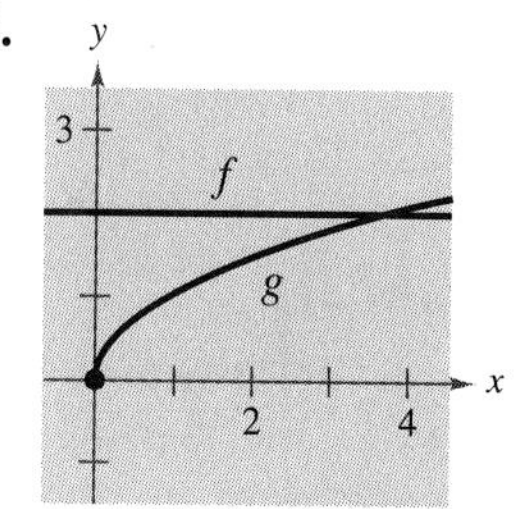

In Exercises 19–26, find (a) $(f + g)(x)$, (b) $(f - g)(x)$, (c) $(fg)(x)$, and (d) $(f/g)(x)$. What is the domain of f/g?

19. $f(x) = x + 1,\quad g(x) = x - 1$

20. $f(x) = 2x - 5,\quad g(x) = 1 - x$

21. $f(x) = x^2,\quad g(x) = 1 - x$

22. $f(x) = 2x - 5,\quad g(x) = 5$

23. $f(x) = x^2 + 5,\quad g(x) = \sqrt{1 - x}$

24. $f(x) = \sqrt{x^2 - 4},\quad g(x) = \dfrac{x^2}{x^2 + 1}$

25. $f(x) = \dfrac{1}{x},\quad g(x) = \dfrac{1}{x^2}$

26. $f(x) = \dfrac{x}{x + 1},\quad g(x) = x^3$

In Exercises 27–38, evaluate the indicated function for $f(x) = x^2 + 1$ and $g(x) = x - 4$.

27. $(f + g)(3)$
28. $(f - g)(-2)$
29. $(f - g)(0)$
30. $(f + g)(1)$
31. $(f - g)(2t)$
32. $(f + g)(t - 1)$
33. $(fg)(4)$
34. $(fg)(-6)$
35. $\left(\frac{f}{g}\right)(5)$
36. $\left(\frac{f}{g}\right)(0)$
37. $\left(\frac{f}{g}\right)(-1) - g(3)$
38. $(2f)(5)$

In Exercises 39–42, graph the functions f, g, and $f + g$ on the same set of coordinate axes.

39. $f(x) = \frac{1}{2}x, \quad g(x) = x - 1$
40. $f(x) = \frac{1}{3}x, \quad g(x) = -x + 4$
41. $f(x) = x^2, \quad g(x) = -2x$
42. $f(x) = 4 - x^2, \quad g(x) = x$

Graphical Reasoning **In Exercises 43 and 44, use a graphing utility to sketch the graphs of f, g, and $f + g$ on the same viewing rectangle. Which function contributes most to the magnitude of the sum when $0 \le x \le 2$? Which function contributes most to the magnitude of the sum when $x > 6$?**

43. $f(x) = 3x, \quad g(x) = -\frac{x^3}{10}$
44. $f(x) = \frac{x}{2}, \quad g(x) = \sqrt{x}$

45. ***Stopping Distance*** While traveling in a car at x miles per hour, you are required to stop quickly to avoid an accident. The distance the car travels during your reaction time is given by $R(x) = \frac{3}{4}x$. The distance traveled while you are braking is given by

$$B(x) = \frac{1}{15}x^2.$$

Find the function giving total stopping distance T. Graph the functions R, B, and T on the same set of coordinate axes for $0 \le x \le 60$.

46. ***Comparing Sales*** You own two restaurants. From 1990 to 1995, the sales R_1 (in thousands of dollars) for one restaurant can be modeled by

$$R_1 = 480 - 8t - 0.8t^2, \qquad t = 0, 1, 2, 3, 4, 5$$

where $t = 0$ represents 1990. During the same 6-year period, the sales R_2 (in thousands of dollars) for the second restaurant can be modeled by

$$R_2 = 254 + 0.78t, \qquad t = 0, 1, 2, 3, 4, 5.$$

(a) Write a function that represents the total sales for the two restaurants. Use a graphing utility to graph the total sales function.

(b) Use the *stacked bar graph* in the figure, which represents the total sales during the 6-year period, to determine whether the total sales have been increasing or decreasing.

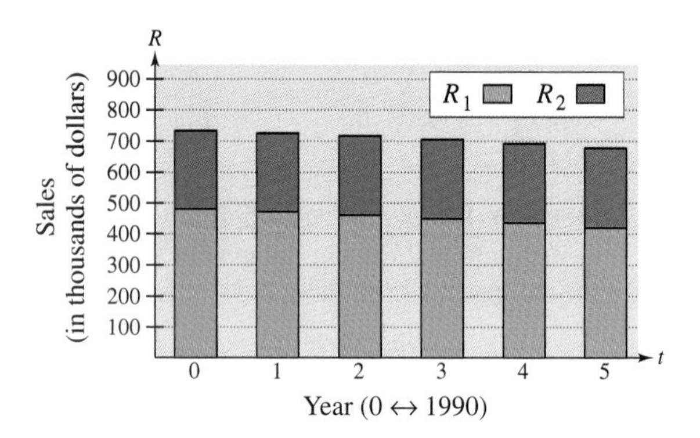

Data Analysis **In Exercises 47 and 48, use the table, which gives the variable costs for operating an automobile in the United States for the years 1985 through 1991. The variables y_1, y_2, and y_3 represent the costs in cents per mile for gas and oil, maintenance, and tires, respectively.** (Source: American Automobile Manufacturers Association)

Year	1985	1986	1987	1988	1989	1990	1991
y_1	6.16	4.48	4.80	5.20	5.20	5.40	6.70
y_2	1.23	1.37	1.60	1.60	1.90	2.10	2.20
y_3	0.65	0.67	0.80	0.80	0.80	0.90	0.90

47. Create a stacked bar graph for the data.

48. Mathematical models for the data are given by

$y_1 = 0.16t^2 - 2.43t + 13.96$

$y_2 = 0.17t + 0.38$

$y_3 = 0.04t + 0.44$

where $t = 5$ represents 1985. Use a graphing utility to graph y_1, y_2, y_3, and $y_1 + y_2 + y_3$ on the same viewing rectangle. Use the model to estimate the total variable cost per mile in 1995.

49. ***Graphical Reasoning*** An electronically controlled thermostat in a home is programmed to automatically lower the temperature during the night (see figure). The temperature in the house T, in degrees Fahrenheit, is given in terms of t, the time in hours on a 24-hour clock.

(a) Explain why T is a function of t.

(b) Approximate $T(4)$ and $T(15)$.

(c) Suppose the thermostat were reprogrammed to produce a temperature H where $H(t) = T(t - 1)$. How would this change the temperature?

(d) Suppose the thermostat were reprogrammed to produce a temperature H where $H(t) = T(t) - 1$. How would this change the temperature in the house?

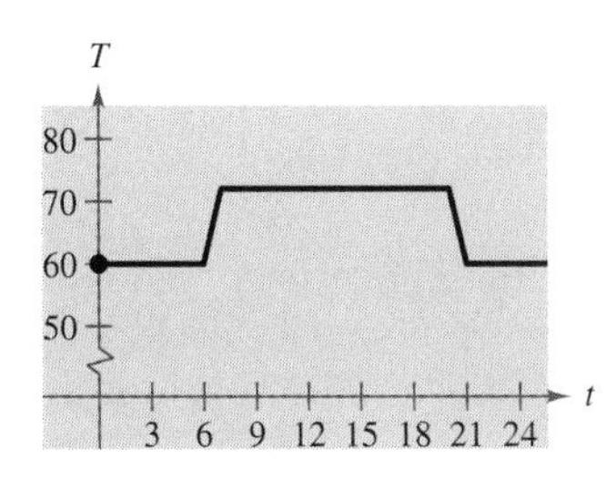

50. ***Think About It*** Write a piecewise-defined function that represents the graph in Exercise 49.

In Exercises 51–54, find (a) $f \circ g$, (b) $g \circ f$, and (c) $f \circ f$.

51. $f(x) = x^2$, $\quad g(x) = x - 1$

52. $f(x) = \sqrt[3]{x - 1}$, $\quad g(x) = x^3 + 1$

53. $f(x) = 3x + 5$, $\quad g(x) = 5 - x$

54. $f(x) = x^3$, $\quad g(x) = \dfrac{1}{x}$

In Exercises 55–62, find (a) $f \circ g$ and (b) $g \circ f$.

55. $f(x) = \sqrt{x + 4}$, $\quad g(x) = x^2$

56. $f(x) = \sqrt[3]{x - 1}$, $\quad g(x) = x^3 + 1$

57. $f(x) = \frac{1}{3}x - 3$, $\quad g(x) = 3x + 1$

58. $f(x) = x^4$, $\quad g(x) = x^4$

59. $f(x) = \sqrt{x}$, $\quad g(x) = \sqrt{x}$

60. $f(x) = 2x - 3$, $\quad g(x) = 2x - 3$

61. $f(x) = |x|$, $\quad g(x) = x + 6$

62. $f(x) = x^{2/3}$, $\quad g(x) = x^6$

In Exercises 63–66, use the graphs of f and g (see figures) to evaluate the functions.

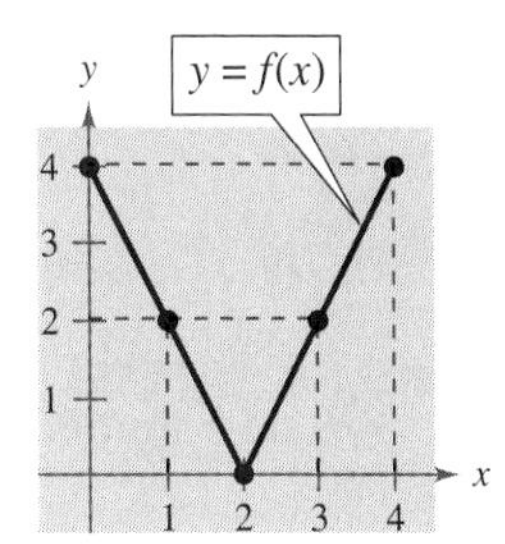

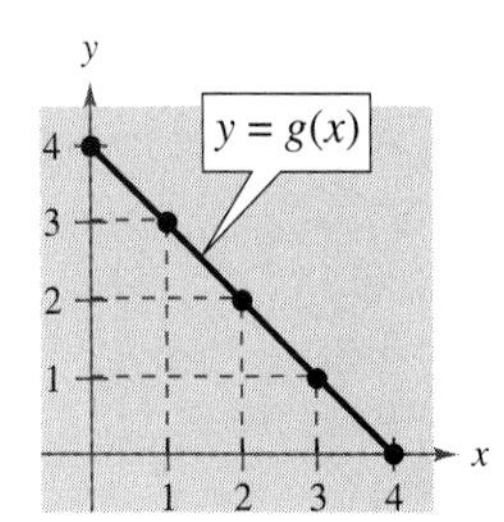

63. (a) $(f + g)(3)$ $\quad$ (b) $\left(\dfrac{f}{g}\right)(2)$

64. (a) $(f - g)(1)$ $\quad$ (b) $(fg)(4)$

65. (a) $(f \circ g)(2)$ $\quad$ (b) $(g \circ f)(2)$

66. (a) $(f \circ g)(1)$ $\quad$ (b) $(g \circ f)(3)$

Graphical Reasoning **In Exercises 67–70, use the graph of f in the figure to sketch the graph of g.**

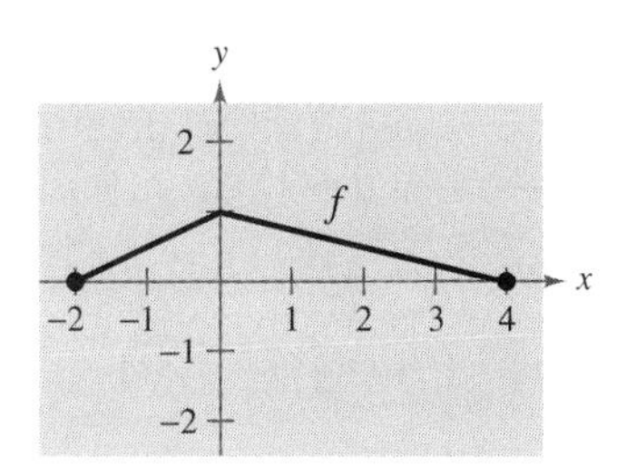

67. $g(x) = f(x) + 2$

68. $g(x) = f(x) - 1$

69. $g(x) = f(-x)$

70. $g(x) = -2f(x)$

In Exercises 71–76, find two functions f and g such that $(f \circ g)(x) = h(x)$. (There are many correct answers.)

71. $h(x) = (2x + 1)^2$

72. $h(x) = (1 - x)^3$

73. $h(x) = \sqrt[3]{x^2 - 4}$

74. $h(x) = \sqrt{9 - x}$

75. $h(x) = \dfrac{1}{x + 2}$

76. $h(x) = \dfrac{4}{(5x + 2)^2}$

In Exercises 77–80, determine the domain of (a) f, (b) g, and (c) $f \circ g$.

77. $f(x) = \sqrt{x}$, $\quad g(x) = x^2 + 1$

78. $f(x) = \dfrac{1}{x}$, $\quad g(x) = x + 3$

79. $f(x) = \dfrac{3}{x^2 - 1}$, $\quad g(x) = x + 1$

80. $f(x) = 2x + 3$, $\quad g(x) = \dfrac{x}{2}$

Average Rate of Change **In Exercises 81–84, find the difference quotient $[f(x + h) - f(x)]/h$ and simplify your answer.**

81. $f(x) = 3x - 4$

82. $f(x) = 1 - x^2$

83. $f(x) = \dfrac{4}{x}$

84. $f(x) = \sqrt{2x + 1}$

85. ***Area*** A square concrete foundation was prepared as a base for a cylindrical tank (see figure).

(a) Express the radius r of the tank as a function of the length x of the sides of the square.

(b) Express the area A of the circular base of the tank as a function of the radius r.

(c) Find and interpret $(A \circ r)(x)$.

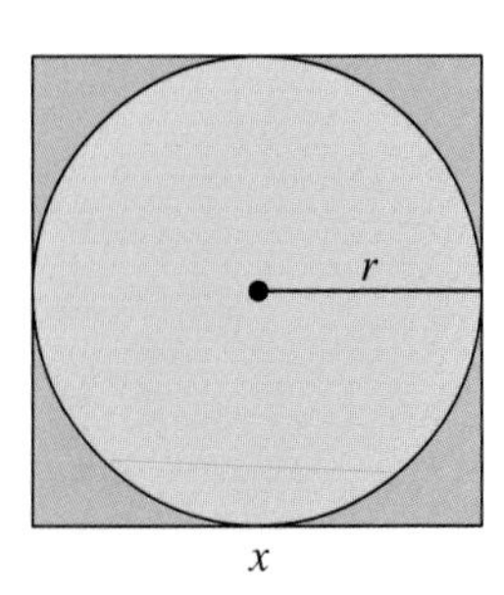

86. ***Ripples*** A pebble is dropped into a calm pond, causing ripples in the form of concentric circles. The radius (in feet) of the outer ripple is given by $r(t) = 0.6t$, where t is the time in seconds after the pebble strikes the water. The area of the circle is given by the function $A(r) = \pi r^2$. Find and interpret $(A \circ r)(t)$.

87. ***Cost*** The weekly cost of producing x units in a manufacturing process is given by the function

$$C(x) = 60x + 750.$$

The number of units produced in t hours is given by $x(t) = 50t$. Find and interpret $(C \circ x)(t)$.

88. ***Think About It*** You are a sales representative for an automobile manufacturer. You are paid an annual salary plus a bonus of 3% of your sales over \$500,000. Consider the two functions

$$f(x) = x - 500{,}000 \quad \text{and} \quad g(x) = 0.03x.$$

If x is greater than \$500,000, which of the following represents your bonus? Explain your reasoning.

(a) $f(g(x))$

(b) $g(f(x))$

89. ***Exploration*** The suggested retail price of a new car is p dollars. The dealership advertised a factory rebate of \$1200 and an 8% discount.

(a) Write a function R in terms of p, giving the cost of the car after receiving the rebate from the factory.

(b) Write a function S in terms of p, giving the cost of the car after receiving the dealership discount.

(c) Form the composite functions $(R \circ S)(p)$ and $(S \circ R)(p)$ and interpret each.

(d) Find $(R \circ S)(18{,}400)$ and $(S \circ R)(18{,}400)$. Which yields the smaller cost for the car? Explain.

90. Prove that the product of two odd functions is an even function, and that the product of two even functions is an even function.

91. ***Conjecture*** Use examples to hypothesize whether the product of an odd function and an even function is even or odd. Then prove your hypothesis.

P.8 Inverse Functions

See Exercise 81 on page 107 for an example of how the inverse of a function can be used to analyze the behavior of a diesel engine.

The Inverse of a Function ▫ *Finding the Inverse of a Function* ▫ *The Graph of the Inverse of a Function*

The Inverse of a Function

Recall from Section P.5 that a function can be represented by a set of ordered pairs. For instance, the function $f(x) = x + 4$ from the set $A = \{1, 2, 3, 4\}$ to the set $B = \{5, 6, 7, 8\}$ can be written as follows.

$$f(x) = x + 4: \{(1, 5), (2, 6), (3, 7), (4, 8)\}$$

By interchanging the first and second coordinates of each of these ordered pairs, you can form the **inverse function** of f, which is denoted by f^{-1}. It is a function from the set B to the set A, and can be written as follows.

$$f^{-1}(x) = x - 4: \{(5, 1), (6, 2), (7, 3), (8, 4)\}$$

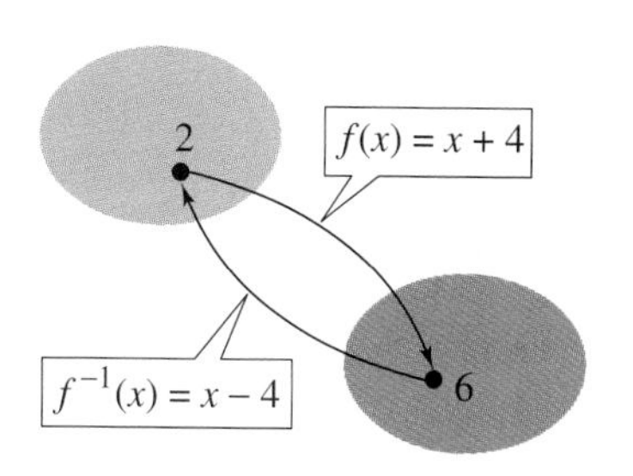

FIGURE P.59

Note that the domain of f is equal to the range of f^{-1}, and vice versa, as shown in Figure P.59. Also note that the functions f and f^{-1} have the effect of "undoing" each other. In other words, when you form the composition of f with f^{-1} or the composition of f^{-1} with f, you obtain the identity function.

$$f(f^{-1}(x)) = f(x - 4) = (x - 4) + 4 = x$$
$$f^{-1}(f(x)) = f^{-1}(x + 4) = (x + 4) - 4 = x$$

NOTE A table of values can help you understand inverse functions. For instance, the following table shows several values of the function f in Example 1.

x	-2	-1	0	1	2
$f(x)$	-8	-4	0	4	8

By interchanging the table's rows, you obtain values of the inverse function f^{-1}.

x	-8	-4	0	4	8
$f^{-1}(x)$	-2	-1	0	1	2

In the first table, each output is 4 times the input, and in the second table, each output is $\frac{1}{4}$ the input.

EXAMPLE 1 *Finding Inverse Functions Informally*

Find the inverse of $f(x) = 4x$. Then verify that both $f(f^{-1}(x))$ and $f^{-1}(f(x))$ are equal to the identity function.

Solution

The given function *multiplies* each input by 4. To "undo" this function, you need to *divide* each input by 4. Thus, the inverse function of $f(x) = 4x$ is

$$f^{-1}(x) = \frac{x}{4}.$$

You can verify that both $f(f^{-1}(x))$ and $f^{-1}(f(x))$ are equal to the identity function as follows.

$$f(f^{-1}(x)) = f\left(\frac{x}{4}\right) = 4\left(\frac{x}{4}\right) = x$$

$$f^{-1}(f(x)) = f^{-1}(4x) = \frac{4x}{4} = x$$

EXAMPLE 2 *Finding Inverse Functions Informally*

Find the inverse of $f(x) = x - 6$. Then verify that both $f(f^{-1}(x))$ and $f^{-1}(f(x))$ are equal to the identity function.

Solution

The given function *subtracts* 6 from each input. To "undo" this function, you need to *add* 6 to each input. Thus, the inverse function of $f(x) = x - 6$ is

$$f^{-1}(x) = x + 6.$$

You can verify that both $f(f^{-1}(x))$ and $f^{-1}(f(x))$ are equal to the identity function as follows.

$$f(f^{-1}(x)) = f(x + 6) = (x + 6) - 6 = x$$
$$f^{-1}(f(x)) = f^{-1}(x - 6) = (x - 6) + 6 = x$$

The formal definition of the inverse of a function is as follows.

DEFINITION OF THE INVERSE OF A FUNCTION

Let f and g be two functions such that

$f(g(x)) = x$ for every x in the domain of g

and

$g(f(x)) = x$ for every x in the domain of f.

Under these conditions, the function g is the **inverse** of the function f. The function g is denoted by f^{-1} (read "f-inverse"). Thus,

$f(f^{-1}(x)) = x$ and $f^{-1}(f(x)) = x.$

The domain of f must be equal to the range of f^{-1}, and the range of f must be equal to the domain of f^{-1}.

NOTE Don't be confused by the use of -1 to denote the inverse function f^{-1}. In this text, whenever we write f^{-1}, we will *always* be referring to the inverse of the function f and *not* to the reciprocal of $f(x)$.

If the function g is the inverse of the function f, it must also be true that the function f is the inverse of the function g. For this reason, you can say that the functions f and g are *inverses of each other.*

EXAMPLE 3 *Verifying Inverse Functions*

Show that the functions are inverses of each other.

$$f(x) = 2x^3 - 1 \quad \text{and} \quad g(x) = \sqrt[3]{\frac{x+1}{2}}$$

Solution

$$\begin{aligned} f(g(x)) = f\left(\sqrt[3]{\frac{x+1}{2}}\right) &= 2\left(\sqrt[3]{\frac{x+1}{2}}\right)^3 - 1 \\ &= 2\left(\frac{x+1}{2}\right) - 1 \\ &= x + 1 - 1 \\ &= x \end{aligned}$$

$$\begin{aligned} g(f(x)) = g(2x^3 - 1) &= \sqrt[3]{\frac{(2x^3 - 1) + 1}{2}} \\ &= \sqrt[3]{\frac{2x^3}{2}} \\ &= \sqrt[3]{x^3} \\ &= x \end{aligned}$$

EXAMPLE 4 *Verifying Inverse Functions*

Which of the functions is the inverse of $f(x) = \dfrac{5}{x-2}$?

$$g(x) = \frac{x-2}{5} \quad \text{and} \quad h(x) = \frac{5}{x} + 2$$

Solution

By forming the composition of f with g, you have

$$f(g(x)) = f\left(\frac{x-2}{5}\right) = \frac{5}{[(x-2)/5] - 2} = \frac{25}{x-12} \neq x.$$

Because this composition is not equal to the identity function x, it follows that g *is not* the inverse of f. By forming the composition of f with h, you have

$$f(h(x)) = f\left(\frac{5}{x} + 2\right) = \frac{5}{(5/x) + 2 - 2} = \frac{5}{5/x} = x.$$

Thus, it appears that h *is* the inverse of f. You can confirm this by showing that the composition of h with f is also equal to the identity function. (Try doing this.)

Finding the Inverse of a Function

For simple functions (such as the ones in Examples 1 and 2) you can find inverse functions by inspection. For more complicated functions, however, it is best to use the following guidelines. The key step in these guidelines is Step 2—interchanging the roles of x and y. This step corresponds to the fact that inverse functions have ordered pairs with the coordinates reversed.

Study Tip

Note in Step 3 of the guidelines for finding the inverse of a function that it is possible that a function has no inverse. For instance, the function $f(x) = x^2$ has no inverse function.

FINDING THE INVERSE OF A FUNCTION

1. In the equation for $f(x)$, replace $f(x)$ by y.
2. Interchange the roles of x and y.
3. If the new equation does not represent y as a function of x, the function f does not have an inverse function. If the new equation does represent y as a function of x, solve the new equation for y.
4. Replace y by $f^{-1}(x)$.
5. Verify that f and f^{-1} are inverses of each other by showing that the domain of f is equal to the range of f^{-1}, the range of f is equal to the domain of f^{-1}, and $f(f^{-1}(x)) = x = f^{-1}(f(x))$.

EXAMPLE 5 *Finding the Inverse of a Function*

Find the inverse of $f(x) = \dfrac{5 - 3x}{2}$.

Solution

$$f(x) = \frac{5 - 3x}{2}$$ Given function

$$y = \frac{5 - 3x}{2}$$ Replace $f(x)$ by y.

$$x = \frac{5 - 3y}{2}$$ Interchange x and y.

$$2x = 5 - 3y$$ Multiply both sides by 2.

$$3y = 5 - 2x$$ Isolate the y-term.

$$y = \frac{5 - 2x}{3}$$ Solve for y.

$$f^{-1}(x) = \frac{5 - 2x}{3}$$ Replace y by $f^{-1}(x)$.

Note that both f and f^{-1} have domains and ranges that consist of the entire set of real numbers. Check that $f(f^{-1}(x)) = x$ and $f^{-1}(f(x)) = x$.

TECHNOLOGY

The following program for the *TI-83* or the *TI-82* will graph a function f and its reflection in the line $y = x$. Programs for other graphing calculator models may be found in the appendix.

```
PROGRAM:REFLECT
:63Xmin/95 → Ymin
:63Xmax/95 → Ymax
:Xscl → Yscl
:"X" → Y2
:DispGraph
:(Xmax - Xmin)/94 → N
:Xmin → X
:While X ≤ Xmax
:Pt-On(Y1, X)
:X + N → X
:End
```

To use this program, enter the function in Y1 and set a viewing rectangle.

The Graph of the Inverse of a Function

The graphs of a function f and its inverse f^{-1} are related to each other in the following way. If the point (a, b) lies on the graph of f, then the point (b, a) must lie on the graph of f^{-1}, and vice versa. This means that the graph of f^{-1} is a *reflection* of the graph of f in the line $y = x$, as shown in Figure P.60.

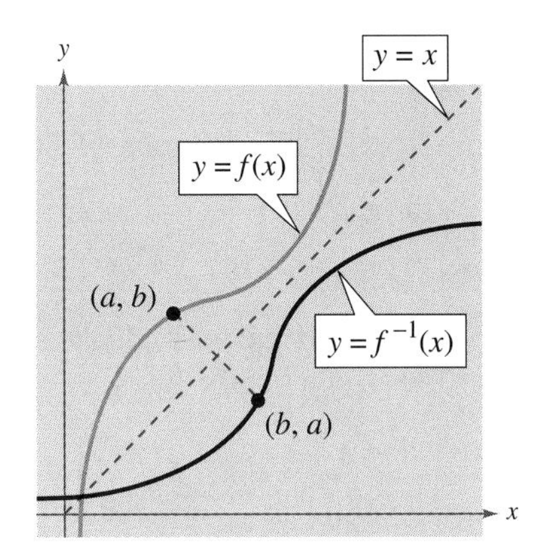

FIGURE P.60

EXAMPLE 6 *The Graphs of f and f^{-1}*

Sketch the graphs of the inverse functions

$$f(x) = 2x - 3 \qquad \text{and} \qquad f^{-1}(x) = \frac{1}{2}(x + 3)$$

on the same rectangular coordinate system and show that the graphs are reflections of each other in the line $y = x$.

Solution

The graphs of f and f^{-1} are shown in Figure P.61. Visually, it appears that the graphs are reflections of each other in the line $y = x$. You can further verify this reflective property by testing a few points on each graph. Note in the following list that if the point (a, b) is on the graph of f, the point (b, a) is on the graph of f^{-1}.

Graph of $f(x) = 2x - 3$	*Graph of $f^{-1}(x) = \frac{1}{2}(x + 3)$*
$(-1, -5)$	$(-5, -1)$
$(0, -3)$	$(-3, 0)$
$(1, -1)$	$(-1, 1)$
$(2, 1)$	$(1, 2)$
$(3, 3)$	$(3, 3)$

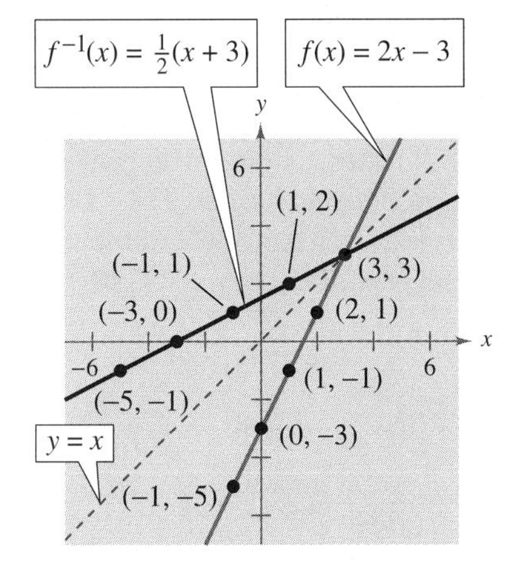

FIGURE P.61

In the study tip on page 100, we mentioned that the function

$$f(x) = x^2$$

has no inverse. What we really meant is that *assuming the domain of f is the entire real line,* the function $f(x) = x^2$ has no inverse. If, however, you restrict the domain of f to the nonnegative real numbers, f does have an inverse, as demonstrated in Example 7.

EXAMPLE 7 *The Graphs of f and f^{-1}*

Sketch the graphs of the inverse functions

$$f(x) = x^2, \quad x \geq 0, \qquad \text{and} \qquad f^{-1}(x) = \sqrt{x}$$

on the same rectangular coordinate system and show that the graphs are reflections of each other in the line $y = x$.

Solution

The graphs of f and f^{-1} are shown in Figure P.62. Visually, it appears that the graphs are reflections of each other in the line $y = x$. You can further verify this reflective property by testing a few points on each graph. Note in the following list that if the point (a, b) is on the graph of f, the point (b, a) is on the graph of f^{-1}.

Graph of $f(x) = x^2, \ x \geq 0$	*Graph of* $f^{-1}(x) = \sqrt{x}$
(0, 0)	(0, 0)
(1, 1)	(1, 1)
(2, 4)	(4, 2)
(3, 9)	(9, 3)

Try showing that $f(f^{-1}(x)) = x$ and $f^{-1}(f(x)) = x$.

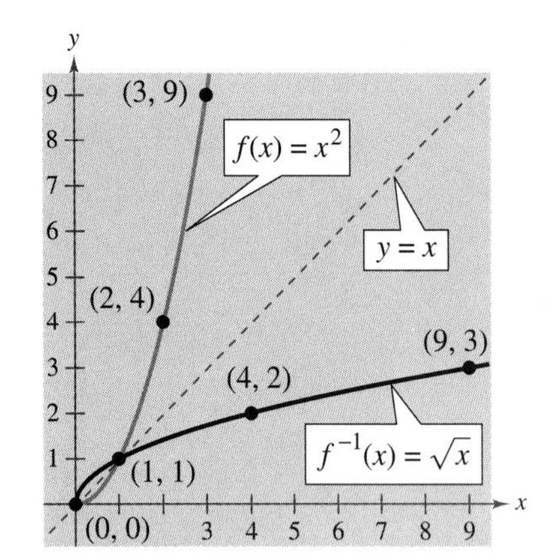

FIGURE P.62

The guidelines for finding the inverse of a function, on page 100, include an *algebraic* test for determining whether a function has an inverse. The reflective property of the graphs of inverse functions gives you a nice *geometric* test for determining whether a function has an inverse. This test is called the **Horizontal Line Test** for inverse functions.

> ***HORIZONTAL LINE TEST FOR INVERSE FUNCTIONS***
>
> A function f has an inverse function if and only if no *horizontal* line intersects the graph of f at more than one point.

EXAMPLE 8 *Applying the Horizontal Line Test*

a. The graph of the function $f(x) = x^3 - 1$ is shown in Figure P.63(a). Because no horizontal line intersects the graph of f at more than one point, you can conclude that f *does* possess an inverse function.

b. The graph of the function $f(x) = x^2 - 1$ is shown in Figure P.63(b). Because it is possible to find a horizontal line that intersects the graph of f at more than one point, you can conclude that f *does not* possess an inverse function.

(a)

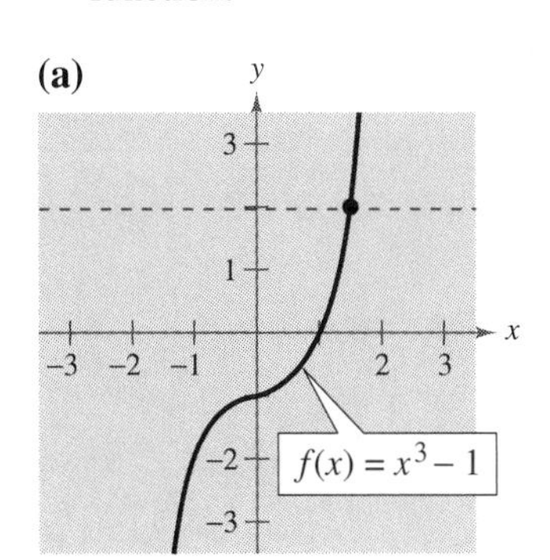

(b)

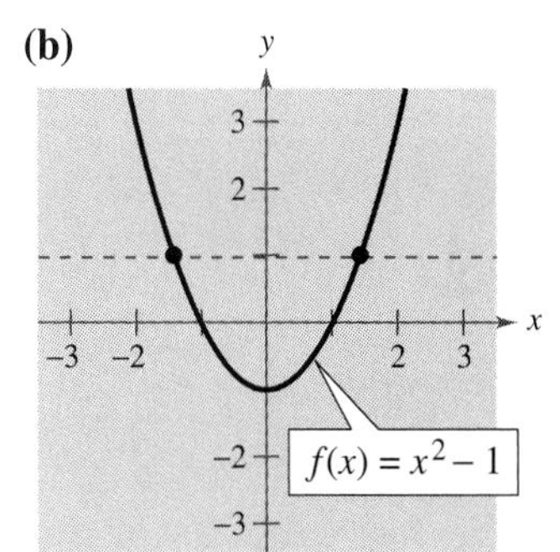

FIGURE P.63

GROUP ACTIVITY

ERROR ANALYSIS

Suppose you are a math instructor, and one of your students has handed in the following quiz. Find the error(s) in each solution and discuss how to explain each error to your student.

1. Find the inverse f^{-1} of $y = \sqrt{2x - 5}$.

$f(x) = \sqrt{2x - 5}$, so

$f^{-1}(x) = \dfrac{1}{\sqrt{2x - 5}}$

2. Find the inverse f^{-1} of $y = \frac{3}{5}x + \frac{1}{3}$.

$f(x) = \frac{3}{5}x + \frac{1}{3}$, so

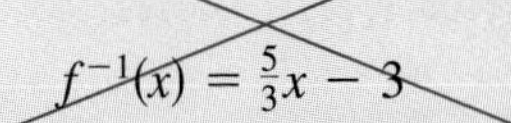

$f^{-1}(x) = \frac{5}{3}x - 3$

WARM UP

Find the domain of the function.

1. $f(x) = \sqrt[3]{x + 1}$

2. $f(x) = \sqrt{x + 1}$

3. $g(x) = \dfrac{2}{x^2 - 2x}$

4. $h(x) = \dfrac{x}{3x + 5}$

Simplify the expression.

5. $2\left(\dfrac{x + 5}{2}\right) - 5$

6. $7 - 10\left(\dfrac{7 - x}{10}\right)$

7. $\sqrt[3]{2\left(\dfrac{x^3}{2} - 2\right) + 4}$

8. $\left(\sqrt[5]{x + 2}\right)^5 - 2$

Solve for x in terms of y.

9. $y = \dfrac{2x - 6}{3}$

10. $y = \sqrt[3]{2x - 4}$

P.8 Exercises

In Exercises 1–4, match the graph of the function with the graph of its inverse. [The graphs of the inverse functions are labeled (a), (b), (c), and (d).]

(a)

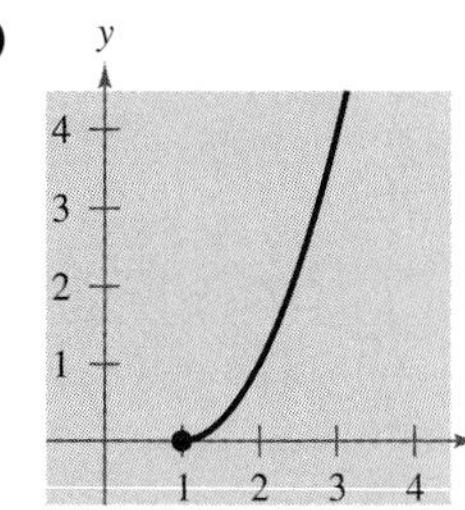

(b)

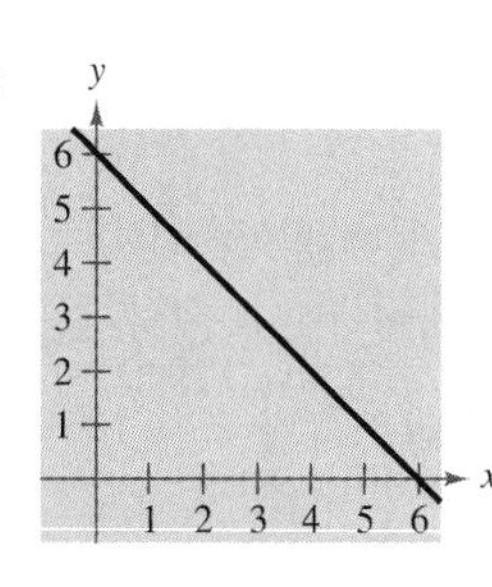

(c)

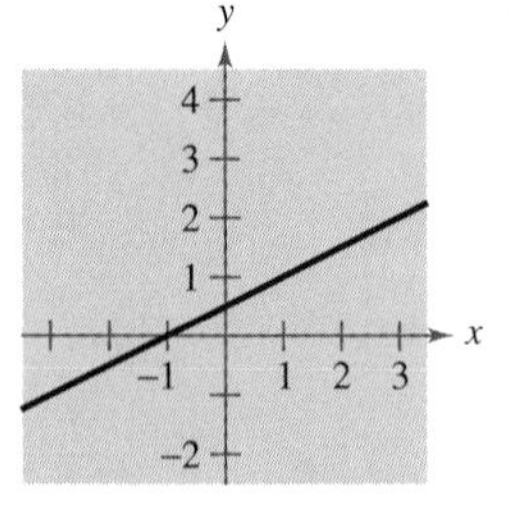

(d)

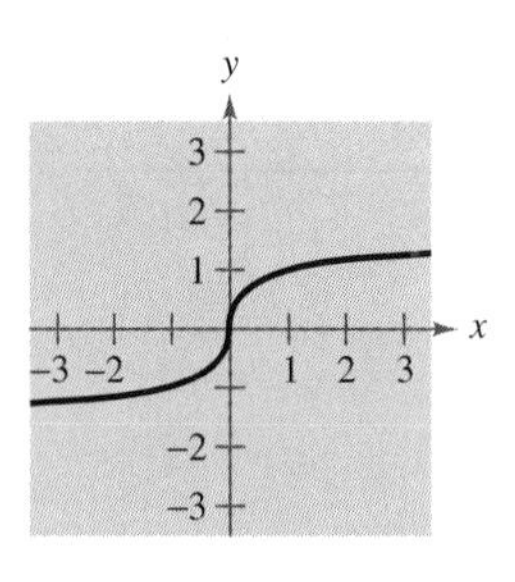

1.

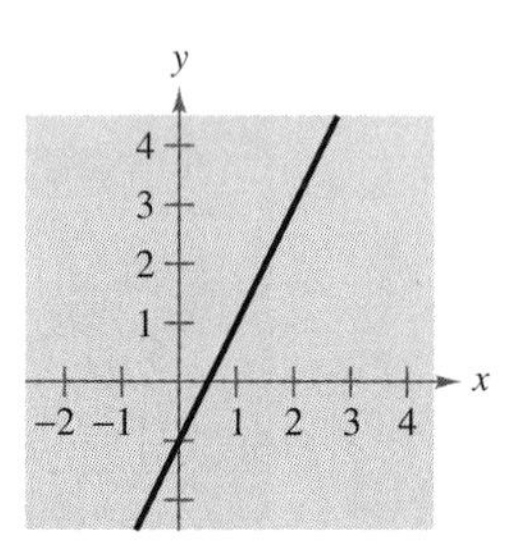

2.

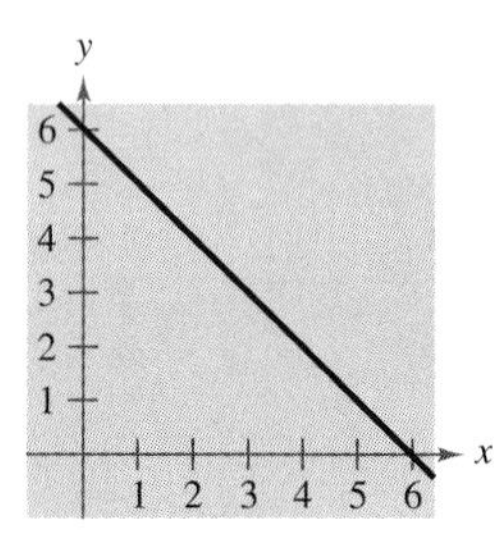

3.

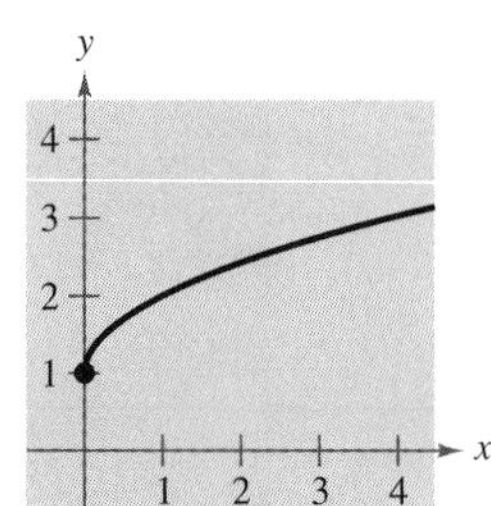

4.

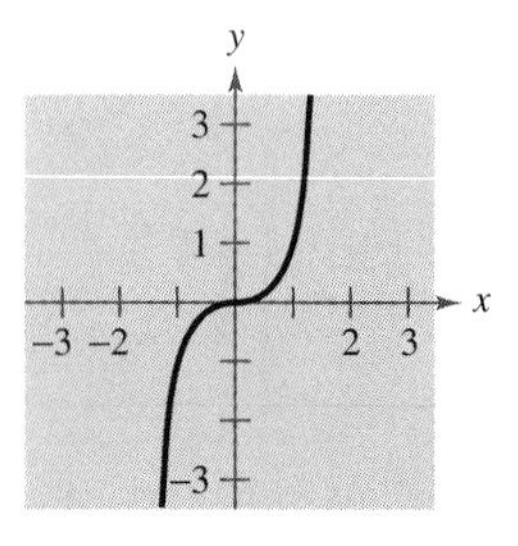

In Exercises 5–10, find the inverse of f informally. Verify that $f(f^{-1}(x)) = x$ and $f^{-1}(f(x)) = x$.

5. $f(x) = 8x$
6. $f(x) = \frac{1}{5}x$
7. $f(x) = x + 10$
8. $f(x) = x - 5$
9. $f(x) = \sqrt[3]{x}$
10. $f(x) = x^5$

In Exercises 11–20, show that f and g are inverse functions (a) algebraically and (b) graphically.

11. $f(x) = 2x, \quad g(x) = \dfrac{x}{2}$
12. $f(x) = x - 5, \quad g(x) = x + 5$
13. $f(x) = 5x + 1, \quad g(x) = \dfrac{x - 1}{5}$
14. $f(x) = 3 - 4x, \quad g(x) = \dfrac{3 - x}{4}$
15. $f(x) = x^3, \quad g(x) = \sqrt[3]{x}$
16. $f(x) = \dfrac{1}{x}, \quad g(x) = \dfrac{1}{x}$
17. $f(x) = \sqrt{x - 4}, \quad g(x) = x^2 + 4, \quad x \geq 0$
18. $f(x) = 1 - x^3, \quad g(x) = \sqrt[3]{1 - x}$
19. $f(x) = 9 - x^2, \quad x \geq 0, \quad g(x) = \sqrt{9 - x}, \quad x \leq 9$
20. $f(x) = \dfrac{1}{1 + x}, \quad x \geq 0$

    $g(x) = \dfrac{1 - x}{x}, \quad 0 < x \leq 1$

In Exercises 21 and 22, does the function have an inverse?

21.

x	-1	0	1	2	3	4
$f(x)$	-2	1	2	1	-2	-6

22.

x	-3	-2	-1	0	2	3
$f(x)$	10	6	4	1	-3	-10

In Exercises 23–26, does the function have an inverse?

23.

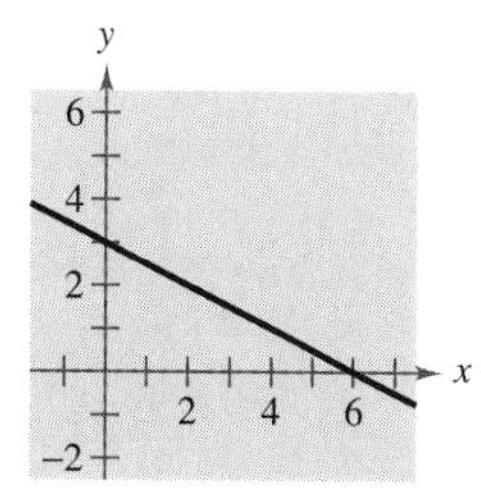

24.

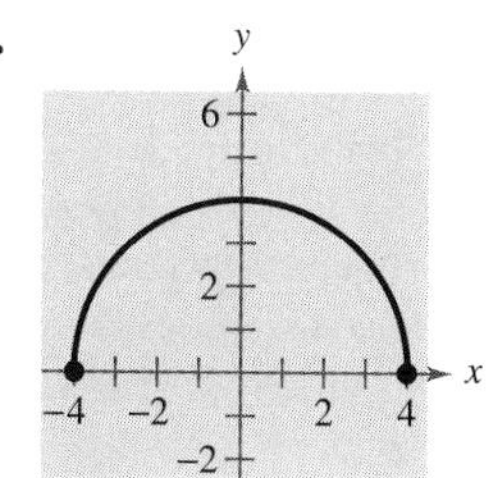

25.

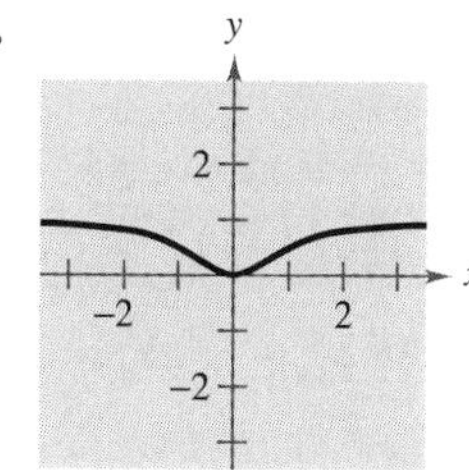

26.

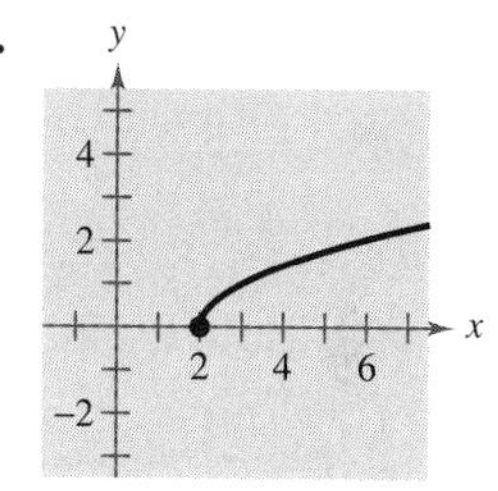

In Exercises 27–32, use a graphing utility to graph the function and use the Horizontal Line Test to determine whether the function has an inverse.

27. $g(x) = \dfrac{4 - x}{6}$
28. $f(x) = 10$
29. $h(x) = |x + 4| - |x - 4|$
30. $g(x) = (x + 5)^3$
31. $f(x) = -2x\sqrt{16 - x^2}$
32. $f(x) = \frac{1}{8}(x + 2)^2 - 1$

In Exercises 33–42, find the inverse of the function f. Then graph both f and f^{-1} on the same coordinate system.

33. $f(x) = 2x - 3$
34. $f(x) = 3x$
35. $f(x) = x^5$
36. $f(x) = x^3 + 1$
37. $f(x) = \sqrt{x}$
38. $f(x) = x^2, \quad x \geq 0$
39. $f(x) = \sqrt{4 - x^2}, \quad 0 \leq x \leq 2$
40. $f(x) = \dfrac{4}{x}$
41. $f(x) = \sqrt[3]{x - 1}$
42. $f(x) = x^{3/5}$

In Exercises 43–58, determine whether the function has an inverse. If it does, find its inverse.

43. $f(x) = x^4$

44. $f(x) = \dfrac{1}{x^2}$

45. $g(x) = \dfrac{x}{8}$

46. $f(x) = 3x + 5$

47. $p(x) = -4$

48. $f(x) = \dfrac{3x + 4}{5}$

49. $f(x) = (x + 3)^2, \quad x \geq -3$

50. $q(x) = (x - 5)^2$

51. $h(x) = \dfrac{1}{x}$

52. $f(x) = |x - 2|, \quad x \leq 2$

53. $f(x) = \sqrt{2x + 3}$

54. $f(x) = \sqrt{x - 2}$

55. $g(x) = x^2 - x^4$

56. $f(x) = \dfrac{x^2}{x^2 + 1}$

57. $f(x) = 25 - x^2, \quad x \leq 0$

58. $f(x) = ax + b, \quad a \neq 0$

In Exercises 59–62, delete part of the graph of the function so that the part that remains has an inverse. Find the inverse and give its domain. (There is more than one correct answer.)

59. $f(x) = (x - 2)^2$

60. $f(x) = 1 - x^4$

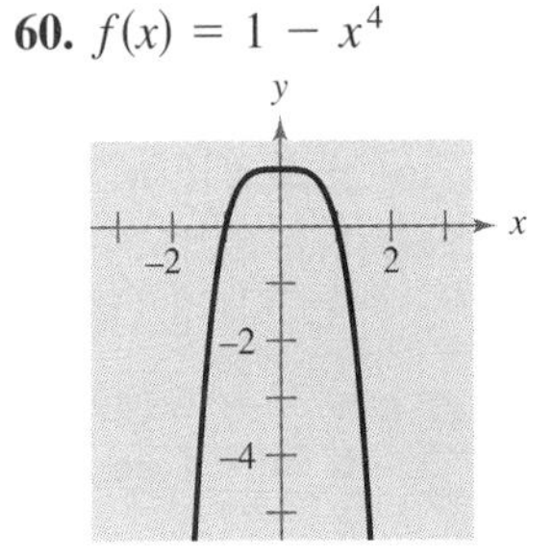

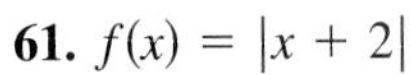

61. $f(x) = |x + 2|$

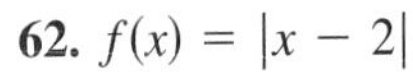

62. $f(x) = |x - 2|$

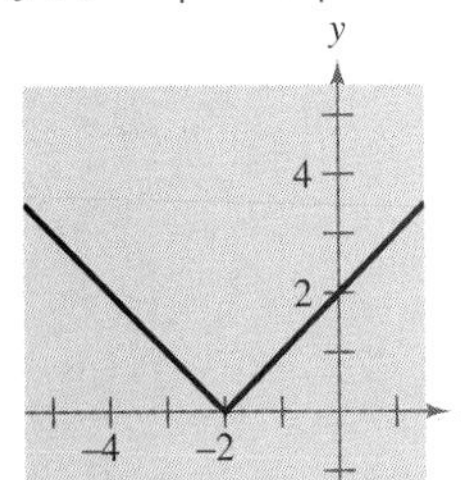

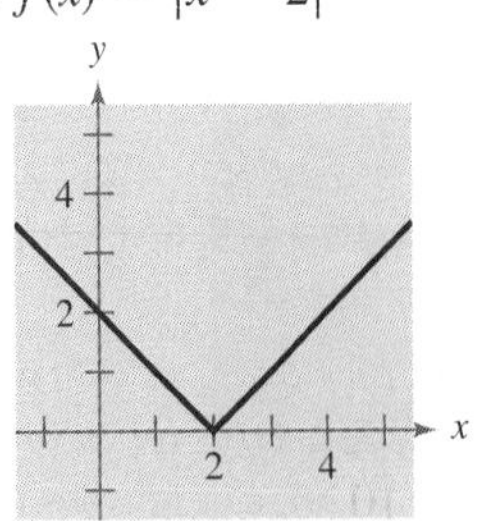

In Exercises 63 and 64, use the graph of the function f to complete the table and sketch the graph of f^{-1}.

63.

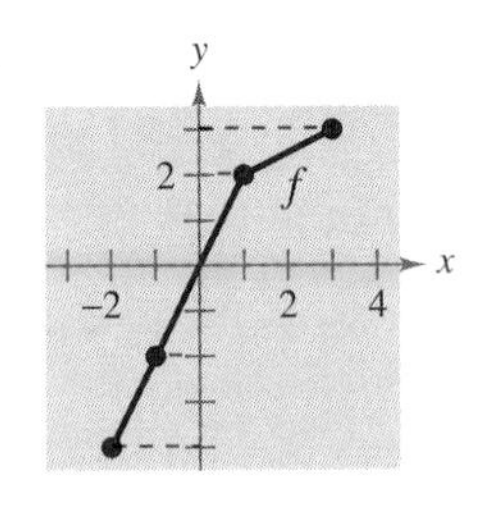

x	$f^{-1}(x)$
-4	
-2	
2	
3	

64.

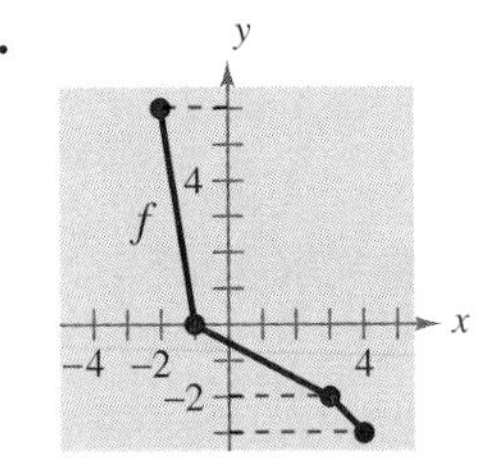

x	$f^{-1}(x)$
-3	
-2	
0	
6	

True or False? **In Exercises 65–68, determine if the statement is true or false. If it is false, give an example that shows why it is false.**

65. If f is an even function, f^{-1} exists.

66. If the inverse of f exists, the y-intercept of f is an x-intercept of f^{-1}.

67. If $f(x) = x^n$ where n is odd, f^{-1} exists.

68. There exists no function f such that $f = f^{-1}$.

In Exercises 69–74, use the functions $f(x) = \frac{1}{8}x - 3$ and $g(x) = x^3$ to find the indicated value or function.

69. $(f^{-1} \circ g^{-1})(1)$

70. $(g^{-1} \circ f^{-1})(-3)$

71. $(f^{-1} \circ f^{-1})(6)$

72. $(g^{-1} \circ g^{-1})(-4)$

73. $(f \circ g)^{-1}$

74. $g^{-1} \circ f^{-1}$

In Exercises 75–78, use the functions $f(x) = x + 4$ and $g(x) = 2x - 5$ to find the specified function.

75. $g^{-1} \circ f^{-1}$

76. $f^{-1} \circ g^{-1}$

77. $(f \circ g)^{-1}$

78. $(g \circ f)^{-1}$

79. ***Hourly Wage*** Your wage is \$8.00 per hour plus \$0.75 for each unit produced per hour. Thus, your hourly wage y in terms of the number of units produced is given by

$$y = 8 + 0.75x.$$

(a) Find the inverse of the function.

(b) What does each variable represent in the inverse function?

(c) Determine the number of units produced when your hourly wage averages \$22.25.

80. ***Cost*** Suppose you need a total of 50 pounds of two commodities costing \$1.25 and \$1.60 per pound, respectively.

(a) Verify that the total cost is

$$y = 1.25x + 1.60(50 - x)$$

where x is the number of pounds of the less expensive commodity.

(b) Find the inverse of the cost function. What does each variable represent in the inverse function?

(c) Use the context of the problem to determine the domain of the inverse function.

(d) Determine the number of pounds of the less expensive commodity purchased if the total cost is \$73.

81. ***Diesel Engine*** The function

$$y = 0.03x^2 + 245.50, \qquad 0 < x < 100$$

approximates the exhaust temperature y in degrees Fahrenheit where x is the percent load for a diesel engine.

(a) Find the inverse of the function. What does each variable represent in the inverse function?

(b) Use a graphing utility to graph the inverse function.

(c) Determine the percent load interval if the exhaust temperature of the engine must not exceed 500 degrees Fahrenheit.

82. ***Think About It*** The function

$$f(x) = k(2 - x - x^3)$$

has an inverse, and $f^{-1}(3) = -2$. Find k.

83. ***Average Miles Per Gallon*** The average miles per gallon f for cars in the United States for 1986 through 1991 is given in the table. The time in years is given by t, with $t = 6$ corresponding to 1986. (Source: U.S. Federal Highway Administration)

t	6	7	8	9	10	11
$f(t)$	18.27	19.20	19.95	20.40	21.02	21.68

(a) Does f^{-1} exist?

(b) If f^{-1} exists, what does it mean in the context of the problem?

(c) If f^{-1} exists, find $f^{-1}(19.95)$.

84. ***Average Miles Per Gallon*** If the table in Exercise 83 were extended to 1992 and if the average number of miles per gallon for that year were 21.02, would f^{-1} exist? Explain.

In Exercises 85–92, solve the equation by any convenient method.

85. $x^2 = 64$

86. $(x - 5)^2 = 8$

87. $4x^2 - 12x + 9 = 0$

88. $9x^2 + 12x + 3 = 0$

89. $x^2 - 6x + 4 = 0$

90. $2x^2 - 4x - 6 = 0$

91. $50 + 5x = 3x^2$

92. $2x^2 + 4x - 9 = 2(x - 1)^2$

93. ***Consecutive Even Integers*** Find two consecutive positive even integers whose product is 288.

94. ***Lawn Mowing*** Two people must mow a rectangular lawn measuring 100 feet by 200 feet. The first person agrees to mow three-fourths of the lawn and starts by mowing around the outside. How wide a strip must the person mow on each of the four sides? If the mower has a 24-inch cut, approximate the required number of trips around the lawn.

95. ***Dimensions of a Triangular Sign*** A triangular sign has a height that is equal to its base. The area of the sign is 10 square feet. Find the base and height of the sign.

96. ***Dimensions of a Triangular Sign*** A triangular sign has a height that is twice its base. The area of the sign is 10 square feet. Find the base and height of the sign.

FOCUS ON CONCEPTS

In this chapter, you studied functions and their graphs. You can use the following questions to check your understanding of several of the basic concepts discussed in this chapter. The answers to these questions are given in the back of the book.

1. With the information given in the graphs, is it possible to determine the slope of the two lines? Is it possible they could have the same slope? Explain.

(a)

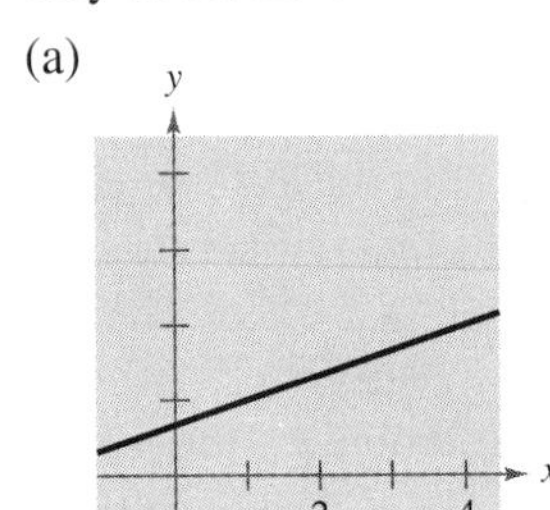

(b)

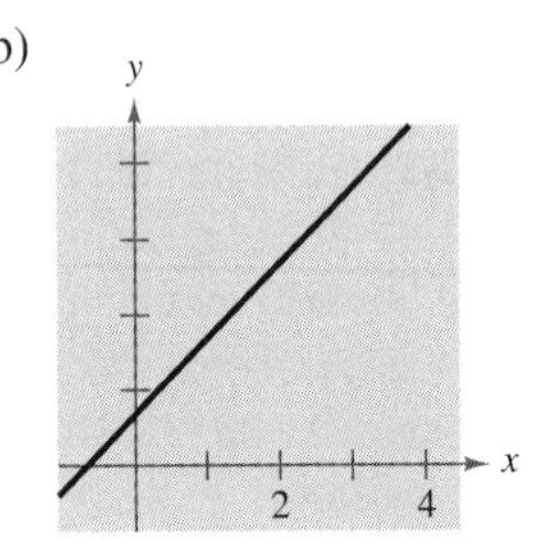

2. The slopes of two lines are -4 and $\frac{5}{2}$. Which is steeper? Explain.

3. The value V of a machine t years after it is purchased is $V = -4000t + 58{,}500$, $0 \le t \le 5$. Explain what the V-intercept and slope measure.

4. Does the relationship shown in the figure represent a function from set A to set B? Explain.

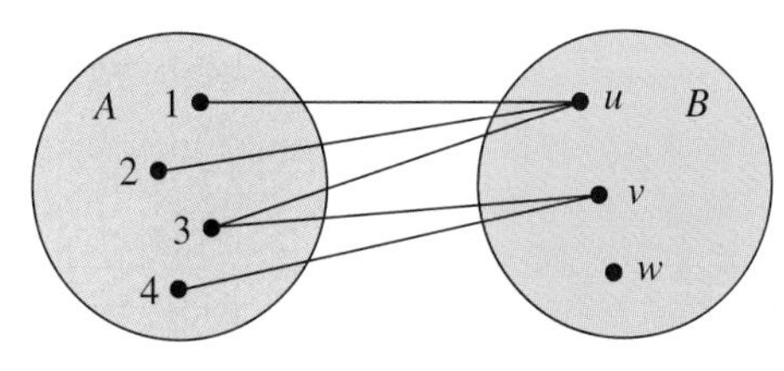

5. Select viewing rectangles using a graphing utility that would show these graphs.

(a)

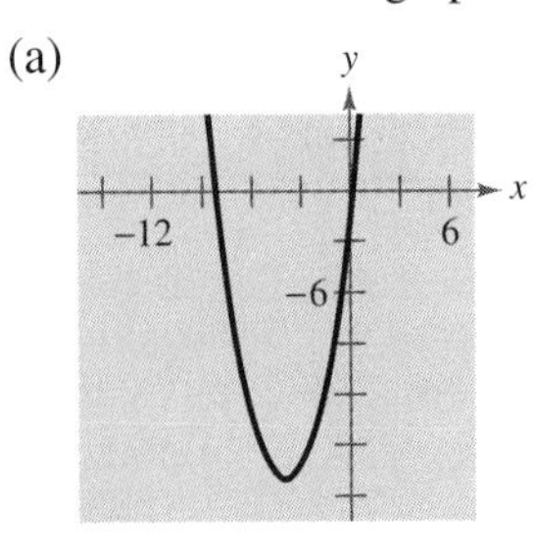

(b)

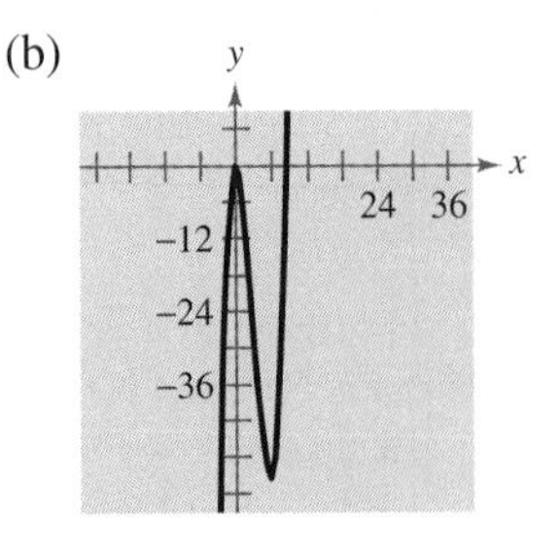

6. If f is an even function, determine if g is even, odd, or neither. Explain.

(a) $g(x) = -f(x)$ (b) $g(x) = f(-x)$

(c) $g(x) = f(x) - 2$ (d) $g(x) = f(x - 2)$

7. Management originally predicted that the profits from the sales of a new product would be approximated by the graph of the function f in the figure. The actual profits are shown by the function g along with a verbal description. Use the concepts of transformations of graphs to write g in terms of f.

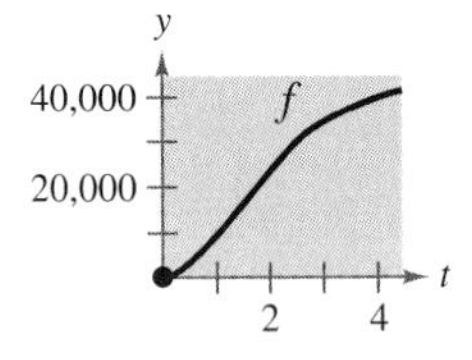

(a) The profits were only three-fourths as large as expected.

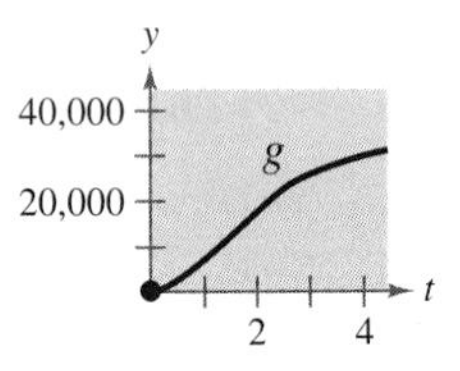

(b) The profits were consistently \$10,000 greater than predicted.

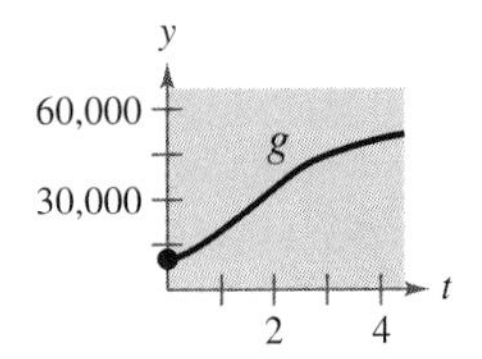

(c) There was a 2-year delay in the introduction of the product. After sales began, profits were as expected.

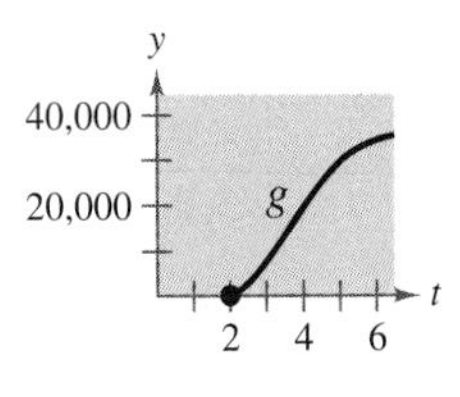

Review Exercises

In Exercises 1 and 2, determine which numbers in the set are (a) natural numbers, (b) integers, (c) rational numbers, and (d) irrational numbers.

1. $\{11, -14, -\frac{8}{9}, \frac{5}{2}, \sqrt{6}, 0.4\}$

2. $\{\sqrt{15}, -22, -\frac{10}{3}, 0, 5.2, \frac{3}{7}\}$

In Exercises 3 and 4, give a verbal description of the subset of real numbers that is represented by the inequality, and sketch the subset on the real number line.

3. $x \leq 7$

4. $x > 1$

In Exercises 5 and 6, use absolute value notation to describe the expression.

5. The distance between x and 7 is at least 4.

6. The distance between y and -30 is less than 5.

In Exercises 7–10, identify the rule of algebra illustrated by the equation.

7. $2x + (3x - 10) = (2x + 3x) - 10$

8. $\dfrac{2}{y+4} \cdot \dfrac{y+4}{2} = 1, \quad y \neq -4$

9. $(t + 4)(2t) = (2t)(t + 4)$

10. $0 + (a - 5) = a - 5$

In Exercises 11 and 12, determine whether the values of x are solutions of the equation.

Equation	*Values*	
11. $3x^2 + 7x = x^2 + 4$	(a) $x = 0$	(b) $x = -4$
	(c) $x = \frac{1}{2}$	(d) $x = -1$
12. $6 + \dfrac{3}{x-4} = 5$	(a) $x = 4$	(b) $x = 0$
	(c) $x = -2$	(d) $x = 1$

In Exercises 13–18, solve the equation (if possible) and check your solution.

13. $4(x + 3) - 3 = 2(4 - 3x) - 4$

14. $\frac{1}{2}(x - 3) - 2(x + 1) = 5$

15. $3\left(1 - \dfrac{1}{5t}\right) = 0$

16. $\dfrac{1}{x-2} = 3$

17. $(x + 4)^2 = 18$

18. $4x^3 - 6x^2 = 0$

19. ***Cost Sharing*** A group agrees to share equally in the cost of a \$48,000 piece of machinery. If they could find two more group members, each member's share of the cost would decrease by \$4000. How many are presently in the group?

20. ***Starting Position*** A fitness center has two running tracks around a rectangular playing floor. The tracks are 1 meter wide and form semicircles at the narrow ends of the rectangular floor (see figure). Determine the distance between the starting positions if two runners must run the same distance to the finish line in one lap around the track.

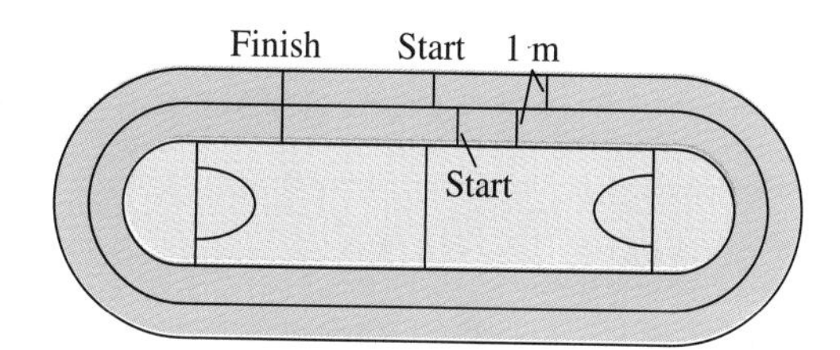

In Exercises 21 and 22, complete the table. Use the resulting solution points to sketch the graph of the equation. Use a graphing utility to verify the graph.

21. $y = -\frac{1}{2}x + 2$

x	-2	0	2	3	4
y					

22. $y = x^2 - 3x$

x	-1	0	1	2	3
y					

In Exercises 23 and 24, (a) plot the points, (b) find the distance between the points, and (c) find the midpoint of the line segment joining the points.

23. $(-3, 8), (1, 5)$ **24.** $(5.6, 0), (0, 8.2)$

In Exercises 25–28, sketch the graph of the equation *by hand.*

25. $y - 2x - 3 = 0$ **26.** $y = 8 - |x|$

27. $y = \sqrt{5 - x}$ **28.** $y = x^2 - 4x$

In Exercises 29–32, use a graphing utility to graph the equation. Approximate any intercepts and identify any symmetry.

29. $y = \frac{1}{4}x^4 - 2x^2$ **30.** $y = 4 - (x - 4)^2$

31. $y = \frac{1}{4}x^3 - 3x$ **32.** $y = x\sqrt{9 - x^2}$

In Exercises 33 and 34, find a setting on a graphing utility such that the graph of the equation agrees with the given graph.

33. $y = 10x^3 - 21x^2$ **34.** $y = 0.002x^2 - 0.06x - 1$

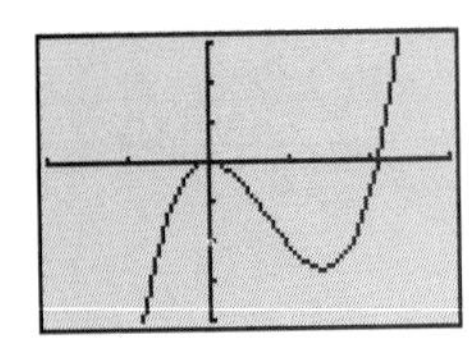

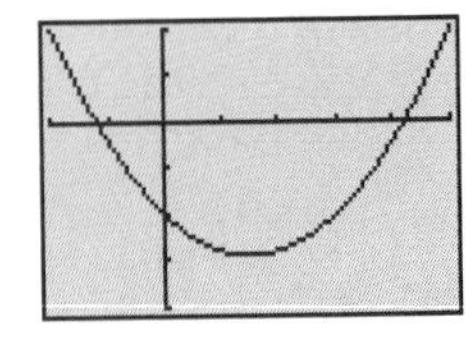

35. Find the center and radius of the circle given by

$(x - 3)^2 + (y + 1)^2 = 9.$

Sketch the graph of the circle.

36. Sketch the standard form of the equation of the circle for which the endpoints of one diameter are $(0, 0)$ and $(4, -6)$.

In Exercises 37 and 38, find an equation of the line that passes through the two points.

37. $(2, 1), (14, 6)$ **38.** $(-2, 2), (3, -10)$

In Exercises 39–42, find an equation of the line that passes through the given point and has the specified slope. Sketch the graph of the line.

	Point	*Slope*
39.	$(0, -5)$	$m = \frac{3}{2}$
40.	$(-2, 6)$	$m = 0$
41.	$(3, 0)$	$m = -\frac{2}{3}$
42.	$(5, 4)$	m is undefined.

Rate of Change **In Exercises 43 and 44, you are given the dollar value of a product in 1996 and the rate at which the value of the item is expected to change during the next 5 years. Use this information to write a linear equation that gives the dollar value V of the product in terms of the year t. (Let $t = 6$ represent 1996.)**

	1996 Value	*Rate*
43.	\$12,500	\$850 increase per year
44.	\$72.95	\$5.15 increase per year

In Exercises 45–48, identify the equations that represent y as a function of x.

45. $16x - y^4 = 0$

46. $2x - y - 3 = 0$

47. $y = \sqrt{1 - x}$

48. $|y| = x + 2$

In Exercises 49 and 50, evaluate the function at the specified values of the independent variable. Simplify your answers.

49. $g(x) = x^{4/3}$

(a) $g(8)$ (b) $g(t + 1)$

(c) $\dfrac{g(8) - g(1)}{8 - 1}$ (d) $g(-x)$

50. $h(x) = 6 - 5x^2$

(a) $h(2)$ (b) $h(x + 3)$

(c) $\dfrac{h(4) - h(2)}{4 - 2}$ (d) $\dfrac{h(x + t) - h(x)}{t}$

In Exercises 51–56, determine the domain of the function. Verify your result with a graphing utility.

51. $f(x) = \sqrt{25 - x^2}$

52. $f(x) = 3x + 4$

53. $g(s) = \dfrac{5}{3s - 9}$

54. $f(x) = \sqrt{x^2 + 8x}$

55. $h(x) = \dfrac{x}{x^2 - x - 6}$

56. $h(t) = |t + 1|$

In Exercises 57 and 58, select the viewing rectangle on a graphing utility that shows the most complete graph of the function.

57. $f(x) = \dfrac{3x}{2(3 - x)}$

Xmin = -4	Xmin = -5	Xmin = 0
Xmax = 4	Xmax = 10	Xmax = 20
Xscl = 1	Xscl = 1	Xscl = 2
Ymin = -3	Ymin = -8	Ymin = 0
Ymax = 3	Ymax = 6	Ymax = 10
Yscl = 1	Yscl = 1	Yscl = 2

58. $f(x) = 4[(0.3x)^3 - 5x]$

Xmin = -200	Xmin = -10	Xmin = -15
Xmax = 200	Xmax = 10	Xmax = 15
Xscl = 50	Xscl = 2	Xscl = 5
Ymin = -500	Ymin = -20	Ymin = -150
Ymax = 500	Ymax = 20	Ymax = 150
Yscl = 50	Yscl = 4	Yscl = 50

In Exercises 59–62, use a graphing utility to graph the function. Use the graph to approximate any x-intercepts of the graph. Set $y = 0$ and solve the resulting equation. Compare the result with the x-intercepts of the graph.

59. $y = 4x^3 - 12x^2 + 8x$

60. $y = 12x^3 - 84x^2 + 120x$

61. $y = \dfrac{1}{x} + \dfrac{1}{x + 1} - 2$

62. $y = \dfrac{4}{x - 3} - \dfrac{4}{x} - 1$

63. Sketch (on the same set of coordinate axes) a graph of f for $c = -2, 0$, and 2.

(a) $f(x) = \sqrt{x} + c$ (b) $f(x) = \sqrt{x - c}$

64. *Cost and Profit* A company produces a product for which the variable cost is \$5.35 per unit and the fixed costs are \$16,000. The company sells the product for \$8.20 and can sell all that it produces.

(a) Find the total cost as a function of x, the number of units produced.

(b) Find the profit as a function of x.

In Exercises 65–68, (a) find f^{-1}, (b) sketch the graphs of f and f^{-1} on the same coordinate system, and (c) verify that $f^{-1}(f(x)) = x = f(f^{-1}(x))$.

65. $f(x) = \frac{1}{2}x - 3$

66. $f(x) = 5x - 7$

67. $f(x) = \sqrt{x + 1}$

68. $f(x) = x^2 - 5, \quad x \geq 0$

In Exercises 69–72, restrict the domain of the function f to an interval over which the function is increasing and determine f^{-1} over that interval.

69. $f(x) = 2(x - 4)^2$

70. $f(x) = |x - 2|$

71. $f(x) = \sqrt{x^2 - 4}$

72. $f(x) = x^{4/3}$

In Exercises 73–80, let $f(x) = 3 - 2x$, $g(x) = \sqrt{x}$, and $h(x) = 3x^2 + 2$ and find the indicated value.

73. $(f - g)(4)$

74. $(f + h)(5)$

75. $(fh)(1)$

76. $\left(\dfrac{g}{h}\right)(1)$

77. $(h \circ g)(7)$

78. $(g \circ f)(-2)$

79. $g^{-1}(3)$

80. $(h \circ f^{-1})(1)$

CHAPTER PROJECT: *A Graphical Approach to Maximization*

In business, a **demand function** gives the price per unit p in terms of the number of units sold x. The demand function whose graph is shown below is

$$p = 40 - 5x^2, \qquad 0 \le x \le \sqrt{8} \qquad \text{Demand function}$$

where x is measured in millions of units. Note that as the price decreases, the number of units sold increases. The **revenue** R (in millions of dollars) is determined by multiplying the number of units sold by the price per unit. Thus,

$$R = xp = x(40 - 5x^2), \qquad 0 \le x \le \sqrt{8} \qquad \text{Revenue function}$$

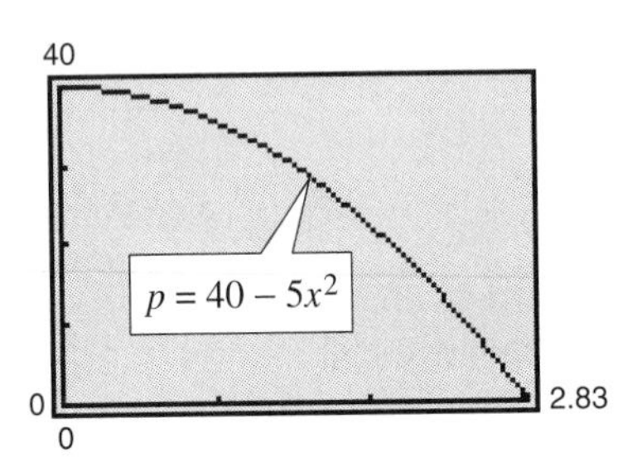

EXAMPLE 1 ***Finding the Maximum Revenue*** — Real Life

Use a graphing utility to sketch the graph of the revenue function

$$R = 40x - 5x^3, \qquad 0 \le x \le \sqrt{8}.$$

How many units should be sold to obtain maximum revenue? What price per unit should be charged to obtain maximum revenue?

Solution

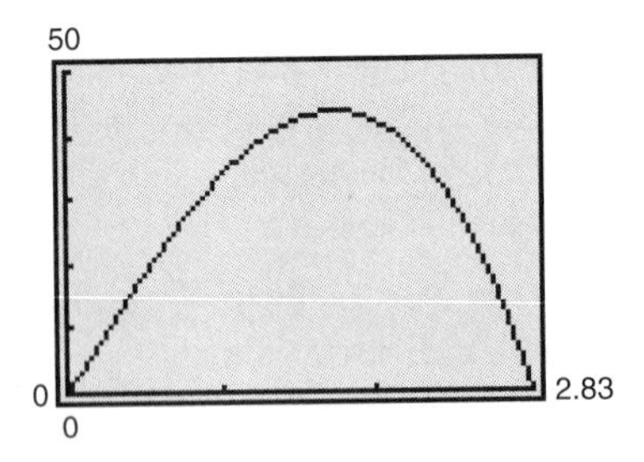

To begin, you need to determine a viewing rectangle that will display the part of the graph that is important to this problem. The domain is given, so you can set the x-boundaries of the graph between 0 and $\sqrt{8}$. To determine the y-boundaries, however, you need to experiment a little. After calculating several values of R, you could decide to use y-boundaries between 0 and 50, as shown in the graph at the left. Next you can use the trace key to find that the maximum revenue of about \$43.5 million occurs when x is approximately 1.64 million units. To find the price per unit that corresponds to this maximum revenue, you can substitute $x = 1.64$ into the demand function to obtain

$$p = 40 - 5(1.64)^2 \approx \$26.55.$$

EXAMPLE 2 *Finding the Maximum Profit*

Real Life

For the revenue function discussed in Example 1, the cost of producing each unit is \$15. How many units should be sold to obtain maximum profit? What price per unit should be charged to obtain maximum profit?

Solution

The total cost C (in millions of dollars) of producing x million units is $C = 15x$. This implies that the profit P (in millions of dollars) obtained by selling x million units is

$$\begin{aligned} P &= R - C \\ &= (40x - 5x^3) - 15x \\ &= -5x^3 + 25x. \end{aligned}$$

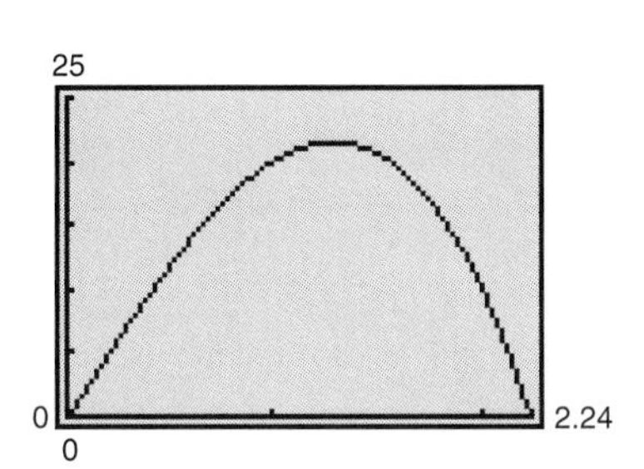

As in Example 1, you can use a graphing utility to graph this function. From the graph, you can approximate that the maximum profit of about \$21.5 million occurs when x is approximately 1.28 million units. The price per unit that corresponds to this maximum profit is

$$p = 40 - 5(1.28)^2 \approx \$31.81.$$

CHAPTER PROJECT INVESTIGATIONS

1. *Can't Give It Away!* For the demand function $p = 40 - 5x^2$, match the points $(0, 40)$ and $(\sqrt{8}, 0)$ with statement (a) or (b). Explain your reasoning.

   (a) No one will buy the product at this price.

   (b) You can't *give* more than this number away.

2. *Exploration* Use a graphing utility to zoom in on the maximum point of the revenue function in Example 1. (Use a setting of $1.62 \le x \le 1.65$ and $43.5 \le y \le 43.6$.) Use the trace feature to improve the accuracy of the approximation obtained in Example 1. Do you think this improved accuracy is appropriate in the context of this particular problem? Does it change the price?

3. *Exploration* Find a setting that allows you to improve graphically the accuracy of the solution in Example 2.

4. *Exploration* In Example 2, suppose that, in addition to the cost of \$15 per unit, there is an initial cost of \$250,000. How does this change the profit function? Does this affect the *price* that corresponds to a maximum profit? Does this affect the *amount* of the maximum profit? Explain.

5. *Maximum Volume of a Box* Consider a box that has a square base and a surface area of 216 square inches. Let x represent the length (in inches) of each side of the base. Show that the volume of the box is given by

   $$V = 54x - \tfrac{1}{2}x^3$$

   where V represents the volume (in cubic inches). Use a graphing utility to approximate the dimensions that produce a box of maximum volume. (Use a graph that yields an accuracy of 0.0001.)

Chapter Test

Take this test as you would take a test in class. After you are done, check your work against the answers given in the back of the book.

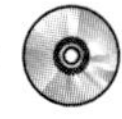

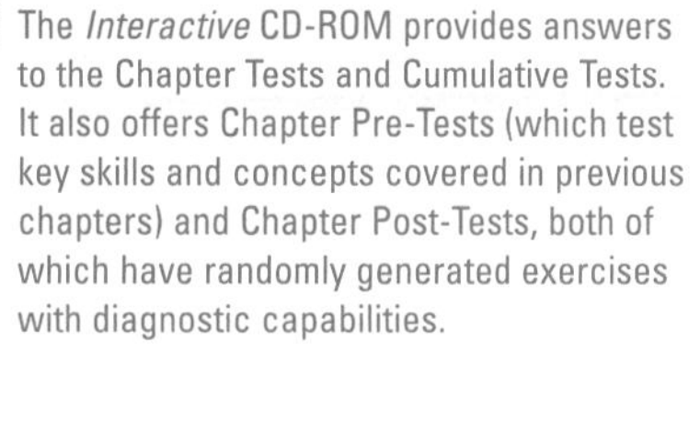
The *Interactive* CD-ROM provides answers to the Chapter Tests and Cumulative Tests. It also offers Chapter Pre-Tests (which test key skills and concepts covered in previous chapters) and Chapter Post-Tests, both of which have randomly generated exercises with diagnostic capabilities.

1. Place the proper inequality symbol (< or >) between the real numbers $-\frac{10}{3}$ and $-|-4|$.

2. Evaluate the quantity without the aid of a calculator.
(a) $\left|\frac{3}{8} + \frac{1}{6} - 2\right|$ (b) $\left(21 \div \frac{3}{4}\right)$

In Exercises 3–8, solve (if possible). Use a graphing utility to verify your answer.

3. $\frac{2}{3}(x - 1) + \frac{1}{4}x = 10$ **4.** $\frac{x-2}{x+2} + \frac{4}{x+2} + 4 = 0$

5. $3x^2 + 6x + 2 = 0$ **6.** $x^4 + x^2 - 6 = 0$

7. $2\sqrt{x} - \sqrt{2x+1} = 1$ **8.** $|3x - 1| = 7$

In Exercises 9–12, use a graphing utility to graph the equation. Check for symmetry and identify any *x*- and *y*-intercepts.

9. $y = 4 - \frac{3}{4}x$ **10.** $y = 4 - \frac{3}{4}|x|$

11. $y = 4 - (x - 2)^2$ **12.** $y = \sqrt{3 - x}$

13. Find the equation of the line through each set of points. Describe the information about the line given by the slope.
(a) $(-3, 6), (3, 2)$ (b) $(4, -2), (7, -2)$

14. The graph of $y^2(4 - x) = x^3$ is shown at the right. Does the graph represent y as a function of x? Explain.

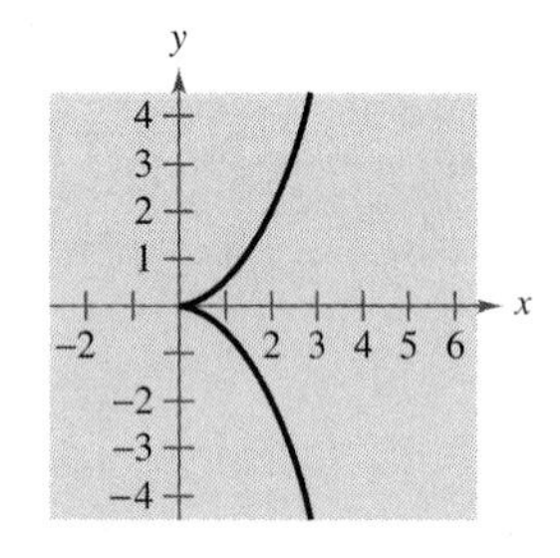

FIGURE FOR 14

15. Given $f(x) = 10 - \sqrt{3 - x}$, find and simplify the function $g(t) = f(t - 3)$. What is the domain of g?

16. The graph of a function g is shown at the right. Sketch a graph of (a) $\frac{1}{2}g(x - 2)$ and (b) $g\left(\frac{1}{2}x\right) - 1$.

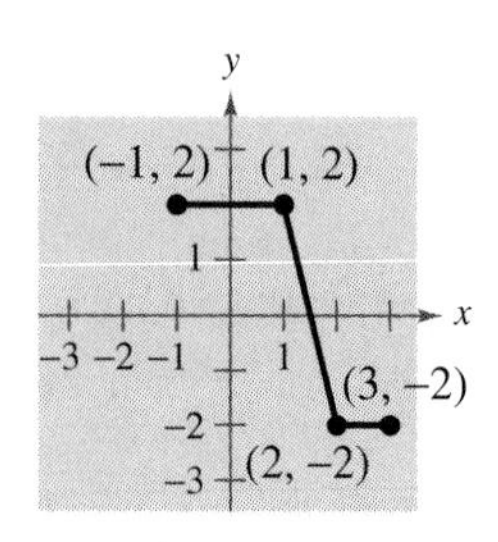

FIGURE FOR 16

In Exercises 17–20, use the functions $f(x) = x^2$ and $g(x) = \sqrt{2 - x}$ to find the specified function and its domain.

17. $(f - g)(x)$ **18.** $\left(\frac{f}{g}\right)(x)$ **19.** $(f \circ g)(x)$ **20.** $g^{-1}(x)$

21. On the first part of a 350-kilometer trip, a salesperson travels for 2 hours and 15 minutes at an average speed of 100 kilometers per hour. Find the average speed required for the remainder of the trip if the salesperson needs to arrive at the destination in another hour and 20 minutes.

1 Trigonometry

In 1995, a huge geological project, called *Project Deep Probe*, set off explosions along a 2100-mile line from northern Canada to the Mexican border. The explosions were recorded by nearly 800 seismographs in Alberta, Montana, and Wyoming.

The goal of the project was to use the seismograph readings to obtain a geological profile of earth's mantle.

The seismographs were located 0.8 miles apart. To find the central angle between two adjacent seismographs, geologists used the formula $s = \theta r$, where θ is measured in radians. Using $s = 0.8$ miles and $r = 4000$ miles, you obtain an angle of 0.0002 radians or 0.0115°.

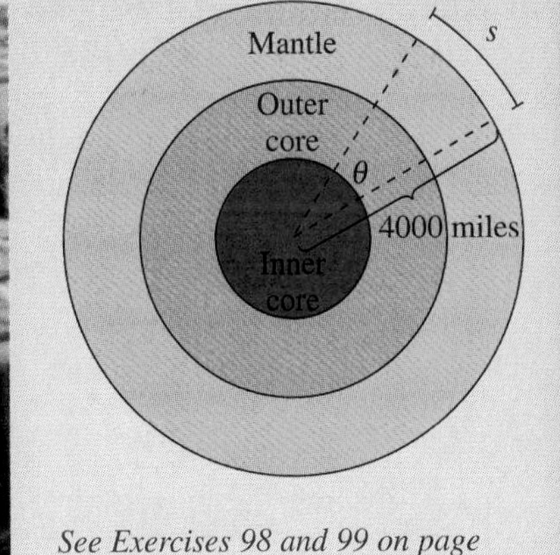

See Exercises 98 and 99 on page 126.

Photos: © Michael Milstein

Project Deep Probe used about 800 portable seismographs from southern Alberta to central Wyoming. Geologists Holger and Reingard Mandler are shown checking a seismograph before burying it in Wyoming.

1.1 Radian and Degree Measure

See Exercises 83–86 on page 125 for examples of how trigonometry can be used to find the distance between two cities of a given longitude.

Angles ▫ *Radian Measure* ▫ *Degree Measure* ▫ *Applications*

Angles

As derived from the Greek language, the word **trigonometry** means "measurement of triangles." Initially, trigonometry dealt with relationships among the sides and angles of triangles and was used in the development of astronomy, navigation, and surveying. With the development of calculus and the physical sciences in the 17th century, a different perspective arose–one that viewed the classic trigonometric relationships as *functions* with the set of real numbers as their domains. Consequently, the applications of trigonometry expanded to include a vast number of physical phenomena involving rotations and vibrations. These phenomena include sound waves, light rays, planetary orbits, vibrating strings, pendulums, and orbits of atomic particles.

The approach in this text incorporates *both* perspectives, starting with angles and their measure.

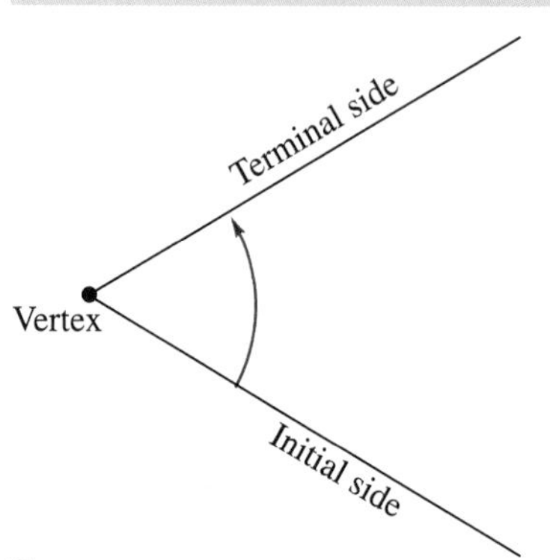

FIGURE 1.1

An **angle** is determined by rotating a ray (half-line) about its endpoint. The starting position of the ray is the **initial side** of the angle, and the position after rotation is the **terminal side,** as shown in Figure 1.1. The endpoint of the ray is the **vertex** of the angle. This perception of an angle fits a coordinate system in which the origin is the vertex and the initial side coincides with the positive x-axis. Such an angle is in **standard position,** as shown in Figure 1.2. **Positive angles** are generated by counterclockwise rotation, and **negative angles** by clockwise rotation, as shown in Figure 1.3. Angles are labeled with Greek letters α (alpha), β (beta), and θ (theta), as well as uppercase letters A, B, and C. In Figure 1.4, note that angles α and β have the same initial and terminal sides. Such angles are **coterminal.**

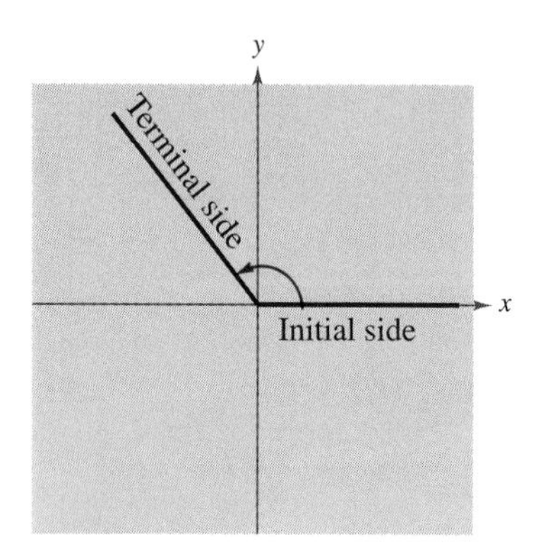

FIGURE 1.2

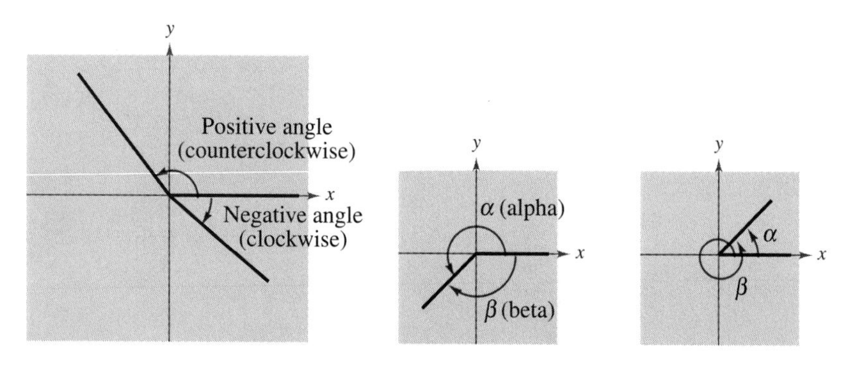

FIGURE 1.3

FIGURE 1.4

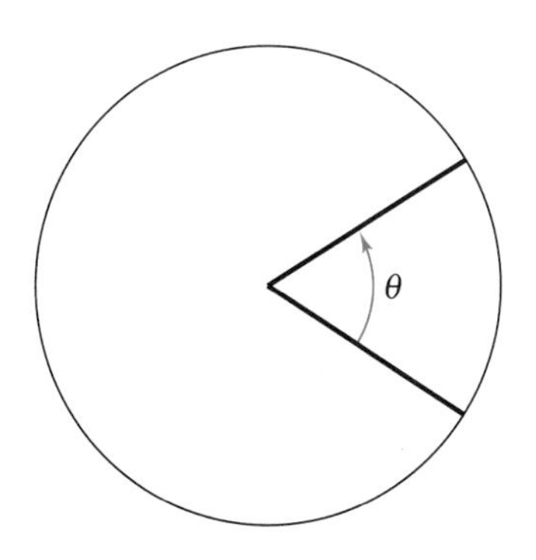

FIGURE 1.5

NOTE One revolution around a circle of radius r corresponds to an angle of 2π radians because

$$\frac{s}{r} = \frac{2\pi r}{r} = 2\pi \text{ radians.}$$

(a)

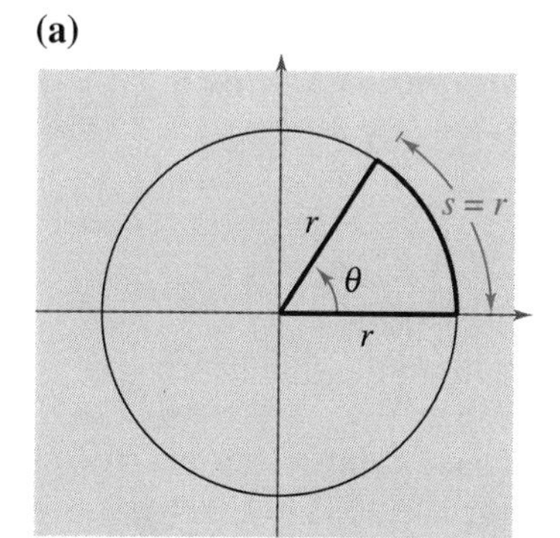

Arc length = radius when $\theta = 1$ radian

(b)

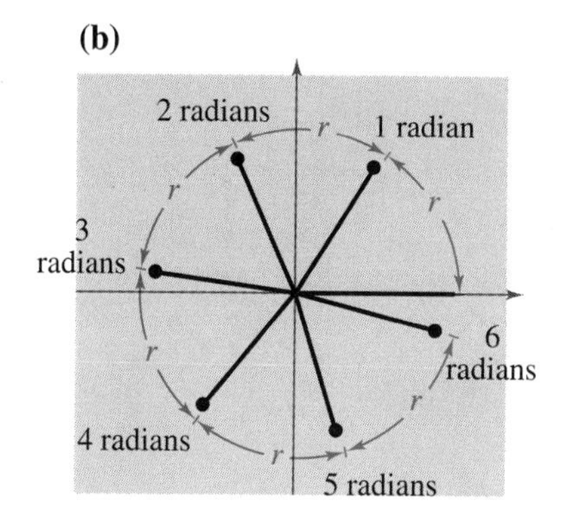

FIGURE 1.6

Radian Measure

The **measure of an angle** is determined by the amount of rotation from the initial side to the terminal side. One way to measure angles is in *radians*. This type of measure is especially useful in calculus. To define a radian, you can use a **central angle** of a circle, one whose vertex is the center of the circle, as shown in Figure 1.5.

DEFINITION OF A RADIAN

One **radian** is the measure of a central angle θ that intercepts an arc s equal in length to the radius r of the circle. See Figure 1.6(a).

Because the circumference of a circle is $2\pi r$, it follows that a central angle of one full revolution (counterclockwise) corresponds to an arc length of $s = 2\pi r$. Moreover, because $2\pi \approx 6.28$, there are just over six radius lengths in a full circle, as shown in Figure 1.6(b). In general, the radian measure of a central angle θ is obtained by dividing the arc length s by r. That is, $s/r = \theta$, where *θ is measured in radians.* Because the units of measure for s and r are the same, this ratio is unitless–it is simply a real number.

Because the radian measure of an angle of one full revolution is 2π, you can obtain the following.

$$\frac{1}{2} \text{ revolution} = \frac{2\pi}{2} = \pi \text{ radians}$$

$$\frac{1}{4} \text{ revolution} = \frac{2\pi}{4} = \frac{\pi}{2} \text{ radians}$$

$$\frac{1}{6} \text{ revolution} = \frac{2\pi}{6} = \frac{\pi}{3} \text{ radians}$$

These and other common angles are shown in Figure 1.7.

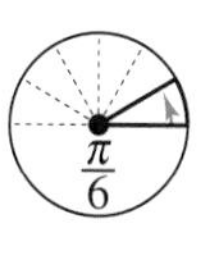

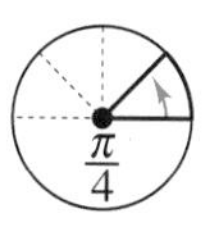

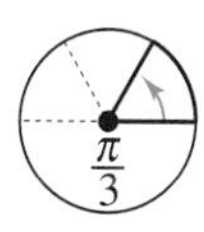

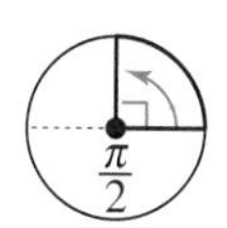

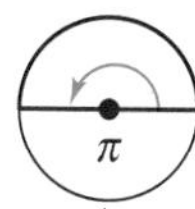

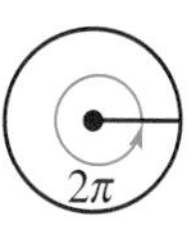

FIGURE 1.7

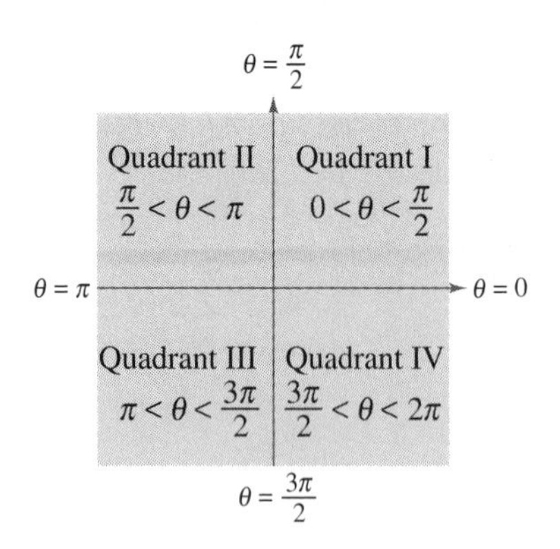

FIGURE 1.8

Recall that the four quadrants in a coordinate system are numbered I, II, III, and IV. Figure 1.8 shows which angles between 0 and 2π lie in each of the four quadrants. Note that angles between 0 and $\pi/2$ are **acute** and that angles between $\pi/2$ and π are **obtuse.**

NOTE The phrase "the terminal side of θ lies in a quadrant" is often abbreviated by simply saying "θ lies in a quadrant." The terminal sides of "quadrant angles" 0, $\pi/2$, π, and $3\pi/2$ do not lie within quadrants ■■

Two angles are coterminal if they have the same initial and terminal sides. For instance, the angles 0 and 2π are coterminal, as are the angles $\pi/6$ and $13\pi/6$. You can find an angle that is coterminal to a given angle θ by adding or subtracting 2π (one revolution), as demonstated in Example 1. A given angle θ has many coterminal angles. For instance, $\theta = \pi/6$ is coterminal with

$$\frac{\pi}{6} + 2n\pi$$

where n is an integer.

EXAMPLE 1 *Sketching and Finding Coterminal Angles*

a. For the positive angle $13\pi/6$, subtract 2π to obtain a coterminal angle

$$\frac{13\pi}{6} - 2\pi = \frac{\pi}{6}.$$ See Figure 1.9(a).

b. For the positive angle $13\pi/4$, subtract 2π to obtain a coterminal angle

$$\frac{3\pi}{4} - 2\pi = -\frac{5\pi}{4}.$$ See Figure 1.9(b).

c. For the negative angle $-2\pi/3$, add 2π to obtain a coterminal angle

$$-\frac{2\pi}{3} + 2\pi = \frac{4\pi}{3}.$$ See Figure 1.9(c).

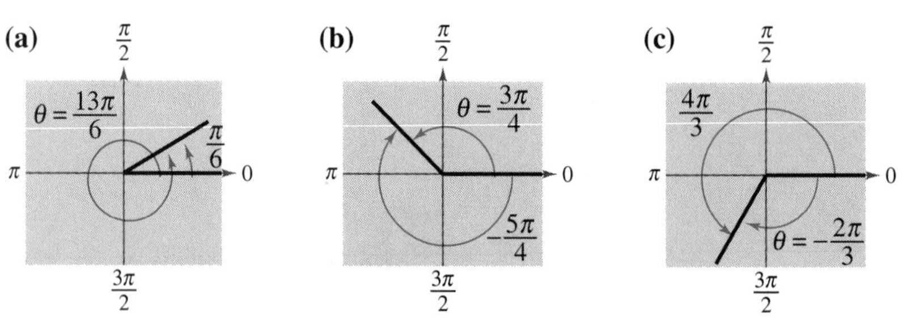

FIGURE 1.9

Two positive angles α and β are **complementary** (complements of each other) if their sum is $\pi/2$. Two positive angles are **supplementary** (supplements of each other) if their sum is π. See Figure 1.10.

(a)

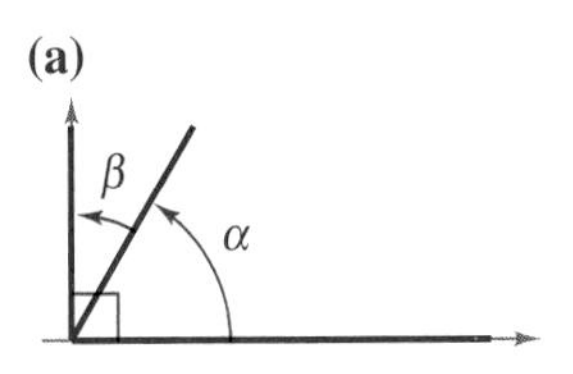

Complementary Angles

(b)

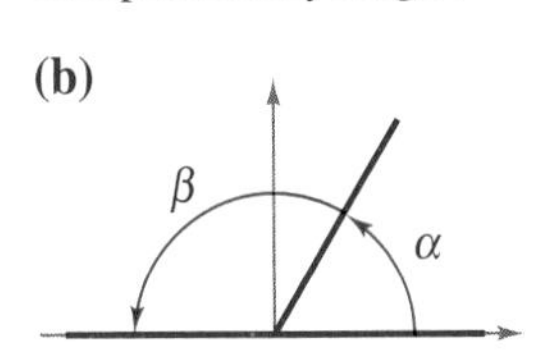

Supplementary Angles

FIGURE 1.10

EXAMPLE 2 *Complementary and Supplementary Angles*

If possible, find the complement and the supplement of (a) $2\pi/5$ and (b) $4\pi/5$.

Solution

a. The complement of $2\pi/5$ is

$$\frac{\pi}{2} - \frac{2\pi}{5} = \frac{5\pi}{10} - \frac{4\pi}{10} = \frac{\pi}{10}.$$

The supplement of $2\pi/5$ is

$$\pi - \frac{2\pi}{5} = \frac{3\pi}{5}.$$

b. Because $4\pi/5$ is greater than $\pi/2$, it has no complement. (Remember to use only *positive* angles for complements.) The supplement is

$$\pi - \frac{4\pi}{5} = \frac{\pi}{5}.$$

Degree Measure

A second way to measure angles is in terms of *degrees*. A measure of **1 degree (1°)** is equivalent to a rotation of $1/360$ of a complete revolution about the vertex. To measure angles, it is convenient to mark degrees on the circumference of a circle, as shown in Figure 1.11. Thus, a full revolution (counterclockwise) corresponds to 360°, a half revolution to 180°, a quarter revolution to 90°, and so on.

Because 2π radians corresponds to one complete revolution, degrees and radians are related by the equations

$$360° = 2\pi \text{ rad}$$

and

$$180° = \pi \text{ rad}.$$

From the latter equation, you obtain

$$1° = \frac{\pi}{180} \text{ rad} \qquad \text{and} \qquad 1 \text{ rad} = \left(\frac{180°}{\pi}\right)$$

which lead to the conversion rules at the top of the next page.

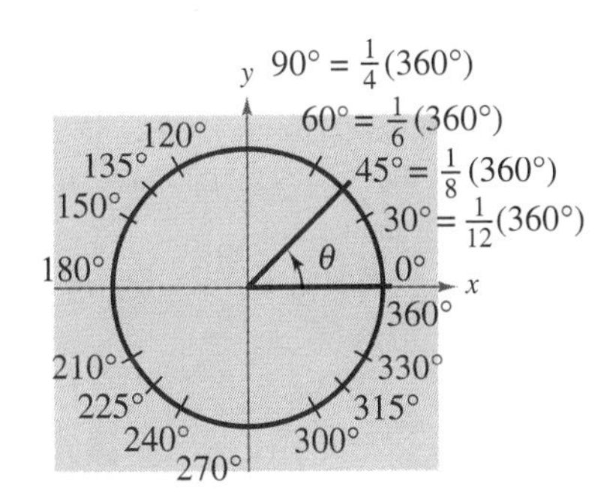

FIGURE 1.11

NOTE Note that when no units of angle measure are specified, *radian measure is implied.* For instance, if you write $\theta = \pi$ or $\theta = 2$, you should mean $\theta = \pi$ radians or $\theta = 2$ radians.

CONVERSIONS BETWEEN DEGREES AND RADIANS

1. To convert degrees to radians, multiply degrees by $\dfrac{\pi \text{ rad}}{180°}$.

2. To convert radians to degrees, multiply radians by $\dfrac{180°}{\pi \text{ rad}}$.

To apply these two conversion rules, use the relationship π rad = 180°.

EXAMPLE 3 *Converting from Degrees to Radians*

a. $135° = (135 \text{ deg})\left(\dfrac{\pi \text{ rad}}{180 \text{ deg}}\right) = \dfrac{3\pi}{4} \text{ rad}$ Multiply by $\pi/180$.

b. $540° = (540 \text{ deg})\left(\dfrac{\pi \text{ rad}}{180 \text{ deg}}\right) = 3\pi \text{ rad}$ Multiply by $\pi/180$.

c. $-270° = (-270 \text{ deg})\left(\dfrac{\pi \text{ rad}}{180 \text{ deg}}\right) = -\dfrac{3\pi}{2} \text{ rad}$ Multiply by $\pi/180$.

EXAMPLE 4 *Converting from Radians to Degrees*

a. $-\dfrac{\pi}{2} \text{ rad} = \left(-\dfrac{\pi}{2} \text{ rad}\right)\left(\dfrac{180 \text{ deg}}{\pi \text{ rad}}\right) = -90°$ Multiply by $180/\pi$.

b. $\dfrac{9\pi}{2} \text{ rad} = \left(\dfrac{9\pi}{2} \text{ rad}\right)\left(\dfrac{180 \text{ deg}}{\pi \text{ rad}}\right) = 810°$ Multiply by $180/\pi$.

c. $2 \text{ rad} = (2 \text{ rad})\left(\dfrac{180 \text{ deg}}{\pi \text{ rad}}\right) = \dfrac{360}{\pi} \approx 114.59°$ Multiply by $180/\pi$.

NOTE If you have a calculator with a "radian-to-degree" conversion key, try using it to verify the result shown in part (c) of Example 4.

TECHNOLOGY

$1' = 1 \text{ minute} = \frac{1}{60}(1°)$

$1'' = 1 \text{ second} = \frac{1}{3600}(1°)$

With calculators it is convenient to use *decimal* degrees to denote fractional parts of degrees. Historically, however, fractional parts of degrees were expressed in *minutes* and *seconds*, using the prime (′) and double prime (″) notations, respectively. Consequently, an angle of 64 degrees, 32 minutes, and 47 seconds, is represented by $\theta = 64° \, 32' \, 47''$. Many calculators have special keys for converting an angle in degrees, minutes, and seconds (D° M′ S″) into decimal degree form, and vice versa.

Applications

The *radian measure* formula, $\theta = s/r$, can be used to measure arc length along a circle. Specifically, for a circle of radius r, a central angle θ intercepts an arc of length s given by

$$s = r\theta \qquad \text{Length of circular arc}$$

where θ is measured in radians.

The *Interactive* CD-ROM shows every example with its solution; clicking on the *Try It!* button brings up similar problems. Guided Examples and Integrated Examples show step-by-step solutions to additional examples. Integrated Examples are related to several concepts in the section.

EXAMPLE 5 *Finding Arc Length*

A circle has a radius of 4 inches. Find the length of the arc intercepted by a central angle of 240°, as shown in Figure 1.12.

Solution

To use the formula $s = r\theta$, first convert 240° to radian measure.

$$240° = (240 \text{ deg})\left(\frac{\pi \text{ rad}}{180 \text{ deg}}\right) \qquad \text{Convert from degrees to radians.}$$

$$= \frac{4\pi}{3} \text{ rad} \qquad \text{Simplify.}$$

Then, using a radius of $r = 4$ inches, you can find the arc length to be

$$s = r\theta \qquad \text{Length of circular arc}$$

$$= 4\left(\frac{4\pi}{3}\right) \qquad \text{Substitute for } r \text{ and } \theta.$$

$$= \frac{16\pi}{3} \qquad \text{Simplify.}$$

$$\approx 16.76 \text{ inches.} \qquad \text{Use a calculator.}$$

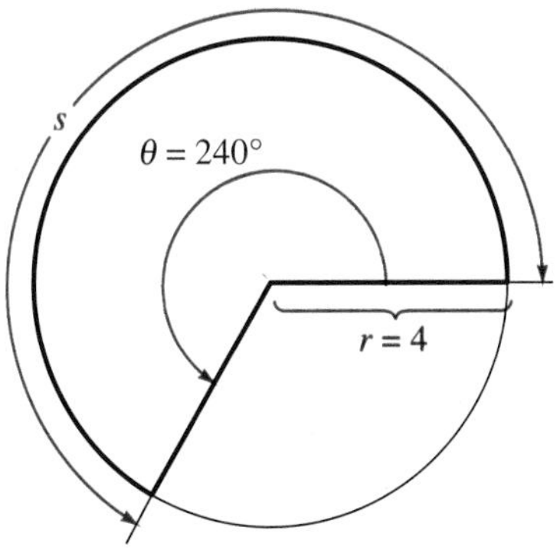

Note that the units for $r\theta$ are determined by the units for r, because θ is given in radian measure and therefore has no units.

FIGURE 1.12

The formula for the length of a circular arc can be used to analyze the motion of a particle moving at a *constant speed* along a circular path. Consider a circle of radius r. If s is the length of the arc traveled in time t, the **speed** of the particle is

$$\text{Speed} = \frac{\text{distance}}{\text{time}} = \frac{s}{t}.$$

Moreover, if θ is the angle (in radian measure) corresponding to the arc length s, the **angular speed** of the particle is

$$\text{Angular speed} = \frac{\theta}{t}.$$

FIGURE 1.13

EXAMPLE 6 Finding the Speed of an Object

The second hand of a clock is 10.2 centimeters long, as shown in Figure 1.13. Find the speed of the tip of this second hand.

Solution

The time required for the second hand to make one full revolution is

$$t = 60 \text{ seconds} = 1 \text{ minute}.$$

The distance traveled by the tip of the second hand in one revolution is

$$s = 2\pi(\text{radius}) = 2\pi(10.2) = 20.4\pi \text{ centimeters}.$$

Therefore, the speed of the tip of the second hand is

$$\text{Speed} = \frac{s}{t} = \frac{20.4\pi \text{ centimeters}}{60 \text{ seconds}} \approx 1.068 \text{ centimeters per second}.$$

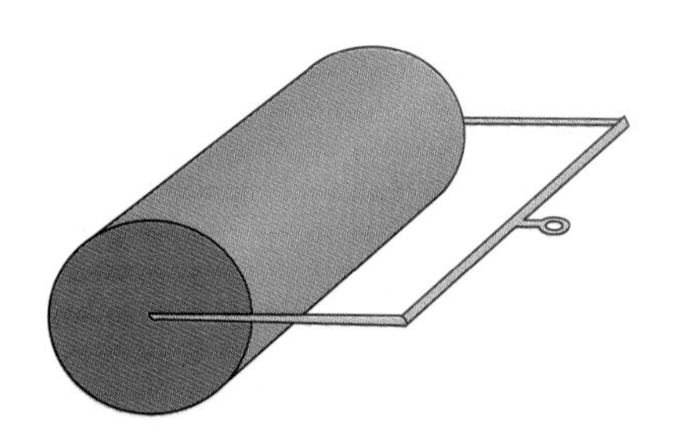

FIGURE 1.14

EXAMPLE 7 Finding Angular Speed

A lawn roller (see Figure 1.14) makes 1.2 revolutions per second. Find the angular speed of the roller in radians per second.

Solution

Because each revolution generates 2π radians, it follows that the roller turns $(1.2)(2\pi) = 2.4\pi$ radians per second. In other words, the angular speed is

$$\text{Angular speed} = \frac{\theta}{t} = \frac{2.4\pi \text{ radians}}{1 \text{ second}} = 2.4\pi \text{ radians per second}.$$

GROUP ACTIVITY

DEGREE AND RADIAN MEASURE

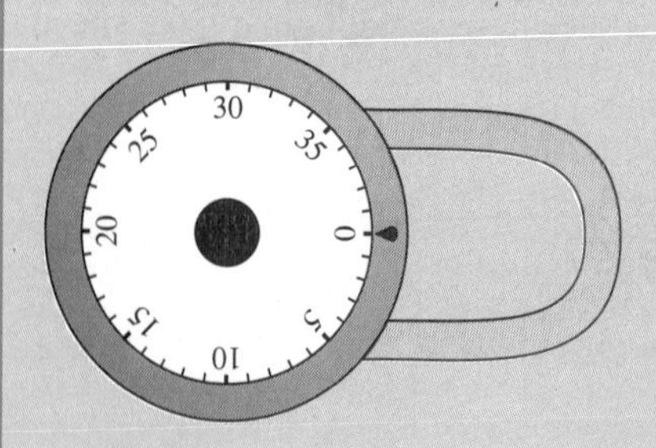

A standard combination lock has 40 numbers (0–39). Suppose the lock is positioned as shown in the figure, with its dial pointing to 0. Choose a three-number combination. Without revealing the combination, describe to your partner how to turn the dial *in terms of degree measure* to open the lock. Each angle measure should be given in standard position from the pointer. Remember that most locks follow a right-left-right pattern. Ask your partner to verify each number of the combination as you go. Switch roles and use radians to describe the combination.

The *Interactive* CD-ROM provides additional help with Warm-Up exercises by providing a hypertext link to the section in which the concept was introduced.

WARM UP

Solve for x.

1. $x + 135 = 180$
2. $790 = 720 + x$
3. $\pi = \dfrac{5\pi}{6} + x$
4. $2\pi - x = \dfrac{5\pi}{3}$
5. $\dfrac{45}{180} = \dfrac{x}{\pi}$
6. $\dfrac{240}{180} = \dfrac{x}{\pi}$
7. $\dfrac{\pi}{180} = \dfrac{x}{20}$
8. $\dfrac{180}{\pi} = \dfrac{330}{x}$
9. $\dfrac{x}{60} = \dfrac{3}{4}$
10. $\dfrac{x}{3600} = 0.0125$

The *Interactive* CD-ROM contains step-by-step solutions to all odd-numbered Section and Review Exercises. It also provides Tutorial Exercises, which link to Guided Examples for additional help.

1.1 Exercises

In Exercises 1–4, estimate the angle to the nearest one-half radian.

1.
3.

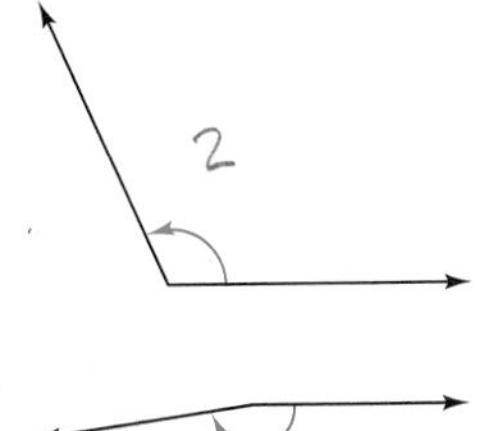

2.
4.

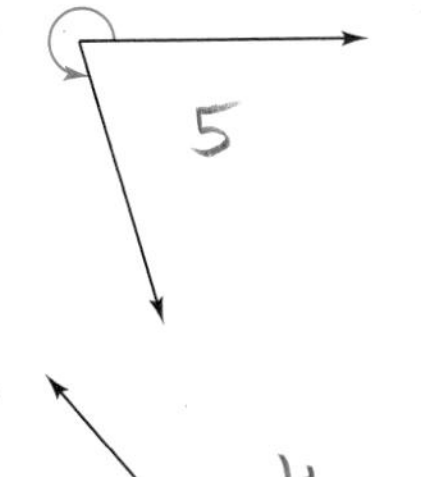

In Exercises 5–10, determine the quadrant in which the angle lies. (The angle measure is given in radians.)

5. (a) $\pi/5$ (b) $7\pi/5$
6. (a) $5\pi/4$ (b) $7\pi/4$
7. (a) $-\pi/12$ (b) $-11\pi/9$
8. (a) -1 (b) -2
9. (a) 3.5 (b) 2.25
10. (a) 5.63 (b) -2.25

In Exercises 11–14, sketch the angle in standard position.

11. (a) $\dfrac{5\pi}{4}$ (b) $\dfrac{2\pi}{3}$
12. (a) $-\dfrac{7\pi}{4}$ (b) $-\dfrac{5\pi}{2}$
13. (a) $\dfrac{11\pi}{6}$ (b) 7π
14. (a) 4 (b) -3

In Exercises 15–18, determine two coterminal angles (one positive and one negative) for the given angle. Give your answers in radians.

15. (a)

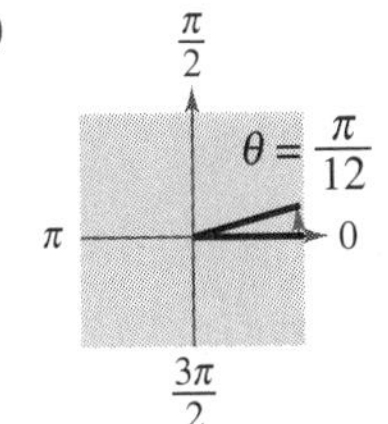

(b)

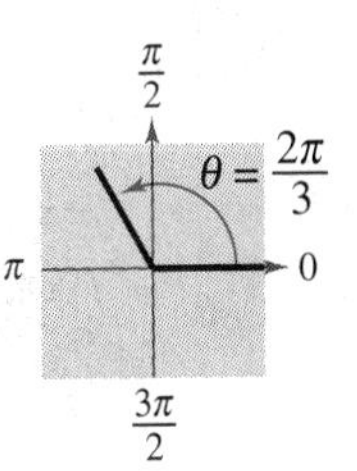

16. (a)

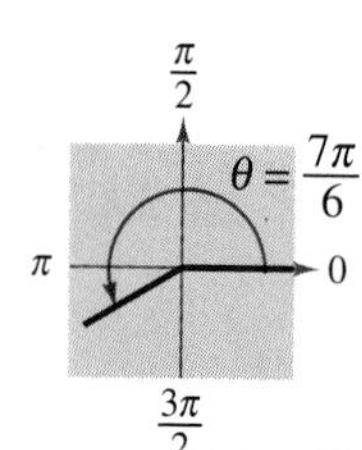

(b)

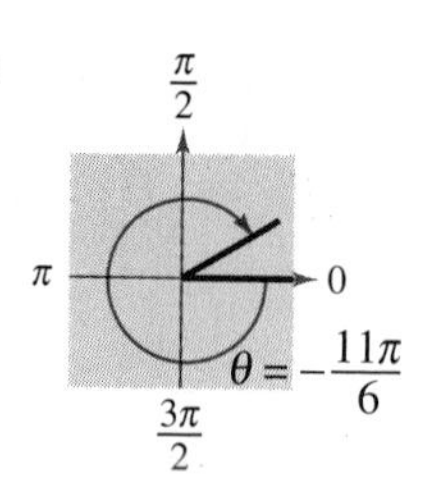

17. (a) $-9\pi/4$ (b) $-2\pi/15$

18. (a) $8\pi/9$ (b) $8\pi/45$

In Exercises 19–20, find (if possible) the complement and supplement of the angle.

19. (a) $\pi/3$ (b) $3\pi/4$

20. (a) 1 (b) 2

In Exercises 21–24, estimate the angle in degrees.

21.

22.

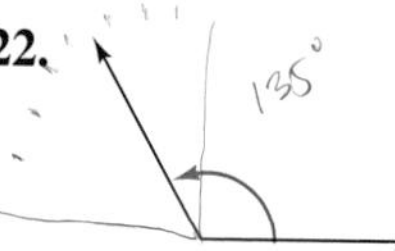

23.

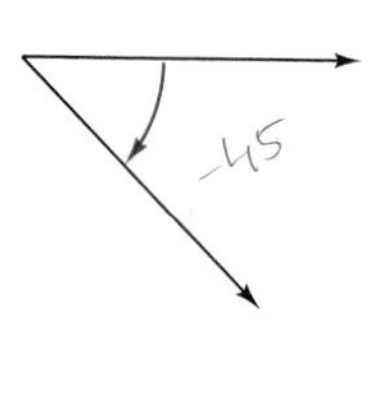

24.

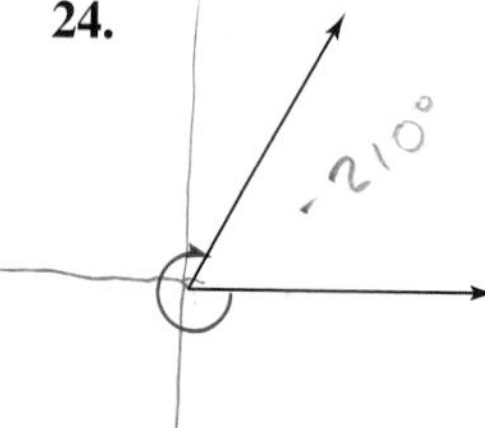

In Exercises 25–28, determine the quadrant in which the angle lies.

25. (a) 130° (b) 285°

26. (a) 8.3° (b) 257° 30′

27. (a) −132° 50′ (b) −336°

28. (a) −260° (b) −3.4°

In Exercises 29–32, sketch the angle in standard position.

29. (a) 30° (b) 150°

30. (a) −270° (b) −120°

31. (a) 405° (b) −480°

32. (a) 750° (b) −600°

In Exercises 33–36, determine two coterminal angles (one positive and one negative) for the given angle. Give your answers in degrees.

33. (a)

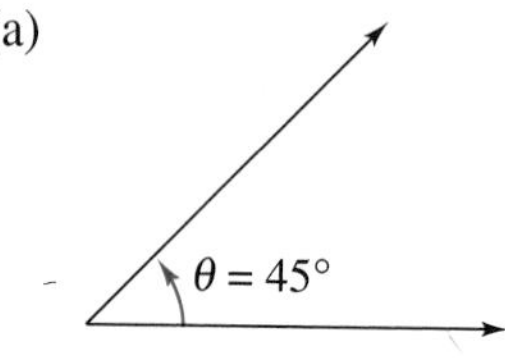

(b)

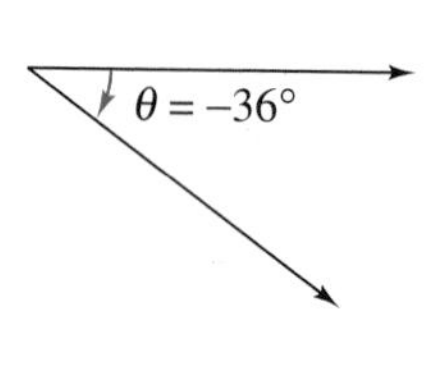

34. (a)

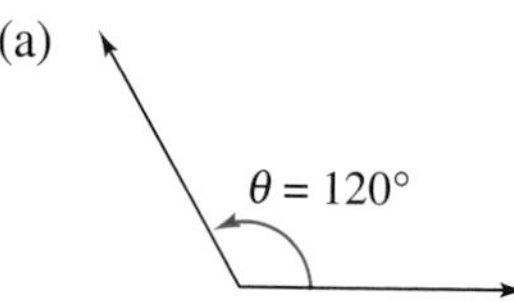

(b)

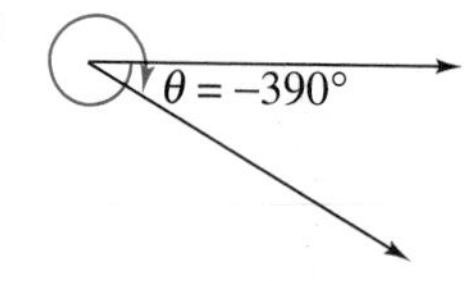

35. (a) 300° (b) 740°

36. (a) −420° (b) 230°

In Exercises 37 and 38, find (if possible) the complement and supplement of the angle.

37. (a) 18° (b) 115°

38. (a) 79° (b) 150°

In Exercises 39–42, express the angle in radian measure as a multiple of π. (Do not use a calculator.)

39. (a) 30° (b) 150°

40. (a) 315° (b) 120°

41. (a) −20° (b) −240°

42. (a) −270° (b) 144°

In Exercises 43–46, express the angle in degree measure. (Do not use a calculator.)

43. (a) $3\pi/2$ (b) $7\pi/6$

44. (a) $-7\pi/12$ (b) $\pi/9$

45. (a) $7\pi/3$ (b) $-11\pi/30$

46. (a) $11\pi/6$ (b) $34\pi/15$

In Exercises 47–54, convert the measure from degrees to radians. Round to three decimal places.

47. $115°$ **48.** $87.4°$

49. $-216.35°$ **50.** $-48.27°$

51. $532°$ **52.** $0.54°$

53. $-0.83°$ **54.** $345°$

In Exercises 55–62, convert the measure from radians to degrees. Round to three decimal places.

55. $\pi/7$ **56.** $5\pi/11$

57. $15\pi/8$ **58.** 6.5π

59. -4.2π **60.** 4.8

61. -2 **62.** -0.57

In Exercises 63–66, convert to decimal degree form.

63. (a) $54° 45'$ (b) $-128° 30'$

64. (a) $245° 10'$ (b) $2° 12'$

65. (a) $85° 18' 30''$ (b) $330° 25''$

66. (a) $-135° 36''$ (b) $-408° 16' 20''$

In Exercises 67–70, convert to D° M′ S″ form.

67. (a) $240.6°$ (b) $-145.8°$

68. (a) $-345.12°$ (b) 0.45

69. (a) 2.5 (b) -3.58

70. (a) -0.355 (b) 0.7865

In Exercises 71–74, find the angle in radians.

71.

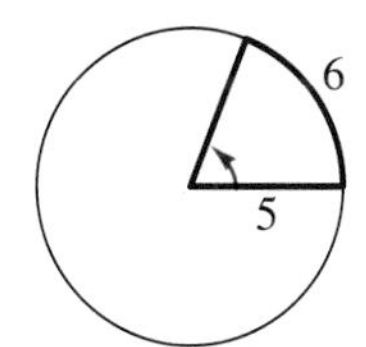

72.

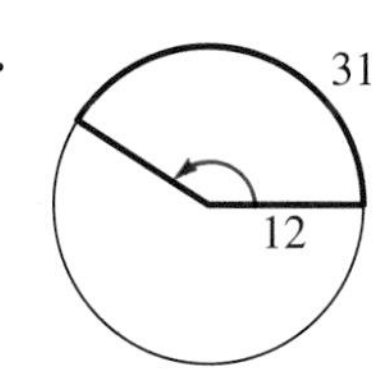

73.

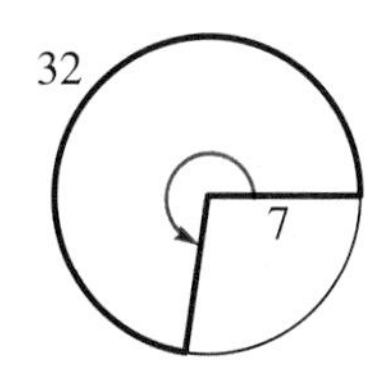

74.

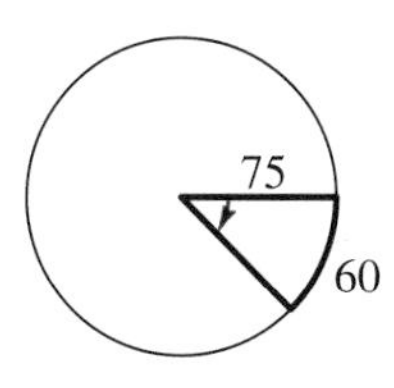

In Exercises 75–78, find the radian measure of the central angle of a circle of the given radius that intercepts an arc of the given length.

	Radius	*Arc Length*
75.	15 inches	4 inches
76.	16 feet	10 feet
77.	14.5 centimeters	25 centimeters
78.	80 kilometers	160 kilometers

In Exercises 79–82, find the length of the arc on a circle of the given radius intercepted by the given central angle.

	Radius	*Central Angle*
79.	15 inches	$180°$
80.	9 feet	$60°$
81.	6 meters	2 radians
82.	40 centimeters	$3\pi/4$ radians

Distance Between Cities **In Exercises 83–86, find the distance between the cities. Assume that earth is a sphere of radius 4000 miles and the cities are on the same meridian (one city is due north of the other).**

	City	*Latitude*
83.	Dallas, Texas	$32° 47' 9''$N
	Omaha, Nebraska	$41° 15' 42''$N
84.	San Francisco, California	$37° 46' 39''$N
	Seattle, Washington	$47° 36' 32''$N
85.	Miami, Florida	$25° 46' 37''$N
	Erie, Pennsylvania	$42° 7' 15''$N
86.	Johannesburg, South Africa	$26° 10'$S
	Jerusalem, Israel	$31° 47'$N

87. *Difference in Latitudes* Assuming that earth is a sphere of radius 6378 kilometers, what is the difference in latitude of two cities, one of which is 600 kilometers due north of the other?

88. *Difference in Latitudes* Assuming that earth is a sphere of radius 6378 kilometers, what is the difference in latitude of two cities, one of which is 800 kilometers due north of the other?

89. ***Instrumentation*** The pointer on a voltmeter is 6 centimeters in length (see figure). Find the angle through which the pointer rotates when it moves 2.5 centimeters on the scale.

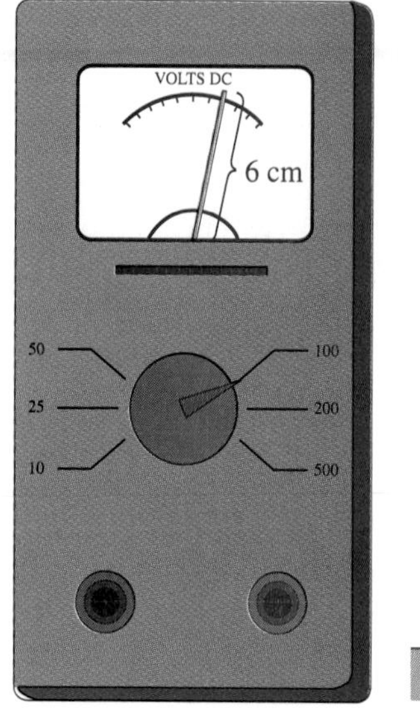

FIGURE FOR 89

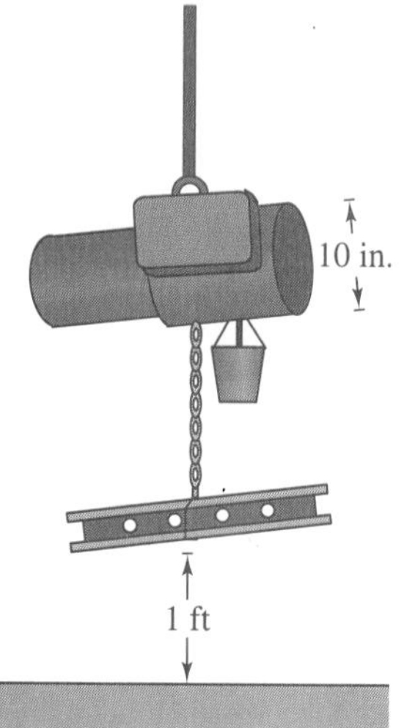

FIGURE FOR 90

90. ***Electric Hoist*** An electric hoist is being used to lift a piece of equipment (see figure). The diameter of the drum on the hoist is 10 inches, and the equipment must be raised 1 foot. Find the number of degrees through which the drum must rotate.

91. ***Angular Speed*** A car is moving at a rate of 50 miles per hour, and the diameter of its wheels is 2.5 feet.

(a) Find the number of revolutions per minute the wheels are rotating.

(b) Find the angular speed of the wheels in radians per minute.

92. ***Angular Speed*** A 2-inch-diameter pulley on an electric motor that runs at 1700 revolutions per minute is connected by a belt to a 4-inch-diameter pulley on a saw arbor.

(a) Find the angular speed (in radians per minute) of each pulley.

(b) Find the revolutions per minute of the saw.

93. ***Think About It*** Is a degree or a radian the larger unit of measure? Explain.

94. ***Think About It*** If the radius of a circle is increasing and the magnitude of a central angle is held constant, how is the length of the intercepted arc changing? Explain.

95. ***Floppy Disk*** The radius of the magnetic disk in a 3.5-inch diskette is 1.68 inches. Find the linear speed of a point on the circumference of the disk if it is rotating at a speed of 360 revolutions per minute.

96. ***Speed of a Bicycle*** The radii of the sprocket assemblies and the wheel of a bicycle (see figure) are 4 inches, 2 inches, and 14 inches, respectively. If the cyclist is pedaling at a rate of 1 revolution per second, find the speed of the bicycle in (a) feet per second and (b) miles per hour.

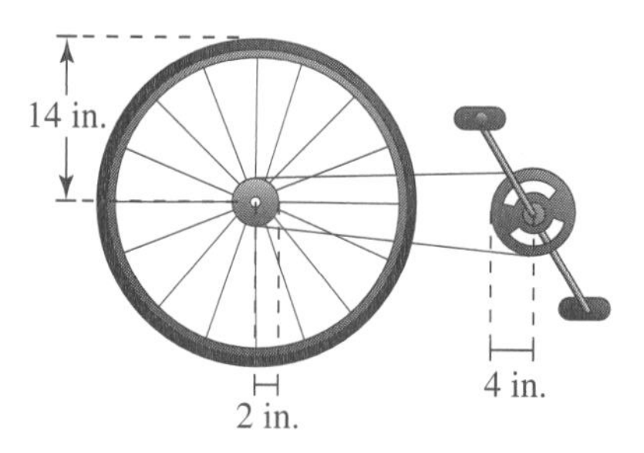

97. ***Geometry*** Prove that the area of a circular sector of radius r with central angle θ is $A = \frac{1}{2}\theta r^2$, where θ is measured in radians.

Chapter Opener **In Exercises 98 and 99, use the information given in the chapter opener on page 115.**

98. Suppose the seismographs were 1.2 miles apart. Find the central angle between the two adjacent seismographs.

99. Suppose the central angle between two adjacent seismographs is 0.031°. Find the distance between the seismographs.

1.2 Trigonometric Functions: The Unit Circle

See Exercise 65 on page 135 for an example of how a trigonometric function can be used to model the current in an electrical circuit.

The Unit Circle ▫ *The Trigonometric Functions* ▫ *Domain and Period of Sine and Cosine* ▫ *Evaluating Trigonometric Functions with a Calculator*

The Unit Circle

The two historical perspectives of trigonometry incorporate different methods for introducing the trigonometric functions. Our first introduction to these functions is based on the unit circle.

Consider the **unit circle** given by

$$x^2 + y^2 = 1 \qquad \text{Unit circle}$$

as shown in Figure 1.15. Imagine that the real number line is wrapped around this circle, with positive numbers corresponding to a counterclockwise wrapping and negative numbers corresponding to a clockwise wrapping, as shown in Figure 1.16.

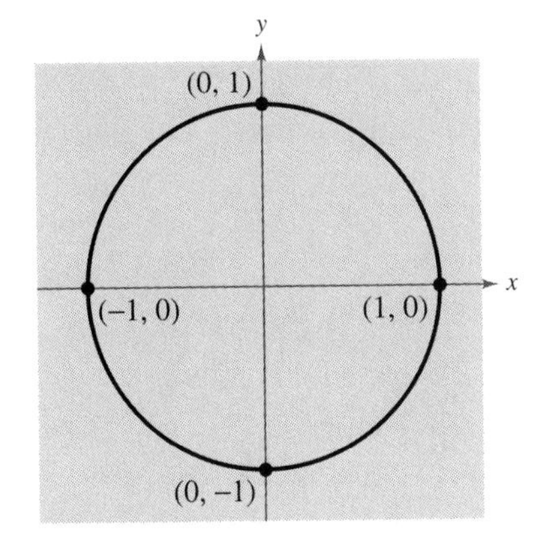

FIGURE 1.15

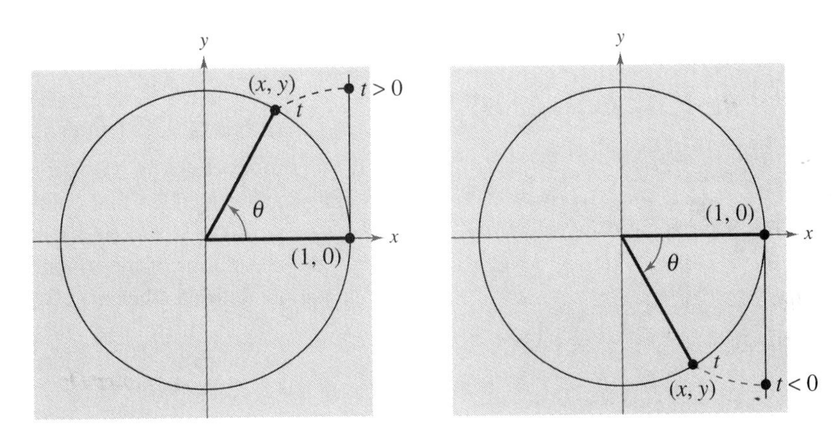

FIGURE 1.16

As the real number line is wrapped around the unit circle, each real number t corresponds to a point (x, y) on the circle. For example, the real number 0 corresponds to the point $(1, 0)$. Moreover, because the unit circle has a circumference of 2π, the real number 2π also corresponds to the point $(1, 0)$.

In general, each real number t also corresponds to a central angle θ (in standard position) whose radian measure is t. With this interpretation of t, the arc length formula $s = r\theta$ (with $r = 1$) indicates the real number t is the length of the arc interpreted by the angle θ, given in radians.

The Trigonometric Functions

From the preceding discussion, it follows that the coordinates x and y are two functions of the real variable t. You can use these coordinates to define the six trigonometric functions of t.

sine	**cosecant**
cosine	**secant**
tangent	**cotangent**

These six functions are normally abbreviated sin, csc, cos, sec, tan, and cot, respectively.

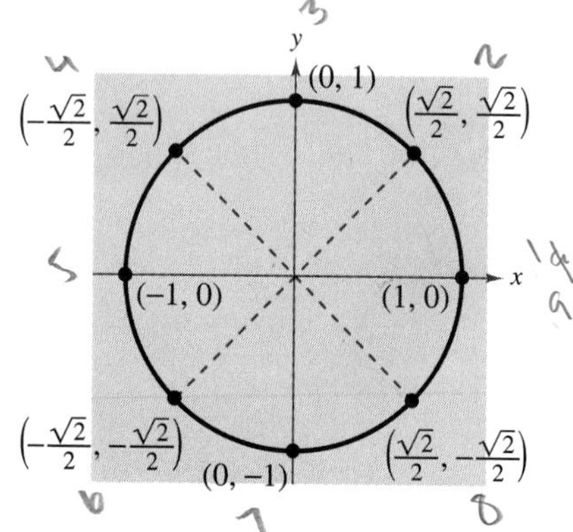

FIGURE 1.17

Definitions of Trigonometric Functions

Let t be a real number and let (x, y) be the point on the unit circle corresponding to t.

$$\sin t = y \qquad \csc t = \frac{1}{y}, \quad y \neq 0$$

$$\cos t = x \qquad \sec t = \frac{1}{x}, \quad x \neq 0$$

$$\tan t = \frac{y}{x}, \quad x \neq 0 \qquad \cot t = \frac{x}{y}, \quad y \neq 0$$

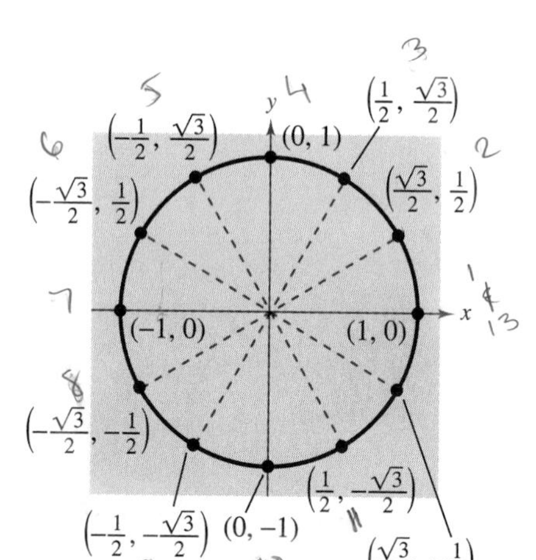

FIGURE 1.18

NOTE The functions in the second column are the *reciprocals* of the corresponding functions in the first column.

In the definitions of the trigonometric functions, note that the tangent and secant are not defined when $x = 0$. For instance, because $t = \pi/2$ corresponds to $(x, y) = (0, 1)$, it follows that $\tan(\pi/2)$ and $\sec(\pi/2)$ are *undefined.* Similarly, the cotangent and cosecant are not defined when $y = 0$. For instance, because $t = 0$ corresponds to $(x, y) = (1, 0)$, cot 0 and csc 0 are *undefined.*

In Figure 1.17, the unit circle has been divided into eight equal arcs, corresponding to t-values of

$$0, \frac{\pi}{4}, \frac{\pi}{2}, \frac{3\pi}{4}, \pi, \frac{5\pi}{4}, \frac{3\pi}{2}, \frac{7\pi}{4}, \text{ and } 2\pi.$$

Similarly, in Figure 1.18, the unit circle has been divided into twelve equal arcs, corresponding to t-values of

$$0, \frac{\pi}{6}, \frac{\pi}{3}, \frac{\pi}{2}, \frac{2\pi}{3}, \frac{5\pi}{6}, \pi, \frac{7\pi}{6}, \frac{4\pi}{3}, \frac{3\pi}{2}, \frac{5\pi}{3}, \frac{11\pi}{6}, \text{ and } 2\pi.$$

Using the (x, y) coordinates in Figures 1.17 and 1.18, you can easily evaluate the trigonometric functions for common t-values. This procedure is demonstrated in Examples 1 and 2.

EXAMPLE 1 *Evaluating Trigonometric Functions*

Evaluate the six trigonometric functions at each real number.

a. $t = \dfrac{\pi}{6}$ **b.** $t = \dfrac{5\pi}{4}$ **c.** $t = 0$ **d.** $t = \pi$

Solution

For each t-value, begin by finding the corresponding point (x, y) on the unit circle. Then use the definitions of trigonometric functions listed on page 128.

a. $t = \pi/6$ corresponds to the point $(x, y) = \left(\sqrt{3}/2, 1/2\right)$.

$$\sin\frac{\pi}{6} = y = \frac{1}{2} \qquad \csc\frac{\pi}{6} = \frac{1}{y} = 2$$

$$\cos\frac{\pi}{6} = x = \frac{\sqrt{3}}{2} \qquad \sec\frac{\pi}{6} = \frac{1}{x} = \frac{2}{\sqrt{3}} = \frac{2\sqrt{3}}{3}$$

$$\tan\frac{\pi}{6} = \frac{y}{x} = \frac{1/2}{\sqrt{3}/2} = \frac{1}{\sqrt{3}} \qquad \cot\frac{\pi}{6} = \frac{x}{y} = \sqrt{3}$$

b. $t = 5\pi/4$ corresponds to the point $(x, y) = \left(-\sqrt{2}/2, -\sqrt{2}/2\right)$.

$$\sin\frac{5\pi}{4} = y = -\frac{\sqrt{2}}{2} \qquad \csc\frac{5\pi}{4} = \frac{1}{y} = -\frac{2}{\sqrt{2}} = -\sqrt{2}$$

$$\cos\frac{5\pi}{4} = x = -\frac{\sqrt{2}}{2} \qquad \sec\frac{5\pi}{4} = \frac{1}{x} = -\frac{2}{\sqrt{2}} = -\sqrt{2}$$

$$\tan\frac{5\pi}{4} = \frac{y}{x} = \frac{-\sqrt{2}/2}{-\sqrt{2}/2} = 1 \qquad \cot\frac{5\pi}{4} = \frac{x}{y} = 1$$

c. $t = 0$ corresponds to the point $(x, y) = (1, 0)$.

$$\sin 0 = y = 0 \qquad \csc 0 = \frac{1}{y} \text{ is undefined.}$$

$$\cos 0 = x = 1 \qquad \sec 0 = \frac{1}{x} = 1$$

$$\tan 0 = \frac{y}{x} = \frac{0}{1} = 0 \qquad \cot 0 = \frac{x}{y} \text{ is undefined.}$$

d. $t = \pi$ corresponds to the point $(x, y) = (-1, 0)$.

$$\sin\pi = y = 0 \qquad \csc\pi = \frac{1}{y} \text{ is undefined.}$$

$$\cos\pi = x = -1 \qquad \sec\pi = \frac{1}{x} = -1$$

$$\tan\pi = \frac{y}{x} = \frac{0}{-1} = 0 \qquad \cot\pi = \frac{x}{y} \text{ is undefined.}$$

EXAMPLE 2 *Evaluating Trigonometric Functions*

Evaluate the six trigonometric functions at $t = -\dfrac{\pi}{3}$.

Solution

Moving *clockwise* around the unit circle, it follows that $t = -\pi/3$ corresponds to the point $(x, y) = (1/2, -\sqrt{3}/2)$.

$$\sin\left(-\frac{\pi}{3}\right) = -\frac{\sqrt{3}}{2} \qquad \csc\left(-\frac{\pi}{3}\right) = -\frac{2}{\sqrt{3}}$$

$$\cos\left(-\frac{\pi}{3}\right) = \frac{1}{2} \qquad \sec\left(-\frac{\pi}{3}\right) = 2$$

$$\tan\left(-\frac{\pi}{3}\right) = -\sqrt{3} \qquad \cot\left(-\frac{\pi}{3}\right) = -\frac{1}{\sqrt{3}}$$

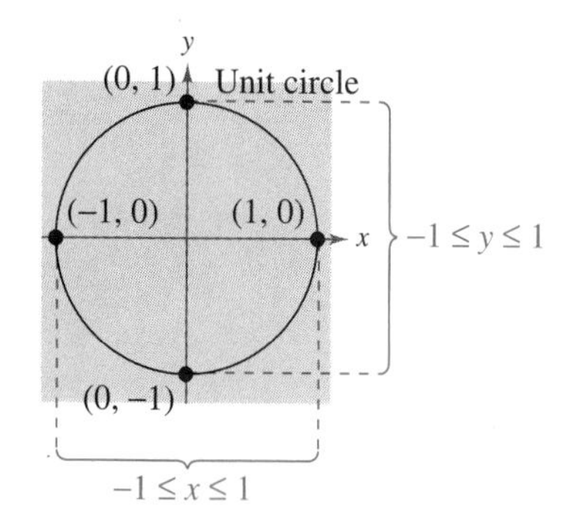

FIGURE 1.19

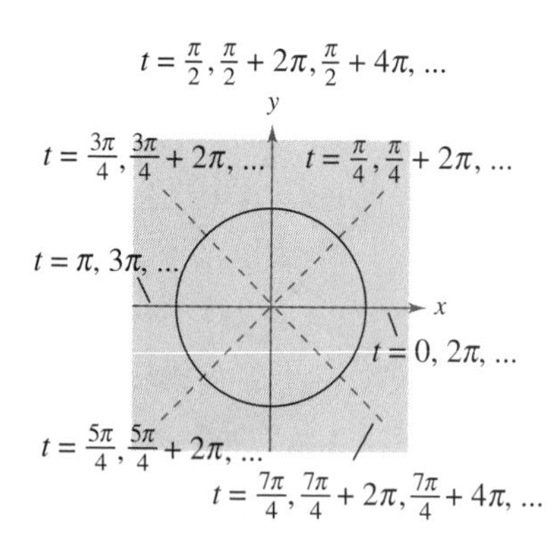

FIGURE 1.20

Domain and Period of Sine and Cosine

The *domain* of the sine and cosine functions is the set of all real numbers. To determine the *range* of these two functions, consider the unit circle shown in Figure 1.19. Because $r = 1$, it follows that $\sin t = y$ and $\cos t = x$. Moreover, because (x, y) is on the unit circle, you know that $-1 \le y \le 1$ and $-1 \le x \le 1$, and it follows that the values of sine and cosine also range between -1 and 1. That is,

$$-1 \le y \le 1 \qquad\qquad -1 \le x \le 1$$
$$-1 \le \sin t \le 1 \quad \text{and} \quad -1 \le \cos t \le 1$$

Suppose you add 2π to each value of t in the interval $[0, 2\pi]$, thus completing a second revolution around the unit circle, as shown in Figure 1.20. The values of $\sin(t + 2\pi)$ and $\cos(t + 2\pi)$ correspond to those of $\sin t$ and $\cos t$. Similar results can be obtained for repeated revolutions (positive or negative) on the unit circle. This leads to the general result

$$\sin(t + 2\pi n) = \sin t \quad \text{and} \quad \cos(t + 2\pi n) = \cos t$$

for any integer n and real number t. Functions that behave in such a repetitive (or cyclic) manner are called **periodic.**

NOTE In Figure 1.20, note that *positive* multiples of 2π are added to the t-values. You could just as well have added *negative* multiples. For instance, $\pi/4 - 2\pi$ and $\pi/4 - 4\pi$ are also coterminal with $\pi/4$.

DEFINITION OF A PERIODIC FUNCTION

A function f is **periodic** if there exists a positive real number c such that

$$f(t + c) = f(t)$$

for all t in the domain of f. The smallest number c for which f is periodic is called the **period** of f.

From this definition it follows that the sine and cosine functions are periodic and have a period of 2π. The other four trigonometric functions are also periodic, and you will study them in more detail in Section 1.6.

Exploration

With your graphing utility in radian and parametric modes, enter

X1T = cos T and Y1T = sin T

and use the following settings.

Tmin = 0
Tmax = 6.3
Tstep = .1
Xmin = -1.5
Xmax = 1.5
Xscl = 1
Ymin = -1 Ymax = 1
Yscl = 1

1. Graph the entered equations and describe the graph.
2. Use the trace key to move the cursor around, the graph. What do the t-values represent? What do the x- and y-values represent?
3. What are the smallest and greatest values for x and y?

EXAMPLE 3 *Using the Period to Evaluate the Sine and Cosine*

a. Because $\dfrac{13\pi}{6} = 2\pi + \dfrac{\pi}{6}$, you have

$$\sin\frac{13\pi}{6} = \sin\left(2\pi + \frac{\pi}{6}\right) = \sin\frac{\pi}{6} = \frac{1}{2}.$$

b. Because $-\dfrac{7\pi}{2} = -4\pi + \dfrac{\pi}{2}$, you have

$$\cos\left(-\frac{7\pi}{2}\right) = \cos\left(-4\pi + \frac{\pi}{2}\right) = \cos\frac{\pi}{2} = 0.$$

Recall from Section P.6 that a function f is *even* if $f(-t) = f(t)$ and is *odd* if $f(-t) = -f(t)$. Of the six trigonometric functions, two are even and four are odd, as stated in the following theorem. Verification of this theorem, using the unit circle, is left as an exercise.

cos t = -cos (π-t)   sec t = -sec (π-t)
tan t = -tan(π-t)   cot t = -cot (π-t)

EVEN AND ODD TRIGONOMETRIC FUNCTIONS

The cosine and secant functions are *even*.

$$\cos(-t) = \cos t \qquad \sec(-t) = \sec t$$

The sine, cosecant, tangent, and cotangent functions are *odd*.

$$\sin(-t) = -\sin t \qquad \csc(-t) = -\csc t$$
$$\tan(-t) = -\tan t \qquad \cot(-t) = -\cot t$$

sin t = sin (π-t)   csc t = csc (π-t)

Evaluating Trigonometric Functions with a Calculator

When evaluating a trigonometric function with a calculator, you need to set the calculator to the desired *mode* of measurement (degrees or radians).

Most calculators do not have keys for the cosecant, secant, and cotangent functions. To evaluate these functions, you can use the $\boxed{x^{-1}}$ key with their respective reciprocal functions sine, cosine, and tangent. For example, to evaluate $\csc(\pi/8)$, use the fact that

$$\csc\frac{\pi}{8} = \frac{1}{\sin(\pi/8)}$$

and enter the following keystroke sequence in radian mode.

[(] [SIN] [(] π [÷] 8 [)] [)] [x^{-1}] [ENTER] Display 2.6131259

EXAMPLE 4 *Using a Calculator*

Use a calculator to evaluate each expression.

a. sin 76.4° **b.** cot 1.5

Solution

Function	*Mode*	*Graphing Calculator Keystrokes*	*Display*
a. sin 76.4°	Degree	[SIN] 76.4 [ENTER]	0.9719610
b. cot 1.5	Radian	[(] [TAN] 1.5 [)] [x^{-1}] [ENTER]	0.0709148

GROUP ACTIVITY

ERROR ANALYSIS

Suppose you are tutoring a student in trigonometry. Your student is asked to evaluate the cosine of 2 radians and, using a calculator, obtains the following.

Keystrokes	*Display*
[COS] 2 [ENTER]	0.999390827

You know that 2 radians lies in the second quadrant. You also know that this implies that the cosine of 2 radians should be negative. What did your student do wrong?

WARM UP

Simplify the expression.

1. $\dfrac{1/2}{-\sqrt{3}/2}$

2. $\dfrac{\sqrt{2}/2}{-\sqrt{2}/2}$

Find a coterminal angle in the interval $[0, 2\pi]$.

3. $\dfrac{8\pi}{3}$

4. $-\dfrac{\pi}{4}$

5. Convert 30° to radians

6. Convert 135° to radians.

7. Convert $\dfrac{\pi}{3}$ to degrees

8. Convert $-\dfrac{3\pi}{2}$ to degrees.

9. Determine the circumference of a circle with radius 1.

10. Determine arc length of a semicircle with radius 1.

1.2 Exercises

In Exercises 1–4, find the six trigonometric functions of t that correspond to the point on the unit circle.

1.
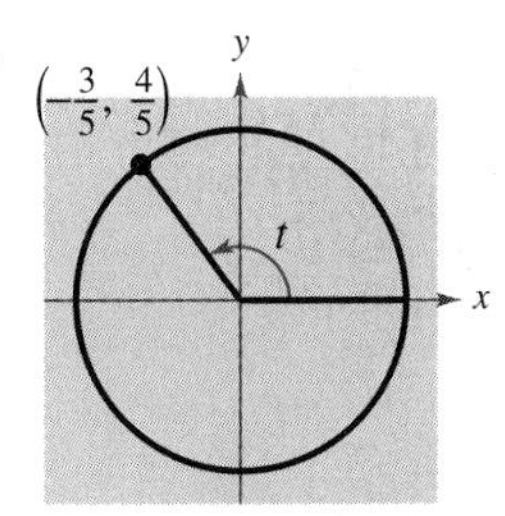

2.
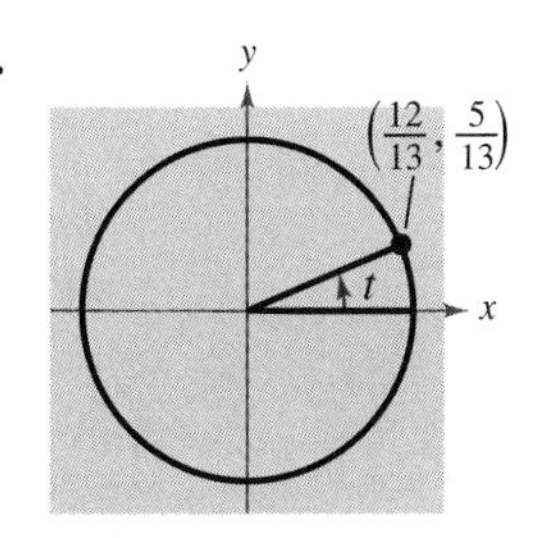

3.
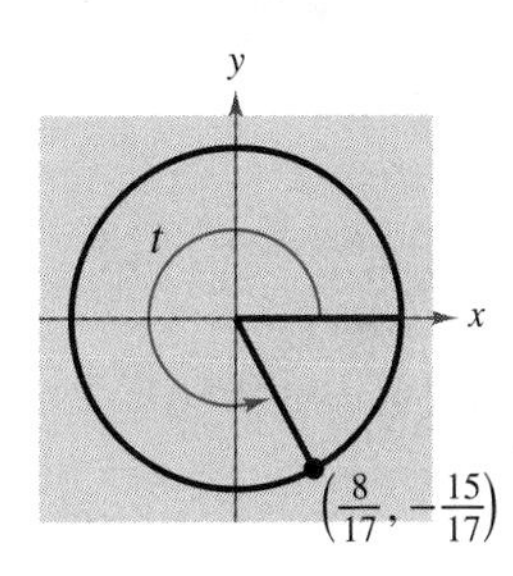

4.
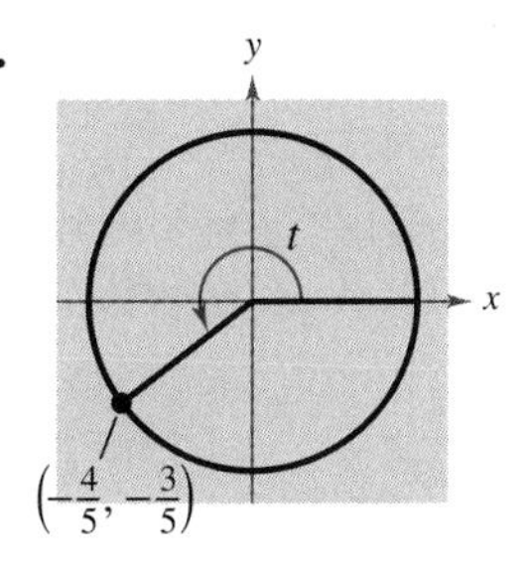

In Exercises 5–12, find the point (x, y) on the unit circle that corresponds to the real number t.

5. $t = \dfrac{\pi}{4}$

6. $t = \dfrac{\pi}{3}$

7. $t = \dfrac{5\pi}{6}$

8. $t = \dfrac{5\pi}{4}$

9. $t = \dfrac{4\pi}{3}$

10. $t = \dfrac{11\pi}{6}$

11. $t = \dfrac{3\pi}{2}$

12. $t = \pi$

In Exercises 13–24, evaluate (if possible) the sine, cosine, and tangent of the real number.

13. $t = \dfrac{\pi}{4}$

14. $t = -\dfrac{\pi}{4}$

15. $t = -\dfrac{\pi}{6}$

16. $t = \dfrac{\pi}{3}$

17. $t = -\frac{5\pi}{4}$

18. $t = -\frac{5\pi}{6}$

19. $t = \frac{11\pi}{6}$

20. $t = \frac{2\pi}{3}$

21. $t = \frac{4\pi}{3}$

22. $t = \frac{7\pi}{4}$

23. $t = -\frac{3\pi}{2}$

24. $t = -2\pi$

In Exercises 25–30, evaluate (if possible) the six trigonometric functions of the real number.

25. $t = \frac{3\pi}{4}$

26. $t = -\frac{2\pi}{3}$

27. $t = \frac{\pi}{2}$

28. $t = \frac{3\pi}{2}$

29. $t = -\frac{4\pi}{3}$

30. $t = -\frac{11\pi}{6}$

In Exercises 31–38, evaluate the trigonometric functions using its period as an aid.

31. $\sin 3\pi$

32. $\cos 3\pi$

33. $\cos \frac{8\pi}{3}$

34. $\sin \frac{9\pi}{4}$

35. $\cos \frac{19\pi}{6}$

36. $\sin\left(-\frac{13\pi}{6}\right)$

37. $\sin\left(-\frac{9\pi}{4}\right)$

38. $\cos\left(-\frac{8\pi}{3}\right)$

In Exercises 39–44, use the value of the trigonometric function to evaluate the indicated functions.

39. $\sin t = \frac{1}{3}$
   (a) $\sin(-t)$
   (b) $\csc(-t)$

40. $\sin(-t) = \frac{2}{5}$
   (a) $\sin t$
   (b) $\csc t$

41. $\cos(-t) = -\frac{7}{8}$
   (a) $\cos t$
   (b) $\sec(-t)$

42. $\cos t = -\frac{3}{4}$
   (a) $\cos(-t)$
   (b) $\sec(-t)$

43. $\sin t = \frac{4}{5}$
   (a) $\sin(\pi - t)$
   (b) $\sin(t + \pi)$

44. $\cos t = \frac{4}{5}$
   (a) $\cos(\pi - t)$
   (b) $\cos(t + \pi)$

In Exercises 45–54, use a calculator to evaluate the expression. Round to four decimal places.

45. $\sin \frac{\pi}{4}$

46. $\tan \pi$

47. $\cos(-3)$

48. $\cot 1$

49. $\cos(-1.7)$

50. $\csc 2.3$

51. $\csc 0.8$

52. $\sec 1.8$

53. $\sec 22.8$

54. $\sin(-0.9)$

In Exercises 55–58, use the figure and a straightedge.

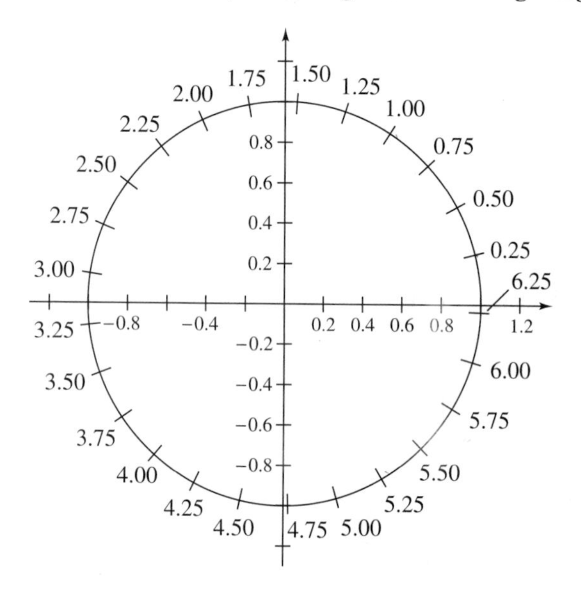

55. Approximate the trigonometric function.
   (a) $\sin 5$
   (b) $\cos 2$

56. Approximate the trigonometric function.
   (a) $\sin 0.75$
   (b) $\cos 2.5$

57. Approximate t where $0 \le t < 2\pi$.
   (a) $\sin t = 0.25$
   (b) $\cos t = -0.25$

58. Approximate t where $0 \le t < 2\pi$.

(a) $\sin t = -0.75$ (b) $\cos t = 0.75$

59. Verify that $\cos 2t \neq 2 \cos t$ by approximating $\cos 1.5$ and $2 \cos 0.75$.

60. Verify that $\sin(t_1 + t_2) \neq \sin t_1 + \sin t_2$ by approximating $\sin 0.25$, $\sin 0.75$, and $\sin 1$.

61. ***Exploration*** Let (x_1, y_1) and (x_2, y_2) be points on the unit circle corresponding to $t = t_1$ and $t = \pi - t_1$, respectively.

(a) Identify the symmetry of the points (x_1, y_1) and (x_2, y_2).

(b) Make a conjecture about any relationship between $\sin t_1$ and $\sin(\pi - t_1)$.

(c) Make a conjecture about any relationship between $\cos t_1$ and $\cos(\pi - t_1)$.

62. ***Exploration*** Let (x_1, y_1) and (x_2, y_2) be points on the unit circle corresponding to $t = t_1$ and $t = t_1 + \pi$, respectively.

(a) Identify the symmetry of the points (x_1, y_1) and (x_2, y_2).

(b) Make a conjecture about any relationship between $\sin t_1$ and $\sin(t_1 + \pi)$.

(c) Make a conjecture about any relationship between $\cos t_1$ and $\cos(t_1 + \pi)$.

63. ***Harmonic Motion*** The displacement from equilibrium of an oscillating weight suspended by a spring is given by

$$y(t) = \frac{1}{4} \cos 6t$$

where y is the displacement in feet and t is the time in seconds. Find the displacement when (a) $t = 0$, (b) $t = \frac{1}{4}$, and (c) $t = \frac{1}{2}$.

64. ***Harmonic Motion*** The displacement from equilibrium of an oscillating weight suspended by a spring and subject to the damping effect of friction is given by

$$y(t) = \frac{1}{4}e^{-t} \cos 6t$$

where y is the displacement in feet and t is the time in seconds. Find the displacement when (a) $t = 0$, (b) $t = \frac{1}{4}$, and (c) $t = \frac{1}{2}$.

65. ***Electrical Circuits*** The initial current and charge in the electrical circuit shown in the figure are zero. When 100 volts is applied to the circuit, the current is given by

$$I = 5e^{-2t} \sin t$$

if the resistance, inductance, and capacitance are 80 ohms, 20 henrys, and 0.01 farad, respectively. Approximate the current when $t = 0.7$ seconds after the voltage is applied.

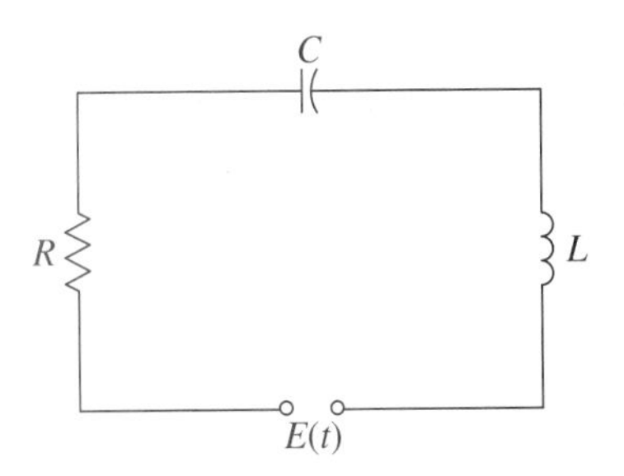

66. Use the unit circle to verify that the cosine and secant functions are even and that the sine, cosecant, tangent, and cotangent functions are odd.

67. ***Think About It*** Because $f(t) = \sin t$ is an odd function and $g(t) = \cos t$ is an even function, what can be said about the function $h(t) = f(t)g(t)$?

68. ***Think About It*** Because $f(t) = \sin t$ and $g(t) = \tan t$ are odd functions, what can be said about the function $h(t) = f(t)g(t)$?

Review **Solve Exercises 69–72 as a review of the skills and problem-solving techniques you learned in previous sections. Find the inverse of the one-to-one function f. Use a graphing utility to graph both f and f^{-1} in the same viewing rectangle.**

69. $f(x) = \frac{1}{2}(3x - 2)$

70. $f(x) = \frac{1}{4}x^3 + 1$

71. $f(x) = \sqrt{x^2 - 4}, \quad x \ge 2$

72. $f(x) = \dfrac{2x}{x + 1}, \quad x > -1$

1.3 Right Triangle Trigonometry

See Exercises 71 and 72 on page 146 for examples of how trigonometry can help find dimensions of machined parts.

The Six Trigonometric Functions ▫ *Trigonometric Identities* ▫ *Evaluating Trigonometric Functions with a Calculator* ▫ *Applications Involving Right Triangles*

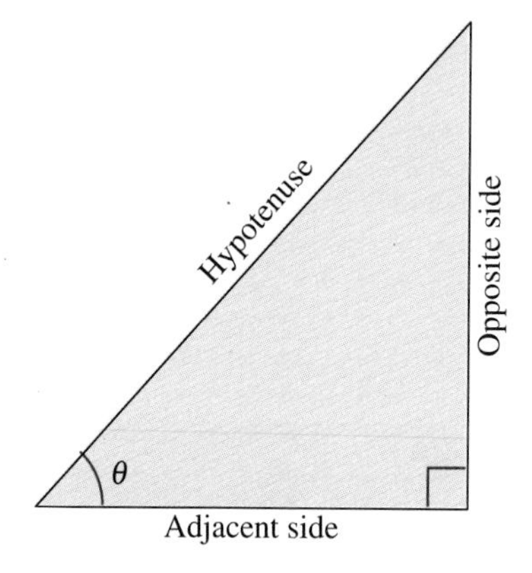

FIGURE 1.21

The Six Trigonometric Functions

Our second look at the trigonometric functions is from a *right triangle* perspective. Consider a right triangle, one of whose acute angles is labeled θ, as shown in Figure 1.21. The three sides of the triangle are the **hypotenuse,** the **opposite side** (the side opposite the angle θ), and the **adjacent side** (the side adjacent to the angle θ). Using the lengths of these three sides, you can form six ratios that define the six trigonometric functions of the acute angle θ.

In the following definition, it is important to see that

$0 < \theta < 90°$ $\qquad$ θ is an acute angle.

and that for such angles the value of each trigonometric function is *positive.*

Right Triangle Definitions of Trigonometric Functions

Let θ be an *acute* angle of a right triangle. The six trigonometric functions *of the angle* θ are defined as follows.

$$\sin\theta = \frac{\text{opp}}{\text{hyp}} \qquad \csc\theta = \frac{\text{hyp}}{\text{opp}}$$

$$\cos\theta = \frac{\text{adj}}{\text{hyp}} \qquad \sec\theta = \frac{\text{hyp}}{\text{adj}}$$

$$\tan\theta = \frac{\text{opp}}{\text{adj}} \qquad \cot\theta = \frac{\text{adj}}{\text{opp}}$$

The abbreviation opp, adj, and hyp represent the lengths of the three sides of a right triangle.

opp = the length of the side *opposite* θ

adj = the length of the side *adjacent* to θ

hyp = the length of the *hypotenuse*

The leading Teutonic mathematical astronomer of the 16th century was Georg Joachim Rhaeticus (1514–1576). He was the first to define the trigonometric functions as ratios of the sides of a right triangle.

NOTE The functions in the second column above are the *reciprocals* of the corresponding functions in the first column. ▪▪

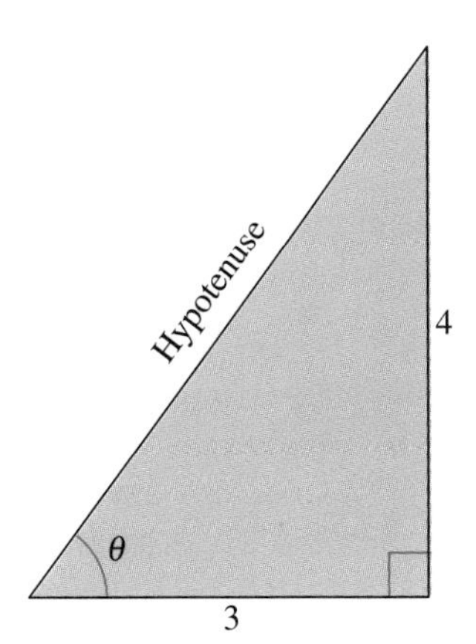

FIGURE 1.22

EXAMPLE 1 *Evaluating Trigonometric Functions*

Find the values of the six trigonometric functions of θ, as shown in Figure 1.22.

Solution

By the Pythagorean Theorem, $(\text{hyp})^2 = (\text{opp})^2 + (\text{adj})^2$, it follows that

$$\text{hyp} = \sqrt{4^2 + 3^2} = \sqrt{25} = 5.$$

Thus, the six trigonometric functions of θ are

$$\sin\theta = \frac{\text{opp}}{\text{hyp}} = \frac{4}{5} \qquad \csc\theta = \frac{\text{hyp}}{\text{opp}} = \frac{5}{4}$$

$$\cos\theta = \frac{\text{adj}}{\text{hyp}} = \frac{3}{5} \qquad \sec\theta = \frac{\text{hyp}}{\text{adj}} = \frac{5}{3}$$

$$\tan\theta = \frac{\text{opp}}{\text{adj}} = \frac{4}{3} \qquad \cot\theta = \frac{\text{adj}}{\text{opp}} = \frac{3}{4}.$$

In Example 1, you were given the lengths of two sides of the right triangle, but not the angle θ. It is more common in trigonometry to be asked to find the trigonometric functions of a *given* acute angle θ. To do this, you can construct a right triangle having θ as one of its angles.

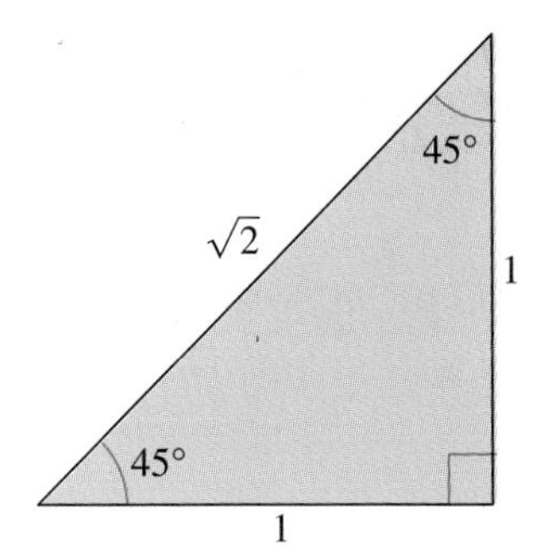

FIGURE 1.23

EXAMPLE 2 *Evaluating Trigonometric Functions of 45°*

Find the values of $\sin 45°$, $\cos 45°$, and $\tan 45°$.

Solution

Construct a right triangle having 45° as one of its acute angles, as shown in Figure 1.23. Choose the length of the adjacent side to be 1. From geometry, you know that the other acute angle is also 45°. Hence, the triangle is isosceles and the length of the opposite side is also 1. Using the Pythagorean Theorem, you find the length of the hypotenuse to be $\sqrt{2}$.

$$\sin 45° = \frac{\text{opp}}{\text{hyp}} = \frac{1}{\sqrt{2}} = \frac{\sqrt{2}}{2}$$

$$\cos 45° = \frac{\text{adj}}{\text{hyp}} = \frac{1}{\sqrt{2}} = \frac{\sqrt{2}}{2}$$

$$\tan 45° = \frac{\text{opp}}{\text{adj}} = \frac{1}{1} = 1$$

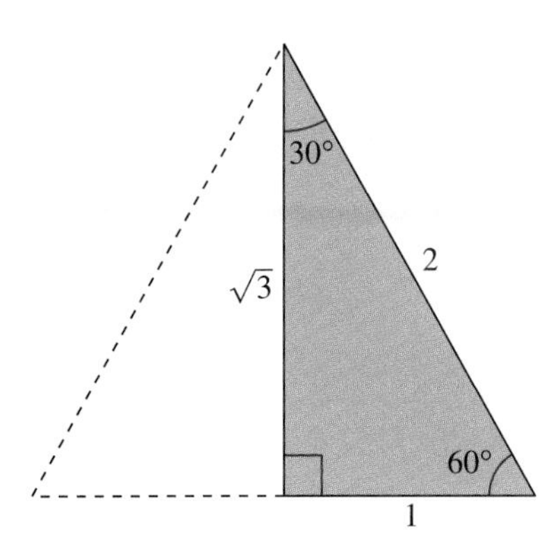

FIGURE 1.24

EXAMPLE 3 Evaluating Trigonometric Functions of 60° and 30°

Use the equilateral triangle shown in Figure 1.24 to find the values of sin 60°, cos 60°, sin 30°, and cos 30°.

Solution

Try using the Pythagorean Theorem and the equilateral triangle to verify the lengths of the sides given in Figure 1.24. For $\theta = 60°$, you have adj $= 1$, opp $= \sqrt{3}$, and hyp $= 2$. Therefore,

$$\sin 60° = \frac{\text{opp}}{\text{hyp}} = \frac{\sqrt{3}}{2} \quad \text{and} \quad \cos 60° = \frac{\text{adj}}{\text{hyp}} = \frac{1}{2}.$$

For $\theta = 30°$, adj $= \sqrt{3}$, opp $= 1$, and hyp $= 2$. Thus,

$$\sin 30° = \frac{\text{opp}}{\text{hyp}} = \frac{1}{2} \quad \text{and} \quad \cos 30° = \frac{\text{adj}}{\text{hyp}} = \frac{\sqrt{3}}{2}.$$

Because the angles 30°, 45°, and 60° ($\pi/6$, $\pi/4$, and $\pi/3$) occur frequently in trigonometry, we suggest that you learn to construct the triangles shown in Figures 1.23 and 1.24.

SINES, COSINES, AND TANGENTS OF SPECIAL ANGLES

$$\sin 30° = \sin\frac{\pi}{6} = \frac{1}{2} \qquad \cos 30° = \cos\frac{\pi}{6} = \frac{\sqrt{3}}{2} \qquad \tan 30° = \tan\frac{\pi}{6} = \frac{\sqrt{3}}{3}$$

$$\sin 45° = \sin\frac{\pi}{4} = \frac{\sqrt{2}}{2} \qquad \cos 45° = \cos\frac{\pi}{4} = \frac{\sqrt{2}}{2} \qquad \tan 45° = \tan\frac{\pi}{4} = 1$$

$$\sin 60° = \sin\frac{\pi}{3} = \frac{\sqrt{3}}{2} \qquad \cos 60° = \cos\frac{\pi}{3} = \frac{1}{2} \qquad \tan 60° = \tan\frac{\pi}{3} = \sqrt{3}$$

In the box, note that $\sin 30° = \frac{1}{2} = \cos 60°$. This occurs because 30° and 60° are complementary angles, and, in general, it can be shown from the right triangle definitions that *cofunctions of complementary angles are equal.* That is, if θ is an acute angle, the following relationships are true.

$$\sin(90° - \theta) = \cos\theta \qquad \cos(90° - \theta) = \sin\theta$$

$$\tan(90° - \theta) = \cot\theta \qquad \cot(90° - \theta) = \tan\theta$$

$$\sec(90° - \theta) = \csc\theta \qquad \csc(90° - \theta) = \sec\theta$$

Trigonometric Identities

In trigonometry, a great deal of time is spent studying relationships between trigonometric functions (identities).

FUNDAMENTAL TRIGONOMETRIC IDENTITIES

Reciprocal Identities

$$\sin\theta = \frac{1}{\csc\theta} \qquad \cos\theta = \frac{1}{\sec\theta} \qquad \tan\theta = \frac{1}{\cot\theta}$$

$$\csc\theta = \frac{1}{\sin\theta} \qquad \sec\theta = \frac{1}{\cos\theta} \qquad \cot\theta = \frac{1}{\tan\theta}$$

Quotient Identities

$$\tan\theta = \frac{\sin\theta}{\cos\theta} \qquad \cot\theta = \frac{\cos\theta}{\sin\theta}$$

Pythagorean Identities

$$\sin^2\theta + \cos^2\theta = 1 \qquad 1 + \tan^2\theta = \sec^2\theta$$

$$1 + \cot^2\theta = \csc^2\theta$$

NOTE Note that $\sin^2\theta$ represents $(\sin\theta)^2$, $\cos^2\theta$ represents $(\cos\theta)^2$, and so on.

EXAMPLE 4 *Applying Trigonometric Identities*

Let θ be an acute angle such that $\sin\theta = 0.6$. Find the values of (a) $\cos\theta$ and (b) $\tan\theta$ using trigonometric identities.

Solution

a. To find the value of $\cos\theta$, use the Pythagorean identity

$$\sin^2\theta + \cos^2\theta = 1.$$

Thus, you have

$$(0.6)^2 + \cos^2\theta = 1$$

$$\cos^2\theta = 1 - (0.6)^2 = 0.64$$

$$\cos\theta = \sqrt{0.64} = 0.8.$$

b. Now, knowing the sine and cosine of θ, you can find the tangent of θ to be

$$\tan\theta = \frac{\sin\theta}{\cos\theta} = \frac{0.6}{0.8} = 0.75.$$

Try using the definitions of $\cos\theta$ and $\tan\theta$, and the triangle shown in Figure 1.25, to check these results.

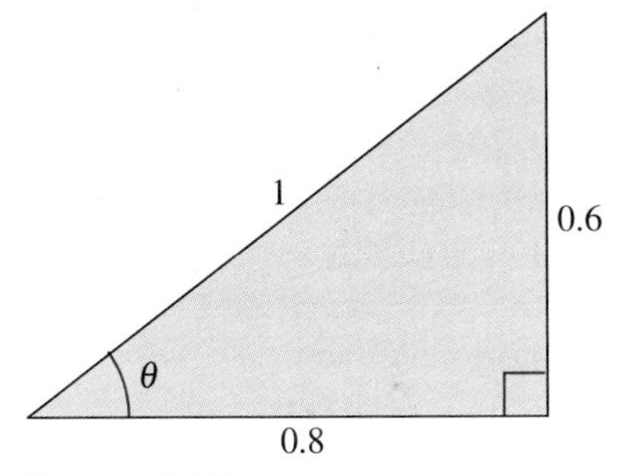

FIGURE 1.25

EXAMPLE 5 *Applying Trigonometric Identities*

Let θ be an acute angle such that $\tan \theta = 3$. Find the values of (a) $\cot \theta$ and (b) $\sec \theta$ using trigonometric identities.

Solution

a. $\cot \theta = \dfrac{1}{\tan \theta}$ Reciprocal identity

$\cot \theta = \dfrac{1}{3}$

b. $\sec^2 \theta = 1 + \tan^2 \theta$ Pythagorean identity

$\sec^2 \theta = 1 + 3^2 = 10$

$\sec \theta = \sqrt{10}$

Try using the definitions of $\cot \theta$ and $\sec \theta$, and the triangle shown in Figure 1.26, to check these results.

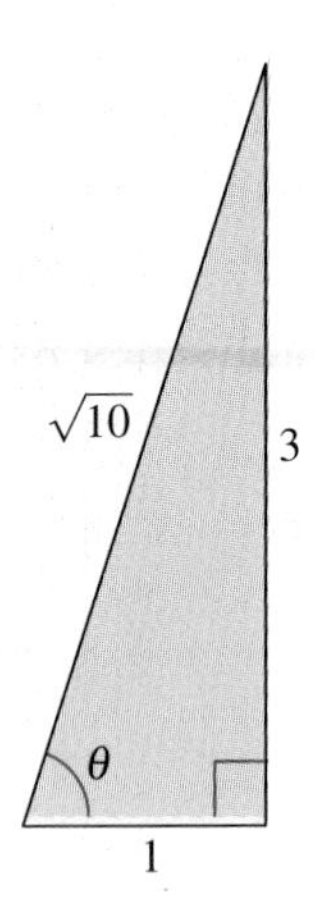

FIGURE 1.26

Evaluating Trigonometric Functions with a Calculator

NOTE Throughout this text, we follow the convention that angles are assumed to be measured in radians unless noted otherwise. For example, sin 1 means the sine of 1 radian and sin 1° means the sine of 1 degree.

To use a calculator to evaluate trigonometric functions of angles measured in degrees, first set the calculator to degree mode and then proceed as demonstrated in Section 1.2. For instance, you can find values of cos 28° and sec 28° as follows.

Function	*Graphing Calculator Keystrokes*	*Display*
cos 28°	[COS] 28 [ENTER]	0.8829476
sec 28°	[(] [COS] 28 [)] [x^{-1}] [ENTER]	1.1325701

EXAMPLE 6 *Using a Calculator*

Use a calculator to evaluate sec(5° 40′ 12″).

Solution

Begin by converting to decimal form.

$$5^\circ\, 40'\, 12'' = 5^\circ + \left(\frac{40}{60}\right)^\circ + \left(\frac{12}{3600}\right)^\circ = 5.67^\circ$$

Then, use a calculator to evaluate sec 5.67°.

$$\sec(5^\circ\, 40'\, 12'') = \sec 5.67^\circ = \frac{1}{\cos 5.67^\circ} \approx 1.00492$$

The *Interactive* CD-ROM offers graphing utility emulators of the *TI-82* and *TI-83*, which can be used with the Examples, Explorations, Technology notes, and Exercises.

Applications Involving Right Triangles

Many applications of trigonometry involve a process called **solving right triangles.** In this type of application, you are usually given one side of a right triangle and one of the acute angles and asked to find one of the other sides, *or* you are given two sides and asked to find one of the acute angles.

EXAMPLE 7 *Solving a Right Triangle*

A surveyor is standing 50 feet from the base of a large tree, as shown in Figure 1.27. The surveyor measures the angle of elevation to the top of the tree as 71.5°. How tall is the tree?

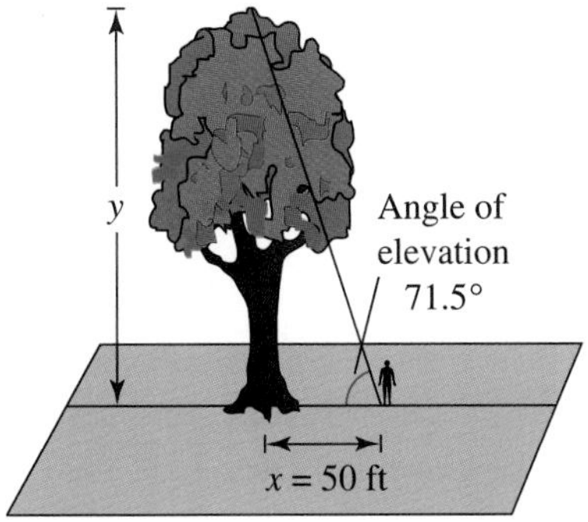

y
Angle of elevation 71.5°
$x = 50$ ft

FIGURE 1.27

Solution

From Figure 1.27, you see that

$$\tan 71.5° = \frac{\text{opp}}{\text{adj}} = \frac{y}{x}$$

where $x = 50$ and y is the height of the tree. Thus, the height of the tree is

$$y = x \tan 71.5° \approx 50(2.98868) \approx 149.4 \text{ feet.}$$

EXAMPLE 8 *Solving a Right Triangle*

A person is 200 yards from a river. Rather than walking directly to the river, the person walks 400 yards along a straight path to the river's edge. Find the acute angle θ between this path and the river's edge, as illustrated in Figure 1.28.

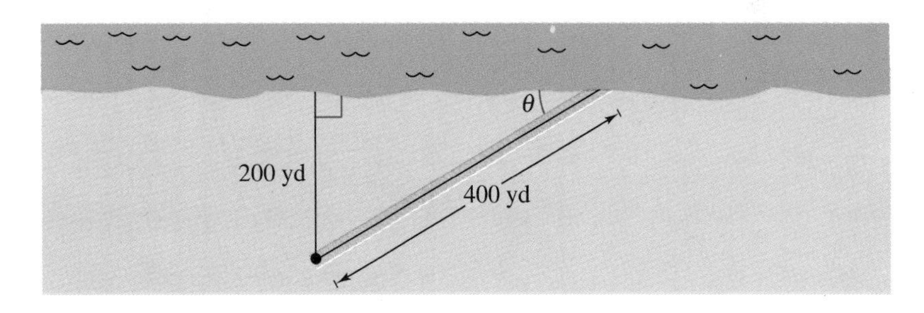

FIGURE 1.28

Solution

From Figure 1.28, you can see that the sine of the angle θ is

$$\sin \theta = \frac{\text{opp}}{\text{hyp}} = \frac{200}{400} = \frac{1}{2}.$$

Now, you recognize that $\theta = 30°$.

Land surveyors use fixed boundaries to find areas of plots of ground. The transit, a small telescope mounted on a tripod, can measure horizontal and vertical angles within small fractions of degrees. *(Photo: "Images © 1995 PhotoDisc, Inc.")*

In Example 8, you were able to recognize that the acute angle that satisfies the equation $\sin \theta = \frac{1}{2}$ is $\theta = 30°$. Suppose, however, that you are given the equation $\sin \theta = 0.6$ and asked to find the acute angle θ. Because

$$\sin 30° = \frac{1}{2} = 0.5000 \qquad \text{and} \qquad \sin 45° = \frac{1}{\sqrt{2}} \approx 0.7071$$

you might guess that θ lies somewhere between 30° and 45°. A more precise value of θ can be found using the *inverse* key on a calculator. To do this, you can use the following keystroke sequence in degree mode.

[SIN⁻¹] .6 [ENTER] Display 36.8699

Thus, you can conclude that if $\sin \theta = 0.6$, $\theta \approx 36.87°$.

EXAMPLE 9 *Solving a Right Triangle*

Real Life

A 12-meter flagpole casts a 9-meter shadow, as shown in Figure 1.29. Find θ, the angle of elevation of the sun.

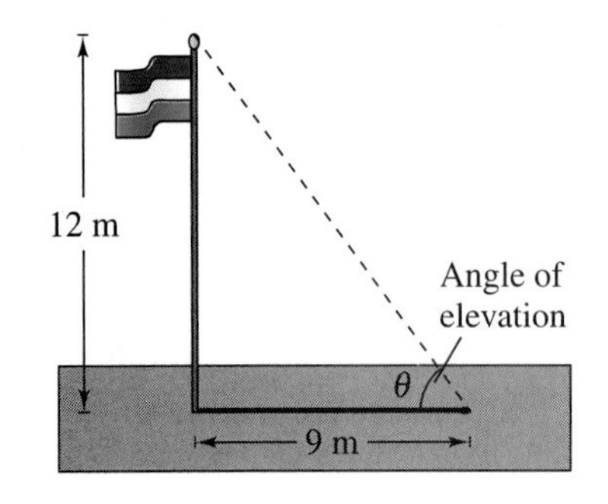

FIGURE 1.29

Solution

Figure 1.29 shows that the *opposite* and *adjacent* sides are known. Thus,

$$\tan \theta = \frac{\text{opp}}{\text{adj}} = \frac{12}{9}.$$

With a calculator in degree mode, you use the keystrokes

[TAN⁻¹] [(] 12 [÷] 9 [)] [ENTER]

to obtain $\theta \approx 53.13°$.

GROUP ACTIVITY

EVALUATING TRIGONOMETRIC FUNCTIONS

Some functions such as $f(x) = 5x$, have the property that

$$f(cx) = cf(x).$$

Do any of the six trigonometric functions have this property? Compare the following values and use the results to justify your answer.

a. $\sin 60°$ and $2 \sin 30°$

b. $\cos 60°$ and $2 \cos 30°$

c. $\tan 60°$ and $2 \tan 30°$

d. $\cot 60°$ and $2 \cot 30°$

e. $\sec 60°$ and $2 \sec 30°$

f. $\csc 60°$ and $2 \csc 30°$

WARM UP

Find the distance between the points.

1. (3, 8), (1, 4)
2. (5, 2), (2, −7)
3. (−4, 0), (2, 8)
4. (−3, −3), (0, 0)

Perform the operations. (Round your answer to two decimal places.)

5. 0.300×4.125
6. 7.30×43.50
7. $\dfrac{151.5}{2.40}$
8. $\dfrac{3740}{28.0}$
9. $\dfrac{19{,}500}{0.007}$
10. $\dfrac{(10.5)(3401)}{1240}$

1.3 Exercises

In Exercises 1–4, find the exact values of the six trigonometric functions of the angle θ given in the figure. (Use the Pythagorean Theorem to find the third side of the triangle.)

1.

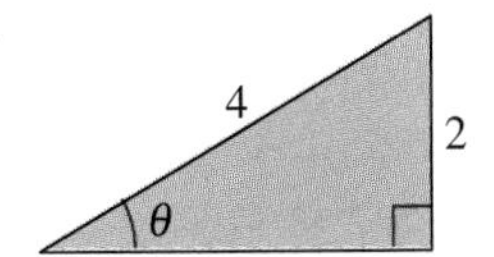

2.

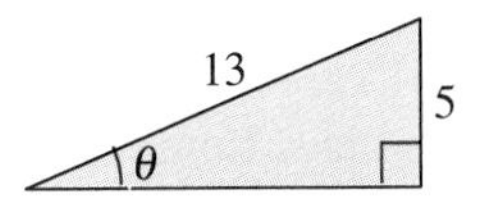

3.

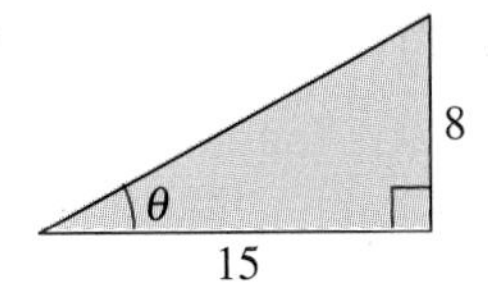

4.

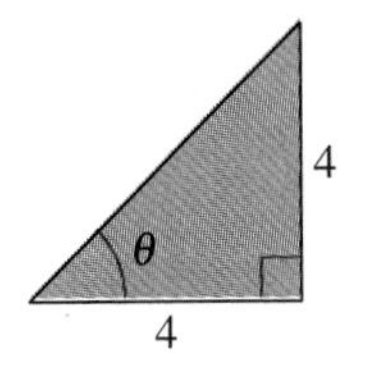

In Exercises 5–8, find the exact values of the six trigonometric functions of the angle θ for each of the triangles. Explain why the function values are the same.

5.

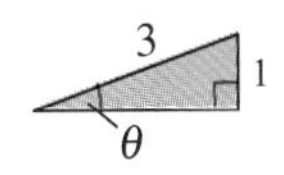

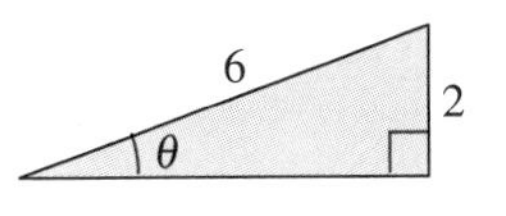

6.

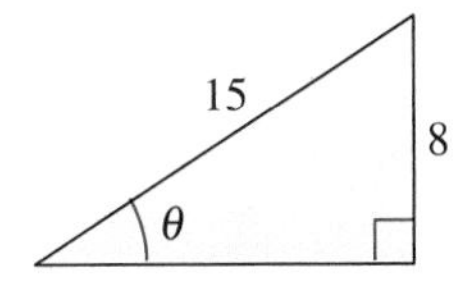

7.5 4 θ

7.

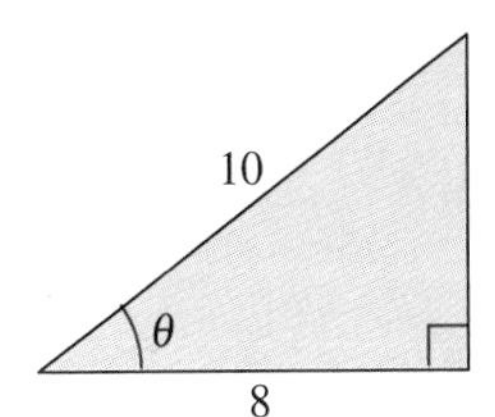

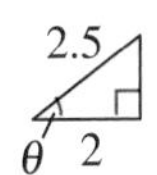

8.

θ 1 2

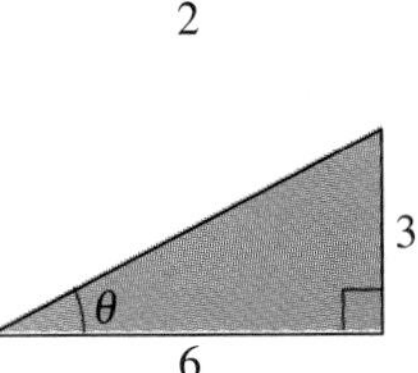

In Exercises 9–16, sketch a right triangle corresponding to the trigonometric function of the acute angle θ. Use the Pythagorean Theorem to determine the third side, and then find the other five trigonometric functions of θ.

9. $\sin\theta = \frac{2}{3}$ **10.** $\cot\theta = 5$

11. $\sec\theta = 2$ **12.** $\cos\theta = \frac{5}{7}$

13. $\tan\theta = 3$ **14.** $\csc\theta = \frac{17}{4}$

15. $\cot\theta = \frac{3}{2}$ **16.** $\sin\theta = \frac{3}{8}$

In Exercises 17–22, use the given function values and the appropriate trigonometric identities (including the relationship between a trigonometric function and its cofunction of a complementary angle) to find the indicated trigonometric functions.

17. $\sin 60° = \dfrac{\sqrt{3}}{2}, \quad \cos 60° = \dfrac{1}{2}$

(a) $\tan 60°$ (b) $\sin 30°$

(c) $\cos 30°$ (d) $\cot 60°$

18. $\sin 30° = \dfrac{1}{2}, \quad \tan 30° = \dfrac{\sqrt{3}}{3}$

(a) $\csc 30°$ (b) $\cot 60°$

(c) $\cos 30°$ (d) $\cot 30°$

19. $\csc\theta = 3, \quad \sec\theta = \dfrac{3\sqrt{2}}{4}$

(a) $\sin\theta$ (b) $\cos\theta$

(c) $\tan\theta$ (d) $\sec(90° - \theta)$

20. $\sec\theta = 5, \quad \tan\theta = 2\sqrt{6}$

(a) $\cos\theta$ (b) $\cot\theta$

(c) $\cot(90° - \theta)$ (d) $\sin\theta$

21. $\cos\alpha = \frac{1}{4}$

(a) $\sec\alpha$ (b) $\sin\alpha$

(c) $\cot\alpha$ (d) $\sin(90° - \alpha)$

22. $\tan\beta = 5$

(a) $\cot\beta$ (b) $\cos\beta$

(c) $\tan(90° - \beta)$ (d) $\csc\beta$

In Exercises 23–32, use trigonometric identities to transform one side of the equation into the other.

23. $\tan\theta\cot\theta = 1$ **24.** $\cos\theta\sec\theta = 1$

25. $\tan\alpha\cos\alpha = \sin\alpha$ **26.** $\cot\alpha\sin\alpha = \cos\alpha$

27. $(1 + \cos\theta)(1 - \cos\theta) = \sin^2\theta$

28. $(1 + \sin\theta)(1 - \sin\theta) = \cos^2\theta$

29. $(\sec\theta + \tan\theta)(\sec\theta - \tan\theta) = 1$

30. $\sin^2\theta - \cos^2\theta = 2\sin^2\theta - 1$

31. $\dfrac{\sin\theta}{\cos\theta} + \dfrac{\cos\theta}{\sin\theta} = \csc\theta\sec\theta$

32. $\dfrac{\tan\beta + \cot\beta}{\tan\beta} = \csc^2\beta$

In Exercises 33–36, evaluate the trigonometric function by memory or by constructing an appropriate triangle for the given special angle.

33. (a) $\cos 60°$ (b) $\tan\dfrac{\pi}{6}$

34. (a) $\csc 30°$ (b) $\sin\dfrac{\pi}{4}$

35. (a) $\cot 45°$ (b) $\cos 45°$

36. (a) $\sin\dfrac{\pi}{3}$ (b) $\csc 45°$

In Exercises 37–46, use a calculator to evaluate each function. Round your answers to four decimal places. (Be sure the calculator is in the correct mode.)

37. (a) $\sin 10°$ (b) $\cos 80°$

38. (a) $\tan 23.5°$ (b) $\cot 66.5°$

39. (a) $\sin 16.35°$ (b) $\csc 16.35°$

40. (a) $\cos 16°\,18'$ (b) $\sin 73°\,56'$

41. (a) $\sec 42°\,12'$ (b) $\csc 48°\,7'$

42. (a) $\cos 4°\,50'\,15''$ (b) $\sec 4°\,50'\,15''$

43. (a) $\cot\dfrac{\pi}{16}$ (b) $\tan\dfrac{\pi}{16}$

44. (a) $\sec 0.75$ (b) $\cos 0.75$

45. (a) $\csc 1$ (b) $\tan\frac{1}{2}$

46. (a) $\sec\left(\dfrac{\pi}{2} - 1\right)$ (b) $\cot\left(\dfrac{\pi}{2} - \dfrac{1}{2}\right)$

In Exercises 47–52, find the values of θ in degrees ($0° < \theta < 90°$) and radians ($0 < \theta < \pi/2$) without the aid of a calculator.

47. (a) $\sin \theta = \frac{1}{2}$ (b) $\csc \theta = 2$

48. (a) $\cos \theta = \frac{\sqrt{2}}{2}$ (b) $\tan \theta = 1$

49. (a) $\sec \theta = 2$ (b) $\cot \theta = 1$

50. (a) $\tan \theta = \sqrt{3}$ (b) $\cos \theta = \frac{1}{2}$

51. (a) $\csc \theta = \frac{2\sqrt{3}}{3}$ (b) $\sin \theta = \frac{\sqrt{2}}{2}$

52. (a) $\cot \theta = \frac{\sqrt{3}}{3}$ (b) $\sec \theta = \sqrt{2}$

In Exercises 53–56, find the values of θ in degrees ($0° < \theta < 90°$) and radians ($0 < \theta < \pi/2$) by using a calculator.

53. (a) $\sin \theta = 0.8191$ (b) $\cos \theta = 0.0175$

54. (a) $\cos \theta = 0.9848$ (b) $\cos \theta = 0.8746$

55. (a) $\tan \theta = 1.1920$ (b) $\tan \theta = 0.4663$

56. (a) $\sin \theta = 0.3746$ (b) $\cos \theta = 0.3746$

In Exercises 57–64, solve for *x, y,* or *r,* as indicated.

57. Solve for y.

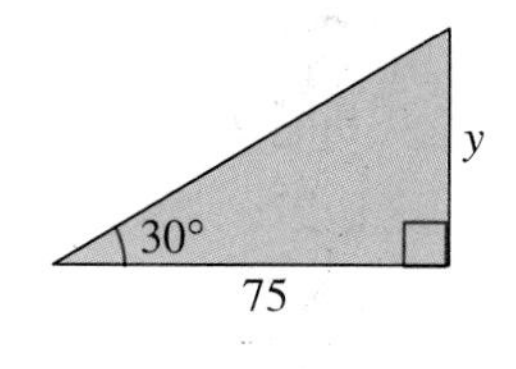

58. Solve for x.

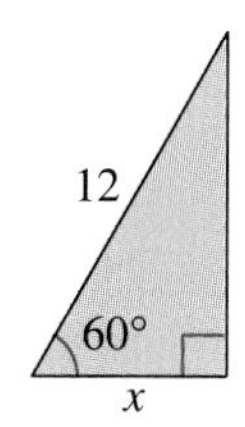

59. Solve for x.

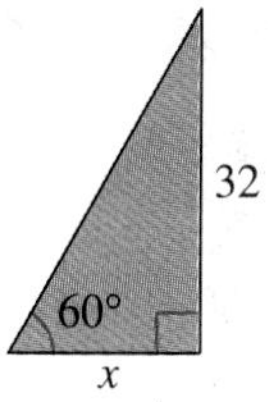

32
60°
x

60. Solve for r.

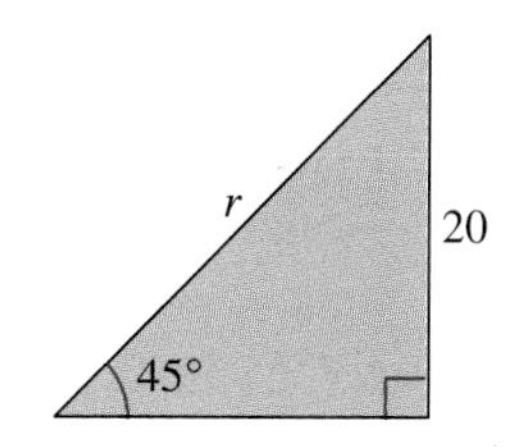

61. Solve for r.

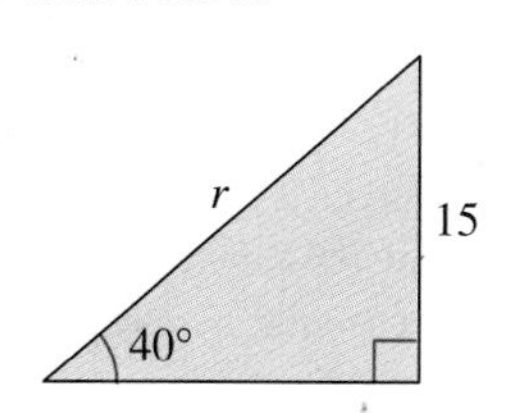

62. Solve for x.

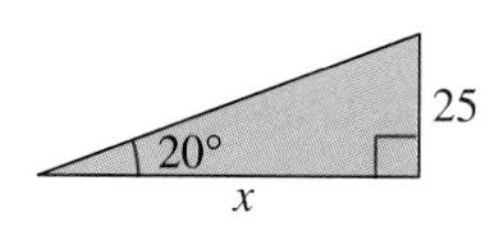

63. Solve for y.

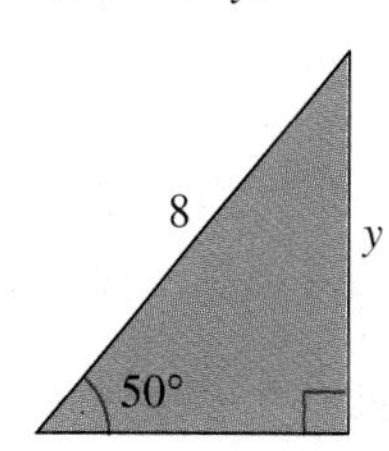

64. Solve for r.

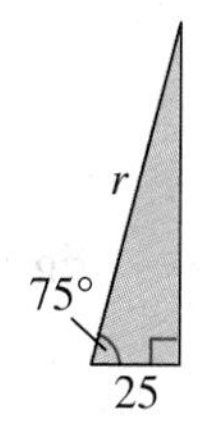

65. ***Height*** A 6-foot person standing 15 feet from a streetlight casts an 8-foot shadow (see figure). What is the height of the streetlight?

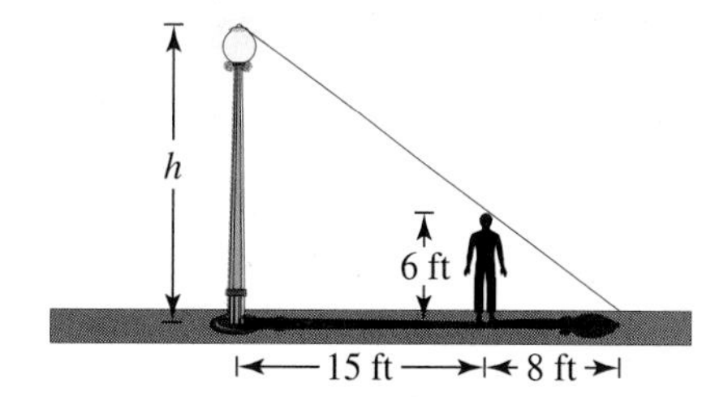

66. ***Height*** A 6-foot person walks from the base of a broadcasting tower directly toward the tip of the shadow cast by the tower. When the person is 132 feet from the tower and 3 feet from the tip of the tower's shadow, the person's shadow starts to appear beyond the tower's shadow.

(a) Draw a right triangle that gives a visual representation of the problem. Show the known quantities on the triangle and use a variable to indicate the height of the tower.

(b) Write an equation involving the unknown.

(c) What is the height of the tower?

67. *Length* A 30-meter line is used to tether a helium-filled balloon. Because of a breeze, the line makes an angle of approximately 75° with the ground.

(a) Draw a right triangle that gives a visual representation of the problem. Show the known quantities on the triangle and use a variable to indicate the height of the balloon.

(b) Write an equation involving the unknown.

(c) What is the height of the balloon?

68. *Width of a River* A biologist wants to know the width w of a river in order to properly set instruments for studying the pollutants in the water. From point A, the biologist walks downstream 100 feet and sights to point C (see figure). From this sighting, it is determined that $\theta = 54°$. How wide is the river?

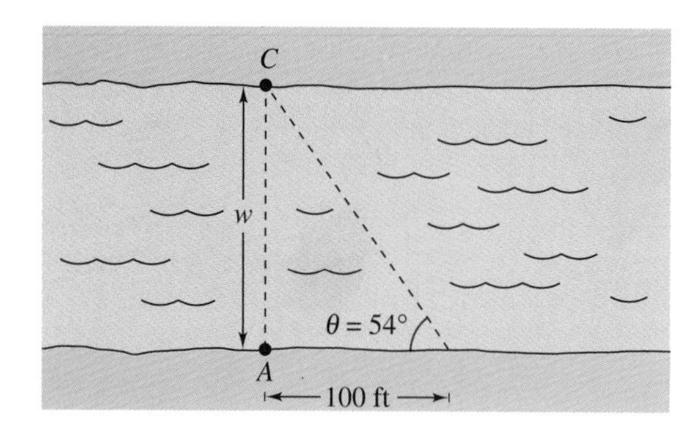

69. *Distance* From a 60-foot observation tower on the coast, a Coast Guard officer sights a boat in difficulty. The angle of depression of the boat is 3° (see figure). How far is the boat from the shoreline?

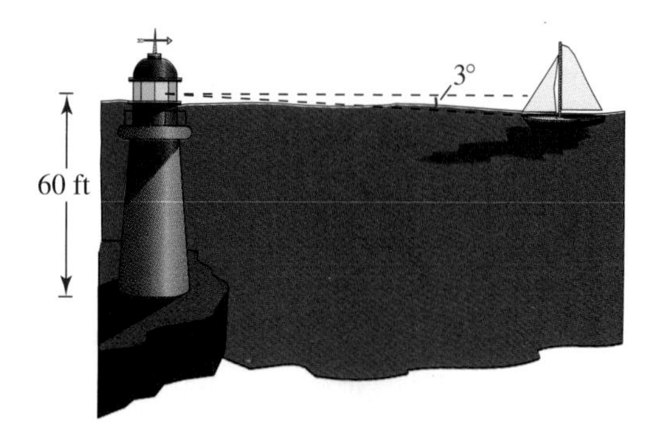

70. *Angle of Elevation* A ramp 20 feet in length rises to a loading platform that is $3\frac{1}{3}$ feet off the ground.

(a) Draw a right triangle that gives a visual representation of the problem. Show the known quantities on the triangle and use a variable to indicate the angle of elevation of the ramp.

(b) Use a trigonometric function to write an equation involving the unknown.

(c) What is the angle of elevation of the ramp?

71. *Machine Shop Calculations* A steel plate has the form of a quarter of a circle with a radius of 60 centimeters. Two 2-centimeter holes are to be drilled in the plate positioned as shown in the figure. Find the coordinates of the center of each hole.

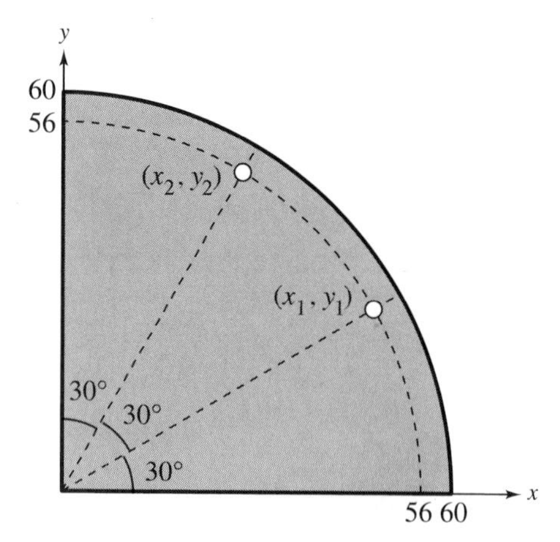

72. *Machine Shop Calculations* A tapered shaft has a diameter of 5 centimeters at the small end and is 15 centimeters long (see figure). If the taper is 3°, find the diameter d of the large end of the shaft.

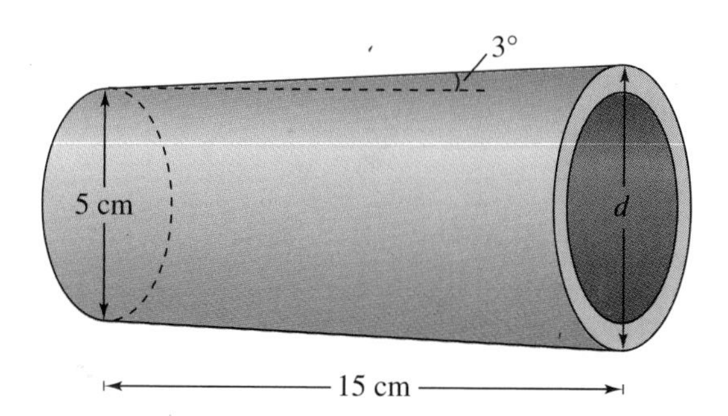

73. ***Geometry*** Use a compass to sketch a quarter of a circle of radius 10 centimeters. Using a protractor, construct an angle of 20° in standard position (see figure). Drop a perpendicular from the point of intersection of the terminal side of the angle and the arc of the circle. By actual measurement, calculate the coordinates (x, y) of the point of intersection and use these measurements to approximate the six trigonometric functions of a 20° angle.

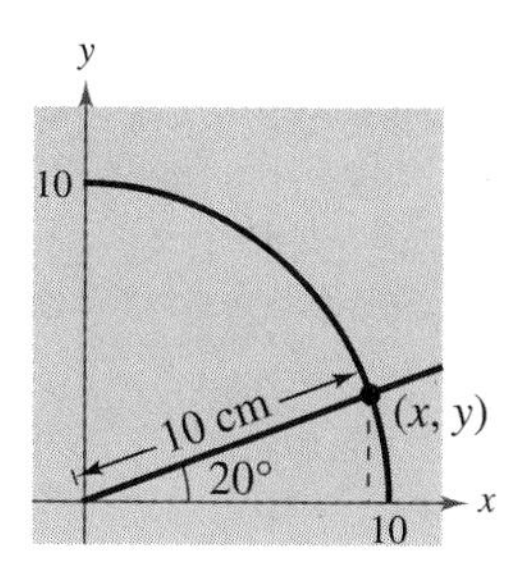

74. ***Geometry*** Repeat Exercise 73 using a 75° angle.

75. ***Exploration***

(a) Complete the table.

θ	0	0.1	0.2	0.3	0.4	0.5
$\sin \theta$						

(b) Is θ or $\sin \theta$ greater for θ in the interval $(0, 0.5]$?

(c) As θ approaches 0, how do θ and $\sin \theta$ compare? Explain.

76. ***Exploration***

(a) Complete the table.

θ	0	0.3	0.6	0.9	1.2	1.5
$\sin \theta$						
$\cos \theta$						

(b) Discuss the behavior of the sine function for θ in the interval $[0, 1.5]$.

(c) Discuss the behavior of the cosine function for θ in the interval $[0, 1.5]$.

(d) Use the definitions of the sine and cosine functions to explain the results of parts (b) and (c).

In Exercises 77–82, determine whether the statement is true or false, and give a reason for your answer.

77. $\sin 60° \csc 60° = 1$

78. $\sec 30° = \csc 60°$

79. $\sin 45° + \cos 45° = 1$

80. $\cot^2 10° - \csc^2 10° = -1$

81. $\dfrac{\sin 60°}{\sin 30°} = \sin 2°$

82. $\tan[(0.8)^2] = \tan^2(0.8)$

Review **Solve Exercises 83–86 as a review of the skills and problem-solving techniques you learned in previous sections. Perform the operations and simplify.**

83. $\dfrac{x^2 - 6x}{x^2 + 4x - 12} \cdot \dfrac{x^2 + 12x + 36}{x^2 - 36}$

84. $\dfrac{2t^2 + 5t - 12}{9 - 4t^2} \div \dfrac{t^2 - 16}{4t^2 + 12t + 9}$

85. $\dfrac{3}{x + 2} - \dfrac{2}{x - 2} + \dfrac{x}{x^2 + 4x + 4}$

86. $\dfrac{\left(\dfrac{3}{x} - \dfrac{1}{4}\right)}{\left(\dfrac{12}{x} - 1\right)}$

1.4 Trigonometric Functions of Any Angle

See Exercise 81 on page 157 for an example of how a trigonometric function can be used to model average daily temperatures.

Introduction ▫ *Reference Angles* ▫ *Trigonometric Functions of Real Numbers*

Introduction

In Section 1.3, the definitions of trigonometric functions were restricted to *acute* angles. In this section, the definitions are extended to cover *any* angle.

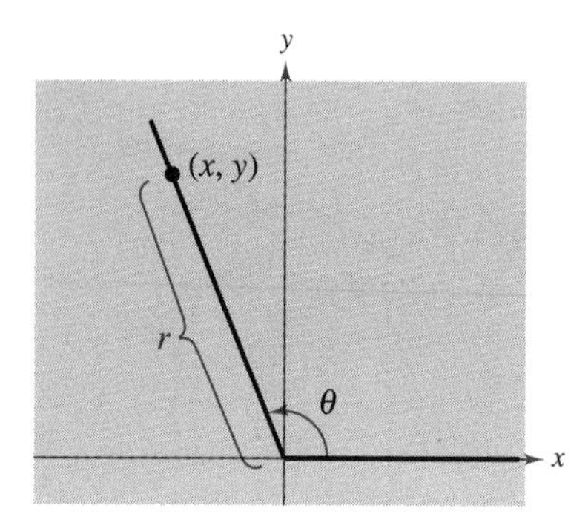

FIGURE 1.30

DEFINITIONS OF TRIGONOMETRIC FUNCTIONS OF ANY ANGLE

Let θ be an angle in standard position with (x, y) a point on the terminal side of θ and $r = \sqrt{x^2 + y^2} \neq 0$, as shown in Figure 1.30.

$$\sin \theta = \frac{y}{r} \qquad \cos \theta = \frac{x}{r}$$

$$\tan \theta = \frac{y}{x}, \quad x \neq 0 \qquad \cot \theta = \frac{x}{y}, \quad y \neq 0$$

$$\sec \theta = \frac{r}{x}, \quad x \neq 0 \qquad \csc \theta = \frac{r}{y}, \quad y \neq 0$$

NOTE If θ is an *acute* angle, these definitions coincide with those given in Section 1.2.

Because $r = \sqrt{x^2 + y^2}$ *cannot* be zero, it follows that the sine and cosine functions are defined for any real value of θ. However, if $x = 0$, the tangent and secant of θ are undefined. For example, the tangent of 90° is undefined. Similarly, if $y = 0$, the cotangent and cosecant of θ are undefined.

EXAMPLE 1 Evaluating Trigonometric Functions

Let $(-3, 4)$ be a point on the terminal side of θ. Find the sine, cosine, and tangent of θ.

Solution

Referring to Figure 1.31, you see that $x = -3$, $y = 4$, and

$$r = \sqrt{x^2 + y^2} = \sqrt{(-3)^2 + 4^2} = \sqrt{25} = 5.$$

Thus, you have the following.

$$\sin \theta = \frac{y}{r} = \frac{4}{5}, \qquad \cos \theta = \frac{x}{r} = -\frac{3}{5}, \qquad \tan \theta = \frac{y}{x} = -\frac{4}{3}$$

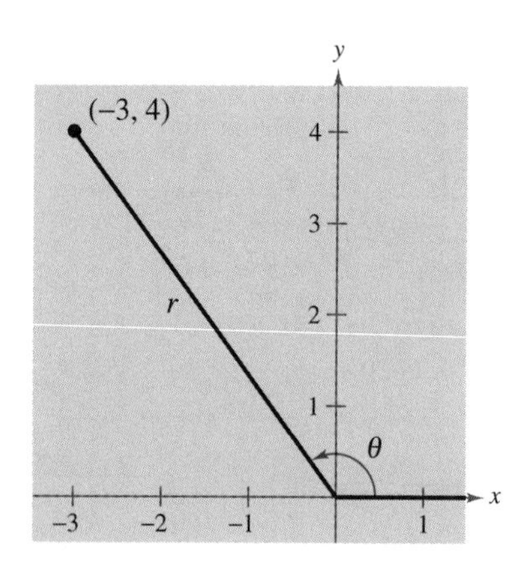

FIGURE 1.31

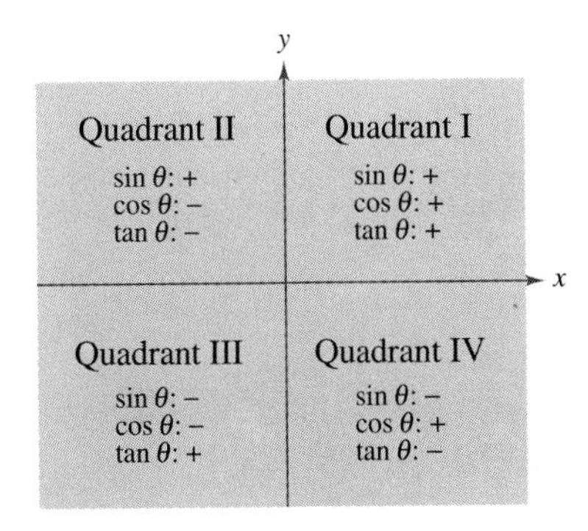

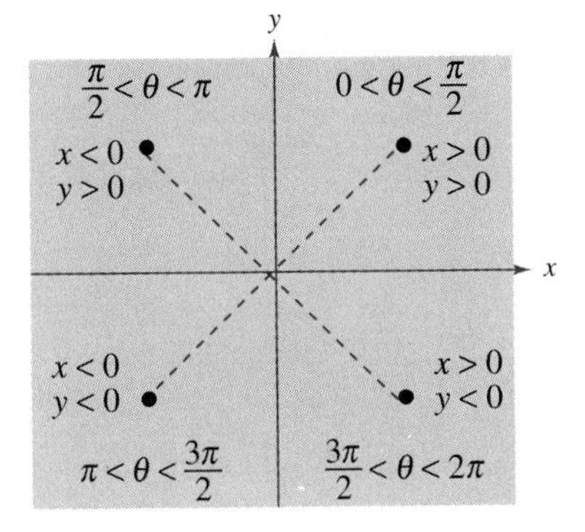

FIGURE 1.32

The *signs* of the trigonometric functions in the four quadrants can be determined easily from the definitions of the functions. For instance, because $\cos\theta = x/r$, it follows that $\cos\theta$ is positive wherever $x > 0$, which occurs in Quadrants I and IV. (Remember, r is always positive.) In a similar manner, you can verify the results shown in Figure 1.32.

EXAMPLE 2 *Evaluating Trigonometric Functions*

Given $\tan\theta = -\frac{5}{4}$ and $\cos\theta > 0$, find $\sin\theta$ and $\sec\theta$.

Solution

Note that θ lies in Quadrant IV because that is the only quadrant in which the tangent is negative and the cosine is positive. Moreover, using

$$\tan\theta = \frac{y}{x} = -\frac{5}{4}$$

and the fact that y is negative in Quadrant IV, you can let $y = -5$ and $x = 4$. Hence, $r = \sqrt{16 + 25} = \sqrt{41}$, and you have the following.

$$\sin\theta = \frac{y}{r} = \frac{-5}{\sqrt{41}} \approx -0.7809$$

$$\sec\theta = \frac{r}{x} = \frac{\sqrt{41}}{4} \approx 1.6008$$

EXAMPLE 3 *Trigonometric Functions of Quadrant Angles*

Evaluate the sine function at the four quadrant angles 0, $\frac{\pi}{2}$, π, and $\frac{3\pi}{2}$.

Solution

To begin, choose a point on the terminal side of each angle, as shown in Figure 1.33. For each of the four given points, $r = 1$, and you have

$$\sin 0 = \frac{y}{r} = \frac{0}{1} = 0 \qquad (x, y) = (1, 0)$$

$$\sin\frac{\pi}{2} = \frac{y}{r} = \frac{1}{1} = 1 \qquad (x, y) = (0, 1)$$

$$\sin\pi = \frac{y}{r} = \frac{0}{1} = 0 \qquad (x, y) = (-1, 0)$$

$$\sin\frac{3\pi}{2} = \frac{y}{r} = \frac{-1}{1} = -1. \qquad (x, y) = (0, -1)$$

Try using Figure 1.33 to evaluate some of the other trigonometric functions at the four quadrant angles.

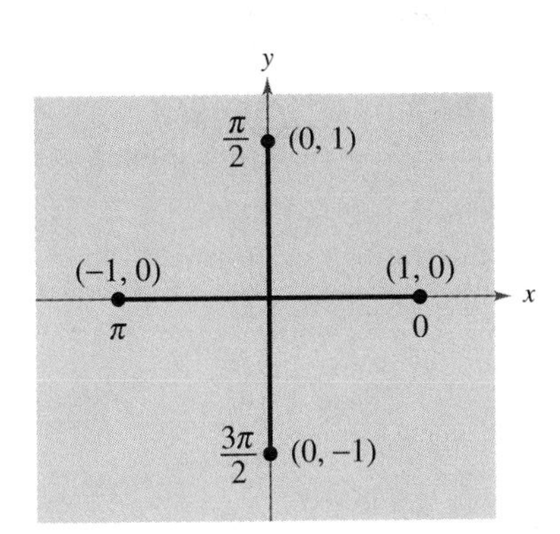

FIGURE 1.33

Reference Angles

The values of the trigonometric functions of angles greater than 90° (or less than 0°) can be determined from their values at corresponding acute angles called **reference angles.**

> **DEFINITION OF REFERENCE ANGLE**
>
> Let θ be an angle in standard position. Its **reference angle** is the acute angle θ' formed by the terminal side of θ and the horizontal axis.

Figure 1.34 shows the reference angle for θ in Quadrants II, III, and IV.

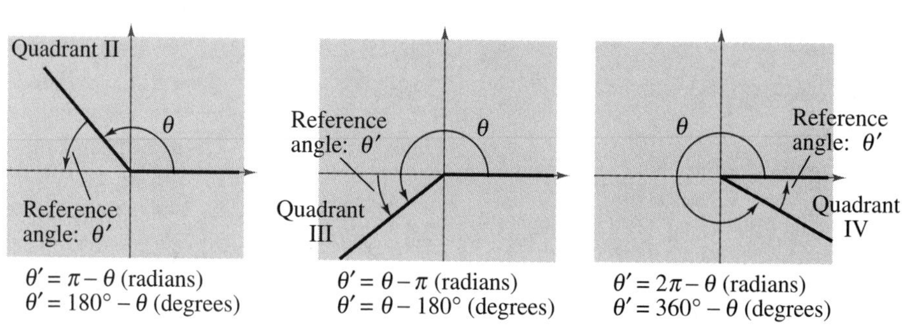

FIGURE 1.34

EXAMPLE 4 *Finding Reference Angles*

Find the reference angle θ'.

a. $\theta = 300°$ **b.** $\theta = 2.3$ **c.** $\theta = -135°$

Solution

a. Because 300° lies in Quadrant IV, the angle it makes with the x-axis is

$$\theta' = 360° - 300° = 60°. \qquad \text{Degrees}$$

b. Because 2.3 lies between $\pi/2 \approx 1.5708$ and $\pi \approx 3.1416$, it follows that it is in Quadrant II and its reference angle is

$$\theta' = \pi - 2.3 \approx 0.8416. \qquad \text{Radians}$$

c. First, determine that $-135°$ is coterminal with 225°, which lies in Quadrant III. Hence, the reference angle is

$$\theta' = 225° - 180° = 45°. \qquad \text{Degrees}$$

Figure 1.35 shows each angle θ and its reference angle θ'.

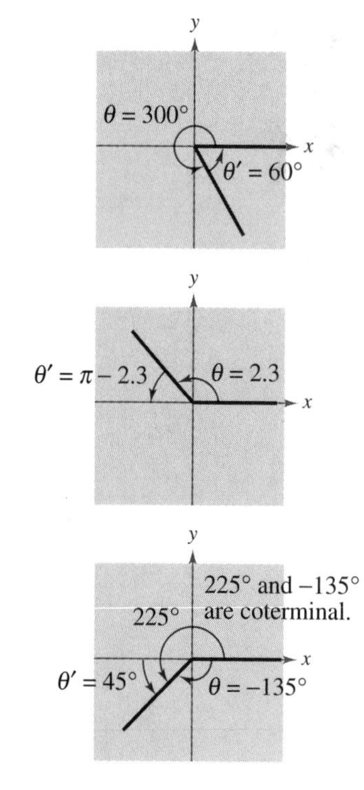

FIGURE 1.35

Trigonometric Functions of Real Numbers

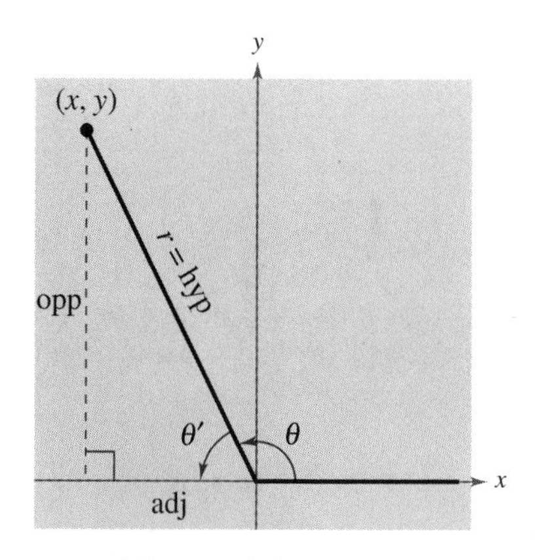

opp = $|y|$, adj = $|x|$

FIGURE 1.36

To see how a reference angle is used to evaluate a trigonometric function, consider the point (x, y) on the terminal side of θ, as shown in Figure 1.36. By definition, you know that

$$\sin \theta = \frac{y}{r} \quad \text{and} \quad \tan \theta = \frac{y}{x}.$$

For the right triangle with acute angle θ' and sides of lengths $|x|$ and $|y|$, you have

$$\sin \theta' = \frac{\text{opp}}{\text{hyp}} = \frac{|y|}{r} \quad \text{and} \quad \tan \theta' = \frac{\text{opp}}{\text{adj}} = \frac{|y|}{|x|}.$$

Thus, it follows that $\sin \theta$ and $\sin \theta'$ are equal, *except possibly in sign.* The same is true for $\tan \theta$ and $\tan \theta'$ *and* for the other four trigonometric functions. In all cases, the sign of the function value can be determined by the quadrant in which θ lies.

Evaluating Trigonometric Functions of Any Angle

To find the value of a trigonometric function of any angle θ:

1. Determine the function value for the associated reference angle θ'.
2. Depending on the quadrant in which θ lies, prefix the appropriate sign to the function value.

By using reference angles and the special angles discussed in the previous section, you can greatly extend the scope of *exact* trigonometric values. For instance, knowing the function values of 30° means that you know the function values of all angles for which 30° is a reference angle. For convenience, the table below gives the exact values of the trigonometric functions of special angles and quadrant angles.

Trigonometric Values of Common Angles

θ (degrees)	0°	30°	45°	60°	90°	180°	270°
θ (radians)	0	$\frac{\pi}{6}$	$\frac{\pi}{4}$	$\frac{\pi}{3}$	$\frac{\pi}{2}$	π	$\frac{3\pi}{2}$
$\sin\theta$	0	$\frac{1}{2}$	$\frac{\sqrt{2}}{2}$	$\frac{\sqrt{3}}{2}$	1	0	-1
$\cos\theta$	1	$\frac{\sqrt{3}}{2}$	$\frac{\sqrt{2}}{2}$	$\frac{1}{2}$	0	-1	0
$\tan\theta$	0	$\frac{\sqrt{3}}{3}$	1	$\sqrt{3}$	Undef.	0	Undef.

EXAMPLE 5 *Trigonometric Functions of Nonacute Angles*

Evaluate the following.

a. $\cos \dfrac{4\pi}{3}$ **b.** $\tan(-210°)$ **c.** $\csc \dfrac{11\pi}{4}$

Solution

a. Because $\theta = 4\pi/3$ lies in Quadrant III, the reference angle is $\theta' = (4\pi/3) - \pi = \pi/3$, as shown in Figure 1.37(a). Because the cosine is negative in Quadrant III, you have

$$\cos \frac{4\pi}{3} = (-) \cos \frac{\pi}{3} = -\frac{1}{2}.$$

b. Because $-210° + 360° = 150°$, it follows that $-210°$ is coterminal with the second-quadrant angle $150°$. Therefore, the reference angle is $\theta' = 180° - 150° = 30°$, as shown in Figure 1.37(b). Finally, because the tangent is negative in Quadrant II, you have

$$\tan(-210°) = (-) \tan 30° = -\frac{\sqrt{3}}{3}.$$

c. Because $(11\pi/4) - 2\pi = 3\pi/4$, it follows that $11\pi/4$ is coterminal with the second-quadrant angle $3\pi/4$. Therefore, the reference angle is $\theta' = \pi - (3\pi/4) = \pi/4$, as shown in Figure 1.37(c). Because the cosecant is positive in Quadrant II, you have

$$\csc \frac{11\pi}{4} = (+) \csc \frac{\pi}{4} = \frac{1}{\sin(\pi/4)} = \sqrt{2}.$$

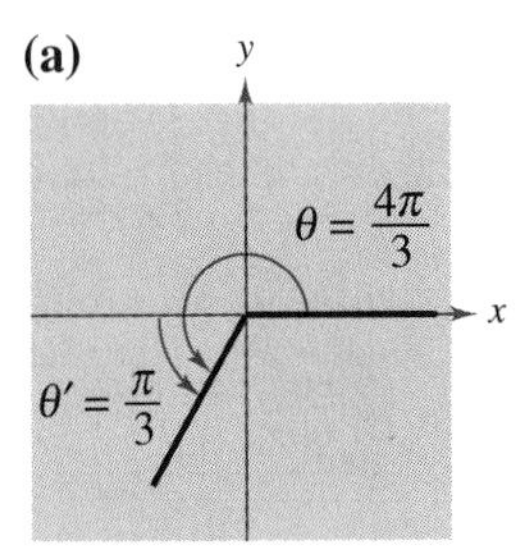

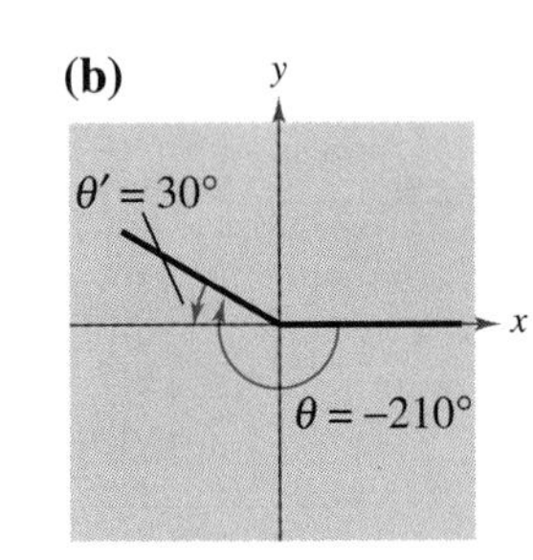

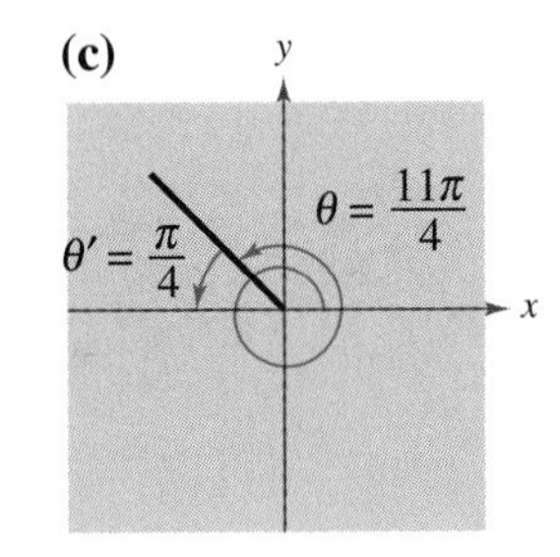

FIGURE 1.37

The fundamental trigonometric identities listed in the previous section (for an acute angle θ) are also valid when θ is any angle in the domain of the function.

EXAMPLE 6 *Using Trigonometric Identities*

Let θ be an angle in Quadrant II such that $\sin\theta = \frac{1}{3}$. Find (a) $\cos\theta$ and (b) $\tan\theta$ by using trigonometric identities.

Solution

a. Using the Pythagorean Identity $\sin^2\theta + \cos^2\theta = 1$, you obtain

$$\left(\frac{1}{3}\right)^2 + \cos^2\theta = 1$$

$$\cos^2\theta = 1 - \frac{1}{9} = \frac{8}{9}.$$

Because $\cos\theta < 0$ lies in Quadrant II, you can use the negative root to obtain

$$\cos\theta = -\frac{\sqrt{8}}{\sqrt{9}} = -\frac{2\sqrt{2}}{3}.$$

b. Using the trigonometric identity $\tan\theta = \sin\theta/\cos\theta$, you obtain

$$\tan\theta = \frac{1/3}{-2\sqrt{2}/3} = -\frac{1}{2\sqrt{2}} = -\frac{\sqrt{2}}{4}.$$

EXAMPLE 7 *Using a Calculator*

a. Use a calculator to evaluate $\cot 410°$ and $\sin(-7)$.

b. Use a calculator to solve $\tan\theta = 4.812$, $0 \le \theta < 2\pi$.

Solution

	Function	*Mode*	*Graphing Calculator Keystrokes*	*Display*
a.	$\cot 410°$	Degree	[( ] [TAN] 410 [ )] [x^{-1}] [ENTER]	0.8390996
	$\sin(-7)$	Radian	[SIN] [( ] [(-)] 7 [ )] [ENTER]	−0.6569866

b. To solve the equation $\tan\theta = 4.812$, you can use the inverse tangent key, as follows.

Equation	*Mode*	*Graphing Calculator Keystrokes*	*Display*
$\tan\theta = 4.812$	Radian	[TAN^{-1}] 4.812 [ENTER]	1.365898912

The angle $\theta \approx 1.366$ lies in Quadrant I. A second value of θ lies in Quadrant III (tangent is positive) and is

$$\theta = \pi + 1.366 \approx 4.507.$$

NOTE For your convenience we have included on the inside cover of this text a summary of basic trigonometry.

At this point, you have completed your introduction to basic trigonometry. You have measured angles in both radians and degrees. You have defined the six trigonometric functions in terms of the unit circle and from a right triangle perspective. In your remaining work with trigonometry you should continue to rely on both perspectives. For instance, in the next two sections on graphing techniques, it helps to think of the trigonometric functions as functions of real numbers. Later, in Section 1.8, you will look at applications involving angles and triangles.

GROUP ACTIVITY

PATTERNS IN TRIGONOMETRIC FUNCTIONS

Complete the following table. Then identify and discuss any inherent patterns in the trigonometric functions. What can you conclude?

Function	Domain	Range	Even/Odd	Period	Zeros
Sine					
Cosine					
Tangent					
Cosecant					
Secant					
Cotangent					

WARM UP

In Exercises 1–6, evaluate the trigonometric function from memory.

1. $\sin 30°$ **2.** $\tan 45°$

3. $\cos \frac{\pi}{4}$ **4.** $\cot \frac{\pi}{3}$

5. $\sec \frac{\pi}{6}$ **6.** $\csc \frac{\pi}{4}$

In Exercises 7–10, use the trigonometric function of an acute angle θ to find the values of the remaining trigonometric functions.

7. $\tan \theta = \frac{3}{2}$ **8.** $\cos \theta = \frac{2}{3}$

9. $\sin \theta = \frac{1}{5}$ **10.** $\sec \theta = 3$

1.4 Exercises

In Exercises 1–4, determine the exact values of the six trigonometric functions of the angle θ.

1. (a)

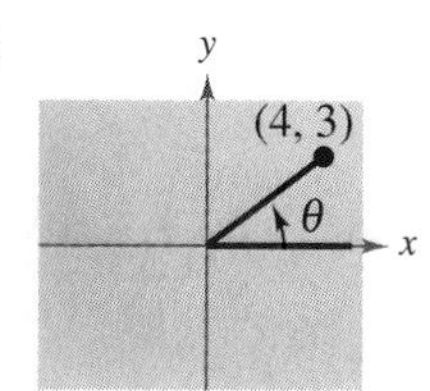

(b)

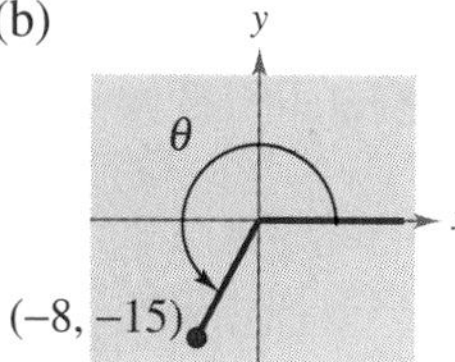

2. (a)

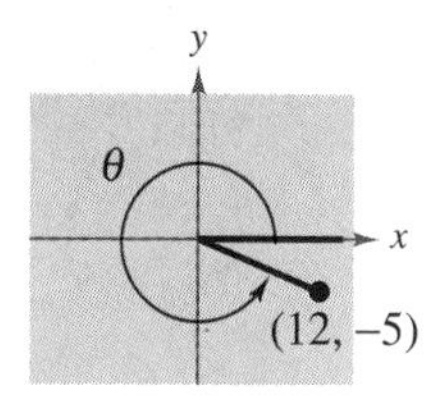

(b)

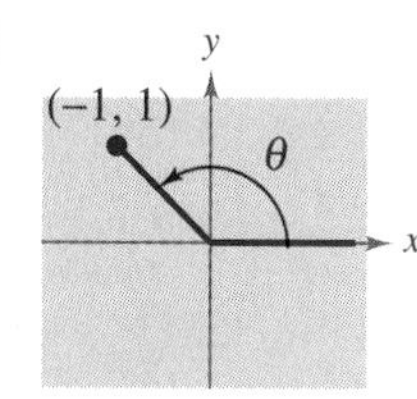

3. (a)

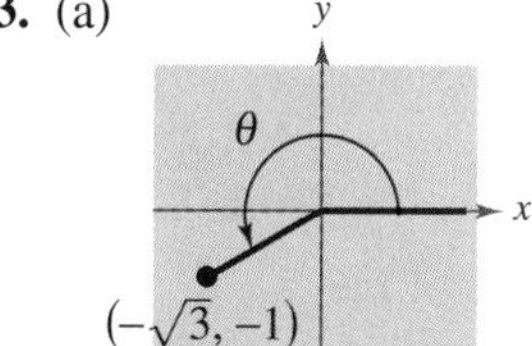

(b)

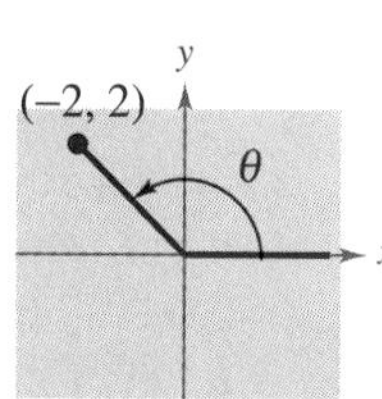

4. (a)

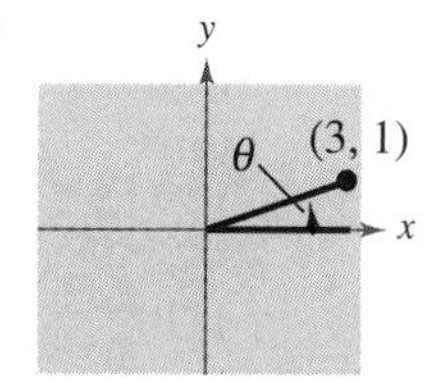

(b)

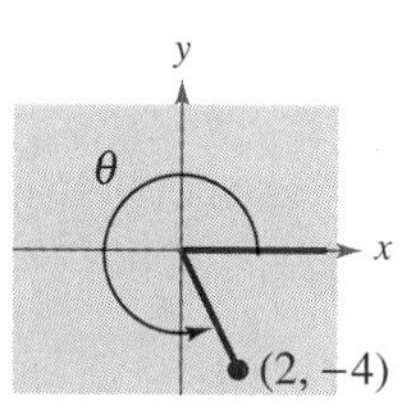

In Exercises 5–8, the point is on the terminal side of an angle in standard position. Determine the exact values of the six trigonometric functions of the angle.

5. (a) $(7, 24)$ (b) $(7, -24)$
6. (a) $(8, 15)$ (b) $(-9, -40)$
7. (a) $(-4, 10)$ (b) $(3, -5)$
8. (a) $(-5, -2)$ (b) $\left(-\frac{3}{2}, 3\right)$

In Exercises 9–12, state the quadrant in which θ lies.

9. (a) $\sin\theta < 0$ and $\cos\theta < 0$
   (b) $\sin\theta > 0$ and $\cos\theta < 0$
10. (a) $\sin\theta > 0$ and $\cos\theta > 0$
    (b) $\sin\theta < 0$ and $\cos\theta > 0$
11. (a) $\sin\theta > 0$ and $\tan\theta < 0$
    (b) $\cos\theta > 0$ and $\tan\theta < 0$
12. (a) $\sec\theta > 0$ and $\cot\theta < 0$
    (b) $\csc\theta < 0$ and $\tan\theta > 0$

In Exercises 13–22, find the values (if possible) of the six trigonometric functions of θ using the functional value and constraint.

	Functional Value	*Constraint*
13.	$\sin\theta = \frac{3}{5}$	θ lies in Quadrant II.
14.	$\cos\theta = -\frac{4}{5}$	θ lies in Quadrant III.
15.	$\tan\theta = -\frac{15}{8}$	$\sin\theta < 0$
16.	$\cos\theta = \frac{8}{17}$	$\tan\theta < 0$
17.	$\cot\theta = -3$	$\cos\theta > 0$
18.	$\csc\theta = 4$	$\cot\theta < 0$
19.	$\sec\theta = -2$	$\sin\theta > 0$
20.	$\cot\theta$ is undefined.	$\frac{\pi}{2} \le \theta \le \frac{3\pi}{2}$
21.	$\sin\theta = 0$	$\sec\theta = -1$
22.	$\tan\theta$ is undefined.	$\pi \le \theta \le 2\pi$

In Exercises 23–26, find the values (if possible) of the six trigonometric functions of θ if the terminal side of θ lies on the given line in the specified quadrant.

	Line	*Quadrant*
23.	$y = -x$	Quadrant II
24.	$y = \frac{1}{3}x$	Quadrant III
25.	$y = 2x$	Quadrant III
26.	$4x + 3y = 0$	Quadrant IV

In Exercises 27–34, find (if possible) the trigonometric function of the quadrant angle.

27. $\cos \pi$
28. $\cos \frac{3\pi}{2}$
29. $\sec \pi$
30. $\sec \frac{3\pi}{2}$
31. $\tan \frac{\pi}{2}$
32. $\tan \pi$
33. $\cot \frac{\pi}{2}$
34. $\csc \pi$

In Exercises 35–42, find the reference angle θ', and sketch θ and θ' in standard position.

35. (a) $\theta = 203°$ (b) $\theta = 127°$
36. (a) $\theta = 309°$ (b) $\theta = 226°$
37. (a) $\theta = -245°$ (b) $\theta = -72°$
38. (a) $\theta = -145°$ (b) $\theta = -239°$
39. (a) $\theta = \frac{2\pi}{3}$ (b) $\theta = \frac{7\pi}{6}$
40. (a) $\theta = \frac{7\pi}{4}$ (b) $\theta = \frac{8\pi}{9}$
41. (a) $\theta = 3.5$ (b) $\theta = 5.8$
42. (a) $\theta = \frac{11\pi}{3}$ (b) $\theta = -\frac{7\pi}{10}$

In Exercises 43–52, evaluate (if possible) the sine, cosine, and tangent of the angles without a calculator.

43. (a) $225°$ (b) $-225°$
44. (a) $300°$ (b) $330°$
45. (a) $750°$ (b) $510°$
46. (a) $-405°$ (b) $-120°$
47. (a) $\frac{4\pi}{3}$ (b) $\frac{2\pi}{3}$
48. (a) $\frac{\pi}{4}$ (b) $\frac{5\pi}{4}$
49. (a) $-\frac{\pi}{6}$ (b) $\frac{5\pi}{6}$
50. (a) $-\frac{\pi}{2}$ (b) $\frac{\pi}{2}$
51. (a) $\frac{11\pi}{4}$ (b) $-\frac{13\pi}{6}$
52. (a) $\frac{10\pi}{3}$ (b) $\frac{17\pi}{3}$

In Exercises 53–62, use a calculator to evaluate the trigonometric function to four decimal places. (Be sure the calculator is set in the correct mode.)

53. (a) $\sin 10°$ (b) $\csc 10°$
54. (a) $\sec 225°$ (b) $\sec 135°$
55. (a) $\cos(-110°)$ (b) $\cos 250°$
56. (a) $\csc 330°$ (b) $\csc 150°$
57. (a) $\tan 240°$ (b) $\cot 210°$
58. (a) $\cot 1.35$ (b) $\tan 1.35$
59. (a) $\tan \frac{\pi}{9}$ (b) $\tan \frac{10\pi}{9}$
60. (a) $\tan\left(-\frac{\pi}{9}\right)$ (b) $\tan\left(-\frac{10\pi}{9}\right)$
61. (a) $\sin 0.65$ (b) $\sin(-5.63)$
62. (a) $\sin(-0.65)$ (b) $\sin 5.63$

In Exercises 63–68, find two values of θ that satisfy the equation. Give your answers in degrees ($0° \le \theta < 360°$) and radians ($0 \le \theta < 2\pi$). Do not use a calculator.

63. (a) $\sin \theta = \frac{1}{2}$ (b) $\sin \theta = -\frac{1}{2}$
64. (a) $\cos \theta = \frac{\sqrt{2}}{2}$ (b) $\cos \theta = -\frac{\sqrt{2}}{2}$
65. (a) $\csc \theta = \frac{2\sqrt{3}}{3}$ (b) $\cot \theta = -1$
66. (a) $\sec \theta = 2$ (b) $\sec \theta = -2$
67. (a) $\tan \theta = 1$ (b) $\cot \theta = -\sqrt{3}$
68. (a) $\sin \theta = \frac{\sqrt{3}}{2}$ (b) $\sin \theta = -\frac{\sqrt{3}}{2}$

In Exercises 69 and 70, use a calculator to approximate two values of θ ($0° \le \theta < 360°$) that satisfy the equation. Round to two decimal places.

69. (a) $\sin \theta = 0.8191$ (b) $\sin \theta = -0.2589$
70. (a) $\cos \theta = 0.8746$ (b) $\cos \theta = -0.2419$

In Exercises 71–74, use a calculator to approximate two values of θ ($0 \le \theta < 2\pi$) that satisfy the equation. Round to three decimal places.

71. (a) $\cos \theta = 0.9848$ (b) $\cos \theta = -0.5890$

72. (a) $\sin \theta = 0.0175$ (b) $\sin \theta = -0.6691$

73. (a) $\tan \theta = 1.192$ (b) $\tan \theta = -8.144$

74. (a) $\cot \theta = 5.671$ (b) $\cot \theta = -1.280$

In Exercises 75–80, find the indicated trigonometric value in the specified quadrant.

	Function	*Quadrant*	*Trigonometric Value*
75.	$\sin \theta = -\frac{3}{5}$	IV	$\cos \theta$
76.	$\cot \theta = -3$	II	$\sin \theta$
77.	$\tan \theta = \frac{3}{2}$	III	$\sec \theta$
78.	$\csc \theta = -2$	IV	$\cot \theta$
79.	$\cos \theta = \frac{5}{8}$	I	$\sec \theta$
80.	$\sec \theta = -\frac{9}{4}$	III	$\tan \theta$

81. ***Average Temperature*** The average daily temperature T (in degrees Fahrenheit) for a certain city is given by

$$T = 45 - 23 \cos\left[\frac{2\pi}{365}(t - 32)\right]$$

where t is the time in days, with $t = 1$ corresponding to January 1. Find the average daily temperatures on the following days.

(a) January 1

(b) July 4 ($t = 185$)

(c) October 18 ($t = 291$)

82. ***Sales*** A company that produces a seasonal product forecasts monthly sales over the next 2 years to be

$$S = 23.1 + 0.442t + 4.3 \sin \frac{\pi t}{6}$$

where S is measured in thousands of units and t is the time in months, with $t = 1$ representing January 1996. Predict sales for the following months.

(a) February 1996 (b) February 1997

(c) September 1996 (d) September 1997

83. ***Distance*** An airplane, flying at an altitude of 6 miles, is on a flight path that passes directly over an observer (see figure). If θ is the angle of elevation from the observer to the plane, find the distance d from the observer to the plane when (a) $\theta = 30°$, (b) $\theta = 90°$, and (c) $\theta = 120°$.

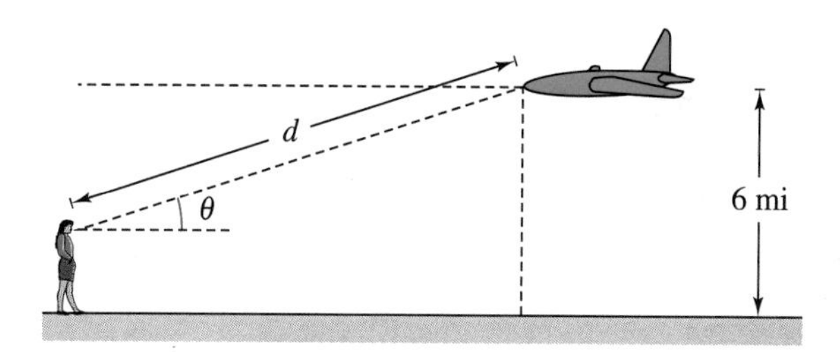

84. ***Essay*** Consider an angle in standard position with $r = 12$ centimeters, as shown in the figure. Write a short paragraph describing the change in the magnitude of x, y, $\sin \theta$, $\cos \theta$, and $\tan \theta$ as θ increases continuously from 0° to 90°.

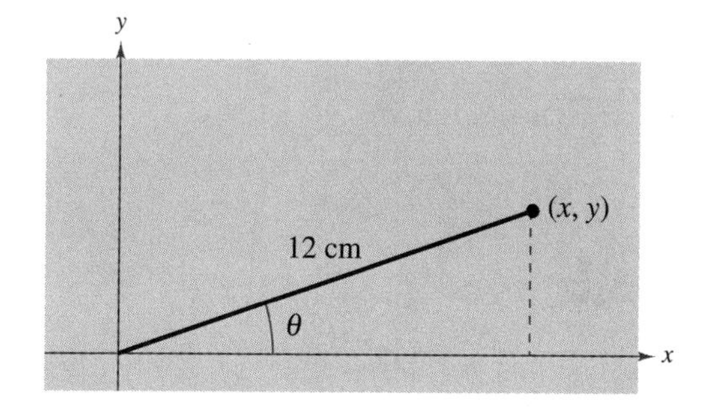

1.5 Graphs of Sine and Cosine Functions

See Exercise 89 on page 169 for an example of how a sine function can be used to model the daily high temperatures in Honolulu and Chicago.

Basic Sine and Cosine Curves ▫ *Amplitude and Period* ▫ *Translations of Sine and Cosine Curves* ▫ *Mathematical Modeling*

Basic Sine and Cosine Curves

In this section you will study techniques for sketching the graphs of the sine and cosine functions. The graph of the sine function is a **sine curve.** In Figure 1.38, the black portion of the graph represents one period of the function and is called **one cycle** of the sine curve. The gray portion of the graph indicates that the basic sine wave repeats indefinitely to the right and left. The graph of the cosine function is shown in Figure 1.39.

Recall from Section 1.2 that the domain of the sine and cosine functions is the set of all real numbers. Moreover, the range of each function is the interval $[-1, 1]$, and each function has a period of 2π. Do you see how this information is consistent with the basic graphs given in Figures 1.38 and 1.39?

NOTE Note in Figures 1.38 and 1.39 that the sine curve is symmetric with respect to the *origin*, whereas the cosine curve is symmetric with respect to the *y-axis*. These properties of symmetry follow from the fact that the sine function is odd whereas the cosine function is even.

x	$\sin x$	$\cos x$
0	0	1
$\frac{\pi}{6}$	$\frac{1}{2}$	$\frac{\sqrt{3}}{2}$
$\frac{\pi}{4}$	$\frac{\sqrt{2}}{2}$	$\frac{\sqrt{2}}{2}$
$\frac{\pi}{3}$	$\frac{\sqrt{3}}{2}$	$\frac{1}{2}$
$\frac{\pi}{2}$	1	0
$\frac{3\pi}{4}$	$\frac{\sqrt{2}}{2}$	$-\frac{\sqrt{2}}{2}$
π	0	-1
$\frac{3\pi}{2}$	-1	0
2π	0	1

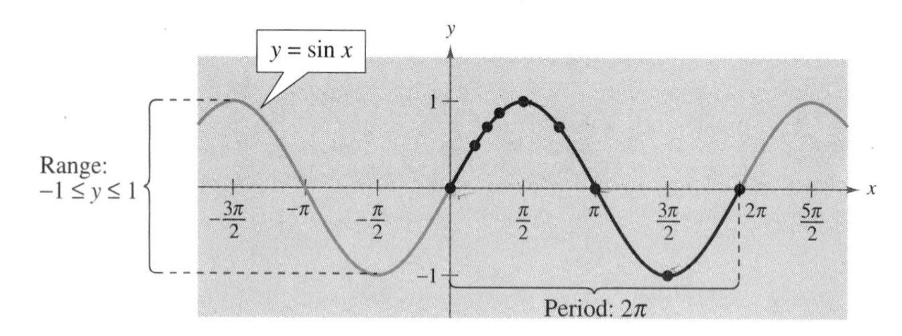

FIGURE 1.38

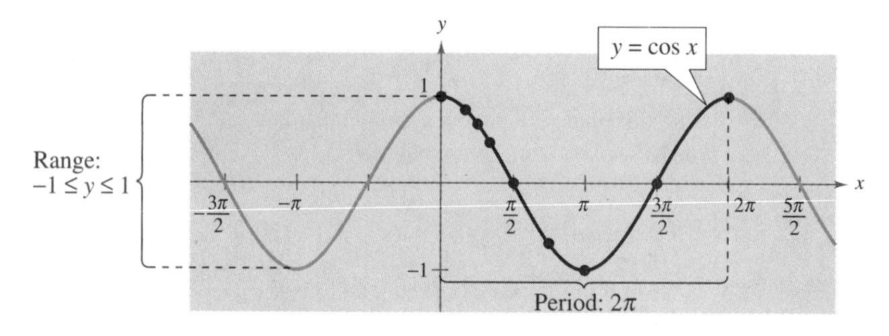

FIGURE 1.39

To sketch the graphs of the basic sine and cosine functions by hand, it helps to note five **key points** in one period of each graph: the *intercepts, maximum points,* and *minimum points.* See Figure 1.40.

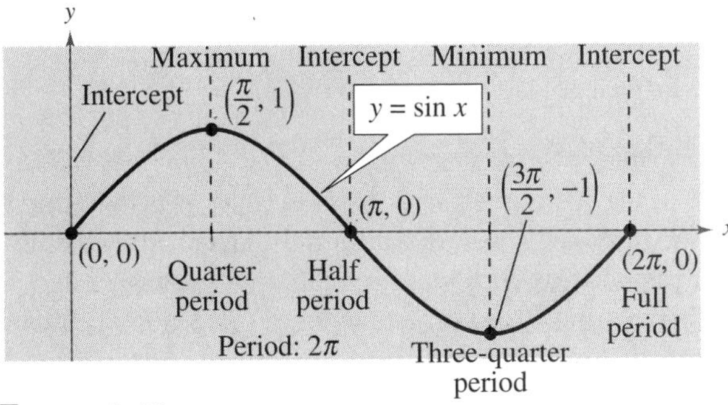

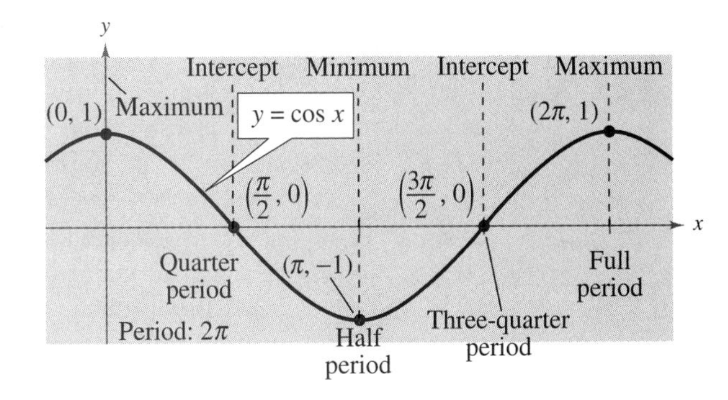

FIGURE 1.40

EXAMPLE 1 *Using Key Points to Sketch a Sine Curve*

Sketch the graph of $y = 2 \sin x$ on the interval $[-\pi, 4\pi]$.

Solution

Note that $y = 2 \sin x = 2(\sin x)$ indicates that the y-values for the key points will have twice the magnitude of the graph of $y = \sin x$. Divide the period 2π into four equal parts to get the following key points for $y = 2 \sin x$.

$$(0, 0), \quad \left(\frac{\pi}{2}, 2\right), \quad (\pi, 0), \quad \left(\frac{3\pi}{2}, -2\right), \quad \text{and} \quad (2\pi, 0)$$

By connecting these key points with a smooth curve and extending the curve in both directions over the interval $[-\pi, 4\pi]$, you obtain the graph shown in Figure 1.41.

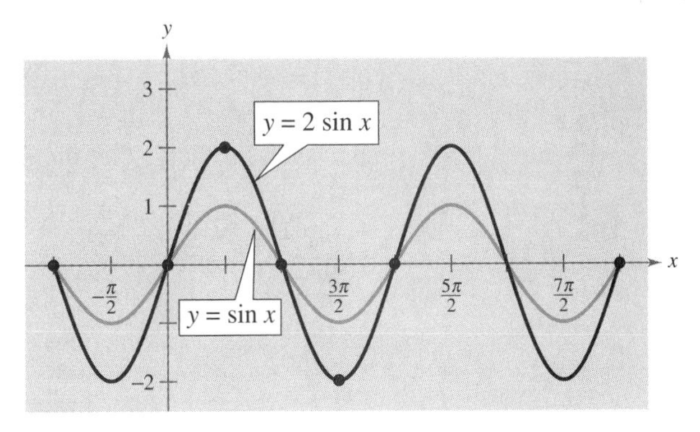

FIGURE 1.41

TECHNOLOGY

When using a graphing utility to graph trigonometric functions, pay special attention to the viewing rectangle you use. For instance, try graphing $y = [\sin(10x)]/10$ on the standard viewing rectangle in radian mode. What do you observe? Use the zoom feature to find a viewing rectangle that displays a good view of the graph.

Amplitude and Period

In the rest of this section you will study the graphic effect of each of the constants *a, b, c,* and *d* in equations of the forms

$$y = d + a\sin(bx - c) \qquad \text{and} \qquad y = d + a\cos(bx - c).$$

A quick review of the transformations studied in Section P.7 should help in this investigation.

The constant factor a in $y = a\sin x$ acts as a *scaling factor*—a *vertical stretch* or *vertical shrink* of the basic sine curve. If $|a| > 1$, the basic sine curve is stretched, and if $|a| < 1$, the basic sine curve is shrunk. The result is that the graph of $y = a\sin x$ ranges between $-a$ and a instead of between -1 and 1. The absolute value of a is the **amplitude** of the function $y = a\sin x$. The range of the function $y = a\sin x$ is $-a \le y \le a$.

> ***Definition of Amplitude of Sine and Cosine Curves***
>
> The **amplitude** of $y = a\sin x$ and $y = a\cos x$ is the largest value of y and is given by
>
> $$\text{Amplitude} = |a|.$$

Example 2 *Scaling: Vertical Shrinking and Stretching*

On the same coordinate axes, sketch the graphs of

$$y = \frac{1}{2}\cos x \qquad \text{and} \qquad y = 3\cos x.$$

Solution

Because the amplitude of $y = \frac{1}{2}\cos x$ is $\frac{1}{2}$, the maximum value is $\frac{1}{2}$ and the minimum value is $-\frac{1}{2}$. Divide one cycle, $0 \le x \le 2\pi$, into four equal parts to get the key points

$$\left(0, \frac{1}{2}\right), \quad \left(\frac{\pi}{2}, 0\right), \quad \left(\pi, -\frac{1}{2}\right), \quad \left(\frac{3\pi}{2}, 0\right), \quad \text{and} \quad \left(2\pi, \frac{1}{2}\right).$$

A similar analysis shows that the amplitude of $y = 3\cos x$ is 3, and the key points are

$$(0, 3), \quad \left(\frac{\pi}{2}, 0\right), \quad (\pi, -3), \quad \left(\frac{3\pi}{2}, 0\right), \quad \text{and} \quad (2\pi, 3).$$

The graphs of these two functions are shown in Figure 1.42.

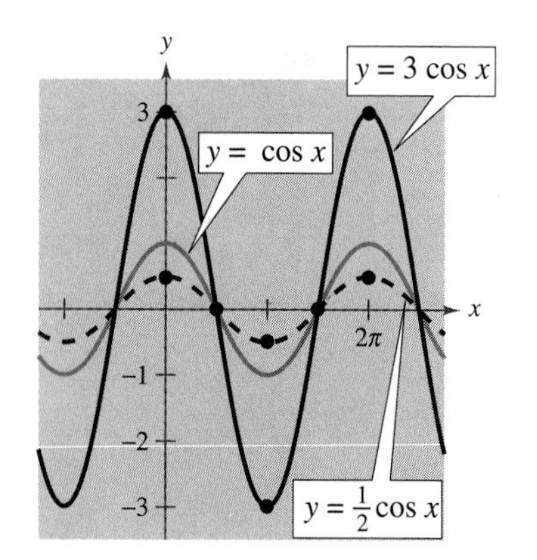

FIGURE 1.42

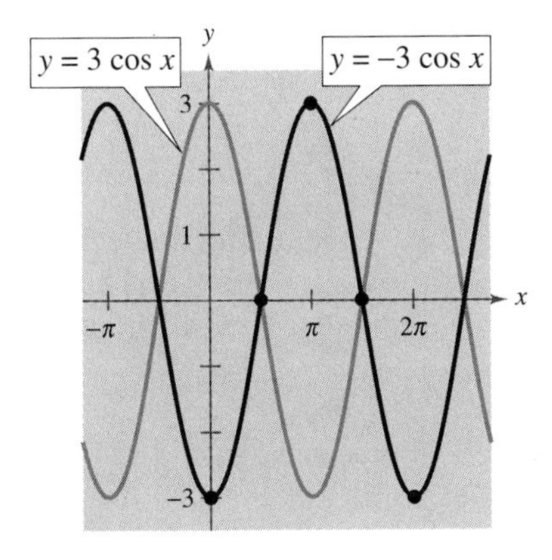

FIGURE 1.43

You know from Section P.7 that the graph of $y = -f(x)$ is a **reflection** in the x-axis of the graph of $y = f(x)$. For instance, the graph of $y = -3 \cos x$ is a reflection of the graph of $y = 3 \cos x$, as shown in Figure 1.43.

Because $y = a \sin x$ completes one cycle from $x = 0$ to $x = 2\pi$, it follows that $y = a \sin bx$ completes one cycle from $x = 0$ to $x = 2\pi/b$.

PERIOD OF SINE AND COSINE FUNCTIONS

Let b be a positive real number. The **period** of $y = a \sin bx$ and $y = a \cos bx$ is $2\pi/b$.

Note that if $0 < b < 1$, the period of $y = a \sin bx$ is greater than 2π and represents a *horizontal stretching* of the graph of $y = a \sin x$. Similarly, if $b > 1$, the period of $y = a \sin bx$ is less than 2π and represents a *horizontal shrinking* of the graph of $y = a \sin x$. If b is negative, we use the identities $\sin(-x) = -\sin x$ and $\cos(-x) = \cos x$ to rewrite the function.

EXAMPLE 3 *Scaling: Horizontal Stretching*

Sketch the graph of $y = \sin \frac{x}{2}$.

Solution

The amplitude is 1. Moreover, because $b = \frac{1}{2}$, the period is

$$\frac{2\pi}{b} = \frac{2\pi}{\frac{1}{2}} = 4\pi.$$

Now, divide the period-interval $[0, 4\pi]$ into four equal parts with the values π, 2π, and 3π to obtain the following key points on the graph.

$$(0, 0), \quad (\pi, 1), \quad (2\pi, 0), \quad (3\pi, -1), \quad \text{and} \quad (4\pi, 0)$$

The graph is shown in Figure 1.44.

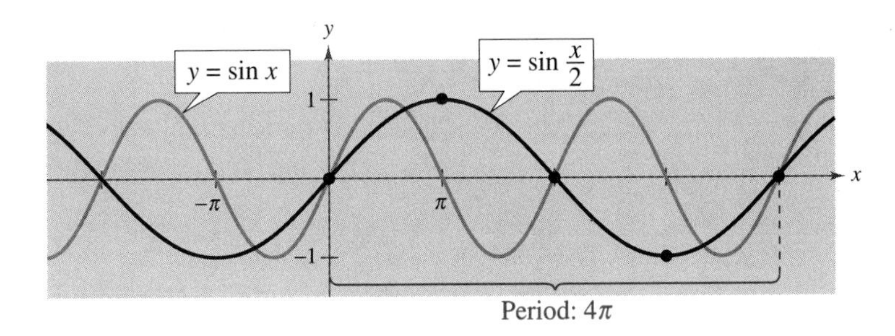

FIGURE 1.44

Study Tip

In general, to divide a period-interval into four equal parts, successively add "period/4," starting with the left endpoint of the interval. For instance, for the period-interval $[-\pi/6, \pi/2]$ of length $2\pi/3$, you would successively add

$$\frac{2\pi/3}{4} = \frac{\pi}{6}$$

to get $-\pi/6, 0, \pi/6, \pi/3$, and $\pi/2$.

Translations of Sine and Cosine Curves

The constant c in the general equations

$$y = a \sin(bx - c) \quad \text{and} \quad y = a \cos(bx - c)$$

creates a *horizontal translation* (shift) of the basic sine and cosine curves. Comparing $y = a \sin bx$ with $y = a \sin(bx - c)$, we find that the graph of $y = a \sin(bx - c)$ completes one cycle from $bx - c = 0$ to $bx - c = 2\pi$. By solving for x, we find the interval for one cycle to be

$$\overbrace{\frac{c}{b}}^{\text{Left endpoint}} \le x \le \overbrace{\frac{c}{b} + \underbrace{\frac{2\pi}{b}}_{\text{Period}}}^{\text{Right endpoint}}.$$

Sketch the graph of $y = \sin(x + c)$, where $c = -\pi/4$, 0, and $\pi/4$. How does the value of c affect the graph?

This implies that the period of $y = a \sin(bx - c)$ is $2\pi/b$, and the graph of $y = a \sin bx$ is shifted by an amount c/b. The number c/b is the **phase shift.**

GRAPHS OF SINE AND COSINE FUNCTIONS

The graphs of $y = a \sin(bx - c)$ and $y = a \cos(bx - c)$ have the following characteristics. (Assume $b > 0$.)

Amplitude $= |a|$ **Period** $= 2\pi/b$

The left and right endpoints of a one-cycle interval can be determined by solving the equations $bx - c = 0$ and $bx - c = 2\pi$.

EXAMPLE 4 Horizontal Translation

Sketch the graph of $y = (1/2) \sin(x - \pi/3)$.

Solution

The amplitude is $\frac{1}{2}$ and the period is 2π. By solving the equations

$$x - \pi/3 = 0 \qquad \text{and} \qquad x - \pi/3 = 2\pi$$
$$x = \pi/3 \qquad\qquad x = 7\pi/3$$

you see that the interval $[\pi/3, 7\pi/3]$ corresponds to one cycle of the graph. Dividing this interval into four equal parts produces the following key points.

$$\left(\frac{\pi}{3}, 0\right), \quad \left(\frac{5\pi}{6}, \frac{1}{2}\right), \quad \left(\frac{4\pi}{3}, 0\right), \quad \left(\frac{11\pi}{6}, -\frac{1}{2}\right), \quad \text{and} \quad \left(\frac{7\pi}{3}, 0\right)$$

The graph is shown in Figure 1.45.

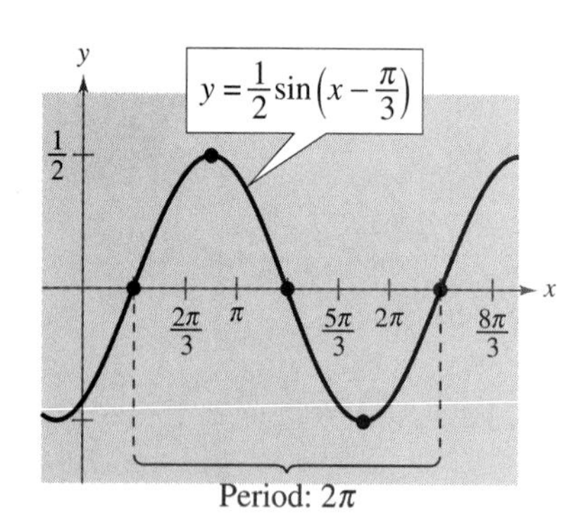

FIGURE 1.45

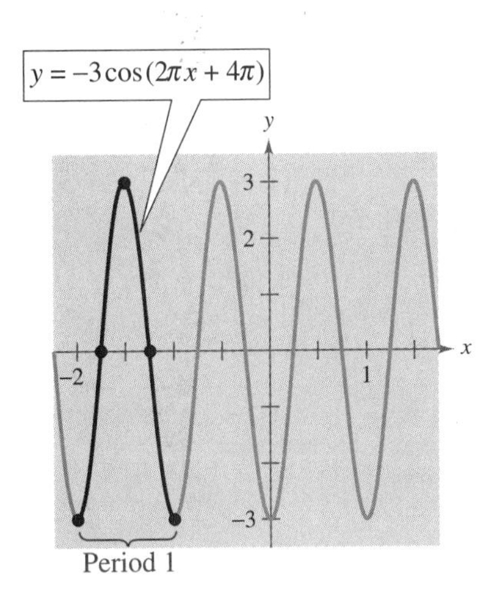

FIGURE 1.46

EXAMPLE 5 *Horizontal Translation*

Sketch the graph of

$$y = -3\cos(2\pi x + 4\pi).$$

Solution

The amplitude is 3 and the period is $2\pi/2\pi = 1$. By solving the equations

$$2\pi x + 4\pi = 0 \qquad \text{and} \qquad 2\pi x + 4\pi = 2\pi$$

$$2\pi x = -4\pi \qquad\qquad 2\pi x = -2\pi$$

$$x = -2 \qquad\qquad x = -1$$

you see that the interval $[-2, -1]$ corresponds to one cycle of the graph. Dividing this interval into four equal parts produces the following key points.

$$(-2, -3), \quad \left(-\frac{7}{4}, 0\right), \quad \left(-\frac{3}{2}, 3\right), \quad \left(-\frac{5}{4}, 0\right), \quad \text{and} \quad (-1, -3)$$

The graph is shown in Figure 1.46.

The final type of transformation is the *vertical translation* caused by the constant d in the equations

$$y = d + a\sin(bx - c) \qquad \text{and} \qquad y = d + a\cos(bx - c).$$

The shift is d units upward for $d > 0$ and downward for $d < 0$. In other words, the graph oscillates about the horizontal line $y = d$ instead of the x-axis.

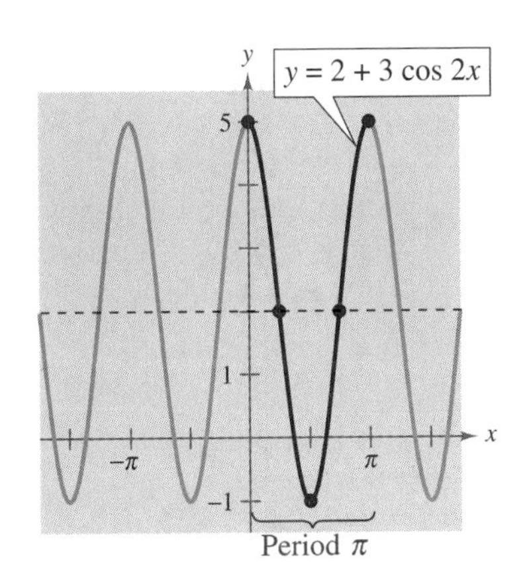

FIGURE 1.47

EXAMPLE 6 *Vertical Translation*

Sketch the graph of

$$y = 2 + 3\cos 2x.$$

Solution

The amplitude is 3 and the period is π. The key points over the interval $[0, \pi]$ are

$$(0, 5), \quad \left(\frac{\pi}{4}, 2\right), \quad \left(\frac{\pi}{2}, -1\right), \quad \left(\frac{3\pi}{4}, 2\right), \quad \text{and} \quad (\pi, 5).$$

The graph is shown in Figure 1.47.

Mathematical Modeling

Sine and cosine functions can be used to model many real-life situations, including electric currents, musical tones, radio waves, tides, sunrises, and weather patterns.

EXAMPLE 7 *Finding a Trigonometric Model*

Throughout the day, the depth of water at the end of a dock varies with the tides. The table shows the depth (in meters) at various times during the morning.

t (time)	Midnight	2 A.M.	4 A.M.	6 A.M.	8 A.M.	10 A.M.	Noon
y (depth)	2.55	3.80	4.40	3.80	2.55	1.80	2.27

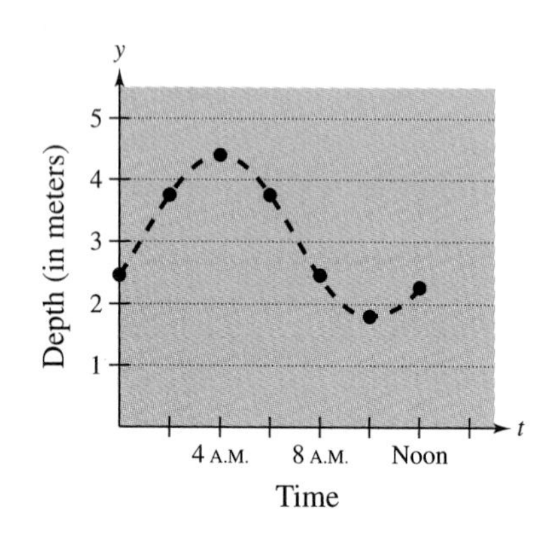

FIGURE 1.48

a. Use a trigonometric function to model this data.

b. Find the depth at 9 A.M. and 3 P.M.

c. A boat needs at least 3 meters of water to moor at the dock. During what times in the afternoon can it safely dock?

Solution

a. Begin by graphing the data, as shown in Figure 1.48. You can use either a sine or cosine model. Suppose you use a cosine model of the form

$$y = a\cos(bt - c) + d.$$

The amplitude is given by

$$a = \frac{1}{2}[(\text{high}) - (\text{low})] = \frac{1}{2}(4.4 - 1.8) = 1.3.$$

The period is

$$p = 2[(\text{low time}) - (\text{high time})] = 2(10 - 4) = 12$$

which implies that $b = 2\pi/p \approx 0.524$. Because high tide occurs 4 hours after midnight, you can conclude that $c/b = 4$, so $c \approx 2.094$. Moreover, because the average depth is $\frac{1}{2}(4.4 + 1.8) = 3.1$, it follows that $d = 3.1$. Thus, you can model the depth with the function

$$y = 1.3\cos(0.542t - 2.094) + 3.1.$$

b. At 9 A.M. and 3 P.M. the depth is as follows.

$$y = 1.3\cos(0.524 \cdot 9 - 2.094) + 3.1 \approx 1.97 \text{ meters} \qquad \text{9 A.M.}$$

$$y = 1.3\cos(0.524 \cdot 15 - 2.094) + 3.1 \approx 4.23 \text{ meters} \qquad \text{3 P.M.}$$

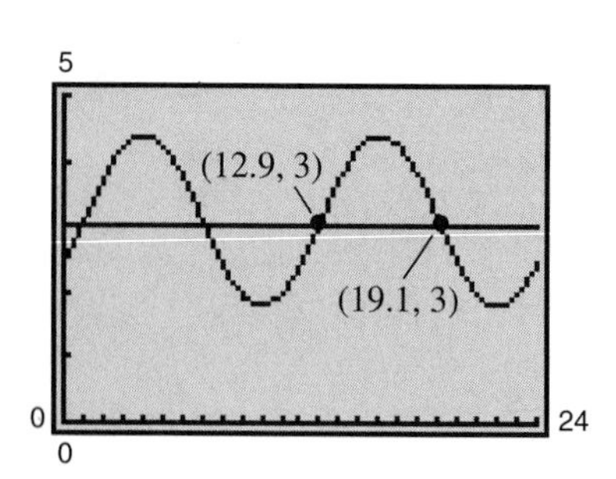

FIGURE 1.49

c. Using a graphing utility, you can graph the model with the line $y = 3$, as shown in Figure 1.49. From the graph, it follows that the depth is at least 3 meters between 12:54 P.M. ($t \approx 12.9$) and 7:06 P.M. ($t \approx 19.1$).

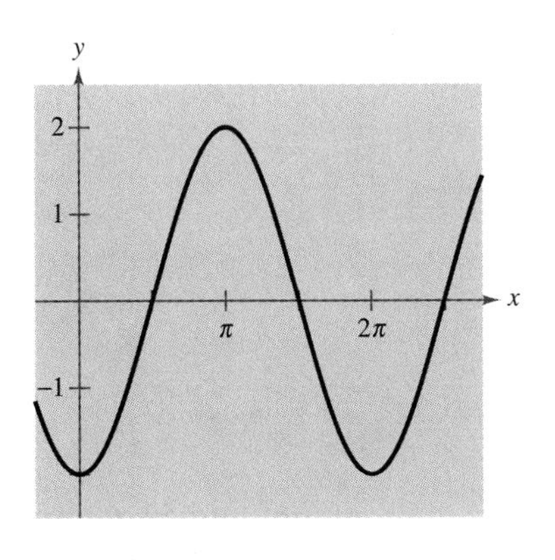

FIGURE 1.50

EXAMPLE 8 Finding an Equation for a Graph

Find the amplitude, period, and phase shift for the sine function whose graph is shown in Figure 1.50. Write an equation for this graph.

Solution

The amplitude for this sine curve is 2, the period is 2π, and there is a right phase shift of $\pi/2$. Thus, you can write

$$y = 2\sin\left(x - \frac{\pi}{2}\right).$$

Try finding a cosine function with this same graph.

GROUP ACTIVITY

A SINE SHOW

Enter the following program in a *TI-83* or a *TI-82* graphing calculator. (Program steps for other graphing calculators are given in the appendix.) This program simultaneously draws the unit circle and the corresponding points on the sine curve. After the circle and sine curve are drawn, you can connect the points on the unit circle with their corresponding points on the sine curve by pressing ENTER. Discuss the relationship that is illustrated.

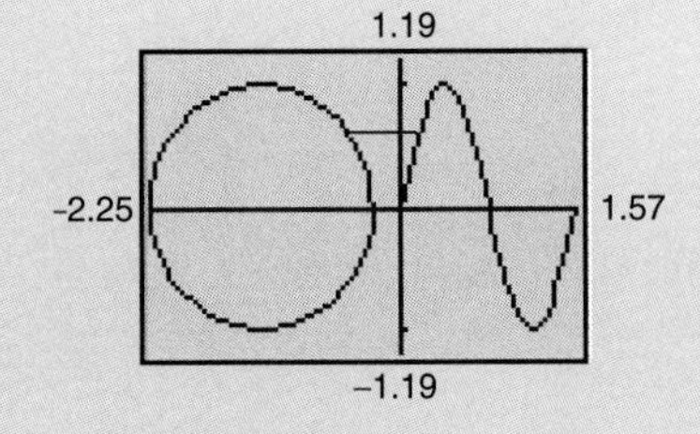


```
PROGRAM:SINESHOW
:Radian
:ClrDraw:FnOff
:Param:Simul
:-2.25→Xmin
:π/2→Xmax
:3→Xscl
:-1.19→Ymin
:1.19→Ymax
:1→Yscl
:0→Tmin
:6.3→Tmax
:.15→Tstep
:"-1.25+cos T"→X1T
:"sin T"→Y1T
:"T/4"→X2T
:"sin T"→Y2T
:DispGraph
:For(N,1,12)
:N*π/6.5→T
:-1.25+cos T→A
:sin T→B
:T/4→C
:Line(A,B,C,B)
:Pause
:End
:Pause:Func
:Sequential:Disp
```

WARM UP

Simplify the expression.

1. $\dfrac{2\pi}{1/3}$
2. $\dfrac{2\pi}{4\pi}$

Solve for x.

3. $2x - \dfrac{\pi}{3} = 0$
4. $2x - \dfrac{\pi}{3} = 2\pi$
5. $3\pi x + 6\pi = 0$
6. $3\pi x + 6\pi = 2\pi$

Evaluate the trigonometric function from memory.

7. $\sin \dfrac{\pi}{2}$
8. $\sin \pi$
9. $\cos 0$
10. $\cos \dfrac{\pi}{2}$

1.5 Exercises

In Exercises 1–14, find the period and amplitude.

1. $y = 3 \sin 2x$

2. $y = 2 \cos 3x$

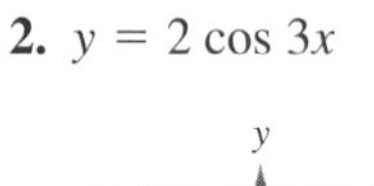

5. $y = \dfrac{2}{3} \sin \pi x$

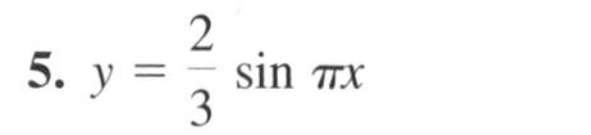

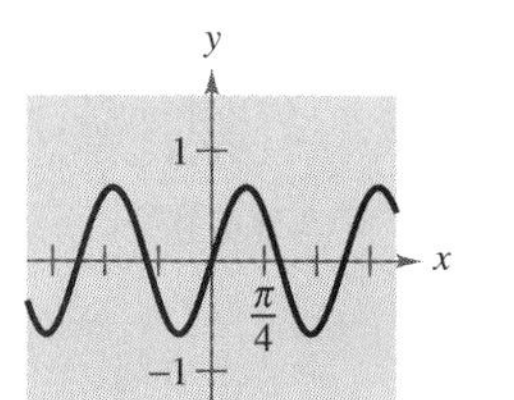

6. $y = \dfrac{3}{2} \cos \dfrac{\pi x}{2}$

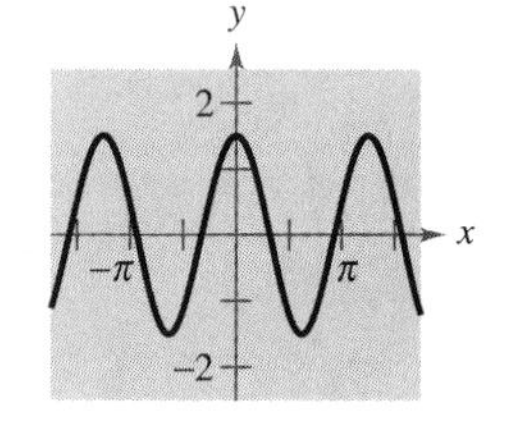

3. $y = \dfrac{5}{2} \cos \dfrac{x}{2}$

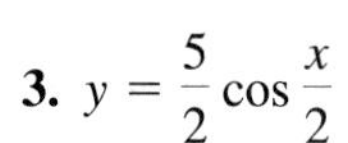

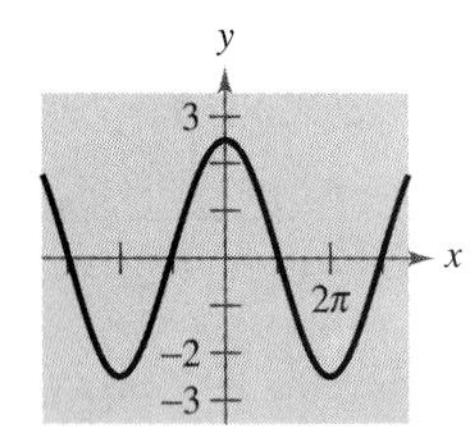

4. $y = -3 \sin \dfrac{x}{3}$

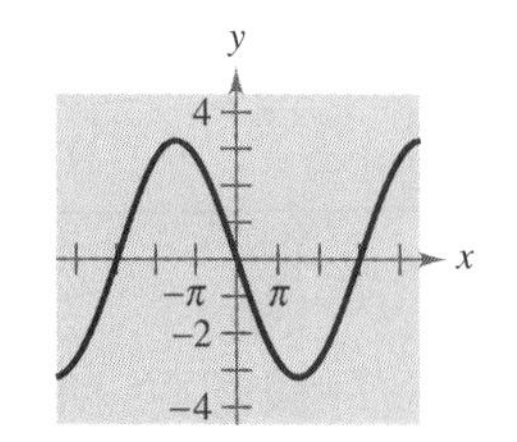

7. $y = -2 \sin x$

8. $y = -\cos \dfrac{2x}{3}$

9. $y = 3 \sin 10x$

10. $y = \frac{1}{3} \sin 8x$

11. $y = \dfrac{1}{2} \cos \dfrac{2x}{3}$

12. $y = \dfrac{5}{2} \cos \dfrac{x}{4}$

13. $y = 3 \sin 4\pi x$

14. $y = \dfrac{2}{3} \cos \dfrac{\pi x}{10}$

In Exercises 15–22, describe the relationship between the graphs of f and g.

15. $f(x) = \sin x$
 $g(x) = \sin(x - \pi)$

16. $f(x) = \cos x$
 $g(x) = \cos(x + \pi)$

17. $f(x) = \cos 2x$
 $g(x) = -\cos 2x$

18. $f(x) = \sin 3x$
 $g(x) = \sin(-3x)$

19. $f(x) = \cos x$
 $g(x) = \cos 2x$

20. $f(x) = \sin x$
 $g(x) = \sin 3x$

21. $f(x) = \sin x$
 $g(x) = 2 + \sin x$

22. $f(x) = \cos 4x$
 $g(x) = -2 + \cos 4x$

In Exercises 23–26, describe the relationship between the graphs of f and g.

23.

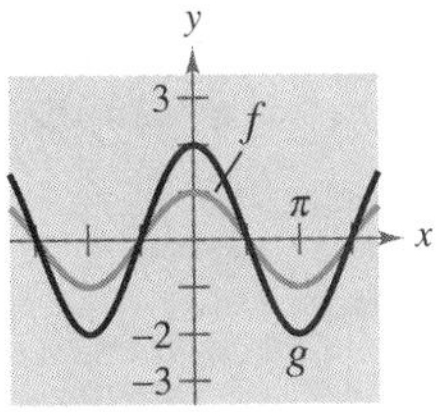

24.

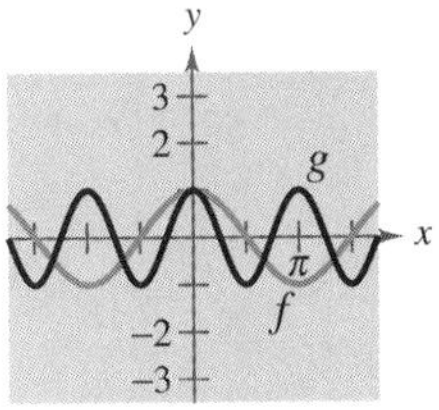

25.

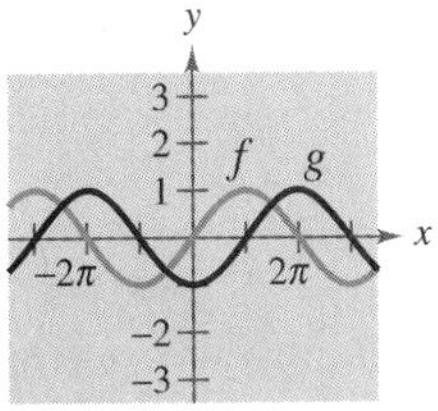

26.

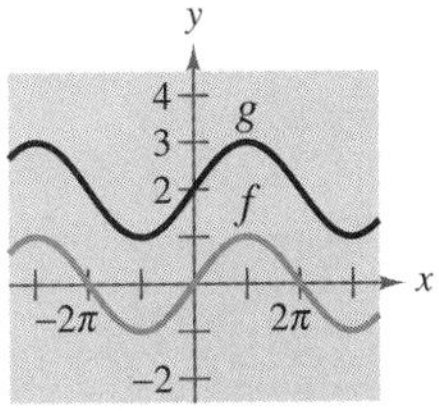

27. *Essay* Use a graphing utility to graph the function $y = a \sin x$ for $a = \frac{1}{2}, a = \frac{3}{2}$, and $a = -3$. Write a paragraph describing the changes in the graph corresponding to the specified changes in a.

28. *Essay* Use a graphing utility to graph the function $y = d + \sin x$ for $d = 2, d = 3.5$, and $d = -2$. Write a paragraph describing the changes in the graph corresponding to the specified changes in d.

29. *Essay* Use a graphing utility to graph the function $y = \sin bx$ for $b = \frac{1}{2}, b = \frac{3}{2}$, and $b = 4$. Write a paragraph describing the changes in the graph corresponding to the specified changes in b.

30. *Essay* Use a graphing utility to graph the function $y = \sin(x - c)$ for $c = 1, c = 3$, and $c = -2$. Write a paragraph describing the changes in the graph corresponding to the specified changes in c.

In Exercises 31–38, graph f and g on the same set of coordinate axes. (Include two full periods.)

31. $f(x) = -2 \sin x$
 $g(x) = 4 \sin x$

32. $f(x) = \sin x$
 $g(x) = \sin \frac{x}{3}$

33. $f(x) = \cos x$
 $g(x) = 1 + \cos x$

34. $f(x) = 2 \cos 2x$
 $g(x) = -\cos 4x$

35. $f(x) = -\frac{1}{2} \sin \frac{x}{2}$
 $g(x) = 3 - \frac{1}{2} \sin \frac{x}{2}$

36. $f(x) = 4 \sin \pi x$
 $g(x) = 4 \sin \pi x - 3$

37. $f(x) = 2 \cos x$
 $g(x) = 2 \cos(x + \pi)$

38. $f(x) = -\cos x$
 $g(x) = -\cos(x - \pi)$

Conjecture **In Exercises 39–42, graph f and g on the same set of coordinate axes. (Include two full periods.) Make a conjecture about the functions.**

39. $f(x) = \sin x, \quad g(x) = \cos\left(x - \frac{\pi}{2}\right)$

40. $f(x) = \sin x, \quad g(x) = -\cos\left(x + \frac{\pi}{2}\right)$

41. $f(x) = \cos x, \quad g(x) = -\sin\left(x - \frac{\pi}{2}\right)$

42. $f(x) = \cos x, \quad g(x) = -\cos(x - \pi)$

In Exercises 43–60, sketch the graph of the function. (Include two full periods.)

43. $y = -2 \sin 6x$

44. $y = -3 \cos 4x$

45. $y = \cos 2\pi x$

46. $y = \frac{3}{2} \sin \frac{\pi x}{4}$

47. $y = -\sin \frac{2\pi x}{3}$

48. $y = 10 \cos \frac{\pi x}{6}$

49. $y = \sin\left(x - \frac{\pi}{4}\right)$

50. $y = \frac{1}{2}\sin(x - \pi)$

51. $y = 3\cos(x + \pi)$

52. $y = 4\cos\left(x + \frac{\pi}{4}\right)$

53. $y = \frac{1}{10}\cos 60\pi x$

54. $y = -3 + 5\cos\frac{\pi t}{12}$

55. $y = 2 - \sin\frac{2\pi x}{3}$

56. $y = 2\cos x - 3$

57. $y = 3\cos(x + \pi) - 3$

58. $y = 4\cos\left(x + \frac{\pi}{4}\right) + 4$

59. $y = \frac{2}{3}\cos\left(\frac{x}{2} - \frac{\pi}{4}\right)$

60. $y = -3\cos(6x + \pi)$

In Exercises 61–68, use a graphing utility to graph the function. (Include two full periods.)

61. $y = -2\sin(4x + \pi)$

62. $y = -4\sin\left(\frac{2}{3}x - \frac{\pi}{3}\right)$

63. $y = \cos\left(2\pi x - \frac{\pi}{2}\right) + 1$

64. $y = 3\cos\left(\frac{\pi x}{2} + \frac{\pi}{2}\right) - 2$

65. $y = -0.1\sin\left(\frac{\pi x}{10} + \pi\right)$

66. $y = 5\sin(\pi - 2x) + 10$

67. $y = 5\cos(\pi - 2x) + 2$

68. $y = \frac{1}{100}\sin 120\pi t$

Graphical Reasoning **In Exercises 69–72, find a and d for the function $f(x) = a\cos x + d$ so that the graph of f matches the figure.**

69.

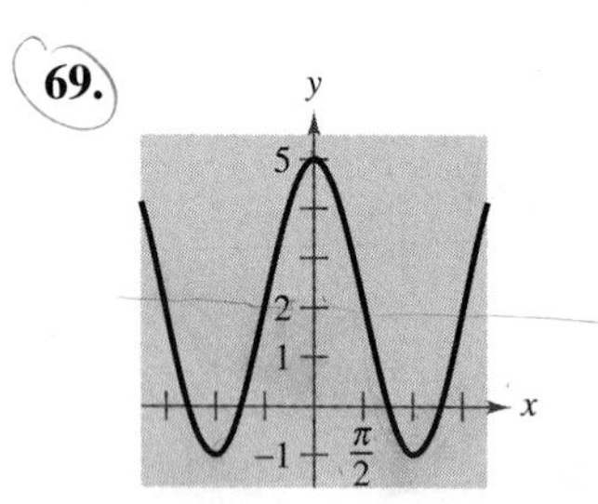

70.

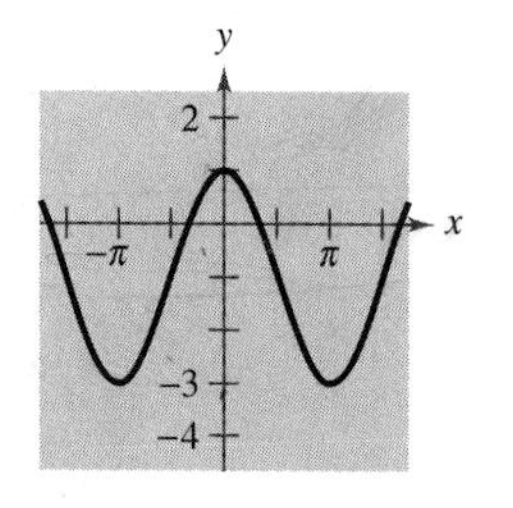

71.

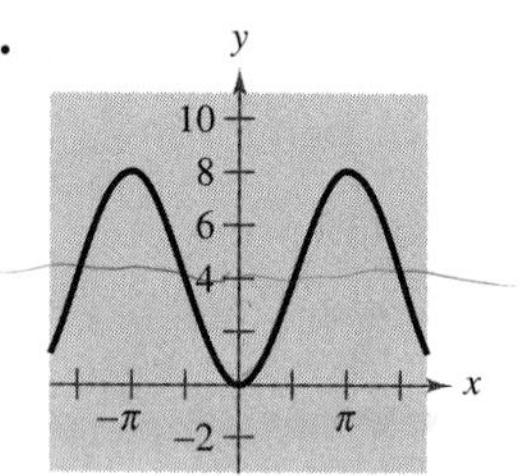

72.

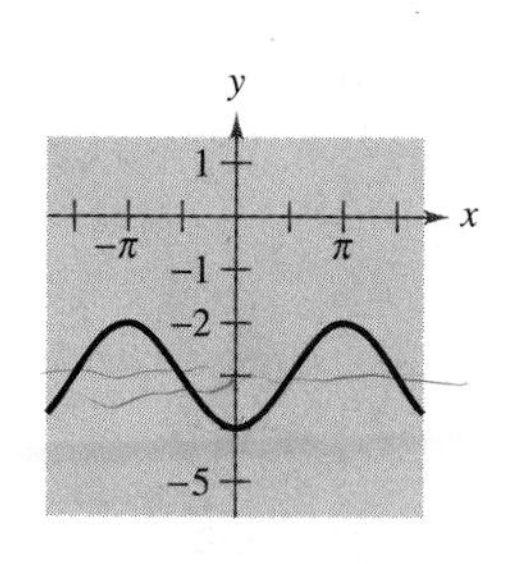

Graphical Reasoning **In Exercises 73–76, find a, b, and c for the function $y = a\sin(bx - c)$ so that the graph of f matches the figure.**

73.

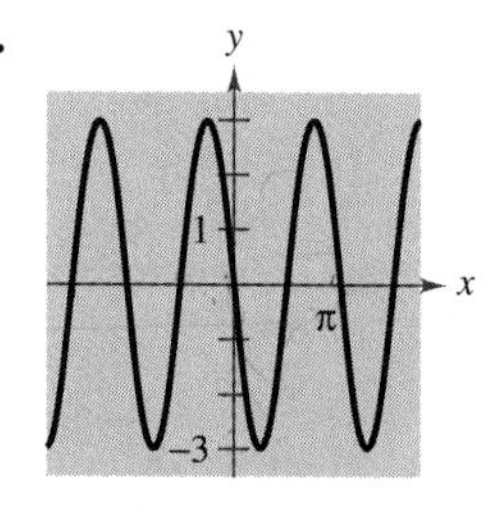

74.

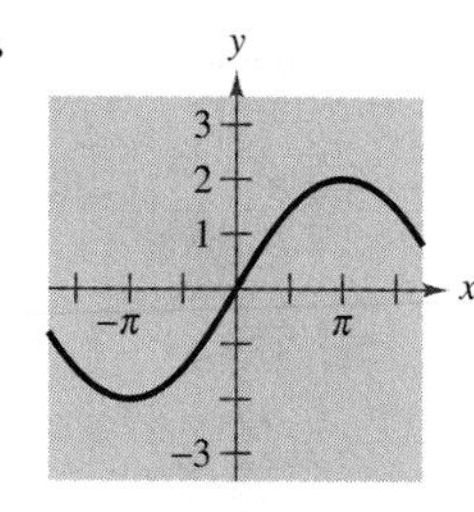

75.

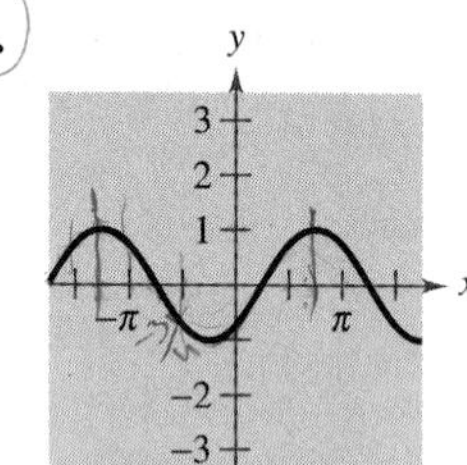

76.

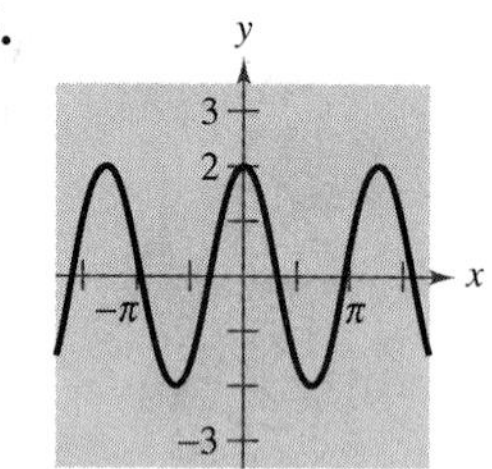

In Exercises 77–80, use a graphing utility to graph y_1 and y_2 in the interval $[-2\pi, 2\pi]$. Use the graphs to find real numbers x such that $y_1 = y_2$.

77. $y_1 = \sin x$
$y_2 = -\frac{1}{2}$

78. $y_1 = \cos x$
$y_2 = -1$

79. $y_1 = \cos x$
$y_2 = \frac{\sqrt{2}}{2}$

80. $y_1 = \sin x$
$y_2 = \frac{\sqrt{3}}{2}$

81. *Exploration* Use a graphing utility to graph h, and use the graph to decide whether h is even, odd, or neither.

(a) $h(x) = \cos^2 x$ (b) $h(x) = \sin^2 x$

82. *Conjecture* If f is an even function and g is an odd function, use the results of Exercise 81 to make a conjecture about h where

(a) $h(x) = [f(x)]^2$ (b) $h(x) = [g(x)]^2$.

83. *Respiratory Cycle* For a person at rest, the velocity v (in liters per second) of air flow during a respiratory cycle is

$$v = 0.85 \sin \frac{\pi t}{3}$$

where t is the time in seconds. (Inhalation occurs when $v > 0$, and exhalation occurs when $v < 0$.)

(a) Find the time for one full respiratory cycle.

(b) Find the number of cycles per minute.

(c) Sketch the graph of the velocity function.

84. *Respiratory Cycle* After exercising for a few minutes, a person has a respiratory cycle for which the velocity of air flow is approximated by

$$v = 1.75 \sin \frac{\pi t}{2}.$$

Use this model to repeat Exercise 83.

85. *Piano Tuning* When tuning a piano, a technician strikes a tuning fork for the A above middle C and sets up wave motion that can be approximated by

$$y = 0.001 \sin 880\pi t$$

where t is the time in seconds.

(a) What is the period of the function?

(b) The frequency f is given by $f = 1/p$. What is the frequency of the note?

86. *Blood Pressure* The function

$$P = 100 - 20 \cos \frac{5\pi t}{3}$$

approximates the blood pressure P in millimeters of mercury at time t in seconds for a person at rest.

(a) Find the period of the function.

(b) Find the number of heartbeats per minute.

Sales **In Exercises 87 and 88, use a graphing utility to graph the sales function over 1 year where S is the sales in thousands of units and t is the time in months, with $t = 1$ corresponding to January.**

87. $S = 22.3 - 3.4 \cos \dfrac{\pi t}{6}$

88. $S = 74.50 + 43.75 \sin \dfrac{\pi t}{6}$

89. *Data Analysis* The table gives the normal daily high temperatures for Honolulu H and Chicago C (in degrees Fahrenheit) for month t with $t = 1$ corresponding to January. (Source: National Oceanic and Atmospheric Association)

t	1	2	3	4	5	6
H	80.1	80.5	81.6	82.8	84.7	86.5
C	29.0	33.5	45.8	58.6	70.1	79.6

t	7	8	9	10	11	12
H	87.5	88.7	88.5	86.9	84.1	81.2
C	83.7	81.8	74.8	63.3	48.4	34.0

(a) A model for Honolulu is given by

$$H(t) = 84.40 + 4.28 \sin\left(\frac{\pi t}{6} + 3.86\right).$$

Find a trigonometric model for Chicago.

(b) Use a graphing utility to graph the data points and the model for the temperatures in Honolulu. How well does the model fit?

(c) Use a graphing utility to graph the data points and the model for the temperatures in Chicago. How well does the model fit?

(d) Use the models to estimate the average annual temperature in each city. Which term of the models did you use? Explain.

(e) What is the period of each model? Are they what you expected? Explain.

(f) Which city has the greater variability in temperature throughout the year? Which factor of the models determines this variability? Explain.

90. *Fuel Consumption* The daily consumption C (in gallons) of diesel fuel on a farm is modeled by

$$C = 30.3 + 21.6 \sin\left(\frac{2\pi t}{365} + 10.9\right)$$

where t is the time in days, with $t = 1$ corresponding to January 1.

(a) What is the period of the model? Is it what you expected? Explain.

(b) What is the average daily fuel consumption? Which term of the model did you use? Explain.

(c) Use a graphing utility to graph the model. Use the graph to approximate the time of the year when consumption exceeds 40 gallons per day.

91. *Exploration* Using calculus, it can be shown that the sine and cosine functions can be approximated by the polynomials

$$\sin x \approx x - \frac{x^3}{3!} + \frac{x^5}{5!} \text{ and } \cos x \approx 1 - \frac{x^2}{2!} + \frac{x^4}{4!}$$

where x is in radians.

(a) Use a graphing utility to graph the sine function and its polynomial approximation on the same viewing rectangle. How do the graphs compare?

(b) Use a graphing utility to graph the cosine function and its polynomial approximation on the same viewing rectangle. How do the graphs compare?

(c) Study the patterns in the polynomial approximations of the sine and cosine functions and guess the next term in each. Then repeat parts (a) and (b). How did the accuracy of the approximations change when additional terms were added?

92. *Exploration* Use the polynomial approximations for the sine and cosine functions given in Exercise 91 to approximate the following functional values. Compare the results with those given by a calculator. Is the error in the approximation the same in each case? Explain.

(a) $\sin \frac{1}{2}$ (b) $\sin 1$ (c) $\sin \frac{\pi}{6}$

(d) $\cos(-0.5)$ (e) $\cos 1$ (f) $\cos \frac{\pi}{4}$

93. *Data Analysis* The percent y of the moon's face that is illuminated on day x of the year 1995, where $x = 300$ represents October 27, is given in the table. (Source: American Museum of Natural History)

x	303	311	319	326	333	341
y	0.5	1.0	0.5	0	0.5	1.0

(a) Create a scatter plot of the data.

(b) Find a trigonometric model that fits the data.

(c) Add the graph of your model in part (b) to the scatter plot. How well does the model fit the data?

(d) Find the moon's percent illumination for December 22, 1995.

1.6 Graphs of Other Trigonometric Functions

See Exercise 68 on page 180 for an example of how a trigonometric function can be used to analyze the distance between a television camera and a parade unit.

Graph of the Tangent Function ▫ *Graph of the Cotangent Function* ▫ *Graphs of the Reciprocal Functions* ▫ *Damped Trigonometric Graphs*

Graph of the Tangent Function

Recall from Section 1.2 that the tangent function is odd. That is,

$$\tan(-x) = -\tan x.$$

Consequently, the graph of

$$y = \tan x$$

is symmetric with respect to the origin. You also know from the identity $\tan x = \sin x/\cos x$ that the tangent is undefined when $\cos x = 0$. Two such values are $x = \pm\pi/2 \approx \pm 1.5708$.

x	$-\frac{\pi}{2}$	-1.57	-1.5	-1	0	1	1.5	1.57	$\frac{\pi}{2}$
$\tan x$	Undef.	-1255.8	-14.1	-1.56	0	1.56	14.1	1255.8	Undef.

tan x approaches $-\infty$ as x approaches $-\pi/2$ from the right

tan x approaches ∞ as x approaches $\pi/2$ from the left

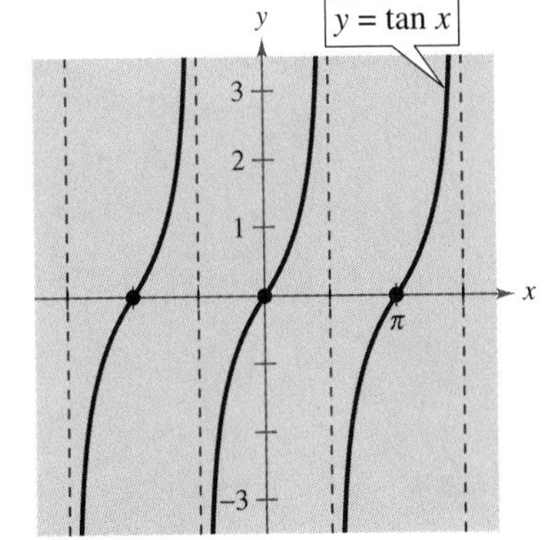

Period: π
Domain: all $x \neq \frac{\pi}{2} + n\pi$
Range: $(-\infty, \infty)$
Vertical asymptotes: $x = \frac{\pi}{2} + n\pi$

FIGURE 1.51

As indicated in the table, tan x increases without bound as x approaches $\pi/2$ from the left, and decreases without bound as x approaches $-\pi/2$ from the right. Thus, the graph of $y = \tan x$ has *vertical asymptotes* at $x = \pi/2$ and $x = -\pi/2$, as shown in Figure 1.51. Moreover, because the period of the tangent function is π, vertical asymptotes also occur when $x = \pi/2 + n\pi$, where n is an integer. The domain of the tangent function is the set of all real numbers other than $x = \pi/2 + n\pi$, and the range is the set of all real numbers.

Sketching the graph of a function of the form $y = a\tan(bx - c)$ is similar to sketching the graph of $y = a\sin(bx - c)$ in that you locate key points that identify the intercepts and asymptotes. Two consecutive asymptotes can be found by solving the equations

$$bx - c = -\frac{\pi}{2} \quad \text{and} \quad bx - c = \frac{\pi}{2}.$$

The midpoint between two consecutive asymptotes is an x-intercept of the graph. After plotting the asymptotes and the x-intercept, plot a few additional points between the two asymptotes and sketch one cycle. Finally sketch one or two additional cycles to the left and right.

NOTE The period of the function $y = a\tan(bx - c)$ is the distance between two consecutive asymptotes. The amplitude of a tangent function is not defined.

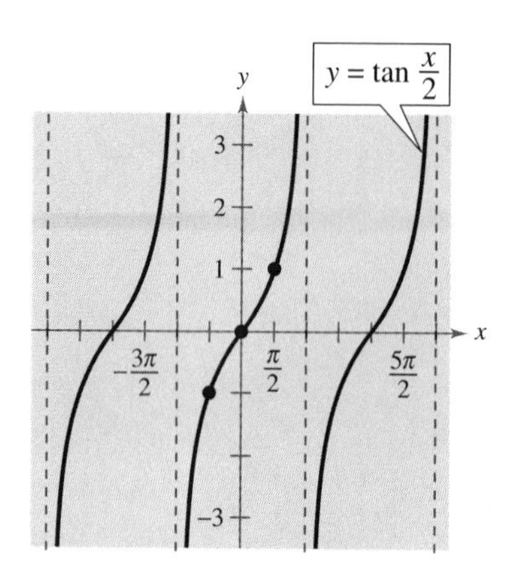

FIGURE 1.52

EXAMPLE 1 Sketching the Graph of a Tangent Function

Sketch the graph of $y = \tan \frac{x}{2}$.

Solution

By solving the equations

$$\frac{x}{2} = -\frac{\pi}{2} \qquad \text{and} \qquad \frac{x}{2} = \frac{\pi}{2}$$

$$x = -\pi \qquad\qquad\qquad x = \pi$$

you can see that two consecutive asymptotes occur at $x = -\pi$ and $x = \pi$. Between these two asymptotes, plot a few points, including the x-intercept, as shown in the table. Three cycles of the graph are shown in Figure 1.52.

x	$-\frac{\pi}{2}$	0	$\frac{\pi}{2}$
$\tan \frac{x}{2}$	-1	0	1

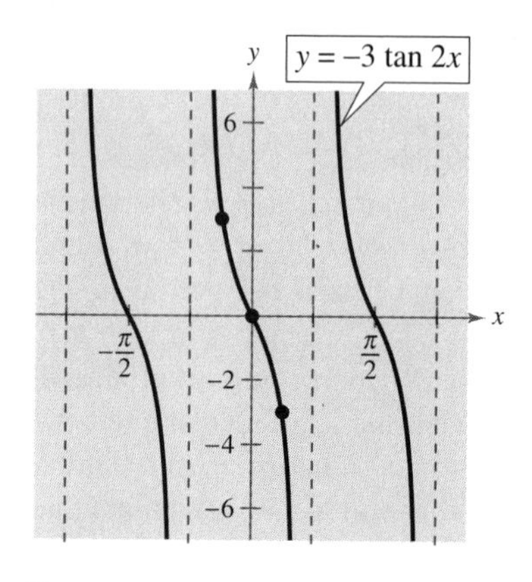

FIGURE 1.53

EXAMPLE 2 Sketching the Graph of a Tangent Function

Sketch the graph of $y = -3 \tan 2x$.

Solution

By solving the equations

$$2x = -\frac{\pi}{2} \qquad \text{and} \qquad 2x = \frac{\pi}{2}$$

$$x = -\frac{\pi}{4} \qquad\qquad\qquad x = \frac{\pi}{4}$$

you can see that two consecutive asymptotes occur at $x = -\pi/4$ and $x = \pi/4$. Between these two asymptotes, plot a few points, including the x-intercept, as shown in the table. Three cycles of the graph are shown in Figure 1.53.

x	$-\frac{\pi}{8}$	0	$\frac{\pi}{8}$
$-3 \tan 2x$	3	0	-3

By comparing the graphs in Examples 1 and 2, you can see that the graph of $y = a\tan(bx - c)$ is increasing between consecutive vertical asymptotes if $a > 0$, and decreasing between consecutive vertical asymptotes if $a < 0$. In other words, the graph for $a < 0$ is a reflection in the x-axis of the graph for $a > 0$.

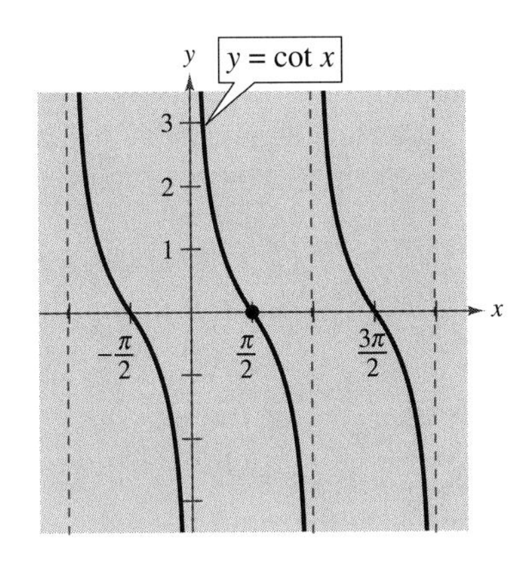

Period: π
Domain: all $x \neq n\pi$
Range: $(-\infty, \infty)$
Vertical asymptotes: $x = n\pi$

FIGURE 1.54

Graph of the Cotangent Function

The graph of the cotangent function is similar to the graph of the tangent function. It also has a period of π. However, from the identity

$$y = \cot x = \frac{\cos x}{\sin x}$$

you can see that the cotangent function has vertical asymptotes at $x = n\pi$ where n is an integer, because $\sin x$ is zero at these x-values. The graph of the cotangent function is shown in Figure 1.54.

EXAMPLE 3 *Sketching the Graph of a Cotangent Function*

Sketch the graph of $y = 2\cot\dfrac{x}{3}$.

Solution

To locate two consecutive vertical asymptotes of the graph, solve the equations $x/3 = 0$ and $x/3 = \pi$, as follows.

$$\frac{x}{3} = 0 \qquad \text{and} \qquad \frac{x}{3} = \pi$$

$$x = 0 \qquad\qquad x = 3\pi$$

Then, between these two asymptotes, plot a few points, including the x-intercept, as shown in the table. Three cycles of the graph are shown in Figure 1.55. (Note that the period is 3π, the distance between consecutive asymptotes.)

x	$\frac{3\pi}{4}$	$\frac{3\pi}{2}$	$\frac{9\pi}{4}$
$2\cot\frac{x}{3}$	2	0	−2

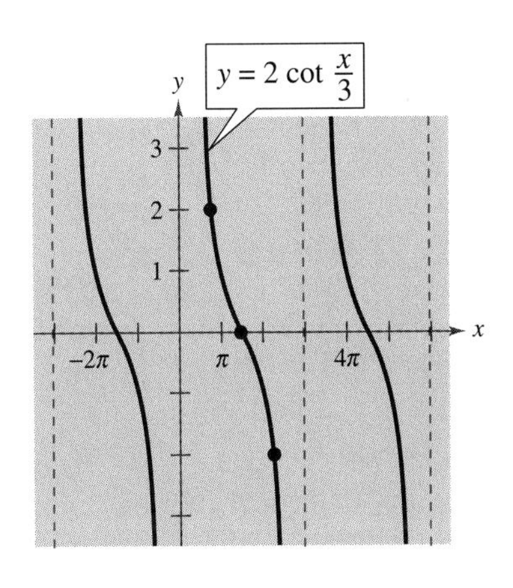

FIGURE 1.55

Graphs of the Reciprocal Functions

The graphs of the two remaining trigonometric functions can be obtained from the graphs of the sine and cosine functions using the reciprocal identities

$$\csc x = 1/\sin x \qquad \text{and} \qquad \sec x = 1/\cos x.$$

For instance, at a given value of x, the y-coordinate of $\sec x$ is the reciprocal of the y-coordinate of $\cos x$. Of course, when $\cos x = 0$, the reciprocal does not exist. Near such values of x, the behavior of the secant function is similar to that of the tangent function. In other words, the graphs of

$$\tan x = \sin x/\cos x \qquad \text{and} \qquad \sec x = 1/\cos x$$

have vertical asymptotes at $x = \pi/2 + n\pi$, where n is an integer and the cosine is zero at these x-values. Similarly,

$$\cot x = \cos x/\sin x \qquad \text{and} \qquad \csc x = 1/\sin x$$

have vertical asymptotes where $\sin x = 0$—that is, at $x = n\pi$.

To sketch the graph of a secant or cosecant function, we suggest that you first make a sketch of its reciprocal function. For instance, the sketch the graph of $y = \csc x$, first sketch the graph of $y = \sin x$. Then take reciprocals of the y-coordinates to obtain points on the graph of $y = \csc x$. We use this procedure to obtain the graphs shown in Figure 1.56.

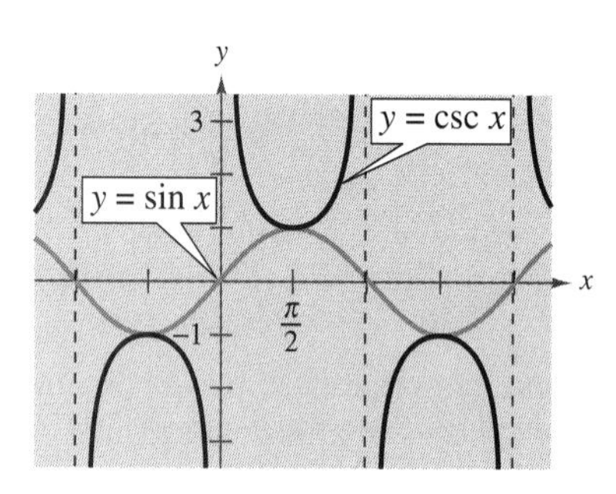

Period: 2π
Domain: all $x \neq n\pi$
Range: $(-\infty, -1]$ and $[1, \infty)$
Vertical asymptotes: $x = n\pi$
Symmetry: origin

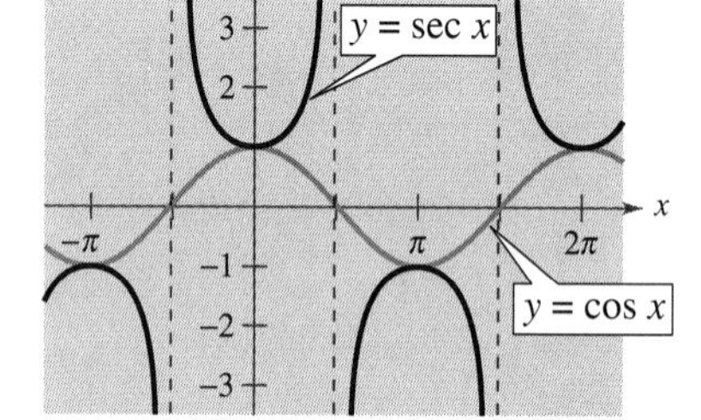

Period: 2π
Domain: all $x \neq \frac{\pi}{2} + n\pi$
Range: $(-\infty, -1]$ and $[1, \infty)$
Vertical asymptotes: $x = \frac{\pi}{2} + n\pi$
Symmetry: y-axis

FIGURE 1.56

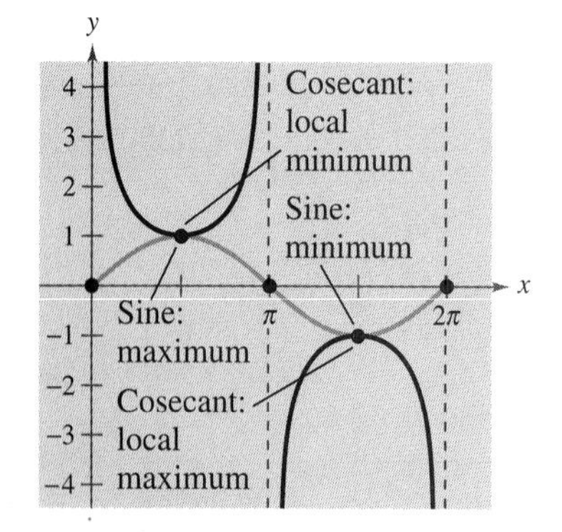

FIGURE 1.57

In comparing the graphs of the secant and cosecant functions with those of the sine and cosine functions, note that the "hills" and "valleys" are interchanged. For example, a hill (or maximum point) on the sine curve corresponds to a valley (a local minimum) on the cosecant curve. Similarly, a valley (or minimum point) on the sine curve corresponds to a hill (a local maximum) on the cosecant curve, as shown in Figure 1.57.

EXAMPLE 4 Sketching the Graph of a Cosecant Function

Sketch the graph of $y = 2\csc\left(x + \dfrac{\pi}{4}\right)$.

Solution

Begin by sketching the graph of

$$y = 2\sin\left(x + \frac{\pi}{4}\right).$$

For this function, the amplitude is 2 and the period is 2π. By solving the equations

$$x + \frac{\pi}{4} = 0 \qquad \text{and} \qquad x + \frac{\pi}{4} = 2\pi$$

$$x = -\frac{\pi}{4} \qquad\qquad x = \frac{7\pi}{4}$$

you can see that one cycle of the sine function corresponds to the interval from $x = -\pi/4$ to $x = 7\pi/4$. The graph of this sine function is represented by the gray curve in Figure 1.58. Because the sine function is zero at the endpoints of this interval, the corresponding cosecant function

$$y = 2\csc\left(x + \frac{\pi}{4}\right) = 2\left(\frac{1}{\sin[x + (\pi/4)]}\right)$$

has vertical asymptotes at $x = -\pi/4$, $x = 3\pi/4$, $x = 7\pi/4$, etc. The graph of the cosecant function is represented by the black curve in Figure 1.58.

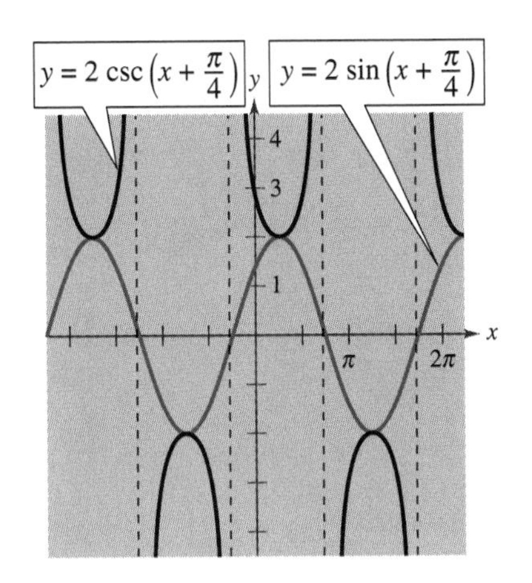

FIGURE 1.58

EXAMPLE 5 Sketching the Graph of a Secant Function

Sketch the graph of $y = \sec 2x$.

Solution

Begin by sketching the graph of $y = \cos 2x$, as indicated by the gray curve in Figure 1.59. Then, form the graph of $y = \sec 2x$ as the black curve in the figure. Note that the x-intercepts of $y = \cos 2x$

$$\left(\frac{\pi}{4}, 0\right),\ \left(\frac{3\pi}{4}, 0\right),\ \left(\frac{5\pi}{4}, 0\right), \ldots$$

correspond to the vertical asymptotes

$$x = \frac{\pi}{4},\quad x = \frac{3\pi}{4},\quad x = \frac{5\pi}{4}, \ldots$$

of the graph of $y = \sec 2x$.

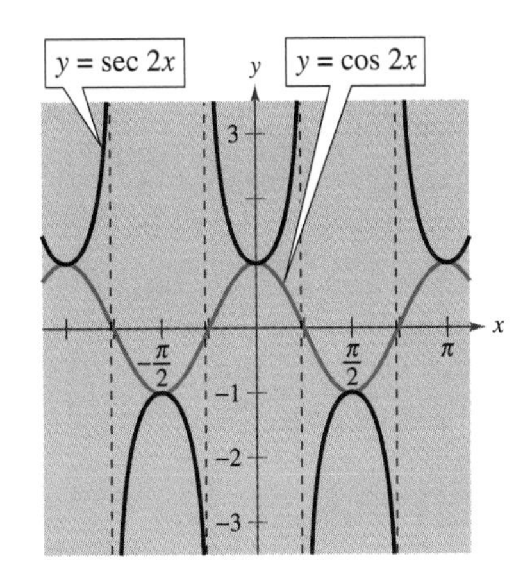

FIGURE 1.59

Damped Trigonometric Graphs

A *product* of two functions can be graphed using properties of the individual functions. For instance, consider the function

$$f(x) = x \sin x$$

as the product of the functions $y = x$ and $y = \sin x$. Using properties of absolute value and the fact that $|\sin x| \le 1$, we have $0 \le |x||\sin x| \le |x|$. Consequently,

$$-|x| \le x \sin x \le |x|$$

which means that the graph of $f(x) = x \sin x$ lies between the lines $y = -x$ and $y = x$. Furthermore, because

$$f(x) = x \sin x = \pm x \quad \text{at} \quad x = \frac{\pi}{2} + n\pi$$

and

$$f(x) = x \sin x = 0 \quad \text{at} \quad x = n\pi$$

the graph of f touches the line $y = -x$ or the line $y = x$ at $x = \pi/2 + n\pi$ and has x-intercepts at $x = n\pi$. A sketch of f is shown in Figure 1.60. In the function $f(x) = x \sin x$, the factor x is called the **damping factor.**

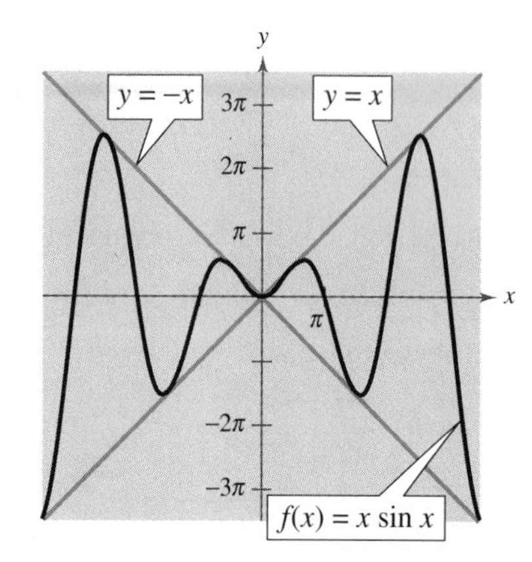

FIGURE 1.60

EXAMPLE 6 *Damped Sine Wave*

Sketch the graph of $f(x) = e^{-x} \sin 3x$.

Solution

Consider $f(x)$ as the product of the two functions

$$y = e^{-x} \quad \text{and} \quad y = \sin 3x$$

each of which has the set of real numbers as its domain. For any real number x, you know that $e^{-x} \ge 0$ and $|\sin 3x| \le 1$. Therefore, $e^{-x}|\sin 3x| \le e^{-x}$, which means that

$$-e^{-x} \le e^{-x} \sin 3x \le e^{-x}.$$

Furthermore, because

$$f(x) = e^{-x} \sin 3x = \pm e^{-x} \quad \text{at} \quad x = \frac{\pi}{6} + \frac{n\pi}{3}$$

and

$$f(x) = e^{-x} \sin 3x = 0 \quad \text{at} \quad x = \frac{n\pi}{3}$$

the graph of f touches the curve $y = -e^{-x}$ and $y = e^{-x}$ at $x = \pi/6 + n\pi/3$ and has intercepts at $x = n\pi/3$. A sketch is shown in Figure 1.61.

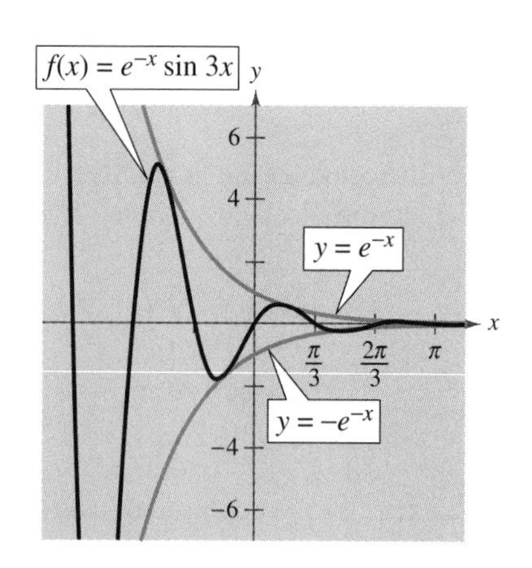

FIGURE 1.61

Figure 1.62 summarizes the six basic trigonometric functions.

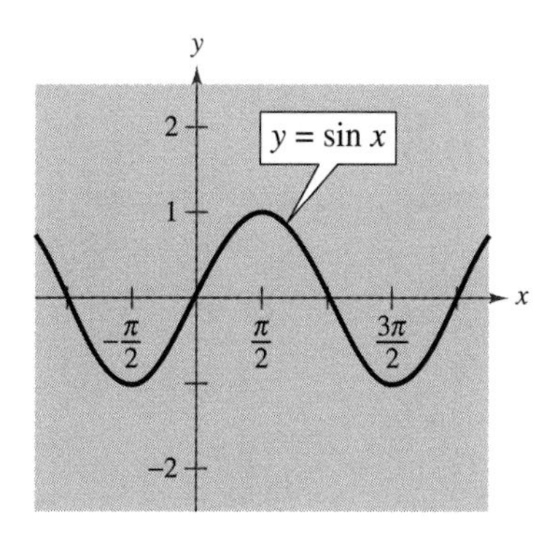

Domain: all reals
Range: $[-1, 1]$
Period: 2π

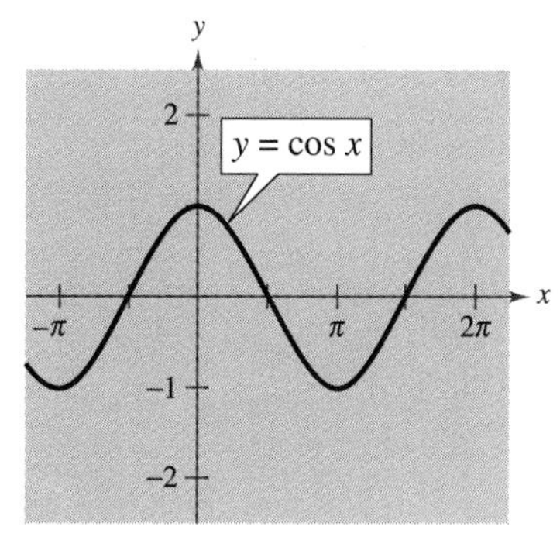

Domain: all reals
Range: $[-1, 1]$
Period: 2π

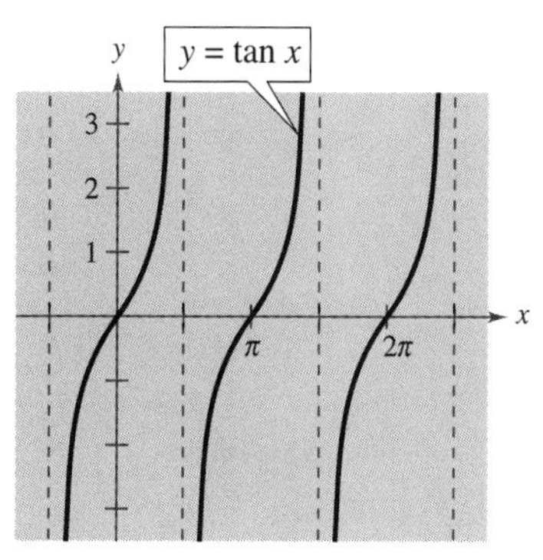

Domain: all $x \neq \frac{\pi}{2} + n\pi$
Range: $(-\infty, \infty)$
Period: π

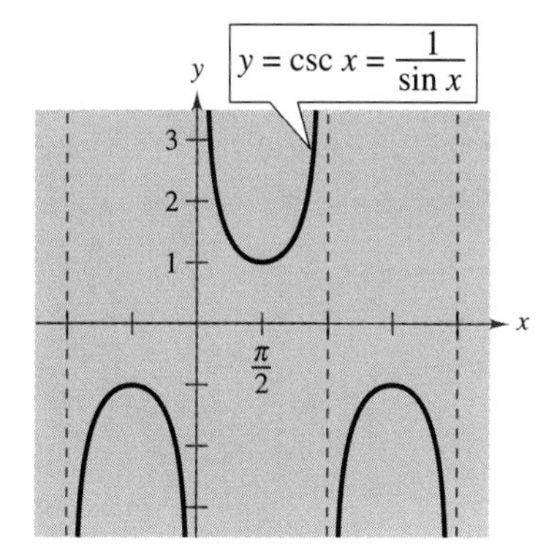

Domain: all $x \neq n\pi$
Range: $(-\infty, -1]$ and $[1, \infty)$
Period: 2π

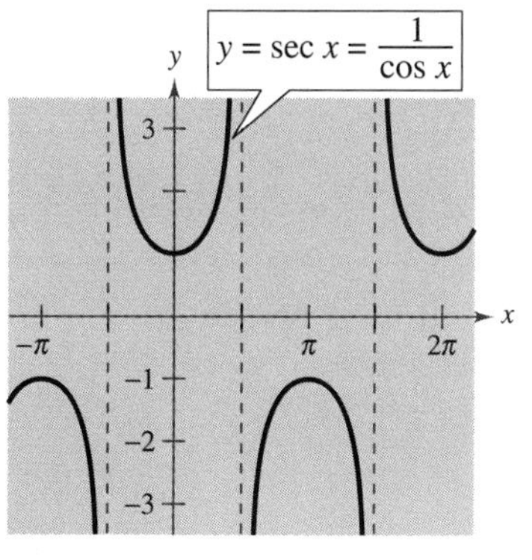

Domain: all $x \neq \frac{\pi}{2} + n\pi$
Range: $(-\infty, -1]$ and $[1, \infty)$
Period: 2π

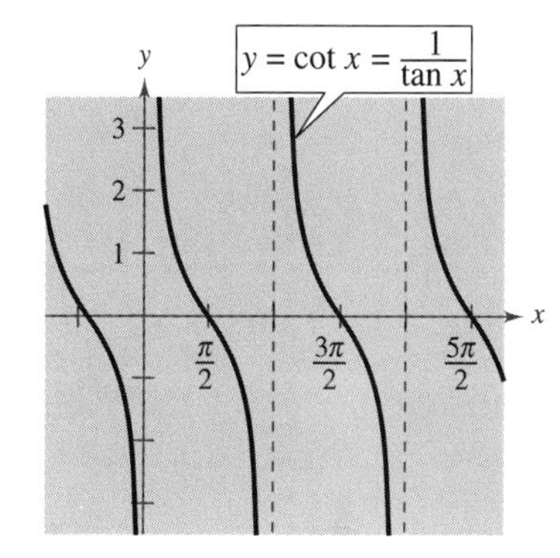

Domain: all $x \neq n\pi$
Range: $(-\infty, \infty)$
Period: π

FIGURE 1.62

GROUP ACTIVITY

COMBINING TRIGONOMETRIC FUNCTIONS

Recall from Section P.7 that functions can be combined arithmetically. This also applies to trigonometric functions. For each of the functions $h(x) = x + \sin x$ and $h(x) = \cos x - \sin 3x$, (a) identify two simpler functions f and g that comprise the combination, (b) use a table to show how to obtain the numerical values of $h(x)$ from the numerical values of $f(x)$ and $g(x)$, and (c) use a graph of f and g to show how h may be formed.

Can you find functions $f(x) = d + a \sin(bx + c)$ and $g(x) = d + a \cos(bx + c)$ such that $f(x) + g(x) = 0$ for all x?

WARM UP

Evaluate the trigonometric function from memory.

1. $\tan 0$
2. $\cos \frac{\pi}{4}$
3. $\tan \frac{\pi}{4}$
4. $\cot \frac{\pi}{2}$
5. $\sin \pi$
6. $\cos \frac{\pi}{2}$

Sketch the graph of the function. (Include two full periods.)

7. $y = -2 \cos 2x$
8. $y = 3 \sin \frac{x}{4}$
9. $y = \frac{3}{2} \sin 2\pi x$
10. $y = -2 \cos \frac{\pi x}{2}$

1.6 Exercises

In Exercises 1–8, match the function with its graph. State the period of the function. [The graphs are labeled (a), (b), (c), (d), (e), (f), (g), and (h).]

(a)

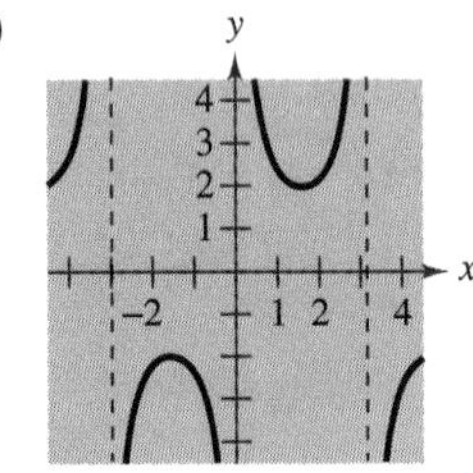

(b)

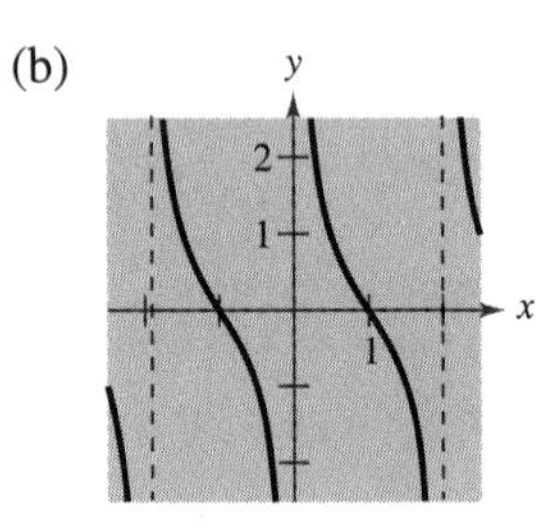

(c)

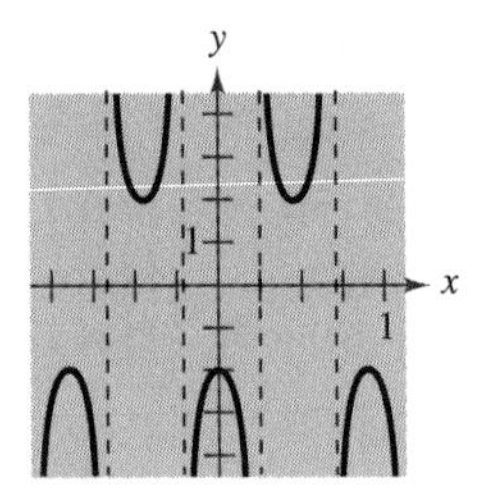

(d)

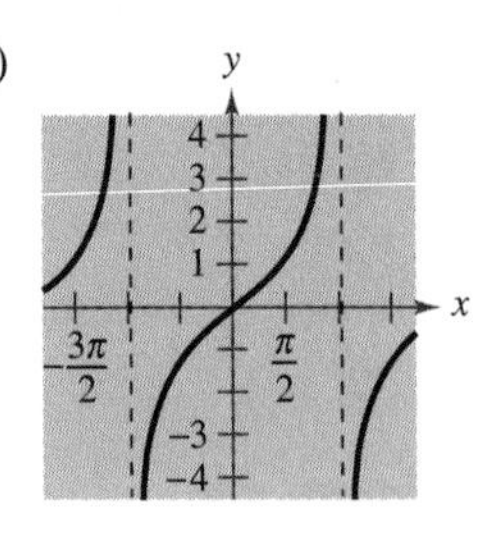

(e)

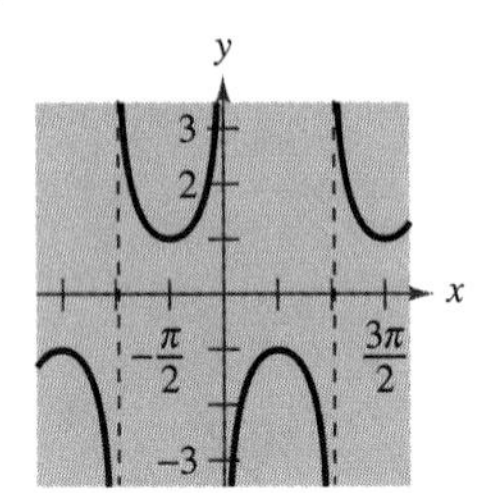

(f)

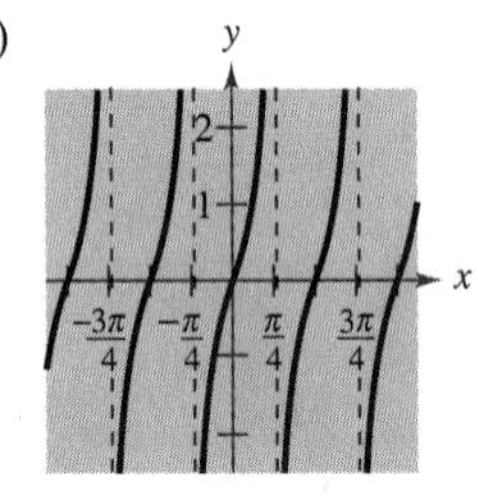

(g)

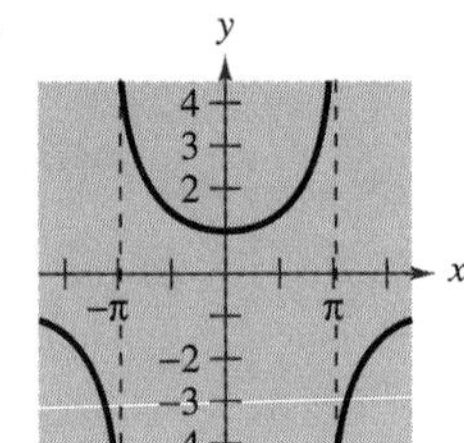

(h)

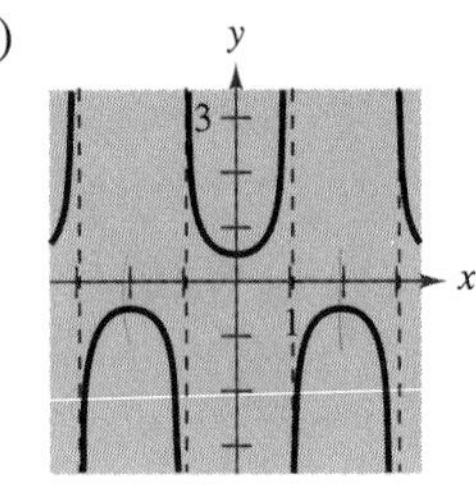

1. $y = \sec \frac{x}{2}$
2. $y = \tan \frac{x}{2}$
3. $y = \tan 2x$
4. $y = 2 \csc x$
5. $y = \cot \frac{\pi x}{2}$
6. $y = \frac{1}{2} \sec \frac{\pi x}{2}$
7. $y = -\csc x$
8. $y = -2 \sec 2\pi x$

In Exercises 9–30, sketch the graph of the function. (Include two full periods.)

9. $y = \frac{1}{3} \tan x$
10. $y = \frac{1}{4} \tan x$
11. $y = \tan 2x$
12. $y = -3 \tan \pi x$
13. $y = -\frac{1}{2} \sec x$
14. $y = \frac{1}{4} \sec x$
15. $y = \sec \pi x$
16. $y = 2 \sec 4x$
17. $y = \sec \pi x - 1$
18. $y = -2 \sec 4x + 2$
19. $y = \csc \frac{x}{2}$
20. $y = \csc \frac{x}{3}$
21. $y = \cot \frac{x}{2}$
22. $y = 3 \cot \frac{\pi x}{2}$
23. $y = \frac{1}{2} \sec 2x$
24. $y = -\frac{1}{2} \tan x$
25. $y = \tan \frac{\pi x}{4}$
26. $y = \sec(x + \pi)$
27. $y = \csc(\pi - x)$
28. $y = \sec(\pi - x)$
29. $y = \frac{1}{4} \csc\left(x + \frac{\pi}{4}\right)$
30. $y = 2 \cot\left(x + \frac{\pi}{2}\right)$

In Exercises 31–40, use a graphing utility to graph the function. (Include two full periods.)

31. $y = \tan \frac{x}{3}$
32. $y = -\tan 2x$
33. $y = -2 \sec 4x$
34. $y = \sec \pi x$
35. $y = \tan\left(x - \frac{\pi}{4}\right)$
36. $y = -\csc(4x - \pi)$
37. $y = \frac{1}{4} \cot\left(x - \frac{\pi}{2}\right)$
38. $y = 0.1 \tan\left(\frac{\pi x}{4} + \frac{\pi}{4}\right)$
39. $y = 2 \sec(2x - \pi)$
40. $y = \frac{1}{3} \sec\left(\frac{\pi x}{2} + \frac{\pi}{2}\right)$

In Exercises 41–44, use a graph to solve the equation on the interval $[-2\pi, 2\pi]$.

41. $\tan x = 1$
42. $\cot x = -\sqrt{3}$
43. $\sec x = -2$
44. $\csc x = \sqrt{2}$

In Exercises 45 and 46, use the graph of the function to determine whether the function is even, odd, or neither.

45. $f(x) = \sec x$
46. $f(x) = \tan x$

47. ***Essay*** Describe the behavior of $f(x) = \tan x$ as x approaches $\pi/2$ from the left and from the right.

48. ***Essay*** Describe the behavior of $f(x) = \csc x$ as x approaches π from the left and from the right.

49. ***Graphical Reasoning*** Consider the functions

$$f(x) = 2 \sin x \quad \text{and} \quad g(x) = \frac{1}{2} \csc x$$

on the interval $(0, \pi)$.

(a) Graph f and g in the same coordinate plane.

(b) Approximate the interval where $f > g$.

(c) Describe the behavior of each of the functions as x approaches π. How is the behavior of g related to the behavior of f as x approaches π?

50. ***Graphical Reasoning*** Consider the functions

$$f(x) = \tan \frac{\pi x}{2} \quad \text{and} \quad g(x) = \frac{1}{2} \sec \frac{\pi x}{2}$$

on the interval $(-1, 1)$.

(a) Use a graphing utility to graph f and g on the same viewing rectangle.

(b) Approximate the interval where $f < g$.

(c) Approximate the interval where $2f < 2g$. How does the result compare with that of part (b)? Explain.

In Exercises 51–54, use a graphing utility to graph the two equations on the same viewing rectangle. Determine analytically whether the expressions are equivalent.

51. $y_1 = \sin x \csc x, \quad y_2 = 1$
52. $y_1 = \sin x \sec x, \quad y_2 = \tan x$
53. $y_1 = \frac{\cos x}{\sin x}, \quad y_2 = \cot x$
54. $y_1 = \sec^2 x - 1, \quad y_2 = \tan^2 x$

In Exercises 55–58, match the function with its graph. Describe the behavior of the function as x approaches zero. [The graphs are labeled (a), (b), (c), and (d).]

(a)

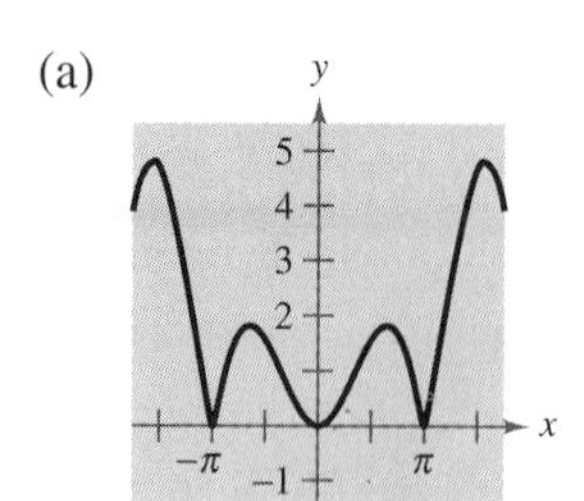

(b)

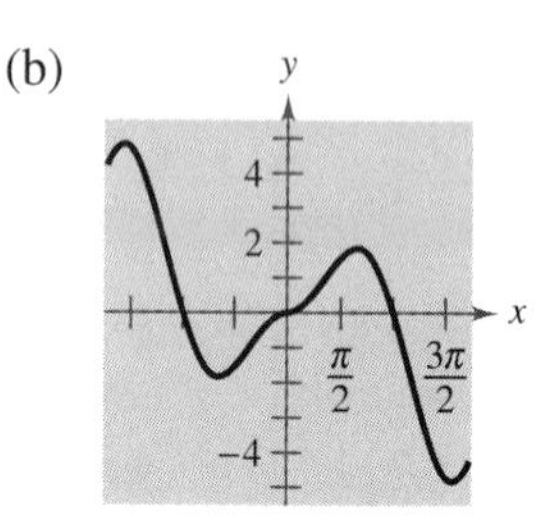

(c)

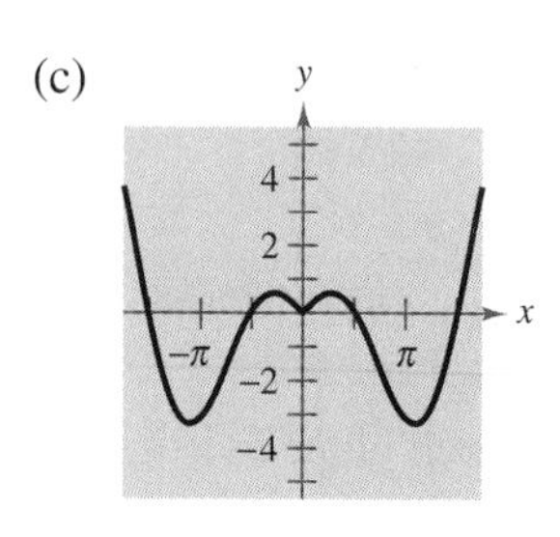

(d)

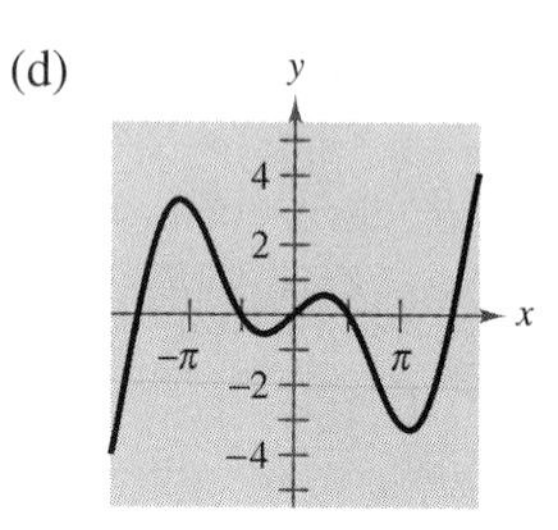

55. $f(x) = x \cos x$

56. $f(x) = |x \sin x|$

57. $g(x) = |x| \sin x$

58. $g(x) = |x| \cos x$

Conjecture **In Exercises 59–62, graph the functions of f and g. Use the graphs to make a conjecture about the relationship between the functions.**

59. $f(x) = \sin x + \cos\left(x + \dfrac{\pi}{2}\right), \quad g(x) = 0$

60. $f(x) = \sin x - \cos\left(x + \dfrac{\pi}{2}\right), \quad g(x) = 2 \sin x$

61. $f(x) = \sin^2 x, \quad g(x) = \frac{1}{2}(1 - \cos 2x)$

62. $f(x) = \cos^2 \dfrac{\pi x}{2}, \quad g(x) = \dfrac{1}{2}(1 + \cos \pi x)$

In Exercises 63–66, use a graphing utility to graph the function and the damping factor of the function on the same viewing rectangle. Describe the behavior of the function as x increases without bound.

63. $f(x) = 2^{-x/4} \cos \pi x$

64. $f(x) = e^{-x} \cos x$

65. $g(x) = e^{-x^2/2} \sin x$

66. $h(x) = 2^{-x^2/4} \sin x$

67. ***Distance*** A plane flying at an altitude of 5 miles over level ground will pass directly over a radar antenna (see figure). Let d be the ground distance from the antenna to the point directly under the plane and let x be the angle of elevation to the plane from the antenna. Write d as a function of x and graph the function over the interval $0 < x < \pi$.

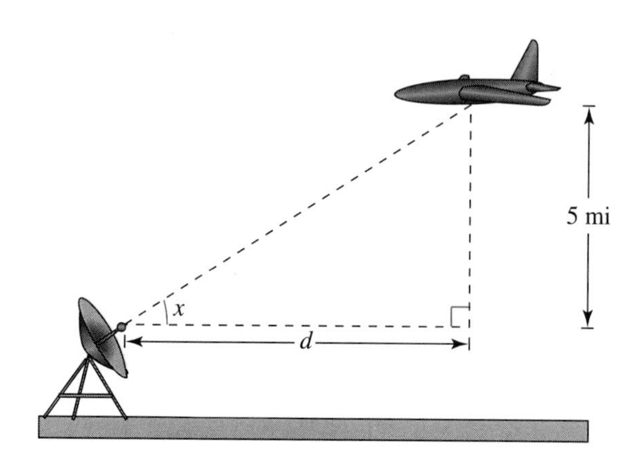

68. ***Television Coverage*** A television camera is on a reviewing platform 36 meters from the street on which a parade will be passing from left to right (see figure). Express the distance d from the camera to a particular unit in the parade as a function of the angle x, and graph the function over the interval $-\pi/2 < x < \pi/2$. (Consider x as negative when a unit in the parade approaches from the left.)

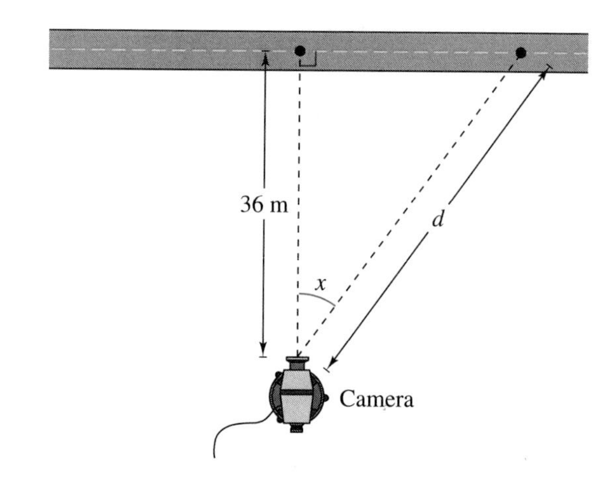

69. ***Predator-Prey Model*** Suppose the population of a certain predator at time t (in months) in a given region is estimated to be

$$P = 10{,}000 + 3000 \sin \frac{2\pi t}{24}.$$

and the population of its primary food source (its prey) is estimated to be

$$p = 15{,}000 + 5000 \cos \frac{2\pi t}{24}.$$

Use the graphs of the models to explain the oscillations in the size of each population.

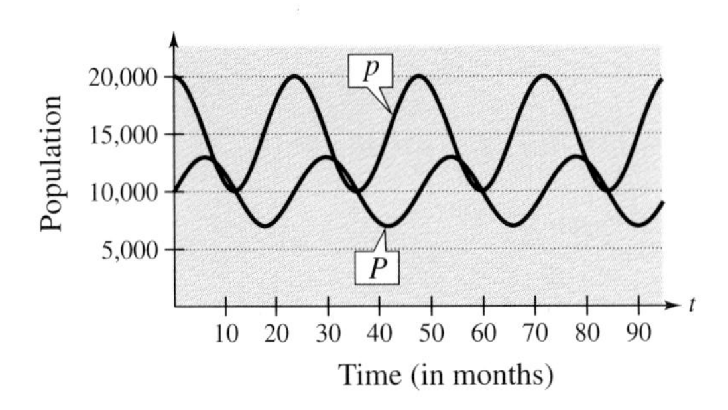

70. ***Normal Temperatures*** The normal monthly high temperatures in degrees Fahrenheit for Erie, Pennsylvania are approximated by

$$H(t) = 54.33 - 20.38 \cos \frac{\pi t}{6} - 15.69 \sin \frac{\pi t}{6}$$

and the normal monthly low temperatures are approximated by

$$L(t) = 39.36 - 15.70 \cos \frac{\pi t}{6} - 14.16 \sin \frac{\pi t}{6}$$

where t is the time in months, with $t = 1$ corresponding to January (see figure). (Source: National Oceanic and Atmospheric Association)

(a) What is the period of each function?

(b) During what part of the year is the difference between the normal high and low temperatures greatest? When is it smallest?

(c) The sun is northernmost in the sky around June 21, but the graph shows the warmest temperatures at a later date. Approximate the lag time of the temperatures relative to the position of the sun.

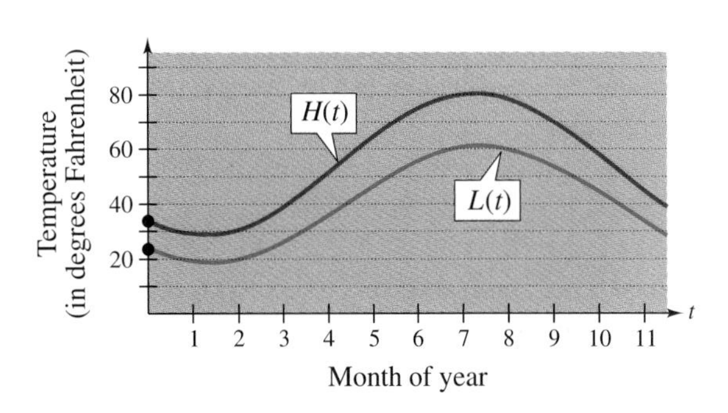

FIGURE FOR 70

71. ***Harmonic Motion*** An object weighing W pounds is suspended from the ceiling by a steel spring (see figure). The weight is pulled downward (positive direction) from its equilibrium position and released. The resulting motion of the weight is described by the function

$$y = \frac{1}{2}e^{-t/4} \cos 4t, \qquad t > 0$$

where y is the distance in feet and t is the time in seconds.

(a) Use a graphing utility to graph the function.

(b) Describe the behavior of the displacement function for increasing values of time t.

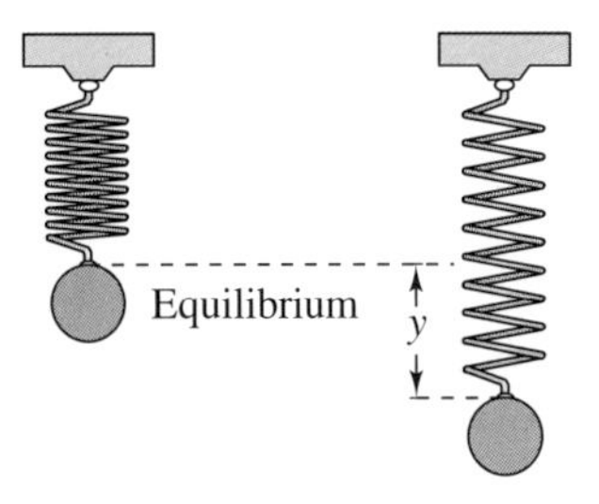

72. ***Exploration*** Consider the function

$$f(x) = x - \cos x.$$

(a) Use a graphing utility to graph the function and verify that there exists a zero between 0 and 1. Use the graph to approximate the zero.

(b) Starting with $x_0 = 1$, generate a sequence $x_1, x_2, x_3, \ldots$ where $x_n = \cos(x_{n-1})$. Verify that the sequence approaches the zero of f.

73. ***Approximation*** Using calculus, it can be shown that the tangent function can be approximated by the polynomial

$$\tan x \approx x + \frac{2x^3}{3!} + \frac{16x^5}{5!}$$

where x is in radians. Use a graphing utility to graph the tangent function and its polynomial approximation on the same viewing rectangle. How do the graphs compare?

74. ***Approximation*** Using calculus, it can be shown that the secant function can be approximated by the polynomial

$$\sec x \approx 1 + \frac{x^2}{2!} + \frac{5x^4}{4!}$$

where x is in radians. Use a graphing utility to graph the secant function and its polynomial approximation on the same viewing rectangle. How do the graphs compare?

75. ***Pattern Recognition***

(a) Use a graphing utility to graph each function.

$$y_1 = \frac{4}{\pi}\left(\sin \pi x + \frac{1}{3}\sin 3\pi x\right)$$

$$y_2 = \frac{4}{\pi}\left(\sin \pi x + \frac{1}{3}\sin 3\pi x + \frac{1}{5}\sin 5\pi x\right)$$

(b) Identify the pattern started in part (a) and find a function y_3 that continues the pattern one more term. Use a graphing utility to graph y_3.

(c) The graphs of parts (a) and (b) approximate the periodic function in the figure. Find a function y_4 that is a better approximation.

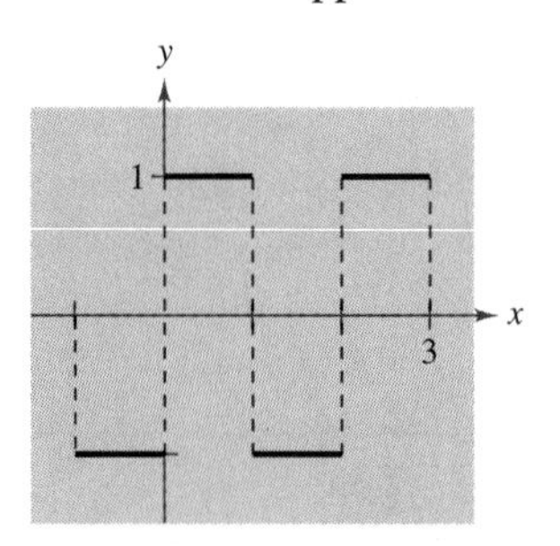

76. ***Sales*** The projected monthly sales S (in thousands of units) of a seasonal product are modeled by

$$S = 74 + 3t + 40 \sin \frac{\pi t}{6}$$

where t is the time in months, with $t = 1$ corresponding to January. Graph the sales function over 1 year.

Exploration **In Exercises 77–82, use a graphing utility to graph the function. Describe the behavior of the function as x approaches zero.**

77. $y = \dfrac{6}{x} + \cos x, \quad x > 0$

78. $y = \dfrac{4}{x} + \sin 2x, \quad x > 0$

79. $g(x) = \dfrac{\sin x}{x}$

80. $f(x) = \dfrac{1 - \cos x}{x}$

81. $f(x) = \sin \dfrac{1}{x}$

82. $h(x) = x \sin \dfrac{1}{x}$

1.7 Inverse Trigonometric Functions

See Exercise 85 on page 192 for an example of how an inverse trigonometric function can be used to analyze a photography setup.

Inverse Sine Function ▫ *Other Inverse Trigonometric Functions* ▫ *Compositions of Functions*

Inverse Sine Function

Recall from Section P.8 that, for a function to have an inverse, it must pass the Horizontal Line Test. From Figure 1.63 it is obvious that $y = \sin x$ does not pass the test because different values of x yield the same y-value. However, if you restrict the domain to the interval $-\pi/2 \le x \le \pi/2$ (corresponding to the black portion of the graph in Figure 1.63), the following properties hold.

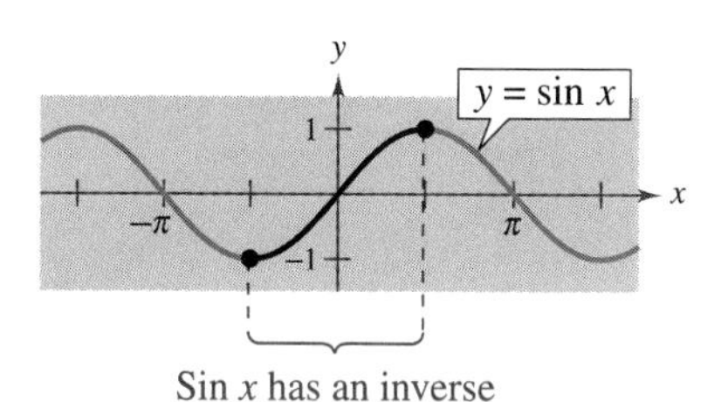

Sin x has an inverse on this interval.

FIGURE 1.63

1. On the interval $[-\pi/2, \pi/2]$, the function $y = \sin x$ is increasing.
2. On the interval $[-\pi/2, \pi/2]$, $y = \sin x$ takes on its full range of values, $-1 \le \sin x \le 1$.
3. On the interval $[-\pi/2, \pi/2]$, $y = \sin x$ passes the Horizontal Line Test.

Thus, on the restricted domain $-\pi/2 \le x \le \pi/2$, $y = \sin x$ has a unique inverse called the **inverse sine function.** It is denoted by

$$y = \arcsin x \qquad \text{or} \qquad y = \sin^{-1} x.$$

The notation $\sin^{-1}x$ is consistent with the inverse function notation $f^{-1}(x)$. The $\arcsin x$ notation (read as "the arcsine of x") comes from the association of a central angle with its subtended *arc length* on a unit circle. Thus, $\arcsin x$ means the angle (or arc) whose sine is x. Both notations, $\arcsin x$ and $\sin^{-1}x$, are commonly used in mathematics, so remember that $\sin^{-1}x$ denotes the *inverse* sine function rather than $1/\sin x$. The values of $\arcsin x$ lie in the interval

$$-\frac{\pi}{2} \le \arcsin x \le \frac{\pi}{2}.$$

The graph of $y = \arcsin x$ is shown in Example 2.

DEFINITION OF INVERSE SINE FUNCTION

The **inverse sine function** is defined by

$$y = \arcsin x \qquad \text{if and only if} \qquad \sin y = x$$

where $-1 \le x \le 1$ and $-\pi/2 \le y \le \pi/2$. The domain of $y = \arcsin x$ is $[-1, 1]$, and the range is $[-\pi/2, \pi/2]$.

NOTE When evaluating the inverse sine function, it helps to remember the phrase "the arcsine of x is the angle (or number) whose sine is x." ▪▪

Study Tip

As with the trigonometric functions, much of the work with the inverse trigonometric functions can be done by *exact* calculations rather than by calculator approximations. Exact calculations help to increase your understanding of the inverse functions by relating them to the triangle definitions of the trigonometric functions.

EXAMPLE 1 *Evaluating the Inverse Sine Function*

If possible, find the exact value.

a. $\arcsin\left(-\frac{1}{2}\right)$ **b.** $\sin^{-1}\frac{\sqrt{3}}{2}$ **c.** $\sin^{-1}2$

Solution

a. Because $\sin(-\pi/6) = -\frac{1}{2}$ for $-\pi/2 \le y \le \pi/2$, it follows that

$$\arcsin\left(-\frac{1}{2}\right) = -\frac{\pi}{6}.$$

b. Because $\sin(\pi/3) = \sqrt{3}/2$ for $-\pi/2 \le y \le \pi/2$, it follows that

$$\sin^{-1}\frac{\sqrt{3}}{2} = \frac{\pi}{3}.$$

c. It is not possible to evaluate $y = \sin^{-1}x$ when $x = 2$ because there is no angle whose sine is 2. Remember that the domain of the inverse sine function is $[-1, 1]$.

EXAMPLE 2 *Graphing the Arcsine Function*

Sketch a graph of $y = \arcsin x$.

Solution

By definition, the equations

$$y = \arcsin x \qquad \text{and} \qquad \sin y = x$$

are equivalent for $-\pi/2 \le y \le \pi/2$. Hence, their graphs are the same. From the interval $[-\pi/2, \pi/2]$, you can assign values to y in the second equation to make a table of values.

y	$-\frac{\pi}{2}$	$-\frac{\pi}{4}$	$-\frac{\pi}{6}$	0	$\frac{\pi}{6}$	$\frac{\pi}{4}$	$\frac{\pi}{2}$
$x = \sin y$	-1	$-\frac{\sqrt{2}}{2}$	$-\frac{1}{2}$	0	$\frac{1}{2}$	$\frac{\sqrt{2}}{2}$	1

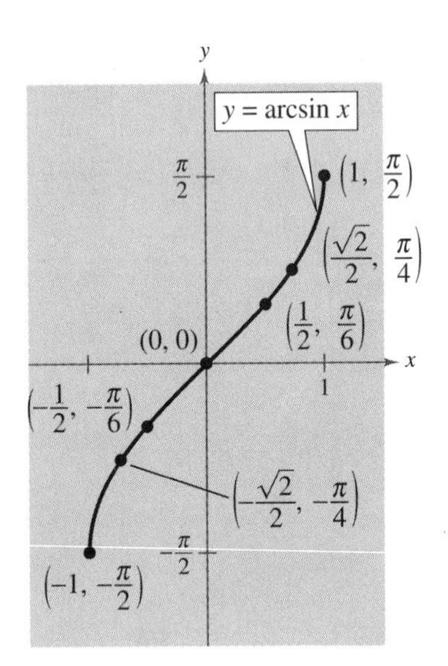

FIGURE 1.64

The resulting graph for $y = \arcsin x$ is shown in Figure 1.64. Note that it is the reflection (in line $y = x$) of the black portion of Figure 1.63. Be sure you see that Figure 1.64 shows the *entire* graph of the inverse sine function. Remember that the range of $y = \arcsin x$ is the closed interval $[-\pi/2, \pi/2]$.

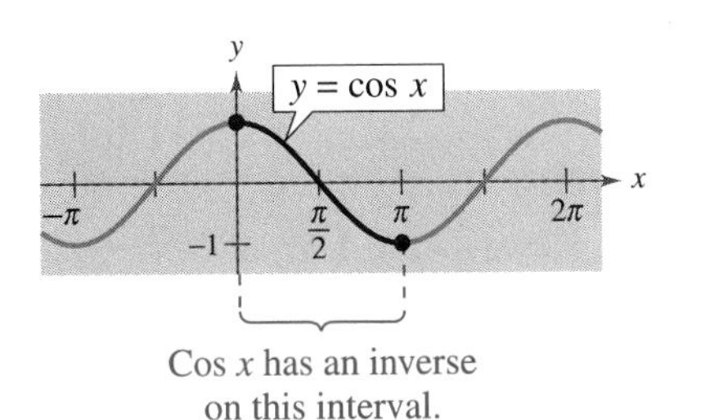

FIGURE 1.65

Other Inverse Trigonometric Functions

The cosine function is decreasing on the interval $0 \le x \le \pi$, as shown in Figure 1.65. Consequently, on this interval the cosine function has an inverse function—the **inverse cosine function**—denoted by

$$y = \arccos x \qquad \text{or} \qquad y = \cos^{-1} x.$$

Similarly, you can define an **inverse tangent function** by restricting the domain of $y = \tan x$ to the interval $(-\pi/2, \pi/2)$. The following list summarizes the definitions of the three most common inverse trigonometric functions. The remaining three are discussed in the exercise set. (The graphs, domains, and ranges of *all six* inverse trigonometric functions are summarized in the appendix.)

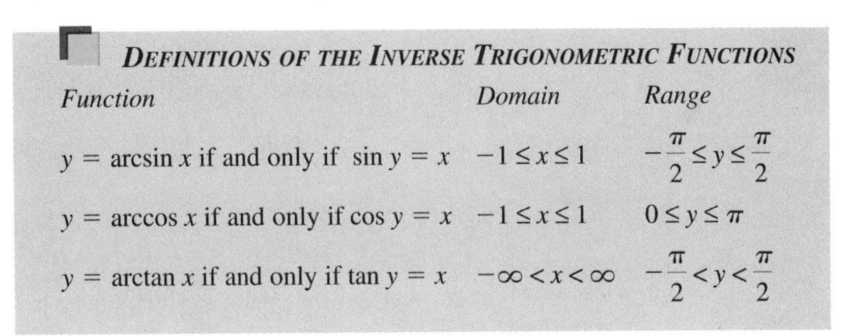

DEFINITIONS OF THE INVERSE TRIGONOMETRIC FUNCTIONS

Function	*Domain*	*Range*
$y = \arcsin x$ if and only if $\sin y = x$	$-1 \le x \le 1$	$-\frac{\pi}{2} \le y \le \frac{\pi}{2}$
$y = \arccos x$ if and only if $\cos y = x$	$-1 \le x \le 1$	$0 \le y \le \pi$
$y = \arctan x$ if and only if $\tan y = x$	$-\infty < x < \infty$	$-\frac{\pi}{2} < y < \frac{\pi}{2}$

The graphs of these three inverse trigonometric functions are shown in Figure 1.66.

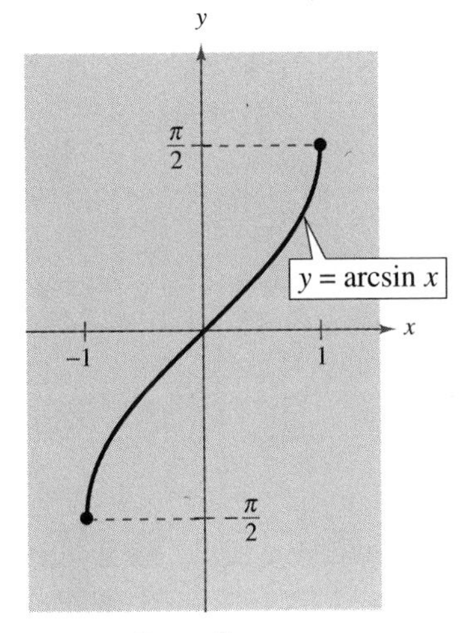

Domain: $[-1, 1]$
Range: $[-\frac{\pi}{2}, \frac{\pi}{2}]$

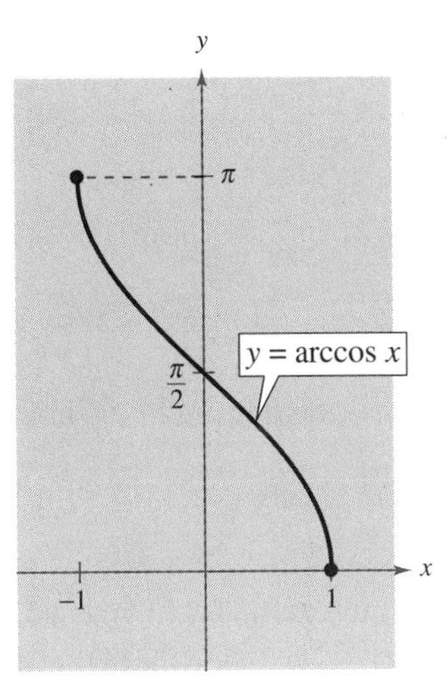

Domain: $[-1, 1]$
Range: $[0, \pi]$

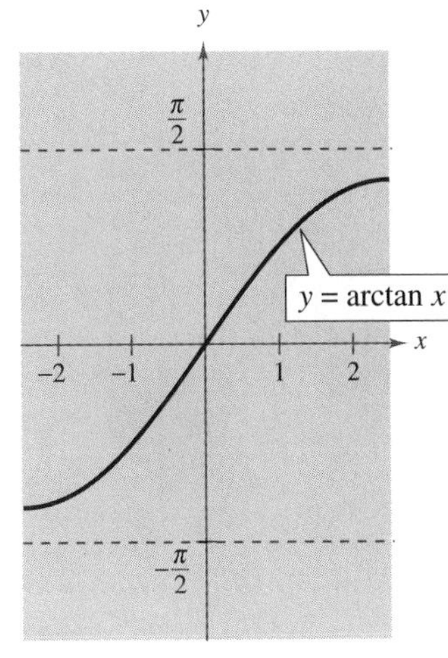

Domain: $(-\infty, \infty)$
Range: $(-\frac{\pi}{2}, \frac{\pi}{2})$

FIGURE 1.66

EXAMPLE 3 *Evaluating Inverse Trigonometric Functions*

Find the exact value.

a. $\arccos \frac{\sqrt{2}}{2}$ **b.** $\arccos(-1)$ **c.** $\arctan 0$

Solution

a. Because $\cos(\pi/4) = \sqrt{2}/2$, and $\pi/4$ lies in $[0, \pi]$, it follows that

$$\arccos \frac{\sqrt{2}}{2} = \frac{\pi}{4}.$$

b. Because $\cos \pi = -1$, and π lies in $[0, \pi]$, it follows that

$$\arccos(-1) = \pi.$$

c. Because $\tan 0 = 0$, and 0 lies in $(-\pi/2, \pi/2)$, it follows that

$$\arctan 0 = 0.$$

EXAMPLE 4 *Calculators and Inverse Trigonometric Functions*

Use a calculator to approximate the value (if possible).

a. $\arctan(-8.45)$ **b.** $\arcsin 0.2447$ **c.** $\arccos 2$

Solution

Function	*Mode*	*Graphing Calculator Keystrokes*
a. $\arctan(-8.45)$	Radian	[TAN⁻¹] [(] [(-)] 8.45 [)] [ENTER]

From the display, it follows that $\arctan(-8.45) \approx -1.453001$.

b. $\arcsin 0.2447$	Radian	[SIN⁻¹] 0.2447 [ENTER]

From the display, it follows that $\arcsin 0.2447 \approx 0.2472103$.

c. $\arccos 2$	Radian	[COS⁻¹] 2 [ENTER]

In real number mode, the calculator should display an *error message* because the domain of the inverse cosine function is $[-1, 1]$.

NOTE In Example 4, if you had set the calculator to degree mode, the display would have been in degrees rather than radians. This convention is peculiar to calculators. By definition, the values of inverse trigonometric functions are always *in radians.*

Compositions of Functions

Recall from Section P.8 that inverse functions possess the properties

$$f(f^{-1}(x)) = x \quad \text{and} \quad f^{-1}(f(x)) = x.$$

The inverse trigonometric versions of these properties are given below.

INVERSE PROPERTIES

If $-1 \le x \le 1$ and $-\pi/2 \le y \le \pi/2$, then

$$\sin(\arcsin x) = x \quad \text{and} \quad \arcsin(\sin y) = y.$$

If $-1 \le x \le 1$ and $0 \le y \le \pi$, then

$$\cos(\arccos x) = x \quad \text{and} \quad \arccos(\cos y) = y.$$

If $-\pi/2 < y < \pi/2$, then

$$\tan(\arctan x) = x \quad \text{and} \quad \arctan(\tan y) = y.$$

NOTE Keep in mind that these inverse properties do not apply for arbitrary values of x and y. For instance,

$$\arcsin\left(\sin\frac{3\pi}{2}\right) = \arcsin(-1) = -\frac{\pi}{2} \ne \frac{3\pi}{2}.$$

In other words, the property

$$\arcsin(\sin y) = y$$

is not valid for values of y outside the interval $[-\pi/2, \pi/2]$.

EXAMPLE 5 *Using Inverse Properties*

If possible, find the exact value.

a. $\tan[\arctan(-5)]$ **b.** $\arcsin\left(\sin\frac{5\pi}{3}\right)$ **c.** $\cos(\cos^{-1}\pi)$

Solution

a. Because -5 lies in the domain of the arctan x, the inverse property applies, and you have

$$\tan[\arctan(-5)] = -5.$$

b. In this case, $5\pi/3$ does not lie within the range of the arcsine function, $-\pi/2 \le y \le \pi/2$. However, $5\pi/3$ is coterminal with

$$\frac{5\pi}{3} - 2\pi = -\frac{\pi}{3}$$

which does lie in the range of the arcsine function, and you have

$$\arcsin\left(\sin\frac{5\pi}{3}\right) = \arcsin\left[\sin\left(-\frac{\pi}{3}\right)\right] = -\frac{\pi}{3}.$$

c. The expression $\cos(\cos^{-1}\pi)$ is not defined because $\cos^{-1}\pi$ is not defined. Remember that the domain of the inverse cosine function is $[-1, 1]$.

Example 6 shows how to use right triangles to find exact values of functions of inverse functions. Then, Example 7 shows how to use triangles to convert a trigonometric expression into an algebraic expression. This conversion technique is used frequently in calculus.

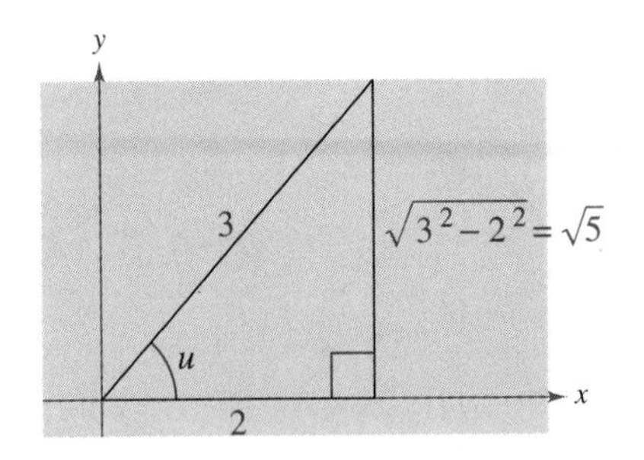

FIGURE 1.67

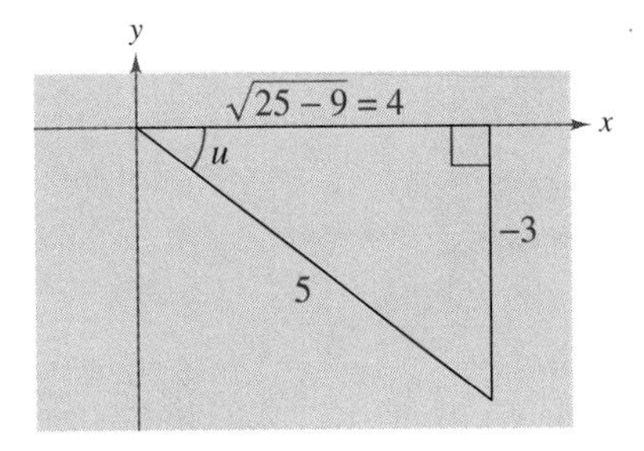

FIGURE 1.68

EXAMPLE 6 *Evaluating Compositions of Functions*

Find the exact value.

a. $\tan\left(\arccos\frac{2}{3}\right)$ **b.** $\cos\left[\arcsin\left(-\frac{3}{5}\right)\right]$

Solution

a. If you let $u = \arccos\frac{2}{3}$, then $\cos u = \frac{2}{3}$. Because $\cos u$ is positive, u is a *first*-quadrant angle. You can sketch and label angle u as shown in Figure 1.67. Consequently,

$$\tan\left(\arccos\frac{2}{3}\right) = \tan u = \frac{\text{opp}}{\text{adj}} = \frac{\sqrt{5}}{2}.$$

b. If you let $u = \arcsin -\frac{3}{5}$, then $\sin u = -\frac{3}{5}$. Because $\sin u$ is negative, u is a *fourth*-quadrant angle. You can sketch and label angle u as shown in Figure 1.68. Consequently,

$$\cos\left[\arcsin\left(-\frac{3}{5}\right)\right] = \cos u = \frac{\text{adj}}{\text{hyp}} = \frac{4}{5}.$$

EXAMPLE 7 *Some Problems from Calculus*

Write each of the following as an algebraic expression in x.

a. $\sin(\arccos 3x),\quad 0 \le x \le \frac{1}{3}$ **b.** $\cot(\arccos 3x),\quad 0 \le x \le \frac{1}{3}$

Solution

If you let $u = \arccos 3x$, then $\cos u = 3x$. Because

$$\cos u = \frac{3x}{1} = \frac{\text{adj}}{\text{hyp}}$$

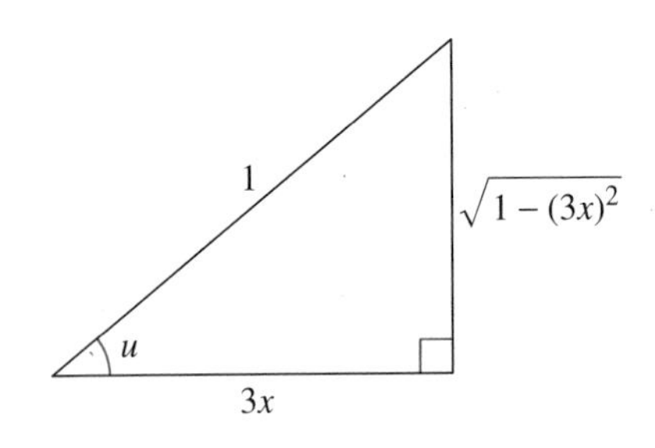

FIGURE 1.69

you can sketch a right triangle with acute angle u, as shown in Figure 1.69. From this triangle, you can easily convert each expression to algebraic form.

a. $\sin(\arccos 3x) = \sin u = \frac{\text{opp}}{\text{hyp}} = \sqrt{1-9x^2},\quad 0 \le x \le \frac{1}{3}$

b. $\cot(\arccos 3x) = \cot u = \frac{\text{adj}}{\text{opp}} = \frac{3x}{\sqrt{1-9x^2}},\quad 0 \le x \le \frac{1}{3}$

NOTE In Example 7, a similar argument can be made for x-values lying in the interval $\left[-\frac{1}{3}, 0\right]$.

GROUP ACTIVITY

INVERSE FUNCTIONS

We have discussed inverse functions for several types of functions. Match the function in the left column with its inverse function in the right column.

1. $f(x) = x$
2. $f(x) = x^2, \quad 0 \le x$
3. $f(x) = x^3$
4. $f(x) = e^x$
5. $f(x) = \ln x$
6. $f(x) = \sin x, \quad -\frac{\pi}{2} \le x \le \frac{\pi}{2}$
7. $f(x) = \cos x, \quad 0 \le x \le \pi$
8. $f(x) = \tan x, \quad -\frac{\pi}{2} < x < \frac{\pi}{2}$

a. $f^{-1}(x) = \arcsin x$
b. $f^{-1}(x) = \ln x$
c. $f^{-1}(x) = \sqrt{x}$
d. $f^{-1}(x) = \arctan x$
e. $f^{-1}(x) = \arccos x$
f. $f^{-1}(x) = \sqrt[3]{x}$
g. $f^{-1}(x) = e^x$
h. $f^{-1}(x) = x$

Discuss reasons for your answers. Verify each pair of inverses algebraically, graphically, and numerically.

WARM UP

Evaluate the trigonometric function from memory.

1. $\sin\left(-\frac{\pi}{2}\right)$ 2. $\cos \pi$ 3. $\tan\left(-\frac{\pi}{4}\right)$ 4. $\sin \frac{\pi}{4}$

Find a real number x in the interval $[-\pi/2, \pi/2]$ that has the same sine value as the given value.

5. $\sin 2\pi$ 6. $\sin \frac{5\pi}{6}$

Find the real number x in the interval $[0, \pi]$ that has the same cosine value as the given value.

7. $\cos 3\pi$ 8. $\cos\left(-\frac{\pi}{4}\right)$

Find a real number x in the interval $(-\pi/2, \pi/2)$ that has the same tangent value as the given value.

9. $\tan 4\pi$ 10. $\tan \frac{3\pi}{4}$

1.7 Exercises

1. ***True or False?*** Explain your reasoning.

$$\sin \frac{5\pi}{6} = \frac{1}{2} \implies \arcsin \frac{1}{2} = \frac{5\pi}{6}$$

2. ***True or False?*** Explain your reasoning.

$$\tan \frac{5\pi}{4} = 1 \implies \arctan 1 = \frac{5\pi}{4}$$

In Exercises 3–18, evaluate the expression without the aid of a calculator.

3. $\arcsin \frac{1}{2}$
4. $\arcsin 0$
5. $\arccos \frac{1}{2}$
6. $\arccos 0$
7. $\arctan \frac{\sqrt{3}}{3}$
8. $\arctan(-1)$
9. $\arccos\left(-\frac{\sqrt{3}}{2}\right)$
10. $\arcsin\left(-\frac{\sqrt{2}}{2}\right)$
11. $\arctan(-\sqrt{3})$
12. $\arctan \sqrt{3}$
13. $\arccos\left(-\frac{1}{2}\right)$
14. $\arcsin \frac{\sqrt{2}}{2}$
15. $\arcsin \frac{\sqrt{3}}{2}$
16. $\arctan\left(-\frac{\sqrt{3}}{3}\right)$
17. $\arctan 0$
18. $\arccos 1$

In Exercises 19–30, use a calculator to approximate the expression. (Round your result to two decimal places.)

19. $\arccos 0.28$
20. $\arcsin 0.45$
21. $\arcsin(-0.75)$
22. $\arccos(-0.7)$
23. $\arctan(-3)$
24. $\arctan 15$
25. $\arcsin 0.31$
26. $\arccos 0.26$
27. $\arccos(-0.41)$
28. $\arcsin(-0.125)$
29. $\arctan 0.92$
30. $\arctan 2.8$

In Exercises 31 and 32, determine the missing coordinates of the points on the graph of the function.

31.

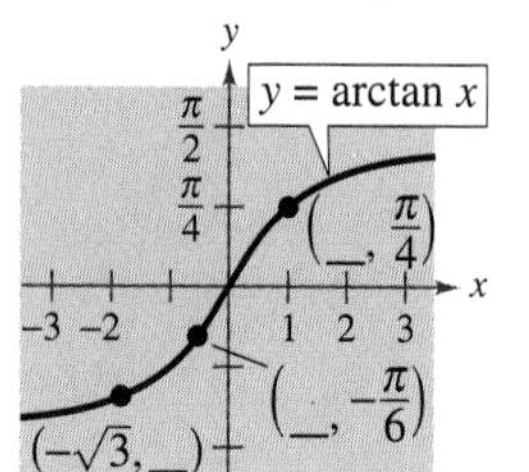

32.

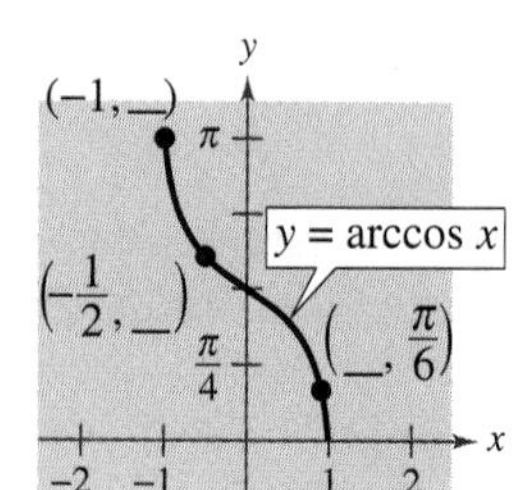

In Exercises 33 and 34, use a graphing utility to graph f, g, and $y = x$ on the same viewing rectangle to verify geometrically that g is the inverse of f. (Be sure to restrict the domain of f properly.)

33. $f(x) = \tan x, \quad g(x) = \arctan x$
34. $f(x) = \sin x, \quad g(x) = \arcsin x$

In Exercises 35–38, use an inverse trigonometric function to write θ as a function of x.

35.

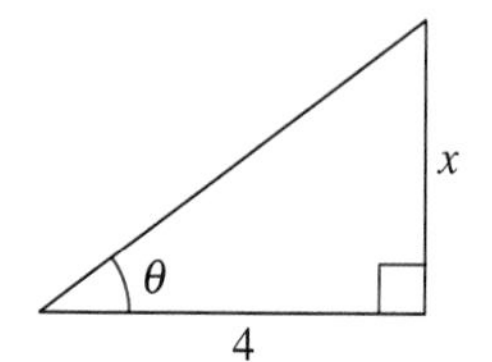

36.

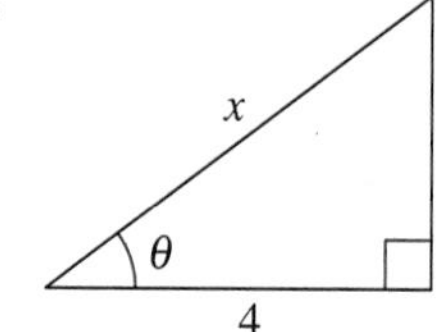

37.

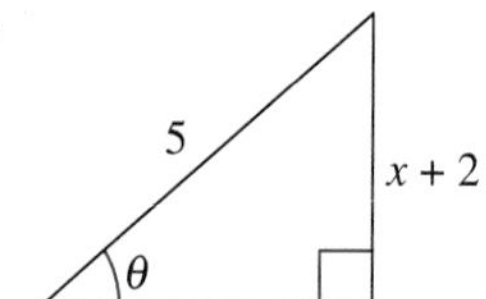

38.

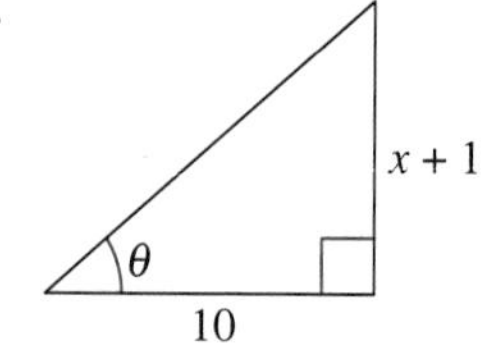

In Exercises 39–44, use the properties of inverse functions to evaluate the expression.

39. $\sin(\arcsin 0.3)$
40. $\tan(\arctan 25)$
41. $\cos[\arccos(-0.1)]$
42. $\sin[\arcsin(-0.2)]$
43. $\arcsin(\sin 3\pi)$
44. $\arccos\left(\cos \dfrac{7\pi}{2}\right)$

In Exercises 45–54, find the exact value of the expression. (*Hint:* Make a sketch of a right triangle.)

45. $\sin\left(\arctan \frac{3}{4}\right)$
46. $\sec\left(\arcsin \frac{4}{5}\right)$
47. $\cos(\arctan 2)$
48. $\sin\left(\arccos \dfrac{\sqrt{5}}{5}\right)$
49. $\cos\left(\arcsin \frac{5}{13}\right)$
50. $\csc\left[\arctan\left(-\frac{5}{12}\right)\right]$
51. $\sec\left[\arctan\left(-\frac{3}{5}\right)\right]$
52. $\tan\left[\arcsin\left(-\frac{3}{4}\right)\right]$
53. $\sin\left[\arccos\left(-\frac{2}{3}\right)\right]$
54. $\cot\left(\arctan \frac{5}{8}\right)$

In Exercises 55–64, write an algebraic expression that is equivalent to the expression. (*Hint:* Sketch a right triangle, as demonstrated in Example 7.)

55. $\cot(\arctan x)$
56. $\sin(\arctan x)$
57. $\cos(\arcsin 2x)$
58. $\sec(\arctan 3x)$
59. $\sin(\arccos x)$
60. $\sec[\arcsin(x - 1)]$
61. $\tan\left(\arccos \dfrac{x}{3}\right)$
62. $\cot\left(\arctan \dfrac{1}{x}\right)$
63. $\csc\left(\arctan \dfrac{x}{\sqrt{2}}\right)$
64. $\cos\left(\arcsin \dfrac{x - h}{r}\right)$

In Exercises 65 and 66, use a graphing utility to graph *f* and *g* on the same viewing rectangle to verify that the two are equal. Explain why they are equal. Identify any asymptotes of the graphs.

65. $f(x) = \sin(\arctan 2x), \quad g(x) = \dfrac{2x}{\sqrt{1 + 4x^2}}$

66. $f(x) = \tan\left(\arccos \dfrac{x}{2}\right), \quad g(x) = \dfrac{\sqrt{4 - x^2}}{x}$

In Exercises 67–70, fill in the blank.

67. $\arctan \dfrac{9}{x} = \arcsin(\quad), \quad x \neq 0$

68. $\arcsin \dfrac{\sqrt{36 - x^2}}{6} = \arccos(\quad), \quad 0 \le x \le 6$

69. $\arccos \dfrac{3}{\sqrt{x^2 - 2x + 10}} = \arcsin(\quad)$

70. $\arccos \dfrac{x - 2}{2} = \arctan(\quad), \quad |x - 2| \le 2$

In Exercises 71–78, sketch a graph of the function.

71. $y = 2 \arccos x$
72. $y = \arcsin \dfrac{x}{2}$
73. $f(x) = \arcsin(x - 1)$
74. $g(t) = \arccos(t + 2)$
75. $f(x) = \arctan 2x$
76. $f(x) = \dfrac{\pi}{2} + \arctan x$
77. $h(v) = \tan(\arccos v)$
78. $f(x) = \arccos \dfrac{x}{4}$

In Exercises 79 and 80, write the given functions in terms of the sine function by using the identity

$$A \cos \omega t + B \sin \omega t = \sqrt{A^2 + B^2} \sin\left(\omega t + \arctan \frac{A}{B}\right).$$

Use a graphing utility to graph both forms of the function. What does the graph imply?

79. $f(t) = 3 \cos 2t + 3 \sin 2t$
80. $f(t) = 4 \cos \pi t + 3 \sin \pi t$

81. *Think About It* Consider the functions

$f(x) = \sin x$ and $f^{-1}(x) = \arcsin x$.

(a) Use a graphing utility to graph the composite functions $f \circ f^{-1}$ and $f^{-1} \circ f$.

(b) Explain why the graphs of part (a) are not the graph of the line $y = x$. Why do the graphs of $f \circ f^{-1}$ and $f^{-1} \circ f$ differ?

82. *Think About It* Use a graphing utility to graph the functions $f(x) = \sqrt{x}$ and $g(x) = 6 \arctan x$. For $x > 0$ it appears that $g > f$. Explain why you know that there exists a positive real number a such that $g < f$ for $x > a$. Approximate the number a.

83. *Docking a Boat* A boat is pulled in by means of a winch located on a dock 10 feet above the deck of the boat (see figure). Let θ be the angle of elevation from the boat to the winch and let s be the length of the rope from the winch to the boat.

(a) Write θ as a function of s.

(b) Find θ when $s = 48$ feet and $s = 24$ feet.

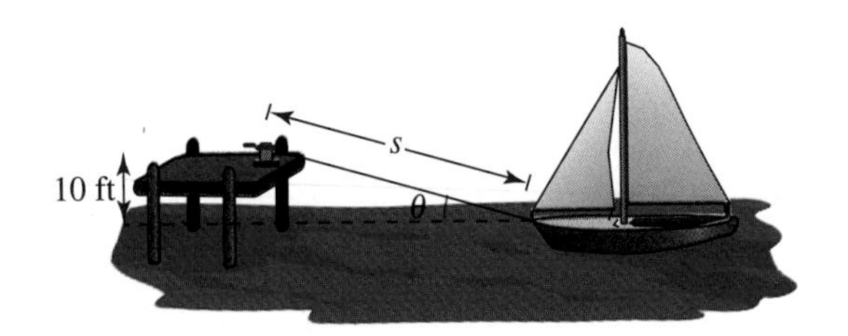

84. *Photography* A television camera at ground level is filming the lift-off of a space shuttle at a point 750 meters from the launch pad (see figure). Let θ be the angle of elevation to the shuttle and let s be the height of the shuttle.

(a) Write θ as a function of s.

(b) Find θ when $s = 300$ meters and $s = 1200$ meters.

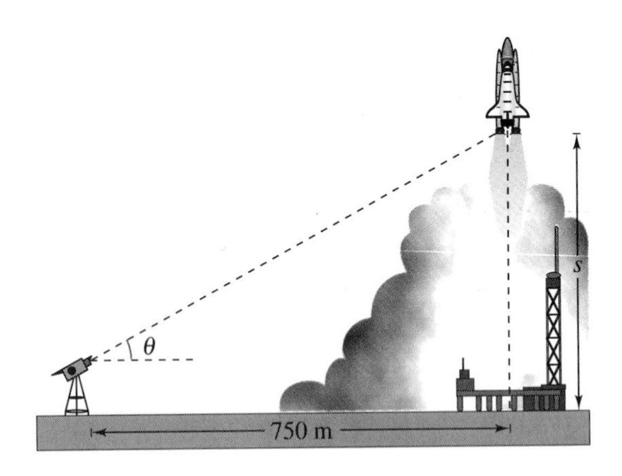

85. *Photography* A photographer is taking a picture of a 3-foot painting hung in an art gallery. The camera lens is 1 foot below the lower edge of the painting (see figure). The angle β subtended by the camera lens x feet from the painting is given by

$$\beta = \arctan \frac{3x}{x^2 + 4}, \qquad x > 0.$$

(a) Use a graphing utility to graph β as a function of x.

(b) Move the cursor along the graph to approximate the distance from the picture when β is maximum.

(c) Identify the asymptote of the graph and discuss its meaning in the context of the problem.

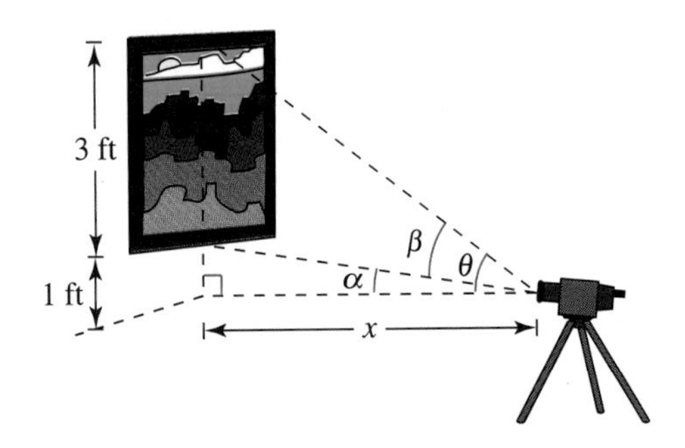

86. *Area* In calculus, it is shown that the area of the region bounded by the graphs of $y = 0$, $y = 1/(x^2 + 1)$, $x = a$, and $x = b$ is given by

Area $= \arctan b - \arctan a$

(see figure). Find the area for the following values of a and b.

(a) $a = 0, b = 1$ (b) $a = -1, b = 1$

(c) $a = 0, b = 3$ (d) $a = -1, b = 3$

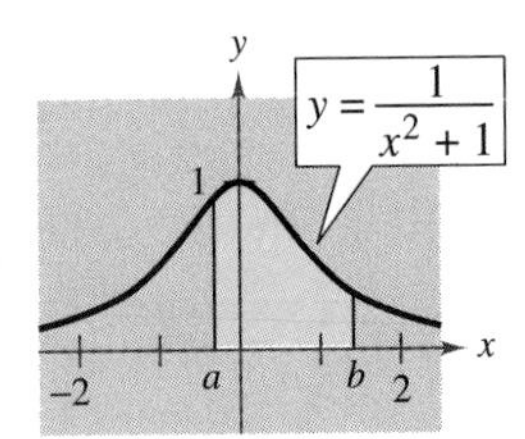

87. ***Angle of Elevation*** An airplane flies at an altitude of 5 miles toward a point directly over an observer. Consider θ and x as shown in the figure.

(a) Write θ as a function of x.

(b) Find θ when $x = 10$ miles and $x = 3$ miles.

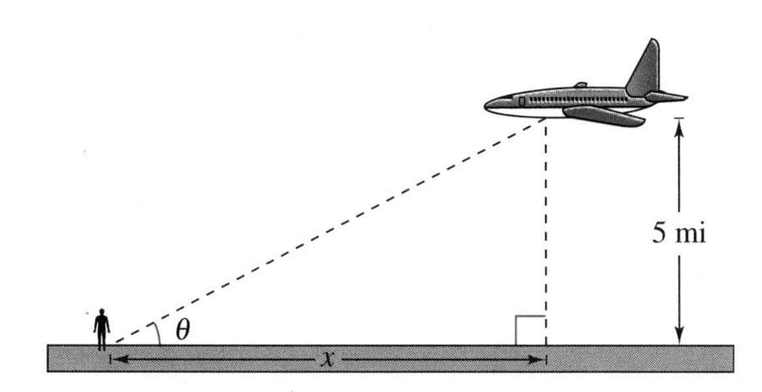

88. ***Security Patrol*** A security car with its spotlight on is parked 20 meters from a long warehouse. Consider θ and x as shown in the figure.

(a) Write θ as a function of x.

(b) Find θ when $x = 5$ meters and $x = 12$ meters.

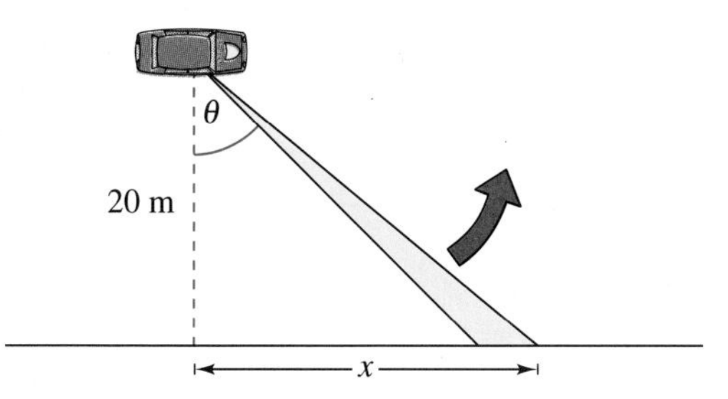

89. Define the inverse cotangent function by restricting the domain of the cotangent function to the interval $(0, \pi)$, and sketch its graph.

90. Define the inverse secant function by restricting the domain of the secant function to the intervals $[0, \pi/2)$ and $(\pi/2, \pi]$, and sketch its graph.

91. Define the inverse cosecant function by restricting the domain of the cosecant function to the intervals $[-\pi/2, 0)$ and $(0, \pi/2]$, and sketch its graph.

92. Use the results of Exercises 89–91 to evaluate the following without using a calculator.

(a) $\operatorname{arcsec} \sqrt{2}$ (b) $\operatorname{arcsec} 1$

(c) $\operatorname{arccot}(-\sqrt{3})$ (d) $\operatorname{arccsc} 2$

In Exercises 93–98, prove the identity.

93. $\arcsin(-x) = -\arcsin x$

94. $\arctan(-x) = -\arctan x$

95. $\arccos(-x) = \pi - \arccos x$

96. $\arctan x + \arctan \dfrac{1}{x} = \dfrac{\pi}{2}, \quad x > 0$

97. $\arcsin x + \arccos x = \dfrac{\pi}{2}$

98. $\arcsin x = \arctan \dfrac{x}{\sqrt{1 - x^2}}$

Review **Solve Exercises 99–102 as a review of the skills and problem-solving techniques you learned in previous sections.**

99. ***Buy Now or Wait?*** A sales representative indicates that if a customer waits another month before purchasing a new car that currently costs \$23,500, the price will increase by 4%. However, the customer will pay an interest penalty of \$725 for the early withdrawal of a certificate of deposit if the car is purchased now. Determine whether the customer should buy now or wait another month.

100. ***Insurance Premium*** The annual insurance premium for a policyholder is normally \$739. However, after having an automobile accident, the policyholder is charged an additional 30%. What is the new annual premium?

101. ***Partnership Costs*** A group of people agree to share equally in the cost of a \$250,000 endowment to a college. If they could find two more people to join the group, each person's share of the cost would decrease by \$6250. How many people are presently in the group?

102. ***Speed*** A boat travels at a speed of 18 miles per hour in still water. It travels 35 miles upstream and then returns to the starting point in a total of 4 hours. Find the speed of the current.

1.8 Applications and Models

See Exercise 58 on page 204 for an example of how a trigonometric function can be used to model harmonic motion.

Applications Involving Right Triangles ▫ *Trigonometry and Bearings* ▫ *Harmonic Motion*

Applications Involving Right Triangles

In keeping with our twofold perspective of trigonometry, this section includes both right triangle applications and applications that emphasize the periodic nature of the trigonometric functions.

NOTE In this section the three angles of a right triangle are denoted by the letters A, B, and C (where C is the right angle), and the lengths of the sides opposite these angles by the letters a, b, and c (where c is the hypotenuse).

EXAMPLE 1 *Solving a Right Triangle*

Solve the right triangle shown in Figure 1.70.

Solution

Because $C = 90°$, it follows that $A + B = 90°$ and $B = 90° - 34.2° = 55.8°$. To solve for a, use the fact that

$$\tan A = \frac{\text{opp}}{\text{adj}} = \frac{a}{b} \quad \Longrightarrow \quad a = b \tan A.$$

Thus, $a = 19.4 \tan 34.2° \approx 13.18$. Similarly, to solve for c, use the fact that

$$\cos A = \frac{\text{adj}}{\text{hyp}} = \frac{b}{c} \quad \Longrightarrow \quad c = \frac{b}{\cos A}.$$

Thus, $c = \dfrac{19.4}{\cos 34.2°} \approx 23.46.$

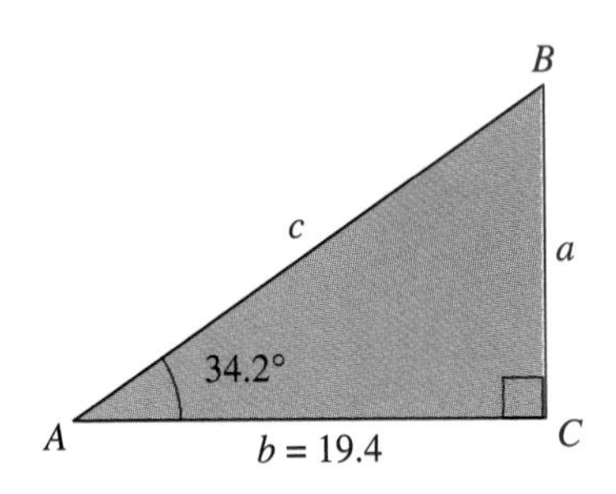

FIGURE 1.70

EXAMPLE 2 *Finding a Side of a Right Triangle*

A safety regulation states that the maximum angle of elevation for a rescue ladder is 72°. If a fire department's longest ladder is 110 feet, what is the maximum safe rescue height?

Solution

A sketch is shown in Figure 1.71. From the equation $\sin A = a/c$, it follows that

$$a = c \sin A = 110 \sin 72° \approx 104.6.$$

Thus, the maximum safe rescue height is about 104.6 feet above the height of the fire truck.

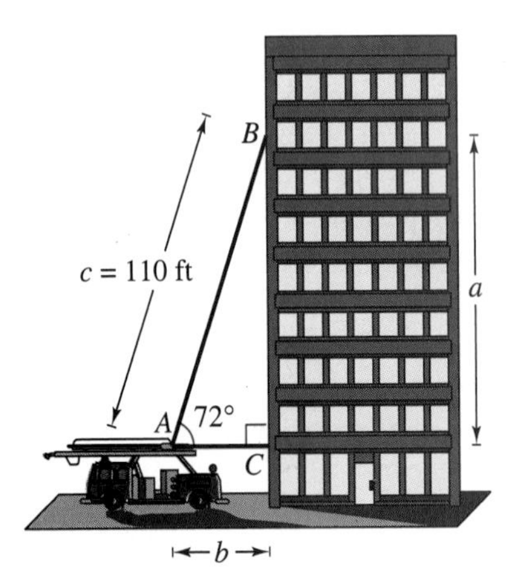

FIGURE 1.71

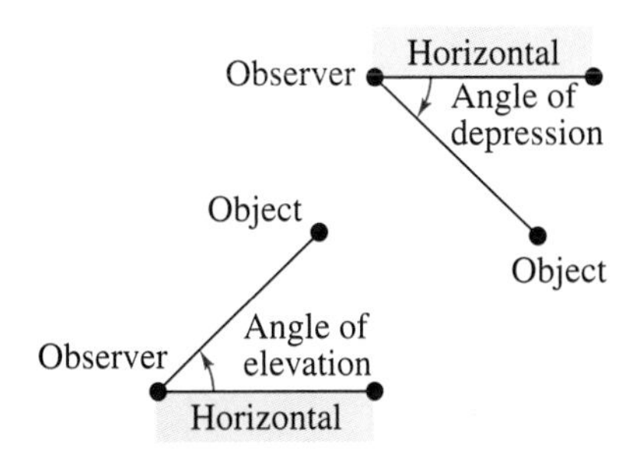

FIGURE 1.72

In Example 2, the term **angle of elevation** represents the angle from the horizontal upward to an object. For objects that lie below the horizontal, it is common to use the term **angle of depression,** as shown in Figure 1.72.

EXAMPLE 3 *Finding a Side of a Right Triangle*

At a point 200 feet from the base of a building, the angle of elevation to the *bottom* of a smokestack is 35°, whereas the angle of elevation to the *top* is 53°, as shown in Figure 1.73. Find the height s of the smokestack alone.

Solution

Note from Figure 1.73 that this problem involves two right triangles. In the smaller right triangle, use the fact that $\tan 35° = a/200$ to conclude that the height of the building is

$$a = 200 \tan 35°.$$

In the larger right triangle, use the equation

$$\tan 53° = \frac{a + s}{200}$$

to conclude that $a + s = 200 \tan 53°$. Hence, the height of the smokestack is

$$\begin{aligned} s &= 200 \tan 53° - a \\ &= 200 \tan 53° - 200 \tan 35° \\ &= 125.4 \text{ feet.} \end{aligned}$$

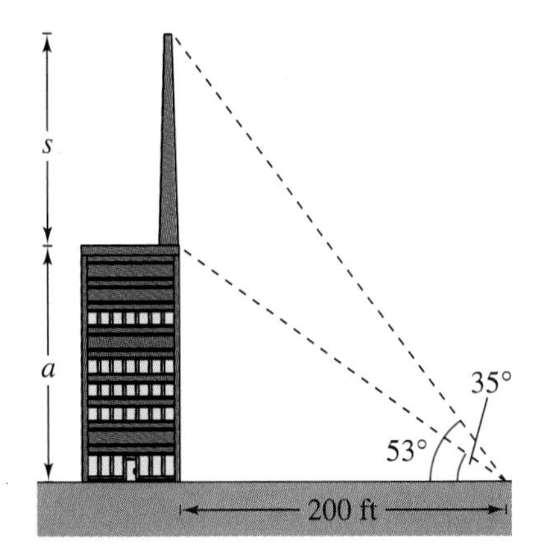

FIGURE 1.73

EXAMPLE 4 *Finding an Acute Angle of a Right Triangle*

A swimming pool is 20 meters long and 12 meters wide. The bottom of the pool is slanted so that the water depth is 1.3 meters at the shallow end and 4 meters at the deep end, as shown in Figure 1.74. Find the angle of depression of the bottom of the pool.

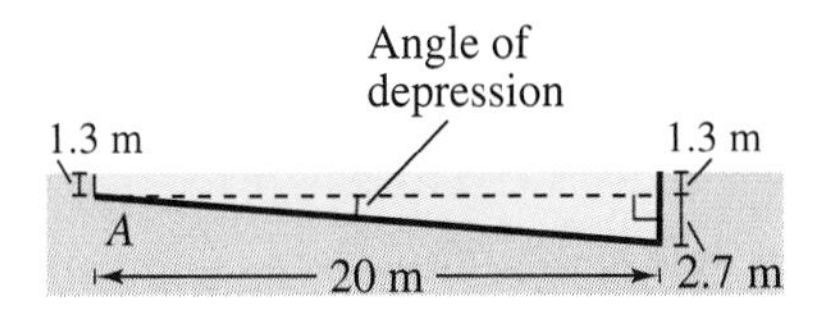

FIGURE 1.74

Solution

Using the tangent function, you see that

$$\tan A = \frac{\text{opp}}{\text{adj}} = \frac{2.7}{20} = 0.135.$$

Thus, the angle of depression is given by

$$\begin{aligned} A &= \arctan 0.135 \\ &\approx 0.13419 \text{ radians} \\ &\approx 7.69°. \end{aligned}$$

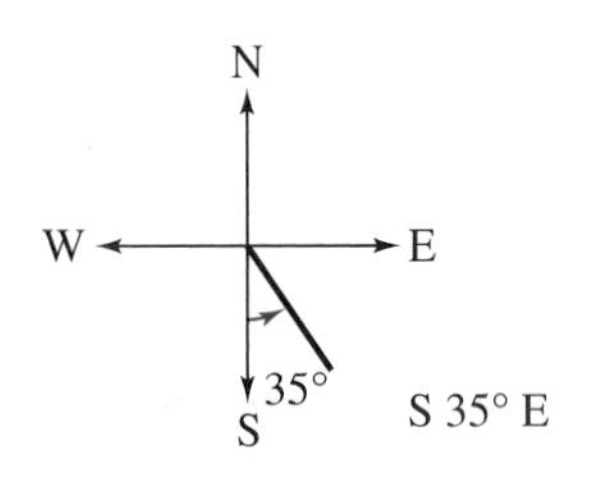

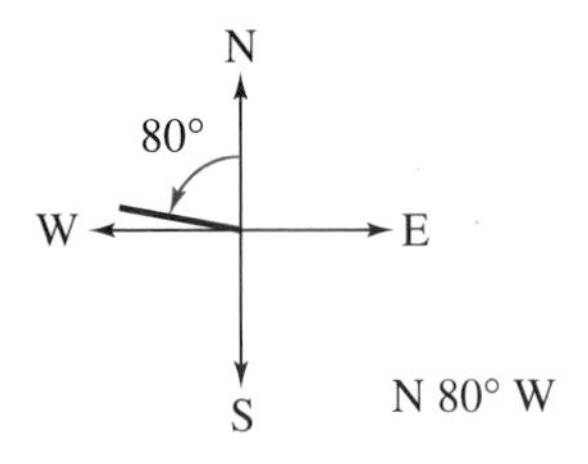

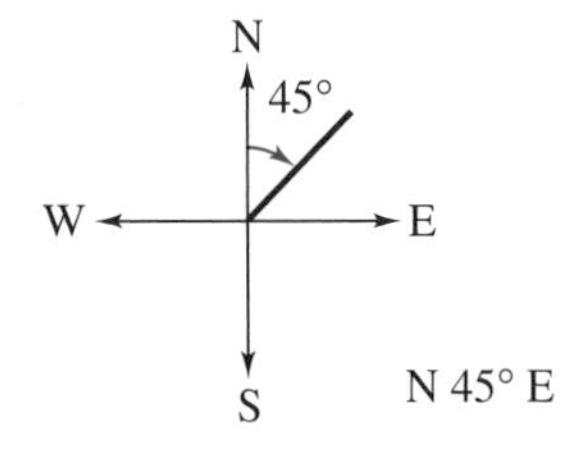

FIGURE 1.75

NOTE The bearing of S 35° E in Figure 1.75 means 35 degrees east of south.

Trigonometry and Bearings

In surveying and navigation, directions are generally given in terms of **bearings.** A bearing measures the acute angle a path or line of sight makes with a fixed north-south line, as shown in Figure 1.75.

EXAMPLE 5 *Finding Directions in Terms of Bearings*

A ship leaves port at noon and heads due west at 20 knots, or 20 nautical miles (nm) per hour. At 2 P.M. the ship changes course to N 54° W, as shown in Figure 1.76. Find the ship's bearing and distance from the port of departure at 3 P.M..

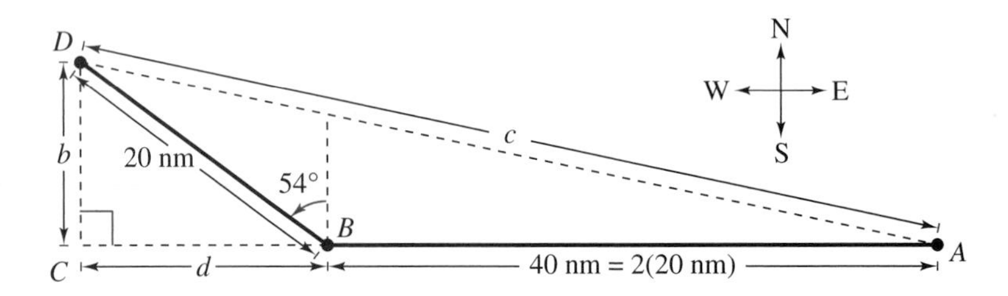

FIGURE 1.76

Solution

In triangle BCD, you have $B = 90° - 54° = 36°$. The two sides of this triangle can be determined to be

$$b = 20 \sin 36° \qquad \text{and} \qquad d = 20 \cos 36°.$$

In triangle ACD, you find angle A as follows.

$$\tan A = \frac{b}{d + 40} = \frac{20 \sin 36°}{20 \cos 36° + 40} \approx 0.2092494$$

$$A \approx \arctan 0.2092494 \approx 0.2062732 \text{ radians} \approx 11.82°$$

The angle with the north-south line is $90° - 11.82° = 78.18°$. Therefore, the bearing of the ship is

N 78.18° W. Bearing

Finally, from triangle ACD, you have $\sin A = b/c$, which yields

$$c = \frac{b}{\sin A} = \frac{20 \sin 36°}{\sin 11.82°}$$

$$\approx 57.4 \text{ nautical miles.}$$ Distance from port

Harmonic Motion

The periodic nature of the trigonometric functions is useful for describing the motion of a point on an object that vibrates, oscillates, rotates, or is moved by wave motion.

For example, consider a ball that is bobbing up and down on the end of a spring, as shown in Figure 1.77. Suppose that 10 centimeters is the maximum distance the ball moves vertically upward or downward from its equilibrium (at rest) position. Suppose further that the time it takes for the ball to move from its maximum displacement above zero to its maximum displacement below zero and back again is $t = 4$ seconds. Assuming the ideal conditions of perfect elasticity and no friction or air resistance, the ball would continue to move up and down in a uniform and regular manner.

From this spring you can conclude that the period (time for one complete cycle) of the motion is

$$\text{Period} = 4 \text{ seconds}$$

and that its amplitude (maximum displacement from equilibrium) is

$$\text{Amplitude} = 10 \text{ centimeters.}$$

Motion of this nature can be described by a sine or cosine function, and is called **simple harmonic motion.**

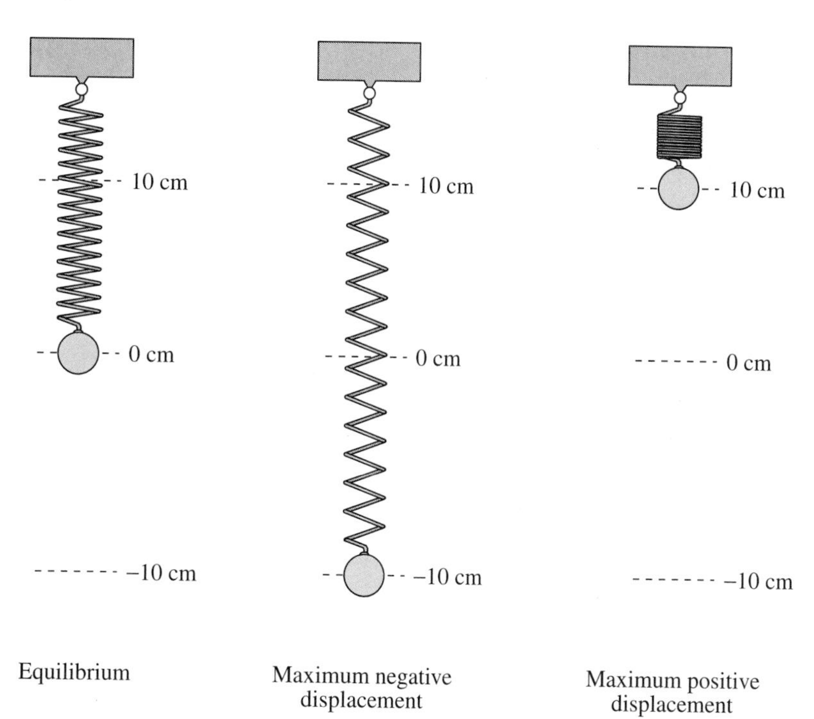

FIGURE 1.77

DEFINITION OF SIMPLE HARMONIC MOTION

A point that moves on a coordinate line is said to be in **simple harmonic motion** if its distance d from the origin at time t is given by either

$$d = a \sin \omega t \qquad \text{or} \qquad d = a \cos \omega t$$

where a and ω are real numbers such that $\omega > 0$. The motion has **amplitude** $|a|$, **period** $2\pi/\omega$, and **frequency** $\omega/2\pi$.

EXAMPLE 6 Simple Harmonic Motion

Real Life

Write the equation for the simple harmonic motion of the ball described in Figure 1.77, where the period is 4 seconds. What is the frequency of this harmonic motion?

Solution

Because the spring is at equilibrium ($d = 0$) when $t = 0$, you use the equation

$$d = a \sin \omega t.$$

Moreover, because the maximum displacement from zero is 10 and the period is 4, you have

$$\text{Amplitude} = |a| = 10$$

$$\text{Period} = \frac{2\pi}{\omega} = 4 \quad \Longrightarrow \quad \omega = \frac{\pi}{2}.$$

Consequently, the equation of motion is

$$d = 10 \sin \frac{\pi}{2} t.$$

Note that the choice of $a = 10$ or $a = -10$ depends on whether the ball initially moves up or down. The frequency is given by

$$\text{Frequency} = \frac{\omega}{2\pi} = \frac{\pi/2}{2\pi} = \frac{1}{4} \text{ cycle per second.}$$

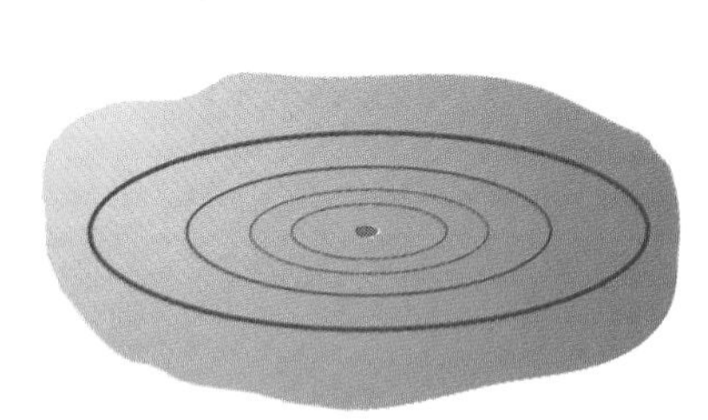

FIGURE 1.78

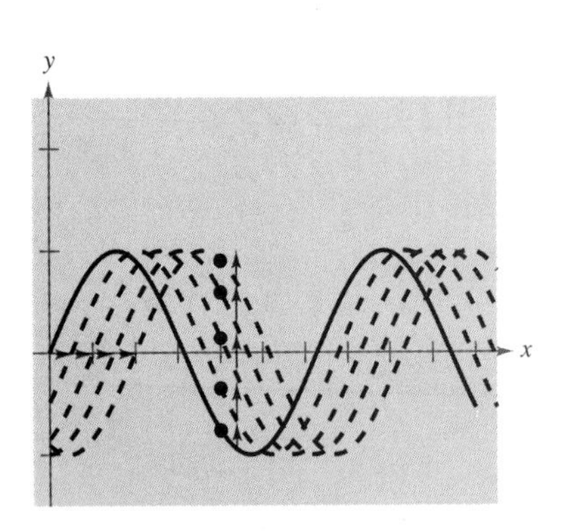

FIGURE 1.79

One illustration of the relationship between sine waves and harmonic motion is seen in the wave motion resulting when a stone is dropped into a calm pool of water. The waves move outward in roughly the shape of sine (or cosine) waves, as shown in Figure 1.78. As an example, suppose you are fishing and your fishing bob is attached so that it does not move horizontally. As the waves move outward from the dropped stone, your fishing bob will move up and down in simple harmonic motion, as shown in Figure 1.79.

TECHNOLOGY

Use the zero or root feature of a graphing utility to verify Example 7(d). To do so using a *TI-83* or *TI-82*, use the following steps.

1. Enter the equation into Y1.
2. Set an appropriate viewing rectangle.
3. Press the keystrokes.
   TI-83
   CALC (2:zero) ENTER
   TI-82
   CALC (2:root) ENTER
4. Cursor to the left of the zero and press ENTER.
5. Cursor to the right of the zero and press ENTER.
6. Press ENTER again and the zero will be displayed.

EXAMPLE 7 *Simple Harmonic Motion*

Given the equation for simple harmonic motion

$$d = 6 \cos \frac{3\pi}{4} t$$

find (a) the maximum displacement, (b) the frequency, (c) the value of d when $t = 4$, and (d) the least positive value of t for which $d = 0$.

Solution

The given equation has the form $d = a \cos \omega t$, with $a = 6$ and $\omega = 3\pi/4$.

a. The maximum displacement (from the point of equilibrium) is given by the amplitude. Thus, the maximum displacement is 6.

b. $\text{Frequency} = \dfrac{\omega}{2\pi} = \dfrac{3\pi/4}{2\pi} = \dfrac{3}{8}$ cycle per unit of time

c. $d = 6 \cos\left[\dfrac{3\pi}{4}(4)\right] = 6 \cos 3\pi = 6(-1) = -6$

d. To find the least positive value of t for which $d = 0$, solve the equation

$$d = 6 \cos \frac{3\pi}{4} t = 0$$

to obtain

$$\frac{3\pi}{4} t = \frac{\pi}{2}, \frac{3\pi}{2}, \frac{5\pi}{2}, \ldots \quad \Longrightarrow \quad t = \frac{2}{3}, 2, \frac{10}{3}, \ldots .$$

Thus, the least positive value of t is $t = \frac{2}{3}$.

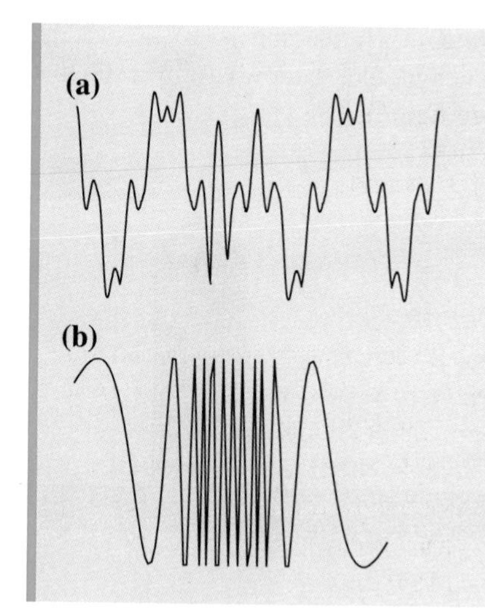

GROUP ACTIVITY

RADIO WAVES

Many different physical phenomena can be characterized by wave motion. These include electromagnetic waves such as radio waves, television waves, and microwaves. Radio waves transmit sound in two different ways. For an AM station, the *amplitude* of the wave is modified to carry sound. The letters AM stand for **amplitude modulation.** An FM radio signal has its *frequency* modified in order to carry sound, hence the term **frequency modulation.** Of the two graphs at the left, one shows an AM wave and the other shows an FM wave. Which is which? Explain your reasoning.

WARM UP

Evaluate the expression and round to two decimal places.

1. $20 \sin 25°$
2. $42 \tan 62°$
3. $\arcsin 0.8723$
4. $\arctan 2.8703$

Solve for x and round to two decimal places.

5. $\cos 22° = \dfrac{x + 13 \sin 22°}{13 \sin 54°}$
6. $\tan 36° = \dfrac{x + 85 \tan 18°}{85}$

Find the amplitude and period of the function.

7. $f(x) = -4 \sin 2x$
8. $f(x) = \frac{1}{2} \sin \pi x$
9. $g(x) = 3 \cos 3\pi x$
10. $g(x) = 0.2 \cot \dfrac{x}{4}$

1.8 Exercises

In Exercises 1–10, solve the right triangle shown in the figure. (Round to two decimal places.)

1. $A = 20°, \quad b = 10$
2. $B = 54°, \quad c = 15$
3. $B = 71°, \quad b = 24$
4. $A = 8.4°, \quad a = 40.5$
5. $a = 6, \quad b = 10$
6. $a = 25, \quad c = 35$
7. $b = 16, \quad c = 52$
8. $b = 1.32, \quad c = 9.45$
9. $A = 12°15', \quad c = 430.5$
10. $B = 65°12', \quad a = 14.2$

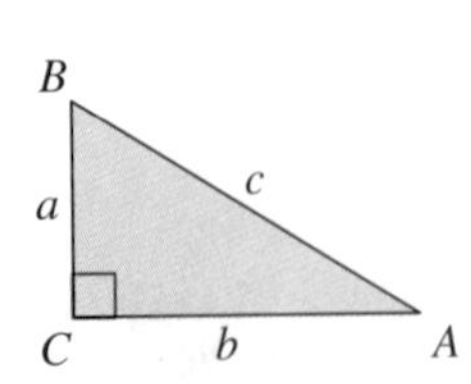

FIGURE FOR 1–10

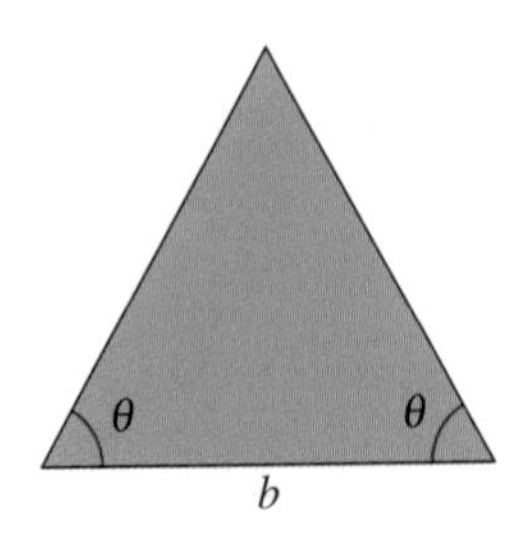

FIGURE FOR 11 AND 12

In Exercises 11 and 12, find the altitude of the isosceles triangle shown in the figure. (Round to two decimal places.)

11. $\theta = 52°, \quad b = 4$ inches
12. $\theta = 18°, \quad b = 10$ meters

13. ***Length of a Shadow*** If the sun is 30° above the horizon, find the length of a shadow cast by a silo that is 60 feet tall (see figure).

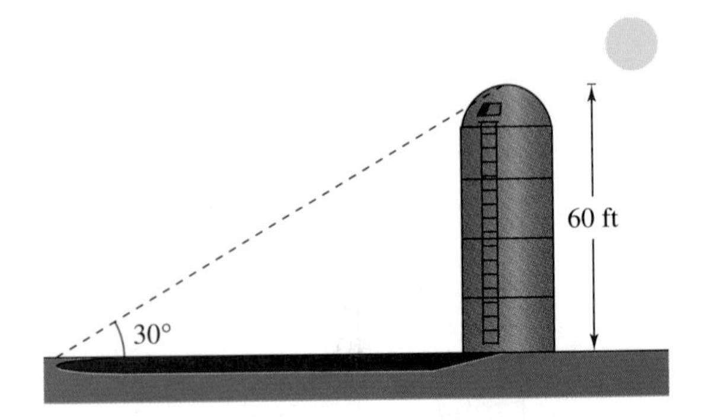

14. ***Length of a Shadow*** If the sun is 20° above the horizon, find the length of a shadow cast by a building that is 600 feet tall.

15. ***Height*** A ladder 16 feet long leans against the side of a house. Find the height h from the top of the ladder to the ground if the angle of elevation of the ladder is 74°.

16. ***Height*** The length of a shadow of a tree is 125 feet when the angle of elevation of the sun is 33°. Approximate the height h of the tree.

17. ***Height*** From a point 50 feet in front of a church, the angles of elevation to the base of the steeple and the top of the steeple are 35° and 47° 40′, respectively.
    (a) Draw right triangles that represent the problem. Label the known and unknown quantities.
    (b) Use a trigonometric function to write an equation involving the unknown.
    (c) Find the height of the steeple.

18. ***Height*** From a point 100 feet in front of the public library, the angles of elevation to the base of the flagpole and the top of the pole are 28° and 39° 45′, respectively. The flagpole is mounted on the front of the library's roof. Find the height of the pole.

19. ***Depth of a Submarine*** The sonar of a navy cruiser detects a submarine that is 4000 feet from the cruiser. The angle between the water line and the submarine is 34° (see figure). How deep is the submarine?

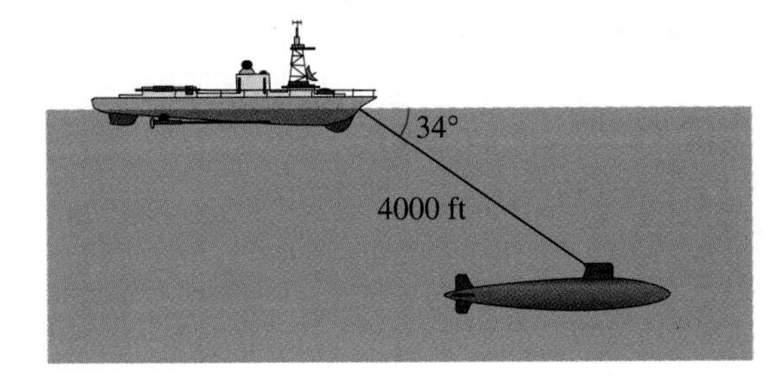

20. ***Height of a Kite*** A 100-foot line is attached to a kite. When the kite has pulled the line taut, the angle of elevation to the kite is approximately 50°. Approximate the height of the kite.

21. ***Angle of Elevation*** An amateur radio operator erects a 75-foot vertical tower for an antenna. Find the angle of elevation to the top of the tower at a point on level ground 50 feet from its base.

22. ***Angle of Elevation*** The height of an outdoor basketball backboard is $12\frac{1}{2}$ feet, and the backboard casts a shadow $17\frac{1}{3}$ feet long.
    (a) Draw a right triangle that represents the problem. Label the known and unknown quantities.
    (b) Use a trigonometric function to write an equation involving the unknown.
    (c) Find the angle of elevation of the sun.

23. ***Angle of Depression*** A spacecraft is traveling in a circular orbit 150 miles above the surface of the earth (see figure). Find the angle of depression from the spacecraft to the horizon. Assume the radius of the earth is 4000 miles.

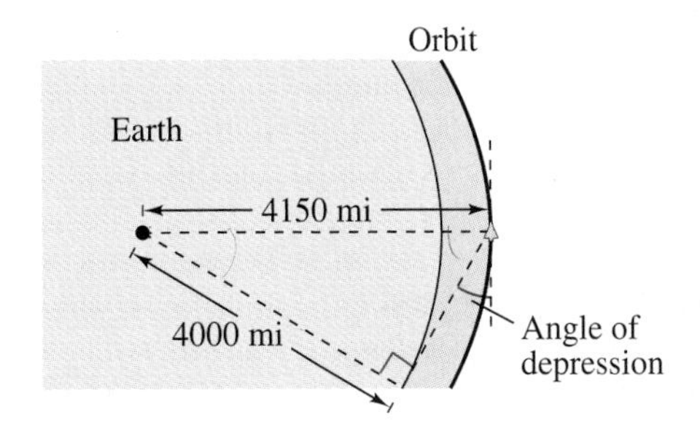

24. ***Angle of Depression*** Find the angle of depression from the top of a lighthouse 250 feet above water level to the water line of a ship 2 miles offshore.

25. ***Airplane Ascent*** When an airplane leaves the runway, its angle of climb is 18° and its speed is 275 feet per second. Find the plane's altitude after one minute.

26. ***Airplane Ascent*** How long will it take the plane in Exercise 25 to climb to an altitude of 10,000 feet?

27. ***Mountain Descent*** A sign on the roadway at the top of a mountain indicates that for the next 4 miles the grade is 10.5° (see figure). Find the change in elevation for a car descending the mountain.

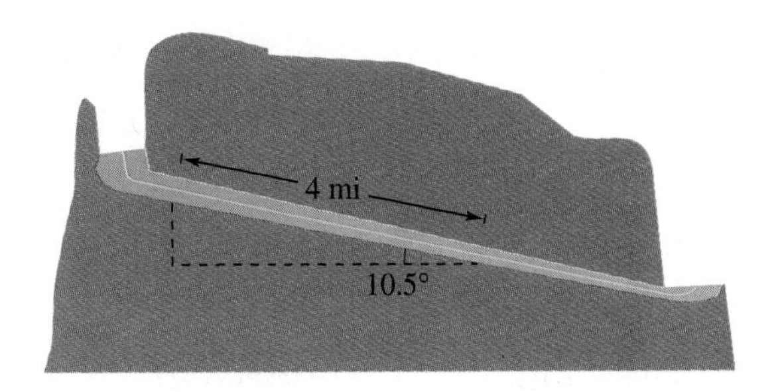

28. ***Mountain Descent*** A sign on the roadway at the top of a mountain indicates that for the next 4 miles the grade is 12%. Find the angle of the grade and the change in elevation for a car descending the mountain.

29. ***Navigation*** An airplane flying at 550 miles per hour has a bearing of N 52° E. After flying 1.5 hours, how far north and how far east will the plane have traveled from its point of departure?

30. ***Navigation*** A ship leaves port at noon and has a bearing of S 27° W. If its speed is 20 knots, how many nautical miles south and how many nautical miles west will the ship have traveled by 6:00 P.M.?

31. ***Surveying*** A surveyor wishes to find the distance across a swamp (see figure). The bearing from A to B is N 32° W. The surveyor walks 50 meters from A, and at the point C the bearing to B is N 68° W. Find (a) the bearing from A to C and (b) the distance from A to B.

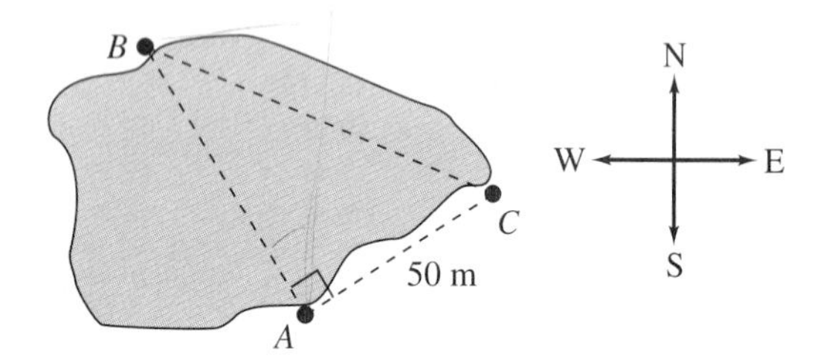

32. ***Location of a Fire*** Two fire towers are 30 kilometers apart, tower A being due west of tower B. A fire is spotted from the towers, and the bearings from A and B are E 14° N and W 34° N, respectively (see figure). Find the distance d of the fire from the line segment AB.

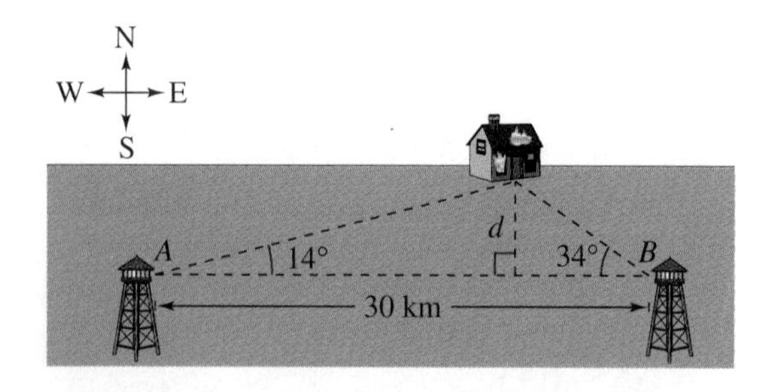

33. ***Navigation*** A ship is 45 miles east and 30 miles south of port. If the captain wants to sail directly to port, what bearing should be taken?

34. ***Navigation*** A plane is 120 miles north and 85 miles east of an airport. If the pilot wants to fly directly to the airport, what bearing should be taken?

35. ***Distance Between Ships*** An observer in a lighthouse 350 feet above sea level observes two ships directly offshore. The angles of depression to the ships are 4° and 6.5° (see figure). How far apart are the ships?

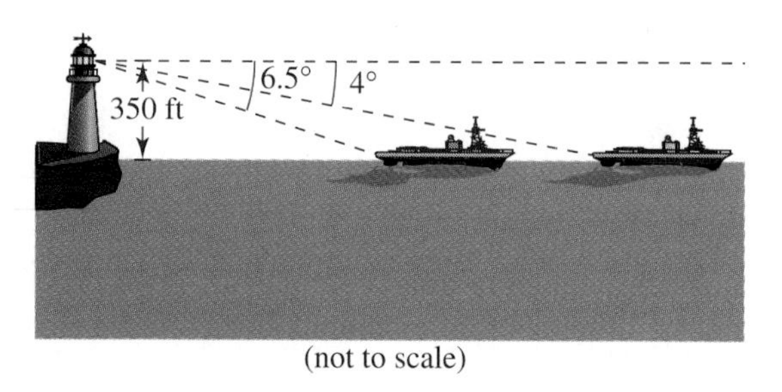

(not to scale)

36. ***Distance Between Towns*** A passenger in an airplane at an altitude of 10 kilometers sees two towns directly to the left of the plane. The angles of depression to the towns are 28° and 55° (see figure). How far apart are the towns?

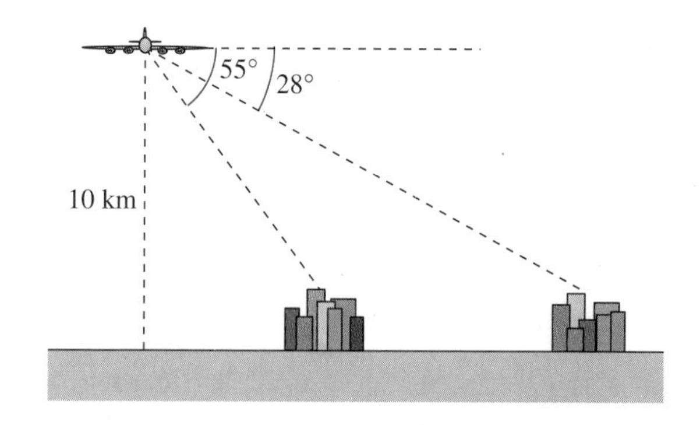

37. ***Altitude of a Plane*** A plane is observed approaching your home and you assume its speed is 550 miles per hour. If the angle of elevation of the plane is 16° at one time and 57° 1 minute later, approximate the altitude of the plane.

38. ***Height of a Mountain*** While traveling across flat land, you notice a mountain directly in front of you. The angle of elevation to the peak is 3.5°. After you drive 13 miles closer to the mountain, the angle of elevation is 9°. Approximate the height of the mountain.

Geometry **In Exercises 39 and 40, find the angle α between two nonvertical lines L_1 and L_2. The angle α satisfies the equation**

$$\tan \alpha = \left| \frac{m_2 - m_1}{1 + m_2 m_1} \right|$$

where m_1 and m_2 are the slopes of L_1 and L_2, respectively. (Assume $m_1 m_2 \neq -1$.)

39. L_1: $3x - 2y = 5$
L_2: $x + y = 1$

40. L_1: $2x + y = 8$
L_2: $x - 5y = -4$

41. ***Geometry*** Determine the angle between the diagonal of the cube and the diagonal of its base, as shown in the figure.

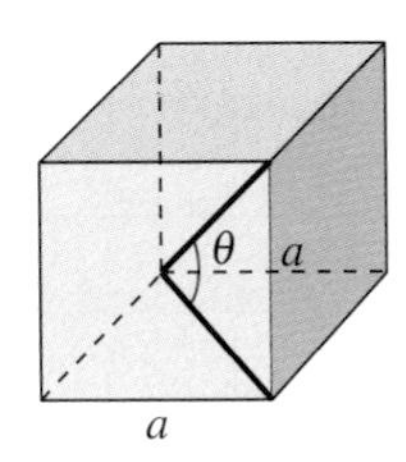

42. ***Geometry*** Determine the angle between the diagonal of the cube and its edge, as shown in the figure.

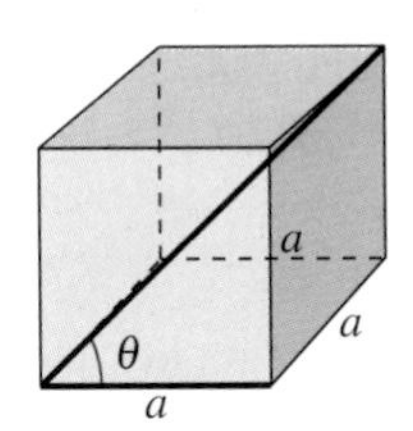

43. ***Wrench Size*** Express the distance y across the flat sides of the hexagonal nut as a function of r, as shown in the figure.

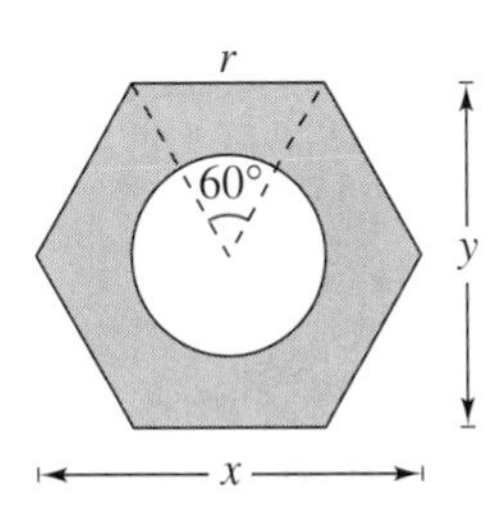

44. ***Bolt Circle*** The figure shows a circular sheet of diameter 40 centimeters, containing 12 equally spaced bolt holes. Determine the straight-line distance between the centers of consecutive bolt holes.

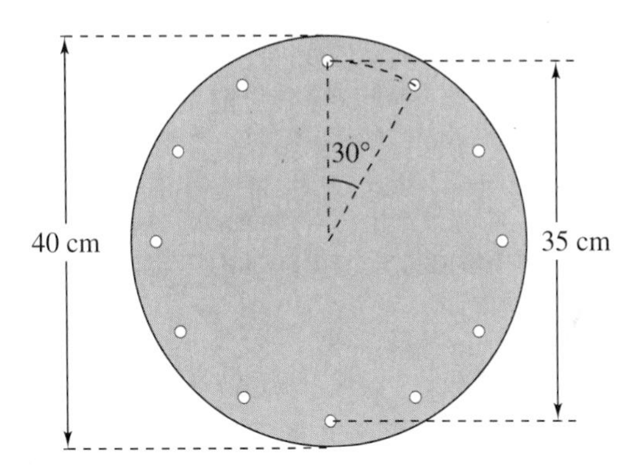

45. ***Geometry*** A regular pentagon is inscribed in a circle of radius 25 inches. Find the length of the sides of the pentagon.

46. ***Geometry*** A regular hexagon is inscribed in a circle of radius 25 inches. Find the length of the sides of the hexagon.

Trusses **In Exercises 47 and 48, find the lengths of all the unknown members of the truss.**

47.

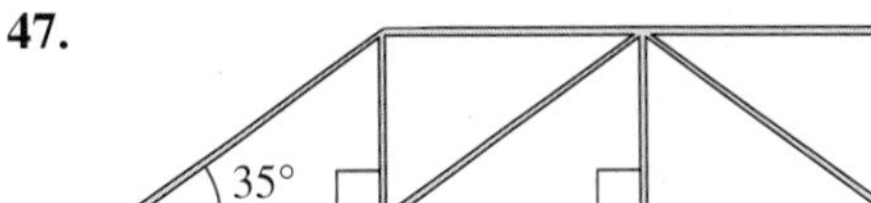

48.

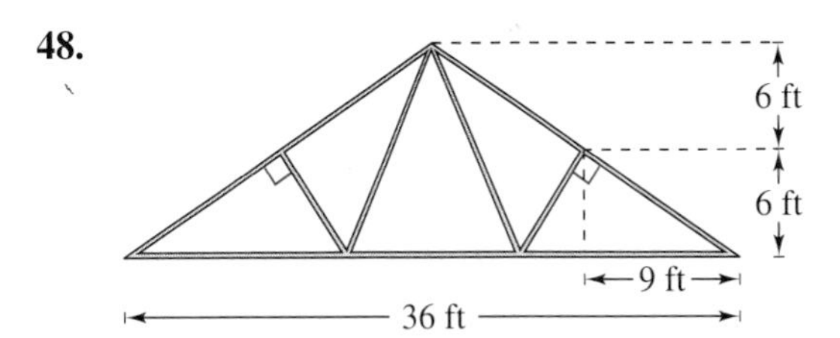

Harmonic Motion **In Exercises 49–52, for the simple harmonic motion described by the trigonometric function, find (a) the maximum displacement, (b) the frequency, and (c) the least positive value of t for which $d = 0$.**

49. $d = 4 \cos 8\pi t$

50. $d = \frac{1}{2} \cos 20\pi t$

51. $d = \frac{1}{16} \sin 120\pi t$

52. $d = \frac{1}{64} \sin 792\pi t$

Harmonic Motion **In Exercises 53–56, find a model for simple harmonic motion satisfying the specified conditions.**

	Displacement ($t = 0$)	*Amplitude*	*Period*
53.	0	4 cm	2 sec
54.	0	3 m	6 sec
55.	3 in.	3 in.	1.5 sec
56.	2 ft	2 ft	10 sec

57. ***Tuning Fork*** A point on the end of a tuning fork moves in simple harmonic motion described by $d = a \sin \omega t$. Find ω given that the tuning fork for middle C has a frequency of 264 vibrations per second.

58. ***Wave Motion*** A buoy oscillates in simple harmonic motion as waves go past. At a given time it is noted that the buoy moves a total of 3.5 feet from its low point to its high point (see figure), and that it returns to its high point every 10 seconds. Write an equation that describes the motion of the buoy if, at $t = 0$, it is at its high point.

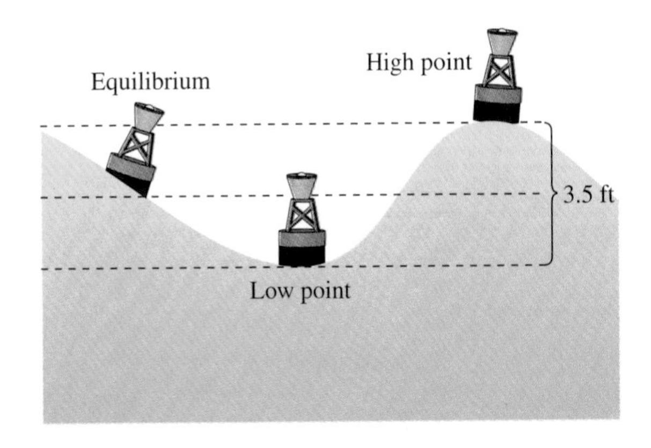

59. ***Springs*** A weight stretches a spring 1.5 inches. The weight is pushed 3 inches above its equilibrium position and released. Its motion is modeled by

$$y = \frac{1}{4} \cos 16t, \qquad t > 0.$$

(a) Graph the function.

(b) What is the period of the oscillations?

(c) Determine the first time the weight passes the point of equilibrium ($y = 0$).

60. ***Numerical and Graphical Analysis*** A 2-meter-high fence is 3 meters from the side of a grain storage bin. A grain elevator must reach from ground level outside the fence to the storage bin (see figure). The objective is to determine the shortest elevator meeting the constraints.

(a) Complete four rows of the table.

θ	L_1	L_2	$L_1 + L_2$
0.1	$\frac{2}{\sin 0.1}$	$\frac{3}{\cos 0.1}$	23.0
0.2	$\frac{2}{\sin 0.2}$	$\frac{3}{\cos 0.2}$	13.1

(b) Use a graphing utility to generate additional rows of the table. Use the table to estimate the minimum length of the elevator.

(c) Write the length $L_1 + L_2$ as a function of θ.

(d) Use a graphing utility to graph the function. Use the graph to estimate the minimum length. How does your estimate compare with that of part (b)?

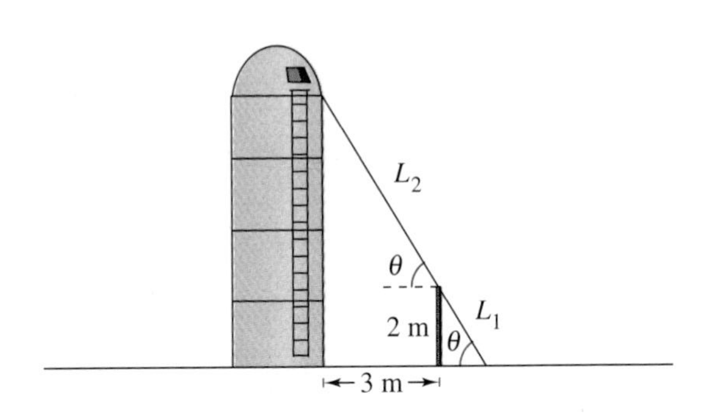

61. ***Numerical and Graphical Analysis*** The cross-sections of an irrigation canal are isosceles trapezoids where the length of three of the sides is 8 feet (see figure). The objective is to find the angle θ that maximizes the area of the cross sections. [*Hint:* The area of a trapezoid is $(h/2)(b_1 + b_2)$.]

(a) Complete six rows of the table.

Base 1	*Base 2*	*Altitude*	*Area*
8	$8 + 16 \cos 10°$	$8 \sin 10°$	22.1
8	$8 + 16 \cos 20°$	$8 \sin 20°$	42.5

(b) Use a graphing utility to generate additional rows of the table. Use the table to estimate the maximum cross-sectional area.

(c) Write the area A as a function of θ.

(d) Use a graphing utility to graph the function. Use the graph to estimate the maximum cross-sectional area. How does your estimate compare with that of part (b)?

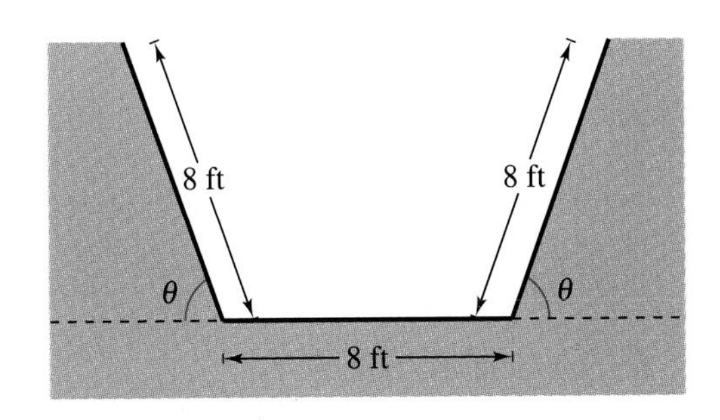

62. ***Data Analysis*** The times S of sunset (Greenwich Mean Time) at 40° north latitude on the 15th of each month are: 1(16:59), 2(17:35), 3(18:06), 4(18:38), 5(19:08), 6(19:30), 7(19:28), 8(18:57), 9(18:09), 10(17:21), 11(16:44), 12(16:36). The month is represented by t, with $t = 1$ corresponding to January. A model (where minutes have been converted to the decimal part of an hour) for this data is

$$S(t) = 18.09 + 1.41 \sin\left(\frac{\pi t}{6} + 4.60\right).$$

(a) Use a graphing utility to graph the data points and the model on the same viewing rectangle.

(b) What is the period of the model? Is it what you expected? Explain.

(c) What is the amplitude of the function? What does it represent in the model? Explain.

63. ***Data Analysis*** The table gives the average sales S (in millions) of an outerwear manufacturer for each month t, where $t = 1$ represents January.

t	1	2	3	4	5	6
S	13.46	11.15	8.00	4.85	2.54	1.70

t	7	8	9	10	11	12
S	2.54	4.85	8.00	11.15	13.46	14.30

(a) Create a scatter plot of the data.

(b) Find a trigonometric model that fits the data. Graph the model on your scatter plot. How well does the model fit?

(c) What is the period of the model? Do you think it is reasonable given the context? Explain your reasoning.

(d) Interpret the meaning of the model's amplitude in the context of the problem.

FOCUS ON CONCEPTS

In this chapter, you studied trigonometry. Use the following questions to check your understanding of several of the basic concepts presented. The answers to these questions are given in the back of the book.

1. In your own words, explain the meaning of (a) an angle in standard position, (b) a negative angle, (c) a coterminal angle, and (d) an obtuse angle.

2. A fan motor turns at a given angular speed. How does the speed of the tips of the blades change if a fan of greater diameter is installed on the motor? Explain.

3. *True or False?* $y = \sin \theta$ is not a function because $\sin 30° = \sin 150°$. Explain.

4. In right triangle trigonometry, $\sin 30° = \frac{1}{2}$ regardless of the size of the triangle. Explain.

5. Describe the behavior of $f(\theta) = \sec \theta$ at the zeros of $g(\theta) = \cos \theta$. Explain.

6. Explain how reference angles are used to find the trigonometric functions of obtuse angles.

In Exercises 7–10, match the function $y = a \sin bx$ with its graph. Base your selection solely on your interpretation of the constants a and b. Explain your reasoning. [The graphs are labeled (a), (b), (c), and (d).]

7. $y = 3 \sin x$

8. $y = -3 \sin x$

9. $y = 2 \sin \pi x$

10. $y = 2 \sin \dfrac{x}{2}$

(a)

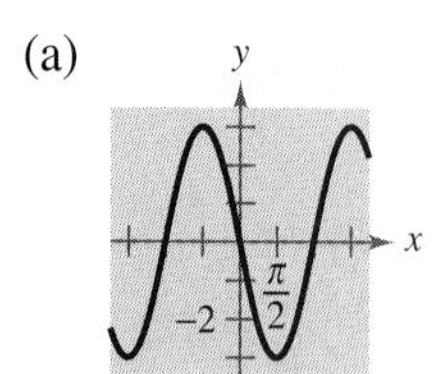

(b)

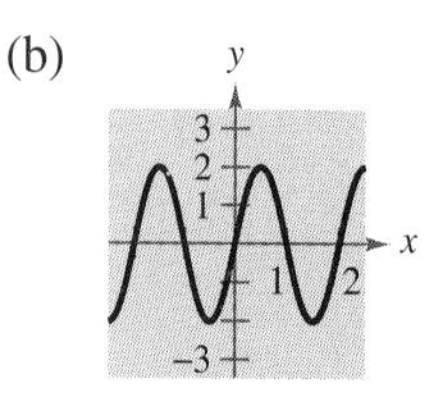

(c)

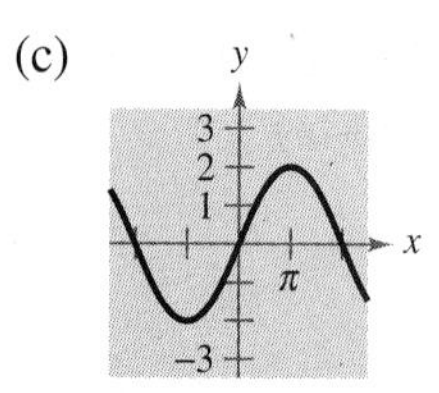

(d)

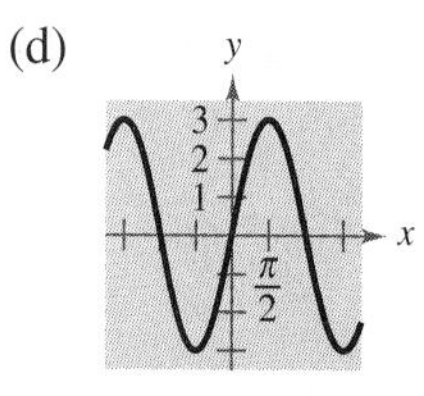

11. The function f is periodic, with period c. Therefore, $f(t + c) = f(t)$. Are the following equal? Explain.

(a) $f(t - 2c) \stackrel{?}{=} f(t)$

(b) $f\left(t + \frac{1}{2}c\right) \stackrel{?}{=} f\left(\frac{1}{2}t\right)$

(c) $f\left(\frac{1}{2}[t + c]\right) \stackrel{?}{=} f\left(\frac{1}{2}t\right)$

12. When graphing the sine and cosine functions, determining the amplitude is part of the analysis. Why is this not true for the other four trigonometric functions?

13. A weight is suspended from a ceiling by a steel spring. The weight is lifted (positive direction) from the equilibrium position and released. The resulting motion of the weight is modeled by

$$y = Ae^{-kt} \cos bt = \frac{1}{5}e^{-t/10} \cos 6t, \qquad t \geq 0$$

where y is the distance in feet from equilibrium and t is the time in seconds. The graph of the function is given in the figure. For each of the following, describe the change in the system without graphing the resulting function.

(a) A is changed from $\frac{1}{5}$ to $\frac{1}{3}$.

(b) k is changed from $\frac{1}{10}$ to $\frac{1}{3}$.

(c) b is changed from 6 to 9.

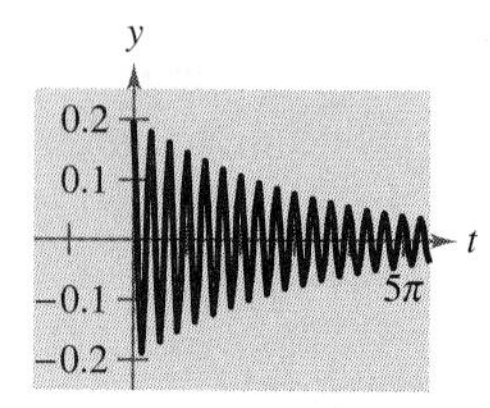

14. *True or False?* Because $\tan 3\pi/4 = -1$, $\arctan(-1) = 3\pi/4$. Explain.

Review Exercises

In Exercises 1–4, sketch the angle in standard position. List one positive and one negative coterminal angle.

1. $\dfrac{11\pi}{4}$
2. $\dfrac{2\pi}{9}$
3. $-110°$
4. $-405°$

In Exercises 5–8, convert the angle measure to decimal form. Round to two decimal places.

5. $135°\ 16'45''$
6. $-234°\ 50''$
7. $5°\ 22'53''$
8. $280°\ 8'50''$

In Exercises 9–12, convert the angle measure to D° M′ S″ form.

9. $135.27°$
10. $25.1°$
11. $-85.15°$
12. $-327.85°$

In Exercises 13–16, convert the angle measure from radians to degrees. Round to two decimal places.

13. $\dfrac{5\pi}{7}$
14. $-\dfrac{3\pi}{5}$
15. -3.5
16. 1.75

In Exercises 17–20, convert the angle measure from degrees to radians. Round to four decimal places.

17. $480°$
18. $-16.5°$
19. $-33°\ 45'$
20. $84°\ 15'$

In Exercises 21–24, find the reference angle for the given angle.

21. $252°$
22. $640°$
23. $-\dfrac{6\pi}{5}$
24. $\dfrac{17\pi}{3}$

In Exercises 25–28, find the point (x, y) on the unit circle that corresponds to the real number t. Then find the sine, cosine, and tangent of t.

25. $t = \dfrac{7\pi}{6}$
26. $t = \dfrac{3\pi}{4}$
27. $t = -\dfrac{\pi}{3}$
28. $t = 2\pi$

In Exercises 29 and 30, find the six trigonometric functions of the angle θ (in standard position) whose terminal side passes through the given point.

29. $(12, 16)$
30. $(4, -8)$

In Exercises 31 and 32, find the remaining five trigonometric functions of θ satisfying the given conditions.

31. $\sec \theta = \frac{6}{5}, \quad \tan \theta < 0$
32. $\tan \theta = -\frac{12}{5}, \quad \sin \theta > 0$

In Exercises 33–36, evaluate the trigonometric function without using a calculator.

33. $\tan \dfrac{\pi}{3}$
34. $\sec \dfrac{\pi}{4}$
35. $\cos 495°$
36. $\csc 270°$

In Exercises 37–40, use a calculator to evaluate the trigonometric function. Round to two decimal places.

37. $\tan 33°$
38. $\csc 105°$
39. $\sec \dfrac{12\pi}{5}$
40. $\sin\left(-\dfrac{\pi}{9}\right)$

In Exercises 41 and 42, find two values of θ in degrees $(0° \le \theta < 360°)$ and in radians $(0 \le \theta < 2\pi)$.

41. $\cos \theta = -\dfrac{\sqrt{2}}{2}$
42. $\sec \theta$ is undefined.

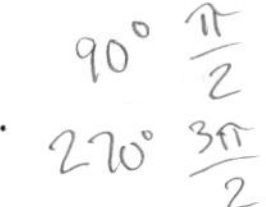

In Exercises 43 and 44, use a calculator to find two values of θ. Express both values in degrees ($0° \le \theta < 360°$) and in radians ($0 \le \theta < 2\pi$).

43. $\sin \theta = 0.8387$ **44.** $\cot \theta = -1.5399$

In Exercises 45 and 46, find the slope of the line with an inclination angle of θ.

45. $\theta = 120°$ **46.** $\theta = 55.8°$

In Exercises 47 and 48, find the angle of inclination of the line.

47. $x + y - 10 = 0$ **48.** $3x - 2y - 4 = 0$

In Exercises 49–60, sketch a graph of the function.

49. $y = 3 \cos 2\pi x$ **50.** $y = -2 \sin \pi x$

51. $f(x) = 5 \sin \dfrac{2x}{5}$ **52.** $f(x) = 8 \cos\left(-\dfrac{x}{4}\right)$

53. $f(x) = -\dfrac{1}{4} \cos \dfrac{\pi x}{4}$ **54.** $f(x) = -\tan \dfrac{\pi x}{4}$

55. $g(t) = \dfrac{5}{2} \sin(t - \pi)$ **56.** $g(t) = 3 \cos(t + \pi)$

57. $h(t) = \tan\left(t - \dfrac{\pi}{4}\right)$ **58.** $h(t) = \sec\left(t - \dfrac{\pi}{4}\right)$

59. $y = \arcsin \dfrac{x}{2}$ **60.** $y = 2 \arccos x$

In Exercises 61–68, use a graphing utility to graph the function. If the function is periodic, find the period.

61. $f(x) = \dfrac{x}{4} - \sin x$ **62.** $y = \dfrac{x}{3} + \cos \pi x$

63. $f(x) = \dfrac{\pi}{2} + \arctan x$ **64.** $y = 4 - \dfrac{x}{4} + \cos \pi x$

65. $h(\theta) = \theta \sin \pi\theta$ **66.** $f(\theta) = \cot \dfrac{\pi\theta}{8}$

67. $f(t) = 2.5e^{-t/4} \sin 2\pi t$ **68.** $f(x) = \arccos(x - \pi)$

In Exercises 69–72, use a graphing utility to graph the function. Use the graph to determine if the function is periodic. If the function is periodic, approximate any relative maximum or minimum points through one period.

69. $f(x) = e^{\sin x}$ **70.** $g(x) = \sin e^x$

71. $g(x) = 2 \sin x \cos^2 x$ **72.** $h(x) = 4 \sin^2 x \cos^2 x$

In Exercises 73–76, write an algebraic expression for the given expression.

73. $\sec[\arcsin(x - 1)]$ **74.** $\tan\left(\arccos \dfrac{x}{2}\right)$

75. $\sin\left(\arccos \dfrac{x^2}{4 - x^2}\right)$ **76.** $\csc(\arcsin 10x)$

77. ***Altitude of a Triangle*** Find the altitude of the triangle in the figure.

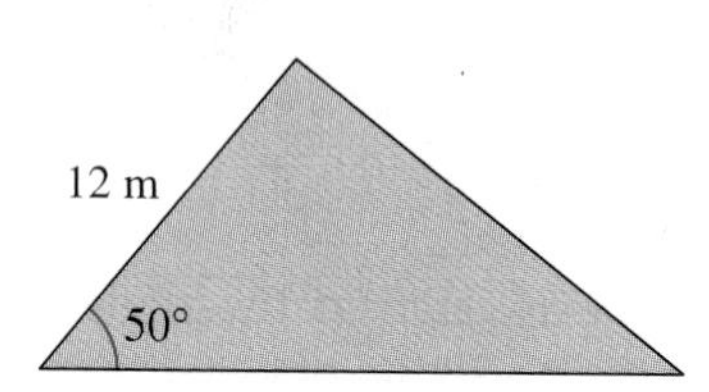

78. ***Angle of Elevation*** The height of a radio transmission tower is 70 meters, and it casts a shadow of length 30 meters (see figure). Find the angle of elevation of the sun.

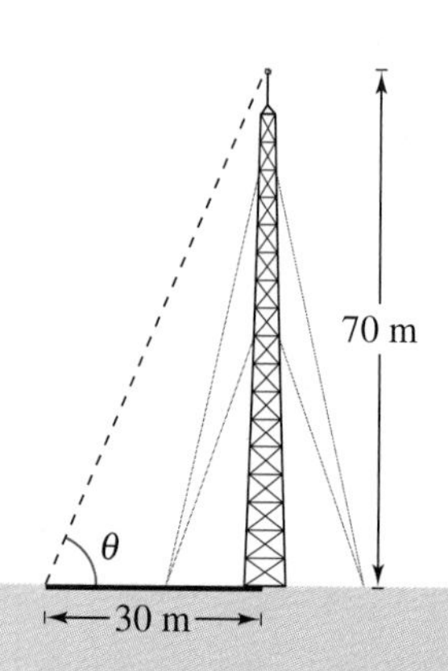

79. ***Shuttle Height*** An observer 2.5 miles from the launch pad of a space shuttle measures the angle of elevation to the base of the shuttle to be 25° soon after lift-off (see figure). How high is the shuttle at that instant? (Assume that the shuttle is still moving vertically.)

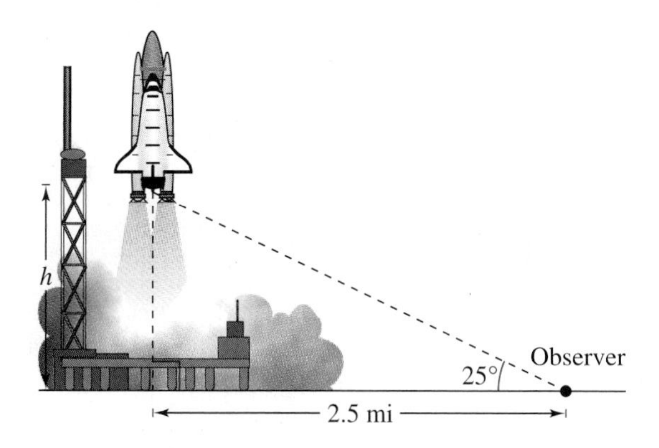

80. ***Distance*** From city A to city B, a plane flies 650 miles at a bearing of N 48° E. From city B to city C, the plane flies 810 miles at a bearing of S 65° E. Find the distance from A to C and the bearing from A to C.

81. ***Railroad Grade*** A train travels 3.5 kilometers on a straight track with a grade of 1° 10′ (see figure). What is the vertical rise of the train in that distance?

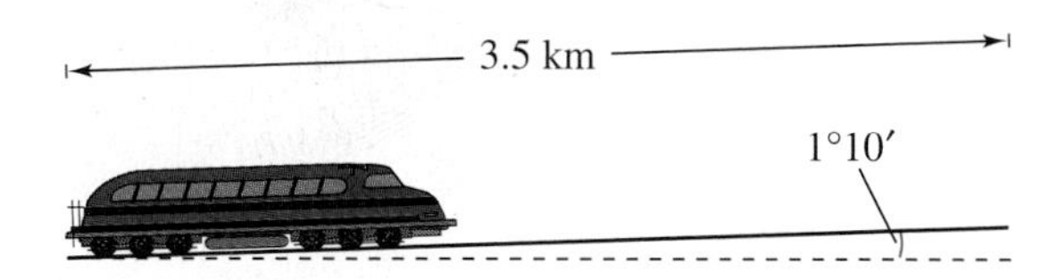

82. ***Distance Between Towns*** A passenger in an airplane flying at 37,000 feet sees two towns directly to the left of the airplane. The angles of depression to the towns are 32° and 76° (see figure). How far apart are the towns?

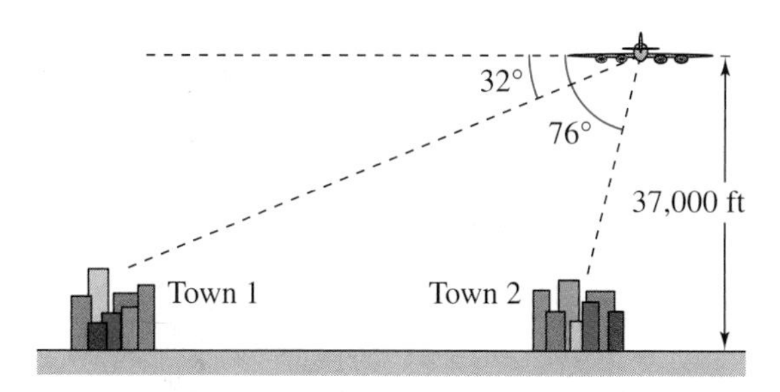

83. ***Exploration*** The base of the triangle in the figure is also the radius of a circular arc.

(a) Find the area A of the shaded region as a function of θ for $0 < \theta < \pi/2$.

(b) Use a graphing utility to graph the area function over the given domain. Interpret the graph in the context of the problem.

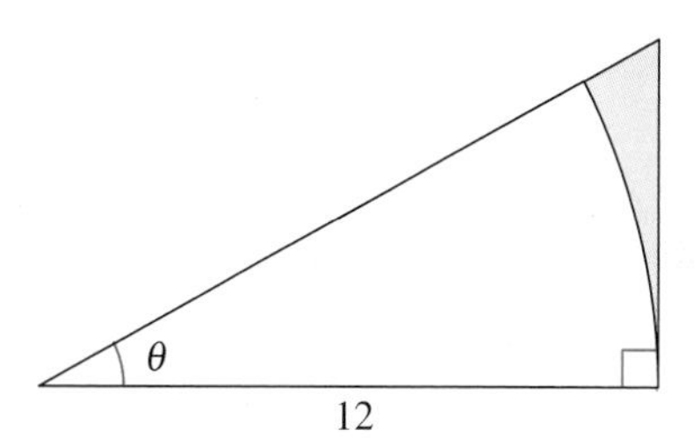

84. ***Exploration*** In calculus it can be shown that the arcsine and arctangent functions can be approximated by the polynomials

$$\arcsin x \approx x + \frac{x^3}{6} + \frac{3x^5}{40} + \frac{5x^7}{112}$$

$$\arctan x \approx x - \frac{x^3}{3} + \frac{x^5}{5} - \frac{x^7}{7}.$$

(a) Use a graphing utility to graph the arcsine function and its polynomial approximation on the same viewing rectangle.

(b) Use a graphing utility to graph the arctangent function and its polynomial approximation on the same viewing rectangle.

(c) Study the pattern in the polynomial approximation of the arctangent function and guess the next term. Then repeat part (b). Do you think your guess was correct? How did the accuracy of the approximation change when additional terms were added?

CHAPTER PROJECT: *Analyzing a Graph*

Graphs of functions that are combinations of algebraic functions and trigonometric functions can be difficult to sketch by hand. For such graphs, a graphing utility is helpful.

EXAMPLE 1 *Sketching the Graph of a Function*

Since 1958, the Mauna Loa Climate Observatory in Hawaii has been collecting data on the carbon dioxide level of earth's atmosphere. A model that closely represents the data is

$$y = 316 + 0.654t + 0.0216t^2 + 2.5 \sin 2\pi t$$

where y represents the monthly average of carbon dioxide concentration (in parts per million) and $t = 0$ represents January 1960, $t = 1$ represents January 1961, etc. Sketch the graph of this function and explain the oscillations in the graph.

Solution

The graph of the function is shown below. From the graph, you can see that the carbon dioxide level fluctuates each year. The low level each year, which occurs toward the end of the summer in the northern hemisphere, is caused by the intake of carbon dioxide in growing plants.

Mauna Loa Climate Observatory
(Photo: NOAA/photo by Bernard G. Mendonca)

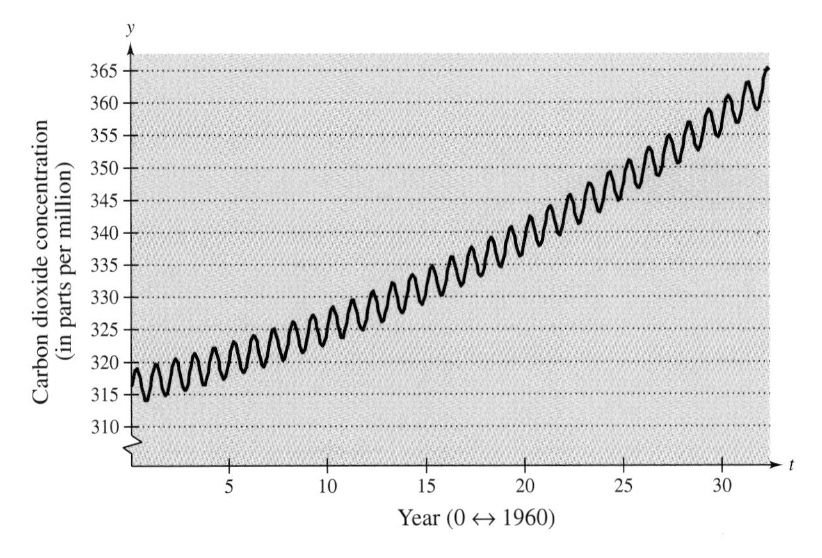

EXAMPLE 2 *Sketching the Graph of a Function*

Sketch the graph of $y = \sin \frac{1}{x}$. Describe the graph near the origin.

Solution

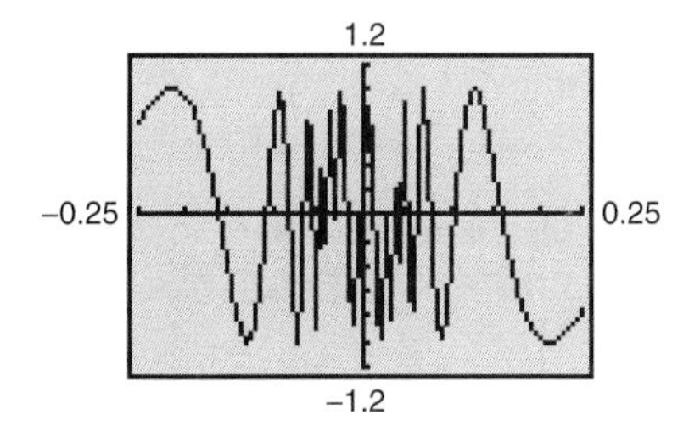

The graph is difficult to sketch, even with a graphing utility. The graph shown at the left was produced with a graphing utility. The low resolution of the utility gives a distorted image of the graph. To obtain a better image, we used high-resolution computer software and obtained the graph below. From the high-resolution graph, you can see that as x approaches the origin from the left or the right, the graph oscillates more and more quickly.

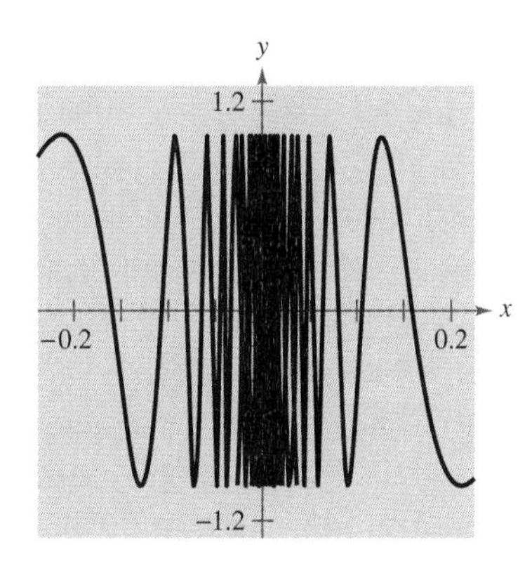

CHAPTER PROJECT INVESTIGATIONS

In Questions 1–6, use a graphing utility to graph the function. Choose a viewing rectangle that you think produces a good representation of the important features of the graph.

1. $y = x^2 + \sin x$
2. $y = x^2 \sin x$
3. $y = |\cos x|$
4. $y = 2 \sin x - \cos 2x$
5. $y = \sin^2 x + \sin x$
6. $y = \dfrac{\sin x}{x}$

7. *Carbon Dioxide Levels* Sketch the graph of the model given in Example 1 for $28 \le t \le 30$. Between January 1988 and January 1990, what were the highest and lowest levels of carbon dioxide? When did each occur?

8. *Throwing a Shot Put* The path of a shot put can be modeled by

$$y = -\frac{16}{v^2 \cos^2 \theta} x^2 + (\tan \theta)x + h$$

where y is the height of the shot put (in feet), x is the horizontal distance (in feet), v is the initial velocity (in feet per second), h is the initial height (in feet), and θ is the angle at which the shot put is thrown.

(a) Choose several values of v, h, and θ and sketch the corresponding graphs. Discuss your results.

(b) Of the graphs you sketched in part (a), which do you think best models a real-life shot-put event? Explain your reasoning.

Chapter Test

Take this test as you would take a test in class. After you are done, check your work against the answers given in the back of the book.

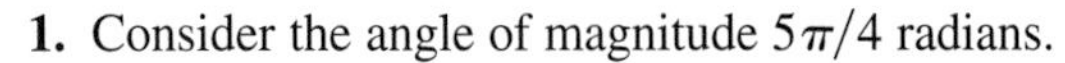

1. Consider the angle of magnitude $5\pi/4$ radians.
   (a) Sketch the angle in standard position.
   (b) Determine two coterminal angles (one positive and one negative).
   (c) Convert the angle to degree measure.
2. A truck is moving at a rate of 90 kilometers per hour, and the diameter of its wheels is 1 meter. Find the angular speed of the wheels in radians per minute.
3. Find the exact values of the six trigonometric functions of the angle θ shown in the figure.

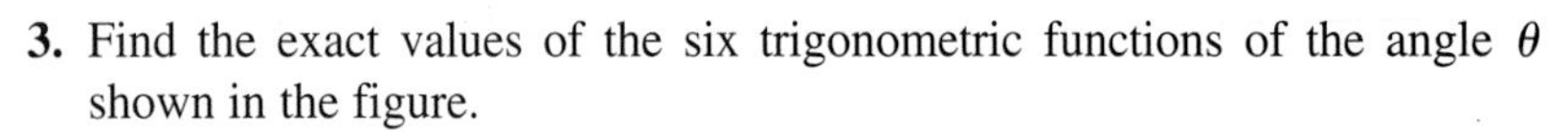

4. Given that $\tan\theta = \frac{3}{2}$, find the other five trigonometric functions of θ.
5. Determine the reference angle θ' of the angle $\theta = 290°$ and sketch θ and θ' in standard position.
6. Determine the quadrant in which θ lies if $\sec\theta < 0$ and $\tan\theta > 0$.
7. Find two values of θ in degrees $(0 \le \theta < 360°)$ if $\cos\theta = -\sqrt{3}/2$. (Do not use a calculator.)
8. Use a calculator to approximate two values of θ in radians $(0 \le \theta < 2\pi)$ if $\csc\theta = 1.030$. Round the result to two decimal places.

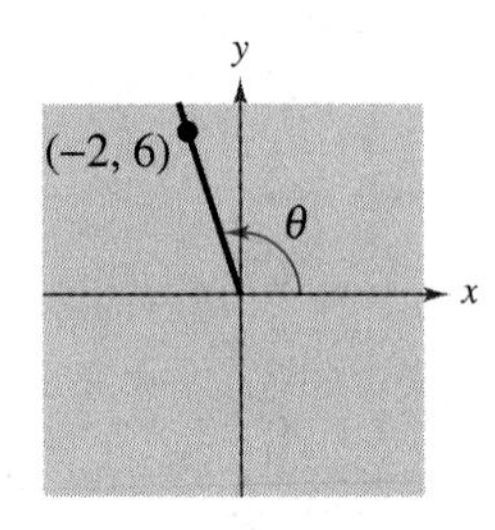

FIGURE FOR 3

The *Interactive* CD-ROM provides answers to the Chapter Tests and Cumulative Tests. It also offers Chapter Pre-Tests (which test key skills and concepts covered in previous chapters) and Chapter Post-Tests, both of which have randomly generated exercises with diagnostic capabilities.

In Exercises 9 and 10, graph the function through two full periods without the aid of a graphing utility.

9. $g(x) = -2\sin\left(x - \frac{\pi}{4}\right)$

10. $f(\alpha) = \frac{1}{2}\tan 2\alpha$

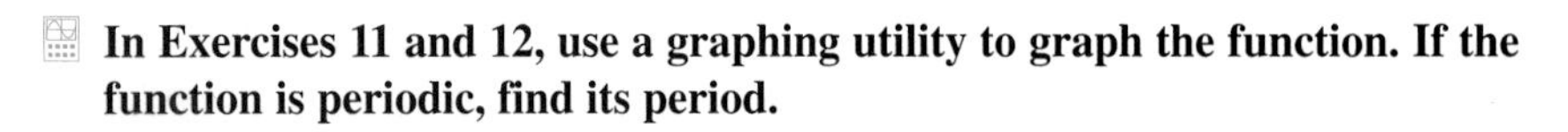

In Exercises 11 and 12, use a graphing utility to graph the function. If the function is periodic, find its period.

11. $y = \sin 2\pi x + 2\cos\pi x$

12. $y = 6e^{-0.12t}\cos(0.25t), \quad 0 \le t \le 32$

13. Find a, b, and c for the function $f(x) = a\sin(bx + c)$ so that the graph of f matches the figure.
14. Find the exact value of $\tan\left(\arccos\frac{2}{3}\right)$ without the aid of a calculator.
15. Graph the function $f(x) = 2\arcsin\left(\frac{1}{2}x\right)$.
16. A ship leaves port at noon and sails at a speed of 18 knots. Its bearing is N 16° W. If the port is positioned at the origin, determine the coordinates of the position of the ship at 3 P.M.

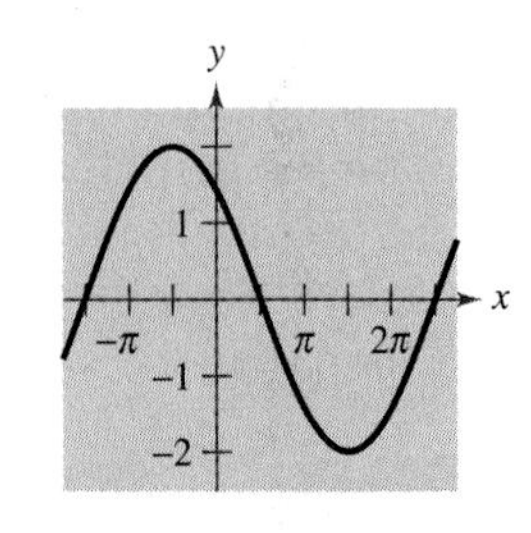

FIGURE FOR 13

2 Analytic Trigonometry

The number of hours of daylight that occur at any location on earth depends on two things: the time of year and the latitude of the location.

Here are models for the number of hours of daylight in Seward, Alaska (60 degrees latitude) and New Orleans, Louisiana (30 degrees latitude).

Seward

$D = 12.2 - 6.4\cos[\pi(t + 0.2)/6]$

New Orleans

$D = 12.2 - 1.9\cos[\pi(t + 0.2)/6]$

In these models, D represents the number of hours of daylight and t represents the month, with $t = 0$ representing January 1.

To find the time of year that both cities receive the same amount of daylight, you can equate the two models and solve for t. When you do that, you obtain $t = 2.8$ (spring equinox) and $t = 8.8$ (fall equinox).

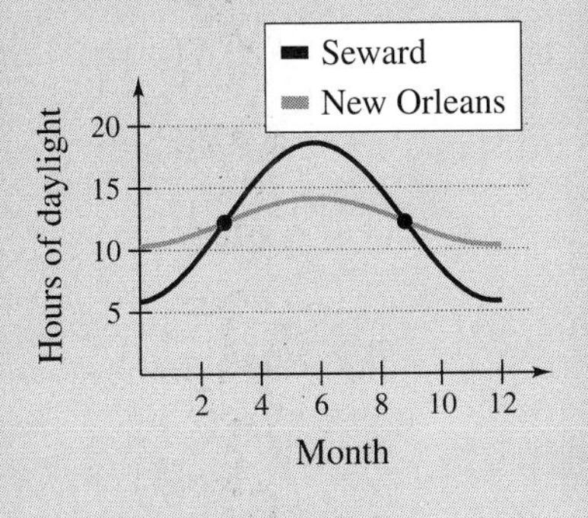

See Exercises 73 and 74 on page 229.

Since 1987, Dr. Warren Washington has directed the Climate Global Dynamics Division of NCAR (National Center for Atmospheric Research) in Boulder, Colorado. He has helped develop innovative computer models to predict long-term weather patterns.

(Photos: University Corporation for Atmospheric Research/National Center for Atmospheric Research/National Science Foundation)

2.1 Using Fundamental Identities

See Exercises 81 and 82 on page 221 for examples of how trigonometric identities can be used to simplify logarithmic expressions.

Introduction ▫ *Using the Fundamental Identities*

Introduction

In Chapter 1, you studied the basic definitions, properties, graphs, and applications of the individual trigonometric functions. In this chapter, you will learn how to use the fundamental identities to

1. evaluate trigonometric functions.
2. simplify trigonometric expressions.
3. develop additional trigonometric identities.
4. solve trigonometric equations.

FUNDAMENTAL TRIGONOMETRIC IDENTITIES

Reciprocal Identities

$$\sin u = \frac{1}{\csc u} \qquad \cos u = \frac{1}{\sec u} \qquad \tan u = \frac{1}{\cot u}$$

$$\csc u = \frac{1}{\sin u} \qquad \sec u = \frac{1}{\cos u} \qquad \cot u = \frac{1}{\tan u}$$

Quotient Identities

$$\tan u = \frac{\sin u}{\cos u} \qquad \cot u = \frac{\cos u}{\sin u}$$

Pythagorean Identities

$$\sin^2 u + \cos^2 u = 1 \qquad 1 + \tan^2 u = \sec^2 u \qquad 1 + \cot^2 u = \csc^2 u$$

Cofunction Identities

$$\sin\left(\frac{\pi}{2} - u\right) = \cos u \qquad \cos\left(\frac{\pi}{2} - u\right) = \sin u$$

$$\tan\left(\frac{\pi}{2} - u\right) = \cot u \qquad \cot\left(\frac{\pi}{2} - u\right) = \tan u$$

$$\sec\left(\frac{\pi}{2} - u\right) = \csc u \qquad \csc\left(\frac{\pi}{2} - u\right) = \sec u$$

Even/Odd Identities

$$\sin(-u) = -\sin u, \qquad \cos(-u) = \cos u, \qquad \tan(-u) = -\tan u$$

$$\csc(-u) = -\csc u, \qquad \sec(-u) = \sec u, \qquad \cot(-u) = -\cot u$$

NOTE Pythagorean identities are sometimes used in radical form such as

$$\sin u = \pm\sqrt{1 - \cos^2 u}$$

or

$$\tan u = \pm\sqrt{\sec^2 u - 1}$$

where the sign depends on the choice of u.

Using the Fundamental Identities

One common use of trigonometric identities is to use given values of trigonometric functions to evaluate other trigonometric functions.

The *Interactive* CD-ROM offers graphing utility emulators of the *TI-82* and *TI-83*, which can be used with the Examples, Explorations, Technology notes, and Exercises.

EXAMPLE 1 *Using Identities to Evaluate a Function*

Use the given values $\sec u = -\frac{3}{2}$ and $\tan u > 0$ to find the values of all six trigonometric functions.

Solution

Using a reciprocal identity, you have

$$\cos u = \frac{1}{\sec u} = \frac{1}{-3/2} = -\frac{2}{3}.$$

Using a Pythagorean identity, you have

$$\sin^2 u = 1 - \cos^2 u = 1 - \left(-\frac{2}{3}\right)^2 = 1 - \frac{4}{9} = \frac{5}{9}.$$

Because $\sec u < 0$ and $\tan u > 0$, it follows that u lies in Quadrant III. Moreover, because $\sin u$ is negative when u is in Quadrant III, you can choose the negative root and obtain $\sin u = -\sqrt{5}/3$. Now, knowing the values of the sine and cosine, you can find the values of all six trigonometric functions.

$$\sin u = -\frac{\sqrt{5}}{3} \qquad \csc u = \frac{1}{\sin u} = -\frac{3}{\sqrt{5}}$$

$$\cos u = -\frac{2}{3} \qquad \sec u = \frac{1}{\cos u} = -\frac{3}{2}$$

$$\tan u = \frac{\sin u}{\cos u} = \frac{-\sqrt{5}/3}{-2/3} = \frac{\sqrt{5}}{2} \qquad \cot u = \frac{1}{\tan u} = \frac{2}{\sqrt{5}}$$

TECHNOLOGY

You can use a graphing utility to check the result of Example 2. To do this, graph

$$y = \sin x \cos^2 x - \sin x$$

and

$$y = -\sin^3 x$$

on the same viewing rectangle, as shown below. Because Example 2 shows the equivalence algebraically and the two graphs appear to coincide, you can conclude that the expressions are equivalent.

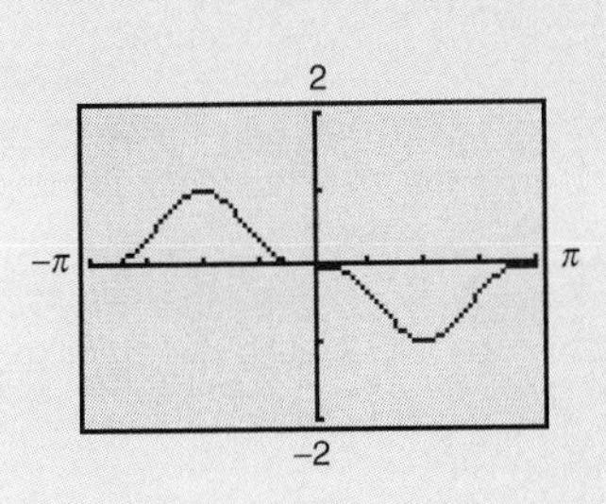

EXAMPLE 2 *Simplifying a Trigonometric Expression*

Simplify $\sin x \cos^2 x - \sin x$.

Solution

Factor out the common monomial factor and then use a fundamental identity.

$$\begin{aligned} \sin x \cos^2 x - \sin x &= \sin x(\cos^2 x - 1) && \text{Monomial factor} \\ &= -\sin x(1 - \cos^2 x) \\ &= -\sin x(\sin^2 x) && \text{Pythagorean identity} \\ &= -\sin^3 x && \text{Multiply.} \end{aligned}$$

The *Interactive* CD-ROM shows every example with its solution; clicking on the *Try It!* button brings up similar problems. Guided Examples and Integrated Examples show step-by-step solutions to additional examples. Integrated Examples are related to several concepts in the section.

EXAMPLE 3 *Factoring Trigonometric Expressions*

Factor each expression.

a. $\sec^2\theta - 1$ **b.** $4\tan^2\theta + \tan\theta - 3$

Solution

a. Here you have the difference of two squares, which factors as

$$\sec^2\theta - 1 = (\sec\theta - 1)(\sec\theta + 1).$$

b. This expression has the polynomial form, $ax^2 + bx + c$, and it factors as

$$4\tan^2\theta + \tan\theta - 3 = (4\tan\theta - 3)(\tan\theta + 1).$$

EXAMPLE 4 *Factoring a Trigonometric Expression*

Factor $\csc^2 x - \cot x - 3$.

Solution

You can use the identity $\csc^2 x = 1 + \cot^2 x$ to rewrite the expression in terms of the cotangent alone.

$$\begin{aligned}\csc^2 x - \cot x - 3 &= (1 + \cot^2 x) - \cot x - 3 && \text{Pythagorean identity}\\ &= \cot^2 x - \cot x - 2 && \text{Combine like terms.}\\ &= (\cot x - 2)(\cot x + 1) && \text{Factor.}\end{aligned}$$

Study Tip

On occasion, factoring or simplifying can best be done by first rewriting the expression in terms of just *one* trigonometric function or in terms of *sine and cosine alone.* These strategies are illustrated in Examples 4 and 5, respectively.

EXAMPLE 5 *Simplifying a Trigonometric Expression*

Simplify $\sin t + \cot t \cos t$.

Solution

Begin by rewriting the expression in terms of sine and cosine.

$$\begin{aligned}\sin t + \cot t\cos t &= \sin t + \left(\frac{\cos t}{\sin t}\right)\cos t && \text{Quotient identity}\\ &= \frac{\sin^2 t + \cos^2 t}{\sin t} && \text{Add fractions.}\\ &= \frac{1}{\sin t} && \text{Pythagorean identity}\\ &= \csc t && \text{Reciprocal identity}\end{aligned}$$

EXAMPLE 6 *Verifying a Trigonometric Identity*

Verify the identity $\dfrac{\sin\theta}{1+\cos\theta}+\dfrac{\cos\theta}{\sin\theta}=\csc\theta$.

Solution

$$\frac{\sin\theta}{1+\cos\theta}+\frac{\cos\theta}{\sin\theta}=\frac{(\sin\theta)(\sin\theta)+(\cos\theta)(1+\cos\theta)}{(1+\cos\theta)(\sin\theta)}$$

$$=\frac{\sin^2\theta+\cos^2\theta+\cos\theta}{(1+\cos\theta)(\sin\theta)} \qquad \text{Multiply.}$$

$$=\frac{\cancel{1+\cos\theta}}{\cancel{(1+\cos\theta)}(\sin\theta)} \qquad \text{Pythagorean identity}$$

$$=\frac{1}{\sin\theta} \qquad \text{Cancel common factor.}$$

$$=\csc\theta \qquad \text{Reciprocal identity}$$

The last two examples in this section involve techniques for rewriting expressions into forms that are useful in calculus.

EXAMPLE 7 *Rewriting a Trigonometric Expression*

Rewrite $\dfrac{1}{1+\sin x}$ so that it is *not* in fractional form.

Solution

From the Pythagorean identity $\cos^2 x = 1-\sin^2 x=(1-\sin x)(1+\sin x)$, you can see that by multiplying both the numerator and the denominator by $(1-\sin x)$ you produce a monomial denominator.

$$\frac{1}{1+\sin x}=\frac{1}{1+\sin x}\cdot\frac{1-\sin x}{1-\sin x} \qquad \text{Multiply numerator and denominator by } (1-\sin x).$$

$$=\frac{1-\sin x}{1-\sin^2 x} \qquad \text{Multiply.}$$

$$=\frac{1-\sin x}{\cos^2 x} \qquad \text{Pythagorean identity}$$

$$=\frac{1}{\cos^2 x}-\frac{\sin x}{\cos^2 x} \qquad \text{Separate fractions.}$$

$$=\frac{1}{\cos^2 x}-\frac{\sin x}{\cos x}\cdot\frac{1}{\cos x} \qquad \text{Separate fractions.}$$

$$=\sec^2 x-\tan x\sec x \qquad \text{Identities}$$

EXAMPLE 8 *Trigonometric Substitution*

Use the substitution $x = 2\tan\theta,\ 0 < \theta < \pi/2$, to express

$$\sqrt{4 + x^2}$$

as a trigonometric function of θ.

Solution

Begin by letting $x = 2\tan\theta$. Then, you can obtain the following.

$$\begin{aligned}\sqrt{4 + x^2} &= \sqrt{4 + (2\tan\theta)^2} && \text{Substitute } 2\tan\theta \text{ for } x.\\ &= \sqrt{4(1 + \tan^2\theta)} \\ &= \sqrt{4\sec^2\theta} && \text{Pythagorean identity}\\ &= 2\sec\theta && \sec\theta > 0 \text{ for } 0 < \theta < \tfrac{\pi}{2}\end{aligned}$$

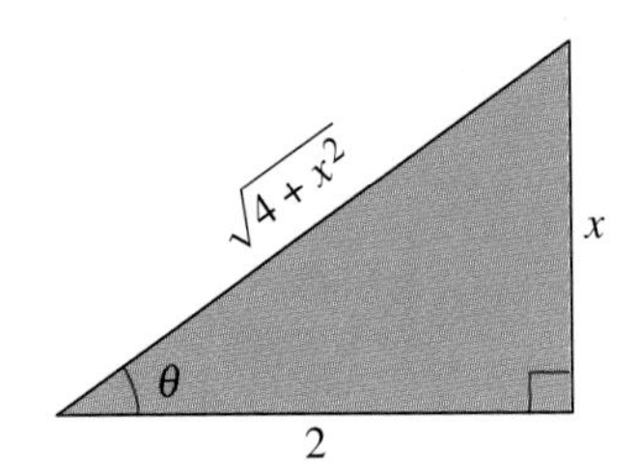

FIGURE 2.1

Figure 2.1 shows the right angle illustration of the trigonometric substitution in Example 8. For $0 < \theta < \pi/2$, you have

$$\text{opp} = x, \quad \text{adj} = 2, \quad \text{and} \quad \text{hyp} = \sqrt{4 + x^2}.$$

With these expressions, you can write the following.

$$\sec\theta = \frac{\sqrt{4 + x^2}}{2}$$

$$2\sec\theta = \sqrt{4 + x^2}$$

GROUP ACTIVITY

REMEMBERING TRIGONOMETRIC IDENTITIES

Most people find the Pythagorean identity involving sine and cosine to be fairly easy to remember: $\sin^2 u + \cos^2 u = 1$. The one involving tangent and secant, however, tends to give some people trouble. They can't remember if the identity is

$$1 + \tan^2 u \stackrel{?}{=} \sec^2 u \qquad \text{or} \qquad 1 + \sec^2 u \stackrel{?}{=} \tan^2 u.$$

Which of these two is the correct Pythagorean identity involving tangent and secant? Discuss how to remember (or derive) this identity. Can you think of easy ways to remember other fundamental trigonometric identities?

SOHCAHTOA

The *Interactive* CD-ROM provides additional help with Warm-Up exercises by providing a hypertext link to the section in which the concept was introduced.

WARM UP

Use a right triangle to evaluate the other five trigonometric functions of the acute angle θ.

1. $\tan\theta = \frac{3}{2}$

2. $\sec\theta = 3$

The point is on the terminal side of an angle θ in standard position. Find the exact values of the six trigonometric functions of θ.

3. $(7, -3)$

4. $(-10, 5)$

Simplify the expression.

5. $\sqrt{1 - \left(\frac{\sqrt{3}}{2}\right)^2}$

6. $\sqrt{\left(\frac{3}{4}\right)^2 + 1}$

7. $\sqrt{1 + \left(\frac{3}{8}\right)^2}$

8. $\sqrt{1 - \left(\frac{\sqrt{5}}{3}\right)^2}$

Perform the operations and simplify.

9. $\frac{4}{1+x} + \frac{x}{4}$

10. $\frac{3}{1-x} - \frac{5}{1+x}$

1 - 13 odd

The *Interactive* CD-ROM contains step-by-step solutions to all odd-numbered Section and Review Exercises. It also provides Tutorial Exercises, which link to Guided Examples for additional help.

2.1 Exercises

In Exercises 1–14, use the given values to evaluate (if possible) the other four trigonometric functions.

1. $\sin x = \frac{1}{2}, \quad \cos x = \frac{\sqrt{3}}{2}$

2. $\tan x = \frac{\sqrt{3}}{3}, \quad \cos x = -\frac{\sqrt{3}}{2}$

3. $\sec\theta = \sqrt{2}, \quad \sin\theta = -\frac{\sqrt{2}}{2}$

4. $\csc\theta = \frac{5}{3}, \quad \tan\theta = \frac{3}{4}$

5. $\tan x = \frac{5}{12}, \quad \sec x = -\frac{13}{12}$

6. $\cot\phi = -3, \quad \sin\phi = \frac{\sqrt{10}}{10}$

7. $\sec\phi = -1, \quad \sin\phi = 0$

8. $\cos\left(\frac{\pi}{2} - x\right) = \frac{3}{5}, \quad \cos x = \frac{4}{5}$

9. $\sin(-x) = -\frac{2}{3}, \quad \tan x = -\frac{2\sqrt{5}}{5}$

10. $\csc x = 5, \quad \cos x > 0$

11. $\tan\theta = 2, \quad \sin\theta < 0$

12. $\sec\theta = -3, \quad \tan\theta < 0$

13. $\sin\theta = -1, \quad \cot\theta = 0$

14. $\tan\theta$ is undefined, $\quad \sin\theta > 0$

In Exercises 15–18, fill in the blanks. (*Note:* The notation $x \to c^+$ indicates that x approaches c from the right and $x \to c^-$ indicates that x approaches c from the left.)

15. As $x \to \frac{\pi}{2}^-$, $\sin x \to$ ____ and $\csc x \to$ ____.

16. As $x \to 0^+$, $\cos x \to$ ____ and $\sec x \to$ ____.

17. As $x \to \frac{\pi}{2}^-$, $\tan x \to$ ____ and $\cot x \to$ ____.

18. As $x \to \pi^+$, $\sin x \to$ ____ and $\csc x \to$ ____.

In Exercises 19–24, match the trigonometric expression with one of the following.

(a) -1 **(b)** $\cos x$ **(c)** $\cot x$

(d) 1 **(e)** $-\tan x$ **(f)** $\sin x$

19. $\sec x \cos x$

20. $\cot x \sin x$

21. $\tan^2 x - \sec^2 x$

22. $(1 - \cos^2 x)(\csc x)$

23. $\dfrac{\sin(-x)}{\cos(-x)}$

24. $\dfrac{\sin[(\pi/2) - x]}{\cos[(\pi/2) - x]}$

In Exercises 25–30, match the trigonometric expression with one of the following.

(a) $\csc x$ **(b)** $\tan x$ **(c)** $\sin^2 x$

(d) $\sin x \tan x$ **(e)** $\sec^2 x$ **(f)** $\sec^2 x + \tan^2 x$

25. $\sin x \sec x$

26. $\cos^2 x(\sec^2 x - 1)$

27. $\sec^4 x - \tan^4 x$

28. $\cot x \sec x$

29. $\dfrac{\sec^2 x - 1}{\sin^2 x}$

30. $\dfrac{\cos^2[(\pi/2) - x]}{\cos x}$

In Exercises 31–44, use the fundamental identities to simplify the expression.

31. $\tan \phi \csc \phi$

32. $\sin \phi(\csc \phi - \sin \phi)$

33. $\cos \beta \tan \beta$

34. $\sec^2 x(1 - \sin^2 x)$

35. $\dfrac{\cot x}{\csc x}$

36. $\dfrac{\csc \theta}{\sec \theta}$

37. $\sec \alpha \cdot \dfrac{\sin \alpha}{\tan \alpha}$

38. $\dfrac{1}{\tan^2 x + 1}$

39. $\dfrac{\sin(-x)}{\cos x}$

40. $\dfrac{\tan^2 \theta}{\sec^2 \theta}$

41. $\cos\left(\dfrac{\pi}{2} - x\right)\sec x$

42. $\cot\left(\dfrac{\pi}{2} - x\right)\cos x$

43. $\dfrac{\cos^2 y}{1 - \sin y}$

44. $\cos t(1 + \tan^2 t)$

In Exercises 45–52, factor the expression and use the fundamental identities to simplify.

45. $\tan^2 x - \tan^2 x \sin^2 x$

46. $\sec^2 x \tan^2 x + \sec^2 x$

47. $\sin^2 x \sec^2 x - \sin^2 x$

48. $\dfrac{\sec^2 x - 1}{\sec x - 1}$

49. $\tan^4 x + 2\tan^2 x + 1$

50. $1 - 2\cos^2 x + \cos^4 x$

51. $\sin^4 x - \cos^4 x$

52. $\csc^3 x - \csc^2 x - \csc x + 1$

In Exercises 53–56, perform the multiplication and use the fundamental identities to simplify.

53. $(\sin x + \cos x)^2$

54. $(\cot x + \csc x)(\cot x - \csc x)$

55. $(\sec x + 1)(\sec x - 1)$

56. $(3 - 3\sin x)(3 + 3\sin x)$

In Exercises 57–60, perform the addition or subtraction and use the fundamental identities to simplify.

57. $\dfrac{1}{1 + \cos x} + \dfrac{1}{1 - \cos x}$

58. $\dfrac{1}{\sec x + 1} - \dfrac{1}{\sec x - 1}$

59. $\dfrac{\cos x}{1 + \sin x} + \dfrac{1 + \sin x}{\cos x}$

60. $\tan x - \dfrac{\sec^2 x}{\tan x}$

In Exercises 61–64, rewrite the expression so that it is *not* in fractional form.

61. $\dfrac{\sin^2 y}{1 - \cos y}$

62. $\dfrac{5}{\tan x + \sec x}$

63. $\dfrac{3}{\sec x - \tan x}$

64. $\dfrac{\tan^2 x}{\csc x + 1}$

Numerical and Graphical Analysis **In Exercises 65–68, use a graphing utility to complete the table and graph the functions. Make a conjecture about y_1 and y_2.**

x	0.2	0.4	0.6	0.8	1.0	1.2	1.4
y_1							
y_2							

65. $y_1 = \cos\left(\dfrac{\pi}{2} - x\right), \quad y_2 = \sin x$

66. $y_1 = \cos x + \sin x \tan x, \quad y_2 = \sec x$

67. $y_1 = \dfrac{\cos x}{1 - \sin x}, \quad y_2 = \dfrac{1 + \sin x}{\cos x}$

68. $y_1 = \sec^4 x - \sec^2 x, \quad y_2 = \tan^2 x + \tan^4 x$

In Exercises 69 and 70, use a graphing utility to determine which of the six trigonometric functions is equal to the expression.

69. $\cos x \cot x + \sin x$

70. $\dfrac{1}{2}\left(\dfrac{1 + \sin \theta}{\cos \theta} + \dfrac{\cos \theta}{1 + \sin \theta}\right)$

In Exercises 71–76, use the trigonometric substitution to write the algebraic expression as a trigonometric function of θ, where $0 < \theta < \pi/2$.

71. $\sqrt{25 - x^2}, \quad x = 5 \sin \theta$

72. $\sqrt{16 - 4x^2}, \quad x = 2 \sin \theta$

73. $\sqrt{x^2 - 9}, \quad x = 3 \sec \theta$

74. $\sqrt{x^2 - 4}, \quad x = 2 \sec \theta$

75. $\sqrt{x^2 + 25}, \quad x = 5 \tan \theta$

76. $\sqrt{x^2 + 100}, \quad x = 10 \tan \theta$

In Exercises 77–80, use a graphing utility to solve the equation for θ, where $0 \le \theta < 2\pi$.

77. $\sin \theta = \sqrt{1 - \cos^2 \theta}$

78. $\cos \theta = -\sqrt{1 - \sin^2 \theta}$

79. $\sec \theta = \sqrt{1 + \tan^2 \theta}$

80. $\tan \theta = \sqrt{\sec^2 \theta - 1}$

In Exercises 81 and 82, rewrite the expression as a single logarithm and simplify the result.

81. $\ln|\cos \theta| - \ln|\sin \theta|$

82. $\ln|\cot t| + \ln(1 + \tan^2 t)$

In Exercises 83–86, determine whether or not the equation is an identity, and give a reason for your answer.

83. $(\sin k\theta)/(\cos k\theta) = \tan \theta$, $\quad k$ is a constant.

84. $1/(5 \cos \theta) = 5 \sec \theta$

85. $\sin \theta \csc \theta = 1$

86. $\sin \theta \csc \phi = 1$

In Exercises 87–90, use a calculator to demonstrate the identity for the given values of θ.

87. $\csc^2 \theta - \cot^2 \theta = 1$, (a) $\theta = 132°$, (b) $\theta = \dfrac{2\pi}{7}$

88. $\tan^2 \theta + 1 = \sec^2 \theta$, (a) $\theta = 346°$, (b) $\theta = 3.1$

89. $\cos\left(\dfrac{\pi}{2} - \theta\right) = \sin \theta$, (a) $\theta = 80°$, (b) $\theta = 0.8$

90. $\sin(-\theta) = -\sin \theta$, (a) $\theta = 250°$, (b) $\theta = \dfrac{1}{2}$

91. Express each of the other trigonometric functions of θ in terms of $\sin \theta$.

92. Express each of the other trigonometric functions of θ in terms of $\cos \theta$.

Review **Solve Exercises 93–96 as a review of the skills and problem-solving techniques you learned in previous sections. Perform the operations and simplify.**

93. $(\sqrt{x} + 5)(\sqrt{x} - 5)$

94. $\sqrt{v}(\sqrt{20} - \sqrt{5})$

95. $(2\sqrt{z} + 3)^2$

96. $50x/(\sqrt{30} - 5)$

2.2 Verifying Trigonometric Identities

See Exercise 72 on page 229 for an example of how trigonometric identities can be used to solve a problem dealing with the coefficient of friction for an object on an inclined plane.

Introduction ❐ *Verifying Trigonometric Identities*

Introduction

In this section, you will study techniques for verifying trigonometric identities. In the next section, you will study techniques for solving trigonometric equations. The key to verifying identities *and* solving equations is the ability to use the fundamental identities and the rules of algebra to rewrite trigonometric expressions.

Remember that a *conditional equation* is an equation that is true for only some of the values in its domain. For example, the conditional equation

$\sin x = 0$ Conditional equation

is true only for $x = n\pi$. When you find these values, you are *solving* the equation. On the other hand, an equation that is true for all real values in the domain of the variable is an *identity*. For example, the familiar equation

$\sin^2 x = 1 - \cos^2 x$ Identity

is true for all real numbers x. Hence, it is an identity.

Although there are similarities, proving that a trigonometric equation is an identity is quite different from solving an equation. There is no well-defined set of rules to follow in verifying trigonometric identities, and the process is best learned by practice.

GUIDELINES FOR VERIFYING TRIGONOMETRIC IDENTITIES

1. Work with one side of the equation at a time. It is often better to work with the more complicated side first.
2. Look for opportunities to factor an expression, add fractions, square a binomial, or create a monomial denominator.
3. Look for opportunities to use the fundamental identities. Note which functions are in the final expression you want. Sines and cosines pair up well, as do secants and tangents, and cosecants and cotangents.
4. If the preceding guidelines do not help, try converting all terms to sines and cosines.
5. Do not just sit and stare at the problem. Try something! Even paths that lead to dead ends give you insights.

Verifying Trigonometric Identities

EXAMPLE 1 *Verifying a Trigonometric Identity*

Verify the identity $\dfrac{\sec^2\theta - 1}{\sec^2\theta} = \sin^2\theta$.

Solution

Because the left side is more complicated, start with it.

$$\frac{\sec^2\theta - 1}{\sec^2\theta} = \frac{(\tan^2\theta + 1) - 1}{\sec^2\theta} \qquad \text{Pythagorean identity}$$

$$= \frac{\tan^2\theta}{\sec^2\theta} \qquad \text{Simplify.}$$

$$= \tan^2\theta(\cos^2\theta) \qquad \text{Reciprocal identity}$$

$$= \frac{\sin^2\theta}{\cancel{\cos^2\theta}}\cancel{(\cos^2\theta)} \qquad \text{Quotient identity}$$

$$= \sin^2\theta \qquad \text{Simplify.}$$

NOTE Here is another way to verify the identity in Example 1.

$$\frac{\sec^2\theta - 1}{\sec^2\theta} = \frac{\sec^2\theta}{\sec^2\theta} - \frac{1}{\sec^2\theta}$$

$$= 1 - \cos^2\theta$$

$$= \sin^2\theta$$

As you can see from the note at the left, there can be more than one way to verify an identity. Your method may differ from that used by your instructor or fellow students. Here is a good chance to be creative and establish your own style, but try to be as efficient as possible.

EXAMPLE 2 *Combining Fractions Before Using Identities*

Verify the identity

$$\frac{1}{1 - \sin\alpha} + \frac{1}{1 + \sin\alpha} = 2\sec^2\alpha.$$

Solution

$$\frac{1}{1 - \sin\alpha} + \frac{1}{1 + \sin\alpha} = \frac{1 + \sin\alpha + 1 - \sin\alpha}{(1 - \sin\alpha)(1 + \sin\alpha)} \qquad \text{Add fractions.}$$

$$= \frac{2}{1 - \sin^2\alpha} \qquad \text{Simplify.}$$

$$= \frac{2}{\cos^2\alpha} \qquad \text{Pythagorean identity}$$

$$= 2\sec^2\alpha \qquad \text{Reciprocal identity}$$

EXAMPLE 3 *Verifying a Trigonometric Identity*

Verify the identity

$$(\tan^2 x + 1)(\cos^2 x - 1) = -\tan^2 x.$$

Solution

By applying identities before multiplying, you obtain the following.

$$\begin{aligned}
(\tan^2 x + 1)(\cos^2 x - 1) &= (\sec^2 x)(-\sin^2 x) && \text{Pythagorean identities} \\
&= -\frac{\sin^2 x}{\cos^2 x} && \text{Reciprocal identity} \\
&= -\left(\frac{\sin x}{\cos x}\right)^2 && \text{Rule of exponents} \\
&= -\tan^2 x && \text{Quotient identity}
\end{aligned}$$

EXAMPLE 4 *Converting to Sines and Cosines*

Verify the identity

$$\tan x + \cot x = \sec x \csc x.$$

Solution

In this case there appear to be no fractions to add, no products to find, and no opportunity to use one of the Pythagorean identities. Hence, try converting the left side into sines and cosines to see what happens.

$$\begin{aligned}
\tan x + \cot x &= \frac{\sin x}{\cos x} + \frac{\cos x}{\sin x} && \text{Quotient identities} \\
&= \frac{\sin^2 x + \cos^2 x}{\cos x \sin x} && \text{Add fractions.} \\
&= \frac{1}{\cos x \sin x} && \text{Pythagorean identity} \\
&= \frac{1}{\cos x} \cdot \frac{1}{\sin x} && \text{Product of fractions} \\
&= \sec x \csc x && \text{Reciprocal identities}
\end{aligned}$$

Recall from algebra that *rationalizing the denominator* is, on occasion, a powerful simplification technique. A related form of this technique works for simplifying trigonometric expressions as well.

EXAMPLE 5 *Verifying Trigonometric Identities*

Verify the identity $\sec y + \tan y = \dfrac{\cos y}{1 - \sin y}$.

Solution

Write with the *right* side. Note that you can create a monomial denominator by multiplying the numerator and denominator by $(1 + \sin y)$.

$$\frac{\cos y}{1 - \sin y} = \frac{\cos y}{1 - \sin y}\left(\frac{1 + \sin y}{1 + \sin y}\right)$$ Multiply numerator and denominator by $(1 + \sin y)$.

$$= \frac{\cos y + \cos y \sin y}{1 - \sin^2 y}$$ Multiply.

$$= \frac{\cos y + \cos y \sin y}{\cos^2 y}$$ Pythagorean identity

$$= \frac{\cos y}{\cos^2 y} + \frac{\cos y \sin y}{\cos^2 y}$$ Separate fractions.

$$= \frac{1}{\cos y} + \frac{\sin y}{\cos y}$$ Simplify.

$$= \sec y + \tan y$$ Identities

EXAMPLE 6 *Working with Each Side Separately*

Verify the identity $\dfrac{\cot^2 \theta}{1 + \csc \theta} = \dfrac{1 - \sin \theta}{\sin \theta}$.

Solution

Working with the left side, you have

$$\frac{\cot^2 \theta}{1 + \csc \theta} = \frac{\csc^2 \theta - 1}{1 + \csc \theta}$$ Pythagorean identity

$$= \frac{(\csc \theta - 1)\cancel{(\csc \theta + 1)}}{\cancel{1 + \csc \theta}}$$ Factor.

$$= \csc \theta - 1.$$ Simplify.

Now, simplifying the right side, you have

$$\frac{1 - \sin \theta}{\sin \theta} = \frac{1}{\sin \theta} - \frac{\sin \theta}{\sin \theta}$$ Separate fractions.

$$= \csc \theta - 1.$$ Reciprocal identity

The identity is verified because both sides are equal to $\csc \theta - 1$.

Study Tip

In Examples 1 through 5, you have been verifying trigonometric identities by working with one side of the equation and converting to the form given on the other side. On occasion it is practical to work with each side *separately,* to obtain one common form equivalent to both sides. This is illustrated in Example 6.

In Example 7, powers of trigonometric functions are rewritten as more complicated sums of products of trigonometric functions. This is a common procedure used in calculus.

EXAMPLE 7 *Two Examples from Calculus*

Verify each identity.

a. $\tan^4 x = \tan^2 x \sec^2 x - \tan^2 x$

b. $\sin^3 x \cos^4 x = (\cos^4 x - \cos^6 x) \sin x$

Solution

a.
$$\begin{aligned} \tan^4 x &= (\tan^2 x)(\tan^2 x) && \text{Separate factors.} \\ &= \tan^2 x(\sec^2 x - 1) && \text{Pythagorean identity} \\ &= \tan^2 x \sec^2 x - \tan^2 x && \text{Multiply.} \end{aligned}$$

b.
$$\begin{aligned} \sin^3 x \cos^4 x &= \sin^2 x \cos^4 x \sin x && \text{Separate factors.} \\ &= (1 - \cos^2 x)\cos^4 x \sin x && \text{Pythagorean identity} \\ &= (\cos^4 x - \cos^6 x)\sin x && \text{Multiply.} \end{aligned}$$

GROUP ACTIVITY

ERROR ANALYSIS

Suppose you are tutoring a student in trigonometry. One of the homework problems your student encounters asks whether the following statement is an identity.

$$\tan^2 x \sin^2 x \stackrel{?}{=} \frac{5}{6} \tan^2 x$$

Your student does not attempt to verify the equivalence algebraically, but mistakenly uses only a graphical approach. Using range settings of Xmin $= -3\pi$, Xmax $= 3\pi$, Xscl $= \pi/2$, Ymin $= -20$, Ymax $= 20$, and Yscl $= 1$, your student graphs both sides of the expression on a graphing utility and concludes that the statement is an identity.

What is wrong with your student's reasoning? Explain.

WARM UP

Factor each expression and, if possible, simplify the result.

1. (a) $x^2 - x^2y^2$
   (b) $\sin^2 x - \sin^2 x \cos^2 x$
2. (a) $x^2 + x^2y^2$
   (b) $\cos^2 x + \cos^2 x \tan^2 x$
3. (a) $x^4 - 1$
   (b) $\tan^4 x - 1$
4. (a) $z^3 + 1$
   (b) $\tan^3 x + 1$
5. (a) $x^3 - x^2 + x - 1$
   (b) $\cot^3 x - \cot^2 x + \cot x - 1$
6. (a) $x^4 - 2x^2 + 1$
   (b) $\sin^4 x - 2\sin^2 x + 1$

Perform the operations and, if possible, simplify the result.

7. (a) $\dfrac{y^2}{x} - x$
   (b) $\dfrac{\csc^2 x}{\cot x} - \cot x$
8. (a) $1 - \dfrac{1}{x^2}$
   (b) $1 - \dfrac{1}{\sec^2 x}$
9. (a) $\dfrac{y}{1+z} + \dfrac{1+z}{y}$
   (b) $\dfrac{\sin x}{1+\cos x} + \dfrac{1+\cos x}{\sin x}$
10. (a) $\dfrac{y}{z} - \dfrac{z}{1+y}$
    (b) $\dfrac{\tan x}{\sec x} - \dfrac{\sec x}{1+\tan x}$

2.2 Exercises

In Exercises 1–44, verify the identity.

1. $\sin t \csc t = 1$
2. $\tan y \cot y = 1$
3. $(1 + \sin\alpha)(1 - \sin\alpha) = \cos^2\alpha$
4. $\cot^2 y(\sec^2 y - 1) = 1$
5. $\cos^2\beta - \sin^2\beta = 1 - 2\sin^2\beta$
6. $\cos^2\beta - \sin^2\beta = 2\cos^2\beta - 1$
7. $\tan^2\theta + 4 = \sec^2\theta + 3$
8. $2 - \sec^2 z = 1 - \tan^2 z$
9. $\sin^2\alpha - \sin^4\alpha = \cos^2\alpha - \cos^4\alpha$
10. $\cos x + \sin x \tan x = \sec x$
11. $\dfrac{\sec^2 x}{\tan x} = \sec x \csc x$
12. $\dfrac{\cot^3 t}{\csc t} = \cos t(\csc^2 t - 1)$
13. $\dfrac{\cot^2 t}{\csc t} = \csc t - \sin t$
14. $\dfrac{1}{\sin x} - \sin x = \dfrac{\cos^2 x}{\sin x}$
15. $\sin^{1/2} x \cos x - \sin^{5/2} x \cos x = \cos^3 x\sqrt{\sin x}$
16. $\sec^6 x(\sec x \tan x) - \sec^4 x(\sec x \tan x) = \sec^5 x \tan^3 x$
17. $\dfrac{1}{\sec x \tan x} = \csc x - \sin x$

18. $\dfrac{\sec\theta - 1}{1 - \cos\theta} = \sec\theta$

19. $\csc x - \sin x = \cos x \cot x$

20. $\sec x - \cos x = \sin x \tan x$

21. $\cos x + \sin x \tan x = \sec x$

22. $\dfrac{\sec x + \tan x}{\sec x - \tan x} = (\sec x + \tan x)^2$

23. $\dfrac{1}{\tan x} + \dfrac{1}{\cot x} = \tan x + \cot x$

24. $\dfrac{1}{\sin x} - \dfrac{1}{\csc x} = \csc x - \sin x$

25. $\dfrac{\cos\theta \cot\theta}{1 - \sin\theta} - 1 = \csc\theta$

26. $\dfrac{1 + \sin\theta}{\cos\theta} + \dfrac{\cos\theta}{1 + \sin\theta} = 2\sec\theta$

27. $\dfrac{1}{\cot x + 1} + \dfrac{1}{\tan x + 1} = 1$

28. $\cos x - \dfrac{\cos x}{1 - \tan x} = \dfrac{\sin x \cos x}{\sin x - \cos x}$

29. $\cos\left(\dfrac{\pi}{2} - x\right)\csc x = 1$

30. $\dfrac{\cos[(\pi/2) - x]}{\sin[(\pi/2) - x]} = \tan x$

31. $\dfrac{\csc(-x)}{\sec(-x)} = -\cot x$

32. $(1 + \sin y)[1 + \sin(-y)] = \cos^2 y$

33. $\dfrac{\cos(-\theta)}{1 + \sin(-\theta)} = \sec\theta + \tan\theta$

34. $\dfrac{1 + \sec(-\theta)}{\sin(-\theta) + \tan(-\theta)} = -\csc\theta$

35. $\dfrac{\sin x \cos y + \cos x \sin y}{\cos x \cos y - \sin x \sin y} = \dfrac{\tan x + \tan y}{1 - \tan x \tan y}$

36. $\dfrac{\tan x + \tan y}{1 - \tan x \tan y} = \dfrac{\cot x + \cot y}{\cot x \cot y - 1}$

37. $\dfrac{\tan x + \cot y}{\tan x \cot y} = \tan y + \cot x$

38. $\dfrac{\cos x - \cos y}{\sin x + \sin y} + \dfrac{\sin x - \sin y}{\cos x + \cos y} = 0$

39. $\sqrt{\dfrac{1 + \sin\theta}{1 - \sin\theta}} = \dfrac{1 + \sin\theta}{|\cos\theta|}$

40. $\sqrt{\dfrac{1 - \cos\theta}{1 + \cos\theta}} = \dfrac{1 - \cos\theta}{|\sin\theta|}$

41. $\sin^2 x + \sin^2\left(\dfrac{\pi}{2} - x\right) = 1$

42. $\sec^2 y - \cot^2\left(\dfrac{\pi}{2} - y\right) = 1$

43. $\csc x \cos\left(\dfrac{\pi}{2} - x\right) = 1$

44. $\sec^2\left(\dfrac{\pi}{2} - x\right) - 1 = \cot^2 x$

In Exercises 45–56, verify the identity algebraically, and use a graphing utility to confirm it graphically.

45. $2\sec^2 x - 2\sec^2 x \sin^2 x - \sin^2 x - \cos^2 x = 1$

46. $\csc x(\csc x - \sin x) + \dfrac{\sin x - \cos x}{\sin x} + \cot x = \csc^2 x$

47. $2 + \cos^2 x - 3\cos^4 x = \sin^2 x(2 + 3\cos^2 x)$

48. $4\tan^4 x + \tan^2 x - 3 = \sec^2 x(4\tan^2 x - 3)$

49. $\csc^4 x - 2\csc^2 x + 1 = \cot^4 x$

50. $\sin x(1 - 2\cos^2 x + \cos^4 x) = \sin^5 x$

51. $\sec^4\theta - \tan^4\theta = 1 + 2\tan^2\theta$

52. $\csc^4\theta - \cot^4\theta = 2\csc^2\theta - 1$

53. $\dfrac{\sin\beta}{1 - \cos\beta} = \dfrac{1 + \cos\beta}{\sin\beta}$

54. $\dfrac{\cot\alpha}{\csc\alpha - 1} = \dfrac{\csc\alpha + 1}{\cot\alpha}$

55. $\dfrac{\tan^3\alpha - 1}{\tan\alpha - 1} = \tan^2\alpha + \tan\alpha + 1$

56. $\dfrac{\sin^3\beta + \cos^3\beta}{\sin\beta + \cos\beta} = 1 - \sin\beta\cos\beta$

In Exercises 57–60, use the properties of logarithms and trigonometric identities to verify the identity.

57. $\ln|\tan \theta| = \ln|\sin \theta| - \ln|\cos \theta|$

58. $\ln|\sec \theta| = -\ln|\cos \theta|$

59. $-\ln(1 + \cos \theta) = \ln(1 - \cos \theta) - 2 \ln|\sin \theta|$

60. $-\ln|\sec \theta + \tan \theta| = \ln|\sec \theta - \tan \theta|$

Think About It **In Exercises 61–64, explain why the equation is *not* an identity and find one value of the variable for which the equation is not true.**

61. $\sin \theta = \sqrt{1 - \cos^2 \theta}$

62. $\tan \theta = \sqrt{\sec^2 \theta - 1}$

63. $\sqrt{\tan^2 x} = \tan x$

64. $\sqrt{\sin^2 x + \cos^2 x} = \sin x + \cos x$

In Exercises 65–68, use the cofunction identities to evaluate the expression without the aid of a calculator.

65. $\sin^2 25° + \sin^2 65°$

66. $\cos^2 18° + \cos^2 72°$

67. $\cos^2 20° + \cos^2 52° + \cos^2 38° + \cos^2 70°$

68. $\sin^2 12° + \sin^2 40° + \sin^2 50° + \sin^2 78°$

69. Verify that for all integers n

$$\cos\left[\frac{(2n + 1)\pi}{2}\right] = 0.$$

70. Verify that for all integers n

$$\sin\left[\frac{(12n + 1)\pi}{6}\right] = \frac{1}{2}.$$

71. ***Rate of Change*** The rate of change of the function

$f(x) = \sin x + \csc x$

with respect to change in the variable x is given by the expression

$\cos x - \csc x \cot x.$

Show that the expression for the rate of change can also be given by

$-\cos x \cot^2 x.$

72. ***Friction*** The forces acting on an object weighing W units on an inclined plane positioned at an angle of θ with the horizontal (see figure) is modeled by

$\mu W \cos \theta = W \sin \theta$

where μ is the coefficient of friction. Solve the equation for μ and simplify the result.

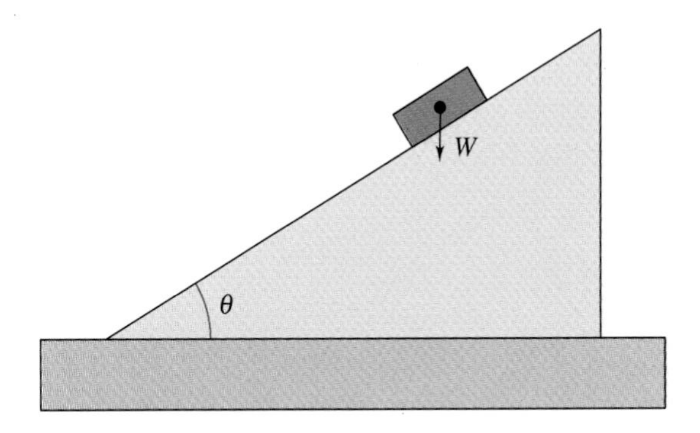

73. ***Chapter Opener*** Which city has the greater variation in the number of daylight hours? Which constant in each model would you use to determine the difference between the greatest and least number of hours of daylight?

74. ***Chapter Opener*** Determine the period of each model.

2.3 Solving Trigonometric Equations

See Exercise 69 on page 240 for an example of how solving a trigonometric equation can help answer questions about the unemployment rate in the United States.

Introduction ▫ *Equations of Quadratic Type* ▫ *Functions Involving Multiple Angles* ▫ *Using Inverse Functions*

Introduction

To solve a trigonometric equation, use standard algebraic techniques such as collecting like terms and factoring. Your preliminary goal is to isolate the trigonometric function involved in the equation.

EXAMPLE 1 *Solving a Trigonometric Equation*

$2 \sin x - 1 = 0$	Original equation
$2 \sin x = 1$	Add 1 to both sides.
$\sin x = \frac{1}{2}$	Divide both sides by 2.

To solve for x, note in Figure 2.2 that the equation $\sin x = \frac{1}{2}$ has solutions $x = \pi/6$ and $x = 5\pi/6$ in the interval $[0, 2\pi)$. Moreover, because $\sin x$ has a period of 2π, there are infinitely many other solutions, which can be written as

$$x = \pi/6 + 2n\pi \qquad \text{and} \qquad x = 5\pi/6 + 2n\pi \qquad \text{General solution}$$

where n is an integer, as shown in Figure 2.2.

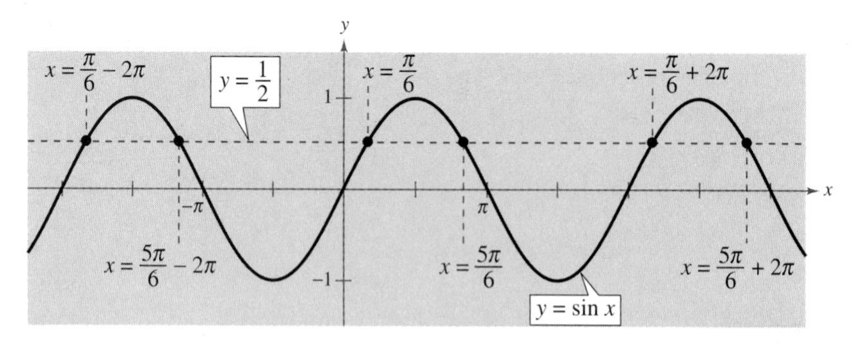

FIGURE 2.2

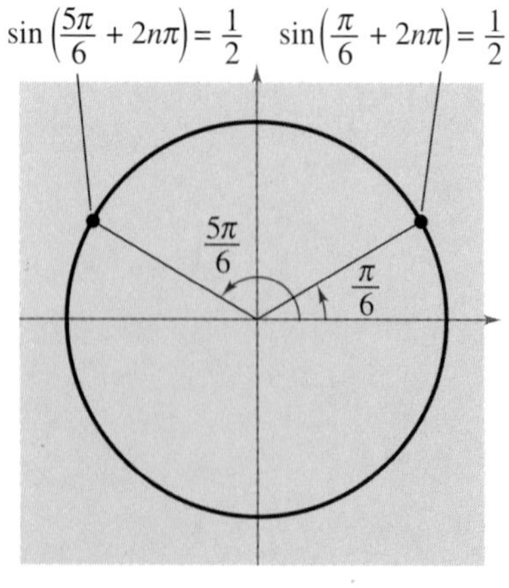

$\sin\left(\frac{5\pi}{6} + 2n\pi\right) = \frac{1}{2}$ $\sin\left(\frac{\pi}{6} + 2n\pi\right) = \frac{1}{2}$

$\frac{5\pi}{6}$ $\frac{\pi}{6}$

FIGURE 2.3

Another way to see that the equation $\sin x = \frac{1}{2}$ has infinitely many solutions is indicated in Figure 2.3. For $0 \le x < 2\pi$, the solutions are $x = \pi/6$ and $x = 5\pi/6$. Any angles that are coterminal with $\pi/6$ or $5\pi/6$ will also be solutions of the equation.

EXAMPLE 2 *Collecting Like Terms*

Solve $\sin x + \sqrt{2} = -\sin x$.

Solution

Your goal is to rewrite the equation so that $\sin x$ is isolated on one side of the equation.

$$\sin x + \sqrt{2} = -\sin x \qquad \text{Original equation}$$
$$\sin x + \sin x = -\sqrt{2} \qquad \text{Add } \sin x \text{ to, and subtract } \sqrt{2} \text{ from, both sides.}$$
$$2 \sin x = -\sqrt{2} \qquad \text{Combine like terms.}$$
$$\sin x = -\frac{\sqrt{2}}{2} \qquad \text{Divide both sides by 2.}$$

Because $\sin x$ has a period of 2π, first find all solutions in the interval $[0, 2\pi)$. These are $x = 5\pi/4$ and $x = 7\pi/4$. Finally, add $2n\pi$ to each of these solutions to get the general form

$$x = \frac{5\pi}{4} + 2n\pi \quad \text{and} \quad x = \frac{7\pi}{4} + 2n\pi \qquad \text{General solution}$$

where n is an integer.

EXAMPLE 3 *Extracting Square Roots*

Solve $3 \tan^2 x - 1 = 0$.

Solution

Your goal is to rewrite the equation so that $\tan x$ is isolated on one side of the equation.

$$3 \tan^2 x - 1 = 0 \qquad \text{Original equation}$$
$$3 \tan^2 x = 1 \qquad \text{Add 1 to both sides.}$$
$$\tan^2 x = \frac{1}{3} \qquad \text{Divide both sides by 3.}$$
$$\tan x = \pm\frac{1}{\sqrt{3}} \qquad \text{Extract square roots.}$$

Because $\tan x$ has a period of π, first find all solutions in the interval $[0, \pi)$. These are $x = \pi/6$ and $x = 5\pi/6$. Finally, add $n\pi$ to each of these solutions to get the general form

$$x = \frac{\pi}{6} + n\pi \quad \text{and} \quad x = \frac{5\pi}{6} + n\pi \qquad \text{General solution}$$

where n is an integer.

TECHNOLOGY

The solutions in Examples 2 and 3 are obtained analytically. You can use a graphing utility to confirm the solutions graphically. For instance, to confirm the solutions found in Example 3, sketch the graph of

$$y = 3 \tan^2 x - 1$$

as shown below.

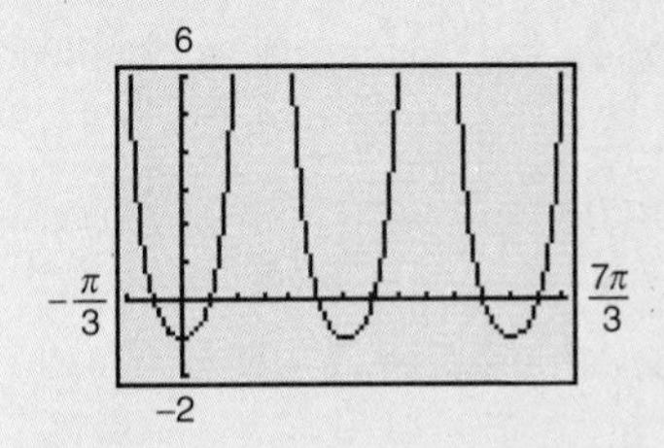

The equations in Examples 1, 2, and 3 involved only one trigonometric function. When two or more functions occur in the same equation, collect all terms on one side and try to separate the functions by factoring or by using appropriate identities. This may produce factors that yield no solutions, as illustrated in Example 4.

EXAMPLE 4 *Factoring*

Solve $\cot x \cos^2 x = 2 \cot x$.

Solution

$$\cot x \cos^2 x = 2 \cot x \qquad \text{Original equation}$$

$$\cot x \cos^2 x - 2 \cot x = 0 \qquad \text{Subtract } 2 \cot x \text{ from both sides.}$$

$$\cot x(\cos^2 x - 2) = 0 \qquad \text{Factor.}$$

By setting each of these factors equal to zero, you obtain the following.

$$\cot x = 0 \quad \text{and} \quad \cos^2 x - 2 = 0$$

$$x = \frac{\pi}{2} \qquad\qquad \cos^2 x = 2$$

$$\cos x = \pm\sqrt{2}$$

No solution is obtained from $\cos x = \pm\sqrt{2}$ because $\pm\sqrt{2}$ are outside the range of the cosine function. Therefore, the general form of the solution is obtained by adding multiples of π to $x = \pi/2$, to get

$$x = \frac{\pi}{2} + n\pi \qquad \text{General solution}$$

where n is an integer. You can confirm this graphically by sketching the graph of $y = \cot x \cos^2 x - 2 \cot x$, as shown in Figure 2.4.

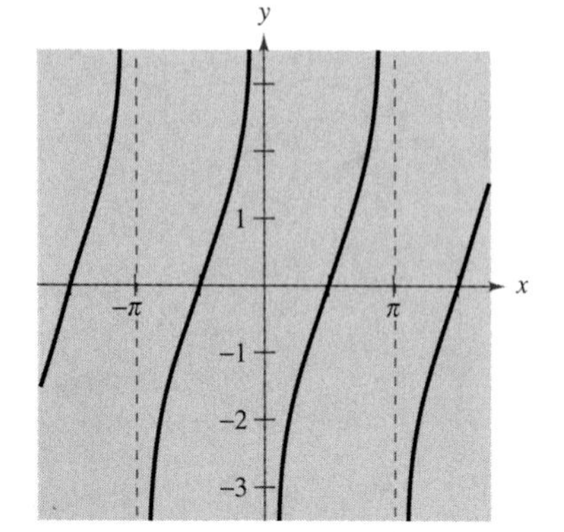

FIGURE 2.4

NOTE In Example 4, don't make the mistake of dividing both sides of the equation by $\cot x$. If you do this, you lose the solutions. Can you see why?

Equations of Quadratic Type

Many trigonometric equations are of quadratic type. Here are a couple of examples.

Quadratic in sin x	*Quadratic in sec x*
$2 \sin^2 x - \sin x - 1 = 0$	$\sec^2 x - 3 \sec x - 2 = 0$
$2(\sin x)^2 - (\sin x) - 1 = 0$	$(\sec x)^2 - 3(\sec x) - 2 = 0$

To solve equations of this type, factor the quadratic or, if this is not possible, use the Quadratic Formula.

EXAMPLE 5 *Factoring an Equation of Quadratic Type*

Find all solutions of $2 \sin^2 x - \sin x - 1 = 0$ in the interval $[0, 2\pi)$.

Solution

Begin by treating the equation as a quadratic in $\sin x$ and factoring.

$$2 \sin^2 x - \sin x - 1 = 0 \qquad \text{Original equation}$$

$$(2 \sin x + 1)(\sin x - 1) = 0 \qquad \text{Factor.}$$

Setting each factor equal to zero, you obtain the following solutions.

$$2 \sin x + 1 = 0 \qquad \text{and} \qquad \sin x - 1 = 0$$

$$\sin x = -\frac{1}{2} \qquad\qquad \sin x = 1$$

$$x = \frac{7\pi}{6}, \frac{11\pi}{6} \qquad\qquad x = \frac{\pi}{2}$$

NOTE In Example 5, the general solution would be

$$x = \frac{7\pi}{6} + 2n\pi$$

$$x = \frac{11\pi}{6} + 2n\pi$$

$$x = \frac{\pi}{2} + 2n\pi$$

where n is an integer.

When working with an equation of quadratic type, be sure that the equation involves a *single* trigonometric function, as shown in the next example.

EXAMPLE 6 *Rewriting with a Single Trigonometric Function*

Solve $2 \sin^2 x + 3 \cos x - 3 = 0$.

Solution

This equation contains both sine and cosine functions. You can rewrite the equation so that it has only cosine functions by using the identity $\sin^2 x = 1 - \cos^2 x$.

$$2 \sin^2 x + 3 \cos x - 3 = 0 \qquad \text{Original equation}$$

$$2(1 - \cos^2 x) + 3 \cos x - 3 = 0 \qquad \text{Pythagorean identity}$$

$$2 \cos^2 x - 3 \cos x + 1 = 0 \qquad \text{Multiply both sides by } -1.$$

$$(2 \cos x - 1)(\cos x - 1) = 0 \qquad \text{Factor.}$$

By setting each factor equal to zero, you can find the solutions in the interval $[0, 2\pi)$ to be $x = 0$, $x = \pi/3$, and $x = 5\pi/3$. The general solution is therefore

$$x = 2n\pi, \qquad x = \frac{\pi}{3} + 2n\pi, \qquad x = \frac{5\pi}{3} + 2n\pi \qquad \text{General solution}$$

where n is an integer.

Sometimes you must square both sides of an equation to obtain a quadratic, as demonstrated in the next example. Because this procedure can introduce extraneous solutions, you should check any solutions in the original equation to see if they are valid or extraneous.

EXAMPLE 7 *Squaring and Converting to Quadratic Type*

Find all solutions of $\cos x + 1 = \sin x$ in the interval $[0, 2\pi)$.

Solution

It is not clear how to rewrite this equation in terms of a single trigonometric function. See what happens when you square both sides of the equation.

$$\cos x + 1 = \sin x \qquad \text{Original equation}$$

$$\cos^2 x + 2\cos x + 1 = \sin^2 x \qquad \text{Square both sides.}$$

$$\cos^2 x + 2\cos x + 1 = 1 - \cos^2 x \qquad \text{Pythagorean identity}$$

$$2\cos^2 x + 2\cos x = 0 \qquad \text{Combine like terms.}$$

$$2\cos x(\cos x + 1) = 0 \qquad \text{Factor.}$$

NOTE In Example 7, the general solution would be

$$x = \frac{\pi}{2} + 2n\pi$$

$$x = \pi + 2n\pi$$

where n is an integer.

Setting each factor equal to zero produces the following.

$2\cos x = 0$ and $\cos x + 1 = 0$

$\cos x = 0$ $\qquad$ $\cos x = -1$

$x = \frac{\pi}{2}, \frac{3\pi}{2}$ $\qquad$ $x = \pi$

Because you squared the original equation, check for extraneous solutions. Of the three possible solutions, $x = 3\pi/2$ is extraneous. (Try checking this.) Thus, in the interval $[0, 2\pi)$, the only two solutions are $x = \pi/2$ and $x = \pi$.

Exploration

Use a graphing utility to confirm the solutions found in Example 7 in two different ways. Do both methods produce the same x-values? Which method do you prefer? Why?

1. Graph both sides of the equation and find the x-coordinates of the points at which the graphs intersect.

   Left side: $y = \cos x + 1$ $\qquad$ *Right side:* $y = \sin x$

2. Graph the equation $y = \cos x + 1 - \sin x$ and find the x-intercepts of the graph.

Functions Involving Multiple Angles

EXAMPLE 8 *Functions of Multiple Angles*

Find all solutions of $2 \cos 3t - 1 = 0$.

Solution

$$2 \cos 3t - 1 = 0 \qquad \text{Original equation}$$

$$2 \cos 3t = 1 \qquad \text{Add 1 to both sides.}$$

$$\cos 3t = \frac{1}{2} \qquad \text{Divide both sides by 2.}$$

In the interval $[0, 2\pi)$, you know that $3t = \pi/3$ and $3t = 5\pi/3$ are the only solutions so that, in general, you have

$$3t = \frac{\pi}{3} + 2n\pi \quad \text{and} \quad 3t = \frac{5\pi}{3} + 2n\pi.$$

Dividing this result by 3, you obtain the general solution

$$t = \frac{\pi}{9} + \frac{2n\pi}{3} \quad \text{and} \quad t = \frac{5\pi}{9} + \frac{2n\pi}{3} \qquad \text{General solution}$$

where n is an integer.

EXAMPLE 9 *Functions of Multiple Angles*

Find all solutions of $3 \tan(x/2) + 3 = 0$.

Solution

$$3 \tan \frac{x}{2} + 3 = 0 \qquad \text{Original equation}$$

$$3 \tan \frac{x}{2} = -3 \qquad \text{Subtract 3 from both sides.}$$

$$\tan \frac{x}{2} = -1 \qquad \text{Divide both sides by 3.}$$

In the interval $[0, \pi)$, you know that $x/2 = 3\pi/4$ is the only solution so that, in general, you have

$$\frac{x}{2} = \frac{3\pi}{4} + n\pi.$$

Multiplying this result by 2, you obtain the general solution

$$x = \frac{3\pi}{2} + 2n\pi \qquad \text{General solution}$$

where n is an integer.

Using Inverse Functions

EXAMPLE 10 *Using Inverse Functions*

Find all solutions of $\sec^2 x - 2\tan x = 4$.

Solution

$$\sec^2 x - 2\tan x = 4 \quad \text{Original equation}$$
$$1 + \tan^2 x - 2\tan x - 4 = 0 \quad \text{Pythagorean identity}$$
$$\tan^2 x - 2\tan x - 3 = 0 \quad \text{Combine like terms.}$$
$$(\tan x - 3)(\tan x + 1) = 0 \quad \text{Factor.}$$

Setting each factor equal to zero, you obtain two solutions in the interval $(-\pi/2, \pi/2)$. [Recall that the range of the inverse tangent function is $(-\pi/2, \pi/2)$.]

$$\tan x = 3, \qquad \tan x = -1$$
$$x = \arctan 3 \qquad x = -\frac{\pi}{4}$$

From 1985 through 1994, the unemployment rate varied between 5.3% and 7.4%. To see the cyclical nature of the unemployment rate during these years, see Exercise 69 on page 240.

Finally, by adding multiples of π, you obtain the general solution

$$x = \arctan 3 + n\pi \quad \text{and} \quad x = -\frac{\pi}{4} + n\pi \quad \text{General solution}$$

where n is an integer.

GROUP ACTIVITY

EQUATIONS WITH NO SOLUTIONS

One of the following equations has solutions and the other two don't. Which two equations do not have solutions?

a. $\sin^2 x - 5\sin x + 6 = 0$

b. $\sin^2 x - 4\sin x + 6 = 0$

c. $\sin^2 x - 5\sin x - 6 = 0$

Can you find conditions involving the constants b and c that will guarantee that the equation

$$\sin^2 x + b\sin x + c = 0$$

has at least one solution on some interval of length 2π?

WARM UP

Solve for θ in the interval $0 \le \theta < 2\pi$.

1. $\cos \theta = -\dfrac{1}{2}$
2. $\sin \theta = \dfrac{\sqrt{3}}{2}$
3. $\cos \theta = \dfrac{\sqrt{2}}{2}$
4. $\sin \theta = -\dfrac{\sqrt{2}}{2}$
5. $\tan \theta = \sqrt{3}$
6. $\tan \theta = -1$

Solve for x.

7. $\dfrac{x}{3} + \dfrac{x}{5} = 1$
8. $2x(x + 3) - 5(x + 3) = 0$
9. $2x^2 - 4x - 5 = 0$
10. $\dfrac{1}{x} = \dfrac{x}{2x + 3}$

2.3 Exercises

In Exercises 1–4, find the x-intercepts of the graph.

1. $y = \sin \dfrac{\pi x}{2} + 1$

2. $y = \sin \pi x + \cos \pi x$

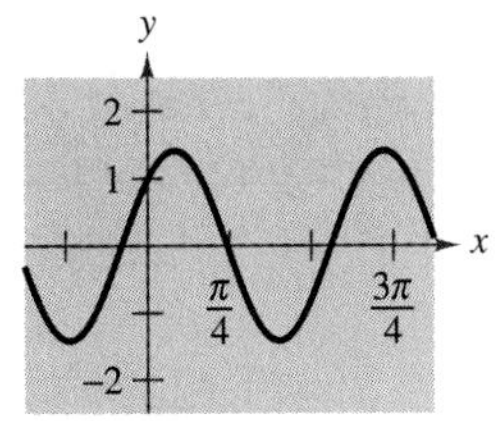

3. $y = \tan^2\left(\dfrac{\pi x}{6}\right) - 3$

4. $y = \sec^4\left(\dfrac{\pi x}{8}\right) - 4$

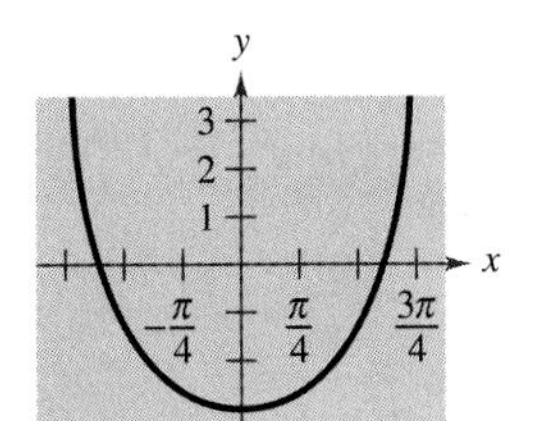

In Exercises 5–10, verify that the x-values are solutions.

5. $2 \cos x - 1 = 0$
   (a) $x = \dfrac{\pi}{3}$ (b) $x = \dfrac{5\pi}{3}$

6. $\csc x - 2 = 0$
   (a) $x = \dfrac{\pi}{6}$ (b) $x = \dfrac{5\pi}{6}$

7. $3 \tan^2 2x - 1 = 0$
   (a) $x = \dfrac{\pi}{12}$ (b) $x = \dfrac{5\pi}{12}$

8. $2 \cos^2 4x - 1 = 0$
   (a) $x = \dfrac{\pi}{16}$ (b) $x = \dfrac{3\pi}{16}$

9. $2 \sin^2 x - \sin x - 1 = 0$
   (a) $x = \dfrac{\pi}{2}$ (b) $x = \dfrac{7\pi}{6}$

10. $\sec^4 x - 4 \sec^2 x = 0$
    (a) $x = \dfrac{2\pi}{3}$ (b) $x = \dfrac{5\pi}{3}$

In Exercises 11–24, solve the equation.

11. $2 \cos x + 1 = 0$

12. $2 \sin x - 1 = 0$

13. $\sqrt{3} \csc x - 2 = 0$

14. $\tan x + 1 = 0$

15. $3 \sec^2 x - 4 = 0$

16. $\csc^2 x - 2 = 0$

17. $2 \sin^2 2x = 1$

18. $\tan^2 3x = 3$

19. $4 \sin^2 x - 3 = 0$

20. $\sin x(\sin x + 1) = 0$

21. $\sin^2 x = 3 \cos^2 x$

22. $\tan 3x(\tan x - 1) = 0$

23. $(3 \tan^2 x - 1)(\tan^2 x - 3) = 0$

24. $\cos 2x(2 \cos x + 1) = 0$

In Exercises 25–40, find all solutions of the equation in the interval $[0, 2\pi)$.

25. $\cos^3 x = \cos x$

26. $\tan^2 x - 1 = 0$

27. $3 \tan^3 x = \tan x$

28. $2 \sin^2 x = 2 + \cos x$

29. $\sec^2 x - \sec x = 2$

30. $\sec x \csc x = 2 \csc x$

31. $2 \sin x + \csc x = 0$

32. $\sin 2x = -\frac{\sqrt{3}}{2}$

33. $\csc x + \cot x = 1$

34. $\tan 3x = 1$

35. $\cos \frac{x}{2} = \frac{\sqrt{2}}{2}$

36. $\sec 4x = 2$

37. $\frac{1 + \cos x}{1 - \cos x} = 0$

38. $2 \sin^2 x + 3 \sin x + 1 = 0$

39. $2 \sec^2 x + \tan^2 x - 3 = 0$

40. $\cos x + \sin x \tan x = 2$

In Exercises 41 and 42, solve both equations. How do the solutions of the algebraic equation compare to the solutions of the trigonometric equation?

41. $6y^2 - 13y + 6 = 0$

$6 \cos^2 x - 13 \cos x + 6 = 0$

42. $y^2 + y - 20 = 0$

$\sin^2 x + \sin x - 20 = 0$

In Exercises 43–56, use a graphing utility to approximate the solutions of the equation in the interval $[0, 2\pi)$.

43. $2 \cos x - \sin x = 0$

44. $4 \sin^3 x + 2 \sin^2 x - 2 \sin x - 1 = 0$

45. $\frac{1 + \sin x}{\cos x} + \frac{\cos x}{1 + \sin x} = 4$

46. $\frac{\cos x \cot x}{1 - \sin x} = 3$

47. $2 \sin x - x = 0$

48. $x \cos x - 1 = 0$

49. $\sec^2 x + 0.5 \tan x - 1 = 0$

50. $\csc^2 x + 0.5 \cot x - 5 = 0$

51. $2 \tan^2 x + 7 \tan x - 15 = 0$

52. $12 \cos^2 x + 5 \cos x - 3 = 0$

53. $12 \sin^2 x - 13 \sin x + 3 = 0$

54. $3 \tan^2 x + 4 \tan x - 4 = 0$

55. $\sin^2 x + 2 \sin x - 1 = 0$

56. $4 \cos^2 x - 4 \cos x - 1 = 0$

In Exercises 57 and 58, (a) use a graphing utility to graph the function and approximate the maximum and minimum points on the graph in the interval $[0, 2\pi)$, and (b) solve the trigonometric equation and demonstrate that its solutions are the x-coordinates of the maximum and minimum points of f. (Calculus is required to find the trigonometric equation.)

	Function	*Trigonometric Equation*
57.	$f(x) = \sin x + \cos x$	$\cos x - \sin x = 0$
58.	$f(x) = 2 \sin x + \cos 2x$	$2 \cos x - 4 \sin x \cos x = 0$

Fixed Point **In Exercises 59 and 60, find the smallest positive fixed point of the function f. [A fixed point of a function f is a real number c such that $f(c) = c$.]**

59. $f(x) = \tan \frac{\pi x}{4}$

60. $f(x) = \cos x$

61. *Graphical Reasoning* Consider the function

$$f(x) = \cos\frac{1}{x}$$

and its graph shown in the figure.

(a) What is the domain of the function?

(b) Identify any symmetry or asymptotes of the graph.

(c) Describe the behavior of the function as $x \to 0$.

(d) How many solutions does the equation

$$\cos\frac{1}{x} = 0$$

have in the interval $[-1, 1]$?

(e) Does the equation $\cos(1/x) = 0$ have a greatest solution? If so, approximate the solution. If not, explain.

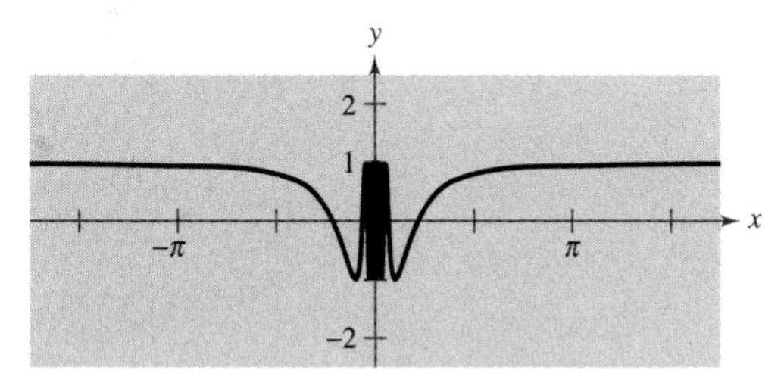

62. *Graphical Reasoning* Consider the function

$$f(x) = \frac{\sin x}{x}$$

and its graph shown in the figure.

(a) What is the domain of the function?

(b) Identify any symmetry or asymptotes of the graph.

(c) Describe the behavior of the function as $x \to 0$.

(d) How many solutions does the equation

$$\frac{\sin x}{x} = 0$$

have in the interval $[-8, 8]$? Find the solutions.

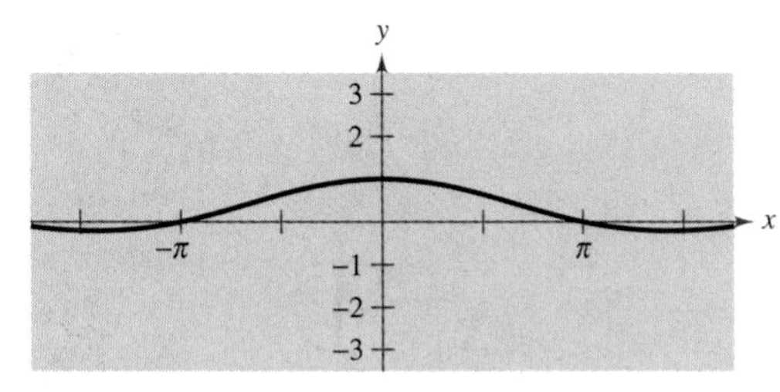

63. *Harmonic Motion* A weight is oscillating on the end of a spring (see figure). The position of the weight relative to the point of equilibrium is given by

$$y = \frac{1}{12}(\cos 8t - 3\sin 8t)$$

where y is the displacement in meters and t is the time in seconds. Find the times when the weight is at the point of equilibrium ($y = 0$) for $0 \leq t \leq 1$.

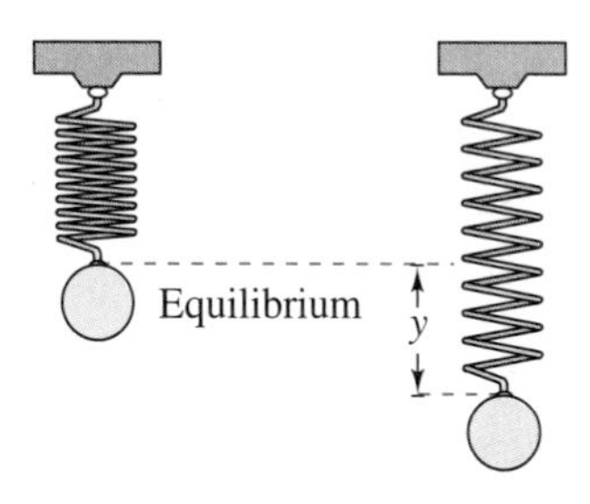

64. *Sales* The monthly sales (in thousands of units) of a seasonal product are approximated by

$$S = 74.50 + 43.75\sin\frac{\pi t}{6}$$

where t is the time in months, with $t = 1$ corresponding to January. Determine the months when sales exceed 100,000 units.

65. *Projectile Motion* A batted baseball leaves the bat at an angle of θ with the horizontal and an initial velocity of $v_0 = 100$ feet per second. The ball is caught by an outfielder 300 feet from home plate (see figure). Find θ if the range r of a projectile is given by

$$f = \frac{1}{32}v_0^2 \sin 2\theta.$$

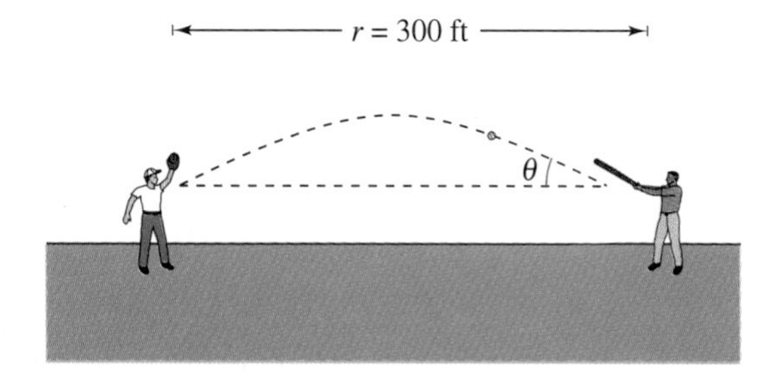

66. *Projectile Motion* A sharpshooter intends to hit a target at a distance of 1000 yards with a gun that has a muzzle velocity of 1200 feet per second (see figure). Neglecting air resistance, determine the minimum angle of elevation of the gun if the range is given by

$$r = \frac{1}{32} v_0^2 \sin 2\theta.$$

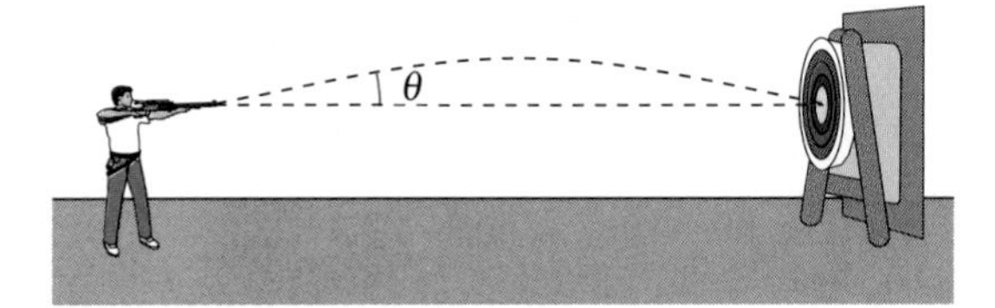

67. *Area* The area of a rectangle (see figure) inscribed in one arch of the graph of $y = \cos x$ is given by

$$A = 2x \cos x, \qquad 0 < x < \frac{\pi}{2}.$$

(a) Use a graphing utility to graph the area function, and approximate the area of the largest inscribed rectangle.

(b) Determine the values of x for which $A \geq 1$.

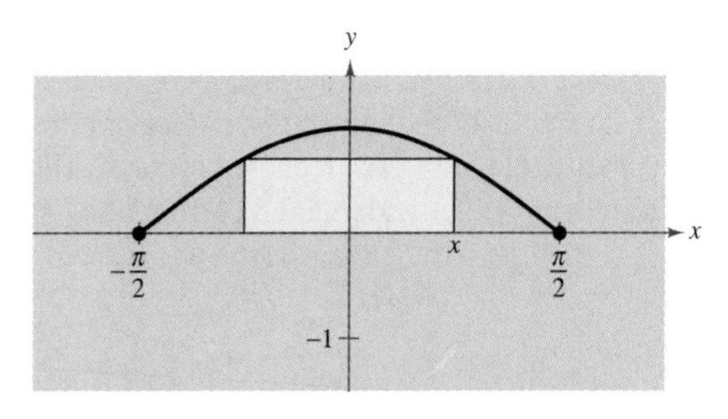

68. *Damped Harmonic Motion* The displacement from equilibrium of a weight oscillating on the end of a spring is given by

$$y = 1.56e^{-0.22t} \cos 4.9t,$$

where y is the displacement in feet and t is the time in seconds. Use a graphing utility to graph the displacement function for $0 \leq t \leq 10$. Find the time beyond which the displacement does not exceed 1 foot from equilibrium.

69. *Data Analysis* The table gives the unemployment rate r for the years 1985 through 1994 in the United States. The time t is measured in years, with $t = 0$ corresponding to 1990. (Source: U.S. Bureau of Labor Statistics)

t	-5	-4	-3	-2	-1
r	7.2	7.0	6.2	5.5	5.3

t	0	1	2	3	4
r	5.5	6.7	7.4	6.8	6.1

(a) Create a scatter plot of the data.

(b) Which of the following models best represents the data? Explain your reasoning.

(1) $r = 1.5 \cos(t + 3.9) + 6.37$

(2) $r = 1.03 \sin(0.9t + 0.44) + 6.19$

(3) $r = 1.05 \sin[0.95(t + 6.32)] + 6.20$

(4) $r = 1.5 \sin[0.5(t + 2.8)] + 6.25$

(c) What term in the model gives the average unemployment rate? What is the rate?

(d) Economists study the lengths of business cycles such as unemployment rates. Based on this short span of time, use the model to give the length of this cycle.

(e) Use the model to estimate the next time the unemployment rate will be 6% or less.

70. *Quadratic Approximation* Consider the function

$$f(x) = 3 \sin(0.6x - 2).$$

(a) Approximate the zero of the function in the interval [0, 6].

(b) A quadratic approximation agreeing with f at $x = 5$ is given by

$$g(x) = -0.45x^2 + 5.52x - 13.70.$$

Use a graphing utility to graph f and g on the same viewing rectangle. Describe the result.

(c) Use the Quadratic Formula to find the zeros of g. Compare the zero in the interval [0, 6] with the result of part (a).

2.4 Sum and Difference Formulas

Introduction ▫ Using Sum and Difference Formulas

See Exercise 79 on page 248 for an example of how a sum formula can be used to help analyze the harmonic motion of a spring.

Introduction

In this and the following section, you will study the derivations and uses of several trigonometric identities and formulas.

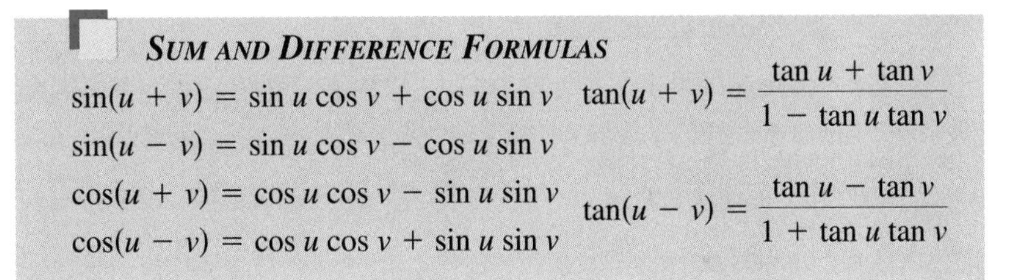

SUM AND DIFFERENCE FORMULAS

$\sin(u + v) = \sin u \cos v + \cos u \sin v$

$\sin(u - v) = \sin u \cos v - \cos u \sin v$

$\cos(u + v) = \cos u \cos v - \sin u \sin v$

$\cos(u - v) = \cos u \cos v + \sin u \sin v$

$\tan(u + v) = \dfrac{\tan u + \tan v}{1 - \tan u \tan v}$

$\tan(u - v) = \dfrac{\tan u - \tan v}{1 + \tan u \tan v}$

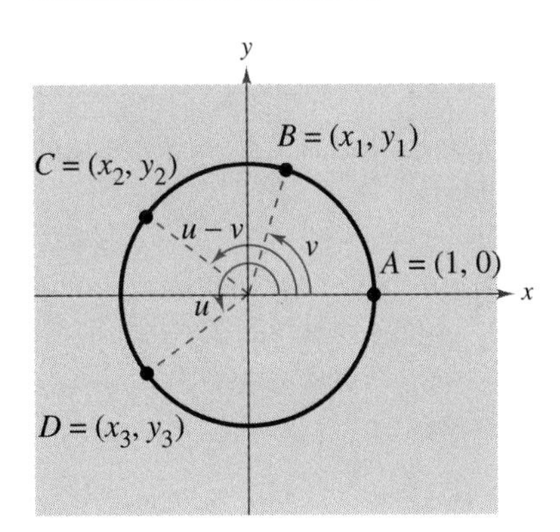

FIGURE 2.5

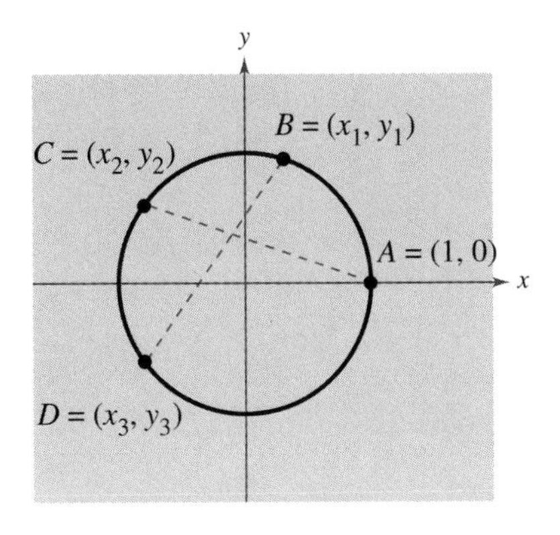

FIGURE 2.6

PROOF ▪▪

Here are proofs for the formulas for $\cos(u \pm v)$. In Figure 2.5, let A be the point $(1, 0)$ and then use u and v to locate the points $B = (x_1, y_1)$, $C = (x_2, y_2)$, and $D = (x_3, y_3)$ on the unit circle. Thus, $x_i^2 + y_i^2 = 1$ for $i = 1, 2, 3$. For convenience, assume that $0 < v < u < 2\pi$. In Figure 2.6, note that arcs AC and BD have the same length. Hence, *line segments* AC and BD are also equal in length, which implies that

$$\begin{aligned}
\sqrt{(x_2 - 1)^2 + (y_2 - 0)^2} &= \sqrt{(x_3 - x_1)^2 + (y_3 - y_1)^2} \\
x_2^2 - 2x_2 + 1 + y_2^2 &= x_3^2 - 2x_1x_3 + x_1^2 + y_3^2 - 2y_1y_3 + y_1^2 \\
(x_2^2 + y_2^2) + 1 - 2x_2 &= (x_3^2 + y_3^2) + (x_1^2 + y_1^2) - 2x_1x_3 - 2y_1y_3 \\
1 + 1 - 2x_2 &= 1 + 1 - 2x_1x_3 - 2y_1y_3 \\
x_2 &= x_3x_1 + y_3y_1.
\end{aligned}$$

Finally, by substituting the values $x_2 = \cos(u - v)$, $x_3 = \cos u$, $x_1 = \cos v$, $y_3 = \sin u$, and $y_1 = \sin v$, you obtain

$$\cos(u - v) = \cos u \cos v + \sin u \sin v.$$

The formula for $\cos(u + v)$ can be established by considering $u + v = u - (-v)$ and using the formula just derived to obtain

$$\begin{aligned}
\cos(u + v) &= \cos[u - (-v)] \\
&= \cos u \cos(-v) + \sin u \sin(-v) \\
&= \cos u \cos v - \sin u \sin v.
\end{aligned}$$ ▪▪

NOTE Note that $\sin(u + v) \neq \sin u + \sin v$. Similar statements can be made for $\cos(u + v)$ and $\tan(u + v)$. ▪▪

Hipparchus, considered the most eminent of Greek astronomers, was born about 160 B.C. in Nicaea. He was credited with the invention of trigonometry. He also derived the sum and difference formulas for $\sin(A \pm B)$ and $\cos(A \pm B)$. *(Illustration: The Granger Collection, New York)*

NOTE Try checking the result obtained in Example 1 on your calculator. You will find that $\cos 75° \approx 0.259$.

Exploration

Graph $y = \cos(x + 2)$ and $y = \cos x + \cos 2$ on the same coordinate plane. What can you conclude about the graphs? Is it true that $\cos(x + 2) = \cos x + \cos 2$?

Graph $y = \sin(x + 4)$ and $y = \sin x + \sin 4$ on the same coordinate plane. What can you conclude about the graphs? Is it true that $\sin(x + 4) = \sin x + \sin 4$?

Using Sum and Difference Formulas

In the remainder of this section, you will study a variety of uses of sum and difference formulas. For instance, Examples 1 and 2 show how sum and difference formulas can be used to find exact values of trigonometric functions involving sums or differences of special angles.

EXAMPLE 1 *Evaluating a Trigonometric Function*

Find the exact value of $\cos 75°$.

Solution

To find the *exact* value of $\cos 75°$, use the fact that $75° = 30° + 45°$. Consequently, the formula for $\cos(u + v)$ yields

$$\begin{aligned}\cos 75° &= \cos(30° + 45°)\\ &= \cos 30° \cos 45° - \sin 30° \sin 45°\\ &= \frac{\sqrt{3}}{2}\left(\frac{\sqrt{2}}{2}\right) - \frac{1}{2}\left(\frac{\sqrt{2}}{2}\right)\\ &= \frac{\sqrt{6} - \sqrt{2}}{4}.\end{aligned}$$

EXAMPLE 2 *Evaluating a Trigonometric Function*

Find the exact value of $\sin \dfrac{\pi}{12}$.

Solution

Using the fact that

$$\frac{\pi}{12} = \frac{\pi}{3} - \frac{\pi}{4}$$

together with the formula for $\sin(u - v)$, you obtain

$$\begin{aligned}\sin \frac{\pi}{12} &= \sin\left(\frac{\pi}{3} - \frac{\pi}{4}\right)\\ &= \sin \frac{\pi}{3} \cos \frac{\pi}{4} - \cos \frac{\pi}{3} \sin \frac{\pi}{4}\\ &= \frac{\sqrt{3}}{2}\left(\frac{\sqrt{2}}{2}\right) - \frac{1}{2}\left(\frac{\sqrt{2}}{2}\right)\\ &= \frac{\sqrt{6} - \sqrt{2}}{4}.\end{aligned}$$

EXAMPLE 3 *Evaluating a Trigonometric Expression*

Find the exact value of $\sin 42° \cos 12° - \cos 42° \sin 12°$.

Solution

Recognizing that this expression fits the formula for $\sin(u - v)$, you can write

$$\begin{aligned}\sin 42° \cos 12° - \cos 42° \sin 12° &= \sin(42° - 12°)\\ &= \sin 30° = \frac{1}{2}.\end{aligned}$$

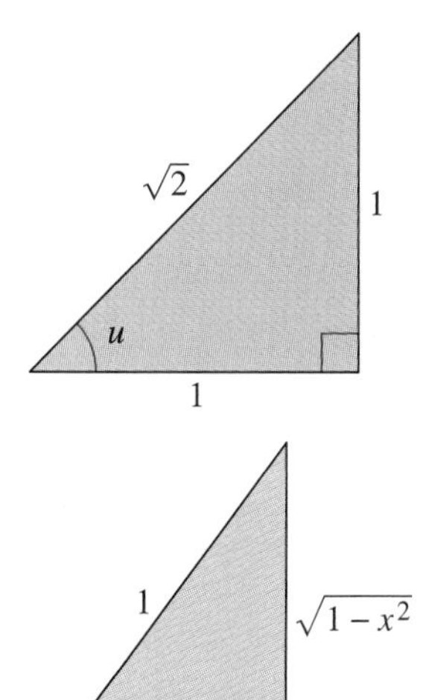

FIGURE 2.7

EXAMPLE 4 *An Application of a Sum Formula*

Evaluate $\cos(\arctan 1 + \arccos x)$.

Solution

This expression fits the formula for $\cos(u + v)$. Angles $u = \arctan 1$ and $v = \arccos x$ are shown in Figure 2.7. Then

$$\begin{aligned}\cos(u + v) &= \cos(\arctan 1)\cos(\arccos x) - \sin(\arctan 1)\sin(\arccos x)\\ &= \frac{1}{\sqrt{2}} \cdot x - \frac{1}{\sqrt{2}} \cdot \sqrt{1 - x^2}\\ &= \frac{x - \sqrt{1 - x^2}}{\sqrt{2}}.\end{aligned}$$

EXAMPLE 5 *Proving a Cofunction Identity*

Prove the cofunction identity $\cos\left(\frac{\pi}{2} - x\right) = \sin x$.

Solution

Using the formula for $\cos(u - v)$, you have

$$\begin{aligned}\cos\left(\frac{\pi}{2} - x\right) &= \cos\frac{\pi}{2}\cos x + \sin\frac{\pi}{2}\sin x\\ &= (0)(\cos x) + (1)(\sin x)\\ &= \sin x.\end{aligned}$$

Sum and difference formulas can be used to derive **reduction formulas** involving expressions such as

$$\sin\left(\theta + \frac{n\pi}{2}\right) \quad \text{and} \quad \cos\left(\theta + \frac{n\pi}{2}\right), \quad \text{where } n \text{ is an integer.}$$

EXAMPLE 6 *Deriving Reduction Formulas*

Simplify each expression.

a. $\cos\left(\theta - \frac{3\pi}{2}\right)$ **b.** $\tan(\theta + 3\pi)$

Solution

a. Using the formula for $\cos(u - v)$, you have

$$\begin{aligned}\cos\left(\theta - \frac{3\pi}{2}\right) &= \cos\theta\cos\frac{3\pi}{2} + \sin\theta\sin\frac{3\pi}{2}\\ &= (\cos\theta)(0) + (\sin\theta)(-1)\\ &= -\sin\theta.\end{aligned}$$

b. Using the formula for $\tan(u + v)$, you have

$$\begin{aligned}\tan(\theta + 3\pi) &= \frac{\tan\theta + \tan 3\pi}{1 - \tan\theta\tan 3\pi}\\ &= \frac{\tan\theta + 0}{1 - (\tan\theta)(0)}\\ &= \tan\theta.\end{aligned}$$

The next example was taken from calculus. It is used to derive the derivative of the sine function.

EXAMPLE 7 *An Application from Calculus*

Verify that

$$\frac{\sin(x + h) - \sin x}{h} = (\cos x)\left(\frac{\sin h}{h}\right) - (\sin x)\left(\frac{1 - \cos h}{h}\right)$$

where $h \neq 0$.

Solution

Using the formula for $\sin(u + v)$, you have

$$\begin{aligned}\frac{\sin(x + h) - \sin x}{h} &= \frac{\sin x\cos h + \cos x\sin h - \sin x}{h}\\ &= \frac{\cos x\sin h - \sin x(1 - \cos h)}{h}\\ &= (\cos x)\left(\frac{\sin h}{h}\right) - (\sin x)\left(\frac{1 - \cos h}{h}\right).\end{aligned}$$

EXAMPLE 8 *Solving a Trigonometric Equation*

Find all solutions of

$$\sin\left(x + \frac{\pi}{4}\right) + \sin\left(x - \frac{\pi}{4}\right) = -1$$

in the interval $[0, 2\pi)$.

Solution

Using sum and difference formulas, rewrite the given equation as

$$\sin x \cos \frac{\pi}{4} + \cos x \sin \frac{\pi}{4} + \sin x \cos \frac{\pi}{4} - \cos x \sin \frac{\pi}{4} = -1$$

$$2 \sin x \cos \frac{\pi}{4} = -1$$

$$2(\sin x)\left(\frac{\sqrt{2}}{2}\right) = -1$$

$$\sin x = -\frac{1}{\sqrt{2}}$$

$$\sin x = -\frac{\sqrt{2}}{2}$$

Therefore, the only solutions in the interval $[0, 2\pi)$ are

$$x = \frac{5\pi}{4} \qquad \text{and} \qquad x = \frac{7\pi}{4}.$$

These solutions are checked graphically in Figure 2.8.

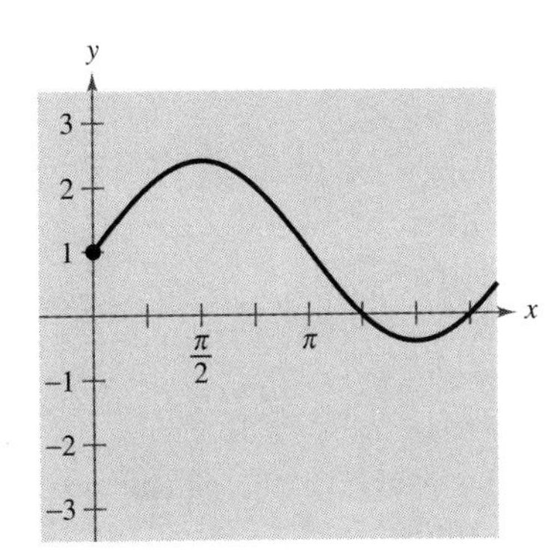

FIGURE 2.8

GROUP ACTIVITY

THE ANGLE BETWEEN TWO LINES

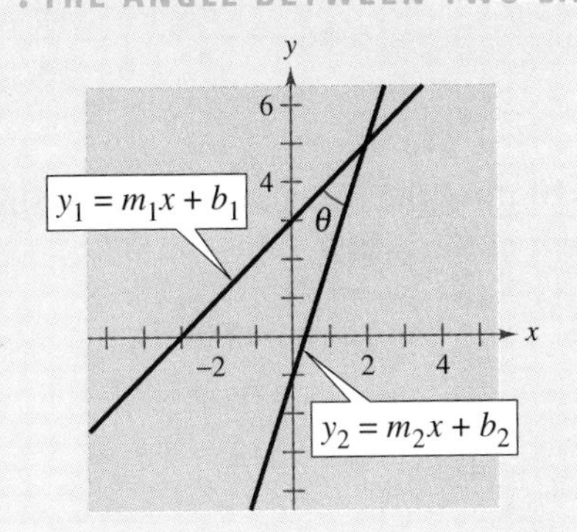

The figure at the left shows two lines whose equations are

$$y_1 = m_1x + b_1 \qquad \text{and} \qquad y_2 = m_2x + b_2.$$

Assume that both lines have positive slopes, as shown in the figure. With others in your group, derive a formula for the angle between the two lines. Then use your formula to find the angle between the following pairs of lines.

a. $y = x$ and $y = \sqrt{3}x$ **b.** $y = x$ and $y = \frac{1}{\sqrt{3}}x$

WARM UP

Use the given information to find $\sin\theta$.

1. $\tan\theta = \frac{1}{3}$, θ in Quadrant I
2. $\cot\theta = \frac{3}{5}$, θ in Quadrant III
3. $\cos\theta = \frac{3}{4}$, θ in Quadrant IV
4. $\sec\theta = -3$, θ in Quadrant II

Find all solutions in the interval $[0, 2\pi)$.

5. $\sin x = \dfrac{\sqrt{2}}{2}$
6. $\cos x = 0$

Simplify the expression.

7. $\tan x \sec^2 x - \tan x$
8. $\dfrac{\cos x \csc x}{\tan x}$
9. $\dfrac{\cos x}{1 - \sin x} - \tan x$
10. $\dfrac{\cos^4 x - \sin^4 x}{\cos^2 x}$

2.4 Exercises

In Exercises 1–4, find the exact value of each expression.

1. (a) $\cos\left(\dfrac{\pi}{4} + \dfrac{\pi}{3}\right)$ (b) $\cos\dfrac{\pi}{4} + \cos\dfrac{\pi}{3}$
2. (a) $\sin\left(\dfrac{3\pi}{4} + \dfrac{5\pi}{6}\right)$ (b) $\sin\dfrac{3\pi}{4} + \sin\dfrac{5\pi}{6}$
3. (a) $\sin\left(\dfrac{7\pi}{6} - \dfrac{\pi}{3}\right)$ (b) $\sin\dfrac{7\pi}{6} - \sin\dfrac{\pi}{3}$
4. (a) $\cos\left(\dfrac{2\pi}{3} - \dfrac{\pi}{6}\right)$ (b) $\cos\dfrac{2\pi}{3} + \cos\dfrac{\pi}{6}$
5. ***Think About It*** Use the results of Exercises 1–4 to determine if the following are true or false. Explain.
   (a) $\sin(u \pm v) = \sin u \pm \sin v$
   (b) $\cos(u \pm v) = \cos u \pm \cos v$
6. ***True or False?*** $\cos\left(x - \dfrac{\pi}{2}\right) = -\sin x$

In Exercises 7–16, find the exact values of the sine, cosine, and tangent of the angle.

7. $75° = 30° + 45°$
8. $15° = 45° - 30°$
9. $105° = 60° + 45°$
10. $165° = 135° + 30°$
11. $195° = 225° - 30°$
12. $255° = 300° - 45°$
13. $\dfrac{11\pi}{12} = \dfrac{3\pi}{4} + \dfrac{\pi}{6}$
14. $\dfrac{7\pi}{12} = \dfrac{\pi}{3} + \dfrac{\pi}{4}$
15. $\dfrac{17\pi}{12} = \dfrac{9\pi}{4} - \dfrac{5\pi}{6}$
16. $-\dfrac{\pi}{12} = \dfrac{\pi}{6} - \dfrac{\pi}{4}$

In Exercises 17–20, find the exact values of the sine, cosine, and tangent of the angle.

17. $285°$
18. $-105°$
19. $-\dfrac{13\pi}{12}$
20. $\dfrac{5\pi}{12}$

In Exercises 21–30, write the expression as the sine, cosine, or tangent of an angle.

21. $\cos 25° \cos 15° - \sin 25° \sin 15°$
22. $\sin 140° \cos 50° + \cos 140° \sin 50°$
23. $\sin 230° \cos 30° - \cos 230° \sin 30°$
24. $\cos 20° \cos 30° + \sin 20° \sin 30°$
25. $\dfrac{\tan 325° - \tan 86°}{1 + \tan 325° \tan 86°}$
26. $\dfrac{\tan 140° - \tan 60°}{1 + \tan 140° \tan 60°}$
27. $\sin 3 \cos 1.2 - \cos 3 \sin 1.2$
28. $\cos \dfrac{\pi}{7} \cos \dfrac{\pi}{5} - \sin \dfrac{\pi}{7} \sin \dfrac{\pi}{5}$
29. $\dfrac{\tan 2x + \tan x}{1 - \tan 2x \tan x}$
30. $\cos 3x \cos 2y + \sin 3x \sin 2y$

In Exercises 31–38, find the exact value of the trigonometric function given that $\sin u = \frac{5}{13}$ and $\cos v = -\frac{3}{5}$. (Both u and v are in Quadrant II.)

31. $\sin(u + v)$
32. $\cos(v - u)$
33. $\cos(u + v)$
34. $\sin(u - v)$
35. $\sec(u + v)$
36. $\csc(u - v)$
37. $\tan(u - v)$
38. $\cot(u + v)$

In Exercises 39–44, find the exact value of the trigonometric function given that $\sin u = -\frac{7}{25}$ and $\cos v = -\frac{4}{5}$. (Both u and v are in Quadrant III.)

39. $\cos(u + v)$
40. $\sin(u + v)$
41. $\sin(v - u)$
42. $\cos(u - v)$
43. $\csc(u + v)$
44. $\sec(v - u)$

In Exercises 45–58, verify the identity.

45. $\sin(3\pi - x) = \sin x$
46. $\sin\left(\dfrac{\pi}{2} + x\right) = \cos x$
47. $\sin\left(\dfrac{\pi}{6} + x\right) = \dfrac{1}{2}(\cos x + \sqrt{3} \sin x)$
48. $\cos\left(\dfrac{5\pi}{4} - x\right) = -\dfrac{\sqrt{2}}{2}(\cos x + \sin x)$
49. $\cos(\pi - \theta) + \sin\left(\dfrac{\pi}{2} + \theta\right) = 0$
50. $\tan\left(\dfrac{\pi}{4} - \theta\right) = \dfrac{1 - \tan \theta}{1 + \tan \theta}$
51. $\cos(x + y)\cos(x - y) = \cos^2 x - \sin^2 y$
52. $\sin(x + y)\sin(x - y) = \sin^2 x - \sin^2 y$
53. $\sin(x + y) + \sin(x - y) = 2 \sin x \cos y$
54. $\cos(x + y) + \cos(x - y) = 2 \cos x \cos y$
55. $\cos(n\pi + \theta) = (-1)^n \cos \theta$, n is an integer.
56. $\sin(n\pi + \theta) = (-1)^n \sin \theta$, n is an integer.
57. $a \sin B\theta + b \cos B\theta = \sqrt{a^2 + b^2} \sin(B\theta + C)$, where $C = \arctan(b/a)$, $a > 0$
58. $a \sin B\theta + b \cos B\theta = \sqrt{a^2 + b^2} \cos(B\theta - C)$, where $C = \arctan(a/b)$, $b > 0$

In Exercises 59–62, verify the identity algebraically and use a graphing utility to confirm it graphically.

59. $\cos\left(\dfrac{3\pi}{2} - x\right) = -\sin x$
60. $\cos(\pi + x) = -\cos x$
61. $\sin\left(\dfrac{3\pi}{2} + \theta\right) + \sin(\pi - \theta) = \sin \theta - \cos \theta$
62. $\tan(\pi + \theta) = \tan \theta$

In Exercises 63–66, use the formulas given in Exercises 57 and 58 to write the trigonometric expression in the following forms.

(a) $\sqrt{a^2 + b^2} \sin(B\theta + C)$
(b) $\sqrt{a^2 + b^2} \cos(B\theta - C)$

63. $\sin \theta + \cos \theta$
64. $3 \sin 2\theta + 4 \cos 2\theta$
65. $12 \sin 3\theta + 5 \cos 3\theta$
66. $\sin 2\theta - \cos 2\theta$

In Exercises 67 and 68, use the formulas given in Exercises 57 and 58 to write the trigonometric expression in the form $a \sin B\theta + b \cos B\theta$.

67. $2 \sin\left(\theta + \dfrac{\pi}{2}\right)$
68. $5 \cos\left(\theta + \dfrac{3\pi}{4}\right)$

In Exercises 69 and 70, write the trigonometric expression as an algebraic expression.

69. $\sin(\arcsin x + \arccos x)$

70. $\sin(\arctan 2x - \arccos x)$

In Exercises 71–74, find all solutions of the equation in the interval $[0, 2\pi)$.

71. $\sin\left(x + \frac{\pi}{3}\right) + \sin\left(x - \frac{\pi}{3}\right) = 1$

72. $\sin\left(x + \frac{\pi}{6}\right) - \sin\left(x - \frac{\pi}{6}\right) = \frac{1}{2}$

73. $\cos\left(x + \frac{\pi}{4}\right) - \cos\left(x - \frac{\pi}{4}\right) = 1$

74. $\tan(x + \pi) + 2\sin(x + \pi) = 0$

In Exercises 75 and 76, use a graphing utility to approximate the solutions in the interval $[0, 2\pi)$.

75. $\cos\left(x + \frac{\pi}{4}\right) + \cos\left(x - \frac{\pi}{4}\right) = 1$

76. $\tan(x + \pi) - \cos\left(x + \frac{\pi}{2}\right) = 0$

77. ***Conjecture*** Consider the function

$$f(\theta) = \sin^2\left(\theta + \frac{\pi}{4}\right) + \sin^2\left(\theta - \frac{\pi}{4}\right).$$

Graph the function and use the graph to create an identity. Prove your conjecture.

78. ***Conjecture*** Three squares of side s are placed side by side (see figure). Make a conjecture about the relationship between the sum $u + v$ and w. Prove your conjecture by using the identity for the tangent of the sum of two angles.

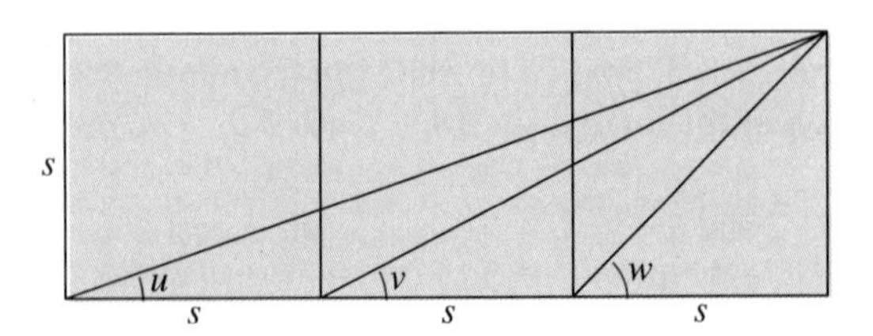

79. ***Harmonic Motion*** A weight is attached to a spring suspended vertically from a ceiling. When a driving force is applied to the system, the weight moves vertically from its equilibrium position, and this motion is modeled by

$$y = \frac{1}{3}\sin 2t + \frac{1}{4}\cos 2t$$

where y is the distance from equilibrium measured in feet and t is the time in seconds.

(a) Write the model in the form

$$y = \sqrt{a^2 + b^2}\sin(Bt + C).$$

(See Exercise 57.)

(b) Find the amplitude of the oscillations of the weight.

(c) Find the frequency of the oscillations of the weight.

80. ***Standing Waves*** The equation of a standing wave is obtained by adding the displacements of two waves traveling in opposite directions (see figure). Assume that each of the waves has amplitude A, period T, and wavelength λ. If the models for these waves are

$$y_1 = A\cos 2\pi\left(\frac{t}{T} - \frac{x}{\lambda}\right) \quad \text{and}$$

$$y_2 = A\cos 2\pi\left(\frac{t}{T} + \frac{x}{\lambda}\right)$$

show that

$$y_1 + y_2 = 2A\cos\frac{2\pi t}{T}\cos\frac{2\pi x}{\lambda}.$$

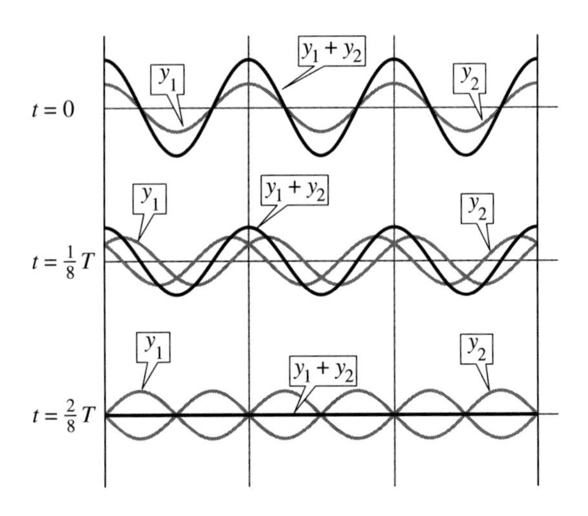

2.5 Multiple-Angle and Product-to-Sum Formulas

See Exercise 116 on page 260 for an example of how a double-angle formula can help analyze the motion of a projectile.

Multiple-Angle Formulas ❐ *Power-Reducing Formulas* ❐ *Half-Angle Formulas* ❐ *Product-to-Sum Formulas*

Multiple-Angle Formulas

In this section you will study four other categories of trigonometric identities.

1. The first category involves functions of multiple angles such as $\sin ku$ or $\cos ku$.
2. The second category involves squares of trigonometric functions such as $\sin^2 u$.
3. The third category involves functions of half-angles such as $\sin(u/2)$.
4. The fourth category involves products of trigonometric functions such as $\sin u \cos v$.

The most commonly used multiple-angle formulas are the **double-angle formulas.** They are used often, so you should learn them.

DOUBLE-ANGLE FORMULAS

$$\sin 2u = 2 \sin u \cos u \qquad \tan 2u = \frac{2 \tan u}{1 - \tan^2 u}$$

$$\begin{aligned} \cos 2u &= \cos^2 u - \sin^2 u \\ &= 2\cos^2 u - 1 \\ &= 1 - 2\sin^2 u \end{aligned}$$

PROOF ▪▪

To prove the first formula, let $v = u$ in the formula for $\sin(u + v)$.

$$\begin{aligned} \sin 2u &= \sin(u + u) \\ &= \sin u \cos u + \cos u \sin u \\ &= 2 \sin u \cos u \end{aligned}$$

To prove the second formula, let $v = u$ in the formula for $\cos(u + v)$.

$$\begin{aligned} \cos 2u &= \cos(u + u) \\ &= \cos u \cos u - \sin u \sin u \\ &= \cos^2 u - \sin^2 u \end{aligned}$$

The tangent double-angle formula can be proven in a similar way. ▪▪

NOTE Note that $\sin 2u \neq 2 \sin u$. Similar statements can be made for $\cos 2u$ and $\tan 2u$. ▪▪

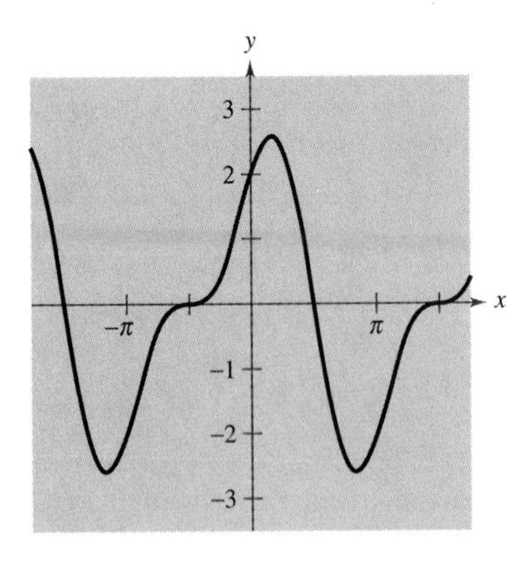

FIGURE 2.9

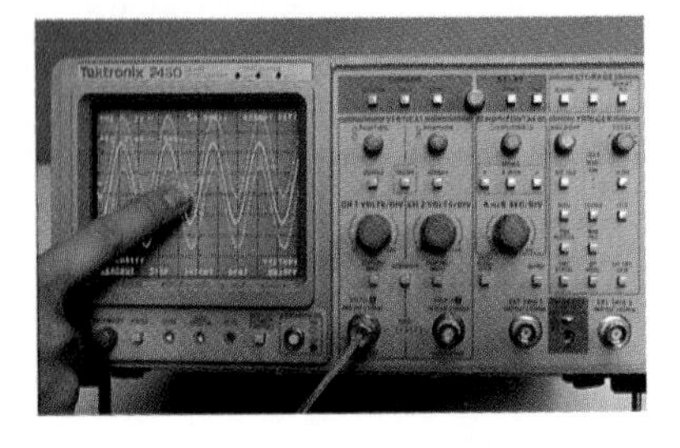

An oscilloscope is an electronic instrument that displays changes in electrical or sound waves on a fluorescent screen. See Exercise 80 on page 248. *(Photo: Peter Aprahamian/ Science Photo Library/Photo Researchers)*

EXAMPLE 1 *Solving a Trigonometric Equation*

Find all solutions of $2 \cos x + \sin 2x = 0$.

Solution

Begin by rewriting the equation so that it involves functions of x (rather than $2x$). Then factor and solve as usual.

$2 \cos x + \sin 2x = 0$	Original equation
$2 \cos x + 2 \sin x \cos x = 0$	Double-angle formula
$2 \cos x(1 + \sin x) = 0$	Factor.
$\cos x = 0, \quad 1 + \sin x = 0$	Set factors equal to zero.
$x = \frac{\pi}{2}, \frac{3\pi}{2} \qquad x = \frac{3\pi}{2}$	Solutions in $[0, 2\pi)$

Therefore, the general solution is

$$x = \frac{\pi}{2} + 2n\pi \qquad \text{and} \qquad x = \frac{3\pi}{2} + 2n\pi$$

where n is an integer. The graph of $y = 2 \cos x + \sin 2x$, as shown in Figure 2.9, allows you to verify these solutions graphically.

EXAMPLE 2 *Using Double-Angle Formulas in Sketching Graphs*

Sketch the graph of $y = 4 \cos^2 x - 2$ over the interval $[0, 2\pi]$.

Solution

Using a double-angle formula, you can rewrite the given function as

$$\begin{aligned} y &= 4 \cos^2 x - 2 \\ &= 2(2 \cos^2 x - 1) \\ &= 2 \cos 2x. \end{aligned}$$

Using the techniques discussed in Section 1.5, you can recognize that the graph of this function has an amplitude of 2 and a period of π. The key points in the interval $[0, \pi]$ are as follows.

Maximum	Intercept	Minimum	Intercept	Maximum
$(0, 2)$	$\left(\frac{\pi}{4}, 0\right)$	$\left(\frac{\pi}{2}, -2\right)$	$\left(\frac{3\pi}{4}, 0\right)$	$(\pi, 2)$

Two cycles of the graph are shown in Figure 2.10.

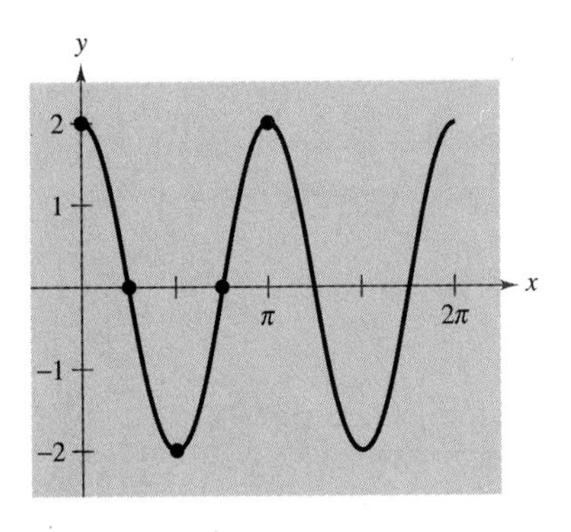

FIGURE 2.10

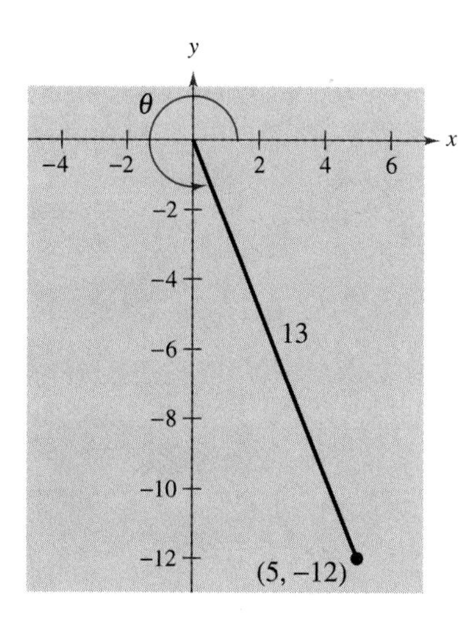

FIGURE 2.11

EXAMPLE 3 *Evaluating Functions Involving Double Angles*

Use the following to find $\sin 2\theta$, $\cos 2\theta$, and $\tan 2\theta$.

$$\cos \theta = \frac{5}{13}, \qquad \frac{3\pi}{2} < \theta < 2\pi$$

Solution

From Figure 2.11, you can see that $\sin \theta = y/r = -12/13$. Consequently, you can write the following.

$$\sin 2\theta = 2 \sin \theta \cos \theta = 2\left(\frac{-12}{13}\right)\left(\frac{5}{13}\right) = -\frac{120}{169}$$

$$\cos 2\theta = 2 \cos^2 \theta - 1 = 2\left(\frac{25}{169}\right) - 1 = -\frac{119}{169}$$

$$\tan 2\theta = \frac{\sin 2\theta}{\cos 2\theta} = \frac{120}{119}$$

The double-angle formulas are not restricted to angles 2θ and θ. Other *double* combinations, such as 4θ and 2θ or 6θ and 3θ, are also valid. Here are two examples.

$$\sin 4\theta = 2 \sin 2\theta \cos 2\theta \qquad \text{and} \qquad \cos 6\theta = \cos^2 3\theta - \sin^2 3\theta$$

By using double-angle formulas together with the sum formulas derived in the preceding section, you can form other multiple-angle formulas.

EXAMPLE 4 *Deriving a Triple-Angle Formula*

Express $\sin 3x$ in terms of $\sin x$.

Solution

$$\begin{aligned}
\sin 3x &= \sin(2x + x) \\
&= \sin 2x \cos x + \cos 2x \sin x \\
&= 2 \sin x \cos x \cos x + (1 - 2 \sin^2 x)\sin x \\
&= 2 \sin x \cos^2 x + \sin x - 2 \sin^3 x \\
&= 2\sin x(1 - \sin^2 x) + \sin x - 2 \sin^3 x \\
&= 2 \sin x - 2 \sin^3 x + \sin x - 2 \sin^3 x \\
&= 3\sin x - 4 \sin^3 x
\end{aligned}$$

Power-Reducing Formulas

The double-angle formulas can be used to obtain the following **power-reducing formulas.**

POWER-REDUCING FORMULAS

$$\sin^2 u = \frac{1 - \cos 2u}{2} \qquad \cos^2 u = \frac{1 + \cos 2u}{2} \qquad \tan^2 u = \frac{1 - \cos 2u}{1 + \cos 2u}$$

PROOF ▪▪

The first two formulas can be verified by solving for $\sin^2 u$ and $\cos^2 u$, respectively, in the double-angle formulas

$$\cos 2u = 1 - 2\sin^2 u \quad \text{and} \quad \cos 2u = 2\cos^2 u - 1.$$

The third formula can be verified using the fact that

$$\tan^2 u = \frac{\sin^2 u}{\cos^2 u}. \quad ▪▪$$

Example 5 shows a typical power reduction that is used in calculus.

EXAMPLE 5 Reducing the Power of a Trigonometric Function

Rewrite $\sin^4 x$ as a sum of first powers of the cosines of multiple angles.

Solution

Note the repeated use of power-reducing formulas.

$$\begin{aligned}
\sin^4 x = (\sin^2 x)^2 &= \left(\frac{1 - \cos 2x}{2}\right)^2 \\
&= \frac{1}{4}(1 - 2\cos 2x + \cos^2 2x) \\
&= \frac{1}{4}\left(1 - 2\cos 2x + \frac{1 + \cos 4x}{2}\right) \\
&= \frac{1}{4} - \frac{1}{2}\cos 2x + \frac{1}{8} + \frac{1}{8}\cos 4x \\
&= \frac{3}{8} - \frac{1}{2}\cos 2x + \frac{1}{8}\cos 4x \\
&= \frac{1}{8}(3 - 4\cos 2x + \cos 4x)
\end{aligned}$$

Half-Angle Formulas

You can derive some useful alternative forms of the power-reducing formulas by replacing u with $u/2$. The results are called **half-angle formulas.**

HALF-ANGLE FORMULAS

$$\sin \frac{u}{2} = \pm\sqrt{\frac{1 - \cos u}{2}}$$

$$\cos \frac{u}{2} = \pm\sqrt{\frac{1 + \cos u}{2}}$$

$$\tan \frac{u}{2} = \frac{1 - \cos u}{\sin u} = \frac{\sin u}{1 + \cos u}$$

The signs of $\sin(u/2)$ and $\cos(u/2)$ depend on the quadrant in which $u/2$ lies.

EXAMPLE 6 *Using a Half-Angle Formula*

Find the exact value of $\sin 105°$.

Solution

Begin by noting that $105°$ is half of $210°$. Then, using the half-angle formula for $\sin(u/2)$ and the fact that $105°$ lies in Quadrant II, you have

$$\begin{aligned}\sin 105° &= \sqrt{\frac{1 - \cos 210°}{2}}\\ &= \sqrt{\frac{1 - (-\cos 30°)}{2}}\\ &= \sqrt{\frac{1 + (\sqrt{3}/2)}{2}}\\ &= \frac{\sqrt{2 + \sqrt{3}}}{2}.\end{aligned}$$

The positive square root is chosen because $\sin \theta$ is positive in Quadrant II.

NOTE Use your calculator to verify the result obtained in Example 6. That is, evaluate $\sin 105°$ and $\left(\sqrt{2 + \sqrt{3}}\right)/2$ and you will see that both values are approximately 0.9659258.

EXAMPLE 7 *Solving a Trigonometric Equation*

Find all solutions of $2 - \sin^2 x = 2\cos^2 \frac{x}{2}$ in the interval $[0, 2\pi]$.

Solution

$$
\begin{aligned}
2 - \sin^2 x &= 2\cos^2 \frac{x}{2} && \text{Original equation} \\
2 - \sin^2 x &= 2\left(\frac{1 + \cos x}{2}\right) && \text{Half-angle formula} \\
2 - \sin^2 x &= 1 + \cos x && \text{Simplify.} \\
2 - (1 - \cos^2 x) &= 1 + \cos x && \text{Pythagorean identity} \\
\cos^2 x - \cos x &= 0 && \text{Simplify.} \\
\cos x(\cos x - 1) &= 0 && \text{Factor.}
\end{aligned}
$$

By setting the factors $\cos x$ and $(\cos x - 1)$ equal to zero, you find that the solutions in the interval $[0, 2\pi]$ are $x = \pi/2$, $x = 3\pi/2$, and $x = 0$.

Product-to-Sum Formulas

Each of the following **product-to-sum formulas** is easily verified using the sum and difference formulas discussed in the preceding section.

PRODUCT-TO-SUM FORMULAS

$$\sin u \sin v = \frac{1}{2}[\cos(u - v) - \cos(u + v)]$$

$$\cos u \cos v = \frac{1}{2}[\cos(u - v) + \cos(u + v)]$$

$$\sin u \cos v = \frac{1}{2}[\sin(u + v) + \sin(u - v)]$$

$$\cos u \sin v = \frac{1}{2}[\sin(u + v) - \sin(u - v)]$$

EXAMPLE 8 *Writing Products as Sums*

Rewrite $\cos 5x \sin 4x$ as a sum or difference.

Solution

$$\cos 5x \sin 4x = \tfrac{1}{2}[\sin(5x + 4x) - \sin(5x - 4x)] = \tfrac{1}{2}\sin 9x - \tfrac{1}{2}\sin x$$

Occasionally, it is useful to reverse the procedure and write a sum of trigonometric functions as a product. This can be accomplished with the following **sum-to-product formulas.**

SUM-TO-PRODUCT FORMULAS

$$\sin x + \sin y = 2 \sin\left(\frac{x+y}{2}\right)\cos\left(\frac{x-y}{2}\right)$$

$$\sin x - \sin y = 2 \cos\left(\frac{x+y}{2}\right)\sin\left(\frac{x-y}{2}\right)$$

$$\cos x + \cos y = 2 \cos\left(\frac{x+y}{2}\right)\cos\left(\frac{x-y}{2}\right)$$

$$\cos x - \cos y = -2 \sin\left(\frac{x+y}{2}\right)\sin\left(\frac{x-y}{2}\right)$$

PROOF ▪▪

To prove the first formula, let $x = u + v$ and $y = u - v$. Then substitute $u = (x + y)/2$ and $v = (x - y)/2$ in the product-to-sum formula.

$$\sin u \cos v = \frac{1}{2}[\sin(u + v) + \sin(u - v)]$$

$$\sin\left(\frac{x+y}{2}\right)\cos\left(\frac{x-y}{2}\right) = \frac{1}{2}(\sin x + \sin y)$$

$$2\sin\left(\frac{x+y}{2}\right)\cos\left(\frac{x-y}{2}\right) = \sin x + \sin y$$ ▪▪

EXAMPLE 9 *Using a Sum-to-Product Formula*

Find the exact value of $\cos 195° + \cos 105°$.

Solution

Using the appropriate sum-to-product formula, you obtain

$$\begin{aligned}\cos 195° + \cos 105° &= 2\cos\left(\frac{195° + 105°}{2}\right)\cos\left(\frac{195° - 105°}{2}\right)\\ &= 2\cos 150° \cos 45°\\ &= 2\left(-\frac{\sqrt{3}}{2}\right)\left(\frac{\sqrt{2}}{2}\right)\\ &= -\frac{\sqrt{6}}{2}.\end{aligned}$$

EXAMPLE 10 *Solving a Trigonometric Equation*

Find all solutions of $\sin 5x + \sin 3x = 0$.

Solution

$$\sin 5x + \sin 3x = 0 \qquad \text{Original equation}$$

$$2 \sin\left(\frac{5x + 3x}{2}\right) \cos\left(\frac{5x - 3x}{2}\right) = 0 \qquad \text{Sum-to-product formula}$$

$$2 \sin 4x \cos x = 0 \qquad \text{Simplify.}$$

By setting the factor $\sin 4x$ equal to zero, you can find that the solutions in the interval $[0, 2\pi)$ are

$$x = 0, \frac{\pi}{4}, \frac{\pi}{2}, \frac{3\pi}{4}, \pi, \frac{5\pi}{4}, \frac{3\pi}{2}, \frac{7\pi}{4}.$$

The equation $\cos x = 0$ yields no additional solutions, and you can conclude that the solutions are of the form

$$x = \frac{n\pi}{4}$$

where n is an integer. These solutions are verified graphically in Figure 2.12.

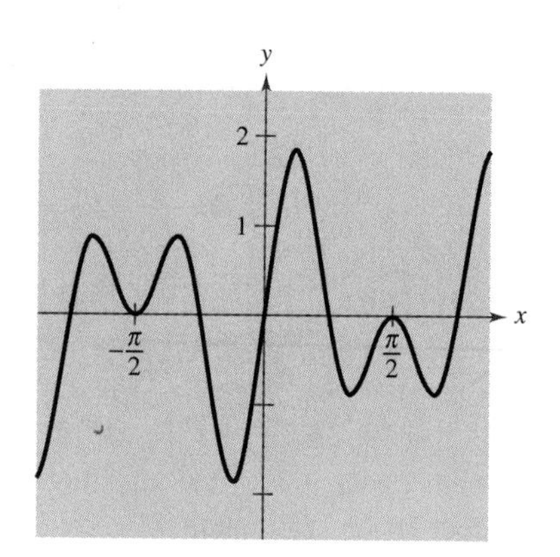

FIGURE 2.12

EXAMPLE 11 *Verifying a Trigonometric Identity*

Verify the identity

$$\frac{\sin t + \sin 3t}{\cos t + \cos 3t} = \tan 2t.$$

Solution

Using appropriate sum-to-product formulas, you have

$$\frac{\sin t + \sin 3t}{\cos t + \cos 3t} = \frac{2 \sin 2t \cos(-t)}{2 \cos 2t \cos(-t)} = \frac{\sin 2t}{\cos 2t} = \tan 2t.$$

GROUP ACTIVITY

DERIVING AN AREA FORMULA

With others in your group, discuss how you can use a double-angle formula or a half-angle formula to derive a formula for the area of an isosceles triangle. Use a labeled sketch to illustrate your derivation. Then write two examples that show how your formula can be used.

WARM UP

Factor the trigonometric expression.

1. $2 \sin x + \sin x \cos x$
2. $\cos^2 x - \cos x - 2$

Find all solutions of the equation in the interval $[0, 2\pi)$.

3. $\sin 2x = 0$
4. $\cos 2x = 0$
5. $\cos \frac{x}{2} = 0$
6. $\sin \frac{x}{2} = 0$

Simplify the expression.

7. $\frac{1 - \cos(\pi/4)}{2}$
8. $\frac{1 + \cos(\pi/3)}{2}$
9. $\frac{2 \sin 3x \cos x}{2 \cos 3x \cos x}$
10. $\frac{(1 - 2\sin^2 x) \cos x}{2 \sin^2 x \cos x}$

2.5 Exercises

In Exercises 1–8, use the figure to find the exact value of the trigonometric function.

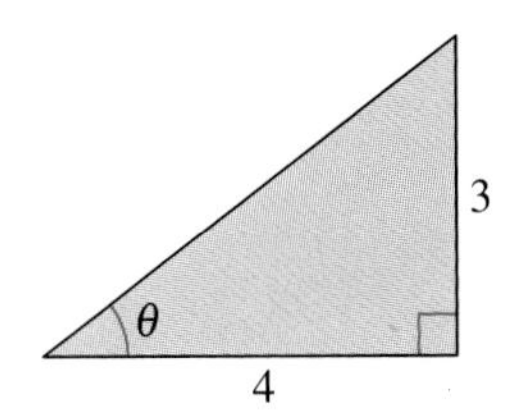

1. $\sin \theta$
2. $\tan \theta$
3. $\cos 2\theta$
4. $\sin 2\theta$
5. $\tan 2\theta$
6. $\sec 2\theta$
7. $\csc 2\theta$
8. $\cot 2\theta$

In Exercises 9–18, find the exact solutions of the equation in the interval $[0, 2\pi)$.

9. $\sin 2x - \sin x = 0$
10. $\sin 2x + \cos x = 0$
11. $4 \sin x \cos x = 1$
12. $\sin 2x \sin x = \cos x$
13. $\cos 2x - \cos x = 0$
14. $\cos 2x + \sin x = 0$
15. $\tan 2x - \cot x = 0$
16. $\tan 2x - 2 \cos x = 0$
17. $\sin 4x = -2 \sin 2x$
18. $(\sin 2x + \cos 2x)^2 = 1$

In Exercises 19–22, use a double-angle formula to rewrite the expression.

19. $6 \sin x \cos x$
20. $4 \sin x \cos x + 2$
21. $4 - 8 \sin^2 x$
22. $(\cos x + \sin x)(\cos x - \sin x)$

In Exercises 23–28, find the exact values of $\sin 2u$, $\cos 2u$, and $\tan 2u$ using the double-angle formulas.

23. $\sin u = \frac{3}{5}, \quad 0 < u < \frac{\pi}{2}$
24. $\cos u = -\frac{2}{3}, \quad \frac{\pi}{2} < u < \pi$

25. $\tan u = \frac{1}{2}, \quad \pi < u < \frac{3\pi}{2}$

26. $\cot u = -4, \quad \frac{3\pi}{2} < u < 2\pi$

27. $\sec u = -\frac{5}{2}, \quad \frac{\pi}{2} < u < \pi$

28. $\csc u = 3, \quad \frac{\pi}{2} < u < \pi$

In Exercises 29–34, use the power-reducing formulas to rewrite the expression in terms of the first power of the cosine.

29. $\cos^4 x$
30. $\sin^4 x$
31. $\sin^2 x \cos^2 x$
32. $\cos^2 x$
33. $\sin^2 x \cos^4 x$
34. $\sin^4 x \cos^2 x$

In Exercises 35–40, use the figure to find the exact value of the trigonometric function.

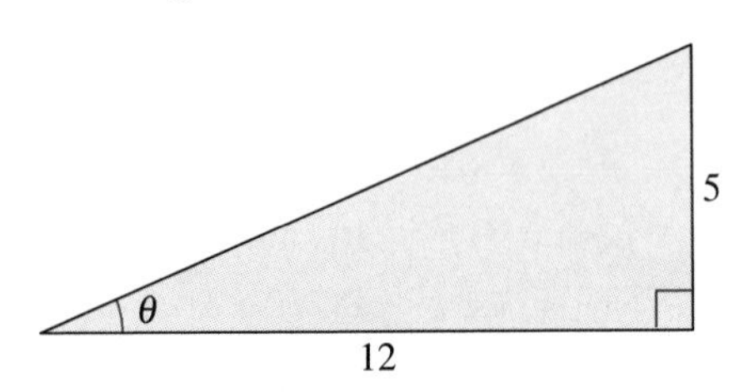

35. $\cos \frac{\theta}{2}$
36. $\sin \frac{\theta}{2}$
37. $\tan \frac{\theta}{2}$
38. $2 \sin \frac{\theta}{2} \cos \frac{\theta}{2}$
39. $\csc \frac{\theta}{2}$
40. $\cot \frac{\theta}{2}$

In Exercises 41–46, use the half-angle formulas to determine the exact values of the sine, cosine, and tangent of the angle.

41. 105°
42. 165°
43. $112^\circ\ 30'$
44. $67^\circ\ 30'$
45. $\frac{\pi}{8}$
46. $\frac{\pi}{12}$

In Exercises 47–52, find the exact values of $\sin(u/2)$, $\cos(u/2)$, and $\tan(u/2)$ using the half-angle formulas.

47. $\sin u = \frac{5}{13}, \quad \frac{\pi}{2} < u < \pi$

48. $\cos u = \frac{3}{5}, \quad 0 < u < \frac{\pi}{2}$

49. $\tan u = -\frac{5}{8}, \quad \frac{3\pi}{2} < u < 2\pi$

50. $\cot u = 3, \quad \pi < u < \frac{3\pi}{2}$

51. $\csc u = -\frac{5}{3}, \quad \pi < u < \frac{3\pi}{2}$

52. $\sec u = -\frac{7}{2}, \quad \frac{\pi}{2} < u < \pi$

In Exercises 53–56, use the half-angle formulas to simplify the expression.

53. $\sqrt{\frac{1 - \cos 6x}{2}}$
54. $\sqrt{\frac{1 + \cos 4x}{2}}$
55. $-\sqrt{\frac{1 - \cos 8x}{1 + \cos 8x}}$
56. $-\sqrt{\frac{1 - \cos(x - 1)}{2}}$

In Exercises 57–60, find the exact zeros of the function in the interval $[0, 2\pi)$. Use the graphing utility to graph the function and verify the zeros.

57. $f(x) = \sin \frac{x}{2} + \cos x$

58. $h(x) = \sin \frac{x}{2} + \cos x - 1$

59. $h(x) = \cos \frac{x}{2} - \sin x$

60. $g(x) = \tan \frac{x}{2} - \sin x$

In Exercises 61–70, use the product-to-sum formulas to write the product as a sum or difference.

61. $6 \sin \frac{\pi}{4} \cos \frac{\pi}{4}$
62. $4 \sin \frac{\pi}{3} \cos \frac{5\pi}{6}$
63. $\sin 5\theta \cos 3\theta$
64. $3 \sin 2\alpha \sin 3\alpha$
65. $5 \cos(-5\beta) \cos 3\beta$
66. $\cos 2\theta \cos 4\theta$

67. $\sin(x + y)\sin(x - y)$

68. $\sin(x + y)\cos(x - y)$

69. $\sin(\theta + \pi)\cos(\theta - \pi)$

70. $10\cos 75° \cos 15°$

In Exercises 71–80, use the sum-to-product formulas to write the sum or difference as a product.

71. $\sin 60° + \sin 30°$

72. $\cos 120° + \cos 30°$

73. $\cos\frac{3\pi}{4} - \cos\frac{\pi}{4}$

74. $\sin 5\theta - \sin 3\theta$

75. $\cos 6x + \cos 2x$

76. $\sin x + \sin 5x$

77. $\sin(\alpha + \beta) - \sin(\alpha - \beta)$

78. $\cos\left(\theta + \frac{\pi}{2}\right) - \cos\left(\theta - \frac{\pi}{2}\right)$

79. $\cos(\phi + 2\pi) + \cos\phi$

80. $\sin\left(x + \frac{\pi}{2}\right) + \sin\left(x - \frac{\pi}{2}\right)$

In Exercises 81–84, find the exact zeros of the function in the interval $[0, 2\pi)$. Use a graphing utility to graph the function and verify the zeros.

81. $g(x) = \sin 6x + \sin 2x$

82. $h(x) = \cos 2x - \cos 6x$

83. $f(x) = \dfrac{\cos 2x}{\sin 3x - \sin x} - 1$

84. $f(x) = \sin^2 3x - \sin^2 x$

In Exercises 85–88, use the figure to find the exact value of the trigonometric function in two ways.

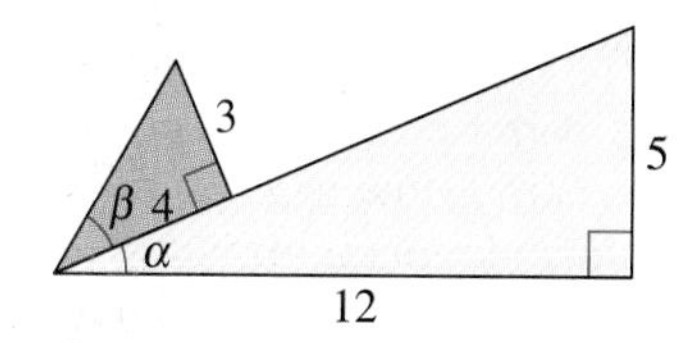

85. $\sin^2\alpha$

86. $\cos^2\alpha$

87. $\sin\alpha\cos\beta$

88. $\cos\alpha\sin\beta$

In Exercises 89–102, verify the identity.

89. $\csc 2\theta = \dfrac{\csc\theta}{2\cos\theta}$

90. $\sec 2\theta = \dfrac{\sec^2\theta}{2 - \sec^2\theta}$

91. $\cos^2 2\alpha - \sin^2 2\alpha = \cos 4\alpha$

92. $\cos^4 x - \sin^4 x = \cos 2x$

93. $(\sin x + \cos x)^2 = 1 + \sin 2x$

94. $\sin\frac{\alpha}{3}\cos\frac{\alpha}{3} = \frac{1}{2}\sin\frac{2\alpha}{3}$

95. $1 + \cos 10y = 2\cos^2 5y$

96. $\dfrac{\cos 3\beta}{\cos\beta} = 1 - 4\sin^2\beta$

97. $\sec\frac{u}{2} = \pm\sqrt{\dfrac{2\tan u}{\tan u + \sin u}}$

98. $\tan\frac{u}{2} = \csc u - \cot u$

99. $\dfrac{\cos 4x + \cos 2x}{\sin 4x + \sin 2x} = \cot 3x$

100. $\dfrac{\sin x \pm \sin y}{\sin x + \cos y} = \tan\dfrac{x \pm y}{2}$

101. $\dfrac{\cos t + \cos 3t}{\sin 3t - \sin t} = \cot t$

102. $\sin\left(\frac{\pi}{6} + x\right) + \sin\left(\frac{\pi}{6} - x\right) = \cos x$

In Exercises 103–106, verify the identity algebraically. Use a graphing utility to confirm the identity.

103. $\cos 3\beta = \cos^3\beta - 3\sin^2\beta\cos\beta$

104. $\sin 4\beta = 4\sin\beta\cos\beta(1 - 2\sin^2\beta)$

105. $(\cos 4x - \cos 2x)/(2\sin 3x) = -\sin x$

106. $(\cos 3x - \cos x)/(\sin 3x - \sin x) = -\tan 2x$

In Exercises 107 and 108, graph the function by using the power-reducing formulas.

107. $f(x) = \sin^2 x$

108. $f(x) = \cos^2 x$

In Exercises 109 and 110, (a) use a graphing utility to graph the function and approximate the maximum and minimum points on the graph in the interval $[0, 2\pi)$. (b) Solve the trigonometric equation and verify that its solutions are the x-coordinates of the maximum and minimum points of f. (Calculus is required to find the trigonometric equation.)

	Function	*Trigonometric Equation*
109.	$f(x) = 4 \sin \frac{x}{2} + \cos x$	$2 \cos \frac{x}{2} - \sin x = 0$
110.	$f(x) = \cos 2x - 2 \sin x$	$-2 \cos x(2 \sin x + 1) = 0$

111. ***Conjecture*** Consider the function

$$f(x) = 2 \sin x[2 \cos^2(x/2) - 1].$$

(a) Use a graphing utility to graph the function.

(b) Make a conjecture about the function that is an identity with f.

(c) Verify your conjecture analytically.

112. ***Exploration*** Consider the function

$$f(x) = \sin^4 x + \cos^4 x.$$

(a) Use the power-reducing formulas to write the function in terms of cosine to the first power.

(b) Determine another way of rewriting the function. Use a graphing utility to rule out incorrectly rewritten functions.

(c) Add a trigonometric term to the function so that it becomes a perfect square trinomial. Rewrite the function as a perfect square trinomial minus the term that you added. Use a graphing utility to rule out incorrectly rewritten functions.

(d) Rewrite the result of part (c) in terms of the sine of a double angle. Use a graphing utility to rule out incorrectly rewritten functions.

(e) When you rewrite a trigonometric expression, the result may not be the same as a friend's. Does this mean that one of you is wrong? Explain.

In Exercises 113 and 114, write the trigonometric expression as an algebraic expression.

113. $\sin(2 \arcsin x)$ **114.** $\cos(2 \arccos x)$

115. ***Area*** The length of each of the two equal sides of an isosceles triangle is 10 meters (see figure). The angle between the two sides is θ.

(a) Express the area of the triangle as a function of $\theta/2$.

(b) Express the area as a function of θ. Determine the value of θ so that the area is a maximum.

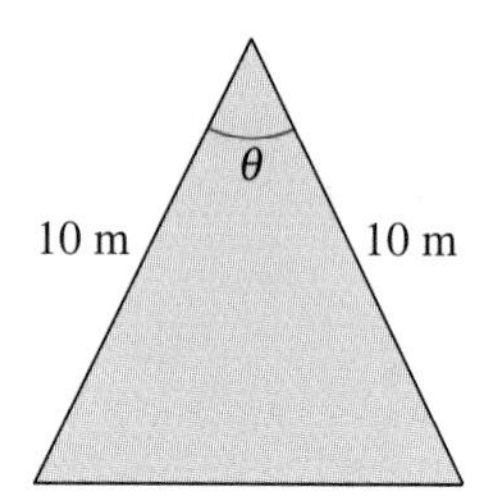

116. ***Projectile Motion*** The range of a projectile fired at an angle θ with the horizontal and with an initial velocity of v_0 feet per second is given by

$$r = \frac{1}{32} v_0^2 \sin 2\theta$$

where r is measured in feet. Determine the expression for the range in terms of θ.

Review **Solve Exercises 117–120 as a review of the skills and problem-solving techniques you learned in previous sections.**

117. The total profit for a company in October was 16% higher than it was in September. The total profit for the two months was \$507,600. Find the profit for each month.

118. Two cars start at a given point and travel in the same direction at average speeds of 48 miles per hour and 56 miles per hour. How much time must elapse before the two cars are 12 miles apart?

119. A 55-gallon barrel contains a mixture with a concentration of 30%. How much of this mixture must be withdrawn and replaced by 100% concentrate to bring the mixture up to 50% concentration?

120. ***Baseball Diamond*** A baseball diamond has the shape of a square where the distance between each of the consecutive bases is 90 feet. Approximate the distance from home plate to second base.

Focus on Concepts

In this chapter, you studied the fundamental identities of trigonometry. Use the following questions to check your understanding of several of these basic concepts presented. The answers to these questions are given in the back of the book.

1. In your own words, describe the difference between an identity and a conditional equation.
2. Describe the difference between verifying an identity and solving an equation.
3. List the reciprocal identities, quotient identities, and Pythagorean identities from memory.
4. Is $\cos\theta = \sqrt{1 - \sin^2\theta}$ an identity? Explain.
5. *True or False?* Usually there is only one correct set of steps to verify an identity. Explain.
6. By observation, determine which of the following is an identity. Explain.
   (a) $\tan(\theta + \pi) \stackrel{?}{=} \tan\theta$
   (b) $\cos(\theta + \pi) \stackrel{?}{=} \cos\theta$
   (c) $\sec\theta\csc\theta \stackrel{?}{=} 1$
   (d) $\tan\theta\cot\theta \stackrel{?}{=} 1$
   (e) $\sin(\theta - \pi) \stackrel{?}{=} -\sin(\pi - \theta)$

In Exercises 7 and 8, use the graph of y_1 and y_2 to determine how to change one function to form the identity $y_1 = y_2$.

7. $y_1 = \sec^2\left(\dfrac{\pi}{2} - x\right)$
   $y_2 = \cot^2 x$

8. $y_1 = \dfrac{\cos 3x}{\cos x}$
   $y_2 = (2\sin x)^2$

In Exercises 9 and 10, use the graph to determine the number of points of intersection of the graphs of y_1 and y_2.

9. $y_1 = 2\sin x$
   $y_2 = 3x + 1$

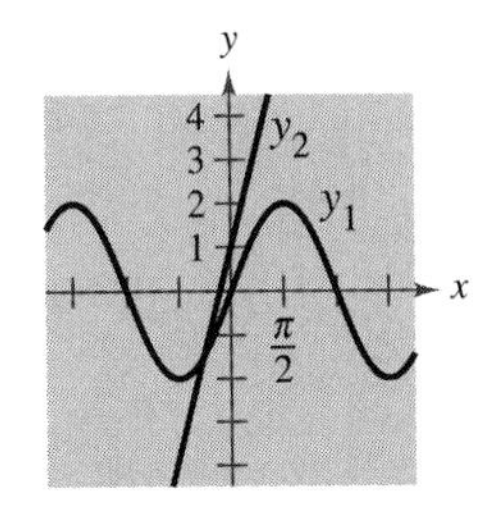

10. $y_1 = 2\sin x$
    $y_2 = \frac{1}{2}x + 1$

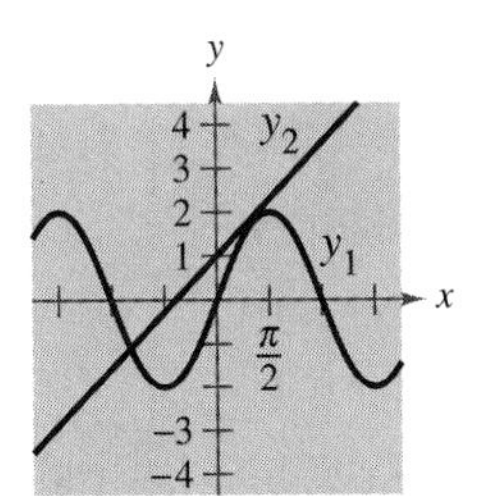

In Exercises 11 and 12, use the graph to determine the number of zeros of the function.

11. $y = \sqrt{x + 3} + 4\cos x$

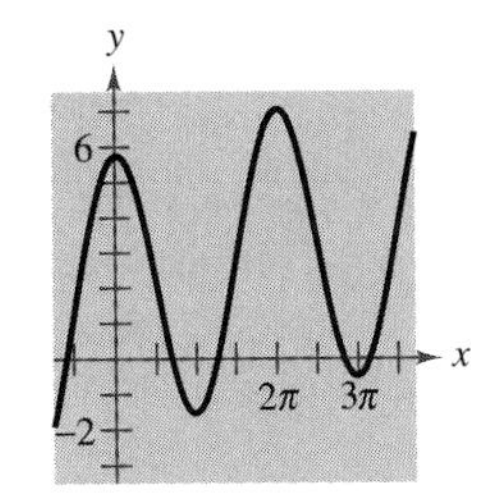

12. $y = 2 - \dfrac{1}{2}x^2 + 3\sin\dfrac{\pi x}{2}$

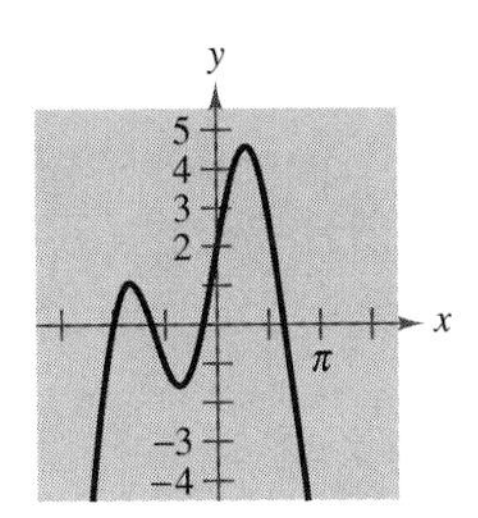

13. Sales of a product are seasonal and can be modeled by the function $y = a + bt + c\sin(dt + e)$, where t is the time in years. What is the value of d?

Review Exercises

In Exercises 1–8, simplify the trigonometric expression.

1. $\dfrac{1}{\cot^2 x + 1}$

2. $\dfrac{\sin 2\alpha}{\cos^2 \alpha - \sin^2 \alpha}$

3. $\dfrac{\sin^2 \alpha - \cos^2 \alpha}{\sin^2 \alpha - \sin \alpha \cos \alpha}$

4. $\dfrac{\sin^3 \beta + \cos^3 \beta}{\sin \beta + \cos \beta}$

5. $\tan^2 \theta(\csc^2 \theta - 1)$

6. $1 - 4 \sin^2 x \cos^2 x$

7. $\dfrac{2 \tan(x + 1)}{1 - \tan^2(x + 1)}$

8. $\sqrt{\dfrac{1 - \cos^2 x}{1 + \cos x}}$

In Exercises 9–26, verify the identity.

9. $\tan x(1 - \sin^2 x) = \frac{1}{2} \sin 2x$

10. $\cos x(\tan^2 x + 1) = \sec x$

11. $\sec^2 x \cot x - \cot x = \tan x$

12. $\sin^3 \theta + \sin \theta \cos^2 \theta = \sin \theta$

13. $\sin^5 x \cos^2 x = (\cos^2 x - 2 \cos^4 x + \cos^6 x)\sin x$

14. $\cos^3 x \sin^2 x = (\sin^2 x - \sin^4 x)\cos x$

15. $\sin 3\theta \sin \theta = \frac{1}{2}(\cos 2\theta - \cos 4\theta)$

16. $\sin 3x \cos 2x = \frac{1}{2}(\sin 5x + \sin x)$

17. $\sqrt{\dfrac{1 - \sin \theta}{1 + \sin \theta}} = \dfrac{1 - \sin\theta}{|\cos\theta|}$

18. $\sqrt{1 - \cos x} = \dfrac{|\sin x|}{\sqrt{1 + \cos x}}$

19. $\cos 3x = 4 \cos^3 x - 3 \cos x$

20. $\cos(x + \pi/2) = -\sin x$

21. $\cot(\pi/2 - x) = \tan x$

22. $\sin(\pi - x) = \sin x$

23. $\dfrac{\sec x - 1}{\tan x} = \tan \dfrac{x}{2}$

24. $\dfrac{2 \cos 3x}{\sin 4x - \sin 2x} = \csc x$

25. $2 \sin y \cos y \sec 2y = \tan 2y$

26. $\dfrac{\sin(\alpha + \beta)}{\cos \alpha \cos \beta} = \tan \alpha + \tan \beta$

In Exercises 27–30, verify the identity algebraically and use a graphing utility to confirm it graphically.

27. $\sin\left(x - \dfrac{3\pi}{2}\right) = \cos x$

28. $\sin 4x = 8 \cos^3 x \sin x - 4 \cos x \sin x$

29. $\tan^2 x = \dfrac{1 - \cos 2x}{1 + \cos 2x}$

30. $\cos^2 5x - \cos^2 x = -\sin 4x \sin 6x$

In Exercises 31–34, find the exact value of the trigonometric function by using the sum, difference, or half-angle formulas.

31. $\sin \dfrac{5\pi}{12} = \sin\left(\dfrac{2\pi}{3} - \dfrac{\pi}{4}\right)$

32. $\cos 285° = \cos(225° + 60°)$

33. $\cos(157° 30') = \cos \dfrac{315°}{2}$

34. $\sin \dfrac{3\pi}{8} = \sin\left[\dfrac{1}{2}\left(\dfrac{3\pi}{4}\right)\right]$

In Exercises 35–40, find the exact value of the trigonometric function given that $\sin u = \frac{3}{4}$, $\cos v = -\frac{5}{13}$, and u and v are in Quadrant II.

35. $\sin(u + v)$

36. $\tan(u + v)$

37. $\cos(u - v)$

38. $\sin 2v$

39. $\cos \dfrac{u}{2}$

40. $\tan 2v$

True or False? **In Exercises 41–44, determine if the statement is true or false. If it is false, make the necessary correction.**

41. If $\dfrac{\pi}{2} < \theta < \pi$, then $\cos \dfrac{\theta}{2} < 0$.

42. $\sin(x + y) = \sin x + \sin y$

43. $4 \sin(-x) \cos(-x) = -2 \sin 2x$

44. $4 \sin 45° \cos 15° = 1 + \sqrt{3}$

In Exercises 45–50, find all solutions of the equation in the interval $[0, 2\pi)$.

45. $\sin x - \tan x = 0$

46. $\csc x - 2 \cot x = 0$

47. $\sin 2x + \sqrt{2} \sin x = 0$

48. $\cos 4x - 7 \cos 2x = 8$

49. $\cos^2 x + \sin x = 1$

50. $\sin 4x - \sin 2x = 0$

In Exercises 51–54, use a graphing utility to graph the function and approximate its zeros in the interval $[0, 2\pi)$. If possible, find the exact values of the zeros algebraically.

51. $y = \dfrac{1 + \sin x}{\cos x} + \dfrac{\cos x}{1 + \sin x} - 4$

52. $y = \cos x - \cos \dfrac{x}{2}$

53. $y = \tan^3 x - \tan^2 x + 3 \tan x - 3$

54. $h(s) = \sin s + \sin 3s + \sin 5s$

55. ***Think About It*** If a trigonometric equation has an infinite number of solutions, is it true that the equation is an identity? Explain.

56. ***Think About It*** Explain why you know from observation that the equation $a \sin x - b = 0$ has no solution if $|a| < |b|$.

In Exercises 57 and 58, write the trigonometric expression as a product.

57. $\cos 3\theta + \cos 2\theta$

58. $\sin(x + \pi/4) - \sin(x - \pi/4)$

In Exercises 59 and 60, write the trigonometric expression as a sum or difference.

59. $\sin 3\alpha \sin 2\alpha$

60. $\cos \dfrac{x}{2} \cos \dfrac{x}{4}$

In Exercises 61 and 62, write the trigonometric expression as an algebraic expression.

61. $\cos(2 \arccos 2x)$

62. $\sin(2 \arctan x)$

63. ***Rate of Change*** The rate of change of the function $f(x) = 2\sqrt{\sin x}$ is given by the expression $\sin^{-1/2} x \cos x$. Show that this expression can also be written as $\cot x \sqrt{\sin x}$.

64. ***Projectile Motion*** A baseball leaves the hand of the person at first base at an angle of θ with the horizontal and an initial velocity of $v_0 = 80$ feet per second. The ball is caught by the person at second base 100 feet away. Find θ if the range r of a projectile is given by

$$r = \frac{1}{32} v_0^2 \sin 2\theta.$$

65. ***Harmonic Motion*** A weight is attached to a spring suspended vertically from a ceiling. When a driving force is applied to the system, the weight moves vertically from its equilibrium position, and this motion is described by the model

$$y = 1.5 \sin 8t - 0.5 \cos 8t$$

where y is the distance from equilibrium measured in feet and t is the time in seconds.

(a) Write the model in the form

$$y = \sqrt{a^2 + b^2} \sin(Bt + C).$$

(b) Find the amplitude of the oscillations of the weight.

(c) Find the frequency of the oscillations of the weight.

66. ***Volume*** A trough for feeding cattle is 4 meters long and its cross sections are isosceles triangles with the two equal sides being $\frac{1}{2}$ meter (see figure). The angle between the two sides is θ.

(a) Express the trough's volume as a function of $\theta/2$.

(b) Express the volume of the trough as a function of θ and determine the value of θ so that the volume is maximum.

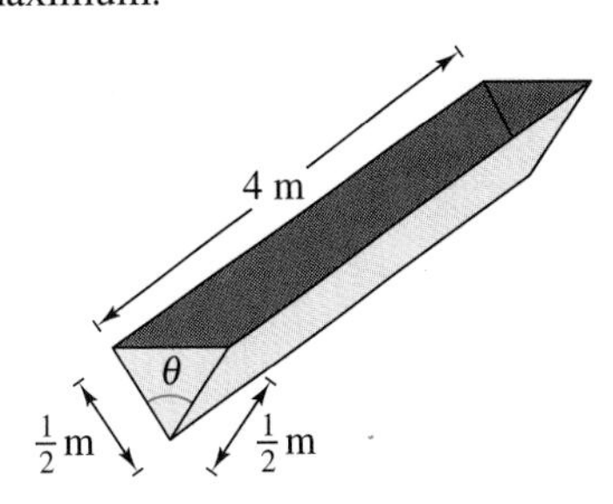

CHAPTER PROJECT: *Solving Equations Graphically*

Equations that involve both algebraic and trigonometric expressions are often difficult to solve analytically. For instance, how would you solve the equation

$$x = \cos x?$$

None of the standard techniques, such as factoring or using a trigonometric identity, can be used to solve this equation. In such cases, a graphing utility can be used to approximate the solution.

EXAMPLE 1 *Approximating Solutions of an Equation*

Approximate all solutions of $x = \cos x$.

Solution

There are several ways that a graphing utility can be used to solve this equation. One way is to graph

$$y = x \quad \text{and} \quad y = \cos x \qquad \text{First method}$$

on the same viewing rectangle and use the trace feature of the graphing utility to approximate the x-coordinate of the point of intersection. From the graph shown below on the left, you can see that the two graphs intersect when x is approximately 0.74. More accuracy can be obtained by using the zoom feature of the graphing utility. Another way to solve the equation is to collect all nonzero terms on one side of the equation, $x - \cos x = 0$. Then, use a graphing utility to graph

$$y = x - \cos x \qquad \text{Second method}$$

and approximate the zeros of the function. By using a viewing rectangle in which $0.735 \le x \le 0.745$ and $-0.01 \le y \le 0.01$, you can obtain the graph shown on the right. From this graph, you can see that the solution $x \approx 0.739$, which is accurate to three decimal places.

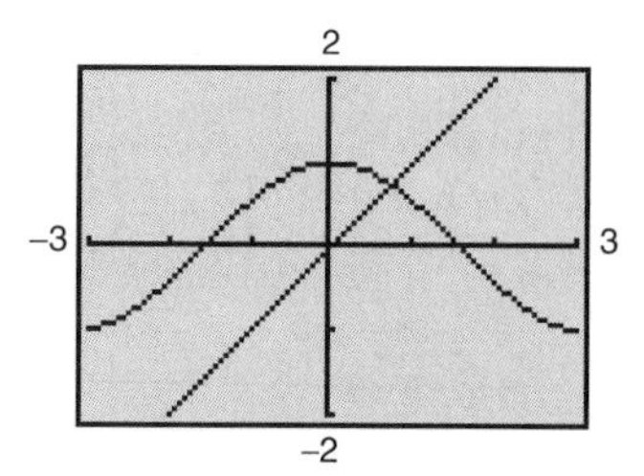

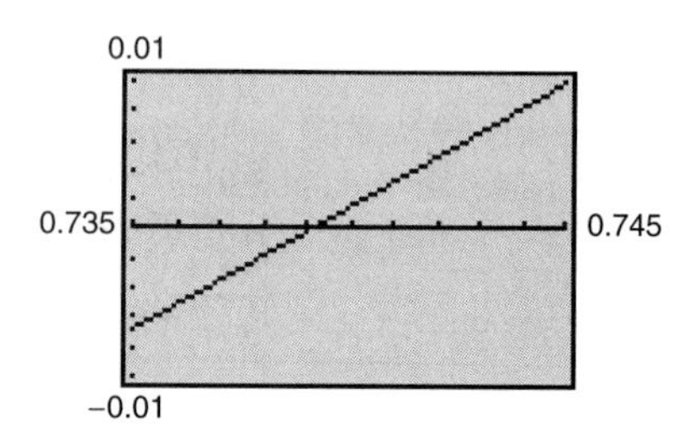

EXAMPLE 2 ***Approximating Solutions of an Equation***

Approximate all solutions of $0.5x + 1.13 = \cos x$.

Solution

When the graphs of $y = 0.5x + 1.13$ and $y = \cos x$ are sketched on the same viewing rectangle, it is clear that one solution occurs when $x \approx -3.8$. From the screen shown below on the left, however, it is unclear whether the two graphs have other points of intersection near $x = -0.5$. To determine whether there are other points of intersection, use the zoom feature. After doing this, you can see that, for x-values near -0.5, the line lies above the cosine curve. Thus, there is only one point of intersection, which occurs when $x \approx -3.819$.

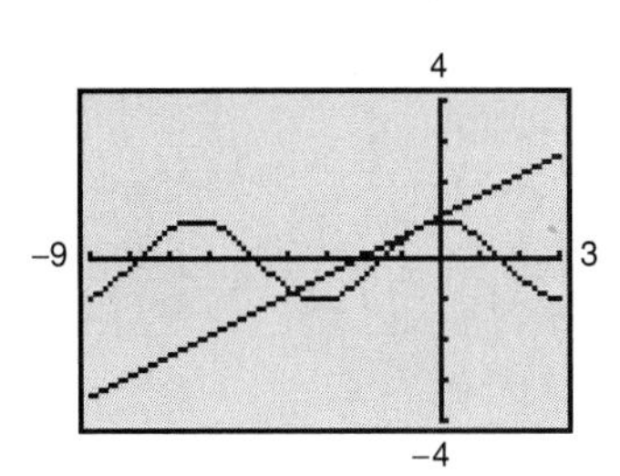

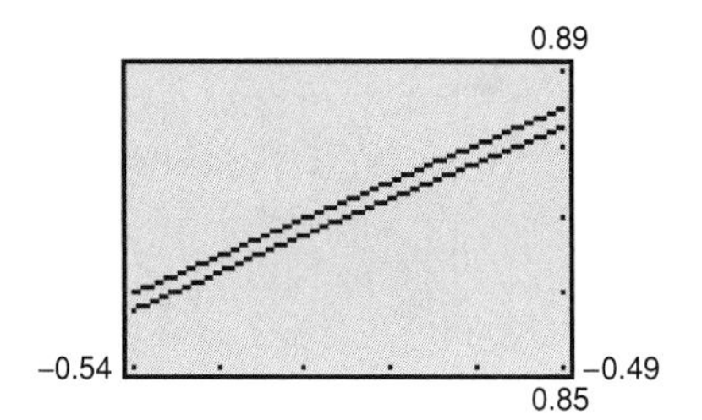

CHAPTER PROJECT INVESTIGATIONS

In Questions 1–4, sketch the graphs of all three functions on the same coordinate plane. Which two graphs are the same? Identify the trigonometric identity that is supported by your discovery.

1. (a) $y = \sin^2 x$
(b) $y = \frac{1}{2}(1 - \cos 2x)$
(c) $y = \frac{1}{2}(1 + \cos 2x)$

2. (a) $y = 2\cos^2 x$
(b) $y = 1 - \cos 2x$
(c) $y = 1 + \cos 2x$

3. (a) $y = 2 \sin x \cos 2x$
(b) $y = \sin 3x + \sin x$
(c) $y = \sin 3x - \sin x$

4. (a) $y = \sin 2x$
(b) $y = 2 \sin x$
(c) $y = 2 \sin x \cos x$

In Exercises 5 and 6, without finding the solutions, decide whether the equation has infinitely many solutions or a finite number of solutions. Explain.

5. $\sin 2x = x$

6. $\tan 2x = x$

In Questions 7–12, use a graphing utility to approximate all solutions of the equation. List the solutions correct to three decimal places.

7. $x + \sin x = 1$

8. $x^2 + \cos x = 2$

9. $5 \cos \dfrac{1}{x^2 + 1} = 3$

10. $|x| + \sec \dfrac{1}{x^2 + 1} = 3$

11. $x + 1.25 = \arctan x$

12. $(x^2 + 1) \cos x = 1$

13. *Seasonal Sales* During 1994, the national monthly sales S (in thousands of units) of a lawn furniture company can be modeled by

$$S = 74.50 + 43.75 \sin \frac{\pi t}{6}$$

where t represents the time in months, with $t = 0$ corresponding to January 1. During which months did sales exceed 100,000 units?

Chapter Test

Take this test as you would take a test in class. After you are done, check your work against the answers given in the back of the book.

The *Interactive* CD-ROM provides answers to the Chapter Tests and Cumulative Tests. It also offers Chapter Pre-Tests (which test key skills and concepts covered in previous chapters) and Chapter Post-Tests, both of which have randomly generated exercises with diagnostic capabilities.

1. If $\tan \theta = \frac{3}{2}$ and $\cos \theta < 0$, use the fundamental identities to evaluate the other five trigonometric functions of θ.
2. Use the fundamental identities to simplify $\csc^2 \beta(1 - \cos^2 \beta)$.
3. Factor and simplify $\dfrac{\sec^4 x - \tan^4 x}{\sec^2 x + \tan^2 x}$.
4. Add and simplify $\dfrac{\cos \theta}{\sin \theta} + \dfrac{\sin \theta}{\cos \theta}$.
5. Determine the values of θ, $0 \le \theta < 2\pi$, for which $\tan \theta = -\sqrt{\sec^2 \theta - 1}$ is true.
6. Use a graphing utility to graph the functions $y_1 = \cos x + \sin x \tan x$ and $y_2 = \sec x$. Make a conjecture about y_1 and y_2. Verify the result analytically.

In Exercises 7–12, verify the identity.

7. $\sin \theta \sec \theta = \tan \theta$
8. $\sec^2 x \tan^2 x + \sec^2 x = \sec^4 x$
9. $\dfrac{\csc \alpha + \sec \alpha}{\sin \alpha + \cos \alpha} = \cot \alpha + \tan \alpha$
10. $\cos\left(x + \dfrac{\pi}{2}\right) = -\sin x$
11. $\sin(n\pi + \theta) = (-1)^n \sin \theta$, n is an integer.
12. $(\sin x + \cos x)^2 = 1 + \sin 2x$

In Exercises 13–16, find all solutions of the equation in the interval $[0, 2\pi)$.

13. $\tan^2 x + \tan x = 0$
14. $\sin 2\alpha - \cos \alpha = 0$
15. $4 \cos^2 x - 3 = 0$
16. $\csc^2 x - \csc x - 2 = 0$
17. Use a graphing utility to approximate the solutions of the equation $3 \cos x - x = 0$ accurate to three decimal places.
18. Explain why the equation $\cos^2 x + \cos x - 6 = 0$ has no solution.
19. Find the exact value of $\cos 105°$ using the fact that $105° = 135° - 30°$.
20. Use the figure to find the exact values of $\sin 2u$ and $\tan 2u$.

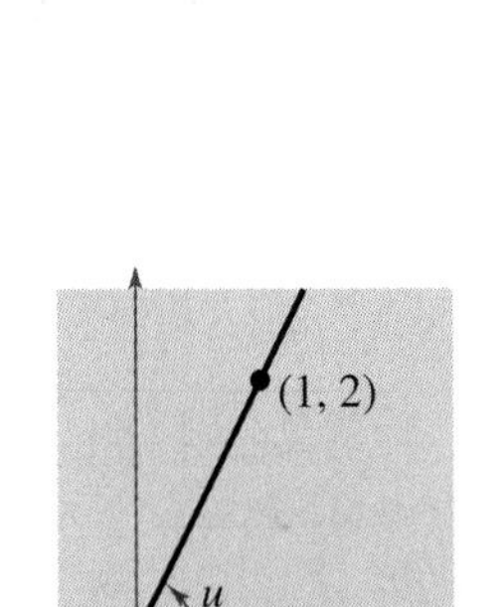

FIGURE FOR 20

3 Additional Topics in Trigonometry

Before a bridge, road, tunnel, or other large structure can be designed and built, the land on which it will be built needs to be surveyed. This is one of the jobs of a civil engineer.

To survey a site, civil engineers rely heavily on trigonometry. One of the primary skills a surveyor must master is that of indirect measurement. For instance, the distances a and c in the diagram below can be found using the Law of Sines.

$$\frac{a}{\sin A} = \frac{b}{\sin B}$$

$$\frac{a}{\sin 69^\circ} = \frac{15.6}{\sin 55^\circ}$$

$$a \approx 17.779$$

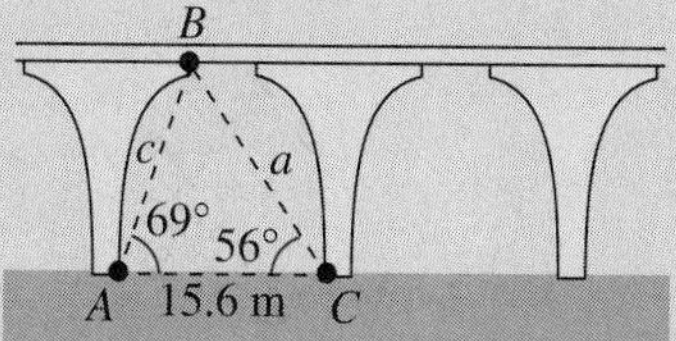

See Exercises 89 and 90 on page 300.

Photos: Courtesy of Keystone Aerial Survey; BET Consultants (inset)

Julio Esquivel is a civil engineer who specializes in surveying. He often uses aerial photographs such as this one to create topographic maps.

3.1 Law of Sines

See Exercise 33 on page 276 for an example of how the Law of Sines can be used to help locate a forest fire.

Introduction ❑ *The Ambiguous Case (SSA)* ❑ *Area of an Oblique Triangle* ❑ *Application*

Introduction

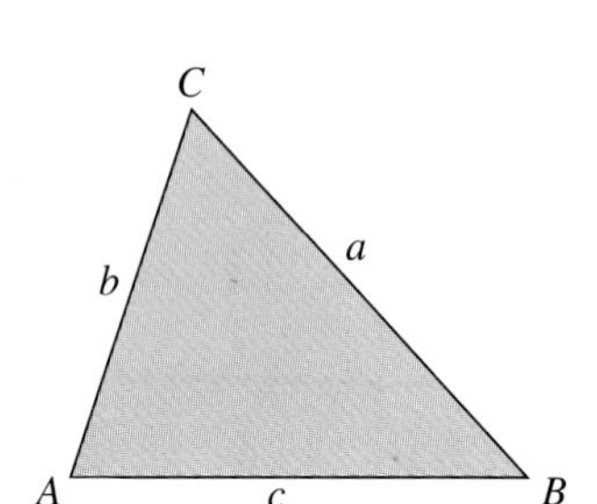

FIGURE 3.1

In Chapter 1 you looked at techniques for solving right triangles. In this section and the next, you will solve **oblique triangles**—triangles that have no right angles. As standard notation, the angles of a triangle are labeled as A, B, and C, and their opposite sides as a, b, and c, as shown in Figure 3.1.

To solve an oblique triangle, you need to know the measure of at least one side and any two other parts of the triangle—either two sides, two angles, or one angle and one side. This breaks down into the following four cases.

1. Two angles and any side (AAS or ASA)
2. Two sides and an angle opposite one of them (SSA)
3. Three sides (SSS)
4. Two sides and their included angle (SAS)

The first two cases can be solved using the **Law of Sines,** whereas the last two cases require the **Law of Cosines** (Section 3.2).

LAW OF SINES

If ABC is a triangle with sides a, b, and c, then

$$\frac{a}{\sin A} = \frac{b}{\sin B} = \frac{c}{\sin C}.$$

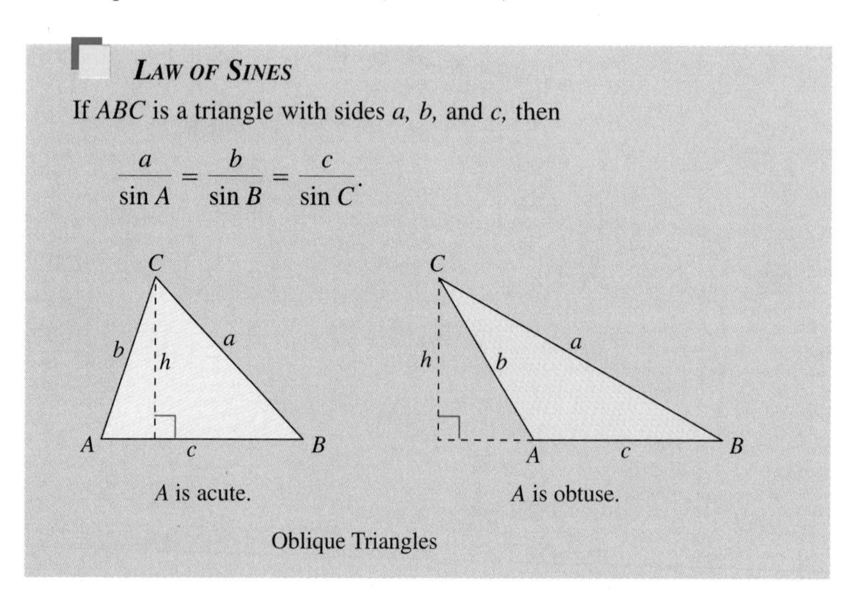

A is acute. A is obtuse.

Oblique Triangles

THINK ABOUT THE PROOF

To prove the Law of Sines, let h be the altitude of either triangle shown in the figure at the right. Then you have

$$\sin A = \frac{h}{b} \text{ or } h = b \sin A$$

$$\sin B = \frac{h}{a} \text{ or } h = a \sin B.$$

By equating the two values of h, you can establish part of the Law of Sines. Can you see how to establish the other part? The details of the proof are given in the appendix.

NOTE The Law of Sines can also be written in the reciprocal form

$$\frac{\sin A}{a} = \frac{\sin B}{b} = \frac{\sin C}{c}.$$

EXAMPLE 1 Given Two Angles and One Side—AAS

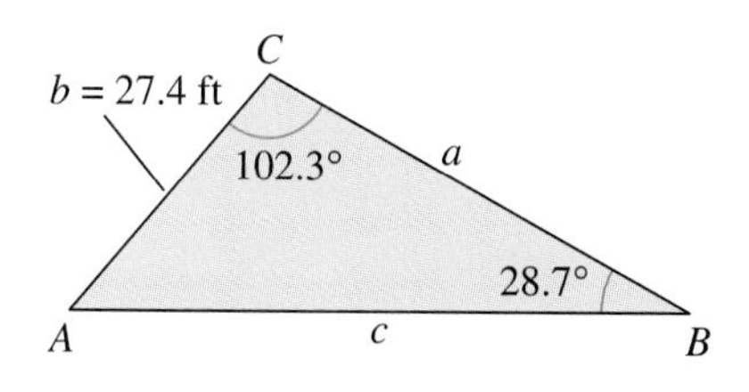

FIGURE 3.2

For the triangle in Figure 3.2, $C = 102.3°$, $B = 28.7°$, and $b = 27.4$ feet. Find the remaining angle and sides.

Solution

The third angle of the triangle is

$$A = 180° - B - C = 180° - 28.7° - 102.3° = 49.0°.$$

By the Law of Sines, you have

$$\frac{a}{\sin 49°} = \frac{b}{\sin 28.7°} = \frac{c}{\sin 102.3°}.$$

Using $b = 27.4$ produces

$$a = \frac{27.4}{\sin 28.7°}(\sin 49°) \approx 43.06 \text{ feet}$$

and

$$c = \frac{27.4}{\sin 28.7°}(\sin 102.3°) \approx 55.75 \text{ feet.}$$

Study Tip

When solving triangles, a careful sketch is useful as a quick test for the feasibility of an answer. Remember that the longest side lies opposite the largest angle, and the shortest side lies opposite the smallest angle.

EXAMPLE 2 Given Two Angles and One Side—ASA

A pole tilts *toward* the sun at an 8° angle from the vertical, and it casts a 22-foot shadow. The angle of elevation from the tip of the shadow to the top of the pole is 43°. How tall is the pole?

Solution

From Figure 3.3, note that $A = 43°$ and $B = 90° + 8° = 98°$. Thus, the third angle is

$$C = 180° - A - B = 180° - 43° - 98° = 39°.$$

By the Law of Sines, you have

$$\frac{a}{\sin 43°} = \frac{c}{\sin 39°}.$$

Because $c = 22$ feet, the length of the pole is

$$a = \frac{22}{\sin 39°}(\sin 43°) \approx 23.84 \text{ feet.}$$

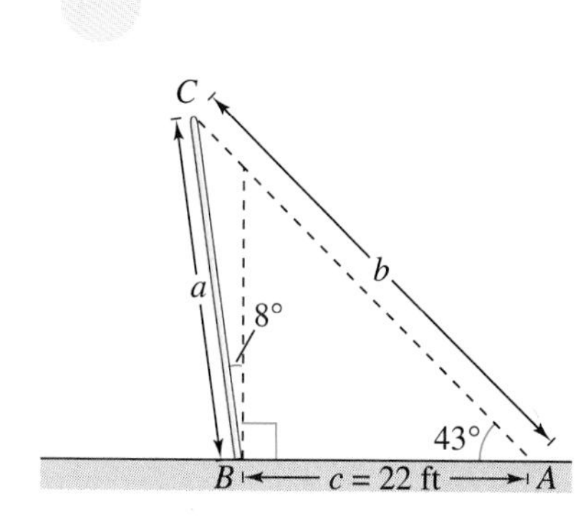

FIGURE 3.3

NOTE For practice, try reworking Example 2 for a pole that tilts *away from* the sun under the same conditions.

The Ambiguous Case (SSA)

In Examples 1 and 2 you saw that two angles and one side determine a unique triangle. However, if two sides and one opposite angle are given, three possible situations can occur: (1) no such triangle exists, (2) one such triangle exists, or (3) two distinct triangles may satisfy the conditions.

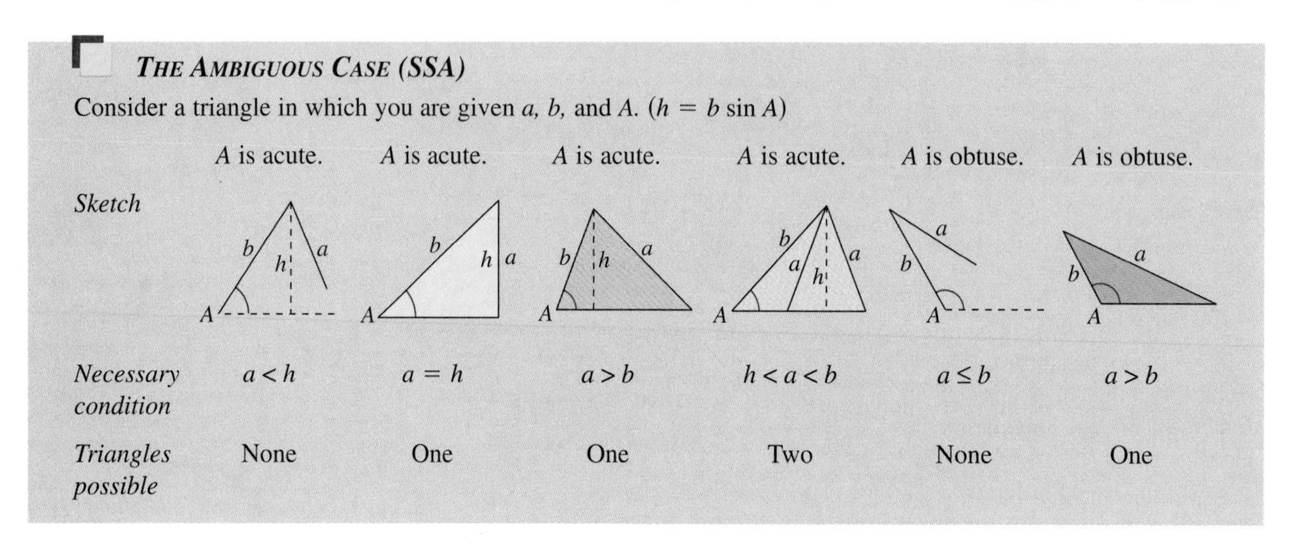

The Ambiguous Case (SSA)

Consider a triangle in which you are given a, b, and A. ($h = b \sin A$)

	A is acute.	A is acute.	A is acute.	A is acute.	A is obtuse.	A is obtuse.
Sketch						
Necessary condition	$a < h$	$a = h$	$a > b$	$h < a < b$	$a \leq b$	$a > b$
Triangles possible	None	One	One	Two	None	One

Example 3 ***Single-Solution Case—SSA***

For the triangle in Figure 3.4, $a = 22$ inches, $b = 12$ inches, and $A = 42°$. Find the remaining side and angles.

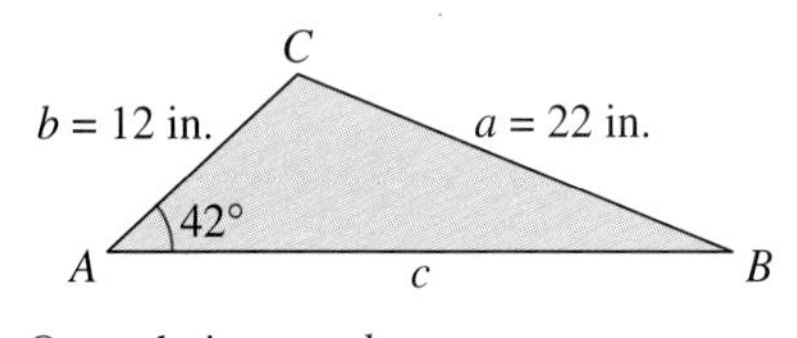

One solution: $a > b$

Figure 3.4

Solution

By the Law of Sines, you have

$$\frac{22}{\sin 42°} = \frac{12}{\sin B}$$

$$\sin B = 12\left(\frac{\sin 42°}{22}\right) \approx 0.3649803$$

$$B \approx 21.41° \qquad B \text{ is acute.}$$

Now, you can determine that $C \approx 180° - 42° - 21.41° = 116.59°$, and the remaining side is given by

$$\frac{c}{\sin 116.59°} = \frac{22}{\sin 42°}$$

$$c = \sin 116.59°\left(\frac{22}{\sin 42°}\right) \approx 29.40 \text{ inches.}$$

EXAMPLE 4 *No-Solution Case—SSA*

Show that there is no triangle for which $a = 15$, $b = 25$, and $A = 85°$.

Solution

Begin by making the sketch shown in Figure 3.5. From this figure it appears that no triangle is formed. You can verify this using the Law of Sines.

$$\frac{a}{\sin A} = \frac{b}{\sin B}$$

$$\frac{15}{\sin 85°} = \frac{25}{\sin B}$$

$$\sin B = 25\left(\frac{\sin 85°}{15}\right) \approx 1.660 > 1$$

This contradicts the fact that $|\sin B| \leq 1$. Hence, no triangle can be formed having sides $a = 15$ and $b = 25$ and an angle of $A = 85°$.

No solution: $a < h$

FIGURE 3.5

EXAMPLE 5 *Two-Solution Case—SSA*

Find two triangles for which $a = 12$ meters, $b = 31$ meters, and $A = 20.5°$.

Solution

By the Law of Sines, you have

$$\frac{a}{\sin A} = \frac{b}{\sin B}$$

$$\sin B = b\left(\frac{\sin A}{a}\right) = 31\left(\frac{\sin 20.5°}{12}\right) \approx 0.9047.$$

There are two angles $B_1 \approx 64.8°$ and $B_2 \approx 115.2°$ between $0°$ and $180°$ whose sine is 0.9047. For $B_1 \approx 64.8°$, you obtain

$$C \approx 180° - 20.5° - 64.8° = 94.7°$$

$$c = \frac{a}{\sin A}(\sin C) = \frac{12}{\sin 20.5°}(\sin 94.7°) \approx 34.15 \text{ meters.}$$

For $B_2 \approx 115.2°$, you obtain

$$C \approx 180° - 20.5° - 115.2° = 44.3°$$

$$c = \frac{a}{\sin A}(\sin C) = \frac{12}{\sin 20.5°}(\sin 44.3°) \approx 23.93 \text{ meters.}$$

The resulting triangles are shown in Figure 3.6.

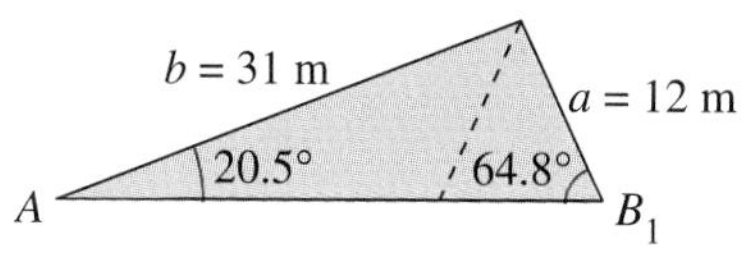

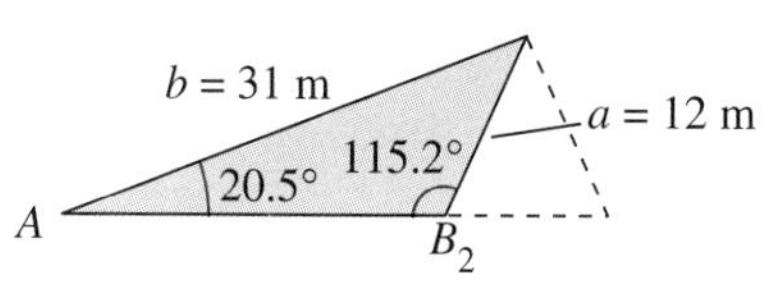

FIGURE 3.6

Area of an Oblique Triangle

The procedure used to prove the Law of Sines leads to a simple formula for the area of an oblique triangle. Referring to Figure 3.7, note that each triangle has a height of $h = b \sin A$. Consequently, the area of each triangle is given by

$$\text{Area} = \frac{1}{2}(\text{base})(\text{height}) = \frac{1}{2}(c)(b \sin A) = \frac{1}{2}bc \sin A.$$

By similar arguments, you can develop the formulas

$$\text{Area} = \frac{1}{2}ab \sin C = \frac{1}{2}ac \sin B.$$

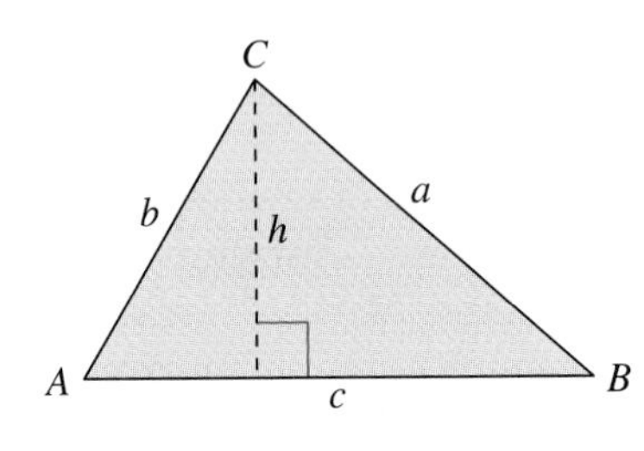

A is acute.

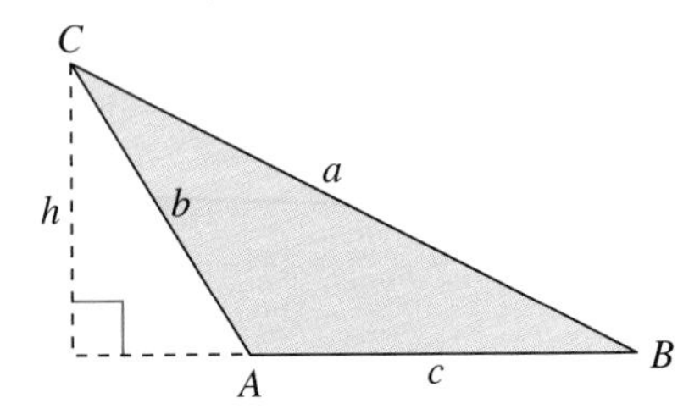

A is obtuse.

FIGURE 3.7

NOTE Note that if angle A is 90°, the formula gives the area for a right triangle:

$$\text{Area} = \frac{1}{2}bc = \frac{1}{2}(\text{base})(\text{height}).$$

Similar results are obtained for angles C and B equal to 90°.

AREA OF AN OBLIQUE TRIANGLE

The area of any triangle is given by one-half the product of the lengths of two sides times the sine of their included angle. That is,

$$\text{Area} = \frac{1}{2}bc \sin A = \frac{1}{2}ab \sin C = \frac{1}{2}ac \sin B.$$

EXAMPLE 6 *Finding the Area of an Oblique Triangle*

Real Life

Find the area of a triangular lot having two sides of lengths 90 meters and 52 meters and an included angle of 102°.

Solution

Consider $a = 90$ m, $b = 52$ m, and angle $C = 102°$, as shown in Figure 3.8. Then the area of the triangle is

$$\text{Area} = \frac{1}{2}ab \sin C = \frac{1}{2}(90)(52)(\sin 102°) \approx 2289 \text{ square meters.}$$

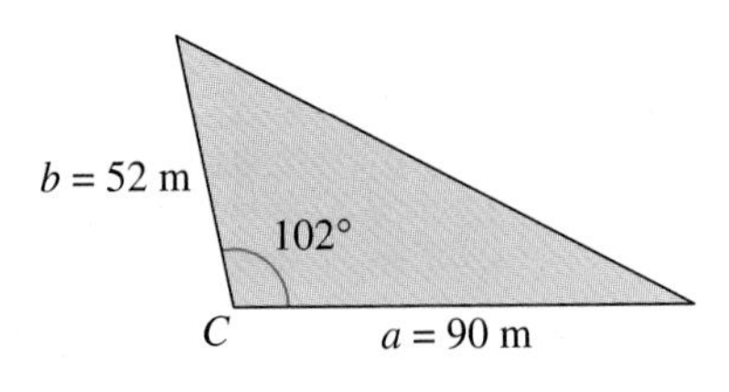

FIGURE 3.8

Application

FIGURE 3.9

EXAMPLE 7 An Application of the Law of Sines

The course for a boat race starts at point A and proceeds in the direction S 52° W to point B, then in the direction S 40° E to point C, and finally back to A, as shown in Figure 3.9. The point C lies 8 kilometers directly south of point A. Approximate the total distance of the race course.

Solution

Because lines BD and AC are parallel, it follows that $\angle BCA \cong \angle DBC$. Consequently, triangle ABC has the measures shown in Figure 3.10. For angle B, you have $B = 180° - 52° - 40° = 88°$. Using the Law of Sines

$$\frac{a}{\sin 52°} = \frac{b}{\sin 88°} = \frac{c}{\sin 40°}$$

you can let $b = 8$ and obtain the following.

$$a = \frac{8}{\sin 88°}(\sin 52°) \approx 6.308 \qquad c = \frac{8}{\sin 88°}(\sin 40°) \approx 5.145$$

The total length of the course is approximately

$$\text{Length} \approx 8 + 6.308 + 5.145 = 19.453 \text{ kilometers.}$$

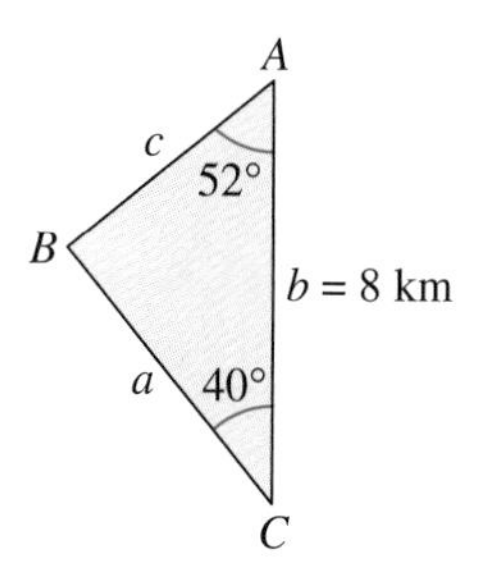

FIGURE 3.10

GROUP ACTIVITY

USING THE LAW OF SINES

In this section, you have been using the Law of Sines to solve *oblique* triangles. Can the Law of Sines also be used to solve a right triangle? If so, write a short paragraph explaining how to use the Law of Sines to solve the following two triangles. Is there an easier way to solve these triangles?

a. (AAS)

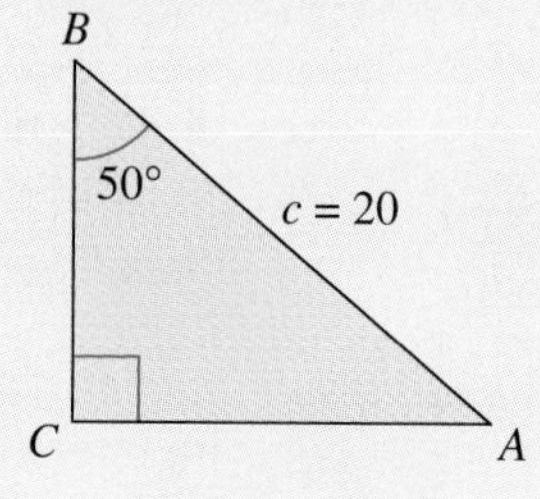

b. (ASA)

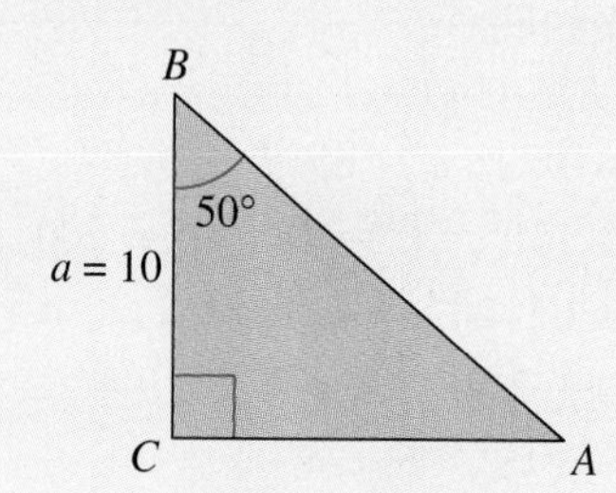

The *Interactive* CD-ROM provides additional help with Warm-Up exercises by providing a hypertext link to the section in which the concept was introduced.

WARM UP

Solve the right triangle. (c is the hypotenuse.)

1. $a = 3,\ c = 6$

2. $a = 5,\ b = 5$

3. $b = 15,\ c = 17$

4. $A = 42°,\ a = 7.5$

5. $B = 10°,\ b = 4$

6. $B = 72°15',\ c = 150$

Find the altitude of the triangle.

7.

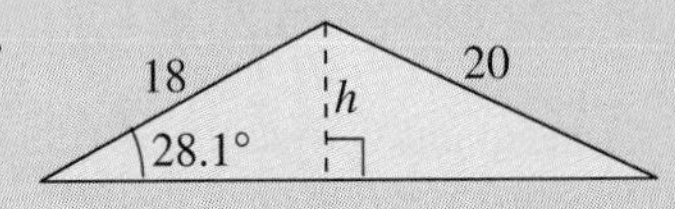

8.

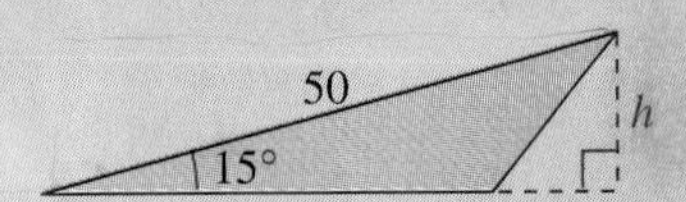

Solve the equation for x.

The *Interactive* CD-ROM contains step-by-step solutions to all odd-numbered Section and Review Exercises. It also provides Tutorial Exercises, which link to Guided Examples for additional help.

9. $\dfrac{2}{\sin 30°} = \dfrac{9}{x}$

10. $\dfrac{100}{\sin 72°} = \dfrac{x}{\sin 60°}$

3.1 Exercises

In Exercises 1–16, use the given information to solve the triangle.

1.

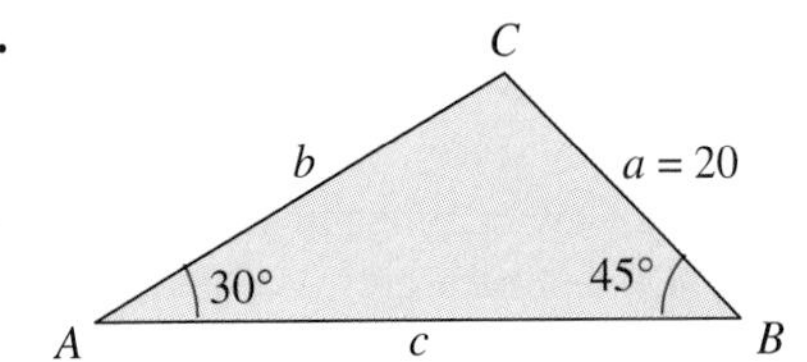

2.

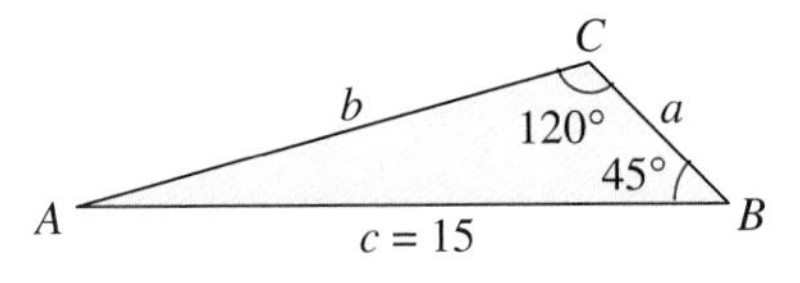

3.

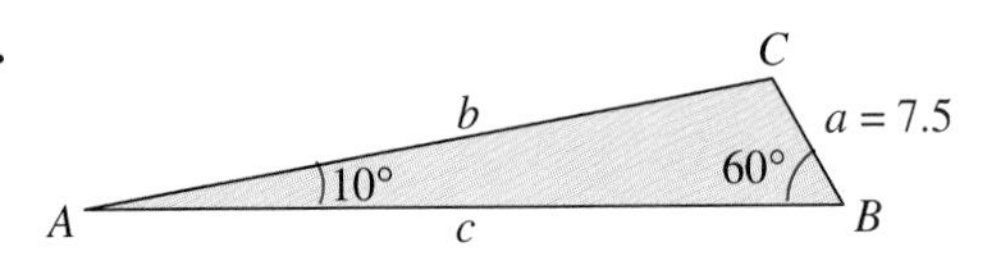

4.

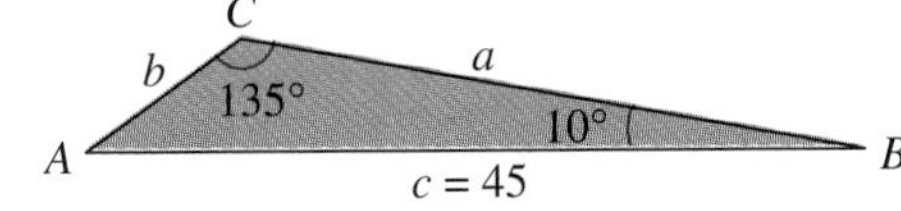

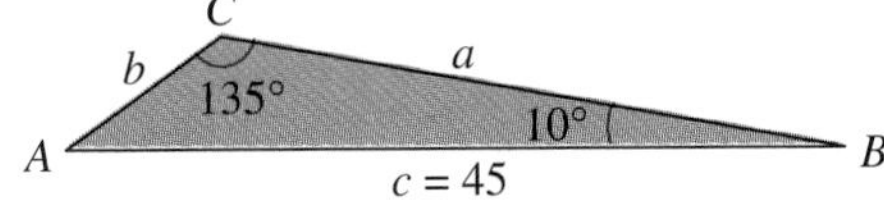

5. $A = 36°,\quad a = 8,\quad b = 5$

6. $A = 60°,\quad a = 9,\quad c = 10$

7. $A = 150°,\quad C = 20°,\quad a = 200$

8. $A = 24.3°,\quad C = 54.6°,\quad c = 2.68$

9. $A = 83°\,20',\quad C = 54.6°,\quad c = 18.1$

10. $A = 5°\,40',\quad B = 8°\,15',\quad b = 4.8$

11. $B = 15°\,30',\quad a = 4.5,\quad b = 6.8$

12. $C = 85°\,20',\quad a = 35,\quad c = 50$

13. $C = 145°,\quad b = 4,\quad c = 14$

14. $A = 100°,\quad a = 125,\quad c = 10$

15. $A = 110°\,15',\quad a = 48,\quad b = 16$

16. $B = 2°\,45',\quad b = 6.2,\quad c = 5.8$

In Exercises 17–22, use the given information to solve (if possible) the triangle. If two solutions exist, find both.

17. $A = 58°, \quad a = 4.5, \quad b = 12.8$

18. $A = 58°, \quad a = 11.4, \quad b = 12.8$

19. $A = 58°, \quad a = 4.5, \quad b = 5$

20. $A = 58°, \quad a = 42.4, \quad b = 50$

21. $A = 110°, \quad a = 125, \quad b = 200$

22. $A = 110°, \quad a = 125, \quad b = 100$

In Exercises 23 and 24, find a value for b such that the triangle has (a) one solution, (b) two solutions, and (c) no solution.

23. $A = 36°, \quad a = 5$ **24.** $A = 60°, \quad a = 10$

25. ***Height*** A flagpole at a right angle to the horizontal is located on a slope that makes an angle of 12° with the horizontal. The pole's shadow is 16 meters long and points directly up the slope. The angle of elevation of the sun is 20°.

(a) Draw a triangle that represents the problem. Show the known quantities on the triangle and use a variable to indicate the height of the flagpole.

(b) Write an equation involving the unknown.

(c) Find the height of the flagpole.

26. ***Height*** Because of prevailing winds, a tree grew so that it was leaning 6° from the vertical. At a point 30 meters from the tree, the angle of elevation to the top of the tree is 22° 50′ (see figure). Find the height h of the tree.

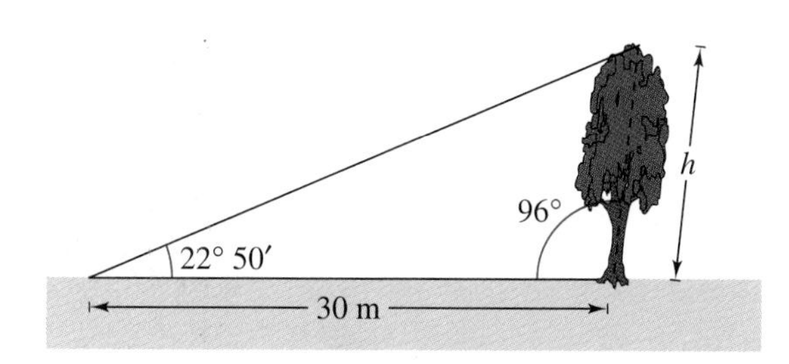

27. ***Angle of Elevation*** A 10-meter telephone pole casts a 17-meter shadow directly down a slope when the angle of elevation of the sun is 42° (see figure). Find, θ, the angle of elevation of the ground.

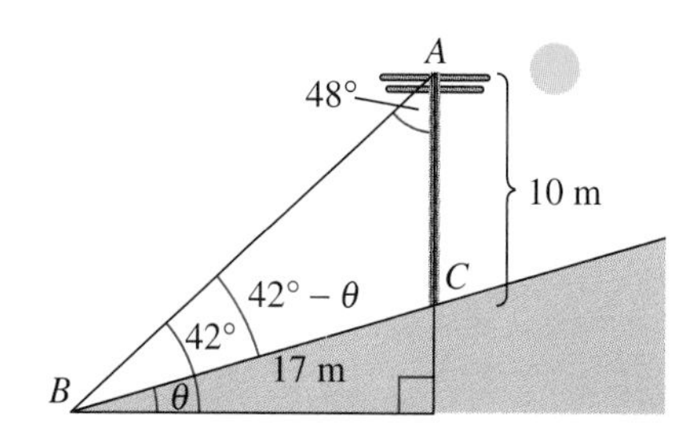

28. ***Flight Path*** A plane flies 500 kilometers with a bearing of N 44° W from B to C (see figure). The plane then flies 720 kilometers from C to A. Find the bearing of the flight from C to A.

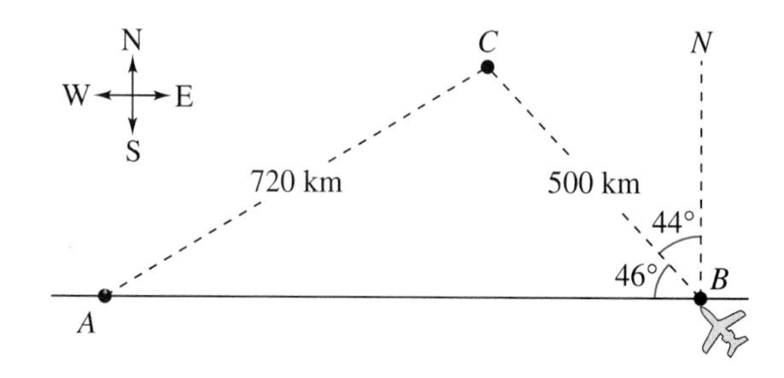

29. ***Bridge Design*** A bridge is to be built across a small lake from B to C (see figure). The bearing from B to C is S 41° W. From a point A, 100 meters from B, the bearings to B and C are S 74° E and S 28° E, respectively. Find the distance from B to C.

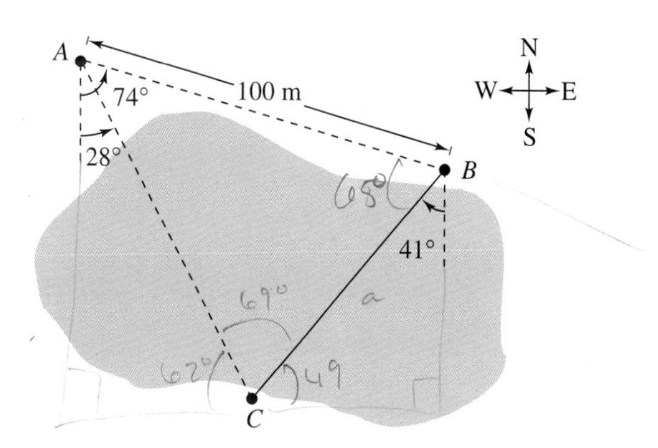

30. ***Railroad Track Design*** The circular arc of a railroad curve has a chord of length 3000 feet and a central angle of 40°.

(a) Draw a figure that visually represents the problem. Show the known quantities on the figure and use variables r and s to represent the radius of the arc and the length of the arc, respectively.

(b) Find the radius r of the circular arc.

(c) Find the length s of the circular arc.

31. ***Glide Path*** A pilot has just started on the glide path for landing at an airport where the length of the runway is 9000 feet. The angles of depression from the plane to the ends of the runway are 17.5° and 18.8°.

(a) Draw a figure that visually represents the problem.

(b) Find the air distance the plane must travel until touching down on the near end of the runway.

(c) Find the ground distance the plane must travel until touching down.

(d) Find the altitude of the plane when the pilot begins the descent.

32. ***Altitude*** The angles of elevation to an airplane from two points A and B on level ground are 51° and 68°, respectively. The points A and B are 2.5 miles apart, and the airplane is east of both points in the same vertical plane. Find the altitude of the plane.

33. ***Locating a Fire*** Two fire towers A and B are 30 kilometers apart. The bearing from A to B is N 65° E. A fire is spotted by a ranger in each tower, and its bearings from A and B are N 28° E and N 16.5° W, respectively (see figure). Find the distance of the fire from each tower.

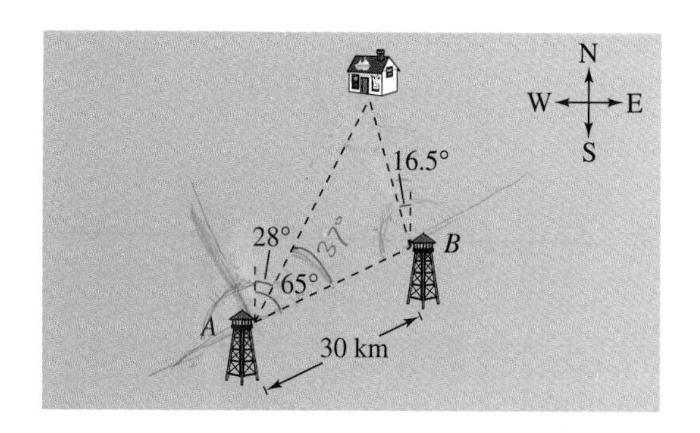

34. ***Distance*** A boat is sailing due east parallel to the shoreline at a speed of 10 miles per hour. At a given time the bearing to the lighthouse is S 70° E, and 15 minutes later the bearing is S 63° E (see figure). Find the distance from the boat to the shoreline if the lighthouse is at the shoreline.

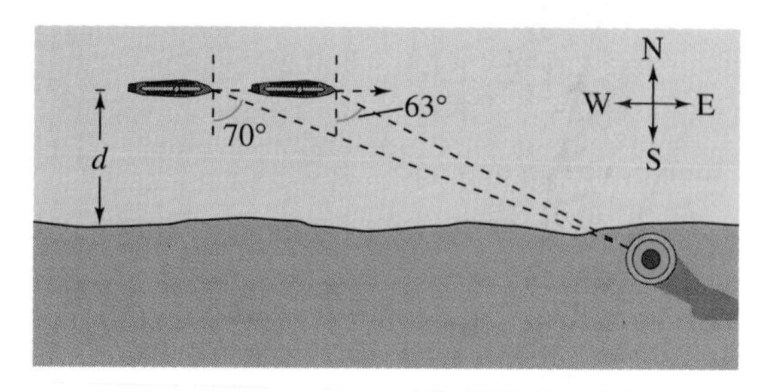

35. ***Distance*** A family is traveling due west on a road that passes a famous landmark. At a given time the bearing to the landmark is N 62° W, and after the family travels 5 miles farther the bearing is N 38° W. What is the closest the family will come to the landmark while on the road?

36. ***Engine Design*** The connecting rod in an engine is 6 inches in length and the radius of the crankshaft is $1\frac{1}{2}$ inches (see figure). Let d be the distance the piston is from the top of its stroke for the angle θ.

(a) Complete the table.

θ	0°	45°	90°	135°	180°
d					

(b) The spark plug fires at $\theta = 5°$ before top dead center. How far is the piston from the top of its stroke at this time?

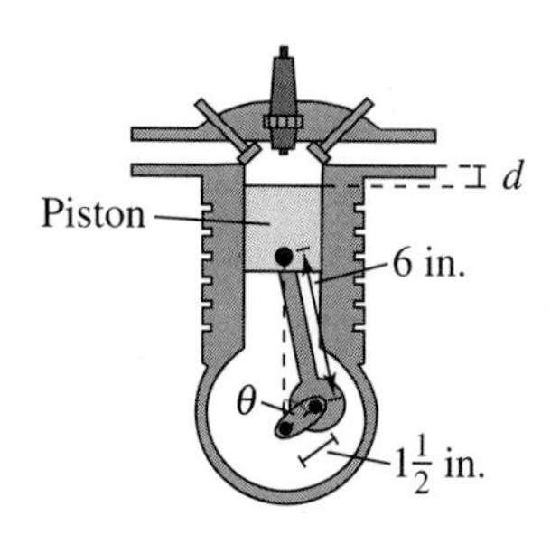

37. ***Graphical and Numerical Analysis*** In the figure, α and β are positive angles.

(a) Write α as a function of β.

(b) Use a graphing utility to graph the function. Determine its domain and range.

(c) Use the result of part (a) to write c as a function of β.

(d) Use a graphing utility to graph the function in part (c). Determine its domain and range.

(e) Complete the table. What can you infer?

β	0	0.4	0.8	1.2	1.6	2.0	2.4	2.8
α								
c								

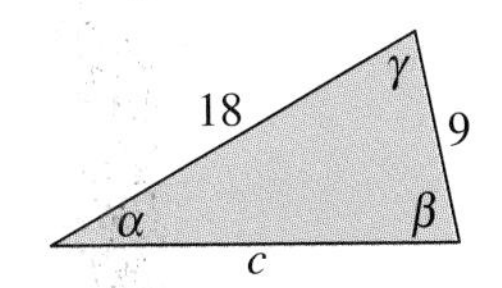

38. ***Distance*** The angles of elevation θ and ϕ to an airplane are being continuously monitored at two observation points A and B that are 2 miles apart (see figure). Write an equation giving the distance d between the plane and point B in terms of θ and ϕ.

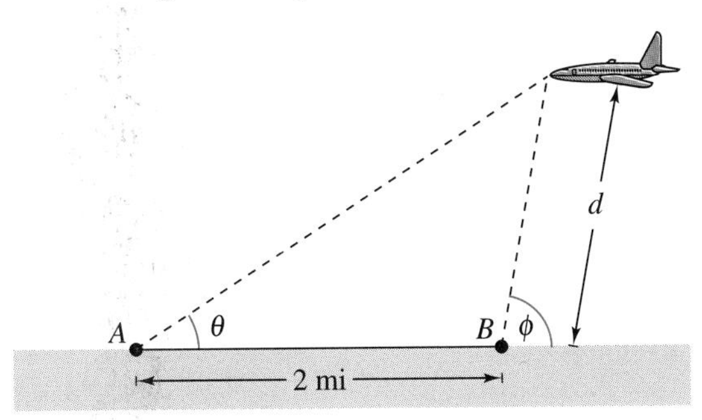

In Exercises 39–44, find the area of the triangle having the indicated sides and angle.

39. $C = 120°, \quad a = 4, \quad b = 6$

40. $B = 72° 30', \quad a = 105, \quad c = 64$

41. $A = 43° 45', \quad b = 57, \quad c = 85$

42. $A = 5° 15', \quad b = 4.5, \quad c = 22$

43. $B = 130°, \quad a = 62, \quad c = 20$

44. $C = 84° 30', \quad a = 16, \quad b = 20$

45. ***Graphical Analysis***

(a) Write the area A of the shaded region in the figure as a function of θ.

(b) Use a graphing utility to graph the area function.

(c) Determine the domain of the area function. Explain how the area of the region and the domain of the function would change if the 8-centimeter line segment were decreased in length.

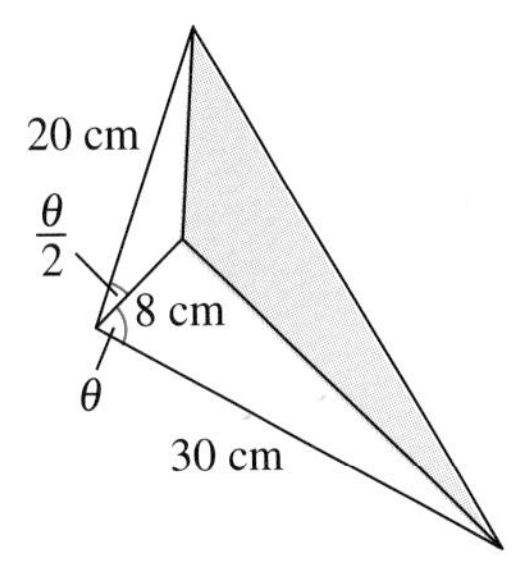

3.2 Law of Cosines

See Exercise 36 on page 286 for an example of how the Law of Cosines can be used to analyze the design of a paper manufacturing machine.

Introduction ▫ *Heron's Formula*

Introduction

Two cases remain in the list of conditions needed to solve an oblique triangle—SSS and SAS. The Law of Sines does not work in either of these cases. To see why, consider the three ratios given in the Law of Sines.

$$\frac{a}{\sin A} = \frac{b}{\sin B} = \frac{c}{\sin C}$$

To use the Law of Sines, you must know at least one side and its opposite angle. If you are given three sides (SSS), or two sides and their included angle (SAS), none of the above ratios would be complete. In such cases you can use the **Law of Cosines.**

LAW OF COSINES

Standard Form	*Alternative Form*
$a^2 = b^2 + c^2 - 2bc\cos A$	$\cos A = \dfrac{b^2 + c^2 - a^2}{2bc}$
$b^2 = a^2 + c^2 - 2ac\cos B$	$\cos B = \dfrac{a^2 + c^2 - b^2}{2ac}$
$c^2 = a^2 + b^2 - 2ab\cos C$	$\cos C = \dfrac{a^2 + b^2 - c^2}{2ab}$

PROOF ■■

Consider a triangle that has three acute angles, as shown in Figure 3.11. In the figure, note that vertex B has coordinates $(c, 0)$. Furthermore, C has coordinates (x, y), where $x = b\cos A$ and $y = b\sin A$. Because a is the distance from vertex C to vertex B, it follows that

$$
\begin{aligned}
a &= \sqrt{(x - c)^2 + (y - 0)^2} \\
a^2 &= (b\cos A - c)^2 + (b\sin A)^2 \\
a^2 &= b^2\cos^2 A - 2bc\cos A + c^2 + b^2\sin^2 A \\
a^2 &= b^2(\sin^2 A + \cos^2 A) + c^2 - 2ab\cos A \\
a^2 &= b^2 + c^2 - 2bc\cos A. \qquad \sin^2 A + \cos^2 A = 1
\end{aligned}
$$

Similar arguments can be used to establish the other two equations. ■■

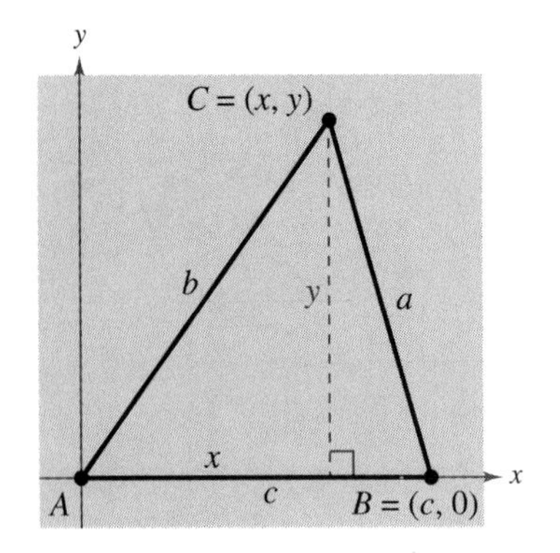

FIGURE 3.11

Note that if $A = 90°$ in Figure 3.11, then $\cos A = 0$ and the first form of the Law of Cosines becomes the Pythagorean Theorem.

$$a^2 = b^2 + c^2$$

Thus, the Pythagorean Theorem is actually just a special case of the more general Law of Cosines.

EXAMPLE 1 *Three Sides of a Triangle—SSS*

Find the three angles of the triangle whose sides have lengths $a = 8$ feet, $b = 19$ feet, and $c = 14$ feet.

Solution

It is a good idea first to find the angle opposite the longest—side b in this case (see Figure 3.12). Using the Law of Cosines, you find that

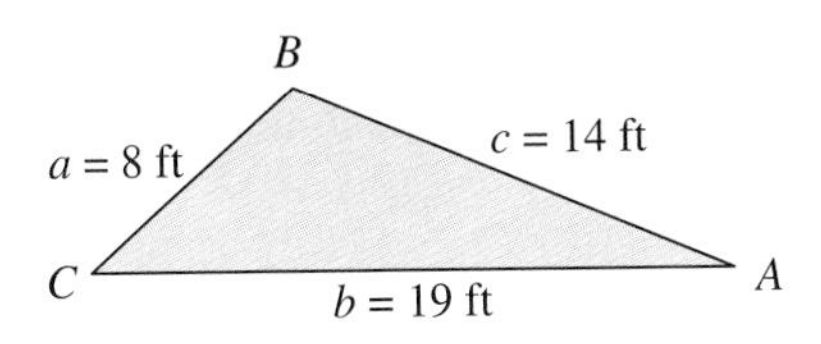

FIGURE 3.12

$$\cos B = \frac{a^2 + c^2 - b^2}{2ac} = \frac{8^2 + 14^2 - 19^2}{2(8)(14)} \approx -0.45089.$$

Because $\cos B$ is negative, you know that B is an *obtuse* angle given by $B \approx 116.80°$. At this point you could use the Law of Cosines to find $\cos A$ and $\cos C$. However, knowing that $B \approx 116.80°$, it is simpler to use the Law of Sines to obtain the following.

$$\frac{b}{\sin B} = \frac{a}{\sin A}$$

$$\sin A = a\left(\frac{\sin B}{b}\right) \approx 8\left(\frac{\sin 116.80°}{19}\right) \approx 0.37582$$

The *Interactive* CD-ROM shows every example with its solution; clicking on the *Try It!* button brings up similar problems. Guided Examples and Integrated Examples show step-by-step solutions to additional examples. Integrated Examples are related to several concepts in the section.

Because B is obtuse, you know that A must be acute, because a triangle can have, at most, one obtuse angle. Thus, $A \approx 22.08°$ and

$$\begin{aligned} C &\approx 180° - 22.08° - 116.80° \\ &= 41.12°. \end{aligned}$$

Do you see why it was wise to find the largest angle *first* in Example 1? Knowing the cosine of an angle, you can determine whether the angle is acute or obtuse. That is,

$\cos\theta > 0$ for $0° < \theta < 90°$ — Acute

$\cos\theta < 0$ for $90° < \theta < 180°.$ — Obtuse

So, in Example 1, once you found that angle B was obtuse, you knew that angles A and C were both acute. If the largest angle is acute, the remaining two angles are acute also.

EXAMPLE 2 *Two Sides and the Included Angle—SAS*

The pitcher's mound on a softball field is 46 feet from home plate and the distance between the bases in 60 feet, as shown in Figure 3.13. (The pitcher's mound is not halfway between home plate and second base.) How far is the pitcher's mound from first base?

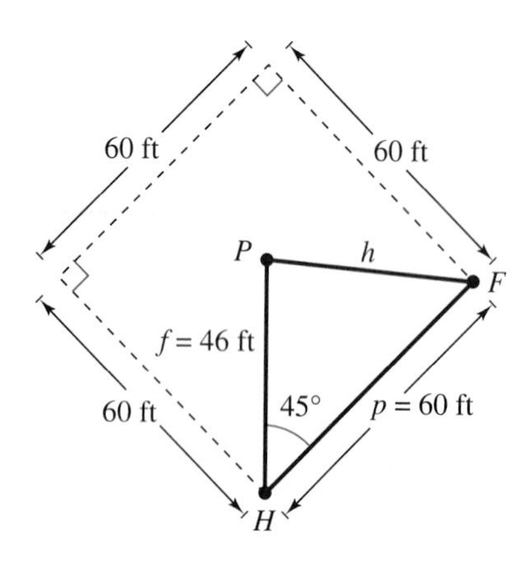

FIGURE 3.13

Solution

In triangle HPF, $H = 45°$ (line HP bisects the right angle at H), $f = 46$, and $p = 60$. Using the Law of Cosines for this SAS case, you have

$$\begin{aligned} h^2 &= f^2 + p^2 - 2fp \cos H \\ &= 46^2 + 60^2 - 2(46)(60) \cos 45° \\ &\approx 1812.8. \end{aligned}$$

Therefore, the approximate distance from the pitcher's mound to first base is

$$h \approx \sqrt{1812.8} \approx 42.58 \text{ feet.}$$

EXAMPLE 3 *Two Sides and the Included Angle—SAS*

A ship travels 60 miles due east, then adjusts its course 15° northward, as shown in Figure 3.14. After traveling 80 miles in that direction, how far is the ship from its point of departure?

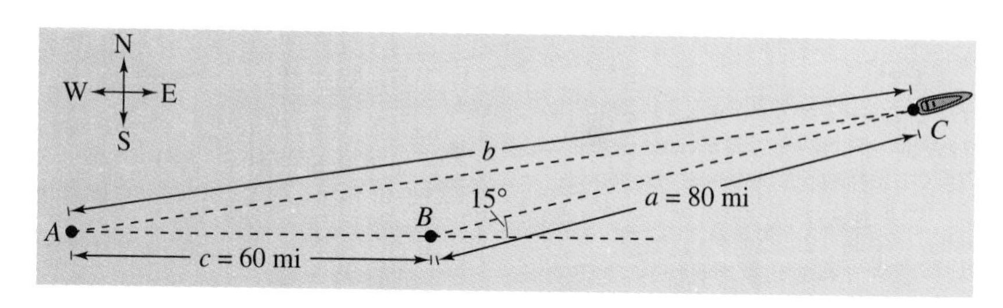

FIGURE 3.14

Solution

You have $c = 60$, $B = 180° - 15° = 165°$, and $a = 80$. Consequently, by the Law of Cosines, it follows that

$$\begin{aligned} b^2 &= a^2 + c^2 - 2ac \cos B \\ &= 80^2 + 60^2 - 2(80)(60) \cos 165° \\ &\approx 19{,}273. \end{aligned}$$

Therefore, the distance b is

$$b \approx \sqrt{19{,}273} \approx 138.8 \text{ miles.}$$

The world's largest ship is the Norwegian oil tanker, Jahre Viking, which is 1503 feet (458 meters) long. It takes 5 minutes to walk the ship's length.

Heron's Formula

The Law of Cosines can be used to establish the following formula for the area of a triangle. This formula is credited to the Greek mathematician Heron (c. 100 B.C.).

HERON'S AREA FORMULA

Given any triangle with sides of lengths a, b, and c, the area of the triangle is

$$\text{Area} = \sqrt{s(s-a)(s-b)(s-c)}$$

where $s = (a+b+c)/2$.

PROOF ▪▪

From the previous section, you know that

$$\begin{aligned}\text{Area} &= \frac{1}{2}bc\sin A \\ &= \sqrt{\frac{1}{4}b^2c^2\sin^2 A} \\ &= \sqrt{\frac{1}{4}b^2c^2(1-\cos^2 A)} \\ &= \sqrt{\left[\frac{1}{2}bc(1+\cos A)\right]\left[\frac{1}{2}bc(1-\cos A)\right]}.\end{aligned}$$

Using the Law of Cosines, you can show that

$$\frac{1}{2}bc(1+\cos A) = \frac{a+b+c}{2} \cdot \frac{-a+b+c}{2}$$

and

$$\frac{1}{2}bc(1-\cos A) = \frac{a-b+c}{2} \cdot \frac{a+b-c}{2}.$$

(See Exercises 49 and 50.) Letting $s = (a+b+c)/2$, these two equations can be rewritten as

$$\frac{1}{2}bc(1+\cos A) = s(s-a)$$

and

$$\frac{1}{2}bc(1-\cos A) = (s-b)(s-c).$$

Thus, you can conclude that

$$\text{Area} = \sqrt{s(s-a)(s-b)(s-c)}. \quad ▪▪$$

The *Interactive* CD-ROM offers graphing utility emulators of the *TI-82* and *TI-83*, which can be used with the Examples, Explorations, Technology notes, and Exercises.

EXAMPLE 4 *Using Heron's Area Formula*

Find the area of the triangular region having sides of lengths $a = 43$ meters, $b = 53$ meters, and $c = 72$ meters.

Solution

Because

$$s = \frac{1}{2}(a + b + c) = \frac{168}{2} = 84,$$

Heron's Formula yields

$$\begin{aligned} \text{Area} &= \sqrt{s(s-a)(s-b)(s-c)} \\ &= \sqrt{84(41)(31)(12)} \\ &\approx 1131.89 \text{ square meters.} \end{aligned}$$

GROUP ACTIVITY

THE AREA OF A TRIANGLE

You have now discussed three different formulas for the area of a triangle.

Standard Formula $\quad \text{Area} = \frac{1}{2}bh$

Oblique Triangle $\quad \text{Area} = \frac{1}{2}bc \sin A = \frac{1}{2}ab \sin C = \frac{1}{2}ac \sin B$

Heron's Formula $\quad \text{Area} = \sqrt{s(s-a)(s-b)(s-c)}$

Use the most appropriate formula to find the area of each triangle. Show your work and give your reasons for choosing each formula.

a.

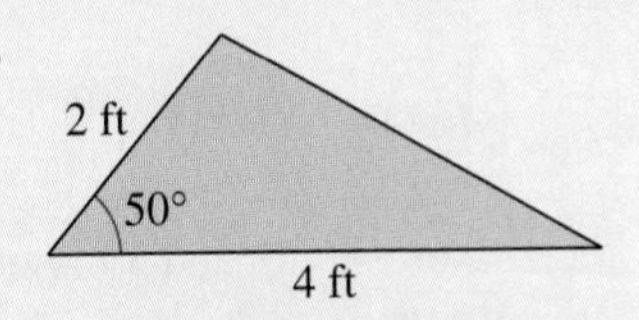

b.

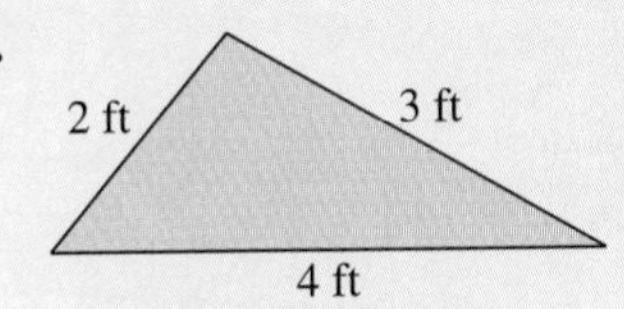

c.

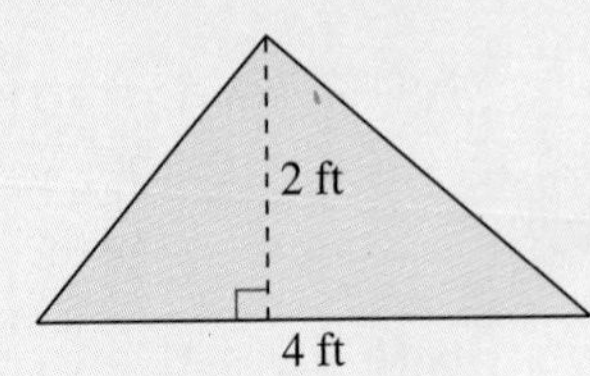

d.

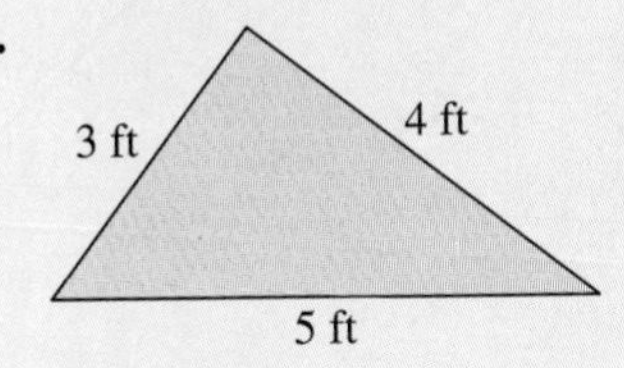

WARM UP

Simplify the expression.

1. $\sqrt{(7-3)^2 + [1-(-5)]^2}$
2. $\sqrt{[-2-(-5)]^2 + (12-6)^2}$

Find the distance between the two points.

3. $(4, -2), (8, 10)$
4. $(1, 3), (7, 12)$

Find the area of the triangle.

5.

6.

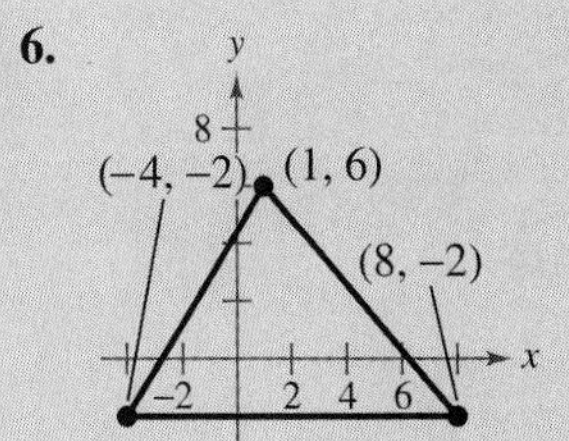

Find (if possible) the remaining sides and angles of the triangle.

7. $A = 10°, C = 100°, b = 25$
8. $A = 20°, C = 90°, c = 100$
9. $B = 30°, b = 6.5, c = 15$
10. $A = 30°, b = 6.5, a = 10$

3.2 Exercises

In Exercises 1–14, use the Law of Cosines to solve the triangle.

1.

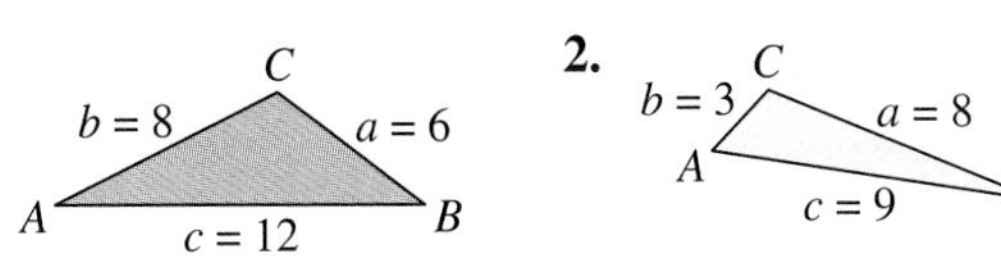

2.

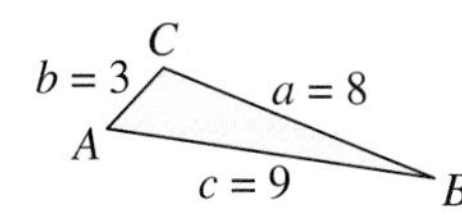

3.

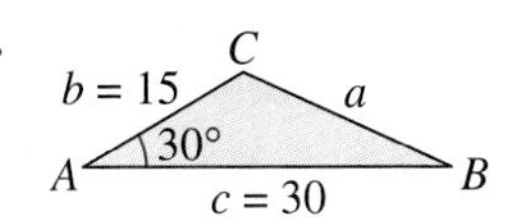

4.

5. $a = 9, \quad b = 12, \quad c = 15$
6. $a = 55, \quad b = 25, \quad c = 72$
7. $a = 75.4, \quad b = 52, \quad c = 52$
8. $a = 1.42, \quad b = 0.75, \quad c = 1.25$
9. $A = 120°, \quad b = 3, \quad c = 10$
10. $A = 55°, \quad b = 3, \quad c = 10$
11. $B = 8° 45', \quad a = 25, \quad c = 15$
12. $B = 75° 20', \quad a = 6.2, \quad c = 9.5$
13. $B = 125° 40', \quad a = 32, \quad c = 32$
14. $C = 15°, \quad a = 6.25, \quad b = 2.15$

In Exercises 15–20, complete the table by solving the parallelogram shown in the figure. (The lengths of the diagonals are given by c and d.)

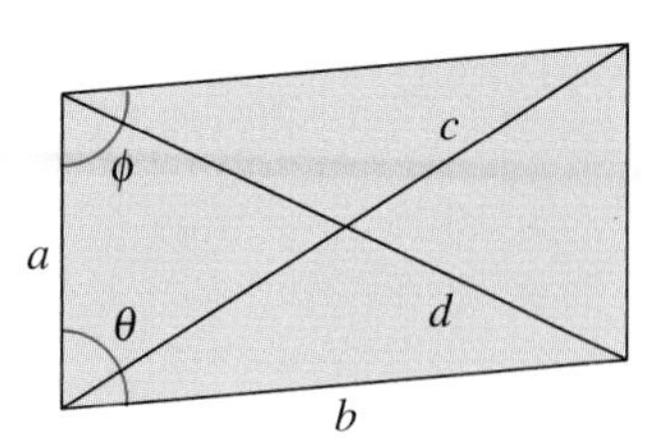

	a	b	c	d	θ	ϕ
15.	4	6			30°	
16.	25	35				120°
17.	10	14	20			
18.	40	60		80		
19.	10		18	12		
20.		25	50	35		

21. ***Navigation*** A boat race runs along a triangular course marked by buoys A, B, and C. The race starts with the boats headed west for 2500 meters. The other two sides of the course lie to the north of the first side, and their lengths are 1100 meters and 2000 meters. Draw a figure that gives a visual representation of the problem, and find the bearings for the last two legs of the race.

22. ***Navigation*** A plane flies 810 miles from A to B with a bearing of N 75° E. Then it flies 648 miles from B to C with a bearing of N 32° E. Draw a figure that visually represents the problem, and find the straight-line distance and bearing from C to A.

23. ***Surveying*** To approximate the length of a marsh, a surveyor walks 300 meters from point A to point B, then turns 80° and walks 250 meters to point C (see figure). Approximate the length AC of the marsh.

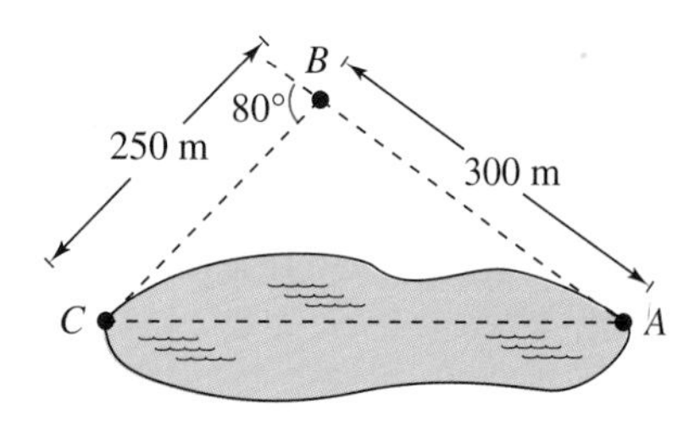

24. ***Surveying*** A triangular parcel of land has 115 meters of frontage, and the other boundaries have lengths of 76 meters and 92 meters. What angles does the frontage make with the two other boundaries?

25. ***Surveying*** A triangular parcel of ground has sides of length 725 feet, 650 feet, and 575 feet. Find the measure of the largest angle.

26. ***Streetlight Design*** Determine the angle θ in the design of the streetlight shown in the figure.

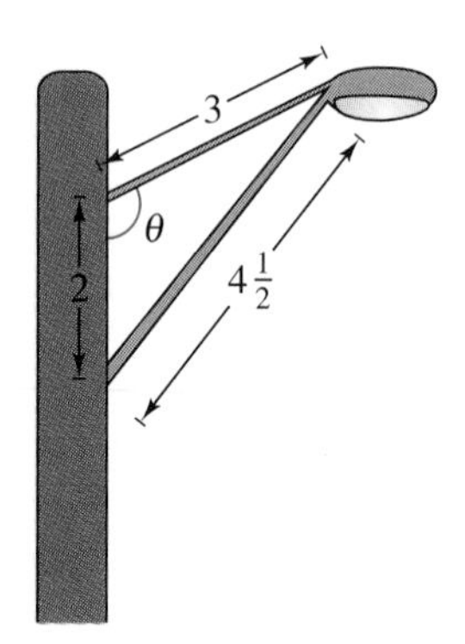

27. ***Distance*** Two ships leave a port at 9 A.M. One travels at a bearing of N 53° W at 12 miles per hour and the other travels at a bearing of S 67° W at 16 miles per hour. Approximate how far apart they are at noon that day.

28. ***Distance*** A 100-foot vertical tower is to be erected on the side of a hill that makes a 6° angle with the horizontal (see figure). Find the length of each of the two guy wires that will be anchored 75 feet uphill and downhill from the base of the tower.

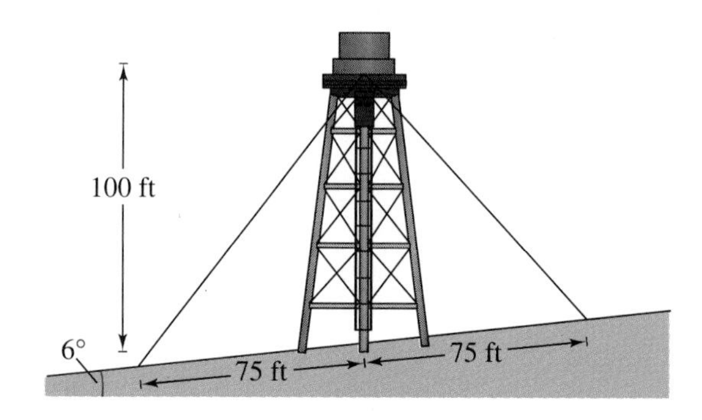

29. *Navigation* On a map, Orlando is 178 millimeters due south of Niagara Falls, Denver is 273 millimeters from Orlando, and Denver is 235 millimeters from Niagara Falls (see figure).

(a) Find the bearing of Denver from Orlando.

(b) Find the bearing of Denver from Niagara Falls.

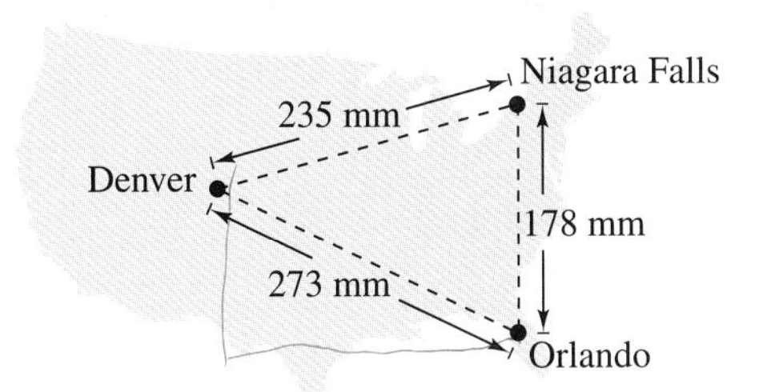

30. *Navigation* On a map, Minneapolis is 165 millimeters due west of Albany, Phoenix is 216 millimeters from Minneapolis, and Phoenix is 368 millimeters from Albany.

(a) Find the bearing of Minneapolis from Phoenix.

(b) Find the bearing of Albany from Phoenix.

31. *Baseball* On a baseball diamond with 90-foot sides, the pitcher's mound is 60.5 feet from home plate. How far is it from the pitcher's mound to third base?

32. *Baseball* The baseball player in center field is playing approximately 330 feet from the television camera that is behind home plate. A batter hits a fly ball that goes to the wall 420 feet from the camera (see figure). Approximate the number of feet that the center fielder has to run to make the catch if the camera turns 8° to follow the play.

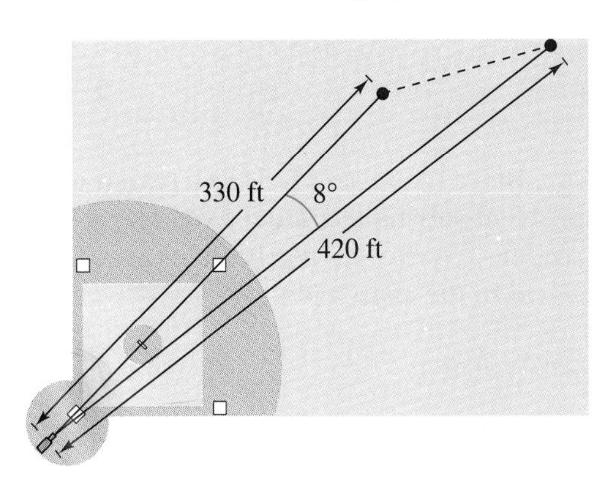

33. *Engineering* If Q is the midpoint of the line segment $\overline{PR}$, find the lengths of the line segments $\overline{PQ}$, $\overline{QS}$, and $\overline{RS}$ on the truss rafter shown in the figure.

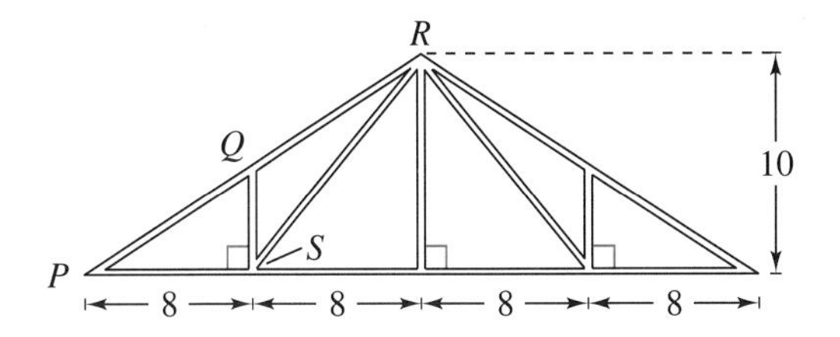

34. *Aircraft Tracking* To determine the distance between two aircraft, a tracking station continuously determines the distance to each aircraft and the angle A between them. Determine the distance a between the planes when $A = 42°$, $b = 35$ miles, and $c = 20$ miles.

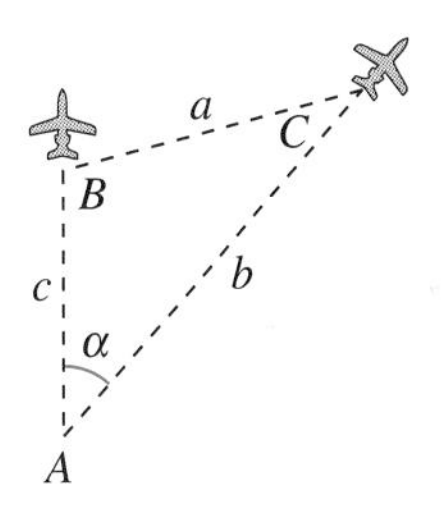

35. *Engine Design* An engine has a 7-inch connecting rod fastened to a crank (see figure).

(a) Use the Law of Cosines to write an equation giving the relationship between x and θ.

(b) Write x as a function of θ. (Select the sign that yields positive values of x.)

(c) Use a graphing utility to graph the function in part (b).

(d) Use the graph in part (c) to determine the maximum distance the piston moves in one cycle.

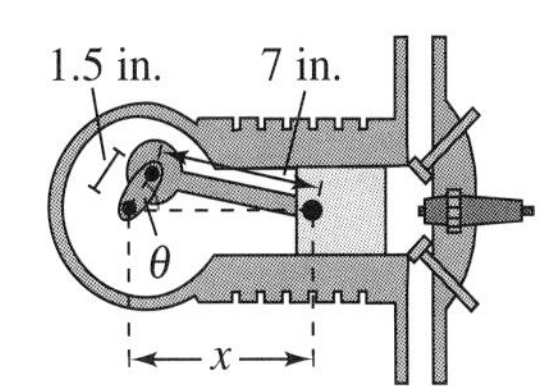

36. ***Paper Manufacturing*** In a certain process with continuous paper, the paper passes across three rollers of radii 3 inches, 4 inches, and 6 inches (see figure). The centers of the 3-inch and 6-inch rollers are d inches apart, and the length of the arc in contact with the paper on the 4-inch roller is s inches. Complete the following table.

d (inches)	9	10	12	13	14	15	16
θ (degrees)							
s (inches)							

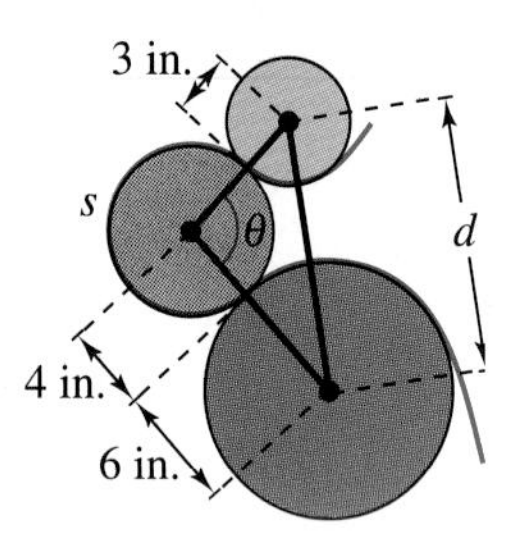

FIGURE FOR 36

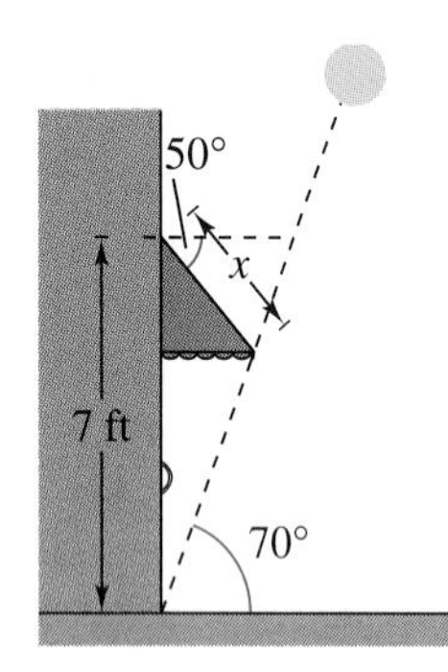

FIGURE FOR 37

37. ***Awning Design*** A retractable awning lowers at an angle of 50° from the top of a patio door that is 7 feet tall (see figure). Find the length x of the awning if no direct sunlight is to enter the door when the angle of elevation of the sun is greater than 70°.

38. ***Circumscribed and Inscribed Circles*** Let R and r be the radii of the circumscribed and inscribed circles of a triangle ABC, respectively, and let $s = (a + b + c)/2$. Prove the following.

(a) $2R = \dfrac{a}{\sin A} = \dfrac{b}{\sin B} = \dfrac{c}{\sin C}$

(b) $r = \sqrt{\dfrac{(s-a)(s-b)(s-c)}{s}}$

Circumscribed and Inscribed Circles **In Exercises 39 and 40, use the results of Exercise 38.**

39. Given the triangle with $a = 25$, $b = 55$, and $c = 72$, find the area of (a) the triangle, (b) the circumscribed circle, and (c) the inscribed circle.

40. Find the length of the largest circular track that can be built on a triangular piece of property whose sides are 200 feet, 250 feet, and 325 feet.

In Exercises 41–46, use Heron's Area Formula to find the area of the triangle.

41. $a = 5, \quad b = 7, \quad c = 10$

42. $a = 2.5, \quad b = 10.2, \quad c = 9$

43. $a = 12, \quad b = 15, \quad c = 9$

44. $a = 75.4, \quad b = 52, \quad c = 52$

45. $a = 20, \quad b = 20, \quad c = 10$

46. $a = 4.25, \quad b = 1.55, \quad c = 3.00$

47. ***Area*** The lengths of the sides of a triangular parcel of land are approximately 200 feet, 500 feet, and 600 feet. Approximate the area of the parcel.

48. ***Area*** The lengths of two adjacent sides of a parallelogram are 4 meters and 6 meters. Find the area of the parallelogram if the angle between the two sides is 30°.

49. Use the Law of Cosines to prove that

$$\frac{1}{2}bc(1 + \cos A) = \frac{a+b+c}{2} \cdot \frac{-a+b+c}{2}.$$

50. Use the Law of Cosines to prove that

$$\frac{1}{2}bc(1 - \cos A) = \frac{a-b+c}{2} \cdot \frac{a+b-c}{2}.$$

Review **Solve Exercises 51–54 as a review of the skills and problem-solving techniques you learned in previous sections. Write an algebraic expression that is equivalent to the expression.**

51. $\sec(\arcsin 2x)$

52. $\tan(\arccos 3x)$

53. $\cot[\arctan(x - 2)]$

54. $\cos\left(\arcsin \dfrac{x-1}{2}\right)$

3.3 Vectors in the Plane

See Exercises 77 and 78 on page 299 for examples of how vectors can be used to analyze the direction and speed of an airplane.

Introduction ▫ *Component Form of a Vector* ▫ *Vector Operations* ▫ *Unit Vectors* ▫ *Direction Angles* ▫ *Applications of Vectors*

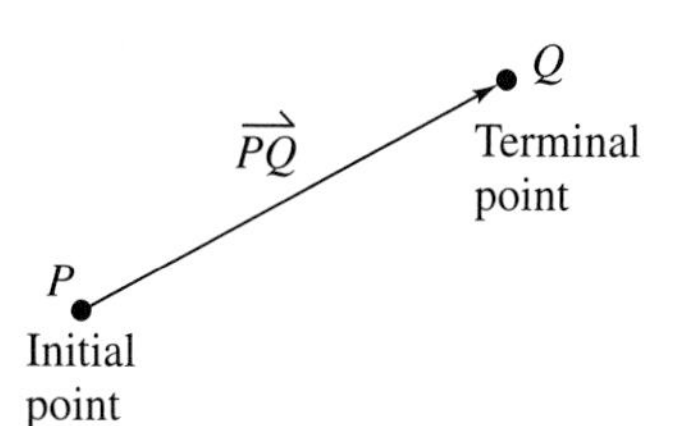

FIGURE 3.15

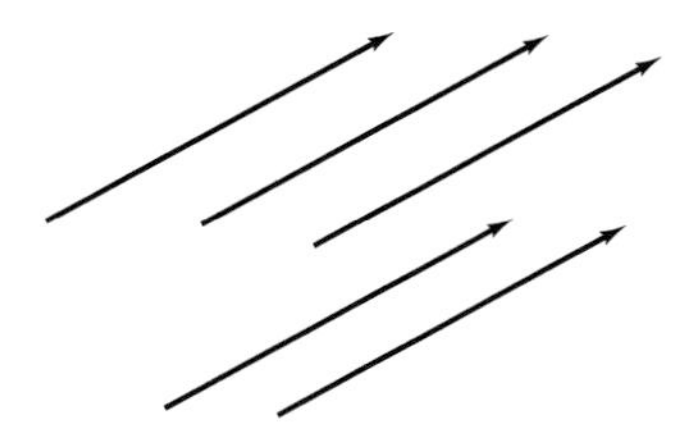

FIGURE 3.16

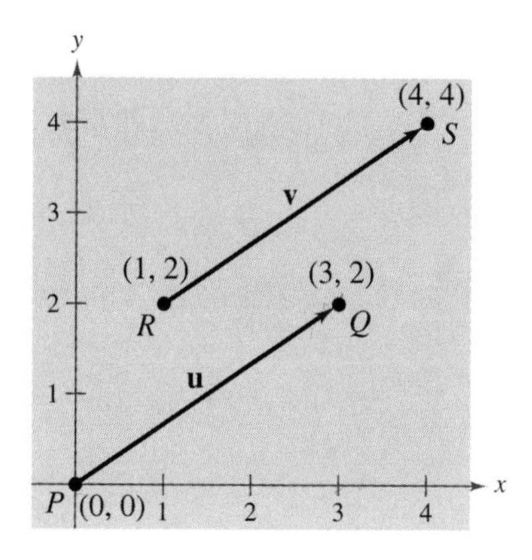

FIGURE 3.17

Introduction

Many quantities in geometry and physics, such as area, time, and temperature, can be represented by a single real number. Other quantities, such as force and velocity, involve both *magnitude* and *direction* and cannot be completely characterized by a single real number. To represent such a quantity, you can use a **directed line segment,** as shown in Figure 3.15. The directed line segment $\overrightarrow{PQ}$ has **initial point** P and **terminal point** Q. Its **length** is denoted by $\|\overrightarrow{PQ}\|$.

Two directed line segments that have the same length (or magnitude) and direction are called **equivalent.** For example, the directed line segments in Figure 3.16 are all equivalent. The set of all directed line segments that are equivalent to given directed line segment $\overrightarrow{PQ}$ is a **vector v in the plane,** written $\mathbf{v} = \overrightarrow{PQ}$. Vectors are denoted by lowercase, boldface letters such as **u**, **v**, and **w**.

Be sure you see that a vector in the plane can be represented by many different directed line segments.

EXAMPLE 1 *Vector Representation by Directed Line Segments*

Let **u** be represented by the directed line segment from $P = (0, 0)$ to $Q = (3, 2)$, and let **v** be represented by the directed line segment from $R = (1, 2)$ to $S = (4, 4)$, as shown in Figure 3.17. Show that $\mathbf{u} = \mathbf{v}$.

Solution

From the distance formula, it follows that $\overrightarrow{PQ}$ and $\overrightarrow{RS}$ have the *same length.*

$$\|\overrightarrow{PQ}\| = \sqrt{(3-0)^2 + (2-0)^2} = \sqrt{13}$$
$$\|\overrightarrow{RS}\| = \sqrt{(4-1)^2 + (4-2)^2} = \sqrt{13}$$

Moreover, both line segments have the *same direction* because they are both directed toward the upper right on lines having a slope of $\frac{2}{3}$. Thus, $\overrightarrow{PQ}$ and $\overrightarrow{RS}$ have the same length and direction, and it follows that $\mathbf{u} = \mathbf{v}$.

Component Form of a Vector

The directed line segment whose initial point is the origin is often the most convenient representative of a set of equivalent directed line segments. This representative of the vector **v** is in **standard position.**

A vector whose initial point is at the origin (0, 0) can be uniquely represented by the coordinates of its terminal point (v_1, v_2). This is the **component form of a vector v,** written

$$\mathbf{v} = \langle v_1, v_2 \rangle.$$

The coordinates v_1 and v_2 are the **components** of **v.** If both the initial point and the terminal point lie at the origin, **v** is the **zero vector** and is denoted by $\mathbf{0} = \langle 0, 0 \rangle$.

NOTE Two vectors $\mathbf{u} = \langle u_1, u_2 \rangle$ and $\mathbf{v} = \langle v_1, v_2 \rangle$ are **equal** if and only if $u_1 = v_1$ and $u_2 = v_2$. For instance, in Example 1, the vector **u** from $P = (0, 0)$ to $Q = (3, 2)$ is

$$\mathbf{u} = \overrightarrow{PQ} = \langle 3 - 0, 2 - 0 \rangle = \langle 3, 2 \rangle$$

and the vector **v** from $R = (1, 2)$ to $S = (4, 4)$ is $\mathbf{v} = \overrightarrow{RS} = \langle 4 - 1, 4 - 2 \rangle = \langle 3, 2 \rangle$.

COMPONENT FORM OF A VECTOR

The component form of the vector with initial point $P = (p_1, p_2)$ and terminal point $Q = (q_1, q_2)$ is

$$\overrightarrow{PQ} = \langle q_1 - p_1, q_2 - p_2 \rangle = \langle v_1, v_2 \rangle = \mathbf{v}.$$

The **length** (or magnitude) of **v** is given by

$$\|\mathbf{v}\| = \sqrt{(q_1 - p_1)^2 + (q_2 - p_2)^2} = \sqrt{v_1^2 + v_2^2}.$$

If $\|\mathbf{v}\| = 1$, **v** is a **unit vector.** Moreover, $\|\mathbf{v}\| = 0$ if and only if **v** is the zero vector **0.**

EXAMPLE 2 *Finding the Component Form of a Vector*

Find the component form and length of the vector **v** that has initial point $(4, -7)$ and terminal point $(-1, 5)$.

Solution

Let $P = (4, -7) = (p_1, p_2)$ and $Q = (-1, 5) = (q_1, q_2)$. Then, the components of $\mathbf{v} = \langle v_1, v_2 \rangle$ are given by

$$v_1 = q_1 - p_1 = -1 - 4 = -5$$
$$v_2 = q_2 - p_2 = 5 - (-7) = 12.$$

Thus, $\mathbf{v} = \langle -5, 12 \rangle$ and the length of **v** is

$$\|\mathbf{v}\| = \sqrt{(-5)^2 + 12^2} = \sqrt{169} = 13,$$

is shown in Figure 3.18.

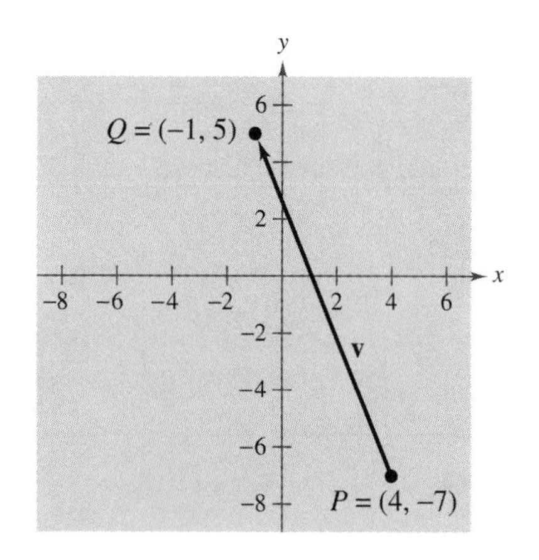

FIGURE 3.18

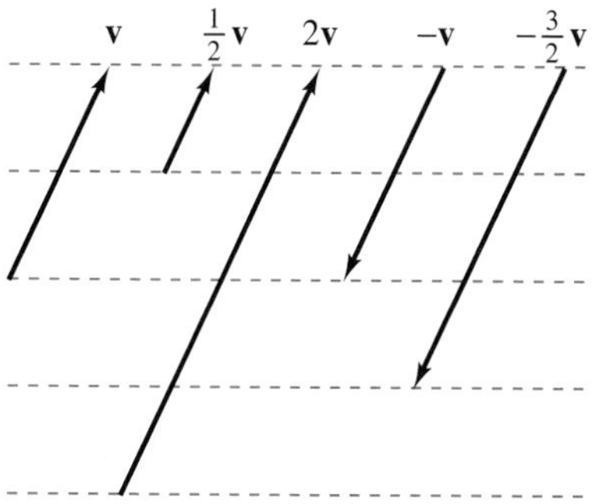

Vector Operations

FIGURE 3.19

The two basic vector operations are **scalar multiplication** and **vector addition.** Geometrically, the product of a vector $\mathbf{v}$ and a scalar k is the vector that is $|k|$ times as long as $\mathbf{v}$. If k is positive, $k\mathbf{v}$ has the same direction as $\mathbf{v}$, and if k is negative, $k\mathbf{v}$ has the direction opposite that of $\mathbf{v}$, as shown in Figure 3.19.

To add two vectors geometrically, position them (without changing length or direction) so that the initial point of one coincides with the terminal point of the other. The sum $\mathbf{u} + \mathbf{v}$ is formed by joining the initial point of the second vector $\mathbf{v}$ with the terminal point of the first vector $\mathbf{u}$, as shown in Figure 3.20. This technique is called the **parallelogram law** for vector addition because the vector $\mathbf{u} + \mathbf{v}$, often called the **resultant** of vector addition, is the diagonal of a parallelogram having $\mathbf{u}$ and $\mathbf{v}$ as its adjacent sides.

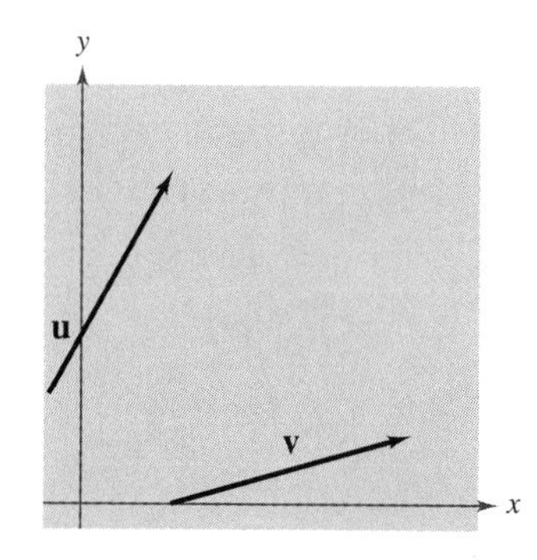

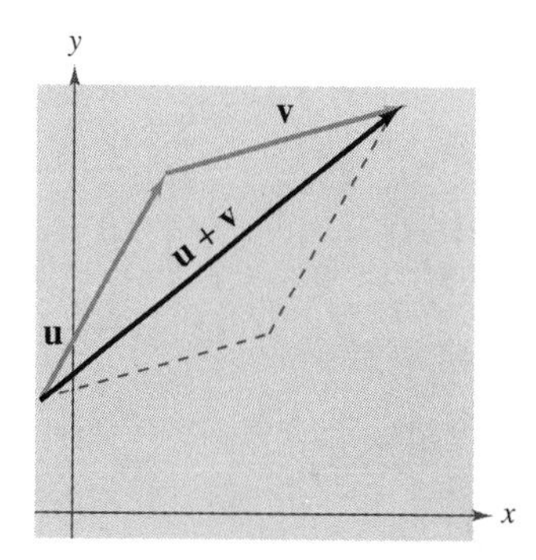

FIGURE 3.20

DEFINITIONS OF VECTOR ADDITION AND SCALAR MULTIPLICATION

Let $\mathbf{u} = \langle u_1, u_2 \rangle$ and $\mathbf{v} = \langle v_1, v_2 \rangle$ be vectors and let k be a scalar (a real number). Then the **sum** of $\mathbf{u}$ and $\mathbf{v}$ is the vector

$$\mathbf{u} + \mathbf{v} = \langle u_1 + v_1, u_2 + v_2 \rangle \qquad \text{Sum}$$

and the **scalar multiple** of k times $\mathbf{u}$ is the vector

$$k\mathbf{u} = k\langle u_1, u_2 \rangle = \langle ku_1, ku_2 \rangle. \qquad \text{Scalar multiple}$$

The **negative** of $\mathbf{v} = \langle v_1, v_2 \rangle$ is

$$-\mathbf{v} = (-1)\mathbf{v} = \langle -v_1, -v_2 \rangle \qquad \text{Negative}$$

and the **difference** of $\mathbf{u}$ and $\mathbf{v}$ is

$$\mathbf{u} - \mathbf{v} = \mathbf{u} + (-\mathbf{v}) = \langle u_1 - v_1, u_2 - v_2 \rangle. \qquad \text{Difference}$$

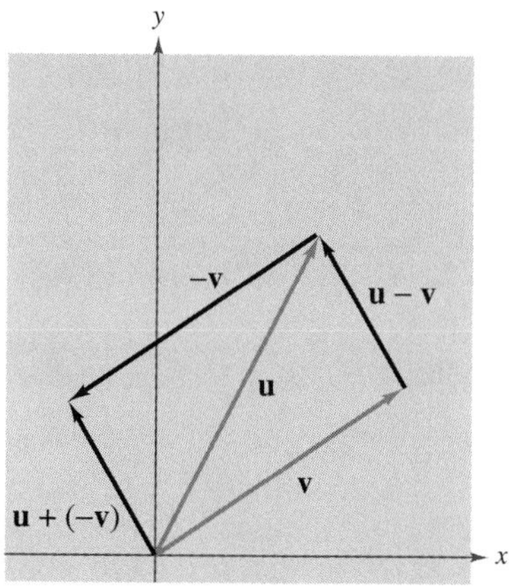

To represent $\mathbf{u} - \mathbf{v}$ graphically, you can use directed line segments with the *same* initial point. The difference $\mathbf{u} - \mathbf{v}$ is the vector from the terminal point of $\mathbf{v}$ to the terminal point of $\mathbf{u}$, as shown in Figure 3.21.

FIGURE 3.21

The component definitions of vector addition and scalar multiplication are illustrated in Example 3. In this example, notice that each of the vector operations can be interpreted geometrically.

EXAMPLE 3 *Vector Operations*

Let $\mathbf{v} = \langle -2, 5 \rangle$ and $\mathbf{w} = \langle 3, 4 \rangle$, and find each of the following vectors.

a. $2\mathbf{v}$ **b.** $\mathbf{w} - \mathbf{v}$ **c.** $\mathbf{v} + 2\mathbf{w}$

Solution

a. Because $\mathbf{v} = \langle -2, 5 \rangle$, you have

$$\begin{aligned} 2\mathbf{v} &= \langle 2(-2), 2(5) \rangle \\ &= \langle -4, 10 \rangle. \end{aligned}$$

A sketch of $2\mathbf{v}$ is shown in Figure 3.22(a).

b. The difference of $\mathbf{w}$ and $\mathbf{v}$ is given by

$$\begin{aligned} \mathbf{w} - \mathbf{v} &= \langle 3 - (-2), 4 - 5 \rangle \\ &= \langle 5, -1 \rangle. \end{aligned}$$

A sketch of $\mathbf{w} - \mathbf{v}$ is shown in Figure 3.22(b).

c. Because $2\mathbf{w} = \langle 6, 8 \rangle$, it follows that

$$\begin{aligned} \mathbf{v} + 2\mathbf{w} &= \langle -2, 5 \rangle + \langle 6, 8 \rangle \\ &= \langle -2 + 6, 5 + 8 \rangle \\ &= \langle 4, 13 \rangle. \end{aligned}$$

A sketch of $\mathbf{v} + 2\mathbf{w}$ is shown in Figure 3.22(c).

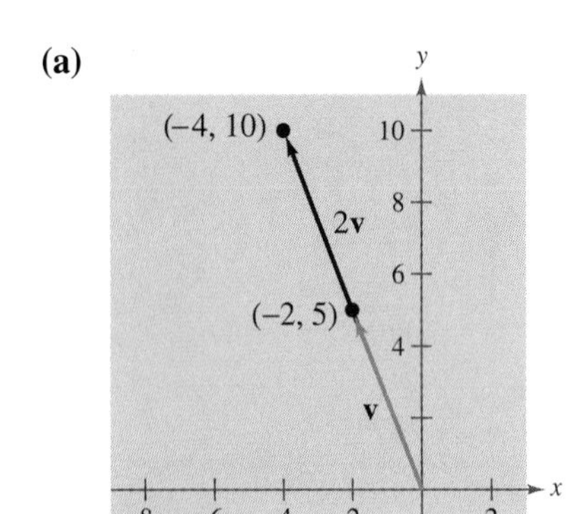

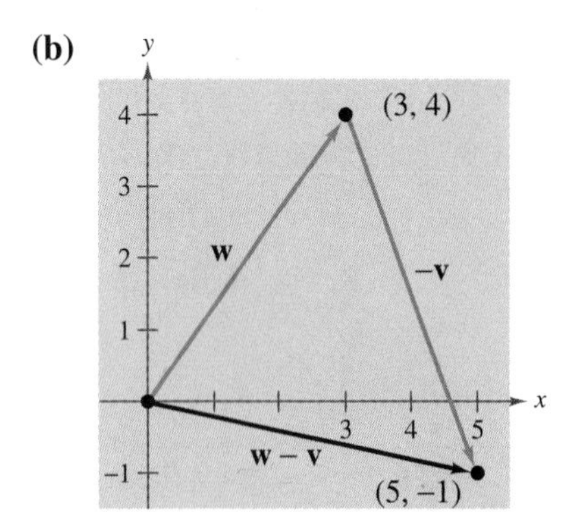

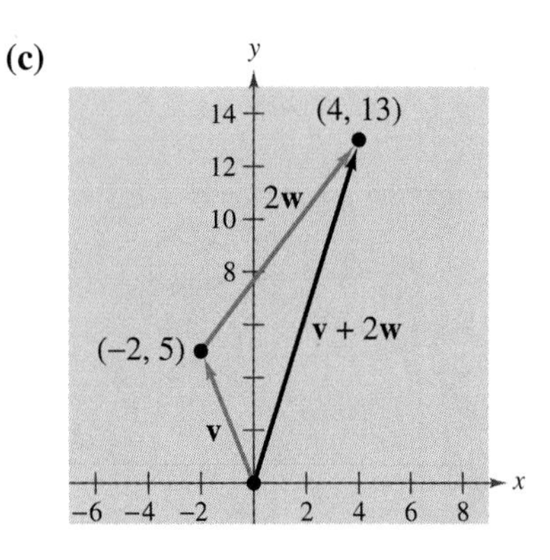

FIGURE 3.22

Vector addition and scalar multiplication share many of the properties of ordinary arithmetic.

NOTE Property 9 can be stated as follows: the length of the vector $c\mathbf{v}$ is the absolute value of c times the length of $\mathbf{v}$.

PROPERTIES OF VECTOR ADDITION AND SCALAR MULTIPLICATION

Let $\mathbf{u}$, $\mathbf{v}$, and $\mathbf{w}$ be vectors and let c and d be scalars. Then the following properties are true.

1. $\mathbf{u} + \mathbf{v} = \mathbf{v} + \mathbf{u}$
2. $(\mathbf{u} + \mathbf{v}) + \mathbf{w} = \mathbf{u} + (\mathbf{v} + \mathbf{w})$
3. $\mathbf{u} + \mathbf{0} = \mathbf{u}$
4. $\mathbf{u} + (-\mathbf{u}) = \mathbf{0}$
5. $c(d\mathbf{u}) = (cd)\mathbf{u}$
6. $(c + d)\mathbf{u} = c\mathbf{u} + d\mathbf{u}$
7. $c(\mathbf{u} + \mathbf{v}) = c\mathbf{u} + c\mathbf{v}$
8. $1(\mathbf{u}) = \mathbf{u}$, $0(\mathbf{u}) = \mathbf{0}$
9. $\|c\mathbf{v}\| = |c|\,\|\mathbf{v}\|$

Unit Vectors

Some of the earliest work with vectors was done by the Irish mathematician William Rowan Hamilton (1805–1865). Hamilton spent many years developing a system of vector-like quantities called quaternions. Although Hamilton was convinced of the benefits of quaternions, the operations he defined did not produce good models for physical phenomena. It wasn't until the latter half of the nineteenth century that the Scottish physicist James Maxwell (1831–1879) restructured Hamilton's quaternions in a form useful for representing physical quantities such as force, velocity, and acceleration.

In many applications of vectors it is useful to find a unit vector that has the same direction as a given nonzero vector $\mathbf{v}$. To do this, you can divide $\mathbf{v}$ by its length to obtain

$$\mathbf{u} = \text{unit vector} = \frac{\mathbf{v}}{\|\mathbf{v}\|} = \left(\frac{1}{\|\mathbf{v}\|}\right)\mathbf{v}.$$

Note that $\mathbf{u}$ is a scalar multiple of $\mathbf{v}$. The vector $\mathbf{u}$ has length 1 and the same direction as $\mathbf{v}$. The vector $\mathbf{u}$ is called a **unit vector in the direction of v.**

EXAMPLE 4 Finding a Unit Vector

Find a unit vector in the direction of $\mathbf{v} = \langle -2, 5\rangle$ and verify that the result has length 1.

Solution

The unit vector in the direction of $\mathbf{v}$ is

$$\frac{\mathbf{v}}{\|\mathbf{v}\|} = \frac{\langle -2, 5\rangle}{\sqrt{(-2)^2 + (5)^2}}$$

$$= \frac{1}{\sqrt{29}}\langle -2, 5\rangle = \left\langle \frac{-2}{\sqrt{29}}, \frac{5}{\sqrt{29}}\right\rangle.$$

This vector has length 1 because

$$\sqrt{\left(\frac{-2}{\sqrt{29}}\right)^2 + \left(\frac{5}{\sqrt{29}}\right)^2} = \sqrt{\frac{4}{29} + \frac{25}{29}} = \sqrt{\frac{29}{29}} = 1.$$

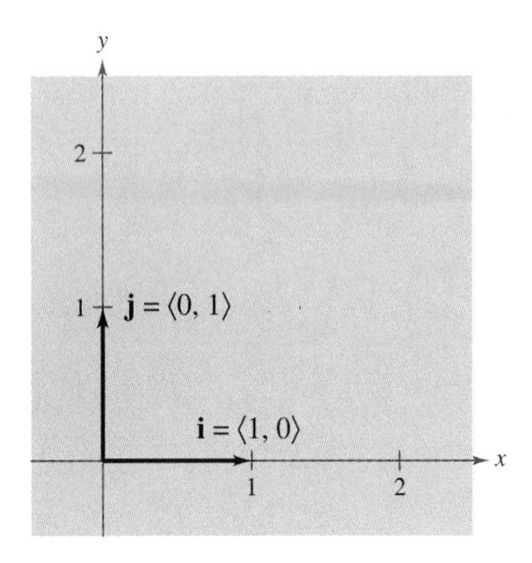

FIGURE 3.23

The unit vectors $\langle 1, 0\rangle$ and $\langle 0, 1\rangle$ are called the **standard unit vectors** and are denoted by

$$\mathbf{i} = \langle 1, 0\rangle \quad \text{and} \quad \mathbf{j} = \langle 0, 1\rangle,$$

as shown in Figure 3.23. (Note that the lowercase letter **i** is written in boldface to distinguish it from the imaginary number $i = \sqrt{-1}$.) These vectors can be used to represent any vector $\mathbf{v} = \langle v_1, v_2\rangle$ as follows.

$$\begin{aligned}\mathbf{v} &= \langle v_1, v_2\rangle \\ &= v_1\langle 1, 0\rangle + v_2\langle 0, 1\rangle \\ &= v_1\mathbf{i} + v_2\mathbf{j}\end{aligned}$$

The scalars v_1 and v_2 are called the **horizontal** and **vertical components of v,** respectively. The vector sum $v_1\mathbf{i} + v_2\mathbf{j}$ is called a **linear combination** of the vectors **i** and **j**. Any vector in the plane can be expressed as a linear combination of the standard unit vectors **i** and **j**.

EXAMPLE 5 *Writing a Linear Combination of Unit Vectors*

Let **u** be the vector with initial point $(2, -5)$ and terminal point $(-1, 3)$. Write **u** as a linear combination of the standard unit vectors **i** and **j**.

Solution

Begin by writing the component form of the vector **u**.

$$\begin{aligned}\mathbf{u} &= \langle -1 - 2, 3 + 5\rangle \\ &= \langle -3, 8\rangle \\ &= -3\mathbf{i} + 8\mathbf{j}\end{aligned}$$

This result is shown graphically in Figure 3.24.

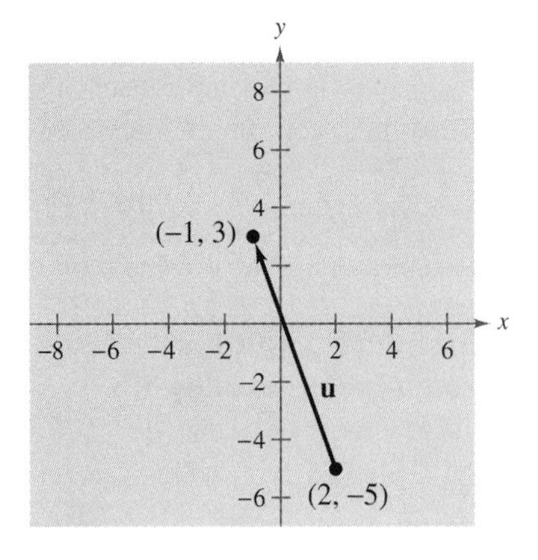

FIGURE 3.24

EXAMPLE 6 *Vector Operations*

Let $\mathbf{u} = -3\mathbf{i} + 8\mathbf{j}$ and $\mathbf{v} = 2\mathbf{i} - \mathbf{j}$. Find $2\mathbf{u} - 3\mathbf{v}$.

Solution

You could solve this problem by converting **u** and **v** to component form. This, however, is not necessary. It is just as easy to perform the operations in unit vector form.

$$\begin{aligned}2\mathbf{u} - 3\mathbf{v} &= 2(-3\mathbf{i} + 8\mathbf{j}) - 3(2\mathbf{i} - \mathbf{j}) \\ &= -6\mathbf{i} + 16\mathbf{j} - 6\mathbf{i} + 3\mathbf{j} \\ &= -12\mathbf{i} + 19\mathbf{j}\end{aligned}$$

Direction Angles

If **u** is a *unit vector* such that θ is the angle (measured counterclockwise) from the positive x-axis to **u**, the terminal point of **u** lies on the unit circle and you have

$$\mathbf{u} = \langle \cos \theta, \sin \theta \rangle = (\cos \theta)\mathbf{i} + (\sin \theta)\mathbf{j}$$

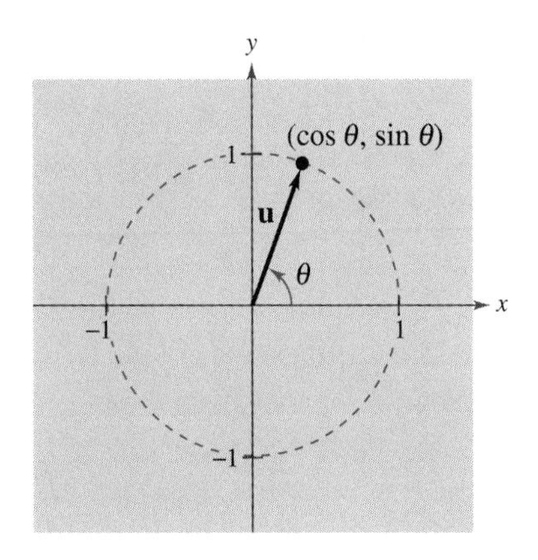

FIGURE 3.25

as shown in Figure 3.25. The angle θ is the **direction angle** of the vector **u**.

Suppose that **u** is a unit vector with direction angle θ. If **v** is any vector that makes an angle θ with the positive x-axis, it has the same direction as **u** and you can write

$$\begin{aligned}\mathbf{v} &= \|\mathbf{v}\| \langle \cos \theta, \sin \theta \rangle \\ &= \|\mathbf{v}\| (\cos \theta)\mathbf{i} + \|\mathbf{v}\| (\sin \theta)\mathbf{j}.\end{aligned}$$

Because $\mathbf{v} = a\mathbf{i} + b\mathbf{j} = \|\mathbf{v}\| (\cos \theta)\, \mathbf{i} + \|\mathbf{v}\| (\sin \theta)\, \mathbf{j}$, it follows that the direction angle θ for **v** is determined from

$$\tan \theta = \frac{\sin \theta}{\cos \theta} = \frac{\|\mathbf{v}\| \sin \theta}{\|\mathbf{v}\| \cos \theta} = \frac{b}{a}.$$

EXAMPLE 7 Finding Direction Angles of Vectors

Find the direction angle of each vector.

a. $\mathbf{u} = 3\mathbf{i} + 3\mathbf{j}$ **b.** $\mathbf{v} = 3\mathbf{i} - 4\mathbf{j}$

Solution

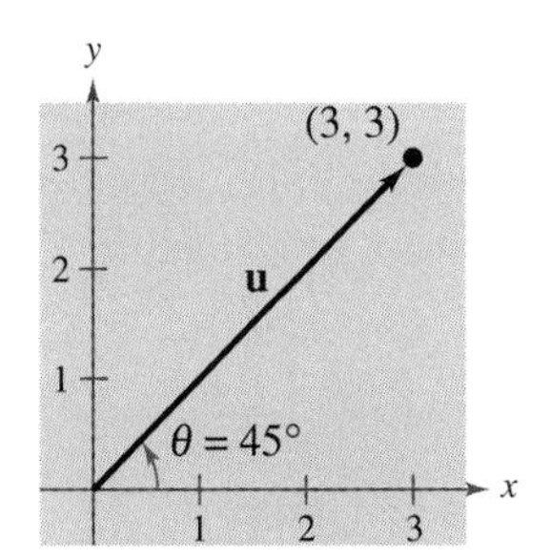

FIGURE 3.26

a. The direction angle is given by

$$\tan \theta = \frac{b}{a} = \frac{3}{3} = 1.$$

Therefore, $\theta = 45°$, as shown in Figure 3.26.

b. The direction angle is given by

$$\tan \theta = \frac{b}{a} = \frac{-4}{3}.$$

Moreover, because $\mathbf{v} = 3\mathbf{i} - 4\mathbf{j}$ lies in Quadrant IV, θ lies in Quadrant IV and its reference angle is

$$\theta = \left| \arctan\left(-\frac{4}{3}\right) \right| \approx |-53.13°| = 53.13°.$$

Therefore, it follows that $\theta \approx 360° - 53.13° = 306.87°$, as shown in Figure 3.27.

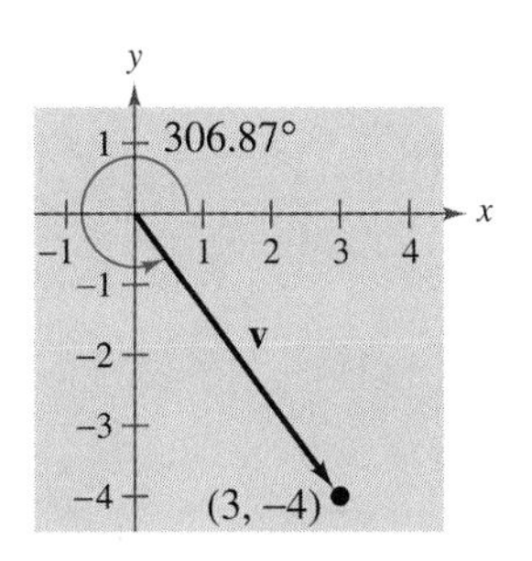

FIGURE 3.27

Applications of Vectors

EXAMPLE 8 *Finding the Component Form of a Vector*

Real Life

Find the component form of the vector that represents the velocity of an airplane descending at a speed of 100 miles per hour at an angle 30° below the horizontal, as shown in Figure 3.28.

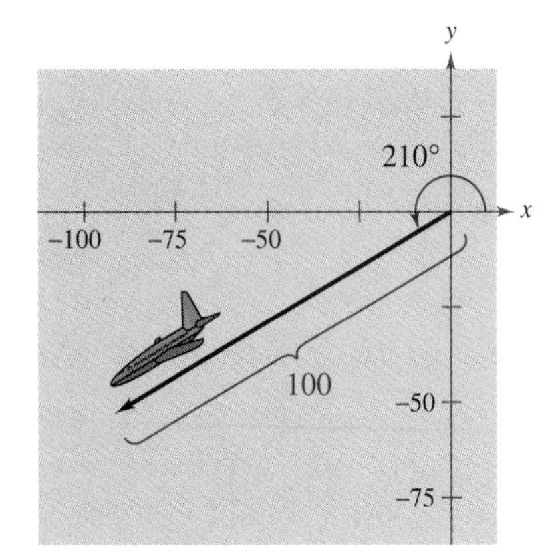

FIGURE 3.28

Solution

The velocity vector **v** has a magnitude of 100 and a direction angle of $\theta = 210°$.

$$\begin{aligned}\mathbf{v} &= \|\mathbf{v}\|(\cos\theta)\mathbf{i} + \|\mathbf{v}\|(\sin\theta)\mathbf{j} \\ &= 100(\cos 210°)\mathbf{i} + 100(\sin 210°)\mathbf{j} \\ &= 100\left(\frac{-\sqrt{3}}{2}\right)\mathbf{i} + 100\left(\frac{-1}{2}\right)\mathbf{j} \\ &= -50\sqrt{3}\,\mathbf{i} - 50\mathbf{j} \\ &= \langle -50\sqrt{3}, -50\rangle\end{aligned}$$

You should check to see that $\|\mathbf{v}\| = 100$.

EXAMPLE 9 *An Application*

A force of 600 pounds is required to pull a boat and trailer up a ramp inclined at 15° from the horizontal. Find the combined weight of the boat and trailer.

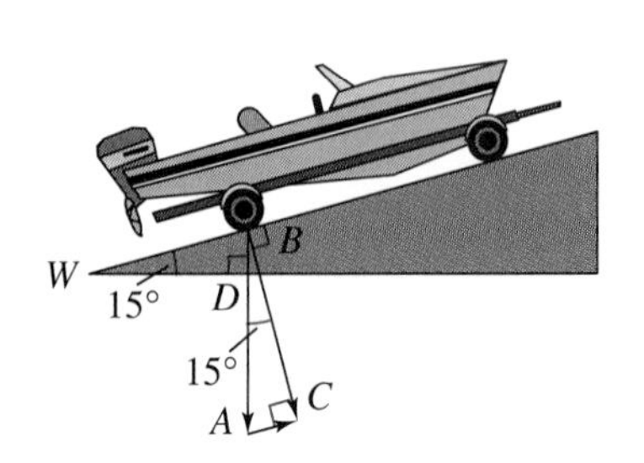

FIGURE 3.29

Solution

Based on Figure 3.29, you can make the following observations.

$\|\overrightarrow{BA}\|$ = force of gravity = combined weight of boat and trailer

$\|\overrightarrow{BC}\|$ = force against ramp

$\|\overrightarrow{AC}\|$ = force required to move boat up ramp = 600 pounds

By construction, triangles BWD and ABC are similar. Hence, angle ABC is 15°. Therefore, in triangle ABC you have

$$\sin 15° = \frac{\|\overrightarrow{AC}\|}{\|\overrightarrow{BA}\|} = \frac{600}{\|\overrightarrow{BA}\|}$$

$$\|BA\| = \frac{600}{\sin 15°} \approx 2318.$$

Consequently, the combined weight is approximately 2318 pounds.

NOTE In Figure 3.29, note that $\overrightarrow{AC}$ is parallel to the ramp.

EXAMPLE 10 An Application

An airplane is traveling at a fixed altitude with a negligible wind factor. The airplane is headed N 30° W at a speed of 500 miles per hour, as shown in Figure 3.30. As the airplane reaches a certain point, it encounters a wind with a velocity of 70 miles per hour in the direction N 45° E. What are the resultant speed and direction of the airplane?

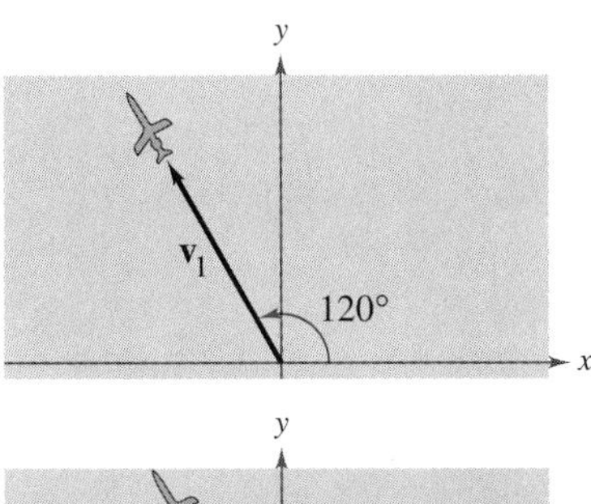

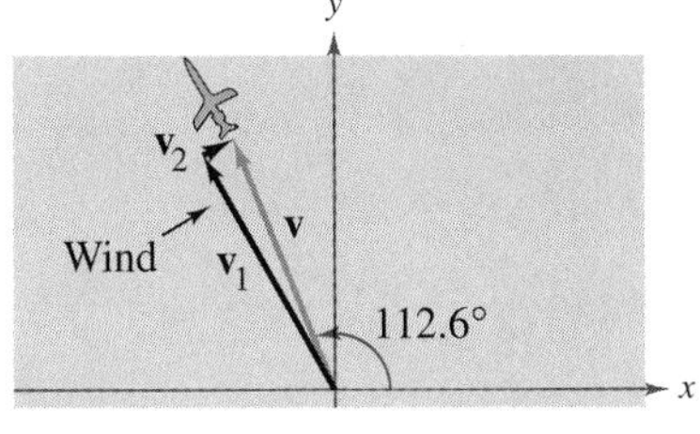

FIGURE 3.30

Solution

Using Figure 3.30, the velocity of the airplane (alone) is given by

$$\mathbf{v}_1 = 500\langle \cos 120°, \sin 120° \rangle = \langle -250, 250\sqrt{3} \rangle$$

and the velocity of the wind is given by

$$\mathbf{v}_2 = 70\langle \cos 45°, \sin 45° \rangle = \langle 35\sqrt{2}, 35\sqrt{2} \rangle.$$

Thus, the velocity of the airplane (in the wind) is given by

$$\mathbf{v} = \mathbf{v}_1 + \mathbf{v}_2 = \langle -250 + 35\sqrt{2}, 250\sqrt{3} + 35\sqrt{2} \rangle \approx \langle -200.5, 482.5 \rangle$$

and the speed of the airplane is

$$\|\mathbf{v}\| = \sqrt{(-200.5)^2 + (482.5)^2} \approx 522.5 \text{ miles per hour.}$$

Finally, if θ if the direction angle of the flight path, you have

$$\tan \theta = \frac{482.5}{-200.5} \approx -2.4065$$

which implies that

$$\theta \approx 180° + \arctan(-2.4065) \approx 180° - 67.4° = 112.6°.$$

GROUP ACTIVITY

VERIFYING THE ASSOCIATIVITY OF VECTOR ADDITION

On page 291, you learned that vector addition is associative—that is, for vectors **u**, **v**, and **w**, **(u + v) + w = u + (v + w)**. Use graph paper and the information in the table to demonstrate geometrically that the resultant vector is the same regardless of whether you add **u** and **v** or **v** and **w** first.

	u	**v**	**w**
Initial Point	$(-3, 2)$	$(1, 6)$	$(2, -2)$
Terminal Point	$(5, -3)$	$(4, -3)$	$(-4, 3)$

WARM UP

Find the distance between the points.

1. $(-2, 6), (5, -15)$ **2.** $(0, 0), (-3, -7)$

Find an equation of the line through the two points.

3. $(3, 1), (-2, 4)$ **4.** $(-2, -3), (4, 5)$

Find an angle θ $(0 \le \theta \le 360°)$ whose vertex is at the origin and whose terminal side passes through the given point.

5. $(-2, 5)$ **6.** $(4, -3)$

Find the sine and cosine of the angle θ.

7. $\theta = 30°$ **8.** $\theta = 120°$

9. $\theta = 300°$ **10.** $\theta = 210°$

3.3 Exercises

In Exercises 1–10, find the component form and the magnitude of the vector v.

1.

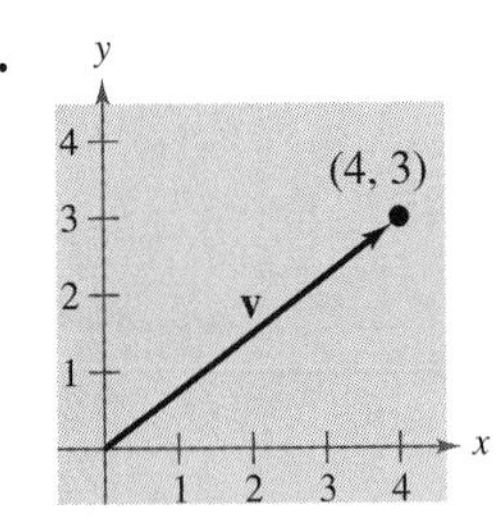

2.

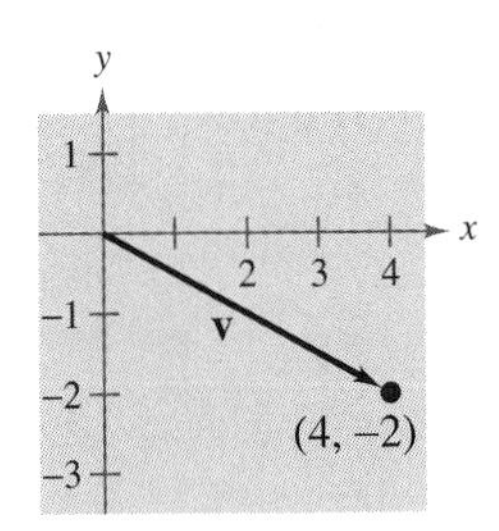

3.

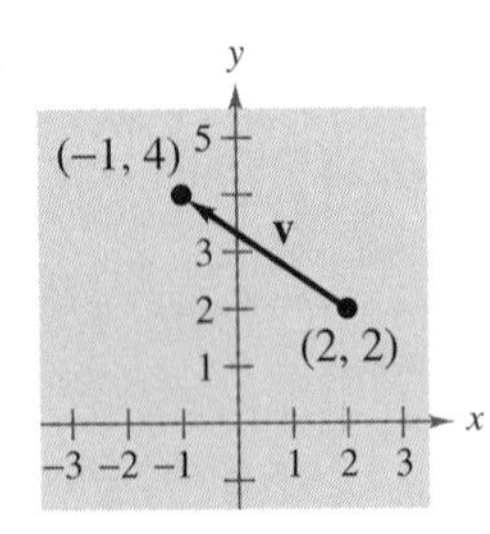

4.

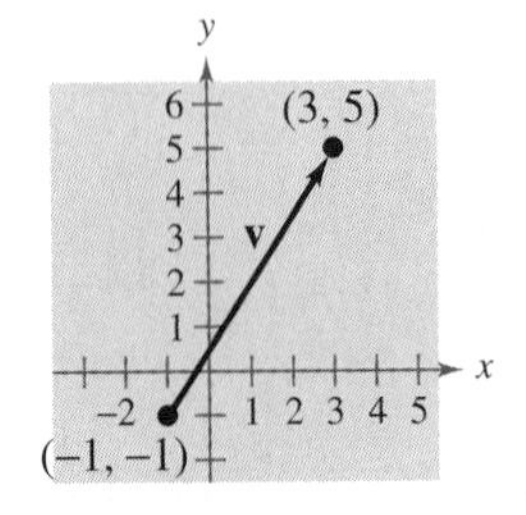

5.

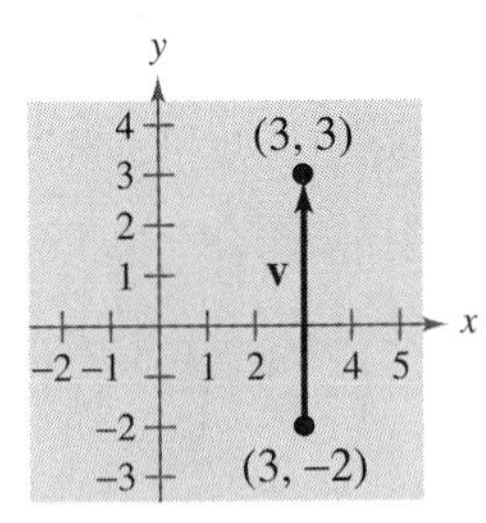

6.

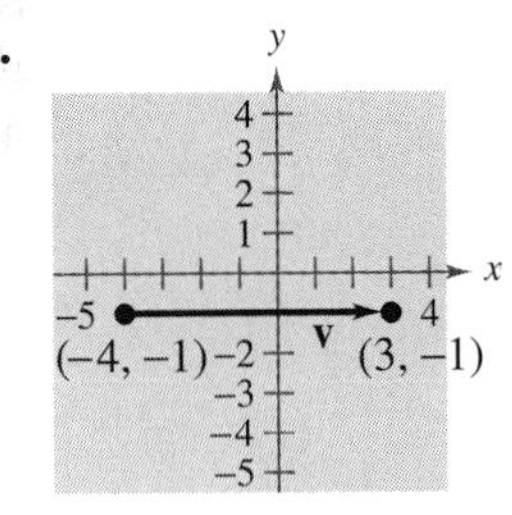

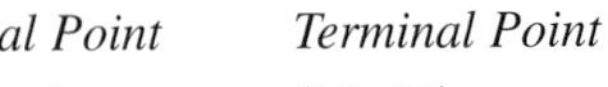

	Initial Point	*Terminal Point*
7.	$(-1, 5)$	$(15, 12)$
8.	$(1, 11)$	$(9, 3)$
9.	$(-3, -5)$	$(5, 1)$
10.	$(-3, 11)$	$(9, 40)$

In Exercises 11–16, use the figure to sketch a graph of the specified vector.

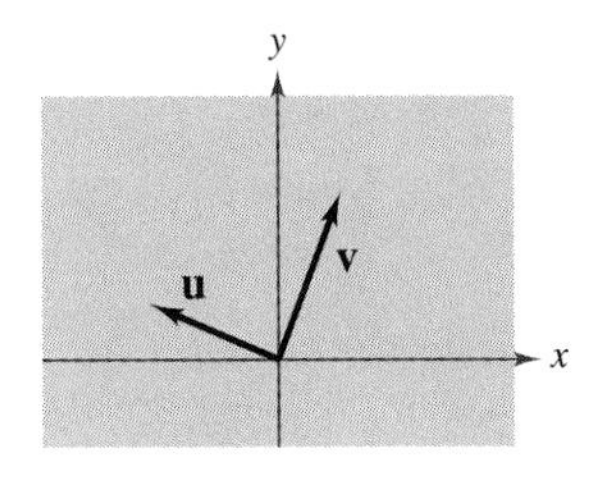

11. $-\mathbf{v}$
12. $3\mathbf{v}$
13. $\mathbf{u} + \mathbf{v}$
14. $\mathbf{u} + 2\mathbf{v}$
15. $\mathbf{u} - \mathbf{v}$
16. $\mathbf{v} - \frac{1}{2}\mathbf{u}$

In Exercises 17–24, find (a) u + v, (b) u − v, and (c) 2u − 3v.

17. $\mathbf{u} = \langle 1, 2 \rangle, \quad \mathbf{v} = \langle 3, 1 \rangle$
18. $\mathbf{u} = \langle 2, 3 \rangle, \quad \mathbf{v} = \langle 4, 0 \rangle$
19. $\mathbf{u} = \langle 4, -2 \rangle, \quad \mathbf{v} = \langle 0, 0 \rangle$
20. $\mathbf{u} = \langle 0, 0 \rangle, \quad \mathbf{v} = \langle 2, 1 \rangle$
21. $\mathbf{u} = \mathbf{i} + \mathbf{j}, \quad \mathbf{v} = 2\mathbf{i} - 3\mathbf{j}$
22. $\mathbf{u} = 2\mathbf{i} - \mathbf{j}, \quad \mathbf{v} = -\mathbf{i} + \mathbf{j}$
23. $\mathbf{u} = 2\mathbf{i}, \quad \mathbf{v} = \mathbf{j}$
24. $\mathbf{u} = 3\mathbf{j}, \quad \mathbf{v} = 2\mathbf{i}$

In Exercises 25–32, find a unit vector in the direction of the given vector.

25. $\mathbf{u} = \langle 5, 0 \rangle$
26. $\mathbf{u} = \langle 0, -3 \rangle$
27. $\mathbf{v} = \langle -2, 2 \rangle$
28. $\mathbf{v} = \langle 5, -12 \rangle$
29. $\mathbf{v} = 4\mathbf{i} - 3\mathbf{j}$
30. $\mathbf{v} = \mathbf{i} + \mathbf{j}$
31. $\mathbf{w} = 2\mathbf{j}$
32. $\mathbf{w} = \mathbf{i} - 2\mathbf{j}$

In Exercises 33–36, find the vector v with the given magnitude and the same direction as u.

	Magnitude	*Direction*
33.	$\|\mathbf{v}\| = 5$	$\mathbf{u} = \langle 3, 3 \rangle$
34.	$\|\mathbf{v}\| = 3$	$\mathbf{u} = \langle 4, -4 \rangle$
35.	$\|\mathbf{v}\| = 7$	$\mathbf{u} = \langle -3, 4 \rangle$
36.	$\|\mathbf{v}\| = 10$	$\mathbf{u} = \langle -10, 0 \rangle$

In Exercises 37–42, find the component form of v and sketch the specified vector operations geometrically, where u = 2i − j and w = i + 2j.

37. $\mathbf{v} = \frac{3}{2}\mathbf{u}$
38. $\mathbf{v} = \mathbf{u} + \mathbf{w}$
39. $\mathbf{v} = \mathbf{u} + 2\mathbf{w}$
40. $\mathbf{v} = -\mathbf{u} + \mathbf{w}$
41. $\mathbf{v} = \frac{1}{2}(3\mathbf{u} + \mathbf{w})$
42. $\mathbf{v} = \mathbf{u} - 2\mathbf{w}$

In Exercises 43–46, find the magnitude and direction angle of the vector v.

43. $\mathbf{v} = 5(\cos 30°\mathbf{i} + \sin 30°\mathbf{j})$
44. $\mathbf{v} = 8(\cos 135°\mathbf{i} + \sin 135°\mathbf{j})$
45. $\mathbf{v} = 6\mathbf{i} - 6\mathbf{j}$
46. $\mathbf{v} = -2\mathbf{i} + 5\mathbf{j}$

In Exercises 47–54, find the component form of v given its magnitude and the angle it makes with the positive *x*-axis. Sketch v.

	Magnitude	*Angle*
47.	$\|\mathbf{v}\| = 3$	$\theta = 0°$
48.	$\|\mathbf{v}\| = 1$	$\theta = 45°$
49.	$\|\mathbf{v}\| = 1$	$\theta = 150°$
50.	$\|\mathbf{v}\| = \frac{5}{2}$	$\theta = 45°$
51.	$\|\mathbf{v}\| = 3\sqrt{2}$	$\theta = 150°$
52.	$\|\mathbf{v}\| = 9$	$\theta = 90°$
53.	$\|\mathbf{v}\| = 2$	$\mathbf{v}$ in the direction $\mathbf{i} + 3\mathbf{j}$
54.	$\|\mathbf{v}\| = 3$	$\mathbf{v}$ in the direction $3\mathbf{i} + 4\mathbf{j}$

In Exercises 55–58, find the component form of the sum of u and v with direction angles $\theta_{\mathbf{u}}$ and $\theta_{\mathbf{v}}$.

	Magnitude	*Angle*
55.	$\|\mathbf{u}\| = 5$	$\theta_{\mathbf{u}} = 0°$
	$\|\mathbf{v}\| = 5$	$\theta_{\mathbf{v}} = 90°$
56.	$\|\mathbf{u}\| = 2$	$\theta_{\mathbf{u}} = 30°$
	$\|\mathbf{v}\| = 2$	$\theta_{\mathbf{v}} = 90°$
57.	$\|\mathbf{u}\| = 20$	$\theta_{\mathbf{u}} = 45°$
	$\|\mathbf{v}\| = 50$	$\theta_{\mathbf{v}} = 180°$
58.	$\|\mathbf{u}\| = 35$	$\theta_{\mathbf{u}} = 25°$
	$\|\mathbf{v}\| = 50$	$\theta_{\mathbf{v}} = 120°$

In Exercises 59–62, use the Law of Cosines to find the angle α between the given vectors. (Assume $0° \leq \alpha \leq 180°$.)

59. $\mathbf{v} = \mathbf{i} + \mathbf{j}$, $\mathbf{w} = 2(\mathbf{i} - \mathbf{j})$

60. $\mathbf{v} = 3\mathbf{i} + \mathbf{j}$, $\mathbf{w} = 2\mathbf{i} - \mathbf{j}$

61. $\mathbf{v} = \mathbf{i} + \mathbf{j}$, $\mathbf{w} = 3\mathbf{i} - \mathbf{j}$

62. $\mathbf{v} = \mathbf{i} + 2\mathbf{j}$, $\mathbf{w} = 2\mathbf{i} - \mathbf{j}$

In Exercises 63 and 64, find the angle between the forces given the magnitude of their resultant. (*Hint:* Write force one as a vector in the direction of the positive x-axis and force two as a vector at an angle θ with the positive x-axis.)

	Force One	*Force Two*	*Resultant Force*
63.	45 pounds	60 pounds	90 pounds
64.	3000 pounds	1000 pounds	3750 pounds

65. ***Think About It*** Consider two forces of equal magnitude acting on a point.

(a) If the magnitude of the resultant is the sum of the magnitudes of the two forces, make a conjecture about the angle between the forces.

(b) If the resultant of the forces is **0**, make a conjecture about the angle between the forces.

(c) Can the magnitude of the resultant be greater than the sum of the magnitudes of the two forces? Explain.

66. ***Graphical Reasoning*** Consider two forces

$\mathbf{F}_1 = \langle 10, 0 \rangle$ and $\mathbf{F}_2 = 5\langle \cos\theta, \sin\theta \rangle$.

(a) Find $\|\mathbf{F}_1 + \mathbf{F}_2\|$.

(b) Determine the magnitude of the resultant as a function of θ. Use a graphing utility to graph the function for $0 \leq \theta < 2\pi$.

(c) Use the graph in part (b) to determine the range of the function. What is its maximum, and for what value of θ does it occur? What is its minimum, and for what value of θ does it occur?

(d) Explain why the magnitude of the resultant is never 0.

67. ***Resultant Force*** Forces with magnitudes of 150 newtons and 220 newtons act on a hook (see figure). The angle between the two forces is 30°. Find the direction and magnitude of the resultant of these forces.

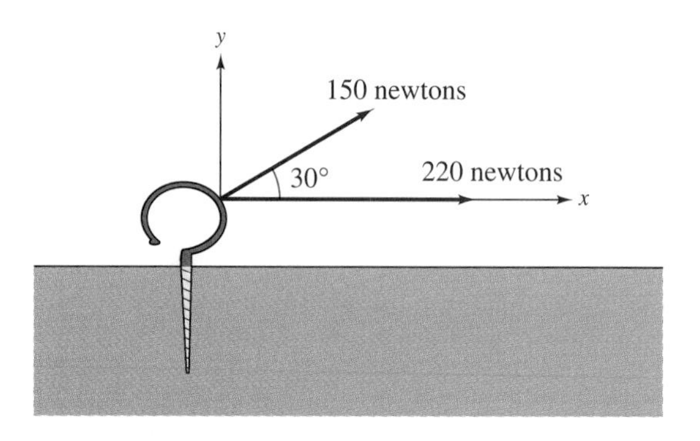

68. ***Resultant Force*** Forces with magnitudes of 2000 newtons and 900 newtons act on a machine part at angles of 30° and −45°, respectively, with the x-axis (see figure). Find the direction and magnitude of the resultant of these forces.

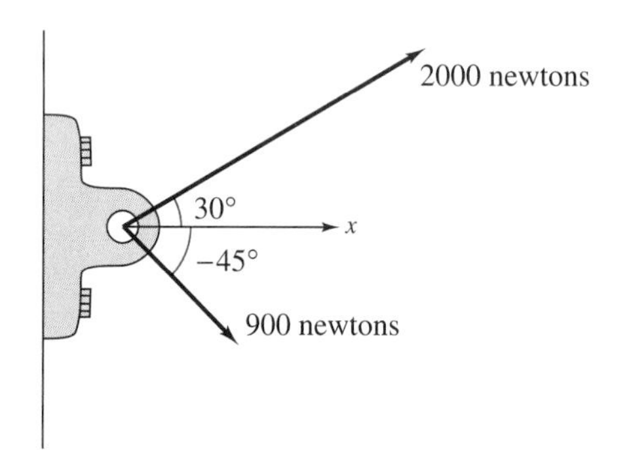

69. ***Resultant Force*** Three forces with magnitudes of 75 pounds, 100 pounds, and 125 pounds act on an object at angles of 30°, 45°, 120°, respectively, with the positive x-axis. Find the direction and magnitude of the resultant of these forces.

70. ***Resultant Force*** Three forces with magnitudes of 70 pounds, 40 pounds, and 60 pounds act on an object at angles of −30°, 45°, and 135°, respectively, with the positive x-axis. Find the direction and magnitude of the resultant of these forces.

71. ***Horizontal and Vertical Components of Velocity*** A ball is thrown with an initial velocity of 80 feet per second, at an angle of 40° with the horizontal (see figure). Find the vertical and horizontal components of the velocity.

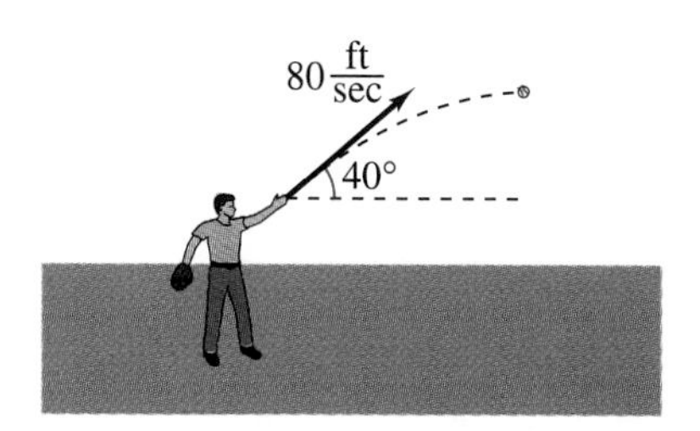

72. ***Horizontal and Vertical Components of Velocity*** A gun with a muzzle velocity of 1200 feet per second is fired at an angle of 6° with the horizontal. Find the vertical and horizontal components of the velocity.

Cable Tension **In Exercises 73 and 74, use the figure to determine the tension in each cable supporting the given load.**

73.

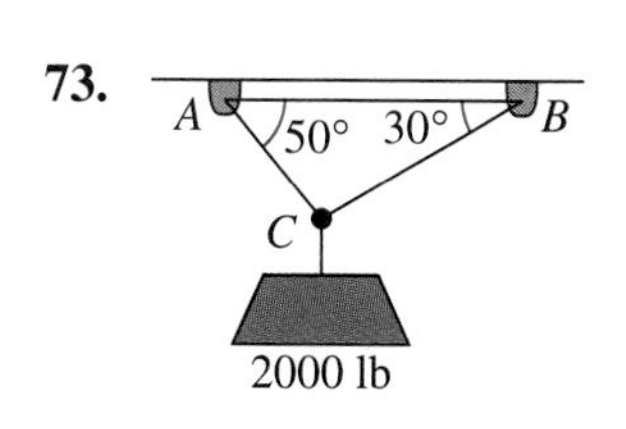

74.

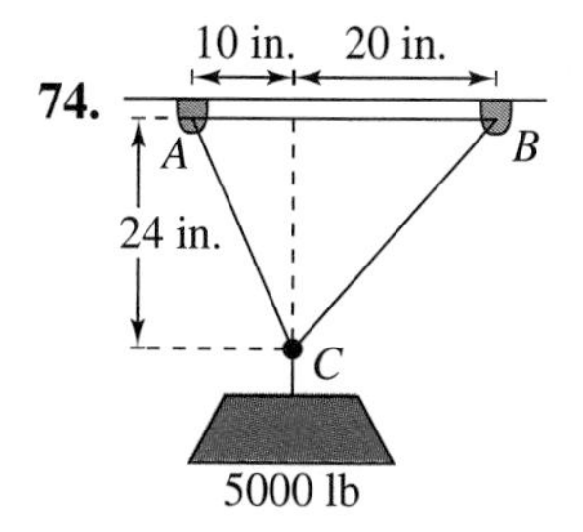

75. ***Barge Towing*** A loaded barge is being towed by two tugboats, and the magnitude of the resultant is 6000 pounds directed along the axis of the barge (see figure). Find the tension in the tow lines if they each make an 18° angle with the axis of the barge.

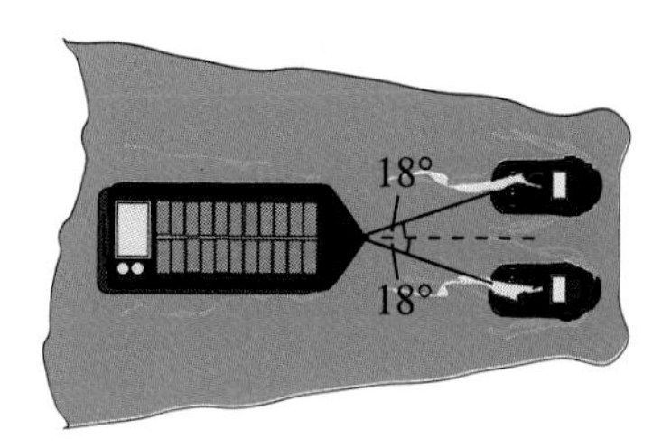

76. ***Shared Load*** To carry a 100-pound cylindrical weight, two people lift on the ends of short ropes that are tied to an eyelet on the top center of the cylinder. Each rope makes a 20° angle with the vertical. Draw a figure that gives a visual representation of the problem, and find the tension in the ropes.

77. ***Navigation*** An airplane is flying in the direction S 32° E, with an airspeed of 875 kilometers per hour. Because of the wind, its ground speed and direction are 800 kilometers per hour and S 40° E, respectively (see figure). Find the direction and speed of the wind.

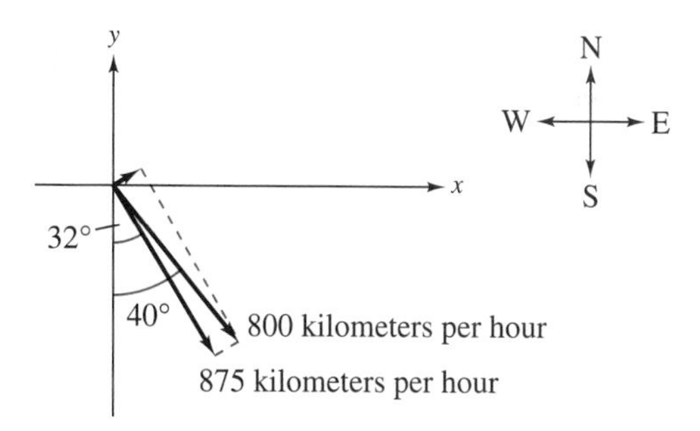

78. ***Navigation*** An airplane's velocity with respect to the air is 580 miles per hour, and it is heading N 60° W. The wind, at the altitude of the plane, is from the southwest and has a velocity of 60 miles per hour. Draw a figure that gives a visual representation of the problem. What is the true direction of the plane, and what is its speed with respect to the ground?

79. ***Work*** A heavy implement is pulled 20 feet across the floor, using a force of 85 pounds. Find the work done if the direction of the force is 60° above the horizontal (see figure). (Use the formula for work, $W = FD$, where F is the component of the force in the direction of motion and D is the distance.)

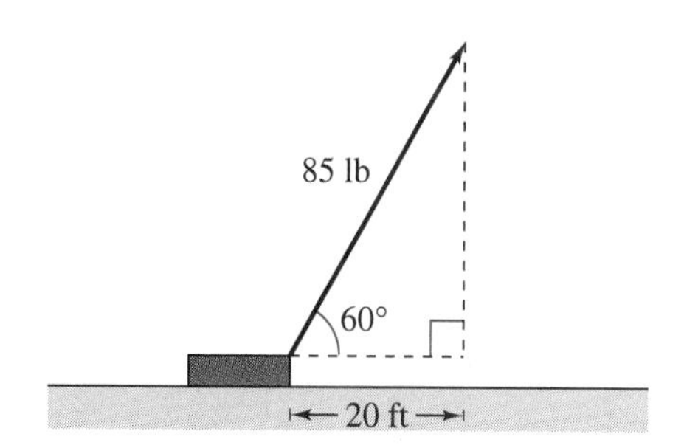

80. ***Tether Ball*** A tether ball weighing 1 pound is pulled outward from the pole by a horizontal force **u** until the rope makes a 45° angle with the pole (see figure). Determine the resulting tension in the rope and the magnitude of **u**.

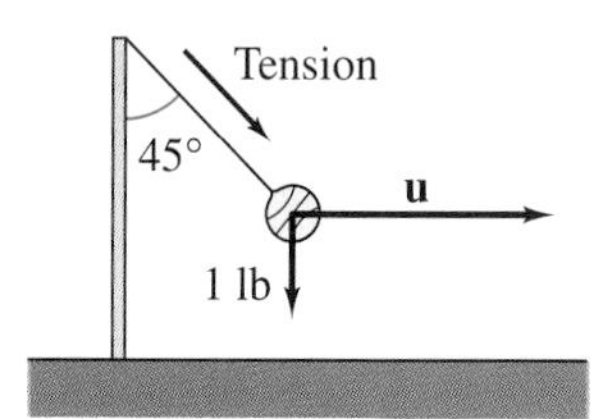

True or False? **In Exercises 81–84, decide whether the statement is true or false. If it is false, explain why or give an example that shows it is false.**

81. If **u** and **v** have the same magnitude and direction, then $\mathbf{u} = \mathbf{v}$.

82. If **u** is a unit vector in the direction of **v**, then $\mathbf{v} = \|\mathbf{v}\|\,\mathbf{u}$.

83. If $\mathbf{v} = a\mathbf{i} + b\mathbf{j} = \mathbf{0}$, then $a = -b$.

84. If $\mathbf{u} = a\mathbf{i} + b\mathbf{j}$ is a unit vector, then $a^2 + b^2 = 1$.

85. Prove that $(\cos\theta)\mathbf{i} + (\sin\theta)\mathbf{j}$ is a unit vector for any value of θ.

86. ***Technology*** Write a program for your graphing utility that graphs two vectors and their difference given the vectors in component form.

In Exercises 87 and 88, use the result of Exercise 86 to find the difference of the vectors shown in the figure.

87.

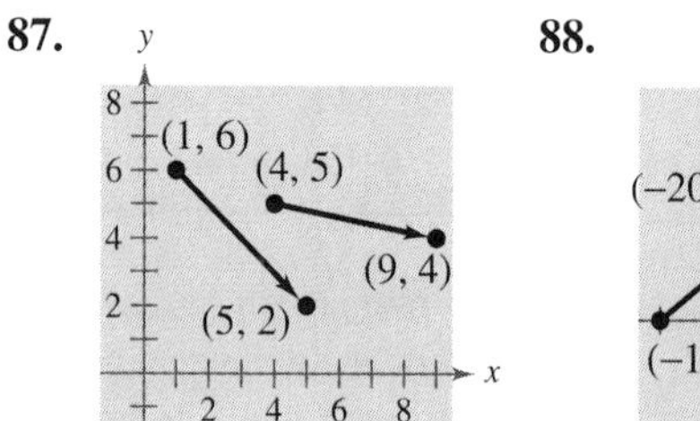

88.

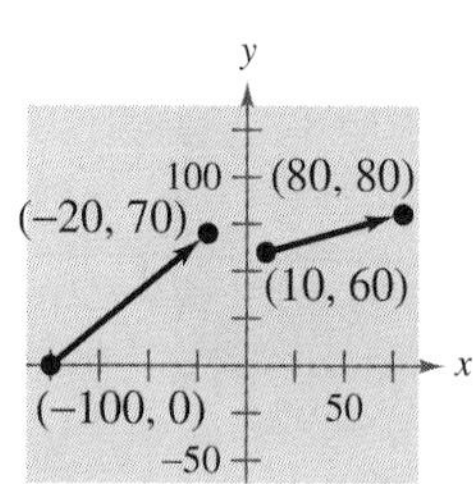

Geodetic surveying is a method of determining the position of points on the earth's surface and the dimension of areas so large that the curvature of the earth must be taken into account. *(Photo: Zigy Kaluzny/ Tony Stone Images)*

89. ***Chapter Opener*** Turn to the figure on page 267 to approximate the distance from the bridge deck to the water level.

90. ***Chapter Opener*** Approximate a in the figure on page 267 if the distance between A and C is 27 meters.

Review **Solve Exercises 91–94 as a review of the skills and problem-solving techniques you learned in previous sections. Use the specified trigonometric substitution to write the algebraic expression as a trigonometric function of θ, where $0 < \theta < \pi/2$.**

91. $\sqrt{x^2 - 64}$, $x = 8\sec\theta$

92. $\sqrt{64 - x^2}$, $x = 8\sin\theta$

93. $\sqrt{x^2 + 36}$, $x = 6\tan\theta$

94. $\sqrt{(x^2 - 25)^3}$, $x = 5\sec\theta$

3.4 Vectors and Dot Products

See Exercise 43 on page 310 for an example of how the dot product can be used to find the force necessary to keep a truck from rolling down a hill.

The Dot Product of Two Vectors ▫ *The Angle Between Two Vectors* ▫ *Finding Vector Components* ▫ *Work*

The Dot Product of Two Vectors

So far you have studied two vector operations—vector addition and multiplication by a scalar—each of which yields another vector. In this section you will study a third vector operation, the **dot product.** This product yields a scalar, rather than a vector.

DEFINITION OF DOT PRODUCT

The **dot product** of $\mathbf{u} = \langle u_1, u_2 \rangle$ and $\mathbf{v} = \langle v_1, v_2 \rangle$ is

$$\mathbf{u} \cdot \mathbf{v} = u_1v_1 + u_2v_2.$$

PROPERTIES OF THE DOT PRODUCT

Let **u, v,** and **w** be vectors in the plane or in space and let c be a scalar.

1. $\mathbf{u} \cdot \mathbf{v} = \mathbf{v} \cdot \mathbf{u}$
2. $\mathbf{0} \cdot \mathbf{v} = 0$
3. $\mathbf{u} \cdot (\mathbf{v} + \mathbf{w}) = \mathbf{u} \cdot \mathbf{v} + \mathbf{u} \cdot \mathbf{w}$
4. $\mathbf{v} \cdot \mathbf{v} = \|\mathbf{v}\|^2$
5. $c(\mathbf{u} \cdot \mathbf{v}) = c\mathbf{u} \cdot \mathbf{v} = \mathbf{u} \cdot c\mathbf{v}$

THINK ABOUT THE PROOF

To prove the second, third, and fifth properties of the dot product, consider the component forms of vectors **u, v,** and **w**. The details of the proof are given in the appendix.

PROOF ▪▪

To prove the first property, let $\mathbf{u} = \langle u_1, u_2 \rangle$ and $\mathbf{v} = \langle v_1\ v_2 \rangle$. Then

$$\begin{aligned}\mathbf{u} \cdot \mathbf{v} &= u_1v_1 + u_2v_2\\ &= v_1u_1 + v_2u_2\\ &= \mathbf{v} \cdot \mathbf{u}.\end{aligned}$$

For the fourth property, let $\mathbf{v} = \langle v_1, v_2 \rangle$. Then

$$\begin{aligned}\mathbf{v} \cdot \mathbf{v} &= v_1^2 + v_2^2\\ &= \left(\sqrt{v_1^2 + v_2^2}\right)^2\\ &= \|\mathbf{v}\|^2.\end{aligned}$$ ▪▪

NOTE In Example 1, be sure you see that the dot product of two vectors is a scalar (a real number), not a vector. Moreover, notice that the dot product can be positive, zero, or negative.

TECHNOLOGY

A graphing utility can be used to find the angle between two vectors. The following program for the *TI-83* or the *TI-82* sketches two vectors $\mathbf{u} = \langle a, b \rangle$ and $\mathbf{v} = \langle c, d \rangle$ in standard position and finds the measure of the angle between them. Use the program to verify Example 4. Before running the program, set an appropriate viewing rectangle. Programs for other graphing calculators may be found in the appendix.

```
:VECANGL
:ClrHome
:Disp "ENTER(A,B)"
:Input "ENTER A",A
:Input "ENTER B",B
:ClrHome
:Disp "ENTER(C,D)"
:Input "ENTER C",C
:Input "ENTER D",D
:Line(0,0,A,B)
:Line(0,0,C,D)
:Pause
:AC+BD→E
:√(A²+B²)→U
:√(C²+D²)→V
:cos⁻¹(E/(UV))→θ
:ClrDraw:ClrHome
:Disp "θ=",θ
:Stop
```

EXAMPLE 1 *Finding Dot Products*

Find each dot product.

a. $\langle 4, 5 \rangle \cdot \langle 2, 3 \rangle$

b. $\langle 2, -1 \rangle \cdot \langle 1, 2 \rangle$

c. $\langle 0, 3 \rangle \cdot \langle 4, -2 \rangle$

Solution

a. $\langle 4, 5 \rangle \cdot \langle 2, 3 \rangle = 4(2) + 5(3) = 8 + 15 = 23$

b. $\langle 2, -1 \rangle \cdot \langle 1, 2 \rangle = 2(1) + (-1)(2) = 2 - 2 = 0$

c. $\langle 0, 3 \rangle \cdot \langle 4, -2 \rangle = 0(4) + 3(-2) = 0 - 6 = -6$

EXAMPLE 2 *Using Properties of Dot Products*

Let $\mathbf{u} = \langle -1, 3 \rangle$, $\mathbf{v} = \langle 2, -4 \rangle$, and $\mathbf{w} = \langle 1, -2 \rangle$. Find each dot product.

a. $(\mathbf{u} \cdot \mathbf{v})\mathbf{w}$

b. $\mathbf{u} \cdot 2\mathbf{v}$

Solution

Begin by finding the dot product of **u** and **v**.

$$\begin{aligned} \mathbf{u} \cdot \mathbf{v} &= \langle -1, 3 \rangle \cdot \langle 2, -4 \rangle \\ &= (-1)(2) + 3(-4) \\ &= -14 \end{aligned}$$

a. $(\mathbf{u} \cdot \mathbf{v})\mathbf{w} = -14\langle 1, -2 \rangle = \langle -14, 28 \rangle$

b. $\mathbf{u} \cdot 2\mathbf{v} = 2(\mathbf{u} \cdot \mathbf{v}) = 2(-14) = -28$

Notice that the first product is a vector, whereas the second is a scalar. Can you see why?

EXAMPLE 3 *Dot Product and Length*

The dot product of **u** with itself is 5. What is the length of **u**?

Solution

Because $\|\mathbf{u}\|^2 = \mathbf{u} \cdot \mathbf{u} = 5$, it follows that

$$\begin{aligned} \|\mathbf{u}\| &= \sqrt{\mathbf{u} \cdot \mathbf{u}} \\ &= \sqrt{5}. \end{aligned}$$

The Angle Between Two Vectors

The **angle between two nonzero vectors** is the angle θ, $0 \le \theta \le \pi$, between their respective standard position vectors, as shown in Figure 3.31. This angle can be found using the dot product. (Note that the angle between the zero vector and another vector is not defined.)

> ***ANGLE BETWEEN TWO VECTORS***
>
> If θ is the angle between two nonzero vectors **u** and **v**, then
>
> $$\cos\theta = \frac{\mathbf{u}\cdot\mathbf{v}}{\|\mathbf{u}\|\,\|\mathbf{v}\|}.$$

PROOF ■■

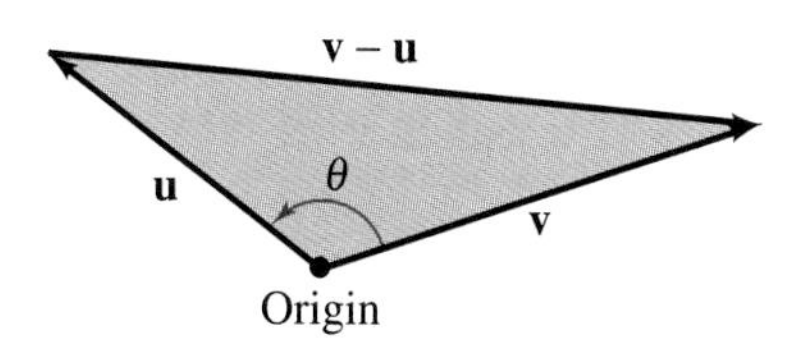

FIGURE 3.31

Consider the triangle determined by vectors **u**, **v**, and **v** − **u**, as shown in Figure 3.31. By the Law of Cosines, you can write

$$\begin{aligned}
\|\mathbf{v}-\mathbf{u}\|^2 &= \|\mathbf{u}\|^2 + \|\mathbf{v}\|^2 - 2\|\mathbf{u}\|\,\|\mathbf{v}\|\cos\theta\\
(\mathbf{v}-\mathbf{u})\cdot(\mathbf{v}-\mathbf{u}) &= \|\mathbf{u}\|^2 + \|\mathbf{v}\|^2 - 2\|\mathbf{u}\|\,\|\mathbf{v}\|\cos\theta\\
(\mathbf{v}-\mathbf{u})\cdot\mathbf{v} - (\mathbf{v}-\mathbf{u})\cdot\mathbf{u} &= \|\mathbf{u}\|^2 + \|\mathbf{v}\|^2 - 2\|\mathbf{u}\|\,\|\mathbf{v}\|\cos\theta\\
\mathbf{v}\cdot\mathbf{v} - \mathbf{u}\cdot\mathbf{v} - \mathbf{v}\cdot\mathbf{u} + \mathbf{u}\cdot\mathbf{u} &= \|\mathbf{u}\|^2 + \|\mathbf{v}\|^2 - 2\|\mathbf{u}\|\,\|\mathbf{v}\|\cos\theta\\
\|\mathbf{v}\|^2 - 2\mathbf{u}\cdot\mathbf{v} + \|\mathbf{u}\|^2 &= \|\mathbf{u}\|^2 + \|\mathbf{v}\|^2 - 2\|\mathbf{u}\|\,\|\mathbf{v}\|\cos\theta\\
-2\mathbf{u}\cdot\mathbf{v} &= -2\|\mathbf{u}\|\,\|\mathbf{v}\|\cos\theta\\
\cos\theta &= \frac{\mathbf{u}\cdot\mathbf{v}}{\|\mathbf{u}\|\,\|\mathbf{v}\|}.
\end{aligned}$$ ■■

EXAMPLE 4 Finding the Angle Between Two Vectors

Find the angle between $\mathbf{u} = \langle 4, 3\rangle$ and $\mathbf{v} = \langle 3, 5\rangle$.

Solution

$$\cos\theta = \frac{\mathbf{u}\cdot\mathbf{v}}{\|\mathbf{u}\|\,\|\mathbf{v}\|} = \frac{\langle 4, 3\rangle \cdot \langle 3, 5\rangle}{\|\langle 4, 3\rangle\|\,\|\langle 3, 5\rangle\|} = \frac{27}{5\sqrt{34}}$$

This implies that the angle between the two vectors is

$$\theta = \arccos\frac{27}{5\sqrt{34}} \approx 22.2°$$

as shown in Figure 3.32.

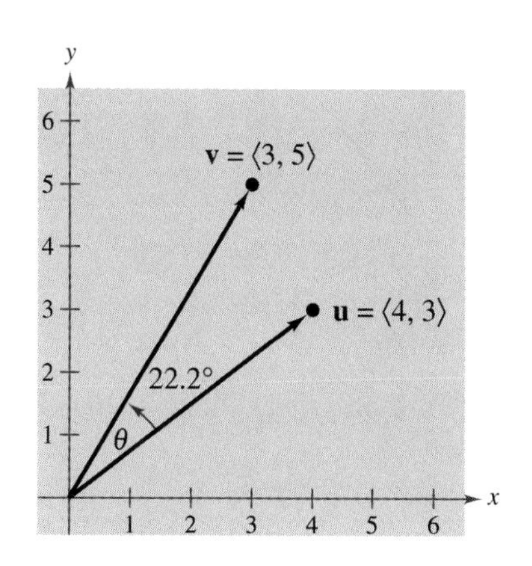

FIGURE 3.32

Rewriting the expression for the angle between two vectors in the form

$$\mathbf{u} \cdot \mathbf{v} = \|\mathbf{u}\| \|\mathbf{v}\| \cos \theta \qquad \text{Alternative form of dot product}$$

produces an alternative way to calculate the dot product. From this form, you can see that because $\|\mathbf{u}\|$ and $\|\mathbf{v}\|$ are always positive, $\mathbf{u} \cdot \mathbf{v}$ and $\cos \theta$ will always have the same sign. Figure 3.33 shows the five possible orientations of two vectors.

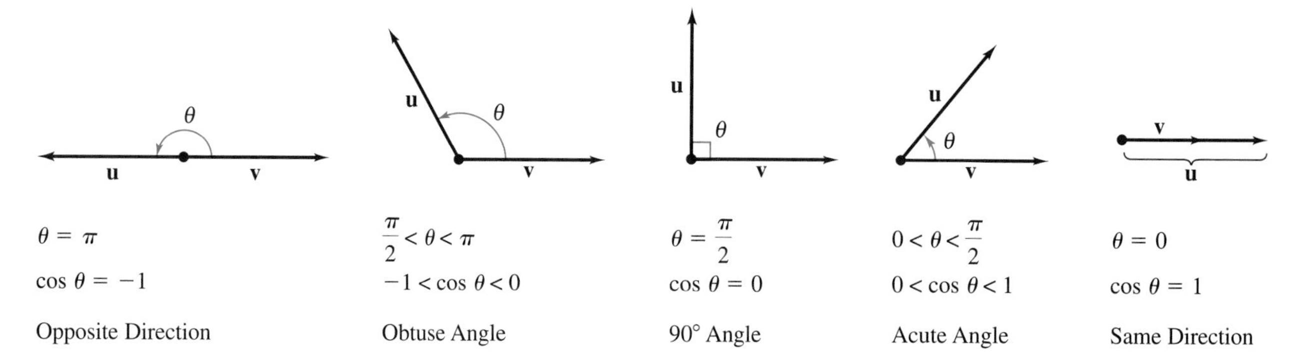

Opposite Direction	Obtuse Angle	90° Angle	Acute Angle	Same Direction
$\theta = \pi$	$\frac{\pi}{2} < \theta < \pi$	$\theta = \frac{\pi}{2}$	$0 < \theta < \frac{\pi}{2}$	$\theta = 0$
$\cos \theta = -1$	$-1 < \cos \theta < 0$	$\cos \theta = 0$	$0 < \cos \theta < 1$	$\cos \theta = 1$

FIGURE 3.33

DEFINITION OF ORTHOGONAL VECTORS

The vectors **u** and **v** are **orthogonal** if $\mathbf{u} \cdot \mathbf{v} = 0$.

The terms "orthogonal" and "perpendicular" mean essentially the same thing—meeting at right angles. By definition, however, the zero vector is orthogonal to every vector **u**, because $\mathbf{0} \cdot \mathbf{u} = 0$.

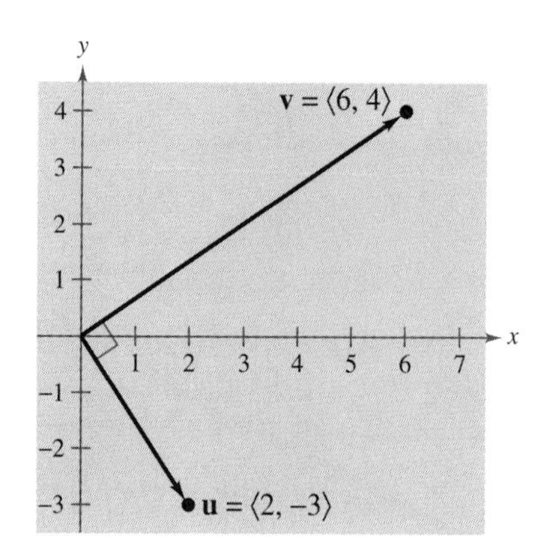

FIGURE 3.34

EXAMPLE 5 *Determining Orthogonal Vectors*

Are the vectors $\mathbf{u} = \langle 2, -3 \rangle$ and $\mathbf{v} = \langle 6, 4 \rangle$ orthogonal?

Solution

Begin by finding the dot product of the two vectors.

$$\begin{aligned} \mathbf{u} \cdot \mathbf{v} &= \langle 2, -3 \rangle \cdot \langle 6, 4 \rangle \\ &= 2(6) + (-3)(4) \\ &= 0 \end{aligned}$$

Because the dot product is 0, the two vectors are orthogonal, as shown in Figure 3.34.

Finding Vector Components

You have already seen applications in which two vectors are added to produce a resultant vector. Many applications in physics and engineering pose the reverse problem—decomposing a given vector into the sum of two **vector components.**

Consider a boat on an inclined ramp, as shown in Figure 3.35. The force $\mathbf{F}$ due to gravity pulls the boat *down* the ramp and *against* the ramp. These two orthogonal forces, $\mathbf{w}_1$ and $\mathbf{w}_2$, are vector components of $\mathbf{F}$. That is,

$$\mathbf{F} = \mathbf{w}_1 + \mathbf{w}_2. \qquad \text{Vector components of } \mathbf{F}$$

The negative of component $\mathbf{w}_1$ represents the force needed to keep the boat from rolling down the ramp, whereas $\mathbf{w}_2$ represents the force that the tires must withstand against the ramp. A procedure for finding $\mathbf{w}_1$ and $\mathbf{w}_2$ is shown below.

FIGURE 3.35

DEFINITION OF VECTOR COMPONENTS

Let $\mathbf{u}$ and $\mathbf{v}$ be nonzero vectors such that

$$\mathbf{u} = \mathbf{w}_1 + \mathbf{w}_2$$

where $\mathbf{w}_1$ and $\mathbf{w}_2$ are orthogonal and $\mathbf{w}_1$ is parallel to $\mathbf{v}$, as shown in Figure 3.36. The vectors $\mathbf{w}_1$ and $\mathbf{w}_2$ are called **vector components** of $\mathbf{u}$. The vector $\mathbf{w}_1$ is the **projection** of $\mathbf{u}$ onto $\mathbf{v}$ and is denoted by

$$\mathbf{w}_1 = \text{proj}_{\mathbf{v}}\mathbf{u}.$$

The vector $\mathbf{w}_2$ is given by $\mathbf{w}_2 = \mathbf{u} - \mathbf{w}_1$.

Study Tip

From the definition of vector components, you can see that it is easy to find the component $\mathbf{w}_2$ once you have found the projection of $\mathbf{u}$ onto $\mathbf{v}$. To find the projection, you can use the dot product.

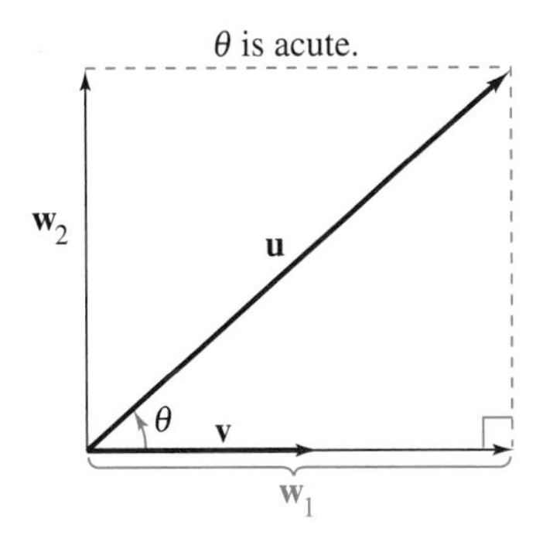

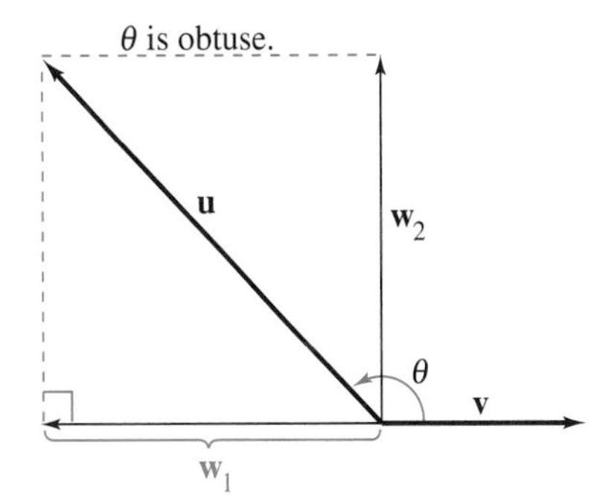

FIGURE 3.36

PROJECTION OF U ONTO V

Let $\mathbf{u}$ and $\mathbf{v}$ be nonzero vectors. The projection of $\mathbf{u}$ onto $\mathbf{v}$ is

$$\text{proj}_{\mathbf{v}}\mathbf{u} = \left(\frac{\mathbf{u} \cdot \mathbf{v}}{\|\mathbf{v}\|^2}\right)\mathbf{v}.$$

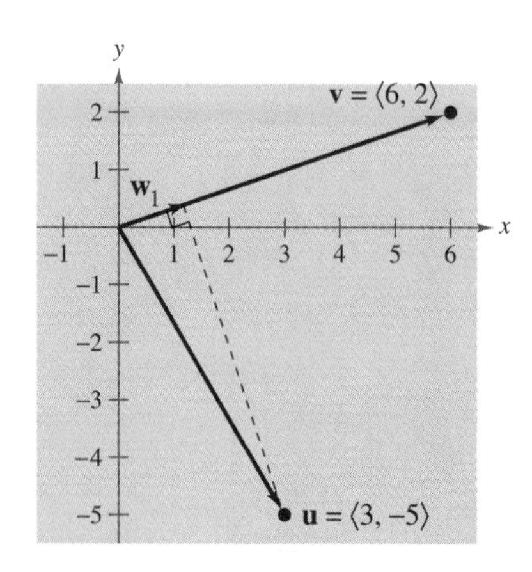

FIGURE 3.37

EXAMPLE 6 *Decomposing a Vector into Components*

Find the projection of $\mathbf{u} = \langle 3, -5 \rangle$ onto $\mathbf{v} = \langle 6, 2 \rangle$. Then write $\mathbf{u}$ as the sum of two orthogonal vectors, one of which is $\text{proj}_{\mathbf{v}}\mathbf{u}$.

Solution

The projection of $\mathbf{u}$ onto $\mathbf{v}$ is

$$\mathbf{w}_1 = \text{proj}_{\mathbf{v}}\mathbf{u} = \left(\frac{\mathbf{u} \cdot \mathbf{v}}{\|\mathbf{v}\|^2}\right)\mathbf{v} = \left(\frac{8}{40}\right)\langle 6, 2 \rangle = \left\langle \frac{6}{5}, \frac{2}{5} \right\rangle,$$

as shown in Figure 3.37. The other component, $\mathbf{w}_2$, is

$$\mathbf{w}_2 = \mathbf{u} - \mathbf{w}_1 = \langle 3, -5 \rangle - \left\langle \frac{6}{5}, \frac{2}{5} \right\rangle = \left\langle \frac{9}{5}, -\frac{27}{5} \right\rangle.$$

Thus, $\mathbf{u} = \mathbf{w}_1 + \mathbf{w}_2 = \left\langle \frac{6}{5}, \frac{2}{5} \right\rangle + \left\langle \frac{9}{5}, -\frac{27}{5} \right\rangle = \langle 3, -5 \rangle.$

EXAMPLE 7 *Finding a Force*

A 600-pound boat sits on a ramp inclined at 30°, as shown in Figure 3.38. What force is required to keep the boat from rolling down the ramp?

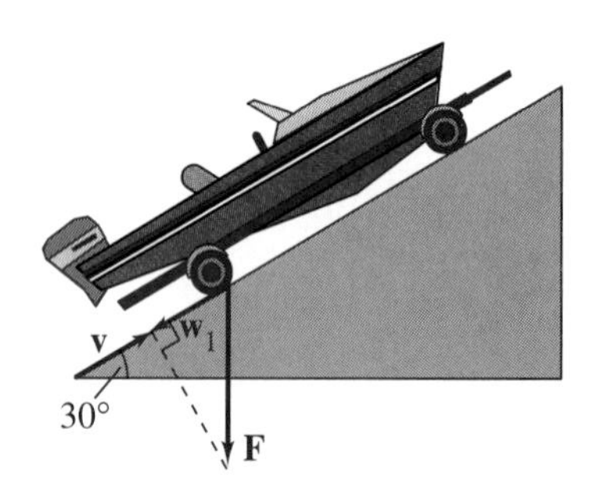

FIGURE 3.38

Solution

Because the force due to gravity is vertical and downward, you can represent the gravitational force by the vector

$$\mathbf{F} = -600\mathbf{j}. \qquad \text{Force due to gravity}$$

To find the force required to keep the boat from rolling down the ramp, project $\mathbf{F}$ onto a unit vector $\mathbf{v}$ in the direction of the ramp, as follows.

$$\mathbf{v} = (\cos 30°)\mathbf{i} + (\sin 30°)\mathbf{j} = \frac{\sqrt{3}}{2}\mathbf{i} + \frac{1}{2}\mathbf{j} \qquad \text{Unit vector along ramp}$$

Therefore, the projection of $\mathbf{F}$ onto $\mathbf{v}$ is given by

$$\begin{aligned}\mathbf{w}_1 &= \text{proj}_{\mathbf{v}}\mathbf{F} \\ &= \left(\frac{\mathbf{F} \cdot \mathbf{v}}{\|\mathbf{v}\|^2}\right)\mathbf{v} \\ &= (\mathbf{F} \cdot \mathbf{v})\mathbf{v} = (-600)\left(\frac{1}{2}\right)\mathbf{v} = -300\left(\frac{\sqrt{3}}{2}\mathbf{i} + \frac{1}{2}\mathbf{j}\right).\end{aligned}$$

The magnitude of this force is 300, and therefore a force of 300 pounds is required to keep the boat from rolling down the ramp.

Work

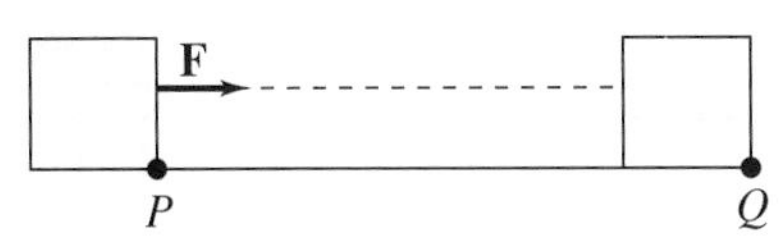

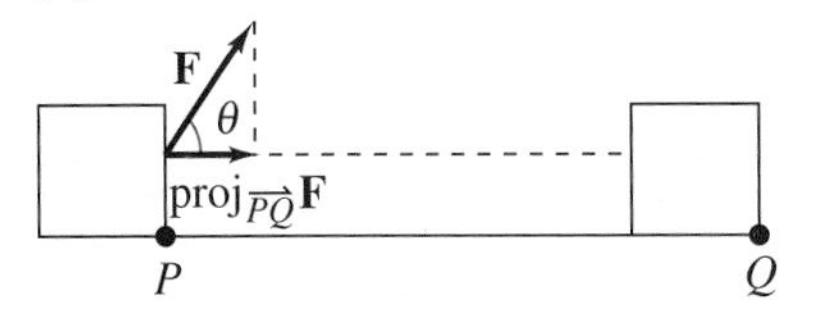

FIGURE 3.39

The work W done by a constant force $\mathbf{F}$ acting along the line of motion of an object is given by

$$\begin{aligned} W &= (\text{magnitude of force})(\text{distance}) \\ &= \|\mathbf{F}\| \, \|\overrightarrow{PQ}\| \end{aligned}$$

as shown in Figure 3.39(a). If the constant force $\mathbf{F}$ is not directed along the line of motion, as shown in Figure 3.39(b), the work W done by the force is

$$W = \|\text{proj}_{\overrightarrow{PQ}} \mathbf{F}\| \, \|\overrightarrow{PQ}\| = (\cos\theta)\|\mathbf{F}\| \, \|\overrightarrow{PQ}\| = \mathbf{F} \cdot PQ.$$

This notion of work is summarized in the following definition.

> ***DEFINITION OF WORK***
>
> The **work** W done by a constant force $\mathbf{F}$ as its point of application moves along the vector $\overrightarrow{PQ}$ is given by either of the following.
>
> **1.** $W = \|\text{proj}_{\overrightarrow{PQ}} \mathbf{F}\| \, \|\overrightarrow{PQ}\|$ Projection form
>
> **2.** $W = \mathbf{F} \cdot \overrightarrow{PQ}$ Dot product form

EXAMPLE 8 Finding Work

To close a sliding door, a person pulls on a rope with a constant force of 50 pounds at a constant angle of 60°, as shown in Figure 3.40. Find the work done in moving the door 12 feet to its closed position.

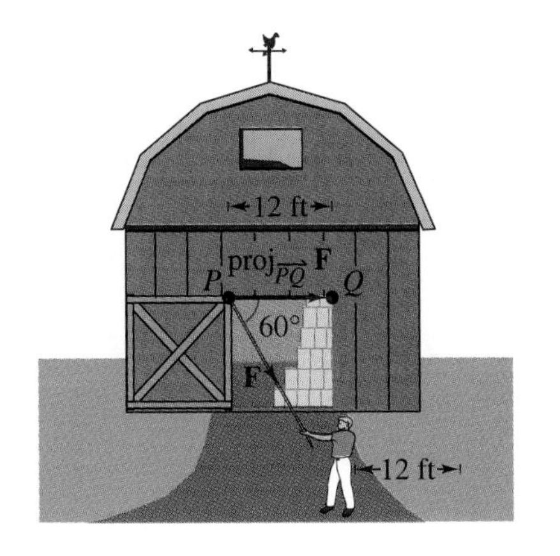

FIGURE 3.40

Solution

Using a projection, you can calculate the work as follows.

$$\begin{aligned} W &= \|\text{proj}_{\overrightarrow{PQ}} \mathbf{F}\| \, \|\overrightarrow{PQ}\| \\ &= (\cos 60°)\|\mathbf{F}\| \, \|\overrightarrow{PQ}\| \\ &= \frac{1}{2}(50)(12) \\ &= 300 \text{ foot–pounds} \end{aligned}$$

Thus, the work done is 300 foot-pounds.

GROUP ACTIVITY

THE SIGN OF THE DOT PRODUCT

On page 304, you were given the alternative form of the dot product of two vectors.

$$\mathbf{u} \cdot \mathbf{v} = \|\mathbf{u}\| \|\mathbf{v}\| \cos \theta \qquad \text{Alternative form of dot product}$$

Use this form to determine the sign of the dot product of **u** and **v** for the vectors shown below. Explain your reasoning.

a.

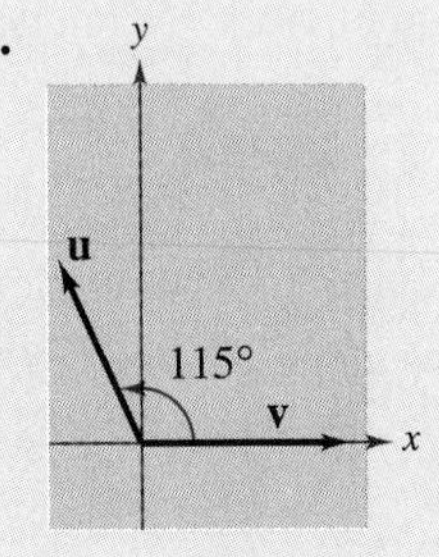

b.

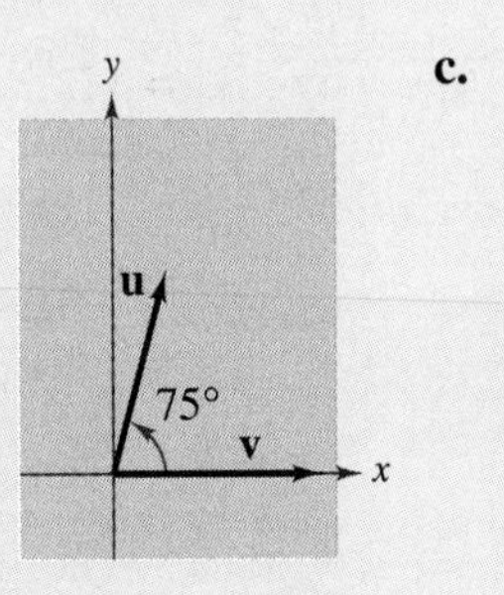

c.

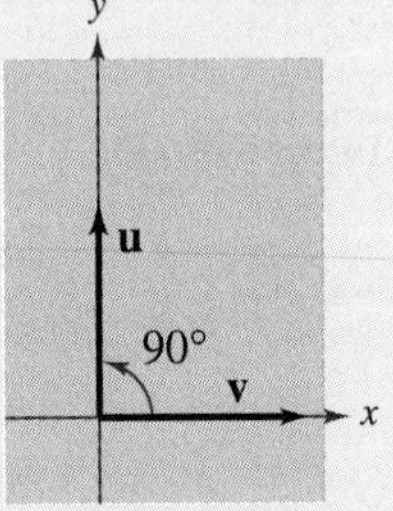

WARM UP

Find (a) $\mathbf{u} + 2\mathbf{v}$ and (b) $\|\mathbf{u}\|$.

1. $\mathbf{u} = \langle 6, -3 \rangle$
$\mathbf{v} = \langle -10, -1 \rangle$

2. $\mathbf{u} = \langle \frac{3}{8}, \frac{4}{5} \rangle$
$\mathbf{v} = \langle \frac{5}{2}, -\frac{1}{10} \rangle$

3. $\mathbf{u} = 4\mathbf{i} - 16\mathbf{j}$
$\mathbf{v} = -5\mathbf{i} + 10\mathbf{j}$

4. $\mathbf{u} = 0.5\mathbf{i} + 1.4\mathbf{j}$
$\mathbf{v} = 4.1\mathbf{i} - 1.8\mathbf{j}$

Find the values of θ in the interval $0 \le \theta \le 2\pi$ that satisfy the equation. Round the result to two decimal places.

5. $\cos \theta = -\frac{1}{2}$

6. $\cos \theta = 0$

7. $\cos \theta = 0.5403$

8. $\cos \theta = -0.9689$

Find a unit vector (a) in the direction of u and (b) in the direction opposite that of u.

9. $\mathbf{u} = \langle 120, -50 \rangle$

10. $\mathbf{u} = \langle \frac{4}{5}, \frac{1}{3} \rangle$

3.4 Exercises

In Exercises 1–4, find the dot product of u and v.

1. $\mathbf{u} = \langle 3, 4\rangle$, $\mathbf{v} = \langle 2, -3\rangle$
2. $\mathbf{u} = \langle 5, 12\rangle$, $\mathbf{v} = \langle -3, 2\rangle$
3. $\mathbf{u} = 4\mathbf{i} - 2\mathbf{j}$, $\mathbf{v} = \mathbf{i} - \mathbf{j}$
4. $\mathbf{u} = 2\mathbf{i} + 5\mathbf{j}$, $\mathbf{v} = 9\mathbf{i} - 3\mathbf{j}$

In Exercises 5–8, use the vectors $\mathbf{u} = \langle 2, 2\rangle$ and $\mathbf{v} = \langle -3, 4\rangle$ to find the indicated quantity. State whether the result is a vector or a scalar.

5. $\mathbf{u} \cdot \mathbf{u}$
6. $\|\mathbf{u}\| - 2$
7. $(\mathbf{u} \cdot \mathbf{v})\mathbf{v}$
8. $\mathbf{u} \cdot 2\mathbf{v}$

In Exercises 9–12, use the dot product to find the length of u.

9. $\mathbf{u} = \langle -5, 12\rangle$
10. $\mathbf{u} = \langle 2, -4\rangle$
11. $\mathbf{u} = 20\mathbf{i} + 25\mathbf{j}$
12. $\mathbf{u} = 6\mathbf{j}$

13. ***Revenue*** The vector $\mathbf{u} = \langle 1245, 2600\rangle$ gives the numbers of units of two products produced by a company. The vector $\mathbf{v} = \langle 12.20, 8.50\rangle$ gives the price (in dollars) of each unit, respectively. Find the dot product, $\mathbf{u} \cdot \mathbf{v}$, and explain what information it gives.
14. ***Revenue*** Repeat Exercise 13 after increasing the prices by 5%. Identify the vector operation used to increase the prices by 5%.

In Exercises 15–20, find the angle θ between the vectors.

15. $\mathbf{u} = \langle 1, 0\rangle$, $\mathbf{v} = \langle 0, -2\rangle$
16. $\mathbf{u} = \langle 4, 4\rangle$, $\mathbf{v} = \langle 2, 0\rangle$
17. $\mathbf{u} = 3\mathbf{i} + 4\mathbf{j}$, $\mathbf{v} = -2\mathbf{j}$
18. $\mathbf{u} = 2\mathbf{i} - 3\mathbf{j}$, $\mathbf{v} = \mathbf{i} - 2\mathbf{j}$
19. $\mathbf{u} = \cos\left(\frac{\pi}{3}\right)\mathbf{i} + \sin\left(\frac{\pi}{3}\right)\mathbf{j}$, $\mathbf{v} = \cos\left(\frac{3\pi}{4}\right)\mathbf{i} + \sin\left(\frac{3\pi}{4}\right)\mathbf{j}$
20. $\mathbf{u} = \cos\left(\frac{\pi}{4}\right)\mathbf{i} + \sin\left(\frac{\pi}{4}\right)\mathbf{j}$, $\mathbf{v} = \cos\left(\frac{\pi}{2}\right)\mathbf{i} + \sin\left(\frac{\pi}{2}\right)\mathbf{j}$

In Exercises 21–24, use a graphing utility to sketch the vectors and find the degree measure of the angle between the vectors.

21. $\mathbf{u} = 3\mathbf{i} + 4\mathbf{j}$, $\mathbf{v} = -7\mathbf{i} + 5\mathbf{j}$
22. $\mathbf{u} = -6\mathbf{i} - 3\mathbf{j}$, $\mathbf{v} = -8\mathbf{i} + 4\mathbf{j}$
23. $\mathbf{u} = 5\mathbf{i} + 5\mathbf{j}$, $\mathbf{v} = -6\mathbf{i} + 6\mathbf{j}$
24. $\mathbf{u} = 2\mathbf{i} - 3\mathbf{j}$, $\mathbf{v} = 4\mathbf{i} + 3\mathbf{j}$

In Exercises 25 and 26, use vectors to find the interior angles of the triangle with the given vertices.

25. $(1, 2), (3, 4), (2, 5)$
26. $(-3, 0), (2, 2), (0, 6)$

In Exercises 27 and 28, find $\mathbf{u} \cdot \mathbf{v}$, where θ is the angle between u and v.

27. $\|\mathbf{u}\| = 4, \|\mathbf{v}\| = 10, \theta = \frac{2\pi}{3}$
28. $\|\mathbf{u}\| = 100, \|\mathbf{v}\| = 250, \theta = \frac{\pi}{6}$

In Exercises 29–34, determine whether u and v are orthogonal, parallel, or neither.

29. $\mathbf{u} = \langle -12, 30\rangle$, $\mathbf{v} = \langle \frac{1}{2}, -\frac{5}{4}\rangle$
30. $\mathbf{u} = \langle 15, 45\rangle$, $\mathbf{v} = \langle -5, 12\rangle$
31. $\mathbf{u} = \frac{1}{4}(3\mathbf{i} - \mathbf{j})$, $\mathbf{v} = 5\mathbf{i} + 6\mathbf{j}$
32. $\mathbf{u} = \mathbf{j}$, $\mathbf{v} = \mathbf{i} - 2\mathbf{j}$
33. $\mathbf{u} = 2\mathbf{i} - 2\mathbf{j}$, $\mathbf{v} = -\mathbf{i} - \mathbf{j}$
34. $\mathbf{u} = \langle \cos\theta, \sin\theta\rangle$, $\mathbf{v} = \langle \sin\theta, -\cos\theta\rangle$

In Exercises 35–38, find the projection of u onto v, and the vector component of u orthogonal to v.

35. $\mathbf{u} = \langle 3, 4\rangle$, $\mathbf{v} = \langle 8, 2\rangle$

36. $\mathbf{u} = \langle 4, 2\rangle$, $\mathbf{v} = \langle 1, -2\rangle$

37. $\mathbf{u} = \langle 0, 3\rangle$, $\mathbf{v} = \langle 2, 15\rangle$

38. $\mathbf{u} = \langle -5, -1\rangle$, $\mathbf{v} = \langle -1, 1\rangle$

In Exercises 39–42, find two vectors in opposite directions that are orthogonal to the vector u. (The answers are not unique.)

39. $\mathbf{u} = \langle 3, 5\rangle$

40. $\mathbf{u} = \langle -8, 3\rangle$

41. $\mathbf{u} = \frac{1}{2}\mathbf{i} - \frac{2}{3}\mathbf{j}$

42. $\mathbf{u} = -\frac{5}{2}\mathbf{i} - 3\mathbf{j}$

43. ***Braking Load*** A truck with a gross weight of 36,000 pounds is parked on a 10° slope (see figure). Assume that the only force to overcome is the force of gravity.

(a) Find the force required to keep the truck from rolling down the hill.

(b) Find the force perpendicular to the hill.

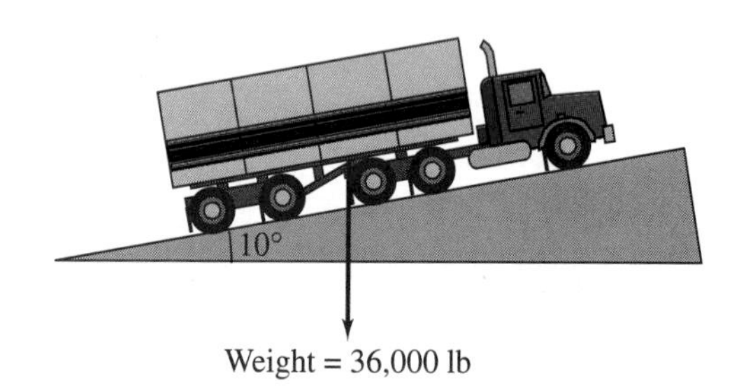

44. ***Braking Load*** Rework Exercise 43 for a truck that is parked on a 12° slope.

45. ***Think About It*** What is known about θ, the angle between two nonzero vectors **u** and **v**, under the following conditions?

(a) $\mathbf{u} \cdot \mathbf{v} = 0$ (b) $\mathbf{u} \cdot \mathbf{v} > 0$ (c) $\mathbf{u} \cdot \mathbf{v} < 0$

46. ***Think About It*** What can be said about the vectors **u** and **v** under the following conditions?

(a) The projection of **u** onto **v** equals **u**.

(b) The projection of **u** onto **v** equals **0**.

47. ***Work*** A 25-kilogram (245-newton) bag of sugar is lifted 3 meters. Determine the work done.

48. ***Work*** Determine the work done by a crane lifting a 2400-pound car 5 feet.

49. ***Work*** A force of 45 pounds in the direction of 30° above the horizontal is required to slide an implement across a floor (see figure). Find the work done if the implement is dragged 20 feet.

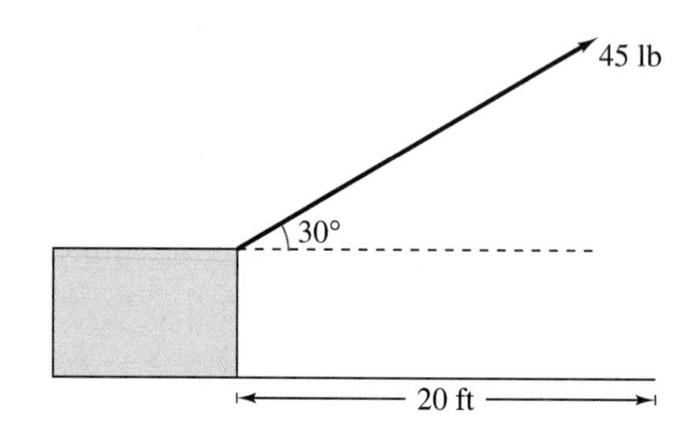

50. ***Work*** A tractor pulls a log 800 meters and the tension in the cable connecting the tractor and log is approximately 1600 kilograms (15,691 newtons). Approximate the work done if the direction of the force is 35° above the horizontal.

Work **In Exercises 51 and 52, find the work done in moving a particle from *P* to *Q* if the magnitude and direction of the force are given by v.**

51. $P = (0, 0)$, $Q = (4, 7)$, $\mathbf{v} = \langle 1, 4\rangle$

52. $P = (1, 3)$, $Q = (-3, 5)$, $\mathbf{v} = -2\mathbf{i} + 3\mathbf{j}$

53. Use vectors to prove that the diagonals of a rhombus are perpendicular.

54. Prove the following.

$$\|\mathbf{u} - \mathbf{v}\|^2 = \|\mathbf{u}\|^2 + \|\mathbf{v}\|^2 - 2\mathbf{u} \cdot \mathbf{v}$$

55. Prove the following properties of the dot product.

(a) $\mathbf{0} \cdot \mathbf{v} = 0$

(b) $\mathbf{u} \cdot (\mathbf{v} + \mathbf{w}) = \mathbf{u} \cdot \mathbf{v} + \mathbf{u} \cdot \mathbf{w}$

(c) $c(\mathbf{u} \cdot \mathbf{v}) = c\mathbf{u} \cdot \mathbf{v} = \mathbf{u} \cdot c\mathbf{v}$

56. Prove that if **u** is orthogonal to **v** and **w**, then **u** is orthogonal to $c\mathbf{v} + d\mathbf{w}$ for any scalars c and d.

Focus on Concepts

In this chapter, you studied methods for solving oblique triangles and vectors in the plane. You can use the following questions to check your understanding of several of the basic concepts presented in the chapter. The answers to these questions are given in the back of the book.

1. State the Law of Sines from memory.

2. State the Law of Cosines from memory.

3. *True or False?* The Law of Sines is true if one of the angles in the triangle is a right angle.

4. If one of the angles in the triangle is a right angle, the Law of Cosines simplifies to what famous theorem?

5. *True or False?* When the Law of Sines is used, the solution is always unique. Explain.

6. What characterizes a vector in the plane?

7. Which vectors in the figure appear to be equivalent?

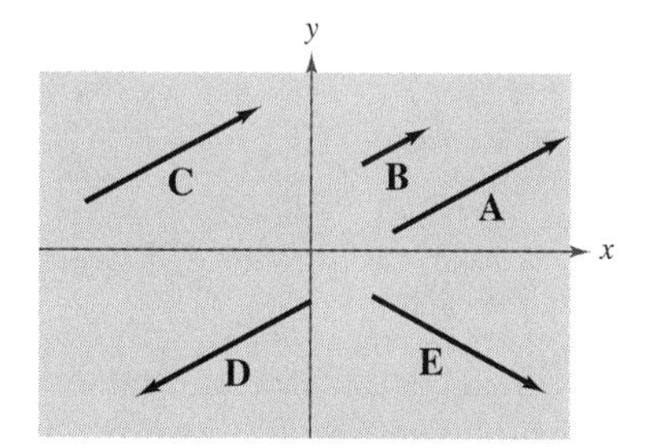

8. The vectors **u** and **v** have the same magnitudes in the two figures. In which figure will the magnitude of the resultant be greater? Give a reason for your answer.

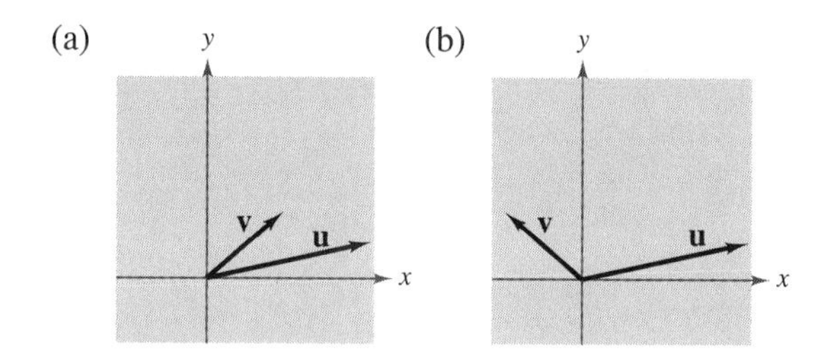

9. Give a geometric description of the scalar multiple $k\mathbf{u}$ of the vector **u**.

10. Give a geometric description of the sum of the vectors **u** and **v**.

11. Which of the two figures shows the difference $\mathbf{u} - \mathbf{v}$? Give a geometric description of the difference and state how you determine its direction.

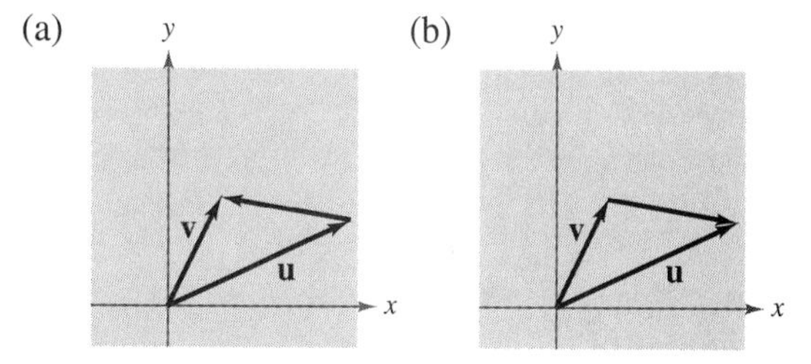

12. Determine whether the dot product of the vectors in each figure is positive, negative, or zero. Explain.

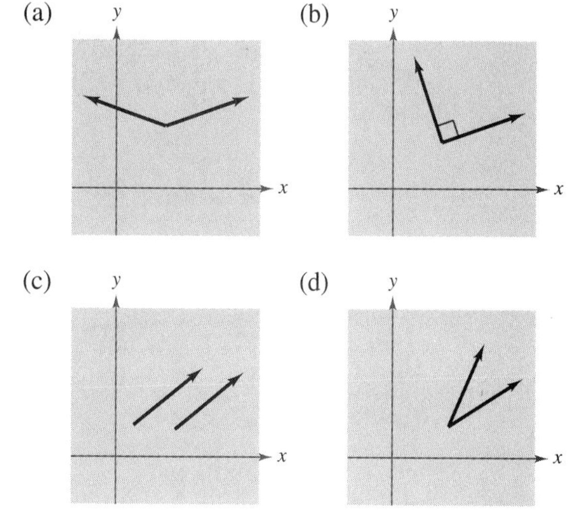

13. Is it possible that $\text{proj}_{\mathbf{v}}(\mathbf{u}) = \mathbf{u}$? If so, give an example.

Review Exercises

In Exercises 1–16, use the given information to solve the triangle (if possible). If two solutions exist, list both.

1. $a = 5, \quad b = 8, \quad c = 10$

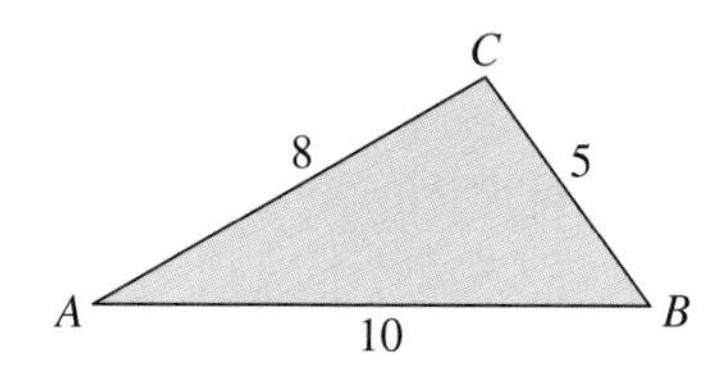

2. $a = 6, \quad b = 9, \quad C = 45°$

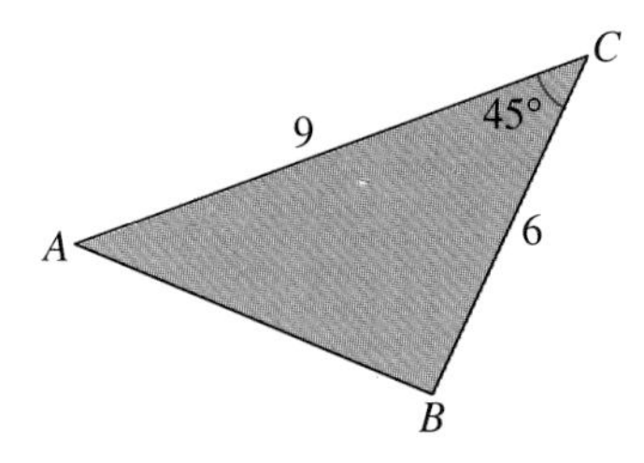

3. $A = 12°, \quad B = 58°, \quad a = 5$

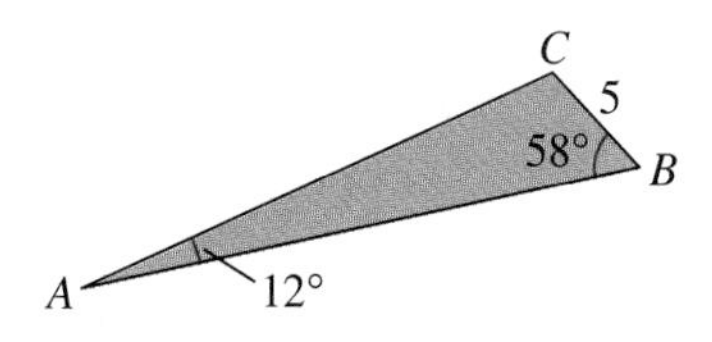

4. $B = 110°, \quad C = 30°, \quad c = 10.5$

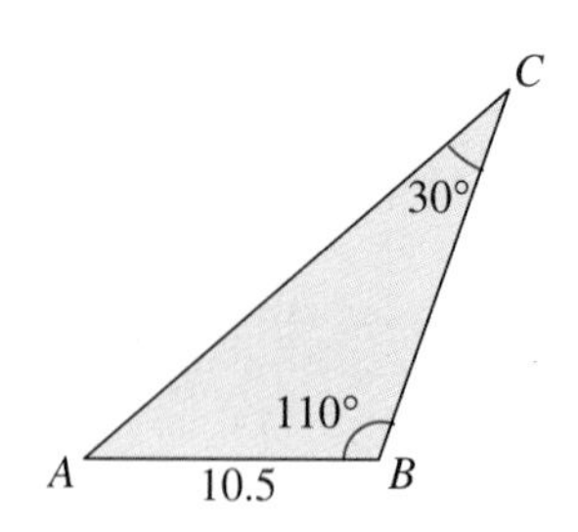

5. $B = 110°, \quad a = 4, \quad c = 4$

6. $a = 80, \quad b = 60, \quad c = 100$

7. $A = 75°, \quad a = 2.5, \quad b = 16.5$

8. $A = 130°, \quad a = 50, \quad b = 30$

9. $B = 115°, \quad a = 7, \quad b = 14.5$

10. $C = 50°, \quad a = 25, \quad c = 22$

11. $A = 15°, \quad a = 5, \quad b = 10$

12. $B = 150°, \quad a = 64, \quad b = 10$

13. $B = 150°, \quad a = 10, \quad c = 20$

14. $a = 2.5, \quad b = 15.0, \quad c = 4.5$

15. $B = 25°, \quad a = 6.2, \quad b = 4$

16. $B = 90°, \quad a = 5, \quad c = 12$

In Exercises 17–20, find the area of the triangle.

17. $a = 4, \quad b = 5, \quad c = 7$

18. $a = 15, \quad b = 8, \quad c = 10$

19. $A = 27°, \quad b = 5, \quad c = 8$

20. $B = 80°, \quad a = 4, \quad c = 8$

21. ***Height*** From a certain distance, the angle of elevation to the top of a building is 17°. At a point 50 meters closer to the building, the angle of elevation is 31°. Approximate the height of the building.

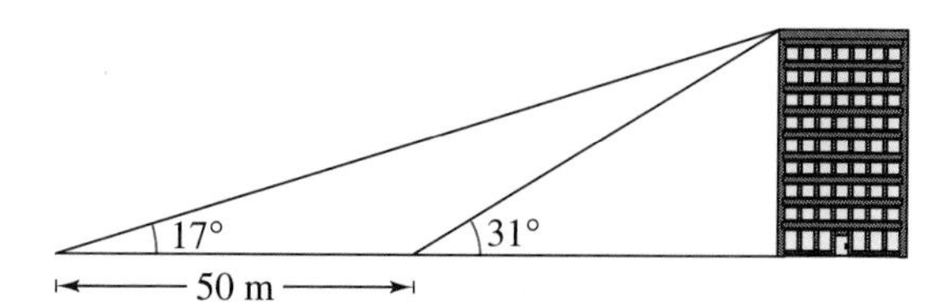

22. ***Geometry*** The lengths of the diagonals of a parallelogram are 10 feet and 16 feet. Find the lengths of the sides of the parallelogram if the diagonals intersect at an angle of 28°.

23. ***Height of a Tree*** Find the height of a tree that stands on a hillside of slope 28° (from the horizontal) if from a point 75 feet down the hill the angle of elevation to the top of the tree is 45° (see figure).

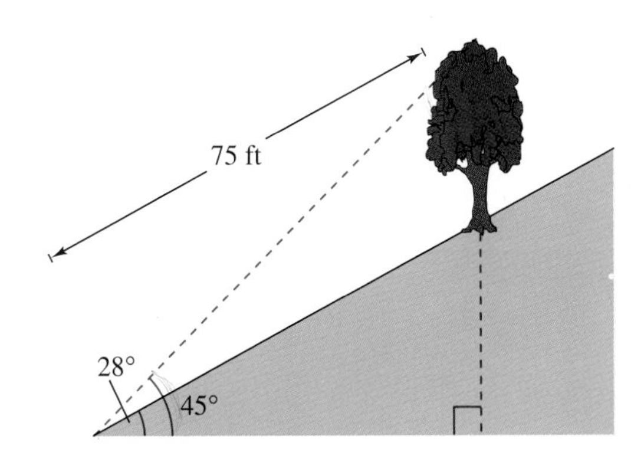

24. ***Surveying*** To approximate the length of a marsh, a surveyor walks 425 meters from point A to point B. Then the surveyor turns 65° and walks 300 meters to point C. Approximate the length AC of the marsh (see figure).

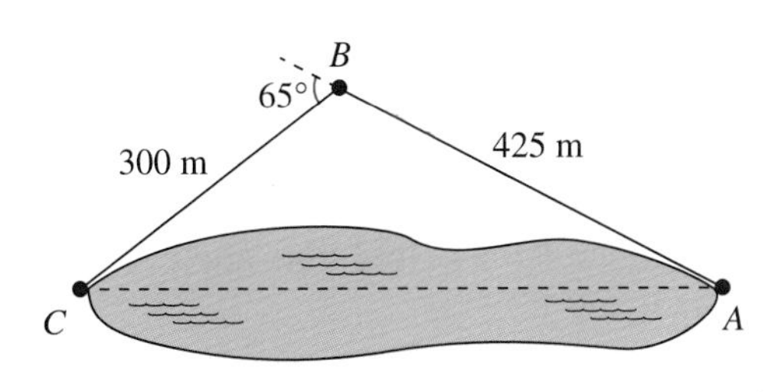

25. ***Navigation*** Two planes leave an airport at approximately the same time. One is flying 425 miles per hour at a bearing of N 5° W, and the other is flying 530 miles per hour at a bearing of N 67° E. Draw a figure that gives a visual representation of the problem and determine the distance between the planes after they have flown for 2 hours.

26. ***River Width*** Determine the width of a river that flows due east, if a tree on the opposite bank has a bearing of N 22° 30′ E and if, after walking 400 feet downstream, a surveyor finds that the tree has a bearing of N 15° W.

In Exercises 27–36, find the component form of the vector v satisfying the given conditions.

27.

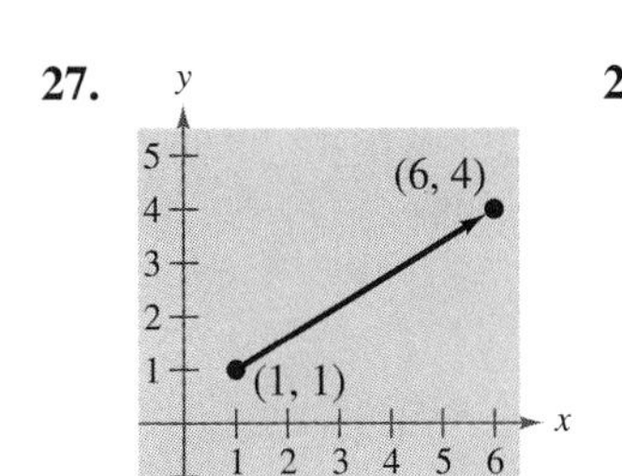

28.

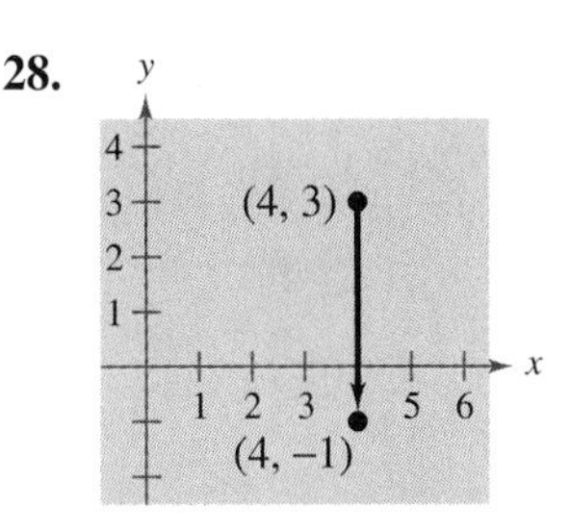

29.

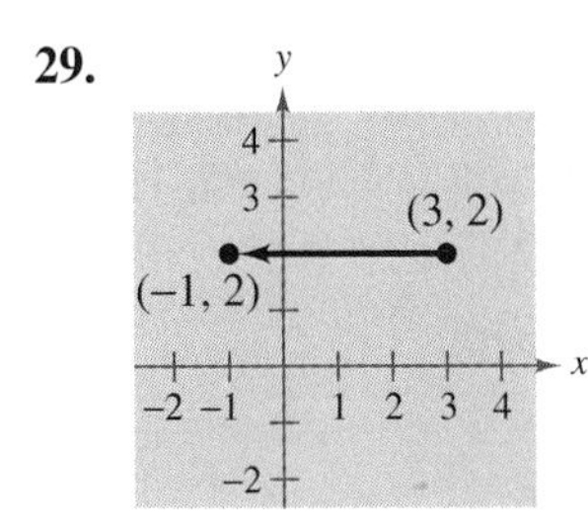

30.

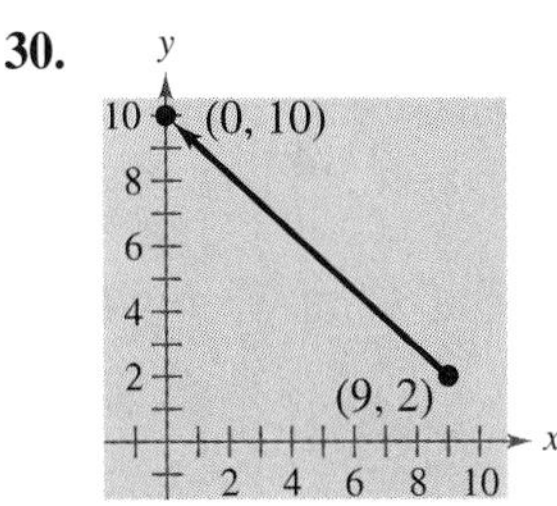

31.

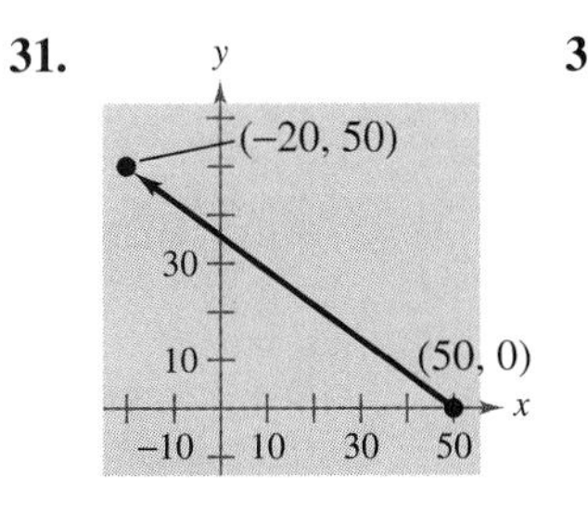

32.

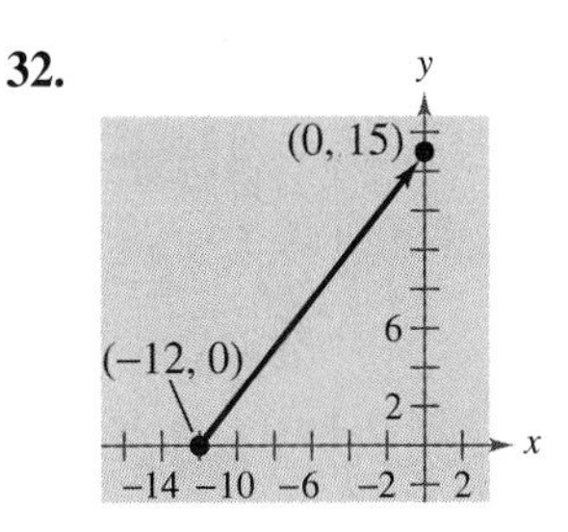

33. Initial point: (0, 10)
Terminal point: (7, 3)

34. Initial point: (1, 5)
Terminal point: (15, 9)

35. $\|\mathbf{v}\| = 8, \quad \theta = 120°$

36. $\|\mathbf{v}\| = \frac{1}{2}, \quad \theta = 225°$

In Exercises 37 and 38, write the vector v in the form $\|\mathbf{v}\|(\mathbf{i} \cos \theta + \mathbf{j} \sin \theta)$.

37. $\mathbf{v} = -10\mathbf{i} + 10\mathbf{j}$

38. $\mathbf{v} = 4\mathbf{i} - \mathbf{j}$

In Exercises 39–42, find the component form of the specified vector and sketch its graph given that $\mathbf{u} = 6\mathbf{i} - 5\mathbf{j}$ and $\mathbf{v} = 10\mathbf{i} + 3\mathbf{j}$.

39. $\dfrac{1}{\|\mathbf{u}\|}\mathbf{u}$

40. $3\mathbf{v}$

41. $4\mathbf{u} - 5\mathbf{v}$

42. $\frac{1}{2}\mathbf{v}$

In Exercises 43 and 44, use a graphing utility to graph the vectors and the resultant of the vectors. Find the magnitude and direction of the resultant.

43.

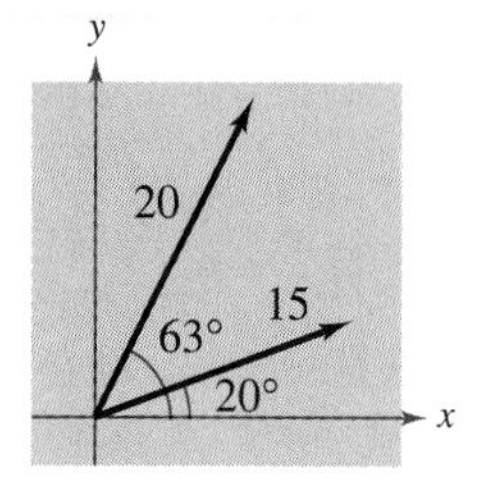

44.

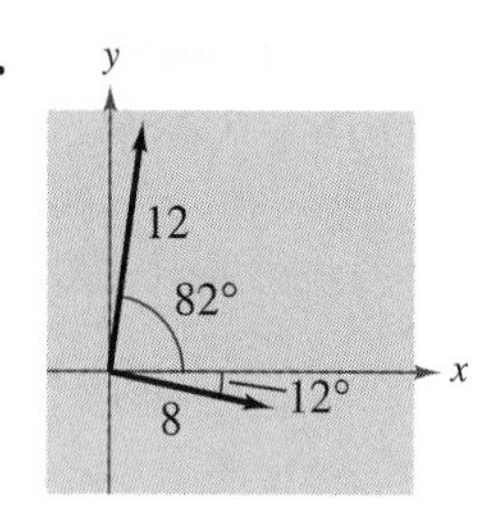

45. ***Resultant Force*** Find the direction and magnitude of the resultant of the three forces shown in the figure.

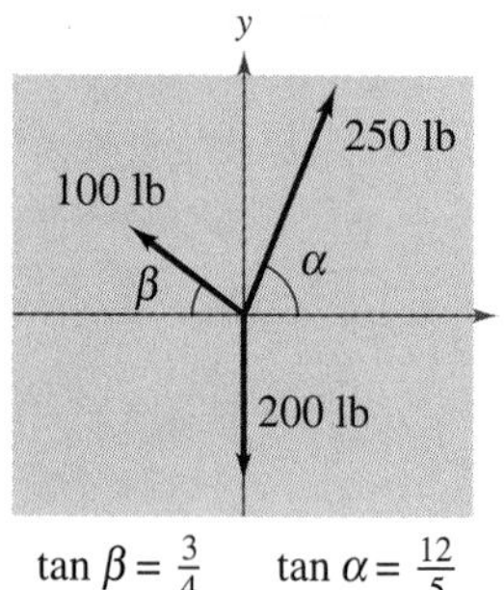

$\tan \beta = \frac{3}{4}$ $\tan \alpha = \frac{12}{5}$

FIGURE FOR 45

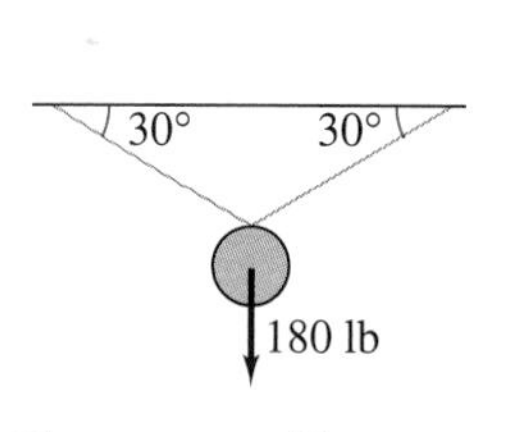

FIGURE FOR 47

46. ***Resultant Force*** Forces of magnitudes 85 pounds and 50 pounds act on a single point. Find the magnitude of the resultant if the angle between the forces is 15°.

47. ***Rope Tension*** A 180-pound weight is supported by two ropes (see figure). Find the tension in each rope.

48. ***Shared Load*** To carry a 40-kilogram cylindrical weight, two people lift on the ends of short ropes that are tied to an eyelet on the top center of the cylinder. Each rope makes a 30° angle with the vertical. Draw a figure that gives a visual representation of the problem, and find the tension in the ropes.

49. ***Braking Force*** A 500-pound motorcycle is headed up a hill inclined at 12°. What force is required to keep the motorcycle from rolling back down the hill when stopped at a red light?

50. ***Navigation*** An airplane has an airspeed of 450 miles per hour at a bearing of N 30° E. If the wind velocity is 20 miles per hour from the west, find the ground-speed and the direction of the plane.

In Exercises 51–54, find the dot product of u and v.

51. $\mathbf{u} = \langle -2, 0 \rangle$, $\mathbf{v} = \langle 4, 8 \rangle$

52. $\mathbf{u} = \langle 8, -2 \rangle$, $\mathbf{v} = \langle -3, -3 \rangle$

53. $\mathbf{u} = -3\mathbf{i} - 7\mathbf{j}$, $\mathbf{v} = 7\mathbf{i} - 3\mathbf{j}$

54. $\mathbf{u} = 6\mathbf{j}$, $\mathbf{v} = 12\mathbf{i} + 8\mathbf{j}$

In Exercises 55 and 56, find u • v.

55.

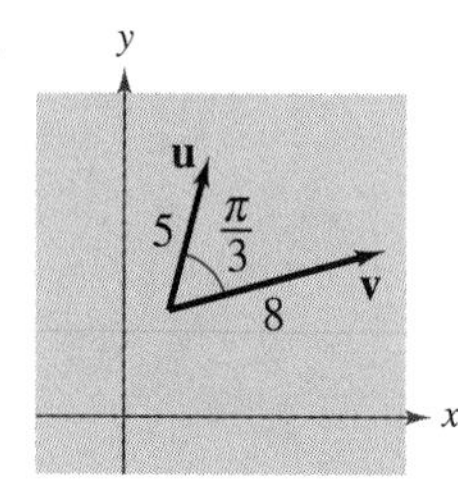

56.

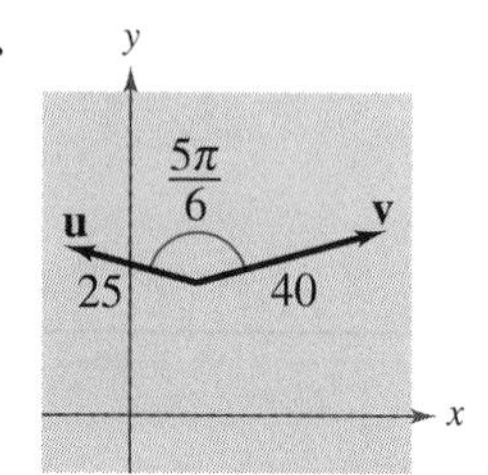

In Exercises 57 and 58, let $\mathbf{u} = \overrightarrow{PQ}$ and $\mathbf{v} = \overrightarrow{PR}$, and find (a) the component forms of u and v, (b) the magnitude of v, (c) u • v, (d) 2u + v, (e) the projection of u onto v, and (f) the vector component of u orthogonal to v.

57. $P = (1, 2),\ Q = (4, 1),\ R = (5, 4)$

58. $P = (-2, -1),\ Q = (5, -1),\ R = (2, 4)$

In Exercises 59–64, find the angle θ between the vectors u and v.

59.

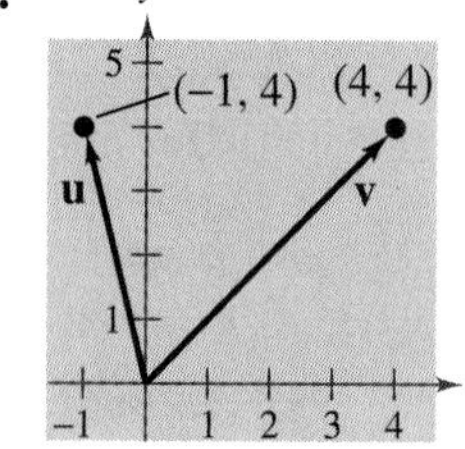

60.

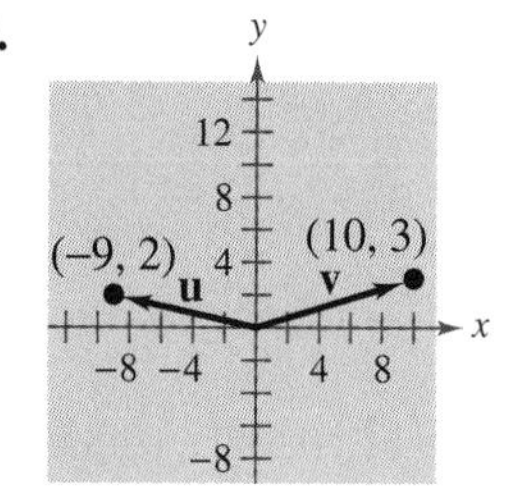

61. $\mathbf{u} = 5\left[\cos\left(\frac{3\pi}{4}\right)\mathbf{i} + \sin\left(\frac{3\pi}{4}\right)\mathbf{j}\right]$

$\mathbf{v} = 2\left[\cos\left(\frac{2\pi}{3}\right)\mathbf{i} + \sin\left(\frac{2\pi}{3}\right)\mathbf{j}\right]$

62. $\mathbf{u} = \langle 4, -1 \rangle,\quad \mathbf{v} = \langle 3, 2 \rangle$

63. $\mathbf{u} = \langle 1, 0 \rangle,\quad \mathbf{v} = \langle -2, 2 \rangle$

64. $\mathbf{u} = \sqrt{3}\mathbf{i} - \mathbf{j},\quad \mathbf{v} = -\mathbf{i} + \mathbf{j}$

65. ***Angle Between Forces*** Forces of 60 pounds and 100 pounds have a resultant force of 125 pounds. Find the angle between the two given forces.

66. ***Angle Between Forces*** Forces of 30 pounds and 40 pounds have a resultant force of 52 pounds. Find the angle between the two given forces.

In Exercises 67–72, find the projection of u onto v, and the vector component of u orthogonal to v.

67. $\mathbf{u} = \langle 3, -2\rangle$, $\mathbf{v} = \langle 4, 1\rangle$

68. $\mathbf{u} = \langle 5, 3\rangle$, $\mathbf{v} = \langle 12, -1\rangle$

69. $\mathbf{u} = \langle 0, 6\rangle$, $\mathbf{v} = \langle 2, 3\rangle$

70. $\mathbf{u} = \langle 2, 6\rangle$, $\mathbf{v} = \langle 7, 0\rangle$

71.

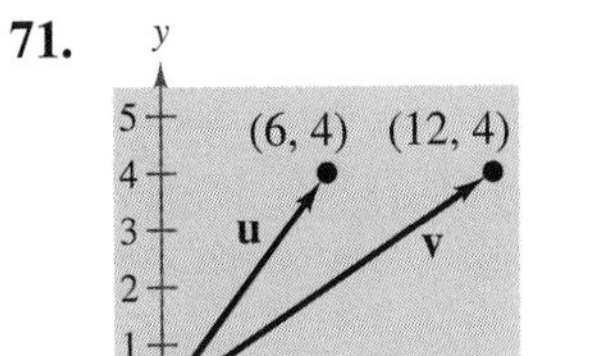

72.

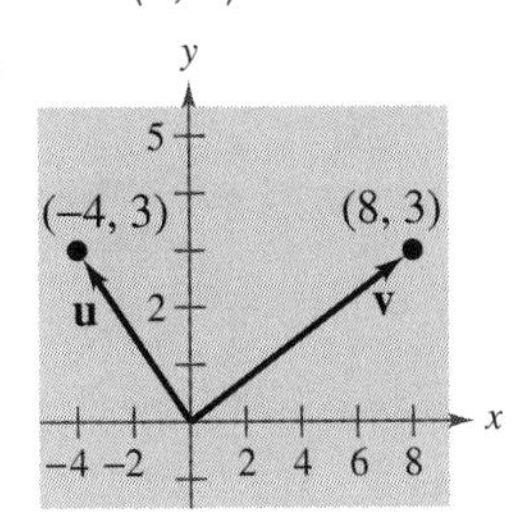

73. ***Work*** A force of 25 pounds in the direction of 30° above the horizontal is required to slide a cement block across a plank (see figure). Find the work done if the length of the plank is 10 feet.

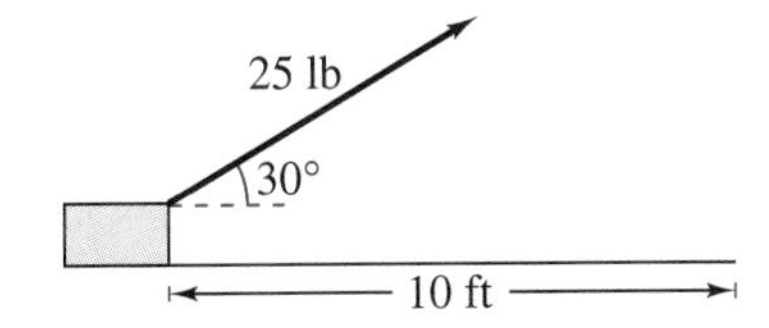

74. ***Work*** Find the work done in moving an object along the vector $\mathbf{v} = 3\mathbf{i} + 2\mathbf{j}$ if the force required is given by $\mathbf{F} = 2\mathbf{i} - \mathbf{j}$.

75. Use vectors to prove that the medians of a triangle pass through a point that is two-thirds the distance from any vertex to the midpoint of the opposite side.

76. Use vectors to prove that the midpoints of *any* quadrilateral are the vertices of a parallelogram.

77. Prove that $\mathbf{w} = (\mathbf{w} \cdot \mathbf{u})\mathbf{u} + (\mathbf{w} \cdot \mathbf{v})\mathbf{v}$ for any vector $\mathbf{w}$ if $\mathbf{u}$ and $\mathbf{v}$ are orthogonal unit vectors.

78. If $\mathbf{u} \neq \mathbf{0}$ and $\mathbf{u} \cdot \mathbf{v} = \mathbf{u} \cdot \mathbf{w}$, does $\mathbf{v} = \mathbf{w}$? Explain your answer.

79. ***Triangle Inequality*** Given any two vectors $\mathbf{u}$ and $\mathbf{v}$, prove the inequality

$$\|\mathbf{u} + \mathbf{v}\| \leq \|\mathbf{u}\| + \|\mathbf{v}\|.$$

(*Hint:* Square both members of the inequality and use the dot product.)

80. ***Triangle Inequality*** Demonstrate the triangle inequality (Exercise 79) for the vectors $\mathbf{u} = \langle 10, 4\rangle$ and $\mathbf{v} = \langle 7, 12\rangle$.

81. ***Cosine of the Difference of Two Angles*** Let $\mathbf{u} = (\cos\alpha)\mathbf{i} + (\sin\alpha)\mathbf{j}$ and $\mathbf{v} = (\cos\beta)\mathbf{i} + (\sin\beta)\mathbf{j}$. Sketch these vectors, and, by interpreting the dot product geometrically, prove that

$$\cos(\alpha - \beta) = \cos\alpha\cos\beta + \sin\alpha\cos\beta.$$

82. Give a geometric argument showing that the reflection of the vector $\mathbf{u}$ through the vector $\mathbf{v}$ is given by

$$\mathbf{w} = \left(\frac{2\mathbf{u} \cdot \mathbf{v}}{\mathbf{v} \cdot \mathbf{v}}\right)\mathbf{v} - \mathbf{u}.$$

In Exercises 83–86, use the result of Exercise 82 to reflect the vector u through the vector v. Show the reflected vector on the graph.

83.

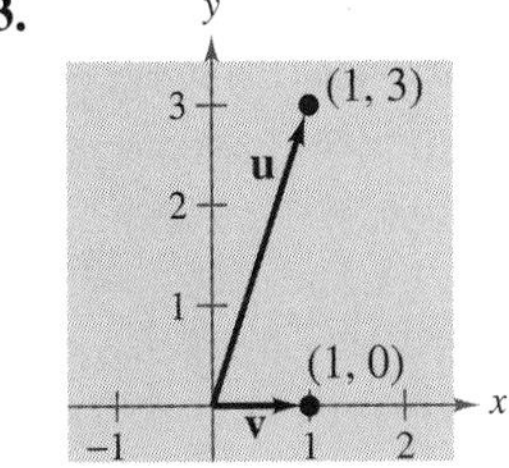

84.

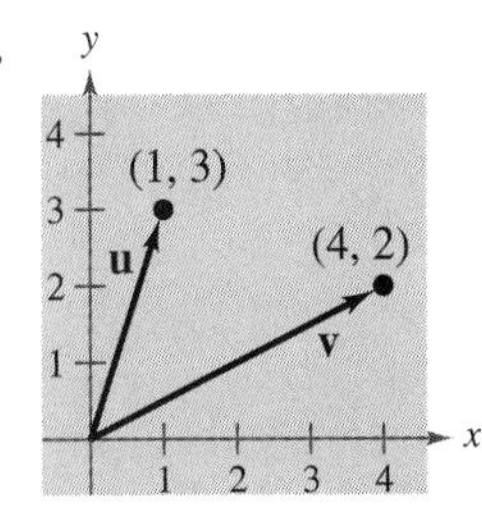

85.

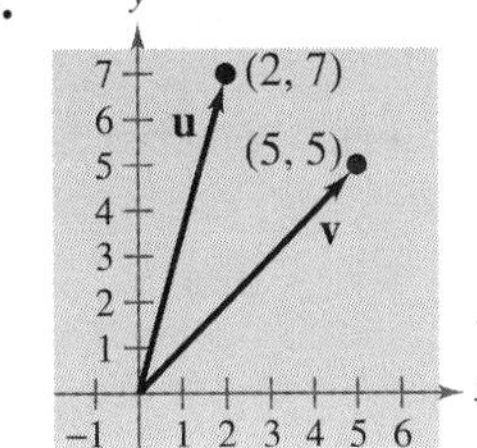

86.

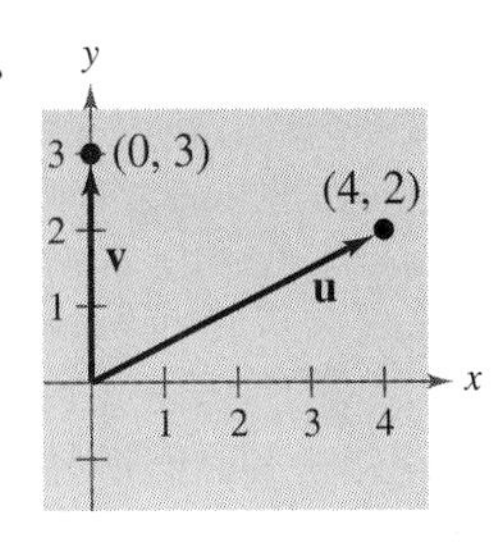

CHAPTER PROJECT: *Adding Vectors Graphically*

The following program is written for a *TI–83* or a *TI–82* graphing calculator. The program sketches two vectors $\mathbf{u} = a\mathbf{i} + b\mathbf{j}$ and $\mathbf{v} = c\mathbf{i} + d\mathbf{j}$ in standard position. Then, using the parallelogram law for vector addition, the program also sketches the vector sum $\mathbf{u} + \mathbf{v}$. *Before* running the program, you should set values that produce an appropriate viewing rectangle. Programs for other graphing calculators may be found in the appendix.

TI–83 or *TI–82* Program

```
PROGRAM:ADDVECT
:Input "ENTER A",A
:Input "ENTER B",B
:Input "ENTER C",C
:Input "ENTER D",D
:Line(0,0,A,B)
:Line(0,0,C,D)
:A+C→E
:B+D→F
:Line(0,0,E,F)
:Line(A,B,E,F)
:Line(C,D,E,F)
:Pause
:ClrDraw
:Stop
```

EXAMPLE 1 *Sketching a Vector Sum*

Use the program listed above to sketch the sum of the vectors $\mathbf{u} = 5\mathbf{i} + 2\mathbf{j}$ and $\mathbf{v} = -4\mathbf{i} + 3\mathbf{j}$.

Solution

To show both vectors and their sum, you can use the viewing rectangle $-6 \le x \le 6$, $-2 \le y \le 6$. Note that this is a "square" setting. That is, the spacing on the horizontal and vertical axes is the same. After running the program and entering $A = 5$, $B = 2$, $C = -4$, and $D = 3$, you should obtain the screen shown below. Note that the vector sum

$$\mathbf{u} + \mathbf{v} = \mathbf{i} + 5\mathbf{j}$$

appears as the diagonal of the parallelogram. The vectors $\mathbf{u}$ and $\mathbf{v}$ appear as two of the sides of the parallelogram.

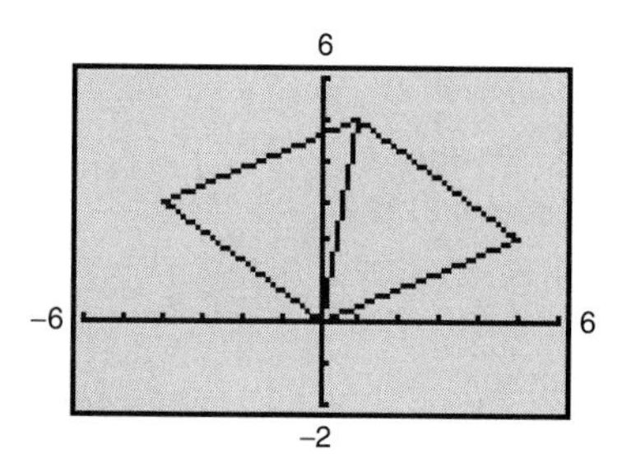

EXAMPLE 2 *Finding an Airplane's Speed and Direction*

Real Life

An airplane is headed N 60° W at a speed of 400 miles per hour. The airplane encounters wind of velocity 75 miles per hour in the direction N 40° E. What are the resultant speed and direction of the airplane? (See Example 10 on page 295.)

Solution

The velocity of the airplane can be represented by the vector

$$\mathbf{v}_1 = 400\langle \cos 150°, \sin 150° \rangle$$

and the velocity of the wind by the vector

$$\mathbf{v}_2 = 75\langle \cos 50°, \sin 50° \rangle.$$

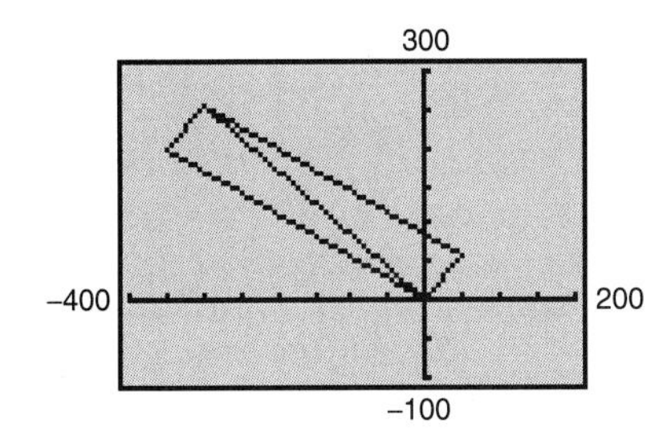

Thus, the resultant velocity of the airplane can be represented by $\mathbf{v}_1 + \mathbf{v}_2$. With the program listed on page 316, you do not need to evaluate the numerical values of the vector coordinates. Simply enter the following.

$A = 400 \cos 150°$ $\qquad B = 400 \sin 150°$

$C = 75 \cos 50°$ $\qquad D = 75 \sin 50°$

Using $-400 \le x \le 200$ and $-100 \le y \le 300$, you should obtain the screen shown at the left.

CHAPTER PROJECT INVESTIGATIONS

In Questions 1–4, use the program on page 316 (or a comparable program on some other graphing utility) to sketch the sum of the vectors. Use the result to estimate graphically the components of the sum. Then check your result analytically. (Use $-9 \le x \le 9$ and $-6 \le y \le 6$.)

1. $\mathbf{u} = 3\mathbf{i} + 4\mathbf{j}, \quad \mathbf{v} = -5\mathbf{i} + \mathbf{j}$

2. $\mathbf{u} = 5\mathbf{i} - 4\mathbf{j}, \quad \mathbf{v} = 3\mathbf{i} + 2\mathbf{j}$

3. $\mathbf{u} = -4\mathbf{i} + 4\mathbf{j}, \quad \mathbf{v} = -2\mathbf{i} - 6\mathbf{j}$

4. $\mathbf{u} = 7\mathbf{i} + 3\mathbf{j}, \quad \mathbf{v} = -2\mathbf{i} - 6\mathbf{j}$

5. *Airplane Speed* After encountering the wind, is the airplane in Example 2 traveling at a higher speed or a lower speed? Explain.

6. *Airplane Speed* Consider the airplane described in Example 2, headed N 60° W at a speed of 400 miles per hour. What wind velocity, in the direction of N 40° E, will produce a resultant direction of N 50° W? Explain how to use the program on page 316 to obtain the answer *experimentally*. Then explain how to obtain the answer analytically.

7. *Airplane Speed* Consider the airplane described in Example 2, headed N 60° W at a speed of 400 miles per hour. What wind direction, at a speed of 75 miles per hour, will produce a resultant direction of N 50° W? Explain how to use the program on page 316 to obtain the answer *experimentally*. Then explain how to obtain the answer analytically.

Cumulative Test for Chapters 1–3

Take this test as you would take a test in class. After you are done, check your work against the answers given in the back of the book.

The *Interactive* CD-ROM provides answers to the Chapter Tests and Cumulative Tests. It also offers Chapter Pre-Tests (which test key skills and concepts covered in previous chapters) and Chapter Post-Tests, both of which have randomly generated exercises with diagnostic capabilities.

1. Consider the angle $\theta = -120°$.

(a) Sketch the angle in standard position.

(b) Determine a coterminal angle in the interval $[0°, 360°)$.

(c) Convert the angle to radian measure.

(d) Find the reference angle θ'.

(e) Find the exact values of the six trigonometric functions of θ.

2. Convert the angle of measure 2.35 radians to degrees. Round the answer to one decimal place.

3. Find $\cos\theta$ if $\tan\theta = -\frac{4}{3}$ and $\sin\theta < 0$.

4. Sketch the graphs of (a) $f(x) = 3 - 2\sin\pi x$ and (b) $g(x) = \dfrac{1}{2}\tan\left(x - \dfrac{\pi}{2}\right)$.

5. Find a, b, and c such that the graph of the function $h(x) = a\cos(bx + c)$ matches the graph in the figure.

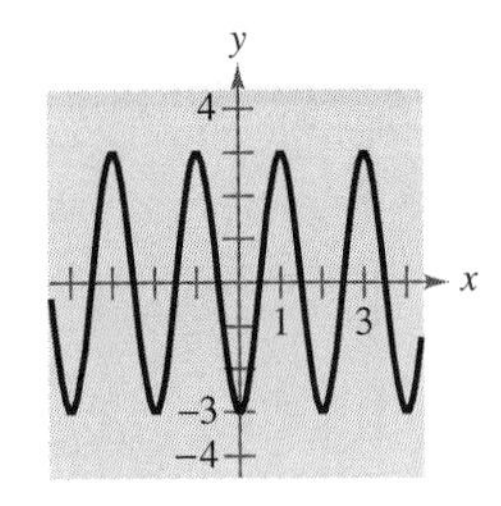

FIGURE FOR 5

6. Write an algebraic expression equivalent to $\sin(\arccos 2x)$.

7. Subtract and simplify: $\dfrac{\sin\theta - 1}{\cos\theta} - \dfrac{\cos\theta}{\sin\theta - 1}$.

8. Prove the identities.

(a) $\cot^2\alpha(\sec^2\alpha - 1) = 1$ (b) $\sin(x + y)\sin(x - y) = \sin^2 x - \sin^2 y$

(c) $\sin^2 x\cos^2 x = \frac{1}{8}(1 - \cos 4x)$

9. Find all solutions of the equations in the interval $[0, 2\pi)$.

(a) $2\cos^2\beta - \cos\beta = 0$ (b) $3\tan\theta - \cot\theta = 0$

10. Find the remaining angles and side of the triangle shown in the figure.

(a) $A = 30°,\ a = 9,\ b = 8$ (b) $A = 30°, b = 8,\ c = 10$

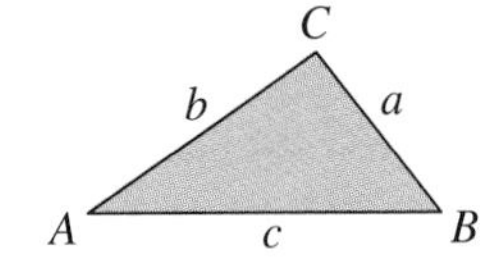

FIGURE FOR 10

11. From a point 200 feet from a flagpole, the angles of elevation to the bottom and top of the flag are $16°\,45'$ and $18°$, respectively. Approximate the height of the flag to the nearest foot.

12. An airplane is flying at an airspeed of 500 kilometers per hour and a bearing of N 30° E. The wind at the altitude of the plane has a velocity of 50 kilometers per hour and a bearing of N 60° E. What is the true direction of the plane, and what is its speed relative to the ground?

13. Find the projection of $\mathbf{u}$ onto $\mathbf{v}$ if $\mathbf{u} = \langle -4, 3\rangle$ and $\mathbf{v} = \langle -1, 5\rangle$.

4 Complex Numbers

The Fundamental Theorem of Algebra implies that an nth-degree polynomial equation has precisely n solutions. This result, however, is true only if repeated and complex solutions are counted. For example, the equation

$$x^4 - 2x^3 + 2x^2 - 2x + 1 = 0$$

has solutions of 1, 1, i, and $-i$.

When first developed, complex numbers were used primarily for theoretical results such as this. Today, however, complex numbers have several other uses.

One use is in creating fractals like that shown in the photograph. To program the fractal with a computer, you plot complex numbers in the complex plane.

In the complex plane, the point (a, b) represents the complex number $a + bi$. For example, the number $2 + 3i$ is plotted below.

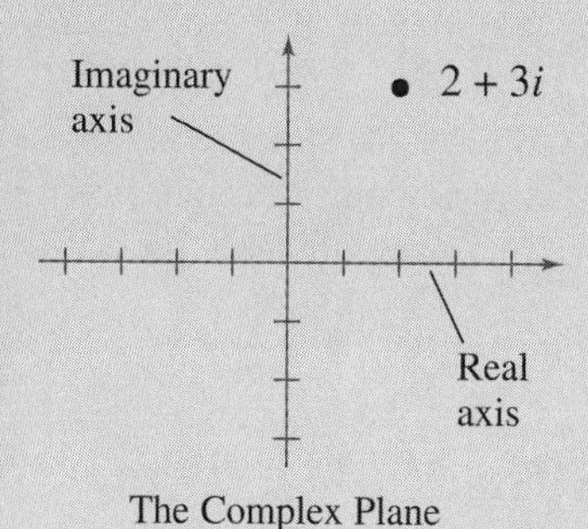

The Complex Plane

See Exercises 1–6 on page 340.

Photos: Jim Zuckerman; Rondi Ballard (inset)

This fractal was created on a computer by photographer Jim Zuckerman. Zuckerman owns and operates a photography and digital imaging company in Northridge, California.

4.1 Complex Numbers

The Imaginary Unit i ▫ *Operations with Complex Numbers* ▫ *Complex Conjugates and Division* ▫ *Complex Solutions of Quadratic Equations*

See Exercises 73–76 on page 327 for examples of how a graphing utility can be used to discover whether a quadratic equation has imaginary solutions.

The Imaginary Unit *i*

Some quadratic equations have no real solutions. For instance, the quadratic equation

$$x^2 + 1 = 0 \qquad \text{Equation with no real solution}$$

has no real solution because there is no real number x that can be squared to produce -1. To overcome this deficiency, mathematicians created an expanded system of numbers using the **imaginary unit *i*,** defined as

$$i = \sqrt{-1} \qquad \text{Imaginary unit}$$

where $i^2 = -1$. By adding real numbers to real multiples of this imaginary unit, we obtain the set of **complex numbers.** Each complex number can be written in the **standard form, $a + bi$.**

Carl Friedrich Gauss (1777–1855) proved that all the roots of any algebraic equation are "numbers" of the form $a + bi$, where a and b are real numbers and i is the square root of -1. These "numbers" were called complex.

DEFINITION OF A COMPLEX NUMBER

For real numbers a and b, the number

$$a + bi$$

is a **complex number.** If $a = 0$ and $b \neq 0$, the complex number bi is an **imaginary number.**

The set of real numbers is a subset of the set of complex numbers, as shown in Figure 4.1. This is true because every real number a can be written as a complex number using $b = 0$. That is, for every real number a, we can write $a = a + 0i$.

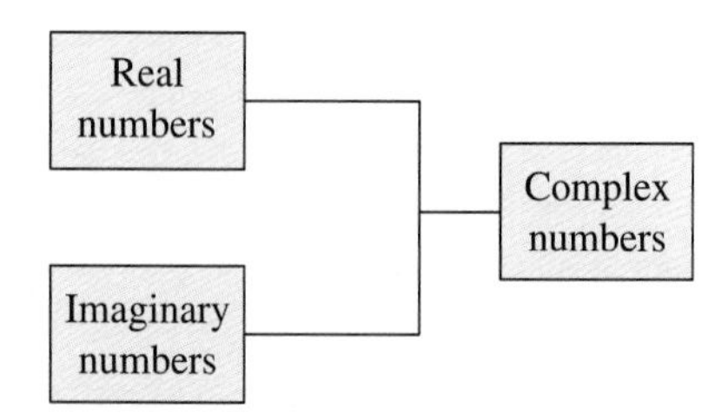

FIGURE 4.1

EQUALITY OF COMPLEX NUMBERS

Two complex numbers $a + bi$ and $c + di$, written in standard form, are **equal** to each other

$$a + bi = c + di \qquad \text{Equality of two complex numbers}$$

if and only if $a = c$ and $b = d$.

Operations with Complex Numbers

To add (or subtract) two complex numbers, you add (or subtract) the real and imaginary parts of the numbers separately.

ADDITION AND SUBTRACTION OF COMPLEX NUMBERS

If $a + bi$ and $c + di$ are two complex numbers written in standard form, their sum and difference are defined as follows.

Sum: $(a + bi) + (c + di) = (a + c) + (b + d)i$

Difference: $(a + bi) - (c + di) = (a - c) + (b - d)i$

The **additive identity** in the complex number system is zero (the same as in the real number system). Furthermore, the **additive inverse** of the complex number $a + bi$ is

$$-(a + bi) = -a - bi. \qquad \text{Additive inverse}$$

Thus, you have

$$(a + bi) + (-a - bi) = 0 + 0i = 0.$$

The *Interactive* CD-ROM shows every example with its solution; clicking on the *Try It!* button brings up similar problems. Guided Examples and Integrated Examples show step-by-step solutions to additional examples. Integrated Examples are related to several concepts in the section.

EXAMPLE 1 *Adding and Subtracting Complex Numbers*

a. $(3 - i) + (2 + 3i) = 3 - i + 2 + 3i$ Remove parentheses.

$= 3 + 2 - i + 3i$ Group like terms.

$= (3 + 2) + (-1 + 3)i$

$= 5 + 2i$ Standard form

b. $2i + (-4 - 2i) = 2i - 4 - 2i$ Remove parentheses.

$= -4 + 2i - 2i$ Group like terms.

$= -4$ Standard form

c. $3 - (-2 + 3i) + (-5 + i) = 3 + 2 - 3i - 5 + i$

$= 3 + 2 - 5 - 3i + i$

$= 0 - 2i$

$= -2i$

NOTE Note in Example 1(b) that the sum of two complex numbers can be a real number.

Exploration

Complete the table:

$i^1 = i$	$i^7 =$
$i^2 = -1$	$i^8 =$
$i^3 = -i$	$i^9 =$
$i^4 = 1$	$i^{10} =$
$i^5 =$	$i^{11} =$
$i^6 =$	$i^{12} =$

What pattern do you see? Write a brief description of how you would find i raised to any positive integer power.

The *Interactive* CD-ROM offers graphing utility emulators of the *TI-82* and *TI-83*, which can be used with the Examples, Explorations, Technology notes, and Exercises.

Many of the properties of real numbers are valid for complex numbers as well. Here are some examples.

Associative Property of Addition and Multiplication

Commutative Property of Addition and Multiplication

Distributive Property of Multiplication Over Addition

Notice below how these properties are used when two complex numbers are multiplied.

$$\begin{aligned}(a+bi)(c+di) &= a(c+di) + bi(c+di) && \text{Distributive}\\ &= ac + (ad)i + (bc)i + (bd)i^2 && \text{Distributive}\\ &= ac + (ad)i + (bc)i + (bd)(-1) && \text{Definition of } i\\ &= ac - bd + (ad)i + (bc)i && \text{Commutative}\\ &= (ac - bd) + (ad + bc)i && \text{Associative}\end{aligned}$$

Rather than trying to memorize this multiplication rule, we suggest that you simply remember how the distributive property is used to multiply two complex numbers. The procedure is similar to multiplying two polynomials and combining like terms (as in the FOIL Method).

EXAMPLE 2 *Multiplying Complex Numbers*

a. $$\begin{aligned}(i)(-3i) &= -3i^2 && \text{Multiply.}\\ &= -3(-1) && i^2 = -1\\ &= 3 && \text{Simplify.}\end{aligned}$$

b. $$\begin{aligned}(2 - i)(4 + 3i) &= 8 + 6i - 4i - 3i^2 && \text{Product of binomials}\\ &= 8 + 6i - 4i - 3(-1) && i^2 = -1\\ &= 8 + 3 + 6i - 4i && \text{Group like terms.}\\ &= 11 + 2i && \text{Standard form}\end{aligned}$$

c. $$\begin{aligned}(3 + 2i)(3 - 2i) &= 9 - 6i + 6i - 4i^2 && \text{Product of binomials}\\ &= 9 - 4(-1) && i^2 = -1\\ &= 9 + 4 && \text{Simplify.}\\ &= 13 && \text{Standard form}\end{aligned}$$

d. $$\begin{aligned}(3 + 2i)^2 &= 9 + 6i + 6i + 4i^2 && \text{Product of binomials}\\ &= 9 + 4(-1) + 12i && i^2 = -1\\ &= 9 - 4 + 12i && \text{Simplify.}\\ &= 5 + 12i && \text{Standard form}\end{aligned}$$

Complex Conjugates and Division

Notice in Example 2(c) that the product of two complex numbers can be a real number. This occurs with pairs of complex numbers of the form $a + bi$ and $a - bi$, called **complex conjugates.**

$$\begin{aligned}(a + bi)(a - bi) &= a^2 - abi + abi - b^2i^2 \\ &= a^2 - b^2(-1) \\ &= a^2 + b^2\end{aligned}$$

To find the quotient of $a + bi$ and $c + di$ where c and d are not both zero, multiply the numerator and denominator by the conjugate of the denominator to obtain

$$\frac{a + bi}{c + di} = \frac{a + bi}{c + di}\left(\frac{c - di}{c - di}\right) = \frac{(ac + bd) + (bc - ad)i}{c^2 + d^2}.$$

Example 3 Dividing Complex Numbers

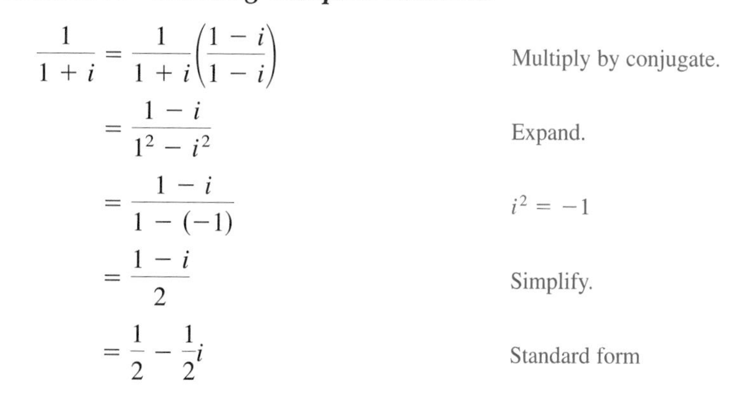

$$\begin{aligned}\frac{1}{1 + i} &= \frac{1}{1 + i}\left(\frac{1 - i}{1 - i}\right) && \text{Multiply by conjugate.} \\ &= \frac{1 - i}{1^2 - i^2} && \text{Expand.} \\ &= \frac{1 - i}{1 - (-1)} && i^2 = -1 \\ &= \frac{1 - i}{2} && \text{Simplify.} \\ &= \frac{1}{2} - \frac{1}{2}i && \text{Standard form}\end{aligned}$$

Some graphing utilities, such as the *TI–92* and the *TI-83* from Texas Instruments, can perform operations with complex numbers. For instance, to divide $2 + 3i$ by $4 - 2i$, enter

[(] 2 [+] 3 [2nd] [i] [)] [÷]
[(] 4 [−] 2 [2nd] [i] [)] [ENTER].

The display is $1/10 + 4/5i$.
(Photo: Courtesy of Texas Instruments)

Example 4 Dividing Complex Numbers

$$\begin{aligned}\frac{2 + 3i}{4 - 2i} &= \frac{2 + 3i}{4 - 2i}\left(\frac{4 + 2i}{4 + 2i}\right) && \text{Multiply by conjugate.} \\ &= \frac{8 + 4i + 12i + 6i^2}{16 - 4i^2} && \text{Expand.} \\ &= \frac{8 - 6 + 16i}{16 + 4} && i^2 = -1 \\ &= \frac{1}{20}(2 + 16i) && \text{Simplify.} \\ &= \frac{1}{10} + \frac{4}{5}i && \text{Standard form}\end{aligned}$$

Complex Solutions of Quadratic Equations

When using the Quadratic Formula to solve a quadratic equation, you often obtain a result such as $\sqrt{-3}$, which you know is not a real number. By factoring out $i = \sqrt{-1}$, you can write this number in standard form.

$$\sqrt{-3} = \sqrt{3(-1)} = \sqrt{3}\sqrt{-1} = \sqrt{3}i$$

The number $\sqrt{3}i$ is called the principal square root of -3.

Study Tip

The definition of principal square root uses the rule

$$\sqrt{ab} = \sqrt{a}\sqrt{b}$$

for $a > 0$ and $b < 0$. This rule is not valid if *both* a and b are negative. For example,

$$\begin{aligned}\sqrt{-5}\sqrt{-5} &= \sqrt{5}i\sqrt{5}i \\ &= \sqrt{25}i^2 \\ &= 5i^2 = -5\end{aligned}$$

whereas

$$\sqrt{(-5)(-5)} = \sqrt{25} = 5.$$

To avoid problems with multiplying square roots of negative numbers, be sure to convert to standard form *before* multiplying.

PRINCIPAL SQUARE ROOT OF A NEGATIVE NUMBER

If a is a positive number, the **principal square root** of the negative number $-a$ is defined as

$$\sqrt{-a} = \sqrt{a}i.$$

EXAMPLE 5 *Writing Complex Numbers in Standard Form*

a. $\sqrt{-3}\sqrt{-12} = \sqrt{3}i\sqrt{12}i = \sqrt{36}i^2 = 6(-1) = -6$

b. $\sqrt{-48} - \sqrt{-27} = \sqrt{48}i - \sqrt{27}i = 4\sqrt{3}i - 3\sqrt{3}i = \sqrt{3}i$

c.
$$\begin{aligned}\left(-1 + \sqrt{-3}\right)^2 &= \left(-1 + \sqrt{3}i\right)^2 \\ &= (-1)^2 - 2\sqrt{3}i + \left(\sqrt{3}\right)^2(i^2) \\ &= 1 - 2\sqrt{3}i + 3(-1) \\ &= -2 - 2\sqrt{3}i\end{aligned}$$

EXAMPLE 6 *Complex Solutions of a Quadratic Equation*

Solve $3x^2 - 2x + 5 = 0$.

Solution

$$x = \frac{-(-2) \pm \sqrt{(-2)^2 - 4(3)(5)}}{2(3)} \qquad \text{Quadratic Formula}$$

$$= \frac{2 \pm \sqrt{-56}}{6} \qquad \text{Simplify.}$$

$$= \frac{2 \pm 2\sqrt{14}i}{6} \qquad \text{Write in } i\text{-form.}$$

$$= \frac{1}{3} \pm \frac{\sqrt{14}}{3}i \qquad \text{Standard form}$$

GROUP ACTIVITY

ERROR ANALYSIS

Suppose you are a math instructor, and one of your students has handed in the following quiz. Find the error(s) in each solution and discuss how to explain each error to your student.

1. Write $\frac{5}{3 - 2i}$ in standard form.

$$\frac{5}{3 - 2i} \cdot \frac{3 + 2i}{3 + 2i} = \frac{15 + 10i}{9 - 4} = 3 + 2i$$

2. Multiply $(\sqrt{-4} + 3)(i - \sqrt{-3})$.

$$\begin{aligned}(\sqrt{-4} + 3)(i - \sqrt{-3}) &= i\sqrt{-4} - \sqrt{-4}\sqrt{-3} + 3i - 3\sqrt{-3} \\ &= -2i - \sqrt{12} + 3i - 3i\sqrt{3} \\ &= (1 - 3\sqrt{3})i - 2\sqrt{3}\end{aligned}$$

3. Sketch the graph of $y = -x^2 + 2$.

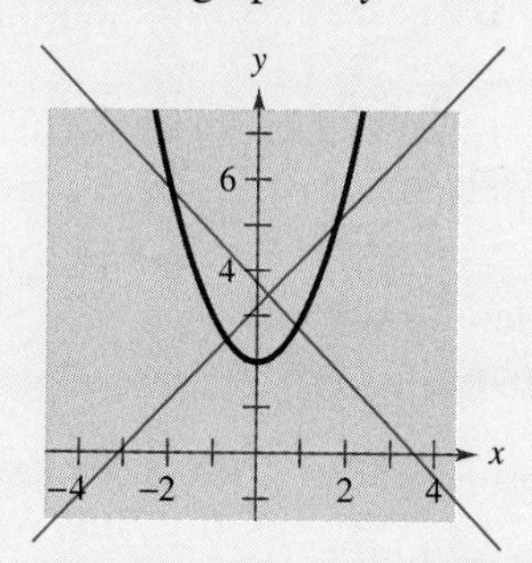

The *Interactive* CD-ROM provides additional help with Warm-Up exercises by providing a hypertext link to the section in which the concept was introduced.

WARM UP

Simplify the expression.

1. $\sqrt{12}$

2. $\sqrt{500}$

3. $\sqrt{20} - \sqrt{5}$

4. $\sqrt{27} - \sqrt{243}$

5. $\sqrt{24}\sqrt{6}$

6. $2\sqrt{18}\sqrt{32}$

7. $\frac{1}{\sqrt{3}}$

8. $\frac{2}{\sqrt{2}}$

Solve the quadratic equation.

9. $x^2 + x - 1 = 0$

10. $x^2 + 2x - 1 = 0$

4.1 Exercises

In Exercises 1–4, find real numbers a and b so that the equation is true.

1. $a + bi = -10 + 6i$
2. $a + bi = 13 + 4i$
3. $(a - 1) + (b + 3)i = 5 + 8i$
4. $(a + 6) + 2bi = 6 - 5i$

In Exercises 5–16, write the complex number in standard form.

5. $4 + \sqrt{-9}$
6. $3 + \sqrt{-16}$
7. $2 - \sqrt{-27}$
8. $1 + \sqrt{-8}$
9. $\sqrt{-75}$
10. 45
11. $-6i + i^2$
12. $-4i^2 + 2i$
13. 8
14. $\left(\sqrt{-4}\right)^2 - 5$
15. $\sqrt{-0.09}$
16. $\sqrt{-0.0004}$

In Exercises 17–26, perform the addition or subtraction and write the result in standard form.

17. $(5 + i) + (6 - 2i)$
18. $(13 - 2i) + (-5 + 6i)$
19. $(8 - i) - (4 - i)$
20. $(3 + 2i) - (6 + 13i)$
21. $\left(-2 + \sqrt{-8}\right) + \left(5 - \sqrt{-50}\right)$
22. $\left(8 + \sqrt{-18}\right) - \left(4 + 3\sqrt{2}i\right)$
23. $13i - (14 - 7i)$
24. $22 + (-5 + 8i) + 10i$
25. $-\left(\frac{3}{2} + \frac{5}{2}i\right) + \left(\frac{5}{3} + \frac{11}{3}i\right)$
26. $(1.6 + 3.2i) + (-5.8 + 4.3i)$

In Exercises 27–40, perform the operation and write the result in standard form.

27. $\sqrt{-6} \cdot \sqrt{-2}$
28. $\sqrt{-5} \cdot \sqrt{-10}$
29. $\left(\sqrt{-10}\right)^2$
30. $\left(\sqrt{-75}\right)^2$
31. $(1 + i)(3 - 2i)$
32. $(6 - 2i)(2 - 3i)$
33. $6i(5 - 2i)$
34. $-8i(9 + 4i)$
35. $\left(\sqrt{14} + \sqrt{10}i\right)\left(\sqrt{14} - \sqrt{10}i\right)$
36. $\left(3 + \sqrt{-5}\right)\left(7 - \sqrt{-10}\right)$
37. $(4 + 5i)^2$
38. $(2 - 3i)^2$
39. $(2 + 3i)^2 + (2 - 3i)^2$
40. $(1 - 2i)^2 - (1 + 2i)^2$

41. ***Error Analysis*** Describe the error.

$$\sqrt{-6}\sqrt{-6} = \sqrt{(-6)(-6)} = \sqrt{36} = 6$$

42. ***Think About It*** **True or False?** There is no complex number that is equal to its conjugate. Explain.

In Exercises 43–50, write the conjugate of the complex number. Multiply the number and its conjugate.

43. $5 + 3i$
44. $9 - 12i$
45. $-2 - \sqrt{5}i$
46. $-4 + \sqrt{2}i$
47. $20i$
48. $\sqrt{-15}$
49. $\sqrt{8}$
50. $1 + \sqrt{8}$

In Exercises 51–64, perform the operation and write the result in standard form.

51. $\dfrac{6}{i}$
52. $-\dfrac{10}{2i}$
53. $\dfrac{4}{4 - 5i}$
54. $\dfrac{3}{1 - i}$
55. $\dfrac{2 + i}{2 - i}$
56. $\dfrac{8 - 7i}{1 - 2i}$
57. $\dfrac{6 - 7i}{i}$
58. $\dfrac{8 + 20i}{2i}$
59. $\dfrac{1}{(4 - 5i)^2}$
60. $\dfrac{(2 - 3i)(5i)}{2 + 3i}$
61. $\dfrac{2}{1 + i} - \dfrac{3}{1 - i}$
62. $\dfrac{2i}{2 + i} + \dfrac{5}{2 - i}$
63. $\dfrac{i}{3 - 2i} + \dfrac{2i}{3 + 8i}$
64. $\dfrac{1 + i}{i} - \dfrac{3}{4 - i}$

In Exercises 65–72, use the Quadratic Formula to solve the quadratic equation.

65. $x^2 - 2x + 2 = 0$

66. $x^2 + 6x + 10 = 0$

67. $4x^2 + 16x + 17 = 0$

68. $9x^2 - 6x + 37 = 0$

69. $4x^2 + 16x + 15 = 0$

70. $9x^2 - 6x - 35 = 0$

71. $16t^2 - 4t + 3 = 0$

72. $5s^2 + 6s + 3 = 0$

Graphical Reasoning **In Exercises 73–76, use a graphing utility to graph the equation. Use the graph to approximate any x-intercepts of the graph. Set $y = 0$ and solve the resulting equation. Compare the result with the x-intercepts of the graph.**

73. $y = \frac{1}{4}(4x^2 - 20x + 25)$

74. $y = -(x^2 - 4x + 3)$

75. $y = -(x^2 - 4x + 5)$

76. $y = \frac{1}{4}(x^2 - 2x + 9)$

77. ***Essay*** Use the results of Exercises 73–76 to describe the relationship between the number of x-intercepts of the graph of $y = ax^2 + bx + c$ and the solutions of the equation $ax^2 + bx + c = 0$.

78. Express each of the following powers of i as i, $-i$, 1, or -1.

(a) i^{40} (b) i^{25}

(c) i^{50} (d) i^{67}

In Exercises 79–86, simplify the complex number and write it in standard form.

79. $-6i^3 + i^2$

80. $4i^2 - 2i^3$

81. $-5i^5$

82. $(-i)^3$

83. $(\sqrt{-75})^3$

84. $(\sqrt{-2})^6$

85. $\dfrac{1}{i^3}$

86. $\dfrac{1}{(2i)^3}$

87. Cube the complex numbers.

$2, \quad -1 + \sqrt{3}i, \quad -1 - \sqrt{3}i$

88. Raise the numbers to the fourth power.

$2, \quad -2, \quad 2i, \quad -2i$

89. Prove that the sum of a complex number $a + bi$ and its conjugate is a real number.

90. Prove that the difference of a complex number $a + bi$ and its conjugate is an imaginary number.

91. Prove that the product of a complex number $a + bi$ and its conjugate is a real number.

92. Prove that the conjugate of the product of two complex numbers $a_1 + b_1i$ and $a_2 + b_2i$ is the product of their conjugates.

93. Prove that the conjugate of the sum of two complex numbers $a_1 + b_1i$ and $a_2 + b_2i$ is the sum of their conjugates.

Review **In Exercises 94–102, review your skills and problem-solving techniques from previous sections.**

94. Subtract: $(x^3 - 3x^2) - (6 - 2x - 4x^2)$

95. Add: $(4 + 3x) + (8 - 6x - x^2)$

96. Multiply: $(3x - \frac{1}{2})(x + 4)$

97. Expand: $(2x - 5)^2$

98. Expand: $[(x + y) + 3]^2$

99. ***Volume of an Oblate Spheroid***

Solve for a: $V = \frac{4}{3}\pi a^2 b$

100. ***Newton's Law of Universal Gravitation***

Solve for r: $F = \alpha\dfrac{m_1m_2}{r^2}$

101. ***Mixture Problem*** A 5-liter container contains a mixture with a concentration of 50%. How much of this mixture must be withdrawn and replaced by 100% concentrate to bring the mixture up to 60% concentration?

102. ***Average Speed*** A business executive traveled at an average speed of 100 kilometers per hour on a 200-kilometer trip. Because of heavy traffic, the average speed on the return trip was 80 kilometers per hour. Find the average speed for the round trip.

4.2 Complex Solutions of Equations

See Exercise 50 on page 334 for an example of how a polynomial can be used to model the path of a baseball.

The Number of Solutions of a Polynomial Equation ▫ *Finding Solutions of Polynomial Equations* ▫ *Finding Zeros of Polynomial Functions*

The Number of Solutions of a Polynomial Equation

The Fundamental Theorem of Algebra implies that a polynomial equation of degree n has precisely n solutions in the complex number system. These solutions can be real or complex and may be repeated.

EXAMPLE 1 *Zeros of Polynomial Functions*

a. The first-degree equation

$$x - 2 = 0 \qquad \text{First-degree equation}$$

has exactly *one* solution: $x = 2$.

b. The second-degree equation

$$x^2 - 6x + 9 = 0 \qquad \text{Second-degree equation}$$
$$(x - 3)(x - 3) = 0 \qquad \text{Factor.}$$

has exactly *two* solutions: $x = 3$ and $x = 3$.

c. The third-degree equation

$$x^3 + 4x = 0 \qquad \text{Third-degree equation}$$
$$x(x - 2i)(x + 2i) = 0 \qquad \text{Factor.}$$

has exactly *three* solutions: $x = 0$, $x = 2i$, and $x = -2i$.

d. The fourth-degree equation

$$x^4 - 1 = 0 \qquad \text{Third-degree equation}$$
$$(x - 1)(x + 1)(x - i)(x + i) = 0 \qquad \text{Factor.}$$

has exactly *four* solutions: $x = 1$, $x = -1$, $x = i$, and $x = -i$.

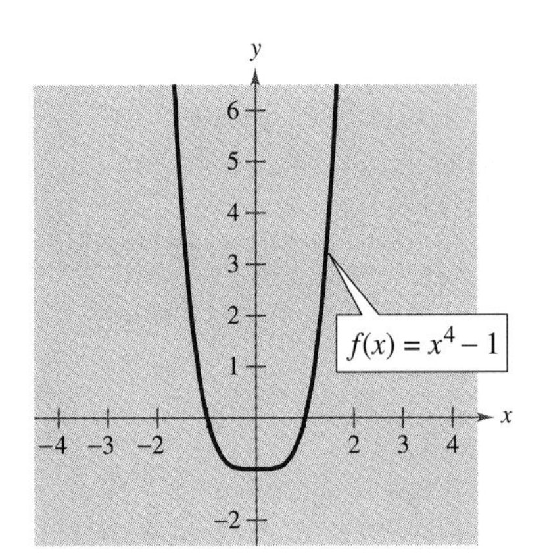

FIGURE 4.2

You can use a graph to check the number of *real* solutions of an equation. For instance, to check the real solutions of $x^4 - 1 = 0$, sketch the graph of $f(x) = x^4 - 1$. As shown in Figure 4.2, the graph has two x-intercepts, which implies that the equation has two real solutions.

Every second-degree equation, $ax^2 + bx + c = 0$, has precisely two solutions given by the Quadratic Formula.

$$x = \frac{-b \pm \sqrt{b^2 - 4ac}}{2a}$$

The expression inside the radical, $b^2 - 4ac$, is called the *discriminant*, and can be used to determine whether the solutions are real, repeated, or complex.

1. If $b^2 - 4ac < 0$, the equation has two complex solutions.
2. If $b^2 - 4ac = 0$, the equation has one repeated real solution.
3. If $b^2 - 4ac > 0$, the equation has two real solutions.

EXAMPLE 2 *Using a Discriminant*

Use the discriminant to find the number of real solutions of each equation.

a. $4x^2 - 20x + 25 = 0$ **b.** $13x^2 + 7x + 1 = 0$ **c.** $5x^2 - 8x = 0$

Solution

a. For this equation, $a = 4$, $b = -20$, and $c = 25$. Thus, the discriminant is

$$b^2 - 4ac = 400 - 4(4)(25) = 400 - 400 = 0.$$

Because the discriminant is zero, there is one repeated solution.

b. For this equation, $a = 13$, $b = 7$, and $c = 1$. Thus, the discriminant is

$$b^2 - 4ac = 49 - 4(13)(1) = 49 - 52 = -3.$$

Because the discriminant is negative, there are two complex solutions.

c. For this equation, $a = 5$, $b = -8$, and $c = 0$. Thus, the discriminant is

$$b^2 - 4ac = 64 - 4(5)(0) = 64.$$

Because the discriminant is positive, there are two real solutions.

NOTE Figure 4.3 shows the graphs of the functions corresponding to the equations in Example 2. Notice that with one repeated solution, the graph touches the x-axis at its x-intercept. With two complex solutions, the graph has no x-intercepts. With two real solutions, the graph crosses the x-axis at its x-intercepts.

(a)

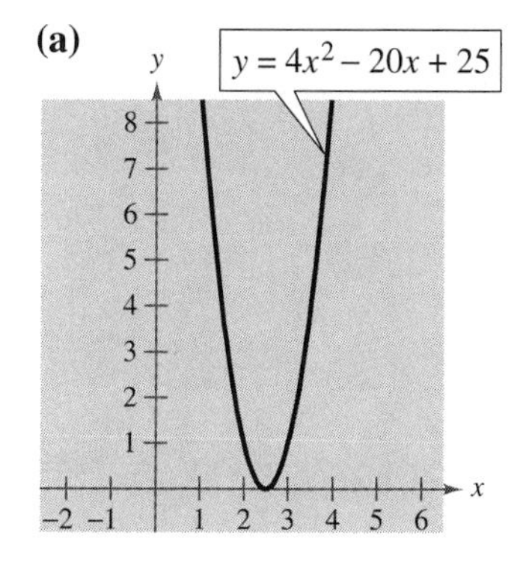

(b)

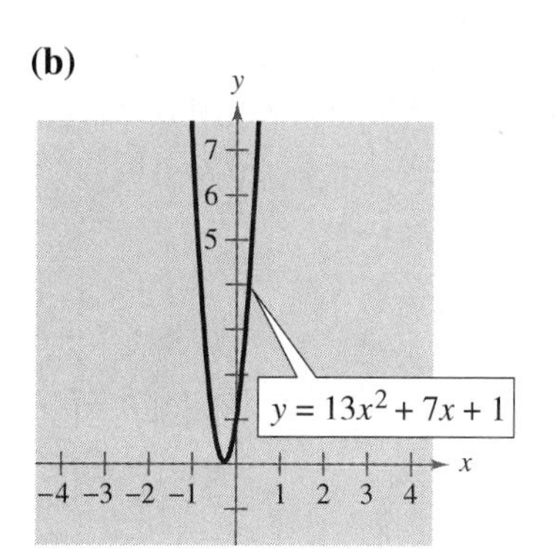

(c)

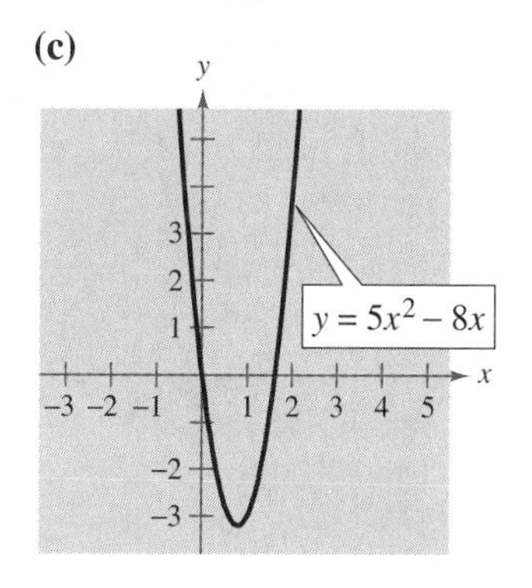

FIGURE 4.3

Finding Solutions of Polynomial Equations

EXAMPLE 3 *Solving a Quadratic Equation*

Solve $x^2 + 2x + 2 = 0$. List any complex solutions in $a + bi$ form.

Solution

Using $a = 1, b = 2$, and $c = 2$, you can apply the Quadratic Formula as follows.

$$\begin{aligned} x &= \frac{-b \pm \sqrt{b^2 - 4ac}}{2a} \\ &= \frac{-2 \pm \sqrt{2^2 - 4(1)(2)}}{2(1)} \\ &= \frac{-2 \pm \sqrt{-4}}{2} \\ &= \frac{-2 \pm 2i}{2} \\ &= -1 \pm i \end{aligned}$$

In Example 3, the two complex solutions are **conjugates.** That is, they are of the form $a \pm bi$. This is not a coincidence, as is indicated by the following theorem.

NOTE Be sure you see that this result is true only if the polynomial has *real* coefficients. For instance, the result applies to the equation $x^2 + 1 = 0$, but not to the equation $x - i = 0$.

COMPLEX SOLUTIONS OCCUR IN CONJUGATE PAIRS

If $a + bi$, $b \neq 0$ is a solution of a polynomial equation with real coefficients, the conjugate $a - bi$ is also a solution of the equation.

EXAMPLE 4 *Solving a Polynomial Equation*

Solve $x^4 - x^2 - 20 = 0$.

Solution

$$\begin{aligned} x^4 - x^2 - 20 &= 0 \\ (x^2 - 5)(x^2 + 4) &= 0 \\ (x + \sqrt{5})(x - \sqrt{5})(x + 2i)(x - 2i) &= 0 \\ x &= -\sqrt{5}, \sqrt{5}, -2i, 2i \end{aligned}$$

Finding Zeros of Polynomial Functions

The problem of finding the zeros of a polynomial function is essentially the same problem as finding the solutions of a polynomial equation. For instance, the zeros of the polynomial function $f(x) = 3x^2 - 4x + 5$ are simply the solutions of the polynomial equation $3x^2 - 4x + 5 = 0$.

EXAMPLE 5 *Finding the Zeros of a Polynomial Function*

Find all zeros of

$$f(x) = x^4 - 3x^3 + 6x^2 + 2x - 60$$

given that $1 + 3i$ is a zero of f.

Solution

Because complex zeros occur in conjugate pairs, you know that $1 - 3i$ is also a solution. This means that both

$$[x - (1 + 3i)] \quad \text{and} \quad [x - (1 - 3i)]$$

are factors of $f(x)$. Multiplying these two factors produces

$$\begin{aligned}[x - (1 + 3i)][x - (1 - 3i)] &= [(x - 1) - 3i][(x - 1) + 3i] \\ &= (x - 1)^2 - (3i)^2 \\ &= x^2 - 2x + 1 - (-9) \\ &= x^2 - 2x + 10.\end{aligned}$$

Using long division, you can divide $x^2 - 2x + 10$ into $f(x)$.

$$\begin{array}{r} x^2 - x - 6 \\ x^2 - 2x + 10 \overline{\big) x^4 - 3x^3 + 6x^2 + 2x - 60} \\ \underline{x^4 - 2x^3 + 10x^2 } \\ -x^3 - 4x^2 + 2x \\ \underline{-x^3 + 2x^2 - 10x } \\ -6x^2 + 12x - 60 \\ \underline{-6x^2 + 12x - 60} \\ 0 \end{array}$$

Therefore,

$$\begin{aligned}f(x) &= (x^2 - 2x + 10)(x^2 - x - 6) \\ &= (x^2 - 2x + 10)(x - 3)(x + 2)\end{aligned}$$

and you can conclude that the zeros of f are $1 + 3i$, $1 - 3i$, 3, and -2.

EXAMPLE 6 *Finding a Polynomial with Given Zeros*

Find a fourth-degree polynomial function with real coefficients that has -1, -1, and $3i$ as zeros.

Solution

Because $3i$ is a zero, you know that $-3i$ is also a zero. Thus, $f(x)$ can be written as

$$f(x) = a(x + 1)(x + 1)(x - 3i)(x + 3i).$$

For simplicity, let $a = 1$ and obtain

$$f(x) = (x^2 + 2x + 1)(x^2 + 9) = x^4 + 2x^3 + 10x^2 + 18x + 9.$$

EXAMPLE 7 *Finding a Polynomial with Given Zeros*

Find a cubic polynomial function f with real coefficients that has 2 and $1 - i$ as zeros, such that $f(1) = 3$.

Solution

Because $1 - i$ is a zero of f, so is $1 + i$. Therefore,

$$\begin{aligned} f(x) &= a(x - 2)[x - (1 - i)][x - (1 + i)] \\ &= a(x - 2)[(x - 1) + i][(x - 1) - i] \\ &= a(x - 2)[(x - 1)^2 - i^2] \\ &= a(x - 2)(x^2 - 2x + 2) \\ &= a(x^3 - 4x^2 + 6x - 4). \end{aligned}$$

To find the value of a, use the fact that $f(1) = 3$ and obtain $f(1) = a(1 - 4 + 6 - 4) = 3$. Thus, $a = -3$ and it follows that

$$f(x) = -3(x^3 - 4x^2 + 6x - 4) = -3x^3 + 12x^2 - 18x + 12.$$

GROUP ACTIVITY

SOLUTIONS, ZEROS, AND INTERCEPTS

Write a paragraph explaining the relationship among the solutions of a polynomial equation, the zeros of a polynomial function, and the x-intercepts of the graph of a polynomial function. Include examples in your paragraph.

The *Interactive* CD-ROM contains step-by-step solutions to all odd-numbered Section and Review Exercises. It also provides Tutorial Exercises, which link to Guided Examples for additional help.

WARM UP

In Exercises 1–4, write each complex number in standard form and give its complex conjugate.

1. $4 - \sqrt{-29}$
2. $-5 - \sqrt{-144}$
3. $-1 + \sqrt{-32}$
4. $6 + \sqrt{-1/4}$

In Exercises 5–10, perform the operations and write the answers in standard form.

5. $(-3 + 6i) - (10 - 3i)$
6. $(12 - 4i) + 20i$
7. $(4 - 2i)(3 + 7i)$
8. $(2 - 5i)(2 + 5i)$
9. $\dfrac{1 + i}{1 - i}$
10. $(3 + 2i)^3$

4.2 Exercises

In Exercises 1–4, determine the number of solutions of the equation in the complex number system.

1. $x^3 - 4x + 5 = 0$
2. $2x^6 + 3x^3 - 10 = 0$
3. $25 - x^4 = 0$
4. $12 - x + 3x^2 - 3x^5 = 0$

In Exercises 5–8, use the discriminant to determine the number of real solutions of the quadratic equation.

5. $2x^2 - 5x + 5 = 0$
6. $2x^2 - x - 1 = 0$
7. $\frac{1}{5}x^2 + \frac{6}{5}x - 8 = 0$
8. $\frac{1}{3}x^2 - 5x + 25 = 0$

In Exercises 9–20, solve the equation. List any complex solutions in the form $a + bi$.

9. $x^2 - 5 = 0$
10. $3x^2 - 1 = 0$
11. $(x + 5)^2 - 6 = 0$
12. $16 - (x - 1)^2 = 0$
13. $x^2 - 8x + 16 = 0$
14. $4x^2 + 4x + 1 = 0$
15. $x^2 + 2x + 5 = 0$
16. $54 + 16x - x^2 = 0$
17. $4x^2 - 4x + 5 = 0$
18. $4x^2 - 4x + 21 = 0$
19. $230 + 20x - 0.5x^2 = 0$
20. $6 - (x - 1)^2 = 0$

Graphical and Analytical Analysis **In Exercises 21–24, find all the zeros of the function. Is there a relationship between the number of real zeros and the number of x-intercepts of the graph? Explain.**

21. $f(x) = x^3 - 4x^2 + x - 4$
22. $f(x) = x^3 - 4x^2 - 4x + 16$

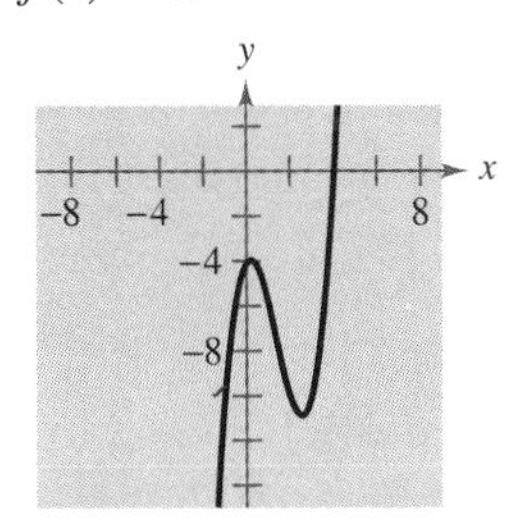

FIGURE FOR 21

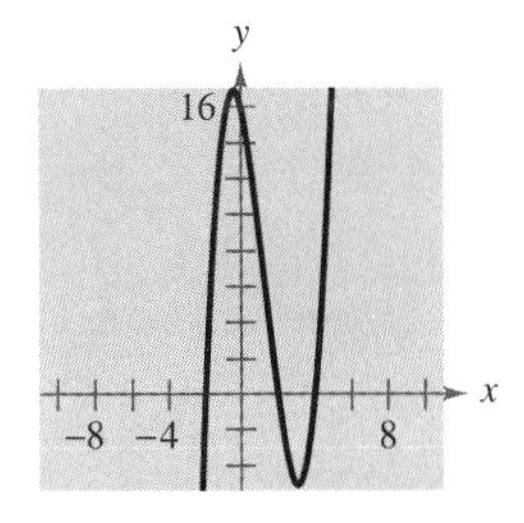

FIGURE FOR 22

23. $f(x) = x^4 + 4x^2 + 4$

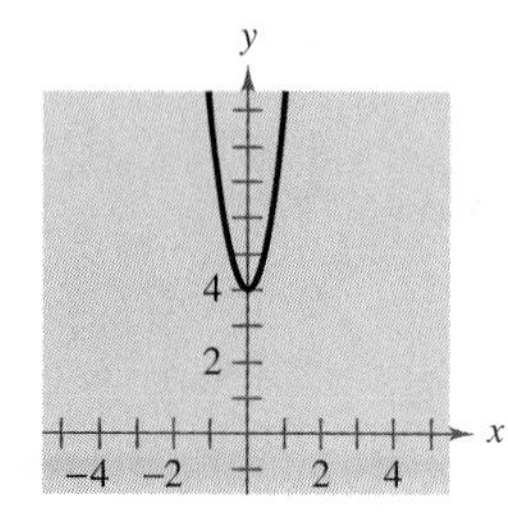

24. $f(x) = x^4 - 3x^2 - 4$

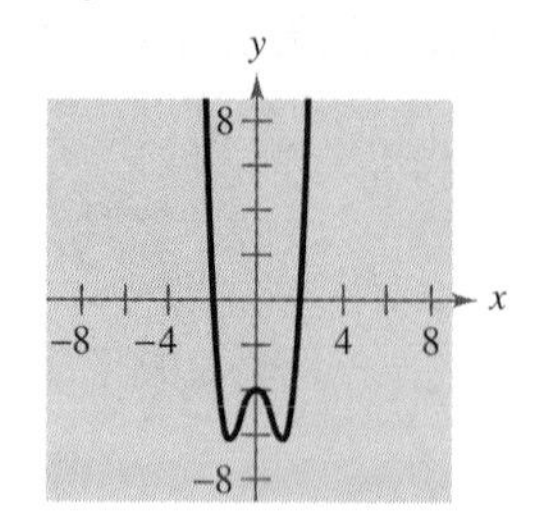

In Exercises 25–28, find all the zeros of the function. Write the polynomial as a product of linear factors.

25. $f(z) = z^2 - 2z + 2$ **26.** $f(x) = x^2 - x + 56$

27. $f(x) = x^4 - 81$ **28.** $f(y) = y^4 - 625$

In Exercises 29–36, use the given zero to find all the zeros of the function.

Function	*Zero*
29. $f(x) = 2x^3 + 3x^2 + 50x + 75$	$5i$
30. $f(x) = 2x^4 - x^3 + 7x^2 - 4x - 4$	$2i$
31. $g(x) = 4x^3 + 23x^2 + 34x - 10$	$-3 + i$
32. $h(x) = 3x^3 - 4x^2 + 8x + 8$	$1 - \sqrt{3}i$
33. $f(x) = x^4 + 3x^3 - 5x^2 - 21x + 22$	$-3 + \sqrt{2}i$
34. $f(x) = x^3 + 4x^2 + 14x + 20$	$-1 - 3i$
35. $h(x) = 8x^3 - 14x^2 + 18x - 9$	$(1 - \sqrt{5}i)/2$
36. $f(x) = 25x^3 - 55x^2 - 54x - 18$	$(-2 + \sqrt{2}i)/5$

In Exercises 37–44, find a polynomial function with integer coefficients that has the given zeros.

37. $1, 5i, -5i$ **38.** $i, -i, 6i, -6i$

39. $2, 4 + i, 4 - i$ **40.** $6, -5 + 2i, -5 - 2i$

41. $-5, -5, 1 + \sqrt{3}i$ **42.** $\frac{2}{3}, -1, 3 + \sqrt{2}i$

43. $\frac{3}{4}, -2, -\frac{1}{2} + i$ **44.** $0, 0, 4, 1 + i$

In Exercises 45 and 46, write the polynomial in completely factored form.

45. $f(x) = x^4 + 6x^2 - 27$

46. $f(x) = x^4 - 2x^3 - 3x^2 + 12x - 18$

47. ***Think About It*** A zero of the function $f(x) = x^3 + ix^2 + ix - 1$ is $x = -i$.

(a) Show that the conjugate $x = i$ is not a zero of f.

(b) Does your result contradict the theorem that complex zeros occur in conjugate pairs? Explain.

48. ***Essay*** Explain whether or not there exists a third-degree polynomial function with integer coefficients that has no real zeros.

49. ***Exploration*** Use a graphing utility to graph the function $f(x) = x^4 - 4x^2 + k$ for different values of k. Find values of k such that the zeros of f satisfy the specified characteristics. (Some parts do not have unique answers.)

(a) Four real zeros

(b) Two real zeros each of multiplicity 2

(c) Two real zeros and two complex zeros

(d) Four complex zeros

50. ***Maximum Height*** A baseball is thrown upward from ground level with an initial velocity of 48 feet per second, and its height h in feet is given by

$$h = 16t^2 + 48t, \quad 0 \le t \le 3$$

where t is the time in seconds. Suppose you are told the ball reaches a height of 64 feet. Is this possible?

51. ***Graphical Reasoning*** Match the graph in the figure with one of the following functions. Explain why each of the others is not the correct function. Use a graphing utility to verify your result.

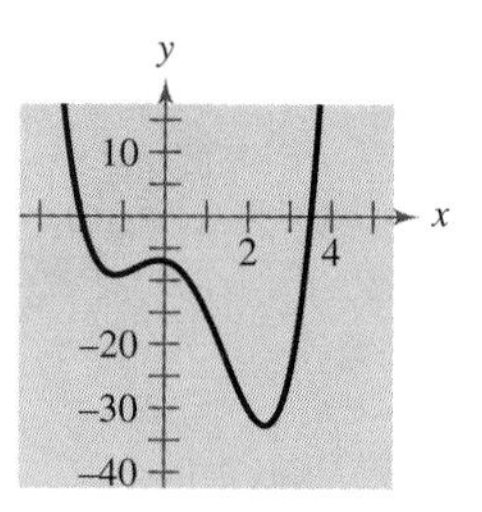

(a) $f(x) = x^2(x + 2)(x - 3.5)$

(b) $g(x) = (x + 2)(x - 3.5)$

(c) $h(x) = (x + 2)(x - 3.5)(x^2 + 1)$

(d) $k(x) = (x + 1)(x + 2)(x - 3.5)$

4.3 Trigonometric Form of a Complex Number

See Exercise 60 on page 341 for an example of the trigonometric form of a complex conjugate.

The Complex Plane ▫ *Trigonometric Form of a Complex Number* ▫ *Multiplication and Division of Complex Numbers*

The Complex Plane

Just as real numbers can be represented by points on the real number line, you can represent a complex number

$$z = a + bi$$

as the point (a, b) in a coordinate plane (the **complex plane**). The horizontal axis is called the **real axis** and the vertical axis is called the **imaginary axis,** as shown in Figure 4.4.

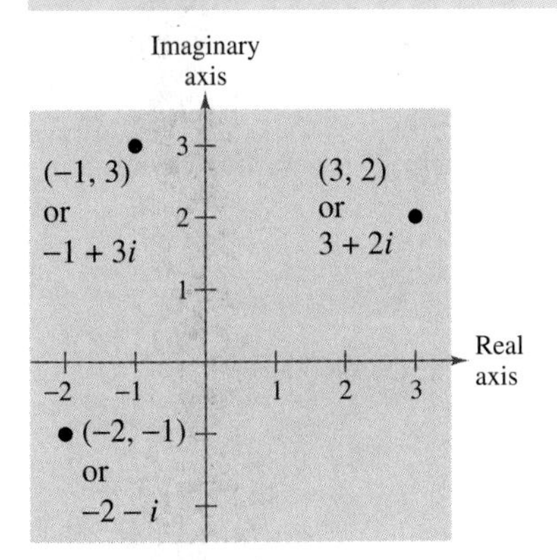

FIGURE 4.4

The **absolute value** of the complex number $a + bi$ is defined as the distance between the origin $(0, 0)$ and the point (a, b).

> **DEFINITION OF THE ABSOLUTE VALUE OF A COMPLEX NUMBER**
>
> The **absolute value** of the complex number $z = a + bi$ is given by
>
> $$|a + bi| = \sqrt{a^2 + b^2}.$$

NOTE If the complex number $a + bi$ is a real number (that is, if $b = 0$), this definition agrees with that given for the absolute value of a real number.

$$|a + 0i| = \sqrt{a^2 + 0^2} = |a|$$

EXAMPLE 1 *Finding the Absolute Value of a Complex Number*

Plot each complex number and find its absolute value.

a. $z = -3i$ **b.** $z = -2 + 5i$

Solution

The points are shown in Figure 4.5.

a. The complex number $z = 0 + (-3)i$ has an absolute value of

$$|z| = \sqrt{0^2 + (-3)^2}$$
$$= 3.$$

b. The complex number $z = -2 + 5i$ has an absolute value of

$$|z| = \sqrt{(-2)^2 + 5^2}$$
$$= \sqrt{29}.$$

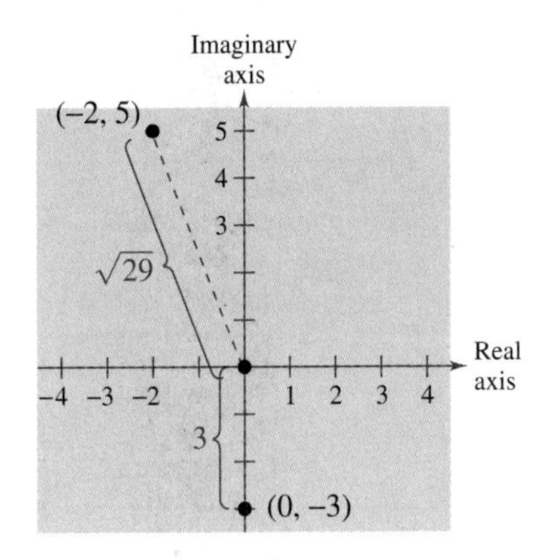

FIGURE 4.5

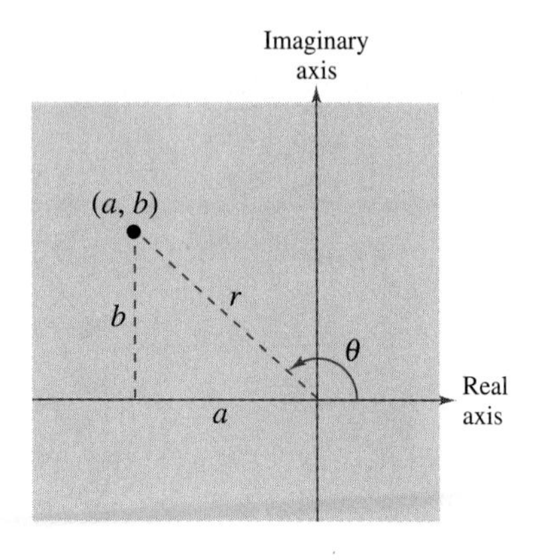

FIGURE 4.6

Trigonometric Form of a Complex Number

In Section 4.1 you learned how to add, subtract, multiply, and divide complex numbers. To work effectively with *powers* and *roots* of complex numbers, it is helpful to write complex numbers in **trigonometric form.** In Figure 4.6, consider the nonzero complex number $a + bi$. By letting θ be the angle from the positive x-axis (measured counterclockwise) to the line segment connecting the origin and the point (a, b), you can write

$$a = r\cos\theta \qquad \text{and} \qquad b = r\sin\theta$$

where $r = \sqrt{a^2 + b^2}$. Consequently, you have

$$a + bi = (r\cos\theta) + (r\sin\theta)i$$

from which you can obtain the **trigonometric form of a complex number.**

NOTE The trigonometric form of a complex number is also called the **polar form.** Because there are infinitely many choices for θ, the trigonometric form of a complex number is not unique. Normally, θ is restricted to the interval $0 \le \theta < 2\pi$, although on occasion it is convenient to use $\theta < 0$.

TRIGONOMETRIC FORM OF A COMPLEX NUMBER

The **trigonometric form** of the complex number $z = a + bi$ is

$$z = r(\cos\theta + i\sin\theta)$$

where $a = r\cos\theta$, $b = r\sin\theta$, $r = \sqrt{a^2 + b^2}$, and $\tan\theta = b/a$. The number r is the **modulus** of z, and θ is called an **argument** of z.

EXAMPLE 2 *Writing a Complex Number in Trigonometric Form*

Write the complex number $z = -2 - 2\sqrt{3}i$ in trigonometric form.

Solution

The absolute value of z is

$$r = \left|-2 - 2\sqrt{3}i\right| = \sqrt{(-2)^2 + \left(-2\sqrt{3}\right)^2} = \sqrt{16} = 4$$

and the angle θ is given by

$$\tan\theta = \frac{b}{a} = \frac{-2\sqrt{3}}{-2} = \sqrt{3}.$$

Because $\tan(\pi/3) = \sqrt{3}$ and $z = -2 - 2\sqrt{3}i$ lies in Quadrant III, you choose θ to be $\theta = \pi + \pi/3 = 4\pi/3$. Thus the trigonometric form is

$$z = r(\cos\theta + i\sin\theta) = 4\left(\cos\frac{4\pi}{3} + i\sin\frac{4\pi}{3}\right).$$

See Figure 4.7.

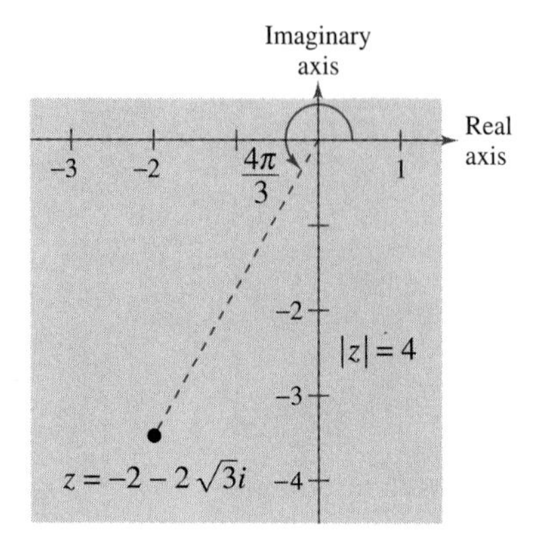

FIGURE 4.7

EXAMPLE 3 *Writing a Complex Number in Trigonometric Form*

Write the complex number $z = 6 + 2i$ in trigonometric form.

Solution

The absolute value of z is

$$r = |6 + 2i| = \sqrt{40} = 2\sqrt{10}$$

and the angle θ is given by

$$\tan \theta = \frac{b}{a} = \frac{2}{6} = \frac{1}{3}.$$

Because θ is in Quadrant I, you can conclude that

$$\theta = \arctan \frac{1}{3} \approx 18.4°.$$

Therefore, the trigonometric form of z is

$$\begin{aligned} z &= r(\cos \theta + i \sin \theta) \\ &= 2\sqrt{10}\left[\cos\left(\arctan \tfrac{1}{3}\right) + i \sin\left(\arctan \tfrac{1}{3}\right)\right] \\ &\approx 2\sqrt{10}(\cos 18.4° + i \sin 18.4°). \end{aligned}$$

This result is illustrated graphically in Figure 4.8.

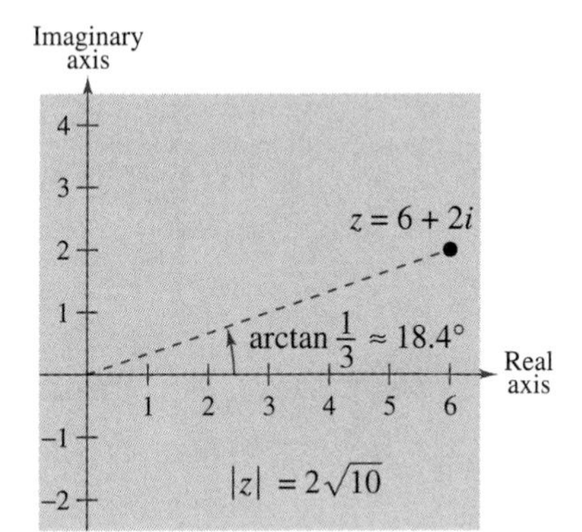

FIGURE 4.8

EXAMPLE 4 *Writing a Complex Number in Standard Form*

Write the complex number in standard form $a + bi$.

$$z = \sqrt{8}\left[\cos\left(-\frac{\pi}{3}\right) + i \sin\left(-\frac{\pi}{3}\right)\right]$$

Solution

Because $\cos(-\pi/3) = 1/2$ and $\sin(-\pi/3) = -\sqrt{3}/2$, you can write

$$\begin{aligned} z &= \sqrt{8}\left[\cos\left(-\frac{\pi}{3}\right) + i \sin\left(-\frac{\pi}{3}\right)\right] \\ &= \sqrt{8}\left(\frac{1}{2} - \frac{\sqrt{3}}{2}i\right) \\ &= 2\sqrt{2}\left(\frac{1}{2} - \frac{\sqrt{3}}{2}i\right) \\ &= \sqrt{2} - \sqrt{6}i. \end{aligned}$$

Multiplication and Division of Complex Numbers

The trigonometric form adapts nicely to multiplication and division of complex numbers. Suppose you are given two complex numbers

$$z_1 = r_1(\cos\theta_1 + i\sin\theta_1) \quad \text{and} \quad z_2 = r_2(\cos\theta_2 + i\sin\theta_2).$$

The product of z_1 and z_2 is

$$\begin{aligned} z_1z_2 &= r_1r_2(\cos\theta_1 + i\sin\theta_1)(\cos\theta_2 + i\sin\theta_2) \\ &= r_1r_2[(\cos\theta_1\cos\theta_2 - \sin\theta_1\sin\theta_2) + i(\sin\theta_1\cos\theta_2 + \cos\theta_1\sin\theta_2)]. \end{aligned}$$

Using the sum and difference formulas for cosine and sine, you can rewrite this equation as

$$z_1z_2 = r_1r_2[\cos(\theta_1 + \theta_2) + i\sin(\theta_1 + \theta_2)].$$

This establishes the first part of the following rule. The second part is left to you (see Exercise 59).

NOTE Note that this rule says that to multiply two complex numbers you multiply moduli and add arguments, whereas to divide two complex numbers you divide moduli and subtract arguments.

PRODUCT AND QUOTIENT OF TWO COMPLEX NUMBERS

Let $z_1 = r_1(\cos\theta_1 + i\sin\theta_1)$ and $z_2 = r_2(\cos\theta_2 + i\sin\theta_2)$ be complex numbers.

$$z_1z_2 = r_1r_2[\cos(\theta_1 + \theta_2) + i\sin(\theta_1 + \theta_2)] \qquad \text{Product}$$

$$\frac{z_1}{z_2} = \frac{r_1}{r_2}[\cos(\theta_1 - \theta_2) + i\sin(\theta_1 - \theta_2)], \quad z_2 \neq 0 \qquad \text{Quotient}$$

EXAMPLE 5 *Dividing Complex Numbers in Trigonometric Form*

Find z_1/z_2, for the following complex numbers.

$$z_1 = 24(\cos 300° + i\sin 300°) \qquad z_2 = 8(\cos 75° + i\sin 75°)$$

Solution

$$\begin{aligned} \frac{z_1}{z_2} &= \frac{24(\cos 300° + i\sin 300°)}{8(\cos 75° + i\sin 75°)} \\ &= \frac{24}{8}[\cos(300° - 75°) + i\sin(300° - 75°)] \\ &= 3(\cos 225° + i\sin 225°) \\ &= 3\left[\left(-\frac{\sqrt{2}}{2}\right) + i\left(-\frac{\sqrt{2}}{2}\right)\right] \\ &= -\frac{3\sqrt{2}}{2} - \frac{3\sqrt{2}}{2}i \end{aligned}$$

EXAMPLE 6 *Multiplying Complex Numbers in Trigonometric Form*

Find the product of the following complex numbers.

$$z_1 = 2\left(\cos\frac{2\pi}{3} + i\sin\frac{2\pi}{3}\right) \qquad z_2 = 8\left(\cos\frac{11\pi}{6} + i\sin\frac{11\pi}{6}\right)$$

Solution

$$\begin{aligned}
z_1z_2 &= 2\left(\cos\frac{2\pi}{3} + i\sin\frac{2\pi}{3}\right) \cdot 8\left(\cos\frac{11\pi}{6} + i\sin\frac{11\pi}{6}\right)\\
&= 16\left[\cos\left(\frac{2\pi}{3} + \frac{11\pi}{6}\right) + i\sin\left(\frac{2\pi}{3} + \frac{11\pi}{6}\right)\right]\\
&= 16\left(\cos\frac{5\pi}{2} + i\sin\frac{5\pi}{2}\right)\\
&= 16\left(\cos\frac{\pi}{2} + i\sin\frac{\pi}{2}\right)\\
&= 16[0 + i(1)]\\
&= 16i
\end{aligned}$$

Check this result by first converting to the standard forms

$$z_1 = -1 + \sqrt{3}i \qquad \text{and} \qquad z_2 = 4\sqrt{3} - 4i$$

and then multiplying algebraically, as in Section 4.1.

GROUP ACTIVITY

MULTIPLYING COMPLEX NUMBERS GRAPHICALLY

With others in your group, discuss how you can graphically approximate the product of the complex numbers. Then, approximate the values of the numbers and check your answers analytically.

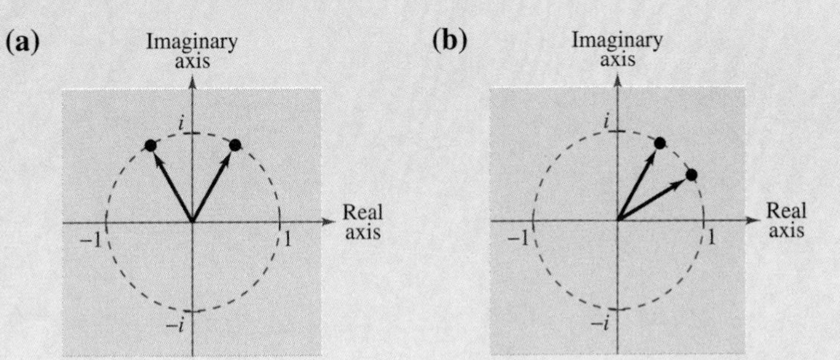

WARM UP

In Exercises 1–4, write the complex number in standard form.

1. $-5 - \sqrt{-100}$
2. $7 + \sqrt{-54}$
3. $-4i + i^2$
4. $3i^3$

In Exercises 5–10, perform the indicated operations and write the answers in standard form.

5. $(3 - 10i) - (-3 + 4i)$
6. $(2 + \sqrt{-50}) + (4 - \sqrt{2}i)$
7. $(4 - 2i)(-6 + i)$
8. $(3 - 2i)(3 + 2i)$
9. $\dfrac{1 + 4i}{1 - i}$
10. $\dfrac{3 - 5i}{2i}$

4.3 Exercises

Chapter Opener **In Exercises 1–6, represent the complex number graphically and find its absolute value.**

1. $-5i$
2. -5
3. $-4 + 4i$
4. $5 - 12i$
5. $6 - 7i$
6. $-8 + 3i$

In Exercises 7–10, express the complex number in trigonometric form.

7.

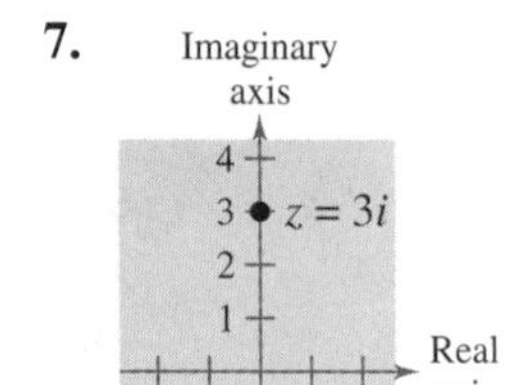

8.

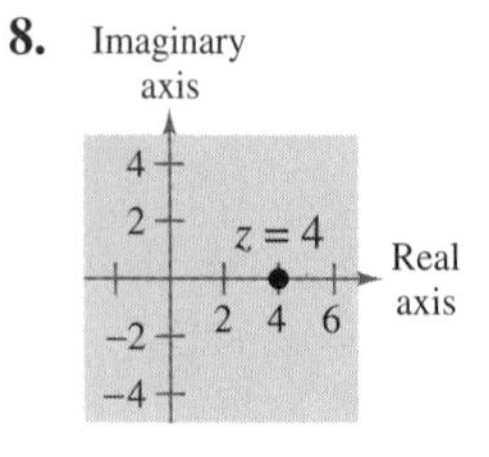

9.

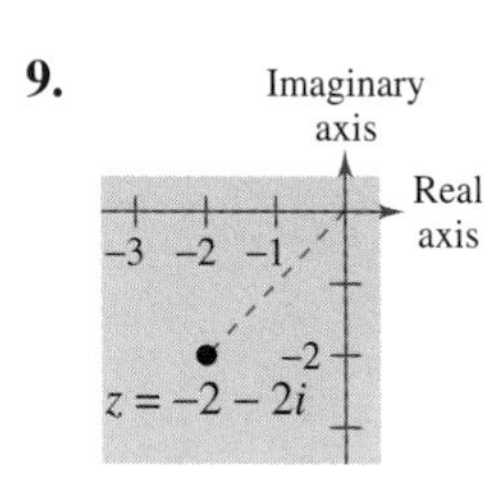

10.

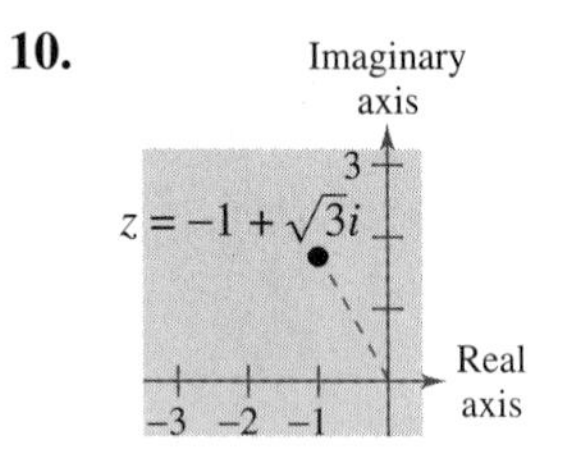

In Exercises 11–24, represent the complex number graphically, and find the trigonometric form of the number.

11. $3 - 3i$
12. $-2 - 2i$
13. $\sqrt{3} + i$
14. $-1 + \sqrt{3}i$
15. $-2(1 + \sqrt{3}i)$
16. $\frac{5}{2}(\sqrt{3} - i)$
17. $6i$
18. 4
19. $-7 + 4i$
20. $3 - i$
21. 7
22. $-2i$
23. $1 + 6i$
24. $2\sqrt{2} - i$

In Exercises 25–28, use a graphing utility to represent the complex number in trigonometric form.

25. $5 + 2i$

26. $-3 + i$

27. $3\sqrt{2} - 7i$

28. $-8 - 5\sqrt{3}i$

In Exercises 29–38, represent the complex number graphically, and find the standard form of the number.

29. $2(\cos 150° + i \sin 150°)$

30. $5(\cos 135° + i \sin 135°)$

31. $\frac{3}{2}(\cos 300° + i \sin 300°)$

32. $\frac{3}{4}(\cos 315° + i \sin 315°)$

33. $3.75\left(\cos \frac{3\pi}{4} + i \sin \frac{3\pi}{4}\right)$

34. $8\left(\cos \frac{\pi}{12} + i \sin \frac{\pi}{12}\right)$

35. $4\left(\cos \frac{3\pi}{2} + i \sin \frac{3\pi}{2}\right)$

36. $7(\cos 0 + i \sin 0)$

37. $3[\cos(18° 45') + i \sin(18° 45')]$

38. $6[\cos(230° 30') + i \sin(230° 30')]$

In Exercises 39–42, use a graphing utility to represent the complex number in standard form.

39. $5\left(\cos \frac{\pi}{9} + i \sin \frac{\pi}{9}\right)$

40. $12\left(\cos \frac{3\pi}{5} + i \sin \frac{3\pi}{5}\right)$

41. $4(\cos 216.5° + i \sin 216.5°)$

42. $9(\cos 58° + i \sin 58°)$

In Exercises 43–52, perform the operation and leave the result in trigonometric form.

43. $\left[3\left(\cos \frac{\pi}{3} + i \sin \frac{\pi}{3}\right)\right]\left[4\left(\cos \frac{\pi}{6} + i \sin \frac{\pi}{6}\right)\right]$

44. $\left[\frac{3}{2}\left(\cos \frac{\pi}{2} + i \sin \frac{\pi}{2}\right)\right]\left[6\left(\cos \frac{\pi}{4} + i \sin \frac{\pi}{4}\right)\right]$

45. $\left[\frac{5}{3}(\cos 140° + i \sin 140°)\right]\left[\frac{2}{3}(\cos 60° + i \sin 60°)\right]$

46. $[0.45(\cos 310° + i \sin 310°)] \times [0.60(\cos 200° + i \sin 200°)]$

47. $\dfrac{2(\cos 120° + i \sin 120°)}{4(\cos 40° + i \sin 40°)}$

48. $\dfrac{\cos 40° + i \sin 40°}{\cos 10° + i \sin 10°}$

49. $\dfrac{\cos(5\pi/3) + i \sin(5\pi/3)}{\cos \pi + i \sin \pi}$

50. $\dfrac{5(\cos 4.3 + i \sin 4.3)}{4(\cos 2.1 + i \sin 2.1)}$

51. $\dfrac{12(\cos 52° + i \sin 52°)}{3(\cos 110° + i \sin 110°)}$

52. $\dfrac{9(\cos 20° + i \sin 20°)}{5(\cos 75° + i \sin 75°)}$

In Exercises 53–58, (a) give the trigonometric form of the complex number, (b) perform the indicated operation using the trigonometric form, and (c) perform the indicated operation using the standard form, and check your result against the answer to part (b).

53. $(2 + 2i)(1 - i)$

54. $(\sqrt{3} + i)(1 + i)$

55. $-2i(1 + i)$

56. $\dfrac{3 + 4i}{1 - \sqrt{3}i}$

57. $\dfrac{5}{2 + 3i}$

58. $\dfrac{4i}{-4 + 2i}$

59. Given two complex numbers $z_1 = r_1(\cos \theta_1 + i \sin \theta_1)$ and $z_2 = r_2(\cos \theta_2 + i \sin \theta_2)$, $z_2 \neq 0$, prove that

$$\frac{z_1}{z_2} = \frac{r_1}{r_2}[\cos(\theta_1 - \theta_2) + i \sin(\theta_1 - \theta_2)].$$

60. Show that the complex conjugate of $z = r(\cos \theta + i \sin \theta)$ is $\bar{z} = r[\cos(-\theta) + i \sin(-\theta)]$.

61. Use the trigonometric forms of z and $\bar{z}$ in Exercise 60 to find

(a) $z\bar{z}$ and (b) $z/\bar{z}$, $\bar{z} \neq 0$.

62. Show that the negative of $z = r(\cos \theta + i \sin \theta)$ is $-z = r[\cos(\theta + \pi) + i \sin(\theta + \pi)]$.

In Exercises 63 and 64, sketch the graph of all complex numbers z satisfying the given condition.

63. $|z| = 2$

64. $\theta = \pi/6$

4.4 DeMoivre's Theorem

See Exercises 37–44 on page 347 for examples of how DeMoivre's Theorem can be used to solve equations.

Powers of Complex Numbers ▫ *Roots of Complex Numbers*

Powers of Complex Numbers

To raise a complex number to a power, consider repeated use of the multiplication rule.

$$
\begin{aligned}
z &= r(\cos\theta + i\sin\theta)\\
z^2 &= r(\cos\theta + i\sin\theta)r(\cos\theta + i\sin\theta) = r^2(\cos 2\theta + i\sin 2\theta)\\
z^3 &= r^2(\cos 2\theta + i\sin 2\theta)r(\cos\theta + i\sin\theta) = r^3(\cos 3\theta + i\sin 3\theta)\\
z^4 &= r^4(\cos 4\theta + i\sin 4\theta)\\
z^5 &= r^5(\cos 5\theta + i\sin 5\theta)
\end{aligned}
$$

This pattern leads to the following important theorem, which is named after the French mathematician Abraham DeMoivre (1667–1754).

DEMOIVRE'S THEOREM

If $z = r(\cos\theta + i\sin\theta)$ is a complex number and n is a positive integer, then

$$z^n = [r(\cos\theta + i\sin\theta)]^n = r^n(\cos n\theta + i\sin n\theta).$$

EXAMPLE 1 *Finding Powers of a Complex Number*

Use DeMoivre's Theorem to find $\left(-1 + \sqrt{3}i\right)^{12}$.

Solution

First convert to trigonometric form. Then apply DeMoivre's Theorem.

$$
\begin{aligned}
\left(-1 + \sqrt{3}i\right)^{12} &= \left[2\left(\cos\frac{2\pi}{3} + i\sin\frac{2\pi}{3}\right)\right]^{12}\\
&= 2^{12}\left[\cos(12)\frac{2\pi}{3} + i\sin(12)\frac{2\pi}{3}\right]\\
&= 4096(\cos 8\pi + i\sin 8\pi)\\
&= 4096(1 + 0)\\
&= 4096
\end{aligned}
$$

Roots of Complex Numbers

Recall that a consequence of the Fundamental Theorem of Algebra is that a polynomial equation of degree n has n solutions in the complex number system. Hence, an equation such as $x^6 = 1$ has six solutions, and in this particular case you can find the six solutions by factoring and using the Quadratic Formula.

$$\begin{aligned} x^6 - 1 &= (x^3 - 1)(x^3 + 1) \\ &= (x - 1)(x^2 + x + 1)(x + 1)(x^2 - x + 1) = 0 \end{aligned}$$

Consequently, the solutions are

$$x = \pm 1, \qquad x = \frac{-1 \pm \sqrt{3}i}{2}, \qquad \text{and} \qquad x = \frac{1 \pm \sqrt{3}i}{2}.$$

Each of these numbers is a sixth root of 1. In general, the ***n*th root** of a complex number is defined as follows.

DEFINITION OF NTH ROOT OF A COMPLEX NUMBER

The complex number $u = a + bi$ is an ***n*th root** of the complex number z if

$$z = u^n = (a + bi)^n.$$

To find a formula for an nth root of a complex number, let u be an nth root of z, where

$$u = s(\cos \beta + i \sin \beta) \qquad \text{and} \qquad z = r(\cos \theta + i \sin \theta).$$

By DeMoivre's Theorem and the fact that $u^n = z$, you have

$$s^n (\cos n\beta + i \sin n\beta) = r(\cos \theta + i \sin \theta).$$

Taking the absolute values of both sides of this equation, it follows that $s^n = r$. Substituting back into the previous equation and dividing by r, you get

$$\cos n\beta + i \sin n\beta = \cos \theta + i \sin \theta.$$

Thus, it follows that

$$\cos n\beta = \cos \theta \qquad \text{and} \qquad \sin n\beta = \sin \theta.$$

Because both sine and cosine have a period of 2π, these last two equations have solutions if and only if the angles differ by a multiple of 2π. Consequently, there must exist an integer k such that

$$\begin{aligned} n\beta &= \theta + 2\pi k \\ \beta &= \frac{\theta + 2\pi k}{n}. \end{aligned}$$

By substituting this value for β into the trigonometric form of u, you get the result stated on the following page.

NOTE When k exceeds $n - 1$, the roots begin to repeat. For instance, if $k = n$, the angle

$$\frac{\theta + 2\pi n}{n} = \frac{\theta}{n} + 2\pi$$

is coterminal with θ/n, which is also obtained when $k = 0$.

NTH ROOTS OF A COMPLEX NUMBER

For a positive integer n, the complex number $z = r(\cos\theta + i\sin\theta)$ has exactly n distinct nth roots given by

$$\sqrt[n]{r}\left(\cos\frac{\theta + 2\pi k}{n} + i\sin\frac{\theta + 2\pi k}{n}\right)$$

where $k = 0, 1, 2, \ldots, n - 1$.

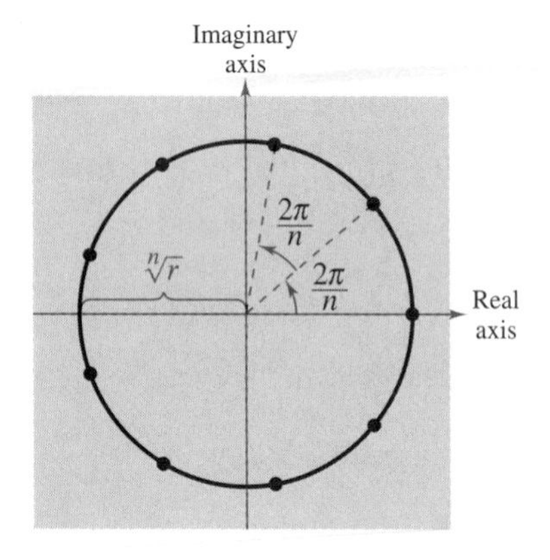

FIGURE 4.9

This formula for the nth roots of a complex number z has a nice geometrical interpretation, as shown in Figure 4.9. Note that because the nth roots of z all have the same magnitude $\sqrt[n]{r}$, they all lie on a circle of radius $\sqrt[n]{r}$ with center at the origin. Furthermore, because successive nth roots have arguments that differ by $2\pi/n$, the n roots are equally spaced around the circle.

You have already found the sixth roots of 1 by factoring and by using the Quadratic Formula. Example 2 shows how you can solve the same problem with the formula for nth roots.

EXAMPLE 2 *Finding nth Roots of a Real Number*

Find all the sixth roots of 1.

Solution

First write 1 in the trigonometric form $1 = 1(\cos 0 + i\sin 0)$. Then, by the nth root formula, with $n = 6$ and $r = 1$, the roots have the form

$$\sqrt[6]{1}\left(\cos\frac{0 + 2\pi k}{6} + i\sin\frac{0 + 2\pi k}{6}\right)$$

or simply $\cos(\pi k/3) + i\sin(\pi k/3)$. Thus, for $k = 0, 1, 2, 3, 4,$ and 5, the sixth roots are as follows. (See Figure 4.10.)

$$\cos 0 + i\sin 0 = 1$$

$$\cos\frac{\pi}{3} + i\sin\frac{\pi}{3} = \frac{1}{2} + \frac{\sqrt{3}}{2}i$$

$$\cos\frac{2\pi}{3} + i\sin\frac{2\pi}{3} = -\frac{1}{2} + \frac{\sqrt{3}}{2}i$$

$$\cos\pi + i\sin\pi = -1$$

$$\cos\frac{4\pi}{3} + i\sin\frac{4\pi}{3} = -\frac{1}{2} - \frac{\sqrt{3}}{2}i$$

$$\cos\frac{5\pi}{3} + i\sin\frac{5\pi}{3} = \frac{1}{2} - \frac{\sqrt{3}}{2}i$$

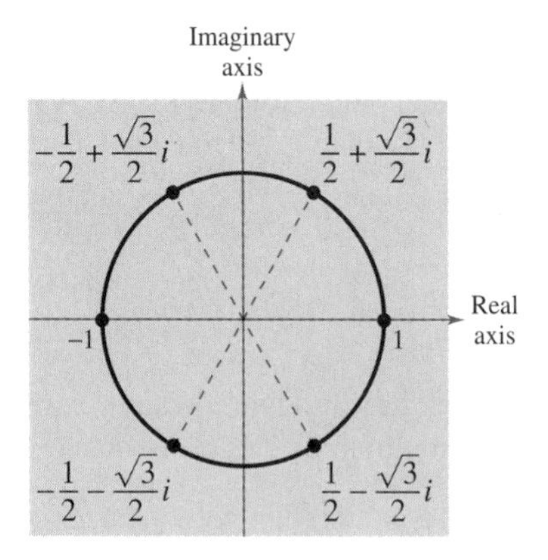

FIGURE 4.10

In Figure 4.10, notice that the roots obtained in Example 2 all have a magnitude of 1 and are equally spaced around the unit circle. Also notice that the complex roots occur in conjugate pairs, as discussed in Section 4.2. The n distinct nth roots of 1 are called the ***n*th roots of unity.**

Exploration

The nth roots of a complex number are useful for solving some polynomial equations. For instance, explain how you can use DeMoivre's Theorem to solve the polynomial equation

$$x^4 + 16 = 0.$$

[*Hint*: Write -16 as $16(\cos \pi + i \sin \pi)$.]

EXAMPLE 3 *Finding the nth Roots of a Complex Number*

Find the three cube roots of $z = -2 + 2i$.

Solution

Because z lies in Quadrant II, the trigonometric form for z is

$$z = -2 + 2i = \sqrt{8}\,(\cos 135° + i \sin 135°).$$

By the formula for nth roots, the cube roots have the form

$$\sqrt[6]{8}\left(\cos \frac{135° + 360°k}{3} + i \sin \frac{135° + 360°k}{3}\right).$$

Finally, for $k = 0, 1$, and 2, you obtain the roots

$$\sqrt{2}(\cos 45° + \text{i} \sin 45°) = 1 + i$$
$$\sqrt{2}(\cos 165° + i \sin 165°) \approx -1.3660 + 0.3660i$$
$$\sqrt{2}(\cos 285° + i \sin 285°) \approx 0.3660 - 1.3660i.$$

GROUP ACTIVITY

A FAMOUS MATHEMATICAL FORMULA

The famous formula

$$e^{a+bi} = e^a(\cos b + i \sin b)$$

is called Euler's Formula, after the German mathematician Leonhard Euler (1707–1783). Although the interpretation of this formula is beyond the scope of this text, we decided to include it because it gives rise to one of the most wonderful equations in mathematics.

$$e^{\pi i} + 1 = 0$$

This elegant equation relates the five most famous numbers in mathematics—0, 1, π, e, and i—in a single equation. Show how Euler's Formula can be used to derive this equation.

WARM UP

In Exercises 1 and 2, simplify the expression.

1. $\sqrt[3]{54}$
2. $\sqrt[4]{16 + 48}$

In Exercises 3–6, write the complex numbers in trigonometric form.

3. $-5 + 5i$
4. $-3i$
5. -12
6. 12

In Exercises 7–10, perform the indicated operation. Leave the result in trigonometric form.

7. $\left(\cos \frac{\pi}{4} + i \sin \frac{\pi}{4}\right)\left(\cos \frac{\pi}{2} + i \sin \frac{\pi}{2}\right)$
8. $\left(\cos \frac{\pi}{12} + i \sin \frac{\pi}{12}\right)\left(\cos \frac{5\pi}{6} + i \sin \frac{5\pi}{6}\right)$
9. $\dfrac{6[\cos(2\pi/3) + i \sin(2\pi/3)]}{3[\cos(\pi/6) + i \sin(\pi/6)]}$
10. $\dfrac{2(\cos 55° + i \sin 55°)}{3(\cos 10° + i \sin 10°)}$

4.4 Exercises

In Exercises 1–12, use DeMoivre's Theorem to find the indicated power of the complex number. Express the result in standard form.

1. $(1 + i)^5$
2. $(2 + 2i)^6$
3. $(-1 + i)^{10}$
4. $(1 - i)^{12}$
5. $2(\sqrt{3} + i)^7$
6. $4(1 - \sqrt{3}i)^3$
7. $[5(\cos 20° + i \sin 20°)]^3$
8. $[3(\cos 150° + i \sin 150°)]^4$
9. $\left(\cos \frac{5\pi}{4} + i \sin \frac{5\pi}{4}\right)^{10}$
10. $\left[2\left(\cos \frac{\pi}{2} + i \sin \frac{\pi}{2}\right)\right]^8$
11. $[5(\cos 3.2 + i \sin 3.2)]^4$
12. $(\cos 0 + i \sin 0)^{20}$

In Exercises 13–16, use a graphing utility and DeMoivre's Theorem to find the indicated power of the complex number. Express the result in standard form.

13. $(3 - 2i)^5$
14. $(\sqrt{5} - 4i)^3$
15. $[3(\cos 15° + i \sin 15°)]^4$
16. $\left[2\left(\cos \frac{\pi}{10} + i \sin \frac{\pi}{10}\right)\right]^5$
17. Show that $-\frac{1}{2}(1 + \sqrt{3}i)$ is a sixth root of 1.
18. Show that $2^{-1/4}(1 - i)$ is a fourth root of -2.

Graphical Reasoning **In Exercises 19 and 20, the graph of the roots of a complex number is given. (a) Write each of the roots in trigonometric form. (b) Identify the complex number whose roots are given. (c) Use a graphing utility to verify the results of part (b).**

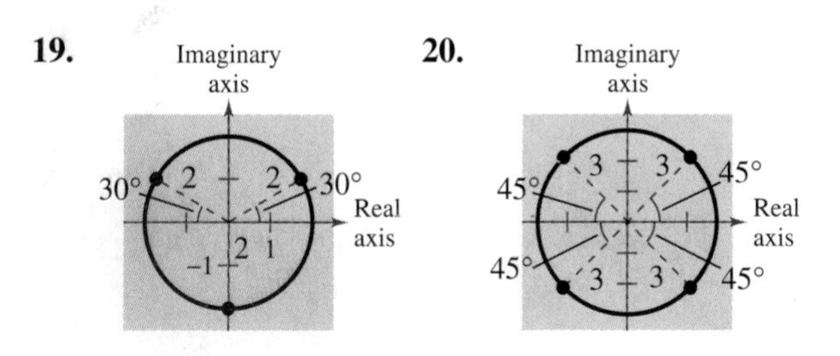

In Exercises 21–32, (a) use DeMoivre's Theorem to find the indicated roots of the complex number, (b) represent each of the roots graphically, and (c) express each of the roots in standard form.

21. Square roots of $5(\cos 120° + i \sin 120°)$
22. Square roots of $16(\cos 60° + i \sin 60°)$
23. Fourth roots of $16\left(\cos \frac{4\pi}{3} + i \sin \frac{4\pi}{3}\right)$
24. Fifth roots of $32\left(\cos \frac{5\pi}{6} + i \sin \frac{5\pi}{6}\right)$
25. Square roots of $-25i$
26. Fourth roots of $625i$
27. Cube roots of $-\frac{125}{2}(1 + \sqrt{3}i)$
28. Cube roots of $-4\sqrt{2}(1 - i)$
29. Cube roots of 8
30. Fourth roots of i
31. Fifth roots of 1
32. Cube roots of 1000

In Exercises 33–36, (a) use DeMoivre's Theorem and a graphing utility to find the indicated roots of the complex number, (b) represent each of the roots graphically, and (c) express each of the roots in standard form.

33. Cube roots of -125
34. Fourth roots of -4
35. Fifth roots of $128(-1 + i)$
36. Sixth roots of $64i$

In Exercises 37–44, find all the solutions of the equation and represent the solutions graphically.

37. $x^4 - i = 0$
38. $x^3 + 1 = 0$
39. $x^5 + 243 = 0$
40. $x^4 - 81 = 0$
41. $x^3 + 64i = 0$
42. $x^6 - 64i = 0$
43. $x^3 - (1 - i) = 0$
44. $x^4 + (1 + i) = 0$

Review **Solve Exercises 45–48, as a review of the skills and problem-solving techniques you learned in previous sections. Use the figure and trigonometric identities to find the exact value of the trigonometric function.**

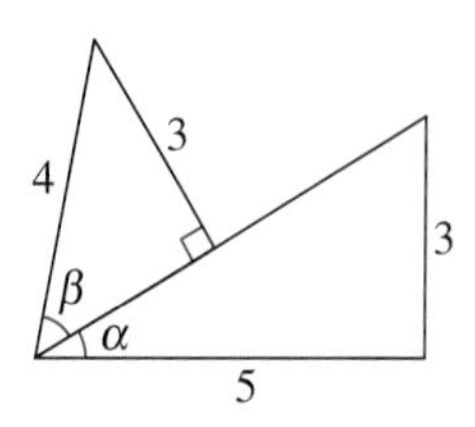

45. $\cos(\alpha + \beta)$
46. $\sin(\alpha - \beta)$
47. $\tan 2\alpha$
48. $\cos \frac{\alpha}{2}$

FOCUS ON CONCEPTS

In this chapter, you studied several concepts that are required in the study of complex numbers and complex solutions of equations. You can use the following questions to check your understanding of several of these basic concepts. The answers to these questions are given in the back of the book.

1. Define the imaginary unit i.
2. Explain why $x^2 + 1 = 0$ does not have real number solutions.
3. Explain why the product of a complex number and its conjugate is a real number.
4. *True or False?* The sum of two complex numbers is never a real number. Explain.
5. The following are graphs of third-degree polynomial functions. Determine the number of real zeros of each function.

(a)

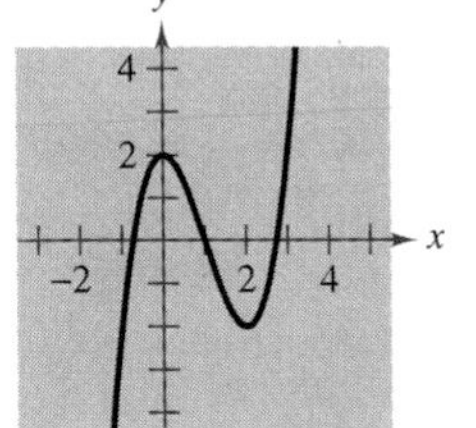

(b)

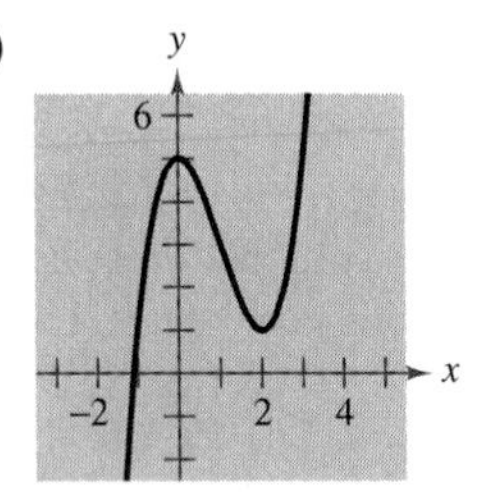

6. The following are graphs of fourth-degree polynomial functions. Determine the number of real zeros of each function.

(a)

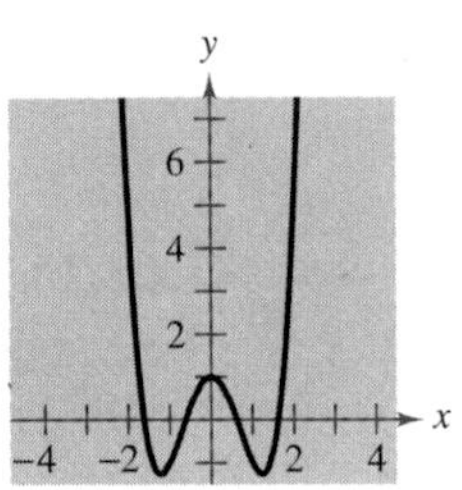

(b)

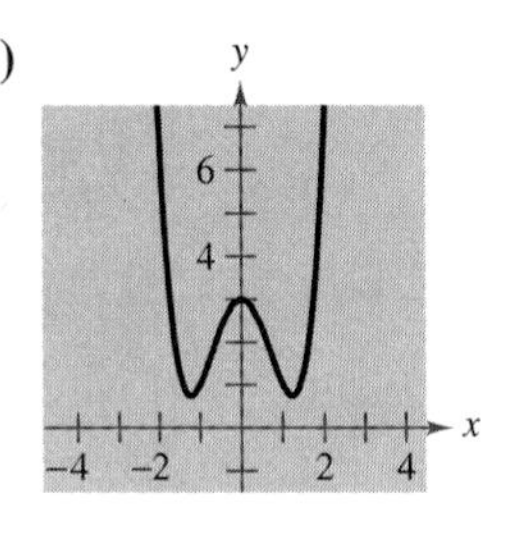

(c)

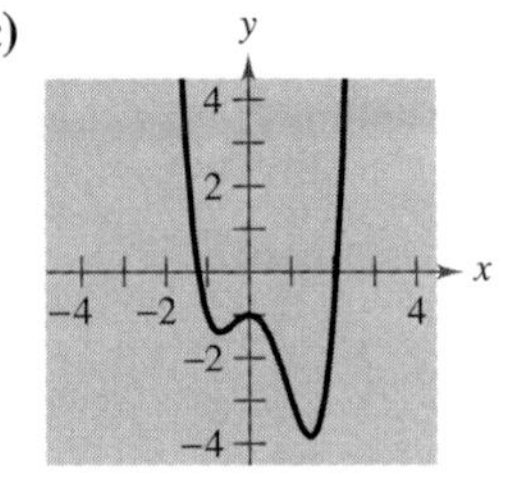

(d)

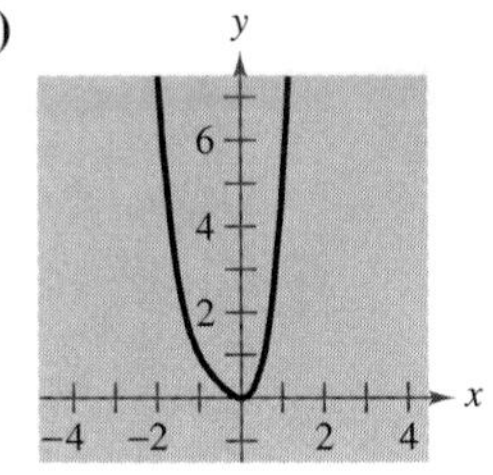

7. The figure below shows z_1 and z_2. Describe $z_1 z_2$ and z_1/z_2.

Imaginary axis

z_2 z_1 θ θ 1 −1 1

Real axis

8. One of the fourth roots of a complex number z is shown in the figure.

(a) How many roots are not shown?

(b) Describe the other roots.

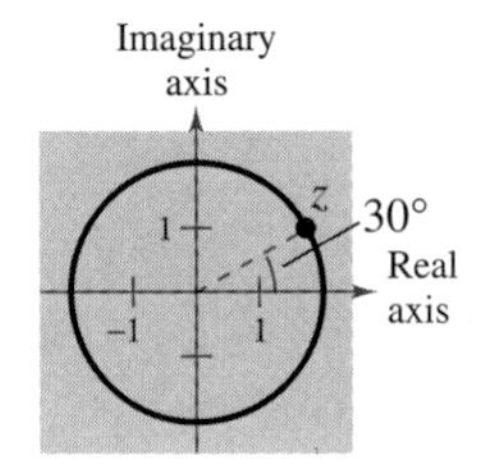

Review Exercises

In Exercises 1–4, write the complex number in standard form.

1. $3 + \sqrt{-25}$ **2.** $5 - \sqrt{-12}$

3. $3i + 6i^3$ **4.** $(\sqrt{-9})^2 + 2$

In Exercises 5–16, perform the operations and write the result in standard form.

5. $(7 + 5i) + (-4 + 2i)$

6. $-(6 - 2i) + (-8 + 3i)$

7. $\left(\frac{\sqrt{2}}{2} - \frac{\sqrt{2}}{2}i\right) - \left(\frac{\sqrt{2}}{2} + \frac{\sqrt{2}}{2}i\right)$

8. $(13 - 8i) - 5i$

9. $5i(13 - 8i)$ **10.** $(1 + 6i)(5 - 2i)$

11. $(10 - 8i)(2 - 3i)$ **12.** $i(6 + i)(3 - 2i)$

13. $\frac{6 + i}{i}$ **14.** $\frac{3 + 2i}{5 + i}$

15. $\frac{4}{-3i}$ **16.** $\frac{1}{(2 + i)^4}$

In Exercises 17–24, use the discriminant to determine the number of real solutions of the equation.

17. $6x^2 + x - 2 = 0$ **18.** $20x^2 - 21x + 4 = 0$

19. $9x^2 - 12x + 4 = 0$ **20.** $3x^2 + 5x + 3 = 0$

21. $0.13x^2 - 0.45x + 0.65 = 0$

22. $4x^2 + \frac{4}{3}x + \frac{1}{9} = 0$

23. $15 + 2x - x^2 = 0$ **24.** $x(x + 12) = -46$

In Exercises 25–30, find all the zeros of the function.

25. $g(x) = x^2 - 2x$ **26.** $f(x) = 6x - x^2$

27. $f(x) = x^2 + 8x + 10$ **28.** $h(x) = 3 + 4x - x^2$

29. $r(x) = 2x^2 + 2x + 3$ **30.** $s(x) = 2x^2 + 5x + 4$

Graphical and Analytical Analysis **In Exercises 31–34, find all the zeros of the function. Is there a relationship between the number of real zeros and the number of x-intercepts of the graph? Explain.**

31. $f(x) = x^3 + 4x^2 - 7x - 10$

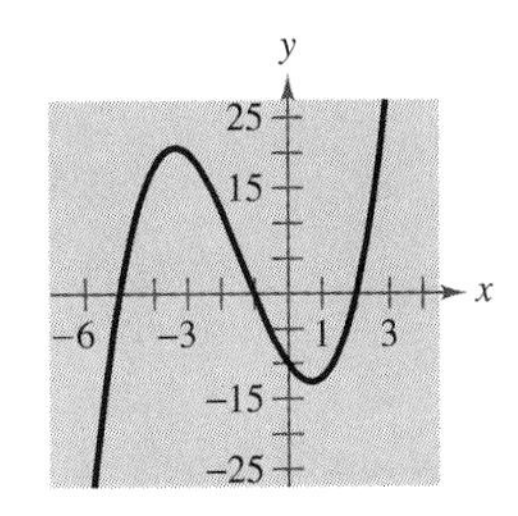

32. $f(x) = x^3 - 3x^2 + 4x - 12$

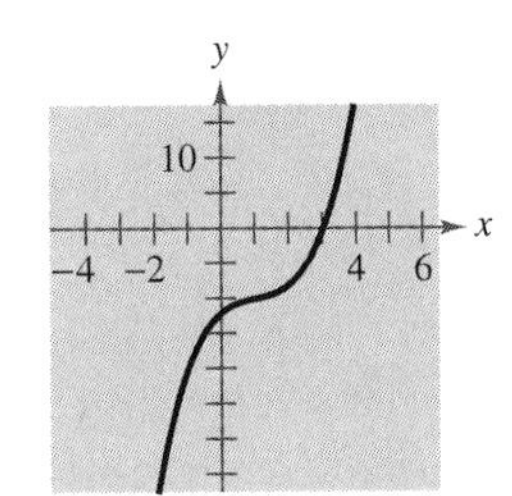

33. $f(x) = x^4 - 8x^3 + 21x^2 - 20x$

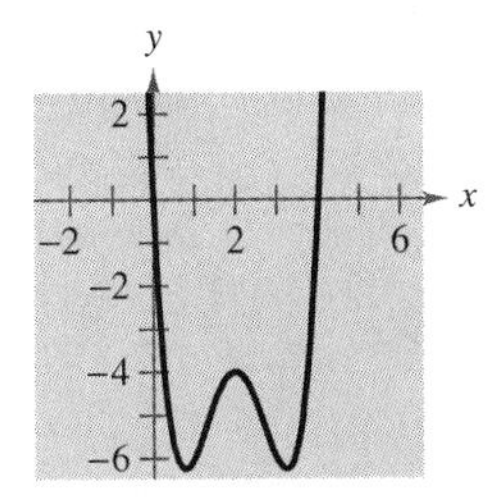

34. $f(x) = x^4 - 4x^3 + 7x^2 - 6x + 2$

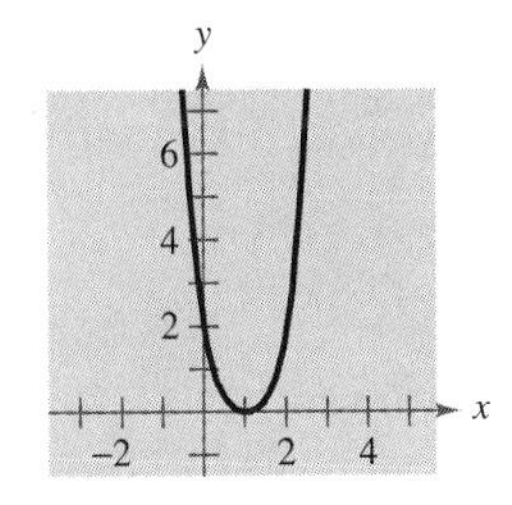

In Exercises 35–42, use the given zero of the function as an aid in finding all its zeros. Write the polynomial as a product of linear factors.

Function	Zero
35. $f(x) = 4x^3 - 11x^2 + 10x - 3$	1
36. $f(x) = 10x^3 + 21x^2 - x - 6$	-2
37. $f(x) = x^3 + 3x^2 - 5x + 25$	-5
38. $g(x) = x^3 - 5x^2 - 9x + 45$	-3
39. $h(x) = x^3 - 18x^2 + 106x - 200$	$7 + i$
40. $f(x) = 5x^3 - 4x^2 + 20x - 16$	$2i$
41. $f(x) = x^4 + 5x^3 + 2x^2 - 50x - 84$	$-3 + \sqrt{5}i$
42. $g(x) = x^4 - 8x^3 + 24x^2 - 36x + 27$	$1 + \sqrt{2}i$

In Exercises 43–50, find a polynomial function with integer coefficients that has the given zeros.

43. $0, 0, 3i, -3i$

44. $1, 1, -2, 2$

45. $-1, -1, \frac{1}{3}, -\frac{1}{2}$

46. $5, 1 - \sqrt{2}, 1 + \sqrt{2}$

47. $\frac{2}{3}, 4, \sqrt{3}i, -\sqrt{3}i$

48. $2, -3, 1 - 2i, 1 + 2i$

49. $\frac{1}{2}, \frac{1}{2}, -1, -1, -1$

50. $-2i, 2i, -4i, 4i$

51. ***Profit*** The demand equation for a certain product is given by $p = 140 - 0.0001x$, where p is the unit price (in dollars) of the product and x is the number of units produced and sold. The cost equation for the product is $C = 80x - 150{,}000$, where C is the total cost (in dollars) and x is the number of units produced. The total profit obtained by producing and selling x units is given by

$$P = xp - C.$$

Suppose you work in the marketing department of the company that produces this product and are asked to determine a price p that would yield a profit of 9 million dollars. Is this possible? Explain.

52. ***Think About It*** Does there exist a fourth-degree polynomial function with integer coefficients that has two real zeros? Explain.

In Exercises 53–56, represent the complex number graphically and find its absolute value.

53. 4

54. $-3i$

55. $3 + 4i$

56. $4 - 3i$

In Exercises 57–60, express the complex number in trigonometric form.

57.

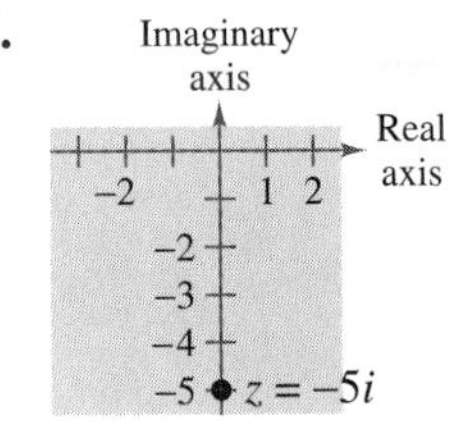

58.

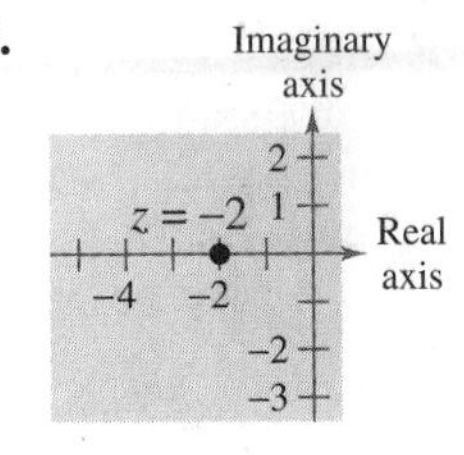

59.

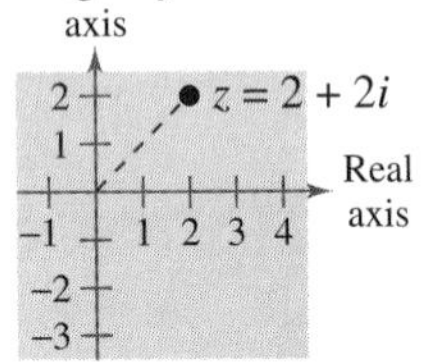

60.

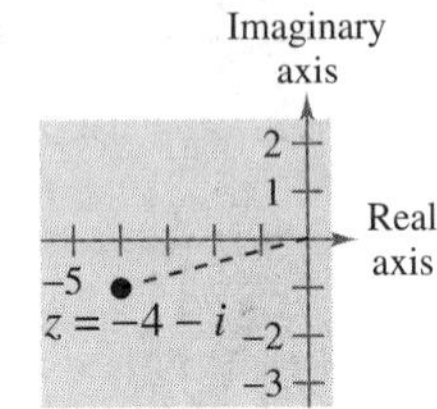

In Exercises 61–64, find the trigonometric form of the complex number.

61. $5 - 5i$

62. $-3\sqrt{3} + 3i$

63. $5 + 12i$

64. -7

In Exercises 65–68, write the complex number in standard form.

65.

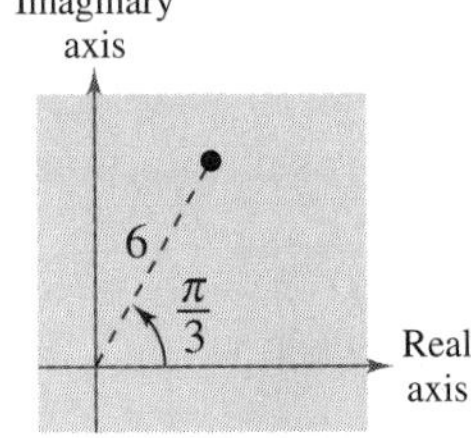

66.

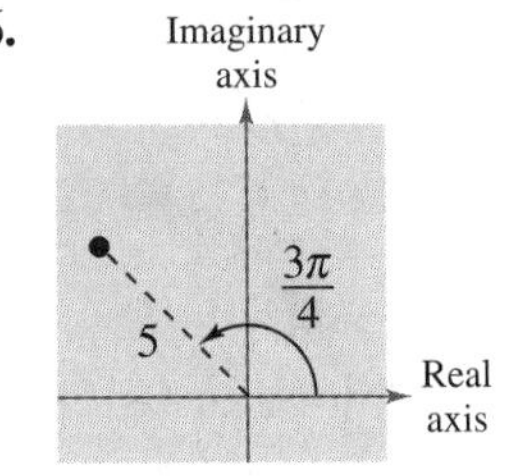

67.

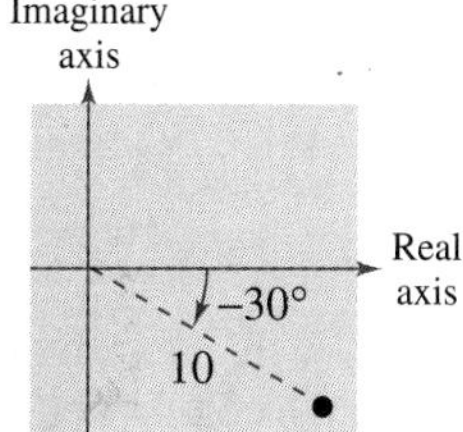

68.

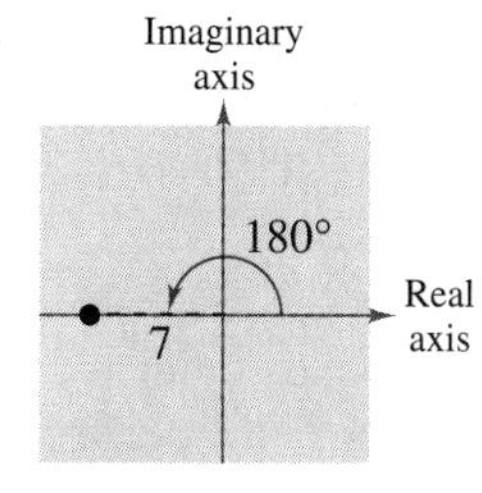

In Exercises 69–72, write the complex number in standard form.

69. $100(\cos 240° + i \sin 240°)$

70. $24(\cos 330° + i \sin 330°)$

71. $13(\cos 0 + i \sin 0)$

72. $8\left(\cos \frac{5\pi}{6} + i \sin \frac{5\pi}{6}\right)$

In Exercises 73–76, (a) express the two complex numbers in trigonometric form, and (b) use the trigonometric form to find $z_1 z_2$ and z_1/z_2.

73. $z_1 = -5,\ z_2 = 5i$

74. $z_1 = 2\sqrt{3} - 2i,\ z_2 = -10i$

75. $z_1 = -3(1 + i),\ z_2 = 2(\sqrt{3} + i)$

76. $z_1 = 5i,\ z_2 = 2(1 - i)$

In Exercises 77–80, use DeMoivre's Theorem to find the indicated power of the complex number. Express the result in standard form.

77. $\left[5\left(\cos \frac{\pi}{12} + i \sin \frac{\pi}{12}\right)\right]^4$

78. $\left[2\left(\cos \frac{4\pi}{15} + i \sin \frac{4\pi}{15}\right)\right]^5$

79. $(2 + 3i)^6$

80. $(1 - i)^8$

In Exercises 81–84, use DeMoivre's Theorem to find the indicated roots of the complex number.

81. Sixth roots of $-729i$

82. Fourth roots of 256

83. Cube roots of -1

84. Fourth roots of $-1 + i$

In Exercises 85–88, find all solutions of the equation and represent the solutions graphically.

85. $x^4 + 81 = 0$

86. $x^5 - 32 = 0$

87. $(x^3 - 1)(x^2 + 1) = 0$

88. $x^3 + 8i = 0$

In Exercises 89–94, find a polynomial function with integer coefficients that has the zeros shown in the graph.

89.

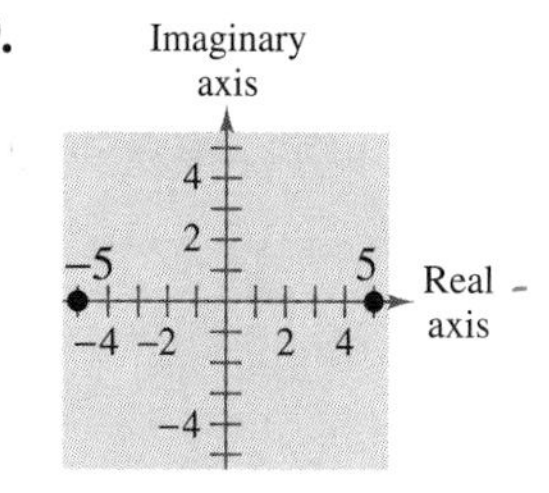

90.

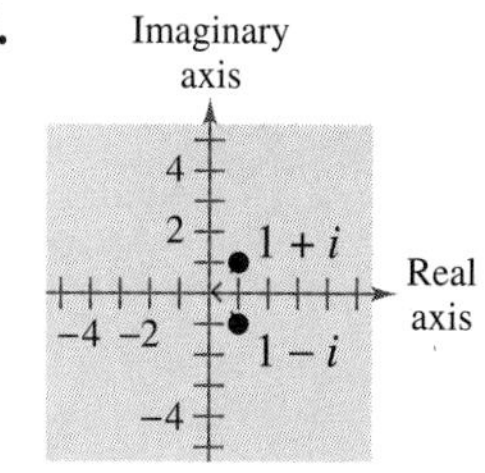

91.

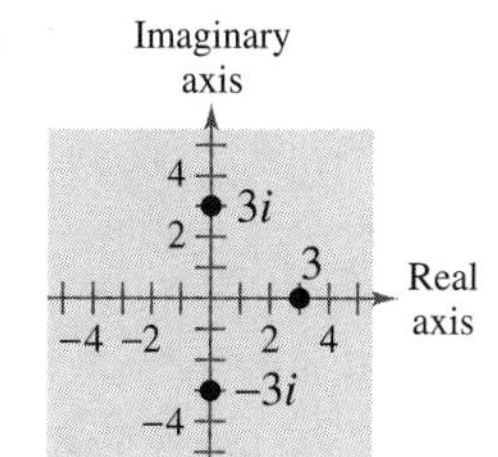

92.

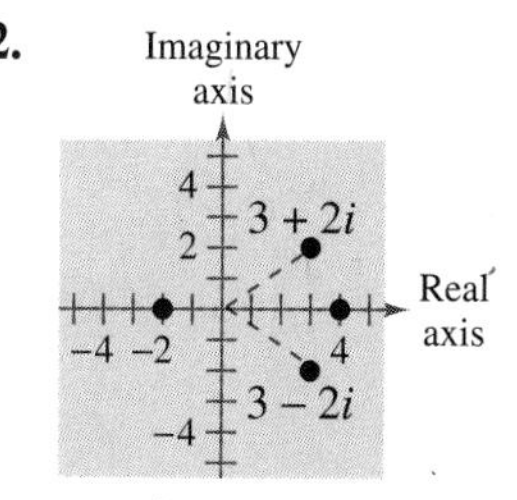

93.

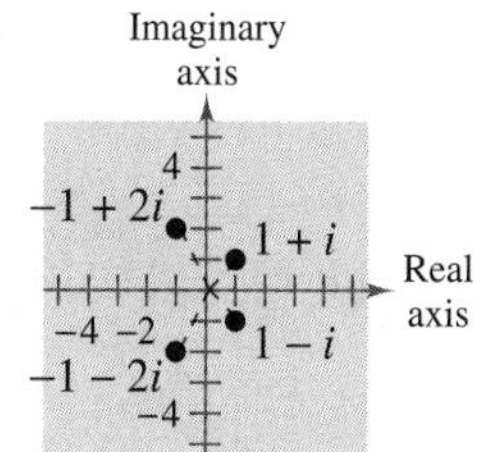

94.

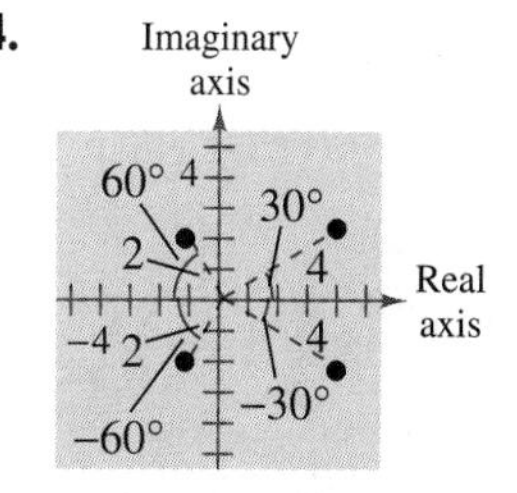

95. For each set of zeros shown graphically in Exercises 89–94, what is the geometric relationship between or among the complex zeros? Explain.

CHAPTER PROJECT: *Using a Graphing Utility to Create Fractals*

Graphing utilities can be used to draw pictures of **fractals** in the complex plane. Although it is possible to obtain crude pictures of fractals with a graphing calculator, you can obtain much better pictures of fractals with a computer that has a color monitor.

The most famous fractal is called the **Mandelbrot Set,** after the Polish-born mathematician Benoit Mandelbrot. To construct the Mandelbrot Set, consider the following sequence of numbers.

$$c, c^2 + c, (c^2 + c)^2 + c, ((c^2 + c)^2 + c)^2 + c, \ldots$$

The behavior of this sequence depends on the value of the complex number c. For some values of c this sequence is **bounded,** and for other values it is **unbounded.** If the sequence is bounded, the complex number c is in the Mandelbrot Set. If the sequence is unbounded, the complex number c is not in the Mandelbrot Set.

EXAMPLE 1 *Members of the Mandelbrot Set*

a. The complex number -2 is in the Mandelbrot Set because for $c = -2$, the corresponding Mandelbrot sequence is

$$-2, 2, 2, 2, 2, 2, \ldots$$

which is bounded. (No number in the sequence has an absolute value greater than 2.)

b. The complex number i is also in the Mandelbrot Set because for $c = i$, the corresponding Mandelbrot sequence is

$$i, -1 + i, -i, -1 + i, -i, -1 + i, \ldots$$

which is bounded. (No number in the sequence has an absolute value greater than $\sqrt{2}$.)

c. The complex number $1 + i$ is *not* in the Mandelbrot Set because for $c = 1 + i$, the corresponding Mandelbrot sequence is

$$1 + i, 1 + 3i, -7 + 7i, 1 - 97i, -9407 - 193i, \ldots$$

which is unbounded. (The absolute values of the numbers in the sequence become arbitrarily large.)

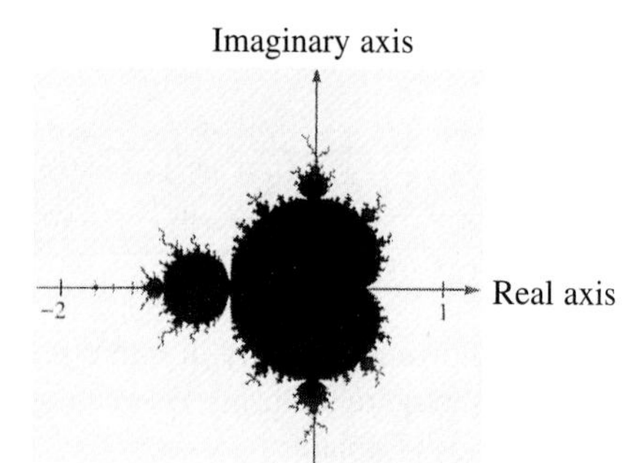

The graph at the left shows a picture of the Mandelbrot Set in the complex plane. In the picture, all the numbers that are in the Mandelbrot Set are plotted as black points, and all numbers that are not in the set are plotted as yellow points.

To add more interest to the picture of the Mandelbrot Set, computer scientists discovered that the points that are not in the set can be assigned a variety of colors, depending on how quickly their sequences diverge. The three pictures at left show different appendages of the Mandelbrot Set.

EXAMPLE 2 Writing a Program to Determine Mandelbrot Points

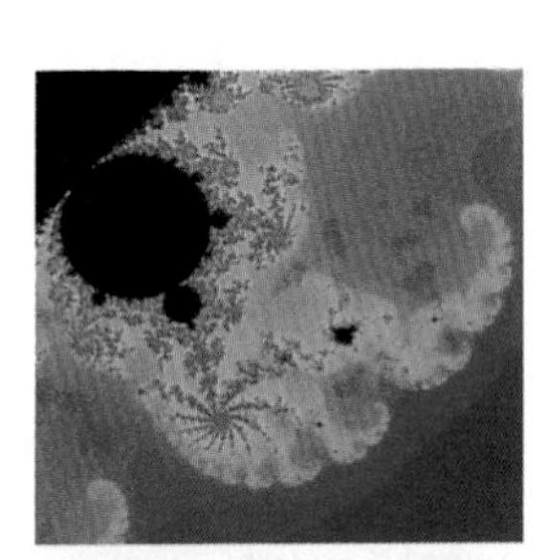

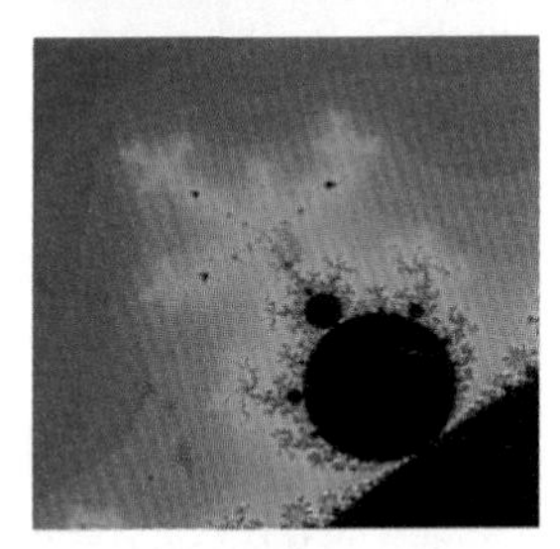

The program shown below is written for a *TI-83* or a *TI-82* graphing calculator. Use the program to determine whether $c = -1 + 0.2i$ is in the Mandelbrot Set. Programs for other graphing calculators may be found in the appendix.

```
:PROGRAM: MANDLBRT
:Input "ENTER REAL PART",A
:Input "ENTER IMAG PART",B
:A→C:B→D
:0→N
:Lbl1
:ClrHome
:N+1→N
:Disp "TERM NUMBER",N
:Disp "REAL PART",A
:Disp "IMAG PART",B
:Pause
:A→F:B→G
:F²–G²+C→A
:2FG+D→B
:Goto 1
```

Solution

To run the program, enter any complex number $c = a + bi$. Press ENTER to see the first term of the sequence. Press ENTER again to see the second term of the sequence. Continue pressing ENTER. If the terms begin to become large, the sequence is unbounded. For the number $c = -1 + 0.2i$, the terms are $-1 + 0.2i$, $-0.04 - 0.2i$, $-1.038 + 0.216i$, $0.032 - 0.249i \ldots$, and thus the sequence is bounded. Therefore, $c = -1 + 0.2i$ is in the Mandelbrot Set.

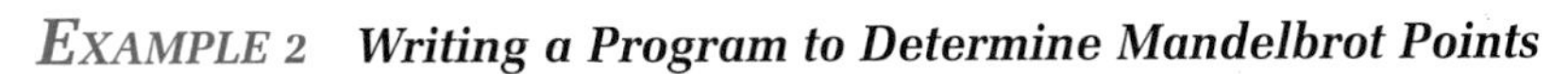

CHAPTER PROJECT INVESTIGATIONS

1. Construct a flowchart for the program given in Example 2. Explain how the program works.
2. Let $c = C + Di$ and let $c^2 + c = F + Gi$. Explain why $(c^2 + c)^2 + c$ is equal to

$$(F^2 - G^2 + C) + (2FG + D)i.$$

In Questions 3–8, use a graphing calculator program or a computer program to determine whether the complex number is in the Mandlebrot Set.

3. $c = 1$
4. $c = -1 + 0.5i$
5. $c = 0.1 + i$
6. $c = -2.1$
7. $c = 0.2 - i$
8. $c = -1 + 0.3i$

Chapter Test

Take this test as you would take a test in class. After you are done, check your work against the answers given in the back of the book.

The *Interactive* CD-ROM provides answers to the Chapter Tests and Cumulative Tests. It also offers Chapter Pre-Tests (which test key skills and concepts covered in previous chapters) and Chapter Post-Tests, both of which have randomly generated exercises with diagnostic capabilities.

1. Write the complex number $-2 + \sqrt{-64}$ in standard form.

2. Perform the operations and write the result in standard form.

(a) $10i - (3 + \sqrt{-25})$ (b) $(3 + 5i)^2$

(c) $(2 + \sqrt{3}i)(2 - \sqrt{3}i)$ (d) $\dfrac{5}{2 + i}$

In Exercises 3 and 4, use the root-finding capabilities of a graphing utility to approximate the real zeros of the function accurate to three decimal places.

3. $f(x) = x^4 - x^3 - 1$ **4.** $f(x) = 3x^5 + 2x^4 - 12x - 8$

In Exercises 5 and 6, use the zero(s) of the function as an aid in finding all its zeros. Write the polynomial as a product of linear factors.

Function	*Zero(s)*
5. $h(x) = x^4 + 3x^2 - 4$	$-1, 1$
6. $g(v) = 2v^3 - 11v^2 + 22v - 15$	$\frac{3}{2}$

In Exercises 7 and 8, find a polynomial function with integer coefficients that has the given zeros.

7. $0, 3, 3 + i, 3 - i$ **8.** $1 + \sqrt{3}i, 1 - \sqrt{3}i, 2, 2$

9. Is it possible for a polynomial function to have exactly one complex zero? Explain.

10. Find the trigonometric form of the complex number.

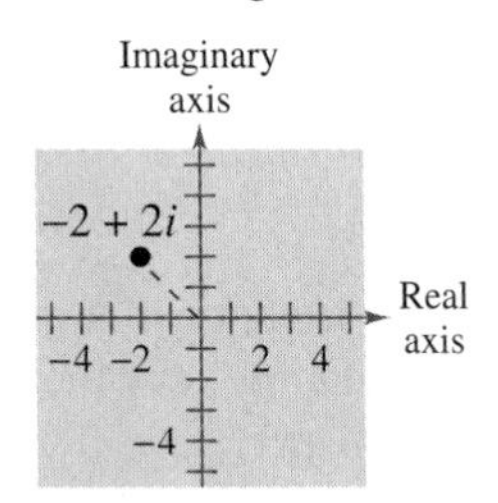

11. Find the product $[4(\cos 30° + i \sin 30°)][6(\cos 120° + i \sin 120°)]$. Write the answer in standard form.

12. Use DeMoivre's Theorem to find the three cube roots of 1.

5 Exponential and Logarithmic Functions

Automobiles are designed with crumple zones that allow the occupants to move short distances when the automobiles come to abrupt stops. The greater the distance moved, the less g's the crash victims experience. (One g is equal to the acceleration due to gravity. For very short periods of time, humans have withstood as much as 40 g's.)

In crash tests with a vehicle moving at 90 kilometers per hour, analysts measured the number of g's that were experienced during deceleration by a crash dummy that was permitted to move x meters during impact.

x	0.2	0.4	0.6	0.8	1.0
g	158	80	53	40	32

A model for this data is

$$g = -3.00 + 11.88 \ln x + \frac{36.94}{x}.$$

The graph of this model is shown below.

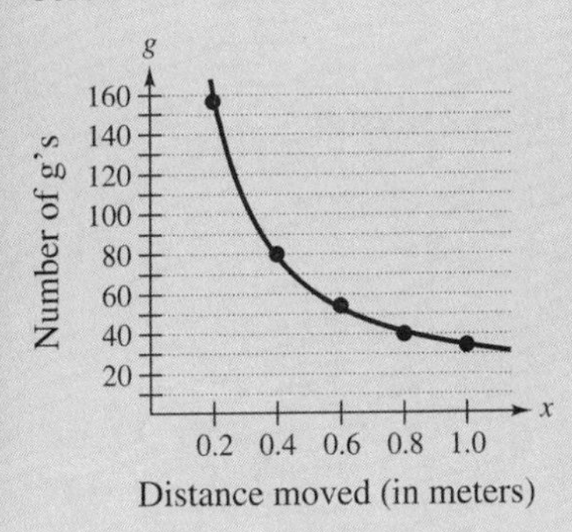

See Exercise 94 on page 396.

Photo: Brad Trent

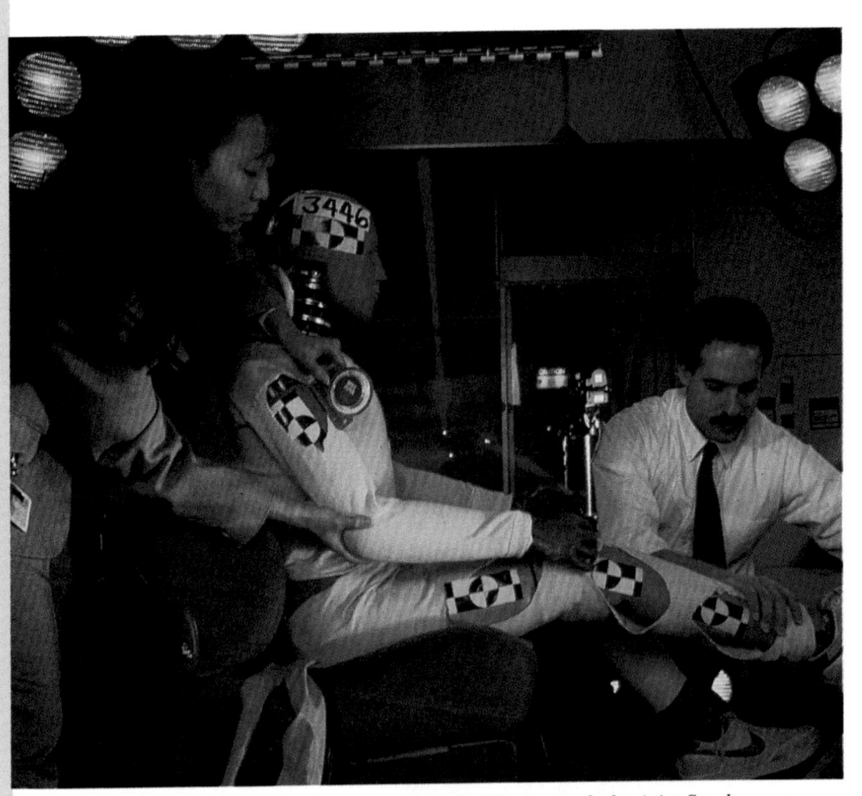

At a General Motors lab, engineer Bonnie Cheung and physicist Stephen Rouhana prepare a dummy for a simulated automobile crash. The laser (in red) helps position the dummy.

5.1 Exponential Functions and Their Graphs

See Exercise 63 on page 367 for an example of how an exponential function can be used to model the meteorological relationship between atmospheric pressure and altitude.

Exponential Functions ▫ *Graphs of Exponential Functions* ▫ *The Natural Base e* ▫ *Applications*

Exponential Functions

In this chapter you will study two types of nonalgebraic functions—*exponential* functions and *logarithmic* functions. These functions are examples of **transcendental functions.**

DEFINITION OF EXPONENTIAL FUNCTION

The **exponential function f with base a** is denoted by

$$f(x) = a^x$$

where $a > 0$, $a \neq 1$, and x is any real number.

NOTE The base $a = 1$ is excluded because it yields

$$f(x) = 1^x = 1.$$

This is a constant function, not an exponential function.

You already know how to evaluate a^x for integer and rational values of x. For example, you know that $4^3 = 64$ and $4^{1/2} = 2$. However, to evaluate 4^x for any real number x, you need to interpret forms with *irrational* exponents. For the purposes of this book, it is sufficient to think of

$$a^{\sqrt{2}} \text{ (where } \sqrt{2} \approx 1.414214)$$

as that value having the successively closer approximations

$$a^{1.4}, a^{1.41}, a^{1.414}, a^{1.4142}, a^{1.41421}, a^{1.414214}, \ldots.$$

Example 1 shows how to use a calculator to evaluate exponential expressions.

EXAMPLE 1 *Evaluating Exponential Expressions*

Use a calculator to evaluate each expression.

a. $2^{-3.1}$ **b.** $2^{-\pi}$

Solution

Number	*Graphing Calculator Keystrokes*	*Display*
a. $2^{-3.1}$	2 [∧] [(-)] 3.1 [ENTER]	0.1166291
b. $2^{-\pi}$	2 [∧] [(-)] π [ENTER]	0.1133147

Graphs of Exponential Functions

The graphs of all exponential functions have similar characteristics, as shown in Examples 2, 3, and 4.

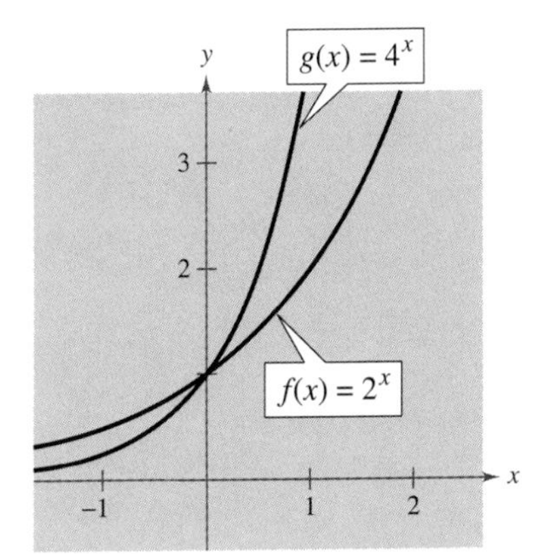

FIGURE 5.1

EXAMPLE 2 *Graphs of $y = a^x$*

In the same coordinate plane, sketch the graph of each function.

a. $f(x) = 2^x$

b. $g(x) = 4^x$

Solution

The table below lists some values for each function, and Figure 5.1 shows their graphs. Note that both graphs are increasing. Moreover, the graph of $g(x) = 4^x$ is increasing more rapidly than the graph of $f(x) = 2^x$.

x	-2	-1	0	1	2	3
2^x	$\frac{1}{4}$	$\frac{1}{2}$	1	2	4	8
4^x	$\frac{1}{16}$	$\frac{1}{4}$	1	4	16	64

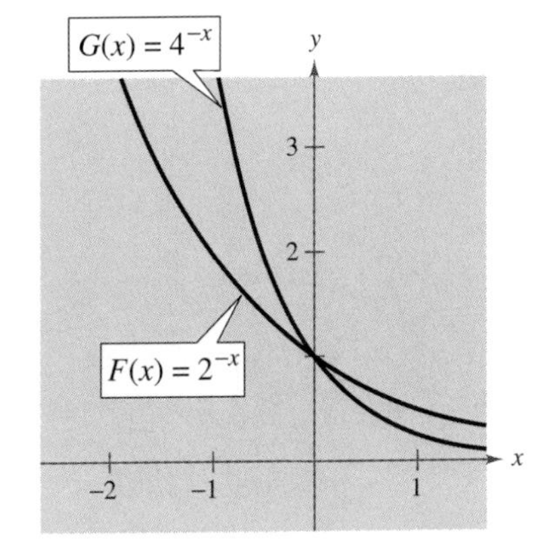

FIGURE 5.2

EXAMPLE 3 *Graphs of $y = a^{-x}$*

In the same coordinate plane, sketch the graph of each function.

a. $F(x) = 2^{-x}$

b. $G(x) = 4^{-x}$

Solution

The table below lists some values for each function, and Figure 5.2 shows their graphs. Note that both graphs are decreasing. Moreover, the graph of $G(x) = 4^{-x}$ is decreasing more rapidly than the graph of $F(x) = 2^{-x}$.

x	-3	-2	-1	0	1	2
2^{-x}	8	4	2	1	$\frac{1}{2}$	$\frac{1}{4}$
4^{-x}	64	16	4	1	$\frac{1}{4}$	$\frac{1}{16}$

NOTE The tables in Examples 2 and 3 were evaluated by hand. You could, of course, use a graphing utility to construct tables with even more values.

Comparing the functions in Examples 2 and 3, observe that

$$F(x) = 2^{-x} = f(-x) \quad \text{and} \quad G(x) = 4^{-x} = g(-x).$$

The *Interactive* CD-ROM offers graphing utility emulators of the *TI-82* and *TI-83*, which can be used with the Examples, Explorations, Technology notes, and Exercises.

Consequently, the graph of F is a reflection (in the y-axis) of the graph of f. The graphs of G and g have the same relationship. The graphs in Figures 5.1 and 5.2 are typical of the exponential functions a^x and a^{-x}. They have one y-intercept and one horizontal asymptote (the x-axis), and they are continuous. The basic characteristics of these exponential functions are summarized in Figure 5.3.

Graph of $y = a^x, a > 1$

- Domain: $(-\infty, \infty)$
- Range: $(0, \infty)$
- Intercept: $(0, 1)$
- Increasing
- x-axis is a horizontal asymptote ($a^x \to 0$ as $x \to -\infty$)
- Continuous

Graph of $y = a^{-x}, a > 1$

- Domain: $(-\infty, \infty)$
- Range: $(0, \infty)$
- Intercept: $(0, 1)$
- Decreasing
- x-axis is a horizontal asymptote ($a^{-x} \to 0$ as $x \to \infty$)
- Continuous

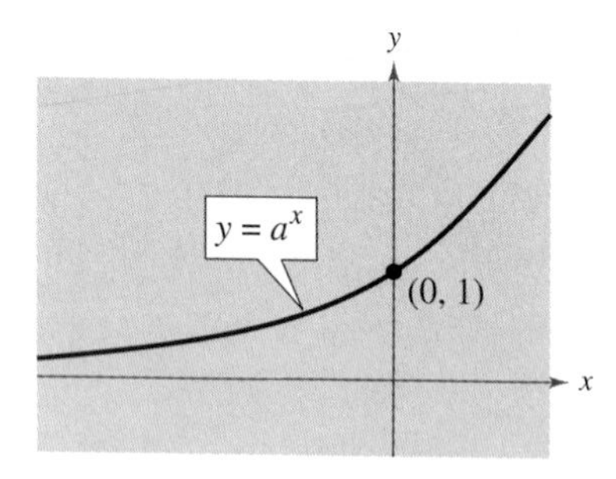

FIGURE 5.3

Exploration

Use a graphing utility to graph $y = a^x$ for $a = 3, 5$, and 7 on the same viewing rectangle. (Use a viewing rectangle in which $-2 \le x \le 1$ and $0 \le y \le 2$.) For instance, the graph of $y = 3^x$ is shown at the left. How do the graphs compare with each other? Which graph is on the top in the interval $(-\infty, 0)$? Which is on the bottom? Which graph is on the top in the interval $(0, \infty)$? Which is on the bottom? Repeat this experiment with the graphs of $y = a^x$ for $a = \frac{1}{3}, \frac{1}{5}$, and $\frac{1}{7}$. (Use a viewing rectangle in which $-1 \le x \le 2$ and $0 \le y \le 2$.) What can you conclude about the relationship between the function's behavior and the value of a?

In the following example, notice how the graph of $y = a^x$ can be used to sketch the graphs of functions of the form $f(x) = b \pm a^{x+c}$.

EXAMPLE 4 Sketching Graphs of Exponential Functions

Each of the following graphs is a transformation of the graph of $f(x) = 3^x$, as shown in Figure 5.4.

a. Because $g(x) = 3^{x+1} = f(x + 1)$, the graph of g can be obtained by shifting the graph of f one unit to the left.

b. Because $h(x) = 3^x - 2 = f(x) - 2$, the graph of h can be obtained by shifting the graph of f down two units.

c. Because $k(x) = -3^x = -f(x)$, the graph of k can be obtained by reflecting the graph of f in the x-axis.

d. Because $j(x) = 3^{-x} = f(-x)$, the graph of j can be obtained by reflecting the graph of f in the y-axis.

NOTE In Figure 5.4, notice that the transformations in parts (a), (c), and (d) keep the x-axis as a horizontal asymptote, but the transformation in part (b) yields a new horizontal asymptote of $y = -2$. Also, be sure to note how the y-intercept is affected by each transformation.

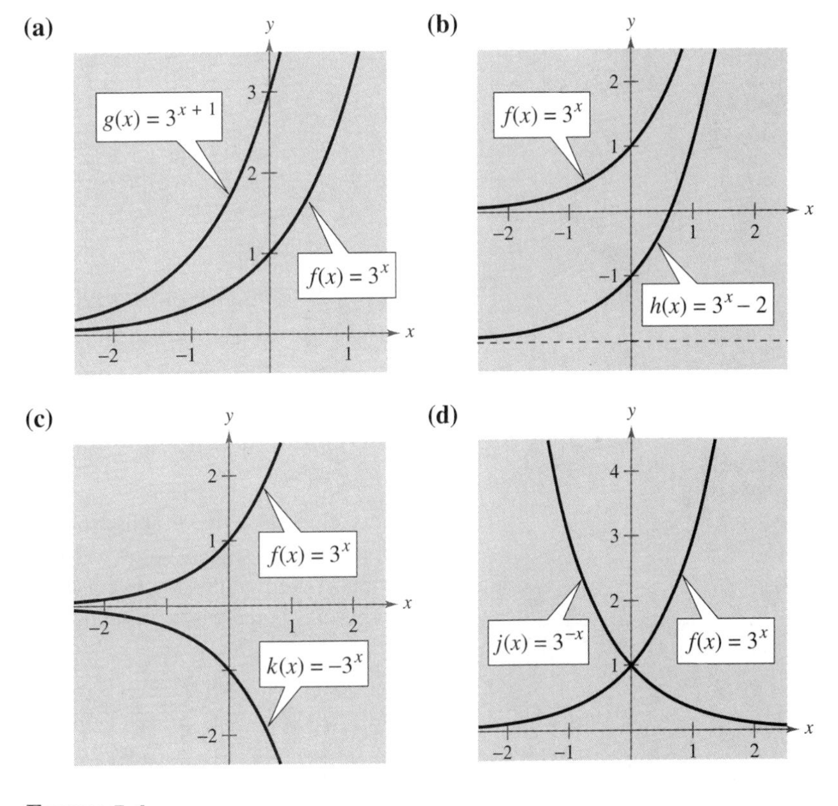

FIGURE 5.4

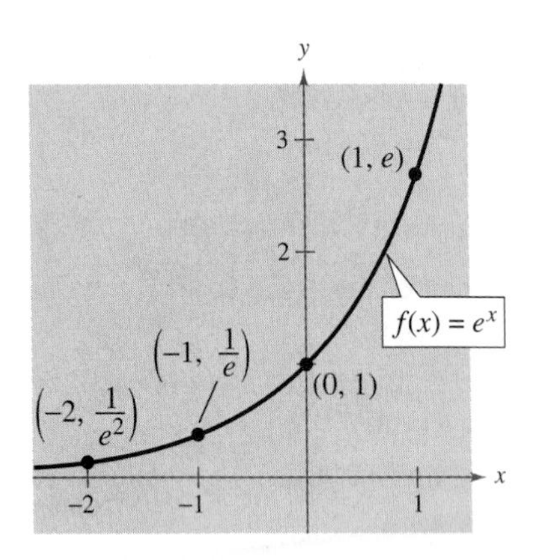

FIGURE 5.5

The Natural Base e

In many applications, the most convenient choice for a base is the irrational number

$$e \approx 2.71828 \ldots .$$

This number is called the **natural base.** The function $f(x) = e^x$ is called the **natural exponential function.** Its graph is shown in Figure 5.5. Be sure you see that for the exponential function $f(x) = e^x$, e is the constant $2.71828 \ldots$, whereas x is the variable.

Example 5 Evaluating the Natural Exponential Function

Use a calculator to evaluate each expression.

a. e^{-2} **b.** e^{-1} **c.** e^1 **d.** e^2

Solution

Number	*Graphing Calculator Keystrokes*	*Display*
a. e^{-2}	e^x (-) 2 ENTER	0.1353353
b. e^{-1}	e^x (-) 1 ENTER	0.3678794
c. e^1	e^x 1 ENTER	2.7182818
d. e^2	e^x 2 ENTER	7.3890561

Example 6 Graphing Natural Exponential Functions

Sketch the graph of each natural exponential function.

a. $f(x) = 2e^{0.24x}$

b. $g(x) = \frac{1}{2}e^{-0.58x}$

Solution

To sketch these two graphs, you can use a graphing utility to construct a table of values, as shown below. After constructing the table, plot the points and connect them with smooth curves, as shown in Figure 5.6. Note that the graph in part (a) is increasing whereas the graph in part (b) is decreasing.

x	-3	-2	-1	0	1	2	3
$f(x)$	0.974	1.238	1.573	2.000	2.543	3.232	4.109
$g(x)$	2.849	1.595	0.893	0.500	0.280	0.157	0.088

(a)

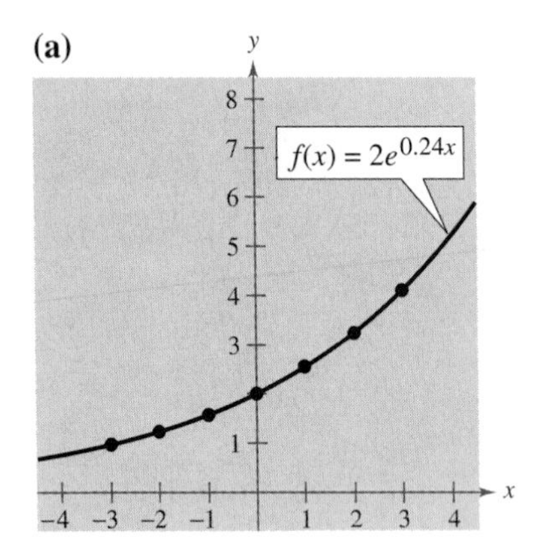

(b)

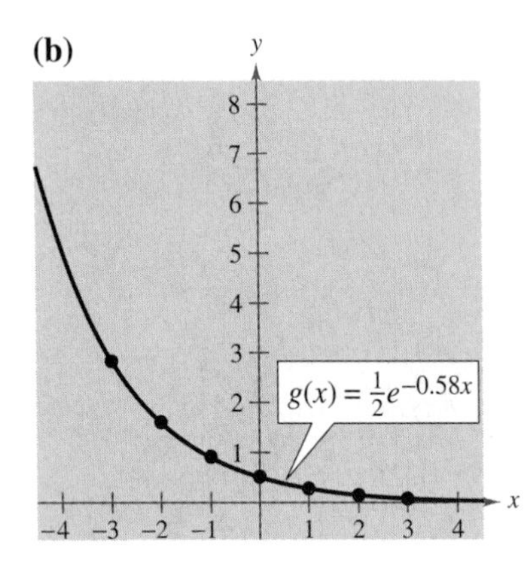

FIGURE 5.6

Applications

One of the most familiar examples of exponential growth is that of an investment earning *continuously compounded interest.*

Suppose a principal P is invested at an annual interest rate r, compounded once a year. If the interest is added to the principal at the end of the year, the balance is

$$P_1 = P + Pr = P(1 + r).$$

This pattern of multiplying the previous principal by $1 + r$ is then repeated each successive year, as shown below.

Year	*Balance After Each Compounding*
0	$P = P$
1	$P_1 = P(1 + r)$
2	$P_2 = P_1(1 + r) = P(1 + r)(1 + r) = P(1 + r)^2$
3	$P_3 = P_2(1 + r) = P(1 + r)^2(1 + r) = P(1 + r)^3$
	$\vdots$
t	$P_t = P(1 + r)^t$

To accommodate more frequent (quarterly, monthly, or daily) compounding of interest, let n be the number of compoundings per year and let t be the number of years. Then the rate per compounding is r/n and the account balance after t years is

$$A = P\left(1 + \frac{r}{n}\right)^{nt}. \qquad \text{Amount with } n \text{ compoundings per year}$$

If you let the number of compoundings n increase without bound, the process approaches what is called **continuous compounding.** In the formula for n compoundings per year, let $m = n/r$. This produces

$$\begin{aligned} A &= P\left(1 + \frac{r}{n}\right)^{nt} \\ &= P\left(1 + \frac{1}{m}\right)^{mrt} \\ &= P\left[\left(1 + \frac{1}{m}\right)^{m}\right]^{rt}. \end{aligned}$$

As m increases without bound, it can be shown that $[1 + (1/m)]^m$ approaches e. (Try the values $m = 10$, 10,000, and 10,000,000.) From this, you can conclude that the formula for continuous compounding is $A = Pe^{rt}$.

Exploration

Use the formula $A = P(1 + r/n)^{nt}$ to calculate the amount in an account when $P = \$3000$, $r = 6\%$, $t = 10$ years, and the number of compoundings is (1) by the day, (2) by the hour, (3) by the minute, and (4) by the second. Use these results to present an argument that increasing the number of compoundings does not mean unlimited growth of the amount in the account.

FORMULAS FOR COMPOUND INTEREST

After t years, the balance A in an account with principal P and annual interest rate r (in decimal form) is given by the following formulas.

1. For n compoundings per year: $A = P\left(1 + \frac{r}{n}\right)^{nt}$
2. For continuous compounding: $A = Pe^{rt}$

NOTE Be sure you see that the annual interest rate must be expressed in decimal form. For instance, 6% should be expressed as 0.06.

EXAMPLE 7 *Compounding n Times and Continuously*

A total of \$12,000 is invested at an annual interest rate of 9%. Find the balance after 5 years if it is compounded

a. quarterly.

b. continuously.

Solution

a. For quarterly compoundings, you have $n = 4$. Thus, in 5 years at 9%, the balance is

$$A = P\left(1 + \frac{r}{n}\right)^{nt} \quad \text{Formula for compound interest}$$

$$= 12{,}000\left(1 + \frac{0.09}{4}\right)^{4(5)} \quad \text{Substitute for } P, r, n, \text{ and } t.$$

$$= \$18{,}726.11. \quad \text{Use a calculator.}$$

b. For continuous compounding, the balance is

$$A = Pe^{rt} \quad \text{Formula for continuous compounding}$$

$$= 12{,}000e^{0.09(5)} \quad \text{Substitute for } P, r, \text{ and } t.$$

$$= \$18{,}819.75. \quad \text{Use a calculator.}$$

Note that continuous compounding yields

$$\$18{,}819.75 - \$18{,}726.11 = \$93.64$$

more than quarterly compounding. This is typical of the two types of compounding. That is, for a given principal, interest rate, and time, continuous compounding will always yield a larger balance than compounding n times a year.

EXAMPLE 8 *Radioactive Decay*

In 1986, a nuclear reactor accident occurred in Chernobyl in what was then the Soviet Union. The explosion spread radioactive chemicals over hundreds of square miles, and the government evacuated the city and the surrounding area. To see why the city is now uninhabited, consider the following model.

$$P = 10e^{-0.00002845t}$$

This model represents the amount of plutonium that remains (from an initial amount of 10 pounds) after t years. Sketch the graph of this function over the interval from $t = 0$ to $t = 100{,}000$. How much of the 10 pounds will remain after 100,000 years?

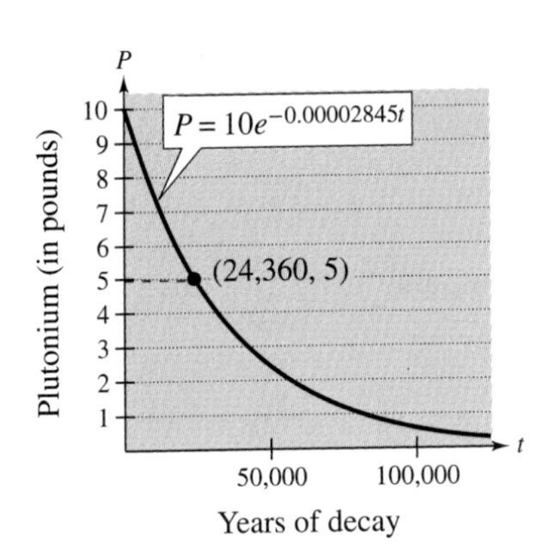

FIGURE 5.7

Solution

The graph of this function is shown in Figure 5.7. Note from this graph that plutonium has a *half-life* of about 24,360 years. That is, after 24,360 years, *half* of the original amount will remain. After another 24,360 years, one-quarter of the original amount will remain, and so on. After 100,000 years, there will still be

$$P = 10e^{-0.00002845(100{,}000)} = 10e^{-2.845} \approx 0.58 \text{ pounds}$$

of plutonium remaining.

GROUP ACTIVITY

IDENTIFYING EXPONENTIAL FUNCTIONS

Which of the following functions generated the two tables below? Discuss how you were able to decide. What do these functions have in common? Are any the same? If so, explain why.

a. $f_1(x) = 2^{(x+3)}$ **b.** $f_2(x) = 8\left(\frac{1}{2}\right)^x$ **c.** $f_3(x) = \left(\frac{1}{2}\right)^{(x-3)}$

d. $f_4(x) = \left(\frac{1}{2}\right)^x + 7$ **e.** $f_5(x) = 7 + 2^x$ **f.** $f_6(x) = (8)2^x$

x	-1	0	1	2	3
$g(x)$	7.5	8	9	11	15

x	-2	-1	0	1	2
$h(x)$	32	16	8	4	2

Create two different exponential functions with y-intercepts of $(0, -3)$. Compare your functions with those of other students in your class.

The *Interactive* CD-ROM provides additional help with Warm-Up exercises by providing a hypertext link to the section in which the concept was introduced.

The *Interactive* CD-ROM contains step-by-step solutions to all odd-numbered Section and Review Exercises. It also provides Tutorial Exercises, which link to Guided Examples for additional help.

WARM UP

Use the properties of exponents to simplify the expression.

1. $5^{2x}(5^{-x})$ **2.** $3^{-x}(3^{3x})$

3. $\dfrac{4^{5x}}{4^{2x}}$ **4.** $\dfrac{10^{2x}}{10^{x}}$

5. $(4^x)^2$ **6.** $(4^{2x})^5$

7. $\left(\dfrac{2^x}{3^x}\right)^{-1}$ **8.** $(4^{6x})^{1/2}$

9. $(2^{3x})^{-1/3}$ **10.** $(16^x)^{1/4}$

5.1 Exercises

In Exercises 1–10, evaluate the expression. Round your result to three decimal places.

1. $(3.4)^{5.6}$ **2.** $5000(2^{-1.5})$

3. $(1.005)^{400}$ **4.** $8^{2\pi}$

5. $5^{-\pi}$ **6.** $\sqrt[3]{4395}$

7. $100^{\sqrt{2}}$ **8.** $e^{1/2}$

9. $e^{-3/4}$ **10.** $e^{3.2}$

Think About It **In Exercises 11–14, use properties of exponents to determine which functions (if any) are the same.**

11. $f(x) = 3^{x-2}$
$g(x) = 3^x - 9$
$h(x) = \frac{1}{9}(3^x)$

12. $f(x) = 4^x + 12$
$g(x) = 2^{2x+6}$
$h(x) = 64(4^x)$

13. $f(x) = 16(4^{-x})$
$g(x) = \left(\frac{1}{4}\right)^{x-2}$
$h(x) = 16(2^{-2x})$

14. $f(x) = 5^{-x} + 3$
$g(x) = 5^{3-x}$
$h(x) = -5^{x-3}$

In Exercises 15–18, match the exponential function with its graph. [The graphs are labeled (a), (b), (c), and (d).]

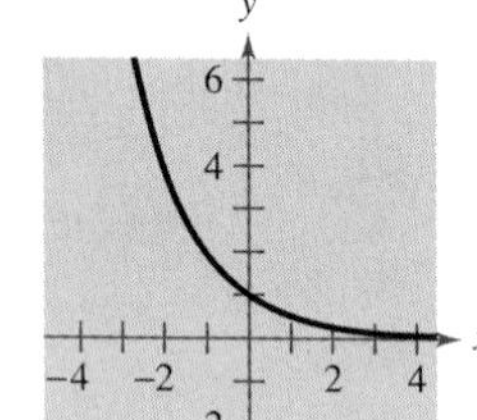

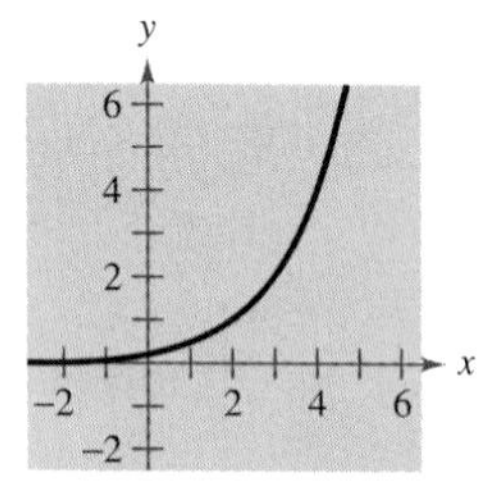

(c)

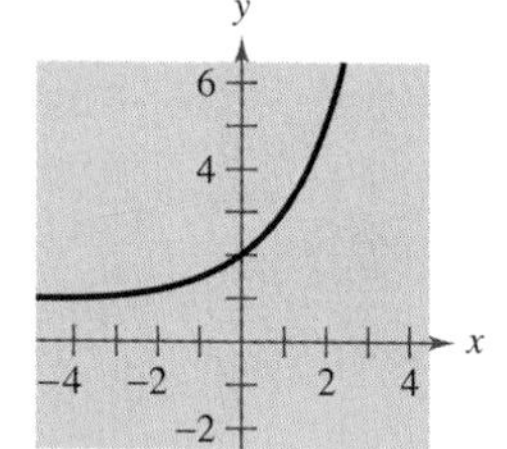

(d)

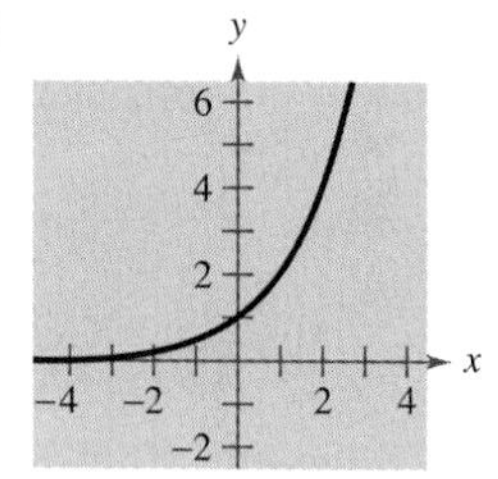

15. $f(x) = 2^x$ **16.** $f(x) = 2^x + 1$

17. $f(x) = 2^{-x}$ **18.** $f(x) = 2^{x-2}$

In Exercises 19–36, graph the exponential function.

19. $g(x) = 5^x$

20. $f(x) = \left(\frac{3}{2}\right)^x$

21. $f(x) = \left(\frac{1}{5}\right)^x = 5^{-x}$

22. $h(x) = \left(\frac{3}{2}\right)^{-x}$

23. $h(x) = 5^{x-2}$

24. $g(x) = \left(\frac{3}{2}\right)^{x+2}$

25. $g(x) = 5^{-x} - 3$

26. $f(x) = \left(\frac{3}{2}\right)^{-x} + 2$

27. $y = 2^{-x^2}$

28. $y = 3^{-|x|}$

29. $y = 3^{x-2} + 1$

30. $y = 4^{x+1} - 2$

31. $y = 1.08^{-5x}$

32. $y = 1.08^{5x}$

33. $s(t) = 2e^{0.12t}$

34. $s(t) = 3e^{-0.2t}$

35. $g(x) = 1 + e^{-x}$

36. $h(x) = e^{x-2}$

37. Graph the functions $y = 3^x$ and $y = 4^x$ and use the graphs to solve the inequalities.

(a) $4^x < 3^x$

(b) $4^x > 3^x$

38. Graph the functions $y = \left(\frac{1}{2}\right)^x$ and $y = \left(\frac{1}{4}\right)^x$ and use the graphs to solve the inequalities.

(a) $\left(\frac{1}{4}\right)^x < \left(\frac{1}{2}\right)^x$

(b) $\left(\frac{1}{4}\right)^x > \left(\frac{1}{2}\right)^x$

39. Use the graph of $f(x) = 3^x$ to graph each of the functions. Identify the transformation.

(a) $g(x) = f(x - 2) = 3^{x-2}$

(b) $h(x) = -\frac{1}{2}f(x) = -\frac{1}{2}(3^x)$

(c) $q(x) = f(-x) + 3 = 3^{-x} + 3$

40. Use a graphing utility to graph each function. Use the graph to find any asymptotes of the function.

(a) $f(x) = \dfrac{8}{1 + e^{-0.5x}}$

(b) $g(x) = \dfrac{8}{1 + e^{-0.5/x}}$

41. Use a graphing utility to graph each function. Use the graph to find where the function is increasing and decreasing, and approximate any relative maximum or minimum values.

(a) $f(x) = x^2e^{-x}$

(b) $g(x) = x2^{3-x}$

42. ***Comparing Functions*** Use a graphing utility to graph $y_1 = e^x$ and each of the functions $y_2 = x^2$, $y_3 = x^3$, $y_4 = \sqrt{x}$, and $y_5 = |x|$. Which function increases at the fastest rate for "large" values of x?

43. ***Conjecture*** Use the result of Exercise 42 to make a conjecture about the rate of growth of $y_1 = e^x$ and $y = x^n$ where n is a natural number and x is "large."

44. ***Essay*** Use the results of Exercises 42 and 43 to describe what is implied when it is stated that a quantity is growing exponentially.

45. ***Graphical Analysis*** Use a graphing utility to graph

$$f(x) = \left(1 + \frac{0.5}{x}\right)^x \quad \text{and} \quad g(x) = e^{0.5}$$

on the same viewing rectangle. What is the relationship between f and g as x increases without bound?

46. ***Conjecture*** Use the result of Exercise 45 to make a conjecture about the value of $[1 + (r/x)]^x$ as x increases without bound. Create a table that illustrates your conjecture for $r = 1$.

Compound Interest **In Exercises 47–50, complete the table to determine the balance A for P dollars invested at rate r for t years and compounded n times per year.**

n	1	2	4	12	365	Continuous
A						

47. $P = \$2500$, $r = 12\%$, $t = 10$ years

48. $P = \$1000$, $r = 10\%$, $t = 10$ years

49. $P = \$2500$, $r = 12\%$, $t = 20$ years

50. $P = \$1000$, $r = 10\%$, $t = 40$ years

Compound Interest **In Exercises 51 and 52, complete the table to determine the amount of money P that should be invested at rate r to produce a final balance of \$100,000 in t years.**

t	1	10	20	30	40	50
P						

51. $r = 9\%$, compounded continuously

52. $r = 12\%$, compounded continuously

53. ***Trust Fund*** On the day of a child's birth, a deposit of \$25,000 is made in a trust fund that pays 8.75% interest, compounded continuously. Determine the balance in this account on the child's 25th birthday.

54. ***Trust Fund*** A deposit of \$5000 is made in a trust fund that pays 7.5% interest, compounded continuously. It is specified that the balance will be given to the college from which the donor graduated after the money has earned interest for 50 years. How much will the college receive?

55. ***Graphical Reasoning*** There are two options for investing \$500. The first earns 7% compounded annually and the second earns 7% simple interest. The figure shows the growth of each investment over a 30-year period.

(a) Identify the two types of investments in the figure. Explain your reasoning.

(b) Verify your answer in part (a) by finding the equations that model the investment growth and graphing the models.

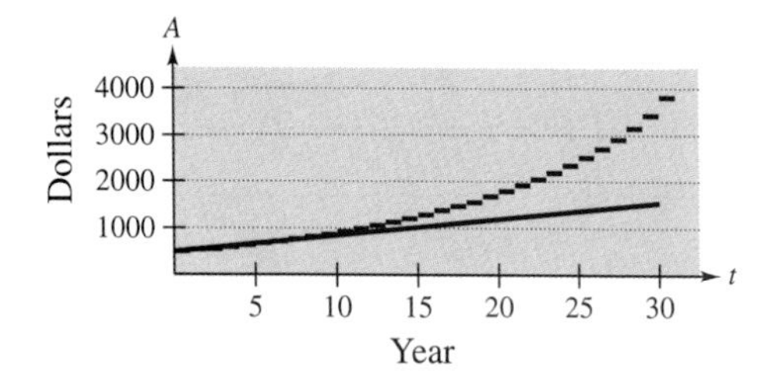

56. ***Depreciation*** After t years, the value of a car that cost \$20,000 is given by

$V(t) = 20{,}000\left(\frac{3}{4}\right)^t.$

Graph the function and determine the value of the car 2 years after it was purchased.

57. ***Inflation*** If the annual rate of inflation averages 4% over the next 10 years, the approximate cost C of goods or services during any year in that decade will be given by

$C(t) = P(1.04)^t$

where t is the time in years and P is the present cost. If the price of an oil change for your car is presently \$23.95, estimate the price 10 years from now.

58. ***Demand Function*** The demand equation for a certain product is given by

$$p = 5000\left(1 - \frac{4}{4 + e^{-0.002x}}\right).$$

(a) Use a graphing utility to graph the demand function for $x > 0$ and $p > 0$.

(b) Find the price p for a demand of $x = 500$ units.

(c) Use the graph in part (a) to approximate the greatest price that will still yield a demand of at least 600 units.

59. ***Bacteria Growth*** A certain type of bacteria increases according to the model

$P(t) = 100e^{0.2197t}$

where t is the time in hours. Find (a) $P(0)$, (b) $P(5)$, and (c) $P(10)$.

60. ***Population Growth*** The population of a town increases according to the model

$P(t) = 2500e^{0.0293t}$

where t is the time in years, with $t = 0$ corresponding to 1990. Use the model to estimate the population in (a) 2000 and (b) 2010.

61. ***Radioactive Decay*** Let Q represent a mass of radium (^{226}Ra), whose half-life is 1620 years. The quantity of radium present after t years is given by

$Q = 25\left(\frac{1}{2}\right)^{t/1620}.$

(a) Determine the initial quantity (when $t = 0$).

(b) Determine the quantity present after 1000 years.

(c) Use a graphing utility to graph the function over the interval $t = 0$ to $t = 5000$.

62. ***Radioactive Decay*** Let Q represent a mass of carbon 14(^{14}C), whose half-life is 5730 years. The quantity of carbon 14 present after t years is given by

$Q = 10\left(\frac{1}{2}\right)^{t/5730}.$

(a) Determine the initial quantity (when $t = 0$).

(b) Determine the quantity present after 2000 years.

(c) Sketch the graph of this function over the interval $t = 0$ to $t = 10{,}000$.

63. ***Data Analysis*** A meteorologist measures the atmospheric pressure P (in kilograms per square meter) at altitude h (in kilometers). The data is shown in the table.

h	0	5	10	15	20
P	10,332	5583	2376	1240	517

A model for the data is given by

$P = 10{,}958e^{-0.15h}$.

(a) Sketch a scatter plot of the data and graph the model on the same set of axes.

(b) Create a table to compare the model with the sample data.

(c) Estimate the atmospheric pressure at a height of 8 kilometers.

(d) Use the graph in part (a) to estimate the altitude at which the atmospheric pressure is 2000 kilograms per square meter.

64. ***Data Analysis*** To estimate the amount of defoliation caused by the gypsy moth during a given year, a forester counts the number x of egg masses on $\frac{1}{40}$ of an acre (circle of radius 18.6 feet) the preceding fall. The percent of defoliation y the next spring is given in the table. (Source: USDA, Forest Service)

x	0	25	50	75	100
y	12	44	81	96	99

(a) A model for the data is given by

$$y = \frac{300}{3 + 17e^{-0.065x}}.$$

Use a graphing utility to create a scatter plot of the data and graph the model on the same viewing rectangle.

(b) Create a table to compare the model with the sample data.

(c) Estimate the percent of defoliation if 36 egg masses are counted on $\frac{1}{40}$ acre.

(d) Use the graph in part (a) to estimate the number of egg masses per $\frac{1}{40}$ acre if you observe that $\frac{2}{3}$ of a forest is defoliated the following spring.

65. ***True or False?*** $e = \dfrac{271{,}801}{99{,}990}$. Explain.

66. ***Think About It***
Which functions are exponential?

(a) $3x$ (b) $3x^2$

(c) 3^x (d) 2^{-x}

67. ***Think About It*** Without using a calculator, why do you know that $2^{\sqrt{2}}$ is greater than 2, but less than 4?

68. ***Exploration*** Use a graphing utility to compare the graph of the function $y = e^x$ with the graphs of the following functions.

(a) $y_1 = 1 + \dfrac{x}{1!}$

(b) $y_2 = 1 + \dfrac{x}{1!} + \dfrac{x^2}{2!}$

(c) $y_3 = 1 + \dfrac{x}{1!} + \dfrac{x^2}{2!} + \dfrac{x^3}{3!}$

69. ***Finding a Pattern*** Identify the pattern of successive polynomials given in Exercise 68. Extend the pattern one more term and compare the graph of the resulting polynomial function with the graph of $y = e^x$. What do you think this pattern implies?

70. Given the exponential function $f(x) = a^x$, show that

(a) $f(u + v) = f(u) \cdot f(v)$.

(b) $f(2x) = [f(x)]^2$.

Review **Solve Exercises 71–74 as a review of the skills and problem-solving techniques you learned in previous sections. Solve for *y*.**

71. $2x - 7y + 14 = 0$

72. $x^2 + 3y = 4$

73. $x^2 + y^2 = 25$

74. $x - |y| = 2$

5.2 Logarithmic Functions and Their Graphs

See Exercises 75 and 76 on page 377 for an example of how a logarithmic function can be used to model the minimum required ventilation rates in public school classrooms.

Logarithmic Functions ▫ Graphs of Logarithmic Functions ▫ The Natural Logarithmic Function ▫ Application

Logarithmic Functions

In Section P.8, you studied the concept of the inverse of a function. There, you learned that if a function has the property that no horizontal line intersects the graph of the function more than once, the function must have an inverse. By looking back at the graphs of the exponential functions introduced in Section 5.1, you will see that every function of the form $f(x) = a^x$ passes the "Horizontal Line Test" and therefore must have an inverse. This inverse function is called the **logarithmic function with base *a*.**

NOTE The equations

$$y = \log_a x \qquad \text{and} \qquad x = a^y$$

are equivalent. The first equation is in logarithmic form and the second is in exponential form.

> **DEFINITION OF LOGARITHMIC FUNCTION**
>
> For $x > 0$ and $0 < a \neq 1$,
>
> $$y = \log_a x \text{ if and only if } x = a^y.$$
>
> The function given by
>
> $$f(x) = \log_a x$$
>
> is called the **logarithmic function with base *a*.**

When evaluating logarithms, remember that *a logarithm is an exponent.* This means that $\log_a x$ is the exponent to which a must be raised to obtain x. For instance, $\log_2 8 = 3$ because 2 must be raised to the third power to get 8.

EXAMPLE 1 *Evaluating Logarithms*

a. $\log_2 32 = 5$ because $2^5 = 32$.

b. $\log_3 27 = 3$ because $3^3 = 27$.

c. $\log_4 2 = \frac{1}{2}$ because $4^{1/2} = \sqrt{4} = 2$.

d. $\log_{10} \frac{1}{100} = -2$ because $10^{-2} = \frac{1}{10^2} = \frac{1}{100}$.

e. $\log_3 1 = 0$ because $3^0 = 1$.

f. $\log_2 2 = 1$ because $2^1 = 2$.

The logarithmic function with base 10 is called the **common logarithmic function.** On most calculators, this function is denoted by LOG.

EXAMPLE 2 *Evaluating Logarithms on a Calculator*

Use a calculator to evaluate each expression.

a. $\log_{10} 10$ **b.** $2\log_{10} 2.5$ **c.** $\log_{10}(-2)$

Solution

Number	*Graphing Calculator Keystrokes*	*Display*
a. $\log_{10} 10$	LOG 10 ENTER	1
b. $2\log_{10} 2.5$	2 × LOG 2.5 ENTER	0.7958800
c. $\log_{10}(-2)$	LOG (-) 2 ENTER	ERROR

Note that the calculator displays an error message when you try to evaluate $\log_{10}(-2)$. The reason for this is that the domain of every logarithmic function is the set of *positive real numbers.*

Study Tip

Because $\log_a x$ is the inverse function of a^x, it follows that the domain of $\log_a x$ is the range of a^x, $(0, \infty)$. In other words, $\log_a x$ is defined only if x is positive.

The following properties follow directly from the definition of the logarithmic function with base a.

PROPERTIES OF LOGARITHMS

1. $\log_a 1 = 0$ because $a^0 = 1$.
2. $\log_a a = 1$ because $a^1 = a$.
3. $\log_a a^x = x$ because $a^x = a^x$.
4. If $\log_a x = \log_a y$, then $x = y$.

EXAMPLE 3 *Using Properties of Logarithms*

Solve each equation for x.

a. $\log_2 x = \log_2 3$

b. $\log_4 4 = x$

Solution

a. Using Property 4, you can conclude that $x = 3$.

b. Using Property 2, you can conclude that $x = 1$.

Graphs of Logarithmic Functions

To sketch the graph of $y = \log_a x$, you can use the fact that the graphs of inverse functions are reflections of each other in the line $y = x$.

EXAMPLE 4 *Graphs of Exponential and Logarithmic Functions*

In the same coordinate plane, sketch the graph of each function.

a. $f(x) = 2^x$ **b.** $g(x) = \log_2 x$

Solution

a. For $f(x) = 2^x$, construct a table of values, as follows.

x	-2	-1	0	1	2	3
$f(x) = 2^x$	$\frac{1}{4}$	$\frac{1}{2}$	1	2	4	8

By plotting these points and connecting them with a smooth curve, you obtain the graph shown in Figure 5.8.

b. Because $g(x) = \log_2 x$ is the inverse of $f(x) = 2^x$, the graph of g is obtained by reflecting the graph of f in the line $y = x$, as shown in Figure 5.8.

FIGURE 5.8

Before you can confirm the result of Example 4 with a graphing utility, you need to know how to enter $\log_2 x$. A procedure using the change-of-base formula is discussed in Section 5.3.

EXAMPLE 5 *Sketching the Graph of a Logarithmic Function*

Sketch the graph of the common logarithmic function $f(x) = \log_{10} x$.

Solution

Begin by constructing a table of values. Note that some of the values can be obtained without a calculator, whereas others require a calculator. Next, plot the points and connect them with a smooth curve, as shown in Figure 5.9.

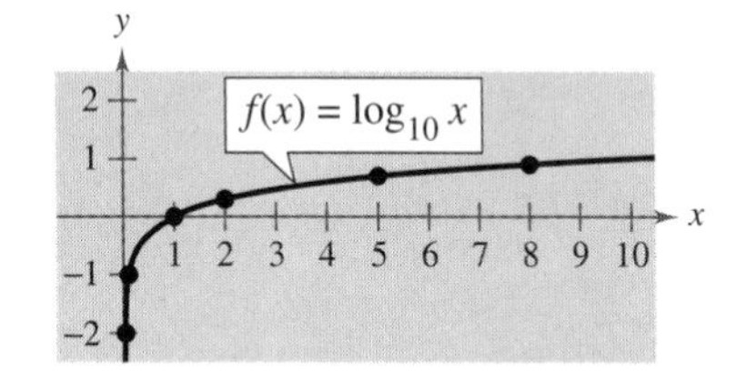

FIGURE 5.9

	Without Calculator				With Calculator		
x	$\frac{1}{100}$	$\frac{1}{10}$	1	10	2	5	8
$\log_{10} x$	-2	-1	0	1	0.301	0.699	0.903

The nature of the graph in Figure 5.9 is typical of functions of the form $f(x) = \log_a x$, $a > 1$. They have one x-intercept and one vertical asymptote. Notice how slowly the graph rises for $x > 1$. In Figure 5.9 you would need to move out to $x = 1000$ before the graph rose to $y = 3$. The basic characteristics of logarithmic graphs are summarized in Figure 5.10.

NOTE In the graph at the right, note that the vertical asymptote occurs at $x = 0$, where $\log_a x$ is *undefined.*

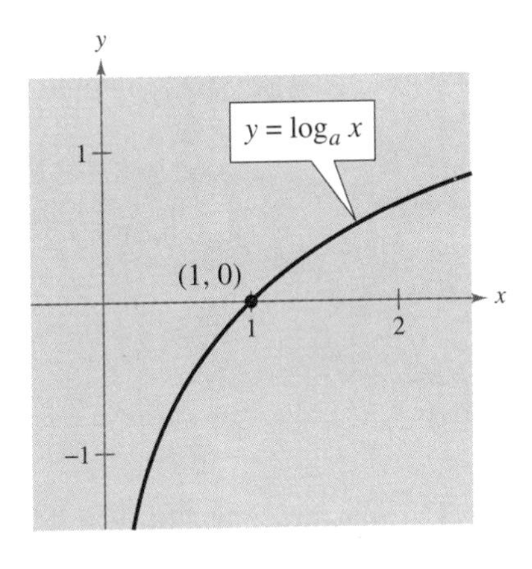

Graph of $y = \log_a x$, $a > 1$

- Domain: $(0, \infty)$
- Range: $(-\infty, \infty)$
- Intercept: $(1, 0)$
- Increasing
- y-axis is a vertical asymptote ($\log_a x \to -\infty$ as $x \to 0^+$)
- Continuous
- Reflection of graph of $y = a^x$ about the line $y = x$

FIGURE 5.10

In the following example, the graph of $\log_a x$ is used to sketch the graphs of functions of the form $y = b \pm \log_a(x + c)$.

EXAMPLE 6 *Sketching the Graphs of Logarithmic Functions*

The graph of each of the following functions is similar to the graph of $f(x) = \log_{10} x$, as shown in Figure 5.11.

a. Because $g(x) = \log_{10}(x - 1) = f(x - 1)$, the graph of g can be obtained by shifting the graph of f one unit to the right.

b. Because $h(x) = 2 + \log_{10} x = 2 + f(x)$, the graph of h can be obtained by shifting the graph of f two units up.

(a)

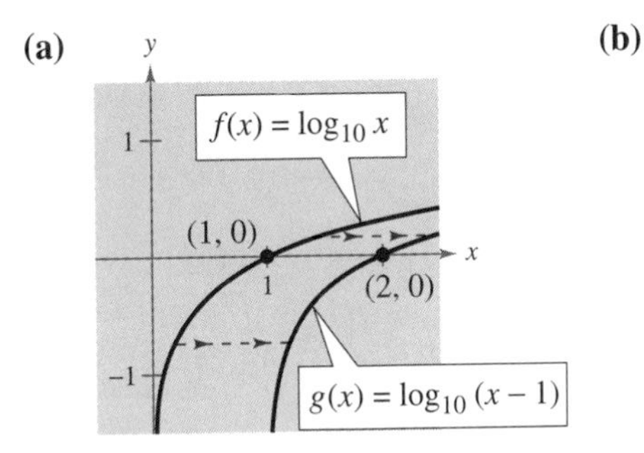

(b)

y
(1, 2)
2
$h(x) = 2 + \log_{10} x$
1
$f(x) = \log_{10} x$
x
(1, 0)
2

FIGURE 5.11

NOTE In Figure 5.11, notice how each transformation affects the vertical asymptote.

The Natural Logarithmic Function

As with exponential functions, the most widely used base for logarithmic functions is the number e, where

$$e \approx 2.718281828 \ldots.$$

The logarithmic function with base e is the **natural logarithmic function** and is denoted by the special symbol $\ln x$, read as "el en of x."

> ***The Natural Logarithmic Function***
>
> The function defined by
>
> $$f(x) = \log_e x = \ln x, \quad x > 0$$
>
> is called the **natural logarithmic function.**

The four properties of logarithms listed on page 369 are also valid for natural logarithms.

> ***Properties of Natural Logarithms***
>
> **1.** $\ln 1 = 0$ because $e^0 = 1$.
> **2.** $\ln e = 1$ because $e^1 = e$.
> **3.** $\ln e^x = x$ because $e^x = e^x$.
> **4.** If $\ln x = \ln y$, then $x = y$.

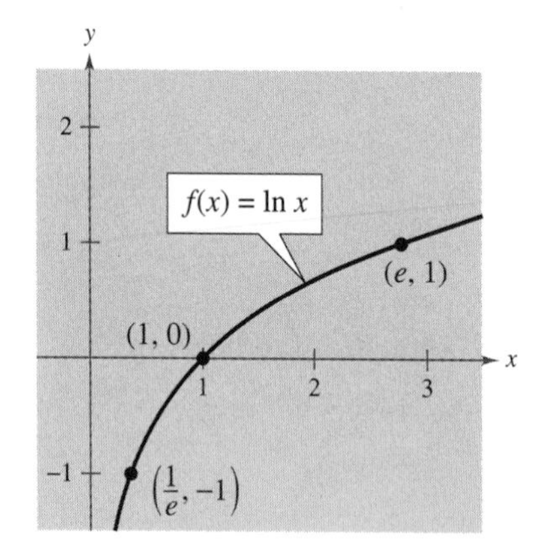

FIGURE 5.12

The graph of the natural logarithmic function is shown in Figure 5.12. Try using a graphing utility to confirm this graph. What is the domain of the natural logarithmic function?

EXAMPLE 7 *Using Properties of Natural Logarithms*

a. $\ln \frac{1}{e} = \ln e^{-1} = -1$	Property 3
b. $\ln e^2 = 2$	Property 3
c. $\ln e^0 = 0$	Property 1
d. $2 \ln e = 2(1) = 2$	Property 2

On most calculators, the natural logarithm is denoted by LN, as illustrated in Example 8.

EXAMPLE 8 *Evaluating the Natural Logarithmic Function*

Use a calculator to evaluate each expression.

a. $\ln 2$ **b.** $\ln 0.3$ **c.** $\ln e^2$ **d.** $\ln(-1)$

Solution

Number	Graphing Calculator Keystrokes	Display
a. $\ln 2$	LN 2 ENTER	0.6931472
b. $\ln 0.3$	LN .3 ENTER	–1.2039728
c. $\ln e^2$	LN e^x 2 ENTER	2
d. $\ln(-1)$	LN (-) 1 ENTER	ERROR

In Example 8, be sure you see that $\ln(-1)$ gives an error message on most calculators. This occurs because the domain of $\ln x$ is the set of positive real numbers (see Figure 5.12). Hence, $\ln(-1)$ is undefined.

EXAMPLE 9 *Finding the Domains of Logarithmic Functions*

Find the domain of each function.

a. $f(x) = \ln(x - 2)$ **b.** $g(x) = \ln(2 - x)$ **c.** $h(x) = \ln x^2$

Solution

a. Because $\ln(x - 2)$ is defined only if $x - 2 > 0$, it follows that the domain of f is $(2, \infty)$.

b. Because $\ln(2 - x)$ is defined only if $2 - x > 0$, it follows that the domain of g is $(-\infty, 2)$. The graph of g is shown in Figure 5.13.

c. Because $\ln x^2$ is defined only if $x^2 > 0$, it follows that the domain of h is all real numbers except $x = 0$.

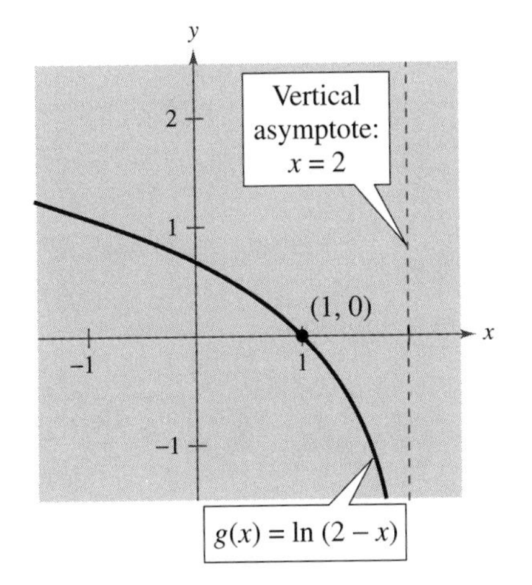

FIGURE 5.13

NOTE In Example 9, suppose you had been asked to analyze the function given by $h(x) = \ln|x - 2|$. How would the domain of this function compare with the domains of the functions given in parts (a) and (b) of the example?

Use a graphing utility to determine the time in months when the average score in Example 10 was 60. Explain your method of solving the problem. Describe another way that you can use a graphing utility to determine the answer.

Application

EXAMPLE 10 *Human Memory Model*

Students participating in a psychological experiment attended several lectures on a subject and were given an exam. Every month for a year after the exam, the students were retested to see how much of the material they remembered. The average scores for the group are given by the *human memory model*

$$f(t) = 75 - 6\ln(t + 1), \quad 0 \le t \le 12$$

where t is the time in months.

a. What was the average score on the original ($t = 0$) exam?

b. What was the average score at the end of $t = 2$ months?

c. What was the average score at the end of $t = 6$ months?

Solution

a. The original average score was

$$f(0) = 75 - 6\ln 1 = 75 - 6(0) = 75.$$

b. After 2 months, the average score was

$$f(2) = 75 - 6\ln 3 \approx 75 - 6(1.0986) \approx 68.4.$$

c. After 6 months, the average score was

$$f(6) = 75 - 6\ln 7 \approx 75 - 6(1.9459) \approx 63.3.$$

The graph of f is shown in Figure 5.14.

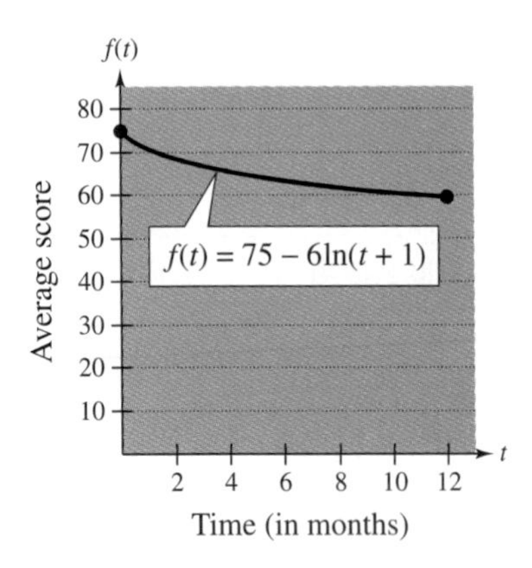

FIGURE 5.14

x	$f_1(x)$	$f_2(x)$
−3	Undefined	Undefined
−2.99999	−11.513	−15.513
−2.9999	−9.210	−13.210
−2	0	−4
0	1.099	−2.901
3	1.792	−2.208
6	2.197	−1.803

GROUP ACTIVITY

TRANSFORMING LOGARITHMIC FUNCTIONS

The table at left gives selected points for two natural logarithmic functions of the form $f(x) = b \pm \ln(x + c)$. Study the table. What can you infer? Compare the given data with data for $f(x) = \ln x$. Try to find a natural logarithmic function that fits each set of data. Explain the method you used. (*Hint:* You might find it easier to find the value of c first.)

WARM UP

Solve for x.

1. $2^x = 8$
2. $4^x = 1$
3. $10^x = 0.1$
4. $e^x = e$

Evaluate the expression. (Round your result to three decimal places.)

5. e^2
6. e^{-1}

Describe how the graph of g is related to the graph of f.

7. $g(x) = f(x + 2)$
8. $g(x) = -f(x)$
9. $g(x) = -1 + f(x)$
10. $g(x) = f(-x)$

5.2 Exercises

In Exercises 1–8, write the logarithmic equation in exponential form. For example, the exponential form of $\log_5 25 = 2$ is $5^2 = 25$.

1. $\log_4 64 = 3$
2. $\log_3 81 = 4$
3. $\log_7 \frac{1}{49} = -2$
4. $\log_{10} \frac{1}{1000} = -3$
5. $\log_{32} 4 = \frac{2}{5}$
6. $\log_{16} 8 = \frac{3}{4}$
7. $\ln 1 = 0$
8. $\ln 4 = 1.386 \ldots$

In Exercises 9–18, write the exponential equation in logarithmic form. For example, the logarithmic form of $2^3 = 8$ is $\log_2 8 = 3$.

9. $5^3 = 125$
10. $8^2 = 64$
11. $81^{1/4} = 3$
12. $9^{3/2} = 27$
13. $6^{-2} = \frac{1}{36}$
14. $10^{-3} = 0.001$
15. $e^3 = 20.0855 \ldots$
16. $e^0 = 1$
17. $e^x = 4$
18. $u^v = w$

In Exercises 19–30, evaluate the expression without using a calculator.

19. $\log_2 16$
20. $\log_2 \frac{1}{8}$
21. $\log_{16} 4$
22. $\log_{27} 9$
23. $\log_7 1$
24. $\log_{10} 1000$
25. $\log_{10} 0.01$
26. $\log_{10} 10$
27. $\log_8 32$
28. $\log_9 243$
29. $\ln e^3$
30. $\log_a a^2$

In Exercises 31–40, use a calculator to evaluate the logarithm. Round to three decimal places.

31. $\log_{10} 345$
32. $\log_{10} \frac{4}{5}$
33. $\log_{10} 145$
34. $\log_{10} 12.5$
35. $\ln 18.42$
36. $\ln\sqrt{42}$
37. $\ln(1 + \sqrt{3})$
38. $\ln(\sqrt{5} - 2)$
39. $\ln 0.32$
40. $\ln 0.75$

In Exercises 41–44, describe the relationship between the graphs of f and g. What is the relationship between the functions f and g?

41. $f(x) = 3^x$
$g(x) = \log_3 x$

42. $f(x) = 5^x$
$g(x) = \log_5 x$

43. $f(x) = e^x$
$g(x) = \ln x$

44. $f(x) = 10^x$
$g(x) = \log_{10} x$

In Exercises 45–50, use the graph of $y = \log_3 x$ to match the given function with its graph. [The graphs are labeled (a), (b), (c), (d), (e), and (f).]

(a)

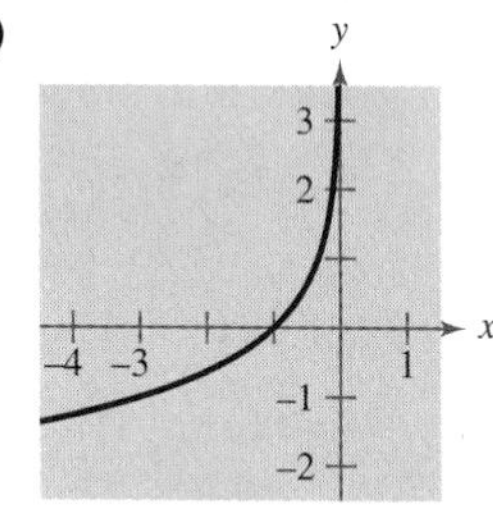

(b)

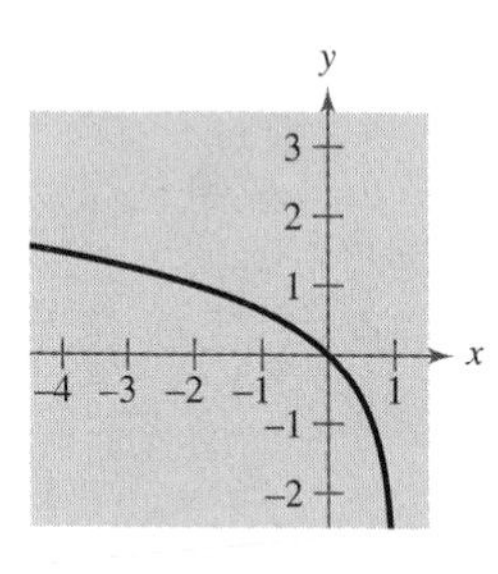

(c)

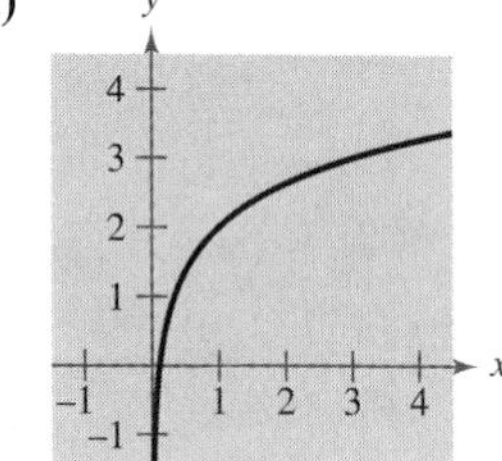

(d)

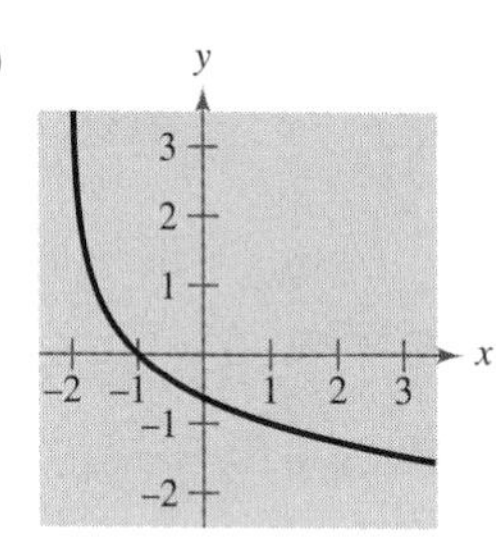

(e)

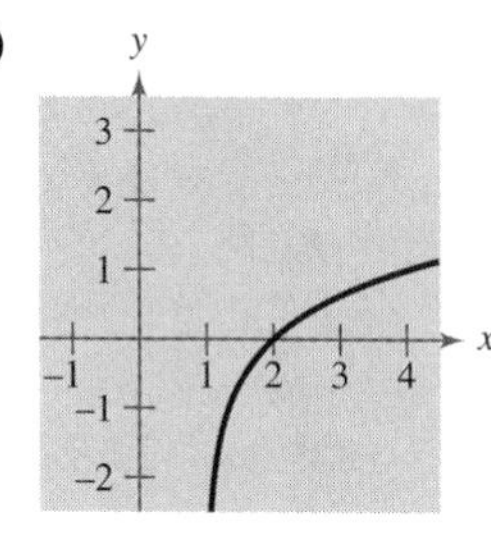

(f)

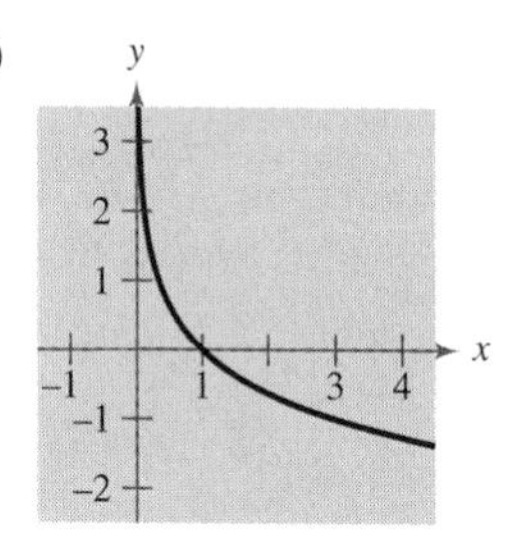

45. $f(x) = \log_3 x + 2$

46. $f(x) = -\log_3 x$

47. $f(x) = -\log_3(x + 2)$

48. $f(x) = \log_3(x - 1)$

49. $f(x) = \log_3(1 - x)$

50. $f(x) = -\log_3(-x)$

In Exercises 51–62, find the domain, vertical asymptote, and x-intercept of the logarithmic function and sketch its graph.

51. $f(x) = \log_4 x$

52. $g(x) = \log_6 x$

53. $y = -\log_3 x + 2$

54. $h(x) = \log_4(x - 3)$

55. $f(x) = -\log_6(x + 2)$

56. $y = \log_5(x - 1) + 4$

57. $y = \log_{10}\left(\frac{x}{5}\right)$

58. $y = \log_{10}(-x)$

59. $f(x) = \ln(x - 2)$

60. $h(x) = \ln(x + 1)$

61. $g(x) = \ln(-x)$

62. $f(x) = \ln(3 - x)$

In Exercises 63–66, use a graphing utility to graph the function. Use the graph to determine the intervals in which the function is increasing and decreasing and approximate any relative maximum or minimum values of the function.

63. $f(x) = |\ln x|$

64. $h(x) = \ln(x^2 + 1)$

65. $f(x) = \frac{x}{2} - \ln \frac{x}{4}$

66. $g(x) = \frac{12 \ln x}{x}$

67. ***Graphical Analysis*** Use a graphing utility to graph f and g on the same viewing rectangle and determine which is increasing at the greater rate for "large" values of x. What can you conclude about the rate of growth of the natural logarithmic function?

(a) $f(x) = \ln x, \quad g(x) = \sqrt{x}$

(b) $f(x) = \ln x, \quad g(x) = \sqrt[4]{x}$

68. ***Exploration*** The table of values was obtained from evaluating a function. Determine which of the statements may be true and which must be false.

x	1	2	8
y	0	1	3

(a) y is an exponential function of x.

(b) y is a logarithmic function of x.

(c) x is an exponential function of y.

(d) y is a linear function of x.

69. ***Exploration*** Use a graphing utility to compare the graph of the function $y = \ln x$ with the graphs of the following functions.

(a) $y = x - 1$

(b) $y = (x - 1) - \frac{1}{2}(x - 1)^2$

(c) $y = (x - 1) - \frac{1}{2}(x - 1)^2 + \frac{1}{3}(x - 1)^3$

70. ***Finding a Pattern*** Identify the pattern of successive polynomials given in Exercise 69. Extend the pattern one more term and compare the graph of the resulting polynomial function with the graph of $y = \ln x$. What do you think the pattern implies?

71. ***Human Memory Model*** Students in a mathematics class were given an exam and then retested monthly with an equivalent exam. The average scores for the class are given by the human memory model

$$f(t) = 80 - 17 \log_{10}(t + 1), \qquad 0 \le t \le 12$$

where t is the time in months.

(a) What was the average score on the original exam $(t = 0)$?

(b) What was the average score after 4 months?

(c) What was the average score after 10 months?

72. ***Population Growth*** The population of a town will double in

$$t = \frac{10 \ln 2}{\ln 67 - \ln 50} \text{ years.}$$

Find t.

73. ***World Population Growth*** The time t in years for the world population to double if it is increasing at a continuous rate of r is given by

$$t = \frac{\ln 2}{r}.$$

(a) Complete the table.

r	0.005	0.01	0.015	0.02	0.025	0.03
t						

(b) Use a reference source to decide which value of r best approximates the actual rate of growth for the world population.

74. ***Investment Time*** A principal P, invested at $9\frac{1}{2}\%$ and compounded continuously, increases to an amount K times the original principal after t years, where t is given by

$$t = \frac{\ln K}{0.095}.$$

(a) Complete the table and interpret your results.

K	1	2	4	6	8	10	12
t							

(b) Sketch a graph of the function.

Ventilation Rates **In Exercises 75 and 76, use the model**

$$y = 80.4 - 11 \ln x, \qquad 100 \le x \le 1500$$

which approximates the minimum required ventilation rate in terms of the air space per child in a public school classroom. In the model, x is the air space per child in cubic feet and y is the ventilation rate in cubic feet per minute.

75. Use a graphing utility to graph the function and approximate the required ventilation rate if there is 300 cubic feet of air space per child.

76. A classroom is designed for 30 students. The air-conditioning system in the room has the capacity of moving 450 cubic feet of air per minute.

(a) Determine the ventilation rate per child, assuming that the room is filled to capacity.

(b) Use the graph of Exercise 75 to estimate the air space required per child.

(c) Determine the minimum number of square feet of floor space required for the room if the ceiling height is 30 feet.

77. ***Work*** The work (in foot-pounds) done in compressing a volume of 9 cubic feet at a pressure of 15 pounds per square inch to a volume of 3 cubic feet is

$$W = 19{,}440(\ln 9 - \ln 3).$$

Find W.

78. ***Sound Intensity*** The relationship between the number of decibels β and the intensity of a sound I in watts per square centimeter is given by

$$\beta = 10 \log_{10}\left(\frac{I}{10^{-16}}\right).$$

(a) Determine the number of decibels of a sound with an intensity of 10^{-4} watts per square centimeter.

(b) Determine the number of decibels of a sound with an intensity of 10^{-6} watts per square centimeter.

(c) The intensity of the sound in part (a) is 100 times as great as that in part (b). Is the number of decibels 100 times as great? Explain.

Monthly Payment **In Exercises 79–82, use the model**

$$t = 10.042 \ln\left(\frac{x}{x - 1250}\right), \qquad 1250 < x$$

which approximates the length of a home mortgage of \$150,000 at 10% in terms of the monthly payment. In the model, t is the length of the mortgage in years and x is the monthly payment in dollars (see figure).

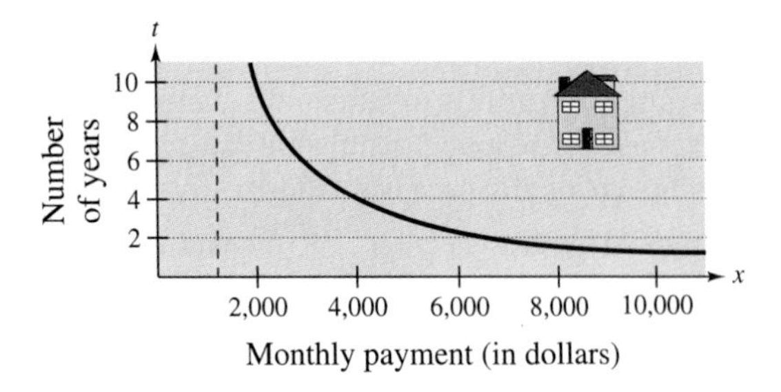

79. Use the model to approximate the length of the mortgage (for \$150,000 at 10%) if the monthly payment is \$1316.35.

80. Use the model to approximate the length of the mortgage (for \$150,000 at 10%) if the monthly payment is \$1982.26.

81. Approximate the total amount paid over the term of the mortgage with a monthly payment of \$1316.35. What is the total interest charge?

82. Approximate the total amount paid over the term of the mortgage with a monthly payment of \$1982.26. What is the total interest charge?

83. (a) Complete the table for the function

$$f(x) = \frac{\ln x}{x}.$$

x	1	5	10	10^2	10^4	10^6
$f(x)$						

(b) Use the table in part (a) to determine what value $f(x)$ approaches as x increases without bound.

(c) Use a graphing utility to confirm the result of part (b).

84. ***Exploration*** Answer the following questions for the function $f(x) = \log_{10} x$. Do not use a calculator.

(a) What is the domain of f?

(b) What is f^{-1}?

(c) If x is a real number between 1000 and 10,000, in which interval will $f(x)$ be found?

(d) In which interval will x be found if $f(x)$ is negative?

(e) If $f(x)$ is increased by one unit, x must have been increased by what factor?

(f) If $f(x_1) = 3n$ and $f(x_2) = n$, what is the ratio of x_1 to x_2?

Review **Solve Exercises 85–88 as a review of the skills and problem-solving techniques you learned in previous sections. Translate the statement into an algebraic expression.**

85. The product of 8 and n is decreased by 3.

86. The total hourly wage for an employee is \$9.25 per hour plus 75 cents for each of q units produced per hour.

87. The total cost for auto repairs if the cost of parts was \$83.95 and there were t hours of labor at \$37.50 per hour

88. The area of a rectangle if the length is 10 units more than the width w

5.3 Properties of Logarithms

See Exercise 89 on page 385 for an example of how a logarithmic function can be used as a human memory model.

Change of Base ▫ *Properties of Logarithms* ▫ *Rewriting Logarithmic Expressions* ▫ *Application*

Change of Base

Most calculators have only two types of log keys, one for common logarithms (base 10) and one for natural logarithms (base e). Although common logs and natural logs are the most frequently used, you may occasionally need to evaluate logarithms to other bases. To do this, you can use the following *change-of-base formula.*

John Napier, a Scottish mathematician, developed logarithms as a way to simplify some of the tedious calculations of his day. Beginning in 1594, Napier worked about 20 years on the invention of logarithms. Napier was only partially successful in his quest to simplify tedious calculations. Nonetheless, the development of logarithms was a step forward and received immediate recognition.

CHANGE-OF-BASE FORMULA

Let a, b, and x be positive real numbers such that $a \neq 1$ and $b \neq 1$. Then $\log_a x$ is given by

$$\log_a x = \frac{\log_b x}{\log_b a}.$$

One way to look at the change-of-base formula is that logarithms to base a are simply *constant multiples* of logarithms to base b. The constant multiplier is $1/(\log_b a)$.

EXAMPLE 1 *Changing Bases Using Common Logarithms*

a. $\log_4 30 = \dfrac{\log_{10} 30}{\log_{10} 4} \approx \dfrac{1.47712}{0.60206} \approx 2.4534$

b. $\log_2 14 = \dfrac{\log_{10} 14}{\log_{10} 2} \approx \dfrac{1.14613}{0.30103} \approx 3.8074$

EXAMPLE 2 *Changing Bases Using Natural Logarithms*

a. $\log_4 30 = \dfrac{\ln 30}{\ln 4} \approx \dfrac{3.40120}{1.38629} \approx 2.4534$

b. $\log_2 14 = \dfrac{\ln 14}{\ln 2} \approx \dfrac{2.63906}{0.693147} \approx 3.8074$

Properties of Logarithms

You know from the previous section that the logarithmic function with base a is the *inverse* of the exponential function with base a. Thus, it makes sense that the properties of exponents should have corresponding properties involving logarithms. For instance, the exponential property $a^0 = 1$ has the corresponding logarithmic property $\log_a 1 = 0$.

NOTE There is no general property that can be used to rewrite $\log_a(u \pm v)$. Specifically, $\log_a(x + y)$ is not equal to $\log_a x + \log_a y$.

PROPERTIES OF LOGARITHMS

Let a be a positive number such that $a \neq 1$, and let n be a real number. If u and v are positive real numbers, the following properties are true.

1. $\log_a(uv) = \log_a u + \log_a v$ **1.** $\ln(uv) = \ln u + \ln v$

2. $\log_a \frac{u}{v} = \log_a u - \log_a v$ **2.** $\ln \frac{u}{v} = \ln u - \ln v$

3. $\log_a u^n = n \log_a u$ **3.** $\ln u^n = n \ln u$

THINK ABOUT THE PROOF

To prove Property 1, let $x = \log_a u$ and $y = \log_a v$. The corresponding exponential forms of these two equations are $a^x = u$ and $a^y = v$. Multiplying u and v produces

$$uv = a^x a^y = a^{x+y}.$$

Can you see how to use this equation to complete the proof? The details of the proof are listed in the appendix.

EXAMPLE 3 *Using Properties of Logarithms*

Write the logarithm in terms of ln 2 and ln 3.

a. $\ln 6$ **b.** $\ln \frac{2}{27}$

Solution

a. $\ln 6 = \ln(2 \cdot 3)$ Rewrite 6 as $2 \cdot 3$.

$= \ln 2 + \ln 3$ Property 1

b. $\ln \frac{2}{27} = \ln 2 - \ln 27$ Property 2

$= \ln 2 - \ln 3^3$ Rewrite 27 as 3^3.

$= \ln 2 - 3 \ln 3$ Property 3

EXAMPLE 4 *Using Properties of Logarithms*

Use the properties of logarithms to verify that $-\ln \frac{1}{2} = \ln 2$.

Solution

$$-\ln \tfrac{1}{2} = -\ln(2^{-1}) = -(-1) \ln 2 = \ln 2$$

Try checking this result on your calculator.

Rewriting Logarithmic Expressions

The properties of logarithms are useful for rewriting logarithmic expressions in forms that simplify the operations of algebra. This is true because they convert complicated products, quotients, and exponential forms into simpler sums, differences, and products, respectively.

EXAMPLE 5 ***Rewriting the Logarithm of a Product***

$$\begin{aligned}\log_{10} 5x^3y &= \log_{10} 5 + \log_{10} x^3y\\ &= \log_{10} 5 + \log_{10} x^3 + \log_{10} y\\ &= \log_{10} 5 + 3\log_{10} x + \log_{10} y\end{aligned}$$

EXAMPLE 6 ***Rewriting the Logarithm of a Quotient***

$$\begin{aligned}\ln\frac{\sqrt{3x-5}}{7} &= \ln(3x-5)^{1/2} - \ln 7\\ &= \frac{1}{2}\ln(3x-5) - \ln 7\end{aligned}$$

In Examples 5 and 6, the properties of logarithms were used to *expand* logarithmic expressions. In Examples 7 and 8, this procedure is reversed and the properties of logarithms are used to *condense* logarithmic expressions.

Exploration

Use a graphing utility to graph the functions

$$y = \ln x - \ln(x-3)$$

and

$$y = \ln\frac{x}{x-3}$$

on the same viewing rectangle. Does the graphing utility show the functions with the same domain? If so, should it? Explain your reasoning.

EXAMPLE 7 ***Condensing a Logarithmic Expression***

$$\begin{aligned}\tfrac{1}{2}\log_{10} x + 3\log_{10}(x+1) &= \log_{10} x^{1/2} + \log_{10}(x+1)^3\\ &= \log_{10}\left[\sqrt{x}\cdot(x+1)^3\right]\end{aligned}$$

EXAMPLE 8 ***Condensing a Logarithmic Expression***

$$\begin{aligned}2\ln(x+2) - \ln x &= \ln(x+2)^2 - \ln x\\ &= \ln\frac{(x+2)^2}{x}\end{aligned}$$

Application

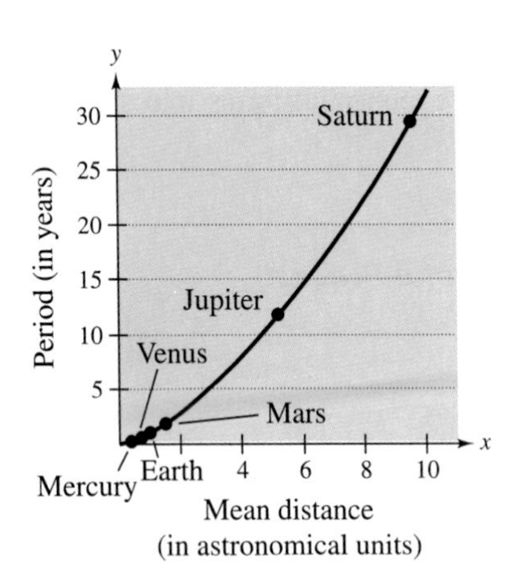

FIGURE 5.15

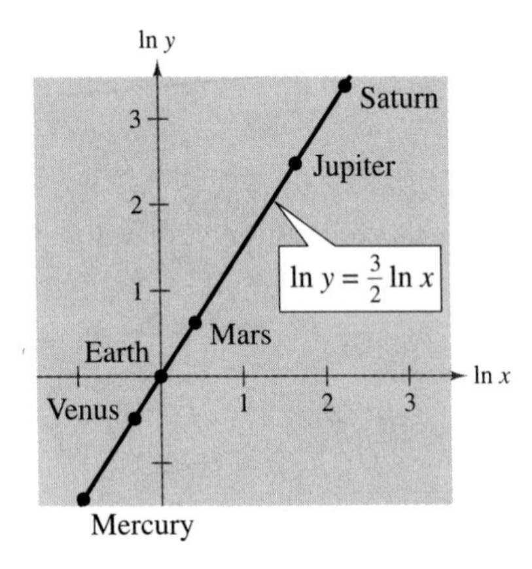

FIGURE 5.16

EXAMPLE 9 *Finding a Mathematical Model*

Real Life

The table gives the mean distance x and the period y of the six planets that are closest to the sun. In the table, the mean distance is given in terms of astronomical units (where the earth's mean distance is defined as 1.0), and the period is given in terms of years. Find an equation that expresses y as a function of x.

Planet	Mercury	Venus	Earth	Mars	Jupiter	Saturn
Period, y	0.241	0.615	1.0	1.881	11.861	29.457
Mean Distance, x	0.387	0.723	1.0	1.523	5.203	9.541

Solution

The points in the table are plotted in Figure 5.15. From this figure it is not clear how to find an equation that relates y and x. To solve this problem, take the natural log of each of the x- and y-values given in the table. This produces the following results.

Planet	Mercury	Venus	Earth	Mars	Jupiter	Saturn
$\ln y$	−1.423	−0.486	0.0	0.632	2.473	3.383
$\ln x$	−0.949	−0.324	0.0	0.421	1.649	2.256

Now, by plotting the points in the second table, you can see that all six of the points appear to lie in a line (see Figure 5.16). You can use a graphical approach or an algebraic approach to find that the slope of this line is $\frac{3}{2}$, and you can therefore conclude that $\ln y = \frac{3}{2} \ln x$. [Try to convert this to $y = f(x)$ form.]

GROUP ACTIVITY

KEPLER'S LAW

The relationship described in Example 9 was first discovered by Johannes Kepler. Use properties of logarithms to rewrite the relationship so that y is expressed as a function of x.

WARM UP

Evaluate the expression without using a calculator.

1. $\log_7 49$
2. $\log_2 \frac{1}{32}$
3. $\ln \dfrac{1}{e^2}$
4. $\log_{10} 0.001$

Simplify the expression.

5. $e^2 e^3$
6. $\dfrac{e^2}{e^3}$
7. $(e^2)^3$
8. $(e^2)^0$

Rewrite the expression in exponential form.

9. $\dfrac{1}{x^2}$
10. $\sqrt{x}$

5.3 Exercises

In Exercises 1 and 2, use a graphing utility to graph the two functions on the same viewing rectangle. Use the graphs to verify that the expressions are equivalent.

1. $f(x) = \log_{10} x$, $g(x) = \dfrac{\ln x}{\ln 10}$
2. $f(x) = \ln x$, $g(x) = \dfrac{\log_{10} x}{\log_{10} e}$

In Exercises 3–6, rewrite the logarithm as a multiple of a common logarithm.

3. $\log_3 5$
4. $\log_4 10$
5. $\log_2 x$
6. $\ln 5$

In Exercises 7–10, rewrite the logarithm as a multiple of a natural logarithm.

7. $\log_3 5$
8. $\log_4 10$
9. $\log_2 x$
10. $\log_{10} 5$

In Exercises 11–18, evaluate the logarithm using the change-of-base formula. Round your result to three decimal places.

11. $\log_3 7$
12. $\log_7 4$
13. $\log_{1/2} 4$
14. $\log_4 0.55$
15. $\log_9 0.4$
16. $\log_{20} 125$
17. $\log_{15} 1250$
18. $\log_{1/3} 0.015$

In Exercises 19–38, use the properties of logarithms to write the expression as a sum, difference, and/or constant multiple of logarithms. (Assume all variables are positive.)

19. $\log_{10} 5x$
20. $\log_{10} 10z$
21. $\log_{10} \dfrac{5}{x}$
22. $\log_{10} \dfrac{y}{2}$
23. $\log_8 x^4$
24. $\log_6 z^{-3}$
25. $\ln \sqrt{z}$
26. $\ln \sqrt[3]{t}$

27. $\ln xyz$

28. $\ln \frac{xy}{z}$

29. $\ln \sqrt{a-1}, \; a > 1$

30. $\ln\left(\frac{x^2-1}{x^3}\right), \; x > 1$

31. $\ln z(z-1)^2, \; z > 1$

32. $\ln \sqrt{\frac{x^2}{y^3}}$

33. $\ln \sqrt[3]{\frac{x}{y}}$

34. $\ln \frac{x}{\sqrt{x^2+1}}$

35. $\ln \frac{x^4\sqrt{y}}{z^5}$

36. $\ln \sqrt{x^2(x+2)}$

37. $\log_b \frac{x^2}{y^2z^3}$

38. $\log_b \frac{\sqrt{x}y^4}{z^4}$

Graphical Analysis **In Exercises 39 and 40, use a graphing utility to graph the two equations on the same viewing rectangle. Use the graphs to verify that the expressions are equivalent.**

39. $y_1 = \ln[x^3(x+4)]$
$y_2 = 3 \ln x + \ln(x+4)$

40. $y_1 = \ln\left(\frac{\sqrt{x}}{x-2}\right)$
$y_2 = \frac{1}{2} \ln x - \ln(x-2)$

In Exercises 41–60, write the expression as the logarithm of a single quantity.

41. $\ln x + \ln 2$

42. $\ln y + \ln z$

43. $\log_4 z - \log_4 y$

44. $\log_5 8 - \log_5 t$

45. $2 \log_2(x+4)$

46. $-4 \log_6 2x$

47. $\frac{1}{3} \log_3 5x$

48. $\frac{3}{2} \log_7(z-2)$

49. $\ln x - 3 \ln(x+1)$

50. $2 \ln 8 + 5 \ln z$

51. $\ln(x-2) - \ln(x+2)$

52. $3 \ln x + 2 \ln y - 4 \ln z$

53. $\ln x - 2[\ln(x+2) + \ln(x-2)]$

54. $4[\ln z + \ln(z+5)] - 2 \ln(z-5)$

55. $\frac{1}{3}[2 \ln(x+3) + \ln x - \ln(x^2-1)]$

56. $2[\ln x - \ln(x+1) - \ln(x-1)]$

57. $\frac{1}{3}[\ln y + 2 \ln(y+4)] - \ln(y-1)$

58. $\frac{1}{2}[\ln(x+1) + 2 \ln(x-1)] + 3 \ln x$

59. $2 \ln 3 - \frac{1}{2} \ln(x^2+1)$

60. $\frac{3}{2} \ln 5t^6 - \frac{3}{4} \ln t^4$

Graphical Analysis **In Exercises 61 and 62, use a graphing utility to graph the two equations on the same viewing rectangle. Use the graphs to verify that the expressions are equivalent.**

61. $y_1 = 2[\ln 8 - \ln(x^2+1)], \; y_2 = \ln\left[\frac{64}{(x^2+1)^2}\right]$

62. $y_1 = \ln x + \frac{1}{3} \ln(x+1), \; y_2 = \ln(x\sqrt[3]{x+1})$

Think About It **In Exercises 63 and 64, use a graphing utility to graph the two equations on the same viewing rectangle. Are the expressions equivalent? Explain.**

63. $y_1 = \ln x^2, \; y_2 = 2 \ln x$

64. $y_1 = \frac{1}{4} \ln[x^4(x^2+1)], \; y_2 = \ln x + \frac{1}{4} \ln(x^2+1)$

Comparing Logarithmic Quantities **In Exercises 65 and 66, compare the logarithmic quantities. If two are equal, explain why.**

65. $\frac{\log_2 32}{\log_2 4}, \quad \log_2 \frac{32}{4}, \quad \log_2 32 - \log_2 4$

66. $\log_7 \sqrt{70}, \quad \log_7 35, \quad \frac{1}{2} + \log_7 \sqrt{10}$

67. ***Think About It*** Graph

$$f(x) = \ln \frac{x}{2}, \quad g(x) = \frac{\ln x}{\ln 2}, \quad h(x) = \ln x - \ln 2$$

on the same set of axes. Which two functions have identical graphs? Explain why.

68. ***Exploration*** Approximate the natural logarithms of as many integers as possible between 1 and 20 given that $\ln 2 \approx 0.6931$, $\ln 3 \approx 1.0986$, and $\ln 5 \approx 1.6094$. (Do not use a calculator.)

In Exercises 69–82, find the exact value of the logarithm without using a calculator. (If this is not possible, state the reason.)

69. $\log_3 9$

70. $\log_6 \sqrt[3]{6}$

71. $\log_4 16^{1.2}$

72. $\log_5 \frac{1}{125}$

73. $\log_3 (-9)$

74. $\log_2 (-16)$

75. $\log_5 75 - \log_5 3$

76. $\log_4 2 + \log_4 32$

77. $\ln e^2 - \ln e^5$

78. $3 \ln e^4$

79. $\log_{10} 0$

80. $\ln 1$

81. $\ln e^{4.5}$

82. $\ln \sqrt[4]{e^3}$

In Exercises 83–88, use the properties of logarithms to simplify the logarithmic expression.

83. $\log_4 8$

84. $\log_2(4^2 \cdot 3^4)$

85. $\log_5 \frac{1}{250}$

86. $\log_{10} \frac{9}{300}$

87. $\ln(5e^6)$

88. $\ln \dfrac{6}{e^2}$

89. ***Human Memory Model*** Students participating in a psychological experiment attended several lectures and were given an exam. Every month for a year after the exam, the students were retested to see how much of the material they remembered. The average scores for the group are given by the memory model

$$f(t) = 90 - 15 \log_{10}(t + 1), \qquad 0 \le t \le 12$$

where t is the time in months.

(a) What was the average score on the original exam $(t = 0)$?

(b) What was the average score after 6 months?

(c) What was the average score after 12 months?

(d) When will the average score decrease to 75?

(e) Use the properties of logarithms to write the function in another form.

(f) Sketch the graph of the function over the specified domain.

90. ***Sound Intensity*** The relationship between the number of decibels β and the intensity of a sound I in watts per square centimeter is given by

$$\beta = 10 \log_{10}\left(\frac{I}{10^{-16}}\right).$$

Use the properties of logarithms to write the formula in simpler form, and determine the number of decibels of a sound with an intensity of 10^{-10} watts per square centimeter.

True or False? **In Exercises 91–96, determine if the statement is true or false given that $f(x) = \ln x$. If the statement is false, state why or give an example to show that it is false.**

91. $f(0) = 0$

92. $f(ax) = f(a) + f(x), \quad a > 0, x > 0$

93. $f(x - 2) = f(x) - f(2), \quad x > 2$

94. $\sqrt{f(x)} = \frac{1}{2} f(x)$

95. If $f(u) = 2f(v)$, then $v = u^2$.

96. If $f(x) < 0$, then $0 < x < 1$.

97. Prove that $\log_b \dfrac{u}{v} = \log_b u - \log_b v$.

98. Prove that $\log_b u^n = n \log_b u$.

Review **Solve Exercises 99–102 as a review of the skills and problem-solving techniques you learned in previous sections. Simplify the expressions.**

99. $\dfrac{24xy^{-2}}{16x^{-3}y}$

100. $\left(\dfrac{2x^2}{3y}\right)^{-3}$

101. $(18x^3y^4)^{-3}(18x^3y^4)^3$

102. $xy(x^{-1} + y^{-1})^{-1}$

5.4 Exponential and Logarithmic Equations

See Exercise 94 on page 396 for an example of how a logarithmic function can be used to model crumple zones for automobile crash tests.

Introduction ❐ *Solving Exponential Equations* ❐ *Solving Logarithmic Equations* ❐ *Applications*

Introduction

So far in this chapter, you have studied the definitions, graphs, and properties of exponential and logarithmic functions. In this section, you will study procedures for *solving equations* involving exponential and logarithmic functions. As a simple example, consider the exponential equation $2^x = 32$. One property of exponential functions states that $a^x = a^y$ if and only if $x = y$. You can solve $2^x = 32$ by rewriting the equation in the form $2^x = 2^5$, which implies that $x = 5$.

Although this method works in some cases, it does not work for an equation as simple as $e^x = 7$. In such a case, solution procedures are based on the fact that the exponential and logarithmic functions are inverses of each other. The following properties are the **inverse properties** of exponential and logarithmic functions.

	Base a	*Base e*
1.	$\log_a a^x = x$	$\ln e^x = x$
2.	$a^{\log_a x} = x$	$e^{\ln x} = x$

Now, to solve $e^x = 7$, you can take the natural logarithms of both sides to obtain

$$e^x = 7 \qquad \text{Original equation}$$
$$\ln e^x = \ln 7 \qquad \text{Take logarithms of both sides.}$$
$$x = \ln 7 \qquad \ln e^x = x \text{ because } e^x = e^x.$$

Here are some guidelines for solving exponential and logarithmic equations.

SOLVING EXPONENTIAL AND LOGARITHMIC EQUATIONS

1. *To solve an exponential equation,* first isolate the exponential expression, then take the logarithms of both sides and solve for the variable.
2. *To solve a logarithmic equation,* rewrite the equation in exponential form and solve for the variable.

Solving Exponential Equations

EXAMPLE 1 *Solving an Exponential Equation*

Solve $e^x = 72$.

Solution

$e^x = 72$	Original equation
$\ln e^x = \ln 72$	Take logarithms of both sides.
$x = \ln 72$	Inverse property of logs and exponents
$x \approx 4.277$	Use a calculator.

The solution is ln 72. Check this in the original equation.

EXAMPLE 2 *Solving an Exponential Equation*

Solve $e^x + 5 = 60$.

Solution

$e^x + 5 = 60$	Original equation
$e^x = 55$	Subtract 5 from both sides.
$\ln e^x = \ln 55$	Take logarithms of both sides.
$x = \ln 55$	Inverse property of logs and exponents
$x \approx 4.007$	Use a calculator.

The solution is ln 55. Check this in the original equation.

TECHNOLOGY

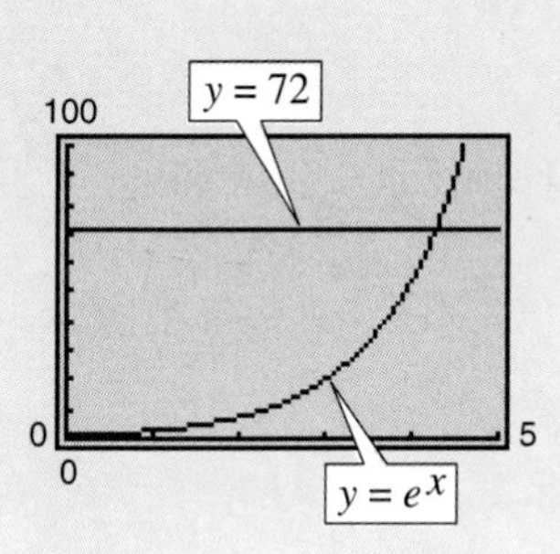

When solving an exponential or logarithmic equation, remember that you can check your solution graphically by "graphing the left and right sides separately" and estimating the x-coordinate of the point of intersection. For instance, to check the solution of the equation in Example 1, you can sketch the graphs of

$$y = e^x \qquad \text{and} \qquad y = 72$$

on the same viewing rectangle, as shown at the left. Notice that the graphs intersect when $x \approx 4.277$, which confirms the solution found in Example 1.

EXAMPLE 3 *Solving an Exponential Equation*

Solve $4e^{2x} = 5$.

Solution

$4e^{2x} = 5$	Original equation
$e^{2x} = \frac{5}{4}$	Divide both sides by 4.
$\ln e^{2x} = \ln \frac{5}{4}$	Take logarithms of both sides.
$2x = \ln \frac{5}{4}$	Inverse property of logs and exponents
$x = \frac{1}{2} \ln \frac{5}{4}$	Divide both sides by 2.
$x \approx 0.112$	Use a calculator.

The solution is $\frac{1}{2} \ln \frac{5}{4}$. Check this in the original equation.

When an equation involves two or more exponential expressions, you can still use a procedure similar to that demonstrated in the first three examples. However, the algebra is a bit more complicated.

EXAMPLE 4 *Solving an Exponential Equation*

Solve $e^{2x} - 3e^x + 2 = 0$.

Solution

$e^{2x} - 3e^x + 2 = 0$		Original equation
$(e^x)^2 - 3e^x + 2 = 0$		Quadratic form
$(e^x - 2)(e^x - 1) = 0$		Factor.
$e^x - 2 = 0$	⇨ $x = \ln 2$	Set 1st factor equal to 0.
$e^x - 1 = 0$	⇨ $x = 0$	Set 2nd factor equal to 0.

The solutions are ln 2 and 0. Check these in the original equation.

NOTE In Example 4, use a graphing utility to graph $y = e^{2x} - 3e^x + 2$. It should have two x-intercepts: one when $x = \ln 2$ and one when $x = 0$.

The *Interactive* CD-ROM shows every example with its solution; clicking on the *Try It!* button brings up similar problems. Guided Examples and Integrated Examples show step-by-step solutions to additional examples. Integrated Examples are related to several concepts in the section.

Solving Logarithmic Equations

To solve a logarithmic equation such as

$\ln x = 3$ Logarithmic form

write the equation in exponential form as follows.

$e^{\ln x} = e^3$ Exponentiate both sides.

$x = e^3$ Exponential form

This procedure is called *exponentiating* both sides of an equation.

EXAMPLE 5 *Solving a Logarithmic Equation*

Solve $\ln x = 2$.

Solution

$\ln x = 2$ Original equation

$e^{\ln x} = e^2$ Exponentiate both sides.

$x = e^2$ Inverse property of exponents and logs

$x \approx 7.389$ Use a calculator.

The solution is e^2. Check this in the original equation.

EXAMPLE 6 *Solving a Logarithmic Equation*

Solve $5 + 2 \ln x = 4$.

Solution

$5 + 2 \ln x = 4$ Original equation

$2 \ln x = -1$ Subtract 5 from both sides.

$\ln x = -\frac{1}{2}$ Divide both sides by 2.

$e^{\ln x} = e^{-1/2}$ Exponentiate both sides.

$x = e^{-1/2}$ Inverse property of exponents and logs

$x \approx 0.607$ Use a calculator.

The solution is $e^{-1/2}$. To check this result graphically, you can sketch the graphs of $y = 5 + 2 \ln x$ and $y = 4$ in the same coordinate plane, as shown in Figure 5.17.

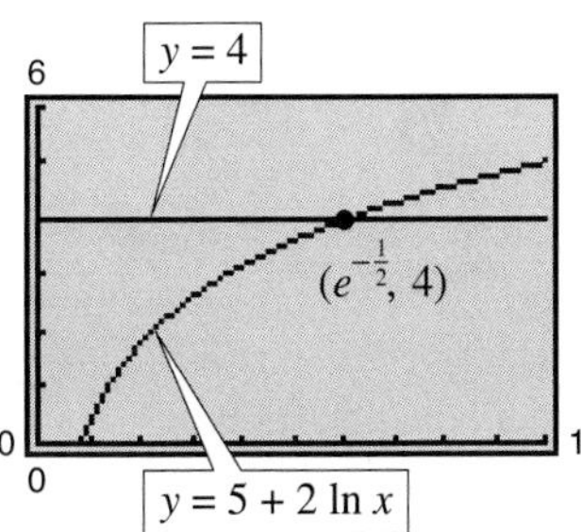

FIGURE 5.17

EXAMPLE 7 *Solving a Logarithmic Equation*

Solve $2 \ln 3x = 4$.

Solution

$2 \ln 3x = 4$	Original equation
$\ln 3x = 2$	Divide both sides by 2.
$e^{\ln 3x} = e^2$	Exponentiate both sides.
$3x = e^2$	Inverse property of exponents and logs
$x = \frac{1}{3}e^2$	Divide both sides by 3.
$x \approx 2.463$	Use a calculator.

The solution is $\frac{1}{3}e^2$. Check this in the original equation.

EXAMPLE 8 *Solving a Logarithmic Equation*

Solve $\ln x - \ln(x - 1) = 1$.

Solution

$\ln x - \ln(x - 1) = 1$	Original equation
$\ln \dfrac{x}{x - 1} = 1$	Property 2 of logarithms
$\dfrac{x}{x - 1} = e^1$	Exponentiate both sides.
$x = ex - e$	Multiply both sides by $x - 1$.
$x - ex = -e$	Subtract ex from both sides.
$x(1 - e) = -e$	Factor.
$x = \dfrac{-e}{1 - e}$	Divide both sides by $1 - e$.
$x = \dfrac{e}{e - 1}$	Simplify.

The solution is $e/(e - 1)$. Check this in the original equation.

In solving exponential or logarithmic equations, the following properties are useful. Can you see where these properties were used in this section?

1. $x = y$ if and only if $\log_a x = \log_a y$.
2. $x = y$ if and only if $a^x = a^y,\ a > 0,\ a \neq 1$.

Applications

EXAMPLE 9 *Doubling an Investment*

You have deposited \$500 in an account that pays 6.75% interest, compounded continuously. How long will it take your money to double?

Solution

Using the formula for continuous compounding, you can find that the balance in the account is given by

$$A = Pe^{rt} = 500e^{0.0675t}.$$

To find the time required for the balance to double, let $A = 1000$, and solve the resulting equation for t.

$500e^{0.0675t} = 1000$	Let $A = 1000$.
$e^{0.0675t} = 2$	Divide both sides by 500.
$\ln e^{0.0675t} = \ln 2$	Take logarithms of both sides.
$0.0675t = \ln 2$	Inverse property of logs and exponents
$t = \dfrac{\ln 2}{0.0675}$	Divide both sides by 0.0675.
$t \approx 10.27$	Use a calculator.

The balance in the account will double after approximately 10.27 years. This result is graphically demonstrated in Figure 5.18.

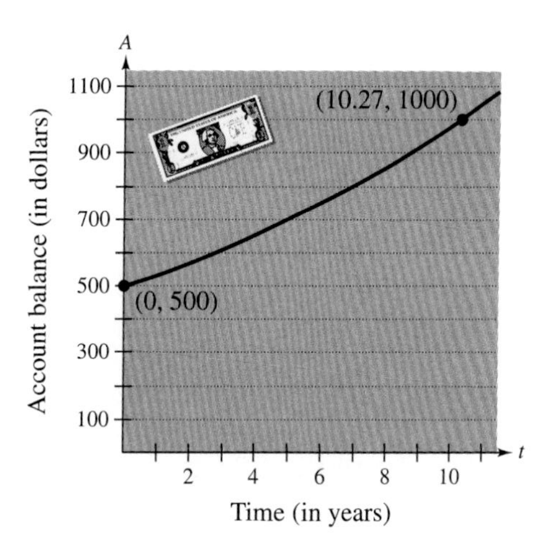

FIGURE 5.18

EXAMPLE 10 *Consumer Price Index for Sugar*

From 1970 to 1993, the Consumer Price Index (CPI) value y for a fixed amount of sugar for the year t can be modeled by the equation

$$y = -169.8 + 86.8 \ln t$$

where $t = 10$ represents 1970 (see Figure 5.19). During which year did the price of sugar reach 4 times its 1970 price of 30.5 on the CPI? (Source: U.S. Bureau of Labor Statistics)

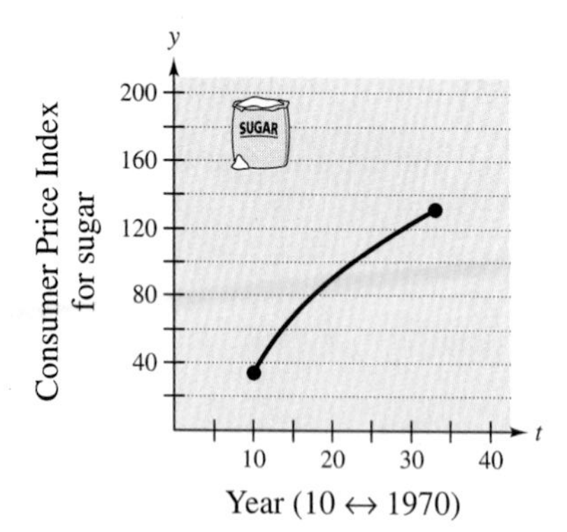

FIGURE 5.19

Solution

$-169.8 + 86.8 \ln t = y$	Given model
$-169.8 + 86.8 \ln t = 122$	Let $y = (4)(30.5) = 122$.
$86.8 \ln t = 291.8$	Add 169.8 to both sides.
$\ln t \approx 3.362$	Divide both sides by 86.8.
$e^{\ln t} \approx e^{3.362}$	Exponentiate both sides.
$t \approx 29$	Inverse property of exponents and logs

The solution is $t \approx 29$ years. Because $t = 10$ represents 1970, it follows that the price of sugar reached 4 times its 1970 price in 1988.

GROUP ACTIVITY

COMPARING MATHEMATICAL MODELS

x	8	9	10	11
y	85	468	2466	5868

x	12	13	14	15
y	12,153	24,016	45,283	60,399

The table gives the monthly amount of traffic y (in millions of packets) on the NSFNET, the backbone of the Internet, for January of each year x from 1988 through 1995, where $x = 8$ represents 1988. (Source: Network Information Center)

(a) Create a scatter plot of the data. Find a linear model for the data, and add its graph to your scatter plot. According to this model, when will monthly network traffic reach 150,000 millions of packets?

(b) Create a new table giving values for $\ln x$ and $\ln y$ and create a scatter plot of this transformed data. Use the method illustrated in Example 9 in Section 5.3 to find a model for the transformed data, and add its graph to your scatter plot. According to this model, when will monthly network traffic reach 150,000 millions of packets?

(c) Solve the model in part (b) for y, and add its graph to your scatter plot in part (a). Which model better fits the original data? Which model will better predict future traffic levels? Explain.

WARM UP

Solve for x.

1. $x \ln 2 = \ln 3$
2. $(x - 1) \ln 4 = 2$
3. $2xe^2 = e^3$
4. $4xe^{-1} = 8$
5. $x^2 - 4x + 5 = 0$
6. $2x^2 - 3x + 1 = 0$

Simplify the expression.

7. $\log_{10} 100^x$
8. $\log_4 64^x$
9. $\ln e^{2x}$
10. $\ln e^{-x^2}$

5.4 Exercises

In Exercises 1–6, determine whether the x-values are solutions of the equation.

1. $4^{2x-7} = 64$
   (a) $x = 5$
   (b) $x = 2$
2. $2^{3x+1} = 32$
   (a) $x = -1$
   (b) $x = 2$
3. $3e^{x+2} = 75$
   (a) $x = -2 + e^{25}$
   (b) $x = -2 + \ln 25$
   (c) $x \approx 1.2189$
4. $5^{2x+3} = 812$
   (a) $x = -1.5 + \log_5 \sqrt{812}$
   (b) $x \approx 0.5813$
   (c) $x = \dfrac{1}{2}\left(-3 + \dfrac{\ln 812}{\ln 5}\right)$
5. $\log_4(3x) = 3$
   (a) $x \approx 20.3560$
   (b) $x = -4$
   (c) $x = \frac{64}{3}$
6. $\ln(x - 1) = 3.8$
   (a) $x = 1 + e^{3.8}$
   (b) $x \approx 45.7012$
   (c) $x = 1 + \ln 3.8$

In Exercises 7–10, approximate the point of intersection of the graphs of f and g. Then solve the equation $f(x) = g(x)$ algebraically.

7. $f(x) = 2^x$
   $g(x) = 8$

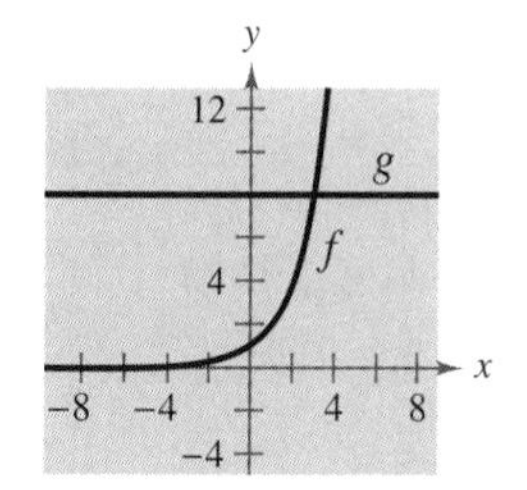

8. $f(x) = 27^x$
   $g(x) = 9$

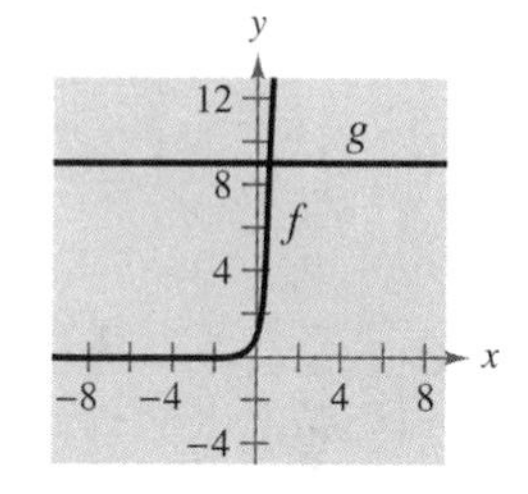

9. $f(x) = \log_3 x$
   $g(x) = 2$

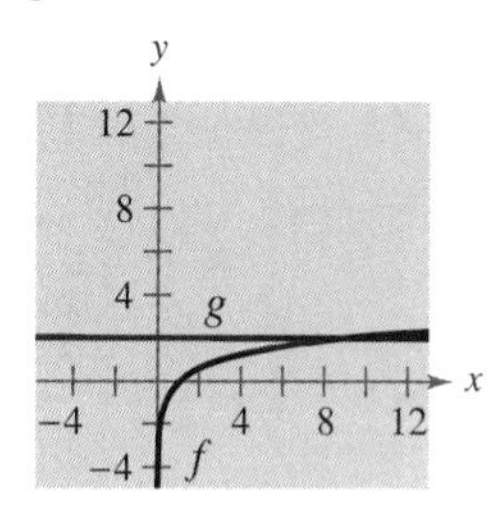

10. $f(x) = \ln (x - 4)$
    $g(x) = 0$

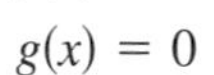

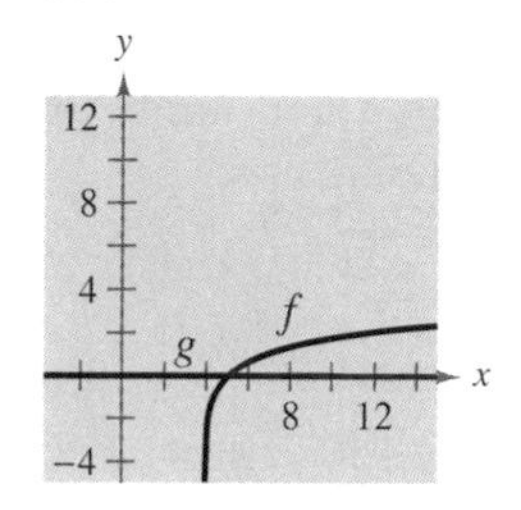

In Exercises 11–20, solve for x.

11. $4^x = 16$
12. $3^x = 243$
13. $7^x = \frac{1}{49}$
14. $8^x = 4$
15. $\left(\frac{3}{4}\right)^x = \frac{27}{64}$
16. $3^{x-1} = 27$
17. $\log_4 x = 3$
18. $\log_x 625 = 4$
19. $\log_{10} x = -1$
20. $\ln(2x - 1) = 0$

In Exercises 21–26, apply the inverse properties of $\ln x$ and e^x to simplify the expression.

21. $\log_{10} 10^{x^2}$
22. $\log_6 6^{2x-1}$
23. $e^{\ln(5x+2)}$
24. $-1 + \ln e^{2x}$
25. $e^{\ln x^2}$
26. $-8 + e^{\ln x^3}$

In Exercises 27–46, solve the exponential equation algebraically. Round the result to three decimal places.

27. $e^x = 10$
28. $4e^x = 91$
29. $7 - 2e^x = 5$
30. $-14 + 3e^x = 11$
31. $e^{3x} = 12$
32. $e^{2x} = 50$
33. $500e^{-x} = 300$
34. $1000e^{-4x} = 75$
35. $e^{2x} - 4e^x - 5 = 0$
36. $e^{2x} - 5e^x + 6 = 0$
37. $20(100 - e^{x/2}) = 500$
38. $\dfrac{400}{1 + e^{-x}} = 350$
39. $10^x = 42$
40. $10^x = 570$
41. $3^{2x} = 80$
42. $6^{5x} = 3000$
43. $5^{-t/2} = 0.20$
44. $4^{-3t} = 0.10$
45. $2^{3-x} = 565$
46. $\left(1 + \dfrac{0.10}{12}\right)^{12t} = 2$

In Exercises 47–50, use a graphing utility to graph the function and approximate its zero accurate to three decimal places.

47. $g(x) = 6e^{1-x} - 25$
48. $f(x) = 3e^{3x/2} - 962$
49. $g(t) = e^{0.09t} - 3$
50. $h(t) = e^{0.125t} - 8$

In Exercises 51–54, solve the exponential equation. Round the result to three decimal places.

51. $8(10^{3x}) = 12$
52. $3(5^{x-1}) = 21$
53. $\left(1 + \dfrac{0.065}{365}\right)^{365t} = 4$
54. $\dfrac{3000}{2 + e^{2x}} = 2$

In Exercises 55–70, solve the logarithmic equation algebraically. Round the result to three decimal places.

55. $\ln x = -3$
56. $\ln x = 2$
57. $\ln 2x = 2.4$
58. $3 \ln 5x = 10$
59. $\ln \sqrt{x + 2} = 1$
60. $\ln(x + 1)^2 = 2$
61. $\log_{10}(z - 3) = 2$
62. $\log_{10} x^2 = 6$
63. $\ln x + \ln(x - 2) = 1$
64. $\ln x + \ln(x + 3) = 1$
65. $\log_{10}(x + 4) - \log_{10} x = \log_{10}(x + 2)$
66. $\log_4 x - \log_4(x - 1) = \frac{1}{2}$
67. $\log_3 x + \log_3(x^2 - 8) = \log_3 8x$
68. $\log_2 x + \log_2(x + 2) = \log_2(x + 6)$
69. $\ln(x + 5) = \ln(x - 1) - \ln(x + 1)$
70. $\ln(x + 1) - \ln(x - 2) = \ln x^2$

In Exercises 71–76, solve the logarithmic equation. Round the result to three decimal places.

71. $6 \log_3(0.5x) = 11$
72. $5 \log_{10}(x - 2) = 11$
73. $2 \ln x = 7$
74. $\ln 4x = 1$
75. $\ln x + \ln(x^2 + 1) = 8$
76. $\log_{10} 8x - \log_{10}\left(1 + \sqrt{x}\right) = 2$

In Exercises 77–80, use a graphing utility to approximate the point of intersection of the graphs. Round the result to three decimal places.

77. $y_1 = 7$
$y_2 = 2^x$

78. $y_1 = 500$
$y_2 = 1500e^{-x/2}$

79. $y_1 = 3$
$y_2 = \ln x$

80. $y_1 = 10$
$y_2 = 4\ln(x - 2)$

Compound Interest **In Exercises 81 and 82, find the time required for a \$1000 investment to double at interest rate *r*, compounded continuously.**

81. $r = 0.085$

82. $r = 0.12$

83. *Think About It* Are the times required for the investments of Exercises 81 and 82 to quadruple twice as long as the times for them to double? Give a reason for your answer and verify your answer algebraically.

84. *Essay* Write a paragraph explaining whether or not the time required for an investment to double is dependent on the size of the investment.

Compound Interest **In Exercises 85 and 86, find the time required for a \$1000 investment to triple at interest rate *r*, compounded continuously.**

85. $r = 0.085$

86. $r = 0.12$

87. *Demand Function* The demand equation for a certain product is given by

$$p = 500 - 0.5(e^{0.004x}).$$

Find the demand x for a price of (a) $p = \$350$ and (b) $p = \$300$.

88. *Demand Function* The demand equation for a certain product is given by

$$p = 5000\left(1 - \frac{4}{4 + e^{-0.002x}}\right).$$

Find the demand x for a price of (a) $p = \$600$ and (b) $p = \$400$.

89. *Forest Yield* The yield V (in millions of cubic feet per acre) for a forest at age t years is given by

$$V = 6.7e^{-48.1/t}.$$

(a) Use a graphing utility to graph the function.

(b) Determine the horizontal asymptote of the function. Interpret its meaning in the context of the problem.

(c) Find the time necessary to obtain a yield of 1.3 million cubic feet.

90. *Trees per Acre* The number of trees per acre N of a certain species is approximated by the model

$$N = 68(10^{-0.04x}), \quad 5 \le x \le 40$$

where x is the average diameter of the trees 3 feet above the ground. Use the model to approximate the average diameter of the trees in a test plot when $N = 21$.

91. *Average Heights* The percent of American males between the ages of 18 and 24 who are no more than x inches tall is given by

$$m(x) = \frac{100}{1 + e^{-0.6114(x-69.71)}}.$$

The percent of American females between the ages of 18 and 24 who are no more than x inches tall is given by

$$f(x) = \frac{100}{1 + e^{-0.66607(x-64.51)}}$$

where m and f are the percents and x is the height in inches. (Source: U.S. National Center for Health Statistics)

(a) Use the graph to determine any horizontal asymptotes of the functions. What do they mean?

(b) What is the median height of each sex?

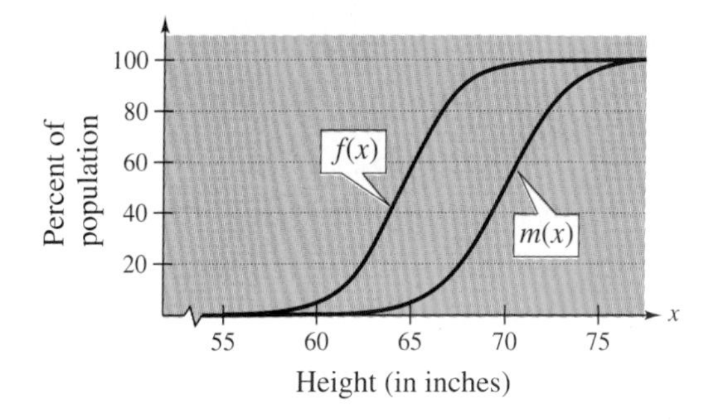

92. ***Human Memory Model*** In a group project in learning theory, a mathematical model for the proportion P of correct responses after n trials was found to be

$$P = \frac{0.83}{1 + e^{-0.2n}}.$$

(a) Use a graphing utility to graph the function.

(b) Use the graph to determine any horizontal asymptotes of the function. Interpret the meaning of the upper asymptote in the context of this problem.

(c) After how many trials will 60% of the responses be correct?

93. ***Data Analysis*** An object at a temperature of 160°C was removed from a furnace and placed in a room at 20°C. The temperature T of the object was measured each hour h and recorded in the table.

h	0	1	2	3	4	5
T	160°	90°	56°	38°	29°	24°

A model for this data is

$$T = 20[1 + 7(2^{-h})].$$

(a) The graph of this model is shown in the figure. Use the graph to identify the horizontal asymptote of the model and interpret the asymptote in the context of the problem.

(b) Approximate the time when the temperature of the object was 100°C.

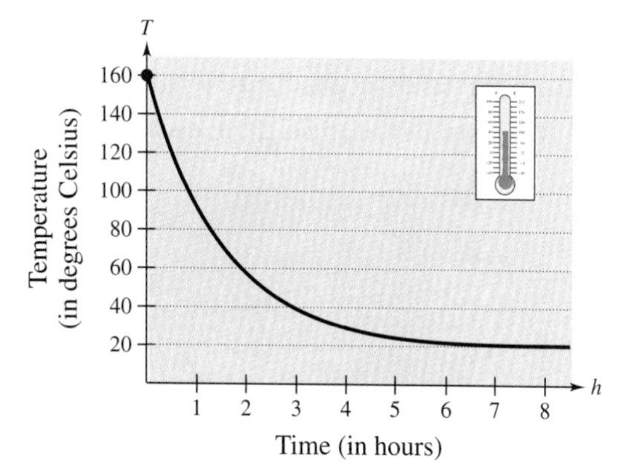

94. ***Chapter Opener*** Automobiles are designed with crumple zones that help protect their occupants in crashes. The crumple zones allow the occupants to move short distances when the automobiles come to abrupt stops. The greater the distance moved, the less g's the crash victims experience. (One g is equal to the acceleration due to gravity. For very short periods of time, humans have withstood as much as 40 g's.) In crash tests with vehicles moving at 90 kilometers per hour, analysts measured the numbers, y, of g's experienced during deceleration by crash dummies that were permitted to move x meters during impact. The data is shown in the table.

x	0.2	0.4	0.6	0.8	1
y	158	80	53	40	32

A model for this data is

$$y = -3.00 + 11.88 \ln x + \frac{36.94}{x}.$$

(a) Use a graphing utility to graph the data points and the model on the same viewing rectangle. How do they compare?

(b) Use the model to estimate the distance traveled during impact if the passenger deceleration must not exceed 30 g's.

(c) Do you think it is practical to lower the number of g's experienced during impact to less than 23? Explain your reasoning.

Review **Solve Exercises 95–98 as a review of the skills and problem-solving techniques you learned in previous sections. Simplify the expression.**

95. $\sqrt{48x^2y^5}$

96. $\sqrt{32} - 2\sqrt{25}$

97. $\sqrt[3]{25} \cdot \sqrt[3]{15}$

98. $\dfrac{3}{\sqrt{10} - 2}$

5.5 Exponential and Logarithmic Models

Introduction ▫ Exponential Growth and Decay ▫ Gaussian Models ▫ Logistics Growth Models ▫ Logarithmic Models

See Exercise 71 on page 409 for an example of comparing an exponential growth model, a linear model, a logarithmic model, and a quadratic model for the same set of data.

Introduction

The five most common types of mathematical models involving exponential functions and logarithmic functions are as follows.

1. Exponential growth model: $y = ae^{bx}, \quad b > 0$
2. Exponential decay model: $y = ae^{-bx}, \quad b > 0$
3. Gaussian model: $y = ae^{-(x-b)^2/c}$
4. Logistics growth model: $y = \dfrac{a}{1 + be^{-(x-c)/d}}$
5. Logarithmic models: $y = a + b \ln x, \quad y = a + b \log_{10} x$

The graphs of the basic forms of these functions are shown in Figure 5.20.

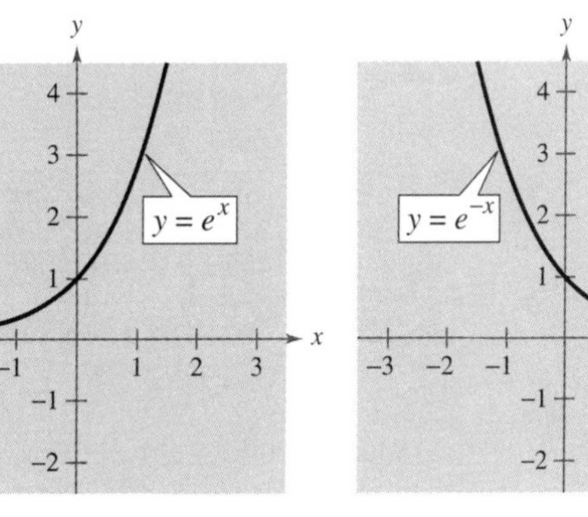

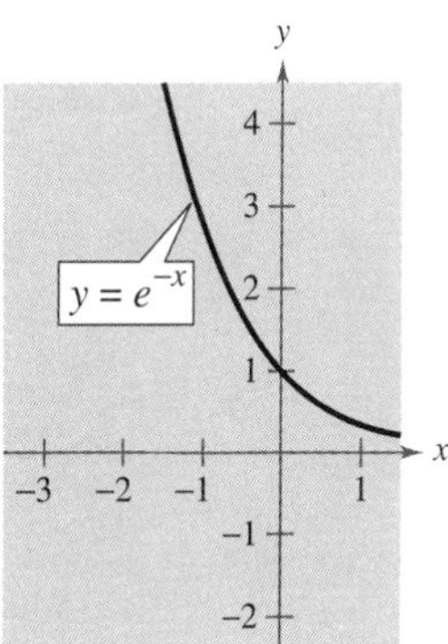

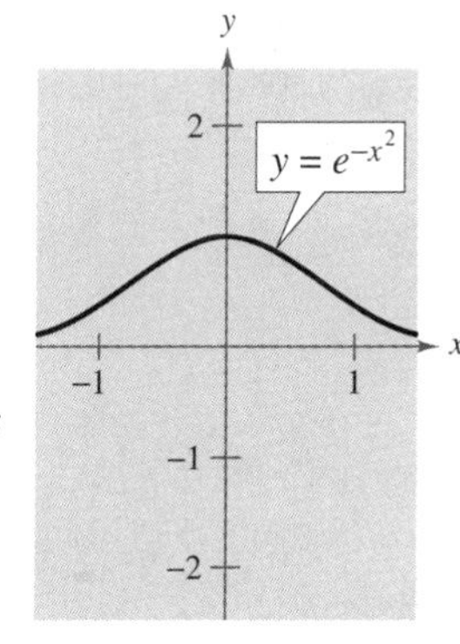

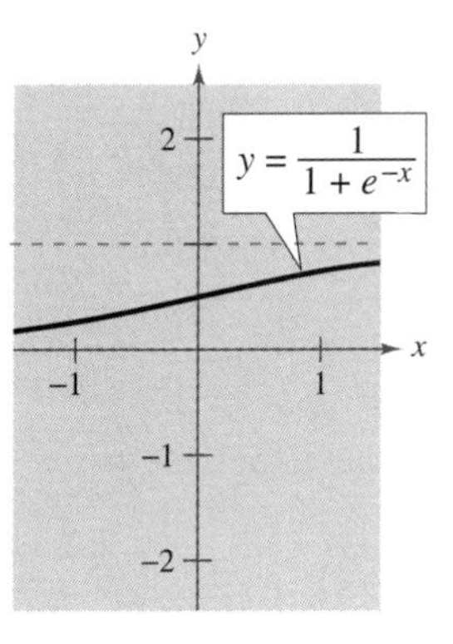

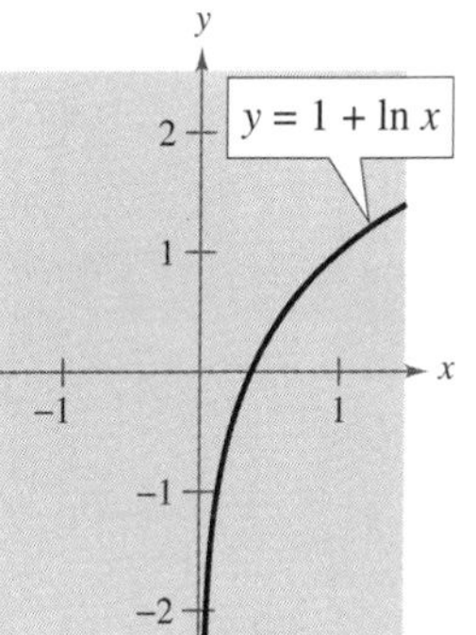

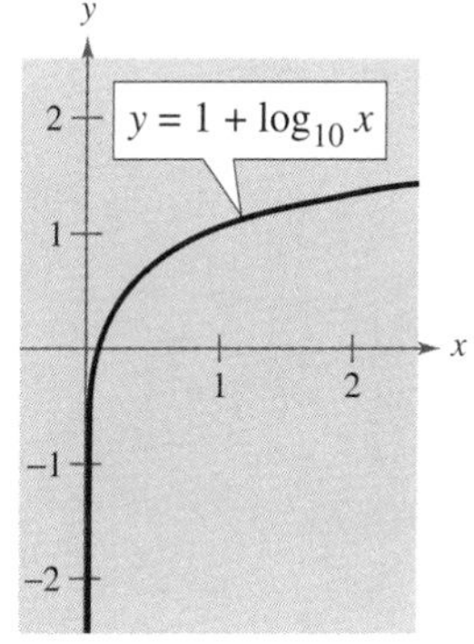

FIGURE 5.20

Study Tip

You can often gain quite a bit of insight into a situation modeled by an exponential or logarithmic function by identifying and interpreting the function's asymptotes. Use the graphs in Figure 5.20 to identify the asymptotes of each function.

Exponential Growth and Decay

EXAMPLE 1 *Population Increase*

Estimates of the world population (in millions) from 1980 through 1992 are shown in the table. The scatter plot of the data is shown in Figure 5.21(a). (Source: Statistical Office of the United Nations)

Year	1980	1985	1986	1987	1988	1989	1990	1991	1992
Population	4453	4850	4936	5024	5112	5202	5294	5384	5478

(a)

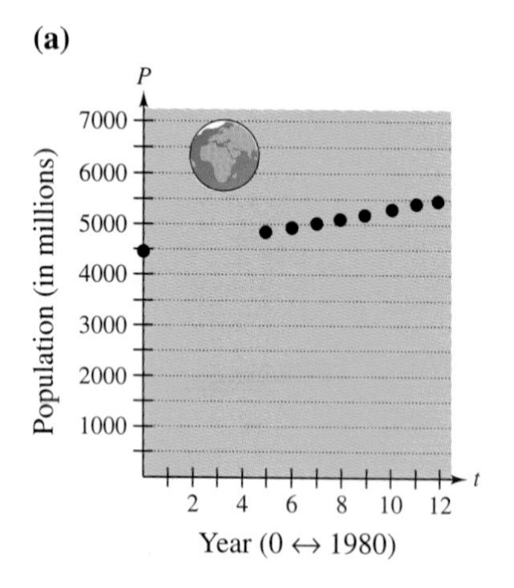

(b)

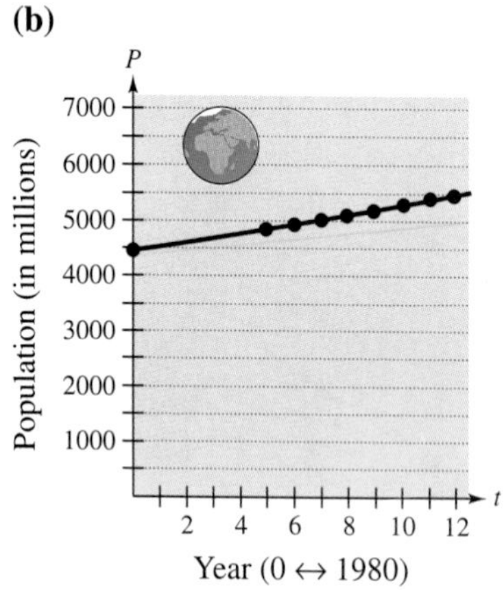

FIGURE 5.21

An exponential growth model that approximates this data is given by

$$P = 4451e^{0.017303t}, \quad 0 \leq t \leq 12$$

where P is the population (in millions) and $t = 0$ represents 1980. Compare the values given by the model with the estimates given by the United Nations. According to this model, when will the world population reach 6 billion?

Solution

The following table compares the two sets of population figures. The graph of the model is shown in Figure 5.21(b).

Year	1980	1985	1986	1987	1988	1989	1990	1991	1992
Population	4453	4850	4936	5024	5112	5202	5294	5384	5478
Model	4451	4853	4938	5024	5112	5201	5292	5384	5478

To find when the world population will reach 6 billion, let $P = 6000$ in the model and solve for t.

$$4451e^{0.017303t} = P \qquad \text{Given model}$$

$$4451e^{0.017303t} = 6000 \qquad \text{Let } P = 6000.$$

$$e^{0.017303t} \approx 1.348 \qquad \text{Divide both sides by 4451.}$$

$$\ln e^{0.017303t} \approx \ln 1.348 \qquad \text{Take logarithms of both sides.}$$

$$0.017303t \approx 0.2986 \qquad \text{Inverse property of logs and exponents}$$

$$t \approx 17.26 \qquad \text{Divide both sides by 0.017303.}$$

According to the model, the world population will reach 6 billion in 1997.

NOTE An exponential model increases (or decreases) by the same percent each year. What is the annual percent increase for the model above?

> **TECHNOLOGY**
>
> Some graphing utilities have curve-fitting capabilities that can be used to find models that represent data. For instance, to find the exponential growth model that approximates the data in the table in Example 1, you can perform the following steps using a *TI-83* or a *TI-82*.
>
> 1. Press STAT ENTER.
> 2. Enter the data in L_1 and L_2.
> 3. Press STAT ▷.
> 4. Select ExpReg and press ENTER ENTER.
>
> Write the exponential growth model you obtain. How does it compare with the model in Example 1?

In Example 1, you were given the exponential growth model. But suppose this model were not given; how could you find such a model? One technique for doing this is demonstrated in Example 2.

EXAMPLE 2 *Finding an Exponential Growth Model*

Find an exponential growth model whose graph passes through the points (0, 4453) and (7, 5024), as shown in Figure 5.22(a).

Solution

The general form of the model is

$$y = ae^{bx}.$$

From the fact that the graph passes through the point (0, 4453), you know that $y = 4453$ when $x = 0$. By substituting these values into the general form of the model, you have

$$4453 = ae^0 \quad \Rightarrow \quad a = 4453.$$

In a similar way, from the fact that the graph passes through the point (7, 5024), you know that $y = 5024$ when $x = 7$. By substituting these values into the model, you have

$$5024 = 4453e^{7b} \quad \Rightarrow \quad b = \frac{1}{7}\ln\frac{5024}{4453} \approx 0.01724.$$

Thus, the exponential growth model is

$$y = 4453e^{0.01724x}.$$

The graph of the model is shown in Figure 5.22(b).

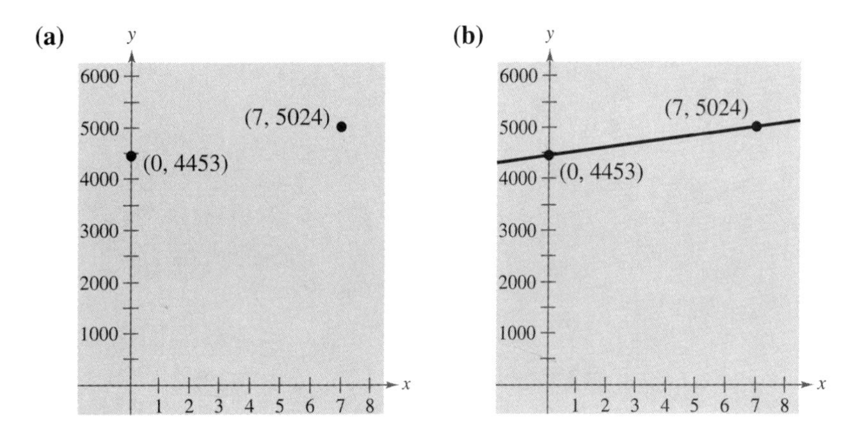

FIGURE 5.22

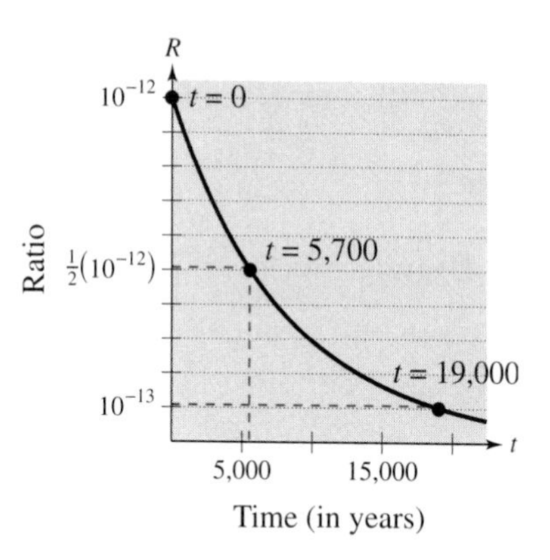

FIGURE 5.23

In living organic material, the ratio of the number of radioactive carbon isotopes (carbon 14) to the number of nonradioactive carbon isotopes (carbon 12) is about 1 to 10^{12}. When organic material dies, its carbon 12 content remains fixed, whereas its radioactive carbon 14 begins to decay with a half-life of about 5700 years. To estimate the age of dead organic material, scientists use the following formula, which denotes the ratio of carbon 14 to carbon 12 present at any time t (in years).

$$R = \frac{1}{10^{12}} e^{-t/8223}$$

The graph of R is shown in Figure 5.23. Note that R decreases as t increases.

EXAMPLE 3 *Carbon Dating*

The ratio of carbon 14 to carbon 12 in a newly discovered fossil is

$$R = \frac{1}{10^{13}}.$$

Estimate the age of the fossil.

Solution

In the carbon dating model, substitute the given value of R to obtain the following.

$$\frac{1}{10^{12}} e^{-t/8223} = R \qquad \text{Given model}$$

$$\frac{e^{-t/8223}}{10^{12}} = \frac{1}{10^{13}} \qquad \text{Let } R = \frac{1}{10^{13}}.$$

$$e^{-t/8223} = \frac{1}{10} \qquad \text{Multiply both sides by } 10^{12}.$$

$$\ln e^{-t/8223} = \ln \frac{1}{10} \qquad \text{Take logarithms of both sides.}$$

$$-\frac{t}{8223} \approx -2.3026 \qquad \text{Inverse property of logs and exponents}$$

$$t \approx 18{,}934 \qquad \text{Multiply both sides by } -8223.$$

Thus, to the nearest thousand years, you can estimate the age of the fossil to be 19,000 years.

NOTE The carbon dating model in Example 3 assumed that the carbon 14/carbon 12 ratio was one part in 10,000,000,000,000. Suppose an error in measurement occurred and the actual ratio was only one part in 8,000,000,000,000. The fossil age corresponding to the actual ratio would then be approximately 17,000 years. Try checking this result.

Gaussian Models

As mentioned at the beginning of this section, Gaussian models are of the form

$$y = ae^{-(x-b)^2/c}.$$

This type of model is commonly used in probability and statistics to represent populations that are **normally distributed.** For *standard* normal distributions, the model takes the form

$$y = \frac{1}{\sigma\sqrt{2\pi}}e^{-x^2/2\sigma^2}$$

where σ is the standard deviation (σ is the lowercase Greek letter sigma). The graph of a Gaussian model is called a **bell-shaped curve.** Try assigning a value to σ and sketching a normal distribution curve with a graphing utility. Can you see why it is called a bell-shaped curve?

The highest possible score for the Scholastic Aptitude Test (SAT) is 1600. The largest number of students ever attaining perfect scores in one year is 13, which has occurred twice, in 1987 and in 1992. *(Photo: Chuck Savage/The Stock Market)*

EXAMPLE 4 *SAT Scores*

Real Life

In 1993, the Scholastic Aptitude Test (SAT) scores for males roughly followed a normal distribution given by

$$y = 0.0026e^{-(x-500)^2/48,000}, \quad 200 \le x \le 800$$

where x is the SAT score for mathematics. Sketch the graph of this function. From the graph, estimate the average SAT score. (Source: College Board)

Solution

The graph of the function is given in Figure 5.24. From the graph, you can see that the average mathematics score for males in 1993 was 500.

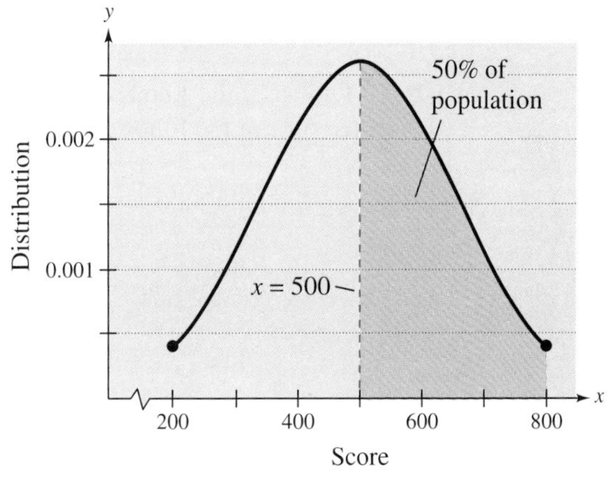

FIGURE 5.24

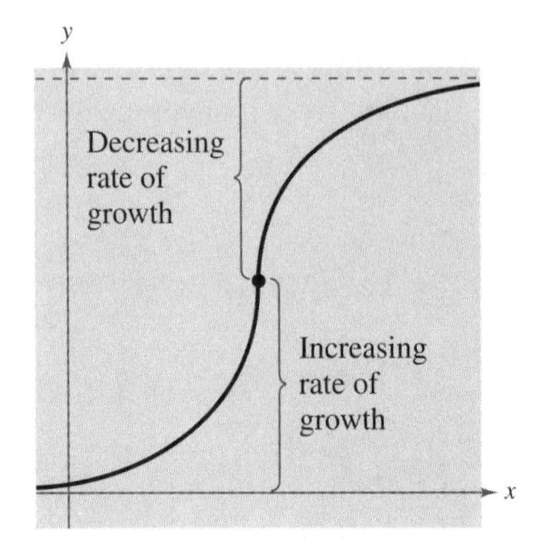

FIGURE 5.25

Logistics Growth Models

Some populations initially have rapid growth, followed by a declining rate of growth, as indicated by the graph in Figure 5.25. One model for describing this type of growth pattern is the **logistics curve** given by the function

$$y = \frac{a}{1 + be^{-(x-c)/d}}$$

where y is the population size and x is the time. An example is a bacteria culture that is initially allowed to grow under ideal conditions, and then under less favorable conditions that inhibit growth. A logistics growth curve is also called a **sigmoidal curve.**

EXAMPLE 5 *Spread of a Virus*

On a college campus of 5000 students, one student returns from vacation with a contagious flu virus. The spread of the virus is modeled by

$$y = \frac{5000}{1 + 4999e^{-0.8t}}, \quad 0 \le t$$

where y is the total number infected after t days. The college will cancel classes when 40% or more of the students are ill.

a. How many are infected after 5 days?

b. After how many days will the college cancel classes?

Solution

a. After 5 days, the number infected is

$$y = \frac{5000}{1 + 4999e^{-0.8(5)}} = \frac{5000}{1 + 4999e^{-4}} \approx 54.$$

b. In this case, the number infected is $(0.40)(5000) = 2000$. Therefore, you solve for t in the following equation.

$$2000 = \frac{5000}{1 + 4999e^{-0.8t}}$$

$$1 + 4999e^{-0.8t} = 2.5$$

$$e^{-0.8t} \approx 0.0003$$

$$\ln e^{-0.8t} \approx \ln 0.0003$$

$$-0.8t \approx -8.1115$$

$$t \approx 10.1$$

Hence, after 10 days, at least 40% of the students will be infected, and the college will cancel classes. The graph of the function is shown in Figure 5.26.

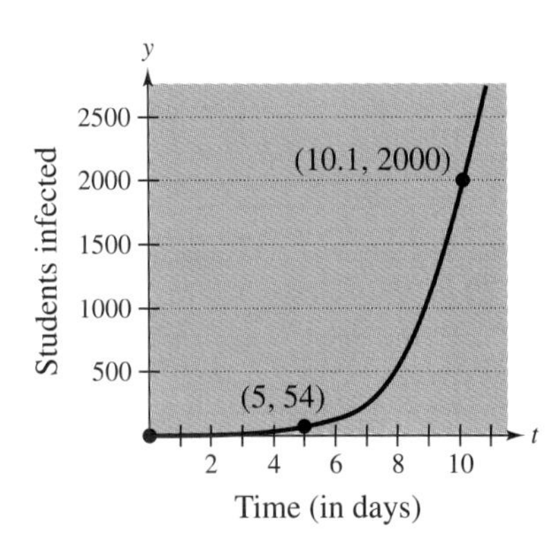

FIGURE 5.26

Twenty seconds of a 7.2 magnitude earthquake in Kobe, Japan, on January 17, 1995, left damage approaching $60 billion. *(Photo: Ben Simmons/The Stock Market)*

Logarithmic Models

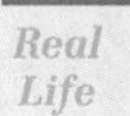

EXAMPLE 6 *Magnitude of Earthquakes*

On the Richter scale, the magnitude R of an earthquake of intensity I is given by

$$R = \log_{10} \frac{I}{I_0}$$

where $I_0 = 1$ is the minimum intensity used for comparison. Find the intensities per unit of area for the following earthquakes. (Intensity is a measure of the wave energy of an earthquake.)

a. Tokyo and Yokohama, Japan in 1923, $R = 8.3$.

b. Kobe, Japan in 1995, $R = 7.2$.

Solution

a. Because $I_0 = 1$ and $R = 8.3$, you have

$$8.3 = \log_{10} I$$
$$I = 10^{8.3} \approx 199{,}526{,}000.$$

b. For $R = 7.2$, you have $7.2 = \log_{10} I$, and

$$I = 10^{7.2} \approx 15{,}849{,}000.$$

Note that an increase of 1.1 units on the Richter scale (from 7.2 to 8.3) represents an increase in intensity by a factor of

$$\frac{199{,}526{,}000}{15{,}849{,}000} \approx 13.$$

In other words, the earthquake in 1923 had a magnitude about 13 times greater than that of the 1995 quake.

GROUP ACTIVITY

COMPARING POPULATION MODELS

The population (in millions) of the United States from 1810 to 1990 is given in the table. (Source: U.S. Bureau of Census) Least squares regression analysis gives the best quadratic model for this data as $P = 0.657t^2 + 0.305t + 6.118$ and the best exponential model for this data as $P = 8.325e^{0.195t}$. Which model better fits the data? Describe the method you used to reach your conclusion.

t	Year	Population
1	1810	7.23
3	1830	12.87
5	1850	23.19
7	1870	39.82
9	1890	62.95
11	1910	91.97
13	1930	122.78
15	1950	151.33
17	1970	203.30
19	1990	250.00

WARM UP

Sketch the graph of the equation.

1. $y = 2^{0.25x}$
2. $y = 2^{-0.25x}$
3. $y = 4 \log_2 x$
4. $y = \ln(x - 3)$
5. $y = e^{-x^2/5}$
6. $y = \dfrac{2}{1 + e^{-x}}$

Solve the equation for x. Round the result to three decimal places.

7. $3e^{2x} = 7$
8. $2e^{-0.2x} = 0.002$
9. $4 \ln 5x = 14$
10. $6 \ln 2x = 12$

5.5 Exercises

In Exercises 1–6, match the function with its graph. [The graphs are labeled (a) through (f).]

1. $y = 2e^{x/4}$
2. $y = 6e^{-x/4}$
3. $y = \frac{1}{16}(x^2 + 8x + 32)$
4. $y = \dfrac{12}{x + 4}$
5. $y = \ln(x + 1)$
6. $y = \sqrt{x}$

(a)

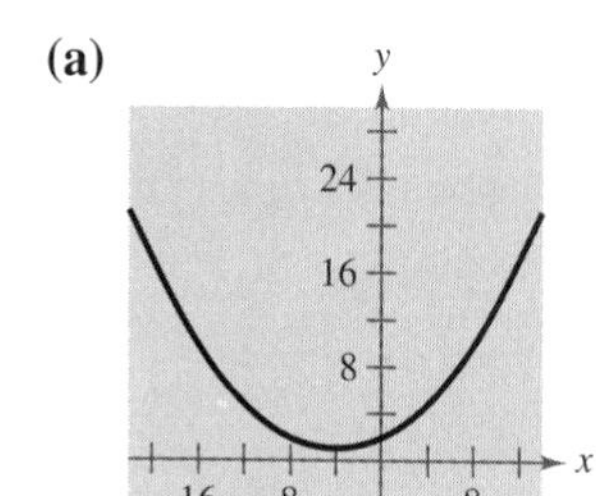

(b)

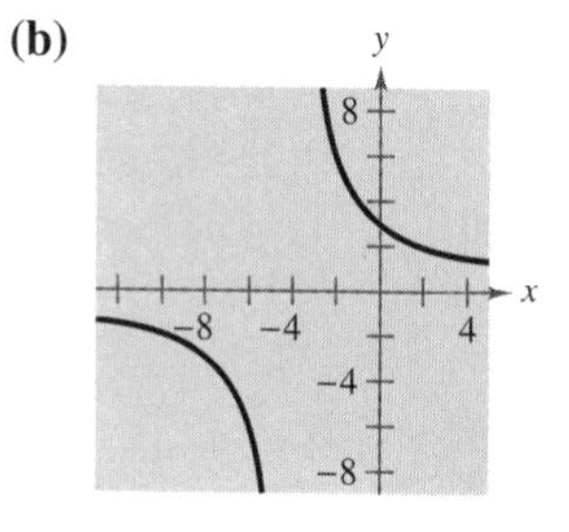

(c)

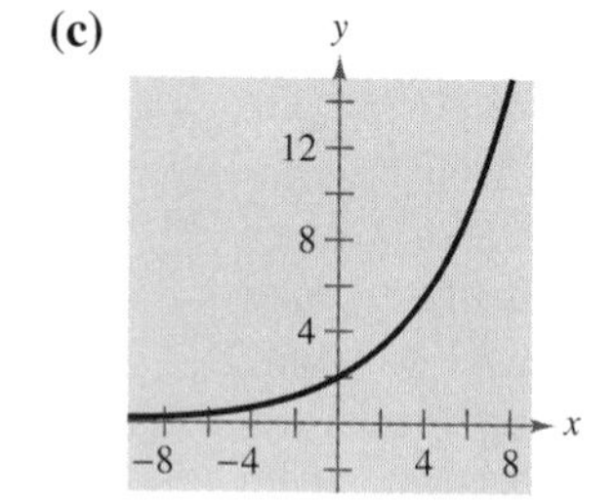

(d)

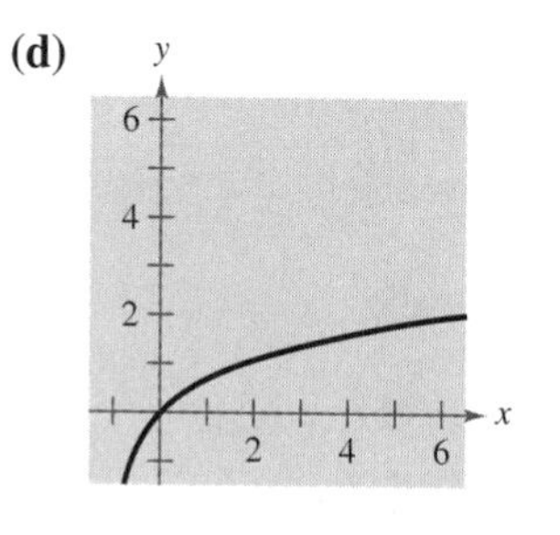

(e)

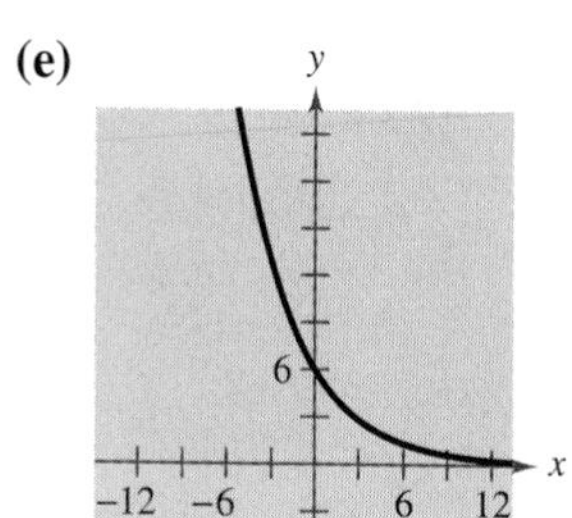

(f)

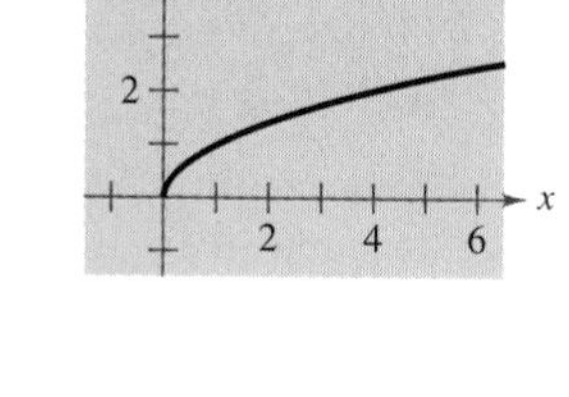

Compound Interest **In Exercises 7–14, complete the table for a savings account in which interest is compounded continuously.**

	Initial Investment	*Annual % Rate*	*Time to Double*	*Amount After 10 Years*
7.	\$1000	12%		
8.	\$20,000	$10\frac{1}{2}$%		
9.	\$750		$7\frac{3}{4}$ yr	
10.	\$10,000		5 yr	
11.	\$500			\$1292.85
12.	\$600			\$19,205.00
13.		4.5%		\$10,000.00
14.		8%		\$20,000.00

Compound Interest **In Exercises 15 and 16, determine the principal P that must be invested at rate r, compounded monthly, so that \$500,000 will be available for retirement in t years.**

15. $r = 7\frac{1}{2}\%$, $t = 20$ **16.** $r = 12\%$, $t = 40$

Compound Interest **In Exercises 17 and 18, determine the time necessary for \$1000 to double if it is invested at interest rate r compounded (a) annually, (b) monthly, (c) daily, and (d) continuously.**

17. $r = 11\%$ **18.** $r = 10\frac{1}{2}\%$

19. ***Compound Interest*** Complete the table for the time t necessary for P dollars to triple if interest is compounded continuously at rate r.

r	2%	4%	6%	8%	10%	12%
t						

20. ***Modeling Data*** Draw a scatter plot of the data in Exercise 19. Use the curve-fitting capabilities of a graphing utility to find a model for the data.

21. ***Compound Interest*** Complete the table for the time t necessary for P dollars to triple if interest is compounded annually at rate r.

r	2%	4%	6%	8%	10%	12%
t						

22. ***Modeling Data*** Draw a scatter plot of the data in Exercise 21. Use the curve-fitting capabilities of a graphing utility to find a model for the data.

23. ***Comparing Investments*** If \$1 is invested in an account over a 10-year period, the amount in the account, where t represents the time in years, is

$$A = 1 + 0.075[\![t]\!] \quad \text{or} \quad A = e^{0.07t}$$

depending on whether the account pays simple interest at $7\frac{1}{2}\%$ or continuous compound interest at 7%. Graph each function on the same set of axes. Which grows at the faster rate?

24. ***Comparing Investments*** If \$1 is invested in an account over a 10-year period, the amount in the account, where t represents the time in years, is

$$A = 1 + 0.06[\![t]\!] \quad \text{or} \quad A = \left(1 + \frac{0.055}{365}\right)^{[\![365t]\!]}$$

depending on whether the account pays simple interest at 6% or compound interest at $5\frac{1}{2}\%$ compounded daily. Use a graphing utility to graph each function on the same viewing rectangle. Which grows at the faster rate?

In Exercises 25–30, complete the table for the given radioactive isotope.

	Isotope	*Half-life (years)*	*Initial Quantity*	*Amount After 1000 Years*
25.	^{226}Ra	1620	10g	
26.	^{226}Ra	1620		1.5g
27.	^{14}C	5730		2g
28.	^{14}C	5730	3g	
29.	^{239}Pu	24,360		2.1g
30.	^{239}Pu	24,360		0.4g

In Exercises 31–34, find the exponential model $y = ae^{bx}$ that fits the points given in the graph or table.

31.

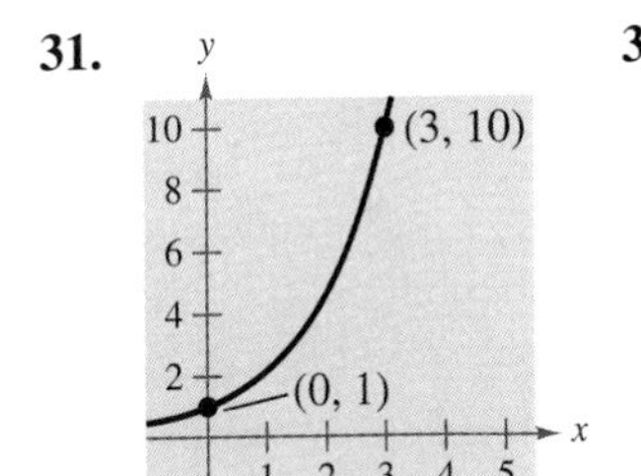

32.

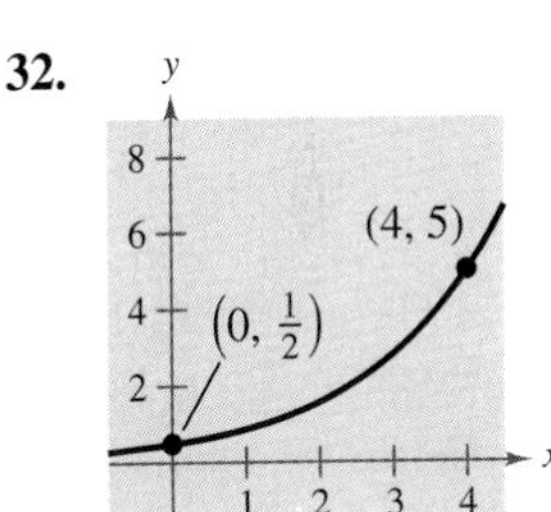

33.

x	0	3
y	1	$\frac{1}{4}$

34.

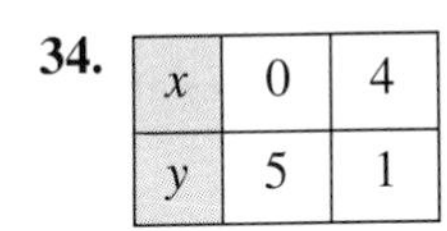

x	0	4
y	5	1

35. ***Population*** The population P of a city is given by

$P = 105{,}300e^{0.015t}$

where $t = 0$ represents 1990. According to this model, when will the population reach 150,000?

36. ***Population*** The population P of a city is given by

$P = 240{,}360e^{0.012t}$

where $t = 0$ represents 1990. According to this model, when will the population reach 275,000?

37. ***Population*** The population P of a city is given by

$P = 2500e^{kt}$

where $t = 0$ represents 1990. In 1945, the population was 1350. Find the value of k, and use this result to predict the population in the year 2010.

38. ***Population*** The population P of a city is given by

$P = 140{,}500e^{kt}$

where $t = 0$ represents 1990. In 1960, the population was 100,250. Find the value of k, and use this result to predict the population in the year 2000.

Population **In Exercises 39–42, the table gives the population (in millions) of a city in 1990 and the projected population (in millions) for the year 2000. Find the exponential growth model $y = ae^{bt}$ for the population by letting $t = 0$ correspond to 1990. Use the model to predict the population of the city in 2010.** (Source: U.S. Bureau of the Census, International Database)

City	*1990*	*2000*
39. Dhaka, Bangladesh	4.22	6.49
40. Houston, Texas	2.30	2.65
41. Detroit, Michigan	3.00	2.74
42. London, United Kingdom	9.17	8.57

43. ***Think About It*** In Exercises 39 and 40, you can see that the populations of Dhaka and Houston are growing at different rates. What constant in the equation $y = ae^{bt}$ is determined by these different growth rates? Discuss the relationship between the different growth rates and the magnitude of the constant.

44. ***Think About It*** In Exercises 39 and 41, you can see that one population is increasing whereas the other is decreasing. What constant in the equation $y = ae^{bt}$ reflects this difference? Explain.

45. ***Bacteria Growth*** The number of bacteria N in a culture is given by the model

$N = 100e^{kt}$

where t is the time in hours. If $N = 300$ when $t = 5$, estimate the time required for the population to double in size.

46. ***Bacteria Growth*** The number of bacteria N in a culture is given by the model

$N = 250e^{kt}$

where t is the time in hours. If $N = 280$ when $t = 10$, estimate the time required for the population to double in size.

47. ***Radioactive Decay*** The half-life of radioactive radium (^{226}Ra) is 1620 years. What percent of a present amount of radioactive radium will remain after 100 years?

48. ***Radioactive Decay*** Carbon 14 dating assumes that the carbon dioxide on earth today has the same radioactive content as it did centuries ago. If this is true, the amount of ^{14}C absorbed by a tree that grew several centuries ago should be the same as the amount of ^{14}C absorbed by a tree growing today. A piece of ancient charcoal contains only 15% as much radioactive carbon as a piece of modern charcoal. How long ago was the tree burned to make the ancient charcoal if the half-life of ^{14}C is 5730 years?

49. ***Comparing Depreciation Models*** A car that cost \$22,000 new has a book value of \$13,000 after 2 years.

(a) Find the straight-line model $V = mt + b$.

(b) Find the exponential model $V = ae^{kt}$.

(c) Use a graphing utility to graph the two models on the same viewing rectangle. Which model depreciates faster in the first 2 years?

(d) Find the book values of the car after 1 year and after 3 years using each model.

(e) Interpret the slope of the straight-line model.

50. ***Depreciation*** A computer that cost \$4600 new has a book value of \$3000 after 2 years. Find the value of the computer after 3 years by using the exponential model $y = ae^{bt}$.

51. ***Sales*** The sales S (in thousands of units) of a new product after it has been on the market t years are given by

$$S(t) = 100(1 - e^{kt}).$$

Fifteen thousand units of the new product were sold the first year.

(a) Complete the model by solving for k.

(b) Sketch the graph of the model.

(c) Use the model to estimate the number of units sold after 5 years.

52. ***Sales and Advertising*** After discontinuing all advertising for a certain product in 1994, the manufacturer noted that sales began to drop according to the model

$$S = \frac{500{,}000}{1 + 0.6e^{kt}}$$

where S represents the number of units sold and $t = 0$ represents 1994. In 1996, the company sold 300,000 units.

(a) Complete the model by solving for k.

(b) Estimate sales in 1999.

53. ***Sales and Advertising*** The sales S (in thousands of units) of a product after x hundred dollars is spent on advertising is given by

$$S = 10(1 - e^{kx}).$$

When \$500 is spent on advertising, 2500 units are sold.

(a) Complete the model by solving for k.

(b) Estimate the number of units that will be sold if advertising expenditures are raised to \$700.

54. ***Profits*** Because of a slump in the economy, a company finds that its annual profits have dropped from \$742,000 in 1994 to \$632,000 in 1996. If the profit follows an exponential pattern of decline, what is the expected profit for 1997? (Let $t = 0$ represent 1994.)

55. ***Learning Curve*** The management at a factory has found that the maximum number of units a worker can produce in a day is 30. The learning curve for the number of units N produced per day after a new employee has worked t days is given by

$$N = 30(1 - e^{kt}).$$

After 20 days on the job, a new employee produces 19 units.

(a) Find the learning curve for this employee (first, find the value of k).

(b) How many days should pass before this employee is producing 25 units per day?

(c) Is the employee's production increasing at a linear rate? Explain your reasoning.

56. ***Endangered Species*** A conservation organization releases 100 animals of an endangered species into a game preserve. The organization believes that the preserve has a carrying capacity of 1000 animals and that the growth of the herd will be modeled by the logistics curve

$$p(t) = \frac{1000}{1 + 9e^{-0.1656t}}$$

where t is measured in months (see figure).

(a) Use a graphing utility to graph the function. Use the graph to determine the horizontal asymptotes, and interpret the meaning of the larger p-value in the context of the problem.

(b) Estimate the population after 5 months.

(c) After how many months will the population be 500?

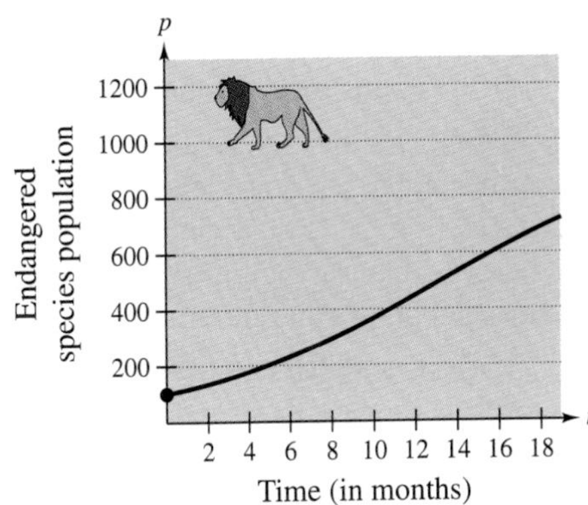

Earthquake Magnitudes **In Exercises 57 and 58, use the Richter scale (see page 403) for measuring the magnitudes of earthquakes.**

57. Find the magnitude R of an earthquake of intensity I (let $I_0 = 1$).

(a) $I = 80{,}500{,}000$

(b) $I = 48{,}275{,}000$

58. Find the intensity I of an earthquake measuring R on the Richter scale (let $I_0 = 1$).

(a) Colombia in 1906, $R = 8.6$

(b) Los Angeles in 1971, $R = 6.7$

Intensity of Sound **In Exercises 59–62, use the following information for determining sound intensity. The level of sound β, in decibels, with an intensity of I is given by**

$$\beta(I) = 10 \log_{10} \frac{I}{I_0}$$

where I_0 is an intensity of 10^{-16} watts per square centimeter, corresponding roughly to the faintest sound that can be heard by the human ear.

59. (a) $I = 10^{-14}$ watts per cm² (faint whisper)

(b) $I = 10^{-9}$ watts per cm² (busy street corner)

(c) $I = 10^{-6.5}$ watts per cm² (air hammer)

(d) $I = 10^{-4}$ watts per cm² (threshold of pain)

60. (a) $I = 10^{-13}$ watts per cm² (whisper)

(b) $I = 10^{-7.5}$ watts per cm² (jet 4 miles from takeoff)

(c) $I = 10^{-7}$ watts per cm² (diesel truck at 25 feet)

(d) $I = 10^{-4.5}$ watts per cm² (auto horn at 3 feet)

61. ***Noise Level*** Due to the installation of noise suppression materials, the noise level in an auditorium was reduced from 93 to 80 decibels. Find the percent decrease in the intensity level of the noise as a result of the installation of these materials.

62. ***Noise Level*** Due to the installation of a muffler, the noise level in an engine was reduced from 88 to 72 decibels. Find the percent decrease in the intensity level of the noise as a result of the installation of the muffler.

Acidity **In Exercises 63–68, use the acidity model given by $\text{pH} = -\log_{10}[H^+]$, where acidity (pH) is a measure of the hydrogen ion concentration $[H^+]$ (measured in moles of hydrogen per liter) of a solution.**

63. Find the pH if $[H^+] = 2.3 \times 10^{-5}$.

64. Find the pH if $[H^+] = 11.3 \times 10^{-6}$.

65. Compute $[H^+]$ for a solution in which $\text{pH} = 5.8$.

66. Compute $[H^+]$ for a solution in which $\text{pH} = 3.2$.

67. A certain fruit has a pH of 2.5 and an antacid tablet has a pH of 9.5. The hydrogen ion concentration of the fruit is how many times the concentration of the tablet?

68. If the pH of a solution is decreased by one unit, the hydrogen ion concentration is increased by what factor?

69. ***Home Mortgage*** A \$120,000 home mortgage for 35 years at $9\frac{1}{2}\%$ has a monthly payment of \$985.93. Part of the monthly payment goes for the interest charge on the unpaid balance, and the remainder of the payment is used to reduce the principal. The amount that goes for interest is given by

$$u = M - \left(M - \frac{Pr}{12}\right)\left(1 + \frac{r}{12}\right)^{12t}$$

and the amount that goes toward reduction of the principal is given by

$$v = \left(M - \frac{Pr}{12}\right)\left(1 + \frac{r}{12}\right)^{12t}.$$

In these formulas, P is the size of the mortgage, r is the interest rate, M is the monthly payment, and t is the time in years.

(a) Use a graphing utility to graph each function on the same viewing rectangle. (The viewing rectangle should show all 35 years of mortgage payments.)

(b) In the early years of the mortgage, the larger part of the monthly payment goes for what purpose? Approximate the time when the monthly payment is evenly divided between interest and principal reduction.

(c) Repeat parts (a) and (b) for a repayment period of 20 years ($M = \$1118.56$). What can you conclude?

70. ***Home Mortgage*** The total interest u paid on a home mortgage of P dollars at interest rate r for t years is given by

$$u = P\left[\frac{rt}{1 - \left(\frac{1}{1 + r/12}\right)^{12t}} - 1\right].$$

Consider a \$120,000 home mortgage at $9\frac{1}{2}\%$.

(a) Use a graphing utility to graph the total interest function.

(b) Approximate the length of the mortgage for which the total interest paid is the same as the size of the mortgage. Is it possible that some people are paying twice as much in interest charges as the size of the mortgage?

71. ***Data Analysis*** The time t (in seconds) required to attain a speed of s miles per hour from a standing start for a 1995 Dodge Avenger is given in the table. (Source: *Road & Track*, March 1995)

s	30	40	50	60	70	80	90
t	3.4	5.0	7.0	9.3	12.0	15.8	20.0

Two models for this data are

$$t_1 = 40.757 + 0.556s - 15.817 \ln s,$$
$$t_2 = 1.2259 + 0.0023s^2.$$

(a) Use a graphing utility to fit a linear model t_3 and an exponential model t_4 to the data.

(b) Use a graphing utility to graph the data points and each model.

(c) Create a table to compare the given data with estimates obtained from each model.

(d) Use the results of part (c) to find the sum of the absolute values of the differences between the data and estimated values given by each model. Based on the four sums, which model do you think better fits the data? Explain.

72. ***Essay*** Use your school's library or some other reference source to write a paper describing John Napier's work with logarithms.

73. ***Essay*** Before the development of electronic calculators and graphing utilities, some computations were done on slide rules. Use your school's library or some other reference source to write a paper describing the use of logarithmic scales on a slide rule.

74. ***Estimating the Time of Death*** At 8:30 A.M., a coroner was called to the home of a person who had died during the night. In order to estimate the time of death, the coroner took the person's temperature twice. At 9:00 A.M. the temperature was 85.7°F and at 9:30 A.M. the temperature was 82.8°F. From these two temperatures the coroner was able to determine that the time elapsed since death and the body temperature were related by the formula

$$t = -2.5 \ln \frac{T - 70}{98.6 - 70}$$

where t is the time in hours elapsed since the person died and T is the temperature (in degrees Fahrenheit) of the person's body. Assume the person had a normal body temperature of 98.6°F at death, and the room temperature was a constant 70°F. (This formula is derived from a general cooling principle called Newton's Law of Cooling.) Use the formula to estimate the time of death of the person.

Review **Solve Exercises 75–78 as a review of the skills and problem-solving techniques you learned in previous sections. Divide by synthetic division.**

75. $\dfrac{4x^3 + 4x^2 - 39x + 36}{x + 4}$

76. $\dfrac{8x^3 - 36x^2 + 54x - 27}{x - \frac{3}{2}}$

77. $(2x^3 - 8x^2 + 3x - 9) \div (x - 4)$

78. $(x^4 - 3x + 1) \div (x + 5)$

FOCUS ON CONCEPTS

In this chapter, you studied several concepts related to exponential and logarithmic functions. Answer the following questions to check your understanding of several of the basic concepts discussed in this chapter. The answers to these questions are given in the back of the book.

1. *Comparing Graphs* The graphs of $y = e^{kt}$ are shown for $k = a, b, c,$ and d. Use the graphs to order $a, b, c,$ and d. Which of the four values are negative? Which are positive?

(a)

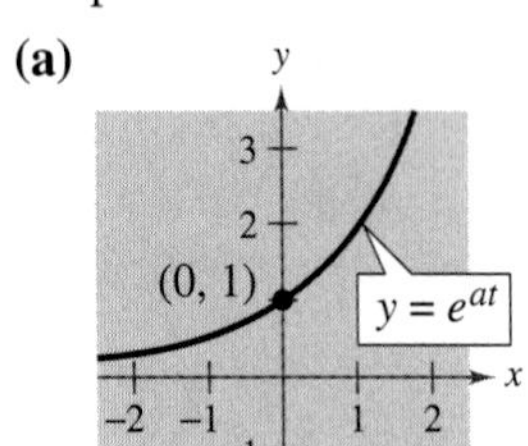

(b)

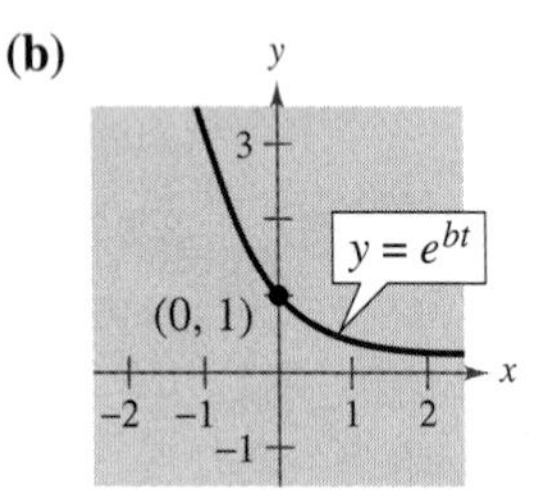

(c)

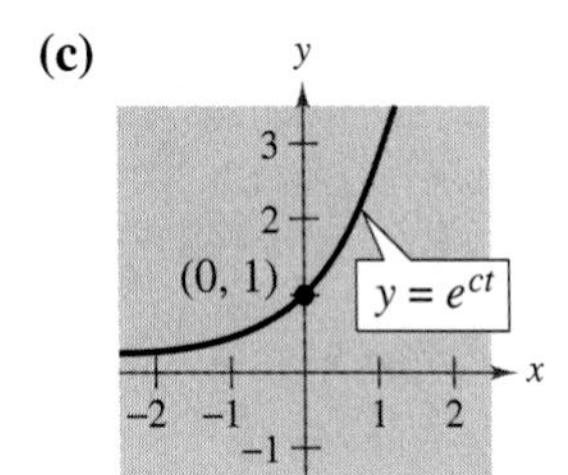

(d)

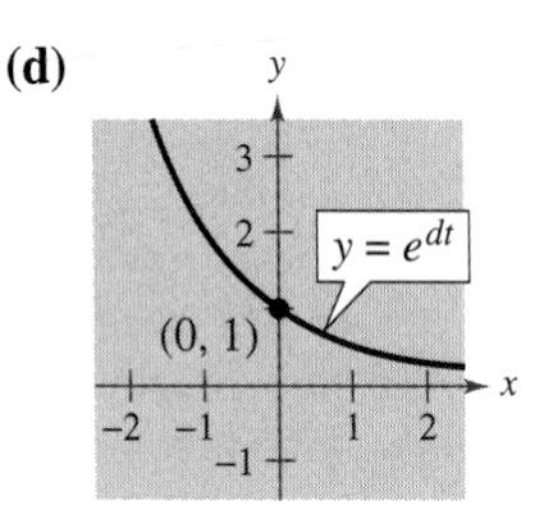

2. *True or False?* Rewrite each verbal statement as an equation. Then decide whether the statement is true or false. If it is false, give an example that shows it is false.

(a) The logarithm of the product of two numbers is equal to the sum of the logarithms of the numbers.

(b) The logarithm of the sum of two numbers is equal to the product of the logarithms of the numbers.

(c) The logarithm of the difference of two numbers is equal to the difference of the logarithms of the numbers.

(d) The logarithm of the quotient of two numbers is equal to the difference of the logarithms of the numbers.

3. *Investing Money* You are investing P dollars at an annual rate of r, compounded continuously, for t years. Which of the following would be most advantageous? Explain your reasoning.

(a) Double the amount you invest.

(b) Double your interest rate.

(c) Double the number of years.

4. Identify the model as linear, logarithmic, exponential, logistic, or none of the above. Explain your reasoning.

(a)

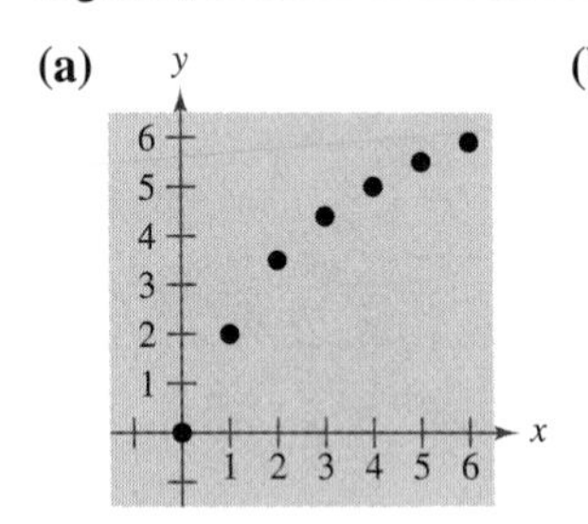

(b)

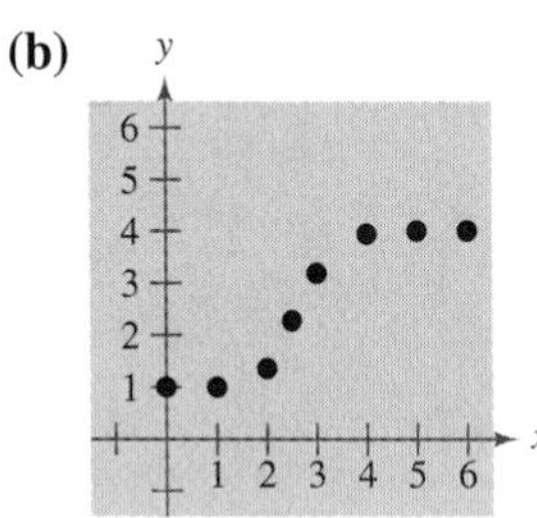

(c)

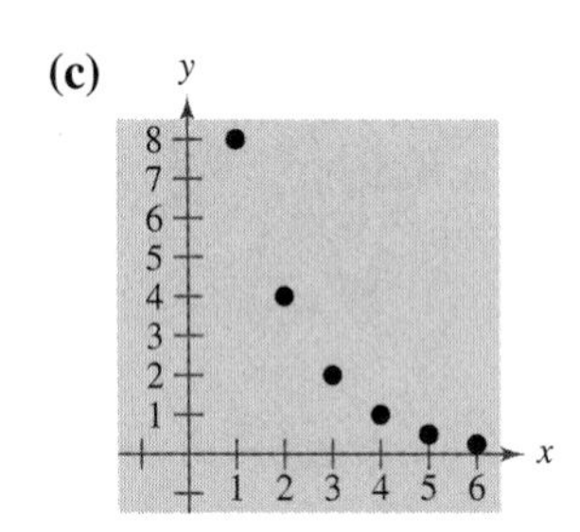

(d)

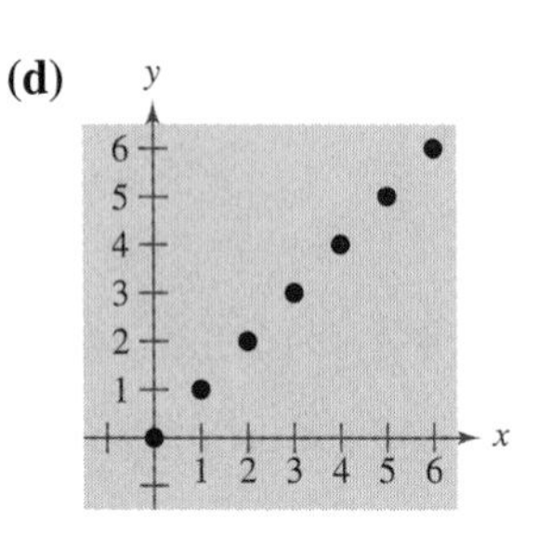

(e)

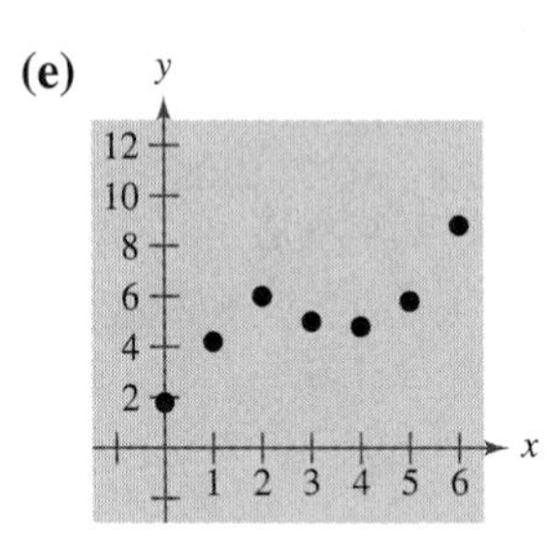

(f)

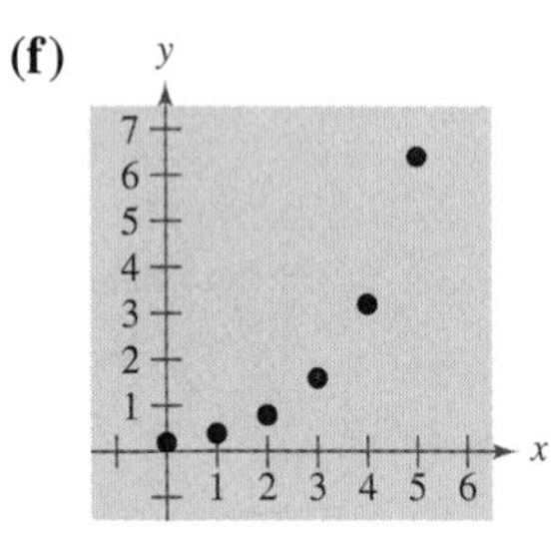

Review Exercises

In Exercises 1–6, match the function with its graph. [The graphs are labeled (a) through (f).]

1. $f(x) = 4^x$

2. $f(x) = 4^{-x}$

3. $f(x) = -4^x$

4. $f(x) = 4^x + 1$

5. $f(x) = \log_4 x$

6. $f(x) = \log_4(x - 1)$

(a)

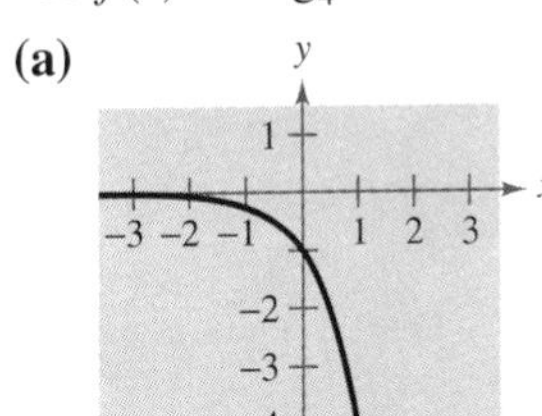

(b)

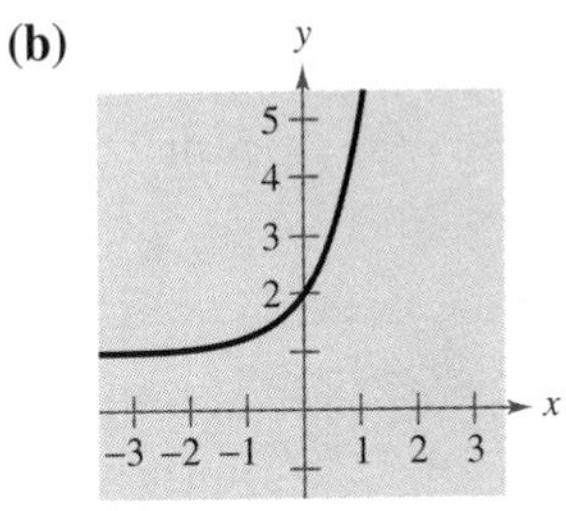

(c)

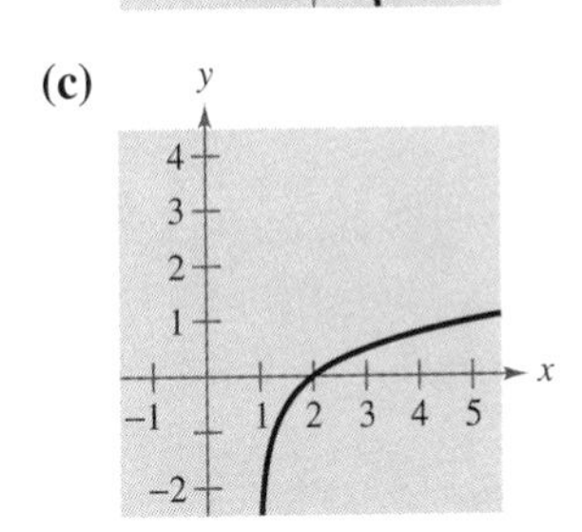

(d)

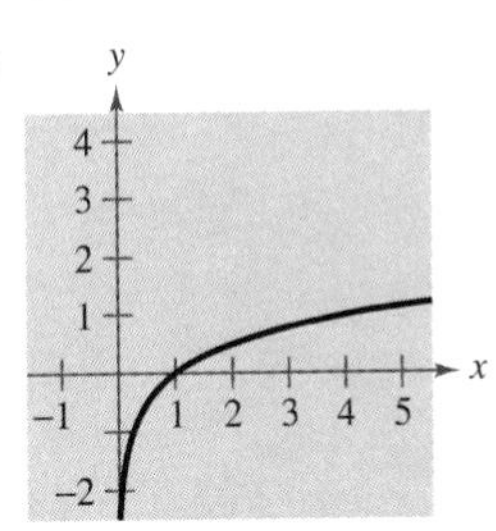

(e)

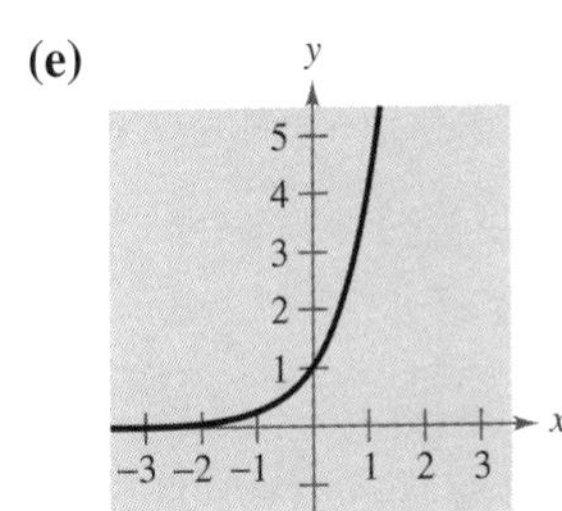

(f)

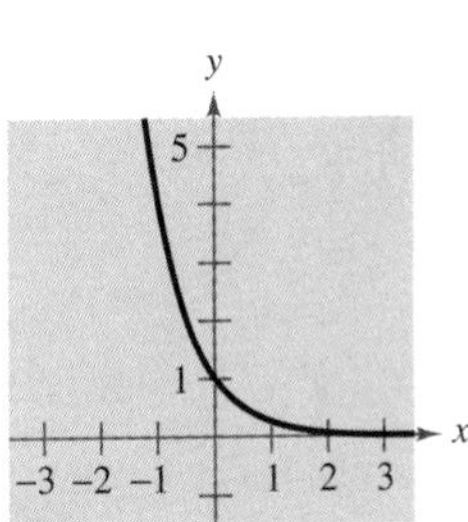

In Exercises 7–12, sketch the graph of the function.

7. $f(x) = 0.3^x$

8. $g(x) = 0.3^{-x}$

9. $h(x) = e^{-x/2}$

10. $h(x) = 2 - e^{-x/2}$

11. $f(x) = e^{x+2}$

12. $s(t) = 4e^{-2/t}, \quad t > 0$

In Exercises 13 and 14, use a graphing utility to graph the function. Identify any asymptotes.

13. $g(x) = 200e^{4/x}$

14. $f(x) = \dfrac{10}{1 + 2^{-0.05x}}$

In Exercises 15 and 16, complete the table to determine the balance A for P dollars invested at rate r for t years and compounded n times per year.

n	1	2	4	12	365	Continuous
A						

15. $P = \$3500,\ r = 10.5\%,\ t = 10$ years

16. $P = \$2000,\ r = 12\%,\ t = 30$ years

In Exercises 17 and 18, complete the table to determine the amount P that should be invested at rate r to produce a balance of \$200,000 in t years.

t	1	10	20	30	40	50
P						

17. $r = 8\%$, compounded continuously

18. $r = 10\%$, compounded monthly

19. ***Waiting Times*** The average time between incoming calls at a switchboard is 3 minutes. The probability of waiting less than t minutes until the next incoming call is approximated by the model

$$F(t) = 1 - e^{-t/3}.$$

If a call has just come in, find the probability that the next call will be within

(a) $\frac{1}{2}$ minute. (b) 2 minutes. (c) 5 minutes.

20. ***Depreciation*** After t years, the value of a car that cost \$14,000 is given by

$$V(t) = 14{,}000\left(\tfrac{3}{4}\right)^t.$$

(a) Use a graphing utility to graph the function.

(b) Find the value of the car 2 years after it was purchased.

(c) According to the model, when does the car depreciate most rapidly? Is this realistic? Explain.

21. ***Trust Fund*** On the day a person was born, a deposit of \$50,000 was made in a trust fund that pays 8.75% interest, compounded continuously.

(a) Find the balance on the person's 35th birthday.

(b) How much longer would the person have to wait to get twice as much?

22. ***Fuel Efficiency*** A certain automobile gets 28 miles per gallon of gasoline for speeds up to 50 miles per hour. Over 50 miles per hour, the number of miles per gallon drops at a rate of 12% for each additional 10 miles per hour. If s is the speed and y is the number of miles per gallon, then

$$y = 28e^{0.6-0.012s}, \qquad s \geq 50.$$

Use this model to complete the table.

s	50	55	60	65	70
y					

In Exercises 23–28, sketch the graph of the function. Identify any asymptotes.

23. $g(x) = \log_2 x$

24. $g(x) = \log_5 x$

25. $f(x) = \ln x + 3$

26. $f(x) = \ln(x - 3)$

27. $h(x) = \ln(e^{x-1})$

28. $f(x) = \frac{1}{4} \ln x$

In Exercises 29 and 30, use a graphing utility to graph the function.

29. $y = \log_{10}(x^2 + 1)$

30. $y = \sqrt{x} \ln(x + 1)$

In Exercises 31 and 32, write the exponential equation in logarithmic form.

31. $4^3 = 64$

32. $25^{3/2} = 125$

In Exercises 33–36, evaluate the expression by hand.

33. $\log_{10} 1000$

34. $\log_9 3$

35. $\ln e^7$

36. $\log_a \dfrac{1}{a}$

In Exercises 37–40, evaluate the logarithm using the change-of-base formula. Do each problem twice, once with common logarithms and once with natural logarithms. Round the result to three decimal places.

37. $\log_4 9$

38. $\log_{1/2} 5$

39. $\log_{12} 200$

40. $\log_3 0.28$

In Exercises 41–44, use the properties of logarithms to write the expression as a sum, difference, and/or multiple of logarithms.

41. $\log_5 5x^2$

42. $\log_7 \dfrac{\sqrt{x}}{4}$

43. $\log_{10} \dfrac{5\sqrt{y}}{x^2}$

44. $\ln\left|\dfrac{x-1}{x+1}\right|$

In Exercises 45–48, write the expression as the logarithm of a single quantity.

45. $\log_2 5 + \log_2 x$

46. $\log_6 y - 2 \log_6 z$

47. $\frac{1}{2} \ln|2x - 1| - 2 \ln|x + 1|$

48. $5 \ln|x - 2| - \ln|x + 2| - 3 \ln|x|$

True or False? **In Exercises 49–54, determine whether the equation or statement is true or false.**

49. $\log_b b^{2x} = 2x$

50. $e^{x-1} = \dfrac{e^x}{e}$

51. $\ln(x + y) = \ln x + \ln y$

52. $\ln(x + y) = \ln(x \cdot y)$

53. $\log\left(\dfrac{10}{x}\right) = 1 - \log x$

54. The domain of the function $f(x) = \ln x$ is the set of all real numbers.

55. ***Snow Removal*** The number of miles s of roads cleared of snow is approximated by the model

$$s = 25 - \frac{13 \ln(h/12)}{\ln 3}, \qquad 2 \leq h \leq 15$$

where h is the depth of the snow in inches. Use this model to find s when $h = 10$ inches.

56. ***Climb Rate*** The time t, in minutes, for a small plane to climb to an altitude of h feet is given by

$$t = 50 \log_{10} \frac{18{,}000}{18{,}000 - h}$$

where 18,000 feet is the plane's absolute ceiling.

(a) Determine the domain of the function appropriate for the context of the problem.

(b) Use a graphing utility to graph the time function and identify any asymptotes.

(c) As the plane approaches its absolute ceiling, what can be said about the time required to further increase its altitude?

(d) Find the time for the plane to climb to an altitude of 4000 feet.

In Exercises 57–62, solve the exponential equation. Round your result to three decimal places.

57. $e^x = 12$

58. $e^{3x} = 25$

59. $3e^{-5x} = 132$

60. $14e^{3x+2} = 560$

61. $e^{2x} - 7e^x + 10 = 0$

62. $e^{2x} - 6e^x + 8 = 0$

In Exercises 63–68, solve the logarithmic equation. Round the result to three decimal places.

63. $\ln 3x = 8.2$

64. $2 \ln 4x = 15$

65. $\ln x - \ln 3 = 2$

66. $\ln\sqrt{x + 1} = 2$

67. $\log(x - 1) = \log(x - 2) - \log(x + 2)$

68. $\log(1 - x) = -1$

In Exercises 69–72, use a graphing utility to solve the equation. Round the result to two decimal places.

69. $2^{0.6x} - 3x = 0$

70. $25e^{-0.3x} = 12$

71. $2 \ln(x + 3) + 3x = 8$

72. $6 \log_{10}(x^2 + 1) - x = 0$

In Exercises 73 and 74, find the exponential function $y = ae^{bx}$ that passes through the points.

73. $(0, 2), (4, 3)$

74. $\left(0, \frac{1}{2}\right), (5, 5)$

75. ***Demand Function*** The demand equation for a certain product is given by

$$p = 500 - 0.5e^{0.004x}.$$

Find the demand x for a price of (a) $p = \$450$ and (b) $p = \$400$.

76. ***Typing Speed*** In a typing class, the average number of words per minute typed after t weeks of lessons was found to be

$$N = \frac{157}{1 + 5.4e^{-0.12t}}.$$

Find the time necessary to type (a) 50 words per minute and (b) 75 words per minute.

77. ***Compound Interest*** A deposit of \$10,000 is made in a savings account for which the interest is compounded continuously. The balance will double in 5 years.

(a) What is the annual interest rate for this account?

(b) Find the balance after 1 year.

78. ***Sound Intensity*** The relationship between the number of decibels β and the intensity of a sound I in watts per square centimeter is given by

$$\beta = 10 \log_{10}\left(\frac{I}{10^{-16}}\right).$$

Determine the intensity of a sound in watts per square centimeter if the decibel level is 125.

79. ***Earthquake Magnitudes*** On the Richter scale, the magnitude R of an earthquake of intensity I is given by

$$R = \log_{10} \frac{I}{I_0}$$

where $I_0 = 1$ is the minimum intensity used for comparison. Find the intensity per unit of area for the following values of R.

(a) $R = 8.4$

(b) $R = 6.85$

(c) $R = 9.1$

CHAPTER PROJECT: *A Graphical Approach to Compound Interest*

A graphing utility can be used to investigate the rates of growth of different types of compound interest.

EXAMPLE 1 *Comparing Balances*

You are depositing \$1000 in a savings account. Which of the following will produce the largest balance?

a. 6% annual interest rate, compounded annually

b. 6% annual interest rate, compounded continuously

c. 6.25% annual interest rate, compounded quarterly

Solution

Option (b) is better than option (a) because, for a given interest rate, continuous compounding yields a larger balance than compounding n times per year. Distinguishing between the second and third options is not as straightforward—the higher interest rate favors option (c), but the "more frequent" compounding favors option (b). One way to compare all three options is to sketch their graphs on the same viewing rectangle.

Option (a)	*Option (b)*	*Option (c)*
$A = 1000(1 + 0.06)^t$	$A = 1000e^{0.06t}$	$A = 1000\left(1 + \frac{0.0625}{4}\right)^{4t}$

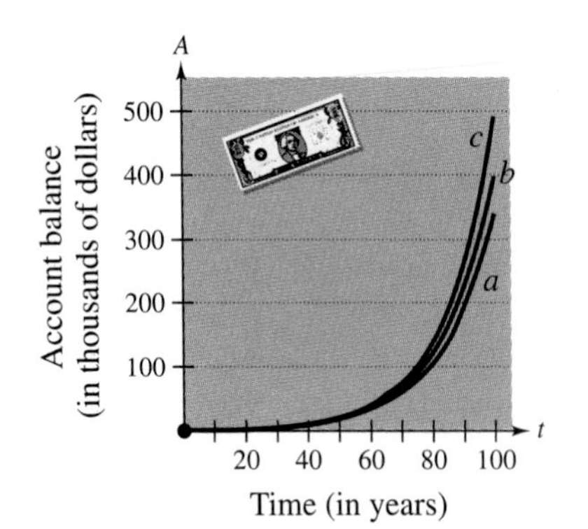

The graphs are shown at the left. On the graph, the t-values vary from 0 years through 100 years. From the graphs, you can conclude that option (c) is better than option (b), and option (b) is better than option (a). Note that for the first 50 years, there is little difference in the graphs. Between 50 and 100 years, however, the balances obtained begin to differ significantly. At the end of 100 years, the balances are (a) \$339,302, (b) \$403,429, and (c) \$493,575.

To help distinguish among different rates and types of compounding, banks use the concept of *effective yield.* The **effective yield** of a savings plan is the percent increase in the balance after *one* year. For instance, in Example 1 the one-year balances are (a) \$1060.00, (b) \$1061.84, and (c) \$1063.98.

Effective Yield (a)	*Effective Yield (b)*	*Effective Yield (c)*
6.000%	6.184%	6.398%

Because option (c) has the greatest effective yield, it is the best option and will yield the highest balance.

If you were to create a retirement plan with a regular savings account, the income tax on the interest would be due each year. With a *tax-deferred* retirement plan, the interest is allowed to build without being taxed until the account reaches maturity.

EXAMPLE 2 *To Defer or Not to Defer*

Real Life

You deposit \$25,000 in an account to accrue interest for 40 years. The account pays 8% compounded annually. Assume that the income tax on the earned interest is 30%. Which of the following plans produces a larger balance after all income tax is paid?

a. *Deferred* The income tax on the interest that is earned is paid in one lump sum at the end of 40 years.

b. *Not Deferred* The income tax on the interest that is earned each year is paid at the end of each year.

Solution

a. The untaxed balance at the end of 40 years is

$$A = 25{,}000(1 + 0.08)^{40} = \$543{,}113.04.$$

The income tax due is $0.3(518{,}113.04) = \$155{,}433.91$, so you are left with a balance of \$387,679.13.

b. You can reason that only 70% of the earned interest will remain in the account each year. The taxed balance at the end of 40 years is

$$A = 25{,}000[1 + 0.08(0.7)]^{40} = \$221{,}053.16.$$

Thus, the tax-deferred plan will produce a significantly greater balance at the end of 40 years. The balances are compared graphically at the left.

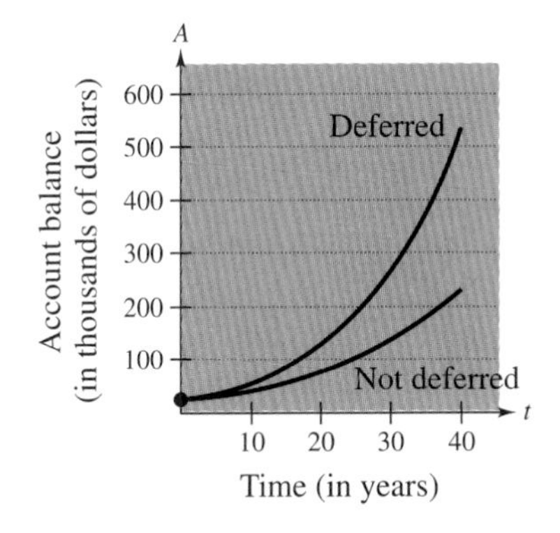

CHAPTER PROJECT INVESTIGATIONS

1. *Comparing Savings Plans* Which would produce a larger balance: an annual interest rate of 8.05% compounded monthly, or an annual interest rate of 8% compounded continuously? Explain.

2. *Exploration* You deposit \$1000 in each of two savings accounts. The interest for the accounts is paid according to the two options described in Question 1. How long would it take for the balance in one of the accounts to exceed the balance in the other account by \$100? By \$100,000?

3. *Comparing Retirement Plans* No income tax is due on the interest earned in some types of investments. You deposit \$25,000 into an account. Which of the following plans is better? Explain.

   (a) *Tax-free* The account pays 5% compounded annually. There is no income tax due on the earned interest.

   (b) *Tax-deferred* The account pays 7% compounded annually. At maturity, the earned interest is taxable at a rate of 40%.

Chapter Test

Take this test as you would take a test in class. After you are done, check your work against the answers given in the back of the book.

The *Interactive* CD-ROM provides answers to the Chapter Tests and Cumulative Tests. It also offers Chapter Pre-Tests (which test key skills and concepts covered in previous chapters) and Chapter Post-Tests, both of which have randomly generated exercises with diagnostic capabilities.

1. Sketch the graph of the function $f(x) = 2^{-x/3}$.
2. Determine the horizontal asymptotes of the function $f(x) = \dfrac{1000}{1 + 4e^{-0.2x}}$.
3. Determine the amount after 30 years if \$5000 is invested at $6\frac{1}{2}\%$ compounded (a) quarterly and (b) continuously.
4. Determine the principle that will yield \$200,000 when invested at 8% compoinded daily for 20 years.
5. Write the logarithmic equation $\log_4 64 = 3$ in exponential form.
6. Write the exponential equation $5^{-2} = \frac{1}{25}$ in logarithmic form.
7. Sketch a graph of the function $g(x) = \log_3(x - 2)$.
8. Use the properties of logarithms to expand $\ln\left(\dfrac{6x^2}{\sqrt{x^2 + 1}}\right)$.
9. Use the properties of logarithms to condense $3 \ln z - [\ln(z + 1) + \ln(z - 1)]$.
10. Use the properties of logarithms to simplify $\log_6 \sqrt{360}$.

In Exercises 11–14, solve the equation. Round the solution to three decimal places.

11. $e^{x/2} = 450$
12. $\left(1 + \dfrac{0.06}{4}\right)^{4t} = 3$
13. $5 \ln(x + 4) = 22$

14. A truck that costs \$28,000 new has a depreciated value of \$20,000 after 1 year. Find the value of the truck when it is 3 years old by using the exponential model $y = Ce^{kt}$.

In Exercises 15–17, the population of a certain species t years after it is introduced into a new habitat is given by $p(t) = 1200/(1 + 3e^{-t/5})$.

15. Determine the population size that was introduced into the habitat.
16. Determine the population after 5 years.
17. After how many years will the population be 800?
18. By observation, identify the equation that corresponds to the graph shown in the figure. Explain your reasoning.

(a) $y = 6e^{-x^2/2}$ (b) $y = \dfrac{6}{1 + e^{-x/2}}$ (c) $y = 6(1 - e^{-x^2/2})$

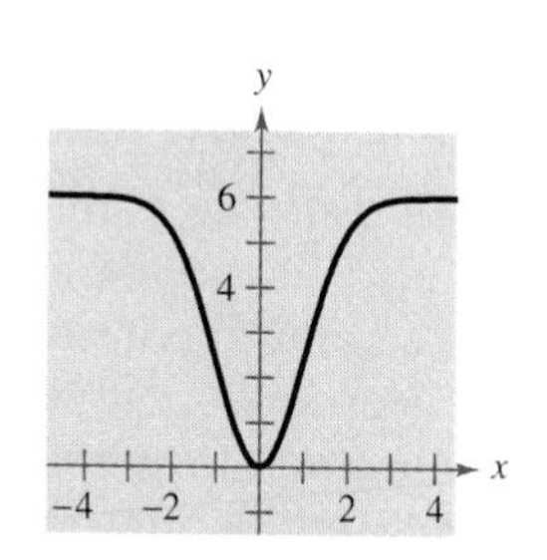

FIGURE FOR 18

6 Topics in Analytic Geometry

On July 16, 1994, the comet Shoemaker-Levy 9 collided with Jupiter, the largest planet in our solar system. The dark spots in the photo of Jupiter, below, show the results of the collision. The impact was filmed from the space probe Galileo, on its way to Jupiter.

All the planets in our solar system travel in elliptical paths about the Sun. Comets, on the other hand, can have varying paths. Some are elliptical, some are parabolic, and some are hyperbolic.

Since the solar system's formation, thousands of comets have collided with the planets and their moons. For evidence of this, all you need to do is look at the moon's surface through a telescope.

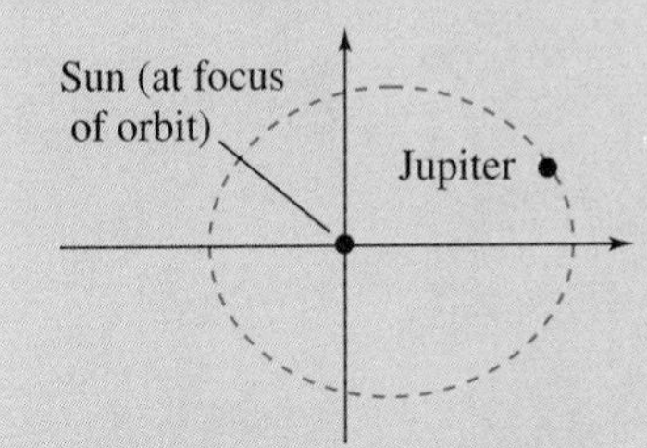

See Exercises 53 and 54 on page 444.

Photos: Alan Levenson; Courtesy of the Space Telescope Science Institute/NASA (inset)

Astronomers David Levy, Carolyn Shoemaker. and Eugene Shoemaker (from left to right) are shown with the 18-inch Schmidt telescope they used to discover the comet that is now named after them.

6.1 Lines

See Exercise 39 on page 424 for an example of how the inclination of a line can be used to determine the height of two mountain peaks.

Inclination of a Line ▫ The Angle Between Two Lines ▫ The Distance Between a Point and a Line

Inclination of a Line

In a previous course, you should have learned that the graph of the linear equation $y = mx + b$ is a nonvertical line with slope m and y-intercept at $(0, b)$. There, the slope of a line was described as the rate of change in y with respect to x. In this section, you will look at the slope of a line in terms of the angle of inclination of the line.

Every nonhorizontal line must intersect the x-axis. The angle formed by such an intersection determines the **inclination** of the line, as specified in the following definition.

THINK ABOUT THE PROOF

If the slope of a line is zero, the line is horizontal and $\theta = 0$. Thus, $m = \tan 0$ is true for horizontal lines because $m = 0 = \tan 0$. If the line has a positive slope, it will intersect the x-axis. Call this point $(x_1, 0)$. If (x_2, y_2) is a second point on the line, the slope is given by

$$m = \frac{y_2 - 0}{x_2 - x_1} = \frac{y_2}{x_2 - x_1} = \tan \theta.$$

Try proving the case in which the line has a negative slope. The details of this proof are listed in the appendix.

DEFINITION OF INCLINATION

The **inclination** of a nonhorizontal line is the positive angle θ (less than 180°) measured counterclockwise from the x-axis to the line. (See Figure 6.1.)

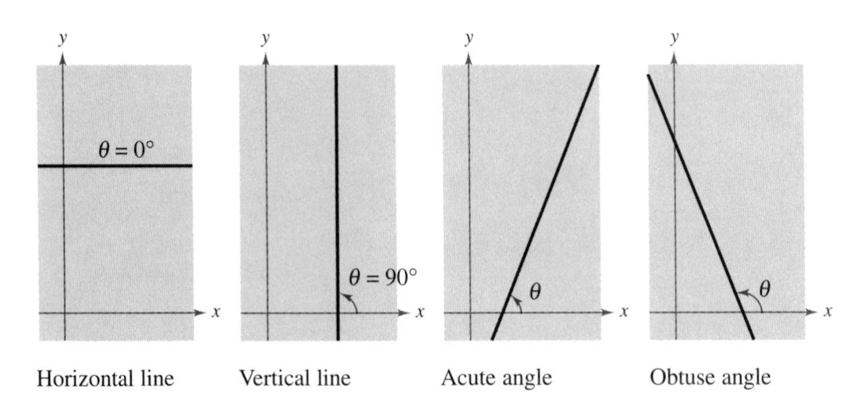

Horizontal line | Vertical line | Acute angle | Obtuse angle

FIGURE 6.1

The inclination of a line is related to its slope in the following manner.

INCLINATION AND SLOPE

If a nonvertical line has inclination θ and slope m, then

$$m = \tan \theta.$$

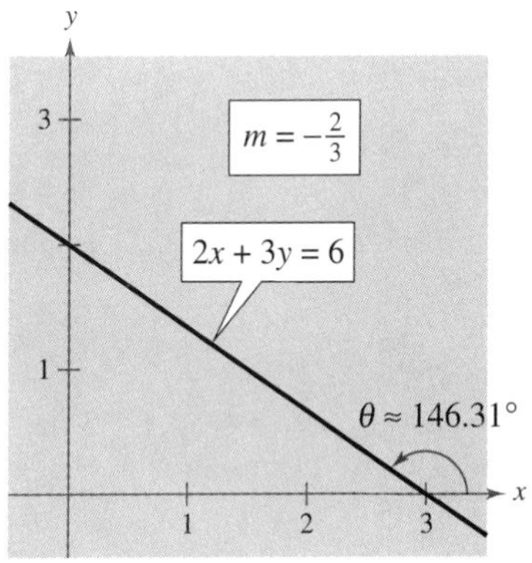

FIGURE 6.2

EXAMPLE 1 Finding the Inclination of a Line

Find the inclination of the line given by $2x + 3y = 6$.

Solution

The slope of this line is $m = -\frac{2}{3}$. Thus, its inclination is determined from the equation

$$\tan \theta = -\frac{2}{3}.$$

From Figure 6.2, it follows that $90° < \theta < 180°$. This means that

$$\begin{aligned}\theta &= 180° + \arctan\left(-\frac{2}{3}\right)\\ &\approx 180° - 33.69°\\ &\approx 146.31°.\end{aligned}$$

The Angle Between Two Lines

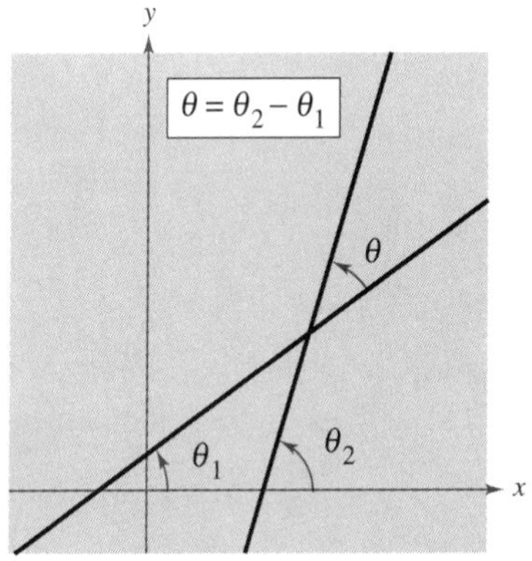

FIGURE 6.3

Two distinct lines in a plane either are parallel or intersecting. If they intersect, their intersection forms two pairs of opposite angles. One pair is acute and the other pair is obtuse. The smaller of these angles is called the **angle between the two lines.** As shown in Figure 6.3, you can use the inclinations of the two lines to find the angle between the two lines. Specifically, if two lines have inclinations θ_1 and θ_2, the angle between the two lines is

$$\theta = \theta_2 - \theta_1$$

where $\theta_1 < \theta_2$. You can use the formula for the tangent of the difference of two angles

$$\begin{aligned}\tan \theta &= \tan(\theta_2 - \theta_1)\\ &= \frac{\tan \theta_2 - \tan \theta_1}{1 + \tan \theta_1 \tan \theta_2}\end{aligned}$$

to obtain the following formula for the angle between two lines.

> ***ANGLE BETWEEN TWO LINES***
>
> If two nonperpendicular lines have slopes m_1 and m_2, the angle between the two lines is given by
>
> $$\tan \theta = \left|\frac{m_2 - m_1}{1 + m_1 m_2}\right|.$$

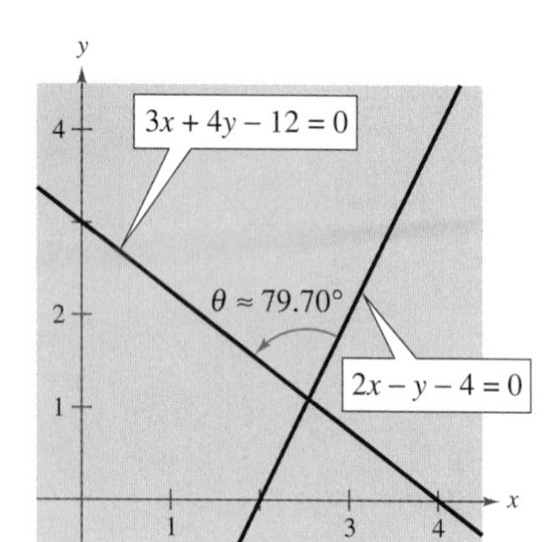

FIGURE 6.4

EXAMPLE 2 *Finding the Angle Between Two Lines*

Find the angle between the following two lines.

Line 1: $2x - y - 4 = 0$
Line 2: $3x + 4y - 12 = 0$

Solution

The two lines have slopes of $m_1 = 2$ and $m_2 = -\frac{3}{4}$, respectively. Thus, the angle between the two lines is given by

$$\begin{aligned}\tan\theta &= \left|\frac{m_2 - m_1}{1 + m_1 m_2}\right| \\ &= \left|\frac{(-3/4) - 2}{1 + (-3/4)(2)}\right| \\ &= \left|\frac{-11/4}{-2/4}\right| \\ &= \frac{11}{2}.\end{aligned}$$

Finally, you can conclude that the angle is

$$\theta = \arctan\frac{11}{2} \approx 79.70^\circ$$

as shown in Figure 6.4.

The Distance Between a Point and a Line

Finding the distance between a line and a point not on the line is an application of perpendicular lines. This distance is defined to be the length of the perpendicular line segment joining the point to the given line, as shown in Figure 6.5.

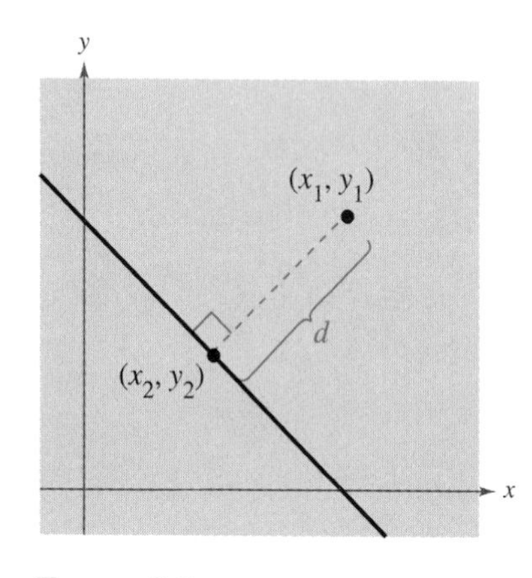

FIGURE 6.5

DISTANCE BETWEEN A POINT AND A LINE

The distance between the point (x_1, y_1) and the line given by $Ax + By + C = 0$ is

$$d = \frac{|Ax_1 + By_1 + C|}{\sqrt{A^2 + B^2}}.$$

NOTE Be sure you see that the values of A, B, and C in this distance formula correspond to the general equation of a line, $Ax + By + C = 0$.

PROOF

For simplicity's sake, assume that the given line is neither horizontal nor vertical. By writing the equation $Ax + By + C = 0$ in slope-intercept form

$$y = -\frac{A}{B}x - \frac{C}{B}$$

you can see that the line has a slope of $m = -A/B$. Thus, the slope of the line passing through (x_1, y_1) and perpendicular to the given line is B/A, and its equation is

$$y - y_1 = \frac{B}{A}(x - x_1).$$

These two lines intersect at the point (x_2, y_2), where

$$x_2 = \frac{B(Bx_1 - Ay_1) - AC}{A^2 + B^2} \quad \text{and} \quad y_2 = \frac{A(-Bx_1 + Ay_1) - BC}{A^2 + B^2}.$$

Finally, the distance between (x_1, y_1) and (x_2, y_2) is

$$\begin{aligned} d &= \sqrt{(x_2 - x_1)^2 + (y_2 - y_1)^2} \\ &= \sqrt{\left(\frac{B^2x_1 - ABy_1 - AC}{A^2 + B^2} - x_1\right)^2 + \left(\frac{-ABx_1 - A^2y_1 - BC}{A^2 + B^2} - y_1\right)^2} \\ &= \sqrt{\frac{A^2(Ax_1 + By_1 + C)^2 + B^2(Ax_1 + By_1 + C)^2}{(A^2 + B^2)^2}} \\ &= \frac{|Ax_1 + By_2 + C|}{\sqrt{A^2 + B^2}}. \end{aligned}$$

EXAMPLE 3 *Finding the Distance Between a Point and a Line*

Find the distance between the point $(4, 1)$ and the line $y = 2x + 1$.

Solution

The general form of the given equation is $-2x + y - 1 = 0$. Hence, the distance between the point and the line is

$$\begin{aligned} d &= \frac{|-2(4) + 1(1) - 1|}{\sqrt{(-2)^2 + 1^2}} \\ &= \frac{8}{\sqrt{5}} \\ &\approx 3.58. \end{aligned}$$

The line and the point are shown in Figure 6.6.

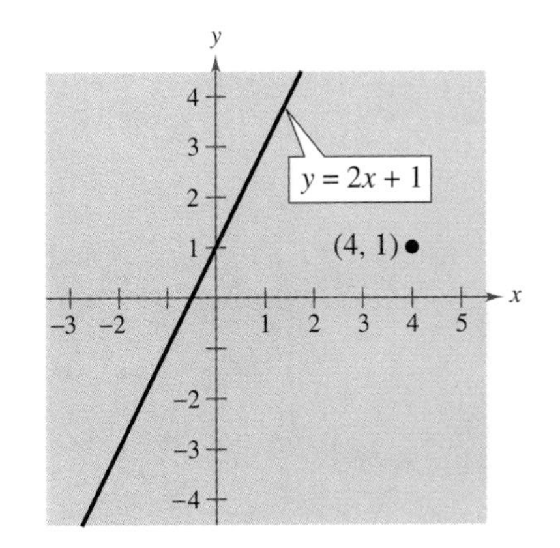

FIGURE 6.6

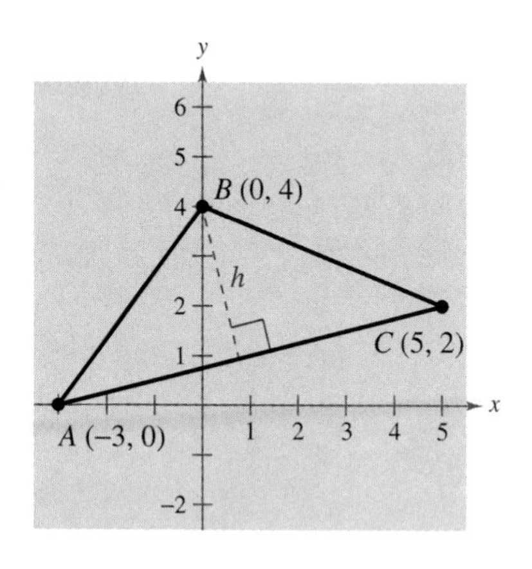

FIGURE 6.7

The *Interactive* CD-ROM shows every example with its solution; clicking on the *Try It!* button brings up similar problems. Guided Examples and Integrated Examples show step-by-step solutions to additional examples. Integrated Examples are related to several concepts in the section.

EXAMPLE 4 *An Application of Two Distance Formulas*

Figure 6.7 shows a triangle with vertices $A(-3, 0)$, $B(0, 4)$, and $C(5, 2)$.

a. Find the altitude from vertex B to side AC.

b. Find the area of the triangle.

Solution

a. To find the altitude, use the formula for the distance between line AC and the point $(0, 4)$. The equation of line AC is obtained as follows.

Slope: $m = \dfrac{2 - 0}{5 + 3} = \dfrac{1}{4}$

Equation:
$$y - 0 = \frac{1}{4}(x + 3)$$
$$4y = x + 3$$
$$x - 4y + 3 = 0$$

Therefore, the distance between this line and the point $(0, 4)$ is

$$\text{Altitude} = h = \frac{|1(0) - 4(4) - 3|}{\sqrt{1^2 + (-4)^2}} = \frac{13}{\sqrt{17}}.$$

b. Using the formula for the distance between two points, you can find the length of the base AC to be

$$b = \sqrt{(5 + 3)^2 + (2 - 0)^2} = \sqrt{68} = 2\sqrt{17}.$$

Finally, the area of the triangle in Figure 6.7 is

$$A = \frac{1}{2}bh = \frac{1}{2}\left(2\sqrt{17}\right)\left(\frac{13}{\sqrt{17}}\right) = 13.$$

GROUP ACTIVITY

INCLINATION AND THE ANGLE BETWEEN TWO LINES

Discuss why the inclination of a line can be an angle that is larger than 90°, but the angle between two lines cannot be larger than 90°. Decide whether the following statement is true or false: "The inclination of a line is the angle between the line and the x-axis." Explain.

The *Interactive* CD-ROM provides additional help with Warm-Up exercises by providing a hypertext link to the section in which the concept was introduced.

The *Interactive* CD-ROM contains step-by-step solutions to all odd-numbered Section and Review Exercises. It also provides Tutorial Exercises, which link to Guided Examples for additional help.

WARM UP

Find the distance between the points.

1. $(-2, 0), (3, 9)$ **2.** $(0, 4), (4, -2)$

Find the slope of the line passing through the points.

3. $(-5, 1), (4, 10)$ **4.** $(0, 2), (7, 8)$

5. $(2, 12), (10, 0)$ **6.** $(-4, 4), (6, -1)$

Find an equation of the line passing through the point with the given slope.

7. $(0, 3),\quad m = \frac{2}{3}$ **8.** $(2, -5),\quad m = \frac{7}{2}$

9. $(3, 20),\quad m = -4$ **10.** $(-6, 4),\quad m = -\frac{3}{4}$

6.1 Exercises

In Exercises 1–8, find the slope of the line with inclination θ.

1.

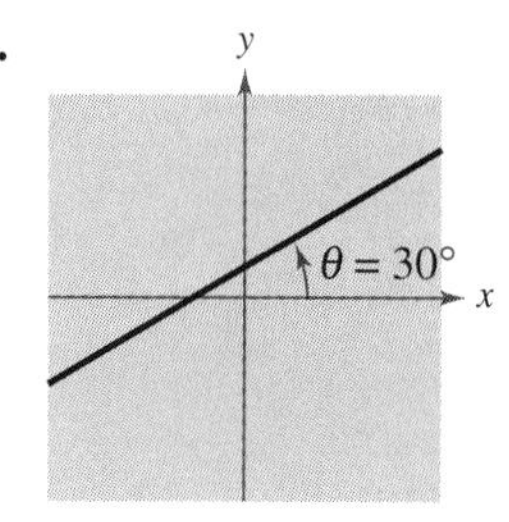

2.

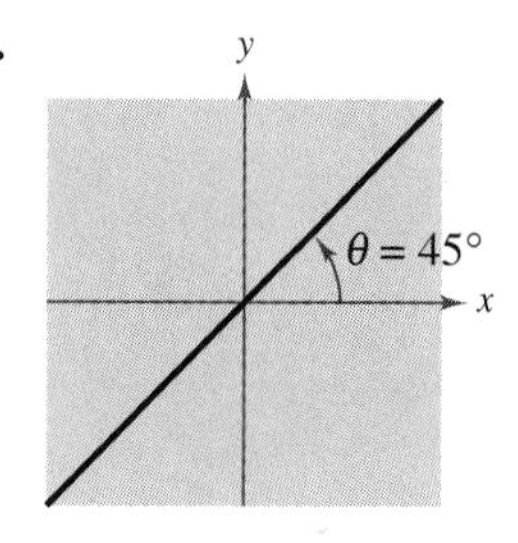

3.

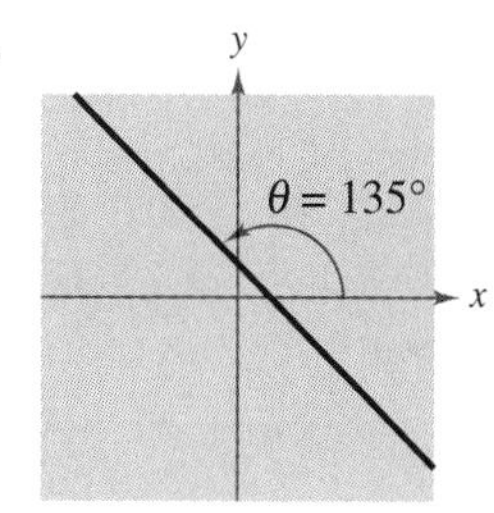

4.

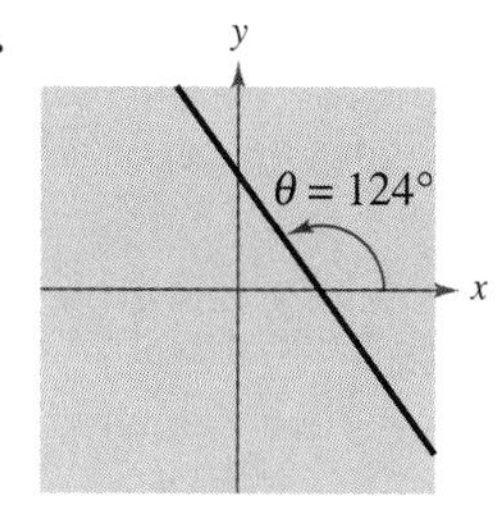

5. $\theta = 38.2°$ **6.** $\theta = 75.4°$

7. $\theta = 110°$ **8.** $\theta = 145.5°$

In Exercises 9–12, find the inclination θ of the line with a slope of m.

9. $m = -1$ **10.** $m = 2$

11. $m = \frac{3}{4}$ **12.** $m = -\frac{5}{2}$

In Exercises 13–16, find the inclination θ of the line passing through the points.

13. $(6, 1), (10, 8)$ **14.** $(-1, -4), (7, 12)$

15. $(-2, 20), (10, 0)$ **16.** $(0, 100), (50, 0)$

In Exercises 17–20, find the inclination of the line.

17. $5x - y + 3 = 0$ **18.** $4x + 5y - 9 = 0$

19. $5x + 3y = 0$ **20.** $x - y - 10 = 0$

21. ***Grade of a Road*** A straight road rises with an inclination of 6.5° from the horizontal. Find the slope of the road and the change in elevation over a 2-mile stretch of the road.

22. ***Grade of a Road*** A straight road rises with an inclination of 4.5° from the horizontal. Find the slope of the road and the change in elevation over a 3-mile stretch of the road.

23. ***Conveyor Design*** A moving conveyor is built so that it rises 1 meter for each 3 meters of horizontal travel.

(a) Find the inclination of the conveyor.

(b) The conveyor runs between two floors in a factory. The distance between the floors is 5 meters. Find the length of the conveyor.

24. ***Pitch of a Roof*** A roof has a rise of 3 feet for every horizontal change of 5 feet. Find the inclination of the roof.

In Exercises 25–34, find the angle θ between the lines.

25. $2x + y = 4$
$x - y = 2$

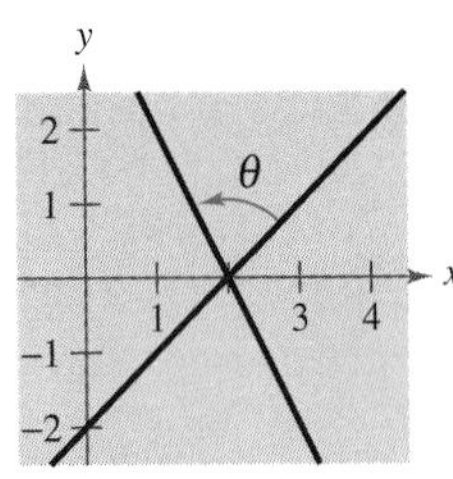

26. $x + 3y = 2$
$x - 2y = -3$

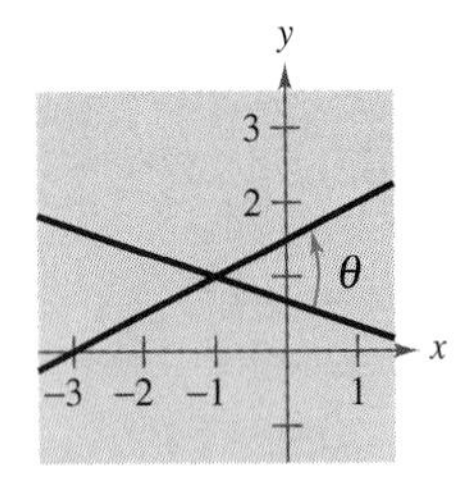

27. $x - y = 0$
$3x - 2y = -1$

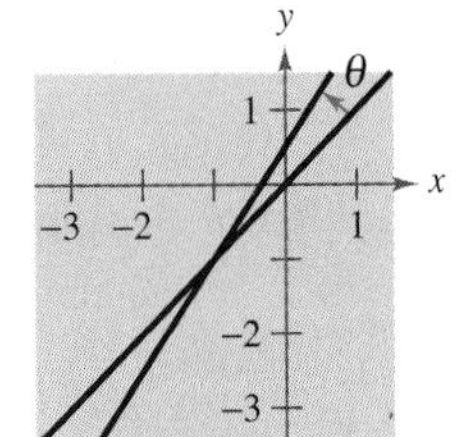

28. $2x - y = 2$
$4x + 3y = 24$

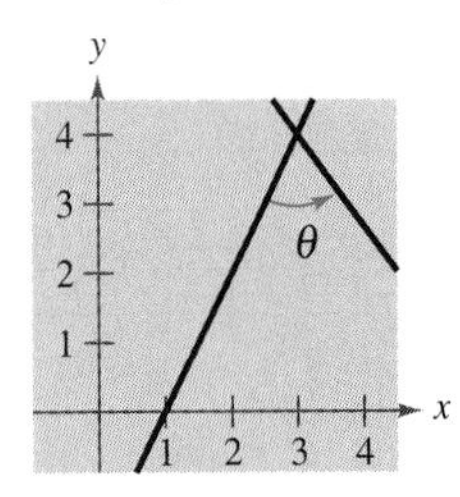

29. $x - 2y = 5$
$6x + 2y = 7$

30. $5x + 3y = 18$
$2x - 6y = -1$

31. $x + 2y = 4$
$x - 2y = 1$

32. $3x - 5y = 2$
$2x + 5y = 13$

33. $0.05x - 0.03y = 0.21$
$0.07x + 0.02y = 0.16$

34. $0.02x - 0.05y = -0.19$
$0.03x + 0.04y = 0.52$

Angle Measurement **In Exercises 35–38, find the slope of each side of the triangle and use the slopes to find the measures of the interior angles.**

35.

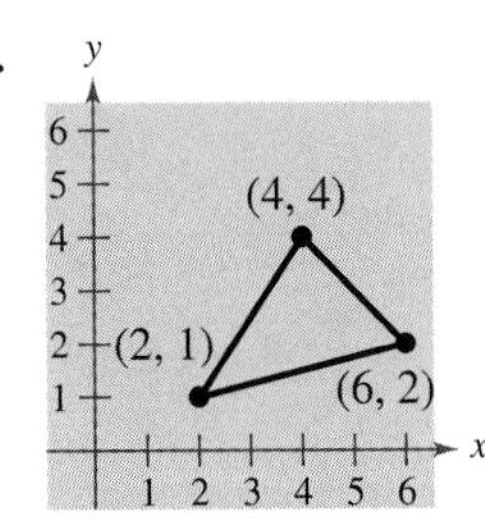

36.

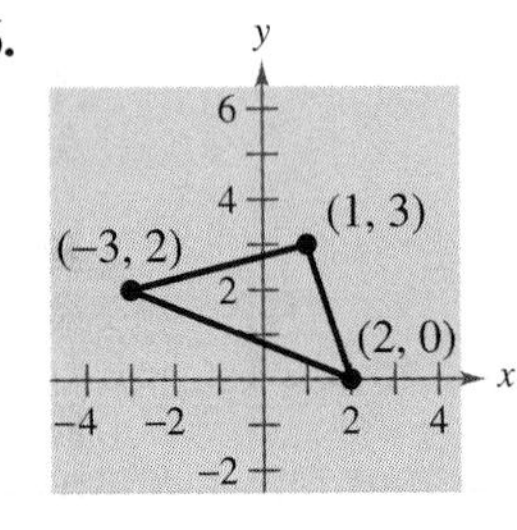

37.

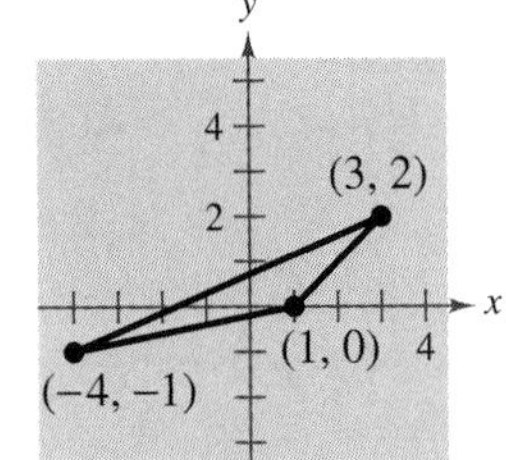

38.

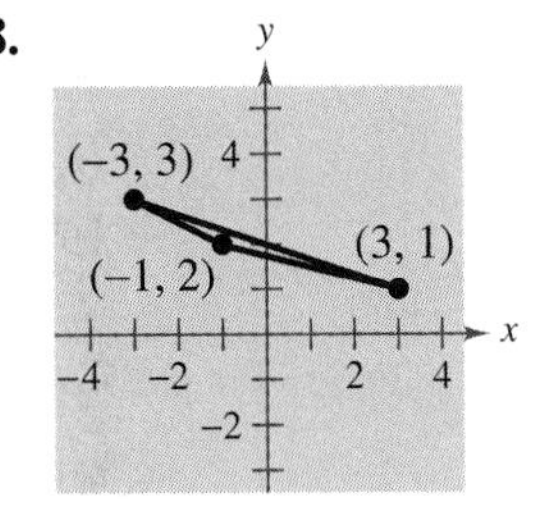

39. ***Mountain Climbing*** Several mountain climbers are located in a mountain pass between two peaks (see figure on page 797). The angles of elevation to the two peaks are 48° and 63°. A range finder shows that the distances to the peaks are 3250 feet and 6700 feet, respectively.

(a) Find the angle between the two lines of sight to the peaks.

(b) Approximate the amount of vertical climb that is necessary to reach the summit of each peak.

FIGURE FOR 39

40. ***Truss*** Find the angles α and β shown in the drawing of the roof truss.

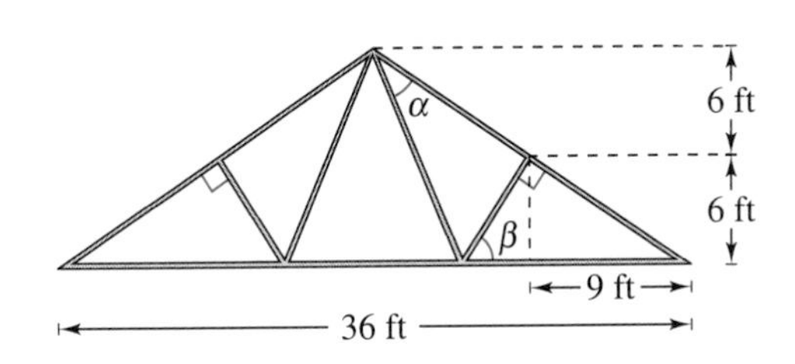

In Exercises 41–48, find the distance between the point and the line.

	Point	*Line*
41.	(0, 0)	$4x + 3y = 0$
42.	(0, 0)	$2x - y = 4$
43.	(2, 3)	$4x + 3y = 10$
44.	(−2, 1)	$x - y = 2$
45.	(6, 2)	$x + 1 = 0$
46.	(10, 8)	$y - 4 = 0$
47.	(0, 8)	$6x - y = 0$
48.	(4, 2)	$x - y = 20$

Area **In Exercises 49–52, (a) find the altitude from vertex B of the triangle to side AC, and (b) find the area of the triangle.**

49. $A = (0, 0), B = (1, 5), C = (3, 1)$

50. $A = (0, 0), B = (4, 5), C = (5, -2)$

51. $A = \left(-\frac{1}{2}, \frac{1}{2}\right), B = (2, 3), C = \left(\frac{5}{2}, 0\right)$

52. $A = (-4, -5), B = (3, 10), C = (6, 10)$

In Exercises 53 and 54, find the distance between the parallel lines.

53. $x + y = 1$
$x + y = 5$

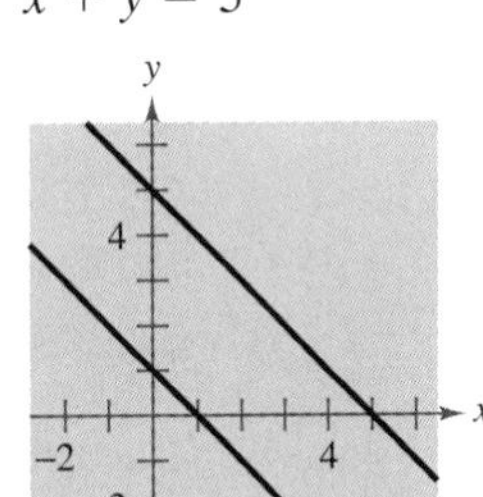

54. $3x - 4y = 1$
$3x - 4y = 10$

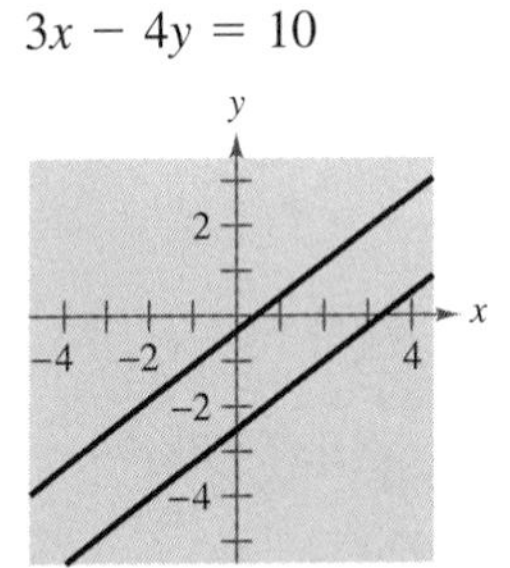

55. ***Exploration*** Consider a line with slope m and y-intercept (0, 4).

(a) Write the distance d between the origin and the line as a function of m.

(b) Graph the function in part (a).

(c) Find the slope that yields the maximum distance between the origin and the line.

(d) Find the asymptote of the graph in part (b) and interpret its meaning in the context of the problem.

56. ***Exploration*** Consider a line with slope m and y-intercept (0, 4).

(a) Write the distance d between the point (3, 1) and the line as a function of m.

(b) Graph the function in part (a).

(c) Find the slope that yields the maximum distance between the origin and the line.

(d) Is it possible for the distance to be 0? If so, what is the slope of the line that yields a distance of 0?

(e) Find the asymptote of the graph in part (b) and interpret its meaning in the context of the problem.

6.2 Introduction to Conics: Parabolas

See Exercise 54 on page 433 for an example of how a parabola can be used to model the cables of a suspension bridge.

Conics ▫ *Parabolas* ▫ *Applications*

Conics

Conic sections were discovered during the classical Greek period, 600 to 300 B.C. The early Greeks were concerned largely with the geometrical properties of conics. It was not until the early 17th century that the broad applicability of conics became apparent, and they then played a prominent role in the early development of calculus.

Each **conic section** (or simply **conic**) is the intersection of a plane and a double-napped cone. Notice in Figure 6.8(a) that in the formation of the four basic conics, the intersecting plane does not pass through the vertex of the cone. When the plane does pass through the vertex, the resulting figure is a **degenerate conic,** as shown in Figure 6.8(b).

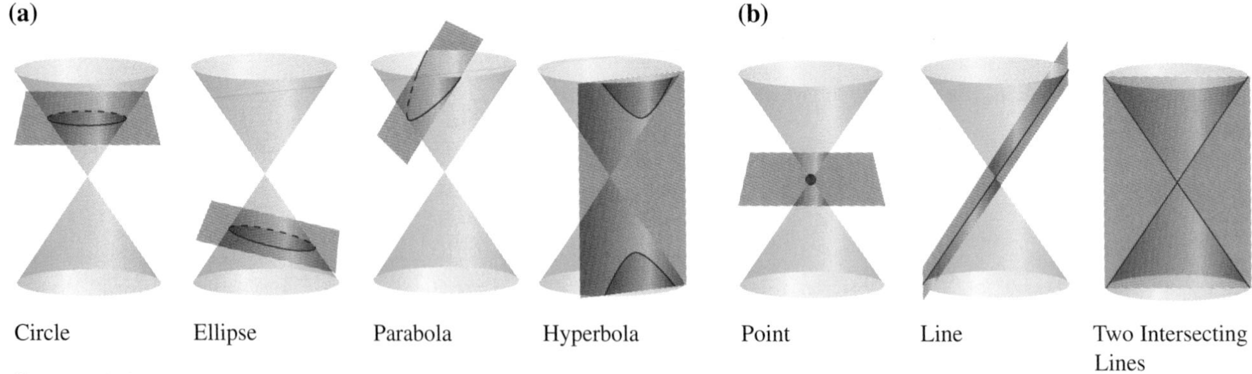

FIGURE 6.8

There are several ways to approach a study of conics. You could begin by defining conics in terms of the intersections of planes and cones, as the Greeks did, or you could define them algebraically in terms of the general second-degree equation

$$Ax^2 + Bxy + Cy^2 + Dx + Ey + F = 0.$$

However, you will study a third approach, in which each of the conics is defined as a **locus** (collection) of points satisfying a geometric property. For example, the definition of a circle as the collection of all points (x, y) that are equidistant from a fixed point (h, k) leads to the standard equation of a circle

$$(x - h)^2 + (y - k)^2 = r^2. \qquad \text{Equation of circle}$$

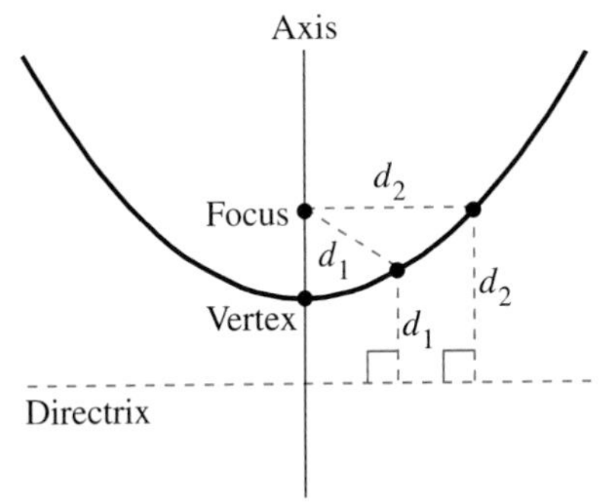

FIGURE 6.9

Parabolas

> ***Definition of Parabola***
>
> A **parabola** is the set of all points (x, y) that are equidistant from a fixed line **(directrix)** and a fixed point **(focus)** not on the line.

The midpoint between the focus and the directrix is called the **vertex,** and the line passing through the focus and the vertex is called the **axis** of the parabola. Note in Figure 6.9 that a parabola is symmetric with respect to its axis. Using the definition of a parabola, you can derive the following **standard form** of the equation of a parabola whose directrix is parallel to the x-axis or to the y-axis.

> ***Standard Equation of a Parabola***
>
> The **standard form** of the equation of a parabola with vertex at (h, k) is as follows.
>
> $(x - h)^2 = 4p(y - k),\ p \neq 0$ Vertical axis, directrix: $y = k - p$
>
> $(y - k)^2 = 4p(x - h),\ p \neq 0$ Horizontal axis, directrix: $x = h - p$
>
> The focus lies on the axis p units (*directed distance*) from the vertex. If the vertex is at the origin $(0, 0)$, the equation takes one of the following forms.
>
> $x^2 = 4py$ Vertical axis
>
> $y^2 = 4px$ Horizontal axis
>
> See Figure 6.10.

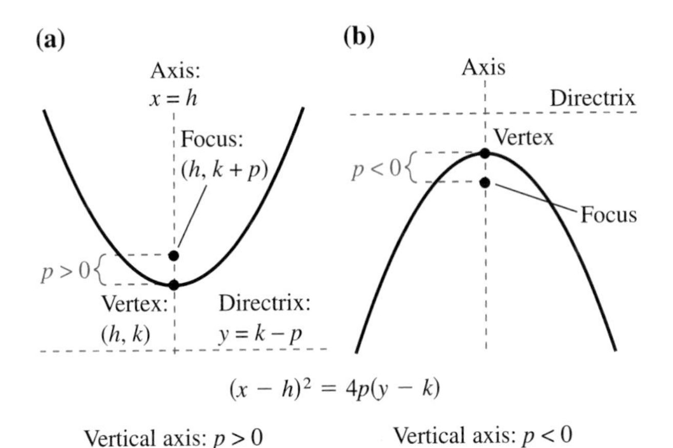

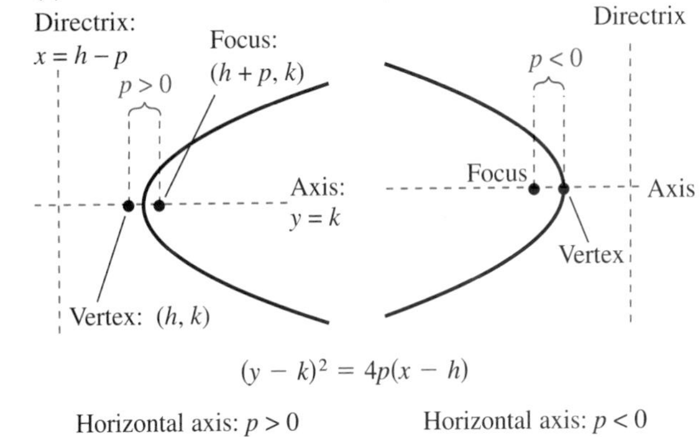

$(x - h)^2 = 4p(y - k)$ $(y - k)^2 = 4p(x - h)$

Vertical axis: $p > 0$ Vertical axis: $p < 0$ Horizontal axis: $p > 0$ Horizontal axis: $p < 0$

FIGURE 6.10

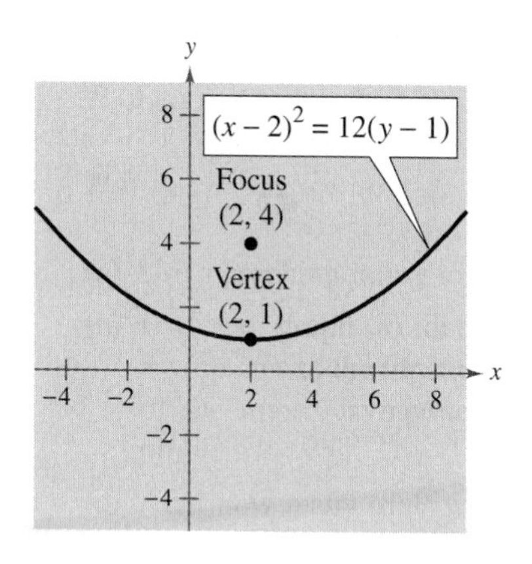

FIGURE 6.11

EXAMPLE 1 *Finding the Standard Equation of a Parabola*

Find the standard form of the equation of the parabola with vertex (2, 1) and focus (2, 4).

Solution

Because the axis of the parabola is vertical, consider the equation

$$(x - h)^2 = 4p(y - k)$$

where $h = 2$, $k = 1$, and $p = 4 - 1 = 3$. Thus, the standard form is

$$(x - 2)^2 = 12(y - 1).$$

The graph of this parabola is shown in Figure 6.11.

NOTE By expanding the standard equation in Example 1, you can obtain the more common quadratic form $y = \frac{1}{12}(x^2 - 4x + 16)$.

EXAMPLE 2 *Finding the Focus of a Parabola*

Find the focus of the parabola given by

$$y = -\frac{1}{2}x^2 - x + \frac{1}{2}.$$

Solution

To find the focus, convert to standard form by completing the square.

$$y = -\frac{1}{2}x^2 - x + \frac{1}{2} \quad \text{Original equation}$$

$$-2y = x^2 + 2x - 1 \quad \text{Multiply both sides by } -2.$$

$$1 - 2y = x^2 + 2x \quad \text{Group terms.}$$

$$2 - 2y = x^2 + 2x + 1 \quad \text{Add 1 to both sides.}$$

$$-2(y - 1) = (x + 1)^2 \quad \text{Standard form}$$

Comparing this equation with $(x - h)^2 = 4p(y - k)$, you can conclude that $h = -1$, $k = 1$, and $p = -\frac{1}{2}$. Because p is negative, the parabola opens downward, as shown in Figure 6.12. Therefore, the focus of the parabola is

$$(h, k + p) = \left(-1, \frac{1}{2}\right). \quad \text{Focus}$$

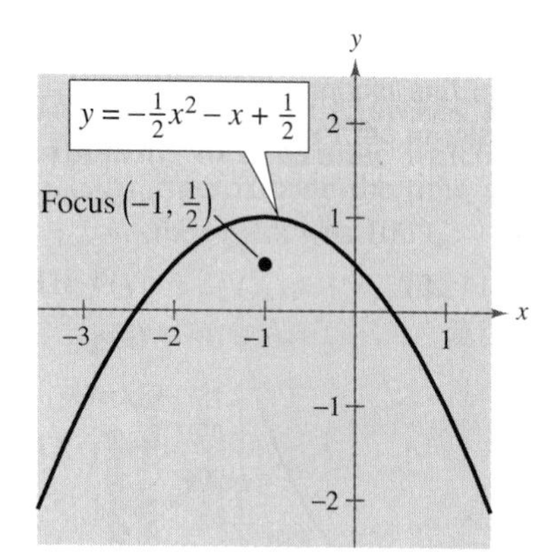

FIGURE 6.12

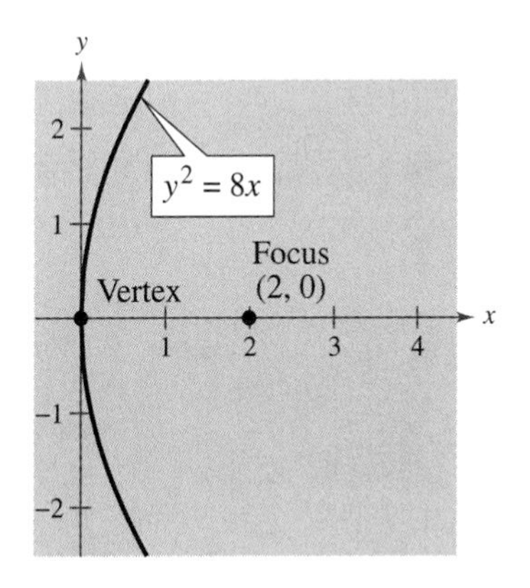

FIGURE 6.13

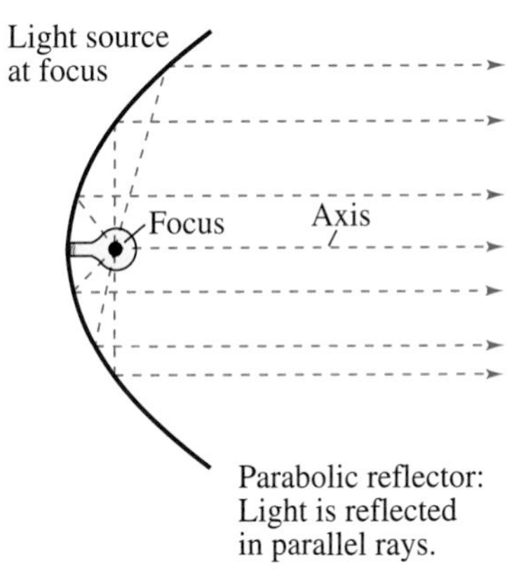

FIGURE 6.14

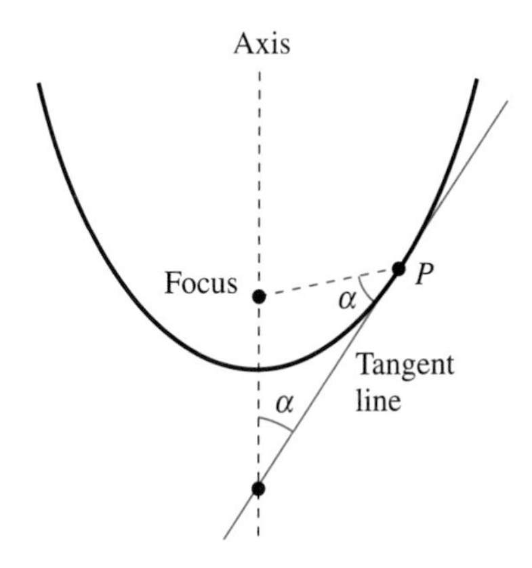

FIGURE 6.15

EXAMPLE 3 *Vertex at the Origin*

Find the standard equation of the parabola with vertex at the origin and focus (2, 0).

Solution

The axis of the parabola is horizontal, passing through (0, 0) and (2, 0), as shown in Figure 6.13. Thus, the standard form is

$$y^2 = 4px$$

where $h = k = 0$ and $p = 2$. Therefore, the equation is

$$y^2 = 8x.$$

NOTE Try using a graphing utility to confirm the equation found in Example 3. To do this, it helps to split the equation into two parts: $y = \sqrt{8x}$ (upper part) and $y = -\sqrt{8x}$ (lower part).

Applications

A line segment that passes through the focus of a parabola and has endpoints on the parabola is called a **focal chord.** The specific focal chord perpendicular to the axis of the parabola is called the **latus rectum.**

Parabolas occur in a wide variety of applications. For instance, a parabolic reflector can be formed by revolving a parabola around its axis. The resulting surface has the property that all incoming rays parallel to the axis are reflected through the focus of the parabola; this is the principle behind the construction of the parabolic mirrors used in reflecting telescopes. Conversely, the light rays emanating from the focus of a parabolic reflector used in a flashlight are all parallel to one another, as shown in Figure 6.14.

A line is **tangent** to a parabola at a point on the parabola if the line intersects, but does not cross, the parabola at the point. Tangent lines to parabolas have special properties related to the use of parabolas in constructing reflective surfaces.

REFLECTIVE PROPERTY OF A PARABOLA

The tangent line to a parabola at a point P makes equal angles with the following two lines (see Figure 6.15).

1. The line passing through P and the focus.
2. The axis of the parabola.

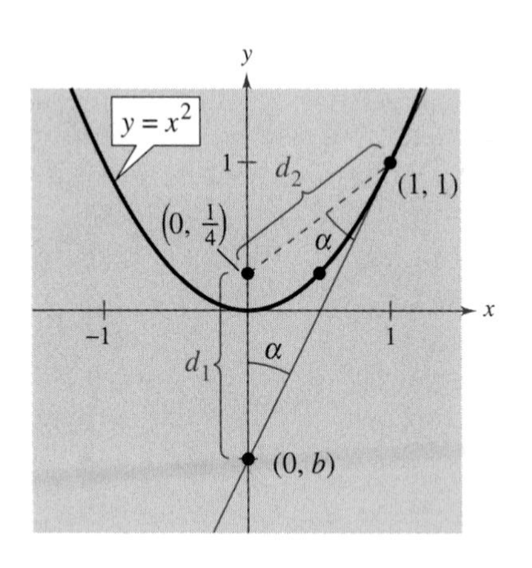

FIGURE 6.16

EXAMPLE 4 *Finding the Tangent Line at a Point on a Parabola*

Find the equation of the tangent line to the parabola given by $y = x^2$ at the point (1, 1).

Solution

For this parabola, $p = \frac{1}{4}$ and the focus is $\left(0, \frac{1}{4}\right)$, as shown in Figure 6.16. You can find the y-intercept $(0, b)$ of the tangent line by equating the lengths of the two sides of the isosceles triangle shown in Figure 6.16:

$$d_1 = \frac{1}{4} - b$$

and

$$d_2 = \sqrt{(1-0)^2 + [1 - (1/4)]^2} = \frac{5}{4}.$$

Setting $d_1 = d_2$ produces

$$\frac{1}{4} - b = \frac{5}{4}$$
$$b = -1.$$

Thus, the slope of the tangent line is

$$m = \frac{1 - (-1)}{1 - 0} = 2$$

and its slope-intercept equation is

$$y = 2x - 1.$$

The *Interactive* CD-ROM offers graphing utility emulators of the *TI-82* and *TI-83*, which can be used with the Examples, Explorations, Technology notes, and Exercises.

NOTE Try using a graphing utility to confirm the result of Example 4. By graphing

$$y = x^2 \qquad \text{and} \qquad y = 2x - 1$$

in the same viewing rectangle, you should be able to see that the line touches the parabola at the point (1, 1).

GROUP ACTIVITY

TELEVISION ANTENNA DISHES

Cross sections of television antenna dishes are parabolic in shape. Write a paragraph describing why these dishes are parabolic.

TECHNOLOGY

You can use a graphing utility to graph the parabola given by

$$(y-2)^2 = 6x$$

by solving for y to get

$$y = 2 + \sqrt{6x}$$

and

$$y = 2 - \sqrt{6x}$$

and then graphing both functions in the same viewing rectangle.

WARM UP

Expand and simplify the expression.

1. $(x-5)^2 - 20$
2. $(x+3)^2 - 1$
3. $10 - (x+4)^2$
4. $4 - (x-2)^2$

Complete the square for the quadratic expression.

5. $x^2 + 6x + 8$
6. $x^2 - 10x + 21$
7. $-x^2 + 2x + 1$
8. $-2x^2 + 4x - 2$

Find an equation of the line passing through the given point with the specified slope.

9. $(1, 6), \quad m = -\frac{2}{3}$
10. $(3, -2), \quad m = \frac{3}{4}$

6.2 Exercises

In Exercises 1–6, match the equation with its graph. [The graphs are labeled (a), (b), (c), (d), (e), and (f).]

1. $y^2 = -4x$
2. $x^2 = 2y$
3. $x^2 = -8y$
4. $y^2 = 12x$
5. $(y-1)^2 = 4(x-3)$
6. $(x+3)^2 = -2(y-1)$

(a)

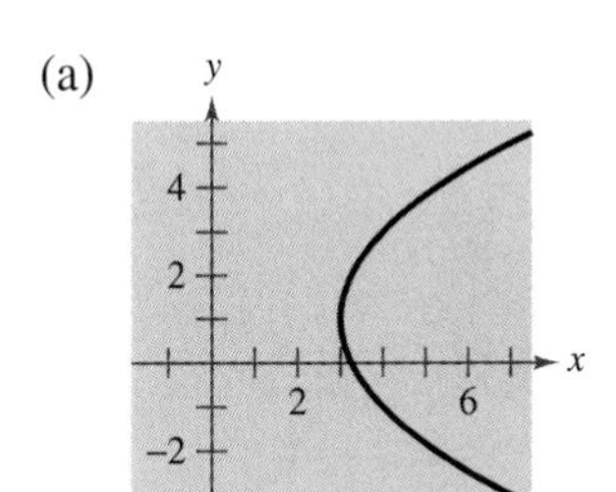

(b)

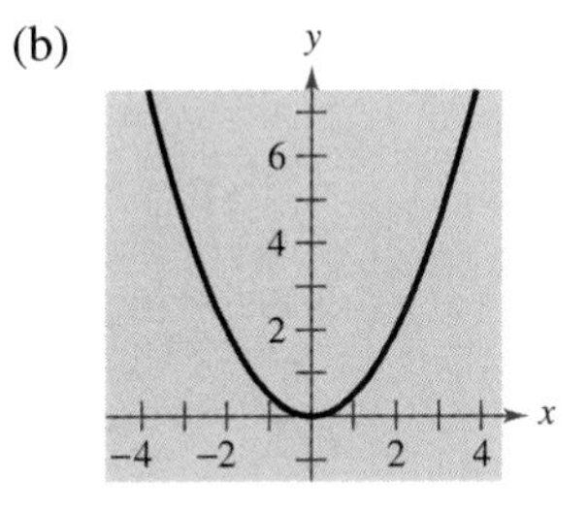

(c)

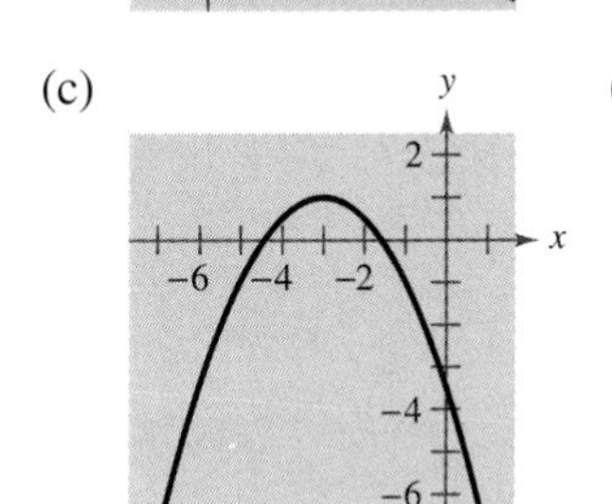

(d)

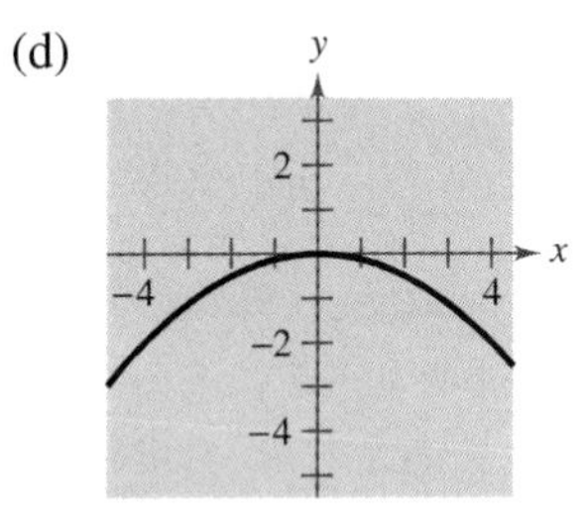

(e)

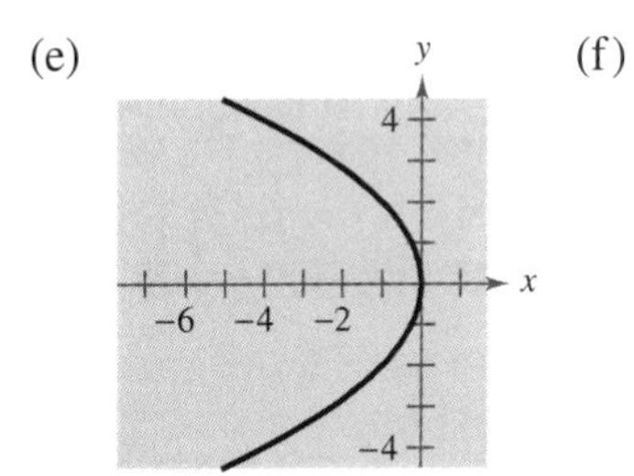

(f)

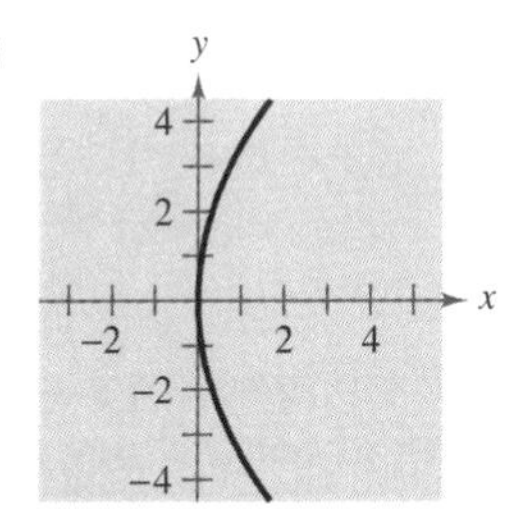

In Exercises 7–20, find the vertex, focus, and directrix of the parabola and sketch its graph.

7. $y = \frac{1}{2}x^2$
8. $y = 2x^2$
9. $y^2 = -6x$
10. $y^2 = 3x$
11. $x^2 + 8y = 0$
12. $x + y^2 = 0$
13. $(x-1)^2 + 8(y+2) = 0$
14. $(x+3) + (y-2)^2 = 0$
15. $\left(y + \frac{1}{2}\right)^2 = 2(x-5)$
16. $\left(x + \frac{1}{2}\right)^2 = 4(y-3)$

17. $y = \frac{1}{4}(x^2 - 2x + 5)$
18. $4x - y^2 - 2y - 33 = 0$
19. $y^2 + 6y + 8x + 25 = 0$
20. $y^2 - 4y - 4x = 0$

In Exercises 21–24, find the vertex, focus, and directrix of the parabola. Use a graphing utility to graph the parabola.

21. $y = -\frac{1}{6}(x^2 + 4x - 2)$
22. $x^2 - 2x + 8y + 9 = 0$
23. $y^2 + x + y = 0$
24. $y^2 - 4x - 4 = 0$

In Exercises 25 and 26, the equations of a parabola and a tangent line to the parabola are given. Use a graphing utility to graph both equations in the same viewing rectangle. Determine the coordinates of the point of tangency.

	Parabola	*Tangent Line*
25.	$y^2 - 8x = 0$	$x - y + 2 = 0$
26.	$x^2 + 12y = 0$	$x + y - 3 = 0$

27. ***Exploration*** Consider the parabola given by

$x^2 = 4py.$

(a) Use a graphing utility to graph the parabola for $p = 1, p = 2, p = 3$, and $p = 4$. Describe the effect on the graph when p increases.

(b) Locate the focus for each parabola in part (a).

(c) For each parabola in part (a), find the length of the chord passing through the focus and parallel to the directrix. How can the length of this chord be determined directly from the standard form of the equation of the parabola?

(d) Explain how the result of part (c) can be used as a sketching aid when graphing parabolas.

28. ***True or False*** Determine whether the following statement is true or false. If it is false, explain why. "It is possible for a parabola to intersect its directrix."

In Exercises 29–40, find an equation of the parabola with its vertex at the origin.

29.

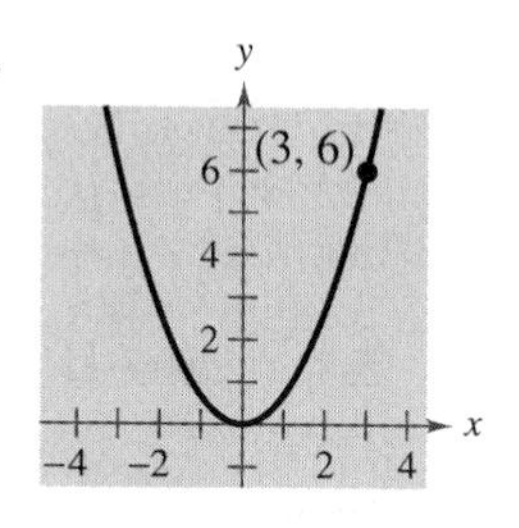

30.

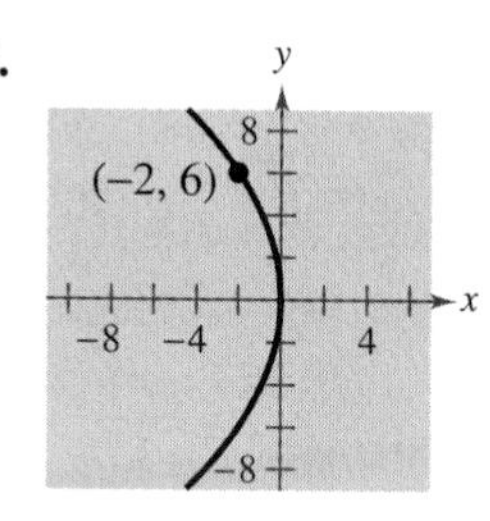

31. Focus: $\left(0, -\frac{3}{2}\right)$
32. Focus: $(2, 0)$
33. Focus: $(-2, 0)$
34. Focus: $(0, -2)$
35. Directrix: $y = -1$
36. Directrix: $x = 3$
37. Directrix: $y = 2$
38. Directrix: $x = -2$
39. Horizontal axis and passes through the point $(4, 6)$
40. Vertical axis and passes through the point $(-2, -2)$

In Exercises 41–50, find an equation of the parabola.

41.

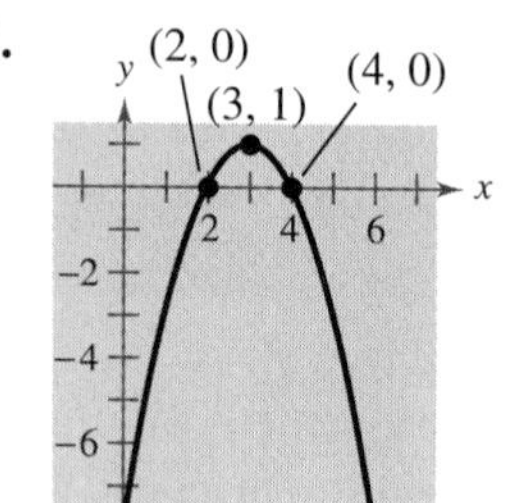

42.

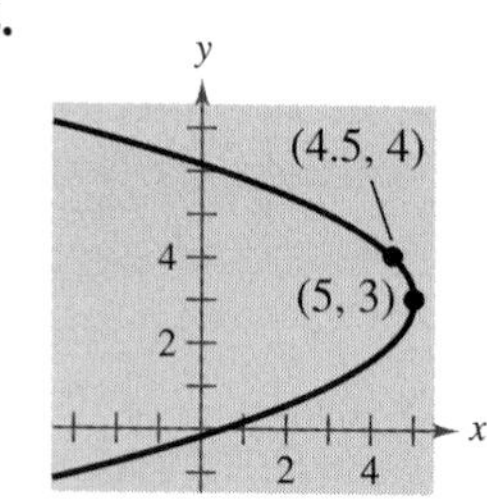

43.

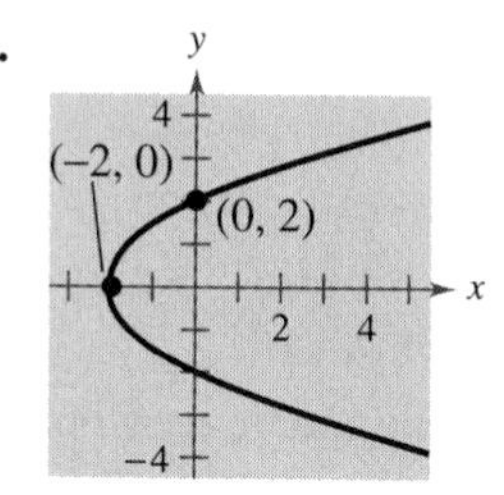

44.

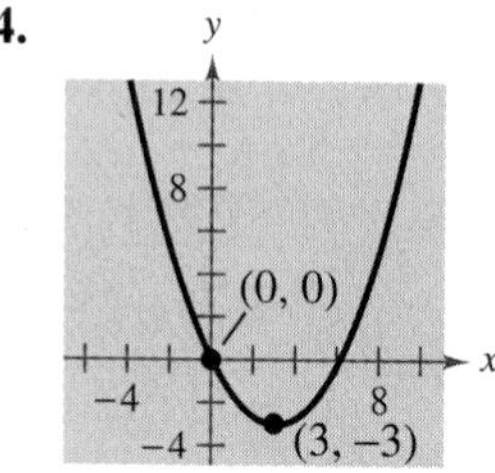

45. Vertex: $(3, 2)$; Focus: $(1, 2)$
46. Vertex: $(-1, 2)$; Focus: $(-1, 0)$
47. Vertex: $(0, 4)$; Directrix: $y = 2$
48. Vertex: $(-2, 1)$; Directrix: $x = 1$
49. Focus: $(2, 2)$; Directrix: $x = -2$
50. Focus: $(0, 0)$; Directrix: $y = 4$

In Exercises 51 and 52, change the equation so that its graph matches the description.

51. $(y - 3)^2 = 6(x + 1)$; upper half of parabola

52. $(y + 1)^2 = 2(x - 2)$; lower half of parabola

53. ***Satellite Antenna*** The receiver in a parabolic television dish antenna is 3.5 feet from the vertex and is located at the focus (see figure). Find an equation of a cross section of the reflector. (Assume that the dish is directed upward and the vertex is at the origin.)

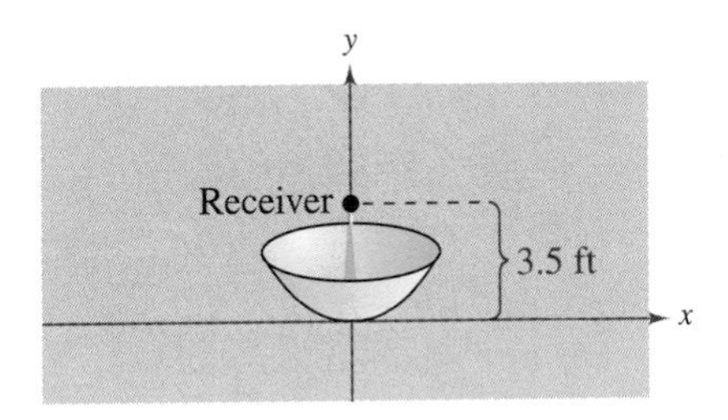

54. ***Suspension Bridge*** Each cable of a suspension bridge is suspended (in the shape of a parabola) between two towers that are 120 meters apart and whose tops are 20 meters above the roadway. The cables touch the roadway midway between the towers.

(a) Create a sketch of the bridge. Draw a rectangular coordinate system on the bridge with the center of the bridge at the origin. Identify the coordinates of the known points.

(b) Find an equation for the parabolic shape of each cable.

(c) Complete the table by finding the heights of the suspension cables over the roadway at distances of x meters from the center of the bridge.

x	0	20	40	60
y				

55. ***Beam Deflection*** A simply supported beam is 16 meters long and has a load at the center (see figure). The deflection of the beam at its center is 3 centimeters. Assume that the shape of the deflected beam is parabolic.

(a) Find an equation of the parabola. (Assume that the origin is at the center of the beam.)

(b) How far from the center of the beam is the deflection 1 centimeter?

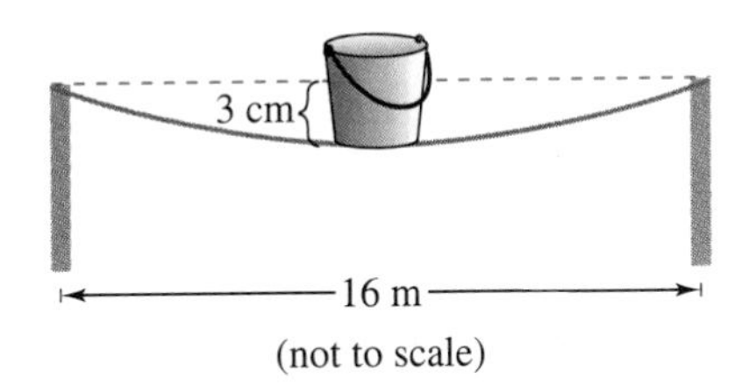

(not to scale)

56. ***Highway Design*** Highway engineers design a parabolic curve for an entrance ramp from a straight street to an interstate highway (see figure). Find an equation of the parabola.

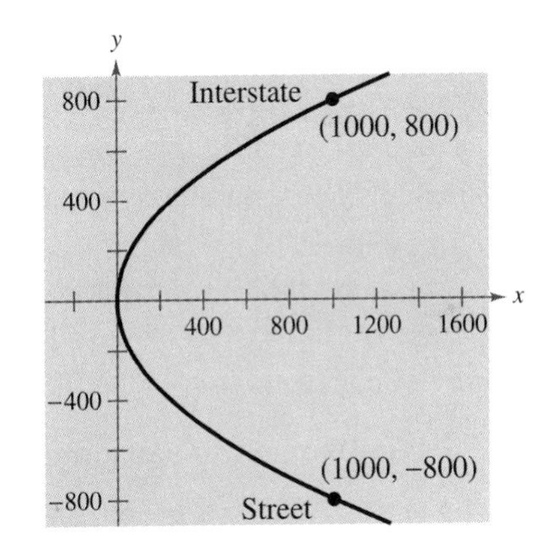

57. ***Revenue*** The revenue R generated by the sale of x units of a product is given by

$$R = 375x - \tfrac{3}{2}x^2.$$

Use a graphing utility to graph the function and approximate the number of sales that will maximize revenue.

58. ***Satellite Orbit*** An earth satellite in a 100-mile-high circular orbit around the earth has a velocity of approximately 17,500 miles per hour. If this velocity is multiplied by $\sqrt{2}$, the satellite will have the minimum velocity necessary to escape the earth's gravity and it will follow a parabolic path with the center of the earth as the focus (see figure).
   (a) Find the escape velocity of the satellite.
   (b) Find an equation of its path (assume that the radius of the earth is 4000 miles).

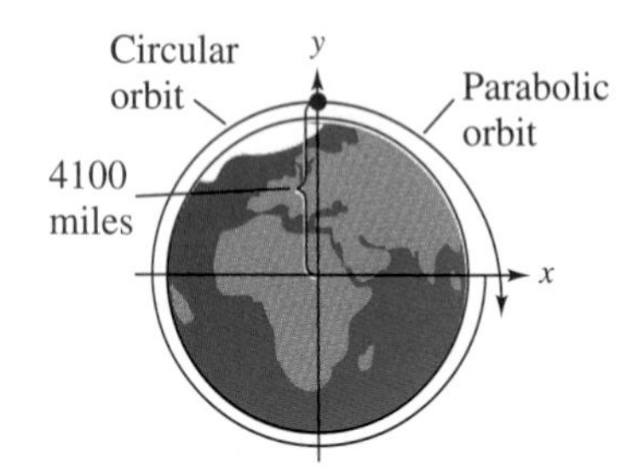

59. ***Path of a Projectile*** The path of a softball is given by the equation

$$y = -0.08x^2 + x + 4.$$

The coordinates x and y are measured in feet, with $x = 0$ corresponding to the position from which the ball was thrown.
   (a) Use a graphing utility to graph the trajectory of the softball.
   (b) Move the cursor along the path to approximate the highest point and the range of the trajectory.

60. ***Projectile Motion*** A bomber is flying at an altitude of 30,000 feet and a speed of 540 miles per hour (792 feet per second). When should a bomb be dropped so that it will hit the target if the path of the bomb is modeled by

$$y = 30{,}000 - \frac{x^2}{39{,}204}?$$

61. ***Exploration*** Let (x_1, y_1) be the coordinates of a point on the parabola $x^2 = 4py$. The equation of the line tangent to the parabola at the point is

$$y - y_1 = \frac{x_1}{2p}(x - x_1).$$

What is the slope of the tangent line?

62. ***Area*** The area of the shaded region in the figure is given by

$$A = \tfrac{4}{3}pb^{3/2}.$$

   (a) Find the area if $p = 2$ and $b = 4$.
   (b) Give a geometric explanation of why the area approaches 0 as p approaches 0.

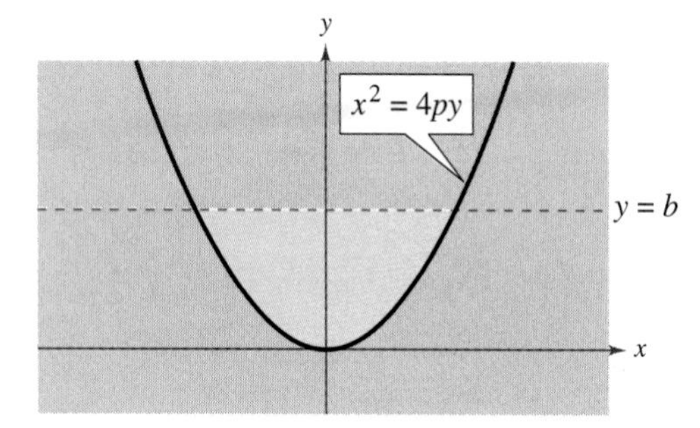

In Exercises 63–66, find an equation of the tangent line to the parabola at the given point, and find the x-intercept of the line.

63. $x^2 = 2y, \quad (4, 8)$
64. $x^2 = 2y, \quad \left(-3, \frac{9}{2}\right)$
65. $y = -2x^2, \quad (-1, -2)$
66. $y = -2x^2, \quad (3, -18)$

Review **Solve Exercises 67–70 as a review of the skills and problem-solving techniques you learned in previous sections.**

67. Find a polynomial with integer coefficients that has the zeros 3, $2 + i$, and $2 - i$.
68. Find all the zeros of

$$f(x) = 2x^3 - 3x^2 + 50x - 75$$

if one of the zeros is $x = \frac{3}{2}$.

69. Find all the zeros of the function

$$g(x) = 6x^4 + 7x^3 - 29x^2 - 28x + 20$$

if two of the zeros are $x = \pm 2$.

70. Use a graphing utility to graph the function

$$h(x) = 2x^4 + x^3 - 19x^2 - 9x + 9.$$

Use the graph to approximate the zeros of h.

6.3 Ellipses

See Exercise 55 on page 444 for an example of how an ellipse can be used to model the orbit of a satellite.

Introduction ❐ Applications ❐ Eccentricity

Introduction

The second type of conic is called an **ellipse,** and is defined as follows.

> **DEFINITION OF ELLIPSE**
> An **ellipse** is the set of all points (x, y) the sum of whose distances from two distinct fixed points (**foci**) is constant. (See Figure 6.17.)

NOTE The line through the foci intersects the ellipse at two points (**vertices**). The chord joining the vertices is the **major axis,** and its midpoint is the **center** of the ellipse. The chord perpendicular to the major axis at the center is the **minor axis** of the ellipse

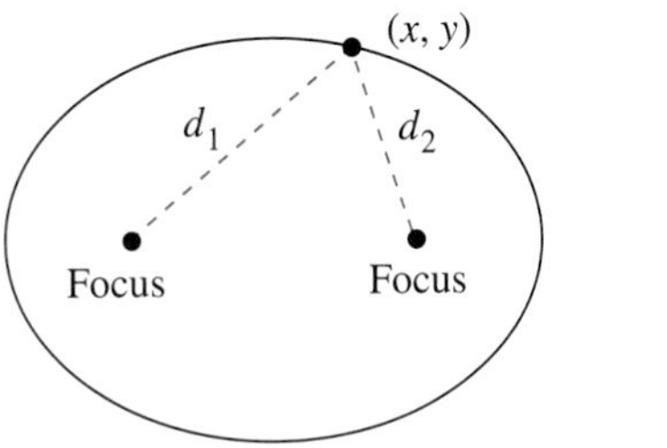

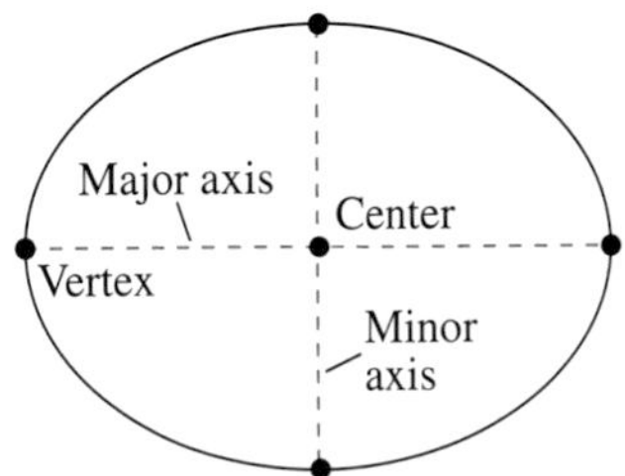

$d_1 + d_2$ is constant.

FIGURE 6.17

To derive the standard form of the equation of an ellipse, consider the ellipse in Figure 6.18 with the following points: center, (h, k); vertices, $(h \pm a, k)$; foci, $(h \pm c, k)$. The sum of the distances from any point on the ellipse to the two foci is constant. Using a vertex point, this constant sum is

$$(a + c) + (a - c) = 2a \qquad \text{Length of major axis}$$

or simply the length of the major axis. Now, if you let (x, y) be *any* point on the ellipse, the sum of the distances between (x, y) and the two foci must also be $2a$. That is,

$$\sqrt{[x - (h - c)]^2 + (y - k)^2} + \sqrt{[x - (h + c)]^2 + (y - k)^2} = 2a.$$

Finally, in Figure 6.18, you can see that $b^2 = a^2 - c^2$, which implies that the equation of the ellipse is

$$b^2(x - h)^2 + a^2(y - k)^2 = a^2b^2$$

$$\frac{(x - h)^2}{a^2} + \frac{(y - k)^2}{b^2} = 1.$$

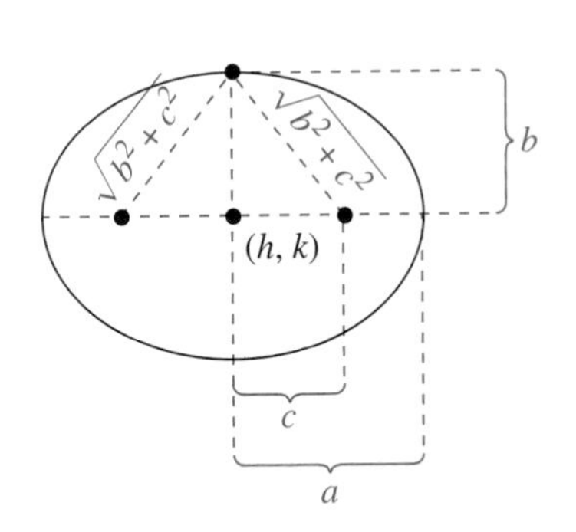

$$2\sqrt{b^2 + c^2} = 2a$$
$$b^2 + c^2 = a^2$$

FIGURE 6.18

In the development on page 435, you would obtain a similar equation if you began with a vertical major axis. Both results are summarized as follows.

STANDARD EQUATION OF AN ELLIPSE

The standard form of the equation of an ellipse, with center (h, k) and major and minor axes of lengths $2a$ and $2b$, where $0 < b < a$, is

$$\frac{(x-h)^2}{a^2} + \frac{(y-k)^2}{b^2} = 1 \qquad \text{Major axis is horizontal.}$$

$$\frac{(x-h)^2}{b^2} + \frac{(y-k)^2}{a^2} = 1 \qquad \text{Major axis is vertical.}$$

The foci lie on the major axis, c units from the center, with $c^2 = a^2 - b^2$. If the center is at the origin $(0, 0)$, the equation takes one of the following forms.

$$\frac{x^2}{a^2} + \frac{y^2}{b^2} = 1 \qquad \text{Major axis is horizontal.}$$

$$\frac{x^2}{b^2} + \frac{y^2}{a^2} = 1 \qquad \text{Major axis is vertical.}$$

A computer animation of this concept appears in the *Interactive* CD-ROM.

Figure 6.19 shows both the vertical and horizontal orientations for an ellipse.

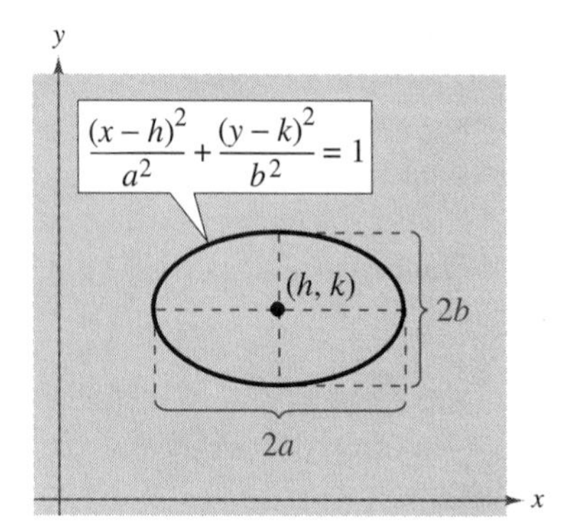

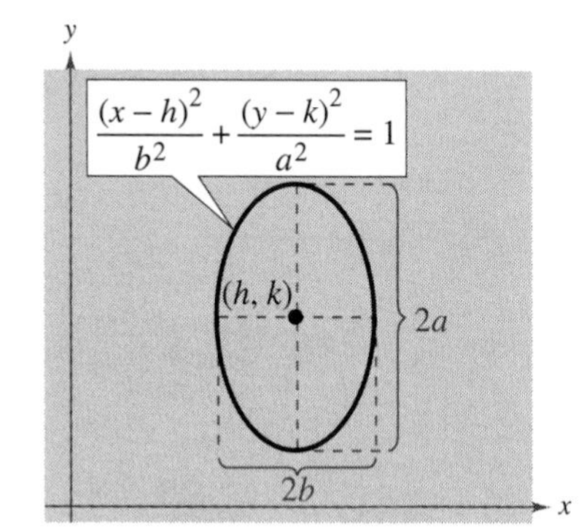

FIGURE 6.19

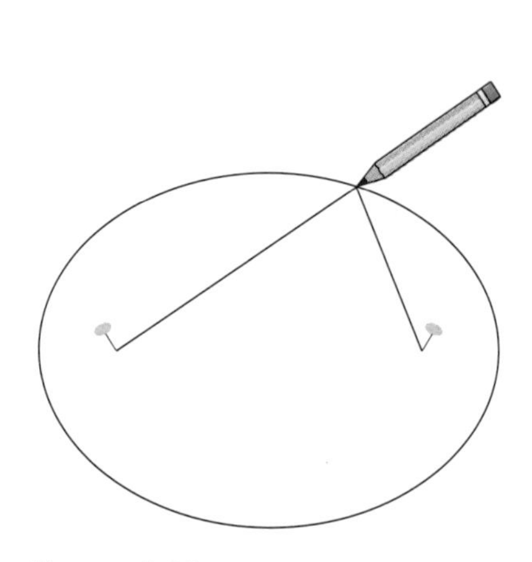

FIGURE 6.20

You can visualize the definition of an ellipse by imagining two thumbtacks placed at the foci, as shown in Figure 6.20. If the ends of a fixed length of string are fastened to the thumbtacks and the string is drawn taut with a pencil, the path traced by the pencil will be an ellipse.

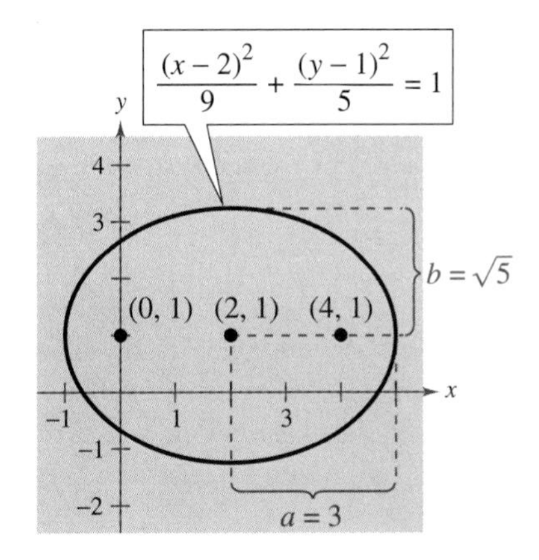

FIGURE 6.21

NOTE In Example 1, note the use of the equation $c^2 = a^2 - b^2$. Don't confuse this equation with the Pythagorean Theorem—there is a sign difference.

EXAMPLE 1 *Finding the Standard Equation of an Ellipse*

Find the standard form of the equation of the ellipse having foci at (0, 1) and (4, 1) and a major axis of length 6, as shown in Figure 6.21.

Solution

Because the foci occur at (0, 1) and (4, 1), the center of the ellipse is (2, 1). This implies that the distance from the center to one of the foci is $c = 2$, and because $2a = 6$ you know that $a = 3$. Now, from $c^2 = a^2 - b^2$, you have

$$\begin{aligned} b &= \sqrt{a^2 - c^2} \\ &= \sqrt{9 - 4} \\ &= \sqrt{5}. \end{aligned}$$

Because the major axis is horizontal, the standard equation is

$$\frac{(x-2)^2}{9} + \frac{(y-1)^2}{5} = 1.$$

EXAMPLE 2 *Writing an Equation in Standard Form*

Sketch the graph of the ellipse whose equation is

$$x^2 + 4y^2 + 6x - 8y + 9 = 0.$$

Solution

Begin by writing the given equation in standard form. In the fourth step, note that 9 and 4 are added to *both* sides of the equation.

$$x^2 + 4y^2 + 6x - 8y + 9 = 0 \qquad \text{Original equation}$$

$$(x^2 + 6x + \square) + (4y^2 - 8y + \square) = -9 \qquad \text{Group terms.}$$

$$(x^2 + 6x + \square) + 4(y^2 - 2y + \square) = -9 \qquad \text{Factor 4 out of } y\text{-terms.}$$

$$(x^2 + 6x + 9) + 4(y^2 - 2y + 1) = -9 + 9 + 4(1)$$

$$(x+3)^2 + 4(y-1)^2 = 4 \qquad \text{Completed square form}$$

$$\frac{(x+3)^2}{4} + \frac{(y-1)^2}{1} = 1 \qquad \text{Standard form}$$

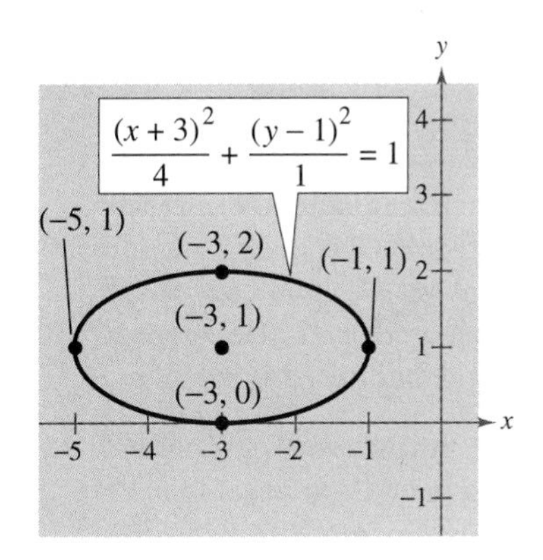

FIGURE 6.22

Now you see that the center occurs at $(h, k) = (-3, 1)$. Because the denominator of the x-term is $a^2 = 2^2$, you can locate the endpoints of the major axis two units to the right and left of the center. Similarly, because the denominator of the y-term is $b^2 = 1^2$, you can locate the endpoints of the minor axis one unit up and down from the center. The graph of this ellipse is shown in Figure 6.22.

EXAMPLE 3 *Analyzing an Ellipse*

Find the center, vertices, and foci of the ellipse given by

$$4x^2 + y^2 - 8x + 4y - 8 = 0.$$

Solution

By completing the square, you can write the given equation in standard form.

$$4x^2 + y^2 - 8x + 4y - 8 = 0$$
$$(4x^2 - 8x + \quad) + (y^2 + 4y + \quad) = 8$$
$$4(x^2 - 2x + \quad) + (y^2 + 4y + \quad) = 8$$
$$4(x^2 - 2x + 1) + (y^2 + 4y + 4) = 8 + 4(1) + 4$$
$$4(x - 1)^2 + (y + 2)^2 = 16$$
$$\frac{(x-1)^2}{4} + \frac{(y+2)^2}{16} = 1$$

Thus, the major axis is vertical, where $h = 1, k = -2, a = 4, b = 2$, and

$$c = \sqrt{16 - 4} = 2\sqrt{3}.$$

Therefore, you have the following.

Center: $(1, -2)$ Vertices: $(1, -6)$, $(1, 2)$ Foci: $(1, -2 - 2\sqrt{3})$, $(1, -2 + 2\sqrt{3})$

The graph of the ellipse is shown in Figure 6.23.

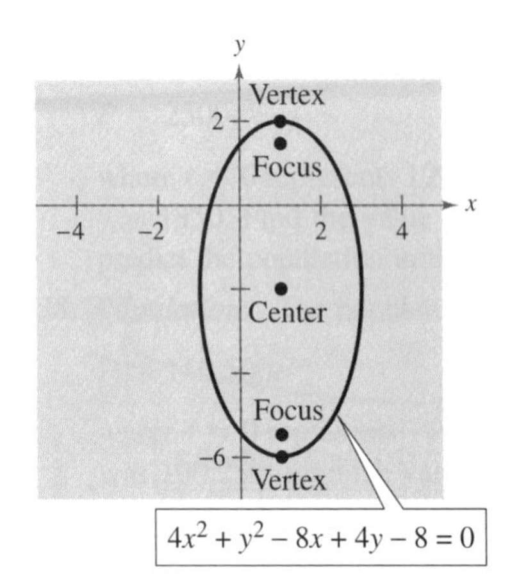

FIGURE 6.23

TECHNOLOGY

You can use a graphing utility to graph an ellipse by graphing the upper and lower portions in the same viewing rectangle. For instance, to graph the ellipse in Example 3, first solve for y to get

$$y_1 = -2 + 4\sqrt{1 - \frac{(x-1)^2}{4}}$$

and

$$y_2 = -2 - 4\sqrt{1 - \frac{(x-1)^2}{4}}.$$

Use a viewing rectangle in which $-6 \le x \le 10$ and $-8 \le y \le 4$. You should obtain the graph shown at the left.

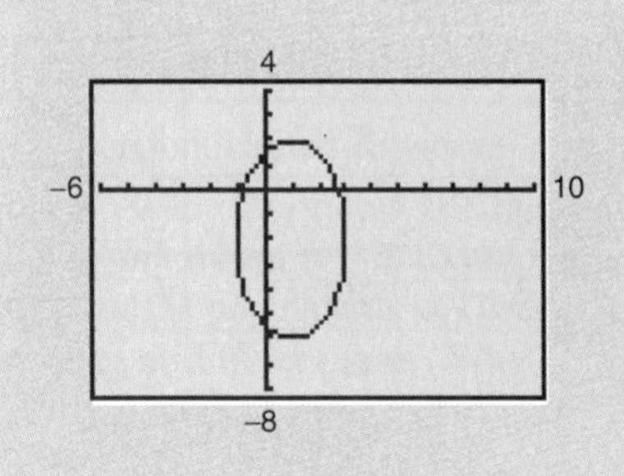

Applications

Ellipses have many practical and aesthetic uses. For instance, machine gears, supporting arches, and acoustical designs often involve elliptical shapes. The orbits of satellites and planets are also ellipses. Example 4 investigates the elliptical orbit of the moon about the earth.

EXAMPLE 4 *An Application Involving an Elliptical Orbit*

The moon travels about the earth in an elliptical orbit with the earth at one focus, as shown in Figure 6.24. The major and minor axes of the orbit have lengths of 768,806 kilometers and 767,746 kilometers, respectively. Find the greatest and smallest distances (the *apogee* and *perigee*) from the earth's center to the moon's center.

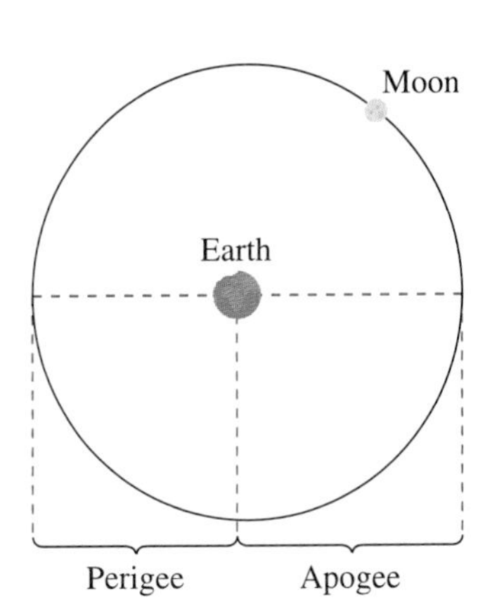

FIGURE 6.24

Solution

Because $2a = 768{,}806$ and $2b = 767{,}746$, you have $a = 384{,}403$ and $b = 383{,}873$, which implies that

$$\begin{aligned} c &= \sqrt{a^2 - b^2} \\ &= \sqrt{384{,}403^2 - 383{,}873^2} \\ &\approx 20{,}179. \end{aligned}$$

Therefore, the greatest distance between the center of the earth and the center of the moon is

$$a + c \approx 404{,}582 \text{ kilometers}$$

and the smallest distance is

$$a - c \approx 364{,}224 \text{ kilometers}.$$

Eccentricity

One of the reasons it was difficult for early astronomers to detect that the orbits of the planets are ellipses is that the foci of the planetary orbits are relatively close to their centers, thus making the orbits nearly circular. To measure the ovalness of an ellipse, you can use the concept of **eccentricity.**

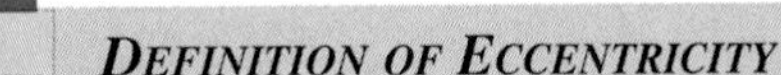

The **eccentricity** e of an ellipse is given by the ratio

$$e = \frac{c}{a}.$$

To see how this ratio is used to describe the shape of an ellipse, note that because the foci of an ellipse are located along the major axis between the vertices and the center, it follows that

$$0 < c < a.$$

For an ellipse that is nearly circular, the foci are close to the center and the ratio c/a is small, as shown in Figure 6.25(a). On the other hand, for an elongated ellipse, the foci are close to the vertices, and the ratio c/a is close to 1, as shown in Figure 6.25(b).

NOTE Note that $0 < e < 1$ for every ellipse.

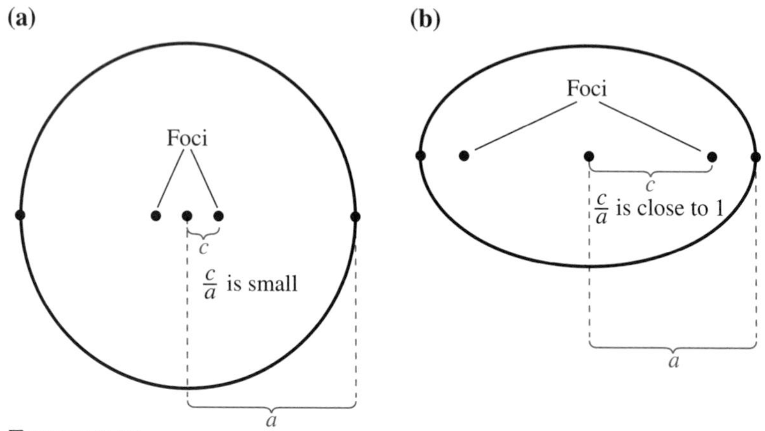

FIGURE 6.25

The time it takes Saturn to orbit the sun is equal to 29.5 earth years. *(Photo: NASA)*

The orbit of the moon has an eccentricity of $e \approx 0.0525$, and the eccentricities of the nine planetary orbits are as follows.

Mercury: $e \approx 0.2056$
Venus: $e \approx 0.0068$
Earth: $e \approx 0.0167$
Mars: $e \approx 0.0934$
Jupiter: $e \approx 0.0484$
Saturn: $e \approx 0.0543$
Uranus: $e \approx 0.0460$
Neptune: $e \approx 0.0082$
Pluto: $e \approx 0.2481$

GROUP ACTIVITY

GRAPHING ELLIPSES

Write an equation of an ellipse and graph it on graph paper. Do not write the equation on your graph. Exchange graphs with another student. Use the graph you receive to reconstruct the equation of the ellipse it represents. Find the eccentricity of the ellipse. Compare and discuss your results.

WARM UP

Sketch the graph of the equation.

1. $x^2 = 9y$
2. $y^2 = 9x$
3. $y^2 = -9x$
4. $x^2 = -9y$

Find the unknown in the equation $c^2 = a^2 - b^2$. (Assume a, b, and c are positive.)

5. $a = 13, b = 5$
6. $a = \sqrt{10}, c = 3$
7. $b = 6, c = 8$
8. $a = 7, b = 5$

Simplify the sum of compound fractions.

9. $\dfrac{x^2}{1/4} + \dfrac{y^2}{1/3}$
10. $\dfrac{(x-1)^2}{4/9} + \dfrac{(y+2)^2}{1/9}$

6.3 Exercises

In Exercises 1–6, match the equation with its graph. [The graphs are labeled (a), (b), (c), (d), (e), and (f).]

1. $\dfrac{x^2}{4} + \dfrac{y^2}{9} = 1$
2. $\dfrac{x^2}{9} + \dfrac{y^2}{4} = 1$
3. $\dfrac{x^2}{4} + \dfrac{y^2}{25} = 1$
4. $\dfrac{y^2}{4} + \dfrac{x^2}{4} = 1$
5. $\dfrac{(x-2)^2}{16} + (y+1)^2 = 1$
6. $\dfrac{(x+2)^2}{4} + \dfrac{(y+2)^2}{16} = 1$

(a)

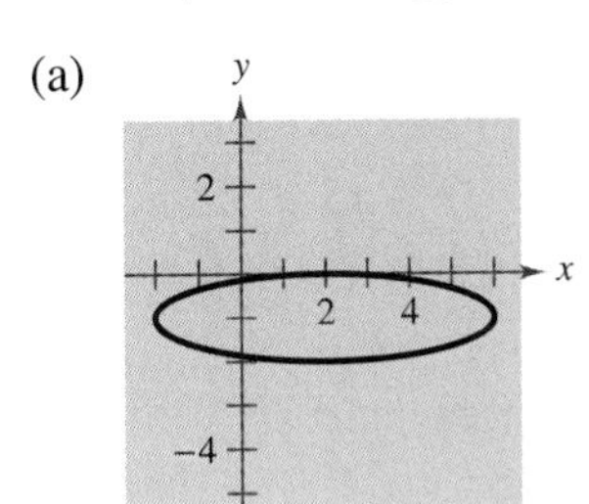

(b)

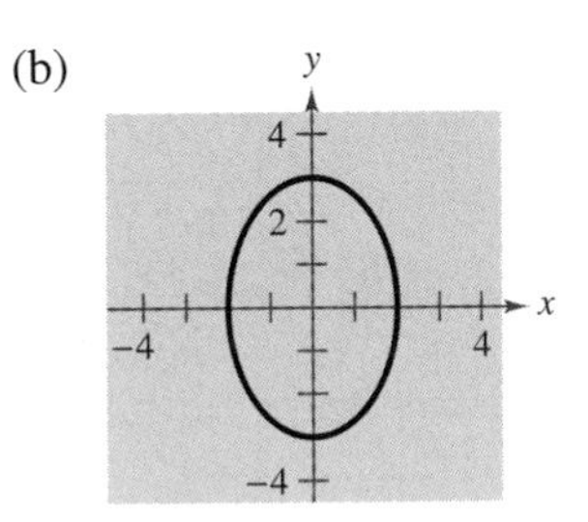

(c)

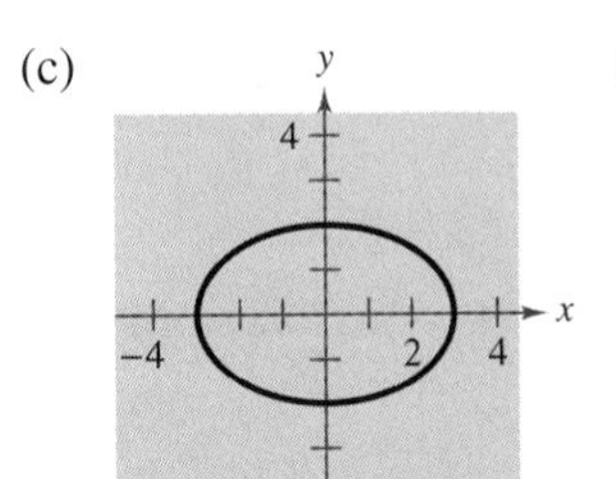

(d)

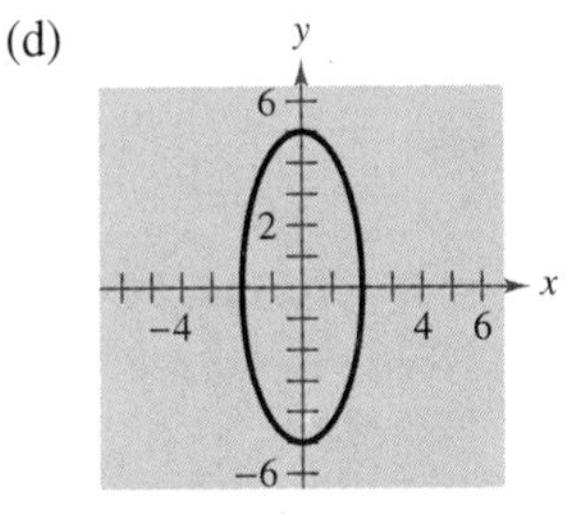

(e)

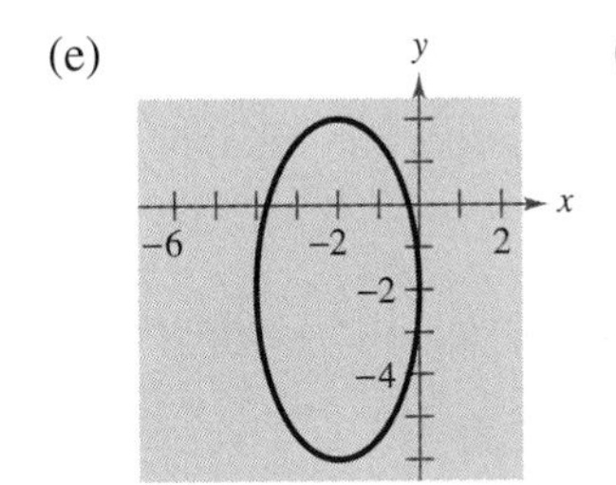

(f)

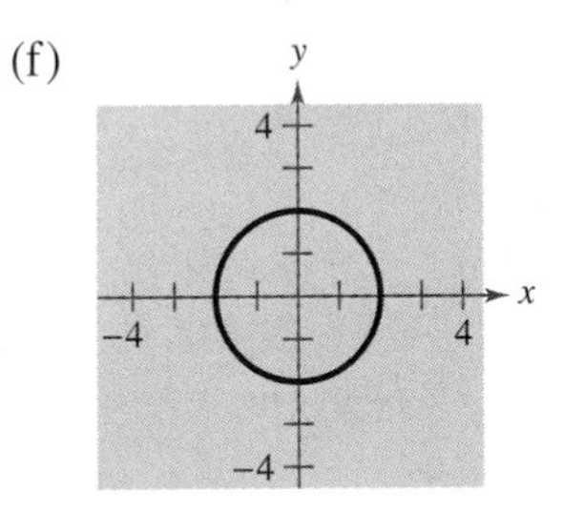

In Exercises 7–18, find the center, vertices, foci, and eccentricity of the ellipse, and sketch its graph.

7. $\dfrac{x^2}{25} + \dfrac{y^2}{16} = 1$
8. $\dfrac{x^2}{144} + \dfrac{y^2}{169} = 1$

9. $\frac{x^2}{16} + \frac{y^2}{25} = 1$

10. $\frac{x^2}{169} + \frac{y^2}{144} = 1$

11. $\frac{x^2}{9} + \frac{y^2}{5} = 1$

12. $\frac{x^2}{28} + \frac{y^2}{64} = 1$

13. $\frac{(x-1)^2}{9} + \frac{(y-5)^2}{25} = 1$

14. $(x+2)^2 + \frac{(y+4)^2}{1/4} = 1$

15. $9x^2 + 4y^2 + 36x - 24y + 36 = 0$

16. $9x^2 + 4y^2 - 36x + 8y + 31 = 0$

17. $16x^2 + 25y^2 - 32x + 50y + 16 = 0$

18. $9x^2 + 25y^2 - 36x - 50y + 61 = 0$

In Exercises 19–22, use a graphing utility to graph the ellipse. Find the center, foci, and vertices. (Recall that it may be necessary to solve the equation for y and obtain two functions.)

19. $5x^2 + 3y^2 = 15$

20. $x^2 + 4y^2 = 4$

21. $12x^2 + 20y^2 - 12x + 40y - 37 = 0$

22. $36x^2 + 9y^2 + 48x - 36y + 43 = 0$

In Exercises 23 and 24, change the equation so that its graph matches the description.

23. $\frac{(x-3)^2}{9} + \frac{y^2}{4} = 1$; right half of ellipse

24. $\frac{(x+1)^2}{16} + \frac{(y-2)^2}{25} = 1$; bottom half of ellipse

In Exercises 25–32, find an equation of the ellipse with its center at the origin.

25.

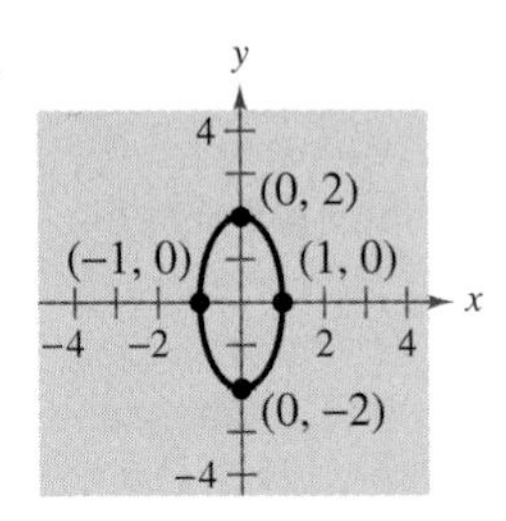

26.

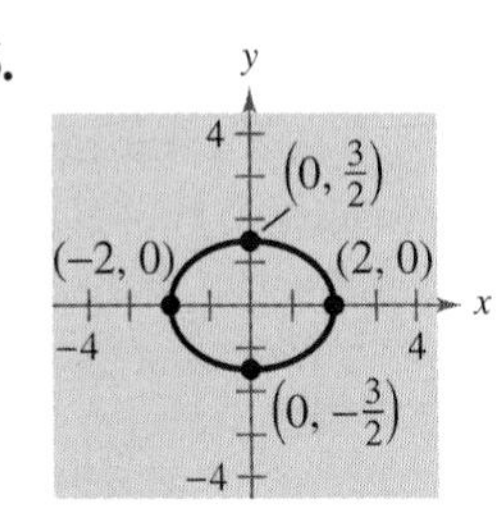

27. Vertices: $(\pm 5, 0)$; Foci: $(\pm 2, 0)$

28. Vertices: $(0, \pm 8)$; Foci: $(0, \pm 4)$

29. Foci: $(\pm 5, 0)$; Major axis of length 12

30. Foci: $(\pm 2, 0)$; Major axis of length 8

31. Vertices: $(0, \pm 5)$; Passes through the point (4, 2)

32. Major axis vertical; Passes through the points (0, 4) and (2, 0)

In Exercises 33–44, find an equation for the specified ellipse.

33.

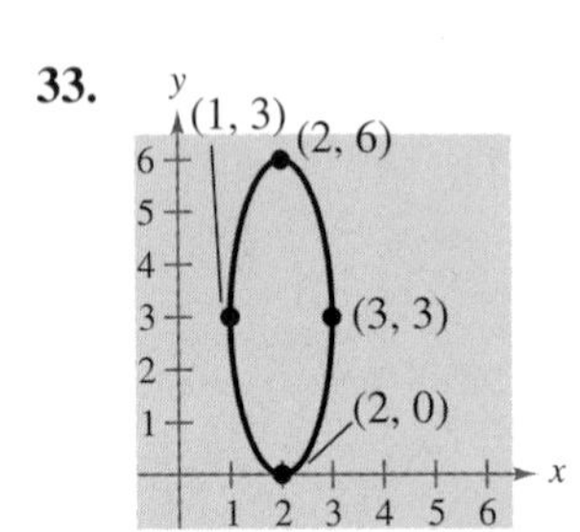

34.

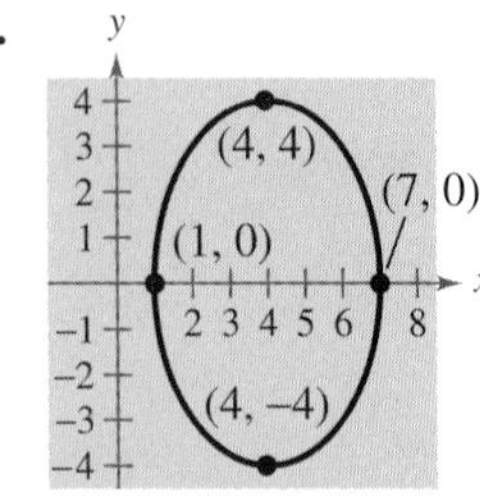

35.

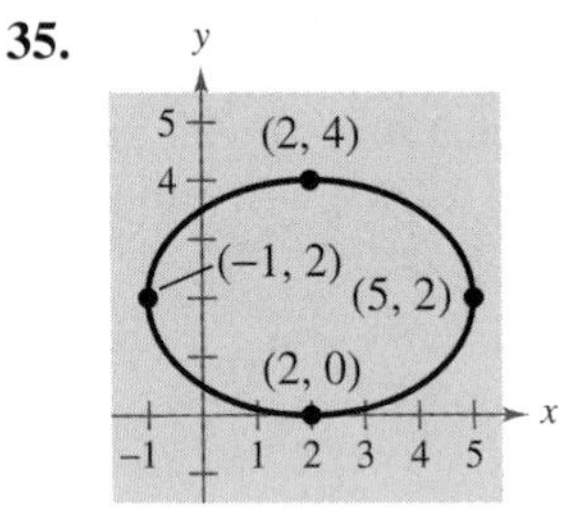

36.

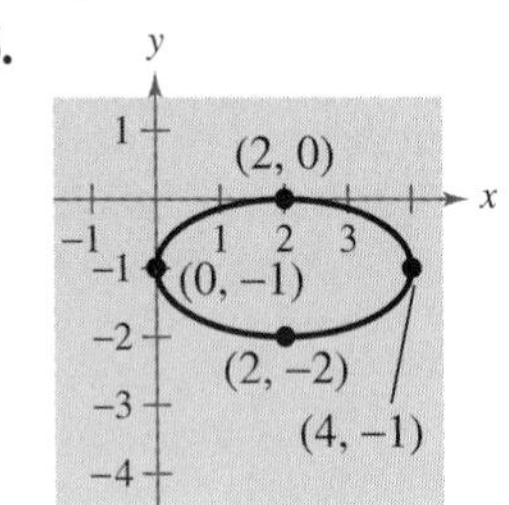

37. Vertices: (0, 2), (4, 2); Minor axis of length 2

38. Foci: (0, 0), (4, 0); Major axis of length 8

39. Foci: (0, 0), (0, 8); Major axis of length 16

40. Center: $(2, -1)$; Vertex: $\left(2, \frac{1}{2}\right)$; Minor axis of length 2

41. Vertices: (3, 1), (3, 9); Minor axis of length 6

42. Center: (3, 2); $a = 3c$; Foci: (1, 2), (5, 2)

43. Center: (0, 4); $a = 2c$; Vertices: $(-4, 4)$, $(4, 4)$

44. Vertices: (5, 0), (5, 12); Endpoints of the minor axis: (0, 6), (10, 6)

45. ***Think About It*** Near the beginning of this section it was noted that an ellipse can be drawn using two thumbtacks, a string of fixed length (greater than the distance between the two tacks), and a pencil. If the ends of the string are fastened at the tacks and the string is drawn taut with a pencil, the path traced by the pencil is an ellipse.

(a) What is the length of the string in terms of a?

(b) Explain why the path is an ellipse.

46. ***Fireplace Arch*** A fireplace arch is to be constructed in the shape of a semiellipse. The opening is to have a height of 2 feet at the center and a width of 6 feet along the base (see figure). The contractor draws the outline of the ellipse by the method described in Exercise 45. Give the required positions of the tacks and the length of the string.

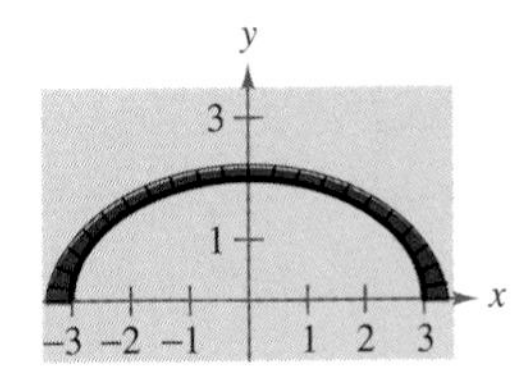

47. ***Geometry*** Sketch a graph of the ellipse that consists of all points (x, y) such that the sum of the distances between (x, y) and two fixed points is 16 units and the foci are located at the centers of the two sets of concentric circles in the figure.

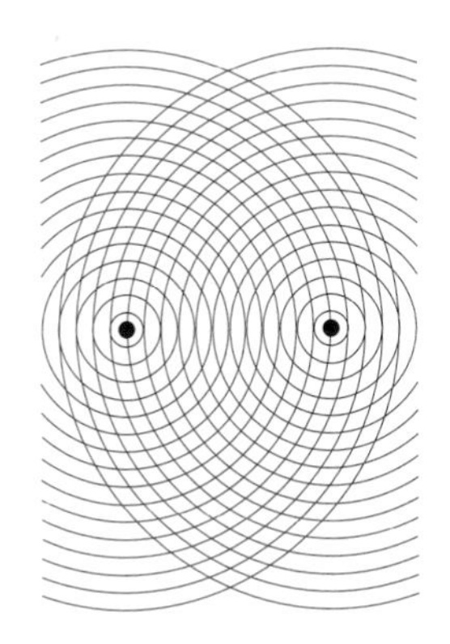

48. ***Mountain Tunnel*** A semielliptical arch over a tunnel for a road through a mountain has a major axis of 100 feet and a height at the center of 30 feet.

(a) Create a sketch for solving the problem. Draw a rectangular coordinate system on the tunnel with the center of the road entering the tunnel at the origin. Identify the coordinates of the known points.

(b) Find an equation of the elliptical tunnel.

(c) Determine the height of the arch 5 feet from the edge of the tunnel.

49. ***Exploration*** The area A of the ellipse

$$\frac{x^2}{a^2} + \frac{y^2}{b^2} = 1$$

is $A = \pi ab$. Let $a + b = 20$.

(a) Write the area of the ellipse as a function of a.

(b) Find the equation of an ellipse with an area of 264 square centimeters.

(c) Complete the table and make a conjecture about the shape of the ellipse with a maximum area.

a	8	9	10	11	12	13
A						

(d) Use a graphing utility to graph the area function, and use the graph to make a conjecture about the shape of the ellipse that yields a maximum area.

50. ***Geometry*** The area of the ellipse in the figure is twice the area of the circle. What is the length of the major axis? (*Hint:* $A = \pi ab$.)

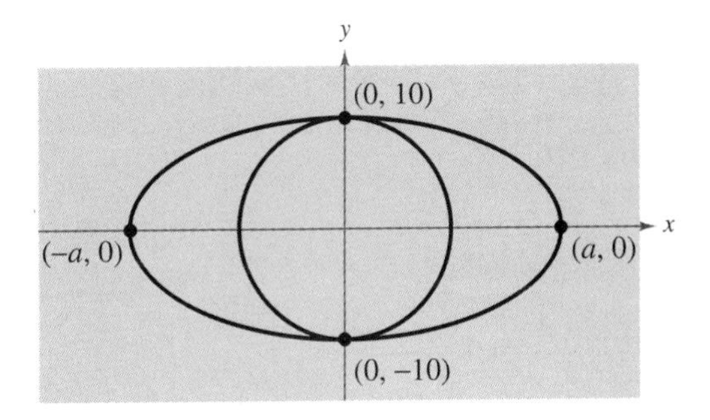

51. Find an equation of the ellipse with vertices $(\pm 5, 0)$ and eccentricity $e = 3/5$.

52. Find an equation of the ellipse with vertices $(0, \pm 8)$ and eccentricity $e = 1/2$.

53. ***Chapter Opener*** Halley's Comet has an elliptical orbit, with the sun at one focus. The eccentricity of the orbit is approximately 0.97. The length of the major axis of the orbit is approximately 36.18 astronomical units. (An astronomical unit is about 93 million miles.) Find an equation for the orbit. Place the center of the orbit at the origin, and place the major axis on the x-axis.

54. ***Chapter Opener*** The comet Encke has an elliptical orbit, with the sun at one focus. Encke ranges from 0.34 to 4.08 astronomical units from the sun. Find an equation of the orbit. Place the center of the orbit at the origin, and place the major axis on the x-axis.

55. ***Satellite Orbit*** The first artificial satellite to orbit earth was Sputnik I (launched by Russia in 1957). Its highest point above earth's surface was 938 kilometers, and its lowest point was 212 kilometers (see figure). The radius of earth is 6378 kilometers. Find the eccentricity of the orbit.

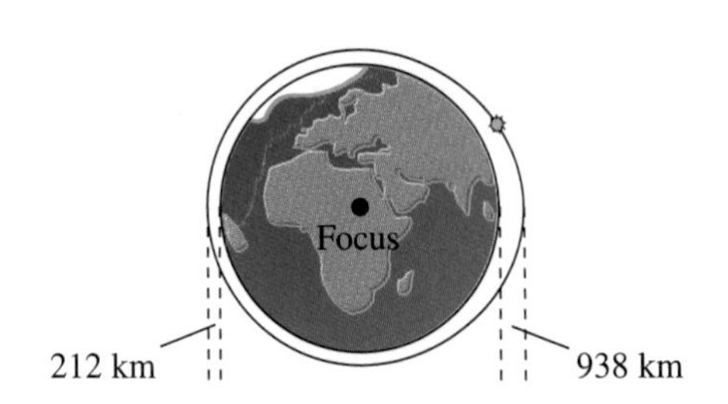

56. ***Exploration***

(a) Show that the equation of an ellipse can be written as

$$\frac{(x-h)^2}{a^2} + \frac{(y-k)^2}{a^2(1-e^2)} = 1.$$

(b) Use a graphing utility to graph the ellipse

$$\frac{(x-2)^2}{4} + \frac{(y-3)^2}{4(1-e^2)} = 1$$

for $e = 0.95$, $e = 0.75$, $e = 0.5$, $e = 0.25$, and $e = 0$.

(c) Use the results of part (b) to make a conjecture about the change in the shape of the ellipse as e approaches 0.

57. ***True or False?*** The graph of $(x^2/4) + y^4 = 1$ is an ellipse. Explain.

58. ***Geometry*** A line segment through a focus with endpoints on the ellipse and perpendicular to the major axis is called a **latus rectum** of the ellipse. Therefore, an ellipse has two latera recta. Knowing the length of the latera recta is helpful in sketching an ellipse because it yields other points on the curve (see figure). Show that the length of each latus rectum is $2b^2/a$.

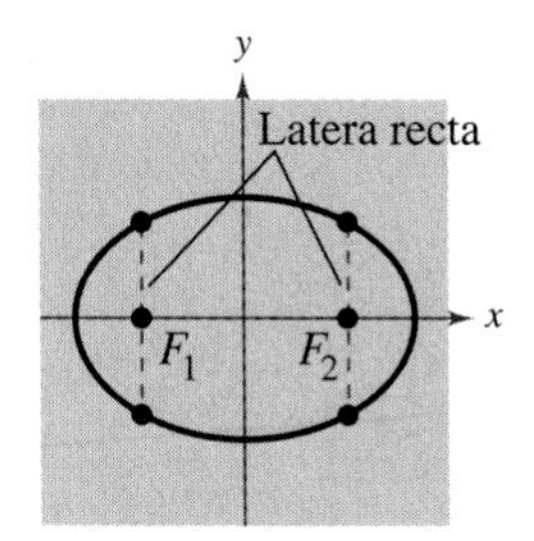

In Exercises 59–62, sketch the graph of the ellipse, making use of the latera recta (see Exercise 58).

59. $\dfrac{x^2}{4} + \dfrac{y^2}{1} = 1$

60. $\dfrac{x^2}{9} + \dfrac{y^2}{16} = 1$

61. $9x^2 + 4y^2 = 36$

62. $5x^2 + 3y^2 = 15$

6.4 Hyperbolas

See Exercise 47 on page 454 for an example of how hyperbolas can be used to locate the position of an explosion that was recorded by three listening stations.

Introduction  *Asymptotes of a Hyperbola* ❐ *Applications*

Introduction

The definition of a hyperbola parallels that of an ellipse. The difference is that for an ellipse the *sum* of the distances between the foci and a point on the ellipse is fixed, whereas for a hyperbola the *difference* of these distances is fixed.

DEFINITION OF HYPERBOLA

A **hyperbola** is the set of all points (x, y) the difference of whose distances from two distinct fixed points (foci) is constant. (See Figure 6.26.)

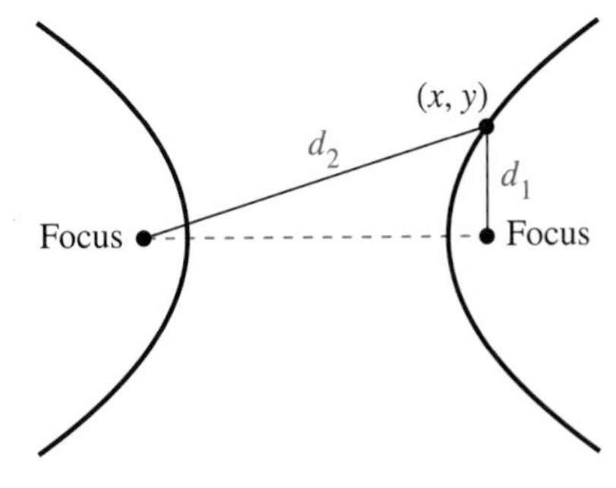

$d_2 - d_1$ is constant.

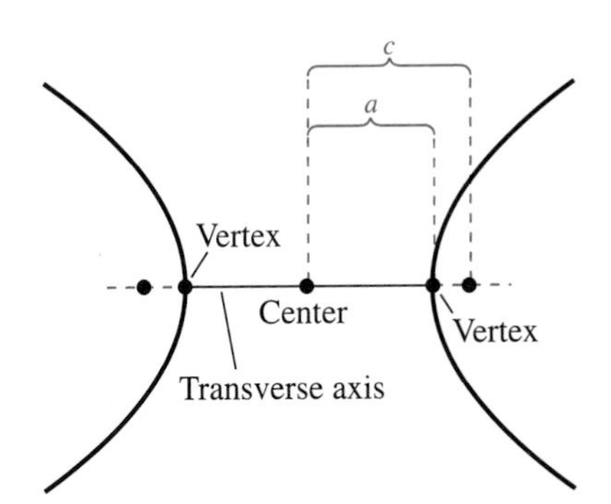

FIGURE 6.26

Every hyperbola has two disconnected **branches.** The line through the two foci intersects a hyperbola at its two **vertices.** The line segment connecting the vertices is called the **transverse axis,** and the midpoint of the transverse axis is called the **center** of the hyperbola. The development of the standard form of the equation of a hyperbola is similar to that of an ellipse.

STANDARD EQUATION OF A HYPERBOLA

The standard form of the equation of a hyperbola with center at (h, k) is

$$\frac{(x-h)^2}{a^2} - \frac{(y-k)^2}{b^2} = 1 \qquad \text{Transverse axis is horizontal.}$$

$$\frac{(y-k)^2}{a^2} - \frac{(x-h)^2}{b^2} = 1. \qquad \text{Transverse axis is vertical.}$$

The vertices are a units from the center, and the foci are c units from the center. Moreover, $c^2 = a^2 + b^2$. If the center of the hyperbola is at the origin $(0, 0)$, the equation takes one of the following forms.

$$\frac{x^2}{a^2} - \frac{y^2}{b^2} = 1 \qquad \text{Transverse axis is horizontal.}$$

$$\frac{y^2}{a^2} - \frac{x^2}{b^2} = 1 \qquad \text{Transverse axis is vertical.}$$

NOTE Note that a, b, and c are related differently for hyperbolas than for ellipses. ▪▪

Figure 6.27 shows the horizontal and vertical orientations for a hyperbola.

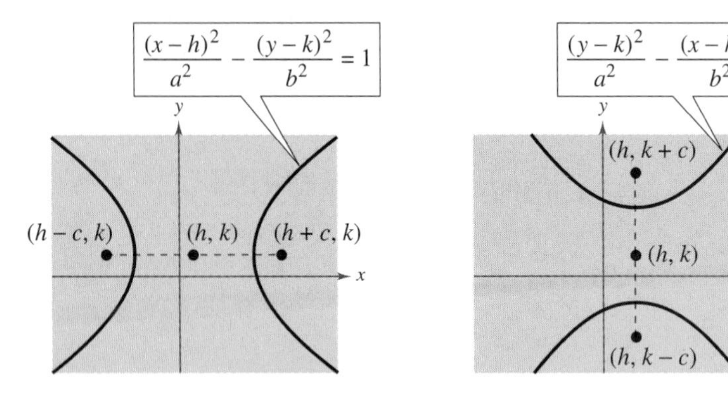

FIGURE 6.27

EXAMPLE 1 *Finding the Standard Equation of a Hyperbola*

Find the standard form of the equation of the hyperbola with foci at $(-1, 2)$ and $(5, 2)$ and vertices at $(0, 2)$ and $(4, 2)$.

Solution

By the Midpoint Formula, the center of the hyperbola occurs at the point $(2, 2)$. Furthermore, $c = 3$ and $a = 2$, and it follows that

$$b^2 = c^2 - a^2 = 3^2 - 2^2 = 9 - 4 = 5.$$

Thus, the equation of the hyperbola is

$$\frac{(x-2)^2}{4} - \frac{(y-2)^2}{5} = 1.$$

Figure 6.28 shows the hyperbola.

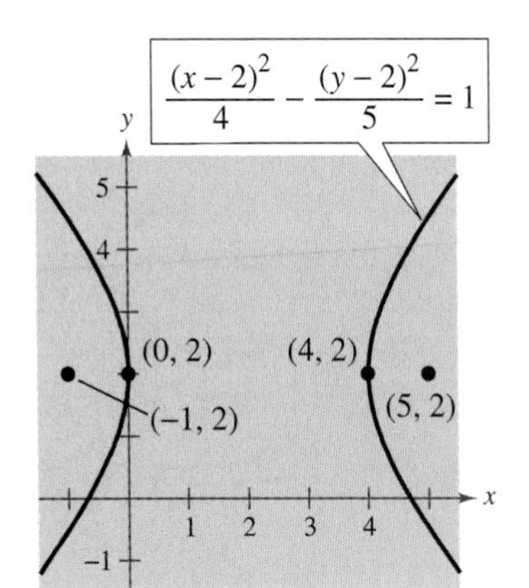

FIGURE 6.28

Exploration

Most graphing utilities have a **parametric** mode. Try using parametric mode to graph the hyperbola given by $x = 2 + 2 \sec t$ and $y = 2 + \sqrt{5} \tan t$. How does the result compare with the graph given in Figure 6.28? (Let the parameter vary from Tmin = 0 to Tmax = 6.28 with Tstep = 0.13.)

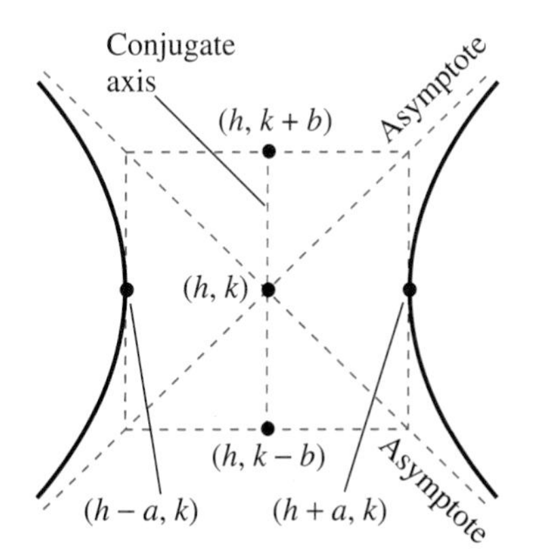

FIGURE 6.29

Asymptotes of a Hyperbola

Each hyperbola has two **asymptotes** that intersect at the center of the hyperbola, as shown in Figure 6.29. The asymptotes pass through the vertices of a rectangle of dimensions $2a$ by $2b$, with its center at (h, k). The line segment of length $2b$ joining $(h, k + b)$ and $(h, k - b)$ [or $(h + b, k)$ and $(h - b, k)$] is the **conjugate axis** of the hyperbola.

ASYMPTOTES OF A HYPERBOLA

$$y = k \pm \frac{b}{a}(x - h)$$ Asymptotes for horizontal transverse axis

$$y = k \pm \frac{a}{b}(x - h)$$ Asymptotes for vertical transverse axis

EXAMPLE 2 *Using Asymptotes to Sketch a Hyperbola*

Sketch the hyperbola whose equation is $4x^2 - y^2 = 16$.

Solution

Divide both sides of the original equation by 16, and rewrite the equation.

$$\frac{x^2}{4} - \frac{y^2}{16} = 1$$ Standard form

From this, you can conclude that the transverse axis is horizontal and the vertices occur at $(-2, 0)$ and $(2, 0)$. Moreover, the ends of the conjugate axis occur at $(0, -4)$ and $(0, 4)$, and you are able to sketch the rectangle shown in Figure 6.30(a). Finally, after drawing the asymptotes through the corners of this rectangle, you can complete the sketch, as shown in Figure 6.30(b).

(a)

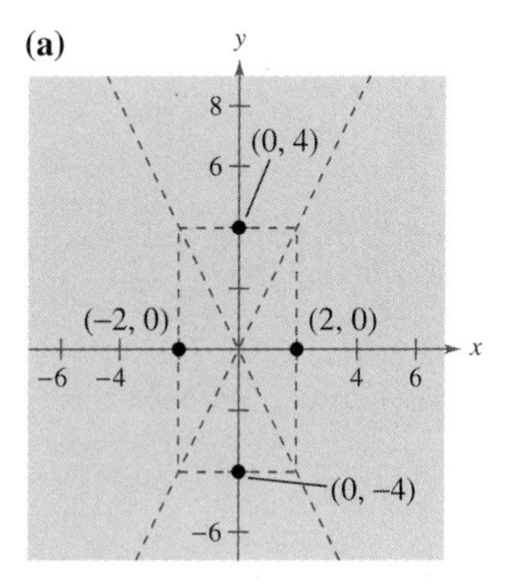

(b)

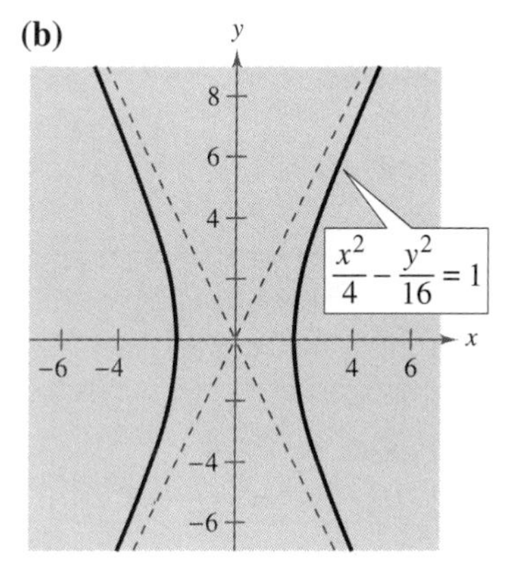

FIGURE 6.30

EXAMPLE 3 *Finding the Asymptotes of a Hyperbola*

Sketch the hyperbola given by $4x^2 - 3y^2 + 8x + 16 = 0$ and find the equations of its asymptotes.

Solution

$$4x^2 - 3y^2 + 8x + 16 = 0 \qquad \text{Original equation}$$
$$4(x^2 + 2x) - 3y^2 = -16$$
$$4(x^2 + 2x + 1) - 3y^2 = -16 + 4$$
$$4(x + 1)^2 - 3y^2 = -12 \qquad \text{Complete the square.}$$
$$\frac{y^2}{4} - \frac{(x + 1)^2}{3} = 1 \qquad \text{Standard form}$$

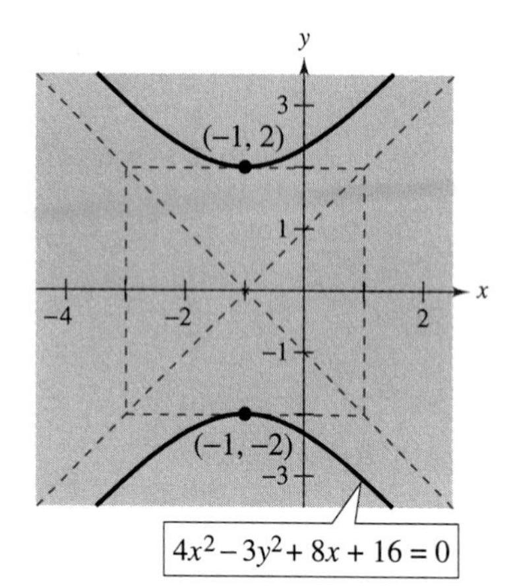

FIGURE 6.31

From this equation you can conclude that the hyperbola is centered at $(-1, 0)$, has vertices at $(-1, 2)$ and $(-1, -2)$, and has a conjugate axis with ends at $(-1 - \sqrt{3}, 0)$ and $(-1 + \sqrt{3}, 0)$. To sketch the hyperbola, draw a rectangle through these four points. The asymptotes are the lines passing through the corners of the rectangle, as shown in Figure 6.31. Finally, using $a = 2$ and $b = \sqrt{3}$, you can conclude that the equations of the asymptotes are

$$y = \frac{2}{\sqrt{3}}(x + 1) \quad \text{and} \quad y = -\frac{2}{\sqrt{3}}(x + 1).$$

NOTE If the constant term F in the equation in Example 3 had been 4 instead of 16, you would have obtained the following degenerate case.

Two Intersecting Lines: $$\frac{y^2}{4} - \frac{(x + 1)^2}{3} = 0$$

TECHNOLOGY

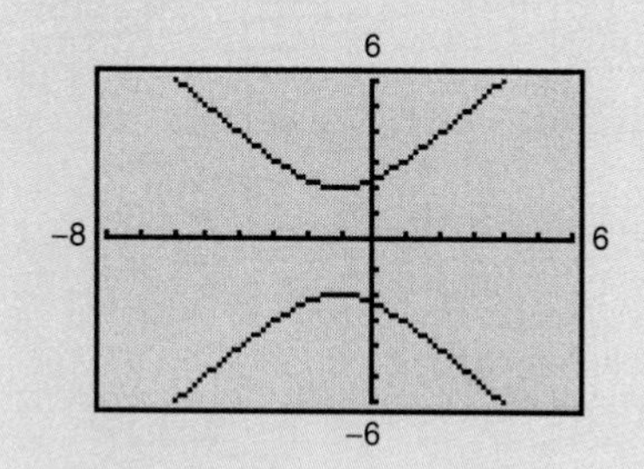

You can use a graphing utility to graph a hyperbola by graphing the upper and lower portions in the same viewing rectangle. For instance, to graph the hyperbola in Example 3, first solve for y to get

$$y_1 = 2\sqrt{1 + \frac{(x + 1)^2}{3}} \quad \text{and} \quad y_2 = -2\sqrt{1 + \frac{(x + 1)^2}{3}}.$$

Use a viewing rectangle in which $-8 \le x \le 6$ and $-6 \le y \le 6$. You should obtain the graph shown at the left.

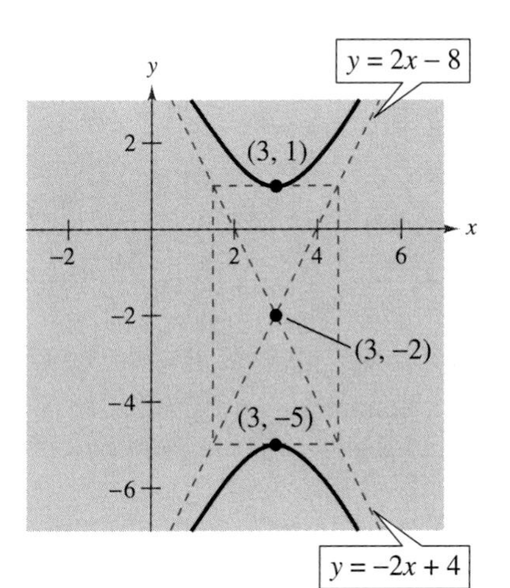

FIGURE 6.32

EXAMPLE 4 *Using Asymptotes to Find the Standard Equation*

Find the standard form of the equation of the hyperbola having vertices at $(3, -5)$ and $(3, 1)$ and having asymptotes $y = 2x - 8$ and $y = -2x + 4$, as shown in Figure 6.32.

Solution

According to the Midpoint Formula, the center of the hyperbola is at $(3, -2)$. Furthermore, the hyperbola has a vertical transverse axis with $a = 3$. From the given equations, you can determine the slopes of the asymptotes to be

$$m_1 = 2 = \frac{a}{b} \qquad \text{and} \qquad m_2 = -2 = -\frac{a}{b}$$

and, because $a = 3$, you can conclude that $b = \frac{3}{2}$. Thus, the standard equation is

$$\frac{(y+2)^2}{9} - \frac{(x-3)^2}{9/4} = 1.$$

As with ellipses, the **eccentricity** of a hyperbola is

$$e = \frac{c}{a} \qquad \text{Eccentricity}$$

and because $c > a$ it follows that $e > 1$. If the eccentricity is large, the branches of the hyperbola are nearly flat, as shown in Figure 6.33(a). If the eccentricity is close to 1, the branches of the hyperbola are more pointed, as shown in Figure 6.33(b).

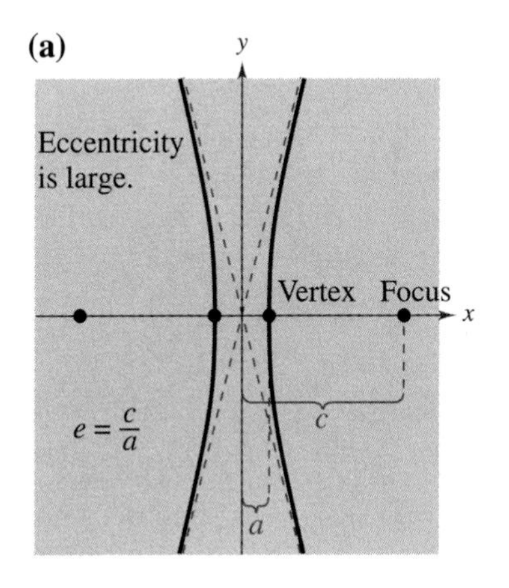

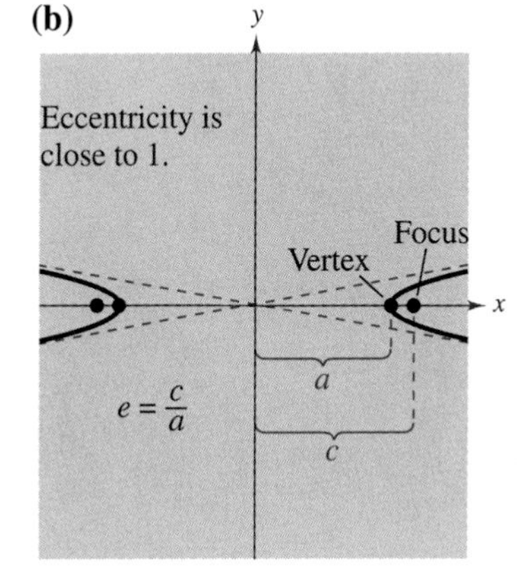

FIGURE 6.33

Applications

The following application was developed during World War II. It shows how the properties of hyperbolas can be used in radar and other detection systems.

EXAMPLE 5 An Application Involving Hyperbolas

Real Life

Two microphones, 1 mile apart, record an explosion. Microphone A receives the sound 2 seconds before microphone B. Where did the explosion occur?

Solution

Assuming sound travels at 1100 feet per second, you know that the explosion took place 2200 feet farther from B than from A, as shown in Figure 6.34. The locus of all points that are 2200 feet closer to A than to B is one branch of the hyperbola $(x^2/a^2) - (y^2/b^2) = 1$, where

$$c = \frac{5280}{2} = 2640 \quad \text{and} \quad a = \frac{2200}{2} = 1100.$$

Thus, $b^2 = c^2 - a^2 = 5{,}759{,}600$, and you conclude that the explosion occurred somewhere on the right branch of the hyperbola given by

$$\frac{x^2}{1{,}210{,}000} - \frac{y^2}{5{,}759{,}600} = 1.$$

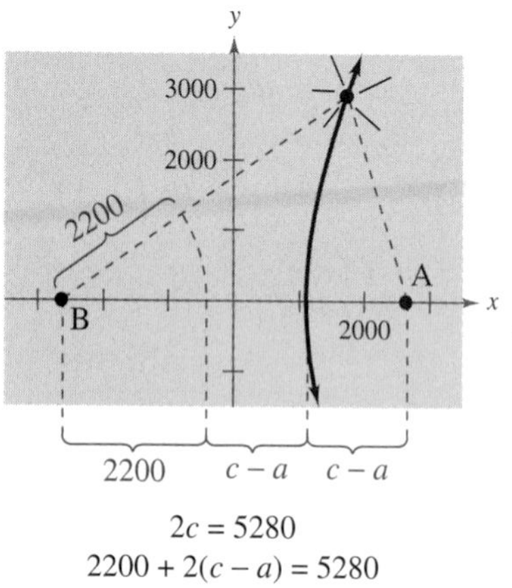

$2c = 5280$

$2200 + 2(c - a) = 5280$

FIGURE 6.34

Another interesting application of conic sections involves the orbits of comets in our solar system. Of the 610 comets identified prior to 1970, 245 have elliptical orbits, 295 have parabolic orbits, and 70 have hyperbolic orbits. The center of the sun is a focus of each of these orbits, and each orbit has a vertex at the point where the comet is closest to the sun, as shown in Figure 6.35. Undoubtedly, there have been many comets with parabolic or hyperbolic orbits that were not identified. We only get to see such comets *once*. Comets with elliptical orbits, such as Halley's comet, are the only ones that remain in our solar system.

If p is the distance between the vertex and the focus in meters, and v is the velocity of the comet at the vertex in meters per second, the type of orbit is determined as follows.

1. *Ellipse:* $v < \sqrt{2GM/p}$
2. *Parabola:* $v = \sqrt{2GM/p}$
3. *Hyperbola:* $v > \sqrt{2GM/p}$

In each of these equations, $M \approx 1.991 \times 10^{30}$ kilograms (the mass of the sun) and $G \approx 6.67 \times 10^{-11}$ cubic meters per gram-second squared.

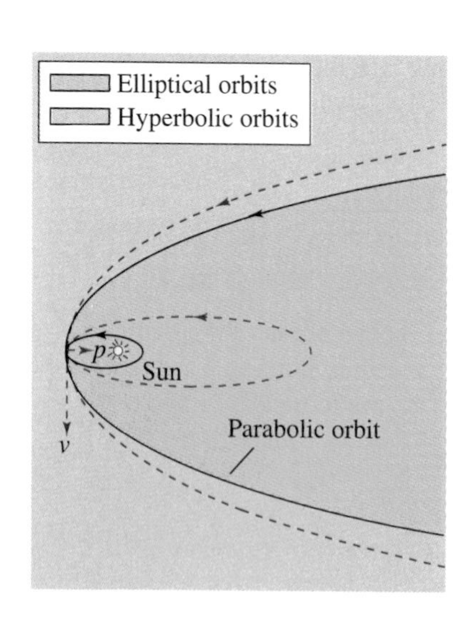

FIGURE 6.35

NOTE The test at the right is valid *if* the graph is a conic. The test does not apply to equations such as $x^2 + y^2 = -1$, which is not a conic.

CLASSIFYING A CONIC FROM ITS GENERAL EQUATION

The graph of $Ax^2 + Cy^2 + Dx + Ey + F = 0$ is one of the following.

1. *Circle:* $A = C$
2. *Parabola:* $AC = 0$ — $A = 0$ or $C = 0$, but not both.
3. *Ellipse:* $AC > 0$ — A and C have like signs.
4. *Hyperbola:* $AC < 0$ — A and C have unlike signs.

EXAMPLE 6 *Classifying Conics from General Equations*

Classify each graph.

a. $4x^2 - 9x + y - 5 = 0$

b. $4x^2 - y^2 + 8x - 6y + 4 = 0$

c. $2x^2 + 4y^2 - 4x + 12y = 0$

Solution

a. For the equation $4x^2 - 9x + y - 5 = 0$, you have

$AC = 4(0) = 0.$ Parabola

Thus, the graph is a parabola.

b. For the equation $4x^2 - y^2 + 8x - 6y + 4 = 0$, you have

$AC = 4(-1) < 0.$ Hyperbola

Thus, the graph is a hyperbola.

c. For the equation $2x^2 + 4y^2 - 4x + 12y = 0$, you have

$AC = 2(4) > 0.$ Ellipse

Thus, the graph is an ellipse.

The first woman to be credited with detecting a new comet was the English astronomer Caroline Herschel (1750–1848). During her long life, Herschel discovered a total of eight new comets.

GROUP ACTIVITY

SKETCHING CONICS

Sketch each of the conics described in Example 6. With others in your group, discuss procedures that allow you to sketch the conics efficiently.

WARM UP

Find the distance between the two points.

1. $(4, 1), (10, 6)$
2. $(-1, 5), (3, -2)$

Sketch the graphs of the lines on the same set of coordinate axes.

3. $y = \pm\frac{1}{2}x$
4. $y = 3 \pm \frac{1}{2}x$
5. $y = 3 \pm \frac{1}{2}(x - 4)$
6. $y = \pm\frac{1}{2}(x - 4)$

Identify the graph of the equation.

7. $x^2 + 4y = 4$
8. $x^2 + 4y^2 = 4$
9. $4x^2 + 4y^2 = 4$
10. $x + 4y^2 = 4$

6.4 Exercises

In Exercises 1–6, match the equation with its graph. [The graphs are labeled (a), (b), (c), (d), (e), and (f).]

1. $\dfrac{x^2}{16} - \dfrac{y^2}{4} = 1$
2. $\dfrac{y^2}{16} - \dfrac{x^2}{4} = 1$
3. $\dfrac{y^2}{9} - \dfrac{x^2}{16} = 1$
4. $\dfrac{y^2}{16} - \dfrac{x^2}{9} = 1$
5. $\dfrac{(x - 1)^2}{16} - \dfrac{y^2}{4} = 1$
6. $\dfrac{(x + 1)^2}{16} - \dfrac{(y - 2)^2}{9} = 1$

(a)

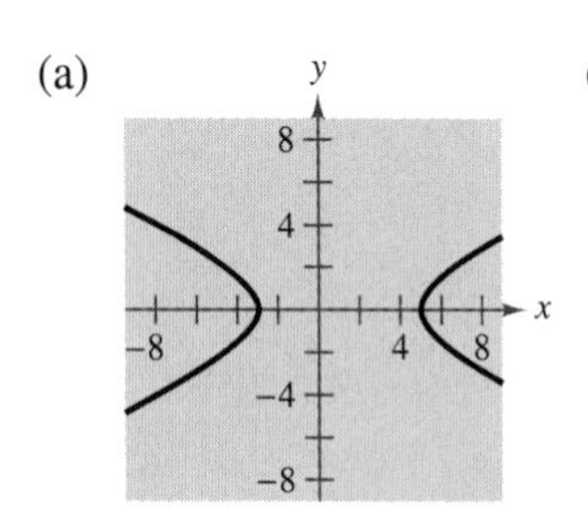

(b)

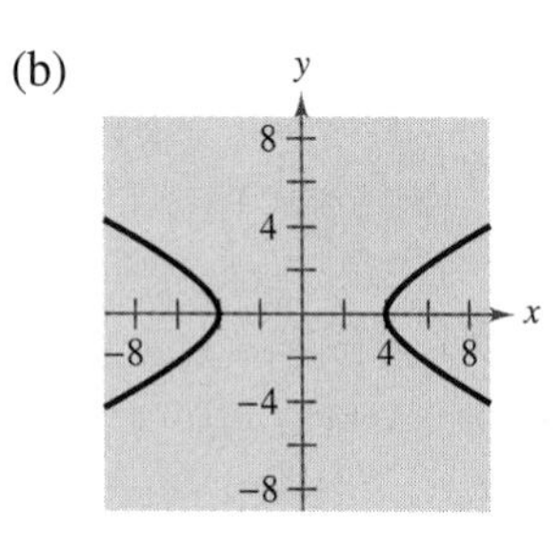

(c)

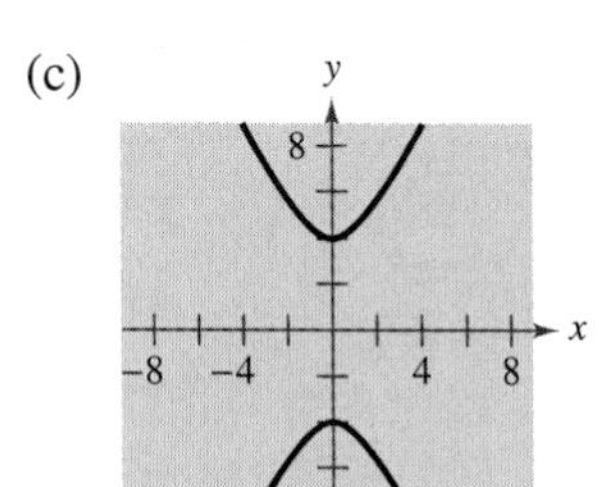

(d)

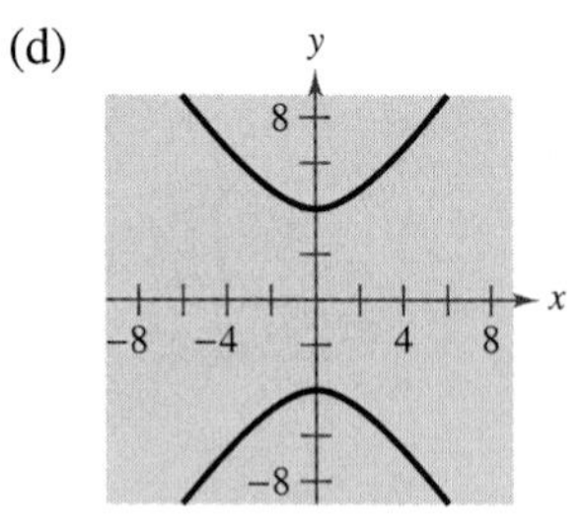

(e)

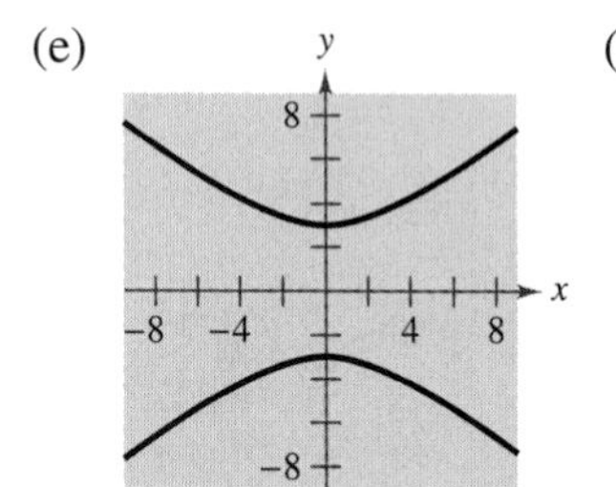

(f)

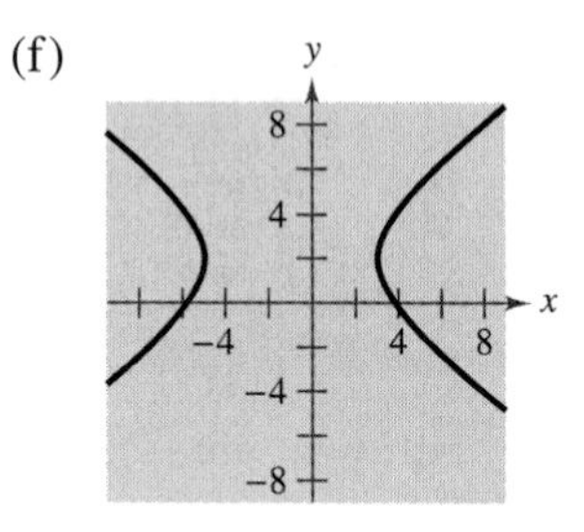

In Exercises 7–20, find the center, vertices, and foci of the hyperbola, and sketch its graph using asymptotes as an aid.

7. $x^2 - y^2 = 1$
8. $\dfrac{x^2}{9} - \dfrac{y^2}{16} = 1$

9. $\dfrac{y^2}{1} - \dfrac{x^2}{4} = 1$

10. $\dfrac{y^2}{9} - \dfrac{x^2}{1} = 1$

11. $\dfrac{y^2}{25} - \dfrac{x^2}{144} = 1$

12. $\dfrac{x^2}{36} - \dfrac{y^2}{4} = 1$

13. $\dfrac{(x-1)^2}{4} - \dfrac{(y+2)^2}{1} = 1$

14. $\dfrac{(x+1)^2}{144} - \dfrac{(y-4)^2}{25} = 1$

15. $(y+6)^2 - (x-2)^2 = 1$

16. $\dfrac{(y-1)^2}{1/4} - \dfrac{(x+3)^2}{1/9} = 1$

17. $9x^2 - y^2 - 36x - 6y + 18 = 0$

18. $x^2 - 9y^2 + 36y - 72 = 0$

19. $x^2 - 9y^2 + 2x - 54y - 80 = 0$

20. $16y^2 - x^2 + 2x + 64y + 63 = 0$

In Exercises 21–24, find the center, vertices, and foci of the hyperbola and graph the hyperbola and its asymptotes with the aid of a graphing utility.

21. $2x^2 - 3y^2 = 6$

22. $3y^2 - 5x^2 = 15$

23. $9y^2 - x^2 + 2x + 54y + 62 = 0$

24. $9x^2 - y^2 + 54x + 10y + 55 = 0$

In Exercises 25–32, find an equation of the specified hyperbola with its center at the origin.

25.

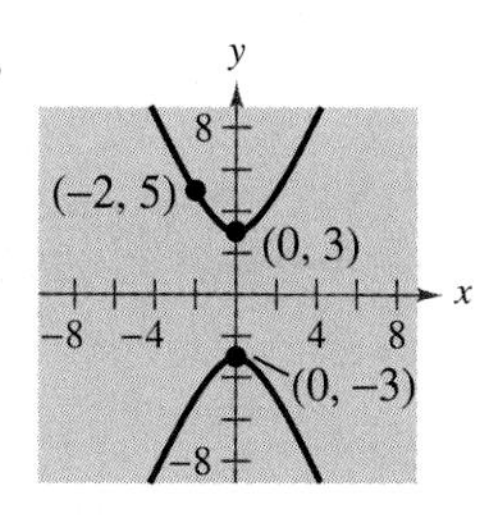

26.

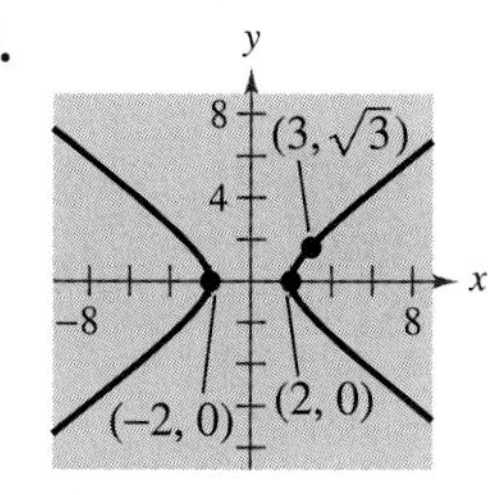

27. Vertices: $(0, \pm 2)$; Foci: $(0, \pm 4)$

28. Vertices: $(\pm 3, 0)$; Foci: $(\pm 5, 0)$

29. Vertices: $(\pm 1, 0)$; Asymptotes: $y = \pm 3x$

30. Vertices: $(0, \pm 3)$; Asymptotes: $y = \pm 3x$

31. Foci: $(0, \pm 8)$; Asymptotes: $y = \pm 4x$

32. Foci: $(\pm 10, 0)$; Asymptotes: $y = \pm \frac{3}{4}x$

In Exercises 33–44, find an equation for the specified hyperbola.

33.

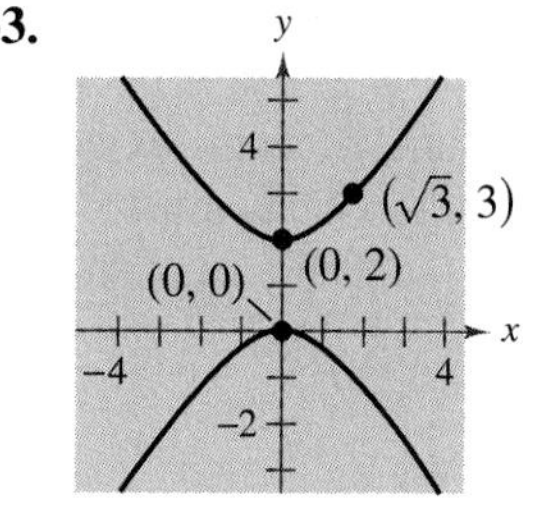

34.

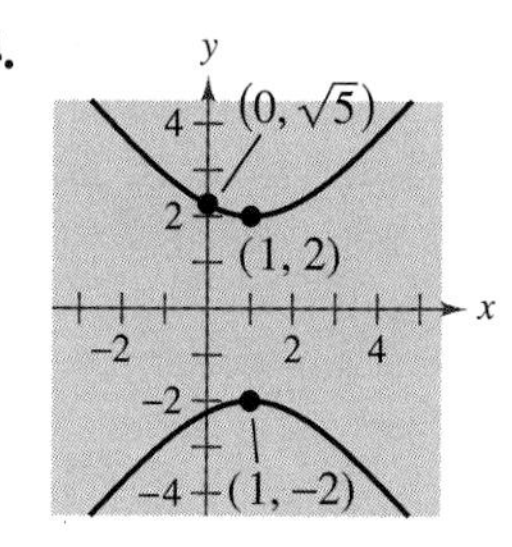

35.

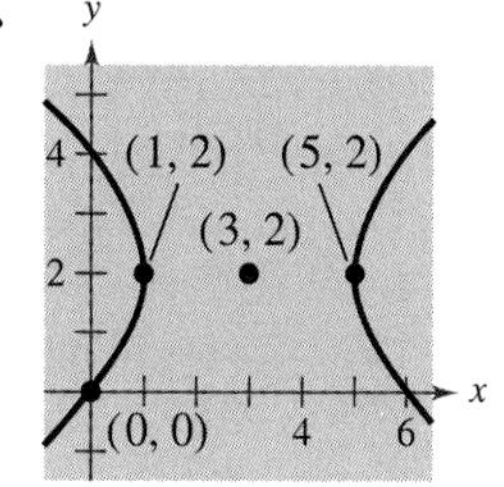

36.

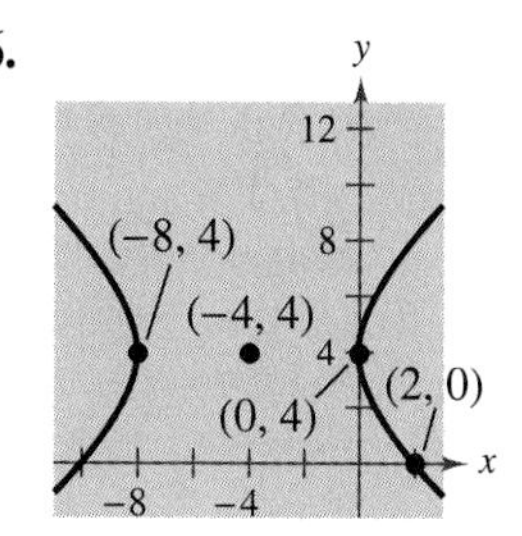

37. Vertices: $(2, 0), (6, 0)$; Foci: $(0, 0), (8, 0)$

38. Vertices: $(2, 3), (2, -3)$; Foci: $(2, 5), (2, -5)$

39. Vertices: $(4, 1), (4, 9)$; Foci: $(4, 0), (4, 10)$

40. Vertices: $(-2, 1), (2, 1)$; Foci: $(-3, 1), (3, 1)$

41. Vertices: $(2, 3), (2, -3)$;
Passes through the point $(0, 5)$

42. Vertices: $(-2, 1), (2, 1)$;
Passes through the point $(4, 3)$

43. Vertices: $(0, 2), (6, 2)$;
Asymptotes: $y = \frac{2}{3}x,\ y = 4 - \frac{2}{3}x$

44. Vertices: $(3, 0), (3, 4)$;
Asymptotes: $y = \frac{2}{3}x,\ y = 4 - \frac{2}{3}x$

In Exercises 45 and 46, determine what part of the graph of a hyperbola

$$\frac{(x-3)^2}{4} - \frac{(y-1)^2}{9} = 1$$

is given by the equation.

45. $x = 3 - \frac{2}{3}\sqrt{1 + (y-1)^2}$

46. $y = 1 + \frac{3}{2}\sqrt{(x-3)^2 - 4}$

47. ***Sound Location*** Three listening stations located at (4400, 0), (4400, 1100), and (−4400, 0) monitor an explosion. If the last two stations detect the explosion 1 second and 5 seconds after the first, respectively, determine the coordinates of the explosion. (Assume that the coordinate system is measured in feet and that sound travels at 1100 feet per second.)

48. ***LORAN*** Long distance radio navigation for aircraft and ships uses synchronized pulses transmitted by widely separated transmitting stations. These pulses travel at the speed of light (186,000 miles per second). The difference in the times of arrival of these pulses at an aircraft or ship is constant on a hyperbola having the transmitting stations as foci. Assume that two stations, 300 miles apart, are positioned on the rectangular coordinate system at points with coordinates (−150, 0) and (150, 0), and that a ship is traveling on a path with coordinates $(x, 75)$ (see figure). Find the x-coordinate of the position of the ship if the time difference between the pulses from the transmitting stations is 1000 microseconds (0.001 second).

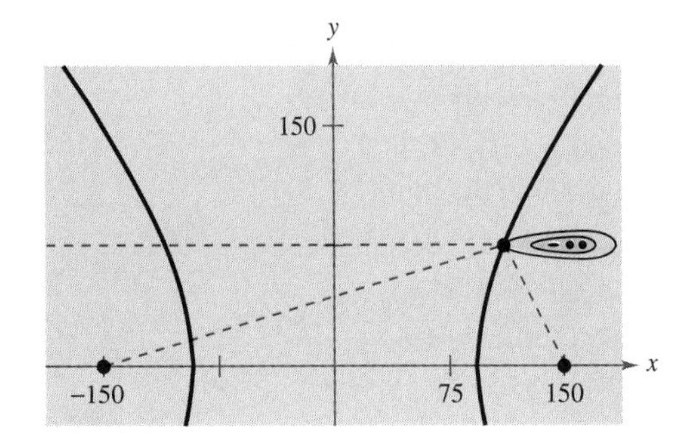

49. ***Hyperbolic Mirror*** A hyperbolic mirror (used in some telescopes) has the property that a light ray directed at a focus will be reflected to the other focus (see figure). The focus of a hyperbolic mirror has coordinates (24, 0). Find the vertex of the mirror if its mount has coordinates (24, 24).

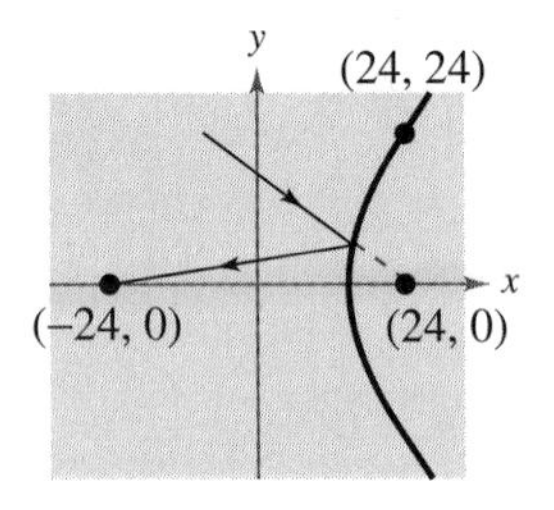

50. Consider a hyperbola centered at the origin with a horizontal transverse axis. Use the definition of a hyperbola to derive its standard form.

In Exercises 51–58, classify the graph of the equation as a circle, a parabola, an ellipse, or a hyperbola.

51. $x^2 + y^2 - 6x + 4y + 9 = 0$

52. $x^2 + 4y^2 - 6x + 16y + 21 = 0$

53. $4x^2 - y^2 - 4x - 3 = 0$

54. $y^2 - 4y - 4x = 0$

55. $4x^2 + 3y^2 + 8x - 24y + 51 = 0$

56. $4y^2 - 2x^2 - 4y - 8x - 15 = 0$

57. $25x^2 - 10x - 200y - 119 = 0$

58. $4x^2 + 4y^2 - 16y + 15 = 0$

Review **Solve Exercises 59–62 as a review of the skills and problem-solving techniques you learned in previous sections.**

59. Subtract: $(x^3 - 3x^2) - (6 - 2x - 4x^2)$

60. Multiply: $\left(3x - \frac{1}{2}\right)(x + 4)$

61. Divide: $\dfrac{x^3 - 3x + 4}{x + 2}$

62. Expand: $[(x + y) + 3]^2$

6.5 Rotation of Conics

See Exercises 5–16 on page 462 for examples of how rotation of the coordinate axes can help you identify the graph of a general second-degree equation.

Rotation ▫ *Invariants Under Rotation*

Rotation

In the previous section you learned that the equation of a conic with axes parallel to one of the coordinate axes has a standard form that can be written in the general form

$$Ax^2 + Cy^2 + Dx + Ey + F = 0. \qquad \text{Horizontal or vertical axis}$$

In this section you will study the equations of conics whose axes are rotated so that they are not parallel to either the x-axis or the y-axis. The general equation for such conics contains an xy-term.

$$Ax^2 + Bxy + Cy^2 + Dx + Ey + F = 0 \qquad \text{Equation in } xy\text{-plane}$$

To eliminate this xy-term, you can use a procedure called **rotation of axes.** The objective is to rotate the x- and y-axes until they are parallel to the axes of the conic. The rotated axes are denoted as the x'-axis and the y'-axis, as shown in Figure 6.36. After the rotation, the equation of the conic in the new $x'y'$-plane will have the form

$$A'(x')^2 + C'(y')^2 + D'x' + E'y' + F' = 0. \qquad \text{Equation in } x'y'\text{-plane}$$

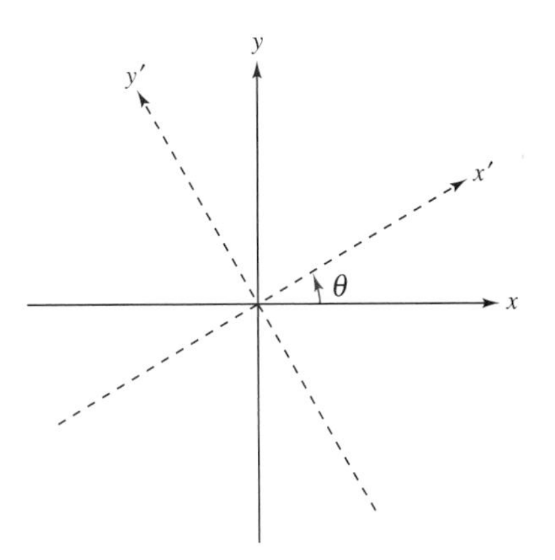

FIGURE 6.36

Because this equation has no xy-term, you can obtain a standard form by completing the square. The following theorem identifies how much to rotate the axes to eliminate the xy-term and also the equations for determining the new coefficients A', C', D', E', and F'.

ROTATION OF AXES TO ELIMINATE AN XY-TERM

The general second-degree equation $Ax^2 + Bxy + Cy^2 + Dx + Ey + F = 0$ can be rewritten as

$$A'(x')^2 + C'(y')^2 + D'x' + E'y' + F' = 0$$

by rotating the coordinate axes through an angle θ, where

$$\cot 2\theta = \frac{A - C}{B}.$$

The coefficients of the new equation are obtained by making the substitutions

$$x = x' \cos \theta - y' \sin \theta \qquad \text{and} \qquad y = x' \sin \theta + y' \cos \theta.$$

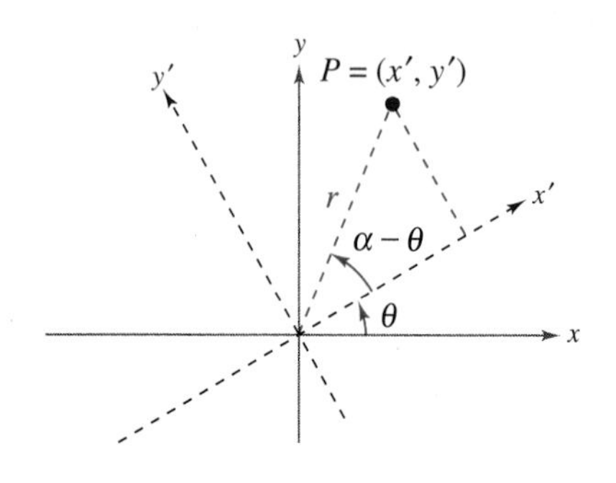

Rotated: $x' = r\cos(\alpha - \theta)$
$y' = r\sin(\alpha - \theta)$

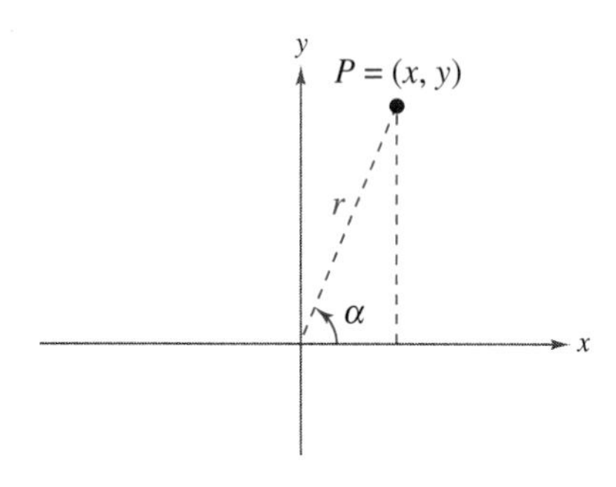

Original: $x = r\cos\alpha$
$y = r\sin\alpha$

FIGURE 6.37

PROOF

You need to discover how the coordinates in the xy-system are related to the coordinates in the $x'y'$-system. To do this, choose a point $P = (x, y)$ in the original system and attempt to find its coordinates (x', y') in the rotated system. In either system the distance r between the point P and the origin is the same. Thus, the equations for x, y, x', and y' are those given in Figure 6.37. Using the formulas for the sine and cosine of the difference of two angles, you have the following.

$$\begin{aligned} x' &= r\cos(\alpha - \theta) \\ &= r(\cos\alpha\cos\theta + \sin\alpha\sin\theta) \\ &= r\cos\alpha\cos\theta + r\sin\alpha\sin\theta \\ &= x\cos\theta + y\sin\theta \\ y' &= r\sin(\alpha - \theta) \\ &= r(\sin\alpha\cos\theta - \cos\alpha\sin\theta) \\ &= r\sin\alpha\cos\theta - r\cos\alpha\sin\theta \\ &= y\cos\theta - x\sin\theta \end{aligned}$$

Solving this system for x and y yields

$$\begin{aligned} x &= x'\cos\theta - y'\sin\theta \\ y &= x'\sin\theta + y'\cos\theta. \end{aligned}$$

Finally, by substituting these values for x and y in the original equation and collecting terms, you obtain

$$\begin{aligned} A' &= A\cos^2\theta + B\cos\theta\sin\theta + C\sin^2\theta \\ C' &= A\sin^2\theta - B\cos\theta\sin\theta + C\cos^2\theta \\ D' &= D\cos\theta + E\sin\theta \\ E' &= -D\sin\theta + E\cos\theta \\ F' &= F. \end{aligned}$$

To eliminate the $x'y'$-term, you must select θ so that $B' = 0$.

$$\begin{aligned} B' &= 2(C - A)\sin\theta\cos\theta + B(\cos^2\theta - \sin^2\theta) \\ &= (C - A)\sin 2\theta + B\cos 2\theta \\ &= B(\sin 2\theta)\left(\frac{C - A}{B} + \cot 2\theta\right) \\ &= 0, \quad \sin 2\theta \neq 0 \end{aligned}$$

If $B = 0$, no rotation is necessary because the xy-term is not present in the original equation. If $B \neq 0$, the only way to make $B' = 0$ is to let

$$\cot 2\theta = \frac{A - C}{B}, \qquad B \neq 0.$$

Thus, you have established the desired results.

EXAMPLE 1 *Rotation of Axes for a Hyperbola*

Write the equation $xy - 1 = 0$ in standard form.

Solution

Because $A = 0$, $B = 1$, and $C = 0$, you have

$$\cot 2\theta = \frac{A - C}{B} = 0 \quad \Rightarrow \quad 2\theta = \frac{\pi}{2} \quad \Rightarrow \quad \theta = \frac{\pi}{4}$$

which implies that

$$x = x' \cos \frac{\pi}{4} - y' \sin \frac{\pi}{4}$$
$$= x'\left(\frac{\sqrt{2}}{2}\right) - y'\left(\frac{\sqrt{2}}{2}\right)$$
$$= \frac{x' - y'}{\sqrt{2}}$$

and

$$y = x' \sin \frac{\pi}{4} + y' \cos \frac{\pi}{4}$$
$$= x'\left(\frac{\sqrt{2}}{2}\right) + y'\left(\frac{\sqrt{2}}{2}\right)$$
$$= \frac{x' + y'}{\sqrt{2}}.$$

The equation in the $x'y'$-system is obtained by substituting these expressions in the equation $xy - 1 = 0$.

$$\left(\frac{x' - y'}{\sqrt{2}}\right)\left(\frac{x' + y'}{\sqrt{2}}\right) - 1 = 0$$
$$\frac{(x')^2 - (y')^2}{2} - 1 = 0$$
$$\frac{(x')^2}{(\sqrt{2})^2} - \frac{(y')^2}{(\sqrt{2})^2} = 1 \qquad \text{Standard form}$$

In the $x'y'$-system, this is a hyperbola centered at the origin with vertices at $(\pm\sqrt{2}, 0)$, as shown in Figure 6.38. To find the coordinates of the vertices in the xy-system, substitute the coordinates $(\pm\sqrt{2}, 0)$ in the equations

$$x = \frac{x' - y'}{\sqrt{2}} \quad \text{and} \quad y = \frac{x' + y'}{\sqrt{2}}.$$

This substitution yields the vertices $(1, 1)$ and $(-1, -1)$ in the xy-system. Note also that the asymptotes of the hyperbola have equations $y' = \pm x'$, which correspond to the original x- and y-axes.

Study Tip

Remember that the substitutions

$$x = x' \cos \theta - y' \sin \theta$$

and

$$y = x' \sin \theta + y' \cos \theta$$

were developed to eliminate the $x'y'$-term in the rotated system. You can use this as a check on your work. In other words, if your final equation contains an $x'y'$-term, you know that you made a mistake.

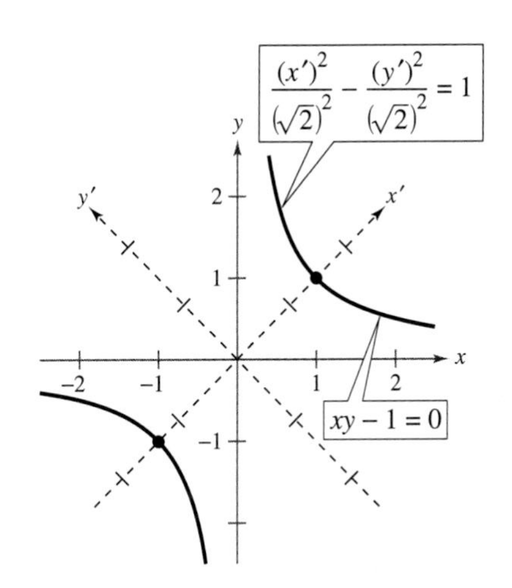

Vertices:
In $x'y'$-system: $(\sqrt{2}, 0)$, $(-\sqrt{2}, 0)$
In xy-system: $(1, 1)$, $(-1, -1)$

FIGURE 6.38

EXAMPLE 2 *Rotation of Axes for an Ellipse*

Sketch the graph of $7x^2 - 6\sqrt{3}xy + 13y^2 - 16 = 0$.

Solution

Because $A = 7$, $B = -6\sqrt{3}$, and $C = 13$, you have

$$\cot 2\theta = \frac{A - C}{B} = \frac{7 - 13}{-6\sqrt{3}} = \frac{1}{\sqrt{3}}$$

which implies that $\theta = \pi/6$. The equation in the $x'y'$-system is obtained by making the substitutions

$$\begin{aligned} x &= x' \cos\frac{\pi}{6} - y' \sin\frac{\pi}{6} \\ &= x'\left(\frac{\sqrt{3}}{2}\right) - y'\left(\frac{1}{2}\right) \\ &= \frac{\sqrt{3}x' - y'}{2} \end{aligned}$$

and

$$\begin{aligned} y &= x' \sin\frac{\pi}{6} + y' \cos\frac{\pi}{6} \\ &= x'\left(\frac{1}{2}\right) + y'\left(\frac{\sqrt{3}}{2}\right) \\ &= \frac{x' + \sqrt{3}y'}{2} \end{aligned}$$

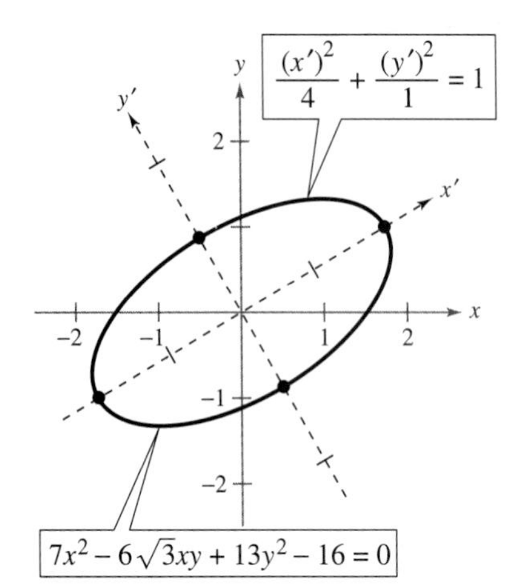

Vertices:
In $x'y'$-system: $(\pm 2, 0)$, $(0, \pm 1)$
In xy-system: $(\pm\sqrt{3}, \pm 1)$, $\left(\pm\frac{1}{2}, \mp\frac{\sqrt{3}}{2}\right)$

FIGURE 6.39

into the original equation. Thus, you have

$$7x^2 - 6\sqrt{3}xy + 13y^2 - 16 = 0$$

$$7\left(\frac{\sqrt{3}x' - y'}{2}\right)^2 - 6\sqrt{3}\left(\frac{\sqrt{3}x' - y'}{2}\right)\left(\frac{x' + \sqrt{3}y'}{2}\right) + 13\left(\frac{x' + \sqrt{3}y'}{2}\right)^2 - 16 = 0$$

which simplifies to

$$\begin{aligned} 4(x')^2 + 16(y') - 16 &= 0 \\ 4(x')^2 + 16(y')^2 &= 16 \\ \frac{(x')^2}{4} + \frac{(y')^2}{1} &= 1. \end{aligned} \qquad \text{Standard form}$$

This is the equation of an ellipse centered at the origin with vertices $(\pm 2, 0)$ in the $x'y'$-system, as shown in Figure 6.39.

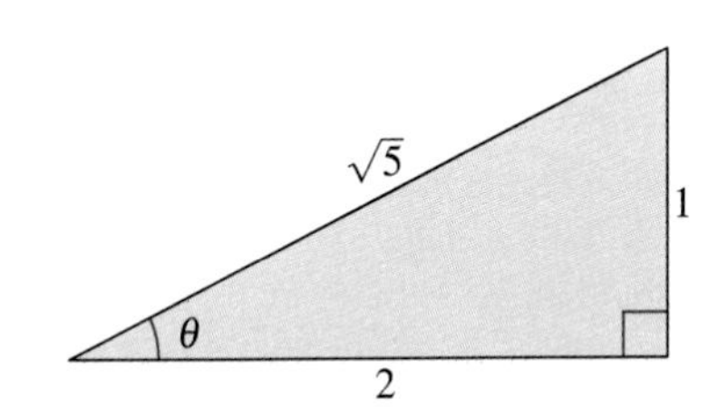

FIGURE 6.40

EXAMPLE 3 *Rotation of Axes for a Parabola*

Sketch the graph of $x^2 - 4xy + 4y^2 + 5\sqrt{5}y + 1 = 0$.

Solution

Because $A = 1$, $B = -4$, and $C = 4$, you have

$$\cot 2\theta = \frac{A - C}{B} = \frac{1 - 4}{-4} = \frac{3}{4}.$$

Using the identity $\cot 2\theta = (\cot^2\theta - 1)/(2\cot\theta)$ produces

$$\cot 2\theta = \frac{3}{4} = \frac{\cot^2\theta - 1}{2\cot\theta}$$

from which you obtain the equation

$$4\cot^2\theta - 4 = 6\cot\theta$$
$$4\cot^2\theta - 6\cot\theta - 4 = 0$$
$$(2\cot\theta - 4)(2\cot\theta + 1) = 0.$$

Considering $0 < \theta < \pi/2$, you have $2\cot\theta = 4$. Thus,

$$\cot\theta = 2 \quad \Rightarrow \quad \theta \approx 26.6°.$$

From the triangle in Figure 6.40, you obtain $\sin\theta = 1/\sqrt{5}$ and $\cos\theta = 2/\sqrt{5}$. Consequently, you use the substitutions

$$x = x'\cos\theta - y'\sin\theta = x'\left(\frac{2}{\sqrt{5}}\right) - y'\left(\frac{1}{\sqrt{5}}\right) = \frac{2x' - y'}{\sqrt{5}}$$

$$y = x'\sin\theta + y'\cos\theta = x'\left(\frac{1}{\sqrt{5}}\right) + y'\left(\frac{2}{\sqrt{5}}\right) = \frac{x' + 2y'}{\sqrt{5}}.$$

Substituting these expressions in the original equation, you have

$$x^2 - 4xy + 4y^2 + 5\sqrt{5}y + 1 = 0$$

$$\left(\frac{2x' - y'}{\sqrt{5}}\right)^2 - 4\left(\frac{2x' - y'}{\sqrt{5}}\right)\left(\frac{x' + 2y'}{\sqrt{5}}\right) + 4\left(\frac{x' + 2y'}{\sqrt{5}}\right)^2 + 5\sqrt{5}\left(\frac{x' + 2y'}{\sqrt{5}}\right) + 1 = 0$$

which simplifies as follows.

$$5(y')^2 + 5x' + 10y' + 1 = 0$$

$$5(y' + 1)^2 = -5x' + 4 \qquad \text{Complete the square.}$$

$$(y' + 1)^2 = (-1)\left(x' - \frac{4}{5}\right) \qquad \text{Standard form}$$

The graph of this equation is a parabola with vertex at $\left(\frac{4}{5}, -1\right)$. Its axis is parallel to the x'-axis in the $x'y'$-system, as shown in Figure 6.41.

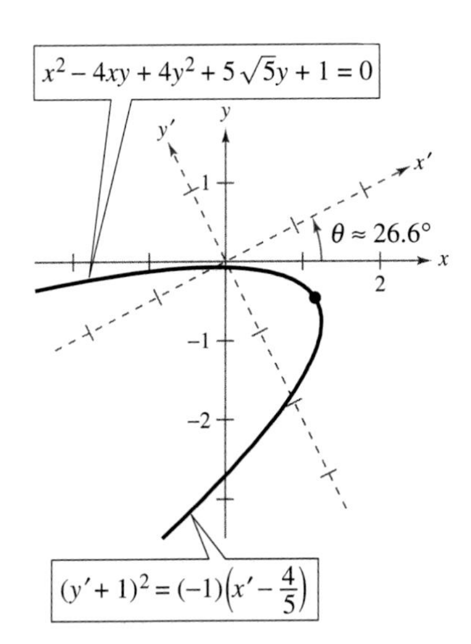

Vertex:
In $x'y'$-system: $\left(\frac{4}{5}, -1\right)$
In xy-system: $\left(\frac{13}{5\sqrt{5}}, -\frac{6}{5\sqrt{5}}\right)$

FIGURE 6.41

Invariants Under Rotation

In the rotation of axes theorem listed at the beginning of this section, note that the constant term is the same in both equations, $F' = F$. Such quantities are **invariant under rotation.** The next theorem lists some other rotation invariants.

ROTATION INVARIANTS

The rotation of the coordinate axes through an angle θ that transforms the equation $Ax^2 + Bxy + Cy^2 + Dx + Ey + F = 0$ into the form

$$A'(x')^2 + C'(y')^2 + D'x' + E'y' + F' = 0$$

has the following rotation invariants.

1. $F = F'$
2. $A + C = A' + C'$
3. $B^2 - 4AC = (B')^2 - 4A'C'$

You can use the results of this theorem to classify the graph of a second-degree equation *with* an xy-term in much the same way you do for a second-degree equation *without* an xy-term. Note that because $B' = 0$, the invariant $B^2 - 4AC$ reduces to

$$B^2 - 4AC = -4A'C'. \qquad \text{Discriminant}$$

This quantity is called the **discriminant** of the equation

$$Ax^2 + Bxy + Cy^2 + Dx + Ey + F = 0.$$

Now, from the classification procedure given in Section 6.4, you know that the sign of $A'C'$ determines the type of graph for the equation

$$A'(x')^2 + C'(y')^2 + D'x' + E'y' + F' = 0.$$

Consequently, the sign of $B^2 - 4AC$ will determine the type of graph for the original equation, as given in the following classification.

CLASSIFICATION OF CONICS BY THE DISCRIMINANT

The graph of the equation $Ax^2 + Bxy + Cy^2 + Dx + Ey + F = 0$ is, except in degenerate cases, determined by its discriminant as follows.

1. *Ellipse or circle:* $B^2 - 4AC < 0$
2. *Parabola:* $B^2 - 4AC = 0$
3. *Hyperbola:* $B^2 - 4AC > 0$

EXAMPLE 4 *Using the Discriminant*

Classify the graph of each of the following equations.

a. $4xy - 9 = 0$

b. $2x^2 - 3xy + 2y^2 - 2x = 0$

c. $x^2 - 6xy + 9y^2 - 2y + 1 = 0$

d. $3x^2 + 8xy + 4y^2 - 7 = 0$

Solution

a. Because $B^2 - 4AC = 16 - 0 > 0$, the graph is a hyperbola.

b. Because $B^2 - 4AC = 9 - 16 < 0$, the graph is a circle or an ellipse.

c. Because $B^2 - 4AC = 36 - 36 = 0$, the graph is a parabola.

d. Because $B^2 - 4AC = 64 - 48 > 0$, the graph is a hyperbola.

GROUP ACTIVITY

CLASSIFYING A GRAPH AS A HYPERBOLA

The graph of $f(x) = 1/x$ is a hyperbola. Discuss how you could use the techniques in this section to verify this, and then do so. Compare your statement with that of another student.

WARM UP

In Exercises 1–6, sketch the graph of the equation.

1. $\dfrac{x^2}{1} + \dfrac{y^2}{9} = 1$

2. $\dfrac{x^2}{1} - \dfrac{y^2}{9} = 1$

3. $x^2 = 6y$

4. $(x - 2)^2 + y^2 = 4$

5. $y^2 = -6x$

6. $x^2 - 6y^2 = -6$

Simplify the expression.

7. $x \cos \dfrac{\pi}{3} - y \sin \dfrac{\pi}{3}$

8. $x \sin\left(-\dfrac{\pi}{6}\right) + y \cos\left(-\dfrac{\pi}{6}\right)$

9. $\left(\dfrac{2x - 3y}{\sqrt{13}}\right)^2$

10. $\left(\dfrac{x - \sqrt{2}y}{\sqrt{3}}\right)^2$

6.5 Exercises

In Exercises 1–4, the $x'y'$-coordinate system has been rotated θ degrees from the xy-coordinate system. The coordinates of a point in the xy-coordinate system are given. Find the coordinates of the point in the rotated coordinate system.

1. $\theta = 90°, (0, 3)$
2. $\theta = 45°, (3, 3)$
3. $\theta = 30°, (1, 4)$
4. $\theta = 60°, (3, 1)$

In Exercises 5–16, rotate the axes to eliminate the xy-term. Sketch the graph of the resulting equation, showing both sets of axes.

5. $xy + 1 = 0$
6. $xy - 4 = 0$
7. $x^2 - 10xy + y^2 + 1 = 0$
8. $xy + x - 2y + 3 = 0$
9. $xy - 2y - 4x = 0$
10. $13x^2 + 6\sqrt{3}xy + 7y^2 - 16 = 0$
11. $5x^2 - 2xy + 5y^2 - 12 = 0$
12. $2x^2 - 3xy - 2y^2 + 10 = 0$
13. $3x^2 - 2\sqrt{3}xy + y^2 + 2x + 2\sqrt{3}y = 0$
14. $16x^2 - 24xy + 9y^2 - 60x - 80y + 100 = 0$
15. $9x^2 + 24xy + 16y^2 + 90x - 130y = 0$
16. $9x^2 + 24xy + 16y^2 + 80x - 60y = 0$

In Exercises 17–22, use a graphing utility to graph the conic. Determine the angle θ through which the axes are rotated. Explain how you used the graphing utility to obtain the graph.

17. $x^2 + xy + y^2 = 10$
18. $x^2 - 4xy + 2y^2 = 6$
19. $17x^2 + 32xy - 7y^2 = 75$
20. $40x^2 + 36xy + 25y^2 = 52$
21. $32x^2 + 50xy + 7y^2 = 52$
22. $4x^2 - 12xy + 9y^2 + (4\sqrt{13} - 12)x - (6\sqrt{13} + 8)y = 91$

In Exercises 23–28, match the graph with its equation. [The graphs are labeled (a), (b), (c), (d), (e), and (f).]

(a)

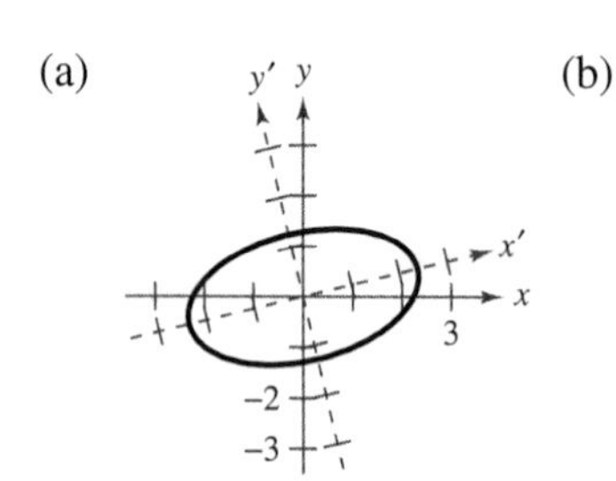

(b)

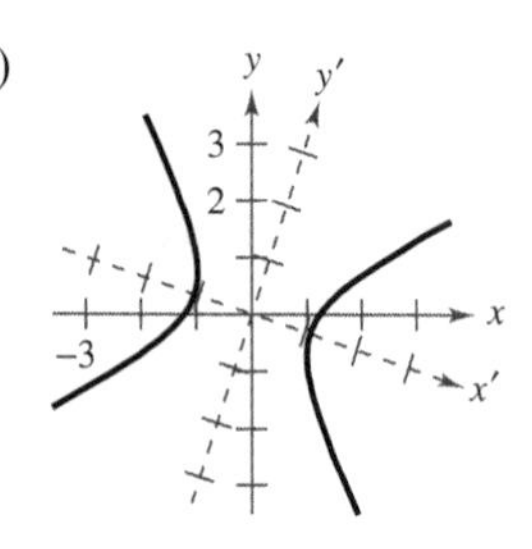

(c)

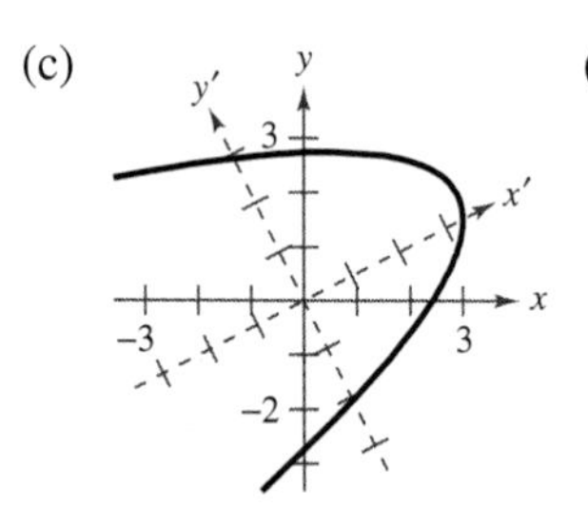

(d)

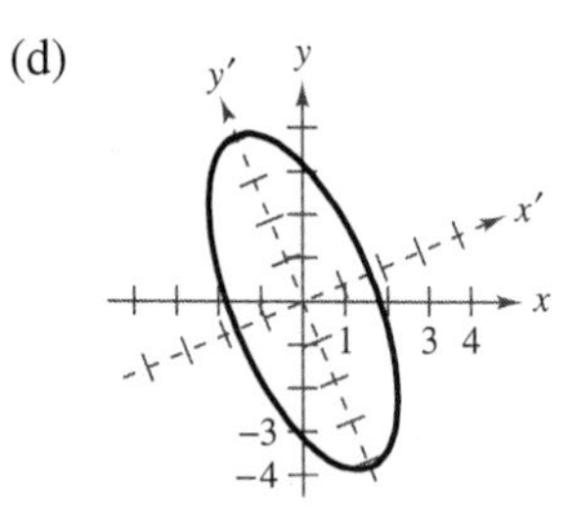

(e)

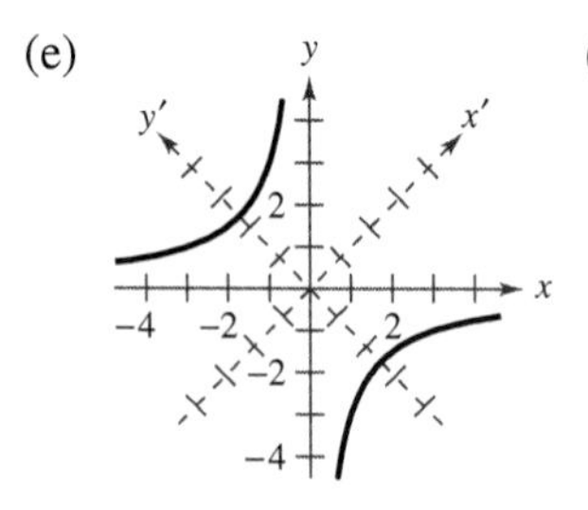

(f)

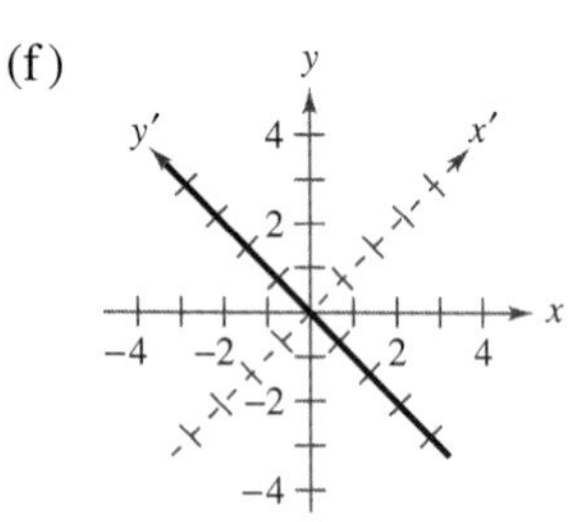

23. $xy + 3 = 0$
24. $x^2 + 2xy + y^2 = 0$
25. $-2x^2 + 3xy + 2y^2 + 3 = 0$
26. $x^2 - xy + 3y^2 - 5 = 0$
27. $3x^2 + 2xy + y^2 - 10 = 0$
28. $x^2 - 4xy + 4y^2 + 10x - 30 = 0$

In Exercises 29–36, use the discriminant to determine whether the graph of the equation is a parabola, an ellipse, or a hyperbola.

29. $16x^2 - 24xy + 9y^2 - 30x - 40y = 0$
30. $x^2 - 4xy - 2y^2 - 6 = 0$

31. $13x^2 - 8xy + 7y^2 - 45 = 0$

32. $2x^2 + 4xy + 5y^2 + 3x - 4y - 20 = 0$

33. $x^2 - 6xy - 5y^2 + 4x - 22 = 0$

34. $36x^2 - 60xy + 25y^2 + 9y = 0$

35. $x^2 + 4xy + 4y^2 - 5x - y - 3 = 0$

36. $x^2 + xy + 4y^2 + x + y - 4 = 0$

In Exercises 37–40, sketch (if possible) the graph of the degenerate conic.

37. $y^2 - 4x^2 = 0$

38. $x^2 + y^2 - 2x + 6y + 10 = 0$

39. $x^2 + 2xy + y^2 - 1 = 0$

40. $x^2 - 10xy + y^2 = 0$

In Exercises 41–48, use a graphing utility to graph the equations and find any points of intersection of the graphs by the method of elimination.

41. $-x^2 + y^2 + 4x - 6y + 4 = 0$
$x^2 + y^2 - 4x - 6y + 12 = 0$

42. $-x^2 - y^2 - 8x + 20y - 7 = 0$
$x^2 + 9y^2 + 8x + 4y + 7 = 0$

43. $-4x^2 - y^2 - 32x + 24y - 64 = 0$
$4x^2 + y^2 + 56x - 24y + 304 = 0$

44. $x^2 - 4y^2 - 20x - 64y - 172 = 0$
$16x^2 + 4y^2 - 320x + 64y + 1600 = 0$

45. $x^2 - y^2 - 12x + 12y - 36 = 0$
$x^2 + y^2 - 12x - 12y + 36 = 0$

46. $x^2 + 4y^2 - 2x - 8y + 1 = 0$
$-x^2 + 2x - 4y - 1 = 0$

47. $-16x^2 - y^2 + 24y - 80 = 0$
$16x^2 + 25y^2 - 400 = 0$

48. $16x^2 - y^2 + 16y - 128 = 0$
$y^2 - 48x - 16y - 32 = 0$

In Exercises 49–54, use a graphing utility to graph the equations and find any points of intersection of the graphs by the method of substitution.

49. $x^2 + y^2 - 25 = 0$
$9x - 4y^2 = 0$

50. $4x^2 + 9y^2 - 36y = 0$
$x^2 + 9y - 27 = 0$

51. $x^2 + 2y^2 - 4x + 6y - 5 = 0$
$x + y + 5 = 0$

52. $x^2 + 2y^2 - 4x + 6y - 5 = 0$
$x^2 - 4x - y + 4 = 0$

53. $xy + x - 2y + 3 = 0$
$x^2 + 4y^2 - 9 = 0$

54. $5x^2 - 2xy + 5y^2 - 12 = 0$
$x + y - 1 = 0$

55. Show that the equation $x^2 + y^2 = r^2$ is invariant under rotation of axes.

56. Find the lengths of the major and minor axes of the ellipse graphed in Exercise 10.

Review **Solve Exercises 57–60 as a review of the skills and problem-solving techniques you learned in previous sections. Graph the rational function.**

57. $g(x) = \dfrac{2}{2 - x}$

58. $f(x) = \dfrac{2x}{2 - x}$

59. $h(t) = \dfrac{t^2}{2 - t}$

60. $g(s) = \dfrac{2}{4 - s^2}$

6.6 Parametric Equations

See Exercise 63 on page 472 for an example of how a set of parametric equations can be used to model the path of a baseball.

Plane Curves ▫ *Sketching a Plane Curve* ▫ *Eliminating the Parameter* ▫ *Finding Parametric Equations for a Graph*

Plane Curves

Up to this point you have been representing a graph by a single equation involving the *two* variables x and y. In this section, you will study situations in which it is useful to introduce a *third* variable to represent a curve in the plane.

To see the usefulness of this procedure, consider the path followed by an object that is propelled into the air at an angle of 45°. If the initial velocity of the object is 48 feet per second, it can be shown that the object follows the parabolic path given by

$$y = -\frac{x^2}{72} + x \qquad \text{Rectangular equation}$$

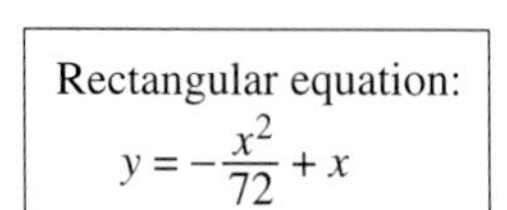

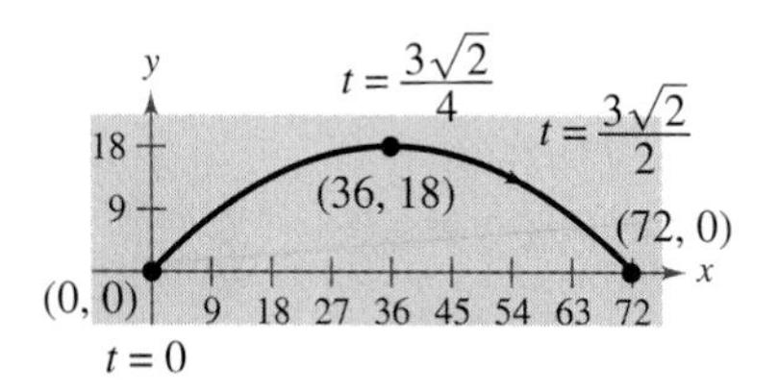

Parametric equations:
$x = 24\sqrt{2}t$
$y = -16t^2 + 24\sqrt{2}t$

Curvilinear Motion:
two variables for position
one variable for time

FIGURE 6.42

as shown in Figure 6.42. However, this equation does not tell the whole story. Although it does tell you *where* the object has been, it doesn't tell you *when* the object was at a given point (x, y) on the path. To determine this time, you can introduce a third variable t, called a **parameter.** It is possible to write both x and y as functions of t to obtain the **parametric equations**

$$x = 24\sqrt{2}t \qquad \text{Parametric equation for } x$$

$$y = -16t^2 + 24\sqrt{2}t. \qquad \text{Parametric equation for } y$$

From this set of equations you can determine that at time $t = 0$, the object is at the point $(0, 0)$. Similarly, at time $t = 1$, the object is at the point $(24\sqrt{2}, 24\sqrt{2} - 16)$, and so on.

For this particular motion problem, x and y are continuous functions of t, and the resulting path is a **plane curve.** (For this text, it is sufficient to think of a *continuous function* as one whose graph can be traced without lifting the pencil from the paper.)

DEFINITION OF PLANE CURVE

If f and g are continuous functions of t on an interval I, the set of ordered pairs $(f(t), g(t))$ is a **plane curve** C. The equations

$$x = f(t) \qquad \text{and} \qquad y = g(t)$$

are **parametric equations** for C, and t is the **parameter.**

Sketching a Plane Curve

When sketching a curve represented by a pair of parametric equations, you still plot points in the xy-plane. Each set of coordinates (x, y) is determined from a value chosen for the parameter t. Plotting the resulting points in the order of *increasing* values of t traces the curve in a specific direction. This is called the **orientation** of the curve.

EXAMPLE 1 Sketching a Curve

Sketch the curve described by the parametric equations

$$x = t^2 - 4 \quad \text{and} \quad y = \frac{t}{2}, \quad -2 \le t \le 3.$$

Solution

Using values of t on the given interval, the parametric equations yield the points (x, y) shown in the table.

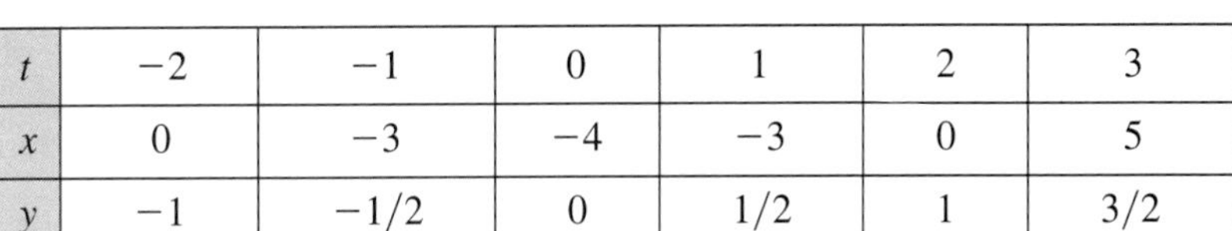

t	-2	-1	0	1	2	3
x	0	-3	-4	-3	0	5
y	-1	$-1/2$	0	$1/2$	1	$3/2$

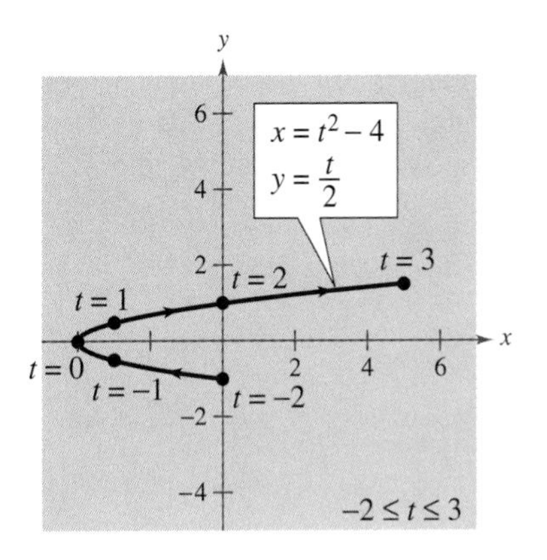

FIGURE 6.43

By plotting these points in the order of increasing t, you obtain the curve C shown in Figure 6.43. Note that the arrows on the curve indicate its orientation as t increases from -2 to 3.

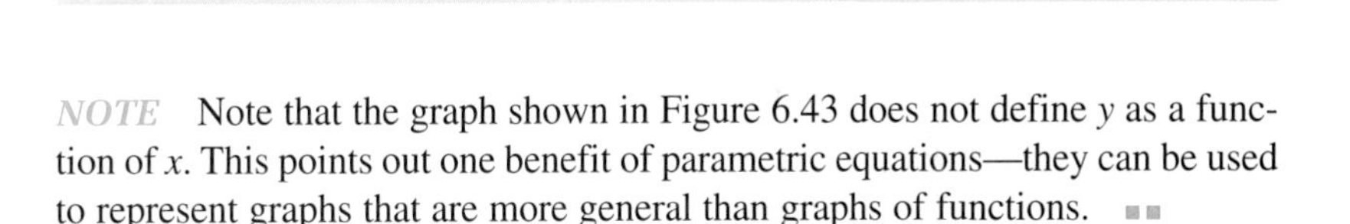

NOTE Note that the graph shown in Figure 6.43 does not define y as a function of x. This points out one benefit of parametric equations—they can be used to represent graphs that are more general than graphs of functions.

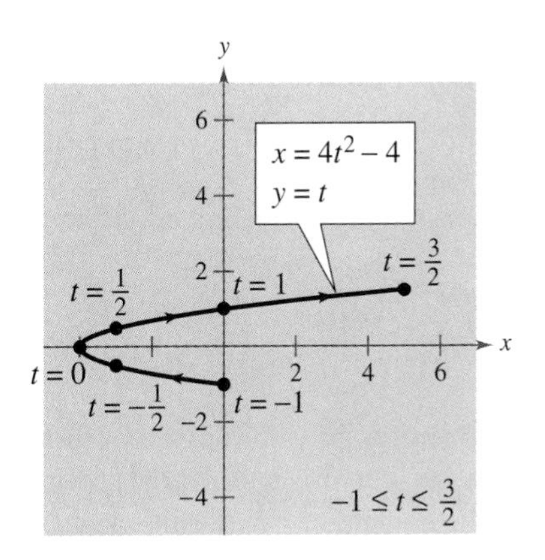

FIGURE 6.44

It often happens that two different sets of parametric equations have the same graph. For example, the set of parametric equations

$$x = 4t^2 - 4 \quad \text{and} \quad y = t, \quad -1 \le t \le \tfrac{3}{2}$$

has the same graph as the set given in Example 1. However, by comparing the values of t in Figures 6.43 and 6.44, you see that this second graph is traced out more *rapidly* (considering t as time) than the first graph. Thus, in applications, different parametric representations can be used to represent various *speeds* at which objects travel along a given path.

Eliminating the Parameter

TECHNOLOGY

Most graphing utilities have a **parametric** graphing mode. If yours does, try entering the parametric equations given in Example 2. Over what values should you let t vary to obtain the graph shown in Figure 6.45?

Example 1 uses simple point plotting to sketch the given curve. This tedious process can sometimes be simplified by finding a rectangular equation (in x and y) that has the same graph. This process is called **eliminating the parameter.**

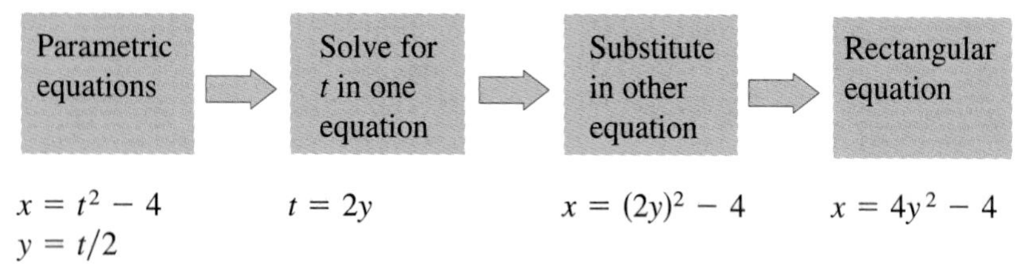

$x = t^2 - 4$ $\quad$ $t = 2y$ $\quad$ $x = (2y)^2 - 4$ $\quad$ $x = 4y^2 - 4$

$y = t/2$

Now you can recognize that the equation $x = 4y^2 - 4$ represents a parabola with a horizontal axis and a vertex at $(-4, 0)$.

When you are converting equations from parametric to rectangular form, you may need to alter the domain of the rectangular equation so that its graph matches the graph of the parametric equations. Such a situation is demonstrated in Example 2.

EXAMPLE 2 Eliminating the Parameter

Sketch the curve represented by the equations

$$x = \frac{1}{\sqrt{t+1}} \quad \text{and} \quad y = \frac{t}{t+1}$$

by eliminating the parameter and adjusting the domain of the resulting rectangular equation.

Solution

Solving for t in the equation for x, you have

$$x = \frac{1}{\sqrt{t+1}} \quad \Longrightarrow \quad x^2 = \frac{1}{t+1}$$

which implies that $t = (1 - x^2)/x^2$. Now, substituting in the equation for y, you obtain

$$y = \frac{t}{t+1} = \frac{(1 - x^2)/x^2}{[(1 - x^2)/x^2] + 1} = 1 - x^2.$$

The rectangular equation, $y = 1 - x^2$, is defined for all values of x, but from the parametric equation for x you can see that the curve is defined only when $t > -1$. This implies that you should restrict the domain of x to positive values, as shown in Figure 6.45.

Parametric equations: $x = \frac{1}{\sqrt{t+1}},\ y = \frac{t}{t+1}$

FIGURE 6.45

It is not necessary for the parameter in a set of parametric equations to represent time. The next example uses an *angle* as the parameter.

EXAMPLE 3 *Eliminating the Parameter*

Sketch the curve represented by

$$x = 3\cos\theta \quad \text{and} \quad y = 4\sin\theta, \quad 0 \le \theta \le 2\pi$$

by eliminating the parameter.

Solution

Begin by solving for $\cos\theta$ and $\sin\theta$ in the given equations.

$$\cos\theta = \frac{x}{3} \quad \text{and} \quad \sin\theta = \frac{y}{4}$$ Solve for $\cos\theta$ and $\sin\theta$.

Make use of the identity $\sin^2\theta + \cos^2\theta = 1$ to form an equation involving only x and y.

$$\cos^2\theta + \sin^2\theta = 1$$ Trigonometric identity

$$\cos^2\theta + \sin^2\theta = \left(\frac{x}{3}\right)^2 + \left(\frac{y}{4}\right)^2 = 1$$ Substitute.

$$\frac{x^2}{9} + \frac{y^2}{16} = 1$$ Rectangular equation

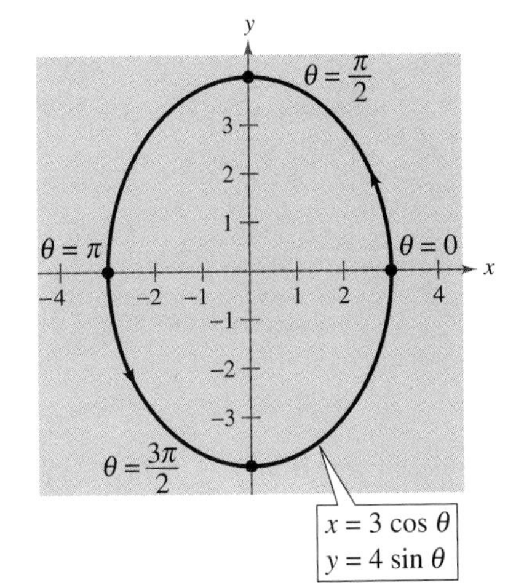

FIGURE 6.46

From this rectangular equation, you can see that the graph is an ellipse centered at $(0, 0)$, with vertices at $(0, 4)$ and $(0, -4)$ and minor axis of length $2b = 6$, as shown in Figure 6.46. Note that the elliptic curve is traced out *counterclockwise* as θ varies from 0 to 2π.

In Examples 2 and 3 it is important to realize that eliminating the parameter is primarily an *aid to curve sketching.* If the parametric equations represent the path of a moving object, the graph alone is not sufficient to describe the object's motion. You still need the parametric equations to tell you the *position, direction,* and *speed* at a given time.

Finding Parametric Equations for a Graph

You have been studying techniques for sketching the graph represented by a set of parametric equations. Now consider the reverse problem—that is, how can you find a set of parametric equations for a given graph or a given physical description? From the discussion following Example 1, you know that such a representation is not unique. This is further demonstrated in Example 4.

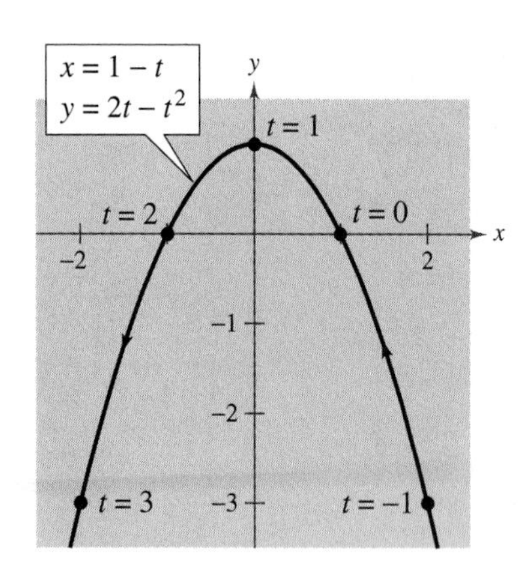

FIGURE 6.47

EXAMPLE 4 *Finding Parametric Equations for a Given Graph*

Find a set of parametric equations to represent the graph of $y = 1 - x^2$, using the following parameters.

a. $t = x$ **b.** $t = 1 - x$

Solution

a. Letting $t = x$, you obtain the parametric equations

$$x = t \quad \text{and} \quad y = 1 - x^2 = 1 - t^2.$$

b. Letting $t = 1 - x$, you obtain

$$x = 1 - t \quad \text{and} \quad y = 1 - (1 - t)^2 = 2t - t^2.$$

In Figure 6.47, note how the resulting curve is oriented by the increasing values of t. For part (a), the curve would have the opposite orientation.

EXAMPLE 5 *Parametric Equations for a Cycloid*

Describe the **cycloid** traced out by a point P on the circumference of a circle of radius a as the circle rolls along a straight line in a plane.

Solution

As the parameter, let θ be the measure of the circle's rotation, and let the point $P = (x, y)$ begin at the origin. When $\theta = 0$, P is at the origin; when $\theta = \pi$, P is at a maximum point $(\pi a, 2a)$; and when $\theta = 2\pi$, P is back on the x-axis at $(2\pi a, 0)$. From Figure 6.48, you can see that $\angle APC = 180° - \theta$. Hence, you have

$$\sin \theta = \sin(180° - \theta) = \sin(\angle APC) = \frac{AC}{a} = \frac{BD}{a}$$

$$\cos \theta = -\cos(180° - \theta) = -\cos(\angle APC) = \frac{AP}{-a}$$

which implies that $AP = -a \cos \theta$ and $BD = a \sin \theta$. Because the circle rolls along the x-axis, you know that $OD = \widehat{PD} = a\theta$. Furthermore, because $BA = DC = a$, you have

NOTE In Example 5, $\widehat{PD}$ represents the arc of the circle between points P and D.

$$x = OD - BD = a\theta - a \sin \theta$$
$$y = BA + AP = a - a \cos \theta.$$

Therefore, the parametric equations are

$$x = a(\theta - \sin \theta) \quad \text{and} \quad y = a(1 - \cos \theta).$$

TECHNOLOGY

Use a graphing utility in parametric mode to obtain a graph similar to Figure 6.48 by graphing the following equations.

$$X_{1T} = T - \sin T$$
$$Y_{1T} = T - \cos T$$

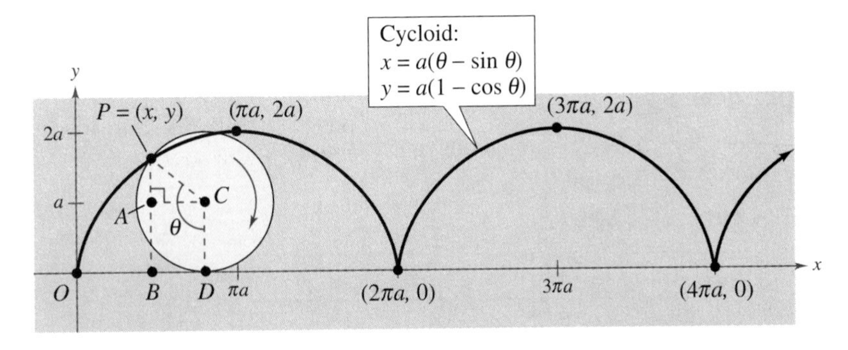

FIGURE 6.48

A computer animation of this concept appears in the *Interactive* CD-ROM.

GROUP ACTIVITY

CHANGING THE ORIENTATION OF A CURVE

The **orientation** of a curve refers to the direction in which the curve is traced as the values of the parameter increase. For instance, as t increases, the ellipse given by

$$x = \cos t \qquad \text{and} \qquad y = 2\sin t$$

is traced out *counterclockwise*. Find a parametric representation for which the same ellipse is traced out *clockwise*.

WARM UP

Sketch the graph of the equation.

1. $y = -\frac{1}{4}x^2$

2. $y = 4 - \frac{1}{4}(x - 2)^2$

3. $16x^2 + y^2 = 16$

4. $-16x^2 + y^2 = 16$

5. $x + y = 4$

6. $x^2 + y^2 = 16$

Simplify the expression.

7. $10\sin^2\theta + 10\cos^2\theta$

8. $5\sec^2\theta - 5$

9. $\sec^4 x - \tan^4 x$

10. $\dfrac{\sin 2\theta}{4\cos\theta}$

6.6 Exercises

1. Consider the parametric equations $x = \sqrt{t}$ and $y = 1 - t$.

(a) Complete the table.

t	0	1	2	3	4
x					
y					

(b) Plot the points (x, y) generated in part (a), and sketch a graph of the parametric equations.

(c) Find the rectangular equation by eliminating the parameter. Sketch its graph. How do the graphs differ?

2. Consider the parametric equations $x = 4\cos^2\theta$ and $y = 2\sin\theta$.

(a) Complete the table.

θ	$-\frac{\pi}{2}$	$-\frac{\pi}{4}$	0	$\frac{\pi}{4}$	$\frac{\pi}{2}$
x					
y					

(b) Plot the points (x, y) generated in part (a), and sketch a graph of the parametric equations.

(c) Find the rectangular equation by eliminating the parameter. Sketch its graph. How do the graphs differ?

In Exercises 3–28, sketch the curve represented by the parametric equations (indicate the direction of the curve). Use a graphing utility to confirm your result. Then eliminate the parameter and write the corresponding rectangular equation whose graph represents the curve.

3. $x = t$, $y = -2t$

4. $x = t$, $y = \frac{1}{2}t$

5. $x = 3t - 1$, $y = 2t + 1$

6. $x = 3 - 2t$, $y = 2 + 3t$

7. $x = \frac{1}{4}t$, $y = t^2$

8. $x = t$, $y = t^3$

9. $x = t + 1$, $y = t^2$

10. $x = \sqrt{t}$, $y = 1 - t$

11. $x = t^3$, $y = \frac{1}{2}t^2$

12. $x = t - 1$, $y = t/(t - 1)$

13. $x = 2t$, $y = |t - 2|$

14. $x = |t - 1|$, $y = t + 2$

15. $x = 3\cos\theta$, $y = 3\sin\theta$

16. $x = \cos\theta$, $y = 3\sin\theta$

17. $x = 4\sin 2\theta$, $y = 2\cos 2\theta$

18. $x = \cos\theta$, $y = 2\sin 2\theta$

19. $x = 4 + 2\cos\theta$, $y = -1 + \sin\theta$

20. $x = 4 + 2\cos\theta$, $y = -1 + 2\sin\theta$

21. $x = 4 + 2\cos\theta$, $y = -1 + 4\sin\theta$

22. $x = \sec\theta$, $y = \tan\theta$

23. $x = 4\sec\theta$, $y = 3\tan\theta$

24. $x = \sec\theta$, $y = \cos\theta$

25. $x = e^{-t}$, $y = e^{3t}$

26. $x = e^{2t}$, $y = e^{t}$

27. $x = t^3$, $y = 3\ln t$

28. $x = \ln 2t$, $y = t^2$

In Exercises 29–32, determine how the plane curves differ from each other.

29. (a) $x = t$, $y = 2t + 1$
(b) $x = \cos\theta$, $y = 2\cos\theta + 1$
(c) $x = e^{-t}$, $y = 2e^{-t} + 1$
(d) $x = e^{t}$, $y = 2e^{t} + 1$

30. (a) $x = t$, $y = t^2 - 1$
(b) $x = t^2$, $y = t^4 - 1$
(c) $x = \sin t$, $y = \sin^2 t - 1$
(d) $x = e^{t}$, $y = e^{2t} - 1$

31. (a) $x = \cos\theta$, $y = \sin\theta$ (b) $x = \sin\theta$, $y = \cos\theta$

(c) $x = \sin^2\theta$, $y = \cos^2\theta$ (d) $x = -\cos\theta$, $y = \sin\theta$

32. (a) $x = t$, $y = t$ (b) $x = t^2$, $y = t^2$

(c) $x = -t$, $y = -t$ (d) $x = t^3$, $y = t^3$

In Exercises 33–36, eliminate the parameter and obtain the standard form of the rectangular equation.

33. Line through (x_1, y_1) and (x_2, y_2):

$x = x_1 + t(x_2 - x_1)$

$y = y_1 + t(y_2 - y_1)$

34. Circle:

$x = h + r\cos\theta$

$y = k + r\sin\theta$

35. Ellipse:

$x = h + a\cos\theta$

$y = k + b\sin\theta$

36. Hyperbola:

$x = h + a\sec\theta$

$y = k + b\tan\theta$

In Exercises 37–44, use the results of Exercises 33–36 to find a set of parametric equations for the line or conic.

37. Line: Passes through $(0, 0)$ and $(5, -2)$

38. Line: Passes through $(1, 4)$ and $(5, -2)$

39. Circle: Center: $(2, 1)$; Radius: 4

40. Circle: Center: $(-3, 1)$; Radius: 3

41. Ellipse: Vertices: $(\pm 5, 0)$
Foci: $(\pm 4, 0)$

42. Ellipse: Vertices: $(4, 7)$, $(4, -3)$
Foci: $(4, 5)$, $(4, -1)$

43. Hyperbola: Vertices: $(\pm 4, 0)$
Foci: $(\pm 5, 0)$

44. Hyperbola: Vertices: $(0, \pm 1)$
Foci: $(0, \pm 2)$

In Exercises 45–48, find two different sets of parametric equations for the given rectangular equation.

45. $y = 3x - 2$

46. $y = \dfrac{1}{x}$

47. $y = x^3$

48. $y = x^2$

In Exercises 49–56, use a graphing utility to obtain a graph of the curve represented by the parametric equations.

49. Cycloid: $x = 2(\theta - \sin\theta)$
$y = 2(1 - \cos\theta)$

50. Cycloid: $x = \theta + \sin\theta$
$y = 1 - \cos\theta$

51. Prolate cycloid: $x = \theta - \frac{3}{2}\sin\theta$
$y = 1 - \frac{3}{2}\cos\theta$

52. Prolate cycloid: $x = 2\theta - 4\sin\theta$
$y = 2 - 4\cos\theta$

53. Hypocycloid: $x = 3\cos^3\theta$
$y = 3\sin^3\theta$

54. Curtate cycloid: $x = 2\theta - \sin\theta$
$y = 2 - \cos\theta$

55. Witch of Agnesi: $x = 2\cot\theta$
$y = 2\sin^2\theta$

56. Folium of Descartes: $x = \dfrac{3t}{1 + t^3}$
$y = \dfrac{3t^2}{1 + t^3}$

In Exercises 57–60, match the parametric equations with the correct graph and describe the domain and range. [The graphs are labeled (a), (b), (c), and (d).]

(a)

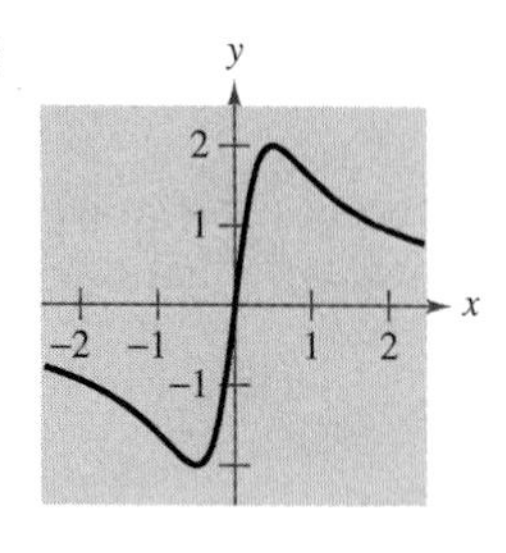

(b)

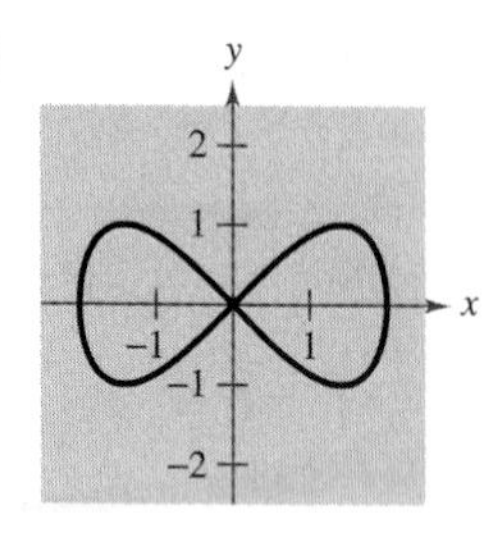

(c)

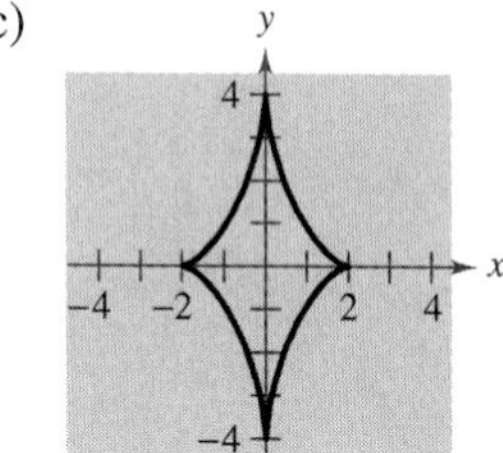

(d)

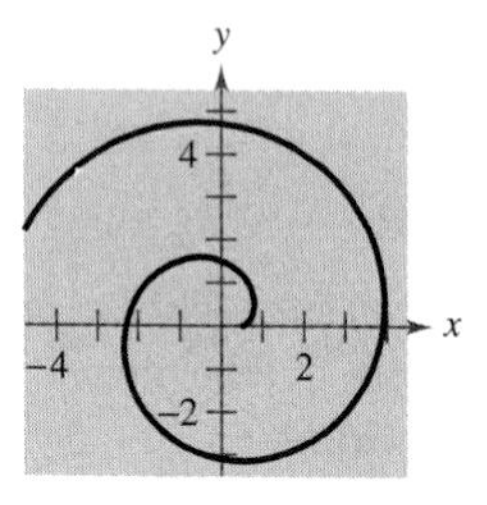

57. Lissajous curve: $x = 2\cos\theta$

$y = \sin 2\theta$

58. Evolute of ellipse: $x = 2\cos^3\theta$

$y = 4\sin^3\theta$

59. Involute of circle: $x = \frac{1}{2}(\cos\theta + \theta\sin\theta)$

$y = \frac{1}{2}(\sin\theta - \theta\cos\theta)$

60. Serpentine curve: $x = \frac{1}{2}\cot\theta$

$y = 4\sin\theta\cos\theta$

Projectile Motion **A projectile is launched at a height of *h* feet above the ground and at an angle θ with the horizontal. If the initial velocity is v_0 feet per second, the path of the projectile is modeled by the parametric equations**

$$x = (v_0\cos\theta)t \quad \text{and} \quad y = h + (v_0\sin\theta)t - 16t^2.$$

In Exercises 61 and 62, use a graphing utility to graph the paths of a projectile launched from ground level at the specified values of θ and v_0. For each case, use the graph to approximate the maximum height and the range of the projectile.

61. (a) $\theta = 20°$, $v_0 = 88$ ft/sec
(b) $\theta = 20°$, $v_0 = 132$ ft/sec
(c) $\theta = 45°$, $v_0 = 88$ ft/sec
(d) $\theta = 45°$, $v_0 = 132$ ft/sec

62. (a) $\theta = 15°$, $v_0 = 60$ ft/sec
(b) $\theta = 15°$, $v_0 = 100$ ft/sec
(c) $\theta = 30°$, $v_0 = 60$ ft/sec
(d) $\theta = 30°$, $v_0 = 100$ ft/sec

63. ***Baseball*** The center-field fence in a ballpark is 10 feet high and 400 feet from home plate. The baseball is hit 3 feet above the ground. It leaves the bat at an angle of θ degrees with the horizontal at a speed of 100 miles per hour (see figure).

(a) Write a set of parametric equations for the path of the baseball.

(b) Use a graphing utility to sketch the path of the baseball if $\theta = 15°$. Is the hit a home run?

(c) Use a graphing utility to sketch the path of the baseball if $\theta = 23°$. Is the hit a home run?

(d) Find the minimum angle required for the hit to be a home run.

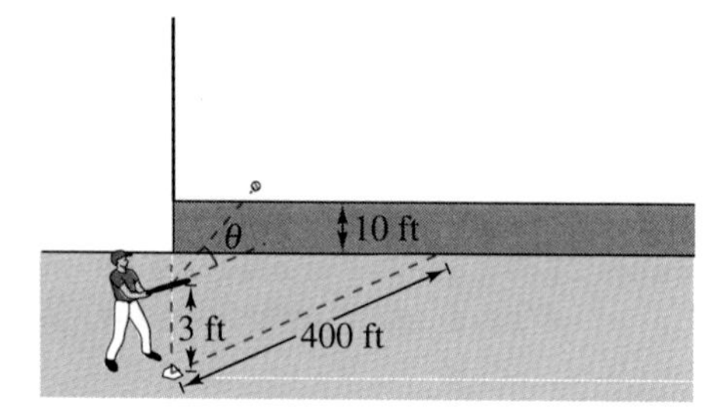

64. ***Football*** The quarterback of a football team releases a pass at a height of 7 feet above the playing field, and the football is caught by a receiver at a height of 4 feet, 30 yards directly downfield. The pass is released at an angle of 35° with the horizontal.

(a) Write a set of parametric equations for the path of the football.

(b) Find the speed of the football at the point of release.

(c) Use a graphing utility to graph the path of the football and approximate its maximum height.

(d) Find the time the receiver has to position himself after the quarterback releases the football.

65. ***Projectile Motion*** Eliminate the parameter t from the position function for the motion of a projectile to show that the rectangular equation is

$$y = -\frac{16 \sec^2 \theta}{v_0^2}x^2 + (\tan \theta)x + h.$$

66. ***Path of a Projectile*** The path of a projectile is given by the rectangular equation

$y = 5 + x - 0.005x^2.$

(a) Use the result of Exercise 65 to find h, v_0, and θ. Find the parametric equations of the path.

(b) Use a graphing utility to graph the rectangular equation for the path of the projectile. Confirm your answer in part (a) by sketching the curve represented by the parametric equations.

(c) Use a graphing utility to approximate the maximum height of the projectile and its range.

67. ***Curtate Cycloid*** A wheel of radius a rolls along a straight line without slipping. The curve traced by a point P that is b units from the center ($b < a$) is called a **curtate cycloid** (see figure). Use the angle θ shown in the figure to find a set of parametric equations for the curve.

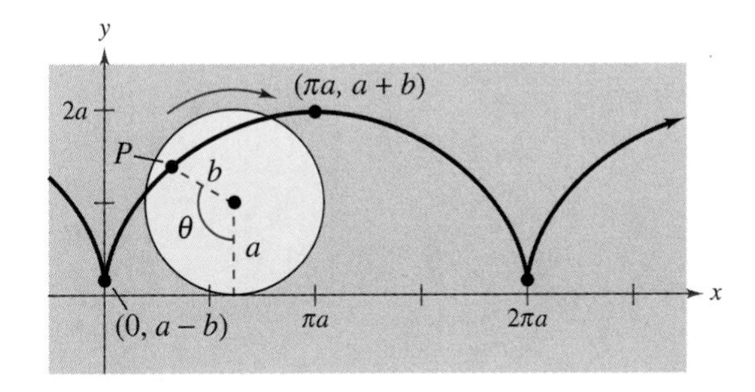

68. ***Epicycloid*** A circle of radius 1 rolls around the outside of a circle of radius 2 without slipping. The curve traced by a point on the circumference of the smaller circle is called an **epicycloid** (see figure). Use the angle θ shown in the figure to find a set of parametric equations for the curve.

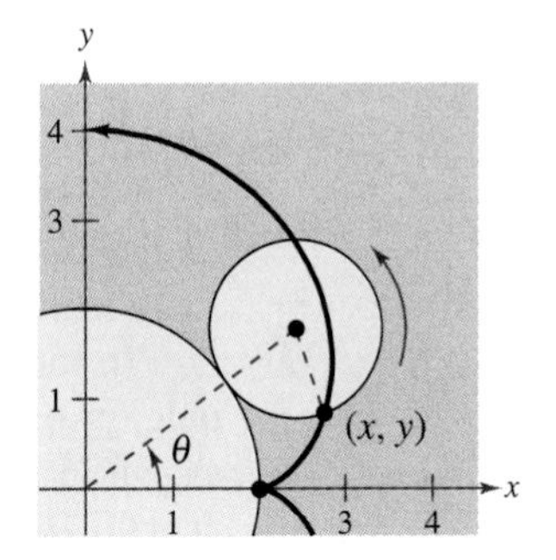

True or False? **In Exercises 69 and 70, determine whether the statement is true or false. If it is false, explain why or give an example that shows it is false.**

69. The two sets of parametric equations $x = t$, $y = t^2 + 1$ and $x = 3t$, $y = 9t^2 + 1$ have the same rectangular equation.

70. The graph of the parametric equations $x = t^2$ and $y = t^2$ is the line $y = x$.

Review **Solve Exercises 71–74 as a review of the skills and problem-solving techniques you learned in previous sections. Solve the system of equations.**

71. $\begin{aligned} 5x - 7y &= 11 \\ -3x + y &= -13 \end{aligned}$

72. $\begin{aligned} 3x + 5y &= 9 \\ 4x - 2y &= -14 \end{aligned}$

73. $\begin{aligned} 3a - 2b + c &= 8 \\ 2a + b - 3c &= -3 \\ a - 3b + 9c &= 16 \end{aligned}$

74. $\begin{aligned} 5u + 7v + 9w &= 4 \\ u - 2v - 3w &= 7 \\ 8u - 2v + w &= 20 \end{aligned}$

6.7 Polar Coordinates

Introduction ▫ *Coordinate Conversion* ▫ *Equation Conversion*

See Exercises 57–62 on page 481 for examples of how to sketch the graph of a polar equation by first converting the equation to rectangular form.

Introduction

So far, you have been representing graphs of equations as collections of points (x, y) on the rectangular coordinate system, where x and y represent the directed distances from the coordinate axes to the point (x, y). In this section you will study a different system called the **polar coordinate system.**

To form the polar coordinate system in the plane, fix a point O, called the **pole** (or **origin**), and construct from O an initial ray called the **polar axis,** as shown in Figure 6.49. Then each point P in the plane can be assigned **polar coordinates** (r, θ) as follows.

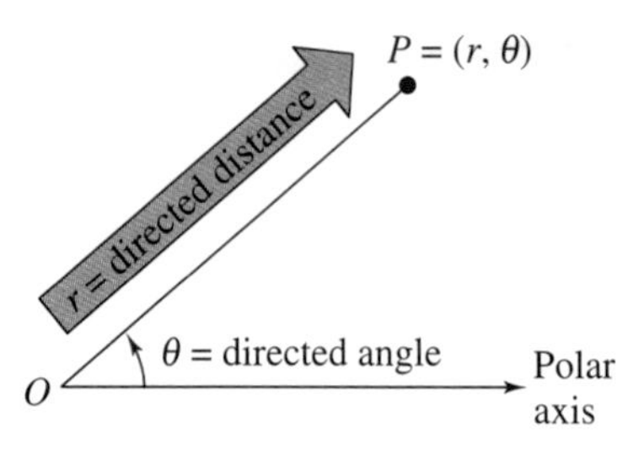

FIGURE 6.49

1. $r =$ *directed distance* from O to P
2. $\theta =$ *directed angle,* counterclockwise from polar axis to segment $\overline{OP}$

EXAMPLE 1 *Plotting Points on the Polar Coordinate System*

a. The point $(r, \theta) = (2, \pi/3)$ lies two units from the pole on the terminal side of the angle $\theta = \pi/3$, as shown in Figure 6.50(a).

b. The point $(r, \theta) = (3, -\pi/6)$ lies three units from the pole on the terminal side of the angle $\theta = -\pi/6$, as shown in Figure 6.50(b).

c. The point $(r, \theta) = (3, 11\pi/6)$ coincides with the point $(3, -\pi/6)$, as shown in Figure 6.50(c).

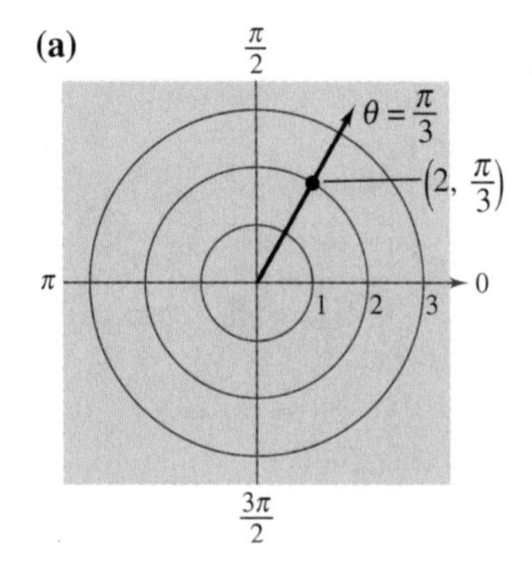

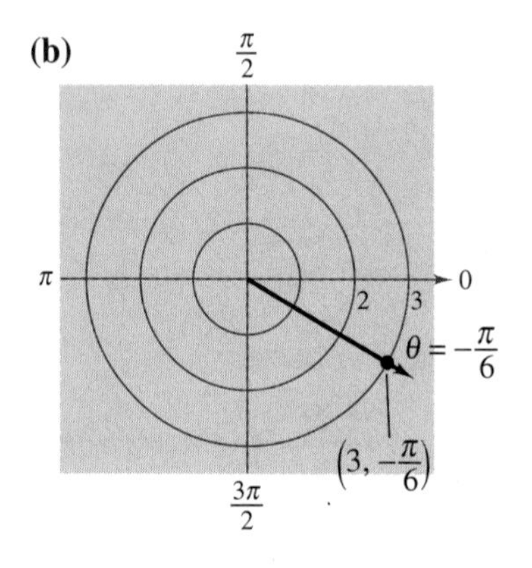

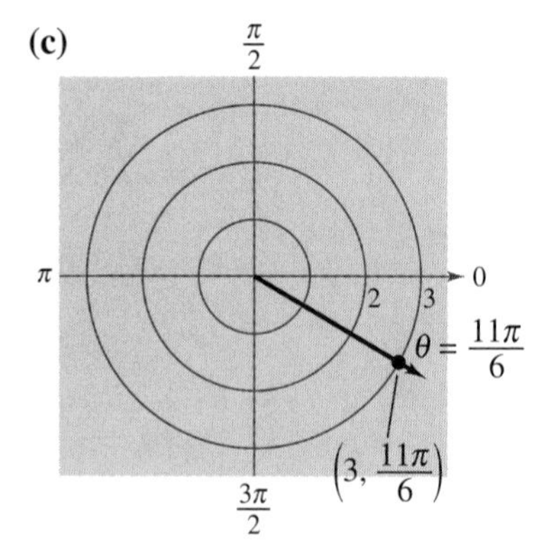

FIGURE 6.50

In rectangular coordinates, each point (x, y) has a unique representation. This is not true for polar coordinates. For instance, the coordinates (r, θ) and $(r, 2\pi + \theta)$ represent the same point, as illustrated in Example 1. Another way to obtain multiple representations of a point is to use negative values for r. Because r is a *directed distance,* the coordinates (r, θ) and $(-r, \theta + \pi)$ represent the same point. In general, the point (r, θ) can be represented as

$$(r, \theta) = (r, \theta \pm 2n\pi)$$

or

$$(r, \theta) = (-r, \theta \pm (2n + 1)\pi)$$

where n is any integer. Moreover, the pole is represented by $(0, \theta)$, where θ is any angle.

EXAMPLE 2 *Multiple Representation of Points*

Plot the point $(3, -3\pi/4)$ and find three additional polar representations of this point, using $-2\pi < \theta < 2\pi$.

Solution

The point is shown in Figure 6.51. Three other representations are as follows.

$$\left(3, -\frac{3\pi}{4} + 2\pi\right) = \left(3, \frac{5\pi}{4}\right) \qquad \text{Add } 2\pi \text{ to } \theta.$$

$$\left(-3, -\frac{3\pi}{4} - \pi\right) = \left(-3, -\frac{7\pi}{4}\right) \qquad \text{Replace } r \text{ by } -r\text{; subtract } \pi \text{ from } \theta.$$

$$\left(-3, -\frac{3\pi}{4} + \pi\right) = \left(-3, \frac{\pi}{4}\right) \qquad \text{Replace } r \text{ by } -r\text{; add } \pi \text{ to } \theta.$$

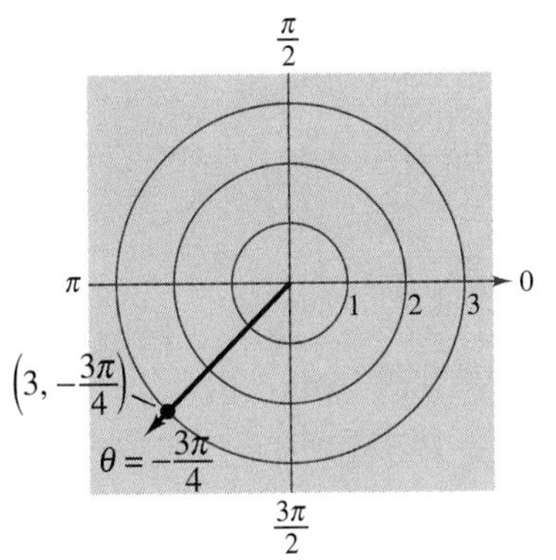

$\left(3, -\frac{3\pi}{4}\right) = \left(3, \frac{5\pi}{4}\right) = \left(-3, -\frac{7\pi}{4}\right) = \left(-3, \frac{\pi}{4}\right) = \ldots$

FIGURE 6.51

Exploration

Most graphing calculators have a **polar** graphing mode. If yours does, try graphing the equation $r = 3$. (Use a setting of $-6 \le x \le 6$ and $-4 \le y \le 4$.) You should obtain a circle of radius 3.

a. Use the trace feature to cursor around the circle. Can you locate the point $(3, 5\pi/4)$?

b. Can you find other polar representations of the point $(3, 5\pi/4)$? If so, explain how you did it.

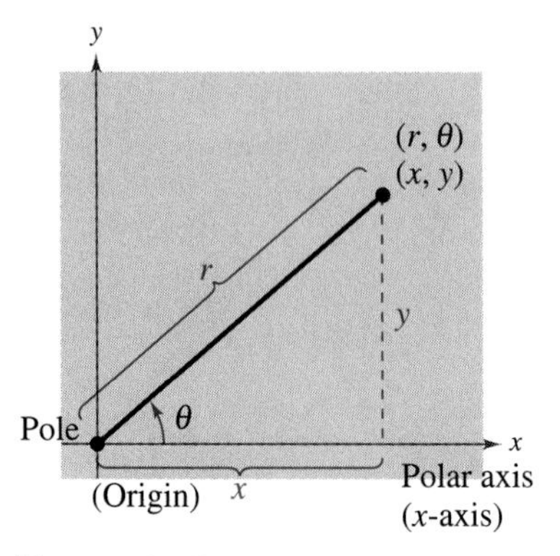

FIGURE 6.52

Coordinate Conversion

To establish the relationship between polar and rectangular coordinates, let the polar axis coincide with the positive x-axis and the pole with the origin, as shown in Figure 6.52. Because (x, y) lies on a circle of radius r, it follows that $r^2 = x^2 + y^2$. Moreover, for $r > 0$, the definitions of the trigonometric functions imply that

$$\tan\theta = \frac{y}{x}, \qquad \cos\theta = \frac{x}{r}, \qquad \text{and} \qquad \sin\theta = \frac{y}{r}.$$

If $r < 0$, you can show that the same relationships hold.

COORDINATE CONVERSION

The polar coordinates (r, θ) are related to the rectangular coordinates (x, y) as follows.

$$x = r\cos\theta \qquad \text{and} \qquad \tan\theta = \frac{y}{x}$$

$$y = r\sin\theta \qquad\qquad r^2 = x^2 + y^2$$

EXAMPLE 3 *Polar-to-Rectangular Conversion*

Convert the points (a) $(2, \pi)$ and (b) $(\sqrt{3}, \pi/6)$ to rectangular coordinates. (See Figure 6.53.)

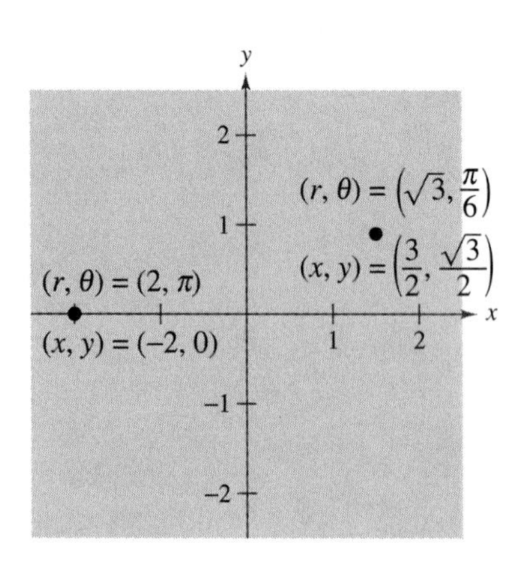

FIGURE 6.53

Solution

a. For the point $(r, \theta) = (2, \pi)$, you have

$$x = r\cos\theta = 2\cos\pi = -2$$

and

$$y = r\sin\theta = 2\sin\pi = 0.$$

The rectangular coordinates are $(x, y) = (-2, 0)$.

b. For the point $(r, \theta) = (\sqrt{3}, \pi/6)$, you have

$$x = \sqrt{3}\cos\frac{\pi}{6} = \sqrt{3}\left(\frac{\sqrt{3}}{2}\right) = \frac{3}{2}$$

and

$$y = \sqrt{3}\sin\frac{\pi}{6} = \sqrt{3}\left(\frac{1}{2}\right) = \frac{\sqrt{3}}{2}.$$

The rectangular coordinates are $(x, y) = (3/2, \sqrt{3}/2)$.

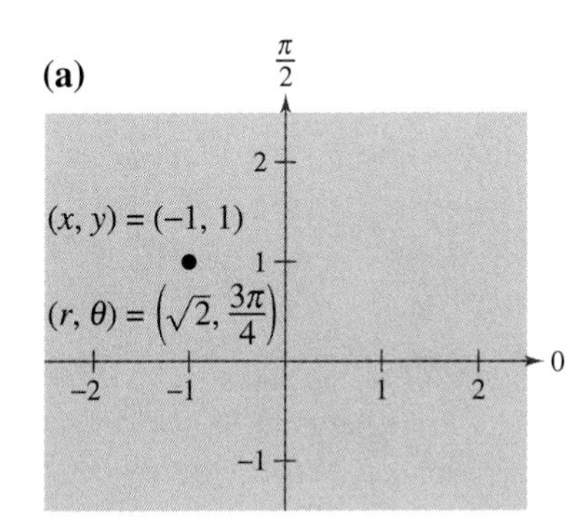

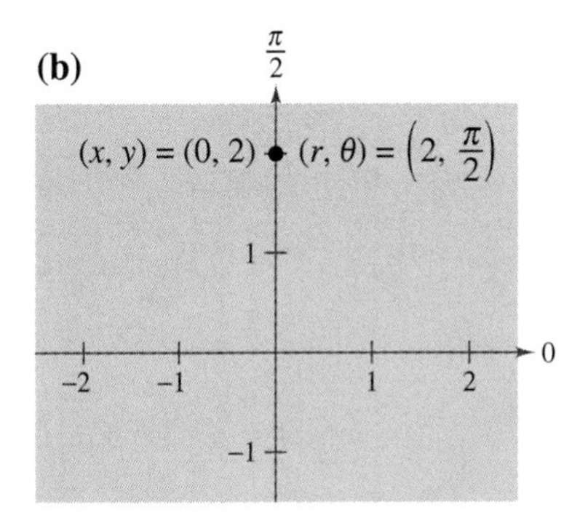

FIGURE 6.54

EXAMPLE 4 Rectangular-to-Polar Conversion

Convert the points (a) $(-1, 1)$ and (b) $(0, 2)$ to polar coordinates.

Solution

a. For the second-quadrant point $(x, y) = (-1, 1)$, you have

$$\tan \theta = \frac{y}{x}$$

$$\tan \theta = -1$$

$$\theta = \frac{3\pi}{4}.$$

Because θ lies in the same quadrant as (x, y), use positive r.

$$r = \sqrt{x^2 + y^2}$$

$$= \sqrt{(-1)^2 + (1)^2}$$

$$= \sqrt{2}$$

Thus, *one* set of polar coordinates is $(r, \theta) = \left(\sqrt{2}, 3\pi/4\right)$, as shown in Figure 6.54(a).

b. Because the point $(x, y) = (0, 2)$ lies on the positive y-axis, choose

$$\theta = \pi/2 \quad \text{and} \quad r = 2.$$

This implies that *one* set of polar coordinates is $(r, \theta) = (2, \pi/2)$, as shown in Figure 6.54(b).

Equation Conversion

By comparing Examples 3 and 4, you can see that point conversion from the polar to the rectangular system is straightforward, whereas point conversion from the rectangular to the polar system is more involved. For equations, the opposite is true. To convert a rectangular equation to polar form, you simply replace x by $r \cos \theta$ and y by $r \sin \theta$. For instance, the rectangular equation $y = x^2$ can be written in polar form as follows.

$$y = x^2 \qquad \text{Rectangular equation}$$

$$r \sin \theta = (r \cos \theta)^2 \qquad \text{Polar equation}$$

On the other hand, converting a polar equation to rectangular form requires considerable ingenuity.

Example 5 demonstrates several polar-to-rectangular conversions that enable you to sketch the graphs of some polar equations.

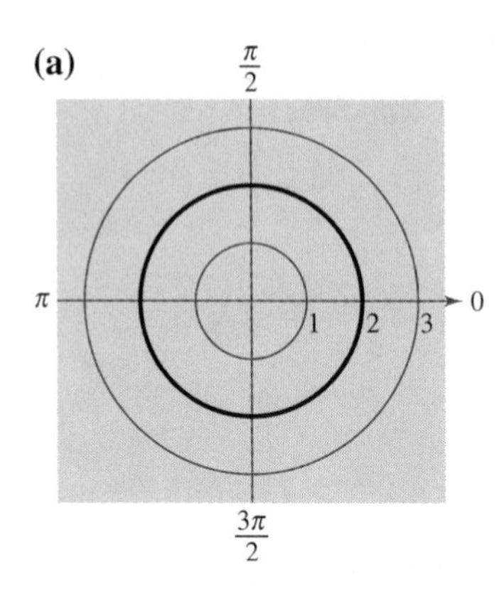

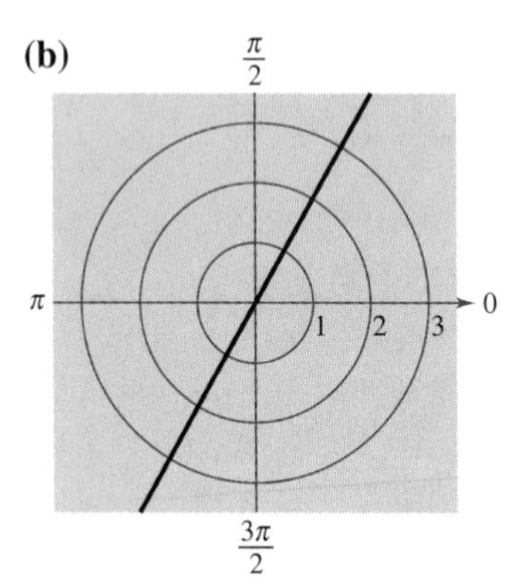

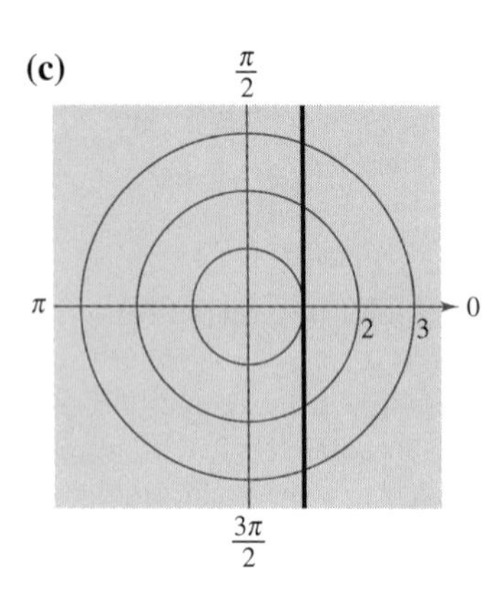

FIGURE 6.55

EXAMPLE 5 *Converting Polar Equations to Rectangular Form*

Describe the graph of each polar equation and find the corresponding rectangular equation.

a. $r = 2$ **b.** $\theta = \dfrac{\pi}{3}$ **c.** $r = \sec \theta$

Solution

a. The graph of the polar equation $r = 2$ consists of all points that are two units from the pole. In other words, this graph is a circle centered at the origin with a radius of 2, as shown in Figure 6.55(a). You can confirm this by converting to rectangular form, using the relationship $r^2 = x^2 + y^2$.

$$\underbrace{r = 2}_{\text{Polar equation}} \quad \Longrightarrow \quad r^2 = 2^2 \quad \Longrightarrow \quad \underbrace{x^2 + y^2 = 2^2}_{\text{Rectangular equation}}$$

b. The graph of the polar equation $\theta = \pi/3$ consists of all points on the line that makes an angle of $\pi/3$ with the positive x-axis, as shown in Figure 6.55(b). To convert to rectangular form, make use of the relationship $\tan \theta = y/x$.

$$\underbrace{\theta = \frac{\pi}{3}}_{\text{Polar equation}} \quad \Longrightarrow \quad \tan \theta = \sqrt{3} \quad \Longrightarrow \quad \underbrace{y = \sqrt{3}x}_{\text{Rectangular equation}}$$

c. The graph of the polar equation $r = \sec \theta$ is not evident by simple inspection, so you convert to rectangular form by using the relationship $r \cos \theta = x$.

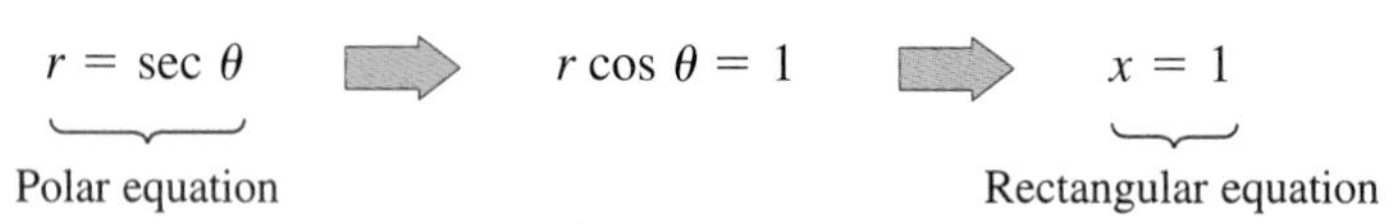

Now you see that the graph is a vertical line, as shown in Figure 6.55(c).

Curve sketching by converting to rectangular form is not always convenient. The next section demonstrates a straightforward point-plotting technique.

GROUP ACTIVITY

USING A GRAPHING UTILITY

Use a graphing utility to confirm the graphs shown in Figure 6.55.

TECHNOLOGY

Use a graphing utility to convert the point $(-2, 0)$ to polar form. The steps for a *TI-83* or a *TI-82* are as follows.

1. Press [ANGLE] and choose R▷Pr(.
2. Enter the values for x and y as (x, y).
3. Repeat steps 1 and 2 choosing R▷Pθ.

Explain how a graphing utility can be used to convert a point in polar form to rectangular form.

WARM UP

Find a positive angle that is coterminal with the given angle.

1. $\dfrac{11\pi}{4}$ **2.** $-\dfrac{5\pi}{6}$

Find the sine and cosine of an angle in standard position of which the terminal side passes through the given point.

3. $(2, 1)$ **4.** $(4, -3)$

Find the magnitude (in radians) of an angle in standard position of which the terminal side passes through the given point.

5. $(-4, 4)$ **6.** $(3, 2)$

Evaluate the trigonometric function without the aid of a calculator.

7. $\sin \dfrac{4\pi}{3}$ **8.** $\cos \dfrac{3\pi}{4}$

Use a calculator to evaluate the trigonometric function.

9. $\cos \dfrac{3\pi}{5}$ **10.** $\sin 1.34$

6.7 Exercises

In Exercises 1–4, a point in polar coordinates is given. Find the corresponding rectangular coordinates of the point.

1. $\left(4, \dfrac{3\pi}{6}\right)$

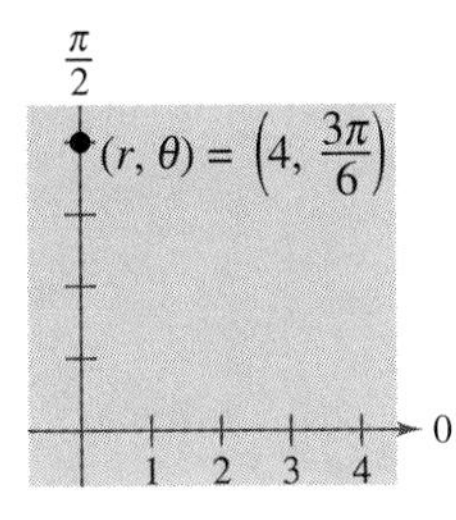

2. $\left(4, \dfrac{3\pi}{2}\right)$

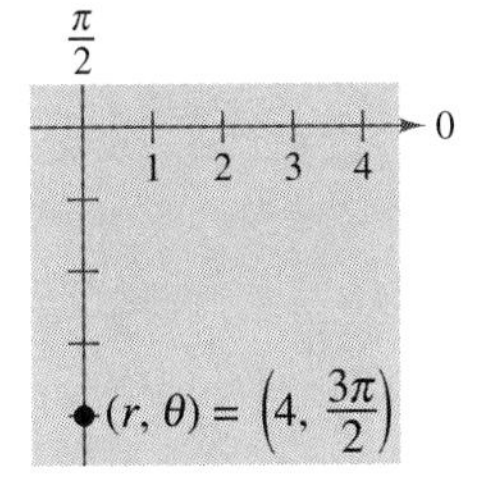

3. $\left(-1, \dfrac{5\pi}{4}\right)$

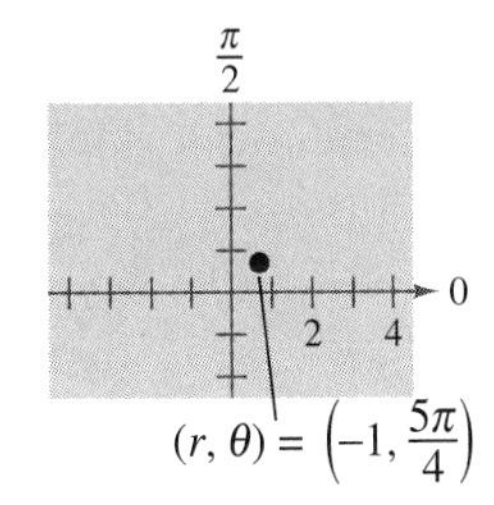

4. $(0, -\pi)$

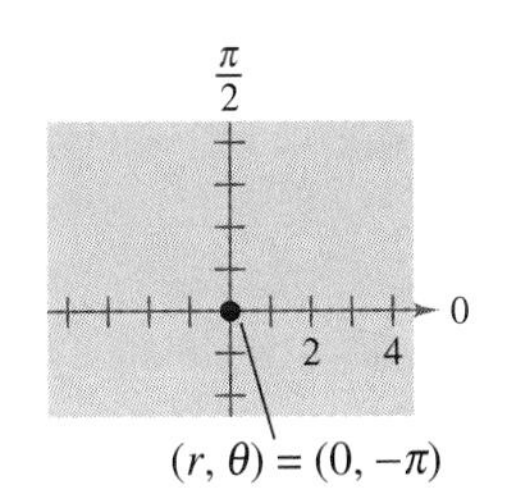

In Exercises 5–10, plot the point given in polar coordinates and find the corresponding rectangular coordinates of the point.

5. $(4, -\pi/3)$
6. $(-1, -3\pi/4)$
7. $(0, -7\pi/6)$
8. $(32, 5\pi/2)$
9. $(\sqrt{2}, 2.36)$
10. $(-3, -1.57)$

In Exercises 11–14, use a graphing utility to find the rectangular coordinates of the point given in polar coordinates.

11. $(2, 3\pi/4)$
12. $(-2, 7\pi/6)$
13. $(-4.5, 1.3)$
14. $(8.25, 3.5)$

In Exercises 15–24, the rectangular coordinates of a point are given. Plot the point and find *two* sets of polar coordinates of the point for $0 \le \theta \le 2\pi$.

15. $(1, 1)$
16. $(0, -5)$
17. $(-6, 0)$
18. $(-3, -3)$
19. $(-3, 4)$
20. $(3, -1)$
21. $(-\sqrt{3}, -\sqrt{3})$
22. $(2, 0)$
23. $(4, 6)$
24. $(5, 12)$

In Exercises 25–30, use a graphing utility to find one set of polar coordinates of the point given in rectangular coordinates.

25. $(3, -2)$
26. $(-4, 1)$
27. $(\sqrt{3}, 2)$
28. $(3\sqrt{2}, 3\sqrt{2})$
29. $\left(\frac{5}{2}, \frac{4}{3}\right)$
30. $(0, -5)$

True or False? **In Exercises 31 and 32, determine whether the statement is true or false. If it is false, explain why or give an example that shows it is false.**

31. If (r_1, θ_1) and (r_2, θ_2) represent the same point on the polar coordinate system, then $|r_1| = |r_2|$.
32. If (r, θ_1) and (r, θ_2) represent the same point on the polar coordinate system, then $\theta_1 = \theta_2 + 2\pi n$, for some integer n.

In Exercises 33–46, convert the rectangular equation to polar form.

33. $x^2 + y^2 = 9$
34. $x^2 + y^2 = a^2$
35. $x^2 + y^2 - 2ax = 0$
36. $x^2 + y^2 - 2ay = 0$
37. $y = 4$
38. $y = b$
39. $x = 10$
40. $x = a$
41. $3x - y + 2 = 0$
42. $4x + 7y - 2 = 0$
43. $xy = 4$
44. $y = x$
45. $(x^2 + y^2)^2 - 9(x^2 - y^2) = 0$
46. $y^2 - 8x - 16 = 0$

In Exercises 47–56, convert the polar equation to rectangular form.

47. $r = 4 \sin \theta$
48. $r = 4 \cos \theta$
49. $\theta = \dfrac{\pi}{6}$
50. $r = 4$
51. $r = 2 \csc \theta$
52. $r^2 = \sin 2\theta$
53. $r = 2 \sin 3\theta$
54. $r = \dfrac{1}{1 - \cos \theta}$
55. $r = \dfrac{6}{2 - 3 \sin \theta}$
56. $r = \dfrac{6}{2 \cos \theta - 3 \sin \theta}$

In Exercises 57–62, convert the polar equation to rectangular form and sketch its graph.

57. $r = 3$

58. $r = 8$

59. $\theta = \frac{\pi}{4}$

60. $\theta = \frac{5\pi}{6}$

61. $r = 3 \sec \theta$

62. $r = 2 \csc \theta$

63. Convert the polar equation

$$r = 2(h \cos \theta + k \sin \theta)$$

to rectangular form and verify that it is the equation of a circle. Find the radius and the rectangular coordinates of the center of the circle.

64. Convert the polar equation $r = \cos \theta + 3 \sin \theta$ to rectangular form and identify the graph.

65. ***Think About It***

(a) Show that the distance between the points (r_1, θ_1) and (r_2, θ_2) is given by

$$\sqrt{r_1^2 + r_2^2 - 2r_1r_2 \cos(\theta_1 - \theta_2)}.$$

(b) Describe the positions of the points relative to each other if $\theta_1 = \theta_2$. Simplify the distance formula for this case. Is the simplification what you expected? Explain.

(c) Simplify the Distance Formula if $\theta_1 - \theta_2 = 90°$. Is the simplification what you expected? Explain.

(d) Choose two points on the polar coordinate system and find the distance between them. Then choose different polar representations of the same two points and apply the Distance Formula again. Discuss the result.

66. ***Exploration***

(a) Set the window format of your graphing utility on rectangular coordinates and locate the cursor at any position off the coordinate axes. Move the cursor horizontally and observe any changes in the displayed coordinates of the points. Explain the changes. Now repeat the process moving the cursor vertically.

(b) Set the window format of your graphing utility on polar coordinates and locate the cursor at any position off the coordinate axes. Move the cursor horizontally and observe any changes in the displayed coordinates of the points. Explain the changes. Now repeat the process moving the cursor vertically.

(c) Explain why the results of parts (a) and (b) are not the same.

Review **Solve Exercises 67–72 as a review of the skills and problem-solving techniques you learned in previous sections. Use determinants to solve the system of equations.**

67.
$$\begin{aligned} 5x - 7y &= -11 \\ -3x + y &= -3 \end{aligned}$$

68.
$$\begin{aligned} 3x + 5y &= 10 \\ 4x - 2y &= -5 \end{aligned}$$

69.
$$\begin{aligned} 3a - 2b + c &= 0 \\ 2a + b - 3c &= 0 \\ a - 3b + 9c &= 8 \end{aligned}$$

70.
$$\begin{aligned} 5u + 7v + 9w &= 15 \\ u - 2v - 3w &= 7 \\ 8u - 2v + w &= 0 \end{aligned}$$

71.
$$\begin{aligned} x + y + z - 3w &= -8 \\ 3x - y - 2z + w &= 7 \\ -x + y - z + 2w &= -2 \\ 2y + w &= -6 \end{aligned}$$

72.
$$\begin{aligned} 2y + 5z + 6w &= 32 \\ 2x + 4y - 5z - w &= -7 \\ 3x - 6y + z + 5w &= 6 \\ 4x - 2y - z &= -12 \end{aligned}$$

6.8 Graphs of Polar Equations

See Exercises 63–66 on page 490 for examples of how a graphing utility can be used to help sketch graphs of polar equations.

Introduction ❐ *Symmetry* ❐ *Zeros and Maximum r-Values* ❐ *Special Polar Graphs*

Introduction

In previous chapters you spent a lot of time learning how to sketch graphs on rectangular coordinates. You began with the basic point-plotting method, which was then enhanced by sketching aids such as symmetry, intercepts, asymptotes, periods, and shifts. This section approaches curve sketching on the polar coordinate system similarly, beginning with a demonstration of point plotting.

EXAMPLE 1 *Graphing a Polar Equation by Point Plotting*

Sketch the graph of the polar equation $r = 4 \sin \theta$.

Solution

The sine function is periodic, so you can get a full range of r-values by considering values of θ in the interval $0 \le \theta \le 2\pi$, shown in the following table.

θ	0	$\frac{\pi}{6}$	$\frac{\pi}{3}$	$\frac{\pi}{2}$	$\frac{2\pi}{3}$	$\frac{5\pi}{6}$	π	$\frac{7\pi}{6}$	$\frac{3\pi}{2}$	$\frac{11\pi}{6}$	2π
r	0	2	$2\sqrt{3}$	4	$2\sqrt{3}$	2	0	-2	-4	-2	0

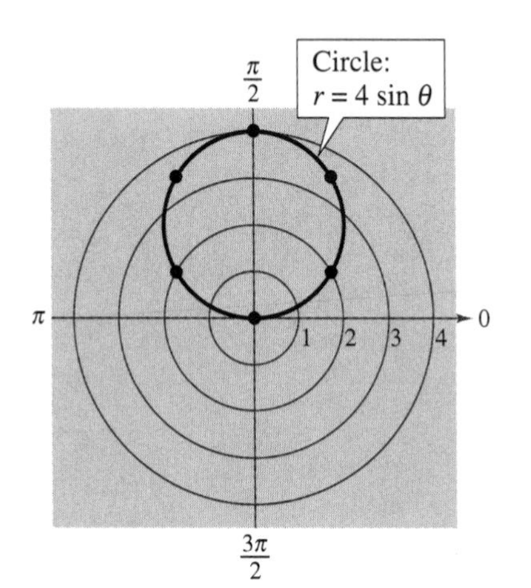

FIGURE 6.56

If you plot these points as shown in Figure 6.56, it appears that the graph is a circle of radius 2 whose center is at the point $(x, y) = (0, 2)$. Try confirming this by squaring both sides of the polar equation and converting the result to rectangular form.

Symmetry

In Figure 6.56, note that as θ increases from 0 to 2π the graph is traced out twice. Moreover, note that the graph is *symmetric with respect to the line* $\theta = \pi/2$. Had you known about this symmetry and retracing ahead of time, you could have used fewer points.

Symmetry with respect to the line $\theta = \pi/2$ is one of three important types of symmetry to consider in polar curve sketching. (See Figure 6.57.)

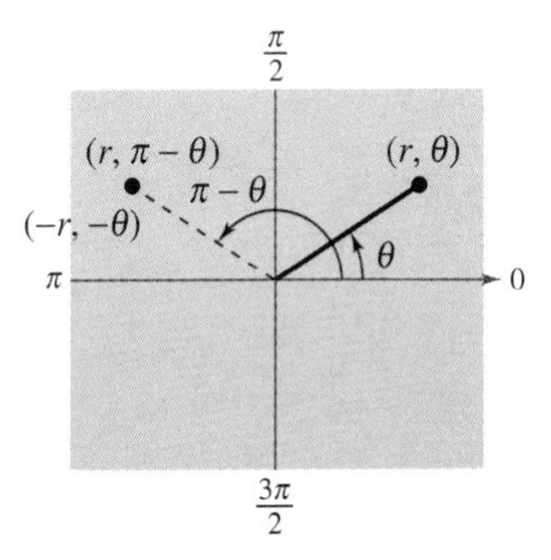

Symmetry with Respect to the Line $\theta = \frac{\pi}{2}$

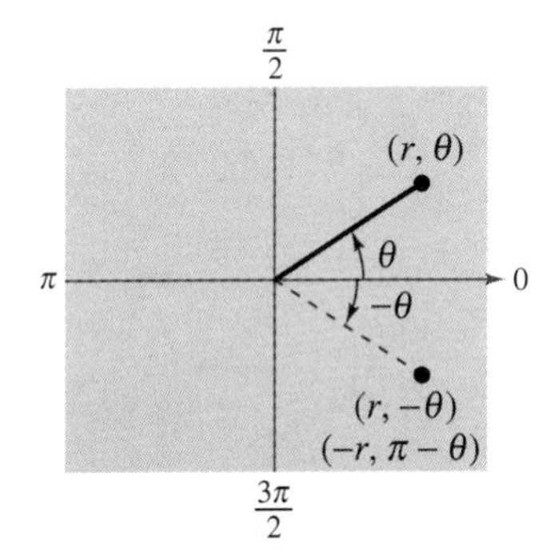

Symmetry with Respect to the Polar Axis

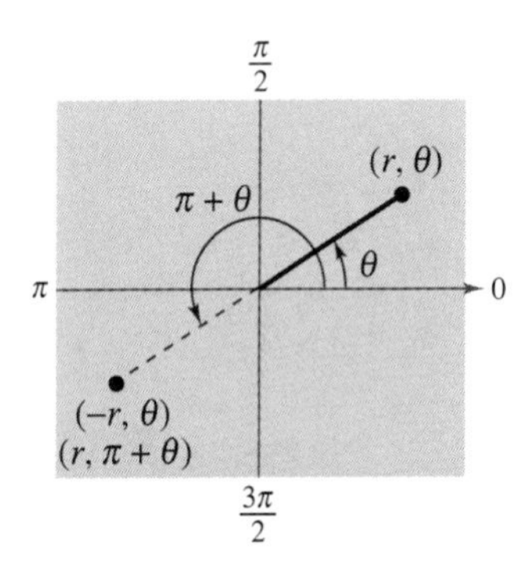

Symmetry with Respect to the Pole

FIGURE 6.57

TEST FOR SYMMETRY IN POLAR COORDINATES

The graph of a polar equation is symmetric with respect to the following if the given substitution yields an equivalent equation.

1. *The line* $\theta = \pi/2$: Replace (r, θ) by $(r, \pi - \theta)$ or $(-r, -\theta)$.
2. *The polar axis:* Replace (r, θ) by $(r, -\theta)$ or $(-r, \pi - \theta)$.
3. *The pole:* Replace (r, θ) by $(r, \pi + \theta)$ or $(-r, \theta)$.

EXAMPLE 2 Using Symmetry to Sketch a Polar Graph

Use symmetry to sketch the graph of $r = 3 + 2 \cos \theta$.

Solution

Replacing (r, θ) by $(r, -\theta)$ produces

$$r = 3 + 2 \cos(-\theta)$$
$$= 3 + 2 \cos \theta.$$

Thus, you can conclude that the curve is symmetric with respect to the polar axis. Plotting the points in the table and using polar axis symmetry, you obtain the graph shown in Figure 6.58.

θ	0	$\pi/3$	$\pi/2$	$2\pi/3$	π
r	5	4	3	2	1

The graph in Figure 6.58 is called a **limaçon.**

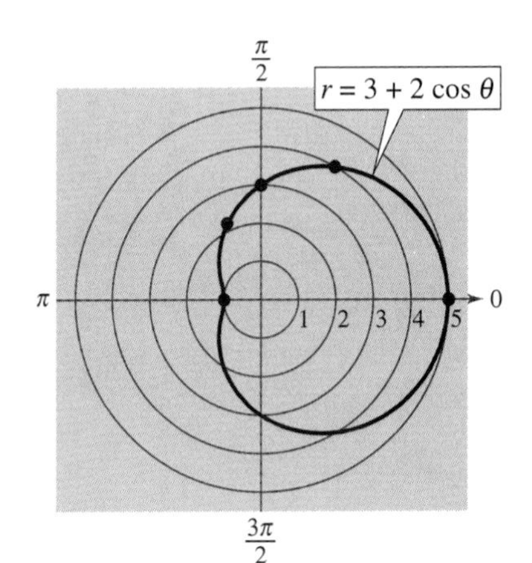

FIGURE 6.58

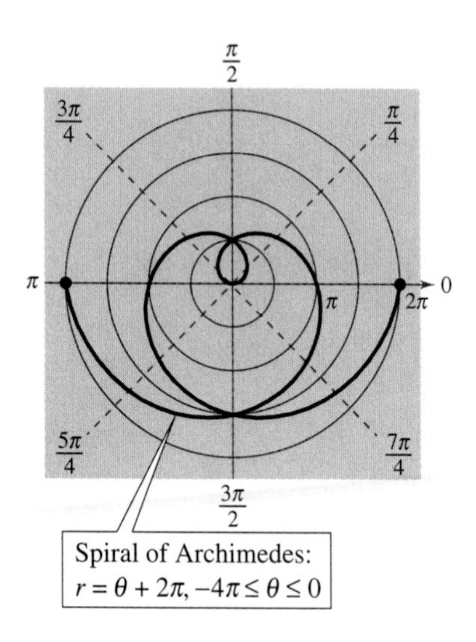

Spiral of Archimedes: $r = \theta + 2\pi, -4\pi \le \theta \le 0$

FIGURE 6.59

The three tests for symmetry in polar coordinates listed on page 483 are sufficient to guarantee symmetry, but they are not necessary. For instance, Figure 6.59 shows the graph of $r = \theta + 2\pi$ to be symmetric with respect to the line $\theta = \pi/2$, and yet the tests on page 483 fail to indicate symmetry.

The equations discussed in Examples 1 and 2 are of the form

$$r = 4 \sin \theta = f(\sin \theta) \qquad \text{and} \qquad r = 3 + 2 \cos \theta = g(\cos \theta).$$

The graph of the first equation is symmetric with respect to the line $\theta = \pi/2$, and the graph of the second equation is symmetric with respect to the polar axis. This observation can be generalized to yield the following *quick test for symmetry.*

1. The graph of $r = f(\sin \theta)$ is symmetric with respect to the line $\theta = \pi/2$.
2. The graph of $r = g(\cos \theta)$ is symmetric with respect to the polar axis.

Zeros and Maximum *r*-Values

Two additional aids to sketching graphs of polar equations involve knowing the θ-values for which $|r|$ is maximum and knowing the θ-values for which $r = 0$. For instance, in Example 1, the maximum value of $|r|$ for $r = 4 \sin \theta$ is $|r| = 4$, and this occurs when $\theta = \pi/2$, as shown in Figure 6.56. Moreover, $r = 0$ when $\theta = 0$.

EXAMPLE 3 *Sketching a Polar Graph*

Sketch the graph of $r = 1 - 2 \cos \theta$.

Solution

From the equation $r = 1 - 2 \cos \theta$, you can obtain the following.

Symmetry: With respect to the polar axis
Maximum Value of $|r|$*:* $r = 3$ when $\theta = \pi$
Zero of r: $r = 0$ when $\theta = \pi/3$

The table shows several θ-values in the interval $[0, \pi]$. By plotting the corresponding points, you can sketch the graph shown in Figure 6.60.

θ	0	$\pi/6$	$\pi/3$	$\pi/2$	$2\pi/3$	$5\pi/6$	π
r	-1	-0.73	0	1	2	2.73	3

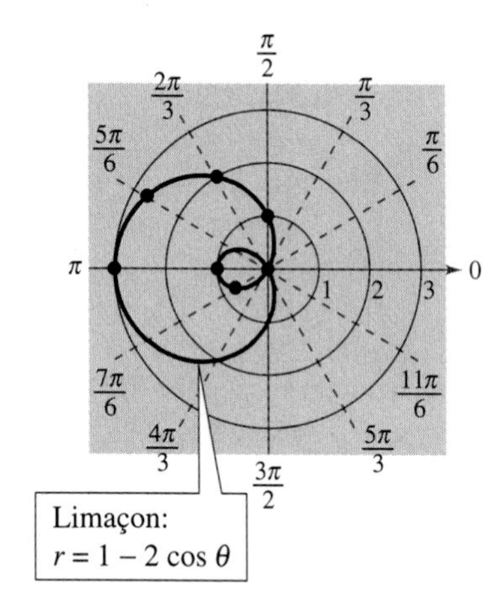

Limaçon: $r = 1 - 2 \cos \theta$

FIGURE 6.60

NOTE Note how the negative r-values determine the *inner loop* of the graph in Figure 6.60. This graph is a limaçon also.

Some curves reach their zeros and maximum r-values at more than one point. Example 4 shows how to handle this situation.

EXAMPLE 4 *Sketching a Polar Graph*

Sketch the graph of $r = 2 \cos 3\theta$.

Solution

Symmetry:	With respect to the polar axis
Maximum Value of $\lvert r \rvert$*:*	$\lvert r \rvert = 2$ when $3\theta = 0, \pi, 2\pi, 3\pi$ or $\theta = 0, \pi/3, 2\pi/3, \pi$
Zeros of r:	$r = 0$ when $3\theta = \pi/2, 3\pi/2, 5\pi/2$ or $\theta = \pi/6, \pi/2, 5\pi/6$

θ	0	$\pi/12$	$\pi/6$	$\pi/4$	$\pi/3$	$5\pi/12$	$\pi/2$
r	2	$\sqrt{2}$	0	$-\sqrt{2}$	-2	$-\sqrt{2}$	0

NOTE The graph shown in Figure 6.61 is called a **rose curve,** and each of the loops on the graph is called a *petal* of the rose curve.

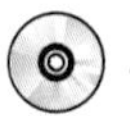

A computer animation of this concept appears in the *Interactive* CD-ROM.

By plotting these points and using the specified symmetry, zeros, and maximum values, you can obtain the graph shown in Figure 6.61. Note how the entire curve is generated as θ increases from 0 to π.

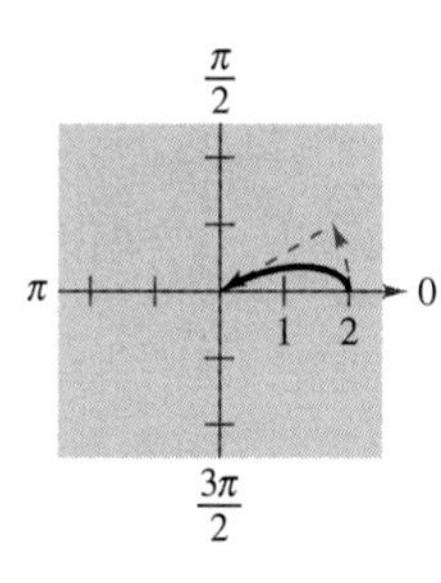

$0 \le \theta \le \frac{\pi}{6}$

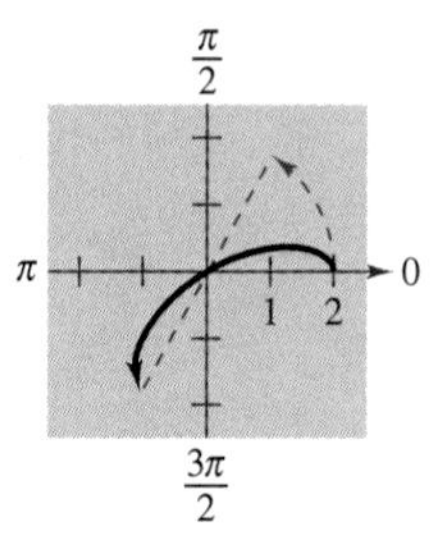

$0 \le \theta \le \frac{\pi}{3}$

$0 \le \theta \le \frac{\pi}{2}$

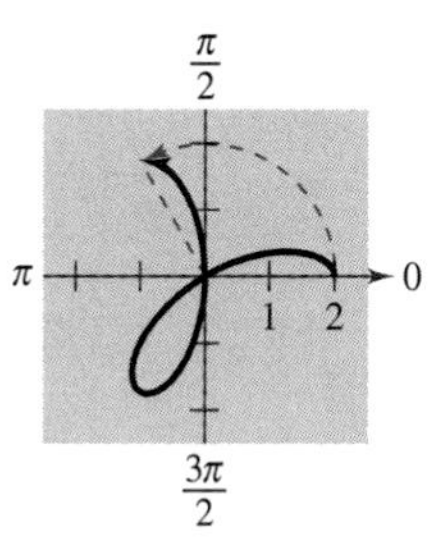

$0 \le \theta \le \frac{2\pi}{3}$

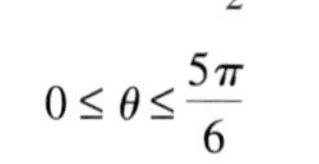

$0 \le \theta \le \frac{5\pi}{6}$

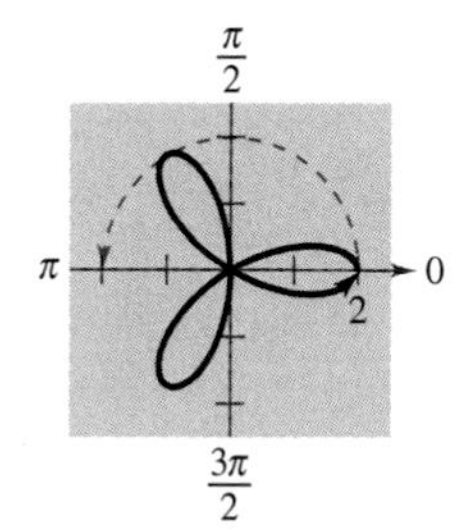

$0 \le \theta \le \pi$

FIGURE 6.61

TECHNOLOGY

Use a graphing utility in polar mode to verify the graph of $r = 2 \cos 3\theta$ shown in Figure 6.61.

Special Polar Graphs

Several important types of graphs have equations that are simpler in polar form than in rectangular form. For example, the circle $r = 4 \sin \theta$ in Example 1 has the more complicated rectangular equation $x^2 + (y - 2)^2 = 4$. Several other types of graphs that have simple polar equations are shown below.

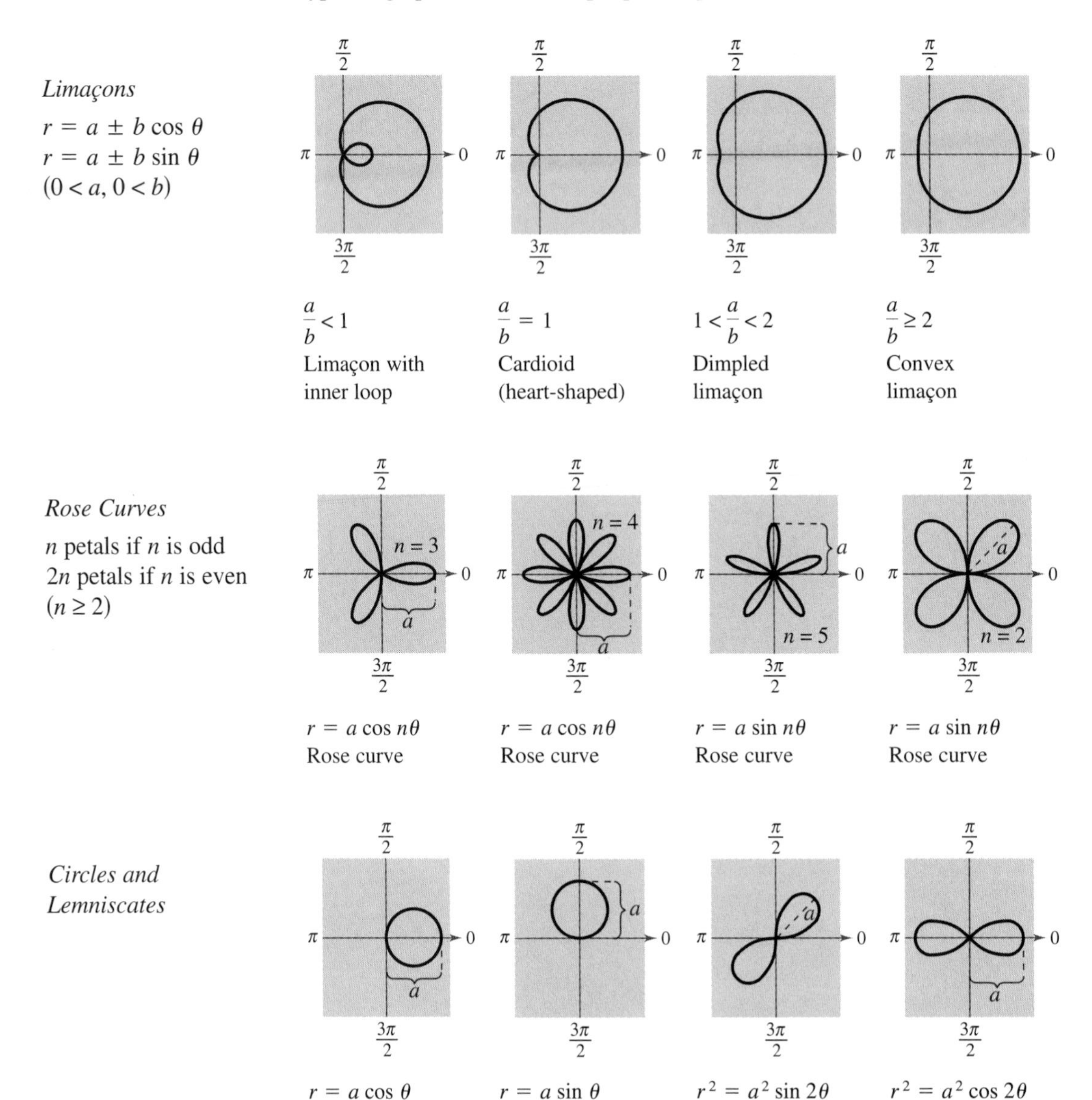

Limaçons

$r = a \pm b \cos \theta$
$r = a \pm b \sin \theta$
$(0 < a, 0 < b)$

$\frac{a}{b} < 1$ Limaçon with inner loop

$\frac{a}{b} = 1$ Cardioid (heart-shaped)

$1 < \frac{a}{b} < 2$ Dimpled limaçon

$\frac{a}{b} \geq 2$ Convex limaçon

Rose Curves

n petals if n is odd
$2n$ petals if n is even
$(n \geq 2)$

$r = a \cos n\theta$ Rose curve

$r = a \cos n\theta$ Rose curve

$r = a \sin n\theta$ Rose curve

$r = a \sin n\theta$ Rose curve

Circles and Lemniscates

$r = a \cos \theta$ Circle

$r = a \sin \theta$ Circle

$r^2 = a^2 \sin 2\theta$ Lemniscate

$r^2 = a^2 \cos 2\theta$ Lemniscate

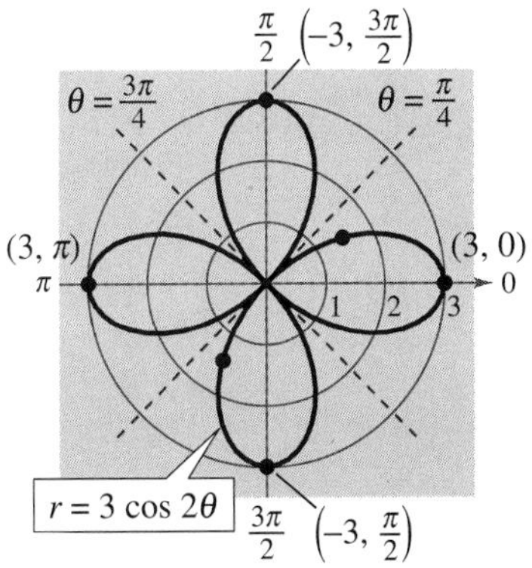

FIGURE 6.62

EXAMPLE 5 *Sketching a Rose Curve*

Sketch the graph of $r = 3 \cos 2\theta$.

Solution

Type of curve: Rose curve with $2n = 4$ petals
Symmetry: With respect to polar axis and the line $\theta = \pi/2$
Maximum Value of $|r|$*:* $|r| = 3$ when $\theta = 0, \pi/2, \pi, 3\pi/2$
Zeros of r*:* $r = 0$ when $\theta = \pi/4, 3\pi/4$

Using this information together with the additional points shown in the following table, you obtain the graph shown in Figure 6.62.

θ	0	$\pi/6$	$\pi/4$	$\pi/3$
r	3	$3/2$	0	$-3/2$

FIGURE 6.63

EXAMPLE 6 *Sketching a Lemniscate*

Sketch the graph of $r^2 = 9 \sin 2\theta$.

Solution

Type of curve: Lemniscate
Symmetry: With respect to the pole
Maximum Value of $|r|$*:* $|r| = 3$ when $\theta = \pi/4$
Zeros of r*:* $r = 0$ when $\theta = 0, \pi/2$

If $\sin 2\theta < 0$, this equation has no solution points. Thus, you restrict the values of θ to those for which $\sin 2\theta \geq 0$.

$$0 \leq \theta \leq \frac{\pi}{2} \quad \text{or} \quad \pi \leq \theta \leq \frac{3\pi}{2}$$

Moreover, using symmetry, you need to consider only the first of these two intervals. By finding a few additional points, you can obtain the graph shown in Figure 6.63.

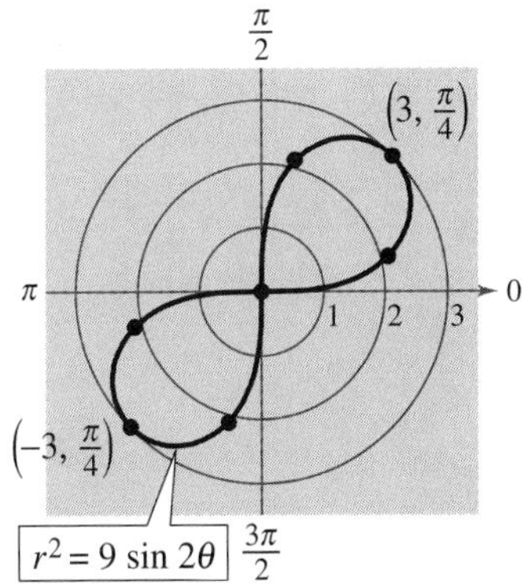

θ	0	$\pi/12$	$\pi/4$	$5\pi/12$	$\pi/2$
$r = \pm 3\sqrt{\sin 2\theta}$	0	$\pm 3/\sqrt{2}$	± 3	$\pm 3/\sqrt{2}$	0

GROUP ACTIVITY

IDENTIFYING GRAPHS

Match each polar equation with the correct graph.

(a)

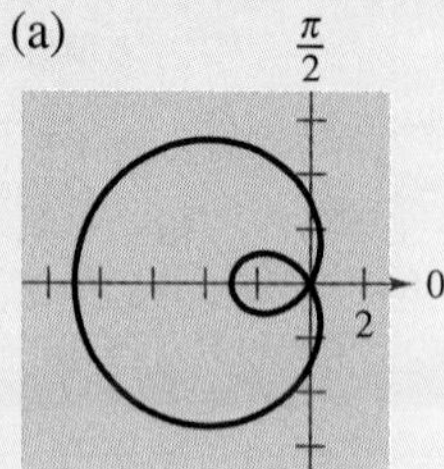

(b)

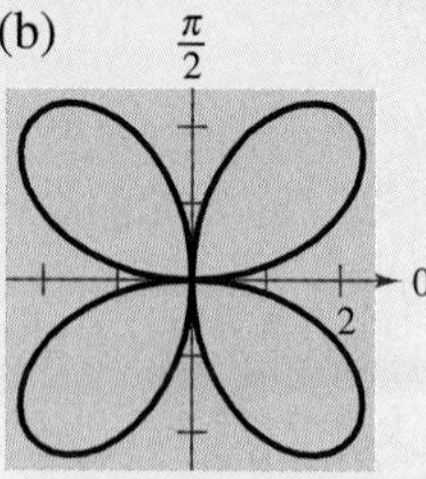

(c)

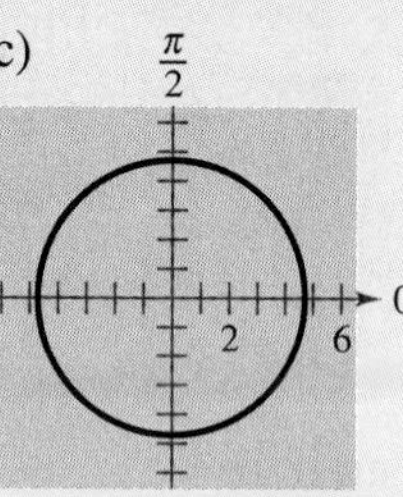

(d)

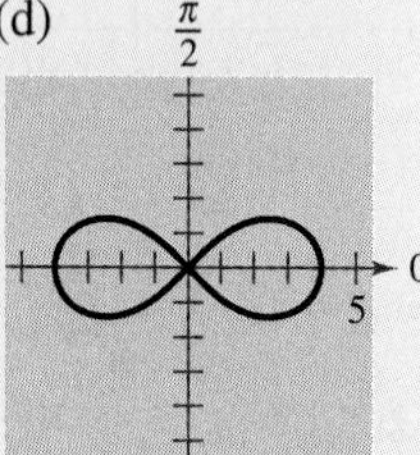

(e)

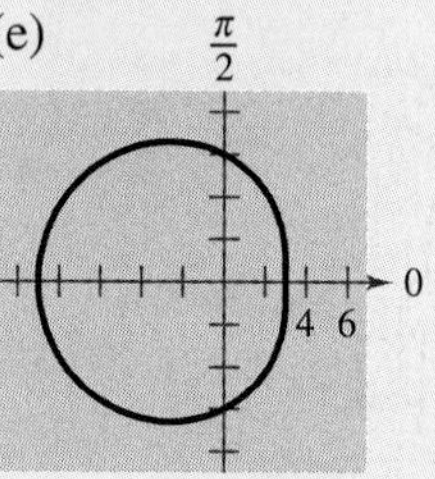

(f)

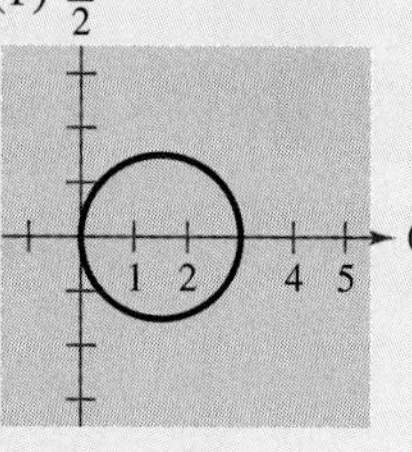

1. $r = 3 \cos \theta$
2. $r = 3 \sin 2\theta$
3. $r = 3(2 - \cos \theta)$
4. $r = 3(1 - 2 \cos \theta)$
5. $r = \dfrac{3\pi}{2}$
6. $r^2 = 16 \cos 2\theta$

WARM UP

Determine the amplitude and period of the function.

1. $y = 5 \sin 4x$
2. $y = 3 \cos 2\pi x$
3. $y = -5 \cos \dfrac{5\pi x}{2}$
4. $y = -\dfrac{1}{2} \sin \dfrac{x}{2}$

Sketch the graph of the function through two periods.

5. $y = 2 \sin x$
6. $y = 3 \cos x$
7. $y = 4 \cos 2x$
8. $y = 2 \sin \pi x$

Use the sum and difference formulas to simplify the expression.

9. $\sin\left(x - \dfrac{\pi}{6}\right)$
10. $\sin\left(x + \dfrac{\pi}{4}\right)$

6.8 Exercises

In Exercises 1–6, identify the type of polar graph.

1.

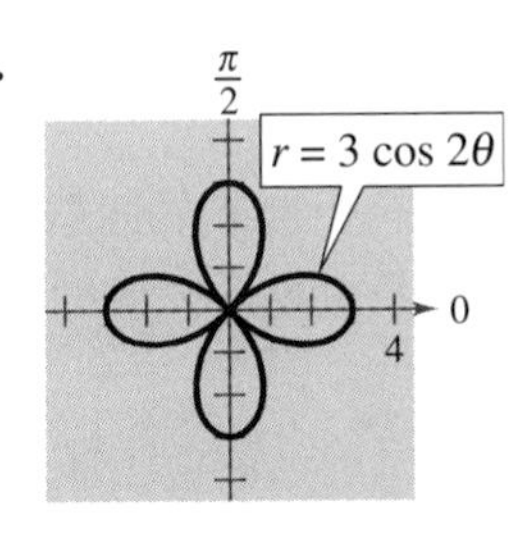

2.

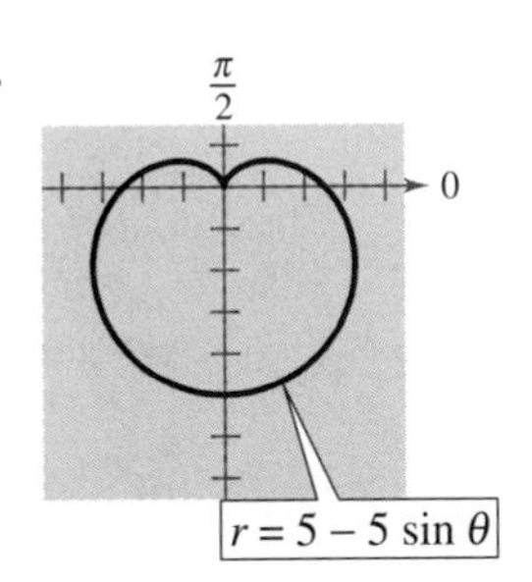

3.

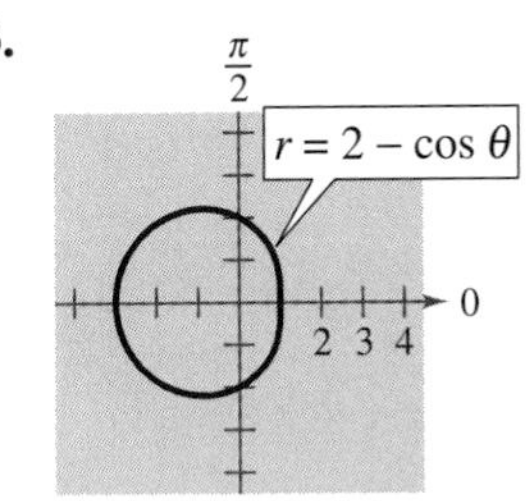

4.

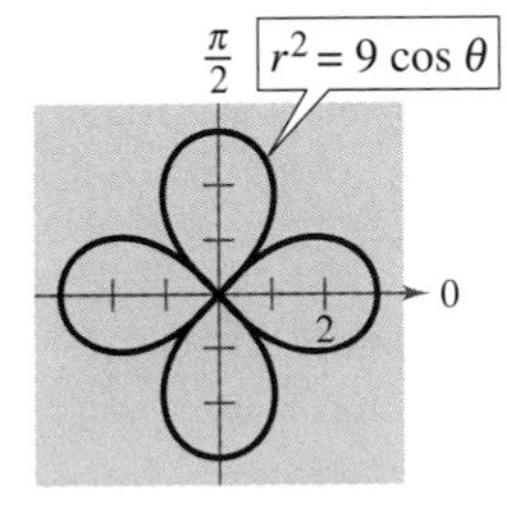

5.

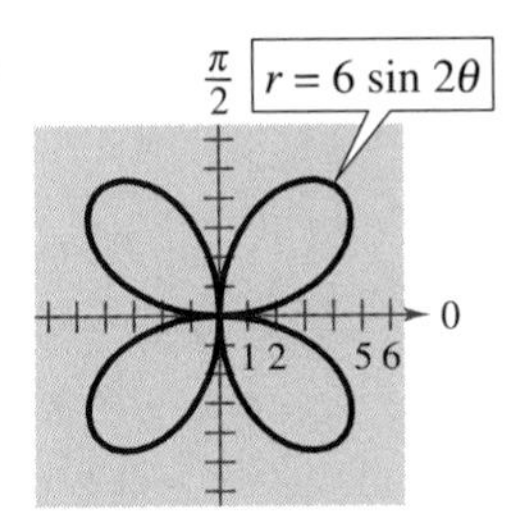

6.

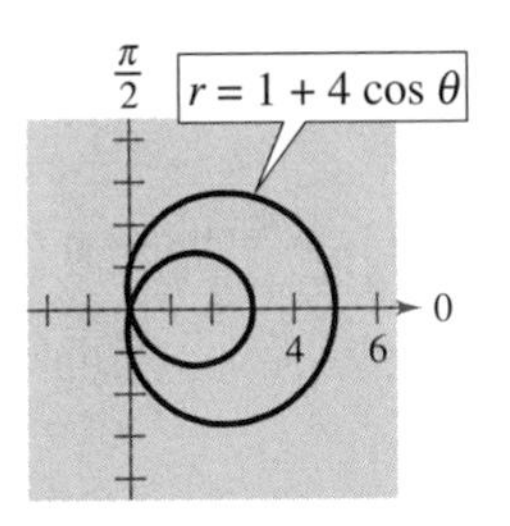

In Exercises 7–12, test for symmetry with respect to $\theta = \pi/2$, the polar axis, and the pole.

7. $r = 10 + 6\cos\theta$

8. $r = 16\cos 3\theta$

9. $r = \dfrac{2}{1 + \sin\theta}$

10. $r = 6\sin\theta$

11. $r = 4\sec\theta\csc\theta$

12. $r^2 = 25\sin 2\theta$

In Exercises 13–16, find the maximum value of $|r|$ and any zeros of r.

13. $r = 10(1 - \sin\theta)$

14. $r = 6 + 12\cos\theta$

15. $r = 4\cos 3\theta$

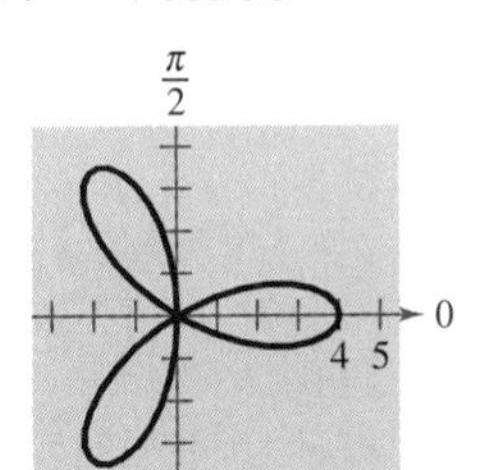

16. $r = 5\sin 2\theta$

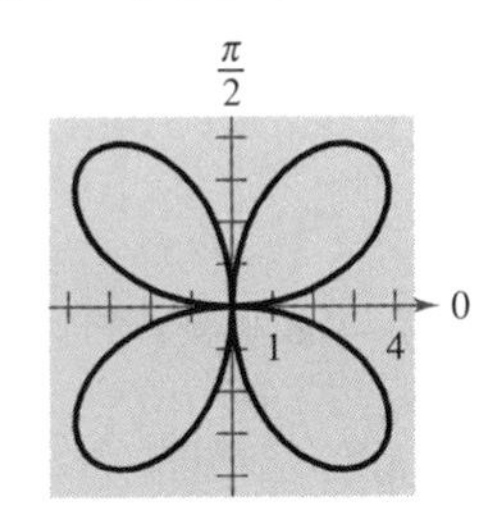

In Exercises 17–46, sketch the graph of the polar equation.

17. $r = 5$

18. $r = 2$

19. $r = \dfrac{\pi}{6}$

20. $r = -\dfrac{\pi}{4}$

21. $r = 3\sin\theta$

22. $r = 3\cos\theta$

23. $r = 3(1 - \cos\theta)$

24. $r = 2(1 - \sin\theta)$

25. $r = 4(1 + \sin\theta)$

26. $r = 1 + \cos\theta$

27. $r = 3 - 2\cos\theta$

28. $r = 5 - 4\sin\theta$

29. $r = 2 + \sin\theta$

30. $r = 4 + 3\cos\theta$

31. $r = 2 + 4\sin\theta$

32. $r = 1 - 2\cos\theta$

33. $r = 3 - 4\cos\theta$

34. $r = 2(1 - 2\sin\theta)$

35. $r = 2\cos 3\theta$

36. $r = -\sin 5\theta$

37. $r = 3\sin 2\theta$

38. $r = 3\cos 2\theta$

39. $r = 2\sec\theta$

40. $r = 3\csc\theta$

41. $r = \dfrac{3}{\sin\theta - 2\cos\theta}$

42. $r = \dfrac{6}{2\sin\theta - 3\cos\theta}$

43. $r^2 = 4\cos 2\theta$

44. $r^2 = 4\sin\theta$

45. $r = \dfrac{\theta}{2}$

46. $r = \theta$

In Exercises 47–54, use a graphing utility to graph the polar equation.

47. $r = 6\cos\theta$

48. $r = \dfrac{\theta}{4}$

49. $r = 3(2 - \sin\theta)$

50. $r = \cos 2\theta$

51. $r = 4 \sin \theta \cos^2 \theta$

52. $r = 2 \cos(3\theta - 2)$

53. $r = 2 \csc \theta + 5$

54. $r = 2 - \sec \theta$

In Exercises 55–62, use a graphing utility to graph the polar equation. Find an interval for θ for which the graph is traced *only once.*

55. $r = 3 - 4 \cos \theta$

56. $r = 2(1 - 2 \sin \theta)$

57. $r = 2 + \sin \theta$

58. $r = 4 + 3 \cos \theta$

59. $r = 2 \cos\left(\dfrac{3\theta}{2}\right)$

60. $r = 3 \sin\left(\dfrac{5\theta}{2}\right)$

61. $r^2 = 4 \sin 2\theta$

62. $r^2 = \dfrac{1}{\theta}$

In Exercises 63–66, use a graphing utility to graph the polar equation and show that the indicated line is an asymptote of the graph.

	Name of Graph	*Polar Equation*	*Asymptote*
63.	Conchoid	$r = 2 - \sec \theta$	$x = -1$
64.	Conchoid	$r = 2 + \csc \theta$	$y = 1$
65.	Hyperbolic spiral	$r = \dfrac{2}{\theta}$	$y = 2$
66.	Strophoid	$r = 2 \cos 2\theta \sec \theta$	$x = -2$

67. ***Exploration*** Sketch the graph of $r = 4 \sin \theta$ over each interval. Describe the part of the graph obtained in each case.

(a) $0 \le \theta \le \dfrac{\pi}{2}$ (b) $\dfrac{\pi}{2} \le \theta \le \pi$

(c) $-\dfrac{\pi}{2} \le \theta \le \dfrac{\pi}{2}$ (d) $\dfrac{\pi}{4} \le \theta \le \dfrac{3\pi}{4}$

68. ***Graphical Reasoning*** Use a graphing utility to graph the polar equation

$$r = 6[1 + \cos(\theta - \phi)]$$

for (a) $\phi = 0$, (b) $\phi = \pi/4$, and (c) $\phi = \pi/2$. Use the graphs to describe the effect of the angle ϕ. Write the equation as a function of $\sin \theta$ for part (c).

69. The graph of $r = f(\theta)$ is rotated about the pole through an angle ϕ. Show that the equation of the rotated graph is $r = f(\theta - \phi)$.

70. Consider the graph of $r = f(\sin \theta)$.

(a) Show that if the graph is rotated counterclockwise $\pi/2$ radians about the pole, the equation of the rotated graph is $r = f(-\cos \theta)$.

(b) Show that if the graph is rotated counterclockwise π radians about the pole, the equation of the rotated graph is $r = f(-\sin \theta)$.

(c) Show that if the graph is rotated counterclockwise $3\pi/2$ radians about the pole, the equation of the rotated graph is $r = f(\cos \theta)$.

In Exercises 71–74, use the results of Exercises 69 and 70.

71. Write an equation for the limaçon $r = 2 - \sin \theta$ after it has been rotated by the given amount.

(a) $\dfrac{\pi}{4}$ (b) $\dfrac{\pi}{2}$ (c) π (d) $\dfrac{3\pi}{2}$

72. Write an equation for the rose curve $r = 2 \sin 2\theta$ after it has been rotated by the given amount.

(a) $\dfrac{\pi}{6}$ (b) $\dfrac{\pi}{2}$ (c) $\dfrac{2\pi}{3}$ (d) π

73. Sketch the graph of each equation.

(a) $r = 1 - \sin \theta$ (b) $r = 1 - \sin\left(\theta - \dfrac{\pi}{4}\right)$

74. Sketch the graph of each equation.

(a) $r = 3 \sec \theta$ (b) $r = 3 \sec\left(\theta - \dfrac{\pi}{4}\right)$

(c) $r = 3 \sec\left(\theta + \dfrac{\pi}{3}\right)$ (d) $r = 3 \sec\left(\theta - \dfrac{\pi}{2}\right)$

75. ***Exploration*** Use a graphing utility to graph and identify $r = 2 + k \cos \theta$ for $k = 0, 1, 2$, and 3.

76. ***Exploration*** Consider the equation $r = 3 \sin k\theta$.

(a) Use a graphing utility to graph the equation for $k = 1.5$. Find the interval for θ for which the graph is traced only once.

(b) Use a graphing utility to graph the equation for $k = 2.5$. Find the interval for θ for which the graph is traced only once.

(c) Is it possible to find an interval for θ for which the graph is traced only once for any rational number k? Explain.

6.9 Polar Equations of Conics

See Exercise 55 on page 499 for an example of how polar coordinates can simplify the equations of planetary orbits.

Alternative Definition of Conics ▫ *Polar Equations of Conics* ▫ *Application*

Alternative Definition of Conics

In Sections 6.3 and 6.4, you learned that the rectangular equations of ellipses and hyperbolas take simple forms when the origin lies at their *centers.* As it happens, there are many important applications of conics in which it is more convenient to use one of the *foci* as the origin for the coordinate system. For example, the sun lies at a focus of the earth's orbit. Similarly, the light source of a parabolic reflector lies at its focus. In this section you will learn that polar equations of conics take simple forms if one of the foci lies at the pole.

To begin, consider the following alternative definition of conic that uses the concept of eccentricity.

> **ALTERNATIVE DEFINITION OF CONIC**
>
> The locus of a point in the plane that moves so that its distance from a fixed point (focus) is in constant ratio to its distance from a fixed line (directrix) is a **conic.** The constant ratio is the **eccentricity** of the conic and is denoted by e. Moreover, the conic is an **ellipse** if $e < 1$, a **parabola** if $e = 1$, and a **hyperbola** if $e > 1$.

In Figure 6.64, note that for each type of conic, the pole corresponds to the fixed point (focus) given in the definition.

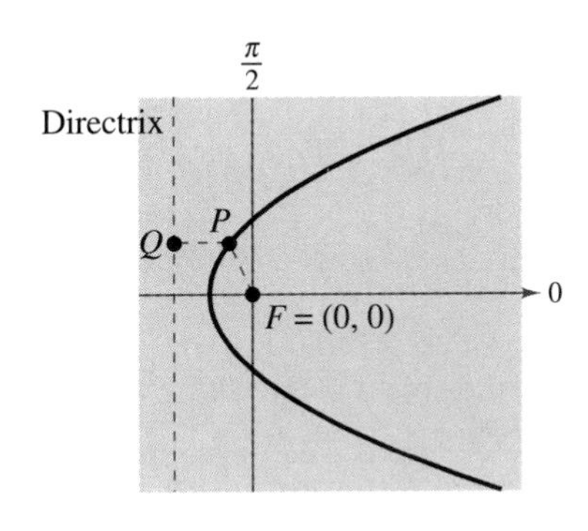

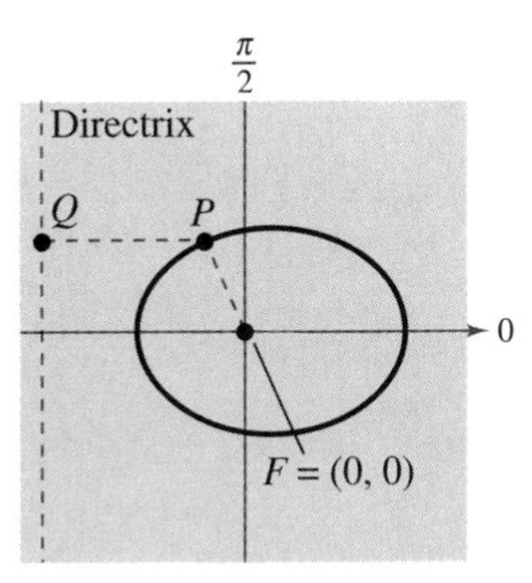

Ellipse: $0 < e < 1$

$\frac{PF}{PQ} < 1$

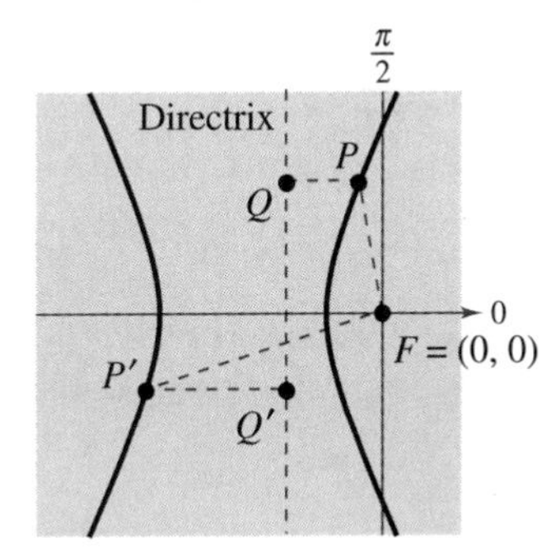

Hyperbola: $e > 1$

$\frac{PF}{PQ} = \frac{P'F}{P'Q'} > 1$

FIGURE 6.64

Polar Equations of Conics

The benefit of locating a focus at the pole can be seen in the proof of the following theorem.

> **POLAR EQUATIONS OF CONICS**
>
> The graph of a polar equation of the form
>
> **1.** $r = \dfrac{ep}{1 \pm e \cos \theta}$
>
> **2.** $r = \dfrac{ep}{1 \pm e \sin \theta}$
>
> is a conic, where $e > 0$ is the eccentricity and $|p|$ is the distance between the focus (pole) and the directrix.

PROOF

A proof for $r = ep/(1 + e \cos \theta)$ with $p > 0$ is listed here. The proofs of the other cases are similar. In Figure 6.65, consider a vertical directrix, p units to the right of the focus $F = (0, 0)$. If $P = (r, \theta)$ is a point on the graph of

$$r = \frac{ep}{1 + e \cos \theta}$$

the distance between P and the directrix is

$$\begin{aligned} PQ &= |p - x| \\ &= |p - r \cos \theta| \\ &= \left| p - \left(\frac{ep}{1 + e \cos \theta} \right) \cos \theta \right| \\ &= \left| p \left(1 - \frac{e \cos \theta}{1 + e \cos \theta} \right) \right| \\ &= \left| \frac{p}{1 + e \cos \theta} \right| \\ &= \left| \frac{r}{e} \right|. \end{aligned}$$

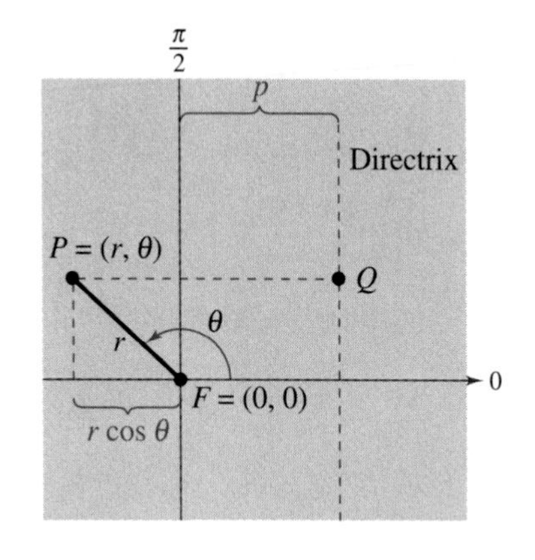

FIGURE 6.65

Moreover, because the distance between P and the pole is simply $PF = |r|$, the ratio of PF to PQ is

$$\begin{aligned} \frac{PF}{PQ} &= \frac{|r|}{|r/e|} \\ &= |e| \\ &= e \end{aligned}$$

and, by definition, the graph of the equation must be a conic.

By completing the proofs of the other three cases, you can see that the equations

$$r = \frac{ep}{1 \pm e \cos \theta}$$ Vertical directrix

correspond to conics with vertical directrices and the equations

$$r = \frac{ep}{1 \pm e \sin \theta}$$ Horizontal directrix

correspond to conics with horizontal directrices. Moreover, the converse is also true—that is, any conic with a focus at the pole and having a horizontal or vertical directrix can be represented by one of the given equations.

EXAMPLE 1 *Determining a Conic from its Equation*

Sketch the graph of the conic given by

$$r = \frac{15}{3 - 2 \cos \theta}.$$

Solution

To determine the type of conic, rewrite the equation as

$$r = \frac{15}{3 - 2 \cos \theta} = \frac{5}{1 - (2/3) \cos \theta}.$$ Divide numerator and denominator by 3.

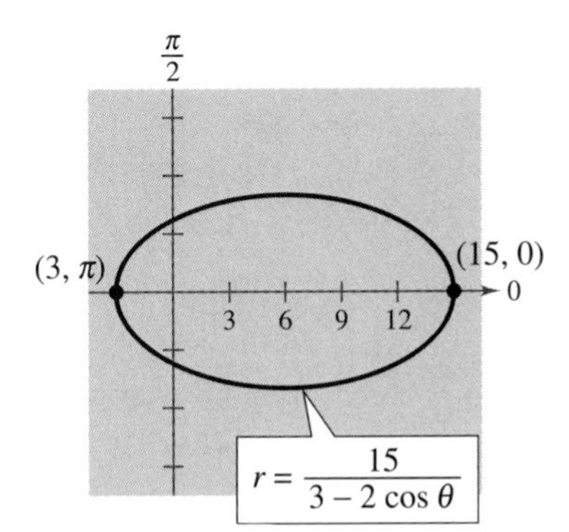

FIGURE 6.66

From this form you can conclude that the graph is an ellipse with $e = \frac{2}{3}$. You can sketch the upper half of the ellipse by plotting points from $\theta = 0$ to $\theta = \pi$, as shown in Figure 6.66. Using symmetry with respect to the polar axis, you can sketch the lower half.

For the ellipse in Figure 6.66, the major axis is horizontal and the vertices lie at $(15, 0)$ and $(3, \pi)$. Thus, the length of the *major* axis is $2a = 18$. To find the length of the *minor* axis, you can use the equations $e = c/a$ and $b^2 = a^2 - c^2$ to conclude that

$$b^2 = a^2 - c^2 = a^2 - (ea)^2 = a^2(1 - e^2).$$ Ellipse

Because $e = \frac{2}{3}$, you have

$$b^2 = 9^2\left[1 - \left(\tfrac{2}{3}\right)^2\right] = 45$$

which implies that $b = \sqrt{45} = 3\sqrt{5}$. Thus, the length of the minor axis is $2b = 6\sqrt{5}$. A similar analysis for hyperbolas yields

$$b^2 = c^2 - a^2 = (ea)^2 - a^2 = a^2(e^2 - 1).$$ Hyperbola

EXAMPLE 2 *Sketching a Conic from its Polar Equation*

Sketch the graph of the polar equation

$$r = \frac{32}{3 + 5 \sin \theta}.$$

Solution

Dividing each term by 3, you have

$$r = \frac{32/3}{1 + (5/3) \sin \theta}.$$

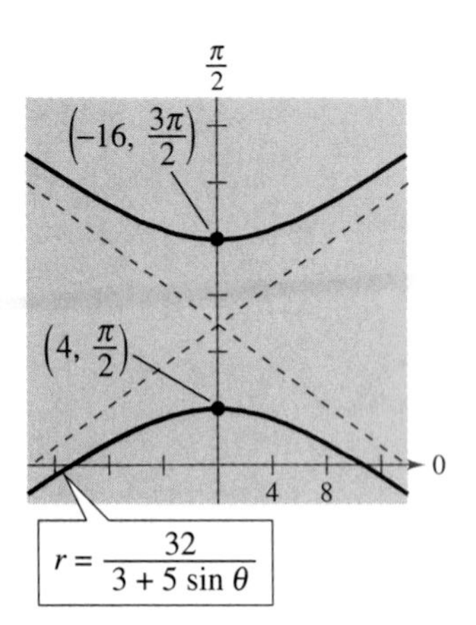

FIGURE 6.67

Because $e = 5/3 > 1$, the graph is a hyperbola. The transverse axis of the hyperbola lies on the line $\theta = \pi/2$ and the vertices occur at $(4, \pi/2)$ and $(-16, 3\pi/2)$. Because the length of the transverse axis is 12, you can see that $a = 6$. To find b, write

$$b^2 = a^2(e^2 - 1) = 6^2\left[\left(\frac{5}{3}\right)^2 - 1\right] = 64.$$

Therefore, $b = 8$. Finally, you can use a and b to determine the asymptotes of the hyperbola and obtain the sketch shown in Figure 6.67.

In the next example you are asked to find a polar equation for a specified conic. To do this, let p be the distance between the pole and the directrix.

1. *Horizontal directrix above the pole:* $r = \dfrac{ep}{1 + e \sin \theta}$

2. *Horizontal directrix below the pole:* $r = \dfrac{ep}{1 - e \sin \theta}$

3. *Vertical directrix to the right of the pole:* $r = \dfrac{ep}{1 + e \cos \theta}$

4. *Vertical directrix to the left of the pole:* $r = \dfrac{ep}{1 - e \cos \theta}$

Try using a graphing utility set in polar mode to verify the four orientations shown above. Remember that e must be positive, but p can be positive or negative.

EXAMPLE 3 *Finding the Polar Equation of a Conic*

Find the polar equation of the parabola whose focus is the pole and whose directrix is the line $y = 3$.

Solution

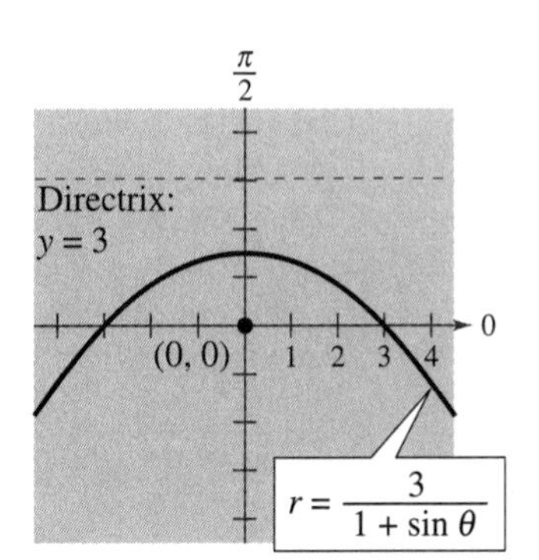

FIGURE 6.68

From Figure 6.68, you can see that the directrix is horizontal and above the pole. Thus, you can choose an equation of the form

$$r = \frac{ep}{1 + e\sin\theta}.$$

Moreover, because the eccentricity of a parabola is $e = 1$ and the distance between the pole and the directrix is $p = 3$, you have the equation

$$r = \frac{3}{1 + \sin\theta}.$$

Applications

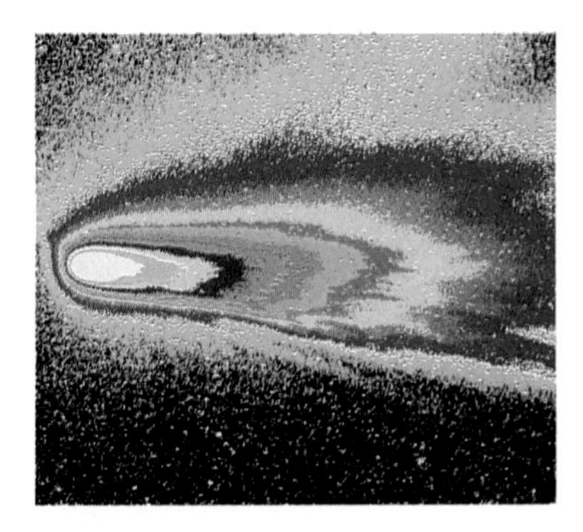

The most distinguishing feature of a comet is its tail, which stretches across space for several tens of millions of kilometers. The most famous is Halley's comet, pictured above, whose sightings have been documented as far back as 240 B.C. in Chinese records. *(Photo: The Royal Observatory, Edinburgh)*

Kepler's Laws (listed below), named after the German astronomer Johannes Kepler (1571–1630), can be used to describe the orbits of the planets about the sun.

1. Each planet moves in an elliptical orbit with the sun at one focus.
2. A ray from the sun to the planet sweeps out equal areas of the ellipse in equal times.
3. The square of the period is proportional to the cube of the mean distance between the planet and the sun.

Although Kepler simply stated these laws on the basis of observation, they were later validated by Isaac Newton (1642–1727). In fact, Newton was able to show that each law can be deduced from a set of universal laws of motion and gravitation that govern the movement of all heavenly bodies, including comets and satellites. This is illustrated in the next example, which involves the comet named after the English mathematician and physicist Edmund Halley (1656–1742).

NOTE If you use earth as a reference with a period of 1 year and a distance of 1 astronomical unit, the proportionality constant in Kepler's third law is 1. For example, because Mars has a mean distance to the sun of $d = 1.523$ astronomical units, its period P is given by $d^3 = P^2$. Thus, the period of Mars is $P = 1.88$ years.

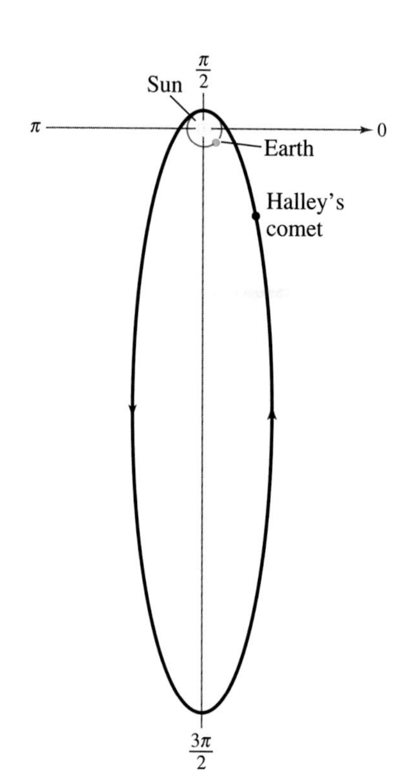

FIGURE 6.69

Real Life

EXAMPLE 4 *Halley's Comet*

Halley's comet has an elliptical orbit with an eccentricity of $e \approx 0.97$. The length of the major axis of the orbit is approximately 36.18 astronomical units. (An *astronomical unit* is defined as the mean distance between the earth and the sun, or 93 million miles.) Find a polar equation for the orbit. How close does Halley's comet come to the sun?

Solution

Using a vertical axis, as shown in Figure 6.69, choose an equation of the form $r = ep/(1 + e \sin\theta)$. Because the vertices of the ellipse occur when $\theta = \pi/2$ and $\theta = 3\pi/2$, you can determine the length of the major axis to be the sum of the r-values of the vertices. That is,

$$2a = \frac{0.97p}{1 + 0.97} + \frac{0.97p}{1 - 0.97} \approx 32.83p \approx 36.18.$$

Thus, $p \approx 1.102$ and $ep \approx (0.97)(1.102) \approx 1.069$. Using this value of ep in the equation, you have

$$r = \frac{1.069}{1 + 0.97 \sin \theta}$$

where r is measured in astronomical units. To find the closest point to the sun (the focus), substitute $\theta = \pi/2$ into this equation to obtain

$$r = \frac{1.069}{1 + 0.97 \sin(\pi/2)}$$

$$\approx 0.54 \text{ astronomical units}$$

$$\approx 50{,}000{,}000 \text{ miles.}$$

GROUP ACTIVITY

COMETS IN OUR SOLAR SYSTEM

Halley's comet is not the only spectacular comet that is periodically visible to viewers on earth. Use your school's library to find information about another comet, and write a paragraph describing some of its characteristics.

WARM UP

Convert the polar point to rectangular form.

1. $(-3, 3\pi/4)$ **2.** $(4, -2\pi/3)$

Convert the rectangular point to polar form.

3. $(0, -3)$ **4.** $(-5, 12)$

Convert the rectangular equation to polar form.

5. $x^2 + y^2 = 25$ **6.** $x^2y = 4$

Convert the polar equation to rectangular form.

7. $r \sin \theta = -4$ **8.** $r = 4 \cos \theta$

Identify and sketch the graph of the polar equation.

9. $r = 1 - \sin \theta$ **10.** $r = 1 + 2 \cos \theta$

6.9 Exercises

Graphical Reasoning **In Exercises 1–4, use a graphing utility to graph the polar equation for (a) $e = 1$, (b) $e = 0.5$, and (c) $e = 1.5$. What can you conclude?**

1. $r = \dfrac{2e}{1 + e \cos \theta}$ **2.** $r = \dfrac{2e}{1 - e \cos \theta}$

3. $r = \dfrac{2e}{1 - e \sin \theta}$ **4.** $r = \dfrac{2e}{1 + e \sin \theta}$

In Exercises 5–8, match the polar equation with its graph. [The graphs are labeled (a), (b), (c), and (d).]

5. $r = \dfrac{4}{1 - \cos \theta}$ **6.** $r = \dfrac{3}{2 - \cos \theta}$

7. $r = \dfrac{3}{1 + 2 \sin \theta}$ **8.** $r = \dfrac{4}{1 + \sin \theta}$

(a)

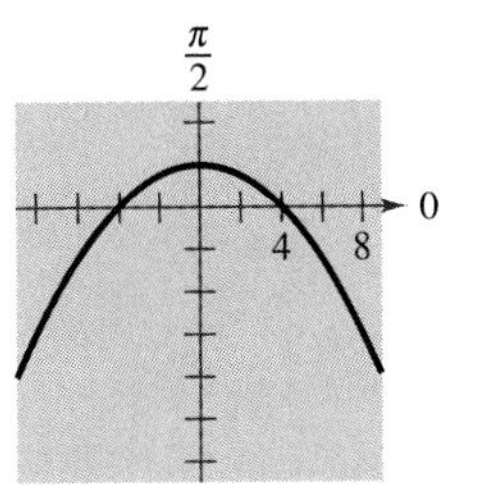

(b)

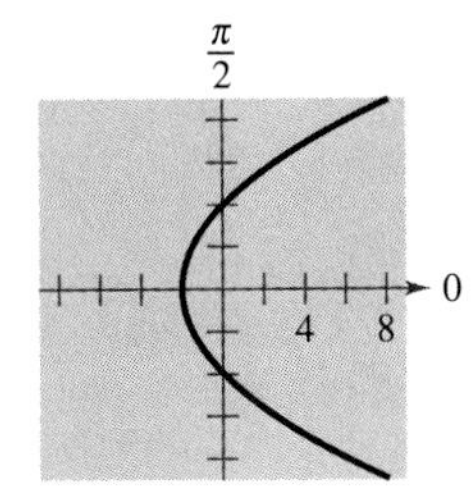

(c)

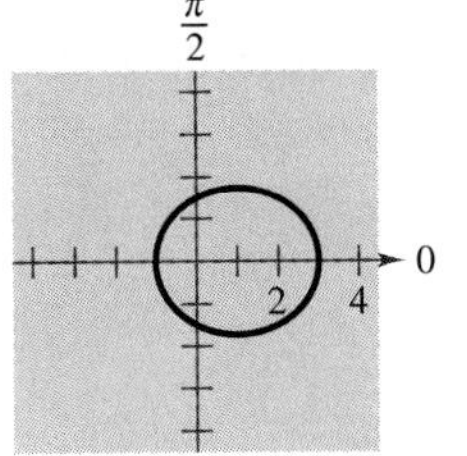

(d)

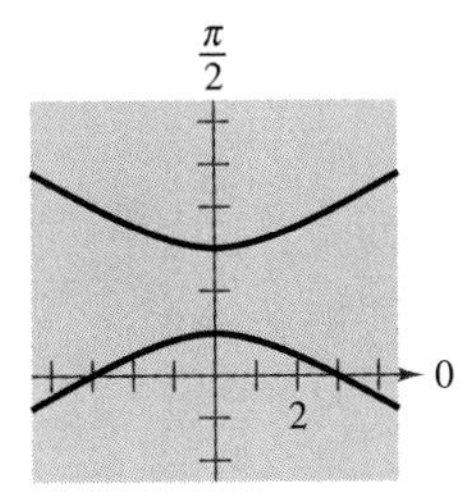

In Exercises 9–22, identify and sketch the graph of the polar equation.

9. $r = \dfrac{2}{1 - \cos\theta}$

10. $r = \dfrac{3}{1 + \sin\theta}$

11. $r = \dfrac{5}{1 + \sin\theta}$

12. $r = \dfrac{6}{1 + \cos\theta}$

13. $r = \dfrac{2}{2 - \cos\theta}$

14. $r = \dfrac{3}{3 + \sin\theta}$

15. $r = \dfrac{4}{2 + \sin\theta}$

16. $r = \dfrac{6}{3 - 2\cos\theta}$

17. $r = \dfrac{3}{2 + 4\sin\theta}$

18. $r = \dfrac{5}{-1 + 2\cos\theta}$

19. $r = \dfrac{3}{2 - 6\cos\theta}$

20. $r = \dfrac{3}{2 + 6\sin\theta}$

21. $r = \dfrac{6}{2 - \cos\theta}$

22. $r = \dfrac{2}{2 + 3\sin\theta}$

In Exercises 23–26, use a graphing utility to graph the polar equation. Identify the graph.

23. $r = \dfrac{-1}{1 - \sin\theta}$

24. $r = \dfrac{-3}{2 + 4\sin\theta}$

25. $r = \dfrac{3}{-4 + 2\cos\theta}$

26. $r = \dfrac{4}{1 - 2\cos\theta}$

In Exercises 27–30, use a graphing utility to graph the rotated conic.

27. $r = \dfrac{2}{1 - \cos(\theta - \pi/4)}$ (See Exercise 9.)

28. $r = \dfrac{3}{3 + \sin(\theta - \pi/3)}$ (See Exercise 14.)

29. $r = \dfrac{4}{2 + \sin(\theta + \pi/6)}$ (See Exercise 15.)

30. $r = \dfrac{5}{-1 + 2\cos(\theta + 2\pi/3)}$ (See Exercise 18.)

In Exercises 31–46, find a polar equation of the conic with its focus at the pole.

	Conic	*Eccentricity*	*Directrix*
31.	Parabola	$e = 1$	$x = -1$
32.	Parabola	$e = 1$	$y = -2$
33.	Ellipse	$e = \frac{1}{2}$	$y = 1$
34.	Ellipse	$e = \frac{3}{4}$	$y = -2$
35.	Hyperbola	$e = 2$	$x = 1$
36.	Hyperbola	$e = \frac{3}{2}$	$x = -1$

	Conic	*Vertex or Vertices*
37.	Parabola	$(1, -\pi/2)$
38.	Parabola	$(4, 0)$
39.	Parabola	$(5, \pi)$
40.	Parabola	$(10, \pi/2)$
41.	Ellipse	$(2, 0), (8, \pi)$
42.	Ellipse	$(2, \pi/2), (4, 3\pi/2)$
43.	Ellipse	$(20, 0), (4, \pi)$
44.	Hyperbola	$(2, 0), (10, 0)$
45.	Hyperbola	$(1, 3\pi/2), (9, 3\pi/2)$
46.	Hyperbola	$(4, \pi/2), (-1, 3\pi/2)$

47. Show that the polar equation of the ellipse

$$\frac{x^2}{a^2} + \frac{y^2}{b^2} = 1 \quad \text{is} \quad r^2 = \frac{b^2}{1 - e^2\cos^2\theta}.$$

48. Show that the polar equation of the hyperbola

$$\frac{x^2}{a^2} - \frac{y^2}{b^2} = 1 \quad \text{is} \quad r^2 = \frac{-b^2}{1 - e^2\cos^2\theta}.$$

In Exercises 49–54, use the results of Exercises 47 and 48 to write the polar form of the equation of the conic.

49. $\dfrac{x^2}{169} + \dfrac{y^2}{144} = 1$

50. $\dfrac{x^2}{25} + \dfrac{y^2}{16} = 1$

51. $\dfrac{x^2}{9} - \dfrac{y^2}{16} = 1$

52. $\dfrac{x^2}{36} - \dfrac{y^2}{4} = 1$

53. Hyperbola One focus: $(5, \pi/2)$
Vertices: $(4, \pi/2), (4, -\pi/2)$

54. Ellipse One focus: $(4, 0)$
Vertices: $(5, 0), (5, \pi)$

55. ***Planetary Motion*** The planets travel in elliptical orbits with the sun at one focus. Assume that the focus is at the pole, the major axis lies on the polar axis, and the length of the major axis is $2a$ (see figure). Show that the polar equation of the orbit is given by

$$r = \frac{(1 - e^2)a}{1 - e\cos\theta}$$

where e is the eccentricity.

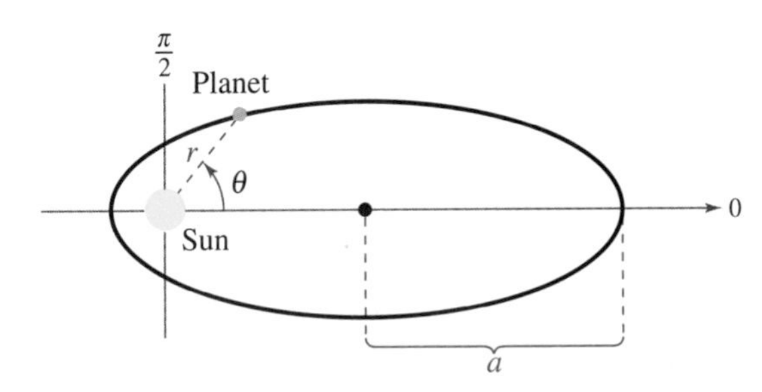

56. ***Planetary Motion*** Use the result of Exercise 55 to show that the minimum distance (*perihelion distance*) from the sun to the planet is $r = a(1 - e)$ and the maximum distance (*aphelion distance*) is $r = a(1 + e)$.

In Exercises 57–60, use the results of Exercises 55 and 56 to find the polar equation of the planet's orbit and the perihelion and aphelion distances.

57. Earth $a = 92.957 \times 10^6$ miles
$e = 0.0167$

58. Saturn $a = 1.427 \times 10^9$ kilometers
$e = 0.0543$

59. Pluto $a = 5.900 \times 10^9$ kilometers
$e = 0.2481$

60. Mercury $a = 36.0 \times 10^6$ miles
$e = 0.2056$

61. ***Satellite Tracking*** A satellite in a 100-mile-high circular orbit around the earth has a velocity of approximately 17,500 miles per hour. If this velocity is multiplied by $\sqrt{2}$, the satellite will have the minimum velocity necessary to escape the earth's gravity and it will follow a parabolic path with the center of the earth as the focus (see figure). Find a polar equation of the parabolic path of the satellite (assume the radius of the earth is 4000 miles). Find the distance between the surface of the earth and the satellite when $\theta = 30°$.

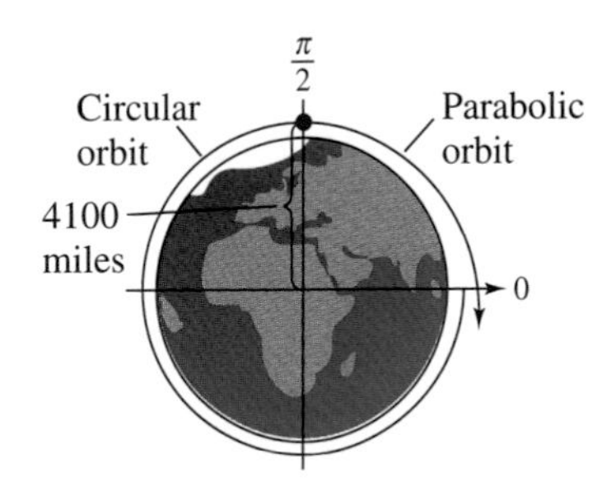

62. ***Explorer 18*** On November 26, 1963, the United States launched Explorer 18. Its low and high points above the surface of the earth were 119 miles and 122,000 miles, respectively (see figure). The center of the earth is at one focus of the orbit. Find the polar equation of the orbit and find the distance between the surface of the earth (assume a radius of 4000 miles) and the satellite when $\theta = 60°$.

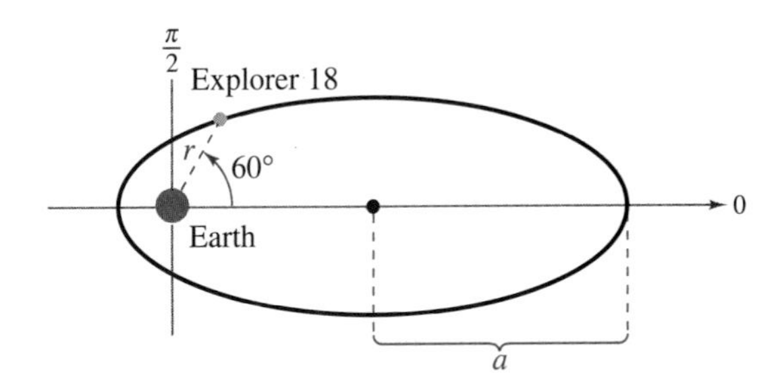

Review **Solve Exercises 63–66 as a review of the skills and problem-solving techniques you learned in previous sections. Graph the function.**

63. $f(x) = 4^{x-2}$ **64.** $g(x) = 4^{2-x}$

65. $h(t) = \log_4(t - 2)$ **66.** $h(s) = \log_4 s^2$

FOCUS ON CONCEPTS

In this chapter, you studied conics, parametric equations, and polar equations. You can use the following questions to check your understanding of several of these basic concepts. The answers to these questions are given in the back of the book.

In Exercises 1 and 2, an equation and four variations are given. In your own words, describe how the graph of each of the variations differs from the graph of the first equation.

1. $y^2 = 8x$

(a) $(y - 2)^2 = 8x$ (b) $y^2 = 8(x + 1)$

(c) $y^2 = -8x$ (d) $y^2 = 4x$

2. $\dfrac{x^2}{4} + \dfrac{y^2}{9} = 1$

(a) $\dfrac{x^2}{9} + \dfrac{y^2}{4} = 1$ (b) $\dfrac{x^2}{4} + \dfrac{y^2}{4} = 1$

(c) $\dfrac{x^2}{4} + \dfrac{y^2}{25} = 1$ (d) $\dfrac{(x - 3)^2}{4} + \dfrac{y^2}{9} = 1$

3. Explain how the central rectangle of a hyperbola can be used to sketch its asymptotes.

4. Consider an ellipse with the major axis horizontal and 10 units in length. The number b in the standard form of the equation of the ellipse must be less than what real number? Explain the change in the shape of the ellipse as b approaches this number.

5. The graph of the parametric equations $x = 2 \sec t$ and $y = 3 \tan t$ is given in the figure. Would the graph change for the equations $x = 2 \sec(-t)$ and $y = 3 \tan(-t)$? If so, how would it change?

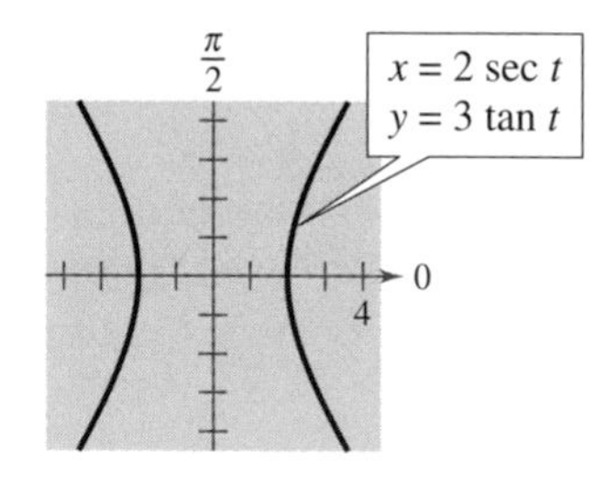

6. A moving object is modeled by the parametric equations $x = 4 \cos t$ and $y = 3 \sin t$, where t is time (see figure). How would the orbit change for

(a) $x = 4 \cos 2t, \quad y = 3 \sin 2t$?

(b) $x = 5 \cos t, \quad y = 3 \sin t$?

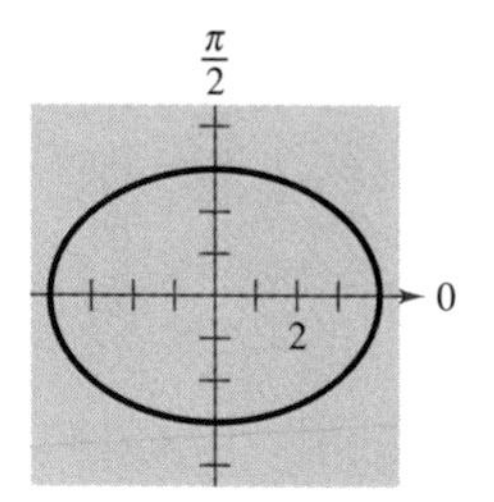

7. *True or False?* Only one set of parametric equations can represent the line $y = 3 - 2x$. Explain.

8. *True or False?* There is a unique polar coordinate representation of each point in the plane. Explain.

9. Identify the type of symmetry each of the following polar points has with the point in the figure.

(a) $\left(-4, \dfrac{\pi}{6}\right)$ (b) $\left(4, -\dfrac{\pi}{6}\right)$ (c) $\left(-4, -\dfrac{\pi}{6}\right)$

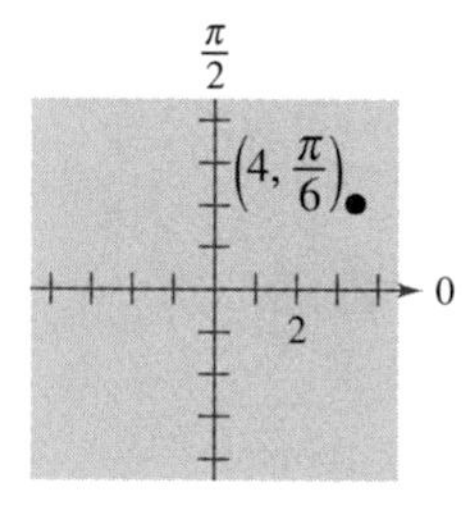

10. What is the relationship between the graphs of the rectangular and polar equations?

(a) $x^2 + y^2 = 25, \quad r = 5$

(b) $x - y = 0, \quad \theta = \dfrac{\pi}{4}$

Review Exercises

In Exercises 1 and 2, find the slope of the line with the given inclination θ.

1. $\theta = 120°$ **2.** $\theta = 55.8°$

In Exercises 3 and 4, find the inclination θ of the given line.

3. $x + y - 10 = 0$ **4.** $3x - 2y - 4 = 0$

In Exercises 5 and 6, find the distance between the point and the line.

	Point	*Line*
5.	$(1, 2)$	$x - y - 3 = 0$
6.	$(0, 4)$	$x + 2y - 2 = 0$

In Exercises 7–16, match the equation with its graph. [The graphs are labeled (a)–(j).]

7. $4x^2 + y^2 = 4$ **8.** $x^2 = 4y$

9. $4x^2 - y^2 = 4$` **10.** $y^2 = -4x$

11. $x^2 + 4y^2 = 4$ **12.** $y^2 - 4x^2 = 4$

13. $x^2 = -6y$ **14.** $x^2 + 5y^2 = 10$

15. $x^2 - 5y^2 = -5$ **16.** $y^2 - 8x = 0$

(a)

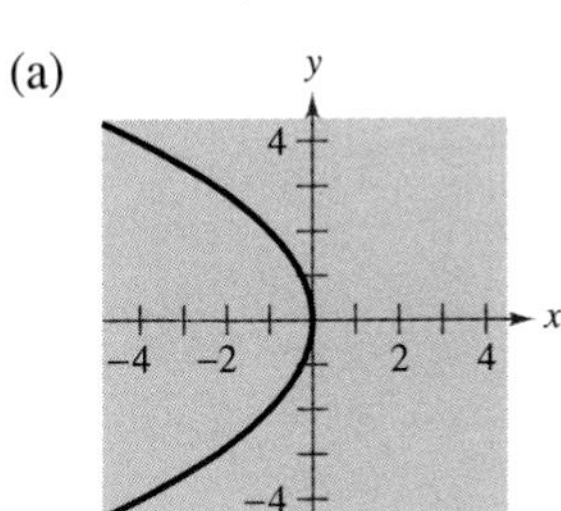

(b)

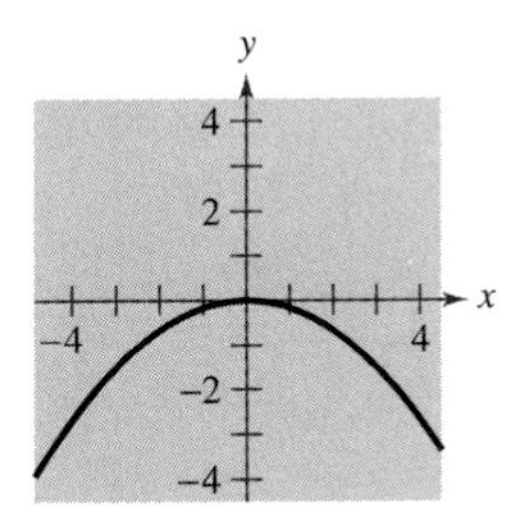

(c)

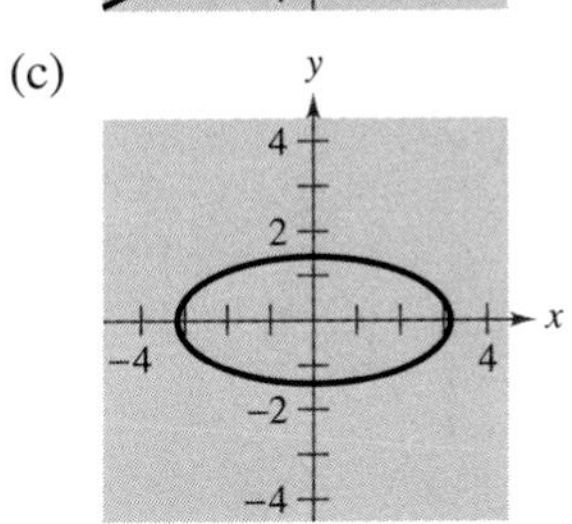

(d)

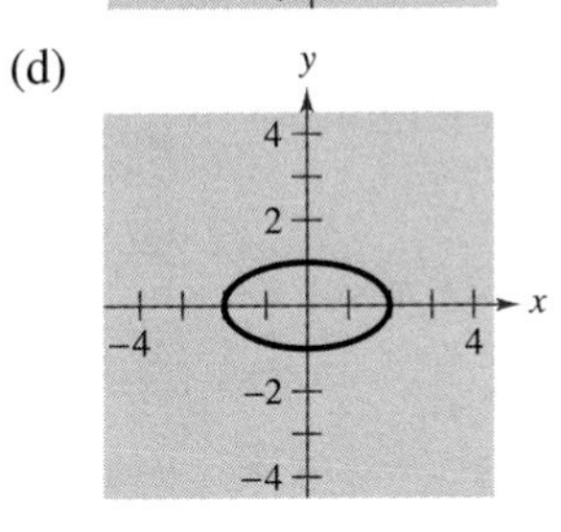

(e)

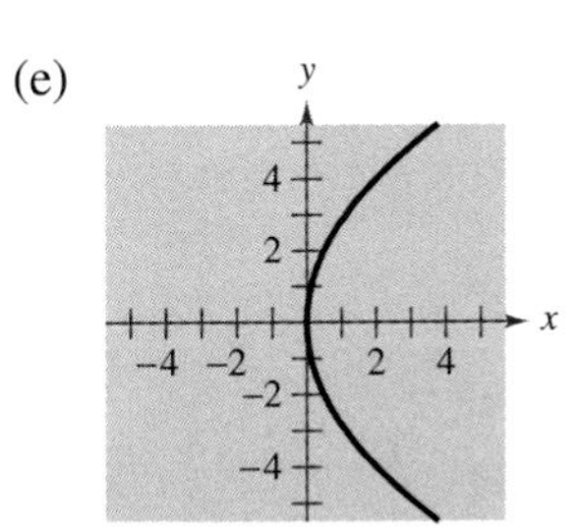

(f)

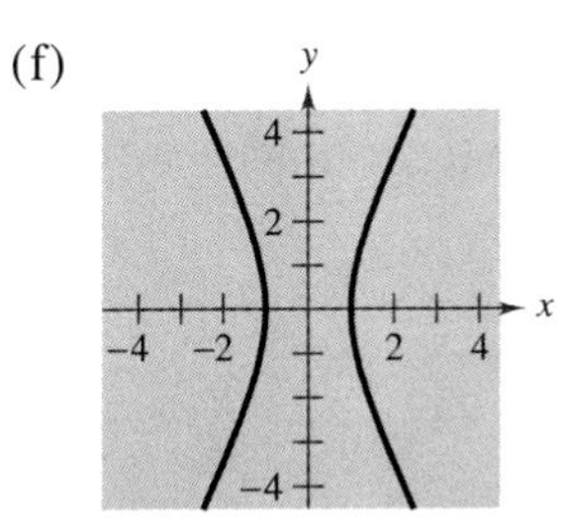

(g)

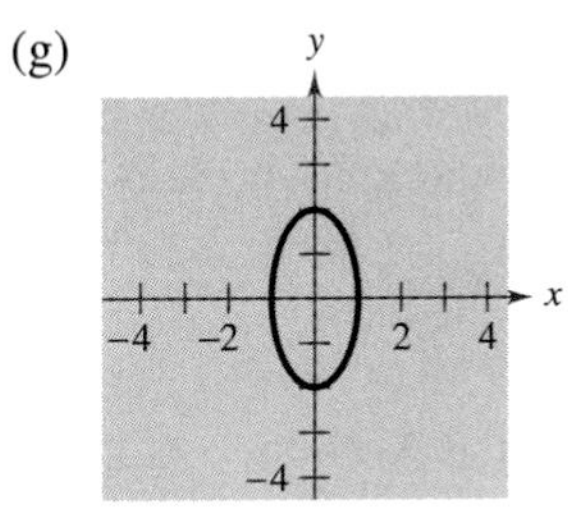

(h)

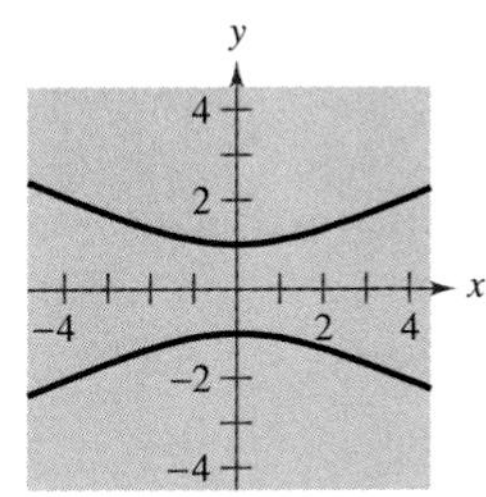

(i)

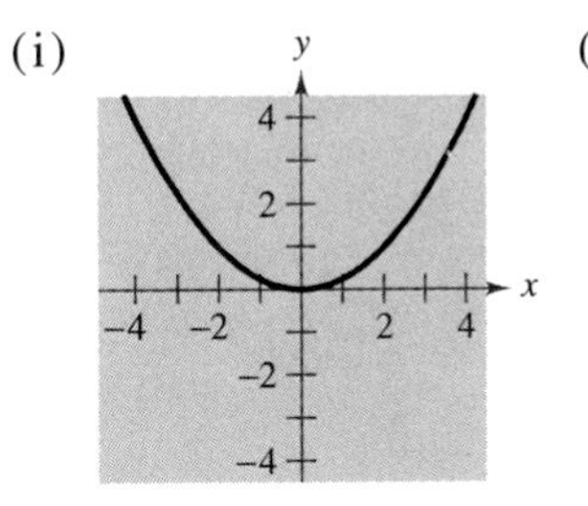

(j)

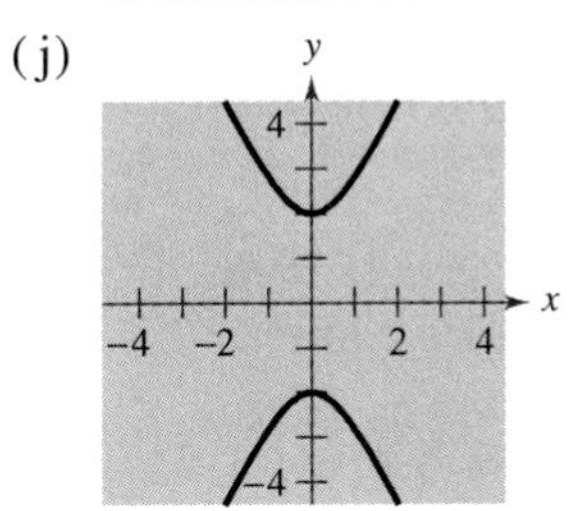

In Exercises 17–30, identify the conic and sketch its graph.

17. $4x - y^2 = 0$

18. $8y + x^2 = 0$

19. $x^2 - 6x + 2y + 9 = 0$

20. $y^2 - 12y - 8x + 20 = 0$

21. $x^2 + y^2 - 2x - 4y + 5 = 0$

22. $16x^2 + 16y^2 - 16x + 24y - 3 = 0$

23. $4x^2 + y^2 = 16$

24. $2x^2 + 6y^2 = 18$

25. $x^2 + 9y^2 + 10x - 18y + 25 = 0$

26. $4x^2 + y^2 - 16x + 15 = 0$

27. $5y^2 - 4x^2 = 20$

28. $x^2 - 9y^2 + 10x + 18y + 7 = 0$

29. $x^2 + y^2 + 2xy + 2\sqrt{2}x - 2\sqrt{2}y + 2 = 0$

30. $9x^2 + 6y^2 + 4xy - 20 = 0$

In Exercises 31–34, solve for y and use a graphing utility to graph the resulting equations. Identify the conic.

31. $3x^2 + 2y^2 - 12x + 12y + 29 = 0$

32. $4x^2 - 4y^2 - 4x + 8y - 11 = 0$

33. $x^2 - 10xy + y^2 + 1 = 0$

34. $40x^2 + 36xy + 25y^2 - 52 = 0$

In Exercises 35–38, find a rectangular equation of the specified parabola.

35. Vertex: $(4, 2)$
Focus: $(4, 0)$

36. Vertex: $(2, 0)$
Focus: $(0, 0)$

37. Vertex: $(0, 2)$
Directrix: $x = -3$

38. Vertex: $(2, 2)$
Directrix: $y = 0$

In Exercises 39–42, find a rectangular equation of the specified ellipse.

39. Vertices: $(-3, 0), (7, 0)$; Foci: $(0, 0), (4, 0)$

40. Vertices: $(2, 0), (2, 4)$; Foci: $(2, 1), (2, 3)$

41. Vertices: $(0, \pm 6)$; Passes through $(2, 2)$

42. Vertices: $(0, 1), (4, 1)$; Foci: $(2, 0), (2, 2)$

In Exercises 43–46, find a rectangular equation of the specified hyperbola.

43. Vertices: $(0, \pm 1)$; Foci: $(0, \pm 3)$

44. Vertices: $(2, 2), (-2, 2)$; Foci: $(4, 2), (-4, 2)$

45. Foci: $(0, 0), (8, 0)$; Asymptotes: $y = \pm 2(x - 4)$

46. Foci: $(3, \pm 2)$; Asymptotes: $y = \pm 2(x - 3)$

47. ***Parabolic Archway*** A parabolic archway is 12 meters high at the vertex. At a height of 10 meters, the width of the archway is 8 meters (see figure). How wide is the archway at ground level?

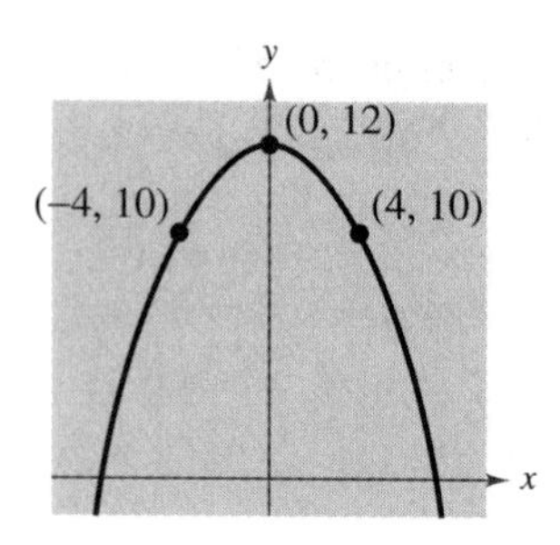

FIGURE FOR 47

48. ***Semielliptical Archway*** A semielliptical archway is set on pillars that are 10 feet apart. Its height (atop the pillars) is 4 feet. Where should the foci be placed in order to sketch the semielliptical arch?

49. ***Locating an Explosion*** Two of your friends live 4 miles apart and on the same "east-west" street, and you live halfway between them. You are talking on a three-way phone call when you hear an explosion. Six seconds later your friend to the east hears the explosion, and your friend to the west hears it 8 seconds after you do. Find equations of two hyperbolas that would locate the explosion. (Sound travels at a rate of 1100 feet per second.)

50. ***True or False?*** The graph of $(x^2/4) - y^4 = 1$ is a hyperbola. Explain.

In Exercises 51–58, sketch the curve represented by the parametric equations and, where possible, write the corresponding rectangular equation by eliminating the parameter. Verify your result with a graphing utility.

51. $x = 2t$
$y = 4t$

52. $x = t^2$
$y = \sqrt{t}$

53. $x = 1 + 4t$
$y = 2 - 3t$

54. $x = t + 4$
$y = t^2$

55. $x = \dfrac{1}{t}$
$y = t^2$

56. $x = \dfrac{1}{t}$
$y = 2t + 3$

57. $x = 6 \cos \theta$
$y = 6 \sin \theta$

58. $x = 3 + 3 \cos \theta$
$y = 2 + 5 \sin \theta$

59. Find a parametric representation of the ellipse with center at $(-3, 4)$, major axis horizontal and eight units in length, and minor axis six units in length.

60. Find a parametric representation of the hyperbola with vertices $(0, \pm 4)$ and foci $(0, \pm 5)$.

61. ***Rotary Engine*** The rotary engine was developed by Felix Wankel in the 1950s. It features a rotor that is basically a modified equilateral triangle. The rotor moves in a chamber that, in two dimensions, is an epitrochoid. Use a graphing utility to graph the chamber modeled by the parametric equations $x = \cos 3\theta + 5\cos\theta$ and $y = \sin 3\theta + 5\sin\theta$.

62. ***Involute of a Circle*** The involute of a circle is described by the endpoint P of a string that is held taut as it is unwound from a spool (see figure). The spool does not rotate.

(a) Show that a parametric representation of the involute of a circle is given by

$$x = r(\cos\theta + \theta\sin\theta)$$
$$y = r(\sin\theta - \theta\cos\theta).$$

(b) Use a graphing utility to graph the involute of a circle given by the parametric equations

$$x = 2(\cos\theta + \theta\sin\theta)$$
$$y = 2(\sin\theta - \theta\cos\theta).$$

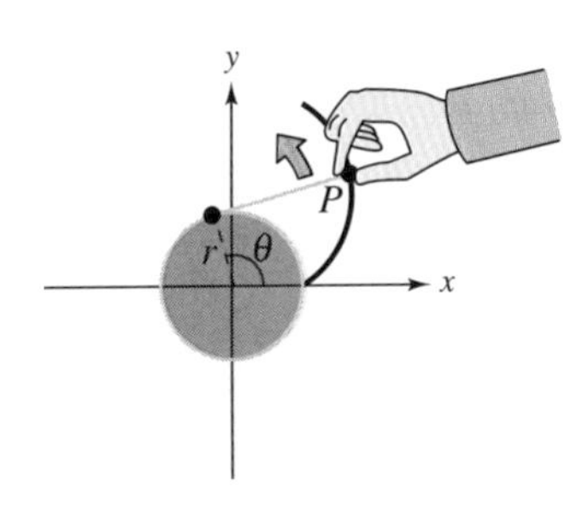

In Exercises 63–66, convert the polar equation to rectangular form.

63. $r = 3\cos\theta$

64. $r = \dfrac{2}{1 + \sin\theta}$

65. $r^2 = \cos 2\theta$

66. $r = 10$

In Exercises 67 and 68, convert the rectangular equation to polar form.

67. $(x^2 + y^2)^2 = ax^2y$

68. $x^2 + y^2 - 4x = 0$

In Exercises 69–80, identify and sketch the graph of the polar equation.

69. $r = 4$

70. $\theta = \dfrac{\pi}{12}$

71. $r = 4\sin 2\theta$

72. $r = 2\theta$

73. $r = -2(1 + \cos\theta)$

74. $r = 3 - 4\cos\theta$

75. $r = 4 - 3\cos\theta$

76. $r = \cos 5\theta$

77. $r = -3\cos 2\theta$

78. $r^2 = \cos 2\theta$

79. $r = \dfrac{2}{1 - \sin\theta}$

80. $r = \dfrac{1}{1 + 2\sin\theta}$

In Exercises 81–84, use a graphing utility to graph the polar equation.

81. $r^2 = 4\sin^2 2\theta$

82. $r = 3\csc\theta$

83. $r = \dfrac{3}{\cos(\theta - \pi/4)}$

84. $r = \dfrac{4}{5 - 3\cos\theta}$

In Exercises 85–90, find a polar equation for the line or conic.

85. Circle — Center: $(5, \pi/2)$; Solution point: $(0, 0)$

86. Line — Solution point: $(0, 0)$; Slope: $\sqrt{3}$

87. Parabola — Vertex: $(2, \pi)$; Focus: $(0, 0)$

88. Parabola — Vertex: $(2, \pi/2)$; Focus: $(0, 0)$

89. Ellipse — Vertices: $(5, 0)$, $(1, \pi)$; One focus: $(0, 0)$

90. Hyperbola — Vertices: $(1, 0)$, $(7, 0)$; One focus: $(0, 0)$

CHAPTER PROJECT: *Graphing in Parametric and Polar Modes*

Most graphing utilities can be set to *parametric mode* to sketch the graph of a curve represented by a pair of parametric equations.

EXAMPLE 1 *Using a Graphing Utility in Parametric Mode*

Use a graphing utility to sketch the curves represented by the parametric equations. For which curve is y a function of x?

a. $x = \cos^3 t, \quad y = \sin^3 t, \quad 0 \le t \le 2\pi$

b. $x = t, \quad y = t^3, \quad -4 \le t \le 4$

Solution

Begin by setting the graphing utility to parametric and radian modes. When choosing a viewing rectangle, you must set not only the minimum and maximum values of x and y, but also the minimum and maximum values of t.

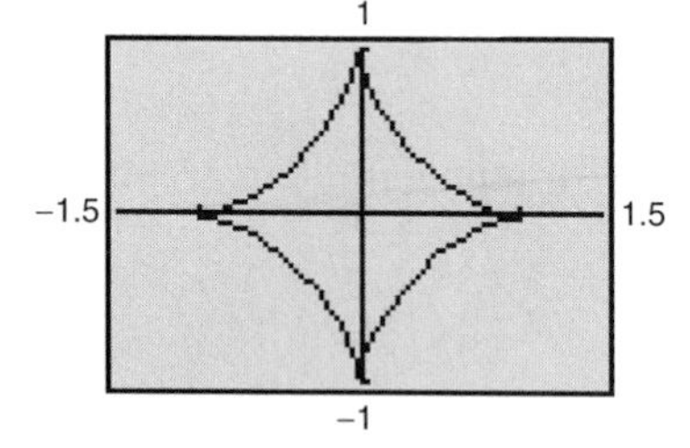

a. Enter the parametric equations for x and y. From the given equations, you can see that the values of x and y vary between -1 and 1. Thus, the following settings are appropriate.

Tmin = 0	Xmin = -1.5	Ymin = -1
Tmax = 6.28	Xmax = 1.5	Ymax = 1
Tstep = .1	Xscl = 1	Yscl = 1

From the graph at the left, you can see that y is not a function of x.

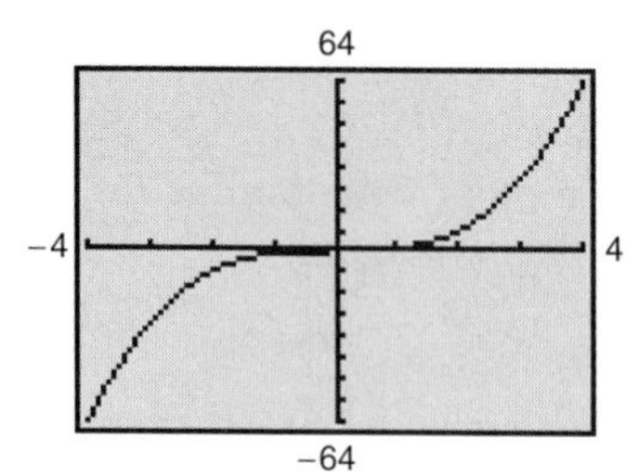

b. Enter the parametric equations for x and y. From the given equations, you can see that the values of x vary between -4 and 4, and the values of y vary between -64 and 64. Thus, the following settings are appropriate.

Tmin = -4	Xmin = -4	Ymin = -64
Tmax = 4	Xmax = 4	Ymax = 64
Tstep = .1	Xscl = 1	Yscl = 8

From the graph at the left, you can see that y is a function of x.

Some graphing utilities have a polar-coordinate mode. If yours doesn't, but does have a parametric mode, you can use the following conversion to sketch the graph of a polar equation.

The graph of the polar equation $r = f(\theta)$ can be written in parametric form, using t as a parameter, as follows.

$$x = f(t)\cos t \qquad \text{and} \qquad y = f(t)\sin t.$$

EXAMPLE 2 *Graphing a Polar Equation*

Use a graphing utility to sketch the graph of $r = 4 \sin \theta$.

Solution

Using Polar Mode: Enter the equation in polar mode and use the following settings.

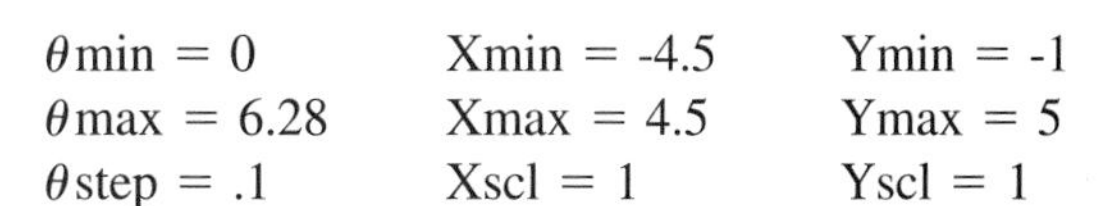

θmin = 0	Xmin = -4.5	Ymin = -1
θmax = 6.28	Xmax = 4.5	Ymax = 5
θstep = .1	Xscl = 1	Yscl = 1

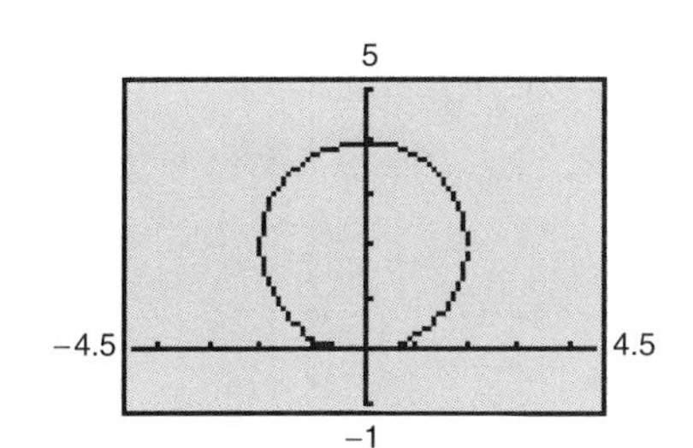

Using Parametric Mode: Begin by rewriting the equation in parametric form.

$$x = 4 \sin t \cos t \quad \text{and} \quad y = 4 \sin^2 t$$

Then enter the parametric equations and use the following settings.

Tmin = 0	Xmin = -4.5	Ymin = -1
Tmax = 6.28	Xmax = 4.5	Ymax = 5
Tstep = .1	Xscl = 1	Yscl = 1

CHAPTER PROJECT INVESTIGATIONS

In Questions 1–6, use a graphing utility to sketch the curve given by the parametric equations. State whether y is a function of x.

1. $x = \cos t$, $y = 3 \sin t$

2. $x = 3 + 4 \cos t$, $y = 1 - 2 \sin t$

3. $x = (t + 1)^2$, $y = \frac{3}{4}t$

4. $x = 2t$, $y = t^3$

5. $x = 2 \cos 2t$, $y = 2 \sin 2t$

6. $x = 4 \cos^2 t$, $y = 2 \sin t$

In Exercises 7–14, use a graphing utility to sketch the graph of the polar equation. Then identify the graph.

7. $r = 5 - 4 \cos \theta$

8. $r = -5 \sec \theta$

9. $r = \dfrac{4}{1 - 3 \cos \theta}$

10. $r = \dfrac{3}{-4 + 2 \cos \theta}$

11. $r = \dfrac{1}{1 - \sin \theta}$

12. $r = 5 \sin 3\theta$

13. $r = 2 \csc \theta$

14. $r = 2 \cos 4\theta$

In Exercises 15 and 16, consider a baseball that is hit at an initial height of 3 feet above the ground, with an initial speed of 100 feet per second, and at an angle of θ with respect to the ground. The path of the baseball at any time t (in seconds) is given by the parametric equations

$$x = 100(\cos \theta)t$$
$$y = 3 + 100(\sin \theta)t - 16t^2.$$

Use a graphing utility to sketch the paths of three baseballs: one with $\theta = 35°$, one with $\theta = 45°$, and one with $\theta = 55°$.

15. Do the three baseballs reach their maximum heights at the same time? Explain.

16. Do the three baseballs reach the same maximum height? Explain.

Cumulative Test for Chapters 4–6

Take this test as you would take a test in class. After you are done, check your work against the answers given in the back of the book.

The *Interactive* CD-ROM provides answers to the Chapter Tests and Cumulative Tests. It also offers Chapter Pre-Tests (which test key skills and concepts covered in previous chapters) and Chapter Post-Tests, both of which have randomly generated exercises with diagnostic capabilities.

1. Write the complex number $3 - \sqrt{-25}$ in standard form.

2. Perform the division $\dfrac{2i}{3 - 4i}$ and write the answer in standard form.

3. Find the zeros of the function $f(x) = x^3 + 4x^2 + 5x$.

4. Write the trigonometric form of the complex number $-12 + 5i$.

5. Evaluate $[3(\cos 30° + i \sin 30°)]^4$. Write the result in standard form.

6. Graph the functions (a) $f(x) = 6(2^{-x})$ and (b) $g(x) = \log_3 x$.

7. Solve the equations (a) $6e^{2x} = 72$ and (b) $\log_2 x + \log_2 5 = 6$.

In Exercises 8–11, graph the conic and identify any vertices and foci.

8. $y^2 - 4x + 4 = 0$

9. $\dfrac{(x-2)^2}{4} + \dfrac{(y+1)^2}{9} = 1$

10. $x^2 - \dfrac{y^2}{4} = 1$

11. $x^2 - 4y^2 - 4x = 0$

12. Find an equation in rectangular coordinates of the parabola with vertex $(3, -2)$, vertical axis, and passing through the point $(0, 4)$.

13. Find an equation in rectangular coordinates of the hyperbola with foci $(0, 0)$ and $(0, 4)$, and asymptotes $y = \pm\frac{1}{2}x + 2$.

14. (a) Determine the number of degrees the axis must be rotated to eliminate the xy-term of the conic $x^2 + 6xy + y^2 - 6 = 0$.

(b) Graph the conic and use a graphing utility to confirm your result.

15. Sketch the curve represented by the parametric equations $x = 2 + 3\cos\theta$ and $y = 2\sin\theta$. Eliminate the parameter and write the corresponding rectangular equation.

16. Find a set of parametric equations of the line passing through the points $(2, -3)$ and $(6, 4)$. (The answer is not unique.)

17. Convert the rectangular equation $x^2 + y^2 - 6y = 0$ to polar form.

18. Sketch the graph of each polar equation.

(a) $r = \dfrac{4}{1 + \cos\theta}$ (b) $r = \dfrac{4}{2 + \cos\theta}$

19. Match the polar equation with its graph at the right.

(a) $r = 2 + 3\sin\theta$ (b) $r = 3\sin\theta$ (c) $r = 3\sin 2\theta$

(i)

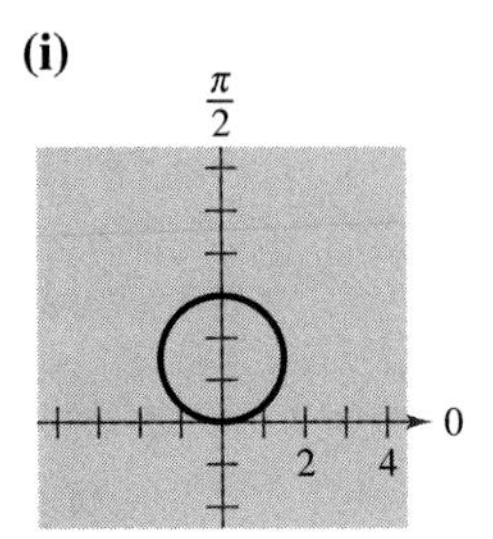

(ii)

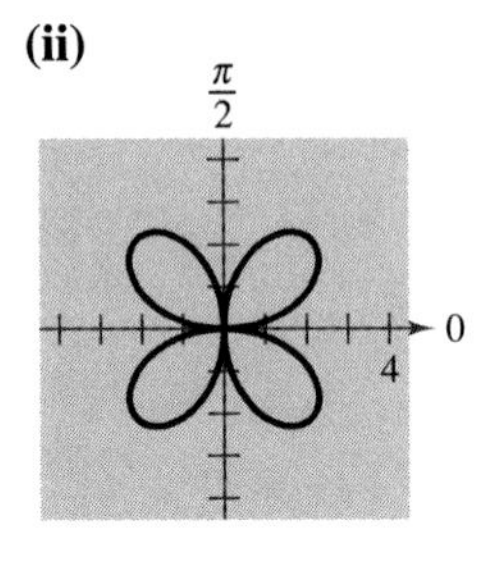

(iii)

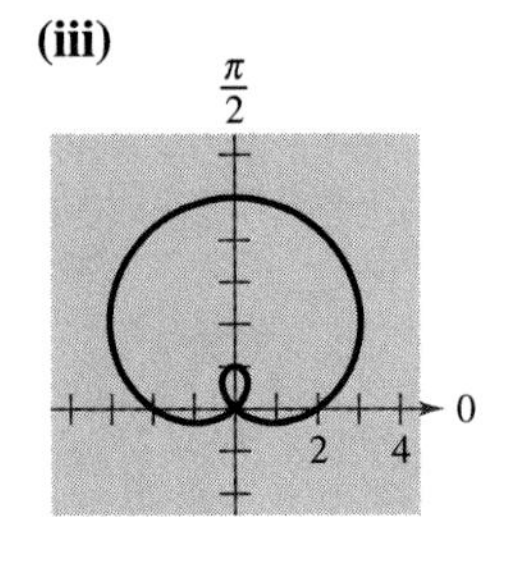

FIGURE FOR 19

APPENDICES

A Further Concepts in Statistics
  A.1 Representing Data and Linear Modeling
  A.2 Measures of Central Tendency and Dispersion
B Think About the Proof
C Programs
D Scientific Calculator Keystrokes

Appendix A Further Concepts in Statistics

Section A.1 Representing Data and Linear Modeling

Stem-and-Leaf Plots ❐ *Histograms and Frequency Distributions* ❐
Scatter Plots ❐ *Fitting a Line to Data*

Stem-and-Leaf Plots

Statistics is the branch of mathematics that studies techniques for collecting, organizing, and interpreting data. In this section, you will study several ways to organize and interpret data.

One type of plot that can be used to organize sets of numbers by hand is a **stem-and-leaf plot.** A set of test scores and the corresponding stem-and-leaf plot are shown below.

Test Scores

93, 70, 76, 58, 86, 93, 82, 78, 83, 86,
64, 78, 76, 66, 83, 83, 96, 74, 69, 76,
64, 74, 79, 76, 88, 76, 81, 82, 74, 70

Stems	*Leaves*
5	8
6	4 4 6 9
7	0 0 4 4 4 6 6 6 6 6 8 8 9
8	1 2 2 3 3 3 6 6 8
9	3 3 6

Note that the *leaves* represent the units digits of the numbers and the *stems* represent the tens digits. Stem-and-leaf plots can also be used to compare two sets of data, as shown in the following example.

EXAMPLE 1 *Comparing Two Sets of Data*

Use a stem-and-leaf plot to compare the test scores given above with the following test scores. Which set of test scores is better?

90, 81, 70, 62, 64, 73, 81, 92, 73, 81, 92, 93, 83, 75, 76,
83, 94, 96, 86, 77, 77, 86, 96, 86, 77, 86, 87, 87, 79, 88

Solution

Begin by ordering the second set of scores.

62, 64, 70, 73, 73, 75, 76, 77, 77, 77, 79, 81, 81, 81, 83,
83, 86, 86, 86, 86, 87, 87, 88, 90, 92, 92, 93, 94, 96, 96

Now that the data has been ordered, you can construct a *double* stem-and-leaf plot by letting the leaves to the right of the stems represent the units digits for the first group of test scores and letting the leaves to the left of the stems represent the units digits for the second group of test scores.

Leaves (2nd Group)	*Stems*	*Leaves (1st Group)*
	5	8
4 2	6	4 4 6 9
9 7 7 7 6 5 3 3 0	7	0 0 4 4 4 6 6 6 6 6 8 8 9
8 7 7 6 6 6 6 3 3 1 1 1	8	1 2 2 3 3 3 6 6 8
6 6 4 3 2 2 0	9	3 3 6

By comparing the two sets of leaves, you can see that the second group of test scores is better than the first group.

EXAMPLE 2 Using a Stem-and-Leaf Plot

Table A.1 shows the percent of the population of each state and the District of Columbia that was at least 65 years old in 1989. Use a stem-and-leaf plot to organize the data. (Source: U.S. Bureau of Census)

TABLE A.1

AK	4.1	AL	12.7	AR	14.8	AZ	13.1	CA	10.6
CO	9.8	CT	13.6	DC	12.5	DE	11.8	FL	18.0
GA	10.1	HI	10.7	IA	15.1	ID	11.9	IL	12.3
IN	12.4	KS	13.7	KY	12.7	LA	11.1	MA	13.8
MD	10.8	ME	13.4	MI	11.9	MN	12.6	MO	13.9
MS	12.4	MT	13.2	NC	12.1	ND	13.9	NE	13.9
NH	11.4	NJ	13.2	NM	10.5	NV	10.9	NY	13.0
OH	12.8	OK	13.3	OR	13.9	PA	15.1	RI	14.8
SC	11.1	SD	14.4	TN	12.6	TX	10.1	UT	8.6
VA	10.8	VT	11.9	WA	11.9	WI	13.4	WV	14.6
WY	9.8								

Solution

Begin by ordering the numbers, as shown below.

4.1, 8.6, 9.8, 9.8, 10.1, 10.1, 10.5, 10.6, 10.7, 10.8, 10.8, 10.9, 11.1, 11.1, 11.4, 11.8, 11.9, 11.9, 11.9, 11.9, 12.1, 12.3, 12.4, 12.4, 12.5, 12.6, 12.6, 12.7, 12.7, 12.8, 13.0, 13.1, 13.2, 13.2, 13.3, 13.4, 13.4, 13.6, 13.7, 13.8, 13.9, 13.9, 13.9, 13.9, 14.4, 14.6, 14.8, 14.8, 15.1, 15.1, 18.0

Next construct the stem-and-leaf plot using the leaves to represent the digits to the right of the decimal points.

Stems	*Leaves*	
4.	1	Alaska has the lowest percent.
5.		
6.		
7.		
8.	6	
9.	8 8	
10.	1 1 5 6 7 8 8 9	
11.	1 1 4 8 9 9 9 9	
12.	1 3 4 4 5 6 6 7 7 8	
13.	0 1 2 2 3 4 4 6 7 8 9 9 9 9	
14.	4 6 8 8	
15.	1 1	
16.		
17.		
18.	0	Florida has the highest percent.

TECHNOLOGY

Try using a computer or graphing calculator to create a histogram for the data at the right. How does the histogram change when the intervals change?

Histograms and Frequency Distributions

With data such as that given in Example 2, it is useful to group the numbers into intervals and plot the frequency of the data in each interval. For instance, the **frequency distribution** and **histogram** shown in Figure A.1 represent the data given in Example 2.

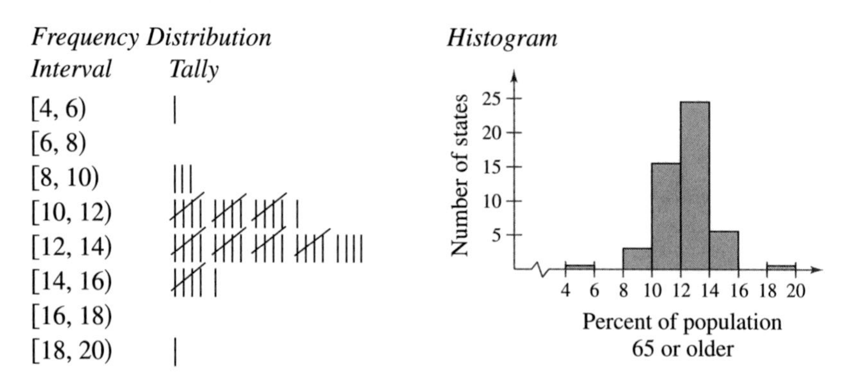

FIGURE A.1

A histogram has a portion of a real number line as its horizontal axis. A **bar graph** is similar to a histogram, except that the rectangles (bars) can be either horizontal or vertical and the labels of the bars are not necessarily numbers.

Another difference between a bar graph and a histogram is that the bars in a bar graph are usually separated by spaces, whereas the bars in a histogram are not separated by spaces.

EXAMPLE 3 Constructing a Bar Graph

The data below shows the average monthly precipitation (in inches) in Houston, Texas. Construct a bar graph for this data. What can you conclude? (Source: PC USA)

January	3.2	February	3.3	March	2.7
April	4.2	May	4.7	June	4.1
July	3.3	August	3.7	September	4.9
October	3.7	November	3.4	December	3.7

Solution

To create a bar graph, begin by drawing a vertical axis to represent the precipitation and a horizontal axis to represent the months. The bar graph is shown in Figure A.2. From the graph, you can see that Houston receives a fairly consistent amount of rain throughout the year, with the driest month tending to be March and the wettest month tending to be September.

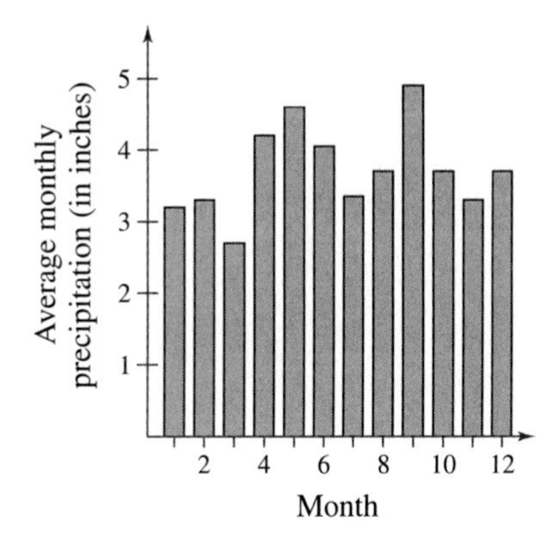

FIGURE A.2

Scatter Plots

Many real-life situations involve finding relationships between two variables, such as the year and the number of people in the labor force. In a typical situation, data is collected and written as a set of ordered pairs. The graph of such a set is called a **scatter plot.**

From the scatter plot in Figure A.3, it appears that the points describe a relationship that is nearly linear. (The relationship is not *exactly* linear because the labor force did not increase by precisely the same amount each year.) A mathematical equation that approximates the relationship between two variables is called a *mathematical model*. When developing a mathematical model, you strive for two (often conflicting) goals—accuracy and simplicity.

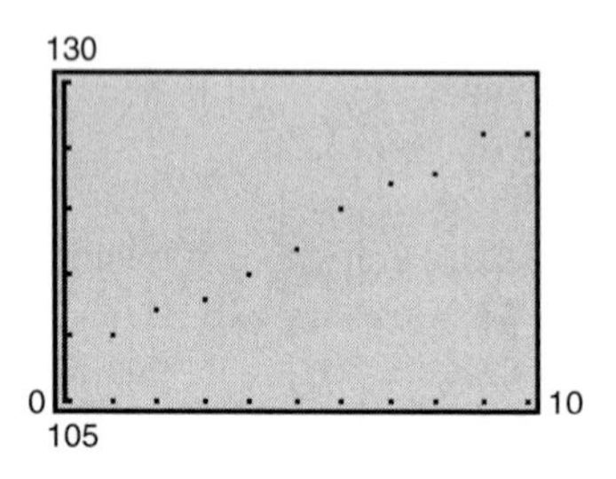

FIGURE A.3

TECHNOLOGY

Most graphing utilities have built-in statistical programs that can create scatter plots. Use your graphing utility to plot the points given in the table below. The data shows the number of people P (in millions) in the United States who were part of the labor force from 1980 through 1990. In the table, t represents the year, with $t = 0$ corresponding to 1980. (Source: U.S. Bureau of Labor Statistics)

t	P
0	109
1	110
2	112
3	113
4	115
5	117

t	P
6	120
7	122
8	123
9	126
10	126

The resulting scatter plot may be seen in Figure A.3.

Consider a collection of ordered pairs of the form (x, y). If y tends to increase as x increases, the collection is said to have a **positive correlation.** If y tends to decrease as x increases, the collection is said to have a **negative correlation.** Figure A.4 shows three examples: one with a positive correlation, one with a negative correlation, and one with no (discernible) correlation.

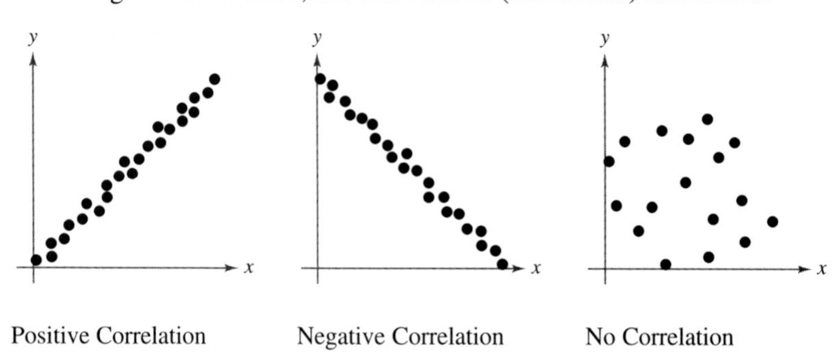

Positive Correlation Negative Correlation No Correlation

FIGURE A.4

Fitting a Line to Data

Finding a linear model that represents the relationship described by a scatter plot is called **fitting a line to data.** You can do this graphically by simply sketching the line that appears to fit the points, finding two points on the line, and then finding the equation of the line that passes through the two points.

EXAMPLE 4 *Fitting a Line to Data*

Find a linear model that relates the year to the number of people P (in millions) who were part of the United States labor force from 1980 through 1990. In Table A.2, t represents the year, with $t = 0$ corresponding to 1980. (Source: U.S. Bureau of Labor Statistics)

TABLE A.2

t	0	1	2	3	4	5	6	7	8	9	10
P	109	110	112	113	115	117	120	122	123	126	126

Solution

After plotting the data from Table A.2, draw the line that you think best represents the data, as shown in Figure A.5. Two points that lie on this line are (0, 109) and (9, 126). Using the point-slope form, you can find the equation of the line to be

$$P = \frac{17}{9}t + 109. \qquad \text{Linear model}$$

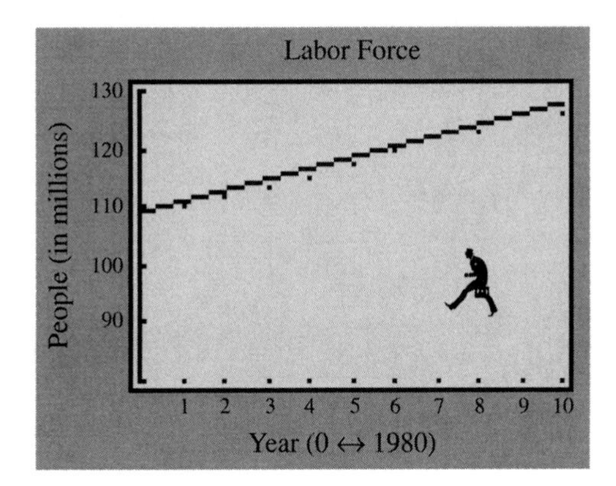

FIGURE A.5

Once you have found a model, you can measure how well the model fits the data by comparing the actual values with the values given by the model, as shown in Table A.3.

TABLE A.3

	t	0	1	2	3	4	5	6	7	8	9	10
Actual →	P	109	110	112	113	115	117	120	122	123	126	126
Model →	P	109	110.9	112.8	114.7	116.6	118.4	120.3	122.2	124.1	126	127.9

The sum of the squares of the differences between the actual values and the model's values is the **sum of the squared differences.** The model that has the least sum is called the **least squares regression line** for the data. For the model in Example 4, the sum of the squared differences is 13.81. The least squares regression line for the data is

$$P = 1.864t + 108.2. \qquad \text{Best-fitting linear model}$$

Its sum of squared differences is 4.7.

LEAST SQUARES REGRESSION LINE

The least squares regression line, $y = ax + b$, for the points (x_1, y_1), (x_2, y_2), (x_3, y_3), . . . , (x_n, y_n) is given by

$$a = \frac{n\sum_{i=1}^{n} x_i y_i - \sum_{i=1}^{n} x_i \sum_{i=1}^{n} y_i}{n\sum_{i=1}^{n} x_i^2 - \left(\sum_{i=1}^{n} x_i\right)^2} \quad \text{and} \quad b = \frac{1}{n}\left(\sum_{i=1}^{n} y_i - a\sum_{i=1}^{n} x_i\right).$$

EXAMPLE 5 ***Finding a Least Squares Regression Line***

Find the least squares regression line for the points $(-3, 0)$, $(-1, 1)$, $(0, 2)$, and $(2, 3)$.

Solution

Begin by constructing a table of values, as shown in Table A.4.

TABLE A.4

x	y	xy	x^2
-3	0	0	9
-1	1	-1	1
0	2	0	0
2	3	6	4
$\sum_{i=1}^{n} x_i = -2$	$\sum_{i=1}^{n} y_i = 6$	$\sum_{i=1}^{n} x_i y_i = 5$	$\sum_{i=1}^{n} x_i^2 = 14$

Applying the formulas for the least squares regression line with $n = 4$ produces

$$a = \frac{n\sum_{i=1}^{n} x_i y_i - \sum_{i=1}^{n} x_i \sum_{i=1}^{n} y_i}{n\sum_{i=1}^{n} x_i^2 - \left(\sum_{i=1}^{n} x_i\right)^2} = \frac{4(5) - (-2)(6)}{4(14) - (-2)^2} = \frac{32}{52} = \frac{8}{13} \quad \text{and}$$

$$b = \frac{1}{n}\left(\sum_{i=1}^{n} y_i - a\sum_{i=1}^{n} x_i\right) = \frac{1}{4}\left[6 - \frac{8}{13}(-2)\right] = \frac{47}{26}.$$

Thus, the least squares regression line is $y = \frac{8}{13}x + \frac{47}{26}$, as shown in Figure A.6.

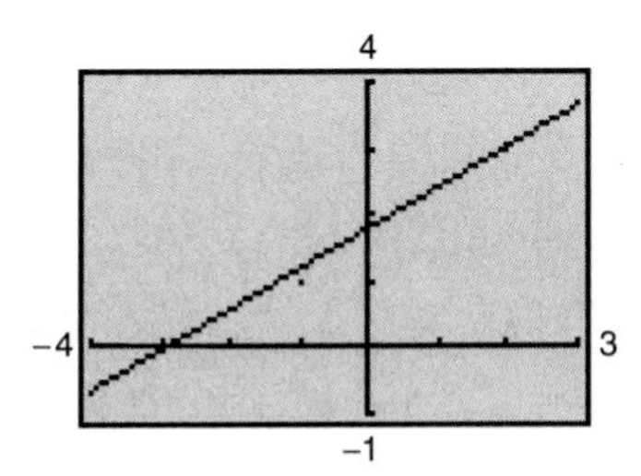

FIGURE A.6

Many calculators have "built-in" least squares regression programs. If your calculator has such a program, try using it to duplicate the results shown in the following example.

EXAMPLE 6 *Finding a Least Squares Regression Line*

The following ordered pairs (w, h) represent the shoe sizes w, and the heights h (in inches), of 25 men. Use a computer program or a statistical calculator to find the least squares regression line for this data.

(10.0, 70.0), (10.5, 71.0), (9.5, 70.0), (11.0, 72.0), (12.0, 74.0),
(8.5, 66.0), (9.0, 68.5), (13.0, 76.0), (10.5, 71.5), (10.5, 70.5),
(10.0, 72.0), (9.5, 70.0), (10.0, 71.0), (10.5, 69.5), (11.0, 71.5),
(12.0, 73.5), (12.5, 74.0), (11.0, 71.5), (9.0, 67.5), (10.0, 70.0),
(13.0, 73.5), (10.5, 72.5), (10.5, 71.0), (11.0, 73.0), (8.5, 68.0)

Solution

A scatter plot for the data is shown in Figure A.7. Note that the plot does not have 25 separate points because some of the ordered pairs graph as the same point. After entering the data into a statistical calculator, you can obtain

$$a = 1.67 \quad \text{and} \quad b = 53.57.$$

Thus, the least squares regression line for the data is

$$h = 1.67w + 53.57.$$

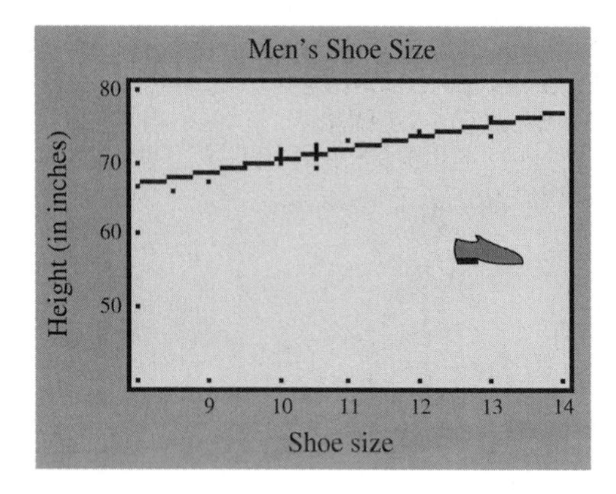

FIGURE A.7

If you use a statistical calculator or computer program to duplicate the results of Example 6, you will notice that the program also outputs a value of $r \approx 0.918$. This number is called the **correlation coefficient** of the data. Correlation coefficients vary between -1 and 1. Basically, the closer $|r|$ is to 1, the better the points can be described by a line. Three examples are shown in Figure A.8.

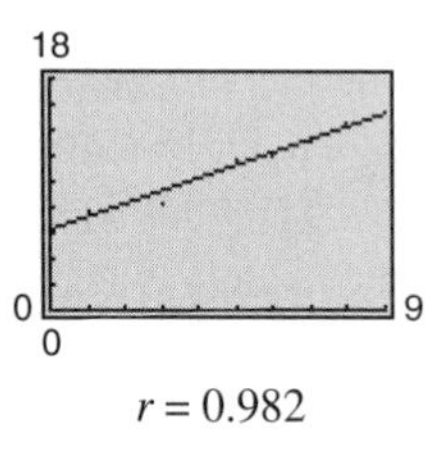

$r = 0.982$

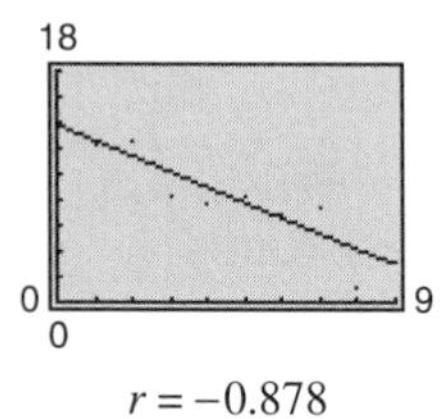

$r = -0.878$

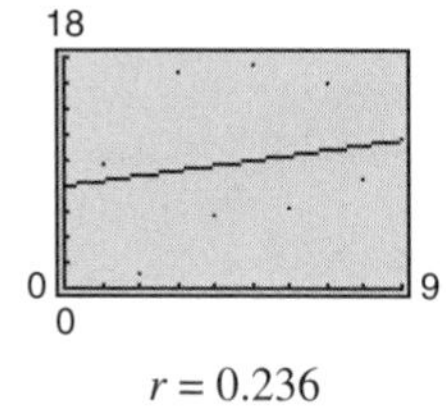

$r = 0.236$

FIGURE A.8

A.1 Exercises

Exam Scores **In Exercises 1 and 2, use the following scores from a math class of 30 students. The scores are given for two 100-point exams.**

Exam #1: **77, 100, 77, 70, 83, 89, 87, 85, 81, 84, 81, 78, 89, 78, 88, 85, 90, 92, 75, 81, 85, 100, 98, 81, 78, 75, 85, 89, 82, 75**

Exam #2: **76, 78, 73, 59, 70, 81, 71, 66, 66, 73, 68, 67, 63, 67, 77, 84, 87, 71, 78, 78, 90, 80, 77, 70, 80, 64, 74, 68, 68, 68**

1. Construct a stem-and-leaf plot for Exam #1.

2. Construct a double stem-and-leaf plot to compare the scores for Exam #1 and Exam #2. Which set of test scores is higher?

3. ***Educational Expenses*** The following table shows the per capita expenditures for public elementary and secondary education in the 50 states and the District of Columbia in 1991. Use a stem-and-leaf plot to organize the data. (Source: National Education Association)

AK	1626	AL	694	AR	668	AZ	892
CA	918	CO	841	CT	1151	DC	1010
DE	891	FL	862	GA	859	HI	784
IA	846	ID	725	IL	788	IN	925
KS	906	KY	725	LA	758	MA	866
MD	944	ME	1062	MI	926	MN	990
MO	742	MS	671	MT	983	NC	813
ND	719	NE	757	NH	881	NJ	1223
NM	915	NV	1004	NY	1186	OH	861
OK	776	OR	925	PA	889	RI	892
SC	835	SD	716	TN	618	TX	905
UT	828	VA	941	VT	992	WA	1095
WI	928	WV	883	WY	1178		

4. ***Snowfall*** The data below shows the seasonal snowfall (in inches) at Erie, Pennsylvania for the years 1960 through 1989 (the amounts are listed in order by year). How would you organize this data? Explain your reasoning. (Source: National Oceanic and Atmospheric Administration)

   69.6, 42.5, 75.9, 115.9, 92.9, 84.8, 68.6, 107.9, 79.7, 85.6, 120.0, 92.3, 53.7, 68.6, 66.7, 66.0, 111.5, 142.8, 76.5, 55.2, 89.4, 71.3, 41.2, 110.0, 106.3, 124.9, 68.2, 103.5, 76.5, 114.9

5. ***Fruit Crops*** The data below shows the cash receipts (in millions of dollars) from fruit crops for farmers in 1990. Construct a bar graph for the data. (Source: U.S. Department of Agriculture)

Apples	1159	Peaches	365
Grapefruit	317	Pears	266
Grapes	1668	Plums and Prunes	293
Lemons	278	Strawberries	560
Oranges	1707		

6. ***Travel to the United States*** The data below gives the places of origin and numbers of travelers (in millions) to the United States in 1991. Construct a horizontal bar graph for this data. (Source: U.S. Travel and Tourism Administration)

Canada	18.9	Mexico	7.0
Europe	7.4	Latin America	2.0
Other	6.8		

Crop Yield **In Exercises 7–10, use the data in the table, where x is the number of units of fertilizer applied to sample plots and y is the yield (in bushels) of a crop.**

x	0	1	2	3	4	5	6	7	8
y	58	60	59	61	63	66	65	67	70

7. Sketch a scatter plot of the data.
8. Determine whether the points are positively correlated, are negatively correlated, or have no discernible correlation.
9. Sketch a linear model that you think best represents the data. Find an equation of the line you sketched. Use the line to predict the yield if 10 units of fertilizer are used.
10. Can the model found in Exercise 9 be used to predict yields for arbitrarily large values of x? Explain.

Speed of Sound **In Exercises 11–14, use the data in the table, where h is the altitude in thousands of feet and v is the speed of sound in feet per second.**

h	0	5	10	15	20	25	30	35
v	1116	1097	1077	1057	1036	1015	995	973

11. Sketch a scatter plot of the data.
12. Determine whether the points are positively correlated, are negatively correlated, or have no discernible correlation.

13. Sketch a linear model that you think best represents the data. Find an equation of the line you sketched. Use the line to predict the speed of sound at an altitude of 27,000 feet.

14. The speed of sound at an altitude of 70,000 feet is approximately 971 feet per second. What does this suggest about the validity of using the model in Exercise 13 to extrapolate beyond the data given in the table?

In Exercises 15 and 16, (a) sketch a scatter plot of the points, (b) find an equation of the linear model you think best represents the data and find the sum of the squared differences, and (c) use the formulas in this section to find the least squares regression line and the sum of the squared differences.

15. $(-1, 0), (0, 1), (1, 3), (2, 3)$

16. $(0, 4), (1, 3), (2, 2), (4, 1)$

In Exercises 17–20, (a) sketch a scatter plot of the points, (b) use the formulas in this section to find the least squares regression line, and (c) sketch the graph of the line.

17. $(-2, 0), (-1, 1), (0, 1), (2, 2)$

18. $(-3, 1), (-1, 2), (0, 2), (1, 3), (3, 5)$

19. $(1, 5), (2, 8), (3, 13), (4, 16), (5, 22), (6, 26)$

20. $(1, 10), (2, 8), (3, 8), (4, 6), (5, 5), (6, 3)$

In Exercises 21–24, use a graphing utility to find the least squares regression line for the data. Sketch a scatter plot and the regression line.

21. $(0, 23), (1, 20), (2, 19), (3, 17), (4, 15), (5, 11), (6, 10)$

22. $(4, 52.8), (5, 54.7), (6, 55.7), (7, 57.8), (8, 60.2), (9, 63.1), (10, 66.5)$

23. $(-10, 5.1), (-5, 9.8), (0, 17.5), (2, 25.4), (4, 32.8), (6, 38.7), (8, 44.2), (10, 50.5)$

24. $(-10, 213.5), (-5, 174.9), (0, 141.7), (5, 119.7), (8, 102.4), (10, 87.6)$

25. ***Advertising*** The management of a department store ran an experiment to determine if a relationship existed between sales S (in thousands of dollars) and the amount spent on advertising x (in thousands of dollars). The following data were collected.

x	1	2	3	4	5	6	7	8
S	405	423	455	466	492	510	525	559

(a) Use a graphing utility to find the least squares regression line. Use the equation to estimate sales if $4500 is spent on advertising.

(b) Make a scatter plot of the data and sketch the graph of the regression line.

(c) Use a computer or calculator to determine the correlation coefficient.

26. ***School Enrollment*** The table gives the preprimary school enrollments y (in millions) for the years 1985 through 1991, where $t = 5$ corresponds to 1985. (Source: U.S. Bureau of the Census)

t	5	6	7	8	9	10	11
y	10.73	10.87	10.87	11.00	11.04	11.21	11.37

(a) Use a computer or calculator to find the least squares regression line. Use the equation to estimate enrollment in 1992.

(b) Make a scatter plot of the data and sketch the graph of the regression line.

(c) Use the computer or calculator to determine the correlation coefficient.

Section A.2 Measures of Central Tendency and Dispersion

Mean, Median, and Mode ▫ *Choosing a Measure of Central Tendency* ▫ *Variance and Standard Deviation*

Mean, Median, and Mode

In many real-life situations, it is helpful to describe data by a single number that is most representative of the entire collection of numbers. Such a number is called a **measure of central tendency.** The most commonly used measures are as follows.

1. The **mean,** or **average,** of n numbers is the sum of the numbers divided by n.
2. The **median** of n numbers is the middle number when the numbers are written in order. If n is even, the median is the average of the two middle numbers.
3. The **mode** of n numbers is the number that occurs most frequently. If two numbers tie for most frequent occurrence, the collection has two modes and is called **bimodal.**

EXAMPLE 1 Comparing Measures of Central Tendency

You are interviewing for a job. The interviewer tells you that the average income of the company's 25 employees is \$60,849. The actual incomes of the 25 employees are shown below. What are the mean, median, and mode of the incomes? Was the person telling the truth?

\$17,305,	\$478,320,	\$45,678,	\$18,980,	\$17,408,
\$25,676,	\$28,906,	\$12,500,	\$24,540,	\$33,450,
\$12,500,	\$33,855,	\$37,450,	\$20,432,	\$28,956,
\$34,983,	\$36,540,	\$250,921,	\$36,853,	\$16,430,
\$32,654,	\$98,213,	\$48,980,	\$94,024,	\$35,671

Solution

The mean of the incomes is

$$\text{Mean} = \frac{17,305 + 478,320 + 45,678 + 18,980 + \cdots + 35,671}{25}$$

$$= \frac{1,521,225}{25}$$

$$= \$60,849.$$

TECHNOLOGY

Statistical calculators have built-in programs for calculating the mean of a collection of numbers. Try using your calculator to find the mean given in Example 1. Then use your calculator to sort the incomes given in Example 1.

To find the median, order the incomes as follows.

$12,500, $12,500, $16,430, $17,305, $17,408,
$18,980, $20,432, $24,540, $25,676, $28,906,
$28,956, $32,654, $33,450, $33,855, $34,983,
$35,671, $36,540, $36,853, $37,450, $45,678,
$48,980, $94,024, $98,213, $250,921, $478,320

From this list, you can see that the median income (the middle number) is $33,450. From the same list, you can see that $12,500 is the only income that occurs more than once. Thus, the mode is $12,500. Technically, the person was telling the truth because the average is (generally) defined to be the mean. However, of the three measures of central tendency

Mean: $60,849 *Median:* $33,450 *Mode:* $12,500

it seems clear that the median is the most representative. The mean is inflated by the two highest salaries.

Choosing a Measure of Central Tendency

Which of the three measures of central tendency is the most representative? The answer is that it depends on the distribution of data *and* the way in which you plan to use the data.

For instance, in Example 1, the mean salary of $60,849 does not seem very representative to a potential employee. To a city income tax collector who wants to estimate 1% of the total income of the 25 employees, however, the mean is precisely the right measure.

EXAMPLE 2 *Choosing a Measure of Central Tendency*

Which measure of central tendency is the most representative of the data given in each of the following frequency distributions?

a.

Number	*Tally*
1	7
2	20
3	15
4	11
5	8
6	3
7	2
8	0
9	15

b.

Number	*Tally*
1	9
2	8
3	7
4	6
5	5
6	6
7	7
8	8
9	9

c.

Number	*Tally*
1	6
2	1
3	2
4	3
5	5
6	5
7	4
8	3
9	0

Solution

a. For these data, the mean is 4.23, the median is 3, and the mode is 2. Of these, the mode is probably the most representative.

b. For these data, the mean and median are each 5 and the modes are 1 and 9 (the distribution is bimodal). Of these, the mean or median is the most representative.

c. For these data, the mean is 4.59, the median is 5, and the mode is 1. Of these, the mean or median is the most representative.

Variance and Standard Deviation

Very different sets of numbers can have the same mean. You will now study two **measures of dispersion,** which give you an idea of how much the numbers in the set differ from the mean of the set. These two measures are called the *variance* of the set and the *standard deviation* of the set.

DEFINITIONS OF VARIANCE AND STANDARD DEVIATION

Consider a set of numbers $\{x_1, x_2, \ldots, x_n\}$ with a mean of $\bar{x}$. The **variance** of the set is

$$v = \frac{(x_1 - \bar{x})^2 + (x_2 - \bar{x})^2 + \cdots + (x_n - \bar{x})^2}{n}$$

and the **standard deviation** of the set is

$$\sigma = \sqrt{v}$$

(σ is the lowercase Greek letter *sigma*).

The standard deviation of a set is a measure of how much a typical number in the set differs from the mean. The greater the standard deviation, the more the numbers in the set *vary* from the mean. For instance, each of the following sets has a mean of 5.

$\{5, 5, 5, 5\}$, $\{4, 4, 6, 6\}$, and $\{3, 3, 7, 7\}$

The standard deviations of the sets are 0, 1, and 2.

$$\sigma_1 = \sqrt{\frac{(5-5)^2 + (5-5)^2 + (5-5)^2 + (5-5)^2}{4}} = 0$$

$$\sigma_2 = \sqrt{\frac{(4-5)^2 + (4-5)^2 + (6-5)^2 + (6-5)^2}{4}} = 1$$

$$\sigma_3 = \sqrt{\frac{(3-5)^2 + (3-5)^2 + (7-5)^2 + (7-5)^2}{4}} = 2$$

EXAMPLE 3 Estimations of Standard Deviation

Consider the three sets of data represented by the following bar graphs. Which set has the smallest standard deviation? Which has the largest?

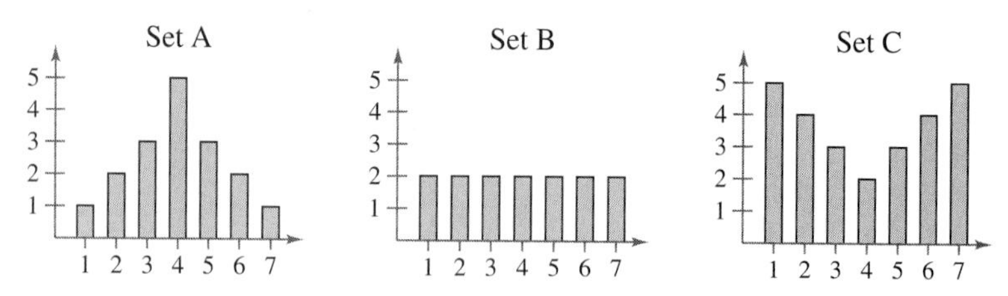

Solution

Of the three sets, the numbers in set A are grouped most closely to the center and the numbers in set C are the most dispersed. Thus, set A has the smallest standard deviation and set C has the largest standard deviation.

EXAMPLE 4 Finding Standard Deviation

Find the standard deviation of each set shown in Example 3.

Solution

Because of the symmetry of each bar graph, you can conclude that each has a mean of $\bar{x} = 4$. The standard deviation of set A is

$$\sigma = \sqrt{\frac{(-3)^2 + 2(-2)^2 + 3(-1)^2 + 5(0)^2 + 3(1)^2 + 2(2)^2 + (3)^2}{17}}$$

$$\approx 1.53.$$

The standard deviation of set B is

$$\sigma = \sqrt{\frac{2(-3)^2 + 2(-2)^2 + 2(-1)^2 + 2(0)^2 + 2(1)^2 + 2(2)^2 + 2(3)^2}{14}}$$

$$= 2.$$

The standard deviation of set C is

$$\sigma = \sqrt{\frac{5(-3)^2 + 4(-2)^2 + 3(-1)^2 + 2(0)^2 + 3(1)^2 + 4(2)^2 + 5(3)^2}{26}}$$

$$\approx 2.22.$$

These values confirm the results of Example 3. That is, set A has the smallest standard deviation and set C has the largest.

TECHNOLOGY

If you have access to a computer or calculator with a standard deviation program, try using it to obtain the results given in Example 4. If you do this, the program will probably output two versions of the standard deviation. In one, the sum of the squared differences is divided by n, and in the other it is divided by $n - 1$. In this text, we always divide by n.

The following alternative formula provides a more efficient way to compute the standard deviation.

ALTERNATIVE FORMULA FOR STANDARD DEVIATION

The standard deviation of $\{x_1, x_2, \ldots, x_n\}$ is

$$\sigma = \sqrt{\frac{x_1^2 + x_2^2 + \cdots + x_n^2}{n} - \bar{x}^2}.$$

Because of messy computations, this formula is difficult to verify. Conceptually, however, the process is straightforward. It consists of showing that the expressions

$$\sqrt{\frac{(x_1 - \bar{x})^2 + (x_2 - \bar{x})^2 + \cdots + (x_n - \bar{x})^2}{n}}$$

and

$$\sqrt{\frac{x_1^2 + x_2^2 + \ldots + x_n^2}{n} - \bar{x}^2}$$

are equivalent. Try verifying this equivalence for the set $\{x_1, x_2, x_3\}$ with $\bar{x} = (x_1 + x_2 + x_3)/3$.

EXAMPLE 5 Using the Alternative Formula

Use the alternative formula for standard deviation to find the standard deviation of the following set of numbers.

5, 6, 6, 7, 7, 8, 8, 8, 9, 10

Solution

Begin by finding the mean of the set, which is 7.4. Thus, the standard deviation is

$$\begin{aligned}\sigma &= \sqrt{\frac{5^2 + 2(6)^2 + 2(7^2) + 3(8^2) + 9^2 + 10^2}{10} - (7.4)^2}\\ &= \sqrt{\frac{568}{10} - 54.76}\\ &= \sqrt{2.04}\\ &\approx 1.43.\end{aligned}$$

You can use the statistical features of a graphing utility to check this result.

A well-known theorem in statistics, called *Chebychev's Theorem*, states that at least $1 - (1/k^2)$ of the numbers in a distribution must lie within k standard deviations of the mean. Thus, 75% of the numbers in the collection must lie within two standard deviations of the mean, and at least 88.9% of the numbers must lie within three standard deviations of the mean. For most distributions, these percentages are low. For instance, in all three distributions shown in Example 3, 100% of the numbers lie within two standard deviations of the mean.

EXAMPLE 6 Describing a Distribution

Table A.5 shows the number of dentists (per 100,000 people) in each state and the District of Columbia. Find the mean and standard deviation of the numbers. What percent of the numbers lie within two standard deviations of the mean? (Source: American Dental Association)

TABLE A.5

AK	66	AL	40	AR	39	AZ	51	CA	62
CO	69	CT	80	DC	94	DE	44	FL	50
GA	46	HI	80	IA	55	ID	53	IL	61
IN	47	KS	51	KY	53	LA	45	MA	74
MD	68	ME	47	MI	62	MN	67	MO	53
MS	37	MT	62	NC	42	ND	47	NE	63
NH	59	NJ	77	NM	45	NV	49	NY	73
OH	55	OK	47	OR	70	PA	61	RI	56
SC	41	SD	49	TN	53	TX	47	UT	66
VA	54	VT	57	WA	68	WI	65	WV	43
WY	52								

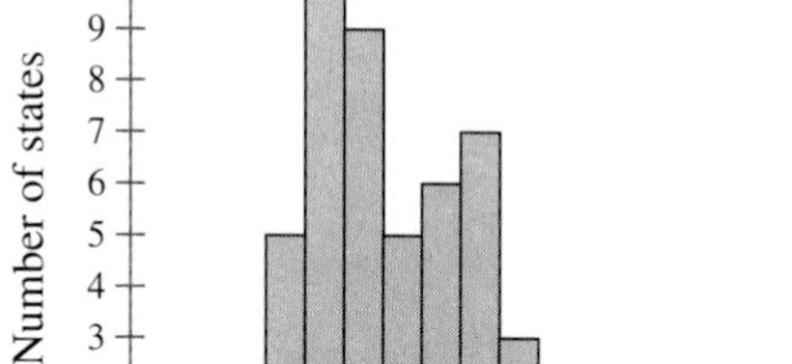

FIGURE A.9

Solution

Begin by entering the numbers into a computer or calculator that has a standard deviation program. After running the program, you should obtain

$$\bar{x} \approx 56.76 \quad \text{and} \quad \sigma \approx 12.14.$$

The interval that contains all numbers that lie with in two standard deviations of the mean is

$$[56.76 - 2(12.14), 56.76 + 2(12.14)] \quad \text{or} \quad [32.48, 81.04].$$

From the histogram in Figure A.9, you can see that all but one of the numbers (98%) lie in this interval—all but the number that corresponds to the number of dentists (per 100,000 people) in the District of Columbia.

A.2 Exercises

In Exercises 1–6, find the mean, median, and mode of the set of measurements.

1. 5, 12, 7, 14, 8, 9, 7

2. 30, 37, 32, 39, 33, 34, 32

3. 5, 12, 7, 24, 8, 9, 7

4. 20, 37, 32, 39, 33, 34, 32

5. 5, 12, 7, 14, 9, 7

6. 30, 37, 32, 39, 34, 32

7. Compare your answers for Exercises 1 and 3 with those for Exercises 2 and 4. Which of the measures of central tendency is sensitive to extreme measurements? Explain your reasoning.

8. (a) Add 6 to each measurement in Exercise 1 and calculate the mean, median, and mode of the revised measurements. How are the measures of central tendency changed?

(b) If a constant k is added to each measurement in a set of data, how will the measures of central tendency change?

9. ***Electric Bills*** A person had the following monthly bills for electricity. What are the mean and median of this collection of bills?

January	\$67.92	February	\$59.84
March	\$52.00	April	\$52.50
May	\$57.99	June	\$65.35
July	\$81.76	August	\$74.98
September	\$87.82	October	\$83.18
November	\$65.35	December	\$57.00

10. ***Car Rental*** A car rental company kept the following record of the number of miles driven by a car that was rented. What are the mean, median, and mode of this set of data?

Monday	410	Tuesday	260
Wednesday	320	Thursday	320
Friday	460	Saturday	150

11. ***Six-Child Families*** A study was done on families having six children. The table gives the number of families in the study with the indicated number of girls. Determine the mean, median, and mode of this set of data.

Number of Girls	0	1	2	3	4	5	6
Frequency	1	24	45	54	50	19	7

12. ***Baseball*** A baseball fan examined the records of a favorite baseball player's performance during his last 50 games. The number of games in which the player had 0, 1, 2, 3, and 4 hits are recorded in the table.

Number of Hits	0	1	2	3	4
Frequency	14	26	7	2	1

(a) Determine the average number of hits per game.

(b) Determine the player's batting average if he had 200 at bats during the 50-game series.

13. Construct a collection of numbers that has the following properties. If this is not possible, explain why it is not.

Mean $= 6$, Median $= 4$, Mode $= 4$

14. Construct a collection of numbers that has the following properties. If this is not possible, explain why it is not.

Mean $= 6$, Median $= 6$, Mode $= 4$

15. ***Test Scores*** A professor records the following scores for a 100-point exam.

99, 64, 80, 77, 59, 72, 87, 79, 92, 88,
90, 42, 20, 89, 42, 100, 98, 84, 78, 91

Which measure of central tendency best describes these test scores?

16. ***Shoe Sales*** A salesman sold eight pairs of a certain style of men's shoes. The sizes of the eight pairs were as follows: $10\frac{1}{2}$, 8, 12, $10\frac{1}{2}$, 10, $9\frac{1}{2}$, 11, and $10\frac{1}{2}$. Which measure (or measures) of central tendency best describes the typical shoe size for this set of data?

In Exercises 17–24, find the mean, variance, and standard deviation of the numbers.

17. 4, 10, 8, 2

18. 3, 15, 6, 9, 2

19. 0, 1, 1, 2, 2, 2, 3, 3, 4

20. 2, 2, 2, 2, 2, 2

21. 1, 2, 3, 4, 5, 6, 7

22. 1, 1, 1, 5, 5, 5

23. 49, 62, 40, 29, 32, 70

24. 1.5, 0.4, 2.1, 0.7, 0.8

In Exercises 25–30, use the alternative formula to find the standard deviation of the numbers.

25. 2, 4, 6, 6, 13, 5

26. 10, 25, 50, 26, 15, 33, 29, 4

27. 246, 336, 473, 167, 219, 359

28. 6.0, 9.1, 4.4, 8.7, 10.4

29. 8.1, 6.9, 3.7, 4.2, 6.1

30. 9.0, 7.5, 3.3, 7.4, 6.0

31. Without calculating the standard deviation, explain why the set $\{4, 4, 20, 20\}$ has a standard deviation of 8.

32. If the standard deviation of a set of numbers is 0, what does this imply about the set?

33. ***Test Scores*** An instructor adds five points to each student's exam score. Will this change the mean or standard deviation of the exam scores? Explain.

34. Consider the four sets of data represented by the histograms. Order the sets from the smallest to the largest variance.

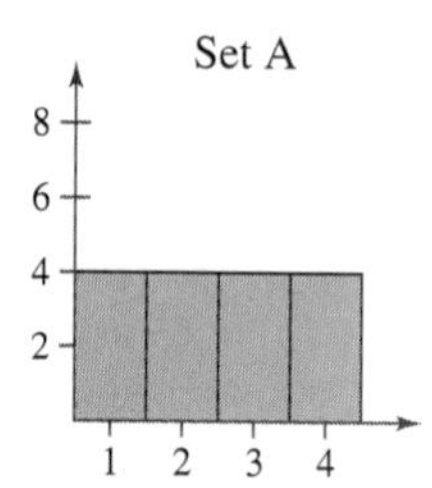

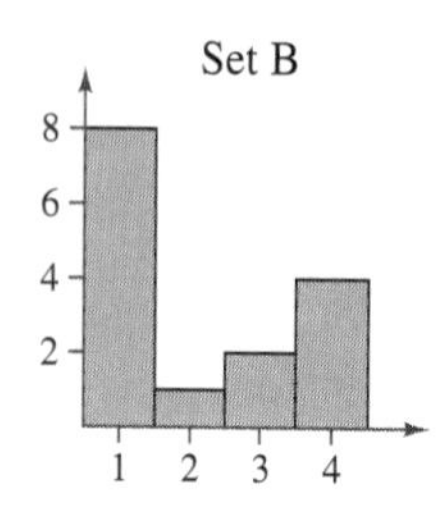

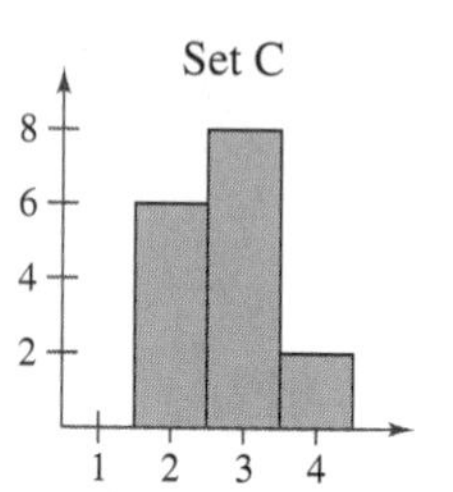

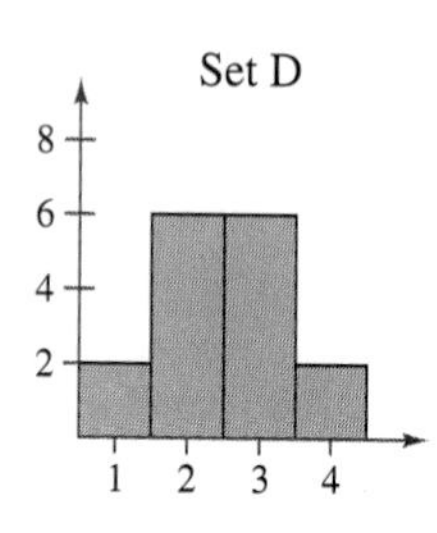

35. ***Test Scores*** The scores on a mathematics exam given to 600 science and engineering students at a college had a mean of 235 and a standard deviation of 28. Use Chebychev's Theorem to determine the intervals containing at least $\frac{3}{4}$ and at least $\frac{8}{9}$ of the scores. How would the intervals change if the standard deviation were 16?

36. ***Precipitation*** The following data represents the annual precipitation (in inches) at Erie, Pennsylvania, for the years 1960 through 1989. Use a computer or calculator to find the mean, variance, and standard deviation of the data. What percent of the data lies within two standard deviations of the mean? (Source: National Oceanic and Atmospheric Administration)

27.41,	36.50,	36.90,	28.11,	36.47,
38.41,	37.74,	37.78,	34.33,	36.58,
41.50,	34.06,	43.55,	38.04,	41.83,
43.03,	43.85,	61.70,	35.04,	55.31,
47.04,	41.97,	41.56,	46.25,	37.79,
45.87,	47.30,	44.86,	38.87,	41.88

Appendix B Think About the Proof

SECTION P.3, PAGE 29

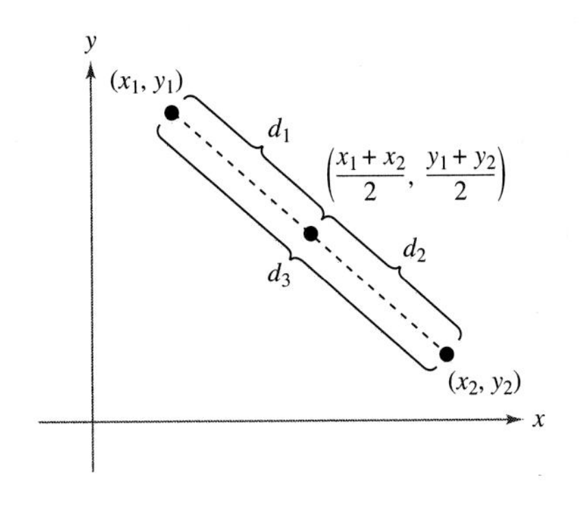

Midpoint Formula

THE MIDPOINT FORMULA

The midpoint of the segment joining the points (x_1, y_1) and (x_2, y_2) is

$$\text{Midpoint} = \left(\frac{x_1 + x_2}{2}, \frac{y_1 + y_2}{2}\right).$$

PROOF

Using the figure, you must show that

$$d_1 = d_2 \quad \text{and} \quad d_1 + d_2 = d_3.$$

By the Distance Formula, you obtain

$$d_1 = \sqrt{\left(\frac{x_1 + x_2}{2} - x_1\right)^2 + \left(\frac{y_1 + y_2}{2} - y_1\right)^2} = \frac{1}{2}\sqrt{(x_2 - x_1)^2 + (y_2 - y_1)^2}$$

$$d_2 = \sqrt{\left(x_2 - \frac{x_1 + x_2}{2}\right)^2 + \left(y_2 - \frac{y_1 + y_2}{2}\right)^2} = \frac{1}{2}\sqrt{(x_2 - x_1)^2 + (y_2 - y_1)^2}$$

$$d_3 = \sqrt{(x_2 - x_1)^2 + (y_2 - y_1)^2}.$$

Thus, it follows that $d_1 = d_2$ and $d_1 + d_2 = d_3$. ■■

SECTION 3.1, PAGE 268

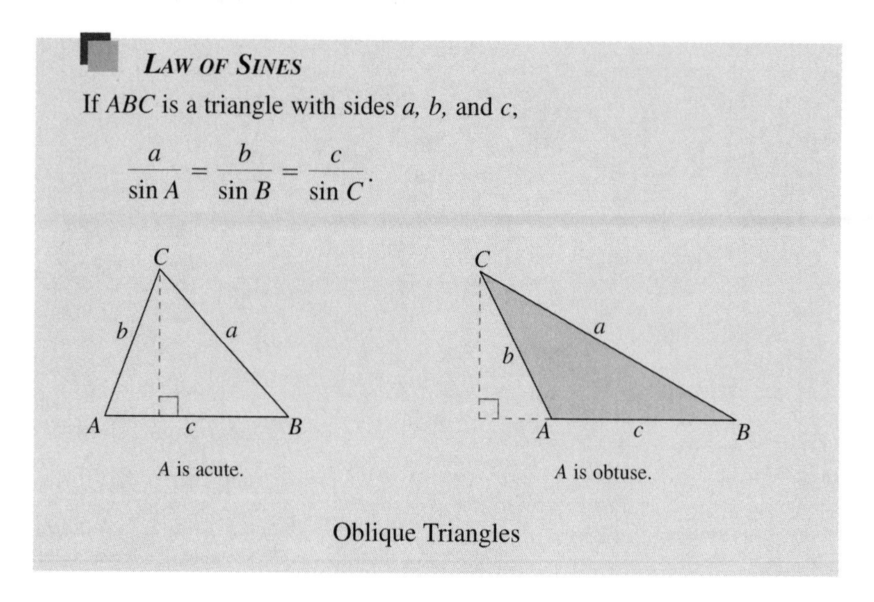

LAW OF SINES

If ABC is a triangle with sides a, b, and c,

$$\frac{a}{\sin A} = \frac{b}{\sin B} = \frac{c}{\sin C}.$$

Oblique Triangles

Proof

Let h be the altitude of either triangle found in the figure showing oblique triangles, above. Then you have

$$\sin A = \frac{h}{b} \quad \text{or} \quad h = b \sin A$$

$$\sin B = \frac{h}{a} \quad \text{or} \quad h = a \sin B.$$

Equating these two values of h, you have

$$a \sin B = b \sin A \quad \text{or} \quad \frac{a}{\sin A} = \frac{b}{\sin B}.$$

Note that $\sin A \neq 0$ and $\sin B \neq 0$ because no angle of a triangle can have a measure of $0°$ or $180°$. In a similar manner, by constructing an altitude from vertex B to side AC (extended), you can show that

$$\frac{a}{\sin A} = \frac{c}{\sin C}.$$

Hence, the Law of Sines is established. ▪▪

SECTION 3.4, PAGE 301

PROPERTIES OF THE DOT PRODUCT

Let **u**, **v**, and **w** be vectors in the plane or in space and let c be a scalar.

1. $\mathbf{u} \cdot \mathbf{v} = \mathbf{v} \cdot \mathbf{u}$
2. $\mathbf{0} \cdot \mathbf{v} = 0$
3. $\mathbf{u} \cdot (\mathbf{v} + \mathbf{w}) = \mathbf{u} \cdot \mathbf{v} + \mathbf{u} \cdot \mathbf{w}$
4. $\mathbf{v} \cdot \mathbf{v} = \|\mathbf{v}\|^2$
5. $c(\mathbf{u} \cdot \mathbf{v}) = c\mathbf{u} \cdot \mathbf{v} = \mathbf{u} \cdot c\mathbf{v}$

Proof

To prove the second property, let $\mathbf{0} = \langle 0, 0 \rangle$ and $\mathbf{v} = \langle v_1, v_2 \rangle$. Then,

$$\begin{aligned}\mathbf{0} \cdot \mathbf{v} &= \langle 0, 0 \rangle \cdot \langle v_1, v_2 \rangle \\ &= 0 \cdot v_1 + 0 \cdot v_2 \\ &= 0 + 0 \\ &= 0.\end{aligned}$$

To prove the third property, let $\mathbf{u} = \langle u_1, u_2 \rangle$, $\mathbf{v} = \langle v_1, v_2 \rangle$, and $\mathbf{w} = \langle w_1, w_2 \rangle$. Then,

$$\begin{aligned}\mathbf{u} \cdot (\mathbf{v} + \mathbf{w}) &= \langle u_1, u_2 \rangle \cdot (\langle v_1, v_2 \rangle + \langle w_1, w_2 \rangle) \\ &= \langle u_1, u_2 \rangle \cdot \langle v_1 + w_1, v_2 + w_2 \rangle \\ &= u_1(v_1 + w_1) + u_2(v_2 + w_2) \\ &= u_1v_1 + u_2v_2 + u_1w_1 + u_2w_2 \\ &= \langle u_1, u_2 \rangle \cdot \langle v_1, v_2 \rangle + \langle u_1, u_2 \rangle \cdot \langle w_1, w_2 \rangle \\ &= \mathbf{u} \cdot \mathbf{v} + \mathbf{u} \cdot \mathbf{w}.\end{aligned}$$

To prove the fifth property, let $\mathbf{u} = \langle u_1, u_2 \rangle$ and $\mathbf{v} = \langle v_1, v_2 \rangle$ and let c be a scalar. Then,

$$\begin{aligned}c(\mathbf{u} \cdot \mathbf{v}) &= c(\langle u_1, u_2 \rangle \cdot \langle v_1, v_2 \rangle) \\ &= c(u_1v_1 + u_2v_2) \\ &= (cu_1)v_1 + (cu_2)v_2 \\ &= \langle cu_1, cu_2 \rangle \cdot \langle v_1, v_2 \rangle = c\mathbf{u} \cdot \mathbf{v} \\ &= u_1(cv_1) + u_2(cv_2) \\ &= \langle u_1, u_2 \rangle \cdot \langle cv_1, cv_2 \rangle = \mathbf{u} \cdot c\mathbf{v}.\end{aligned}$$

■■

SECTION 5.3, PAGE 380

PROPERTIES OF LOGARITHMS

Let a be a positive number such that $a \neq 1$, and let n be a real number. If u and v are positive real numbers, the following properties are true.

1. $\log_a(uv) = \log_a u + \log_a v$	**1.** $\ln(uv) = \ln u + \ln v$
2. $\log_a \frac{u}{v} = \log_a u - \log_a v$	**2.** $\ln \frac{u}{v} = \ln u - \ln v$
3. $\log_a u^n = n \log_a u$	**3.** $\ln u^n = n \ln u$

PROOF

To prove Property 1, let

$$x = \log_a u \quad \text{and} \quad y = \log_a v.$$

The corresponding exponential forms of these two equations are

$$a^x = u \quad \text{and} \quad a^y = v.$$

Multiplying u and v produces $uv = a^x a^y = a^{x+y}$. The corresponding logarithmic form of $uv = a^{x+y}$ is $\log_a(uv) = x + y$. Hence, $\log_a(uv) = \log_a u + \log_a v$. ■■

SECTION 6.1, PAGE 418

INCLINATION AND SLOPE

If a nonvertical line has inclination θ and slope m,

$$m = \tan \theta.$$

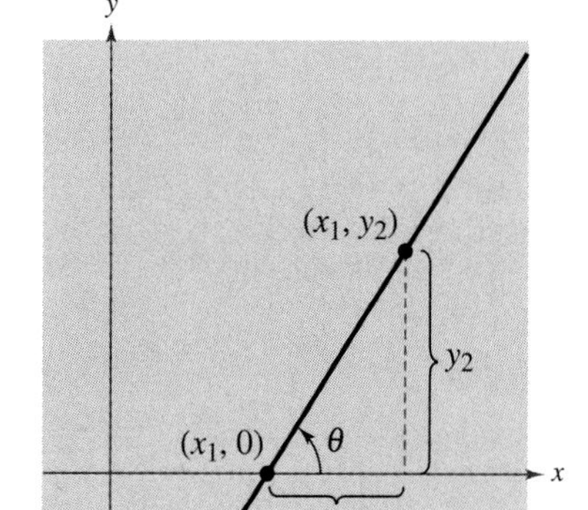

PROOF

If $m = 0$, the line is horizontal and $\theta = 0$. Thus, the result is true for horizontal lines because $m = 0 = \tan 0$.

If the line has a positive slope, it will intersect the x-axis. Label this point $(x_1, 0)$, as shown in the figure. If (x_2, y_2) is a second point on the line, the slope is given by

$$m = \frac{y_2 - 0}{x_2 - x_1} = \frac{y_2}{x_2 - x_1} = \tan \theta.$$

The case in which the line has a negative slope is left for you to prove. ■■

Appendix C Programs

Reflections and Shifts Program (Section P.7)

This program, referenced in the Technology note on page 85, will sketch a graph of the function $y = R(x + H)^2 + V$, where $R = \pm 1$, H is an integer between -6 and 6, and V is an integer between -3 and 3. This program gives you practice working with reflections, horizontal shifts, and vertical shifts.

Note: On the *TI-83* and the *TI-82*, the "int" and "rand" commands may be entered through the "NUM" and "PRB" menus, respectively, accessed by pressing the MATH key. The "=" and "<" symbols may be entered through the "TEST" menu accessed by pressing the TEST key. Other commands, such as "If," "Then," "Else," and "End," may be entered through the "CTL" menu accessed by pressing the PRGM key. The commands "Xmin," "Xmax," "Xscl," "Ymin," "Ymax," and "Yscl" may be entered through the "Window" menu accessed by pressing the VARS key. The commands "DispGraph" and "Pause" may be entered through the "I/O" and "CTL" menus, respectively, accessed by pressing the PRGM key. For additional keystroke instructions, see previous programs in this appendix. Keystroke sequences for similar commands on other calculators will vary. Consult the user manual for your calculator.

TI-80

```
PROGRAM:PARABOL
:-6+INT (12RAND)→H
:-3+INT (6RAND)→V
:RAND→R
:IF R <.5
:THEN
:-1→R
:ELSE
:1→R
:END
:"R(X+H)²+V"→Y1
:-9→XMIN
:9→XMAX
:1→XSCL
:-6→YMIN
:6→YMAX
:1→YSCL
:DISPGRAPH
:PAUSE
:DISP "Y=R(X+H)²+V²"
:DISP "R=",R
:DISP "H=",H
:DISP "V=",V
:PAUSE
```

Press ENTER after viewing the graph to display the values of the integers.

TI-81

```
Prgm2:PARABOLA
:Rand→H
:-6+Int (12H)→H
:Rand→V
:-3+Int (6V)→V
:Rand→R
:If R <.5
:-1→R
:If R >.49
:1→R
:"R(X+H)²+V"→Y1
:-9→Xmin
:9→Xmax
:1→Xscl
:-6→Ymin
:6→Ymax
:1→Yscl
:DispGraph
:Pause
:Disp "Y=R(X+H)²+V"
:Disp "R="
:Disp R
:Disp "H="
:Disp H
:Disp "V="
:Disp V
:End
```

Press ENTER after viewing the graph to display the values of the integers.

TI-83
TI-82

```
PROGRAM:PARABOLA
:-6+int (12rand)→H
:-3+int (6rand)→V
:rand→R
:If R < .5
:Then
:-1→R
:Else
:1→R
:End
:"R(X+H)²+V"→Y1
:-9→Xmin
:9→Xmax
:1→Xscl
:-6→Ymin
:6→Ymax
:1→Yscl
:DispGraph
:Pause
:Disp "Y=R(X+H)²+V"
:Disp "R=",R
:Disp "H=",H
:Disp "V=",V
:Pause
```

Press ENTER after viewing the graph to display the values of the integers.

TI-85

```
PROGRAM:PARABOLA
:rand→H
:-6+int (12H)→H
:rand→V
:-3+int (6V)→V
:rand→R
:If R < .5
:-1→R
:If R > .49
:1→R
:y1=R(x+H)²+V
:-9→xMin
:9→xMax
:1→xScl
:-6→yMin
:6→yMax
:1→yScl
:DispG
:Pause
:Disp "Y=R(X+H)²+V"
:Disp "R=",R
:Disp "H=",H
:Disp "V=",V
:Pause
```

Press ENTER after viewing the graph to display the values of the integers.

TI-92

```
Parabola( )
Prgm
ClrHome
ClrIO
setMode("Split Screen",
   "Left-Right")
setMode("Split 1 App","Home")
setMode("Split 2 App","Graph")
-6+int (12rand( ))→h
-3+int (6rand( ))→v
rand( )→r
If r < .5 Then
   -1→r
      Else
   1→r
EndIf
r*(x+h)^2+v→y1(x)
-9→xmin
9→xmax
1→xscl
-6→ymin
6→ymax
1→yscl
DispG
Disp "y1(x)=r(x+h)^2+v"
Output 20,1, "r=":Output 20,11,r
Output 40,1, "h=":Output 40,11,h
Output 60,1, "v=":Output 60,11,v
Pause
setMode("Split Screen","Full")
EndPrgm
```

Casio fx-7700G

```
PARABOLA
-6+INT (12Ran#)→H
-3+INT (6Ran#)→V
-1→R:Ran#<0.5 ⇒1→R
Range -9,9,1,-6,6,1
Graph Y=R(X+H)²+V ◢
"Y=R(X+H)²+V"
"R=":R ◢
"H=":H ◢
"V=":V
```

Press [EXE] after viewing the graph to display the values of the integers.

Casio fx-7700GE
Casio fx-9700GE
Casio CFX-9800G

```
PARABOLA↵
-6+Int (12Ran#)→H↵
-3+Int (6Ran#)→V↵
Ran#→R↵
R< .5⇒-1→R↵
R≥ .5⇒1→R↵
Range -9,9,1,-6,6,1↵
Graph Y=R(X+H)²+V ◢
"Y=R(X+H)²+V"↵
"R=":R◢
"H=":H◢
"V=":V
```

Press [EXE] after viewing the graph to display the values of the integers.

Sharp EL-9200C
Sharp EL-9300C

```
parabola
-----------REAL
h=int (random*12) -6
v=int (random*6) -3
s=(random*2) -1
r=s/abs s
Range -9,9,1,-6,6,1
Graph r(X+h)²+v
Wait
Print "y=r(X+h)²+v
Print r
Print h
Print v
End
```

Press [ENTER] after viewing the graph to display the values of the integers.

HP-38G

PARABOLA

```
PARABOLA PROGRAM
-6+INT(12RANDOM)▶H:
-3+INT(6RANDOM)▶V:
RANDOM ▶R:
IF R>.5
   THEN -1▶R:
   ELSE 1▶R:
END:
'R*(X+H)²+V'▶F1(X):
CHECK 1:
```

```
PARANS PROGRAM
ERASE:
DISP 2;"Y=R(X+H)²+V":
DISP 3;"R="R:
DISP 4;"H="H:
DISP 5;"V="V:
FREEZE:
```

```
PARABOLA.SV PROGRAM
SETVIEWS "RUN
PARABOLA";PARABOLA;1;
"ANSWER";PARANS;1;
"  ";PARABOLA.SV;0:
```

1. Press [LIB]. Highlight the Function aplet. Press {{SAVE}}. Enter the name PARABOLA for the new aplet and press {{OK}}.
2. Press ■ [SETUP-PLOT] and set XRNG: from −12 to 12, YRNG: from −6 to 6, and XTICK: and YTICK: to 1.
3. Enter the 3 programs PARABOLA, PARANS, PARABOLA.SV.
4. Run the program PARABOLA.SV.
5. Enter the PARABOLA aplet.
6. Press ■ [VIEWS]. Highlight RUN PARABOLA and press {{OK}}.
7. After viewing the graph, press ■ [VIEWS]. Highlight ANSWER and press {{OK}} to see the values of the integers.
8. Press {{OK}} to return to the graph.
9. Repeat steps 6, 7, and 8 for a new parabola.

Graph Reflection Program (Section P.8)

This program, shown in the Technology note on page 101, will graph a function f and its reflection in the line $y = x$.

Note: On the *TI-83* and the *TI-82*, the "While" command may be entered through the "CTL" menu accessed by pressing the PRGM key. The "Pt-On(" command may be entered through the "POINTS" menu accessed by pressing the DRAW key. For additional keystroke instructions, see previous programs in this appendix. Keystroke sequences required for similar commands on other calculators will vary. Consult the user manual for your calculator.

TI-80

```
PROGRAM:REFLECT
:47XMIN/63→YMIN
:47XMAX/63→YMAX
:XSCL→YSCL
:"X"→Y2
:DISPGRAPH
:(XMAX−XMIN)/62→I
:XMIN→X
:LBL A
:PT-ON(Y1,X)
:X+I→X
:If X>XMAX
:STOP
:GOTO A
```

To use this program, enter the function in Y1 and set a viewing rectangle.

TI-81

```
Prgm3:REFLECT
:2Xmin/3→Ymin
:2Xmax/3→Ymax
:Xscl→Yscl
:"X"→Y2
:DispGraph
:(Xmax−Xmin)/95→I
:Xmin→X
:Lbl 1
:Pt-On(Y1,X)
:X+I→X
:If X>Xmax
:End
:Goto 1
```

To use this program, enter the function in Y_1 and set a viewing rectangle.

TI-83
TI-82

```
PROGRAM:REFLECT
:63Xmin/95→Ymin
:63Xmax/95→Ymax
:Xscl→Yscl
:"X"→Y2
:DispGraph
:(Xmax−Xmin)/94→I
:Xmin→X
:While X≤Xmax
:Pt-On(Y1,X)
:X+I→X
:End
```

To use this program, enter the function in Y_1 and set a viewing rectangle.

TI-85

```
PROGRAM:REFLECT
:63*xMin/127→yMin
:63*xMax/127→yMax
:xScl→yScl
:y2=x
:DispG
:(xMax−xMin)/126→I
:xMin→x
:Lbl A
:PtOn(y1,x)
:x+I→x
:If x>xMax
:Stop
:Goto A
```

To use this program, enter the function in y1 and set a viewing rectangle.

TI-92

```
Prgm
103xmin/239→ymin
103xmax/239→ymax
xscl→yscl
x→y2(x)
DispG
(xmax−xmin)/238→n
xmin→x
While x<xmax
   PtOn y1(x),x
   x+n→x
EndWhile
EndPrgm
```

To use this program, enter the function in y1 and set an appropriate viewing window.

Casio fx-7700G

```
REFLECTION
"GRAPH -A TO A"
"A="?→A
Range -A,A,1,-2A÷3,2A÷3,1
Graph Y=f1
-A→B
Lbl 1
B→X
Plot f1,B
B+A÷32→B
B≤A⇒Goto1 :Graph Y=X
```

To use this program, enter the function in f_1.

Casio fx-7700GE

```
REFLECTION
"GRAPH -A TO A"↵
"A="?→A↵
Range -A,A,1,-2A÷3,2A÷3,1↵
Graph Y=f1↵
-A→B↵
Lbl 1↵
B→X↵
Plot f1,B↵
B+A÷32→B↵
B≤A⇒Goto1:Graph Y=X
```

To use this program, enter the function in f1.

Casio fx-9700GE

```
REFLECTION↵
63Xmin÷127→A↵
63Xmax÷127→B↵
Xscl→C↵
Range , , , A, B, C↵
(Xmax−Xmin)÷126→I↵
Xmax→M↵
Xmin→D↵
Graph Y=f1↵
Lbl 1↵
D→X↵
Plot f1,D↵
D+I→D↵
D≤M⇒Goto 1:Graph Y=X
```

To use this program, enter the function in f1 and set a viewing rectangle.

Casio CFX-9800G

```
REFLECTION↵
63Xmin÷95→A↵
63Xmax÷95→B↵
Xscl→C↵
Range , , , A, B, C↵
(Xmax−Xmin)÷94→I↵
Xmax→M↵
Xmin→D↵
Graph Y=f1↵
```

Casio CFX-9800G

(Continued)

```
Lbl 1↵
D→X↵
Plot f1,D↵
D+I→D↵
D≤M⇒Goto 1:Graph Y=X
```

To use this program, enter the function in f1 and set a viewing rectangle.

Sharp EL-9200C
Sharp EL-9300C

```
reflection
————REAL
Goto top
Label equation
Y=f(X)
Return
Label rng
xmin=-10
xmax=10
xstp=(xmax−xmin)/10
ymin=2xmin/3
ymax=2xmax/3
ystp=xstp
Range xmin,xmax,xstp,ymin,
   ymax,ystp
Return
Label top
Gosub rng
Graph X
step=(xmax−xmin)/(94*2)
X=xmin
Label 1
Gosub equation
Plot X,Y
Plot Y, X
X=X+step
If X<=xmax Goto 1
End
```

To use this program, replace f(X) with your expression in X.

Graphing a Sine Function (Section 1.5)

The program, shown in the Group Activity on page 165, will simultaneously draw a unit circle and the corresponding points on the sine curve. After the circle and sine curve are drawn, you can connect the points on the unit circle with their corresponding points on the sine curve by pressing ENTER or EXE.

TI-80

```
PROGRAM:SINESHO
:RADIAN
:CLRDRAW:FNOFF
:PARAM:SIMUL
:-2.25→XMIN
:π/2→XMAX
:3→XSCL
:-1.5→YMIN
:1.5→YMAX
:1→YSCL
:0→TMIN
:6.3→TMAX
:.15→TSTEP
:"-1.25+COS T"→X1T
:"SIN T"→Y1T
:"T/4"→X2T
:"SIN T"→Y2T
:DISPGRAPH
:FOR(N,1,12)
:Nπ/6.5→T
:"-1.25+COS T"→A
:SIN T→B
:T/4→C
:LINE(A,B,C,B)
:PAUSE
:END
:PAUSE:FUNC
:SEQUENTIAL:DISP
```

TI-81

```
PrgmA:SINESHOW
:Rad
:ClrDraw
:Param
:Simul
:-2.25→Xmin
:π/2→Xmax
:3→Xscl
:-1.19→Ymin
:1.19→Ymax
:1→Yscl
:0→Tmin
:6.3→Tmax
:.15→Tstep
:"-1.25+cos T"→X1T
:"sin T"→Y1T
:"T/4"→X2T
:"sin T"→Y2T
:DispGraph
:1→N
:Lbl 1
:IS>(N,12)
:Goto 2
:Pause
:Function
:Sequence
:Disp " "
:End
:Lbl 2
:Nπ/6.5→T
:-1.25+cos T→A
:sin T→B
:T/4→C
:Line(A,B,C,B)
:Pause
:Goto 1
```

TI-83
TI-82

```
PROGRAM:SINESHOW
:Radian
:ClrDraw:FnOff
:Param:Simul
:-2.25→Xmin
:π/2→Xmax
:3→Xscl
:-1.19→Ymin
:1.19→Ymax
:1→Yscl
:0→Tmin
:6.3→Tmax
:.15→Tstep
:"-1.25+cos (T)"→X1T
:"sin (T)"→Y1T
:"T/4"→X2T
:"sin (T)"→Y2T
:DispGraph
:For(N,1,12)
:Nπ/6.5→T
:-1.25+cos (T)→A
:sin(T)→B
:T/4→C
:Line(A,B,C,B)
:Pause
:End
:Pause :Func
:Sequential:Disp
```

TI-85

```
PROGRAM:SINESHOW
:Radian
:ClDrw:FnOff
:Param:SimulG
:-2.25→xMin
:π/2→xMax
:3→xScl
:-1.1→yMin
:1.1→yMax
:1→yScl
:0→tMin
:6.3→tMax
:.15→tStep
:xt1=-1.25+cos t
:yt1=sin t
:xt2=t/4
:yt2=sin t
:For(N,1,12)
:N*π/6.5→t
:-1.25+cos t→A
:sin t→B
:t/4→C
:Line(A,B,C,B)
:Pause
:End
:Pause :Func
:SeqG:Disp
```

TI-92

```
sineshow()
Prgm
Disp
ClrDraw:FnOff
setMode("Graph", "Parametric")
setGraph("Graph Order",
    "Simul")
-2.9→xmin
3π/4→xmax
3→xscl
-1.1→ymin
1.1→ymax
1→yscl
0→tmin
6.3→tmax
.15→tstep
-1.25+cos(t)→xt1(t)
sin(t)→yt1(t)
t/4→xt2(t)
sin(t)→yt2(t)
DispG
For N,1,12
N*π/6.5→t
-1.25+cos(t)→A
sin(t)→B
t/4→C
Line A,B,C,B
Pause
EndFor
Pause
setMode("Graph", "Function")
setGraph("Graph order",
    "Seq")
setMode("Split 1 App",
    "Home")
EndPrgm
```

Casio fx-7700G

```
SINESHOW
Rad
Range -2.25,π÷2,3,-1.19,1.19,
    10,6.3,.15
Graph(X,Y)=(-1.25+cos T,sinT)
Graph(X,Y)=(T÷4,sinT)
0→N
Lbl 1
N+1→N
Nπ÷6.5→T
-1.25+cos T→A
sin T→B
T÷4→C
Plot A,B
Plot C,B
Line◢
N<12⇒Goto 1
```

Press MODE SHIFT × to change to parametric mode when starting to write this program.

Casio fx-7700GE
Casio fx-9700GE
Casio CFX-9800G

```
SINESHOW↵
Rad↵
Range -2.25,π÷2,3,-1.19,1.19,
    1,0,6.3,.15↵
Graph(X,Y)=(-1.25+cos T,sin T)↵
Graph(X,Y)=(T÷4,sin T)↵
0→N↵
Lbl 1↵
N+1→N↵
Nπ÷6.5→T↵
-1.25+cosT→A↵
sin T→B↵
T÷4→C↵
Plot A,B↵
Plot C,B↵
Line◢
N<12⇒Goto 1↵
Cls
```

When starting to write this program, press SHIFT SET UP and select PRM or PARM for the GRAPH TYPE to change to parametric mode.

Sharp EL-9200C
Sharp EL-9300C

```
sineshow
————————REAL
m=sin⁻¹ 1/(π/2)
Range -2.25,π/2,3,-1.19,1.19,1
step=π/15
θ=0
xco=-.25
xso=0
yo=0
Label 1
θ=θ+step
xc=cos(mθ)−1.25
xs=θ/4
y=sin (mθ)
Line xco,yo,xc,y
Line xso,yo,xs,y
xco=xc
xso=xs
yo=y
If θ<(2π) Goto 1
step=π/6
θ=0
Label 2
θ=θ+step
xc=cos (mθ)−1.25
xs=θ/4
y=sin (mθ)
Line xc,y,xs,y
Wait
If θ<2π Goto 2
End
```

HP-38G Programs

```
SINESHOW PROGRAM
ASIN(1)/(π/2)►M:
0►T:
-.25►A:
0►B:
0►C:
LINE -3;0;π/2;0:
LINE 0;-1.1;0;1.1:
FOR T=0 TO 31π/15
   STEP π/15;
   RUN "DRAW.SINE":
END:
0►T:
FOR T=0 TO 2π
   STEP π/6;
   RUN "DRAW.LINE":
END
```

```
DRAW.SINE PROGRAM
COS(MT)−1.25►D:
T/4►E:
SIN(MT)►F:
LINE A;C;D;F:
LINE B;C;E;F:
D►A:
E►B:
F►C:
```

```
DRAW.LINE PROGRAM
COS(MT)−1.25►D:
T/4►E:
SIN(MT)►F:
LINE D;F;E;F:
FREEZE
```

1. Enter the 3 programs SINESHOW, DRAW.SINE, and DRAW.LINE.
2. Set the plot range in the Function aplet to $-3 \le x \le \pi/2$ and $-1.1 \le y \le 1.1$. Set the angle measure to radians.
3. Run the SINESHOW program.

Finding the Angle between Two Vectors (Section 3.4)

The program, shown in the Technology note on page 302, will sketch two vectors and calculate the measure of the angle between the vectors. Be sure to set an appropriate viewing rectangle.

TI-80

```
:PROGRAM:VECANGL
:CLRHOME
:DEGREE
:DISP "ENTER (A,B)"
:INPUT "ENTER A",A
:INPUT "ENTER B",B
:CLRHOME
:DISP "ENTER (C,D)"
:INPUT "ENTER C",C
:INPUT "ENTER D",D
:LINE(0,0,A,B)
:LINE(0,0,C,D)
:PAUSE
:AC+BD→E
:√ (A²+B²)→U
:√ (C²+D²)→V
:COS⁻¹(E/(UV))→θ
:DISP "θ=", θ
:CLRDRAW
```

TI-81

```
:PrgmB:VECANGL
:ClrHome
:Deg
:Disp "ENTER (A,B)"
:Disp "ENTER A"
:Input A
:Disp "ENTER B"
:Input B
:ClrHome
:Disp "ENTER (C,D)"
:Disp "ENTER C"
:Input C
:Disp "ENTER D"
:Input D
:Line(0,0,A,B)
:Line(0,0,C,D)
:Pause
:AC+BD→E
:√ (A²+B²)→U
:√ (C²+D²)→V
:cos⁻¹(E/(UV))→θ
:Disp "θ="
:Disp θ
:ClrDraw
:End
```

TI-83
TI-82

```
:PROGRAM:VECANGL
:ClrHome
:Degree
:Disp "ENTER (A,B)"
:Input "ENTER A",A
:Input "ENTER B",B
:ClrHome
:Disp "ENTER (C,D)"
:Input "ENTER C",C
:Input "ENTER D",D
:Line(0,0,A,B)
:Line(0,0,C,D)
:Pause
:AC+BD→E
:√ (A²+B²)→U
:√ (C²+D²)→V
:cos⁻¹(E/(UV))→θ
:ClrDraw:ClrHome
:Disp "θ=",θ
:Stop
```

TI-85

```
:PROGRAM:VECANGL
:ClLCD
:Radian
:Disp "enter (A,B)"
:Input "enter A",A
:Input "enter B",B
:ClLCD
:Disp "enter (C,D)"
:Input "enter C",C
:Input "enter D",D
:Line(0,0,A,B)
:Line(0,0,C,D)
:Pause
:A*C+B*D→E
:√ (A²+B²)→U
:√ (C²+D²)→V
:cos⁻¹(E/(U*V))→T
:T*180/π→T
:Disp "T=",T
:ClDrw
```

TI-92

```
vecangl( )
Prgm
FnOff
ClrHome:ClrDraw
SetMode("Split Screen",
    "Left-Right")
SetMode("Split 1 App", "Home")
SetMode("Split 2 App", "Graph")
SetMode("Exact/Approx",
    "Approximate")
ClrIO
Disp "ENTER (A,B)"
Input "ENTER A", A
Input "ENTER B", B
Line(0,0,A,B)
Pause
ClrIO
Disp "ENTER (C,D)"
Input "ENTER C",C
Input "ENTER D",D
Line(0,0,C,D)
Pause
ClrIO
A*C+B*D→E
√ ((A^2+B^2))→U
√ (C^2+D^2)→V
cos⁻¹ (E/(U*V))→θ
Disp "θ=",θ
Pause
SetMode("Exact/Approx", "Auto")
SetMode("Split Screen", "Full")
SetMode("Split 1 App", "Home")
Stop
EndPrgm
```

Casio fx-7700G

```
VECANGL
Cls
Deg
"ENTER (A,B)"
"A="?→A
"B="?→B
"ENTER (C,D)"
"C="?→C
"D="?→D
Plot 0,0
Plot A,B
Line
Plot 0,0
Plot C,D
Line◢
AC+BD→E
√ (A²+B²)→U
√ (C²+D²)→V
cos⁻¹(E÷UV)→θ
"θ="
θ
```

Casio fx-7700GE
Casio fx-9700GE
Casio CFX-9800G

```
VECANGL↵
Cls↵
Deg↵
"ENTER (A,B)"↵
"A="?→A↵
"B="?→B↵
"ENTER (C,D)"↵
"C="?→C↵
"D="?→D↵
Plot 0,0↵
Plot A,B↵
Line↵
Plot 0,0↵
Plot C,D↵
Line◢
AC+BD→E↵
√(A²+B²)→U↵
√(C²+D²)→V↵
cos⁻¹(E÷UV)→θ↵
"θ="↵
θ
```

Sharp EL-9200C
Sharp EL-9300C

```
vecangl
————————REAL
ClrG
ClrT
Print"enter (a,b)"
Input a
Input b
ClrT
Print"enter (c,d)"
Input c
Input d
Line 0,0,a,b
Line 0,0,c,d
Wait
e=a*c+b*d
u=√(a²+b²)
v=√(c²+d²)
t=cos⁻¹(e/(u*v))
Print t
End
```

Set the calculator to degree mode before running the program.

HP-38G

```
VECANGL PROGRAM
INPUT A; "ENTER (A,B)";
    "ENTER A";;1:
INPUT B; "ENTER (A,B)";
    "ENTER B";;1:
INPUT C; "ENTER (C,D)";
    "ENTER C";;1:
INPUT D; "ENTER (C,D)";
    "ENTER D";;1:
ERASE:
LINE−10;0;10;0:
LINE 0;−10;0;10:
LINE 0;0;A;B:
LINE 0;0;C;D:
FREEZE:
AC+BD► E
√ (A²+B²)►U:
√ (C²+D²)►V:
ACOS(E/(UV))►T:
ERASE:
DISP 3; "ANGLE= "T:
FREEZE
```

The Function aplet should have a plot range of $-10 \le x \le 10$ and $-10 \le y \le 10$. Set the MODE to degrees before running the program.

Adding Vectors Graphically (Chapter 3 Project)

The program, shown in the Chapter 3 Project on page 316, will sketch two vectors in standard position. Using the parallelogram law for the vector addition, the program also sketches the vector sum. Be sure to set an appropriate viewing rectangle.

TI-80

```
:PROGRAM:ADDVECT
:CLRDRAW
:DISP "ENTER(A,B)"
:INPUT "ENTER A",A
:INPUT "ENTER B",B
:DISP "ENTER (C,D)"
:INPUT "ENTER C",C
:INPUT "ENTER D",D
:LINE(0,0,A,B)
:LINE(0,0,C,D)
:A+C→E
:B+D→F
:LINE(0,0,E,F)
:LINE(A,B,E,F)
:LINE(C,D,E,F)
:PAUSE
```

TI-81

```
:PrgmC:ADDVECT
:ClrDraw
:Disp "ENTER(A,B)"
:Disp "ENTER A"
:Input A
:Disp "ENTER B"
:Input B
:Disp "ENTER (C,D)"
:Disp "ENTER C"
:Input C
:Disp "ENTER D"
:Input D
:Line(0,0,A,B)
:Line(0,0,C,D)
:A+C→E
:B+D→F
:Line(0,0,E,F)
:Line(A,B,E,F)
:Line(C,D,E,F)
:Pause
:End
```

TI-83
TI-82

```
:PROGRAM:ADDVECT
:ClrDraw
:Input "ENTER A",A
:Input "ENTER B",B
:Input "ENTER C",C
:Input "ENTER D",D
:Line(0,0,A,B)
:Line(0,0,C,D)
:A+C→E
:B+D→F
:Line(0,0,E,F)
:Line(A,B,E,F)
:Line(C,D,E,F)
:Pause
:Stop
```

TI-85

```
:PROGRAM:ADDVECT
:ClrDraw
:Input "enter A",A
:Input "enter B",B
:Input "enter C",C
:Input "enter D",D
:Line(0,0,A,B)
:Line(0,0,C,D)
:A+C→E
:B+D→F
:Line(0,0,E,F)
:Line(A,B,E,F)
:Line(C,D,E,F)
:Pause
:Disp
```

TI-92

```
addvect( )
Prgm
ClrIO
Input "ENTER a ",a
Input "ENTER b ",b
Input "ENTER c ",c
Input "ENTER d ",d
ClrDraw
Line(0,0,a,b)
Line(0,0,c,d)
a+c→e
b+d→f
Line 0,0,e,f
Line a,b,e,f
Line c,d,e,f
Pause
setMode("Split 1 App","Home")
Stop
EndPrgm
```

Casio fx-7700G

```
ADDVECT
Cls
"A="?→A
"B="?→B
"C="?→C
"D="?→D
Plot 0,0
Plot A,B
Line
Plot 0,0
Plot C,D
Line ◢
A+C→E
B+D→F
Plot 0,0
Plot E,F
Line
Plot A,B
Plot E,F
Line
Plot C,D
Plot E,F
Line ◢
```

Casio fx-7700GE
Casio fx-9700GE
Casio CFX-9800G

```
ADDVECT↵
Cls↵
"A="?→A↵
"B="?→B↵
"C="?→C↵
"D="?→D↵
Plot 0,0↵
Plot A,B↵
Line↵
Plot 0,0↵
Plot C,D↵
Line◢
A+C→E↵
B+D→F↵
Plot 0,0↵
Plot E,F↵
Line↵
Plot A,B↵
Plot E,F↵
Line↵
Plot C,D↵
Plot E,F↵
Line◢
```

Sharp El-9200C
Sharp EL-9300C

```
addvect
——————REAL
ClrG
Input a
Input b
Input c
Input d
Line 0,0,a,b
Line 0,0,c,d
e=a+c
f=b+d
Line 0,0,e,f
Line a,b,e,f
Line c,d,e,f
Wait
End
```

HP-38G PROGRAMS

```
ADDVECT PROGRAM
INPUT A;; "ENTER A";;1:
INPUT B;; "ENTER B";;1:
INPUT C;; "ENTER C";;1:
INPUT D;; "ENTER D";;1:
ERASE:
LINE−10;0;10;0:
LINE 0;−10;0;10:
LINE 0;0;A;B:
LINE 0;0;C;D:
FREEZE:
A+C▶ E
B+D▶ F
LINE 0;0;E;F:
LINE A;B;E;F:
LINE C;D;E;F:
FREEZE
```

The Function aplet should have a plot range of $-10 \le x \le 10$ and $-10 \le y \le 10$.

Mandelbrot Set Program (Chapter 4 Project)

The program, shown in the Chapter 4 Project on page 353, can be used to determine if a complex number is in the Mandelbrot Set. The number is entered in two parts, the real part of the complex number and the imaginary part of the complex number. After entering the number, press [ENTER] or [EXE] to see the next number in the sequence. If the terms of the sequence become very large, the sequence is unbounded and the complex number is not in the Mandelbrot Set.

TI-80

```
PROGRAM:MANDLBR
:INPUT "ENTER REAL PART",A
:INPUT "ENTER IMAG PART",B
:A→C:B→D
:0→N
:LBL 1
:CLRHOME
:N+1→N
:DISP "TERM NUMBER",N
:DISP "REAL PART",A
:DISP "IMAG PART",B
:PAUSE
:A→F:B→G
:F²–G²+C→A
:2FG+D→B
:GOTO 1
```

Press [ON] [2] to quit the program. The program will give an error message when a number exceeds the capacity of the calculator. Press [2].

TI-81

```
Prgm4:MANDLBRT
:Disp "ENTER REAL PART"
:Input A
:Disp "ENTER IMAG PART"
:Input B
:A→C
:B→D
:0→N
:Lbl 1
:ClrHome
:N+1→N
:Disp "TERM NUMBER"
:Disp N
:Disp "REAL PART"
:Disp A
:Disp "IMAG PART"
:Disp B
:Pause
:A→F
:B→G
:F²–G²+C→A
:2FG+D→B
:Goto 1
```

Press [ON] [2] to quit the program. The program will give an error message when a number exceeds the capacity of the calculator. Press [2].

TI-82
TI-83

```
PROGRAM:MANDLBRT
:Input "ENTER REAL PART",A
:Input "ENTER IMAG PART",B
:A→C:B→D
:0→N
:Lbl 1
:ClrHome
:N+1→N
:Disp "TERM NUMBER",N
:Disp "REAL PART",A
:Disp "IMAG PART",B
:Pause
:A→F:B→G
:F²–G²+C→A
:2FG+D→B
:Goto 1
```

For the *TI-82,* press [ON] [2] to quit the program. For the *TI-83,* press [ON] [1] to quit the program. The program will give an error message when a number exceeds the capacity of the calculator. For the *TI-82,* press [2]. For the *TI-83,* press [1].

TI-85

```
PROGRAM:MANDLBRT
:Input "ENTER REAL PART",A
:Input "ENTER IMAG PART",B
:A→C:B→D
:0→N
:Lbl Z
:ClLCD
:N+1→N
:Disp "TERM NUMBER",N
:Disp "REAL PART",A
:Disp "IMAG PART",B
:Pause
:A→F:B→G
:F²–G²+C→A
:2*F*G+D→B
:Goto Z
```

Press [ON] [F5] to quit the program. The program will give an error message when a number exceeds the capacity of the calculator. Press [F5].

TI-92

```
:mandlbrt( )
:Prgm
:Input "enter real part",a
:Input "enter imaginary part",b
:a→c:b→d
:0→n
:Lbl top
:ClrIO
:n+1→n
:Disp "term number",n
:Disp "real part",a
:Disp "imaginary part",b
:Pause
:a→f:b→g
:f^2–g^2+c→a
:2*f*g+d→b
:Goto top
:EndPrgm
```

Press ON ENTER to quit the program. The ∞ symbol will be displayed when a number has exceeded the capacity of the calculator.

Casio fx-7700G

```
MANDLBRT
"ENTER REAL PART":?→A
"ENTER IMAG PART":?→B
A→C:B→D
0→N
Lbl 1
Cls
N+1→N
"TERM NUMBER":N◢
"REAL PART":A◢
"IMAG PART":B◢
:A→F:B→G
:F²–G²+C→A
:2FG+D→B
:GoTo 1
```

When running the program, press EXE after each variable is displayed to see the value of the next variable. To quit the program, press AC AC. The program will give an error message when a number exceeds the capacity of the calculator. Press AC.

Casio fx-7700GE
Casio fx-9700GE
Casio CFX-9800G

```
MANDLBRT↵
"ENTER REAL PART":?→A↵
"ENTER IMAG PART":?→B↵
A→C:B→D↵
0→N↵
Lbl 1↵
Cls↵
N+1→N↵
"TERM NUMBER":N◢
"REAL PART":A◢
"IMAG PART":B◢
A→F:B→G↵
F²–G²+C→A↵
2FG+D→B↵
Goto 1
```

When running the program, press [EXE] after each variable is displayed to see the value of the next variable. To quit the program, press [AC/ON] [AC/ON]. The program will give an error message when a number exceeds the capacity of the calculator. Press [AC/ON].

Sharp EL-9300C

```
mandlbrt
————————REAL
Print "enter real part"
Input A
Print "enter imag part"
Input B
c=a
d=b
n=0
Label 1
ClrT
n=n+1
Print "term number"
Print n
Print "real part"
Print a
Print "imag part"
Print b
Wait
f=a
g=b
a=f²–g²+c
b=2f*g+d
Goto 1
```

To quit the program, press [ON] [CL]. The program will give an error message when a number exceeds the capacity of the calculator. Press [CL].

HP-38G

```
MANDLBRT PROGRAM
INPUT A;"ENTER REAL
    PART ";"ENTER A";"";1:
INPUT B;"ENTER IMAG
    PART ";"ENTER B";"";0:
A►C:
B►D:
0►N:
WHILE N<20 REPEAT
    ERASE:
    N+1►N:
    DISP 1;"TERM NUMBER"N:
    DISP 3;"REAL PART":
    DISP 4;A:
    DISP 6;"IMAG PART":
    DISP 7;B:
    FREEZE:
    A►F:
    B►G:
    F²–G²+C►A:
    2FG+D►B:
END:
```

This program will give the first twenty terms of the sequence.

Appendix D Scientific Calculator Keystrokes

Chapter 1

PAGE 132, EVALUATING TRIGONOMETRIC FUNCTIONS WITH A CALCULATOR

π [÷] 8 [=] [sin] [1/x]   In radian mode

PAGE 132, EXAMPLE 4 USING A CALCULATOR

a. 76.4 [sin]   **b.** 1.5 [tan] [1/x]

PAGE 132, GROUP ACTIVITY

30 [cos]

PAGE 140, EVALUATING TRIGONOMETRIC FUNCTIONS WITH A CALCULATOR

28 [cos]   In degree mode

28 [cos] [1/x]   In degree mode

PAGE 142 (TOP OF PAGE)

.6 [INV] [sin]

PAGE 142, EXAMPLE 9 USING TRIGONOMETRY TO SOLVE A RIGHT TRIANGLE

12 [÷] 9 [=] [INV] [tan]

PAGE 153, EXAMPLE 7 USING A CALCULATOR

a. $\cot 410°$ 410 [tan] [1/x]

$\sin(-7)$ 7 [+/−] [sin]

b. 4.812 [INV] [tan]

PAGE 186, EXAMPLE 4 CALCULATORS AND INVERSE TRIGONOMETRIC FUNCTIONS

a. 8.45 [+/−] [INV] [tan]

b. 0.2447 [INV] [sin]

c. 2 [INV] [cos]

Chapter 5

PAGE 356, EXAMPLE 1 EVALUATING EXPONENTIAL EXPRESSIONS

a. 2 [y^x] 3.1 [+/−] [=]

b. 2 [y^x] π [+/−] [=]

PAGE 360, EXAMPLE 5 EVALUATING THE NATURAL EXPONENTIAL FUNCTIONS

a. 2 [+/−] [e^x] or 2 [+/−] [INV] [ln x]

b. 1 [+/−] [e^x] or 1 [+/−] [INV] [ln x]

c. 1 [e^x] or 1 [INV] [ln x]

d. 2 [e^x] or 2 [INV] [ln x]

PAGE 369, EXAMPLE 2 EVALUATING LOGARITHMS ON A CALCULATOR

a. 10 [log]

b. 2.5 [log] [×] 2 [=]

c. 2 [+/−] [log]

PAGE 373, EXAMPLE 8 EVALUATING THE NATURAL LOGARITHMIC FUNCTION

a. 2 [ln x]

b. .3 [ln x]

c. 2 [e^x] [ln x]

d. 1 [+/−] [ln x]

ANSWERS TO WARM UPS, ODD-NUMBERED EXERCISES, FOCUS ON CONCEPTS, AND TESTS

CHAPTER P

Section P.1 *(page 10)*

1. (a) 5, 1 (b) $-9, 5, 0, 1$ (c) $-9, -\frac{7}{2}, 5, \frac{2}{3}, 0, 1$
(d) $\sqrt{2}$

3. (a) None (b) -13 (c) $2.01, 0.666\ldots, -13$
(d) $0.010110111\ldots$

5. (a) $\frac{6}{3}$ (b) $\frac{6}{3}$ (c) $-\frac{1}{3}, \frac{6}{3}, -7.5$ (d) $-\pi, \frac{1}{2}\sqrt{2}$

7. 0.625 **9.** $0.\overline{123}$ **11.** $-1 < 2.5$

13. $\frac{3}{2} < 7$ **15.** $-4 > -8$

17. $\frac{5}{6} > \frac{2}{3}$

19. $x \le 5$ is the set of all real numbers less than or equal to 5. Unbounded

21. $x < 0$ is the set of all negative real numbers. Unbounded

23. $x \ge 4$ is the set of all real numbers greater than or equal to 4. Unbounded

25. $-2 < x < 2$ is the set of all real numbers greater than -2 and less than 2. Bounded

27. $-1 \le x < 0$ is the set of all negative real numbers greater than or equal to -1. Bounded

29. $\frac{127}{90}, \frac{584}{413}, \frac{7071}{5000}, \sqrt{2}, \frac{47}{33}$ **31.** $x < 0$ **33.** $y \ge 0$

35. $A \ge 30$ **37.** 10 **39.** $\pi - 3 \approx 0.1416$

41. -1 **43.** -9 **45.** 3.75 **47.** $|-3| > -|-3|$

49. $-5 = -|5|$ **51.** $-|-2| = -|2|$ **53.** 4

55. $\frac{5}{2}$ **57.** 51 **59.** $\frac{128}{75}$ **61.** $|x - 5| \le 3$

63. $|7 - 18| = 11$ miles **65.** $|y| \ge 6$

67. $|\$113{,}356 - \$112{,}700| = \$656 > \$500.00$
$0.05(\$112{,}700) = \$5635$
Because the actual expenses differ from the budget by more than \$500.00, there is failure to meet the "budget variance test."

69. $|\$37{,}335 - \$37{,}640| = \$305 < \$500$
$0.05(\$37{,}640) = \$1882$
Because the difference between the actual expenses and the budget is less than \$500 and less than 5% of the budgeted amount, there is compliance with the "budget variance test."

71. $|77.8 - 92.2| = 14.4$
There was a deficit of \$14.4 billion.

73. $|1031.3 - 1252.7| = 221.4$
There was a deficit of \$221.4 billion.

75. (a) No. If one is negative while the other is positive, they are unequal.
(b) $|u + v| \le |u| + |v|$

77. $7x, 4$ **79.** $4x^3, x, -5$

81. (a) -10 (b) -6 **83.** (a) 14 (b) 2

85. (a) Division by 0 is undefined. (b) 0

87. Commutative Property of Addition

89. Multiplicative Inverse Property

91. Distributive Property

93. Multiplicative Identity Property

95. Associative and Commutative Properties of Multiplication

97. 0 **99.** Division by 0 is undefined. **101.** 6 **103.** $\frac{1}{2}$

105. $\frac{3}{8}$ **107.** $\frac{3}{10}$ **109.** 48 **111.** -2.57 **113.** 1.56

115.

n	1	0.5	0.01	0.0001	0.000001
$5/n$	5	10	500	50,000	5,000,000

117.

n	1	10	100	10,000	100,000
$5/n$	5	0.5	0.05	0.0005	0.00005

Section P.2 *(page 22)*

Warm Up *(page 22)*

1. $-3x - 10$ **2.** $5x - 12$ **3.** x **4.** $x + 26$

5. $\frac{8x}{15}$ **6.** $\frac{3x}{4}$ **7.** $-\frac{1}{x(x+1)}$ **8.** $\frac{5}{x}$

9. $\frac{7x-8}{x(x-2)}$ **10.** $-\frac{2}{x^2-1}$

1. (a) No (b) No (c) Yes (d) No

3. (a) Yes (b) Yes (c) No (d) No

5. (a) Yes (b) No (c) No (d) No

7. Identity **9.** Conditional **11.** Identity

13. (a) The equations have the same solutions and the one is derived from the other by the steps for generating equivalent equations given in this section.

$2x = 5,\ 2x + 3 = 8$

(b) Remove symbols of grouping, combine like terms, reduce fractions.
Add (or subtract) the same quantity to (from) both sides of the equation.
Multiply (or divide) both sides of the equation by the same nonzero quantity.
Interchange the two sides of the equation.

15. 9 **17.** -4 **19.** 9 **21.** No solution

23. 10 **25.** 4 **27.** 3 **29.** 5 **31.** No solution

33. $\frac{11}{6}$ **35.** 0 **37.** All real numbers

39. $\frac{1+4b}{2+a},\ a \neq -2$

41. (a)

x	-1	0	1	2	3	4
$3.2x - 5.8$	-9	-5.8	-2.6	0.6	3.8	7

(b) $1 < x < 2$; The expression changes from negative to positive in this interval.

(c)

x	1.5	1.6	1.7	1.8	1.9	2
$3.2x - 5.8$	-1	-0.68	-0.36	-0.04	0.28	0.6

(d) $1.8 < x < 1.9$. To improve accuracy, evaluate the expression in this interval and determine where the sign changes.

43. $0, -\frac{1}{2}$ **45.** $4, -2$ **47.** $3, -\frac{1}{2}$ **49.** $-\frac{3}{2}, 11$

51. $0, \pm 3$ **53.** $0, 3, -1$ **55.** $0, \frac{3}{2}, 6$

57. ± 4; ± 4.00 **59.** $\pm 2\sqrt{3}$; ± 3.46

61. $12 \pm 3\sqrt{2}$; 16.24, 7.76 **63.** $-2 \pm 2\sqrt{3}$; 1.46, -5.46

65. 0, 2 **67.** $-3 \pm \sqrt{7}$ **69.** $2 \pm 2\sqrt{3}$

71. $\frac{1}{2}, -1$ **73.** $-4 \pm 2\sqrt{5}$ **75.** $\frac{2}{3} \pm \frac{\sqrt{7}}{3}$

77. $-\frac{3}{2} \pm \frac{\sqrt{13}}{2}$ **79.** $\frac{2}{7}$ **81.** $2 \pm \frac{\sqrt{6}}{2}$

83. False. The product must equal zero for the Zero-Factor Property to be used.

85. (a) $0, -\frac{b}{a}$ (b) 0, 1 **87.** $\pm\sqrt{3}, \pm 1$

89. $-\frac{1}{5}, -\frac{1}{3}$ **91.** $\frac{1}{4}$ **93.** $1, -\frac{125}{8}$ **95.** 26

97. -16 **99.** 5, 6 **101.** 0 **103.** 9

105. 3 **107.** -59, 69 **109.** 500 units

111. $h = \frac{1}{\pi r}\sqrt{S^2 - \pi^2 r^4}$ **113.** $a = 4,\ b = 24$

115. $1, -3$ **117.** $3, -2$ **119.** $\pm 3, -6$

121. $x^2 - 2x - 15 = 0$

123. $x = 7.5$ and $y = 23\frac{1}{3}$ or
$x = 17.5$ and $y = 10$

Section P.3 *(page 37)*

Warm Up *(page 37)*

1. $14x - 42$ **2.** $-17s$ **3.** $-24y^7$

4. $2(t+1)(t-1)(t+2)$ **5.** $5x^2\sqrt{6}$

6. $2\sqrt{x+3}$ **7.** $y = x^3 + 4x$ **8.** $x^2 + y^2 = 4$

9. $y = 4x^2 + 8$ **10.** $y^2 = -3x + 4$

1.

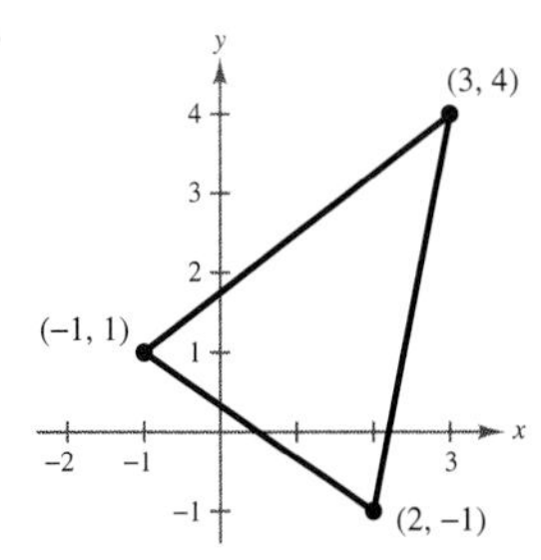

3. A: $(2, 6)$, B: $(-6, -2)$, C: $(4, -4)$, D: $(-3, 2)$

5. $(-3, 4)$ **7.** $(-5, -5)$

9. Quadrant IV **11.** Quadrant III

13. Point on x-axis: $y = 0$; point on y-axis: $x = 0$

15.

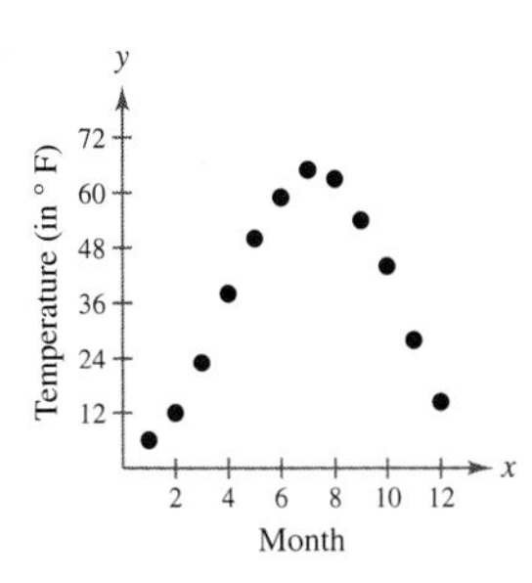

17. $4^2 + 3^2 = 5^2$ **19.** $10^2 + 3^2 = (\sqrt{109})^2$

21. (a) (b) 10 (c) $(5, 4)$

23. (a)

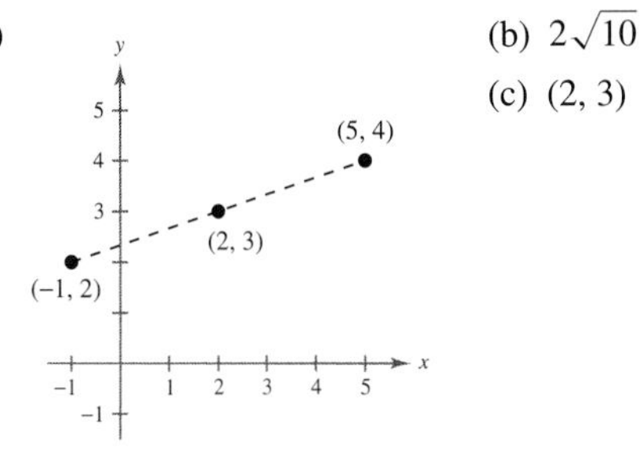

(b) $2\sqrt{10}$ (c) $(2, 3)$

25. (a)

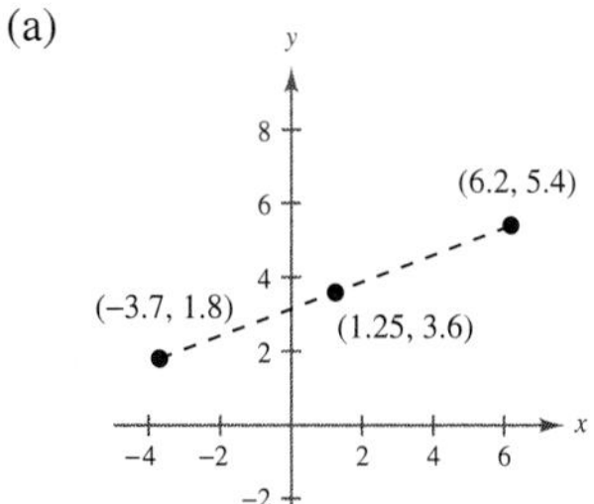

(b) $\sqrt{110.97}$ (c) $(1.25, 3.6)$

27. (1993, \$630,000)

29.

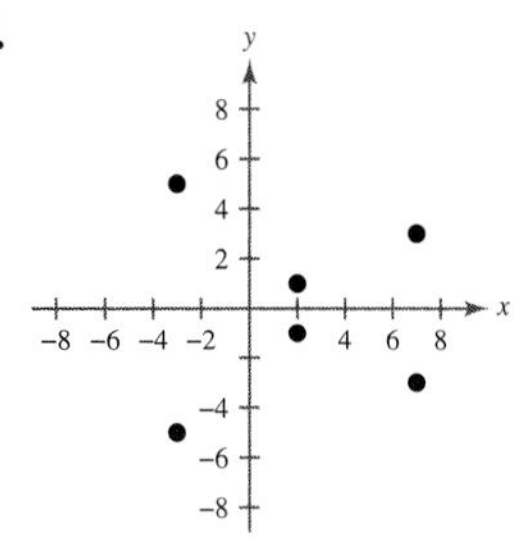

Reflection in the x-axis

31. (a) Yes (b) Yes **33.** (a) Yes (b) Yes

35.

x	-2	0	$\frac{2}{3}$	1	2
y	-4	-1	0	$\frac{1}{2}$	2

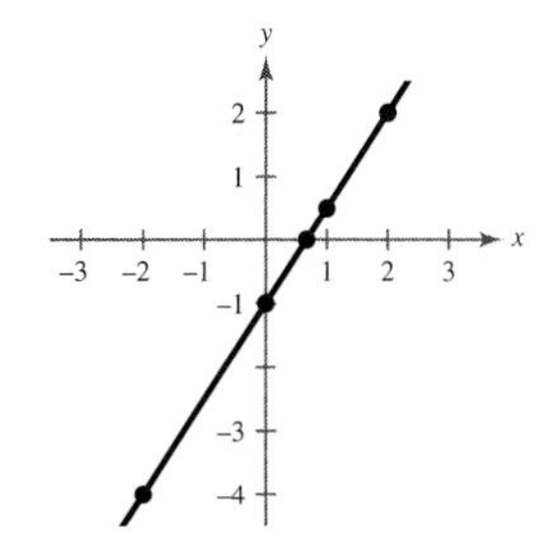

37.

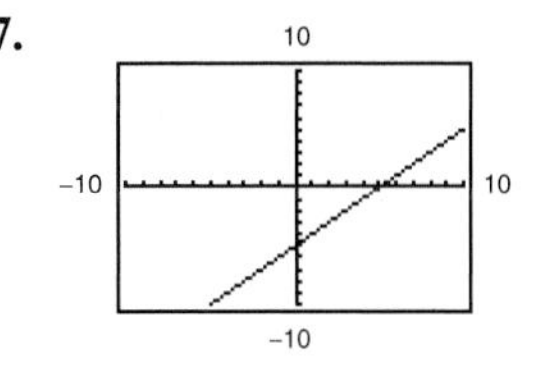

$(5, 0)$, $(0, -5)$

39.

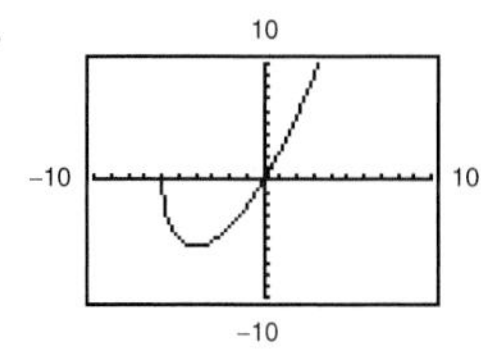

$(0, 0)$, $(-6, 0)$

41. y-axis symmetry **43.** y-axis symmetry

45. Origin symmetry

47.

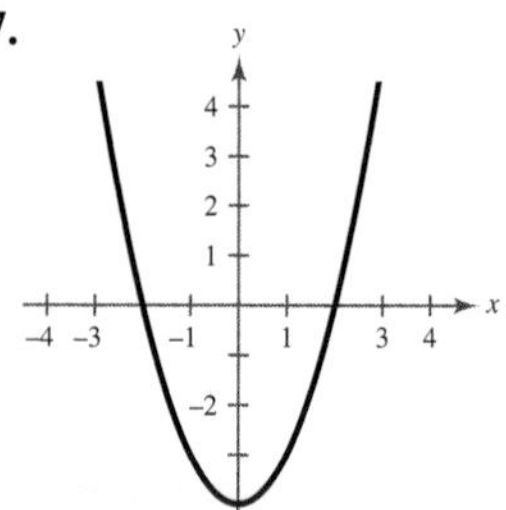

49.

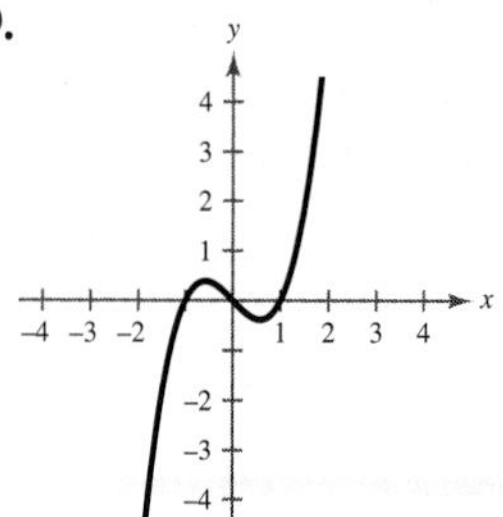

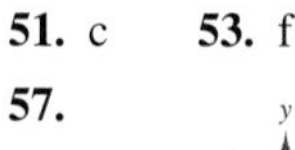

51. c **53.** f **55.** b

57.

No symmetry

59.

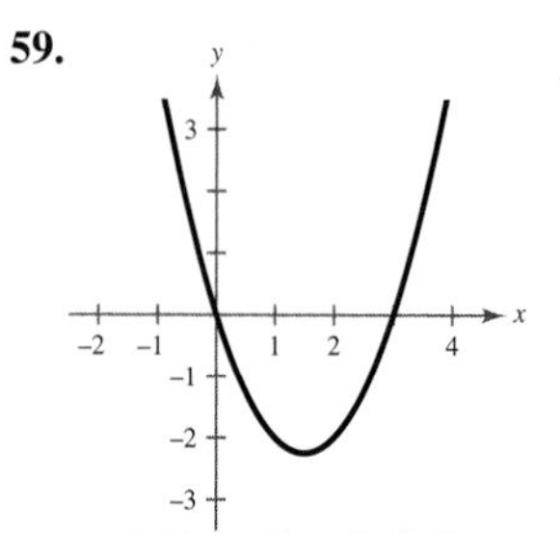

No symmetry

61.

No symmetry

63.

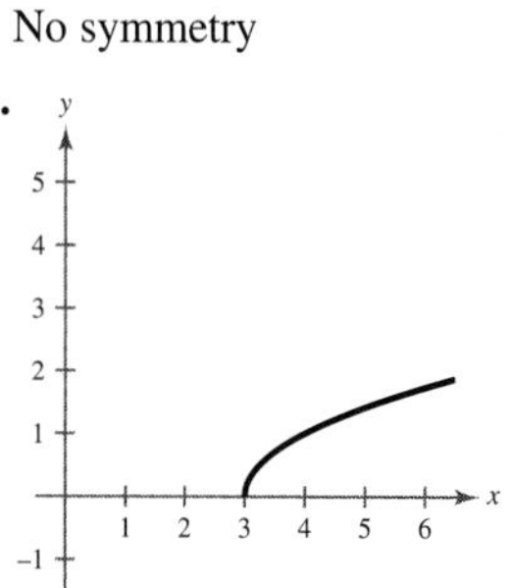

No symmetry

65.

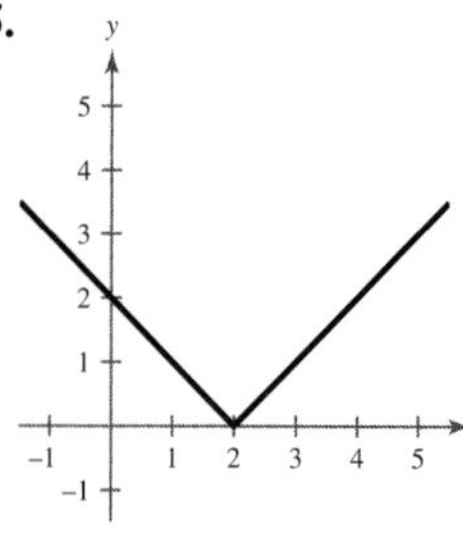

No symmetry

67.

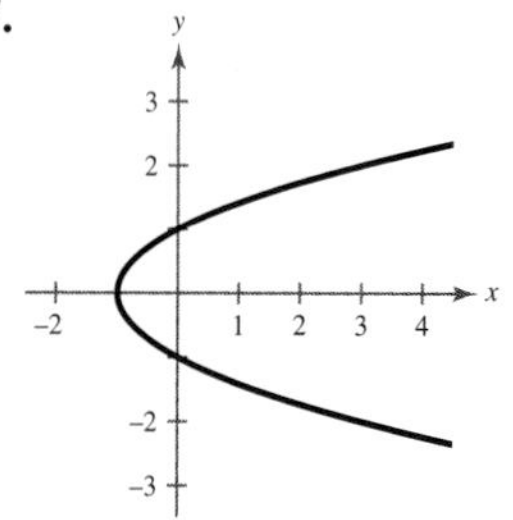

Symmetry: x-axis

69.

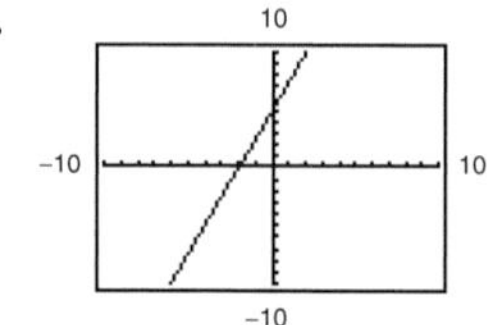

The standard setting gives a more complete graph.

71.

Xmin = -5
Xmax = 5
Xscl = 1
Ymin = -30
Ymax = 30
Yscl = 10

73.

Xmin = -10
Xmax = 20
Xscl = 5
Ymin = -5
Ymax = 30
Yscl = 5

75.

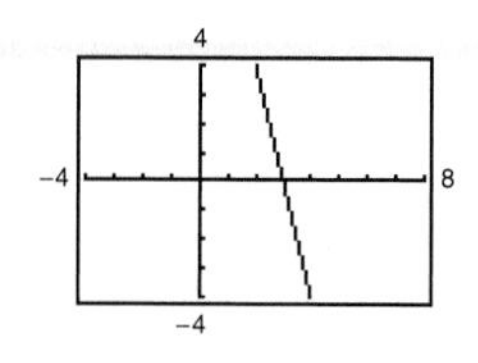

3

77.

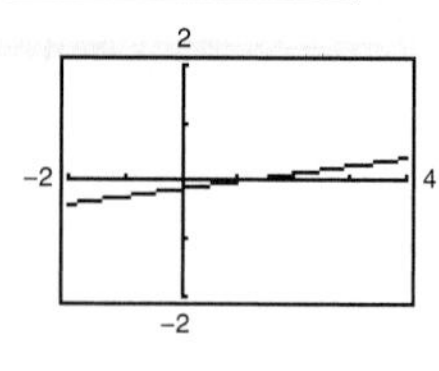

1

79.

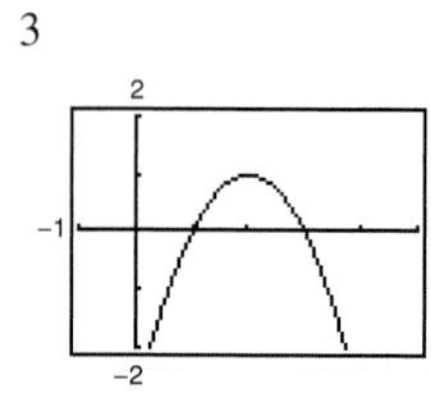

1, 3

81.

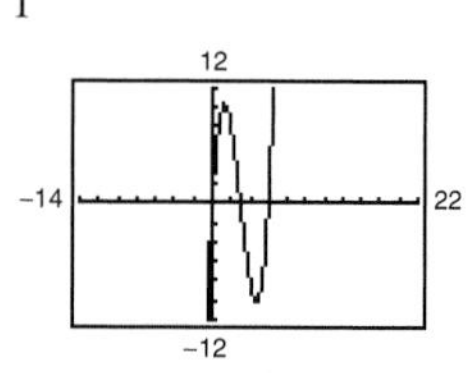

0, 3, 6

83. $x^2 + y^2 = 9$ **85.** $(x - 2)^2 + (y + 1)^2 = 16$

87. $(x + 1)^2 + (y - 2)^2 = 5$

89. $(x - 3)^2 + (y - 4)^2 = 25$

91.

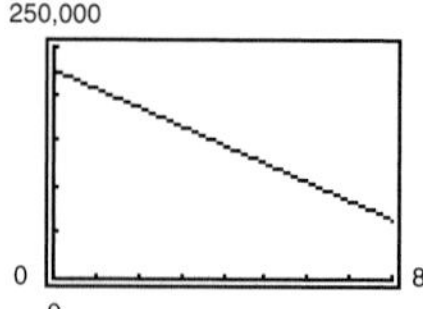

93. (a)

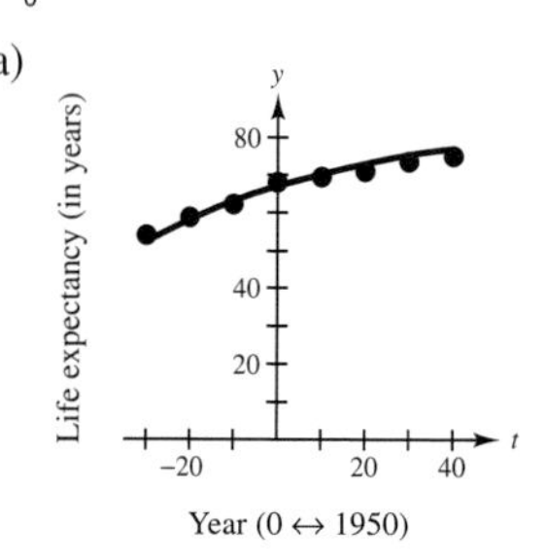

(b) 1998: 77.7; 2000: 78.0

Section P.4 *(page 51)*

Warm Up *(page 51)*

1. $-\frac{9}{2}$ **2.** $-\frac{13}{3}$ **3.** $-\frac{5}{4}$ **4.** $\frac{1}{2}$

5. $y = \frac{2}{3}x - \frac{5}{3}$ **6.** $y = -2x$ **7.** $y = 3x - 1$

8. $y = \frac{2}{3}x + 5$ **9.** $y = -2x + 7$ **10.** $y = x + 3$

1. (a) L_2 (b) L_3 (c) L_1

3.

5. $\frac{8}{5}$ **7.** 0 **9.** -4

11. $m = 2$

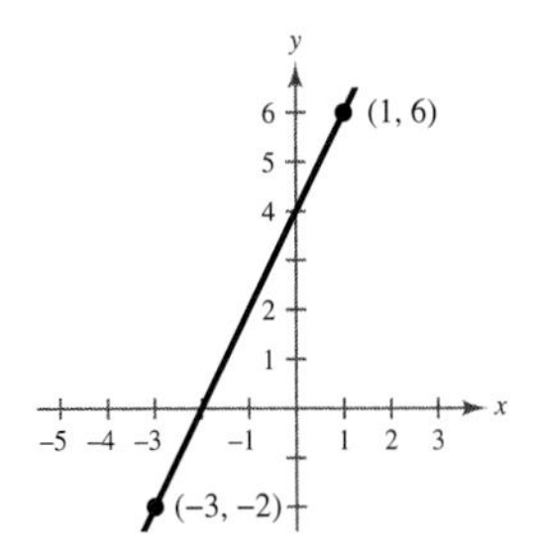

13. m is undefined.

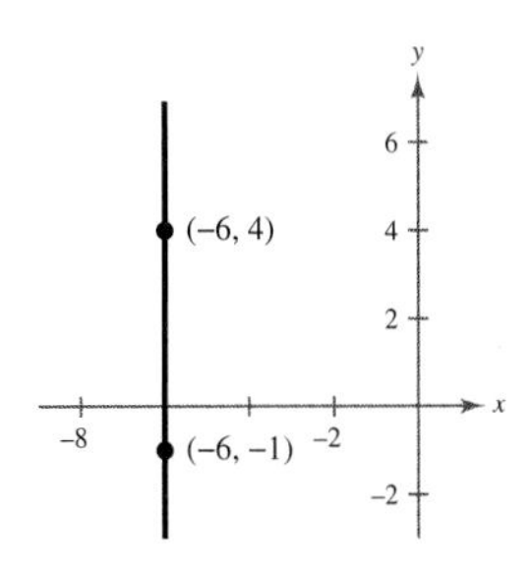

15. $m = \frac{4}{3}$

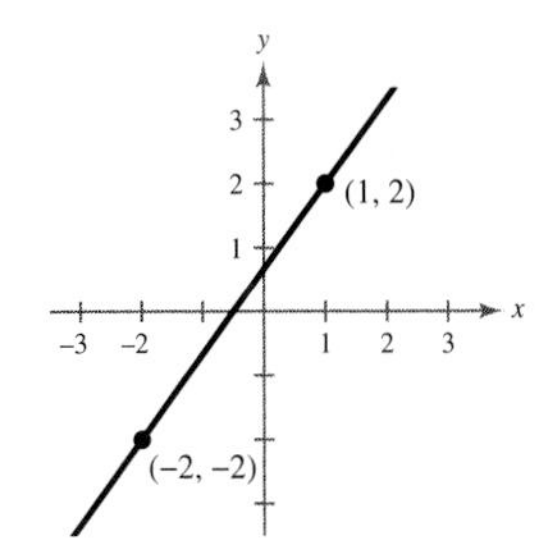

17. $(0, 1), (3, 1), (-1, 1)$ **19.** $(6, -5), (7, -4), (8, -3)$

21. $(-8, 0), (-8, 2), (-8, 3)$ **23.** Perpendicular

25. Parallel

27. Yes. The rate of change remains the same on a line.

29. (a) Sales increasing 135 units per year.

(b) No change in sales.

(c) Sales decreasing 40 units per year.

31. (a) 1989–1990 (b) 1988–1989

33. $16{,}666\frac{2}{3}$ feet $\approx$ 3.16 miles

35. $m = 5$; Intercept: $(0, 3)$

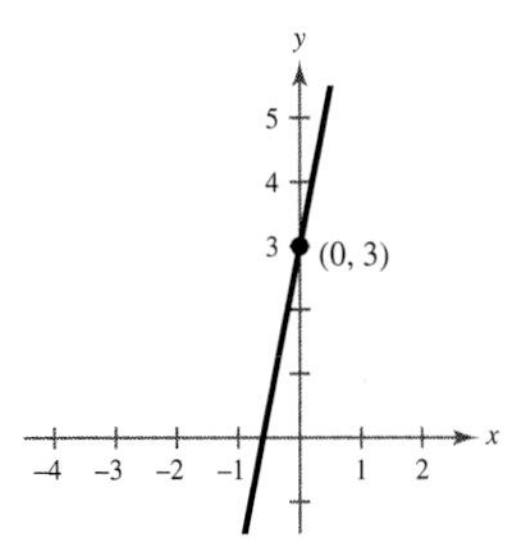

37. m is undefined. There is no y-intercept.

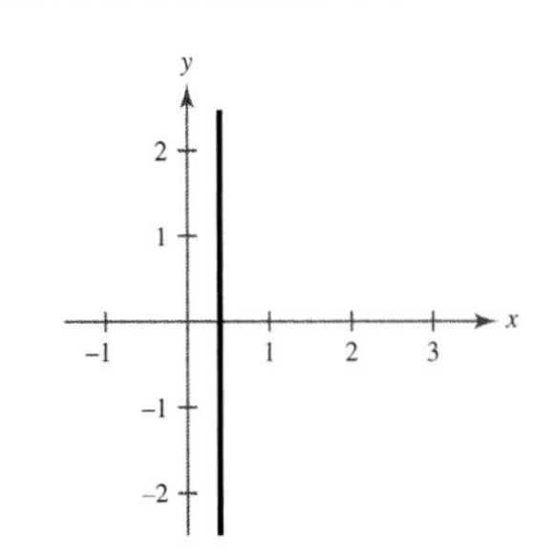

39. $m = -\frac{7}{6}$; Intercept: $(0, 5)$

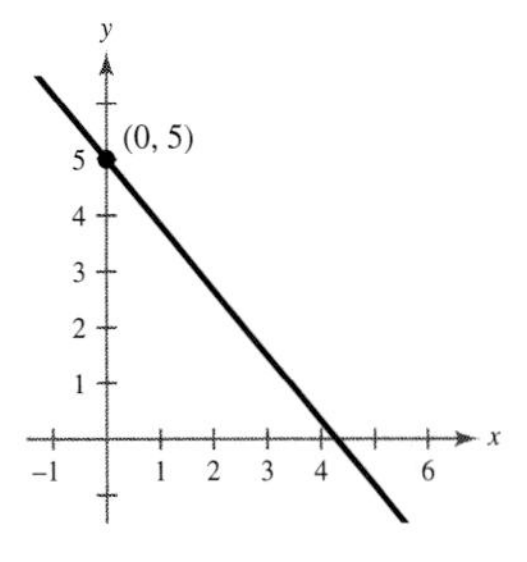

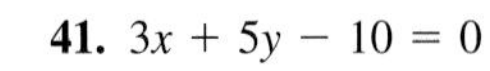

41. $3x + 5y - 10 = 0$

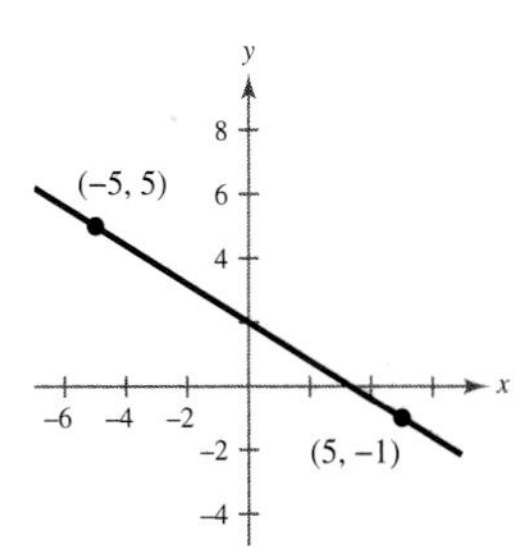

43. $x + 2y - 3 = 0$

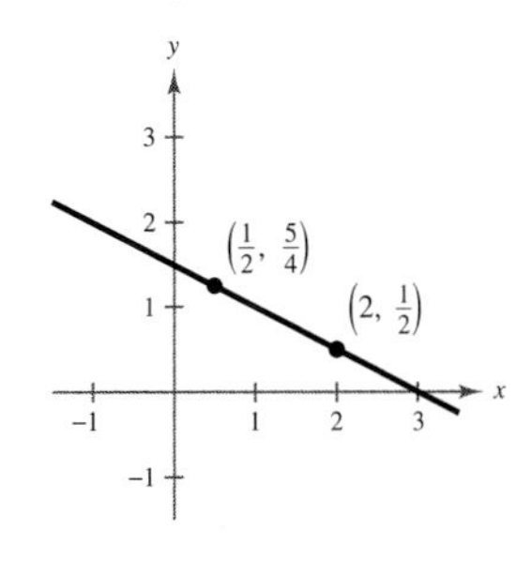

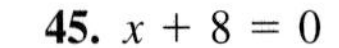

45. $x + 8 = 0$

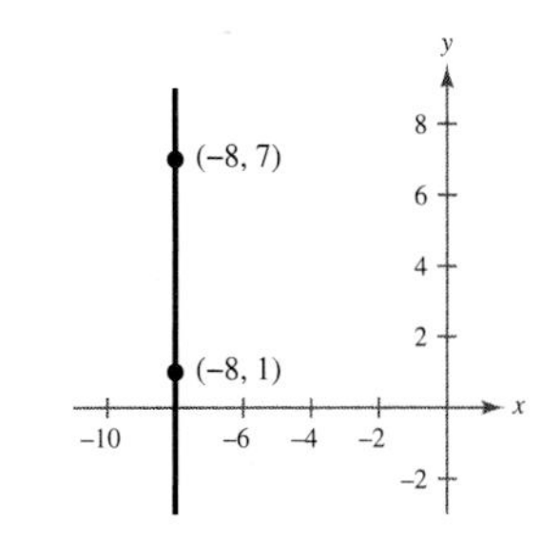

47. $2x - 5y + 1 = 0$

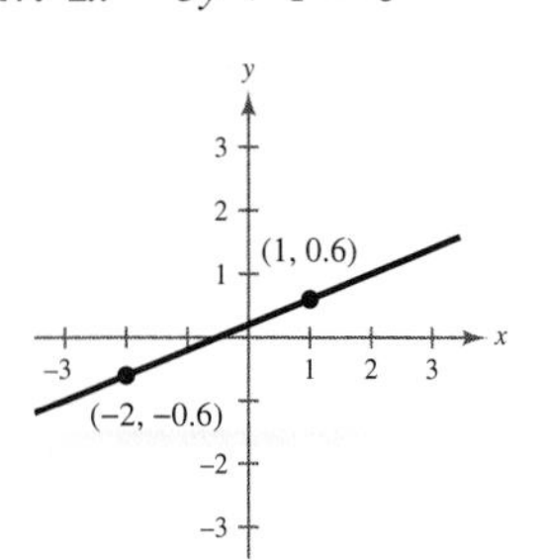

49. $3x - y - 2 = 0$

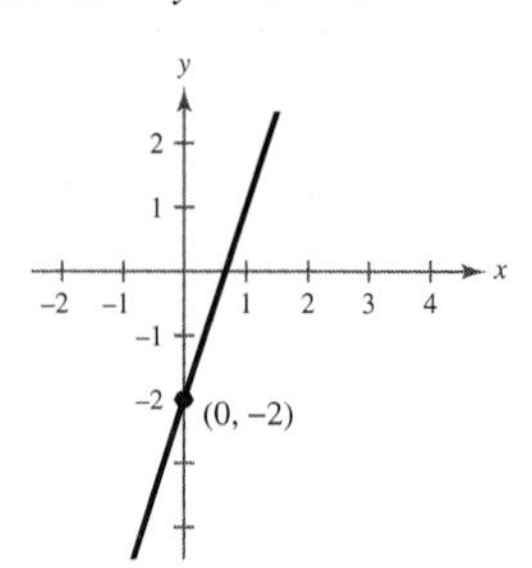

51. $2x + y = 0$

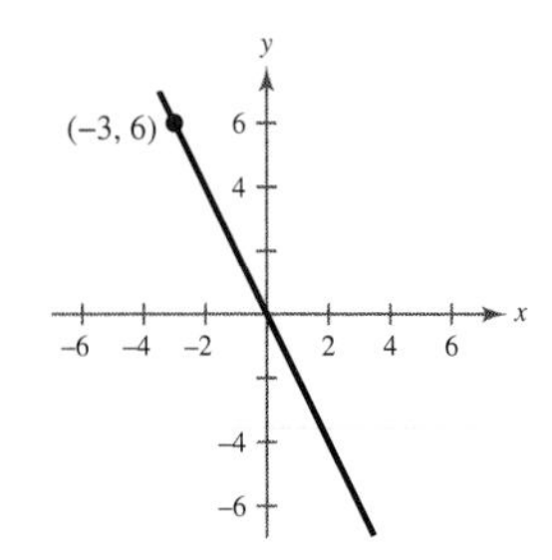

53. $x + 3y - 4 = 0$

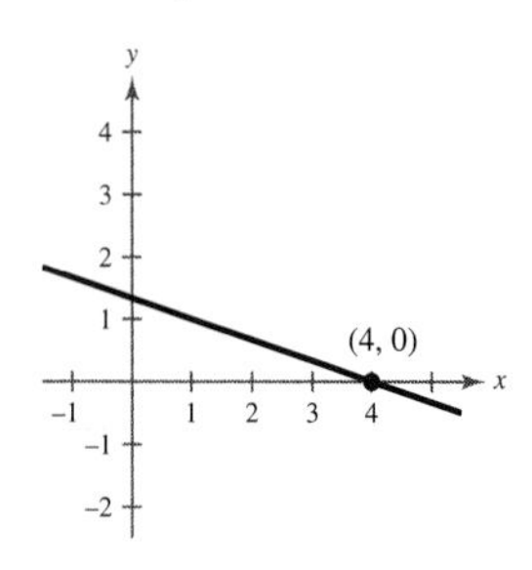

55. $x - 6 = 0$

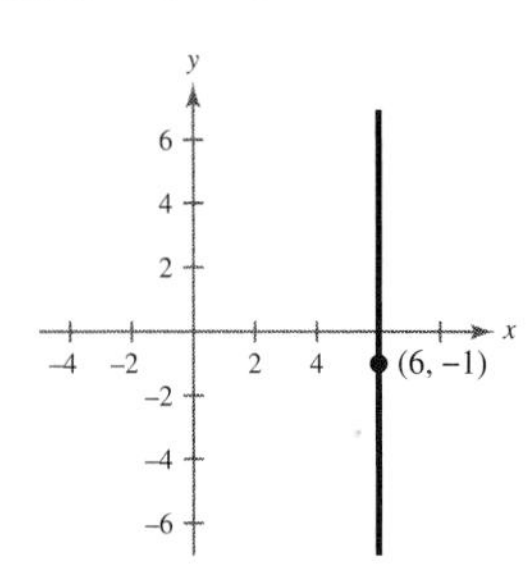

57. $8x - 6y - 17 = 0$

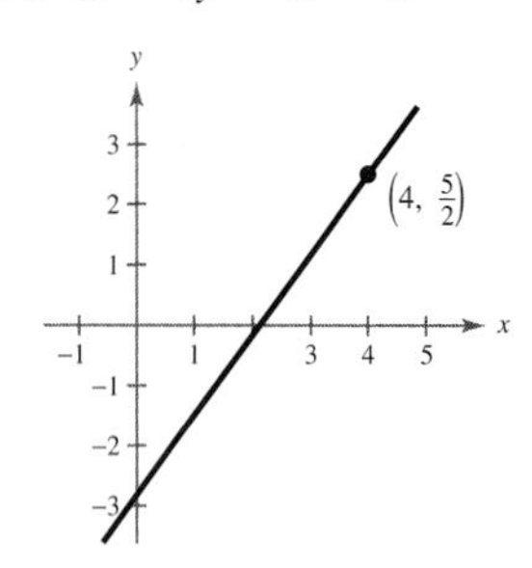

59. $3x + 2y - 6 = 0$ **61.** $12x + 3y + 2 = 0$

63. $x + y - 3 = 0$

65. (a) $2x - y - 3 = 0$ (b) $x + 2y - 4 = 0$

67. (a) $3x + 4y + 2 = 0$ (b) $4x - 3y + 36 = 0$

69. (a) $y = 0$ (b) $x + 1 = 0$

71. Parallel

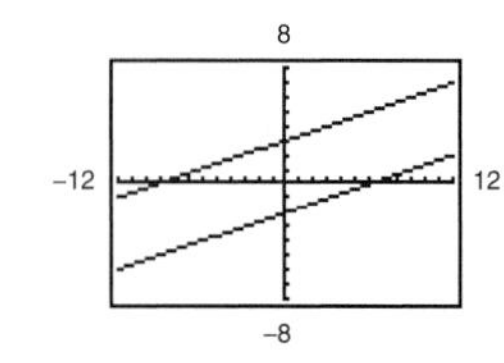

73. Neither

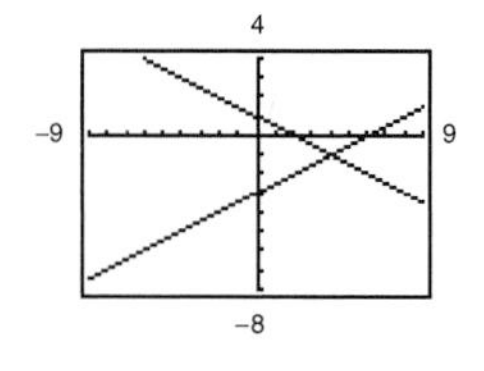

75. Perpendicular

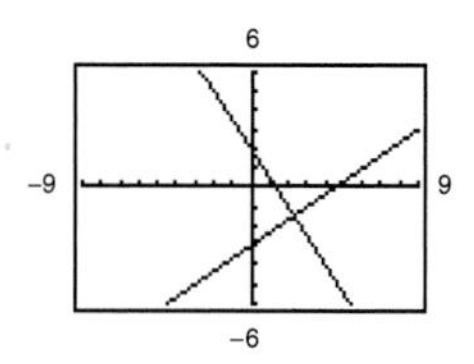

77.

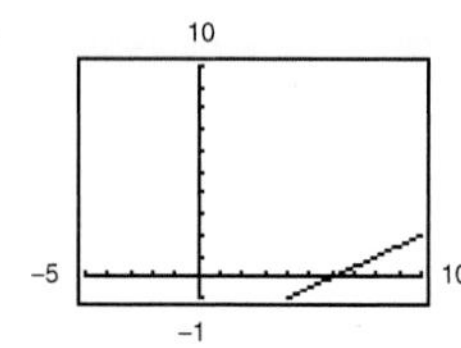

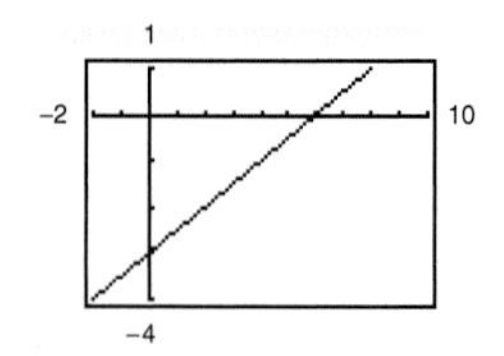

The second setting shows the x- and y-intercepts more clearly.

79.

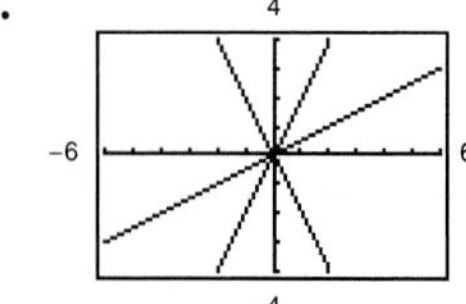

(b) is perpendicular to (c).

81.

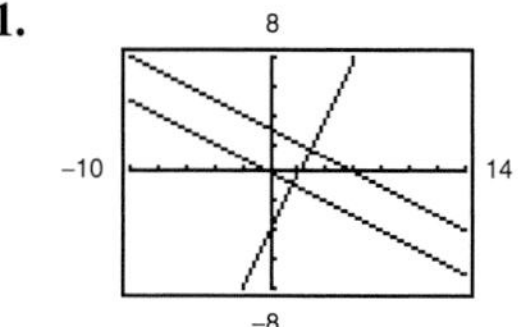

(a) is parallel to (b).

(c) is perpendicular to (a) and (b).

83. $V = 1790 + 125t$ **85.** b **87.** a

89. $3x - 2y - 1 = 0$

91. $F = \frac{9}{5}C + 32$ **93.** $39,500

95. $V = -175t + 875$ **97.** $S = 0.85L$

99. (a) $C = 16.75t + 36{,}500$ (b) $R = 27t$
(c) $P = 10.25t - 36{,}500$ (d) $t \approx 3561$ hours

101. (a)

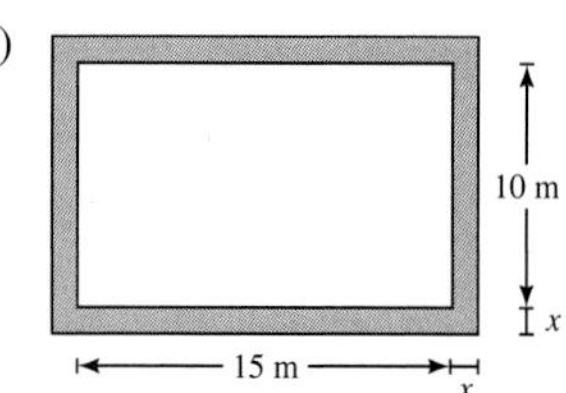

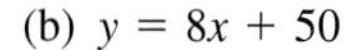

(b) $y = 8x + 50$

(c) (d) 8 meters

103. $C = 120 + 0.26x$ **105.** $y = 91x + 164$

107. d **109.** a

Section P.5 *(page 65)*

Warm Up *(page 65)*

1. -73 **2.** 13 **3.** $2(x + 2)$ **4.** $-8(x - 2)$
5. $y = \frac{7}{5} - \frac{2}{5}x$ **6.** $y = \pm x$ **7.** $x \le \frac{9}{2}$
8. $x \ge -\frac{2}{3}$ **9.** $-3 \le x \le 3$ **10.** $x \le 1,\ x \ge 2$

1. Yes **3.** No **5.** Yes **7.** No

9. (a) Function
(b) Not a function, because the element 1 in A corresponds to two elements, -2 and 1, in B.
(c) Function
(d) Not a function, because not every element in A is matched with an element in B.

11. Each is a function. For each year there corresponds one and only one circulation.

13. Not a function **15.** Function **17.** Function

19. Not a function **21.** Function

23. (a) 4 (b) 0 (c) $4x$ (d) $(x + c)$

25. (a) -1 (b) -9 (c) $2x - 5$

27. (a) 0 (b) -0.75 (c) $x^2 + 2x$

29. (a) 1 (b) 2.5 (c) $3 - 2|x|$

31. (a) $-\frac{1}{9}$ (b) Undefined (c) $\dfrac{1}{y^2 + 6y}$

33. (a) 1 (b) -1 (c) $\dfrac{|x - 1|}{x - 1}$

35. (a) -1 (b) 2 (c) 6

37.

x	-2	-1	0	1	2
$f(x)$	1	-2	-3	-2	1

39.

t	-5	-4	-3	-2	-1
$h(t)$	1	$\frac{1}{2}$	0	$\frac{1}{2}$	1

41.

x	-2	-1	0	1	2
$f(x)$	5	$\frac{9}{2}$	4	1	0

43. 5 **45.** ± 3 **47.** $2, -1$ **49.** $3, 0$

51. All real numbers x **53.** All real numbers $t \ne 0$

55. $y \ge 10$ **57.** $-1 \le x \le 1$

59. All real numbers $x \ne 0, -2$

61. $\{(-2, 4), (-1, 1), (0, 0), (1, 1), (2, 4)\}$

63. $\{(-2, 0), (-1, 1), (0, \sqrt{2}), (1, \sqrt{3}), (2, 2)\}$

65. The domain is the set of inputs of the function, and the range is the set of outputs.

67. $g(x) = -2x^2$ **69.** $r(x) = \dfrac{32}{x}$ **71.** $3 + h, h \ne 0$

73. $3x^2 + 3xc + c^2, c \ne 0$ **75.** $3, x \ne 0$ **77.** $A = \dfrac{C^2}{4\pi}$

79. (a)

Height x	Width	Volume V
1	$24 - 2(1)$	$1[24 - 2(1)]^2 = 484$
2	$24 - 2(2)$	$2[24 - 2(2)]^2 = 800$
3	$24 - 2(3)$	$3[24 - 2(3)]^2 = 972$
4	$24 - 2(4)$	$4[24 - 2(4)]^2 = 1024$
5	$24 - 2(5)$	$5[24 - 2(5)]^2 = 980$
6	$24 - 2(6)$	$6[24 - 2(6)]^2 = 864$

Maximum when $x = 4$

(b)

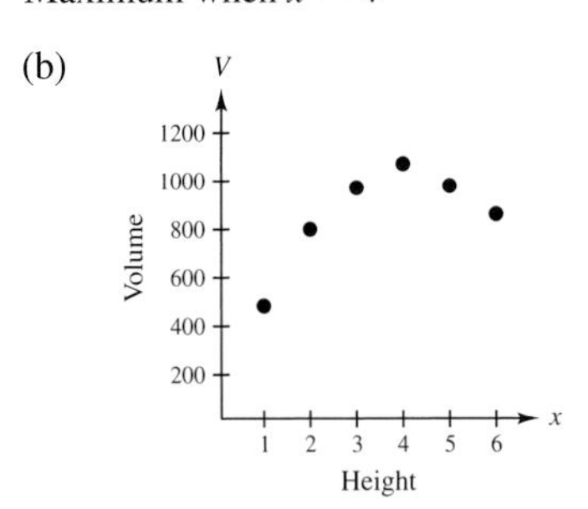

V is a function of x.

(c) $V = x(24 - 2x)^2,\ 0 < x < 12$

81. $A = \dfrac{x^2}{2(x - 2)},\ x > 2$

83. $V = x^2y = x^2(108 - 4x) = 108x^2 - 4x^3$, $0 < x < 27$

85. (a) $C = 12.30x + 98{,}000$
(b) $R = 17.98x$
(c) $P = 5.68x - 98{,}000$

87. (a) $R = \dfrac{240n - n^2}{20}, n \geq 80$

(b)

n	90	100	110	120	130	140	150
$R(n)$	\$675	\$700	\$715	\$720	\$715	\$700	\$675

The revenue is maximum when 120 people take the trip.

89. (a)

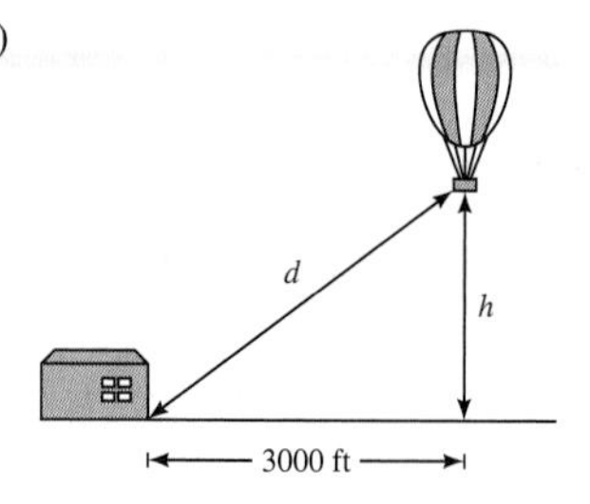

(b) $h = \sqrt{d^2 - 3000^2}$, $[3000, \infty)$

91. $\frac{15}{8}$ **93.** $-\frac{1}{5}$

Section P.6 *(page 78)*

Warm Up *(page 78)*

1. -8 **2.** 0 **3.** $-\dfrac{3}{x}$ **4.** $x^2 + 3$ **5.** $0, \pm 4$
6. $\frac{1}{2}, 1$ **7.** All real numbers $x \neq 4$
8. All real numbers $x \neq 4, 5$ **9.** $t \leq \frac{5}{3}$
10. All real numbers

1. Domain: all real numbers
Range: $(-\infty, 1]$

3. Domain: $(-\infty, -1], [1, \infty)$
Range: $[0, \infty)$

5. Domain: $[-4, 4]$
Range: $[0, 4]$

7. Function **9.** Not a function **11.** Function

13. Yes. For each value of y there corresponds one and only one value of x.

15. Second setting **17.** First setting

19. (a) Increasing on $(-\infty, \infty)$ (b) Odd function

21. (a) Increasing on $(-\infty, 0)$, $(2, \infty)$
Decreasing on $(0, 2)$
(b) Neither even nor odd

23. (a)

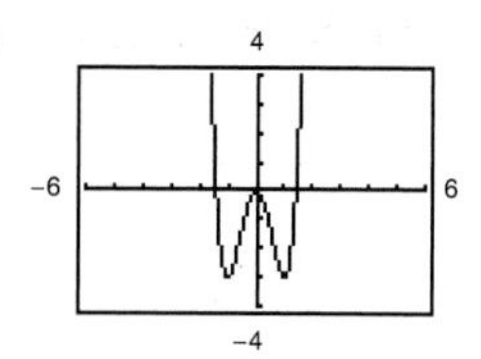

(b) Increasing on $(-1, 0)$, $(1, \infty)$
Decreasing on $(-\infty, -1)$, $(0, 1)$
(c) Even function

25. (a)

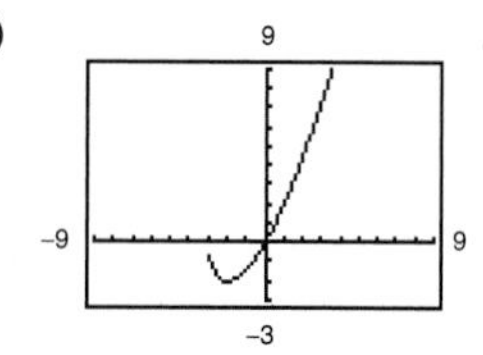

(b) Increasing on $(-2, \infty)$
Decreasing on $(-3, -2)$
(c) Neither even nor odd

27. Even **29.** Odd **31.** Neither even nor odd

33. (a) $\left(\frac{3}{2}, 4\right)$ (b) $\left(\frac{3}{2}, -4\right)$

35. Even

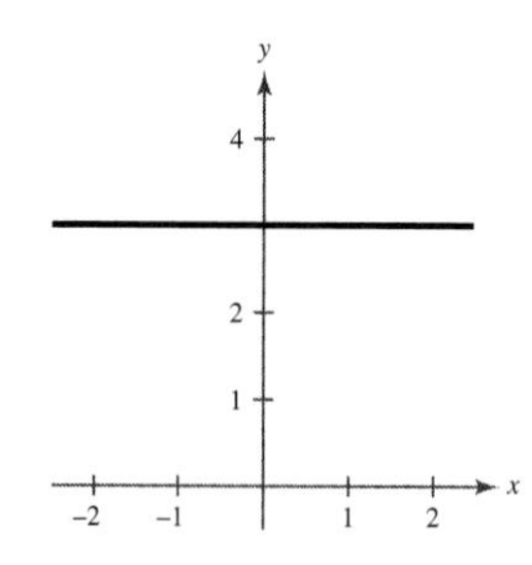

37. Neither even nor odd

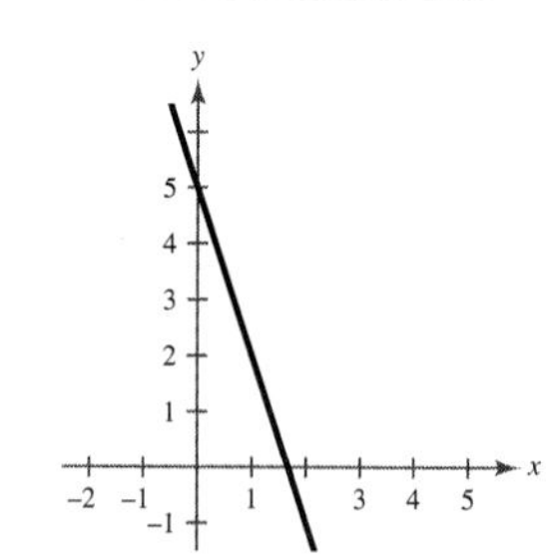

39. Even

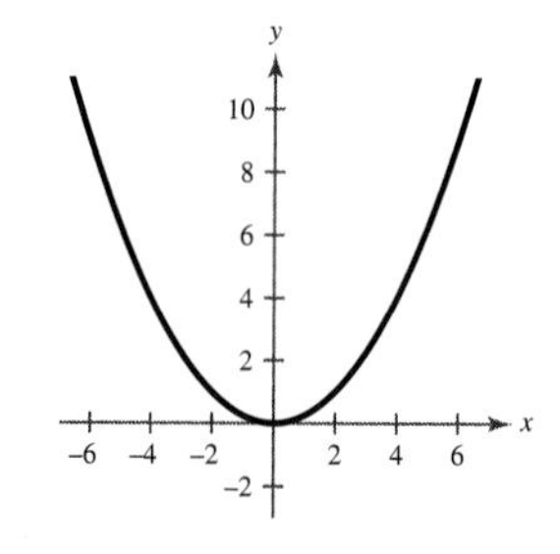

41. Neither even nor odd

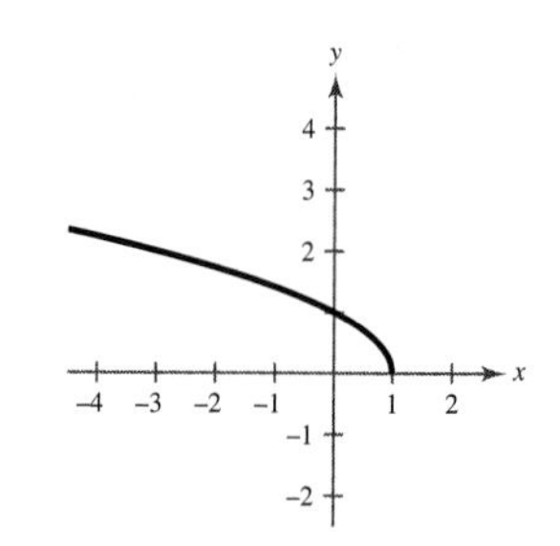

43. Neither even nor odd

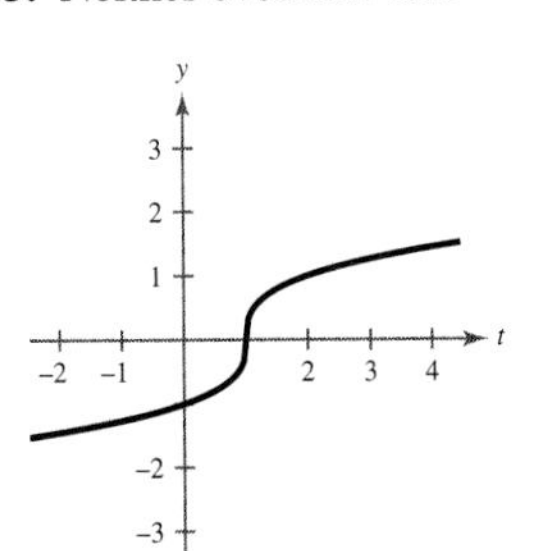

45. Neither even nor odd

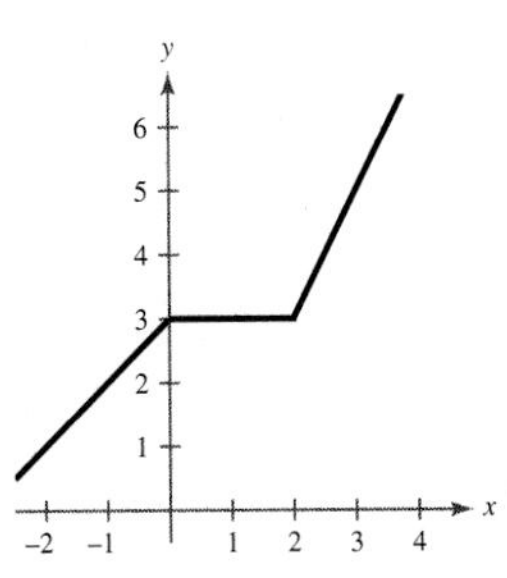

47. $(-\infty, 4]$

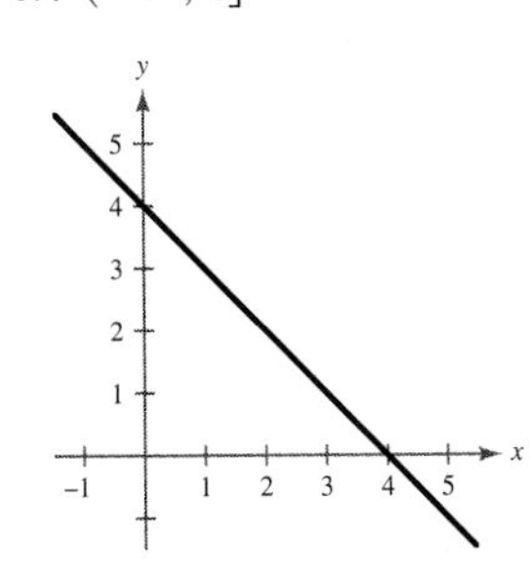

49. $(-\infty, -3]$, $[3, \infty)$

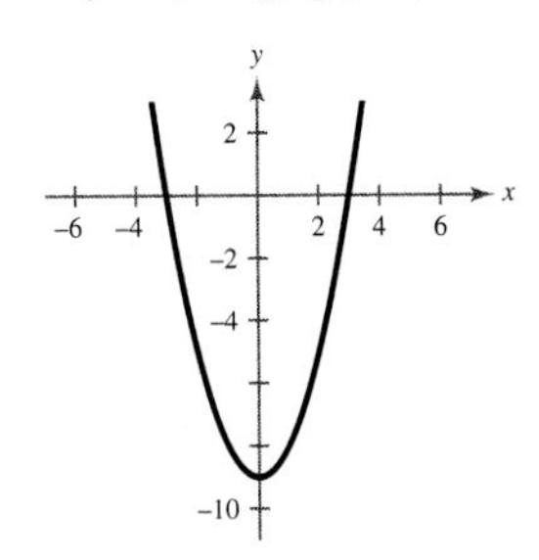

51. $[-1, 1]$

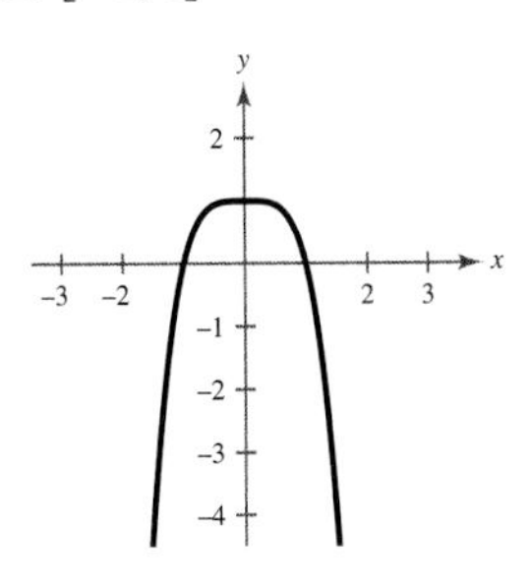

53. $(-\infty, \infty)$

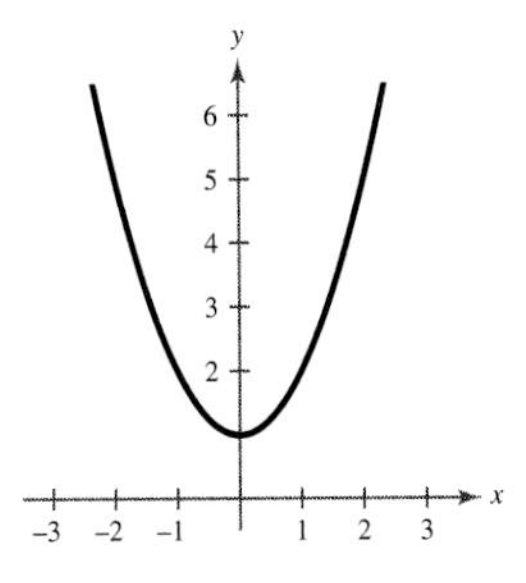

55. $f(x) < 0$ for all x

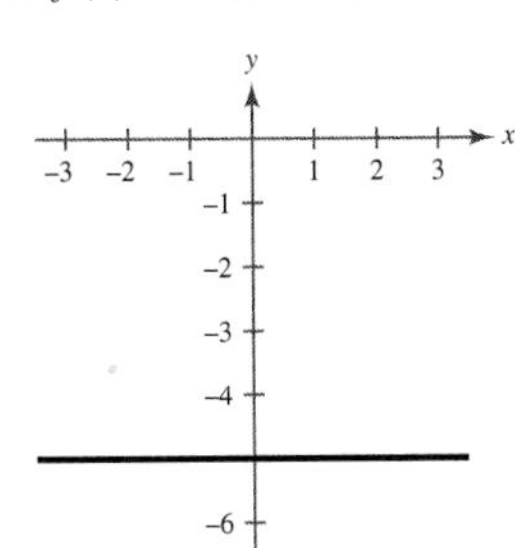

57.

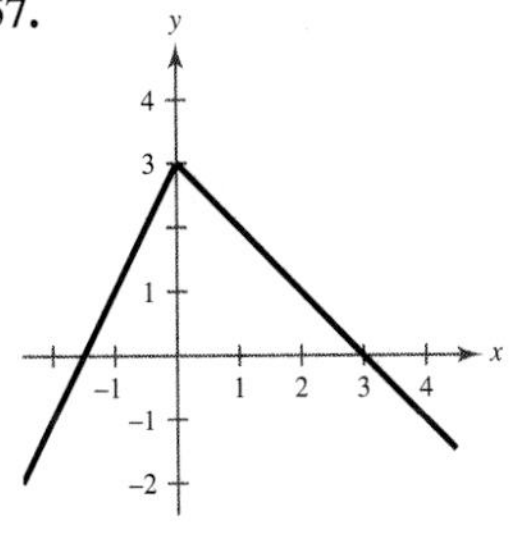

59.

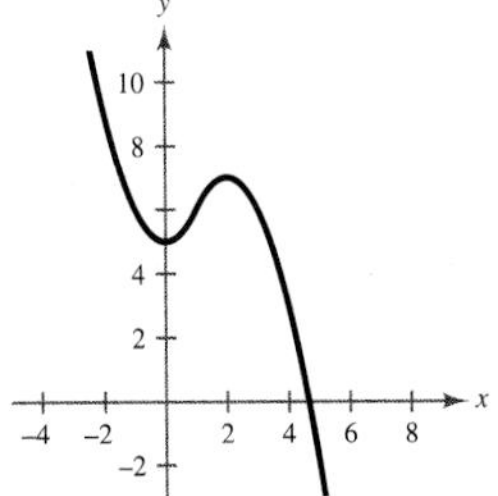

61.

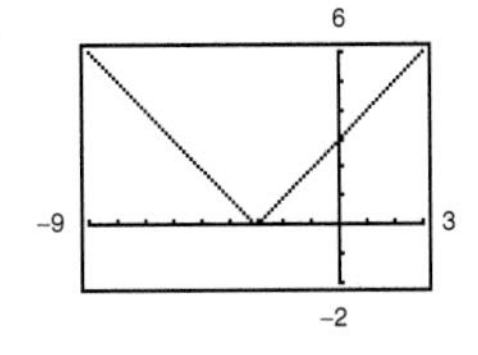

Domain: all real numbers

Range: $[0, \infty)$

63.

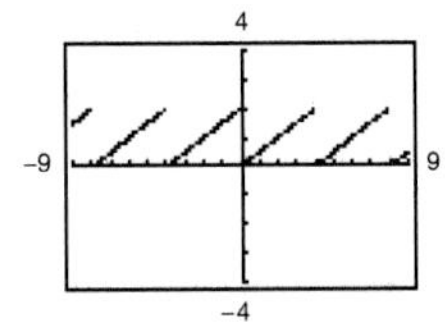

Domain: $(-\infty, \infty)$

Range: $[0, 2)$

Sawtooth pattern

65. (a)

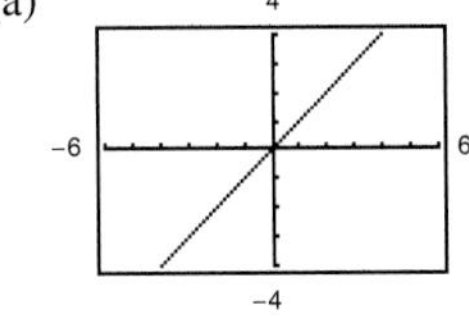

(b)

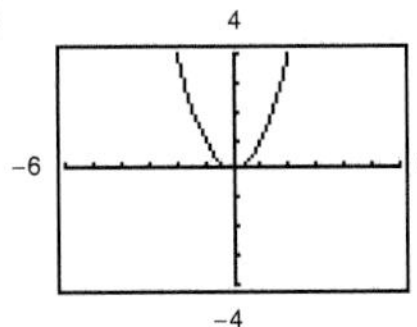

(c)

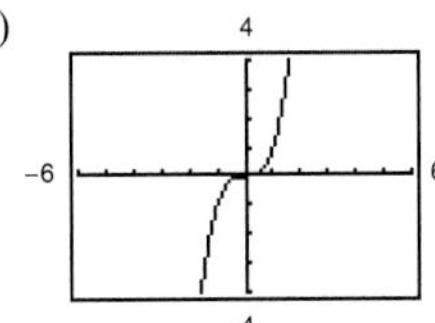

(d)

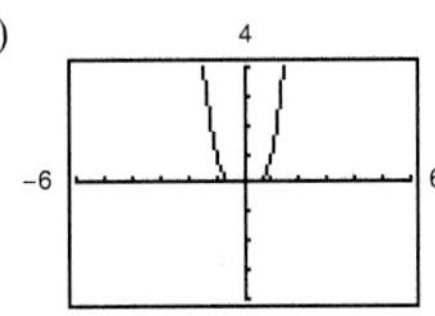

(e)

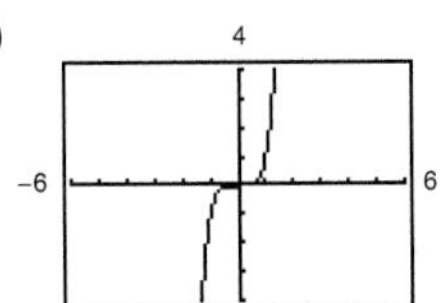

(f)

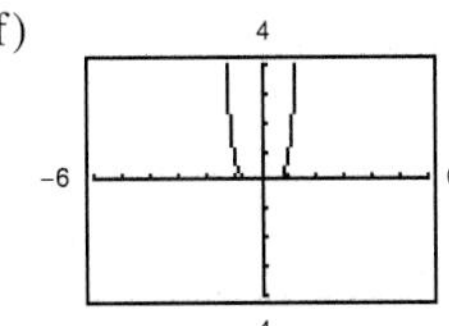

All the graphs pass through the origin. The graphs of the odd powers of x are symmetric with respect to the origin, and the graphs of the even powers are symmetric with respect to the y-axis. As the powers increase, the graphs become flatter in the interval $-1 < x < 1$.

67. (a) C_2 is the appropriate model, because the cost does not increase until after the next minute of conversation has started.

(b) $7.85

69. 350,000 units

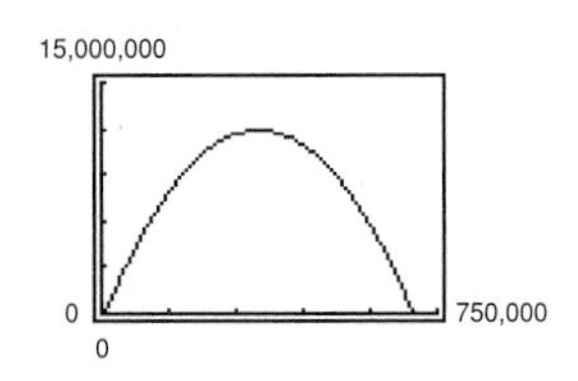

71. $h = -x^2 + 4x - 3$ **73.** $h = 2x - x^2$

75. $L = \frac{1}{2}y^2$ **77.** $L = 4 - y^2$

79. (a) Domain: $-4 \le t \le 3$

(b)

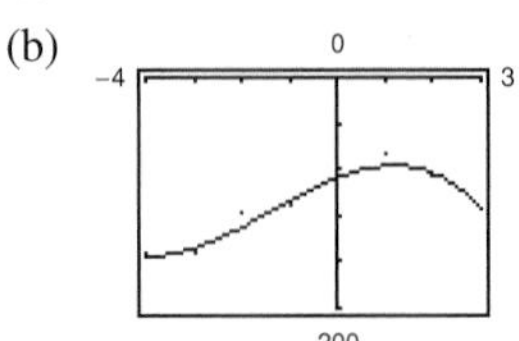

(c) 1986, 1990

(d) Because the balance would continue to decrease.

81. (a) \$10,000 (b) 50,000,000 (c) 1%

83. Answers will vary. **85.** 0, 10 **87.** $0, \pm i$

Section P.7 *(page 92)*

Warm Up *(page 92)*

1. $\dfrac{1}{x(1-x)}$ 2. $-\dfrac{12}{(x+3)(x-3)}$ 3. $\dfrac{3x-2}{x(x-2)}$

4. $\dfrac{4x-5}{3(x-5)}$ 5. $\dfrac{\sqrt{x^2-1}}{x+1}, x \ne 1$ 6. $\dfrac{x+1}{x(x+2)}, x \ne 2$

7. $5(x-2), x \ne -2$ 8. $\dfrac{x+1}{(x-2)(x+3)}, x \ne -5, -1, 0$

9. $\dfrac{1+5x}{3x-1}, x \ne 0$ 10. $\dfrac{x+4}{4x}, x \ne 4$

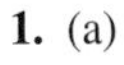

1. (a)

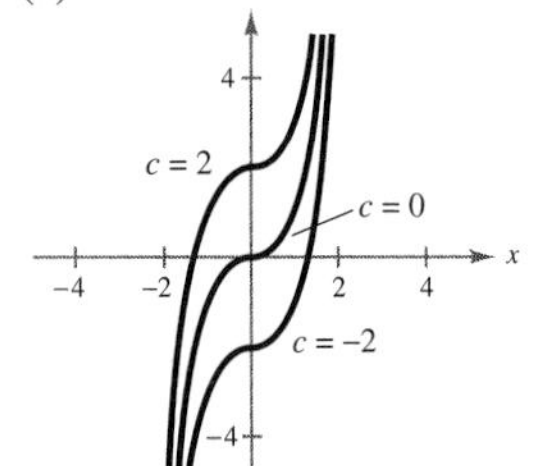

(b)

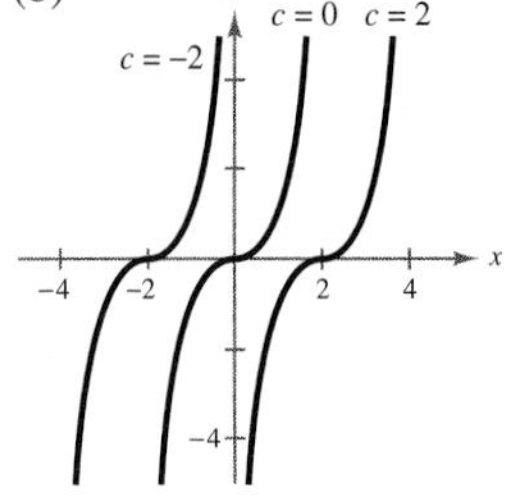

3. (a)

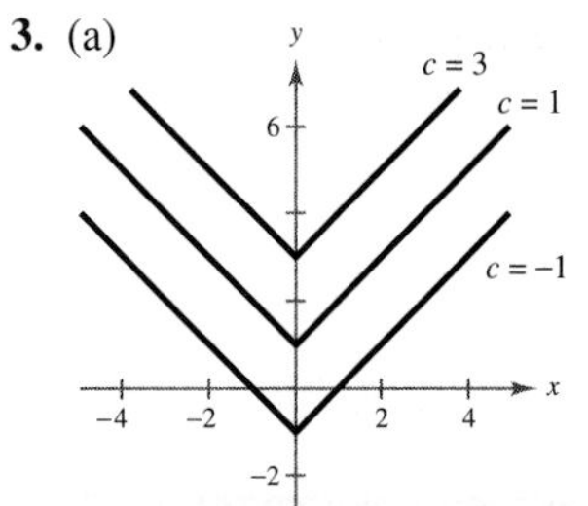

(b)

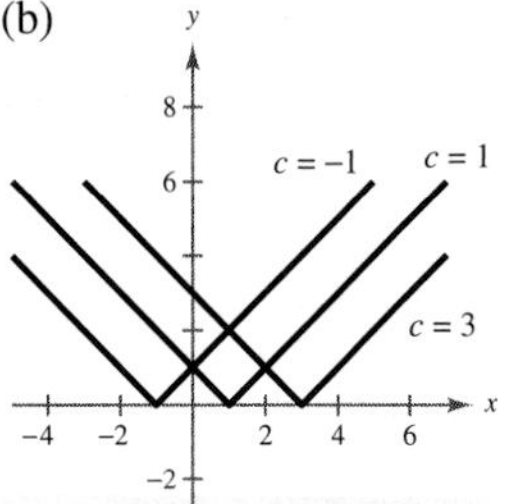

(c)

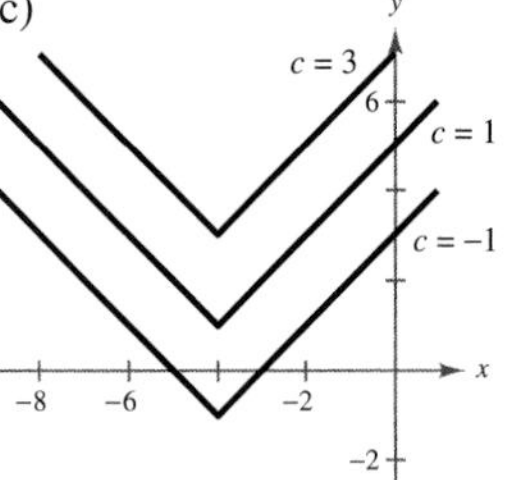

5. (a)

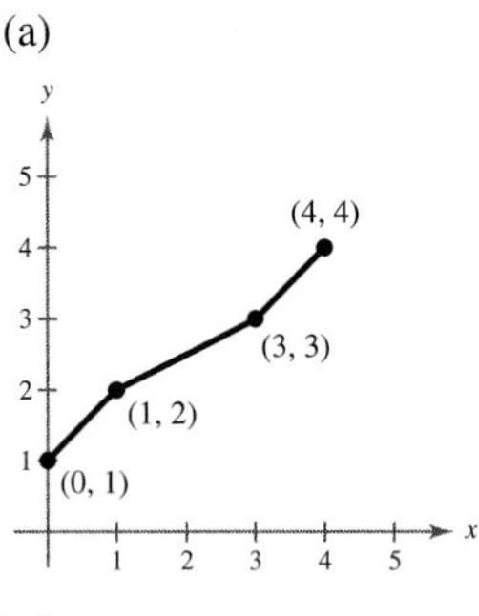

(b)

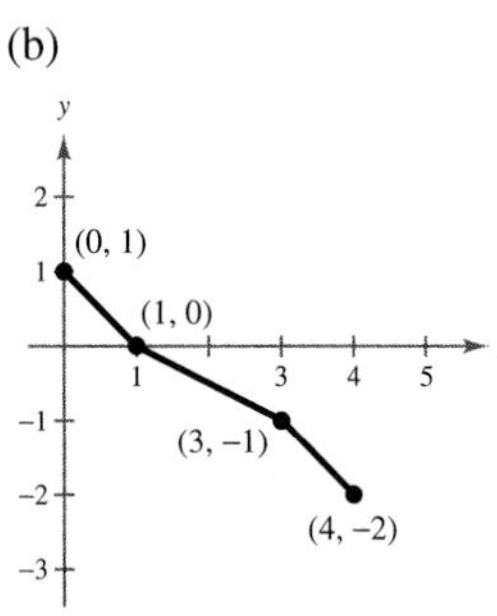

(c)

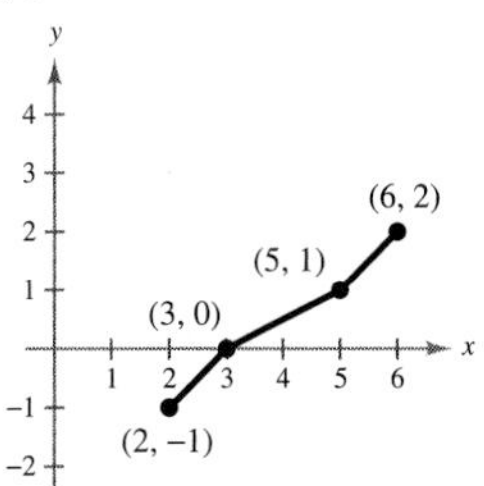

(d)

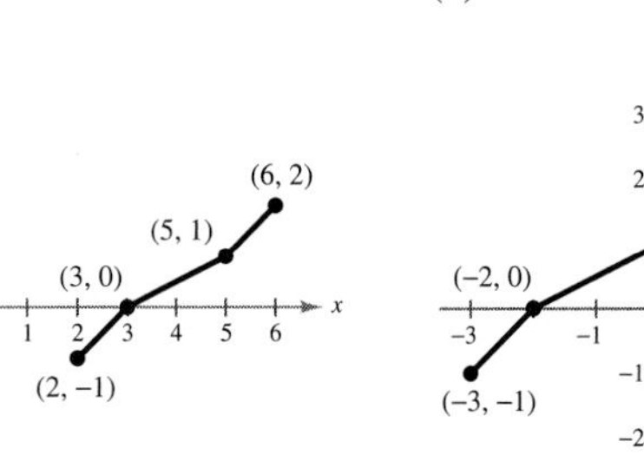

(e)

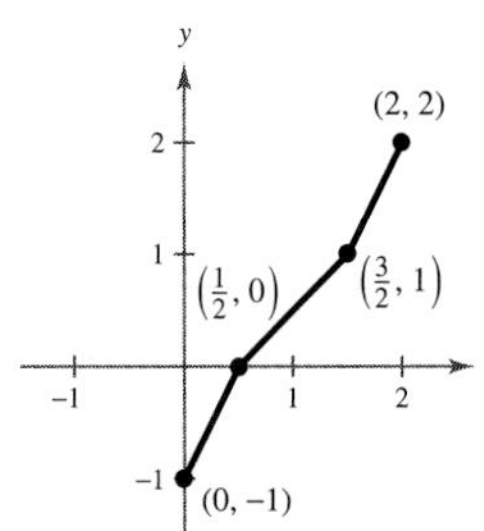

(f)

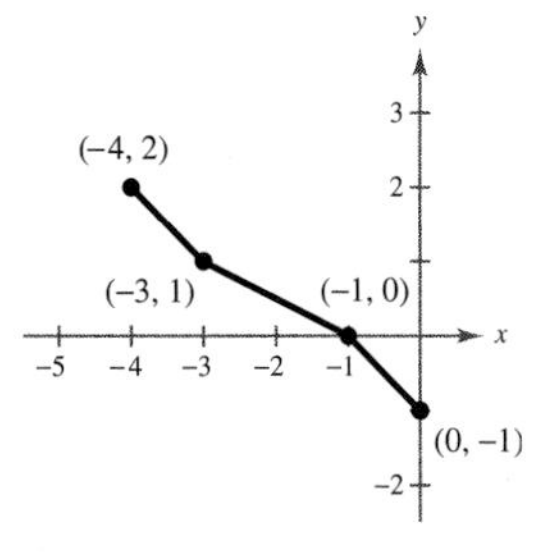

7. (a) $y = x^2 - 1$ (b) $y = 1 - (x + 1)^2$

9. Horizontal shift of $y = x^3$

$y = (x - 2)^3$

11. Reflection in the x-axis of $y = x^2$

$y = -x^2$

13. Reflection in the x-axis and a vertical shift of $y = \sqrt{x}$

$y = 1 - \sqrt{x}$

15.

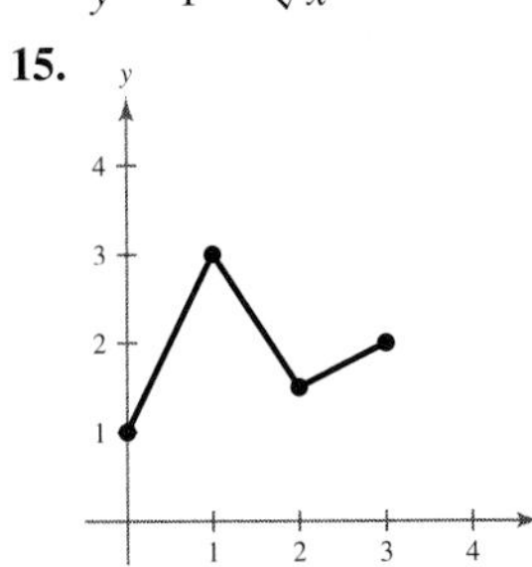

17.

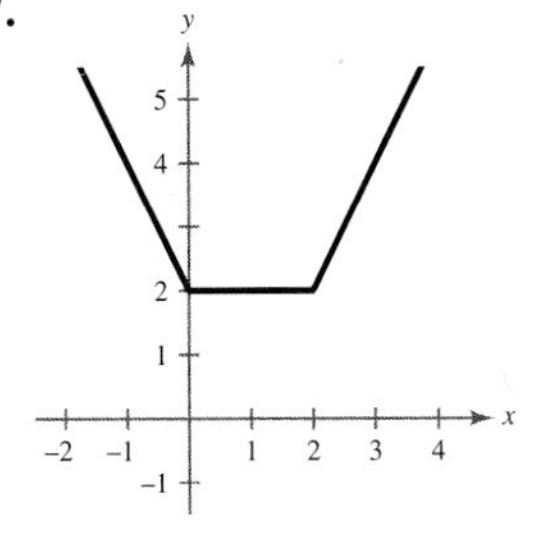

19. (a) $2x$ (b) 2 (c) $x^2 - 1$ (d) $\dfrac{x+1}{x-1}$, $x \neq 1$

21. (a) $x^2 - x + 1$ (b) $x^2 + x - 1$

(c) $x^2 - x^3$ (d) $\dfrac{x^2}{1-x}$, $x \neq 1$

23. (a) $x^2 + 5 + \sqrt{1-x}$ (b) $x^2 + 5 - \sqrt{1-x}$

(c) $(x^2 + 5)\sqrt{1-x}$ (d) $\dfrac{x^2+5}{\sqrt{1-x}}$, $x < 1$

25. (a) $\dfrac{x+1}{x^2}$ (b) $\dfrac{x-1}{x^2}$ (c) $\dfrac{1}{x^3}$ (d) x, $x \neq 0$

27. 9 **29.** 5 **31.** $4t^2 - 2t + 5$ **33.** 0

35. 26 **37.** $\frac{3}{5}$

39.

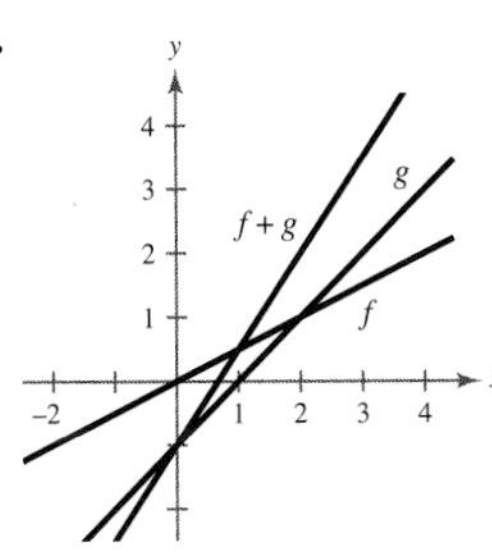

41.

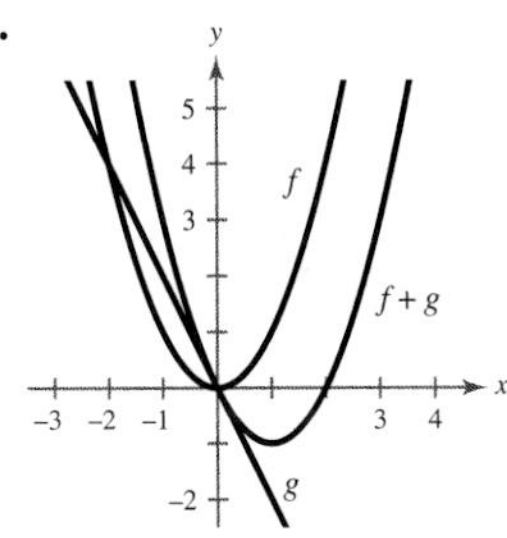

43.

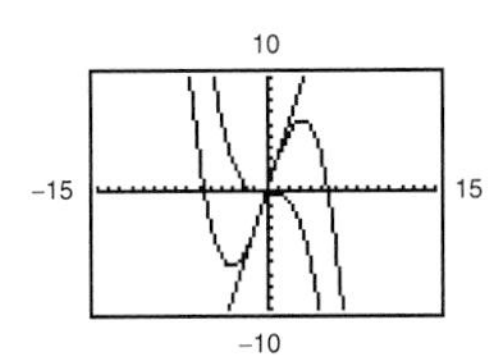

$f(x)$, $g(x)$

45. $T = \frac{3}{4}x + \frac{1}{15}x^2$

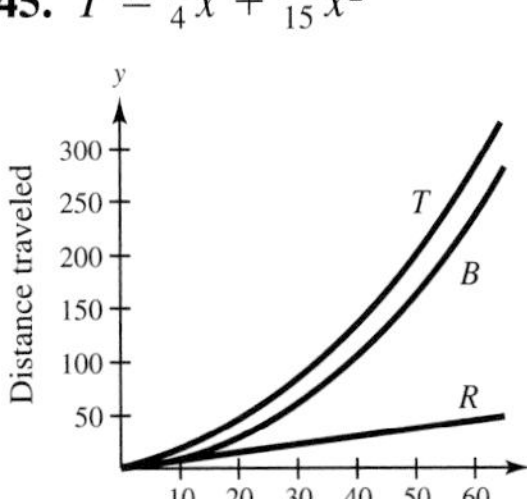

47.

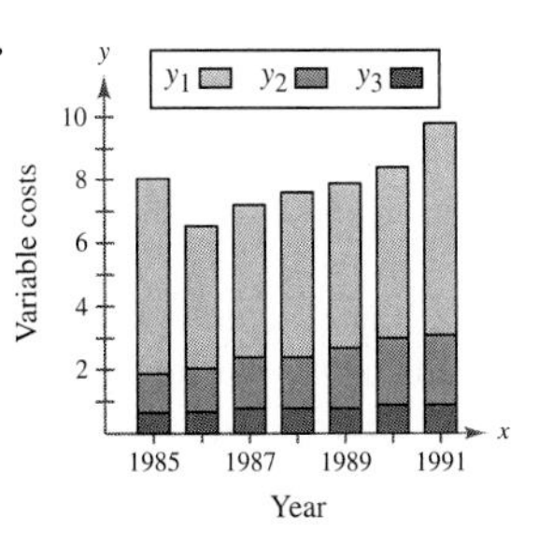

49. (a) For each time t there corresponds one and only one temperature T.

(b) 60°, 72°

(c) All the temperature changes would be 1 hour later.

(d) The temperature would be decreased by 1 degree.

51. (a) $(x - 1)^2$ (b) $x^2 - 1$ (c) x^4

53. (a) $20 - 3x$ (b) $-3x$ (c) $9x + 20$

55. (a) $\sqrt{x^2 + 4}$ (b) $x + 4, x \geq -4$

57. (a) $x - \frac{8}{3}$ (b) $x - 8$ **59.** (a) $\sqrt[4]{x}$ (b) $\sqrt[4]{x}$

61. (a) $|x + 6|$ (b) $|x| + 6$

63. (a) 3 (b) 0 **65.** (a) 0 (b) 4

67.

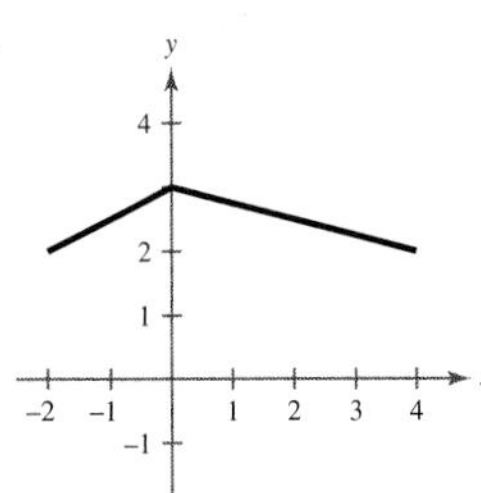

69.

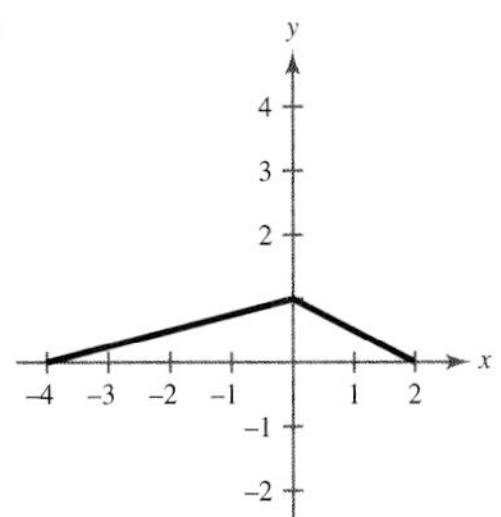

71. $f(x) = x^2$, $g(x) = 2x + 1$

73. $f(x) = \sqrt[3]{x}$, $g(x) = x^2 - 4$

75. $f(x) = \dfrac{1}{x}$, $g(x) = x + 2$

77. (a) $x \geq 0$ (b) All real numbers (c) All real numbers

79. (a) All real numbers $x \neq \pm 1$ (b) All real numbers

(c) All real numbers $x \neq -2, 0$

81. 3 **83.** $\dfrac{-4}{x(x+h)}$

85. (a) $r(x) = \dfrac{x}{2}$ (b) $A(r) = \pi r^2$

(c) $(A \circ r)(x) = \pi\left(\dfrac{x}{2}\right)^2$;

$(A \circ r)(x)$ represents the area of the circular base of the tank on the square foundation with side length y.

87. $(C \circ x)(t) = 3000t + 750$

$(C \circ x)(t)$ represents the cost after t production hours.

89. (a) $R = p - 1200$ (b) $S = 0.92p$

(c) $(R \circ S)(p) = 0.92p - 1200$

$(S \circ R)(p) = 0.92(p - 1200)$

(d) $(R \circ S)(18{,}400) = 15{,}728$

$(S \circ R)(18{,}400) = 15{,}824$

$(R \circ S)(18{,}400)$ is smaller, because all of the \$1200 is taken off for the rebate.

91. Odd

Section P.8 *(page 104)*

Warm Up *(page 104)*

1. All real numbers **2.** $[-1, \infty)$
3. All real numbers $x \neq 0, 2$
4. All real numbers $x \neq -\frac{5}{3}$ **5.** x **6.** x
7. x **8.** x **9.** $x = \dfrac{3}{2}y + 3$ **10.** $x = \dfrac{y^3}{2} + 2$

1. c **3.** a **5.** $f^{-1}(x) = \frac{1}{8}x$
7. $f^{-1}(x) = x - 10$ **9.** $f^{-1}(x) = x^3$

11. (a) $f(g(x)) = f\left(\dfrac{x}{2}\right) = 2\left(\dfrac{x}{2}\right) = x$

$g(f(x)) = g(2x) = \dfrac{(2x)}{2} = x$

(b)

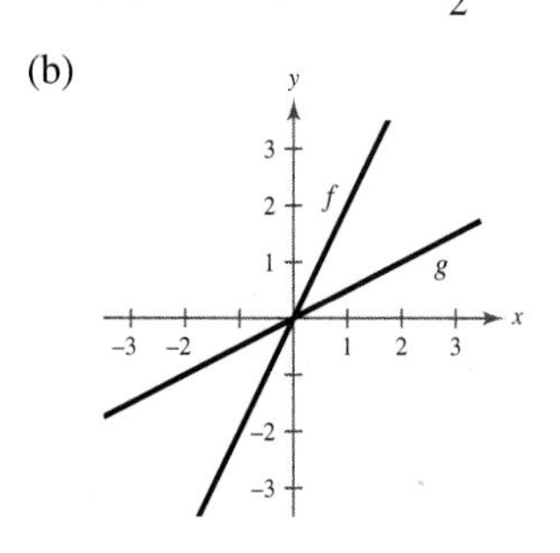

13. (a) $f(g(x)) = f\left(\dfrac{x-1}{5}\right) = 5\left(\dfrac{x-1}{5}\right) + 1 = x$

$g(f(x)) = g(5x + 1) = \dfrac{(5x + 1) - 1}{5} = x$

(b)

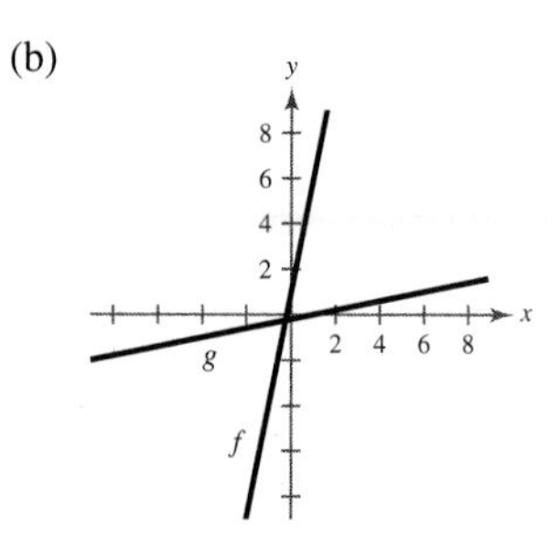

15. (a) $f(g(x)) = f(\sqrt[3]{x}) = (\sqrt[3]{x})^3 = x$

$g(f(x)) = g(x^3) = \sqrt[3]{x^3} = x$

(b)

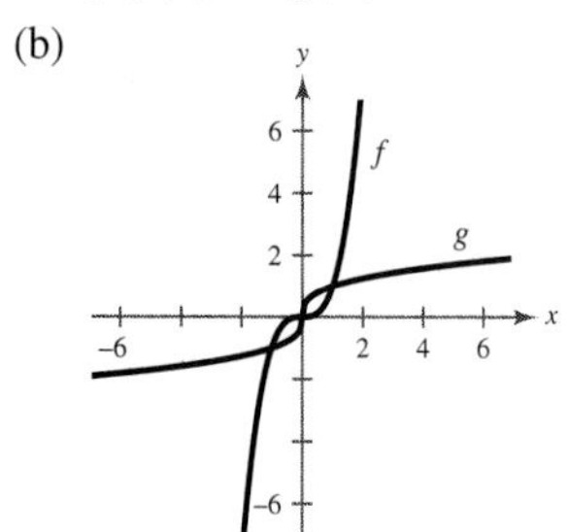

17. (a) $f(g(x)) = f(x^2 + 4),\ x \geq 0$

$= \sqrt{(x^2 + 4) - 4} = x$

$g(f(x)) = g\left(\sqrt{x - 4}\right)$

$= \left(\sqrt{x - 4}\right)^2 + 4 = x$

(b)

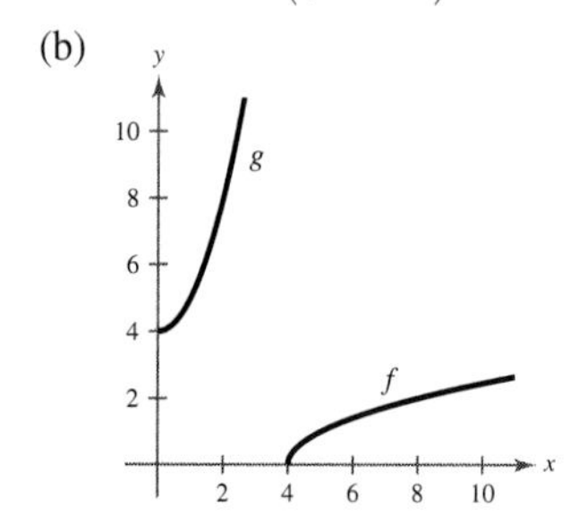

19. (a) $f(g(x)) = f\left(\sqrt{9 - x}\right),\ x \leq 9$

$= 9 - \left(\sqrt{9 - x}\right)^2 = x$

$g(f(x)) = g(9 - x^2),\ x \geq 0$

$= \sqrt{9 - (9 - x^2)} = x$

(b)

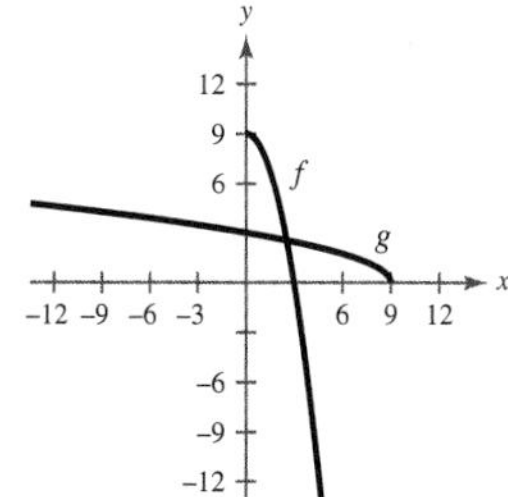

21. No **23.** Yes **25.** No

27.

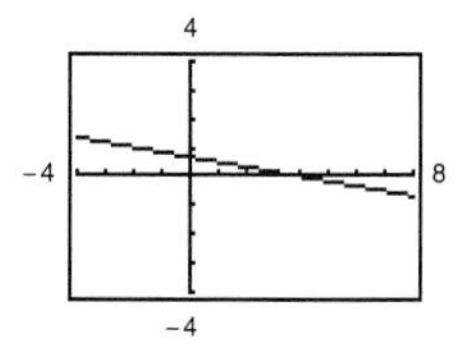

Yes

29.

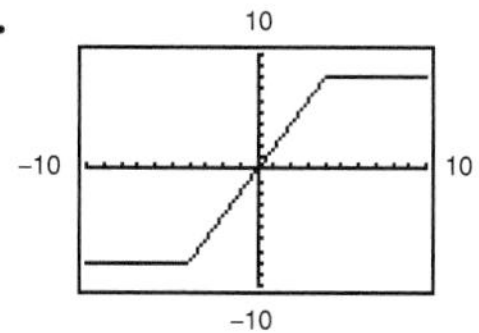

No

31.

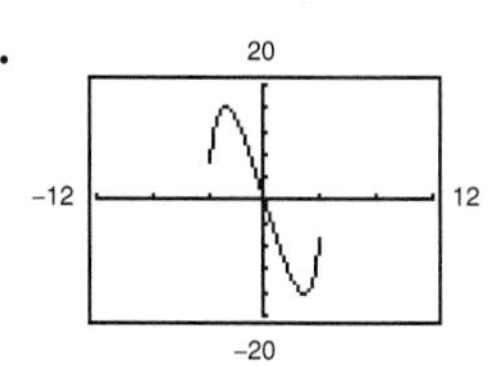

No

33. $f^{-1}(x) = \dfrac{x + 3}{2}$

35. $f^{-1}(x) = \sqrt[5]{x}$

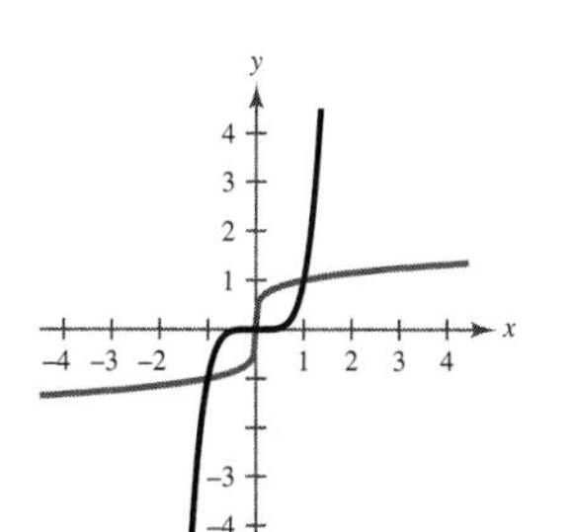

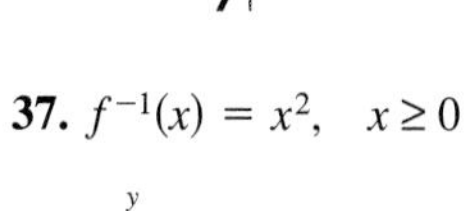
37. $f^{-1}(x) = x^2, \quad x \geq 0$

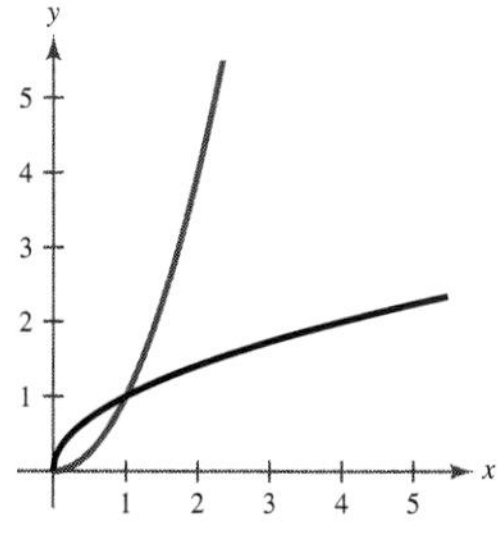

39. $f^{-1}(x) = \sqrt{4 - x^2}, \quad 0 \leq x \leq 2$

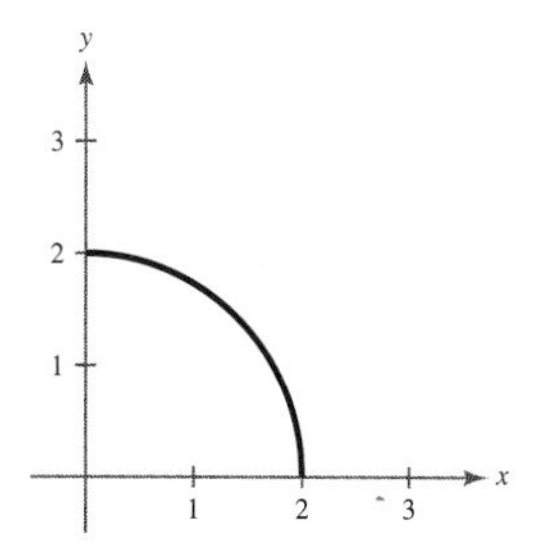

41. $f^{-1}(x) = x^3 + 1$

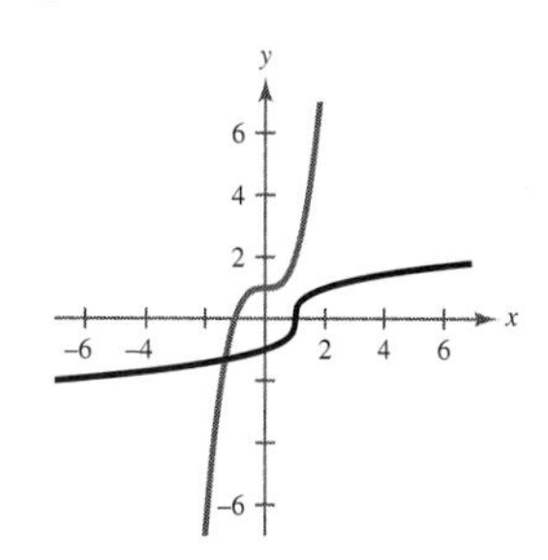

43. No inverse **45.** $g^{-1}(x) = 8x$ **47.** No inverse

49. $f^{-1}(x) = \sqrt{x} - 3, \quad x \geq 0$ **51.** $h^{-1}(x) = \dfrac{1}{x}$

53. $f^{-1}(x) = \dfrac{x^2 - 3}{2}, \quad x \geq 0$ **55.** No inverse

57. $f^{-1}(x) = -\sqrt{25 - x}, \quad x \leq 25$

59. $y = \sqrt{x} + 2, \quad x \geq 0$ **61.** $y = x - 2, \quad x \geq 0$

63.

x	-4	-2	2	3
$f^{-1}(x)$	-2	-1	1	3

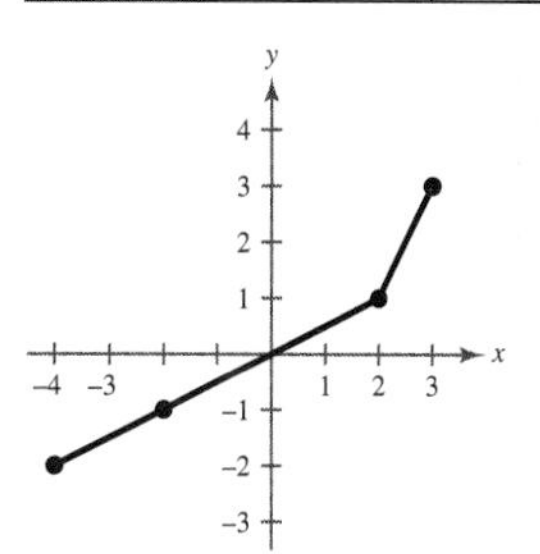

65. False. $f(x) = x^2$ **67.** True **69.** 32

71. 600 **73.** $2\sqrt[3]{x + 3}$ **75.** $\dfrac{x + 1}{2}$ **77.** $\dfrac{x + 1}{2}$

79. (a) $y = \dfrac{x - 8}{0.75}$

(b) y = number of units produced; x = hourly wage

(c) 19

81. (a) $y = \sqrt{\dfrac{x - 245.50}{0.03}}$, $245.5 < x < 545.5$

x = degrees Fahrenheit

y = % load

(b)

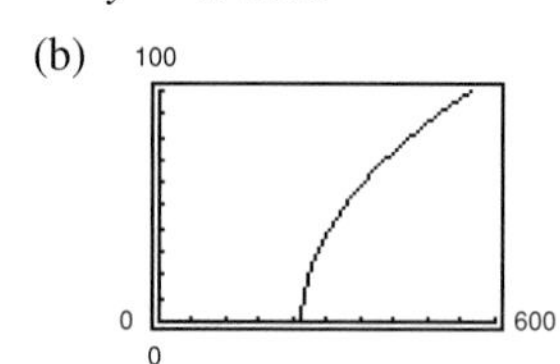

(c) $0 < x < 92.11$

83. (a) Yes

(b) f^{-1} yields the year for a given average fuel consumption.

(c) 8

85. ± 8 **87.** $\frac{3}{2}$ **89.** $3 \pm \sqrt{5}$ **91.** $5, -\frac{10}{3}$

93. 16, 18 **95.** $b = h = 2\sqrt{5}$

Focus on Concepts *(page 108)*

1. No. The slope cannot be determined without knowing the scale on the y-axis. The slopes could be the same.
2. -4. The slope with the greatest magnitude corresponds to the steepest line.
3. V-intercept: Initial cost; Slope: Annual depreciation
4. No. The element 3 in the domain corresponds to two elements in the range.
5. (a)

Xmin = -15
Xmax = 6
Xscl = 3
Ymin = -18
Ymax = 6
Yscl = 3

(b)

Xmin = -24
Xmax = 36
Xscl = 6
Ymin = -54
Ymax = 12
Yscl = 6

6. (a) Even. The graph is a reflection in the x-axis.
   (b) Even. The graph is a reflection in the y-axis.
   (c) Even. The graph is a vertical translation of f.
   (d) Neither. The graph is a horizontal translation of f.
7. (a) $g(t) = \frac{3}{4} f(t)$ (b) $g(t) = f(t) + 10{,}000$
   (c) $g(t) = f(t - 2)$

Review Exercises *(page 109)*

1. (a) 11 (b) $11, -14$
(c) $11, -14, -\frac{8}{9}, \frac{5}{2}, 0.4$ (d) $\sqrt{6}$

3. The set consists of all real numbers less than or equal to 7.

5. $|x - 7| \geq 4$ **7.** Associative Property of Addition

9. Commutative Property of Multiplication

11. (a) No (b) Yes (c) Yes (d) No

13. $-\frac{1}{2}$ **15.** $\frac{1}{5}$ **17.** $-4 \pm 3\sqrt{2}$ **19.** 4

21.

x	-2	0	2	3	4
y	3	2	1	$\frac{1}{2}$	0

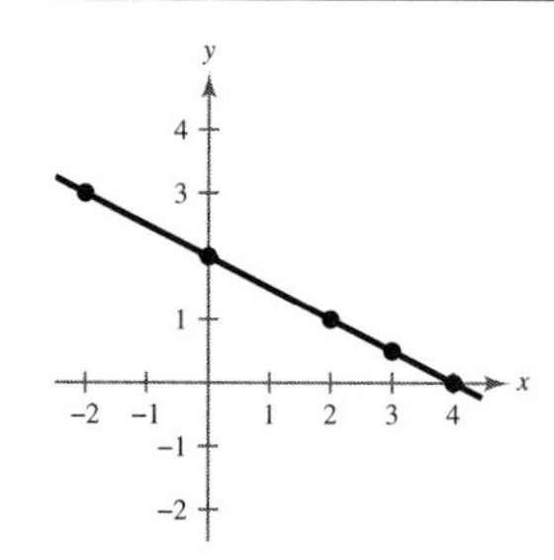

23. (a)

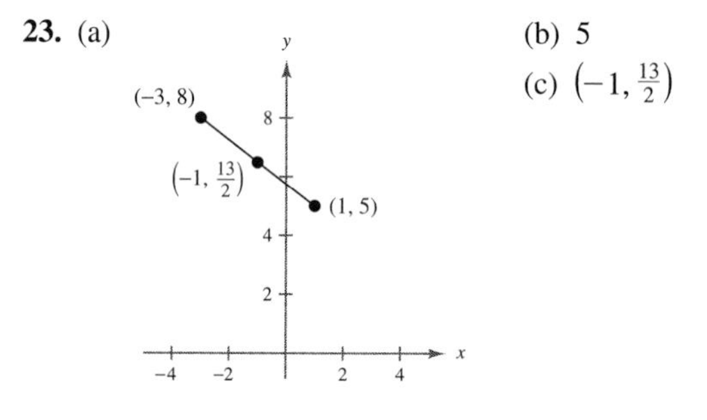

(b) 5
(c) $\left(-1, \frac{13}{2}\right)$

25. **27.**

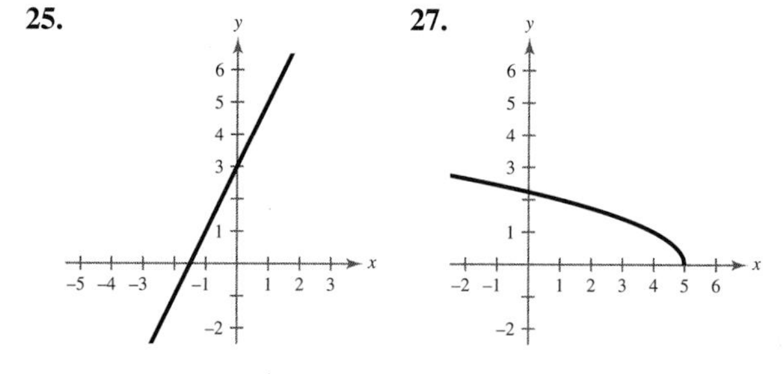

29.

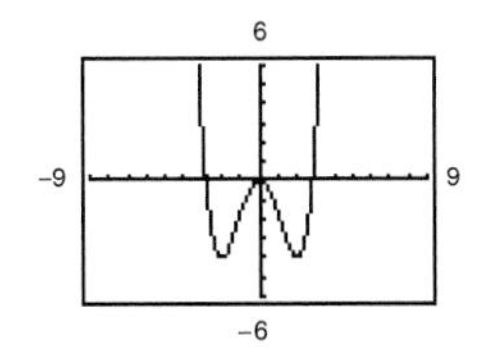

Intercepts: $(0, 0)$, $(\pm 2\sqrt{2}, 0)$

31.

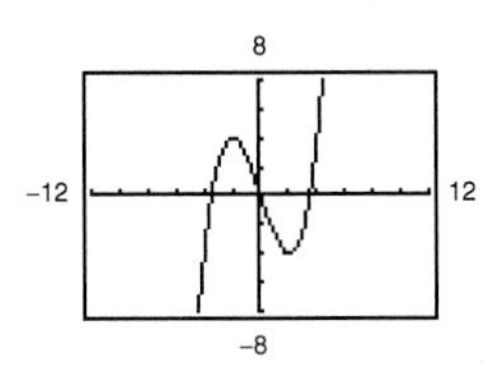

Intercepts: $(0, 0)$, $(\pm 2\sqrt{3}, 0)$

33.

Xmin = -2
Xmax = 3
Xscl = 1
Ymin = -20
Ymax = 15
Yscl = 5

35. Center: $(3, -1)$; radius $= 3$

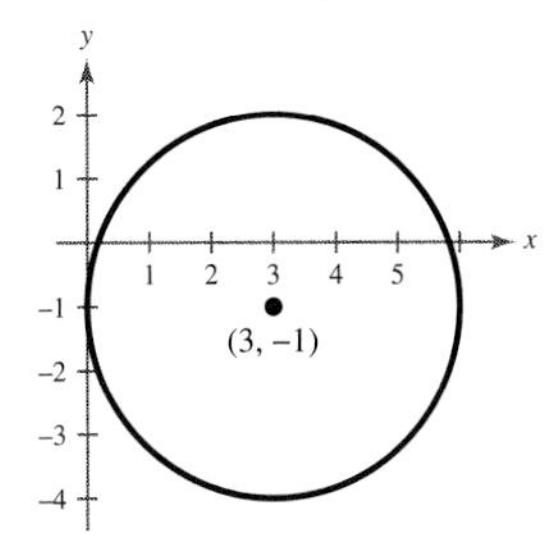

37. $5x - 12y + 2 = 0$

39.

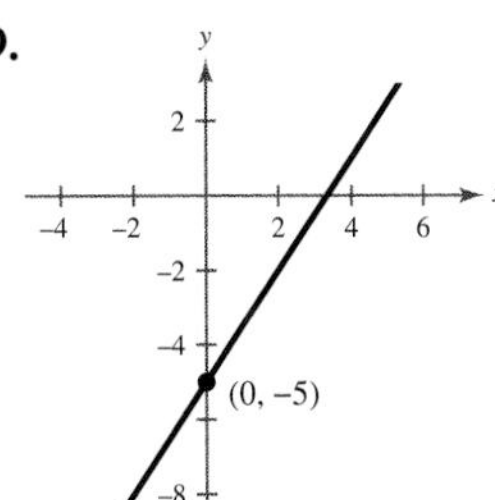

$3x - 2y - 10 = 0$

41.

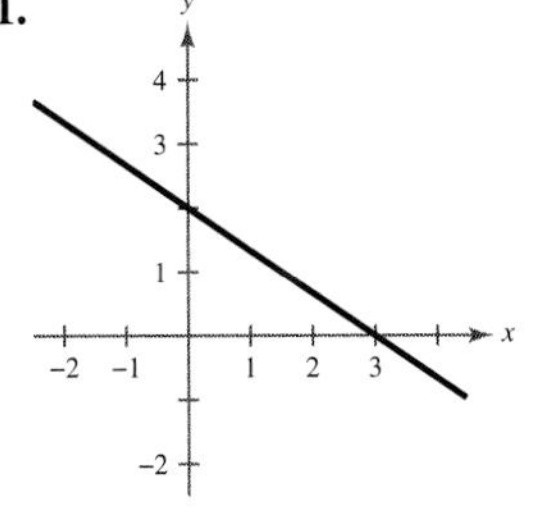

$2x + 3y - 6 = 0$

43. $y = 850t + 7400$ **45.** No **47.** Yes

49. (a) 16 (b) $(t + 1)^{4/3}$ (c) $\frac{15}{7}$ (d) $x^{4/3}$

51. $[-5, 5]$ **53.** All real numbers except $s = 3$

55. All real numbers except $x = -2, 3$

57. Second setting

59.

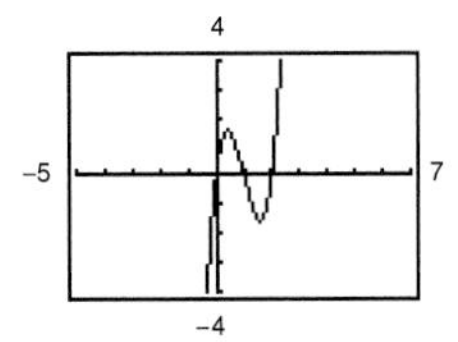

$x = 0, 1, 2$

61.

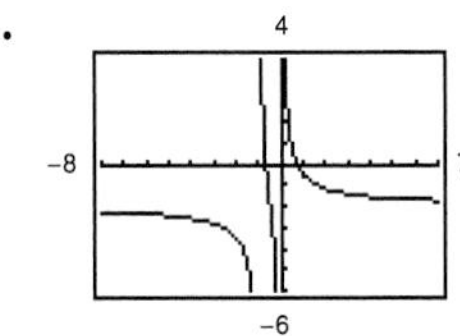

$x = \pm\frac{\sqrt{2}}{2}$

63. (a)

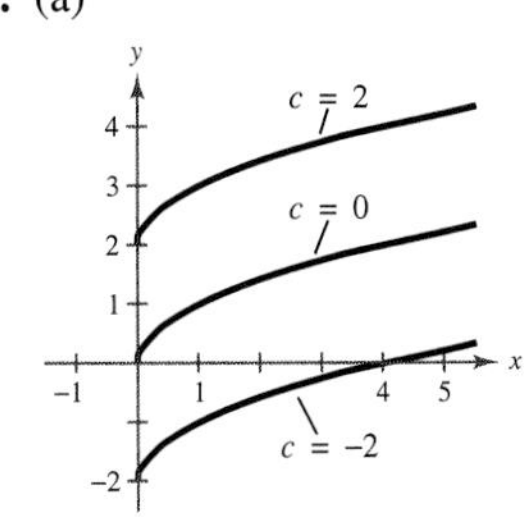

(b)

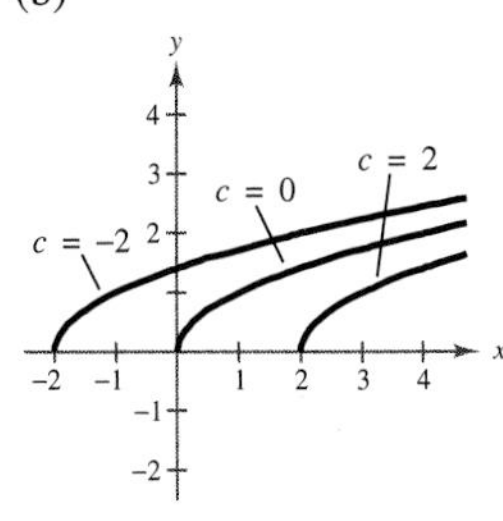

65. (a) $f^{-1}(x) = 2x + 6$

(b)

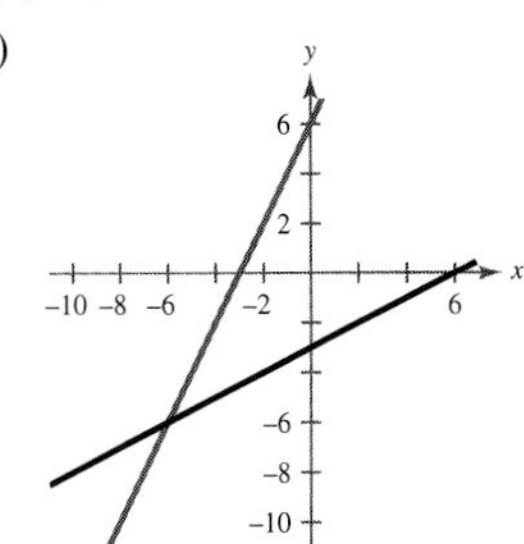

(c)
$$f^{-1}(f(x)) = f^{-1}\left(\tfrac{1}{2}x - 3\right)$$
$$= 2\left(\tfrac{1}{2}x - 3\right) + 6$$
$$= x$$
$$f(f^{-1}(x)) = f(2x + 6)$$
$$= \tfrac{1}{2}(2x + 6) - 3$$
$$= x$$

67. (a) $f^{-1}(x) = x^2 - 1, \quad x \geq 0$

(b)

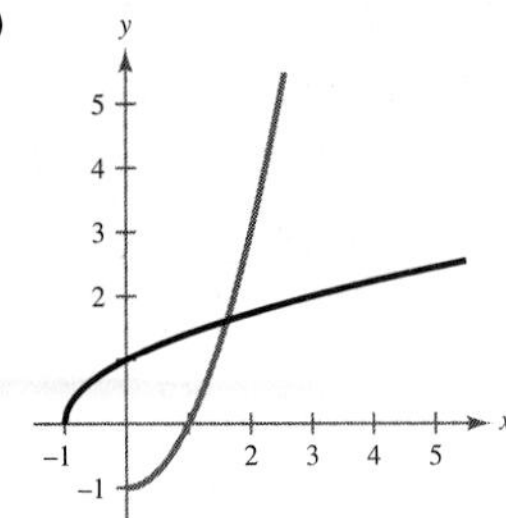

(c) $f^{-1}(f(x)) = f^{-1}(\sqrt{x+1})$
$= (x + 1) - 1$
$= x$

$f(f^{-1}(x)) = f(x^2 - 1), \quad x \geq 0$
$= \sqrt{x^2 - 1 + 1}$
$= x$

69. $x \geq 4, \ f^{-1}(x) = \sqrt{\dfrac{x}{2}} + 4$

71. $x \geq 2, \ f^{-1}(x) = \sqrt{x^2 + 4}, \ x \geq 0$

73. -7 **75.** 5 **77.** 23 **79.** 9

Chapter Test *(page 114)*

1. $-\frac{10}{3} > -|-4|$ **2.** (a) $\frac{35}{24}$ (b) 28 **3.** $\frac{128}{11}$

4. No solution **5.** $\dfrac{-3 \pm \sqrt{3}}{3}$ **6.** $\pm\sqrt{2}$

7. 4 **8.** $-2, \frac{8}{3}$

9.

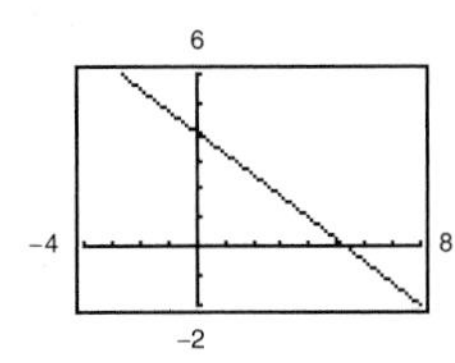

No symmetry

$(0, 4), \left(\frac{16}{3}, 0\right)$

10.

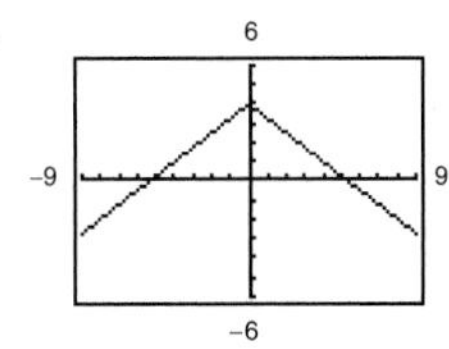

y-axis symmetry

$(0, 4), \left(\pm\frac{16}{3}, 0\right)$

11.

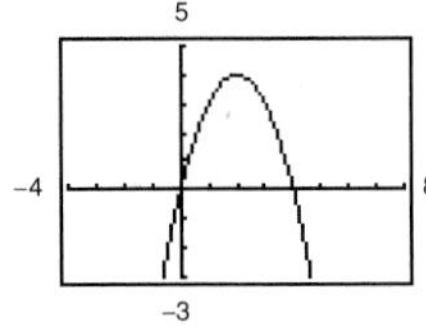

No symmetry

$(0, 0), (4, 0)$

12.

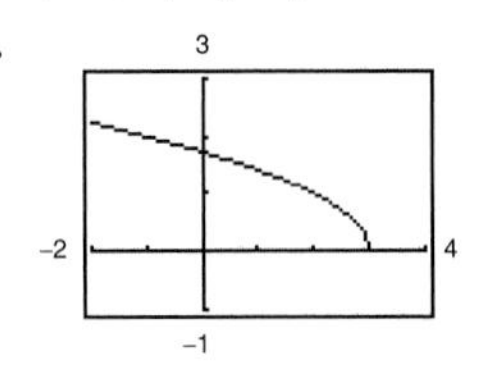

No symmetry

$(0, \sqrt{3}), (3, 0)$

13. (a) $2x + 3y - 12 = 0$ (b) $y + 2 = 0$

14. No. For some values of x there correspond more than one value of y.

15. $g(t) = 10 - \sqrt{6 - t}, \ (-\infty, 6]$

16. (a) (b)

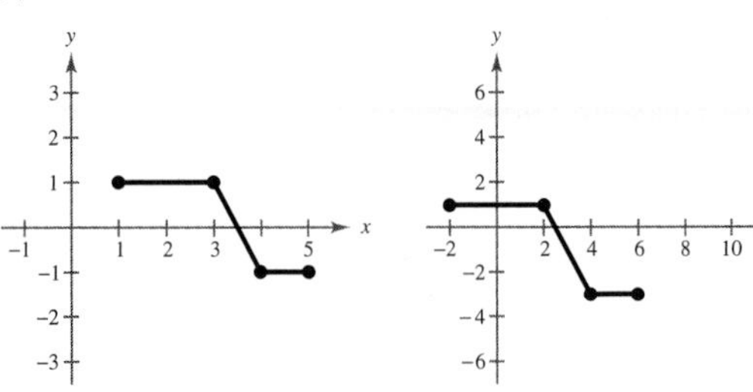

17. $x^2 - \sqrt{2 - x}$ **18.** $\dfrac{x^2}{\sqrt{2 - x}}$ **19.** $2 - x, \ x \leq 2$

20. $2 - x^2$ **21.** $93\frac{3}{4}$ kilometers per hour

CHAPTER 1

Section 1.1 *(page 123)*

Warm Up *(page 123)*

1. 45 **2.** 70 **3.** $\dfrac{\pi}{6}$ **4.** $\dfrac{\pi}{3}$ **5.** $\dfrac{\pi}{4}$ **6.** $\dfrac{4\pi}{3}$

7. $\dfrac{\pi}{9}$ **8.** $\dfrac{11\pi}{6}$ **9.** 45 **10.** 45

1. 2 **3.** -3 **5.** (a) Quadrant I (b) Quadrant III

7. (a) Quadrant IV (b) Quadrant II

9. (a) Quadrant III (b) Quadrant II

11. (a) (b)

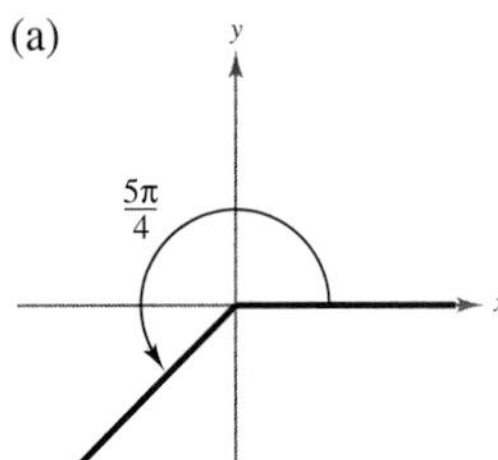

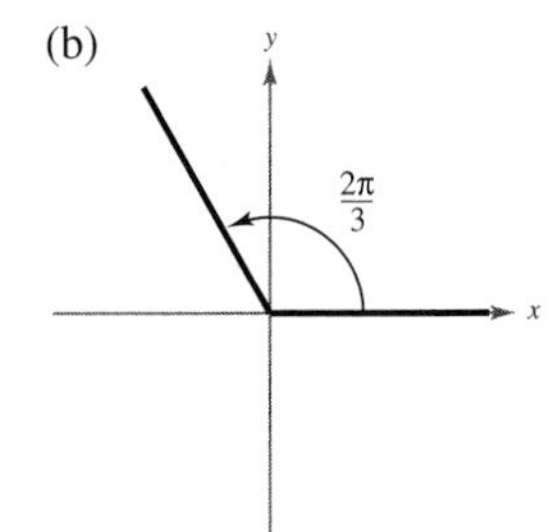

13. (a) (b)

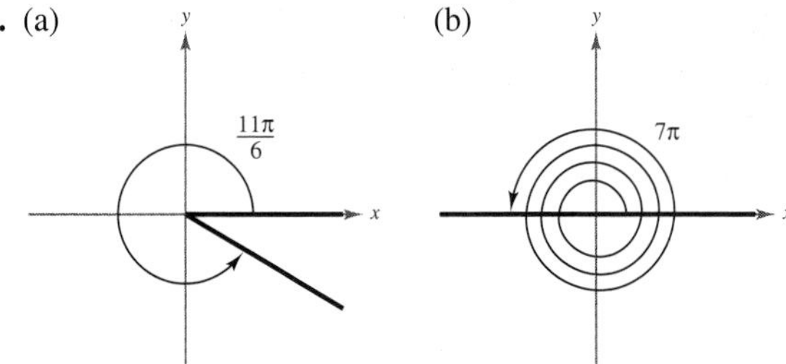

15. (a) $\frac{25\pi}{12}, -\frac{23\pi}{12}$ (b) $\frac{8\pi}{3}, -\frac{4\pi}{3}$

17. (a) $\frac{7\pi}{4}, -\frac{\pi}{4}$ (b) $\frac{28\pi}{15}, -\frac{32\pi}{15}$

19. (a) Complement: $\frac{\pi}{6}$; Supplement: $\frac{2\pi}{3}$

(b) Complement: none; Supplement: $\frac{\pi}{4}$

21. 210° **23.** −45°

25. (a) Quadrant II (b) Quadrant IV

27. (a) Quadrant III (b) Quadrant I

29. (a) (b)

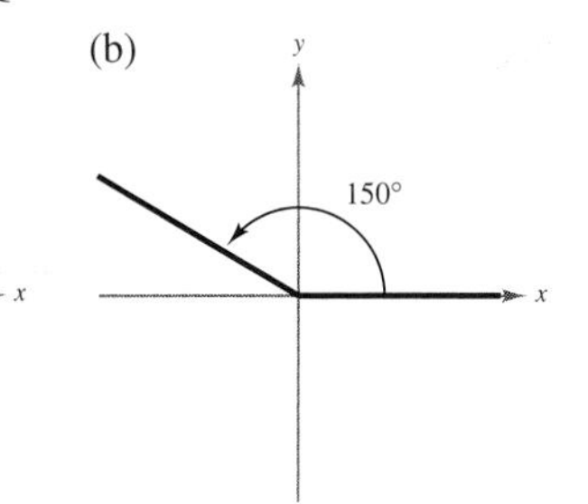

31. (a) (b)

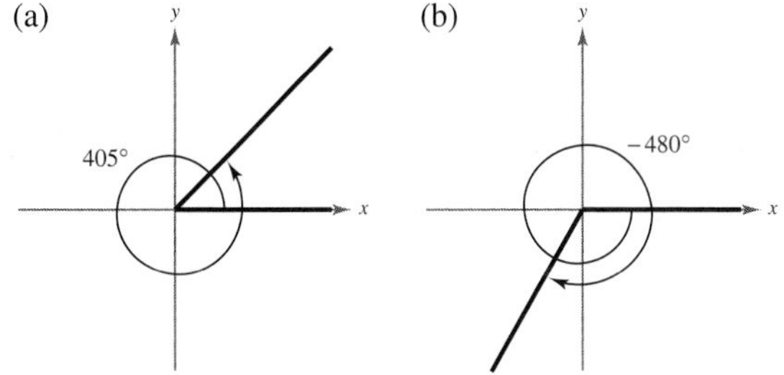

33. (a) 405°, −315° (b) 324°, −396°

35. (a) 660°, −60° (b) 20°, −340°

37. (a) Complement: 72°; Supplement: 162°

(b) Complement: none; Supplement: 65°

39. (a) $\frac{\pi}{6}$ (b) $\frac{5\pi}{6}$ **41.** (a) $-\frac{\pi}{9}$ (b) $-\frac{4\pi}{3}$

43. (a) 270° (b) 210° **45.** (a) 420° (b) −66°

47. 2.007 **49.** −3.776 **51.** 9.285 **53.** −0.014

55. 25.714° **57.** 337.5° **59.** −756°

61. −114.592° **63.** (a) 54.75° (b) −128.5°

65. (a) 85.308° (b) 330.007°

67. (a) 240° 36′ (b) −145° 48′

69. (a) 143° 14′ 22″ (b) −205° 7′ 8″

71. $\frac{6}{5}$ rad **73.** $4\frac{4}{7}$ rad **75.** $\frac{4}{15}$ rad **77.** $\frac{50}{29}$ rad

79. 15π inches ≈ 47.12 inches **81.** 12 meters

83. 591.72 miles **85.** 1141.02 miles

87. 0.094 rad ≈ 5.39° **89.** $\frac{5}{12}$ rad

91. (a) 560.2 revolutions per minute

(b) 3520 rad per minute

93. Radian. 1 rad ≈ 57.3°

95. 20.16π inches per second

97. Answers will vary. **99.** ≈ 2.16 miles

Section 1.2 *(page 133)*

Warm Up *(page 133)*

1. $-\frac{\sqrt{3}}{3}$ **2.** −1 **3.** $\frac{2\pi}{3}$ **4.** $\frac{7\pi}{4}$ **5.** $\frac{\pi}{6}$

6. $\frac{3\pi}{4}$ **7.** 60° **8.** −270° **9.** 2π **10.** π

1. $\sin t = \frac{4}{5}$
$\cos t = -\frac{3}{5}$
$\tan t = -\frac{4}{3}$
$\csc t = \frac{5}{4}$
$\sec t = -\frac{5}{3}$
$\cot t = -\frac{3}{4}$

3. $\sin t = -\frac{15}{17}$
$\cos t = \frac{8}{17}$
$\tan t = -\frac{15}{8}$
$\csc t = -\frac{17}{15}$
$\sec t = \frac{17}{8}$
$\cot t = -\frac{8}{15}$

5. $\left(\frac{\sqrt{2}}{2}, \frac{\sqrt{2}}{2}\right)$ **7.** $\left(-\frac{\sqrt{3}}{2}, \frac{1}{2}\right)$ **9.** $\left(-\frac{1}{2}, -\frac{\sqrt{3}}{2}\right)$

11. (0, −1)

13. $\sin\frac{\pi}{4} = \frac{\sqrt{2}}{2}$

$\cos\frac{\pi}{4} = \frac{\sqrt{2}}{2}$

$\tan\frac{\pi}{4} = 1$

15. $\sin\left(-\frac{\pi}{6}\right) = -\frac{1}{2}$

$\cos\left(-\frac{\pi}{6}\right) = \frac{\sqrt{3}}{2}$

$\tan\left(-\frac{\pi}{6}\right) = -\frac{\sqrt{3}}{3}$

17. $\sin\left(-\frac{5\pi}{4}\right) = \frac{\sqrt{2}}{2}$

$\cos\left(-\frac{5\pi}{4}\right) = -\frac{\sqrt{2}}{2}$

$\tan\left(-\frac{5\pi}{4}\right) = -1$

19. $\sin\frac{11\pi}{6} = -\frac{1}{2}$

$\cos\frac{11\pi}{6} = \frac{\sqrt{3}}{2}$

$\tan\frac{11\pi}{6} = -\frac{\sqrt{3}}{3}$

21. $\sin\frac{4\pi}{3} = -\frac{\sqrt{3}}{2}$

$\cos\frac{4\pi}{3} = -\frac{1}{2}$

$\tan\frac{4\pi}{3} = \sqrt{3}$

23. $\sin\left(-\frac{3\pi}{2}\right) = 1$

$\cos\left(-\frac{3\pi}{2}\right) = 0$

$\tan\left(-\frac{3\pi}{2}\right)$ is undefined.

25. $\sin\frac{3\pi}{4} = \frac{\sqrt{2}}{2}$

$\cos\frac{3\pi}{4} = -\frac{\sqrt{2}}{2}$

$\tan\frac{3\pi}{4} = -1$

$\csc\frac{3\pi}{4} = \sqrt{2}$

$\sec\frac{3\pi}{4} = -\sqrt{2}$

$\cot\frac{3\pi}{4} = -1$

27. $\sin\frac{\pi}{2} = 1$

$\cos\frac{\pi}{2} = 0$

$\tan\frac{\pi}{2}$ is undefined.

$\csc\frac{\pi}{2} = 1$

$\sec\frac{\pi}{2}$ is undefined.

$\cot\frac{\pi}{2} = 0$

29. $\sin\left(-\frac{4\pi}{3}\right) = \frac{\sqrt{3}}{2}$

$\cos\left(-\frac{4\pi}{3}\right) = -\frac{1}{2}$

$\tan\left(-\frac{4\pi}{3}\right) = -\sqrt{3}$

$\csc\left(-\frac{4\pi}{3}\right) = \frac{2\sqrt{3}}{3}$

$\sec\left(-\frac{4\pi}{3}\right) = -2$

$\cot\left(-\frac{4\pi}{3}\right) = -\frac{\sqrt{3}}{3}$

31. $\sin 3\pi = \sin \pi = 0$

33. $\cos\frac{8\pi}{3} = \cos\frac{2\pi}{3} = -\frac{1}{2}$

35. $\cos\frac{19\pi}{6} = \cos\frac{7\pi}{6} = -\frac{\sqrt{3}}{2}$

37. $\sin\left(-\frac{9\pi}{4}\right) = \sin\frac{7\pi}{4} = -\frac{\sqrt{2}}{2}$ **39.** (a) $-\frac{1}{3}$ (b) -3

41. (a) $-\frac{7}{8}$ (b) $-\frac{8}{7}$ **43.** (a) $\frac{4}{5}$ (b) $-\frac{4}{5}$

45. 0.7071 **47.** -0.9900 **49.** -0.1288

51. 1.3940 **53.** -1.4486 **55.** (a) -1 (b) -0.4

57. (a) 0.25, 2.89 (b) 1.82, 4.46

59. $0.0707 = \cos 1.5 \neq 2\cos 0.75 = 1.4634$

61. (a) y-axis (b) $\sin t_1 = \sin(\pi - t_1)$
(c) $\cos(\pi - t_1) = -\cos t_1$

63. (a) 0.2500 foot (b) 0.0177 foot (c) -0.2475 foot

65. 0.794 **67.** Odd

69. $f^{-1}(x) = \frac{2}{3}(x + 1)$
$f(x) = \frac{1}{2}(3x - 2)$

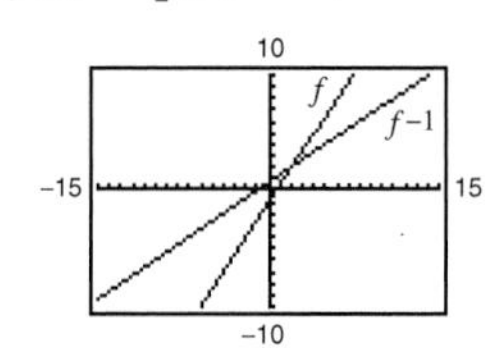

71. $f^{-1}(x) = \sqrt{x^2 + 4}, \quad x \geq 0$

$f(x) = \sqrt{x^2 - 4}, \quad x \geq 2$

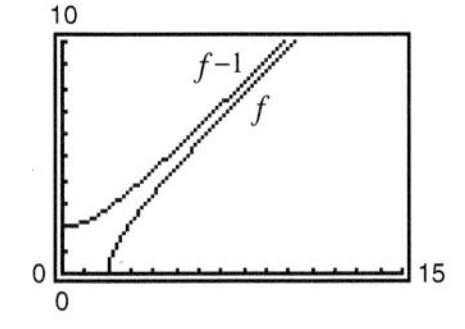

Section 1.3 *(page 143)*

Warm Up *(page 143)*

1. $2\sqrt{5}$ **2.** $3\sqrt{10}$ **3.** 10 **4.** $3\sqrt{2}$ **5.** 1.24
6. 317.55 **7.** 63.13 **8.** 133.57
9. 2,785,714.29 **10.** 28.80

1. $\sin\theta = \frac{1}{2}$, $\cos\theta = \frac{\sqrt{3}}{2}$, $\tan\theta = \frac{\sqrt{3}}{3}$, $\csc\theta = 2$, $\sec\theta = \frac{2\sqrt{3}}{3}$, $\cot\theta = \sqrt{3}$

3. $\sin\theta = \frac{8}{17}$, $\cos\theta = \frac{15}{17}$, $\tan\theta = \frac{8}{15}$, $\csc\theta = \frac{17}{8}$, $\sec\theta = \frac{17}{15}$, $\cot\theta = \frac{15}{8}$

5. $\sin\theta = \frac{1}{3}$, $\cos\theta = \frac{2\sqrt{2}}{3}$, $\tan\theta = \frac{\sqrt{2}}{4}$, $\csc\theta = 3$, $\sec\theta = \frac{3\sqrt{2}}{4}$, $\cot\theta = 2\sqrt{2}$

The triangles are similar, and corresponding sides are proportional.

7. $\sin\theta = \frac{3}{5}$, $\cos\theta = \frac{4}{5}$, $\tan\theta = \frac{3}{4}$, $\csc\theta = \frac{5}{3}$, $\sec\theta = \frac{5}{4}$, $\cot\theta = \frac{4}{3}$

The triangles are similar, and corresponding sides are proportional.

9.

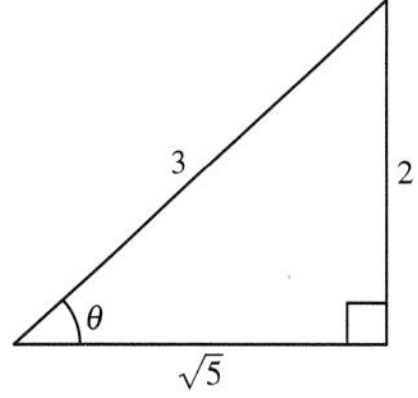

$\cos\theta = \frac{\sqrt{5}}{3}$, $\tan\theta = \frac{2\sqrt{5}}{5}$, $\csc\theta = \frac{3}{2}$, $\sec\theta = \frac{3\sqrt{5}}{5}$, $\cot\theta = \frac{\sqrt{5}}{2}$

11.

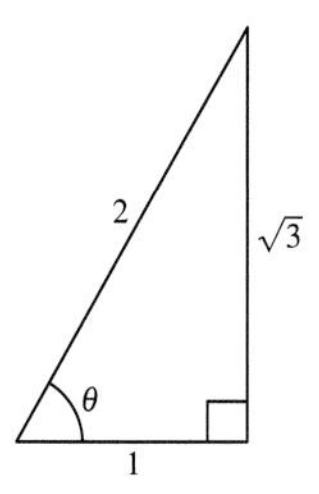

$\sin\theta = \frac{\sqrt{3}}{2}$, $\cos\theta = \frac{1}{2}$, $\tan\theta = \sqrt{3}$, $\csc\theta = \frac{2\sqrt{3}}{3}$, $\cot\theta = \frac{\sqrt{3}}{3}$

13.

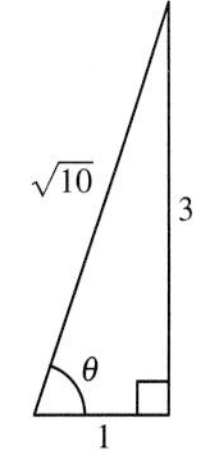

$\sin\theta = \frac{3\sqrt{10}}{10}$, $\cos\theta = \frac{\sqrt{10}}{10}$, $\csc\theta = \frac{\sqrt{10}}{3}$, $\sec\theta = \sqrt{10}$, $\cot\theta = \frac{1}{3}$

15.

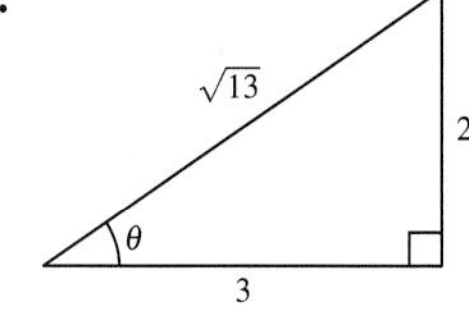

$\sin\theta = \frac{2\sqrt{13}}{13}$, $\cos\theta = \frac{3\sqrt{13}}{13}$, $\tan\theta = \frac{2}{3}$, $\csc\theta = \frac{\sqrt{13}}{2}$, $\sec\theta = \frac{\sqrt{13}}{3}$

17. (a) $\sqrt{3}$ (b) $\frac{1}{2}$ (c) $\frac{\sqrt{3}}{2}$ (d) $\frac{\sqrt{3}}{3}$

19. (a) $\frac{1}{3}$ (b) $\frac{2\sqrt{2}}{3}$ (c) $\frac{\sqrt{2}}{4}$ (d) 3

21. (a) 4 (b) $\frac{\sqrt{15}}{4}$ (c) $\frac{\sqrt{15}}{15}$ (d) $\frac{1}{4}$

23.–31. Answers will vary.

33. (a) $\frac{1}{2}$ (b) $\frac{\sqrt{3}}{3}$ **35.** (a) 1 (b) $\frac{\sqrt{2}}{2}$

37. (a) 0.1736 (b) 0.1736 **39.** (a) 0.2815 (b) 3.5523

41. (a) 1.3499 (b) 1.3432 **43.** (a) 5.0273 (b) 0.1989

45. (a) 1.1884 (b) 0.5463

47. (a) $30° = \frac{\pi}{6}$ (b) $30° = \frac{\pi}{6}$

49. (a) $60° = \frac{\pi}{3}$ (b) $45° = \frac{\pi}{4}$

51. (a) $60° = \frac{\pi}{3}$ (b) $45° = \frac{\pi}{4}$

53. (a) $55° \approx 0.960$ (b) $89° \approx 1.553$

55. (a) $50° \approx 0.873$ (b) $25° \approx 0.436$ **57.** $25\sqrt{3}$

59. $\frac{32\sqrt{3}}{3}$ **61.** 23.3 **63.** 6.1 **65.** $17\frac{1}{4}$ feet

67. (a)

30
x
75°

(b) $\sin 75° = \frac{x}{30}$

(c) 29.0 meters

69. 1144.9 feet **71.** $(x_1, y_1) = (28\sqrt{3}, 28)$
$(x_2, y_2) = (28, 28\sqrt{3})$

73. $\sin 20° \approx 0.34$
$\cos 20° \approx 0.94$
$\tan 20° \approx 0.36$
$\csc 20° \approx 2.92$
$\sec 20° \approx 1.06$
$\cot 20° \approx 2.75$

75. (a)

θ	0	0.1	0.2	0.3	0.4	0.5
$\sin\theta$	0	0.0998	0.1987	0.2955	0.3894	0.4794

(b) θ

(c) $\sin\theta$ approaches θ as θ approaches 0.

77. True, $\csc x = \frac{1}{\sin x}$ **79.** False, $\frac{\sqrt{2}}{2} + \frac{\sqrt{2}}{2} \neq 1$

81. False, $1.7321 \neq 0.0349$ **83.** $\frac{x}{x-2}$

85. $\frac{2(x^2 - 5x - 10)}{(x-2)(x+2)^2}$

Section 1.4 *(page 154)*

Warm Up *(page 154)*

1. $\frac{1}{2}$ **2.** 1 **3.** $\frac{\sqrt{2}}{2}$ **4.** $\frac{\sqrt{3}}{3}$ **5.** $\frac{2\sqrt{3}}{3}$ **6.** $\sqrt{2}$

7. $\sin\theta = \frac{3\sqrt{13}}{13}$
$\cos\theta = \frac{2\sqrt{13}}{13}$
$\csc\theta = \frac{\sqrt{13}}{3}$
$\sec\theta = \frac{\sqrt{13}}{2}$
$\cot\theta = \frac{2}{3}$

8. $\sin\theta = \frac{\sqrt{5}}{3}$
$\tan\theta = \frac{\sqrt{5}}{2}$
$\csc\theta = \frac{3\sqrt{5}}{5}$
$\sec\theta = \frac{3}{2}$
$\cot\theta = \frac{2\sqrt{5}}{5}$

9. $\cos\theta = \frac{2\sqrt{6}}{5}$
$\tan\theta = \frac{\sqrt{6}}{12}$
$\csc\theta = 5$
$\sec\theta = \frac{5\sqrt{6}}{12}$
$\cot\theta = 2\sqrt{6}$

10. $\sin\theta = \frac{2\sqrt{2}}{3}$
$\cos\theta = \frac{1}{3}$
$\tan\theta = 2\sqrt{2}$
$\csc\theta = \frac{3\sqrt{2}}{4}$
$\cot\theta = \frac{\sqrt{2}}{4}$

1. (a) $\sin\theta = \frac{3}{5}$, $\cos\theta = \frac{4}{5}$, $\tan\theta = \frac{3}{4}$, $\csc\theta = \frac{5}{3}$, $\sec\theta = \frac{5}{4}$, $\cot\theta = \frac{4}{3}$
(b) $\sin\theta = -\frac{15}{17}$, $\cos\theta = -\frac{8}{17}$, $\tan\theta = \frac{15}{8}$, $\csc\theta = -\frac{17}{15}$, $\sec\theta = -\frac{17}{8}$, $\cot\theta = \frac{8}{15}$

3. (a) $\sin\theta = -\frac{1}{2}$, $\cos\theta = -\frac{\sqrt{3}}{2}$, $\tan\theta = \frac{\sqrt{3}}{3}$, $\csc\theta = -2$, $\sec\theta = -\frac{2\sqrt{3}}{3}$, $\cot\theta = \sqrt{3}$
(b) $\sin\theta = \frac{\sqrt{2}}{2}$, $\cos\theta = -\frac{\sqrt{2}}{2}$, $\tan\theta = -1$, $\csc\theta = \sqrt{2}$, $\sec\theta = -\sqrt{2}$, $\cot\theta = -1$

5. (a) $\sin\theta = \frac{24}{25}$, $\cos\theta = \frac{7}{25}$, $\tan\theta = \frac{24}{7}$, $\csc\theta = \frac{25}{24}$, $\sec\theta = \frac{25}{7}$, $\cot\theta = \frac{7}{24}$
(b) $\sin\theta = -\frac{24}{25}$, $\cos\theta = \frac{7}{25}$, $\tan\theta = -\frac{24}{7}$, $\csc\theta = -\frac{25}{24}$, $\sec\theta = \frac{25}{7}$, $\cot\theta = -\frac{7}{24}$

7. (a) $\sin\theta = \frac{5\sqrt{29}}{29}$, $\cos\theta = -\frac{2\sqrt{29}}{29}$, $\tan\theta = -\frac{5}{2}$, $\csc\theta = \frac{\sqrt{29}}{5}$, $\sec\theta = -\frac{\sqrt{29}}{2}$, $\cot\theta = -\frac{2}{5}$
(b) $\sin\theta = -\frac{5\sqrt{34}}{34}$, $\cos\theta = \frac{3\sqrt{34}}{34}$, $\tan\theta = -\frac{5}{3}$, $\csc\theta = -\frac{\sqrt{34}}{5}$, $\sec\theta = \frac{\sqrt{34}}{3}$, $\cot\theta = -\frac{3}{5}$

9. (a) Quadrant III (b) Quadrant II

11. (a) Quadrant II (b) Quadrant IV

13. $\sin\theta = \frac{3}{5}$, $\cos\theta = -\frac{4}{5}$, $\tan\theta = -\frac{3}{4}$, $\csc\theta = \frac{5}{3}$, $\sec\theta = -\frac{5}{4}$, $\cot\theta = -\frac{4}{3}$

15. $\sin\theta = -\frac{15}{17}$, $\cos\theta = \frac{8}{17}$, $\tan\theta = -\frac{15}{8}$, $\csc\theta = -\frac{17}{15}$, $\sec\theta = \frac{17}{8}$, $\cot\theta = -\frac{8}{15}$

17. $\sin\theta = -\frac{\sqrt{10}}{10}$, $\cos\theta = \frac{3\sqrt{10}}{10}$, $\tan\theta = -\frac{1}{3}$, $\csc\theta = -\sqrt{10}$, $\sec\theta = \frac{\sqrt{10}}{3}$, $\cot\theta = -3$

19. $\sin\theta = \frac{\sqrt{3}}{2}$, $\cos\theta = -\frac{1}{2}$, $\tan\theta = -\sqrt{3}$, $\csc\theta = \frac{2\sqrt{3}}{3}$, $\sec\theta = -2$, $\cot\theta = -\frac{\sqrt{3}}{3}$

21. $\sin\theta = 0$, $\cos\theta = -1$, $\tan\theta = 0$, $\csc\theta$ is undefined. $\sec\theta = -1$, $\cot\theta$ is undefined.

23. $\sin\theta = \frac{\sqrt{2}}{2}$, $\cos\theta = -\frac{\sqrt{2}}{2}$, $\tan\theta = -1$, $\csc\theta = \sqrt{2}$, $\sec\theta = -\sqrt{2}$, $\cot\theta = -1$

25. $\sin\theta = -\frac{2\sqrt{5}}{5}$, $\cos\theta = -\frac{\sqrt{5}}{5}$, $\tan\theta = 2$, $\csc\theta = -\frac{\sqrt{5}}{2}$, $\sec\theta = -\sqrt{5}$, $\cot\theta = \frac{1}{2}$

27. -1 **29.** -1 **31.** Undefined **33.** 0

35. (a) $\theta' = 23°$

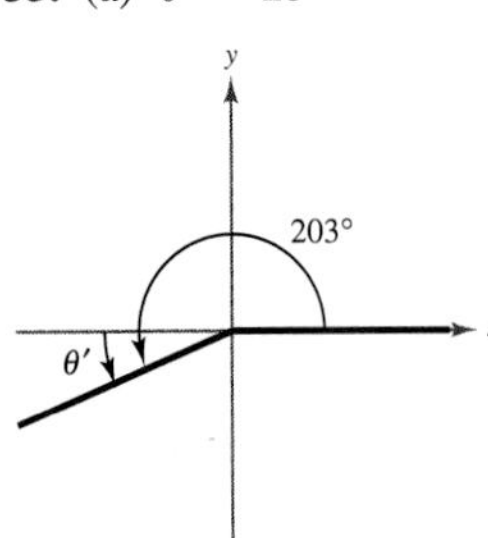

(b) $\theta' = 53°$

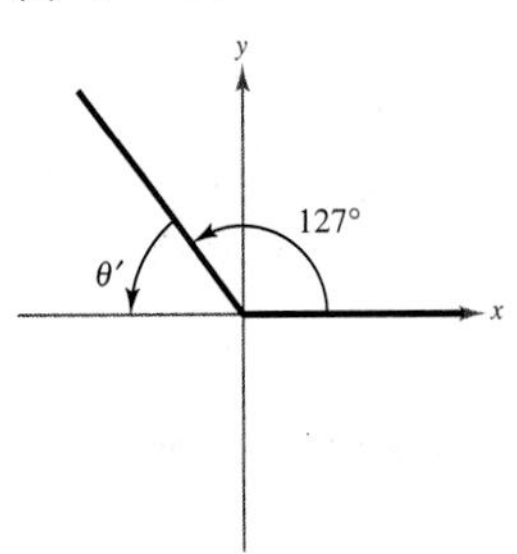

37. (a) $\theta' = 65°$

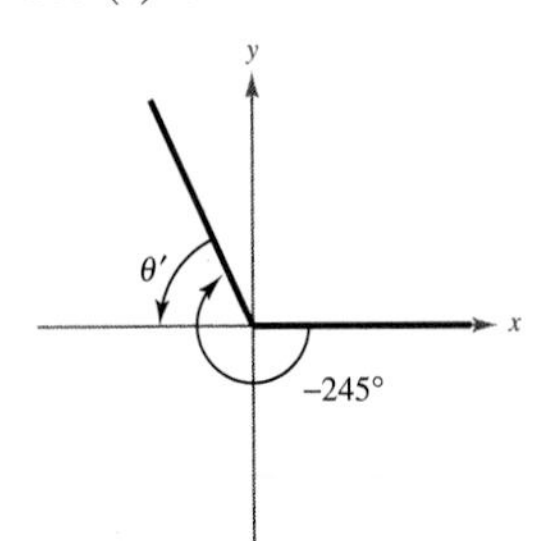

(b) $\theta' = 72°$

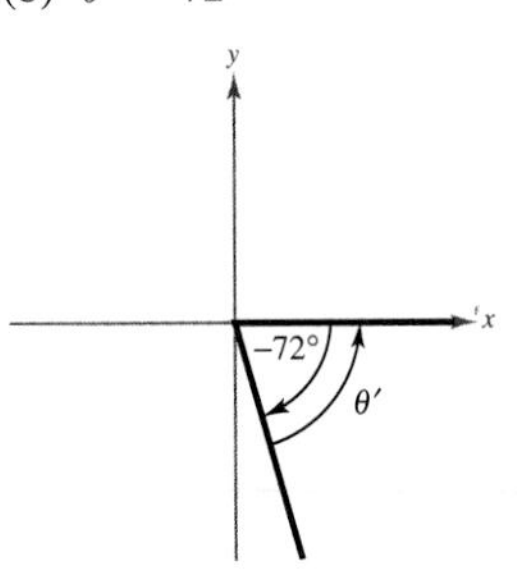

39. (a) $\theta' = \frac{\pi}{3}$

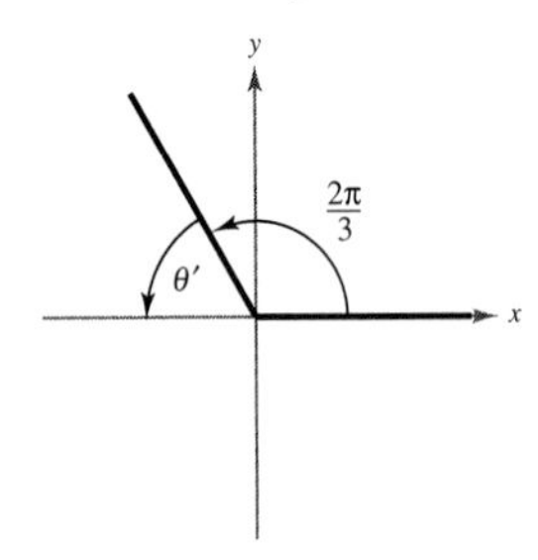

(b) $\theta' = \frac{\pi}{6}$

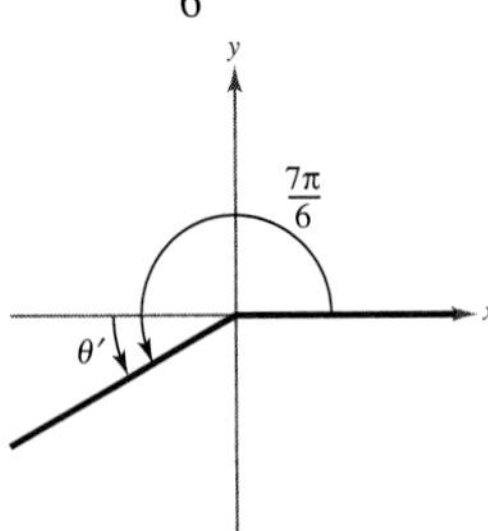

41. (a) $\theta' = 3.5 - \pi$

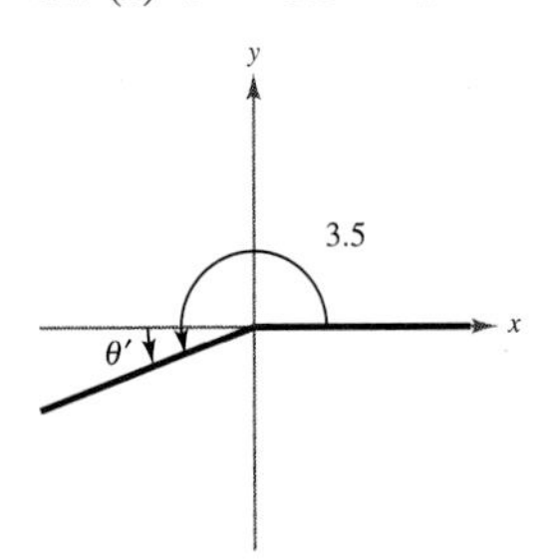

(b) $\theta' = 2\pi - 5.8$

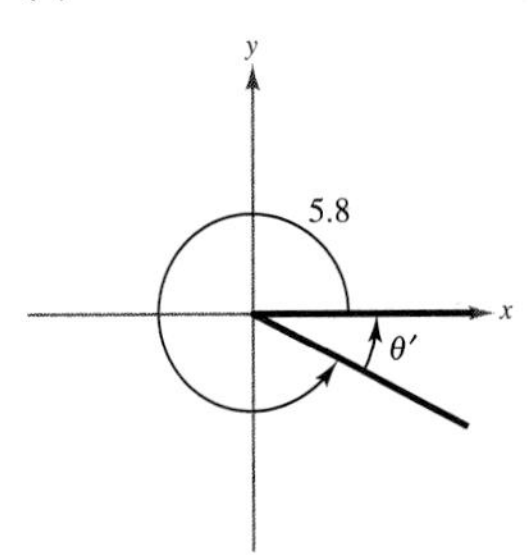

43. (a) $\sin 225° = -\frac{\sqrt{2}}{2}$ (b) $\sin(-225°) = \frac{\sqrt{2}}{2}$

$\cos 225° = -\frac{\sqrt{2}}{2}$ $\cos(-225°) = -\frac{\sqrt{2}}{2}$

$\tan 225° = 1$ $\tan(-225°) = -1$

45. (a) $\sin 750° = \frac{1}{2}$ (b) $\sin 510° = \frac{1}{2}$

$\cos 750° = \frac{\sqrt{3}}{2}$ $\cos 510° = -\frac{\sqrt{3}}{2}$

$\tan 750° = \frac{\sqrt{3}}{3}$ $\tan 510° = -\frac{\sqrt{3}}{3}$

47. (a) $\sin\frac{4\pi}{3} = -\frac{\sqrt{3}}{2}$ (b) $\sin\frac{2\pi}{3} = \frac{\sqrt{3}}{2}$

$\cos\frac{4\pi}{3} = -\frac{1}{2}$ $\cos\frac{2\pi}{3} = -\frac{1}{2}$

$\tan\frac{4\pi}{3} = \sqrt{3}$ $\tan\frac{2\pi}{3} = -\sqrt{3}$

49. (a) $\sin\left(-\frac{\pi}{6}\right) = -\frac{1}{2}$ (b) $\sin\left(\frac{5\pi}{6}\right) = \frac{1}{2}$

$\cos\left(-\frac{\pi}{6}\right) = \frac{\sqrt{3}}{2}$ $\cos\left(\frac{5\pi}{6}\right) = -\frac{\sqrt{3}}{2}$

$\tan\left(-\frac{\pi}{6}\right) = -\frac{\sqrt{3}}{3}$ $\tan\left(\frac{5\pi}{6}\right) = -\frac{\sqrt{3}}{3}$

51. (a) $\sin\frac{11\pi}{4} = \frac{\sqrt{2}}{2}$ (b) $\sin\left(-\frac{13\pi}{6}\right) = -\frac{1}{2}$

$\cos\frac{11\pi}{4} = -\frac{\sqrt{2}}{2}$ $\cos\left(-\frac{13\pi}{6}\right) = \frac{\sqrt{3}}{2}$

$\tan\frac{11\pi}{4} = -1$ $\tan\left(-\frac{13\pi}{6}\right) = -\frac{\sqrt{3}}{3}$

53. (a) 0.1736 (b) 5.7588

55. (a) −0.3420 (b) −0.3420

57. (a) 1.7321 (b) 1.7321

59. (a) 0.3640 (b) 0.3640

61. (a) 0.6052 (b) 0.6077

63. (a) $30° = \frac{\pi}{6}$, $150° = \frac{5\pi}{6}$ (b) $210° = \frac{7\pi}{6}$, $330° = \frac{11\pi}{6}$

65. (a) $60° = \frac{\pi}{3}$, $120° = \frac{2\pi}{3}$ (b) $135° = \frac{3\pi}{4}$, $315° = \frac{7\pi}{4}$

67. (a) $45° = \frac{\pi}{4}$, $225° = \frac{5\pi}{4}$ (b) $150° = \frac{5\pi}{6}$, $330° = \frac{11\pi}{6}$

69. (a) 54.99°, 125.01° (b) 195.00°, 345.00°

71. (a) 0.175, 6.109 (b) 2.201, 4.083

73. (a) 0.873, 4.014 (b) 1.693, 4.835

75. $\frac{4}{5}$ **77.** $-\frac{\sqrt{13}}{2}$ **79.** $\frac{8}{5}$

81. (a) 25.2°F (b) 65.1°F (c) 50.8°F

83. (a) 12 miles (b) 6 miles (c) 6.9 miles

Section 1.5 *(page 166)*

Warm Up *(page 166)*

1. 6π **2.** $\frac{1}{2}$ **3.** $\frac{\pi}{6}$ **4.** $\frac{7\pi}{6}$ **5.** -2 **6.** $-\frac{4}{3}$

7. 1 **8.** 0 **9.** 1 **10.** 0

1. Period: π
Amplitude: 3

3. Period: 4π
Amplitude: $\frac{5}{2}$

5. Period: 2
Amplitude: $\frac{2}{3}$

7. Period: 2π
Amplitude: 2

9. Period: $\frac{\pi}{5}$
Amplitude: 3

11. Period: 3π
Amplitude: $\frac{1}{2}$

13. Period: $\frac{1}{2}$
Amplitude: 3

15. g is a shift of f π units to the right.

17. g is a reflection of f about the x-axis.

19. The period of f is twice the period of g.

21. Shift the graph of f two units up to obtain the graph of g.

23. The graph of g has twice the amplitude of the graph of f.

25. The graph of g is a horizontal shift of the graph of f π units to the right.

27.

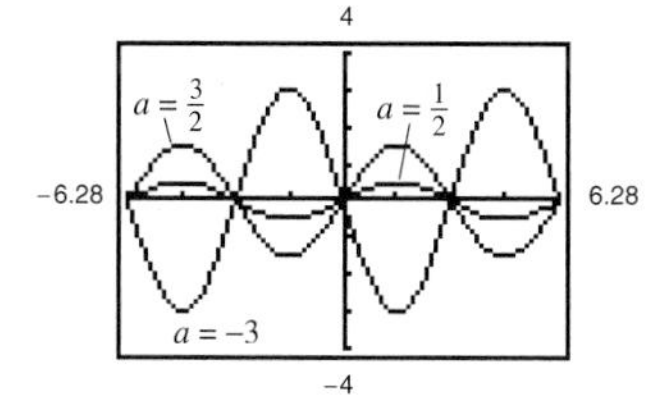

Amplitude changes

29.

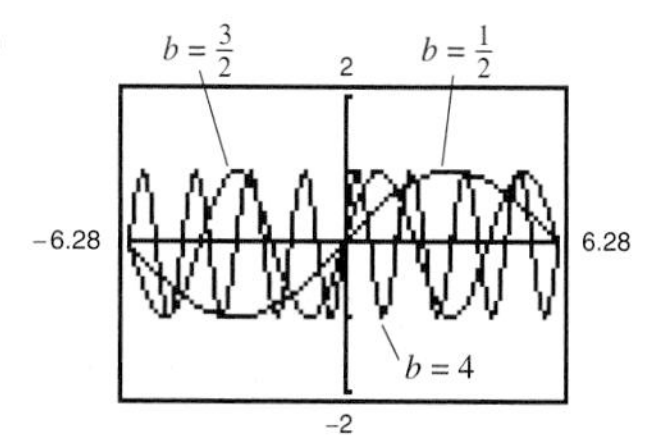

Period changes

31.

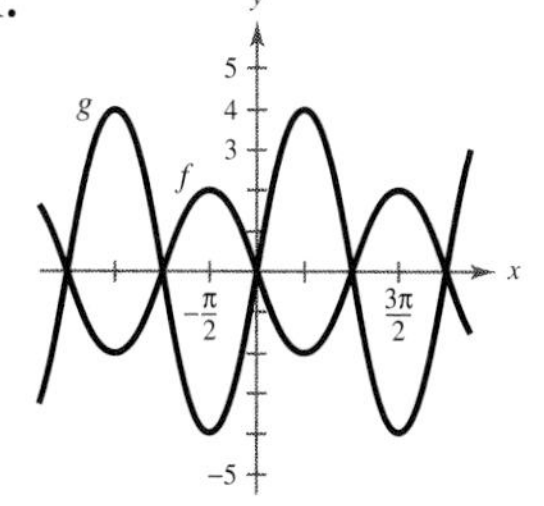

33.

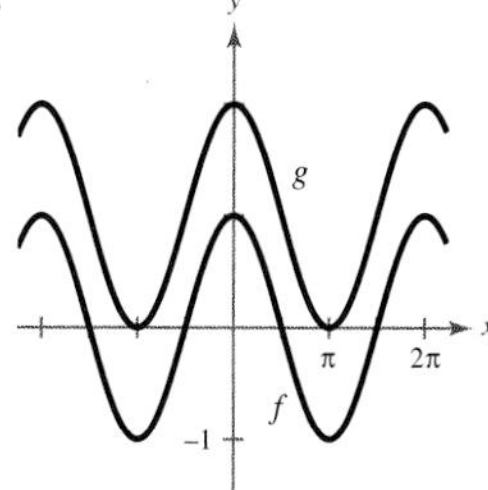

35.

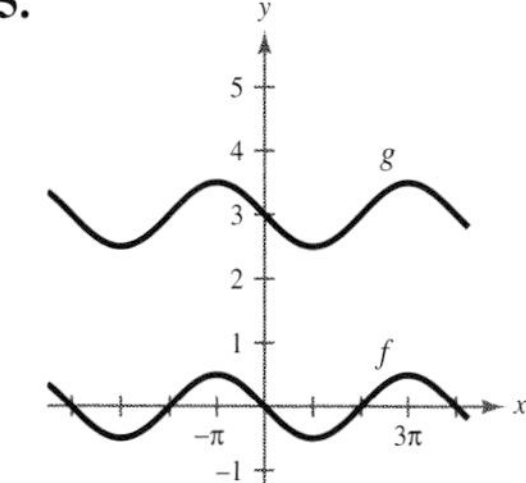

37.

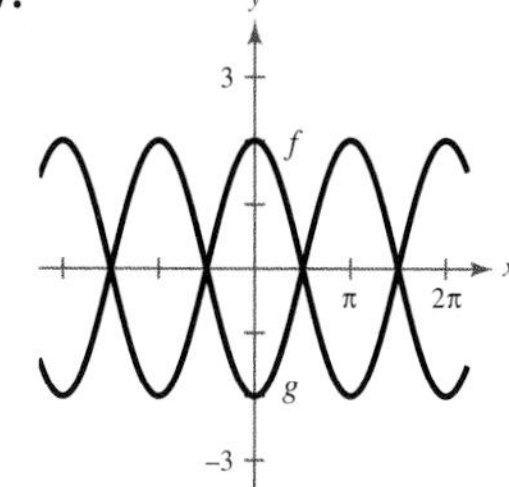

39.

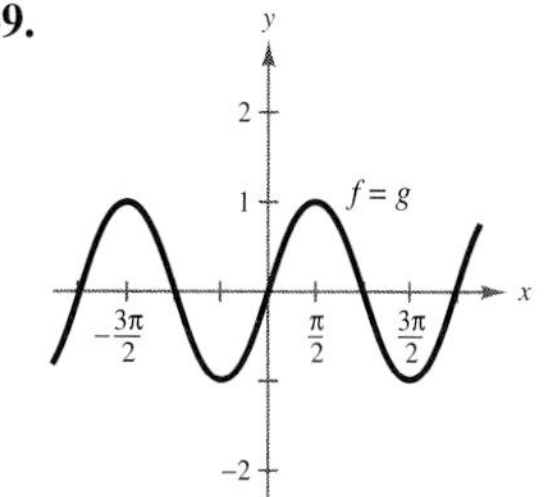

41.

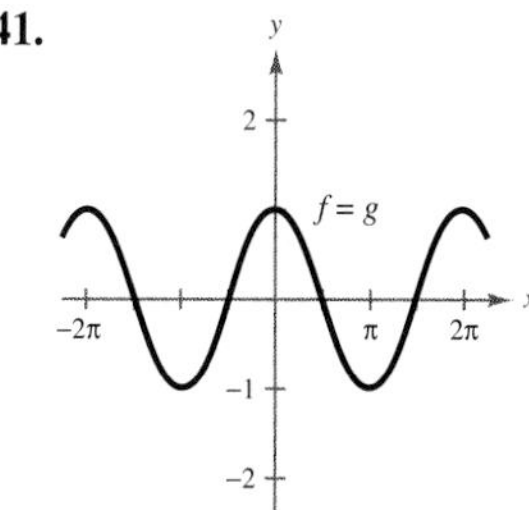

43.

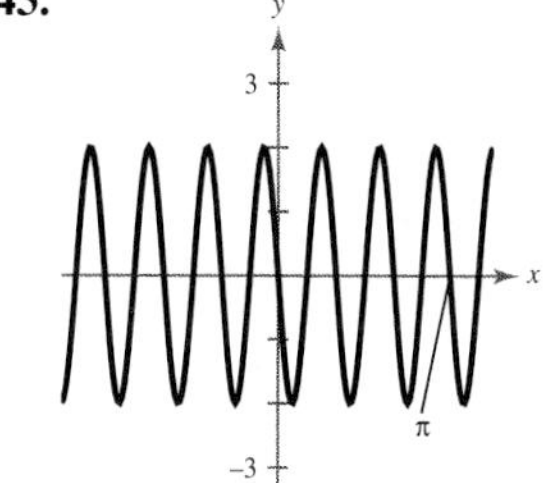

45.

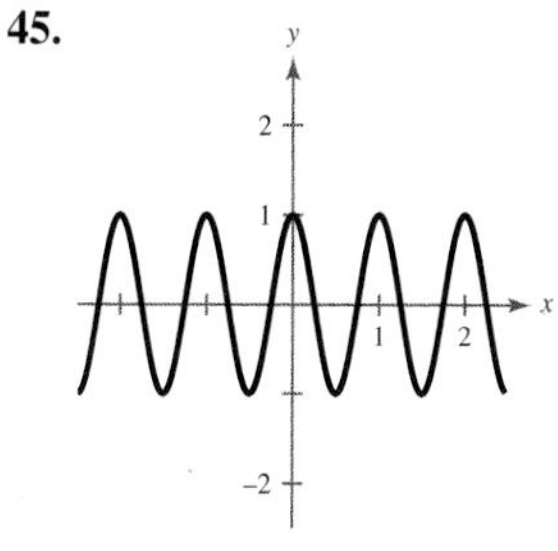

47.

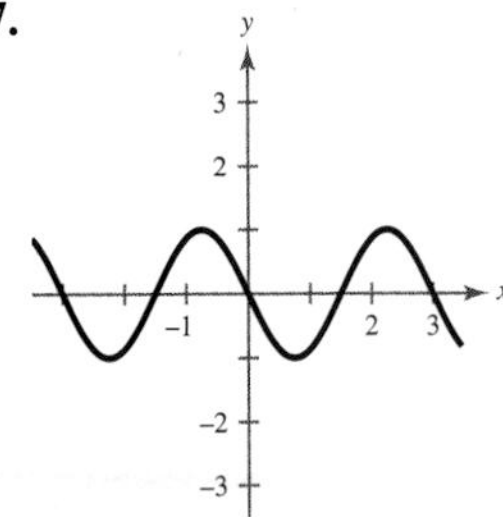

49.

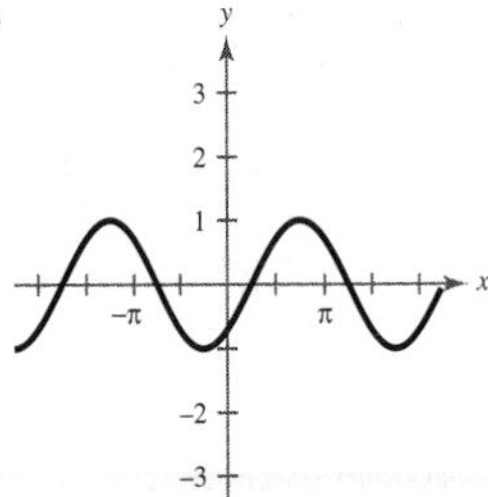

51.

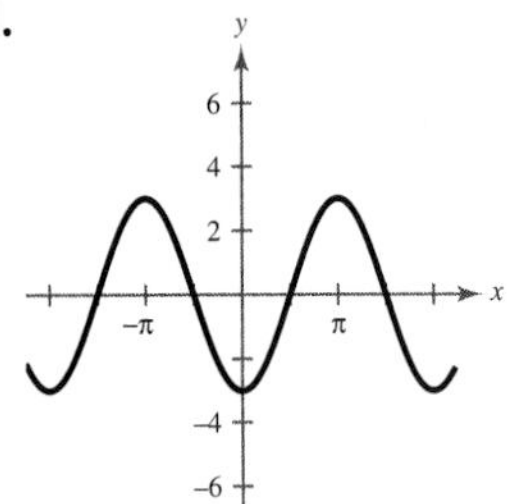

53.

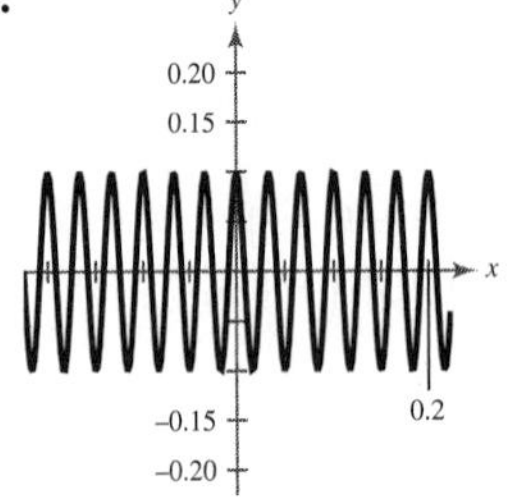

55.

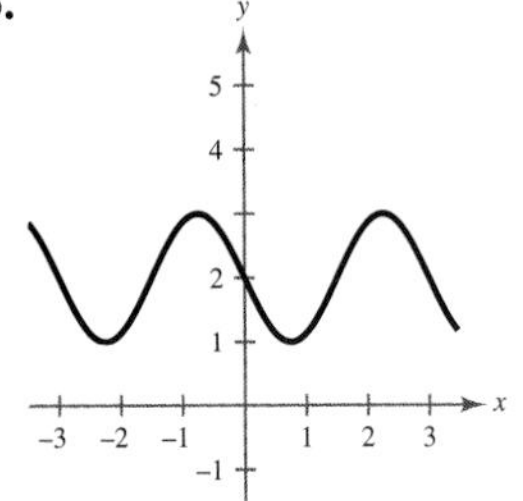

57.

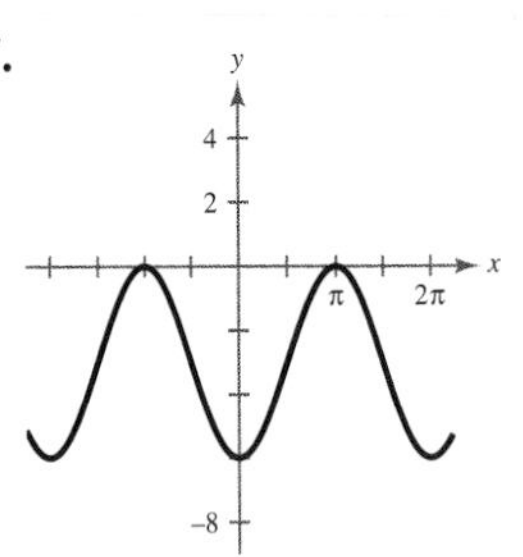

59.

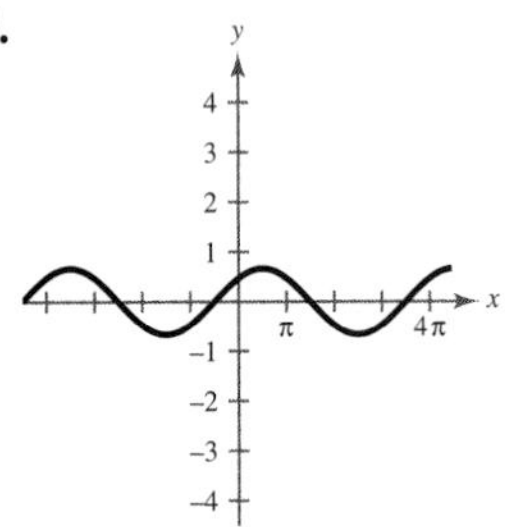

61.

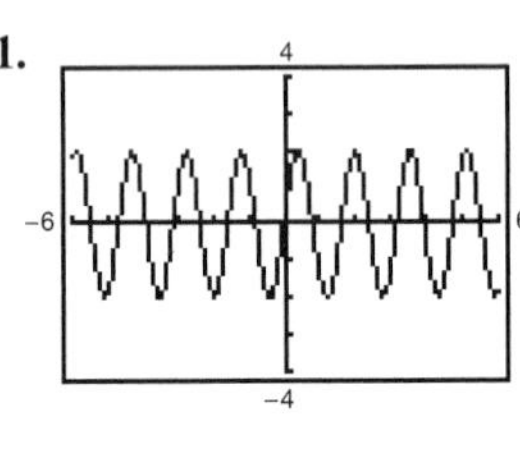

63.

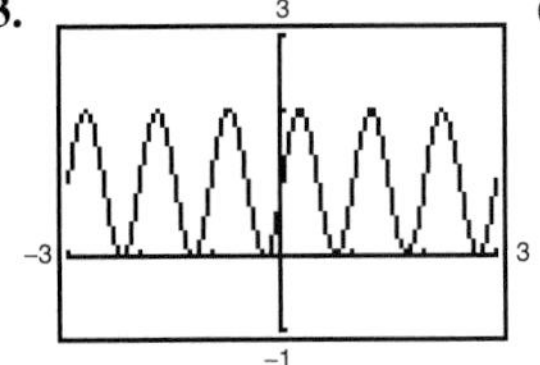

65.

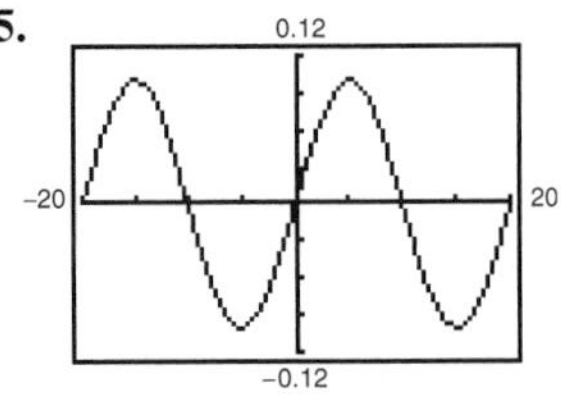

67.

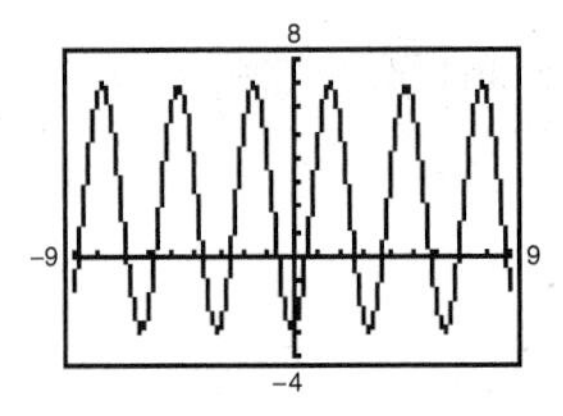

69. $y = 2 + 3\cos x$ **71.** $y = -4\cos x + 4$

73. $y = -3\sin(2x)$ **75.** $y = \sin\left(x - \frac{\pi}{4}\right)$

77.

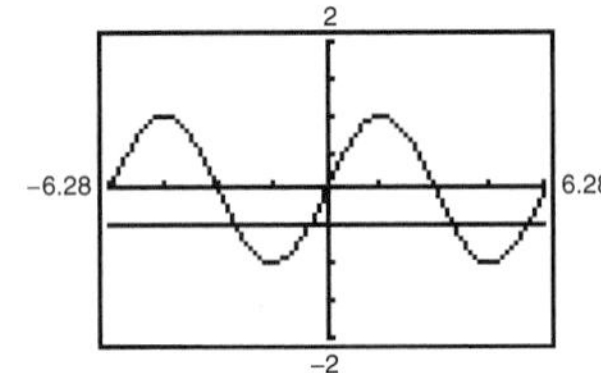

$x = -\frac{\pi}{6}, -\frac{5\pi}{6}, \frac{7\pi}{6}, \frac{11\pi}{6}$

79.

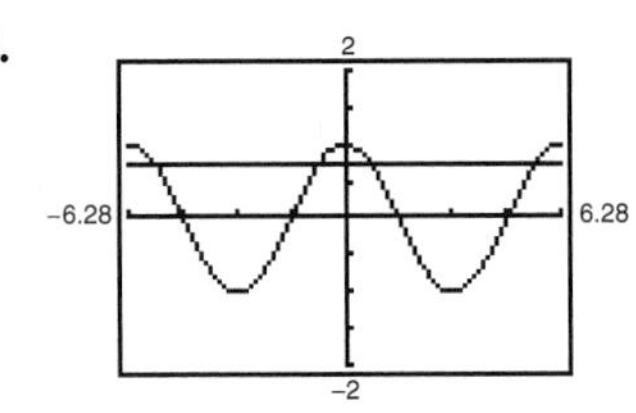

$x = \pm\frac{\pi}{4}, \pm\frac{7\pi}{4}$

81. (a) Even (b) Even

83. (a) 6 seconds (b) 10 cycles per minute

(c)

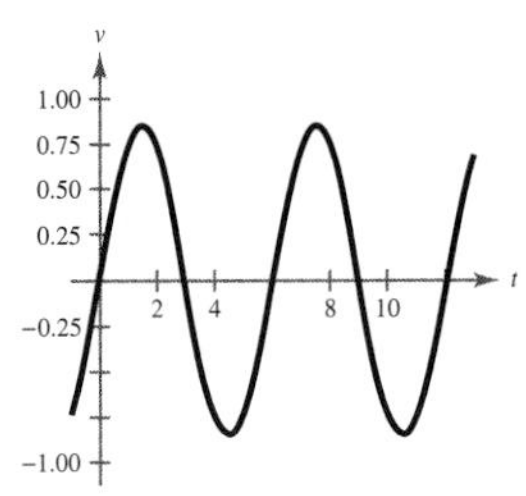

85 (a) $\frac{1}{440}$ second (b) 440 cycles per second

87.

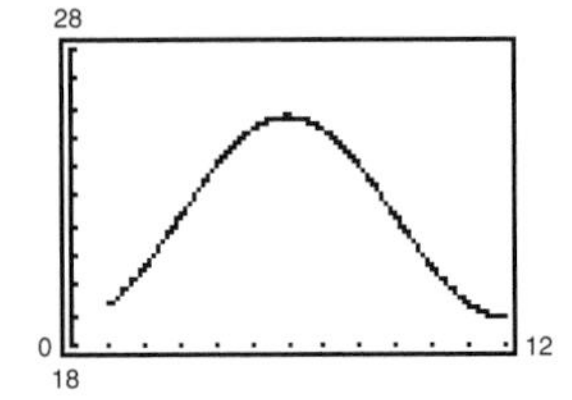

89. (a) $C(t) = 56.35 + 27.35 \sin\left(\frac{\pi t}{6} + 4.19\right)$

(b)

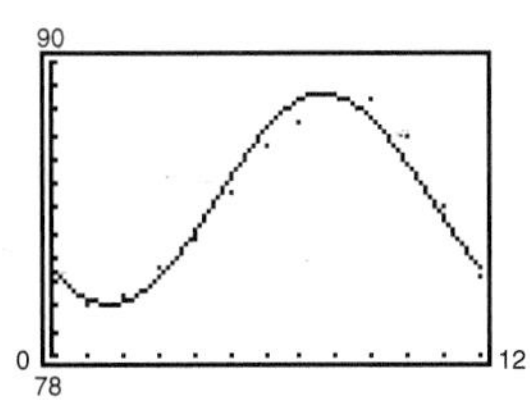

(c)

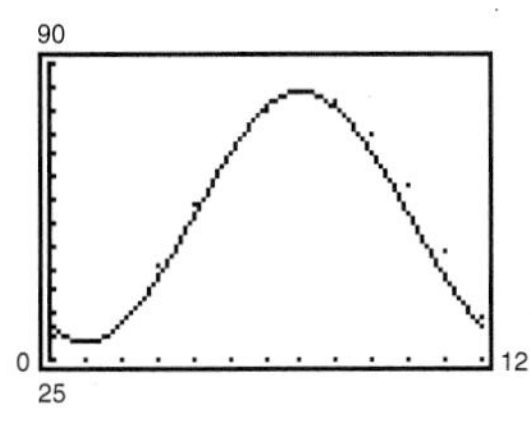

(d) Honolulu: 84.40; Chicago: 56.35. Vertical translation (d)

(e) 12. Yes. One full period is 1 year.

(f) Chicago, amplitude

91. (a)

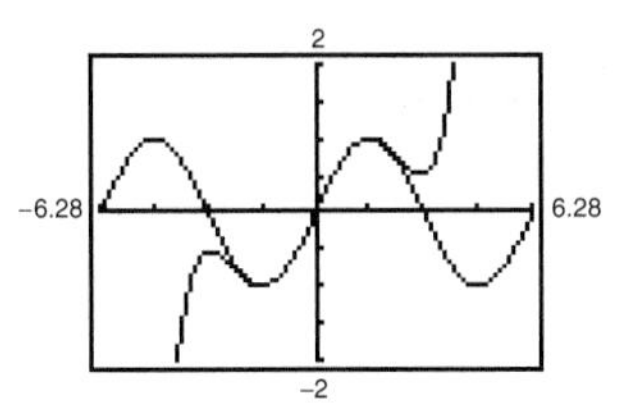

They appear to coincide from $-\frac{\pi}{2}$ to $\frac{\pi}{2}$.

(b)

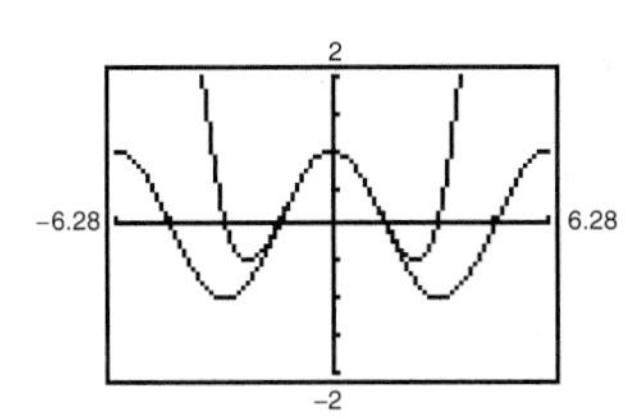

They appear to coincide from $-\frac{\pi}{2}$ to $\frac{\pi}{2}$.

(c) $-\frac{x^7}{7!}, -\frac{x^6}{6!}$

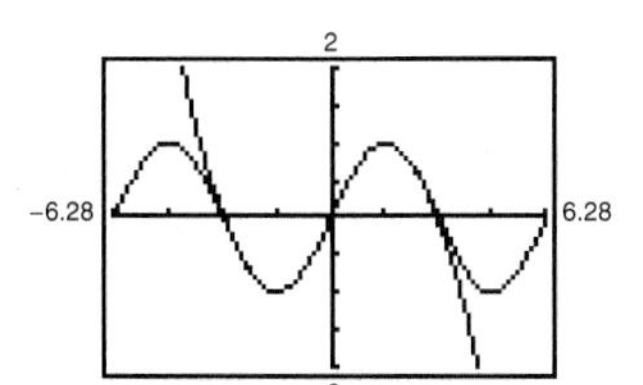

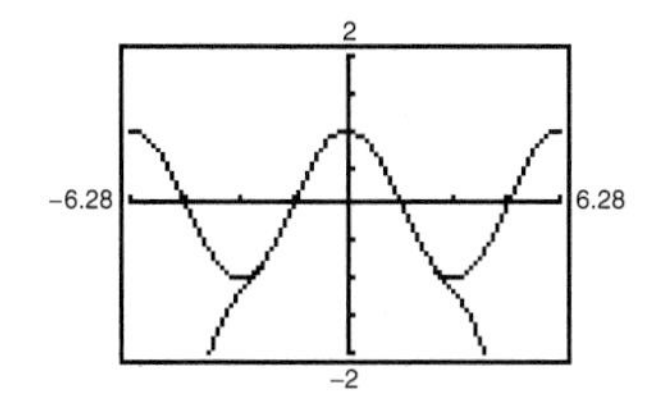

The accuracy increased.

93. (a)

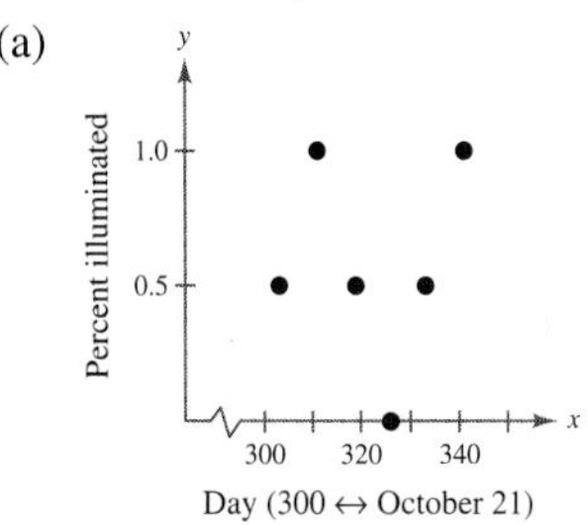

(b) $y = \frac{1}{2} + \frac{1}{2}\sin\left[\frac{\pi}{15}(t - 303)\right]$

(c)

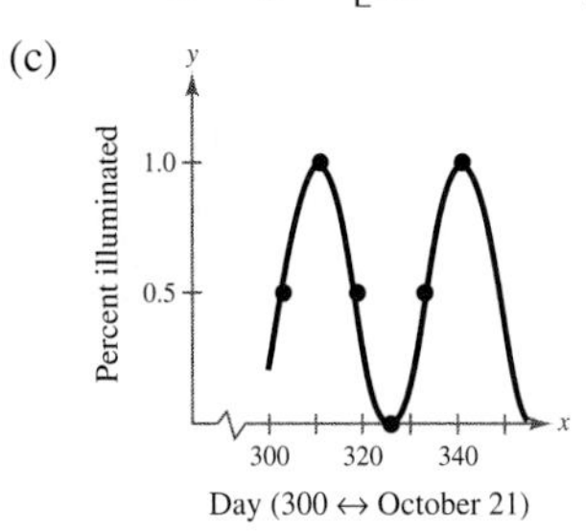

(d) 0

Section 1.6 *(page 171)*

Warm Up *(page 171)*

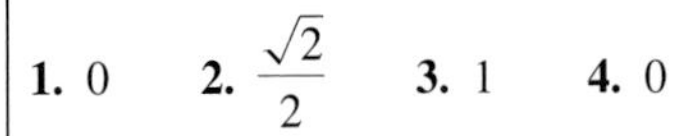

1. 0 **2.** $\frac{\sqrt{2}}{2}$ **3.** 1 **4.** 0 **5.** 0 **6.** 0

7.

8.

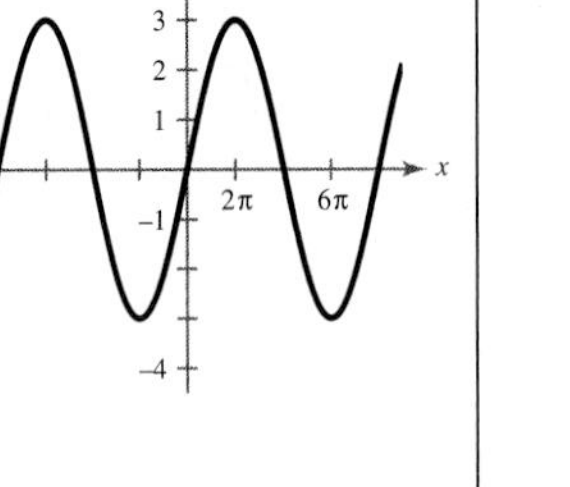

9.

10.

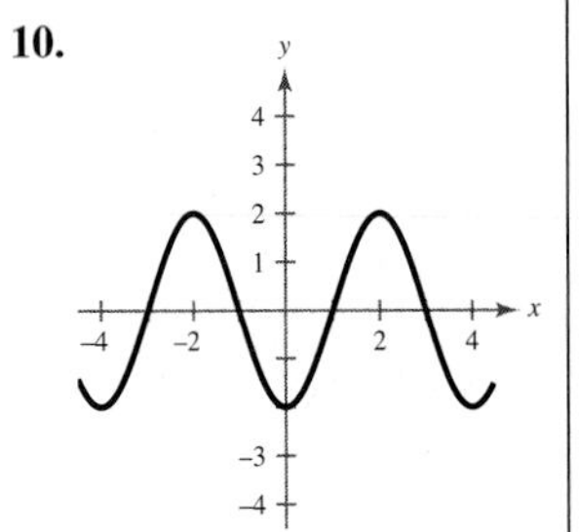

1. g, 4π **3.** f, $\frac{\pi}{2}$ **5.** b, 2 **7.** e, 2π

9.

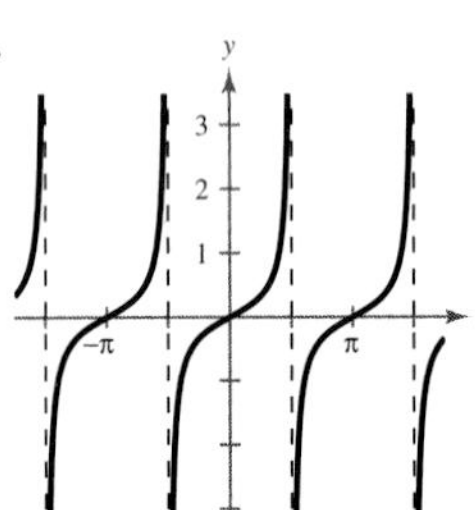

11.

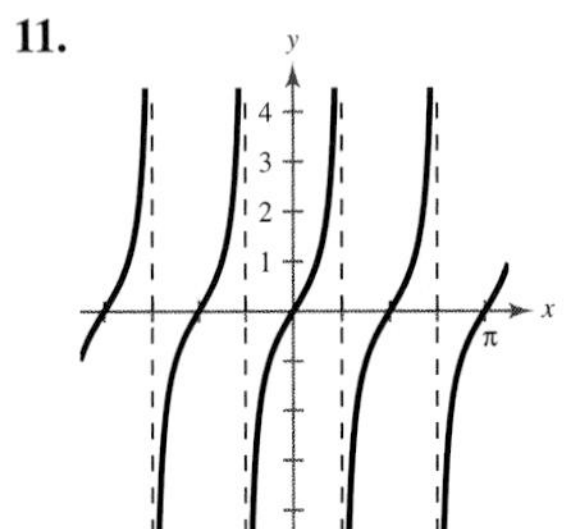

13.

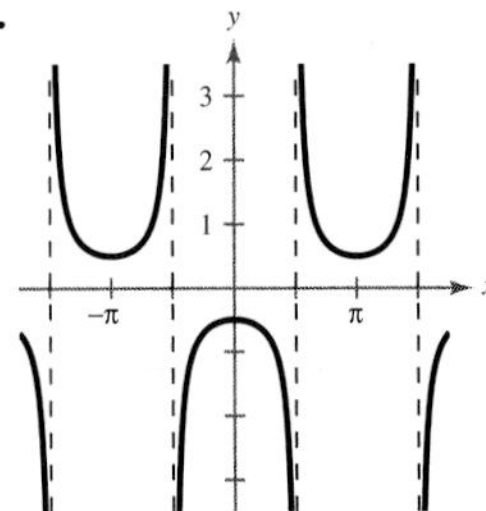

15.

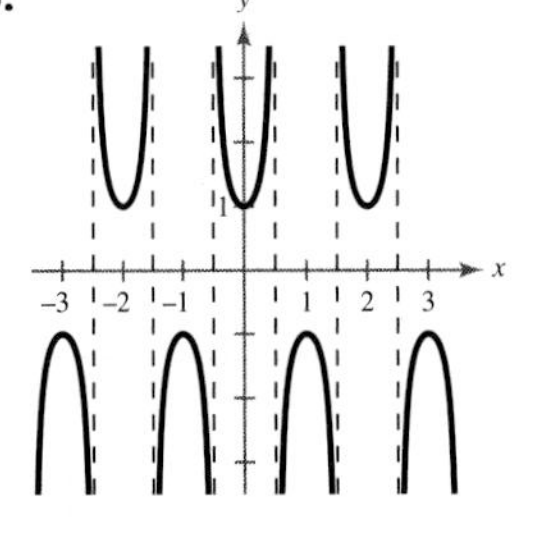

17.

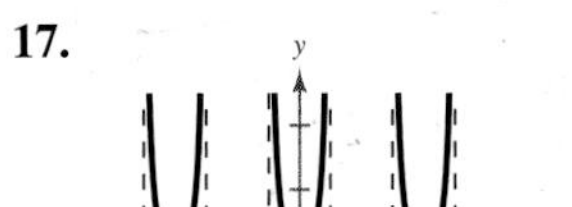

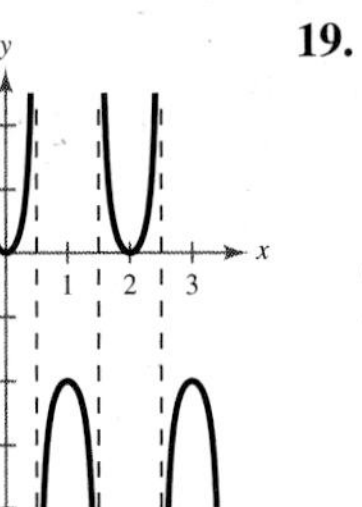

19.

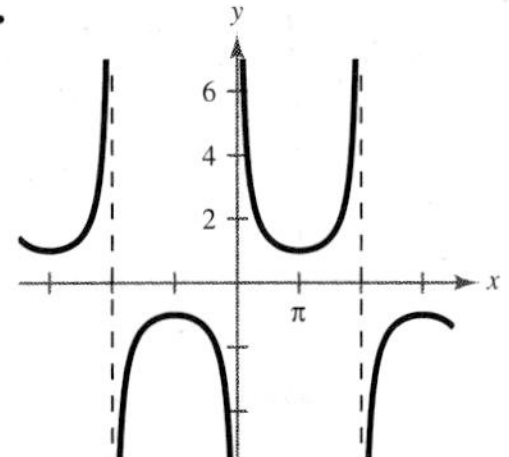

21.

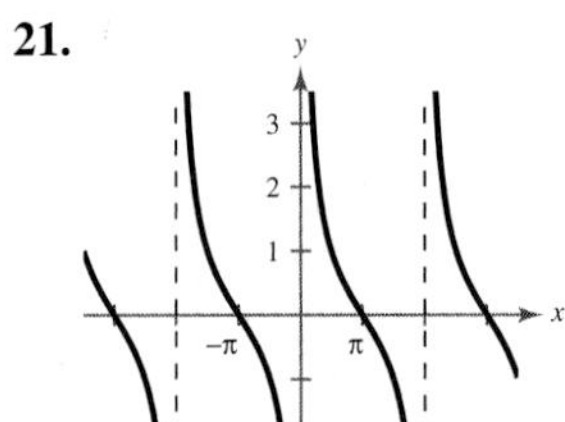

23.

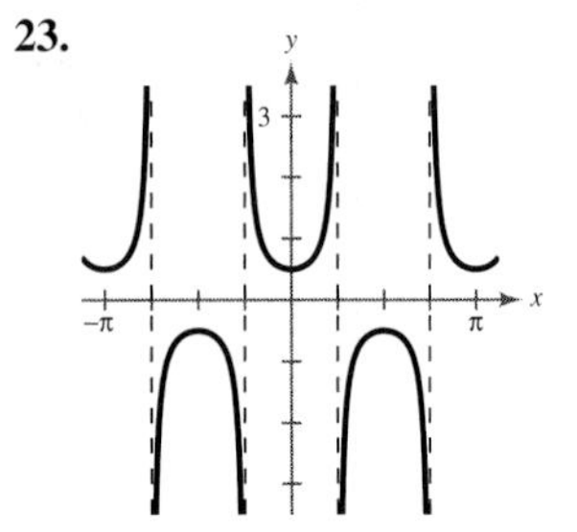

25.

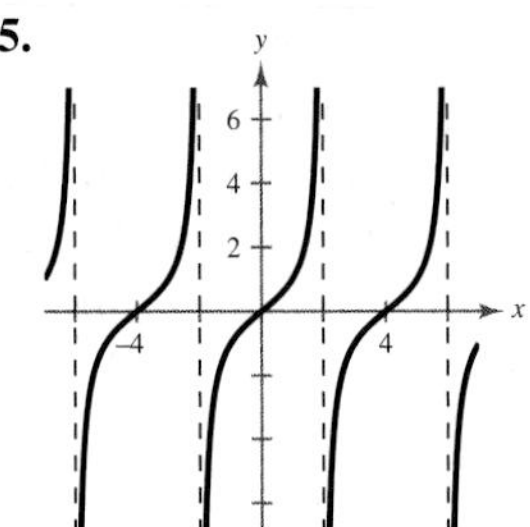

27.

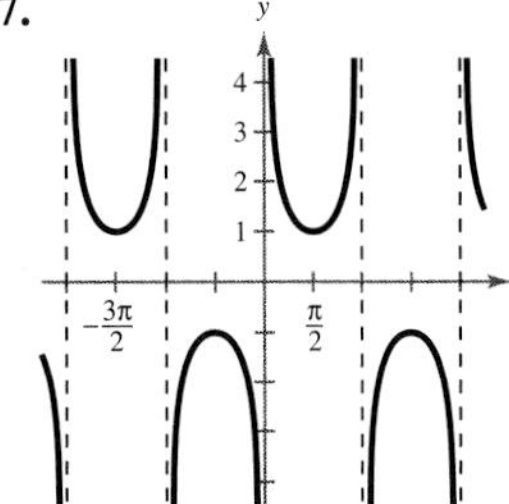

29.

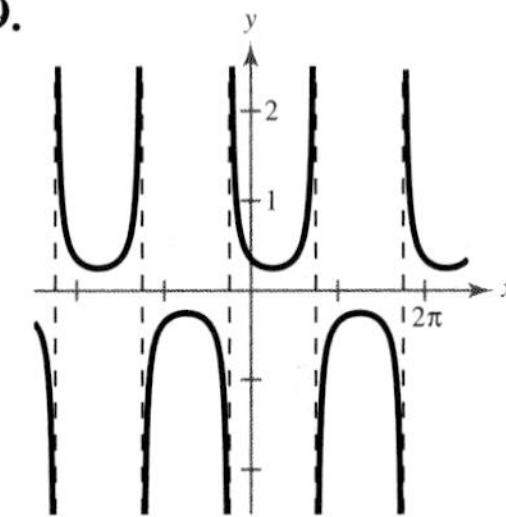

31.

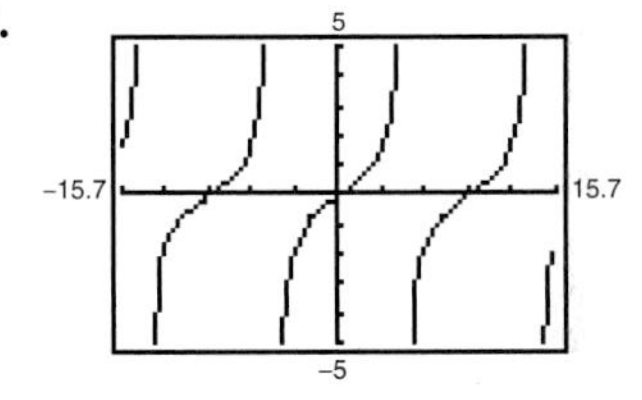

33.

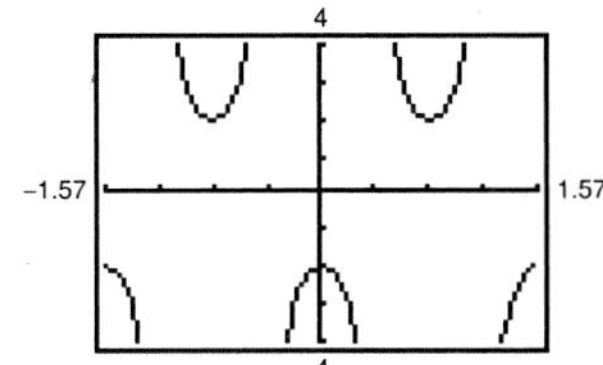

35.

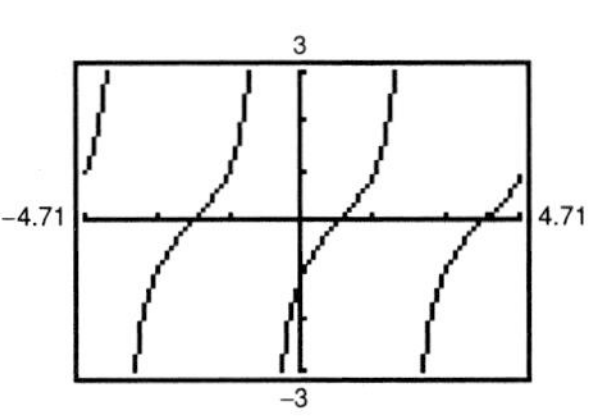

37.

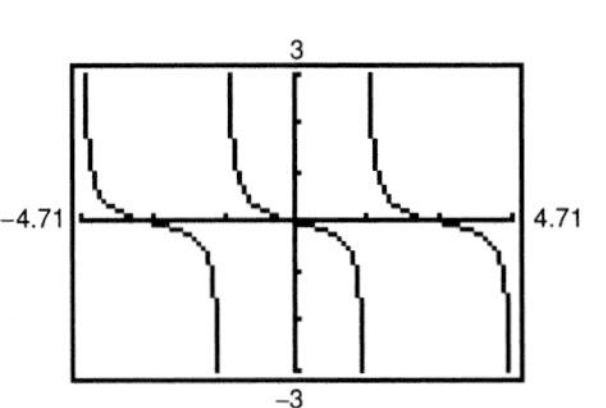

39.

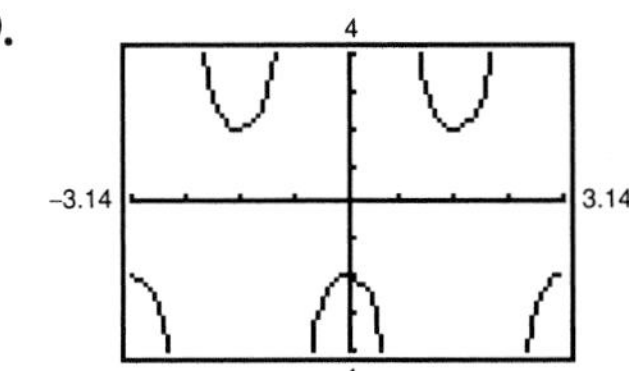

41. $-\frac{7\pi}{4}, -\frac{3\pi}{4}, \frac{\pi}{4}, \frac{5\pi}{4}$ **43.** $-\frac{4\pi}{3}, -\frac{2\pi}{3}, \frac{2\pi}{3}, \frac{4\pi}{3}$

45. Even **47.** As x approaches $\pi/2$ from the left, f approaches ∞. As x approaches $\pi/2$ from the right, f approaches $-\infty$.

49. (a)

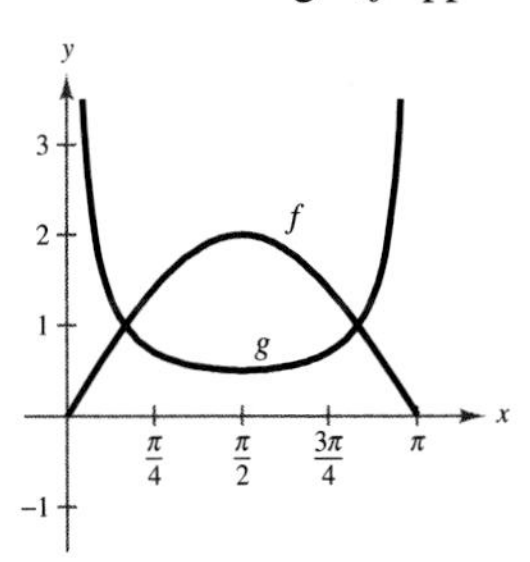

(b) $\frac{\pi}{6} < x < \frac{5\pi}{6}$

(c) Sine approaches 0 and cosecant approaches $\pm\infty$ because the cosecant is the reciprocal of the sine.

51.

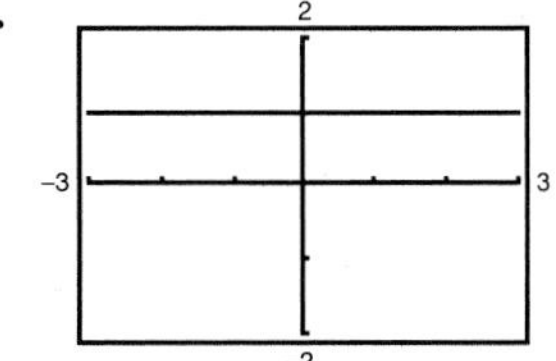

Not equivalent

53.

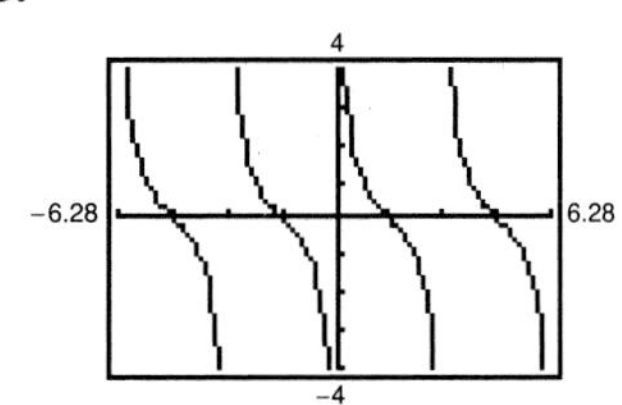

Equivalent

55. d, 0 **57.** b, 0

59.

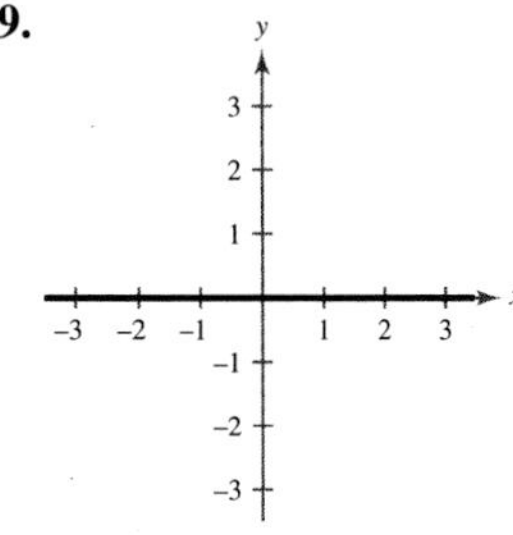

Equal

61.

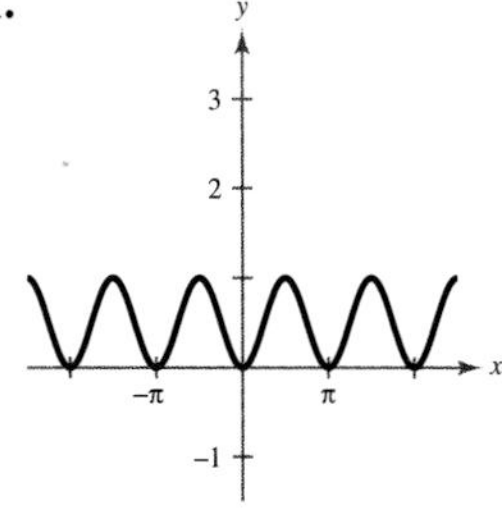

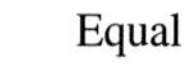

Equal

63.

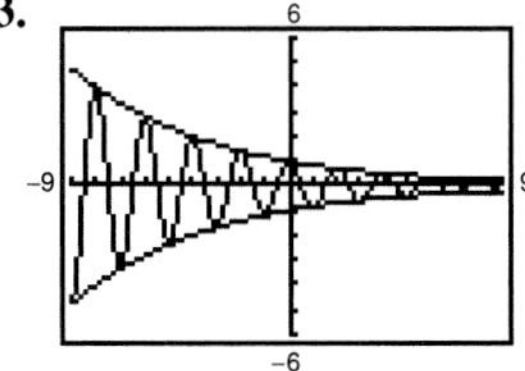

0

65.

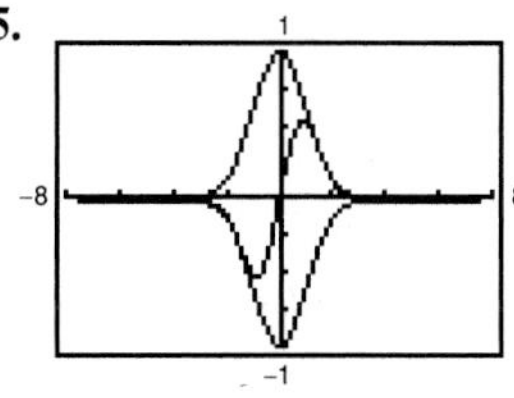

0

67. $d = 5 \cot x$

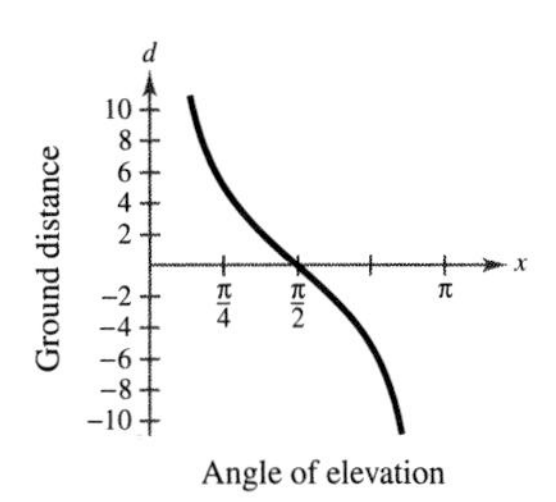

69. As the predator population increases, the number of prey decreases. When the number of prey is small, the number of predators decreases.

71. (a)

(b) Damped sine wave; goes to 0 as t increases.

73.

75. (a)

(b)

$$y_3 = \frac{4}{\pi}\left[\sin(\pi x) + \frac{1}{3}\sin(3\pi x) + \frac{1}{5}\sin(5\pi x) + \frac{1}{7}\sin(7\pi x)\right]$$

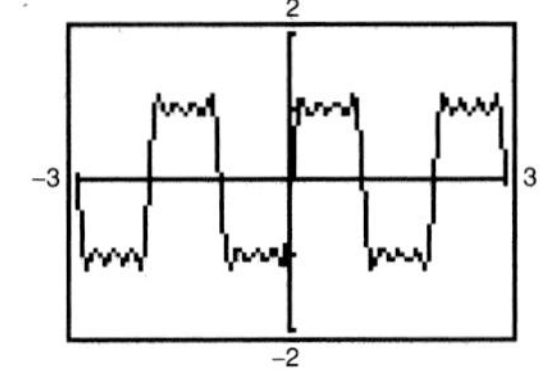

(c)

$$y_4 = \frac{4}{\pi}\left[\sin(\pi x) + \frac{1}{3}\sin(3\pi x) + \frac{1}{5}\sin(5\pi x) + \frac{1}{7}\sin(7\pi x) + \frac{1}{9}\sin(9\pi x)\right]$$

77.

∞

79.

1

81.

Oscillates

Section 1.7 *(page 189)*

Warm Up *(page 189)*

1. -1 **2.** -1 **3.** -1 **4.** $\frac{\sqrt{2}}{2}$ **5.** 0 **6.** $\frac{\pi}{6}$
7. π **8.** $\frac{\pi}{4}$ **9.** 0 **10.** $-\frac{\pi}{4}$

1. False. $\frac{5\pi}{6}$ is not in the range of the arcsine.

3. $\frac{\pi}{6}$ **5.** $\frac{\pi}{3}$ **7.** $\frac{\pi}{6}$ **9.** $\frac{5\pi}{6}$ **11.** $-\frac{\pi}{3}$ **13.** $\frac{2\pi}{3}$

15. $\frac{\pi}{3}$ **17.** 0 **19.** 1.29 **21.** -0.85 **23.** -1.25

25. 0.32 **27.** 1.99 **29.** 0.74 **31.** $-\frac{\pi}{3}, -\frac{1}{\sqrt{3}}, 1$

33.

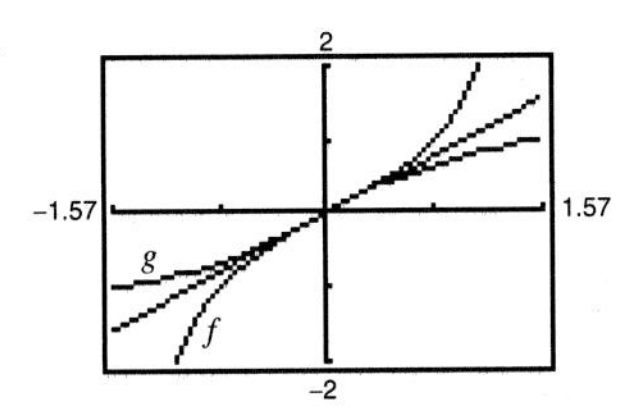

35. $\theta = \arctan \frac{x}{4}$ **37.** $\theta = \arcsin \frac{x+2}{5}$ **39.** 0.3

41. -0.1 **43.** 0 **45.** $\frac{3}{5}$ **47.** $\frac{\sqrt{5}}{5}$ **49.** $\frac{12}{13}$

51. $\frac{\sqrt{34}}{5}$ **53.** $\frac{\sqrt{5}}{3}$ **55.** $\frac{1}{x}$ **57.** $\sqrt{1-4x^2}$

59. $\sqrt{1-x^2}$ **61.** $\frac{\sqrt{9-x^2}}{x}$ **63.** $\frac{\sqrt{x^2+2}}{x}$

65.

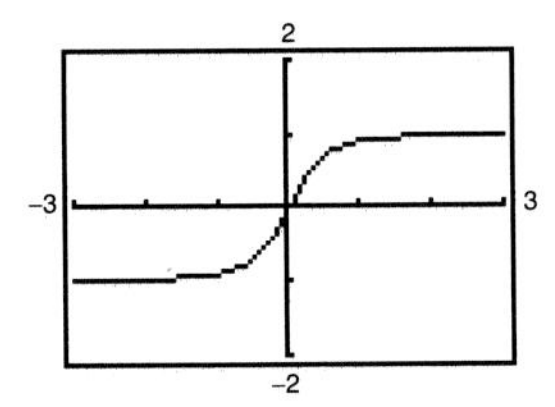

$y = \pm 1$

67. $\frac{9}{\sqrt{x^2+81}}, x > 0; \frac{-9}{\sqrt{x^2+81}}, x < 0$

69. $\frac{|x-1|}{\sqrt{x^2-2x+10}}$

71.

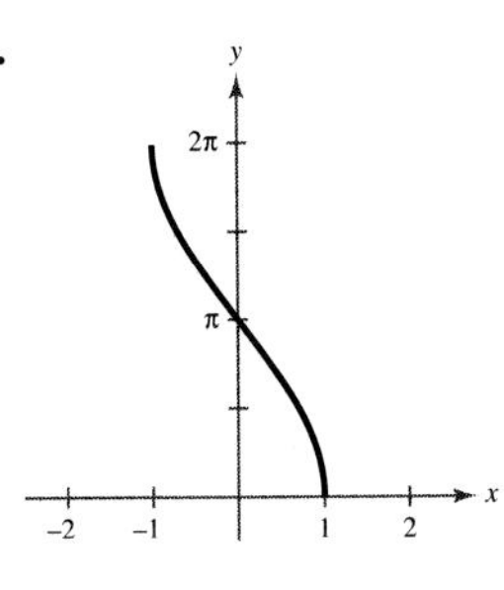

73.

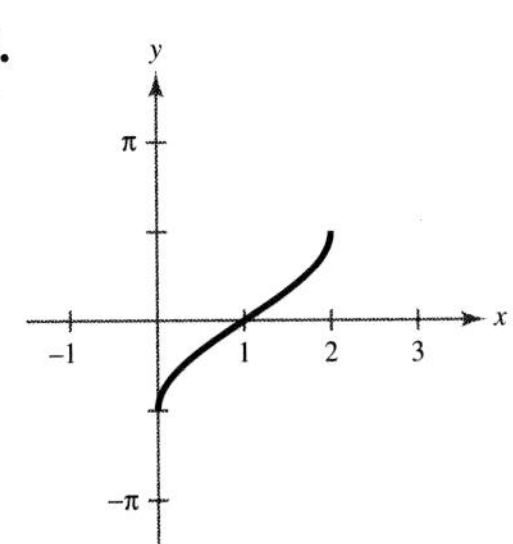

75.

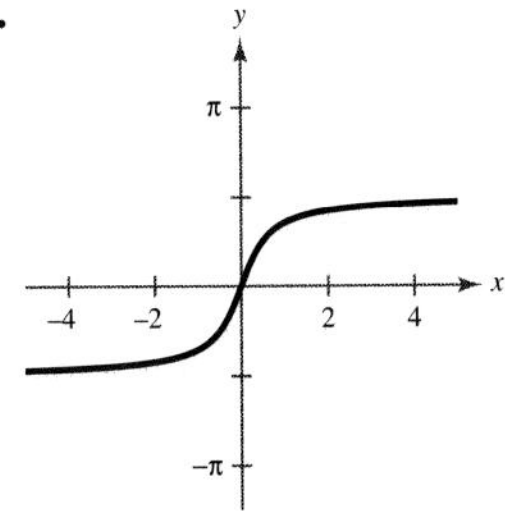

77.

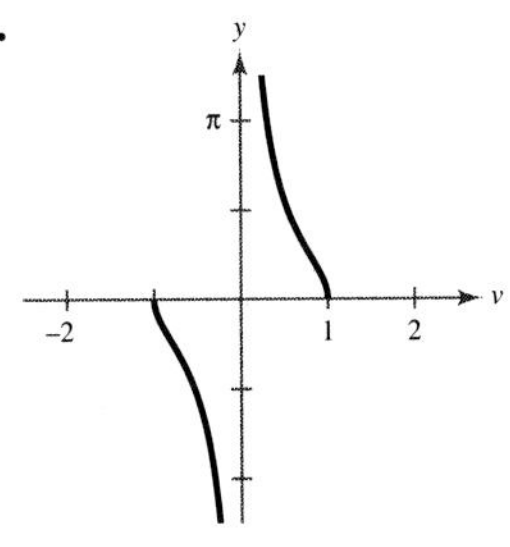

79. $3\sqrt{2}\sin\left(2t + \frac{\pi}{4}\right)$

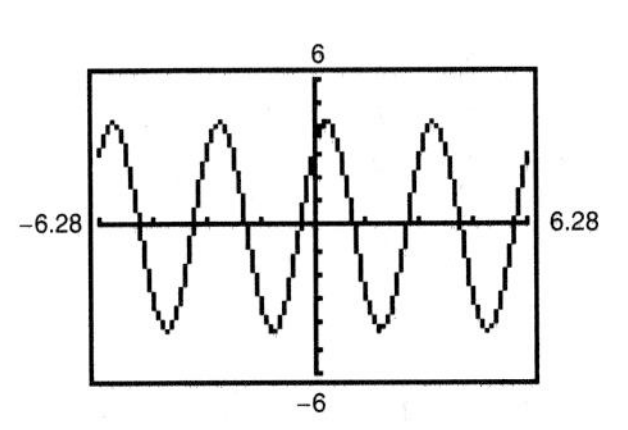

81. (a) $f \circ f^{-1}$

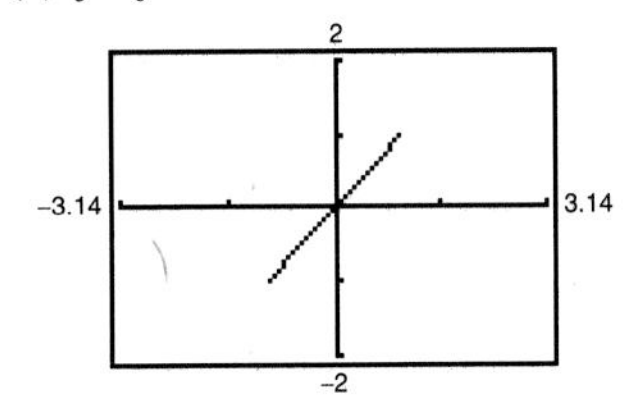

$f^{-1} \circ f$

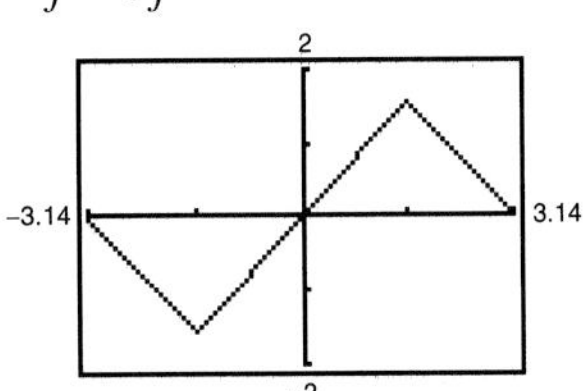

(b) The domain and range of the functions are restricted. The graphs of $f \circ f^{-1}$ and $f^{-1} \circ f$ differ because of the domains and ranges of f and f^{-1}.

83. (a) $\theta = \arcsin \frac{10}{s}$ (b) 0.21, 0.43

85. (a)

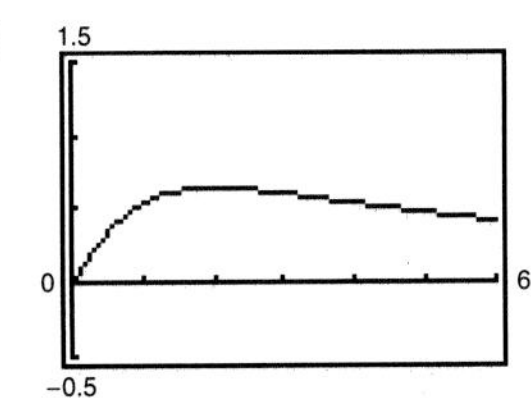

(b) 2 feet
(c) $\beta = 0$

87. (a) $\theta = \arctan \frac{5}{x}$ (b) 26.6°, 59.0°

89. Domain: $(-\infty, \infty)$; Range: $(0, \pi)$

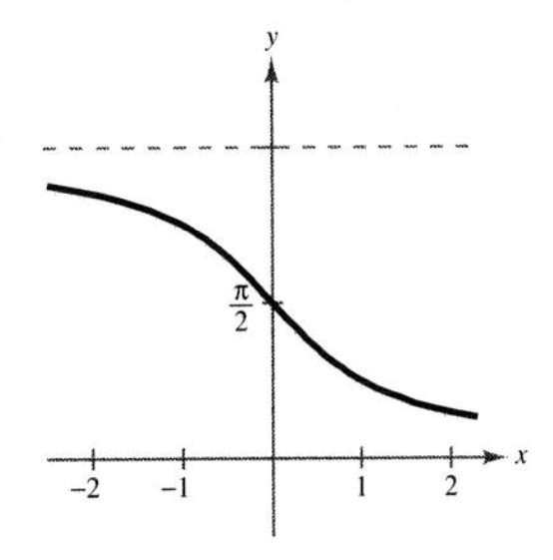

91. Domain: $(-\infty, -1] \cup [1, \infty)$
Range: $[-\pi/2, 0) \cup (0, \pi/2]$

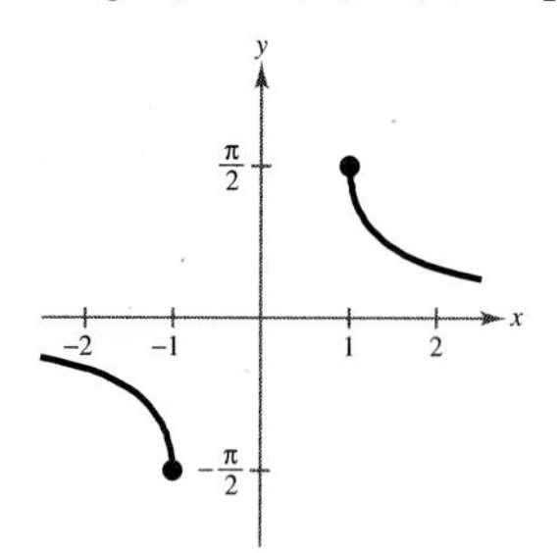

93.–97. Answers will vary. **99.** Buy now. **101.** 8

Section 1.8 *(page 200)*

Warm Up *(page 200)*

1. 8.45	**2.** 78.99	**3.** 1.06	**4.** 1.24	**5.** 4.88
6. 34.14	**7.** 4; π	**8.** $\frac{1}{2}$; 2	**9.** 3; $\frac{2}{3}$	**10.** 0.2; 8π

1. $a \approx 3.64$, $c \approx 10.64$, $B = 70°$
3. $a \approx 8.26$, $c \approx 25.38$, $A = 19°$
5. $c \approx 11.66$, $A \approx 30.96°$, $B \approx 59.04°$
7. $a \approx 49.48$, $A \approx 72.08°$, $B = 17.92°$
9. $a \approx 91.34$, $b \approx 420.70$, $B = 77°45'$

11. 2.56 inches **13.** 103.9 feet **15.** 15.4 feet

17. (a)

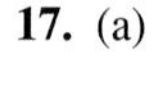

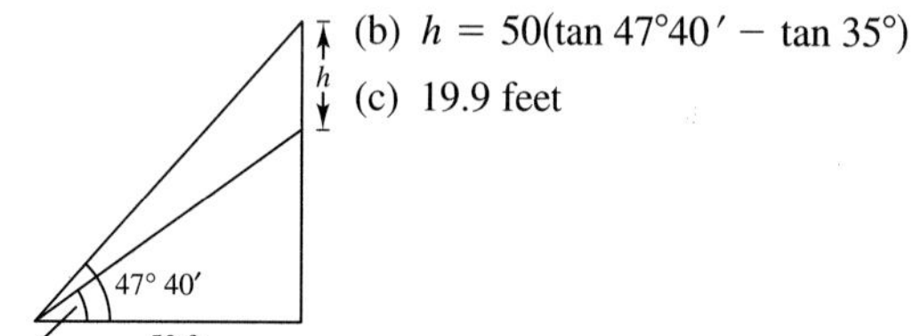

(b) $h = 50(\tan 47°40' - \tan 35°)$
(c) 19.9 feet

19. 2236.8 feet **21.** 56.3° **23.** 15.5° **25.** 5099 feet
27. 0.73 mile **29.** 508 miles north; 650 miles east
31. (a) N 58° E (b) 68.82 meters **33.** N 56.3° W
35. 1933.3 feet **37.** ≈ 3.23 miles or $\approx 17{,}054$ feet
39. 78.7° **41.** 35.3° **43.** $y = \sqrt{3}\,r$ **45.** 29.4 inches
47. $a \approx 7$, $c \approx 12.2$ **49.** (a) 4 (b) 4 (c) $\frac{1}{16}$
51. (a) $\frac{1}{16}$ (b) 60 (c) $\frac{1}{120}$ **53.** $y = 4\sin(\pi t)$

55. $y = 3\cos\left(\dfrac{4\pi t}{3}\right)$ **57.** $\omega = 528\pi$

59. (a)

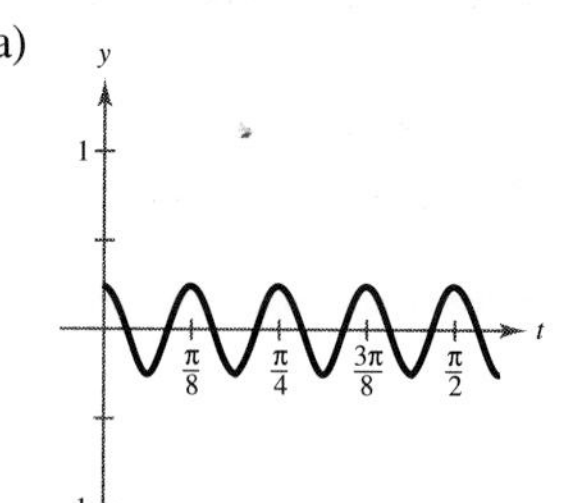

(b) $\dfrac{\pi}{8}$ seconds
(c) $\dfrac{\pi}{32}$ seconds

61. (a) and (b)

Base 1	*Base 2*	*Altitude*	*Area*
8	8 + 16 cos 10°	8 sin 10°	22.1
8	8 + 16 cos 20°	8 sin 20°	42.5
8	8 + 16 cos 30°	8 sin 30°	59.7
8	8 + 16 cos 40°	8 sin 40°	72.7
8	8 + 16 cos 50°	8 sin 50°	80.5
8	8 + 16 cos 60°	8 sin 60°	83.1
8	8 + 16 cos 70°	8 sin 70°	80.7

83.1 (maximum cross-sectional area)

(c) $A = 64(1 + \cos\theta)(\sin\theta)$

(d)

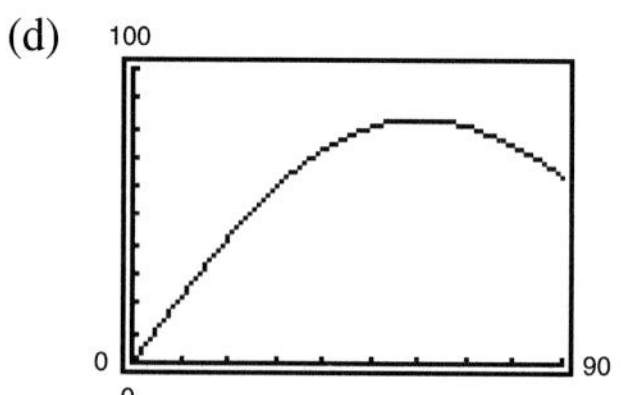

83.1

63. (a)

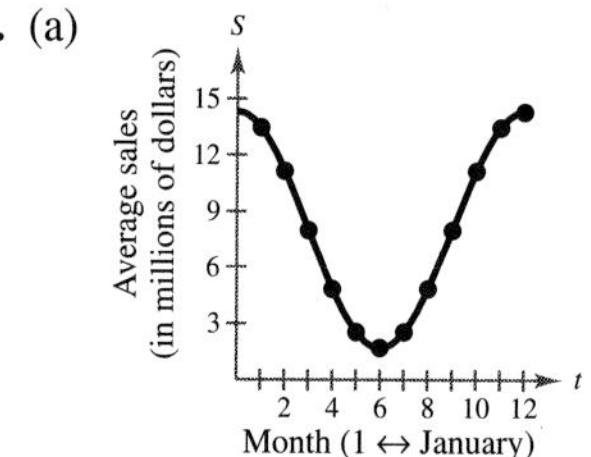

(b) $S = 8 + 6.3\cos\left(\dfrac{\pi t}{6}\right)$

(c) 12. Yes, sales of outerwear are seasonal.

(d) Maximum displacement from average sales of \$8 million

Focus on Concepts *(page 206)*

1. (a) The vertex is at the origin and the initial side is on the positive x-axis.
(b) Clockwise rotation of the terminal side.
(c) Two angles in standard position where the terminal sides coincide.
(d) The magnitude of the angle is between 90° and 180°.

2. Increases. The linear velocity is proportional to the radius.

3. False. For each θ there corresponds exactly one value of y.

4. Corresponding sides of similar triangles are proportional.

5. Undefined because $\sec\theta = 1/\cos\theta$.

6. Determine the trigonometric function of the reference angle and prefix the appropriate sign.

7. d; the period is 2π and the amplitude is 3.

8. a; the period is 2π and, because $a < 0$, the graph is reflected about the x-axis.

9. b; the period is 2 and the amplitude is 2.

10. c; the period is 4π and the amplitude is 2.

11. (a) Equal; two-period shift
(b) Not equal; $f\left(t + \frac{1}{2}c\right)$ is a horizontal translation and $f\left(\frac{1}{2}t\right)$ is a period change.
(c) Not equal; for example, $\sin\left[\frac{1}{2}(\pi + 2\pi)\right] \neq \sin\left(\frac{1}{2}\pi\right)$.

12. Their range is $(-\infty, \infty)$ or $(-\infty, -1] \cup [1, \infty)$.

13. (a) The displacement is increased.
(b) The friction damps the oscillations more quickly.
(c) The frequency of the oscillations increases.

14. False. $3\pi/4$ is not in the range of the arctangent function.

Review Exercises *(page 207)*

1.

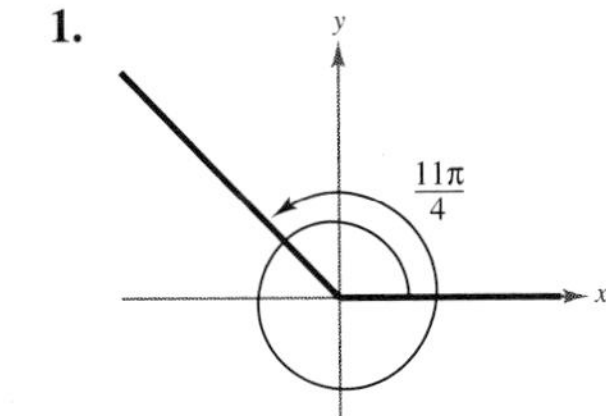

$\frac{3\pi}{4}, -\frac{5\pi}{4}$

3.

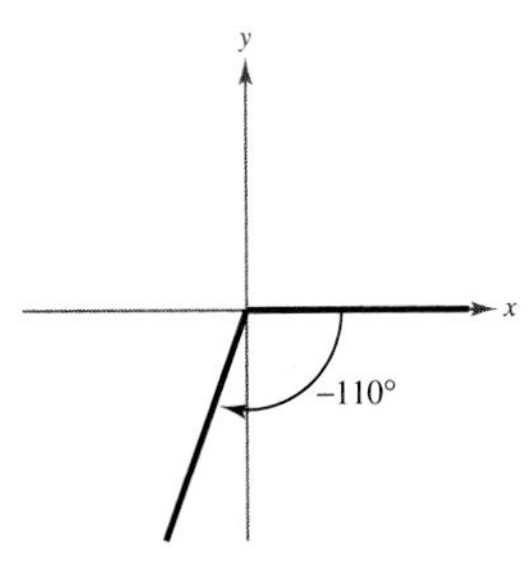

250°, −470°

5. 135.28° **7.** 5.38° **9.** 135°16′12″ **11.** −85°9′

13. 128.57° **15.** −200.54° **17.** 8.3776

19. −0.5890 **21.** 72° **23.** $\frac{\pi}{5}$

25. $\sin\theta = -\frac{1}{2}$, $\cos\theta = -\frac{\sqrt{3}}{2}$, $\tan\theta = \frac{1}{\sqrt{3}}$

27. $\sin\theta = -\frac{\sqrt{3}}{2}$, $\cos\theta = \frac{1}{2}$, $\tan\theta = -\sqrt{3}$

29. $\sin\theta = \frac{4}{5}$, $\cos\theta = \frac{3}{5}$, $\tan\theta = \frac{4}{3}$, $\csc\theta = \frac{5}{4}$, $\sec\theta = \frac{5}{3}$, $\cot\theta = \frac{3}{4}$

31. $\sin\theta = -\frac{\sqrt{11}}{6}$, $\cos\theta = \frac{5}{6}$, $\tan\theta = -\frac{\sqrt{11}}{5}$, $\csc\theta = -\frac{6\sqrt{11}}{11}$, $\cot\theta = -\frac{5\sqrt{11}}{11}$

33. $\sqrt{3}$ **35.** $-\frac{\sqrt{2}}{2}$ **37.** 0.65 **39.** 3.24

41. $135° = \frac{3\pi}{4}$, $225° = \frac{5\pi}{4}$

43. $\approx 57°$; ≈ 0.9949; $\approx 123°$; ≈ 2.1467

45. $-\sqrt{3}$ **47.** 135°

49.

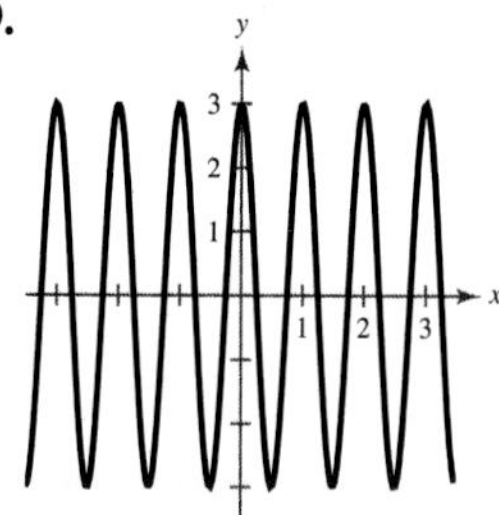

51.

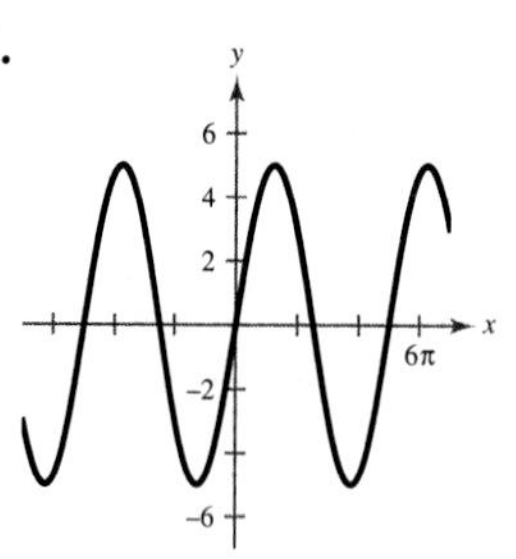

53.

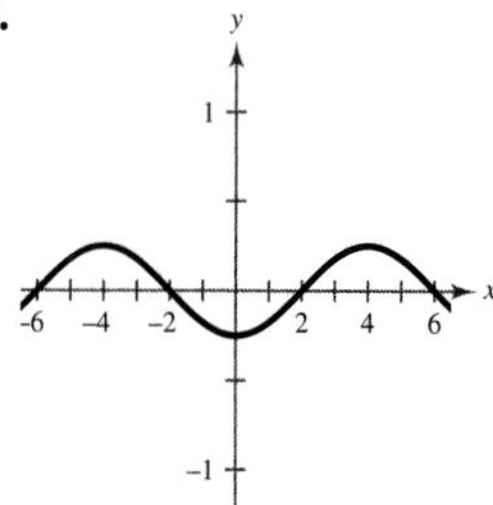

55.

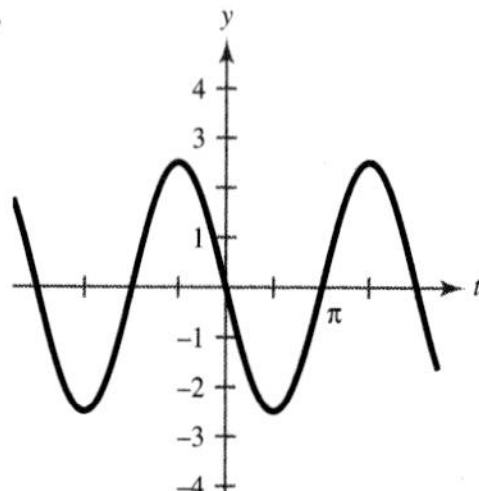

57.

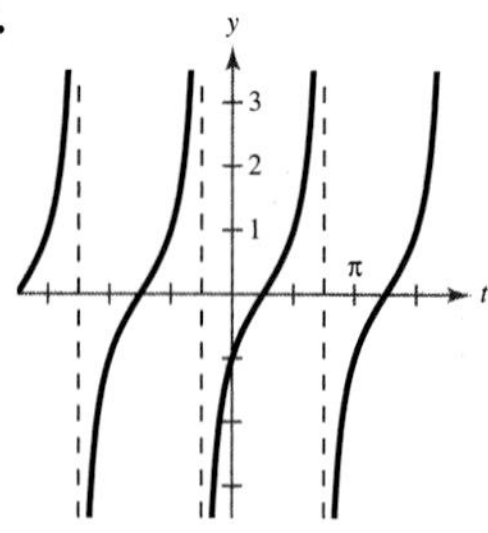

59.

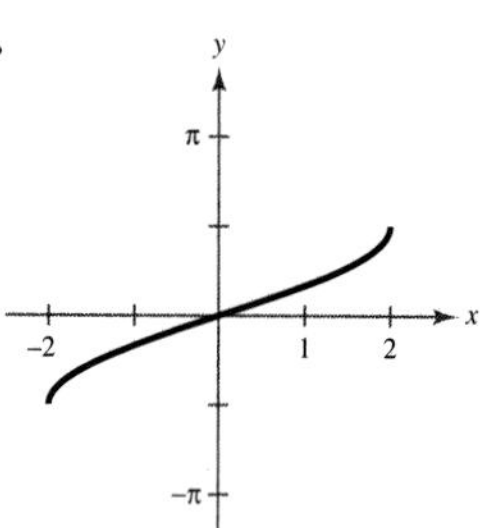

61.

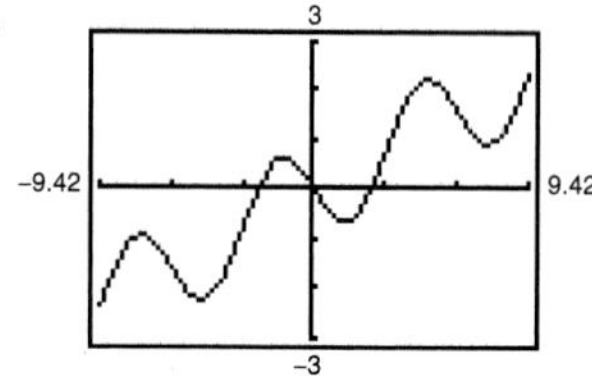

Not periodic

63.

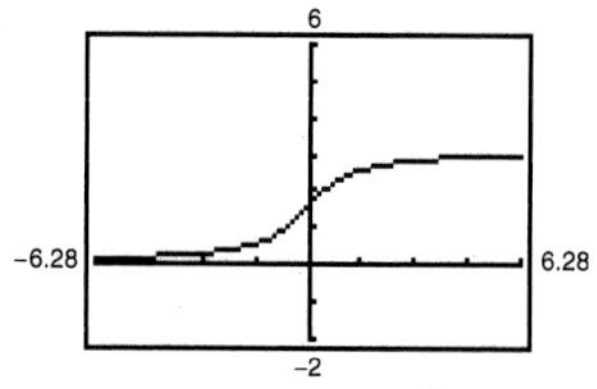

Not periodic

65.

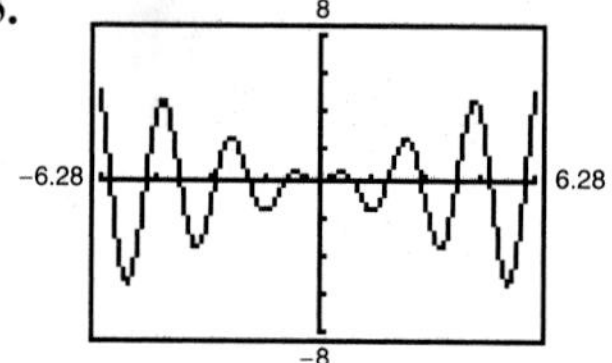

Not periodic

67.

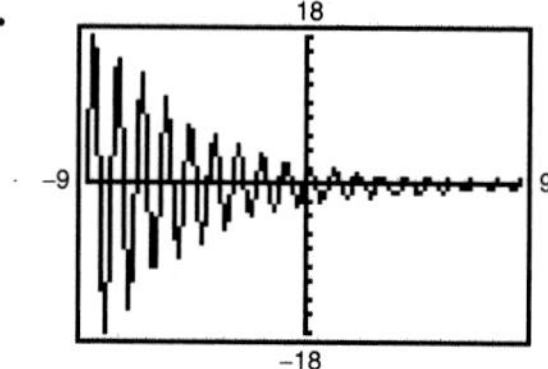

Not periodic

69.

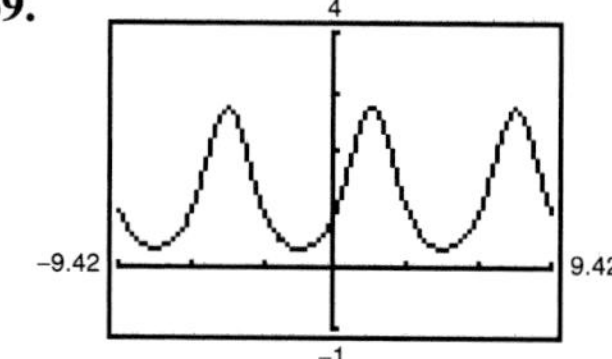

Periodic; $\left(\frac{\pi}{2}, e\right)$, $\left(\frac{3\pi}{2}, e^{-1}\right)$

71.

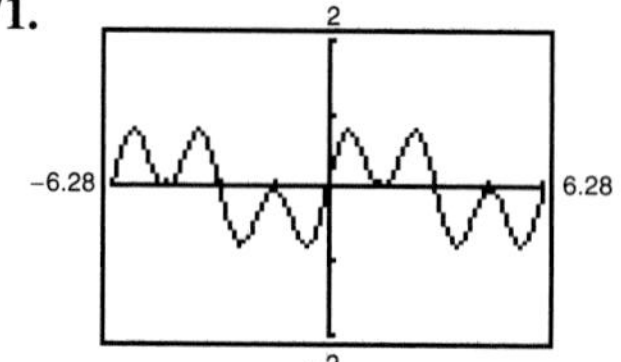

Periodic; $\left(\frac{\pi}{2}, 0\right)$, $\left(\frac{3\pi}{2}, 0\right)$, (0.61, 0.77), (2.53, 0.77), (3.76, −0.77), (5.67, −0.77)

73. $\frac{\sqrt{-x^2 + 2x}}{-x^2 + 2x}$ **75.** $\frac{2\sqrt{4 - 2x^2}}{4 - x^2}$ **77.** 9.2 meters

79. 1.2 miles **81.** 0.07 kilometer

83. (a) $A = 72(\tan \theta - \theta)$

(b)

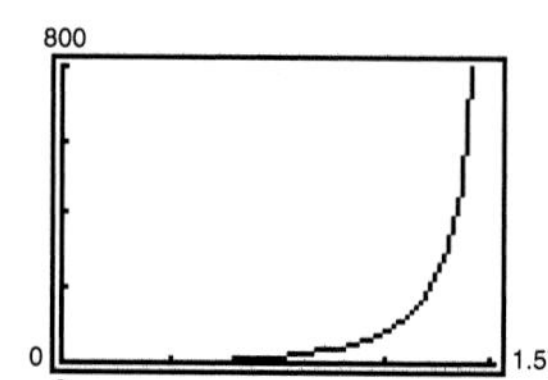

Area increases without bound as θ approaches $\pi/2$.

Chapter Test (page 212)

1. (a)

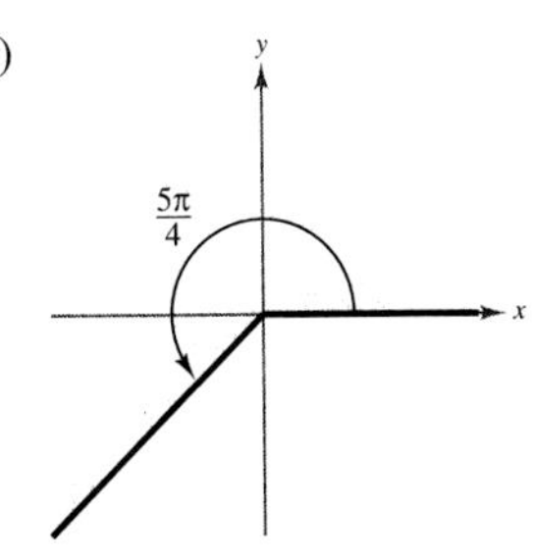

(b) $\frac{13\pi}{4}, -\frac{3\pi}{4}$

(c) 225°

2. 3000 radians per minute

3. $\sin\theta = \frac{3\sqrt{10}}{10}$

$\cos\theta = -\frac{\sqrt{10}}{10}$

$\tan\theta = -3$

$\csc\theta = \frac{\sqrt{10}}{3}$

$\sec\theta = -\sqrt{10}$

$\cot\theta = -\frac{1}{3}$

4. For $0 \le \theta \le \frac{\pi}{2}$:

$\sin\theta = \frac{3\sqrt{13}}{13}$

$\cos\theta = \frac{2\sqrt{13}}{13}$

$\csc\theta = \frac{\sqrt{13}}{3}$

$\sec\theta = \frac{\sqrt{13}}{2}$

$\cot\theta = \frac{2}{3}$

For $\pi \le \theta \le \frac{3\pi}{2}$:

$\sin\theta = -\frac{3\sqrt{13}}{13}$

$\cos\theta = -\frac{2\sqrt{13}}{13}$

$\csc\theta = -\frac{\sqrt{13}}{3}$

$\sec\theta = -\frac{\sqrt{13}}{2}$

$\cot\theta = \frac{2}{3}$

5. 70°

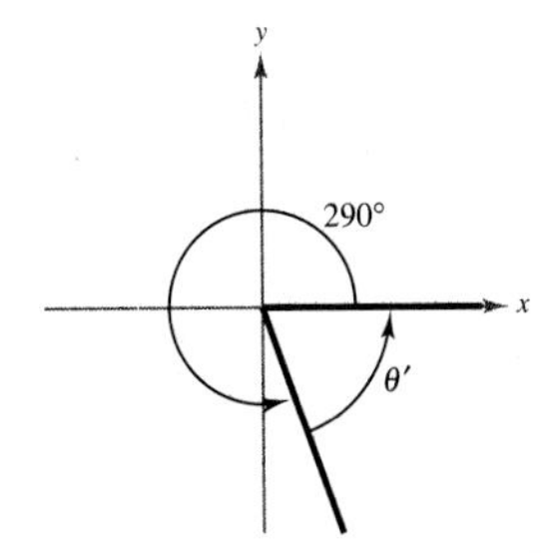

6. III **7.** 150°, 210° **8.** 1.33, 1.81

9.

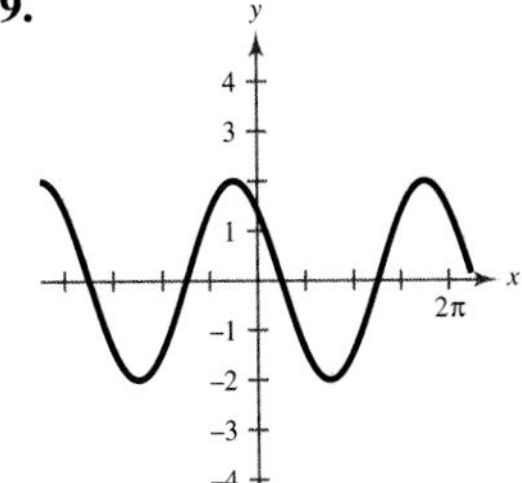

10.

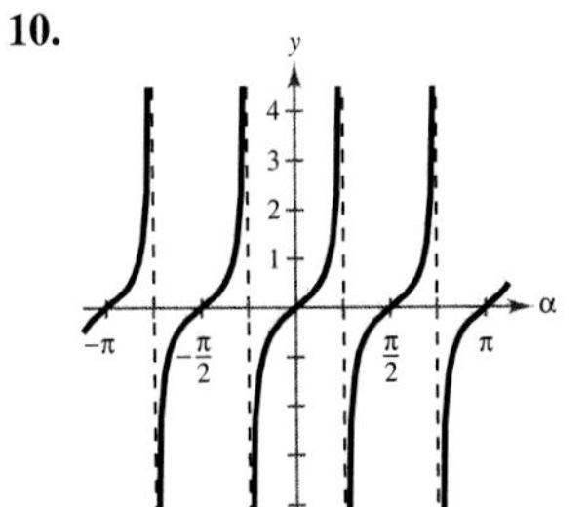

11.

Period: 2

12.

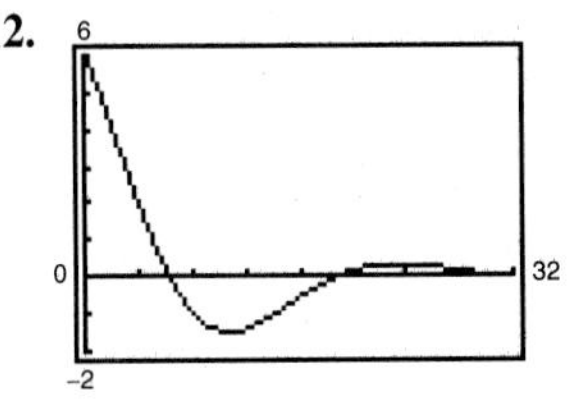

13. $a = -2, b = \frac{1}{2}, c = -\frac{\pi}{4}$ **14.** $\frac{\sqrt{5}}{2}$

15.

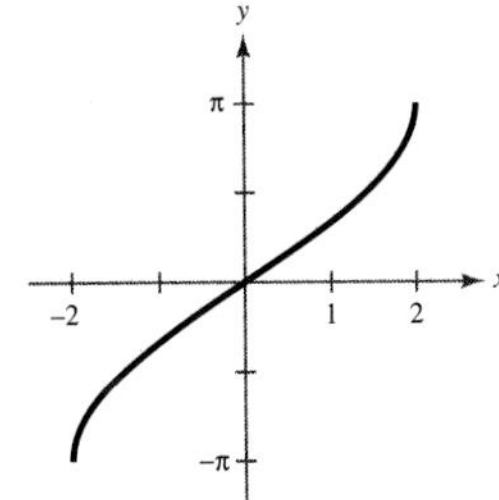

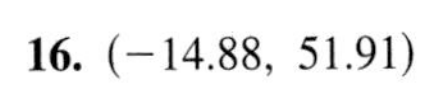

16. (−14.88, 51.91)

CHAPTER 2

Section 2.1 *(page 219)*

Warm Up *(page 219)*

1. $\sin\theta = \frac{3\sqrt{13}}{13}$, $\cos\theta = \frac{2\sqrt{13}}{13}$, $\tan\theta = \frac{3}{2}$, $\csc\theta = \frac{\sqrt{13}}{3}$, $\sec\theta = \frac{\sqrt{13}}{2}$, $\cot\theta = \frac{2}{3}$

2. $\sin\theta = \frac{2\sqrt{2}}{3}$, $\cos\theta = \frac{1}{3}$, $\tan\theta = 2\sqrt{2}$, $\csc\theta = \frac{3\sqrt{2}}{4}$, $\sec\theta = 3$, $\cot\theta = \frac{\sqrt{2}}{4}$

3. $\sin\theta = -\frac{3\sqrt{58}}{58}$, $\cos\theta = \frac{7\sqrt{58}}{58}$, $\tan\theta = -\frac{3}{7}$, $\csc\theta = -\frac{\sqrt{58}}{3}$, $\sec\theta = \frac{\sqrt{58}}{7}$, $\cot\theta = -\frac{7}{3}$

4. $\sin\theta = \frac{\sqrt{5}}{5}$, $\cos\theta = -\frac{2\sqrt{5}}{5}$, $\tan\theta = -\frac{1}{2}$, $\csc\theta = \sqrt{5}$, $\sec\theta = -\frac{\sqrt{5}}{2}$, $\cot\theta = -2$

5. $\frac{1}{2}$ **6.** $\frac{5}{4}$ **7.** $\frac{\sqrt{73}}{8}$ **8.** $\frac{2}{3}$ **9.** $\frac{x^2 + x + 16}{4(x + 1)}$

10. $\frac{8x - 2}{1 - x^2}$

1. $\sin x = \frac{1}{2}$, $\cos x = \frac{\sqrt{3}}{2}$, $\tan x = \frac{\sqrt{3}}{3}$, $\csc x = 2$, $\sec x = \frac{2\sqrt{3}}{3}$, $\cot x = \sqrt{3}$

3. $\sin\theta = -\frac{\sqrt{2}}{2}$, $\cos\theta = \frac{\sqrt{2}}{2}$, $\tan\theta = -1$, $\csc\theta = -\sqrt{2}$, $\sec\theta = \sqrt{2}$, $\cot\theta = -1$

5. $\sin x = -\frac{5}{13}$, $\cos x = -\frac{12}{13}$, $\tan x = \frac{5}{12}$, $\csc x = -\frac{13}{5}$, $\sec x = -\frac{13}{12}$, $\cot x = \frac{12}{5}$

7. $\sin\phi = 0$, $\cos\phi = -1$, $\tan\phi = 0$, $\csc\phi$ is undefined. $\sec\phi = -1$, $\cot\phi$ is undefined.

9. $\sin x = \frac{2}{3}$, $\cos x = -\frac{\sqrt{5}}{3}$, $\tan x = -\frac{2\sqrt{5}}{5}$, $\csc x = \frac{3}{2}$, $\sec x = -\frac{3\sqrt{5}}{5}$, $\cot x = -\frac{\sqrt{5}}{2}$

11. $\sin\theta = -\frac{2\sqrt{5}}{5}$, $\cos\theta = -\frac{\sqrt{5}}{5}$, $\tan\theta = 2$, $\csc\theta = -\frac{\sqrt{5}}{2}$, $\sec\theta = -\sqrt{5}$, $\cot\theta = \frac{1}{2}$

13. $\sin\theta = -1$, $\cos\theta = 0$, $\tan\theta$ is undefined. $\csc\theta = -1$, $\sec\theta$ is undefined. $\cot\theta = 0$

15. 1, 1 **17.** ∞, 0 **19.** d **21.** a **23.** e
25. b **27.** f **29.** e **31.** $\sec\phi$ **33.** $\sin\beta$
35. $\cos x$ **37.** 1 **39.** $-\tan x$ **41.** $\tan x$
43. $1 + \sin y$ **45.** $\sin^2 x$ **47.** $\sin^2 x \tan^2 x$

49. $\sec^4 x$ **51.** $\sin^2 x - \cos^2 x$ **53.** $1 + 2 \sin x \cos x$

55. $\tan^2 x$ **57.** $2 \csc^2 x$ **59.** $2 \sec x$ **61.** $1 + \cos y$

63. $3(\sec x + \tan x)$

65.

x	0.2	0.4	0.6	0.8	1.0
y_1	0.1987	0.3894	0.5646	0.7174	0.8415
y_2	0.1987	0.3894	0.5646	0.7174	0.8415

x	1.2	1.4
y_1	0.9320	0.9854
y_2	0.9320	0.9854

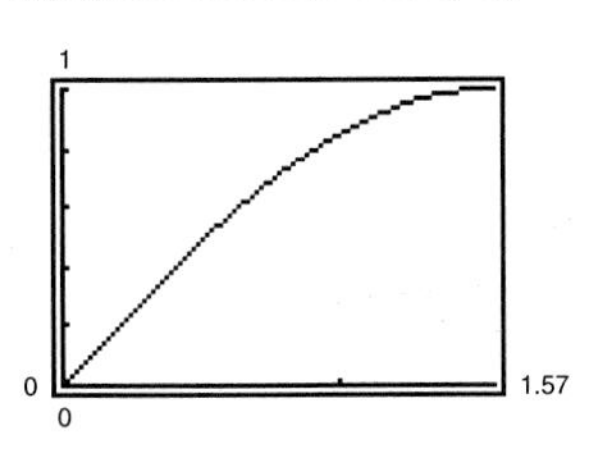

$y_1 = y_2$

67.

x	0.2	0.4	0.6	0.8	1.0
y_1	1.2230	1.5085	1.8958	2.4650	3.4082
y_2	1.2230	1.5085	1.8958	2.4650	3.4082

x	1.2	1.4
y_1	5.3319	11.6814
y_2	5.3319	11.6814

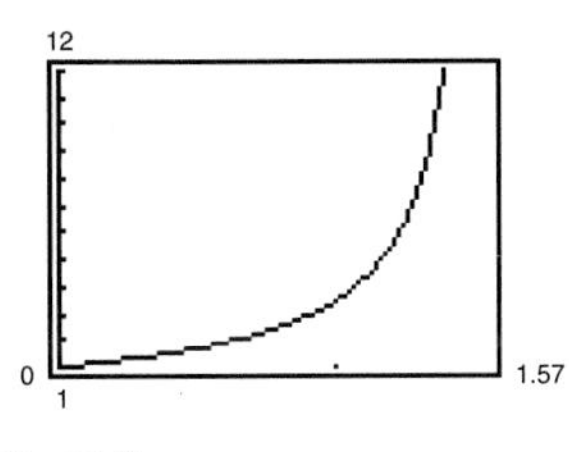

$y_1 = y_2$

69. $\csc x$ **71.** $5 \cos \theta$ **73.** $3 \tan \theta$ **75.** $5 \sec \theta$

77. $0 \le \theta \le \pi$ **79.** $0 \le \theta < \frac{\pi}{2}, \frac{3\pi}{2} < \theta < 2\pi$

81. $\ln|\cot \theta|$ **83.** Not an identity because $\frac{\sin k\theta}{\cos k\theta} = \tan k\theta$

85. Identity because $\sin \theta \cdot \frac{1}{\sin \theta} = 1$

87. (a) $\csc^2 132° - \cot^2 132° \approx 1.8107 - 0.8107 = 1$

(b) $\csc^2 \frac{2\pi}{7} - \cot^2 \frac{2\pi}{7} \approx 1.6360 - 0.6360 = 1$

89. (a) $\cos(90° - 80°) = \sin 80° \approx 0.9848$

(b) $\cos\left(\frac{\pi}{2} - 0.8\right) = \sin 0.8 \approx 0.7174$

91. $\cos \theta = \pm\sqrt{1 - \sin^2\theta}$

$\tan \theta = \pm\frac{\sin \theta}{\sqrt{1 - \sin^2\theta}}$

$\csc \theta = \frac{1}{\sin \theta}$

$\sec \theta = \pm\frac{1}{\sqrt{1 - \sin^2\theta}}$

$\cot \theta = \pm\frac{\sqrt{1 - \sin^2\theta}}{\sin \theta}$

93. $x - 25$ **95.** $4z + 12\sqrt{z} + 9$

Section 2.2 *(page 227)*

Warm Up *(page 227)*

1. (a) $x^2(1 + y)(1 - y)$ (b) $\sin^4 x$

2. (a) $x^2(1 + y^2)$ (b) 1

3. (a) $(x^2 + 1)(x + 1)(x - 1)$

(b) $\sec^2 x(\tan x + 1)(\tan x - 1)$

4. (a) $(z + 1)(z^2 - z + 1)$

(b) $(\tan x + 1)(\sec^2 x - \tan x)$ or

$(\tan x + 1)(\sec x - \sin x) \sec x$

5. (a) $(x - 1)(x^2 + 1)$ (b) $(\cot x - 1)\csc^2 x$

6. (a) $(x + 1)^2(x - 1)^2$ (b) $\cos^4 x$

7. (a) $\frac{y^2 - x^2}{x}$ (b) $\tan x$ **8.** (a) $\frac{x^2 - 1}{x^2}$ (b) $\sin^2 x$

9. (a) $\frac{y^2 + (1 + z)^2}{y(1 + z)}$ (b) $2 \csc x$

10. (a) $\frac{y(1 + y) - z^2}{z(1 + y)}$ (b) $\frac{\tan x - 1}{\sec x(1 + \tan x)}$

1.–59. Answers will vary.

61. $\sin\theta = \pm\sqrt{1-\cos^2\theta}$; $\dfrac{7\pi}{4}$

63. $\sqrt{\tan^2 x} = |\tan x|$; $\dfrac{3\pi}{4}$ **65.** 1 **67.** 2

69.–71. Answers will vary.

73. Seward; 6.4 and 1.9

Section 2.3 *(page 237)*

Warm Up *(page 237)*

1. $\dfrac{2\pi}{3}, \dfrac{4\pi}{3}$ **2.** $\dfrac{\pi}{3}, \dfrac{2\pi}{3}$ **3.** $\dfrac{\pi}{4}, \dfrac{7\pi}{4}$ **4.** $\dfrac{7\pi}{4}, \dfrac{5\pi}{4}$

5. $\dfrac{\pi}{3}, \dfrac{4\pi}{3}$ **6.** $\dfrac{3\pi}{4}, \dfrac{7\pi}{4}$ **7.** $\dfrac{15}{8}$ **8.** $-3, \dfrac{5}{2}$

9. $\dfrac{2 \pm \sqrt{14}}{2}$ **10.** $-1, 3$

1. $x = -1, 3$ **3.** $x = \pm 2$ **5.–9.** Answers will vary.

11. $\dfrac{2\pi}{3} + 2n\pi, \dfrac{4\pi}{3} + 2n\pi$ **13.** $\dfrac{\pi}{3} + 2n\pi, \dfrac{2\pi}{3} + 2n\pi$

15. $\dfrac{\pi}{6} + n\pi, \dfrac{5\pi}{6} + n\pi$

17. $\dfrac{\pi}{8} + n\pi, \dfrac{3\pi}{8} + n\pi, \dfrac{5\pi}{8} + n\pi, \dfrac{7\pi}{8} + n\pi$

19. $\dfrac{\pi}{3} + n\pi, \dfrac{2\pi}{3} + n\pi$ **21.** $\dfrac{\pi}{3} + n\pi, \dfrac{2\pi}{3} + n\pi$

23. $\dfrac{\pi}{6} + n\pi, \dfrac{5\pi}{6} + n\pi, \dfrac{\pi}{3} + n\pi, \dfrac{2\pi}{3} + n\pi$

25. $0, \dfrac{\pi}{2}, \pi, \dfrac{3\pi}{2}$ **27.** $0, \pi, \dfrac{\pi}{6}, \dfrac{5\pi}{6}, \dfrac{7\pi}{6}, \dfrac{11\pi}{6}$

29. $\dfrac{\pi}{3}, \dfrac{5\pi}{3}, \pi$ **31.** No solution **33.** $\dfrac{\pi}{2}$ **35.** $\dfrac{\pi}{2}$

37. π **39.** $\dfrac{\pi}{6}, \dfrac{5\pi}{6}, \dfrac{7\pi}{6}, \dfrac{11\pi}{6}$

41. $\dfrac{2}{3}, \dfrac{3}{2}$; $0.8411 + 2n\pi, 5.4421 + 2n\pi$

43. 1.1071, 4.2487 **45.** 1.0472, 5.2360 **47.** 0, 1.8955

49. 0, 2.6779, 3.1416, 5.8195

51. 0.9828, 1.7682, 4.1244, 4.9098

53. 0.3398, 0.8481, 2.2935, 2.8018 **55.** 0.4271, 2.7145

57.

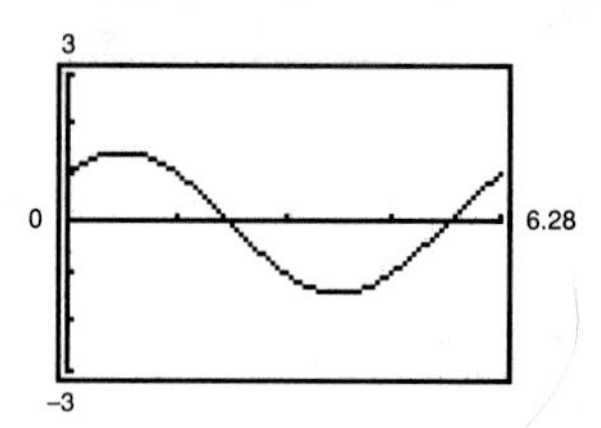

Maximum: $\left(\dfrac{\pi}{4}, \sqrt{2}\right)$

Minimum: $\left(\dfrac{5\pi}{4}, -\sqrt{2}\right)$

59. 1

61. (a) All real numbers except $x = 0$

(b) y-axis symmetry; horizontal asymptote: $y = 1$

(c) Oscillates

(d) Infinitely many solutions

(e) Yes, 0.6366

63. 0.04, 0.43, 0.83 **65.** 37°, 53°

67. (a)

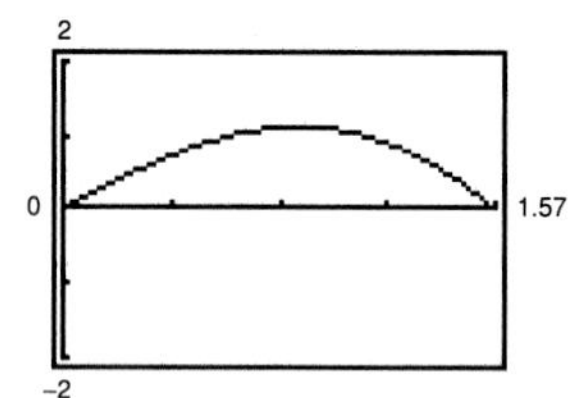

$x \approx 0.86,\ A \approx 1.12$

(b) $0.6 < x < 1.1$

69. (a)

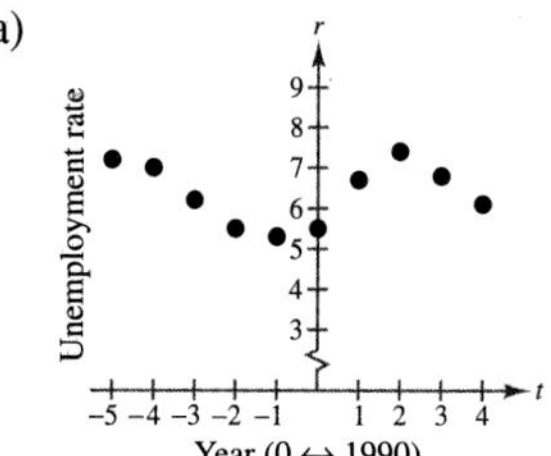

(b) (3)

(c) Constant: 6.20%

(d) 7 years

(e) 2000

Section 2.4 *(page 246)*

Warm Up *(page 246)*

1. $\frac{\sqrt{10}}{10}$	**2.** $\frac{-5\sqrt{34}}{34}$	**3.** $-\frac{\sqrt{7}}{4}$	**4.** $\frac{2\sqrt{2}}{3}$
5. $\frac{\pi}{4}, \frac{3\pi}{4}$	**6.** $\frac{\pi}{2}, \frac{3\pi}{2}$	**7.** $\tan^3 x$	**8.** $\cot^2 x$
9. $\sec x$	**10.** $1 - \tan^2 x$		

1. (a) $\frac{\sqrt{2} - \sqrt{6}}{4}$ (b) $\frac{\sqrt{2} + 1}{2}$

3. (a) $\frac{1}{2}$ (b) $\frac{-\sqrt{3} - 1}{2}$

5. False. Parts (a) and (b) are unequal in Exercises 1–4.

7. $\sin 75° = \frac{\sqrt{2}}{4}(1 + \sqrt{3})$

$\cos 75° = \frac{\sqrt{2}}{4}(\sqrt{3} - 1)$

$\tan 75° = \sqrt{3} + 2$

9. $\sin 105° = \frac{\sqrt{2}}{4}(\sqrt{3} + 1)$

$\cos 105° = \frac{\sqrt{2}}{4}(1 - \sqrt{3})$

$\tan 105° = -2 - \sqrt{3}$

11. $\sin 195° = \frac{\sqrt{2}}{4}(1 - \sqrt{3})$

$\cos 195° = -\frac{\sqrt{2}}{4}(\sqrt{3} + 1)$

$\tan 195° = 2 - \sqrt{3}$

13. $\sin \frac{11\pi}{12} = \frac{\sqrt{2}}{4}(\sqrt{3} - 1)$

$\cos \frac{11\pi}{12} = -\frac{\sqrt{2}}{4}(\sqrt{3} + 1)$

$\tan \frac{11\pi}{12} = -2 + \sqrt{3}$

15. $\sin \frac{17\pi}{12} = -\frac{\sqrt{2}}{4}(\sqrt{3} + 1)$

$\cos \frac{17\pi}{12} = \frac{\sqrt{2}}{4}(1 - \sqrt{3})$

$\tan \frac{17\pi}{12} = 2 + \sqrt{3}$

17. $\sin 285° = -\frac{\sqrt{2}}{4}(\sqrt{3} + 1)$

$\cos 285° = \frac{\sqrt{2}}{4}(\sqrt{3} - 1)$

$\tan 285° = -(2 + \sqrt{3})$

19. $\sin\left(-\frac{13\pi}{12}\right) = \frac{\sqrt{2}}{4}(\sqrt{3} - 1)$

$\cos\left(-\frac{13\pi}{12}\right) = -\frac{\sqrt{2}}{4}(\sqrt{3} + 1)$

$\tan\left(-\frac{13\pi}{12}\right) = -2 + \sqrt{3}$

21. $\cos 40°$ **23.** $\sin 200°$ **25.** $\tan 239°$ **27.** $\sin 1.8$

29. $\tan 3x$ **31.** $-\frac{63}{65}$ **33.** $\frac{16}{65}$ **35.** $\frac{65}{16}$

37. $\frac{33}{56}$ **39.** $\frac{3}{5}$ **41.** $\frac{44}{125}$ **43.** $\frac{5}{4}$

45.–61. Answers will vary.

63. (a) $\sqrt{2}\sin\left(\theta + \frac{\pi}{4}\right)$ (b) $\sqrt{2}\cos\left(\theta - \frac{\pi}{4}\right)$

65. (a) $13\sin(3\theta + 0.3948)$ (b) $13\cos(3\theta - 1.1760)$

67. $2\cos\theta$ **69.** 1 **71.** $\frac{\pi}{2}$ **73.** $\frac{5\pi}{4}, \frac{7\pi}{4}$

75. $\frac{\pi}{4}, \frac{7\pi}{4}$

77.

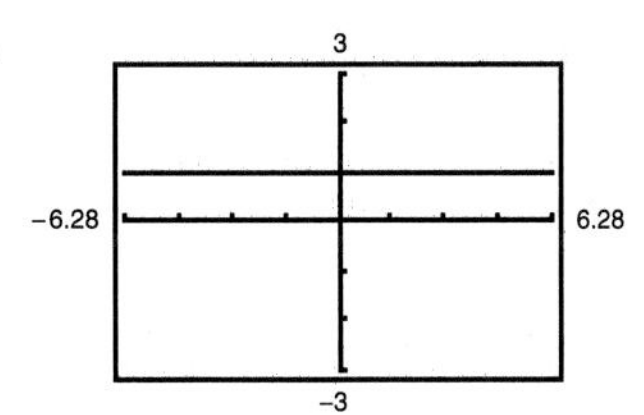

$\sin^2\left(\theta + \frac{\pi}{4}\right) + \sin^2\left(\theta - \frac{\pi}{4}\right) = 1$

79. (a) $y = \frac{5}{12}\sin(2t + 0.6435)$

(b) $\frac{5}{12}$ foot (c) $\frac{1}{\pi}$ cycle per second

Section 2.5 *(page 257)*

Warm Up *(page 257)*

1. $\sin x(2 + \cos x)$ **2.** $(\cos x - 2)(\cos x + 1)$

3. $0, \frac{\pi}{2}, \pi, \frac{3\pi}{2}$ **4.** $\frac{\pi}{4}, \frac{3\pi}{4}, \frac{5\pi}{4}, \frac{7\pi}{4}$

5. π **6.** 0 **7.** $\frac{2 - \sqrt{2}}{4}$ **8.** $\frac{3}{4}$

9. $\tan 3x$ **10.** $\frac{1}{2}\csc^2 x - 1$

1. $\frac{3}{5}$ **3.** $\frac{7}{25}$ **5.** $\frac{24}{7}$ **7.** $\frac{25}{24}$ **9.** $0, \frac{\pi}{3}, \pi, \frac{5\pi}{3}$

11. $\frac{\pi}{12}, \frac{5\pi}{12}, \frac{13\pi}{12}, \frac{17\pi}{12}$ **13.** $0, \frac{2\pi}{3}, \frac{4\pi}{3}$

15. $\frac{\pi}{2}, \frac{\pi}{6}, \frac{5\pi}{6}, \frac{7\pi}{6}, \frac{3\pi}{2}, \frac{11\pi}{6}$ **17.** $0, \frac{\pi}{2}, \pi, \frac{3\pi}{2}$

19. $3 \sin 2x$ **21.** $4 \cos 2x$

23. $\sin 2u = \frac{24}{25}$
$\cos 2u = \frac{7}{25}$
$\tan 2u = \frac{24}{7}$

25. $\sin 2u = \frac{4}{5}$
$\cos 2u = \frac{3}{5}$
$\tan 2u = \frac{4}{3}$

27. $\sin 2u = -\frac{4\sqrt{21}}{25}$
$\cos 2u = -\frac{17}{25}$
$\tan 2u = \frac{4\sqrt{21}}{17}$

29. $\frac{1}{8}(3 + 4\cos 2x + \cos 4x)$ **31.** $\frac{1}{8}(1 - \cos 4x)$

33. $\frac{1}{32}(2 + \cos 2x - 2\cos 4x - \cos 6x)$

35. $\frac{5}{\sqrt{26}}$ **37.** $\frac{1}{5}$ **39.** $\sqrt{26}$

41. $\sin 105° = \frac{1}{2}\sqrt{2 + \sqrt{3}}$
$\cos 105° = -\frac{1}{2}\sqrt{2 - \sqrt{3}}$
$\tan 105° = -2 - \sqrt{3}$

43. $\sin 112° 30' = \frac{1}{2}\sqrt{2 + \sqrt{2}}$
$\cos 112° 30' = -\frac{1}{2}\sqrt{2 - \sqrt{2}}$
$\tan 112° 30' = -1 - \sqrt{2}$

45. $\sin \frac{\pi}{8} = \frac{1}{2}\sqrt{2 - \sqrt{2}}$
$\cos \frac{\pi}{8} = \frac{1}{2}\sqrt{2 + \sqrt{2}}$
$\tan \frac{\pi}{8} = \sqrt{2} - 1$

47. $\sin \frac{u}{2} = \frac{5\sqrt{26}}{26}$
$\cos \frac{u}{2} = \frac{\sqrt{26}}{26}$
$\tan \frac{u}{2} = 5$

49. $\sin \frac{u}{2} = \sqrt{\frac{89 - 8\sqrt{89}}{178}}$
$\cos \frac{u}{2} = -\sqrt{\frac{89 + 8\sqrt{89}}{178}}$
$\tan \frac{u}{2} = \frac{8 - \sqrt{89}}{5}$

51. $\sin \frac{u}{2} = \frac{3\sqrt{10}}{10}$
$\cos \frac{u}{2} = -\frac{\sqrt{10}}{10}$
$\tan \frac{u}{2} = -3$

53. $|\sin 3x|$ **55.** $-|\tan 4x|$ **57.** π

59. $\frac{\pi}{3}, \pi, \frac{5\pi}{3}$ **61.** $3\left(\sin \frac{\pi}{2} + \sin 0\right)$

63. $\frac{1}{2}(\sin 8\theta + \sin 2\theta)$ **65.** $\frac{5}{2}(\cos 8\beta + \cos 2\beta)$

67. $\frac{1}{2}(\cos 2y - \cos 2x)$ **69.** $\frac{1}{2}(\sin 2\theta + \sin 2\pi)$

71. $2 \sin 45° \cos 15°$ **73.** $-2 \sin \frac{\pi}{2} \sin \frac{\pi}{4}$

75. $2 \cos 4x \cos 2x$ **77.** $2 \cos \alpha \sin \beta$

79. $2 \cos(\phi + \pi) \cos \pi$

81. $0, \frac{\pi}{4}, \frac{\pi}{2}, \frac{3\pi}{4}, \pi, \frac{5\pi}{4}, \frac{3\pi}{2}, \frac{7\pi}{4}$

83. $\frac{\pi}{6}, \frac{5\pi}{6}$ **85.** $\frac{25}{169}$ **87.** $\frac{4}{13}$

89.–101. Answers will vary.

103.

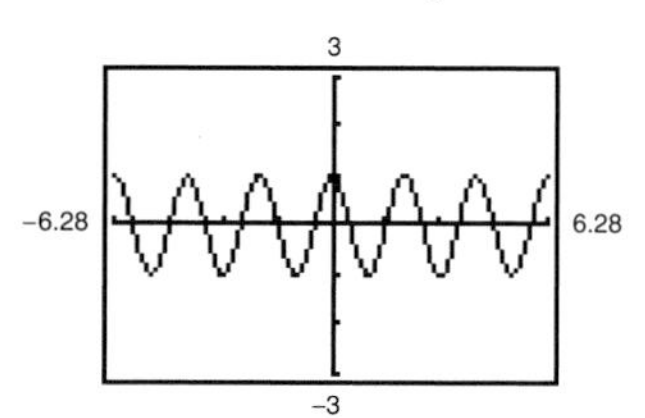

105.

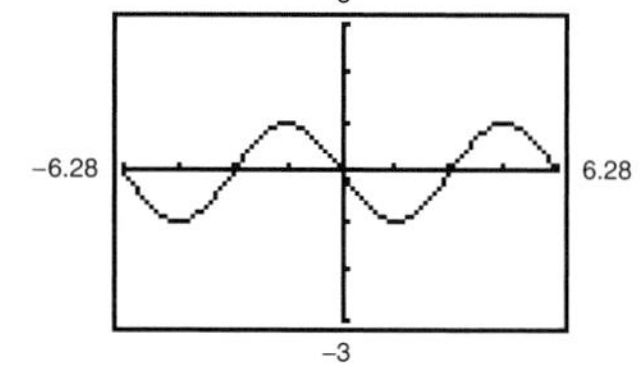

107.

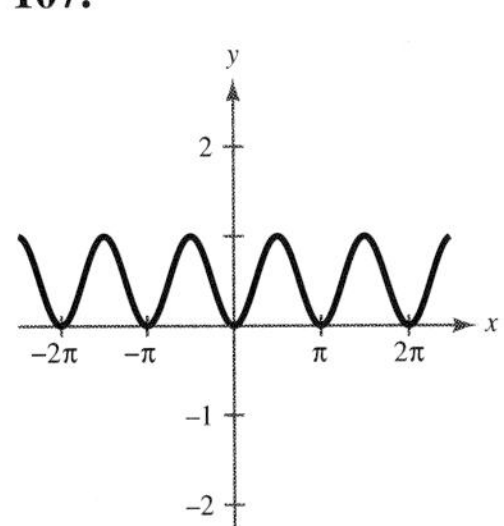

109.

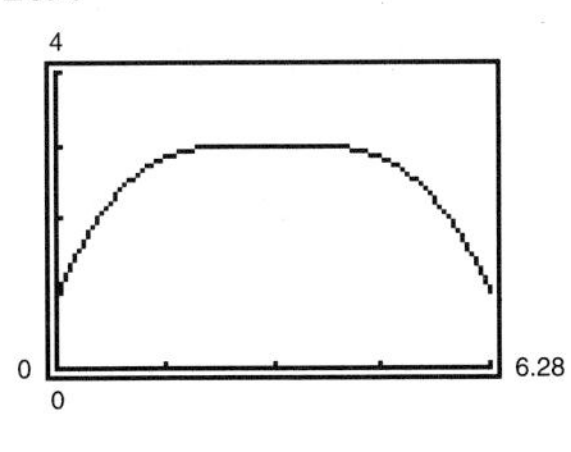

Maximum: $(\pi, 3)$

111. (a)

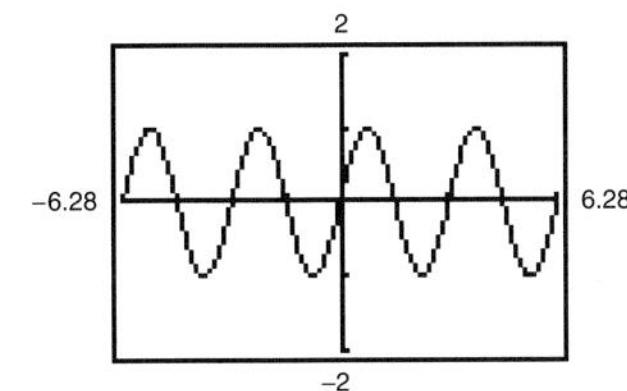

(b) $g(x) = \sin 2x$

(c) Answers will vary.

113. $2x\sqrt{1 - x^2}$

115. (a) $A = 100 \sin \dfrac{\theta}{2} \cos \dfrac{\theta}{2}$

(b) $A = 50 \sin \theta$

The area is maximum when $\theta = \dfrac{\pi}{2}$.

117. September: $235,000   **119.** 15.7 gallons

October: $272,600

Focus on Concepts *(page 261)*

1. An identity is true for all values of the variable and a conditional equation is true for some values of the variable.
2. When proving an identity you use the fundamental identities and rules of algebra to transform one expression into another. To solve a trigonometric equation, use standard algebraic techniques and identities to isolate a trigonometric function involved in the equation. Find the value of the variable by using the inverse of the trigonometric function.
3. Reciprocal identities: $\csc \theta = \dfrac{1}{\sin \theta}$, $\sec \theta = \dfrac{1}{\cos \theta}$, $\cot \theta = \dfrac{1}{\tan \theta}$

   Quotient identities: $\tan \theta = \dfrac{\sin \theta}{\cos \theta}$, $\cot \theta = \dfrac{\cos \theta}{\sin \theta}$

   Pythagorean identities: $\sin^2\theta + \cos^2\theta = 1$, $\tan^2\theta + 1 = \sec^2\theta$, $1 + \cot^2\theta = \csc^2\theta$
4. No. $\cos \theta = \pm\sqrt{1 - \sin^2\theta}$
5. False. The order in which algebraic operations and fundamental identities are done may vary.
6. (a) True. The period of tangent is π.

   (b) False. The period of cosine is 2π.

   (c) False. $\sec \theta \cos \theta = 1$

   (d) True.

   (e) True. $\sin(-\alpha) = -\sin \alpha$

7. $y_1 = y_2 + 1$ **8.** $y_1 = 1 - y_2$ **9.** 1   **10.** 3

11. 5   **12.** 4   **13.** 2π

Review Exercises *(page 262)*

1. $\sin^2 x$ **3.** $1 + \cot \alpha$ **5.** 1   **7.** $\tan(2x + 2)$

9.–25. Answers will vary.

27.

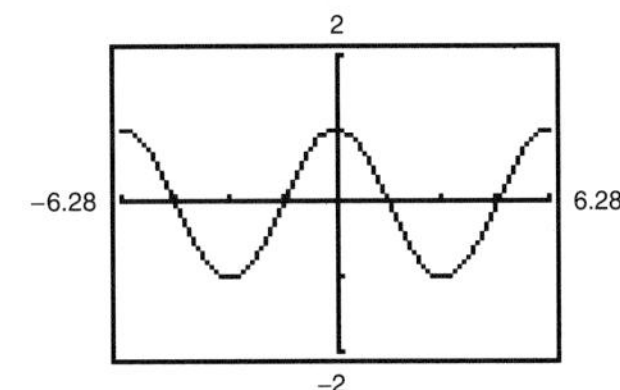

29.

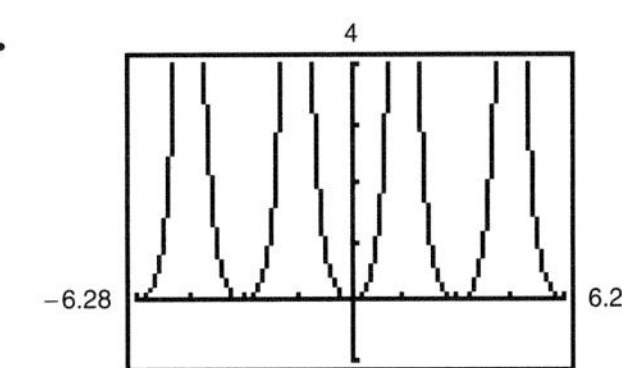

31. $\dfrac{\sqrt{2}}{4}(\sqrt{3} + 1)$ **33.** $-\dfrac{1}{2}\sqrt{2 + \sqrt{2}}$

35. $-\dfrac{3}{52}(5 + 4\sqrt{7})$ **37.** $\dfrac{1}{52}(36 + 5\sqrt{7})$

39. $\dfrac{1}{4}\sqrt{2(4 - \sqrt{7})}$

41. False. If $\dfrac{\pi}{2} < \theta < \pi$, then $\cos \dfrac{\theta}{2} > 0$.

43. True   **45.** $0, \pi$ **47.** $0, \dfrac{3\pi}{4}, \pi, \dfrac{5\pi}{4}$

49. $0, \dfrac{\pi}{2}, \pi$ **51.** $\dfrac{\pi}{3}, \dfrac{5\pi}{3}$ **53.** $\dfrac{\pi}{4}, \dfrac{5\pi}{4}$

55. False. $\sin \theta = \frac{1}{2}$ has an infinite number of solutions but is not an identity.

57. $2\cos\frac{5\theta}{2}\cos\frac{\theta}{2}$ **59.** $\frac{1}{2}(\cos\alpha - \cos 5\alpha)$

61. $8x^2 - 1$ **63.** Answers will vary.

65. (a) $y = \frac{1}{2}\sqrt{10}\sin\left(8t - \arctan\frac{1}{3}\right)$

(b) $\frac{1}{2}\sqrt{10}$ feet

(c) $\frac{4}{\pi}$ cycles per second

Chapter Test *(page 266)*

1. $\sin\theta = -\frac{3\sqrt{13}}{13}$ **2.** 1 **3.** 1

$\cos\theta = -\frac{2\sqrt{13}}{13}$

$\csc\theta = -\frac{\sqrt{13}}{3}$

$\sec\theta = -\frac{\sqrt{13}}{2}$

$\cot\theta = \frac{2}{3}$

4. $\csc\theta\sec\theta$ **5.** $\theta = 0, \frac{\pi}{2} < \theta < \pi, \frac{3\pi}{2} < \theta < 2\pi$

6.

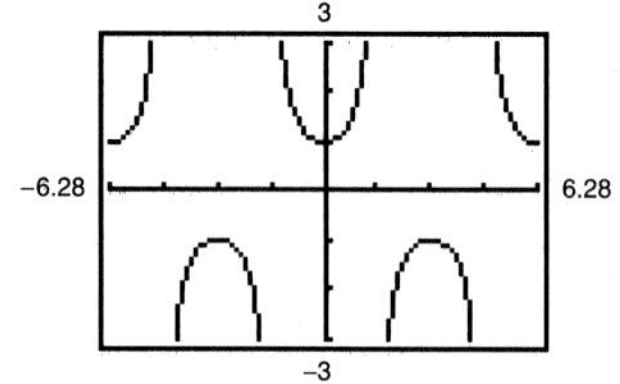

$y_1 = y_2$

7.–12. Answers will vary. **13.** $0, \frac{3\pi}{4}, \pi, \frac{7\pi}{4}$

14. $\frac{\pi}{6}, \frac{\pi}{2}, \frac{5\pi}{6}, \frac{3\pi}{2}$ **15.** $\frac{\pi}{6}, \frac{5\pi}{6}, \frac{7\pi}{6}, \frac{11\pi}{6}$

16. $\frac{\pi}{6}, \frac{5\pi}{6}, \frac{3\pi}{2}$ **17.** $-2.938, -2.663, 1.170$

18. $|\cos^2 x + \cos x| \le 2$ for all x

19. $\frac{\sqrt{2} - \sqrt{6}}{4}$ **20.** $\sin 2u = \frac{4}{5}, \tan 2u = -\frac{4}{3}$

CHAPTER 3

Section 3.1 *(page 274)*

Warm Up *(page 274)*

1. $b = 3\sqrt{3}, A = 30°, B = 60°$

2. $c = 5\sqrt{2}, A = 45°, B = 45°$

3. $a = 8, A \approx 28.07°, B \approx 61.93°$

4. $b \approx 8.33, c \approx 11.21, B = 48°$

5. $a \approx 22.69, c \approx 23.04, A = 80°$

6. $a \approx 45.73, b \approx 142.86, A = 17°45'$ **7.** 8.48

8. 12.94 **9.** 2.25 **10.** 91.06

1. $C = 105°, b \approx 28.28, c \approx 38.64$

3. $C = 110°, b \approx 37.40, c \approx 40.59$

5. $B \approx 21.55°, C \approx 122.45°, c \approx 11.49$

7. $B = 10°, b \approx 69.46, c \approx 136.81$

9. $B = 42°4', a \approx 22.05, b \approx 14.88$

11. $A \approx 10°11', C \approx 154°19', c \approx 11.03$

13. $A \approx 25.57°, B \approx 9.43°, a \approx 10.5$

15. $B \approx 18°13', C \approx 51°32', c \approx 40.06$

17. No solution

19. Two solutions

$B \approx 70.4°, C \approx 51.6°, c \approx 4.16$

$B \approx 109.6°, C \approx 12.4°, c \approx 1.14$

21. No solution

23. (a) $b \le 5, b = \frac{5}{\sin 36°}$

(b) $5 < b < \frac{5}{\sin 36°}$

(c) $b > \frac{5}{\sin 36°}$

25. (a)

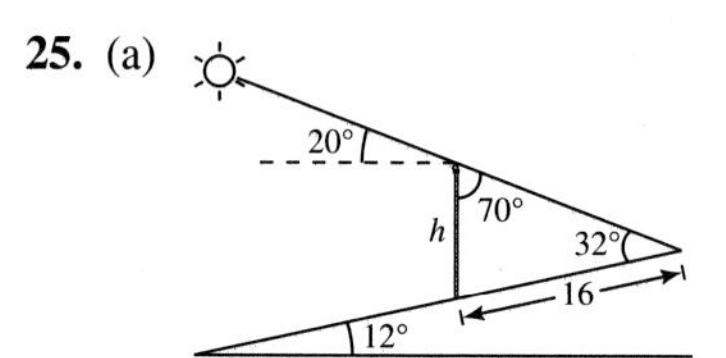

(b) $\frac{16}{\sin 70°} = \frac{h}{\sin 32°}$

(c) 9 meters

27. 16.1° **29.** 77 meters

31. (a)

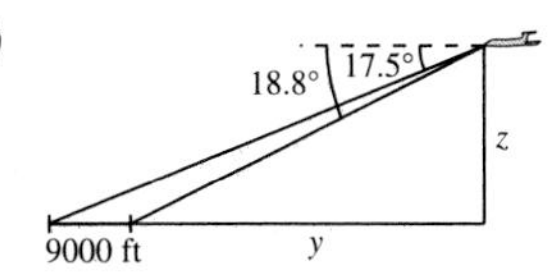

(b) 22.6 miles
(c) 21.4 miles
(d) 38,443 feet

33. 42.3 kilometers, 25.8 kilometers **35.** 4.55 miles

37. (a) $\alpha = \arcsin(0.5 \sin \beta)$

(b)

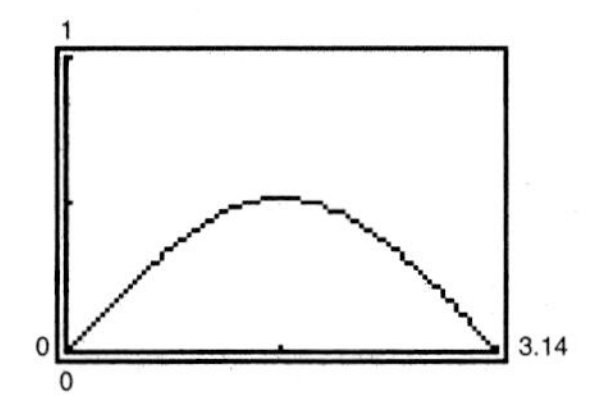

Domain: $0 < \beta < \pi$

Range: $0 < \alpha \le \frac{\pi}{6}$

(c) $c = \dfrac{18 \sin[\pi - \beta - \arcsin(0.5 \sin \beta)]}{\sin \beta}$

(d)

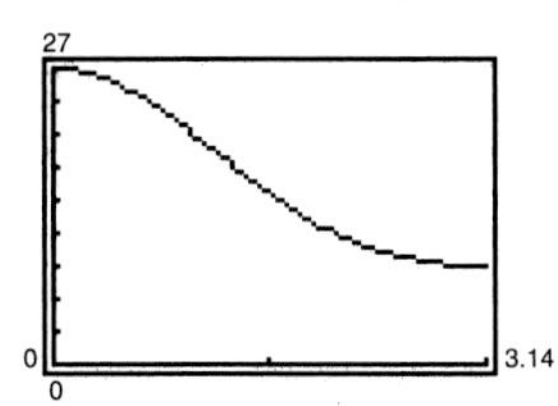

Domain: $0 < \beta < \pi$

Range: $9 < c < 27$

(e)

β	0	0.4	0.8	1.2	1.6	2.0
α	0	0.1960	0.3669	0.4848	0.5234	0.4720
c	27	25.95	27.07	19.19	15.33	12.29

β	2.4	2.8
α	0.3445	0.1683
c	10.31	9.27

39. 10.4 **41.** 1675.2 **43.** 474.9

45. (a) $20\left[15 \sin \dfrac{3\theta}{2} - 4 \sin \dfrac{\theta}{2} - 6 \sin \theta\right]$

(b)

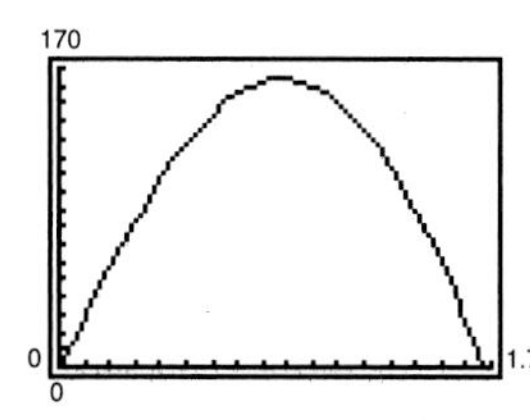

(c) Domain: $0 \le \theta \le 1.6690$

The domain would increase in length and the area would increase.

Section 3.2 *(page 283)*

Warm Up *(page 283)*

1. $2\sqrt{13}$ **2.** $3\sqrt{5}$ **3.** $4\sqrt{10}$ **4.** $3\sqrt{13}$
5. 20 **6.** 48 **7.** $a \approx 4.62,\ c \approx 26.20,\ B = 70°$
8. $a \approx 34.20,\ b \approx 93.97,\ B = 70°$
9. No solution **10.** $a \approx 15.09, B \approx 18.97°, C \approx 131.03°$

1. $A \approx 26.4°,\ B \approx 36.3°,\ C \approx 117.3°$
3. $B \approx 23.8°,\ C \approx 126.2°,\ a \approx 18.6$
5. $A \approx 36.9°,\ B \approx 53.1°,\ C \approx 90°$
7. $A \approx 92.94°,\ B \approx 43.53°,\ C \approx 43.53°$
9. $a \approx 11.79,\ B \approx 12.7°,\ C \approx 47.3°$
11. $A \approx 158°37',\ C \approx 12°38',\ b \approx 10.4$
13. $A = 27°10',\ B = 27°10',\ c \approx 56.9$

	a	b	c	d	θ	ϕ
15.	4	6	9.67	3.23	30°	150°
17.	10	14	20	13.86	68.2°	111.8°
19.	10	11.58	18	12	67.1°	112.9°

21.

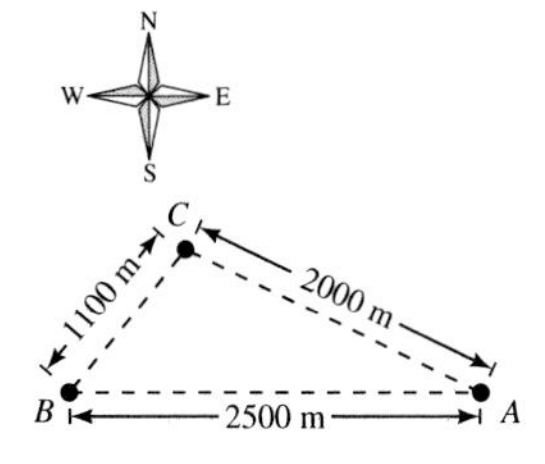

N 39° E, S 64.7° E

23. 422.5 meters **25.** 72.3° **27.** 43.3 miles

29. (a) N 58.4° W (b) S 81.5° W **31.** 63.7 feet

33. $\overline{PQ} \approx 9.4,\ \overline{QS} = 5,\ \overline{RS} \approx 12.8$

35. (a) $49 = 2.25 + x^2 - 3x \cos \theta$

(b) $x = \frac{1}{2}\left(3 \cos \theta + \sqrt{9 \cos^2\theta + 187}\right)$

(c)

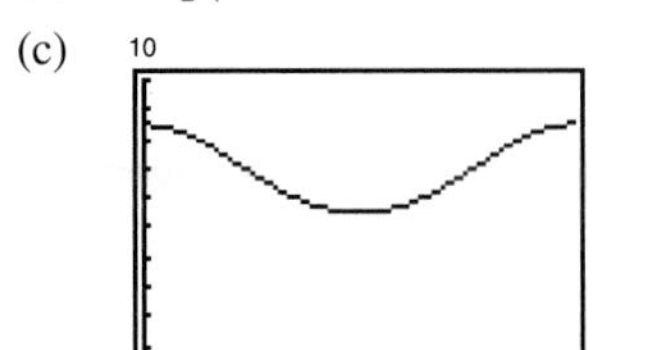

(d) 6 inches

37. 2.76 feet **39.** (a) 570.60 (b) 5910.68 (c) 177.09

41. 16.25 **43.** 54 **45.** 96.82

47. 46,837.5 square feet **49.** Answers will vary.

51. $\dfrac{1}{\sqrt{1 - 4x^2}}$ **53.** $\dfrac{1}{x - 2}$

Section 3.3 *(page 296)*

Warm Up *(page 296)*

1. $7\sqrt{10}$ **2.** $\sqrt{58}$ **3.** $3x + 5y - 14 = 0$

4. $4x - 3y - 1 = 0$ **5.** 111.8° **6.** 323.1°

7. $\frac{1}{2}, \frac{\sqrt{3}}{2}$ **8.** $\frac{\sqrt{3}}{2}, -\frac{1}{2}$ **9.** $-\frac{\sqrt{3}}{2}, \frac{1}{2}$

10. $-\frac{1}{2}, -\frac{\sqrt{3}}{2}$

1. $\langle 4, 3\rangle, \|\mathbf{v}\| = 5$ **3.** $\langle -3, 2\rangle, \|\mathbf{v}\| = \sqrt{13}$

5. $\langle 0, 5\rangle, \|\mathbf{v}\| = 5$ **7.** $\mathbf{v} = \langle 16, 7\rangle, \|\mathbf{v}\| = \sqrt{305}$

9. $\mathbf{v} = \langle 8, 6\rangle, \|\mathbf{v}\| = 10$

11.

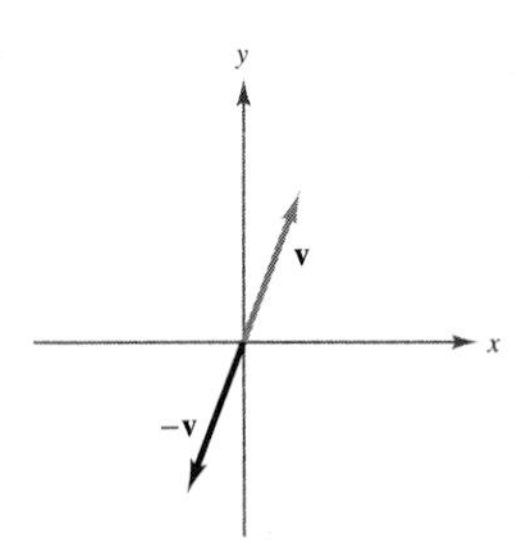

13.

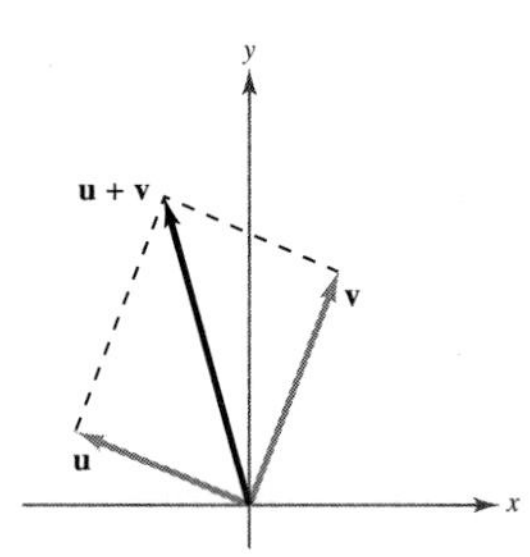

15.

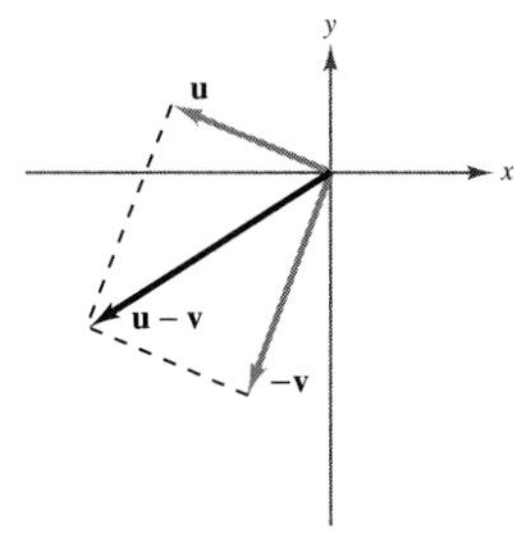

17. (a) $\langle 4, 3\rangle$ (b) $\langle -2, 1\rangle$ (c) $\langle -7, 1\rangle$

19. (a) $\langle 4, -2\rangle$ (b) $\langle 4, -2\rangle$ (c) $\langle 8, -4\rangle$

21. (a) $3\mathbf{i} - 2\mathbf{j}$ (b) $-\mathbf{i} + 4\mathbf{j}$ (c) $-4\mathbf{i} + 11\mathbf{j}$

23. (a) $2\mathbf{i} + \mathbf{j}$ (b) $2\mathbf{i} - \mathbf{j}$ (c) $4\mathbf{i} - 3\mathbf{j}$

25. $\langle 1, 0\rangle$ **27.** $\left\langle -\frac{1}{\sqrt{2}}, \frac{1}{\sqrt{2}}\right\rangle$ **29.** $\frac{4}{5}\mathbf{i} - \frac{3}{5}\mathbf{j}$

31. $\mathbf{j}$ **33.** $\left\langle \frac{5}{\sqrt{2}}, \frac{5}{\sqrt{2}}\right\rangle$ **35.** $\left\langle -\frac{21}{5}, \frac{28}{5}\right\rangle$

37. $\mathbf{v} = \langle 3, -\frac{3}{2}\rangle$

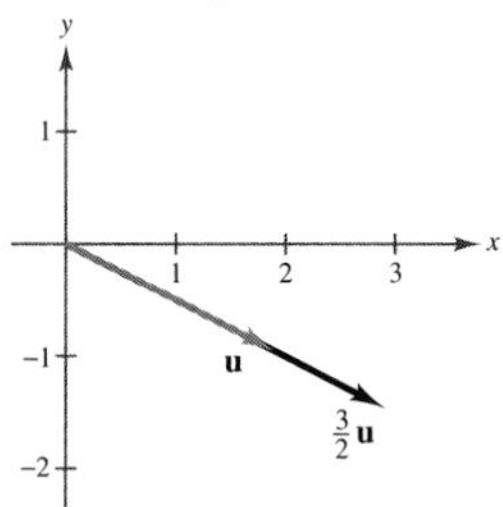

39. $\mathbf{v} = \langle 4, 3\rangle$

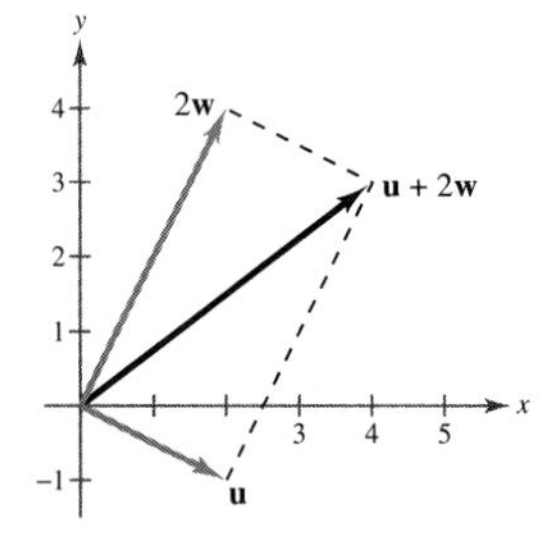

41. $\mathbf{v} = \langle \frac{7}{2}, -\frac{1}{2} \rangle$

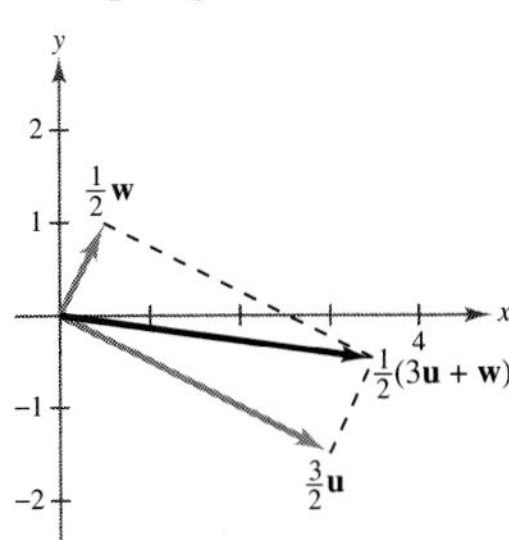

43. $\|\mathbf{v}\| = 5,\ \theta = 30°$ **45.** $\|\mathbf{v}\| = 6\sqrt{2},\ \theta = 315°$

47. $\mathbf{v} = \langle 3, 0 \rangle$

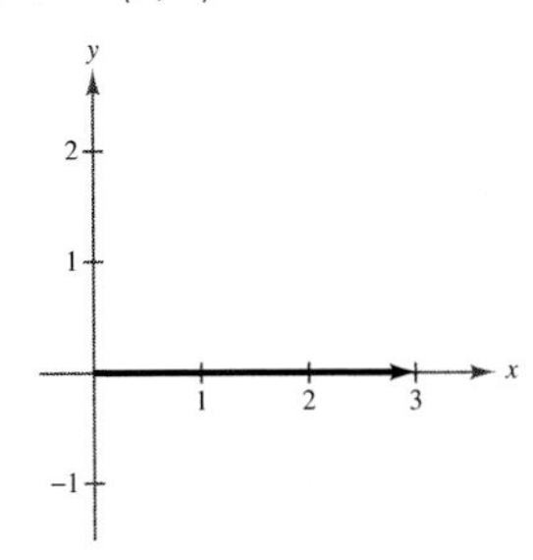

49. $\mathbf{v} = \left\langle -\frac{\sqrt{3}}{2}, \frac{1}{2} \right\rangle$

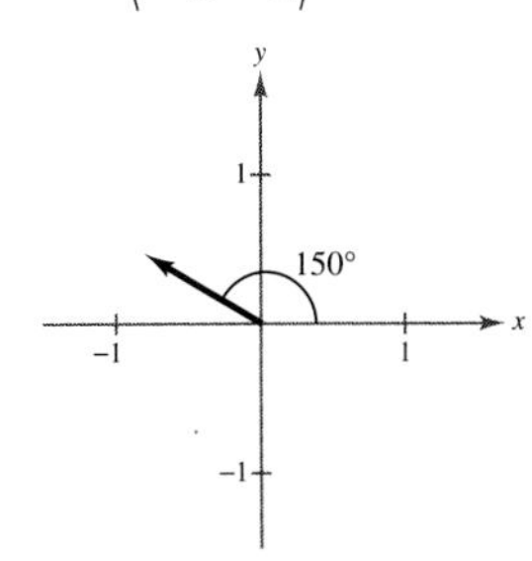

51. $\mathbf{v} = \left\langle -\frac{3\sqrt{6}}{2}, \frac{3\sqrt{2}}{2} \right\rangle$

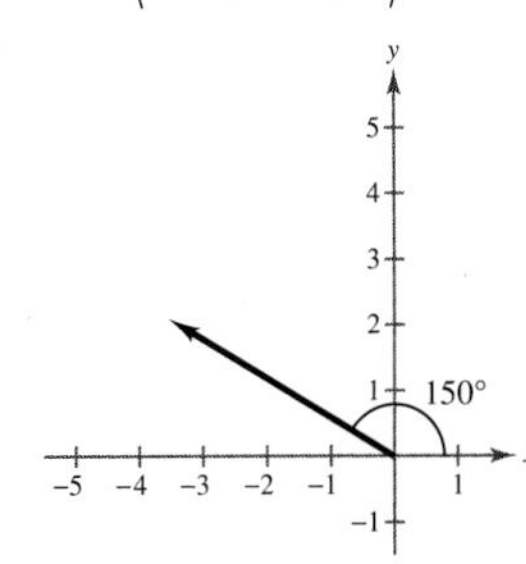

53. $\mathbf{v} = \left\langle \frac{\sqrt{10}}{5}, \frac{3\sqrt{10}}{5} \right\rangle$

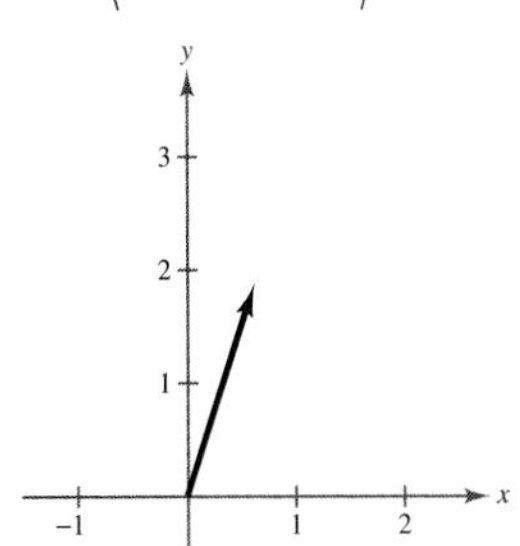

55. $\langle 5, 5 \rangle$ **57.** $\langle 10\sqrt{2} - 50,\ 10\sqrt{2} \rangle$

59. 90° **61.** 63.4° **63.** 62.7°

65. (a) 0° (b) 180°

(c) No. The magnitude is at most equal to the sum when the angle between the vectors is 0°.

67. 12.1°, 357.85 newtons **69.** 71.3°, 228.5 pounds

71. Horizontal component: $80 \cos 40° \approx 61.28$ feet per second

Vertical component: $80 \sin 40° \approx 51.42$ feet per second

73. $T_{AC} \approx 1758.8$ pounds **75.** 3154.4 pounds

$T_{BC} \approx 1305.4$ pounds

77. N 21.4° E, 138.7 kilometers per hour

79. 850 foot-pounds **81.** True **83.** False. $a = b = 0$

85. Answers will vary. **87.** $\langle 1, 3 \rangle$ or $\langle -1, -3 \rangle$

89. 14.7 meters **91.** $8 \tan \theta$ **93.** $6 \sec \theta$

Section 3.4 *(page 308)*

Warm Up *(page 308)*

1. (a) $\langle -14, -5 \rangle$ (b) $3\sqrt{5}$

2. (a) $\left\langle \frac{43}{8}, \frac{3}{5} \right\rangle$ (b) $\frac{\sqrt{1249}}{40}$

3. (a) $-6\mathbf{i} + 4\mathbf{j}$ (b) $4\sqrt{17}$

4. (a) $8.7\mathbf{i} - 2.2\mathbf{j}$ (b) 1.5 **5.** 2.09, 4.19

6. 1.57, 4.71 **7.** 1, 5.28 **8.** 2.89, 3.39

9. (a) $\langle \frac{12}{13}, -\frac{5}{13} \rangle$ (b) $\langle -\frac{12}{13}, \frac{5}{13} \rangle$

10. (a) $\langle \frac{12}{13}, \frac{5}{13} \rangle$ (b) $\langle -\frac{12}{13}, -\frac{5}{13} \rangle$

1. -6 **3.** 6 **5.** 8, scalar **7.** $\langle -6, 8 \rangle$, vector

9. 13 **11.** $5\sqrt{41}$ **13.** \$37,289 total revenue **15.** 90°

17. 143.13° **19.** $\frac{5\pi}{12}$

21.

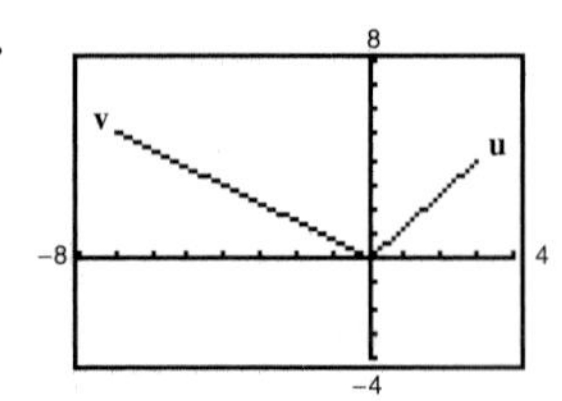

91.33°

23.

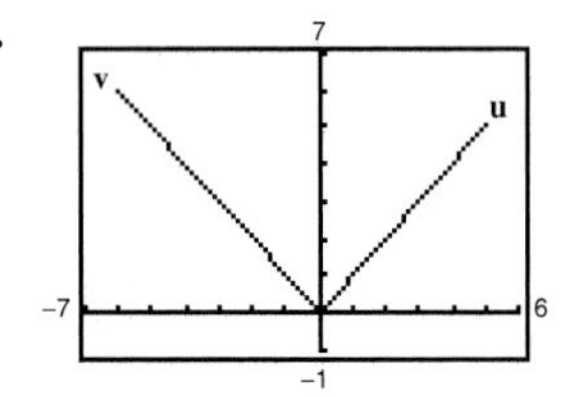

90°

25. 26.6°, 63.4°, 90° **27.** −20 **29.** Parallel

31. Neither **33.** Orthogonal **35.** $\frac{16}{17}\langle 4, 1\rangle, \frac{13}{17}\langle -1, 4\rangle$

37. $\frac{45}{229}\langle 2, 15\rangle, \frac{6}{229}\langle -15, 2\rangle$ **39.** $\langle -5, 3\rangle, \langle 5, -3\rangle$

41. $\frac{2}{3}\mathbf{i} + \frac{1}{2}\mathbf{j}, -\frac{2}{3}\mathbf{i} - \frac{1}{2}\mathbf{j}$

43. (a) 6251.3 pounds (b) 35,453.1 pounds

45. (a) $\theta = \frac{\pi}{2}$ (b) $0 \le \theta < \frac{\pi}{2}$ (c) $\frac{\pi}{2} < \theta \le \pi$

47. 735 newton-meters **49.** 779.4 foot-pounds **51.** 32

53.–55. Answers may vary.

Focus on Concepts *(page 311)*

1. $\frac{a}{\sin A} = \frac{b}{\sin B} = \frac{c}{\sin C}$

2. $a^2 = b^2 + c^2 - 2bc \cos A,\ b^2 = a^2 + c^2 - 2ac \cos B,$
$c^2 = a^2 + b^2 - 2ab \cos C$

3. True **4.** Pythagorean Theorem

5. False. There may be no solution, one solution, or two solutions.

6. Direction and magnitude **7.** A, C

8. a. The angle between the vectors is acute.

9. If $k > 0$, the direction is the same and the magnitude is k times as great.

If $k < 0$, the result is a vector in the opposite direction and the magnitude is k times as great.

10. The diagonal of the parallelogram with **u** and **v** as its adjacent sides

11. b. Visualize the sum of **u** and −**v**.

12. $\mathbf{u} \cdot \mathbf{v} = \|\mathbf{u}\|\,\|\mathbf{v}\| \cos \theta$

(a) Negative because $\frac{\pi}{2} < \theta < \pi$

(b) Zero because $\theta = \frac{\pi}{2}$

(c) Positive because $\theta = 0$

(d) Positive because $0 < \theta < \frac{\pi}{2}$

13. Yes. $\mathbf{u} = \langle 3, 3\rangle,\ \mathbf{v} = \langle 1, 1\rangle$

Review Exercises *(page 312)*

1. $A \approx 29.7°,\ B \approx 52.4°,\ C \approx 97.9°$

3. $C = 110°,\ b \approx 20.4,\ c \approx 22.6$

5. $A = 35°,\ C = 35°,\ b \approx 6.6$

7. No solution **9.** $A \approx 25.9°,\ C \approx 39.1°,\ c \approx 10.1$

11. $B \approx 31.2°,\ C \approx 133.8°,\ c \approx 13.9$
$B \approx 148.8°,\ C \approx 16.2°,\ c \approx 5.39$

13. $A \approx 9.9°,\ C \approx 20.1°,\ b \approx 29.1$

15. $A \approx 40.9°,\ C \approx 114.1°,\ c \approx 8.6$
$A \approx 139.1°,\ C \approx 15.9°,\ c \approx 2.6$

17. 9.798 **19.** 9.08 **21.** 31.1 meters **23.** 31.0 feet

25.

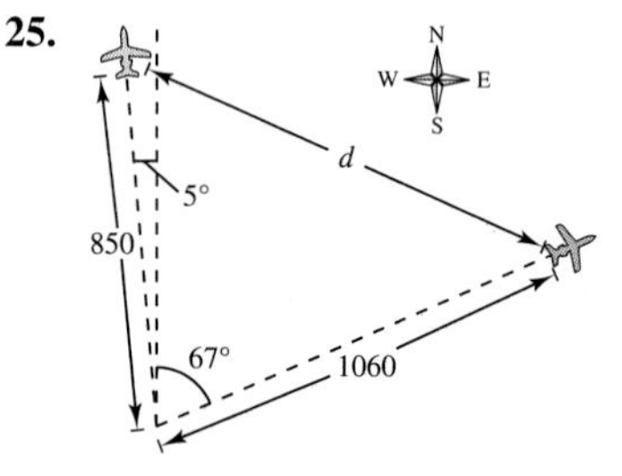

1135 miles

27. $\langle 5, 3\rangle$ **29.** $\langle -4, 0\rangle$ **31.** $\langle -70, 50\rangle$

33. $\langle 7, -7\rangle$ **35.** $\langle -4, 4\sqrt{3}\rangle$

37. $10\sqrt{2}(\mathbf{i} \cos 135° + \mathbf{j} \sin 135°)$

39. $\left\langle \frac{6}{\sqrt{61}}, -\frac{5}{\sqrt{61}} \right\rangle$

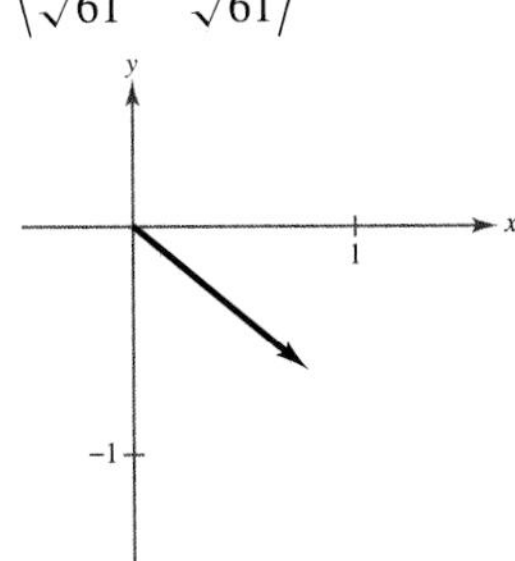

41. $\langle -26, -35 \rangle$

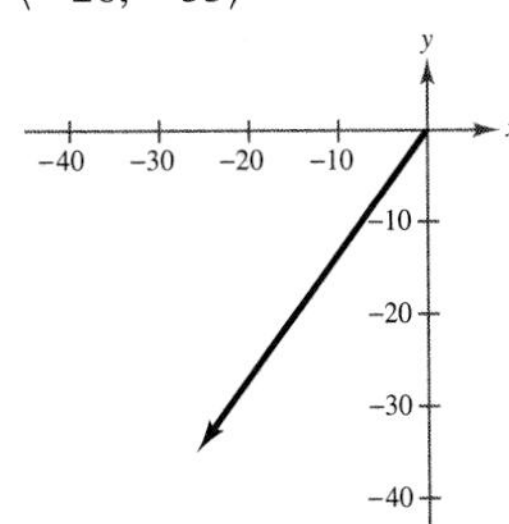

43.

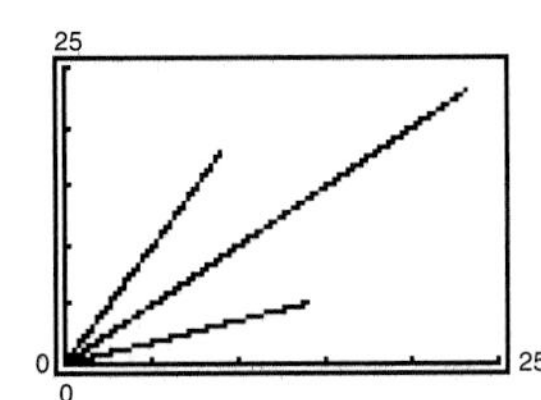

45. 92.2 pounds, 79.9° **47.** 180 pounds

49. 104 pounds **51.** -8 **53.** 0 **55.** 20

57. (a) $\mathbf{u} = \langle 3, -1 \rangle$, $\mathbf{v} = \langle 4, 2 \rangle$ (b) $2\sqrt{5}$ (c) 10
(d) $\langle 10, 0 \rangle$ (e) $\langle 2, 1 \rangle$ (f) $\langle 1, -2 \rangle$

59. 59.0° **61.** 15° **63.** 135° **65.** 80.3°

67. $\frac{10}{17}\langle 4, 1 \rangle$ **69.** $\frac{18}{13}\langle 2, 3 \rangle$ **71.** $\frac{11}{5}\langle 3, 1 \rangle$

73. $125\sqrt{3}$ foot-pounds **75.–81.** Answers will vary.

83. $\langle 1, -3 \rangle$ **85.** $\langle 7, 2 \rangle$

Cumulative Test for Chapters 1–3
(page 318)

1. (a)

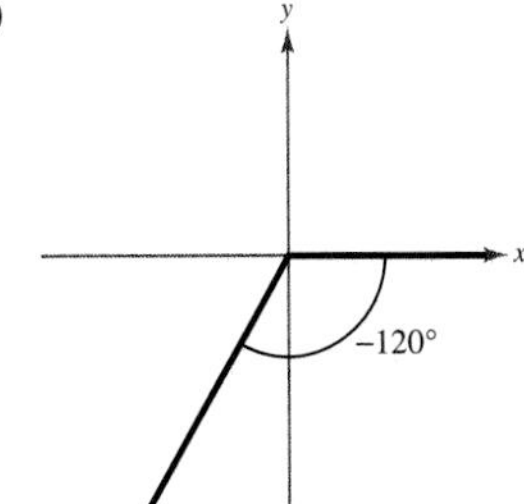

(b) 240°

(c) $-\frac{2\pi}{3}$

(d) 60°

(e) $\sin(-120°) = -\frac{\sqrt{3}}{2}$

$\cos(-120°) = -\frac{1}{2}$

$\tan(-120°) = \sqrt{3}$

$\csc(-120°) = -\frac{2\sqrt{3}}{3}$

$\sec(-120°) = -2$

$\cot(-120°) = \frac{\sqrt{3}}{3}$

2. 134.6° **3.** $\frac{3}{5}$

4. (a)

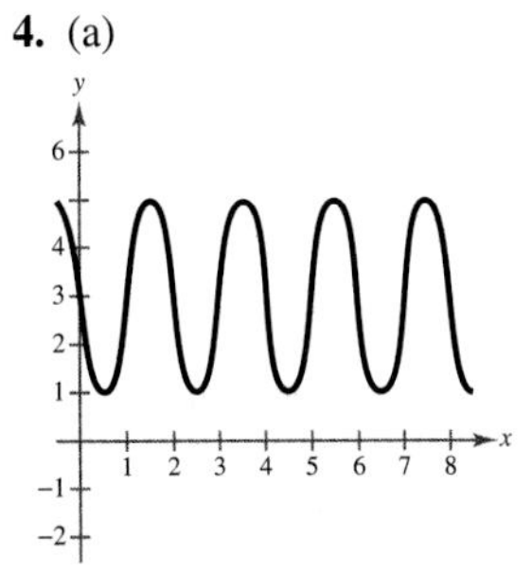

(b)

5. $h(x) = -3\cos(\pi x)$ **6.** $\sqrt{1 - 4x^2}$ **7.** $2\tan\theta$

8. Answers will vary.

9. (a) $\frac{\pi}{3}, \frac{\pi}{2}, \frac{3\pi}{2}, \frac{5\pi}{3}$ (b) $\frac{\pi}{6}, \frac{5\pi}{6}, \frac{7\pi}{6}, \frac{11\pi}{6}$

10. (a) $B \approx 26.4°$, $C \approx 123.6°$, $c \approx 15$
(b) $a \approx 5.0$, $B \approx 52.5°$, $C \approx 97.5°$

11. 5 feet **12.** N 32.6° E, 543.9 kilometers per hour

13. $\frac{19}{26}\langle -1, 5 \rangle$

CHAPTER 4

Section 4.1 *(page 325)*

Warm Up *(page 325)*

1. $2\sqrt{3}$ **2.** $10\sqrt{5}$ **3.** $\sqrt{5}$ **4.** $-6\sqrt{3}$

5. 12 **6.** 48 **7.** $\dfrac{\sqrt{3}}{3}$ **8.** $\sqrt{2}$

9. $-\dfrac{1}{2} \pm \dfrac{\sqrt{5}}{2}$ **10.** $-1 \pm \sqrt{2}$

1. $a = -10,\ b = 6$ **3.** $a = 6,\ b = 5$
5. $4 + 3i$ **7.** $2 - 3\sqrt{3}i$ **9.** $5\sqrt{3}i$
11. $-1 - 6i$ **13.** 8 **15.** $0.3i$ **17.** $11 - i$
19. 4 **21.** $3 - 3\sqrt{2}i$ **23.** $-14 + 20i$
25. $\frac{1}{6} + \frac{7}{6}i$ **27.** $-2\sqrt{3}$ **29.** -10 **31.** $5 + i$
33. $12 + 30i$ **35.** 24 **37.** $-9 + 40i$ **39.** -10
41. $\sqrt{-6}\sqrt{-6} = \sqrt{6}i\sqrt{6}i = 6i^2 = -6$
43. $5 - 3i,\ 34$ **45.** $-2 + \sqrt{5}i,\ 9$
47. $-20i,\ 400$ **49.** $\sqrt{8},\ 8$ **51.** $-6i$
53. $\frac{16}{41} + \frac{20}{41}i$ **55.** $\frac{3}{5} + \frac{4}{5}i$
57. $-7 - 6i$ **59.** $-\frac{9}{1681} + \frac{40}{1681}i$ **61.** $-\frac{1}{2} - \frac{5}{2}i$
63. $\frac{62}{949} + \frac{297}{949}i$ **65.** $1 \pm i$ **67.** $-2 \pm \frac{1}{2}i$
69. $-\dfrac{3}{2}, -\dfrac{5}{2}$ **71.** $\dfrac{1}{8} \pm \dfrac{\sqrt{11}}{8}i$

73.

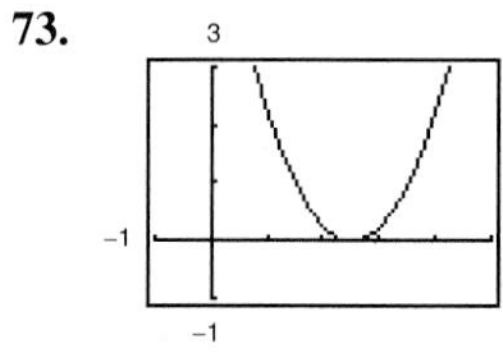

$x = \frac{5}{2}$

75.

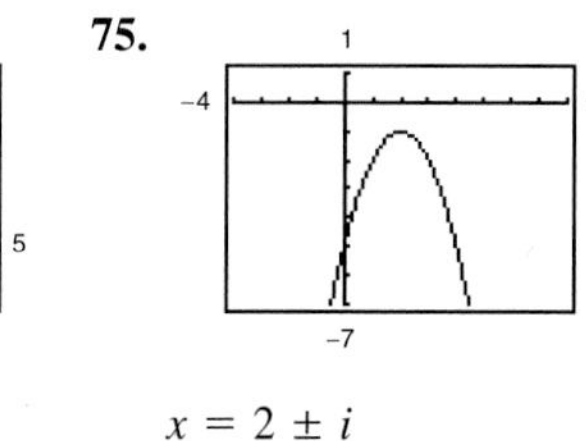

$x = 2 \pm i$

77. The number of x-intercepts of the graph corresponds to the number of real solutions of the equation. If there are no x-intercepts, the quadratic equation has two complex solutions.

79. $-1 + 6i$ **81.** $-5i$ **83.** $-375\sqrt{3}i$
85. i **87.** 8, 8, 8 **89.–93.** Answers will vary.
95. $-x^2 - 3x + 12$ **97.** $4x^2 - 20x + 25$
99. $a = \dfrac{1}{2}\sqrt{\dfrac{3V}{\pi b}}$ **101.** 1 liter

Section 4.2 *(page 333)*

Warm Up *(page 333)*

1. $4 - \sqrt{29}i, 4 + \sqrt{29}i$ **2.** $-5 - 12i, -5 + 12i$
3. $-1 + 4\sqrt{2}i, -1 - 4\sqrt{2}i$ **4.** $6 + \frac{1}{2}i, 6 - \frac{1}{2}i$
5. $-13 + 9i$ **6.** $12 + 16i$ **7.** $26 + 22i$ **8.** 29
9. i **10.** $-9 + 46i$

1. Three solutions **3.** Four solutions
5. No real solutions **7.** Two real solutions
9. $\pm\sqrt{5}$ **11.** $-5 \pm \sqrt{6}$ **13.** 4 **15.** $-1 \pm 2i$
17. $\frac{1}{2} \pm i$ **19.** $20 \pm 2\sqrt{215}$ **21.** (4, 0); same
23. No real zeros, no x-intercepts
25. $1 \pm i$; $(z - 1 + i)(z - 1 - i)$
27. $\pm 3, \pm 3i$; $(x + 3)(x - 3)(x + 3i)(x - 3i)$
29. $-\frac{3}{2}, \pm 5i$ **31.** $-3 \pm i, \frac{1}{4}$ **33.** $1, 2, -3 \pm \sqrt{2}i$
35. $\dfrac{3}{4}, \dfrac{1}{2} \pm \dfrac{\sqrt{5}}{2}i$ **37.** $x^3 - x^2 + 25x - 25$
39. $x^3 - 10x^2 + 33x - 34$
41. $x^4 + 8x^3 + 9x^2 - 10x + 100$
43. $16x^4 + 36x^3 + 16x^2 + x - 30$
45. $(x + 3i)(x - 3i)(x + \sqrt{3})(x - \sqrt{3})$
47. (a) Answers will vary.
(b) f does not have real coefficients.
49. (a) $0 < k < 4$ (b) $k = 4$ (c) $k < 0$ (d) $k > 4$
51. (a) No. f has (0, 0) as an intercept.
(b) No. The function must be at least a fourth-degree polynomial.
(c) Yes
(d) No. k has $(-1, 0)$ as an intercept.

Section 4.3 *(page 340)*

Warm Up *(page 340)*

1. $-5 - 10i$ **2.** $7 + 3\sqrt{6}i$ **3.** $-1 - 4i$ **4.** $-3i$
5. $6 - 14i$ **6.** $6 + 4\sqrt{2}i$ **7.** $-22 + 16i$ **8.** 13
9. $-\frac{3}{2} + \frac{5}{2}i$ **10.** $-\frac{5}{2} - \frac{3}{2}i$

1. 5

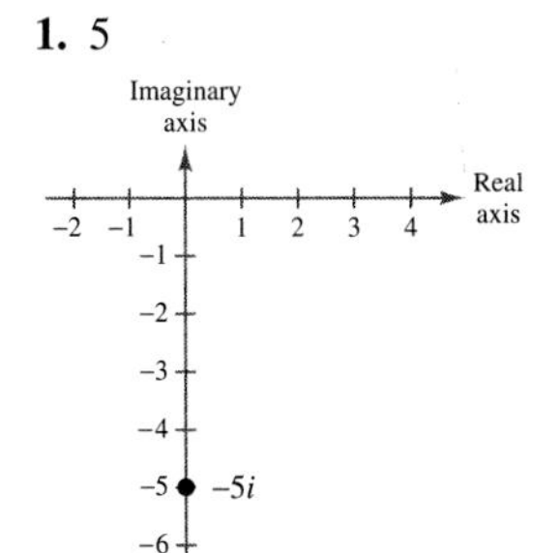

3. $4\sqrt{2}$

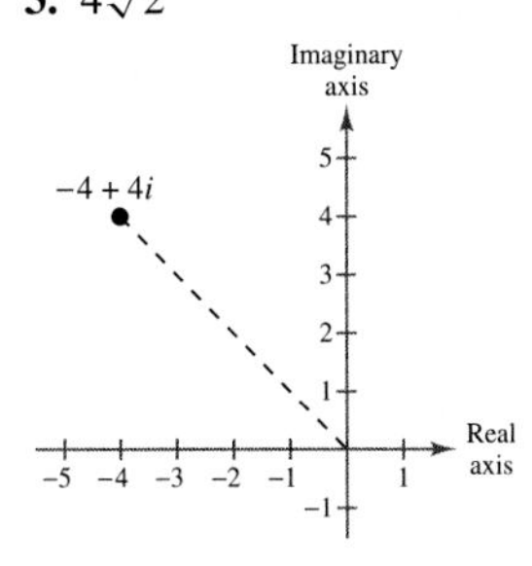

5. $\sqrt{85}$

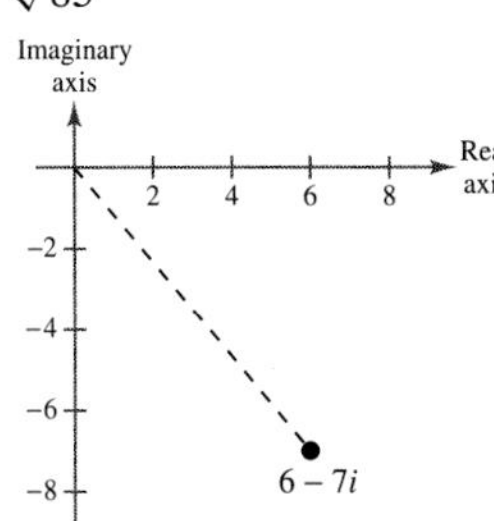

7. $3\left(\cos\frac{\pi}{2} + i\sin\frac{\pi}{2}\right)$ **9.** $2\sqrt{2}\left(\cos\frac{5\pi}{4} + i\sin\frac{5\pi}{4}\right)$

11. $3\sqrt{2}\left(\cos\frac{7\pi}{4} + i\sin\frac{7\pi}{4}\right)$

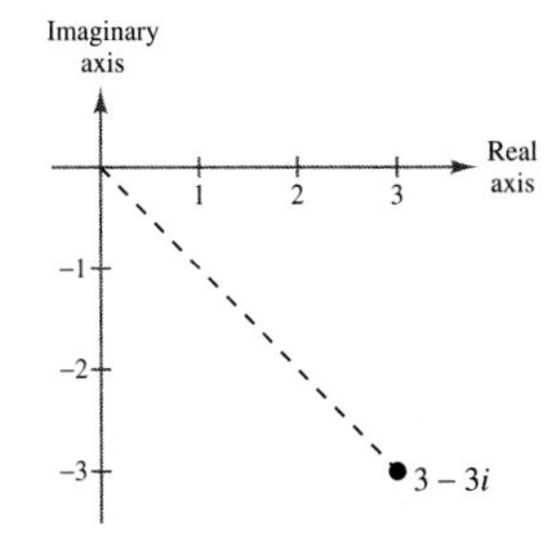

13. $2\left(\cos\frac{\pi}{6} + i\sin\frac{\pi}{6}\right)$

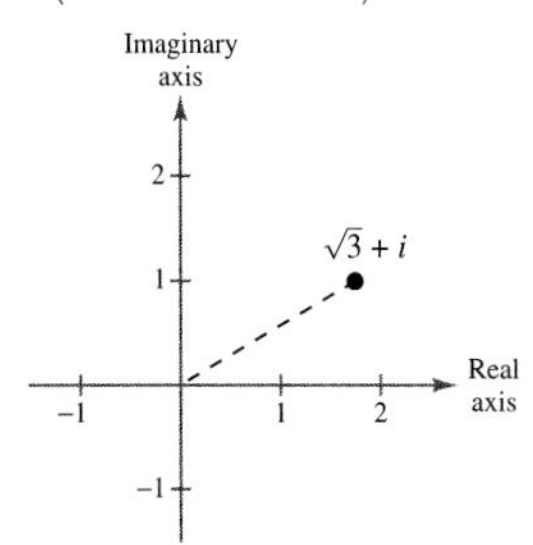

15. $4\left(\cos\frac{4\pi}{3} + i\sin\frac{4\pi}{3}\right)$

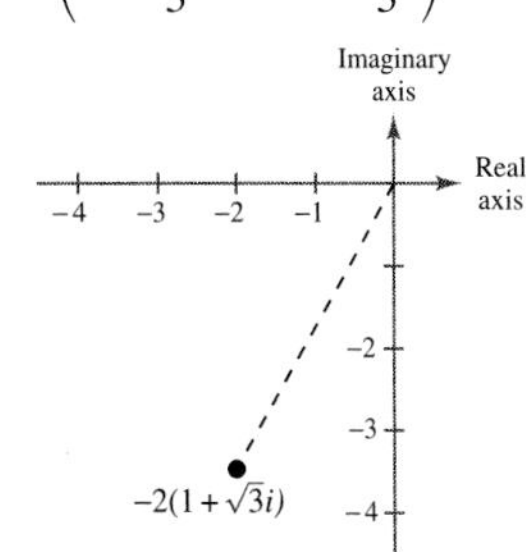

17. $6\left(\cos\frac{\pi}{2} + i\sin\frac{\pi}{2}\right)$

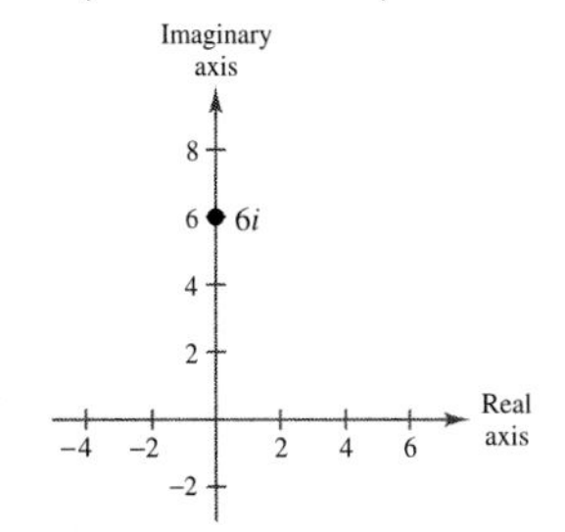

19. $\sqrt{65}(\cos 2.62 + i\sin 2.62)$

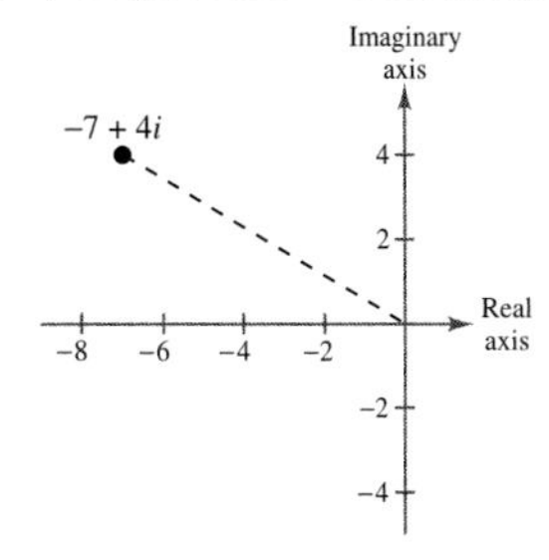

21. $7(\cos 0 + i \sin 0)$

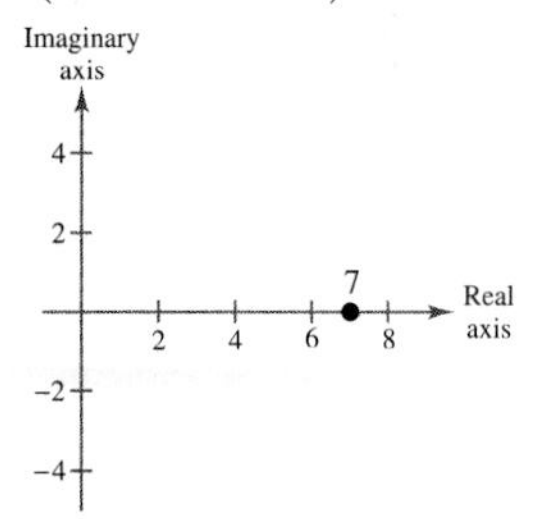

23. $\sqrt{37}(\cos 1.41 + i \sin 1.41)$

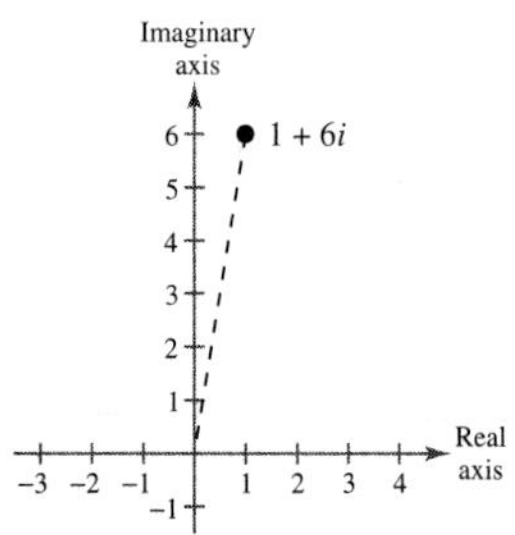

25. $5.39(\cos 0.38 + i \sin 0.38)$

27. $8.19(\cos 5.26 + i \sin 5.26)$

29. $-\sqrt{3} + i$

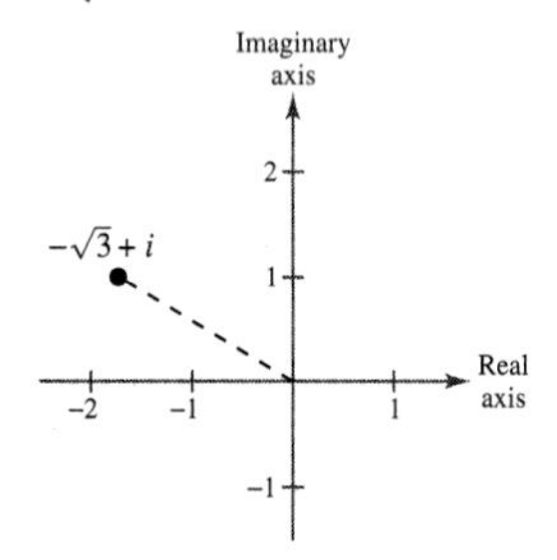

31. $\frac{3}{4} - \frac{3\sqrt{3}}{4}i$

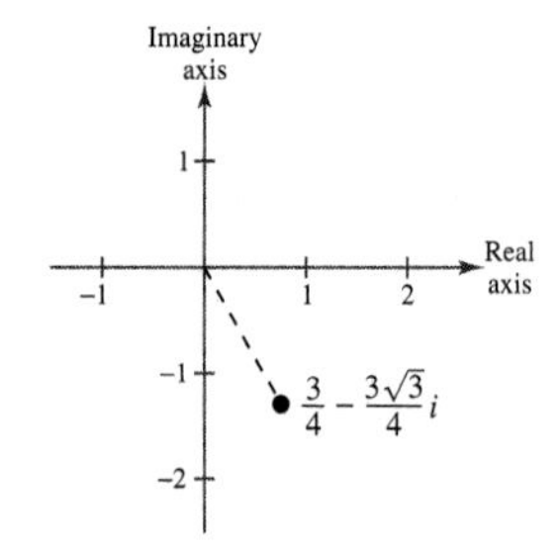

33. $-\frac{15\sqrt{2}}{8} + \frac{15\sqrt{2}}{8}i$

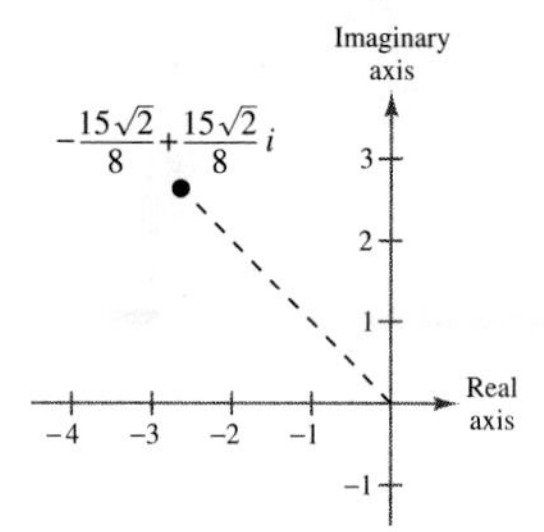

35. $-4i$

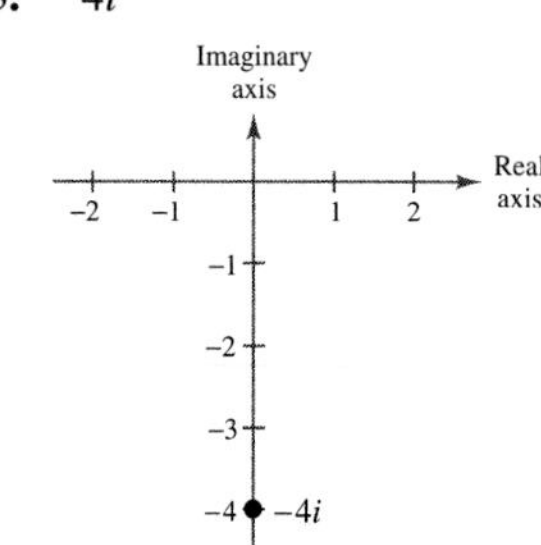

37. $2.8408 + 0.9643i$

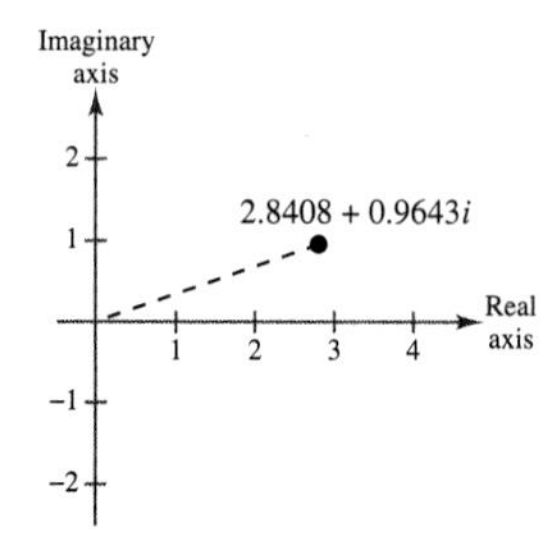

39. $4.70 + 1.71i$ **41.** $-3.22 - 2.38i$

43. $12\left(\cos \frac{\pi}{2} + i \sin \frac{\pi}{2}\right)$ **45.** $\frac{10}{9}(\cos 200° + i \sin 200°)$

47. $\frac{1}{2}(\cos 80° + i \sin 80°)$ **49.** $\cos \frac{2\pi}{3} + i \sin \frac{2\pi}{3}$

51. $4[\cos(-58°) + i \sin(-58°)]$

53. (a) $2\sqrt{2}(\cos 45° + i \sin 45°)$; $\sqrt{2}[\cos(-45°) + i \sin(-45°)]$

(b) $4(\cos 0° + i \sin 0°) = 4$

(c) 4

55. (a) $2[\cos(-90°) + i\sin(-90°)]$;
$\sqrt{2}(\cos 45° + i\sin 45°)$
(b) $2\sqrt{2}[\cos(-45°) + i\sin(-45°)] = 2 - 2i$
(c) $2 - 2i$

57. (a) $5(\cos 0° + i\sin 0°)$;
$\sqrt{13}(\cos 56.31° + i\sin 56.31°)$
(b) $\frac{5}{\sqrt{13}}[\cos(-56.31°) + i\sin(-56.31°)]$
$\approx 0.7692 - 1.154i$
(c) $\frac{10}{13} - \frac{15}{13}i \approx 0.7692 - 1.154i$

59. Answers will vary.

61. (a) r^2 (b) $\cos(2\theta) + i\sin(2\theta)$

63.

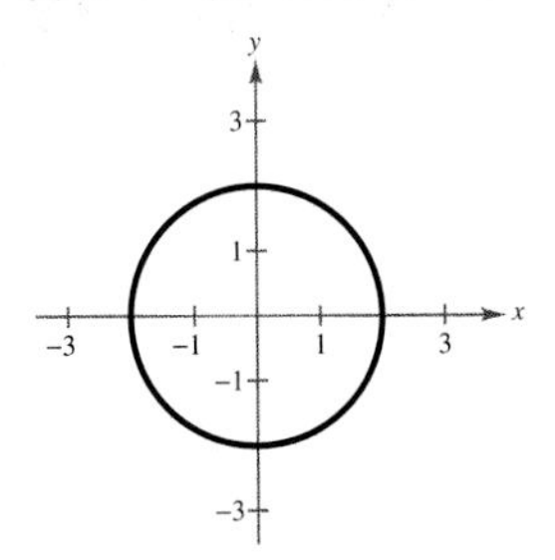

Section 4.4 *(page 346)*

Warm Up *(page 346)*

1. $3\sqrt[3]{2}$ **2.** $2\sqrt{2}$ **3.** $5\sqrt{2}(\cos 135° + i\sin 135°)$
4. $3(\cos 270° + i\sin 270°)$ **5.** $12(\cos 180° + i\sin 180°)$
6. $12(\cos 0° + i\sin 0°)$ **7.** $\cos\frac{3\pi}{4} + i\sin\frac{3\pi}{4}$
8. $\cos\frac{11\pi}{12} + i\sin\frac{11\pi}{12}$ **9.** $2\left(\cos\frac{\pi}{2} + i\sin\frac{\pi}{2}\right)$
10. $\frac{2}{3}(\cos 45° + i\sin 45°)$

1. $-4 - 4i$ **3.** $-32i$ **5.** $-128\sqrt{3} - 128i$

7. $\frac{125}{2} + \frac{125\sqrt{3}}{2}i$ **9.** i **11.** $608.02 + 144.69i$

13. $-597 - 122i$ **15.** $\frac{81}{2} + \frac{81\sqrt{3}}{2}i$

17. Answers will vary.

19. (a) $2(\cos 30° + i\sin 30°)$
$2(\cos 150° + i\sin 150°)$
$2(\cos 270° + i\sin 270°)$
(b) $8i$

21. (a) $\sqrt{5}(\cos 60° + i\sin 60°)$
$\sqrt{5}(\cos 240° + i\sin 240°)$
(b)

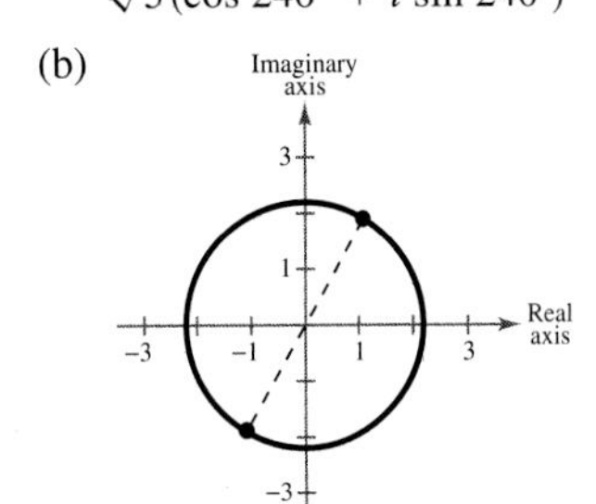

(c) $\frac{\sqrt{5}}{2} + \frac{\sqrt{15}}{2}i,\ -\frac{\sqrt{5}}{2} - \frac{\sqrt{15}}{2}i$

23. (a) $2\left(\cos\frac{\pi}{3} + i\sin\frac{\pi}{3}\right)$
$2\left(\cos\frac{5\pi}{6} + i\sin\frac{5\pi}{6}\right)$
$2\left(\cos\frac{4\pi}{3} + i\sin\frac{4\pi}{3}\right)$
$2\left(\cos\frac{11\pi}{6} + i\sin\frac{11\pi}{6}\right)$
(b)

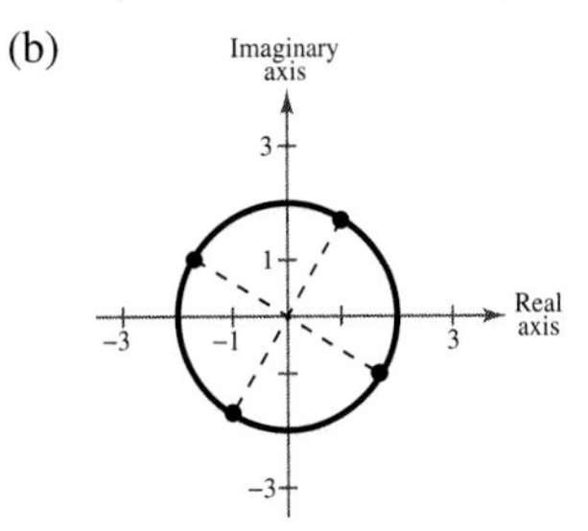

(c) $1 + \sqrt{3}i,\ -\sqrt{3} + i,\ -1 - \sqrt{3}i,\ \sqrt{3} - i$

25. (a) $5\left(\cos \frac{3\pi}{4} + i \sin \frac{3\pi}{4}\right)$

$5\left(\cos \frac{7\pi}{4} + i \sin \frac{7\pi}{4}\right)$

(b)

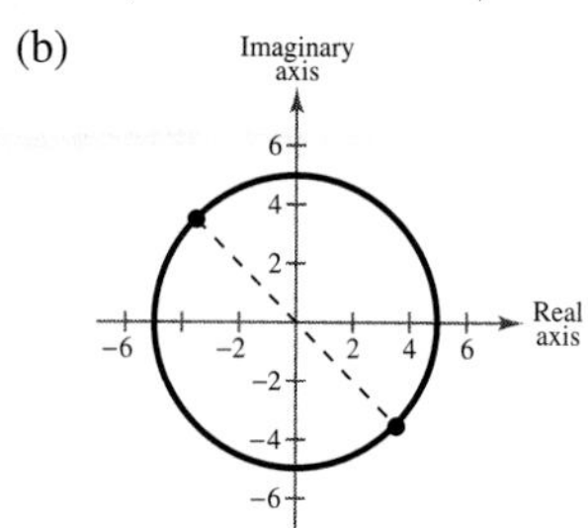

(c) $-\frac{5\sqrt{2}}{2} + \frac{5\sqrt{2}}{2}i,\ \frac{5\sqrt{2}}{2} - \frac{5\sqrt{2}}{2}i$

27. (a) $5\left(\cos \frac{4\pi}{9} + i \sin \frac{4\pi}{9}\right)$

$5\left(\cos \frac{10\pi}{9} + i \sin \frac{10\pi}{9}\right)$

$5\left(\cos \frac{16\pi}{9} + i \sin \frac{16\pi}{9}\right)$

(b)

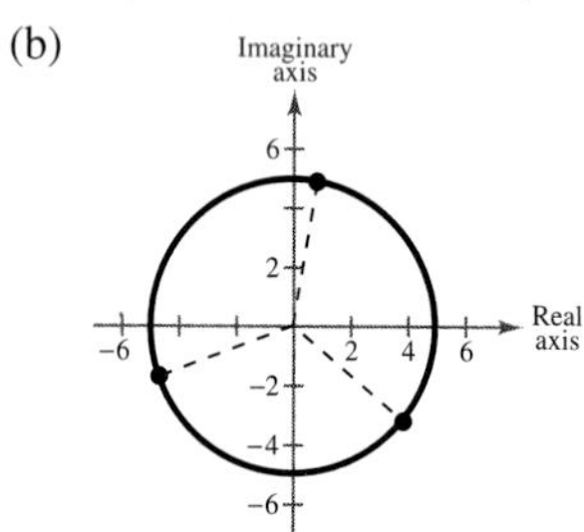

(c) $0.8682 + 4.924i,\ -4.698 - 1.710i,\ 3.830 - 3.214i$

29. (a) $2(\cos 0 + i \sin 0)$

$2\left(\cos \frac{2\pi}{3} + i \sin \frac{2\pi}{3}\right)$

$2\left(\cos \frac{4\pi}{3} + i \sin \frac{4\pi}{3}\right)$

(b)

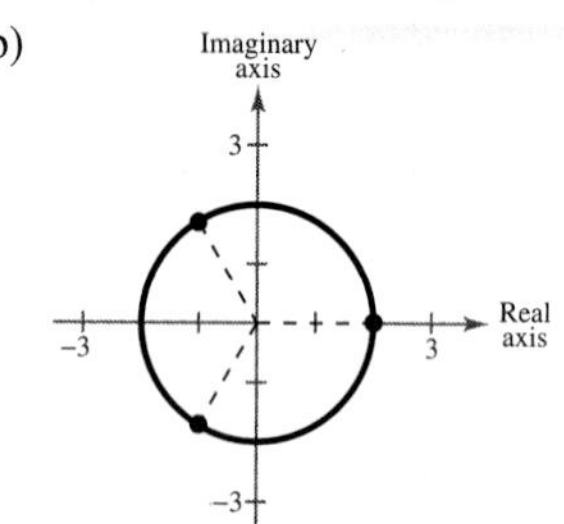

(c) $2,\ -1 + \sqrt{3}i,\ -1 - \sqrt{3}i$

31. (a) $\cos 0 + i \sin 0$

$\cos \frac{2\pi}{5} + i \sin \frac{2\pi}{5}$

$\cos \frac{4\pi}{5} + i \sin \frac{4\pi}{5}$

$\cos \frac{6\pi}{5} + i \sin \frac{6\pi}{5}$

$\cos \frac{8\pi}{5} + i \sin \frac{8\pi}{5}$

(b)

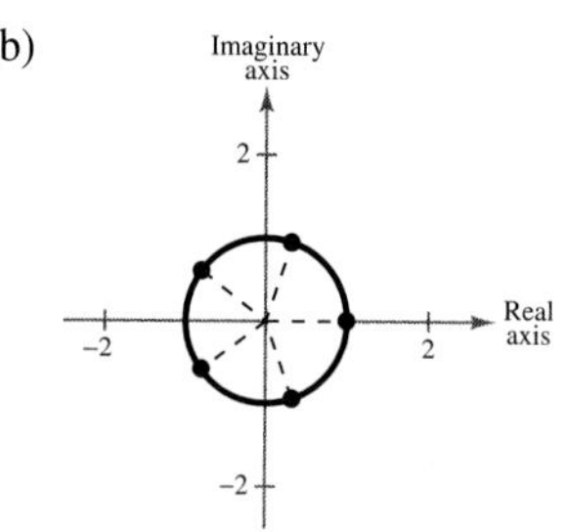

(c) $1, 0.3090 + 0.9511i,\ -0.8090 + 0.5878i,$
$-0.8090 - 0.5878i, 0.3090 - 0.9511i$

33. (a) $5(\cos 60° + i \sin 60°)$

$5(\cos 180° + i \sin 180°)$

$5(\cos 300° + i \sin 300°)$

(b)

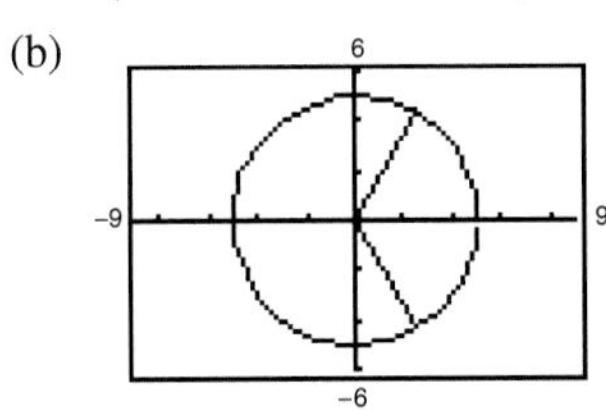

(c) $-5, \ \dfrac{5}{2} \pm \dfrac{5\sqrt{3}}{2} i$

35. (a) $2\sqrt[5]{4\sqrt{2}}\,(\cos 27° + i \sin 27°)$

$2\sqrt[5]{4\sqrt{2}}\,(\cos 99° + i \sin 99°)$

$2\sqrt[5]{4\sqrt{2}}\,(\cos 171° + i \sin 171°)$

$2\sqrt[5]{4\sqrt{2}}\,(\cos 243° + i \sin 243°)$

$2\sqrt[5]{4\sqrt{2}}\,(\cos 315° + i \sin 315°)$

(b)

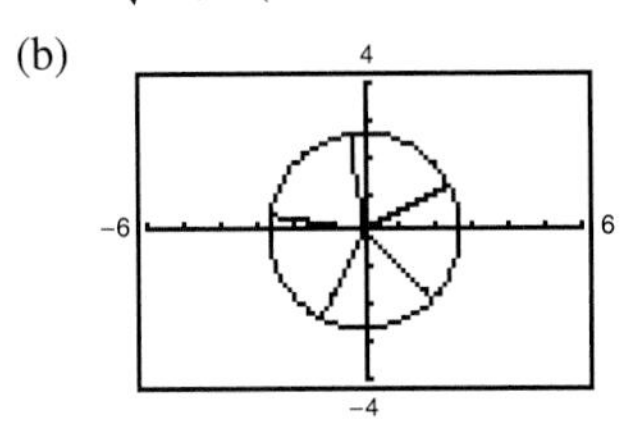

(c) $2.52 + 1.28i, \ -0.44 + 2.79i, \ -2.79 + 0.44i,$
$-1.28 - 2.52i, \ 2 - 2i$

37. $\cos \dfrac{\pi}{8} + i \sin \dfrac{\pi}{8}$

$\cos \dfrac{5\pi}{8} + i \sin \dfrac{5\pi}{8}$

$\cos \dfrac{9\pi}{8} + i \sin \dfrac{9\pi}{8}$

$\cos \dfrac{13\pi}{8} + i \sin \dfrac{13\pi}{8}$

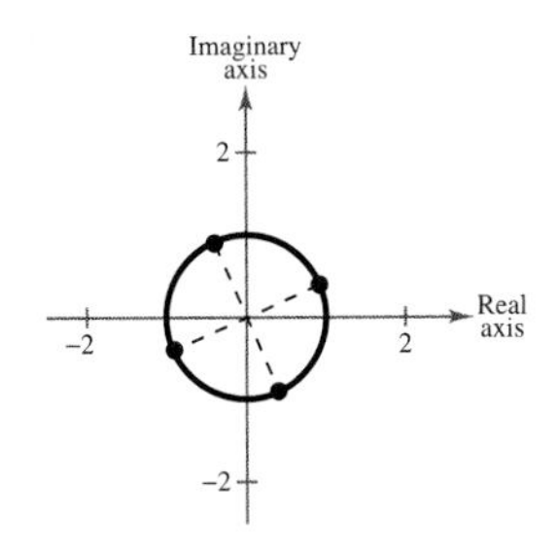

39. $3\left(\cos \dfrac{\pi}{5} + i \sin \dfrac{\pi}{5}\right)$

$3\left(\cos \dfrac{3\pi}{5} + i \sin \dfrac{3\pi}{5}\right)$

$3(\cos \pi + i \sin \pi)$

$3\left(\cos \dfrac{7\pi}{5} + i \sin \dfrac{7\pi}{5}\right)$

$3\left(\cos \dfrac{9\pi}{5} + i \sin \dfrac{9\pi}{5}\right)$

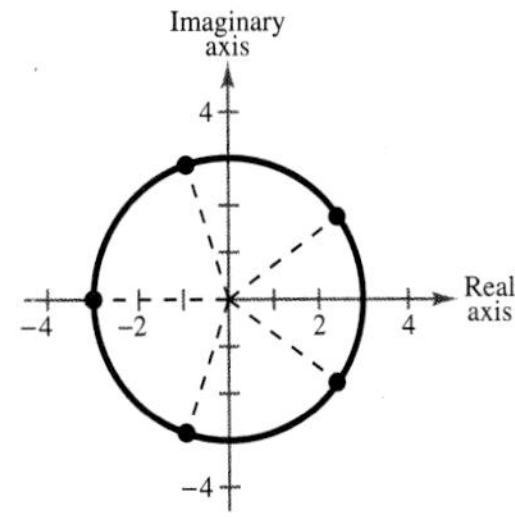

41.

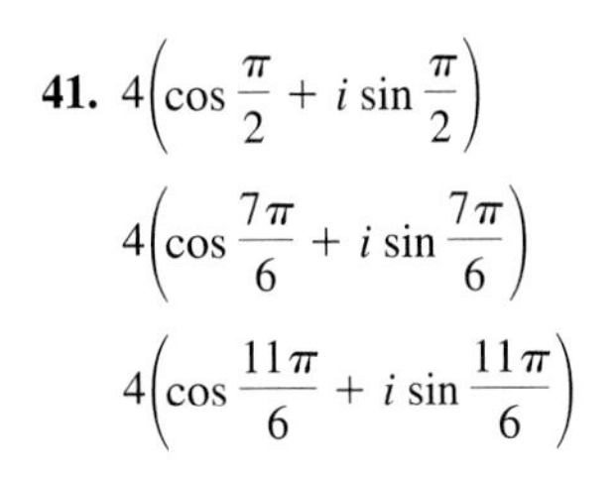

$4\left(\cos \dfrac{\pi}{2} + i \sin \dfrac{\pi}{2}\right)$

$4\left(\cos \dfrac{7\pi}{6} + i \sin \dfrac{7\pi}{6}\right)$

$4\left(\cos \dfrac{11\pi}{6} + i \sin \dfrac{11\pi}{6}\right)$

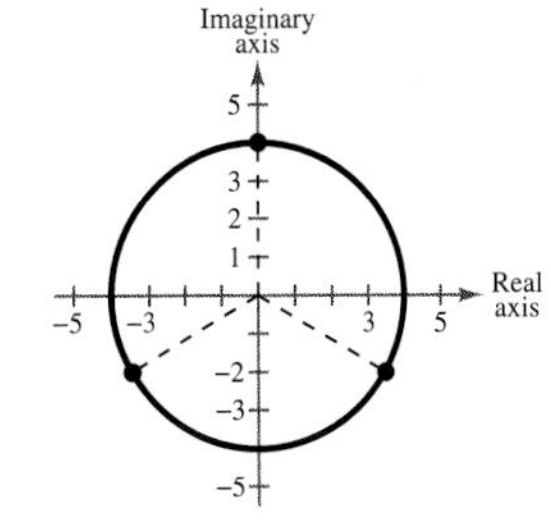

43. $\sqrt[6]{2}(\cos 105° + i \sin 105°)$
$\sqrt[6]{2}(\cos 225° + i \sin 225°)$
$\sqrt[6]{2}(\cos 345° + i \sin 345°)$

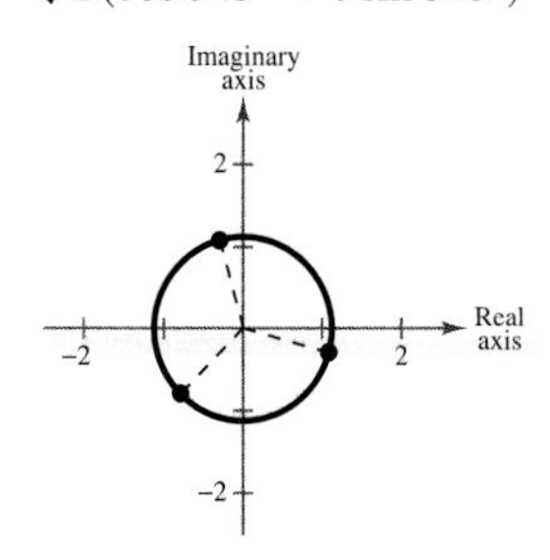

45. $\dfrac{5\sqrt{7} - 9}{4\sqrt{34}}$ **47.** $\dfrac{15}{8}$

Focus on Concepts *(page 348)*

1. $i = \sqrt{-1}$

2. The sum of 1 and the square of a real number is always greater than or equal to 1.

3. $(a + bi)(a - bi) = a^2 - (bi)^2 = a^2 - b^2i^2 = a^2 + b^2$

4. False. $(a + bi) + (a - bi) = 2a$

5. (a) Three real zeros (b) One real zero

6. (a) Four real zeros (b) No real zeros
(c) Two real zeros (d) One real zero

7. $z_1 z_2 = -4, \; \dfrac{z_1}{z_2} = -\dfrac{1}{4}z_1^2.$

8. (a) 3
(b) On the circle 120°, 210°, and 300° from the positive x-axis

Review Exercises *(page 349)*

1. $3 + 5i$ **3.** $-3i$ **5.** $3 + 7i$ **7.** $-\sqrt{2}i$

9. $40 + 65i$ **11.** $-4 - 46i$ **13.** $1 - 6i$

15. $\frac{4}{3}i$ **17.** Two real solutions **19.** One real solution

21. No real solutions **23.** Two real solutions

25. 0, 2 **27.** $-4 \pm \sqrt{6}$ **29.** $-\dfrac{1}{2} \pm \dfrac{\sqrt{5}}{2}i$

31. $-5, -1, 2$; same **33.** 0, 4; same

35. $\frac{3}{4}, 1; f(x) = (4x - 3)(x - 1)^2$

37. $1 \pm 2i, -5; f(x) = (x + 5)(x - 1 - 2i)(x - 1 + 2i)$

39. $7 \pm i, \; 4; h(x) = (x - 4)(x - 7 - i)(x - 7 + i)$

41. $-3 \pm \sqrt{5}i, \; 3, -2;$
$f(x) = (x + 2)(x - 3)(x + 3 - \sqrt{5}i)(x + 3 + \sqrt{5}i)$

43. $f(x) = x^4 + 9x^2$

45. $f(x) = 6x^4 + 13x^3 + 7x^2 - x - 1$

47. $f(x) = 3x^4 - 14x^3 + 17x^2 - 42x + 24$

49. $f(x) = 4x^5 + 8x^4 + x^3 - 5x^2 - x + 1$

51. No

53. **55.**

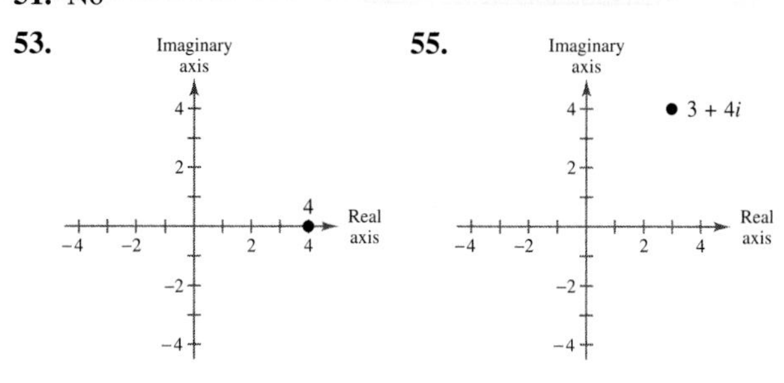

4 5

57. $5(\cos 270° + i \sin 270°)$

59. $2\sqrt{2}(\cos 45° + i \sin 45°)$

61. $5\sqrt{2}(\cos 315° + i \sin 315°)$

63. $13(\cos 67.38° + i \sin 67.38°)$ **65.** $3 + 3\sqrt{3}i$

67. $5\sqrt{3} - 5i$ **69.** $-50 - 50\sqrt{3}i$ **71.** 13

73. (a) $z_1 = 5(\cos 180° + i \sin 180°)$
$z_2 = 5(\cos 90° + i \sin 90°)$
(b) $z_1 z_2 = 25(\cos 270° + i \sin 270°)$
$\dfrac{z_1}{z_2} = \cos 90° + i \sin 90°$

75. (a) $z_1 = 3\sqrt{2}\left(\cos \dfrac{5\pi}{4} + i \sin \dfrac{5\pi}{4}\right)$
$z_2 = 4\left(\cos \dfrac{\pi}{6} + i \sin \dfrac{\pi}{6}\right)$
(b) $z_1 z_2 = 12\sqrt{2}\left(\cos \dfrac{17\pi}{12} + i \sin \dfrac{17\pi}{12}\right)$
$\dfrac{z_1}{z_2} = \dfrac{3\sqrt{2}}{4}\left(\cos \dfrac{13\pi}{12} + i \sin \dfrac{13\pi}{12}\right)$

77. $\dfrac{625}{2} + \dfrac{625\sqrt{3}}{2}i$ **79.** $2035 - 828i$

81. $3\left(\cos\frac{\pi}{4} + i\sin\frac{\pi}{4}\right)$

$3\left(\cos\frac{7\pi}{12} + i\sin\frac{7\pi}{12}\right)$

$3\left(\cos\frac{11\pi}{12} + i\sin\frac{11\pi}{12}\right)$

$3\left(\cos\frac{5\pi}{4} + i\sin\frac{5\pi}{4}\right)$

$3\left(\cos\frac{19\pi}{12} + i\sin\frac{19\pi}{12}\right)$

$3\left(\cos\frac{23\pi}{12} + i\sin\frac{23\pi}{12}\right)$

83. $\cos\frac{\pi}{3} + i\sin\frac{\pi}{3}$

$\cos\pi + i\sin\pi$

$\cos\frac{5\pi}{3} + i\sin\frac{5\pi}{3}$

85. $3\left(\cos\frac{\pi}{4} + i\sin\frac{\pi}{4}\right)$

$3\left(\cos\frac{3\pi}{4} + i\sin\frac{3\pi}{4}\right)$

$3\left(\cos\frac{5\pi}{4} + i\sin\frac{5\pi}{4}\right)$

$3\left(\cos\frac{7\pi}{4} + i\sin\frac{7\pi}{4}\right)$

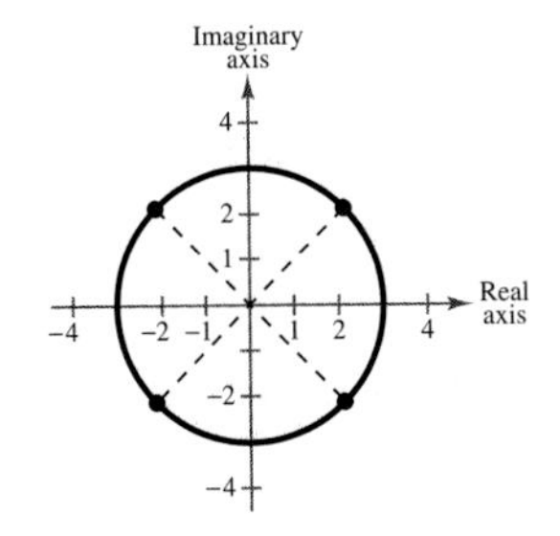

87. $\cos 0 + i\sin 0$

$\cos\frac{\pi}{2} + i\sin\frac{\pi}{2}$

$\cos\frac{2\pi}{3} + i\sin\frac{2\pi}{3}$

$\cos\frac{4\pi}{3} + i\sin\frac{4\pi}{3}$

$\cos\frac{3\pi}{2} + i\sin\frac{3\pi}{2}$

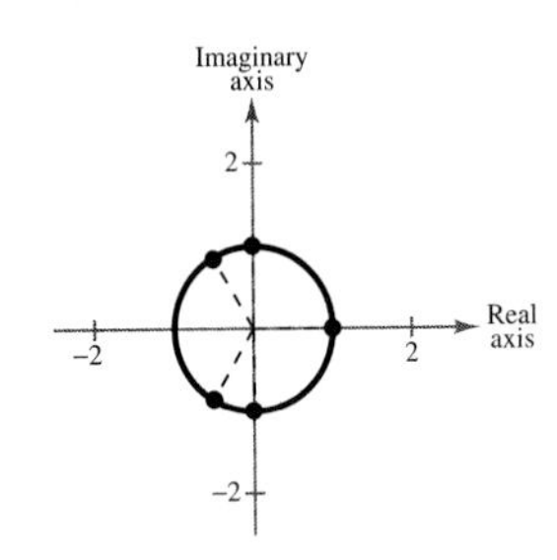

89. $f(x) = x^2 - 25$ **91.** $f(x) = x^3 - 3x^2 + 9x - 27$

93. $f(x) = x^4 + 3x^2 - 6x + 10$

95. The complex conjugates are reflections in the real axis.

Chapter Test *(page 354)*

1. $-2 + 8i$

2. (a) $-3 + 5i$ (b) $-16 + 30i$ (c) 7 (d) $2 - i$

3. $-0.819,\ 1.380$ **4.** $-1.414,\ -0.667,\ 1.414$

5. $\pm 1, \pm 2i;\ h(x) = (x + 1)(x - 1)(x + 2i)(x - 2i)$

6. $\frac{3}{2},\ 2 \pm i;\ g(v) = (2v - 3)(v - 2 + i)(v - 2 - i)$

7. $f(x) = x^4 - 9x^3 + 28x^2 - 30x$

8. $f(x) = x^4 - 6x^3 + 16x^2 - 24x + 16$

9. No. If $a + bi$ is a zero, its conjugate $a - bi$ is also a zero.

10. $2\sqrt{2}(\cos 135° + i\sin 135°)$

11. $-12\sqrt{3} + 12i$

12. $1,\ \cos 120° + i\sin 120° = -\frac{1}{2} + \frac{\sqrt{3}}{2}i,$

$\cos 240° + i\sin 240° = -\frac{1}{2} - \frac{\sqrt{3}}{2}i$

CHAPTER 5

Section 5.1 *(page 364)*

Warm Up *(page 364)*

1. 5^x **2.** 3^{2x} **3.** 4^{3x} **4.** 10^x **5.** 4^{2x} **6.** 4^{10x}
7. $\left(\frac{3}{2}\right)^x$ **8.** 4^{3x} **9.** 2^{-x} **10.** $16^{x/4}$

1. 946.852 **3.** 7.352 **5.** 0.006 **7.** 673.639
9. 0.472 **11.** $f(x) = h(x)$ **13.** $f(x) = g(x) = h(x)$
15. d **17.** a

19.

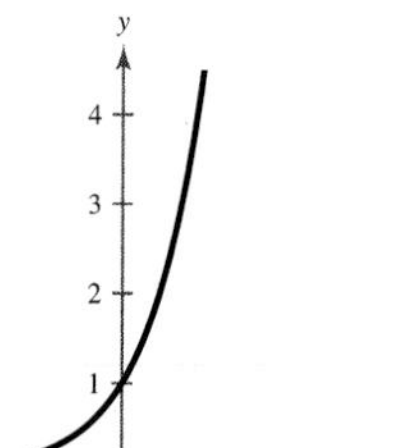

21.

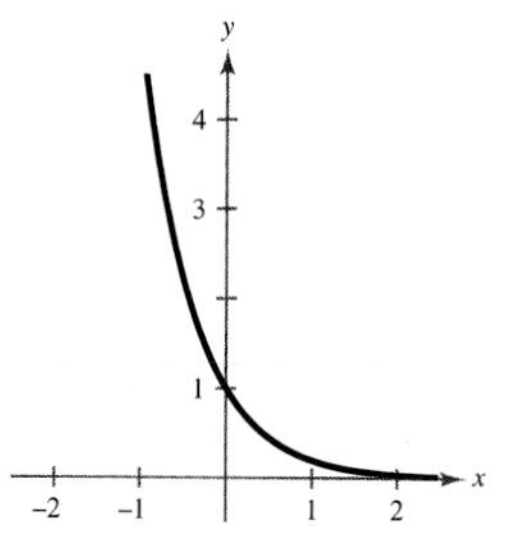

23.

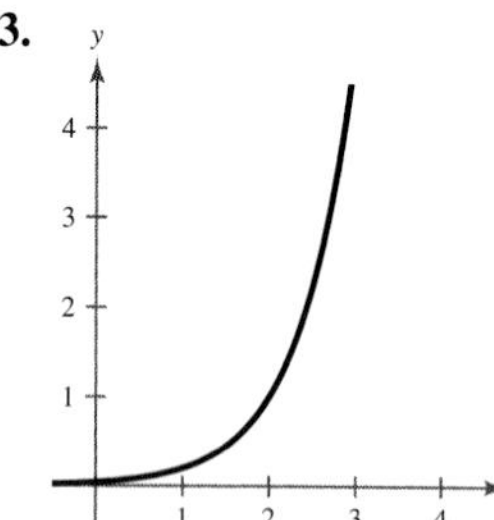

25.

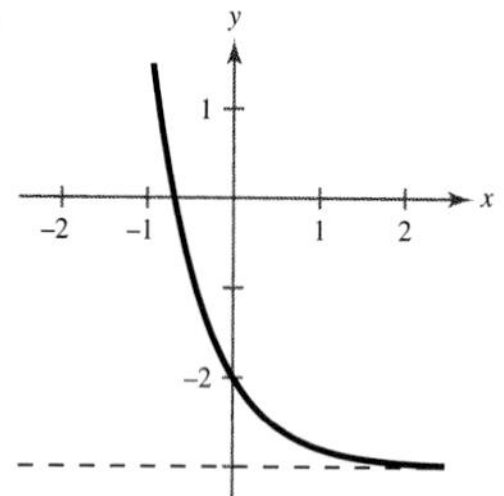

27.

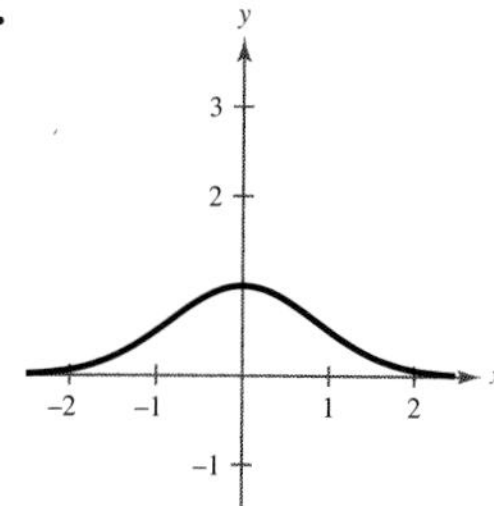

29.

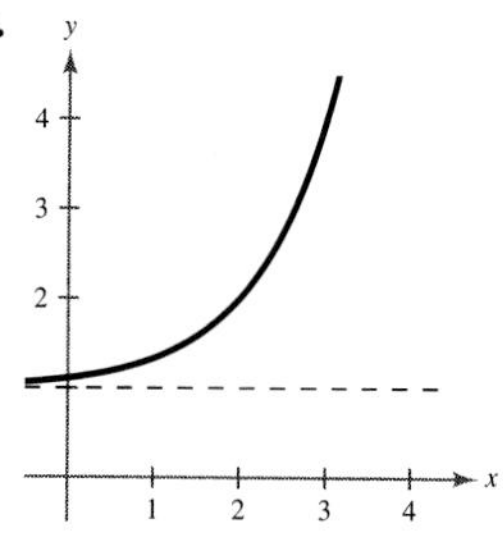

31.

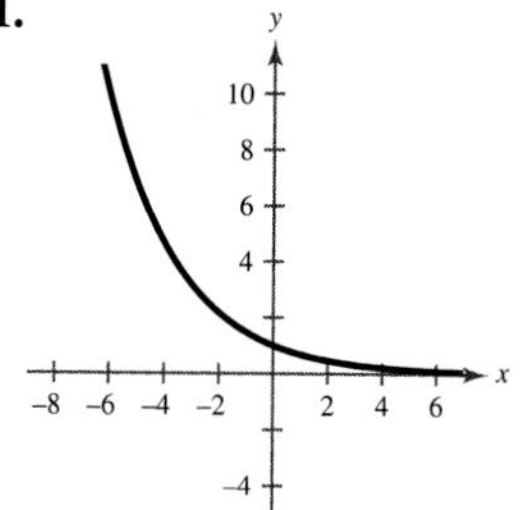

33.

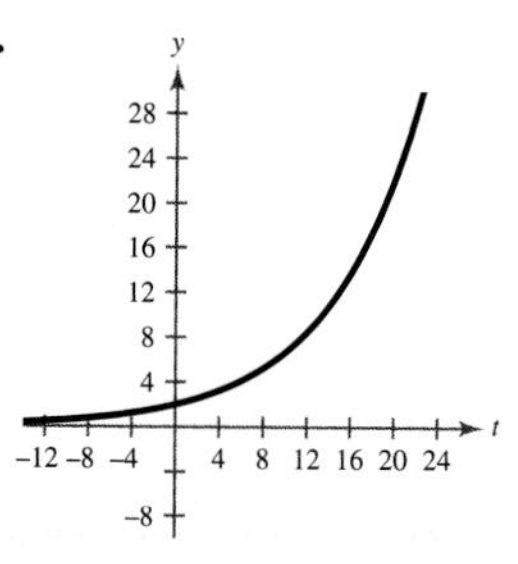

35.

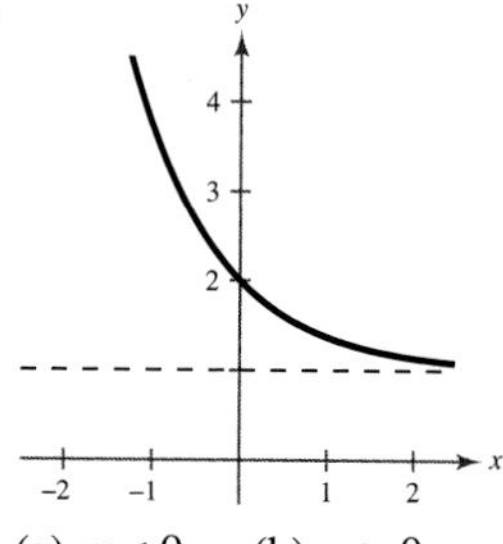

37. (a) $x < 0$ (b) $x > 0$

39. (a)

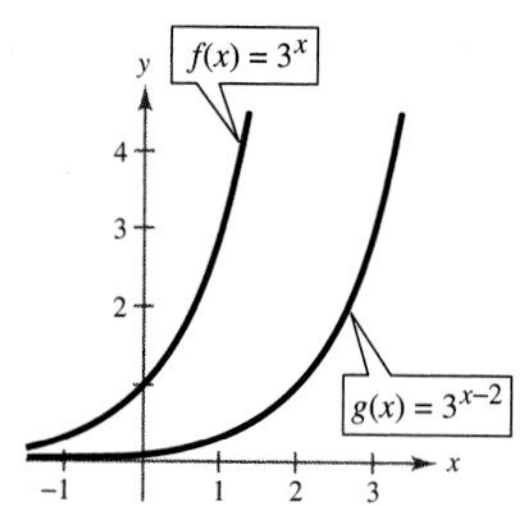

Horizontal shift two units to the right

(b)

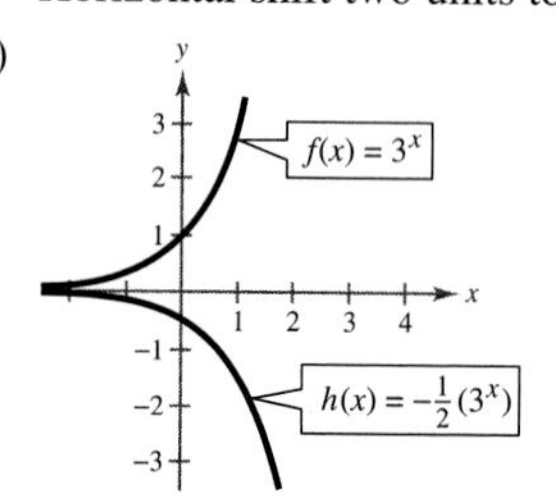

Vertical shrink and a reflection about the x-axis

(c)

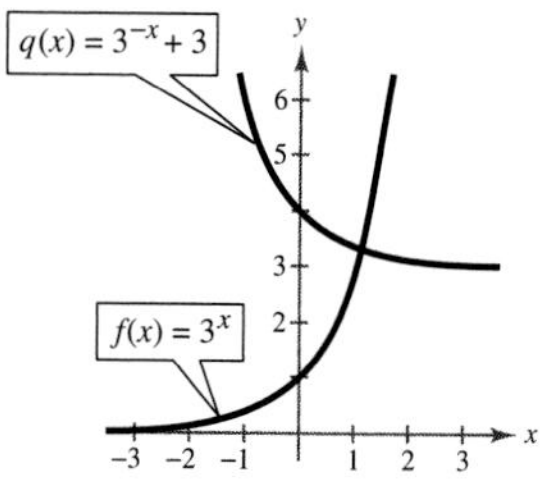

Reflection about the y-axis and a vertical translation

41. (a)

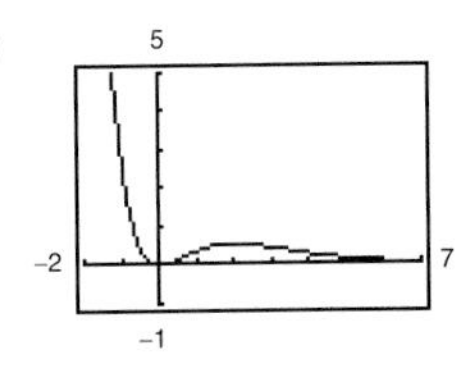

Decreasing: $(-\infty, 0)$, $(2, \infty)$

Increasing: $(0, 2)$

Relative maximum: $(2, 4e^{-2})$

Relative minimum: $(0, 0)$

(b)

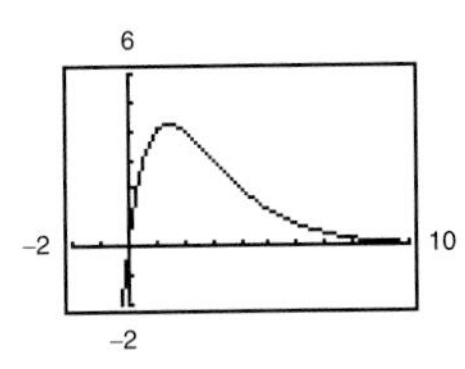

Decreasing: $(1.44, \infty)$

Increasing: $(-\infty, 1.44)$

Relative maximum: $(1.44, 4.25)$

43. The exponential function increases at a faster rate.

45.

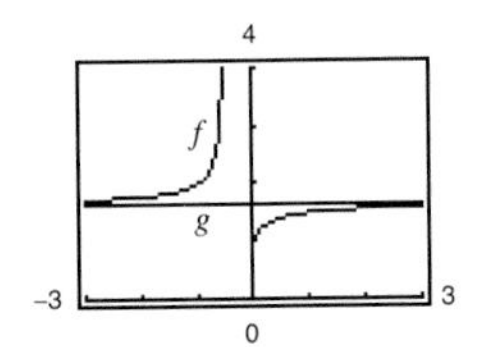

As $x \to \infty$, $f(x) \to g(x)$.

47.

n	1	2	4
A	$7764.62	$8017.84	$8155.09

n	12	365	Continuous
A	$8250.97	$8298.66	$8300.29

49.

n	1	2	4
A	$24,115.73	$25,714.29	$26,602.23

n	12	365	Continuous
A	$27,231.38	$27,547.07	$27,557.94

51.

t	1	10	20
P	$91,393.12	$40,656.97	$16,529.89

t	30	40	50
P	$6720.55	$2732.37	$1110.90

53. $222,822.57

55. (a) The steeper curve represents the investment earning compounded interest, because the compound interest earns more than simple interest.

(b) $A = 500(1.07)^t$

$A = 500(0.07)t + 500$

57. $35.45 **59.** (a) 100 (b) 300 (c) 900

61. (a) 25 units (b) 16.30 units

(c)

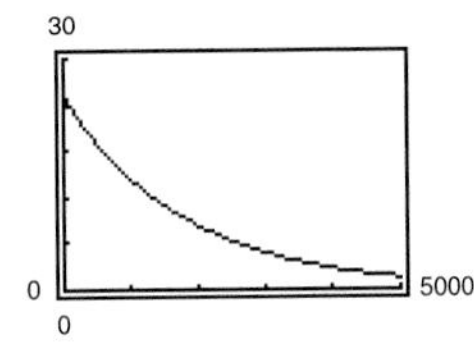

63. (a)

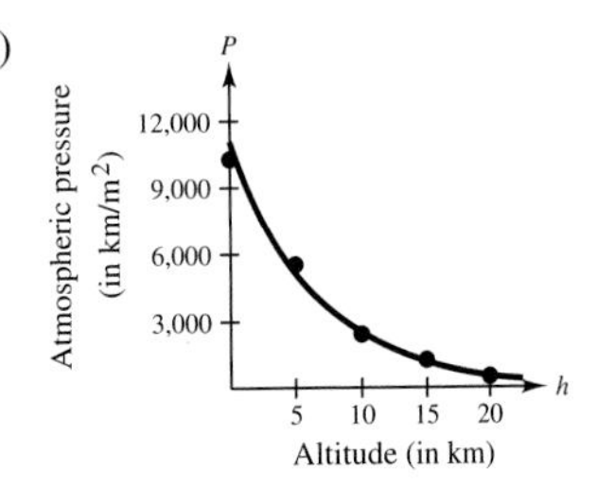

(b)

h	0	5	10	15	20
P	10,958	5176	2445	1155	546

(c) 3300 kilograms per square meter

(d) 11.3 kilometers

65. False. e is an irrational number.

67. $1 < \sqrt{2} < 2$

$2^1 < 2^{\sqrt{2}} < 2^2$

69. $y_4 = 1 + \frac{x}{1!} + \frac{x^2}{2!} + \frac{x^3}{3!} + \frac{x^4}{4!}$

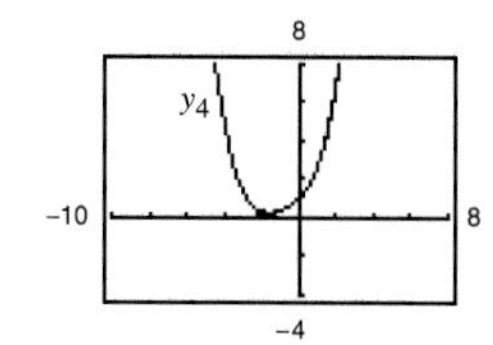

As more terms are added, the polynomial approaches e^x.

71. $y = \frac{1}{7}(2x + 14)$ **73.** $y = \pm\sqrt{25 - x^2}$

Section 5.2 *(page 375)*

Warm Up *(page 375)*

1. 3 **2.** 0 **3.** −1 **4.** 1 **5.** 7.389
6. 0.368 **7.** Shifted two units to the left
8. Reflection about the x-axis **9.** Shifted one unit downward
10. Reflection about the y-axis

1. $4^3 = 64$ **3.** $7^{-2} = \frac{1}{49}$ **5.** $32^{2/5} = 4$
7. $e^0 = 1$ **9.** $\log_5 125 = 3$ **11.** $\log_{81} 3 = \frac{1}{4}$
13. $\log_6 \frac{1}{36} = -2$ **15.** $\ln 20.0855 = 3$
17. $\ln 4 = x$ **19.** 4 **21.** $\frac{1}{2}$ **23.** 0
25. −2 **27.** $\frac{5}{3}$ **29.** 3 **31.** 2.538
33. 2.161 **35.** 2.913 **37.** 1.005 **39.** −1.139

41.

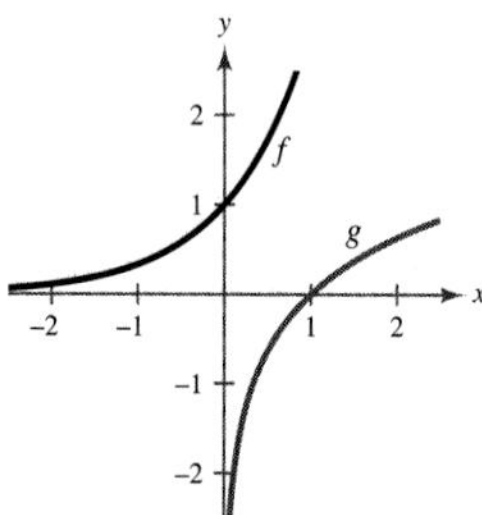

$g = f^{-1}$

43.

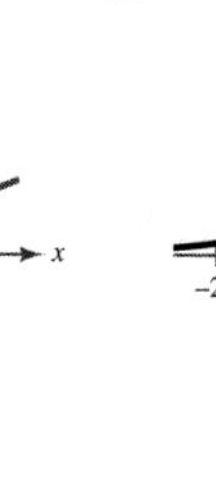

$g = f^{-1}$

45. c **47.** d **49.** b

51. Domain: $(0, \infty)$
Vertical asymptote: $x = 0$
Intercept: $(1, 0)$

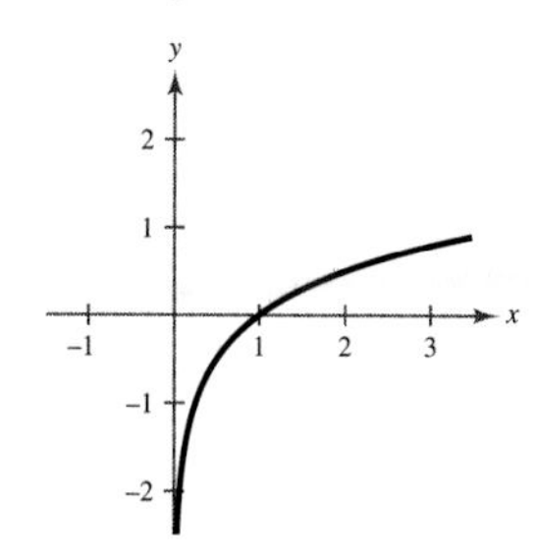

53. Domain: $(0, \infty)$
Vertical asymptote: $x = 0$
Intercept: $(9, 0)$

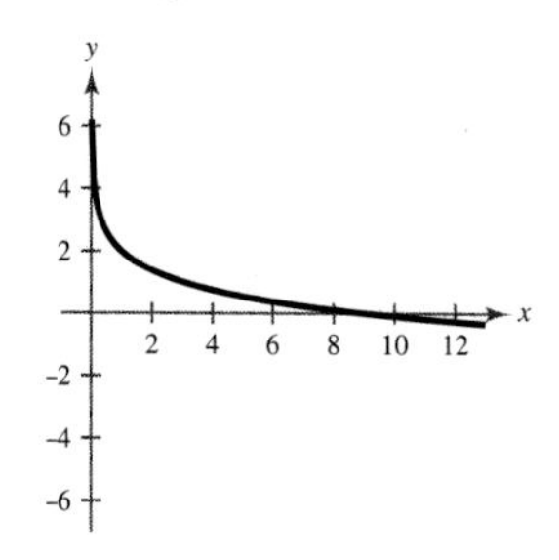

55. Domain: $(-2, \infty)$
Vertical asymptote: $x = -2$
Intercept: $(-1, 0)$

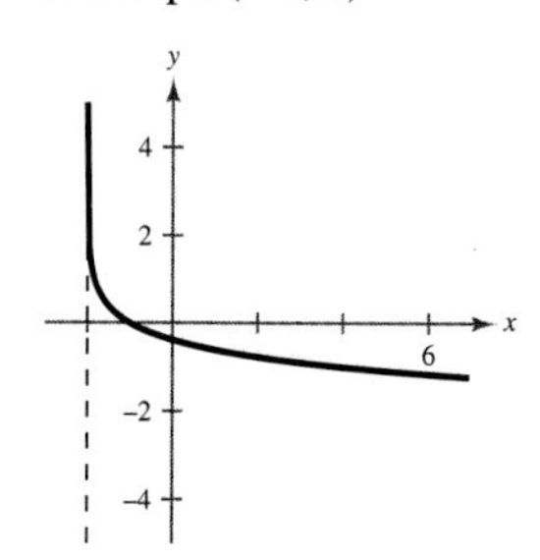

57. Domain: $(0, \infty)$

Vertical asymptote: $x = 0$

Intercept: $(5, 0)$

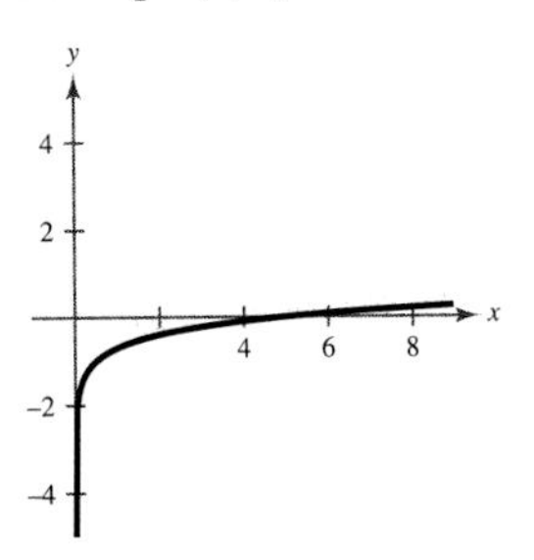

59. Domain: $(2, \infty)$

Vertical asymptote: $x = 2$

Intercept: $(3, 0)$

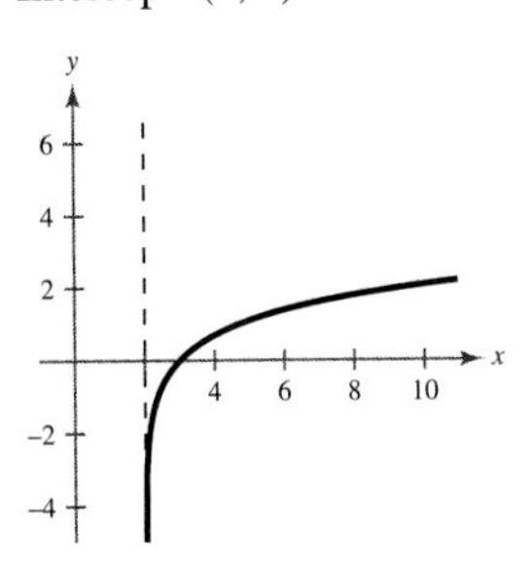

61. Domain: $(-\infty, 0)$

Vertical asymptote: $x = 0$

Intercept: $(-1, 0)$

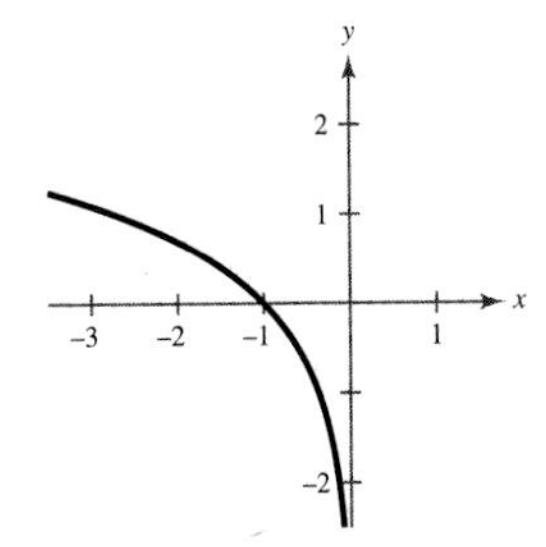

63. Decreasing: $(0, 1)$

Increasing: $(1, \infty)$

Relative minimum: $(1, 0)$

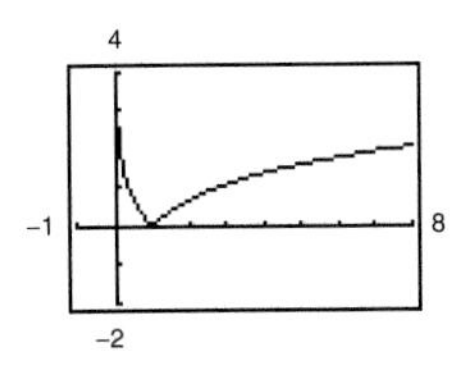

65. Decreasing: $(0, 2)$

Increasing: $(2, \infty)$

Relative minimum: $\left(2, 1 - \ln\frac{1}{2}\right)$

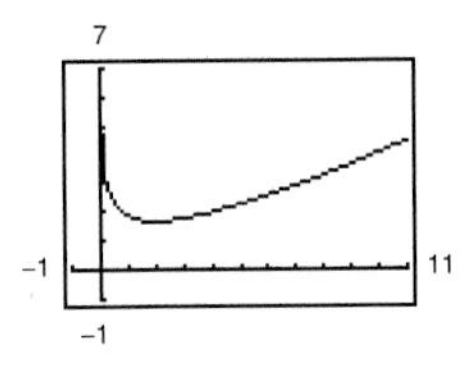

67. (a)

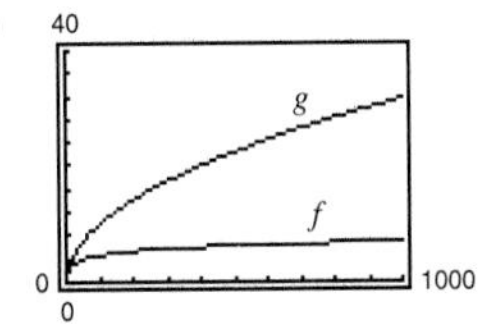

$g(x)$; the natural log function grows at a slower rate than the square root function.

(b)

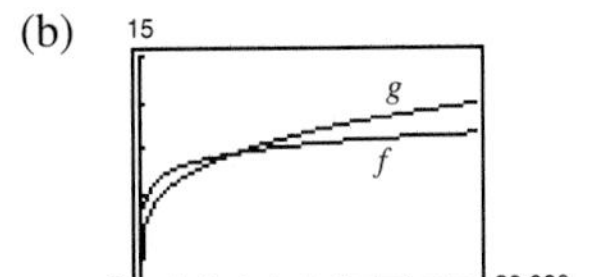

$g(x)$; the natural log function grows at a slower rate than the fourth root function.

69.

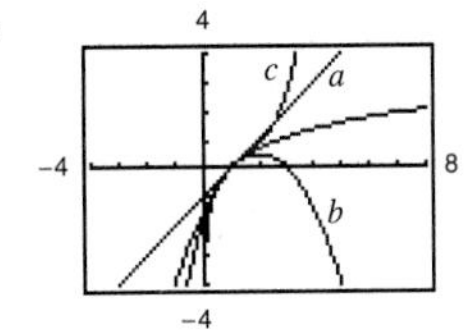

71. (a) 80 (b) 68.1 (c) 62.3

73.

r	0.005	0.010	0.015
t	138.6 yr	69.3 yr	46.2 yr

r	0.020	0.025	0.030
t	34.7 yr	27.7 yr	23.1 yr

75.

17.66 cubic feet per minute

77. 21,357 foot-pounds **79.** 30 years

81. Total amount: \$473,886
Interest: \$323,886

83. (a)

x	1	5	10	10^2
$f(x)$	0	0.322	0.230	0.046

x	10^4	10^6
$f(x)$	0.00092	0.0000138

(b) 0

(c)

85. $8n - 3$ **87.** $83.95 + 37.50t$

Section 5.3 *(page 383)*

Warm Up *(page 383)*

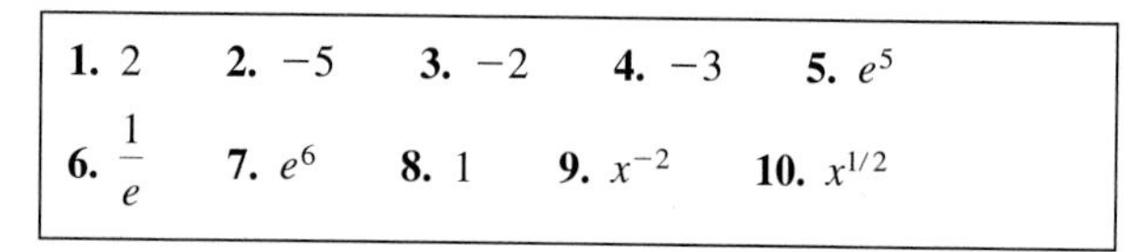

1. 2 **2.** -5 **3.** -2 **4.** -3 **5.** e^5
6. $\frac{1}{e}$ **7.** e^6 **8.** 1 **9.** x^{-2} **10.** $x^{1/2}$

1.

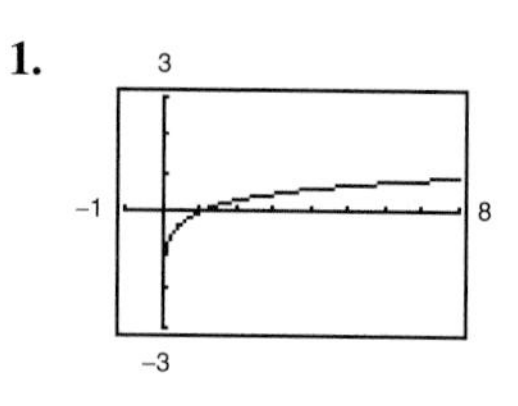

3. $\frac{\log_{10} 5}{\log_{10} 3}$ **5.** $\frac{\log_{10} x}{\log_{10} 2}$ **7.** $\frac{\ln 5}{\ln 3}$ **9.** $\frac{\ln x}{\ln 2}$ **11.** 1.771

13. -2.000 **15.** -0.417 **17.** 2.633

19. $\log_{10} 5 + \log_{10} x$ **21.** $\log_{10} 5 - \log_{10} x$ **23.** $4 \log_8 x$

25. $\frac{1}{2} \ln z$ **27.** $\ln x + \ln y + \ln z$ **29.** $\frac{1}{2} \ln(a - 1)$

31. $\ln z + 2 \ln(z - 1)$ **33.** $\frac{1}{3} \ln x - \frac{1}{3} \ln y$

35. $4 \ln x + \frac{1}{2} \ln y - 5 \ln z$

37. $2 \log_b x - 2 \log_b y - 3 \log_b z$

39.

41. $\ln 2x$ **43.** $\log_4 \frac{z}{y}$ **45.** $\log_2(x + 4)^2$

47. $\log_3 \sqrt[3]{5x}$ **49.** $\ln \frac{x}{(x + 1)^3}$ **51.** $\ln \frac{x - 2}{x + 2}$

53. $\ln \frac{x}{(x^2 - 4)^2}$ **55.** $\ln \sqrt[3]{\frac{x(x + 3)^2}{x^2 - 1}}$

57. $\ln \frac{\sqrt[3]{y(y + 4)^2}}{y - 1}$ **59.** $\ln \frac{9}{\sqrt{x^2 + 1}}$

61.

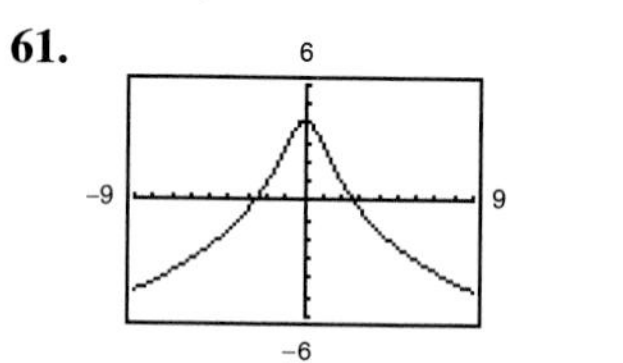

63.

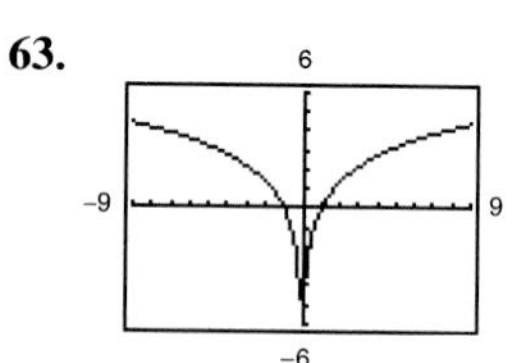

No. The domains differ.

65. $\log_2 \frac{32}{4} = \log_2 32 - \log_2 4$

67.

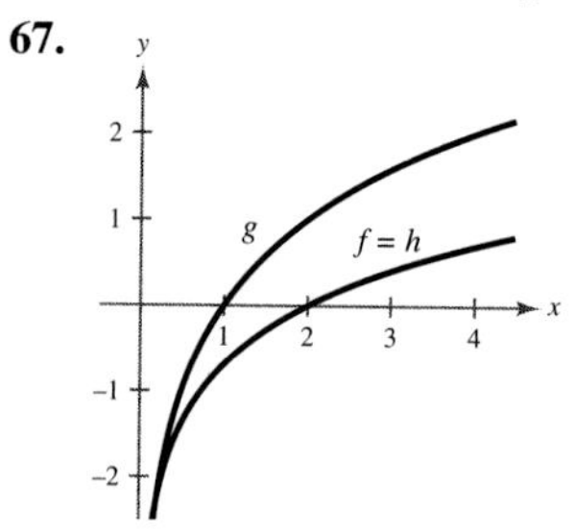

$f(x) = h(x)$

69. 2 **71.** 2.4 **73.** -9 is not in the domain of $\log_3 x$.

75. 2 **77.** -3 **79.** 0 is not in the domain of $\log_{10} x$.

81. 4.5 **83.** $\frac{3}{2}$ **85.** $-3 - \log_5 2$ **87.** $6 + \ln 5$

89. (a) 90 (b) 77 (c) 73 (d) 9 months
(e) $90 - \log_{10}(t + 1)^{15}$
(f)

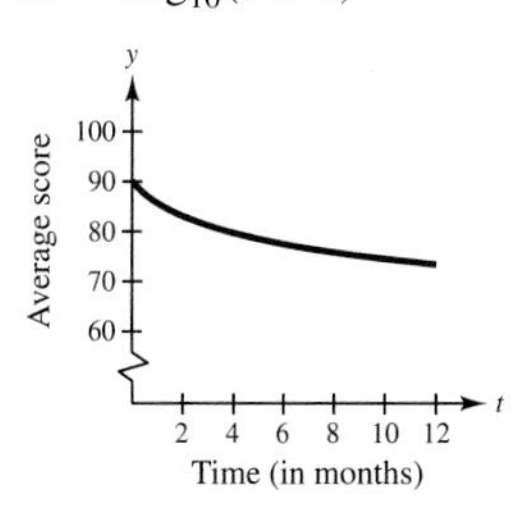

91. False. $\ln 1 = 0$ **93.** False. $\ln(x - 2) \neq \ln x - \ln 2$

95. False. $u = v^2$ **97.** Answers will vary.

99. $\dfrac{3x^4}{2y^3}, x \neq 0$ **101.** 1

Section 5.4 *(page 393)*

Warm Up *(page 393)*

1. $\dfrac{\ln 3}{\ln 2}$	**2.** $1 + \dfrac{2}{\ln 4}$	**3.** $\dfrac{e}{2}$	**4.** $2e$	**5.** $2 \pm i$
6. $\frac{1}{2}, 1$	**7.** $2x$	**8.** $3x$	**9.** $2x$	**10.** $-x^2$

1. (a) Yes (b) No **3.** (a) No (b) Yes (c) Yes
5. (a) No (b) No (c) Yes **7.** (3, 8) **9.** (9, 2)
11. 2 **13.** -2 **15.** 3 **17.** 64 **19.** $\frac{1}{10}$
21. x^2 **23.** $5x + 2$ **25.** x^2 **27.** $\ln 10 \approx 2.303$

29. 0 **31.** $\dfrac{\ln 12}{3} \approx 0.828$ **33.** $\ln \dfrac{5}{3} \approx 0.511$

35. $\ln 5 \approx 1.609$ **37.** $2 \ln 75 \approx 8.635$

39. $\log_{10} 42 \approx 1.623$ **41.** $\dfrac{\ln 80}{2 \ln 3} \approx 1.994$

43. 2 **45.** $\dfrac{\ln 8 - \ln 565}{\ln 2} \approx -6.142$

47.

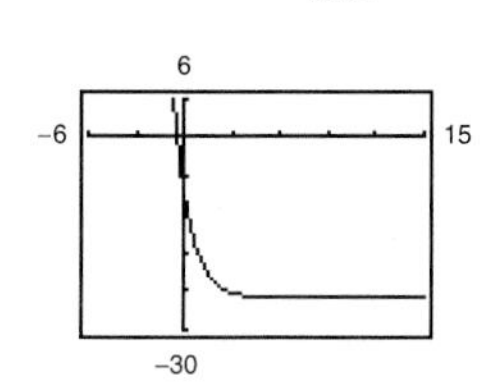

-0.427

49.

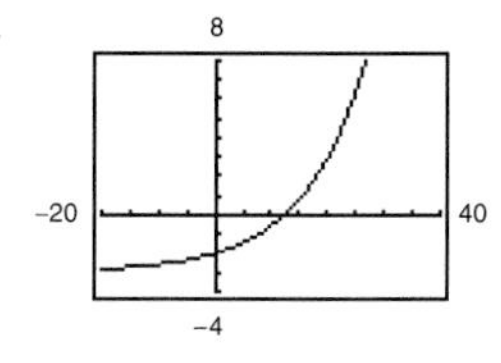

12.207

51. 0.059 **53.** 21.330 **55.** $e^{-3} \approx 0.050$

57. $\dfrac{e^{2.4}}{2} \approx 5.512$ **59.** $e^2 - 2 \approx 5.389$ **61.** 103

63. $1 + \sqrt{1 + e} \approx 2.928$ **65.** $\dfrac{-1 + \sqrt{17}}{2} \approx 1.562$

67. 4 **69.** No solution **71.** 14.988
73. 33.115 **75.** 14.369

77.

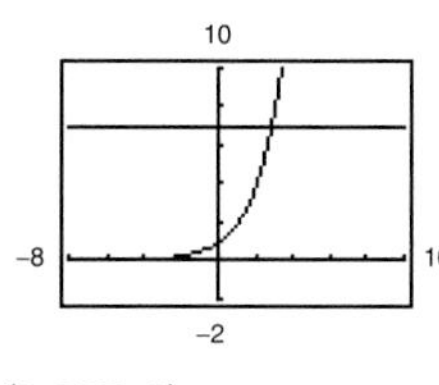

(2.807, 7)

79.

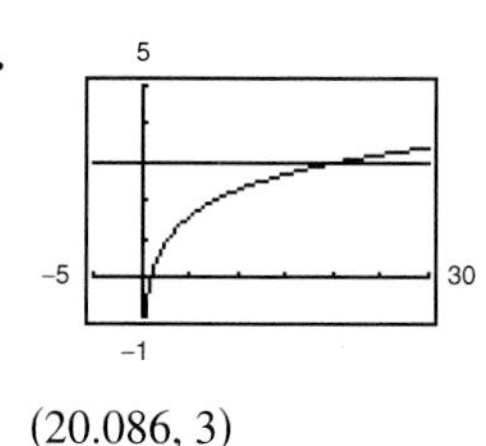

(20.086, 3)

81. 8.2 years

83. Yes. Time to double: $t = \dfrac{\ln 2}{r}$;

Time to quadruple: $t = \dfrac{\ln 4}{r} = 2\left(\dfrac{\ln 2}{r}\right)$

85. 12.9 years **87.** (a) 1426 units (b) 1498 units

89. (a)

(b) $y = 6.7$. Yield will approach 6.7 million cubic feet per acre.
(c) 29.3 years

91. (a) $y = 100$ and $y = 0$; the range falls between 0% and 100%
(b) Males: 69.71 inches Females: 64.51 inches

93. (a) $y = 20$; Room temperature (b) 0.81 hour

95. $4|x|y^2\sqrt{3y}$ **97.** $5\sqrt[3]{3}$

Section 5.5 *(page 404)*

Warm Up *(page 404)*

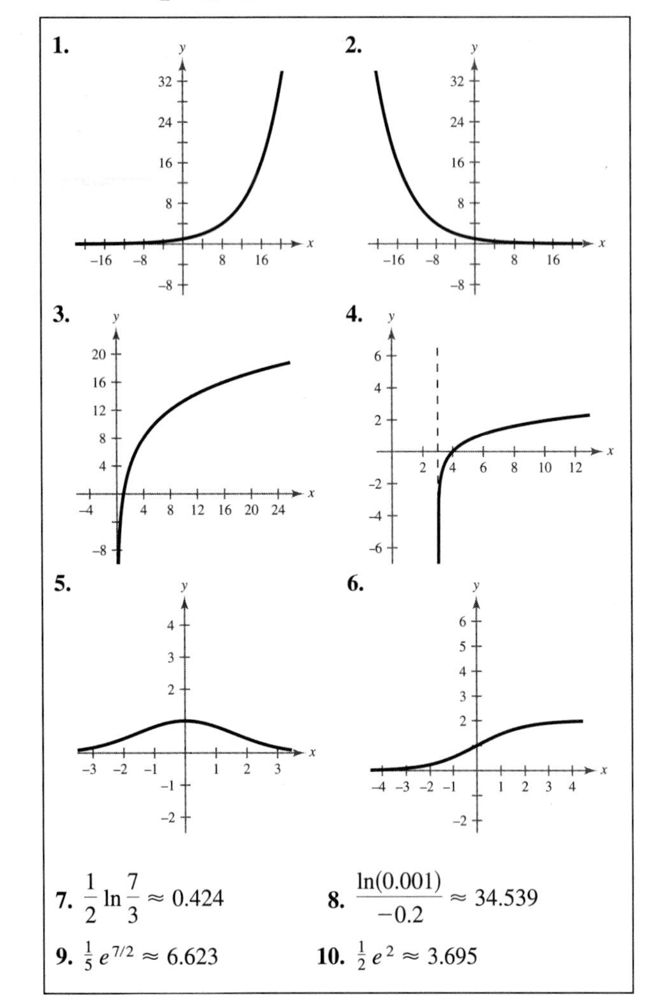

7. $\frac{1}{2} \ln \frac{7}{3} \approx 0.424$

8. $\frac{\ln(0.001)}{-0.2} \approx 34.539$

9. $\frac{1}{5} e^{7/2} \approx 6.623$

10. $\frac{1}{2} e^{2} \approx 3.695$

1. c **3.** a **5.** d

	Initial Investment	*Annual % Rate*	*Time to Double*	*Amount After 10 years*
7.	$1000	12%	5.78 yr	$3,320.12
9.	$750	8.94%	7.75 yr	$1,833.67
11.	$500	9.5%	7.30 yr	$1,292.85
13.	$6376.28	4.5%	15.4 yr	$10,000.00

15. $112,087.09

17. (a) 6.642 years (b) 6.330 years (c) 6.302 years (d) 6.301 years

19.

r	2%	4%	6%	8%	10%	12%
t	54.93	27.47	18.31	13.73	10.99	9.16

21.

r	2%	4%	6%	8%	10%	12%
t	55.48	28.01	18.85	14.27	11.53	9.69

23.

Continuous compounding

	Isotope	*Half-Life (Years)*	*Initial Quantity*	*Amount After 1000 years*
25.	^{226}Ra	1620	10 g	6.52 g
27.	^{14}C	5730	2.26 g	2 g
29.	^{239}Pu	24,360	2.16 g	2.1 g

31. $y = e^{0.7675x}$ **33.** $y = e^{-0.4621x}$

35. 2013 **37.** $k = 0.0137$, 3288

39. $y = 4.22e^{0.0430t}$, 9.97 million

41. $y = 3e^{-0.0091t}$, 2.50 million

43. The greater rate of growth, the greater the value of b.

45. 3.15 hours **47.** 95.8%

49. (a) $V = -4500t + 22,000$ (b) $V = 22,000e^{-0.263t}$

(c)

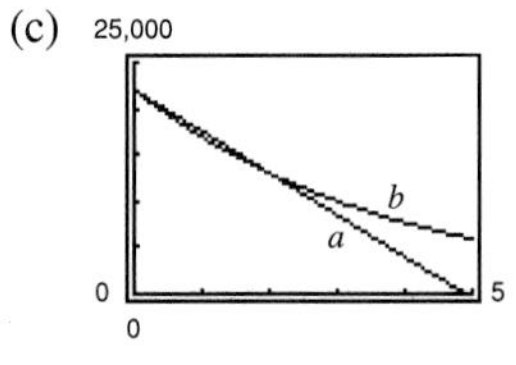

Exponential

(d) 1 year. Straight-line: $17,500;
Exponential: $16,912
3 years. Straight-line: $8500;
Exponential: $9995

(e) Decreases $4500 per year

51. (a) $S(t) = 100(1 - e^{-0.1625t})$

(b)

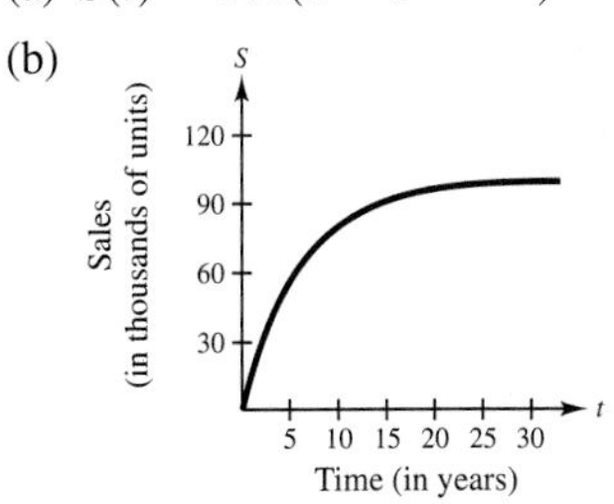

(c) 55,625

53. (a) $S = 10(1 - e^{-0.0575x})$ (b) 3314

55. (a) $N = 30(1 - e^{-0.050t})$ (b) 36 days

(c) No. It is not a linear function.

57. (a) 7.91 (b) 7.68

59. (a) 20 (b) 70 (c) 95 (d) 120

61. 95%

63. 4.64

65. 1.58×10^{-6} moles per liter

67. 10^7

69. (a)

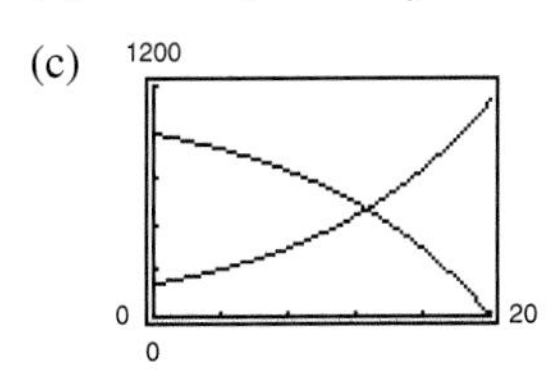

(b) Interest; $t \approx 28$ years

(c)

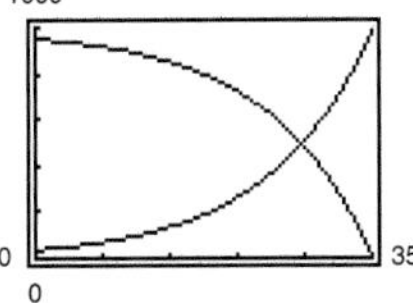

Interest; $t \approx 12.7$ years

71. (a) $t_3 = 0.2729s - 6.0143$

$t_4 = 1.5385e^{0.0291s}$

(b)

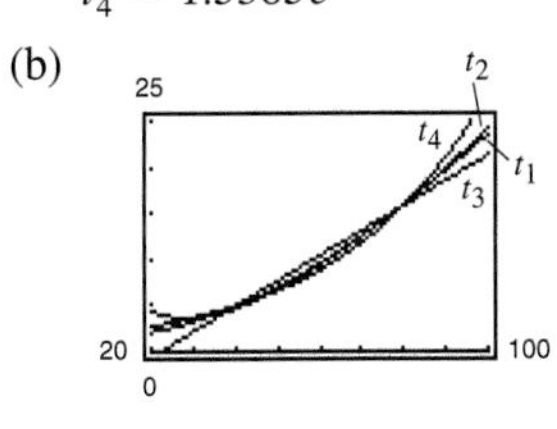

(c)

s	30	40	50	60	70	80	90
t_1	3.6	4.6	6.7	9.4	12.5	15.9	19.6
t_2	3.3	4.9	7.0	9.5	12.5	15.9	19.9
t_3	2.2	4.9	7.6	10.4	13.1	15.8	18.5
t_4	3.7	4.9	6.6	8.8	11.8	15.8	21.2

(d) Model: t_1; Sum = 2.0

Model: t_2; Sum = 1.1

Model: t_3; Sum = 5.6

Model: t_4; Sum = 2.7

Quadratic model fits best.

73. Answers will vary.

75. $4x^2 - 12x + 9$

77. $2x^2 + 3 + \dfrac{3}{x - 4}$

Focus on Concepts *(page 410)*

1. $b < d < a < c$

b and d are negative.

2. (a) True. $\log_b uv = \log_b u + \log_b v,\ u, v > 0$

(b) False.

$2.04 \approx \log_{10}(10 + 100) \neq (\log_{10} 10)(\log_{10} 100) = 2$

(c) False.

$1.95 \approx \log_{10}(100 - 10) \neq \log_{10} 100 - \log_{10} 10 = 1$

(d) True. $\log_b \dfrac{u}{v} = \log_b u - \log_b v,\ u, v > 0$

3. For $t < 14$ years double the amount invested. For $t \geq 14$ years double the interest rate or time because it doubles the exponent in the exponential function.

4. (a) Logarithmic (b) Logistic (c) Exponential (d) Linear (e) None of the above (f) Exponential

Review Exercises *(page 411)*

1. e **3.** a **5.** d

7.

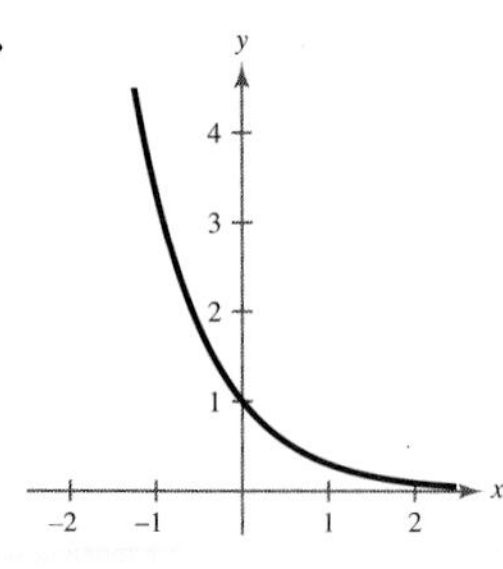

9.

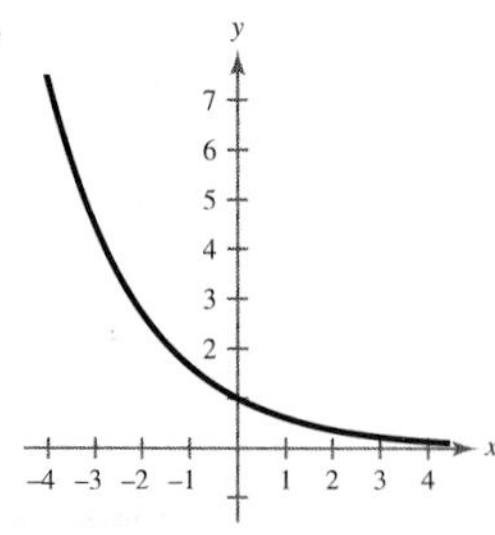

11.

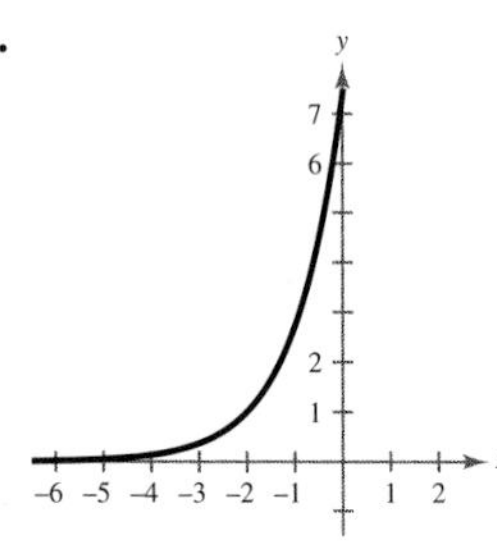

13.

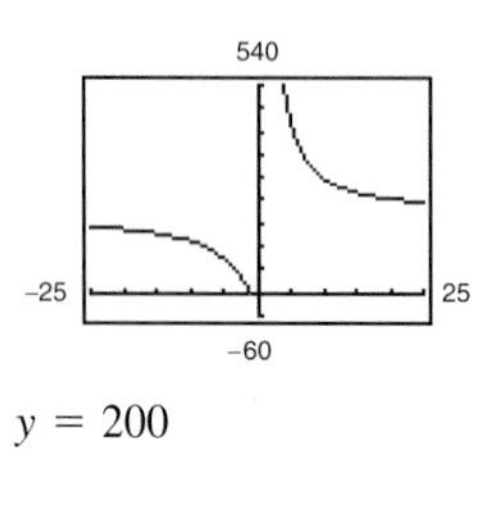

$y = 200$

15.

n	1	2	4	12
A	\$9499.28	\$9738.91	\$9867.22	\$9956.20

n	365	Continuous
A	\$10,000.27	\$10,001.78

17.

t	1	10	20
P	\$184,623.27	\$89,865.79	\$40,379.30

t	30	40	50
P	\$18,143.59	\$8152.44	\$3663.13

19. (a) 0.154 (b) 0.487 (c) 0.811

21. (a) \$1,069,047.14 (b) 7.9 years

23.

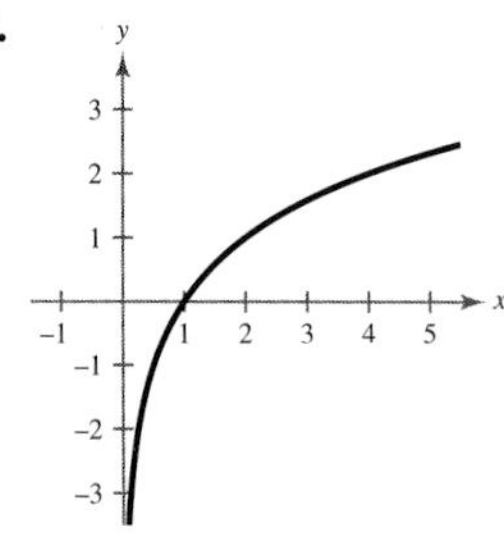

25.

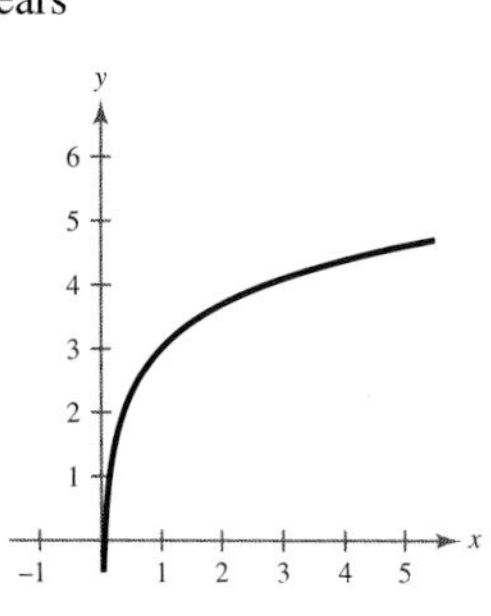

27.

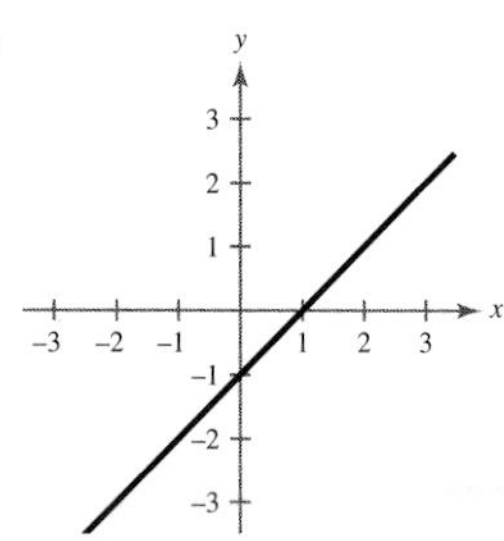

29.

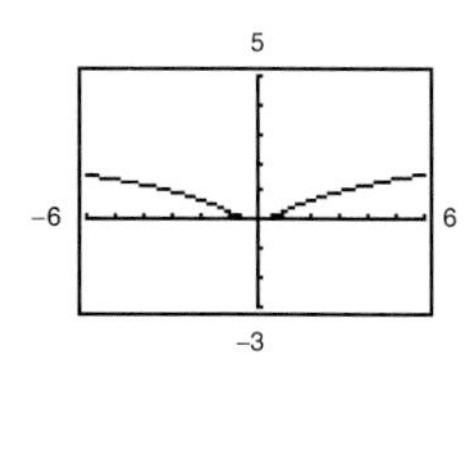

31. $\log_4 64 = 3$ **33.** 3 **35.** 7 **37.** 1.585

39. 2.132 **41.** $1 + 2\log_5 |x|$

43. $\log_{10} 5 + \frac{1}{2}\log_{10} y - 2\log_{10} |x|$ **45.** $\log_2 5x$

47. $\ln \frac{\sqrt{|2x - 1|}}{(x + 1)^2}$ **49.** True **51.** False **53.** True

55. 27.16 miles **57.** $\ln 12 \approx 2.485$

59. $-\frac{\ln 44}{5} \approx -0.757$ **61.** $\ln 2 \approx 0.693$, $\ln 5 \approx 1.609$

63. $\frac{1}{3}e^{8.2} \approx 1213.650$ **65.** $3e^2 \approx 22.167$

67. No solution **69.** 0.39, 7.48 **71.** 1.64

73. $y = 2e^{0.1014x}$ **75.** (a) 1151 units (b) 1325 units

77. (a) 13.86% (b) \$11,486.65

79. (a) $10^{8.4}$ (b) $10^{6.85}$ (c) $10^{9.1}$

Chapter Test *(page 416)*

1.

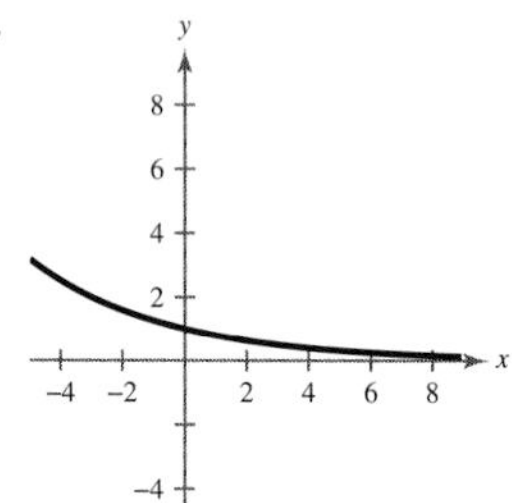

2. $y = 1000$

3. (a) \$34,596.89 (b) \$35,143.44

4. \$40,386.38 **5.** $4^3 = 64$ **6.** $\log_5 \frac{1}{25} = -2$

7.

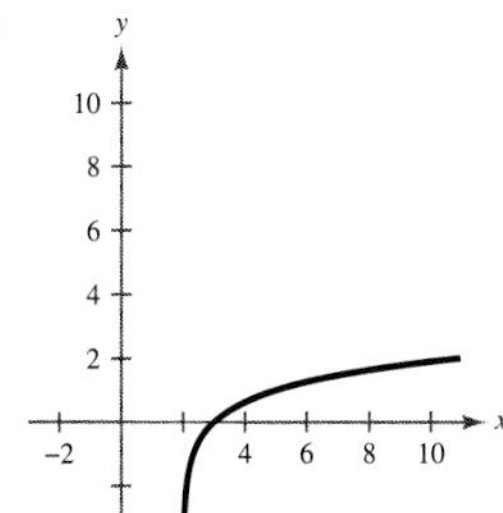

8. $\ln 6 + 2\ln x - \frac{1}{2}\ln(x^2 + 1)$ **9.** $\ln \dfrac{z^3}{z^2 - 1}$

10. $1 + \frac{1}{2}\log_6 10$ **11.** 12.218 **12.** 18.447

13. 77.451 **14.** $10,204 **15.** 300 **16.** 570

17. 9 years **18.** c

55. (a) $d = \dfrac{4}{\sqrt{m^2 + 1}}$

(b)

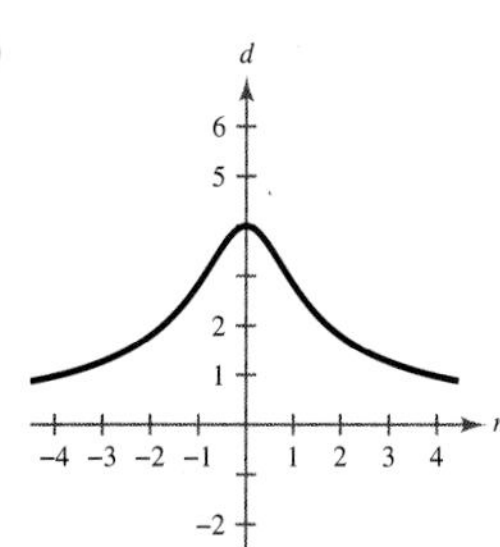

(c) $m = 0$

(d) $d = 0$. As the line approaches the vertical, the distance approaches 0.

CHAPTER 6

Section 6.1 *(page 423)*

Warm Up *(page 423)*

1. $\sqrt{106}$	**2.** $2\sqrt{13}$	**3.** 1	**4.** $\frac{6}{7}$	**5.** $-\frac{3}{2}$
6. $-\frac{1}{2}$	**7.** $2x - 3y + 9 = 0$		**8.** $7x - 2y - 24 = 0$	
9. $4x + y - 32 = 0$	**10.** $3x + 4y + 2 = 0$			

1. $\dfrac{\sqrt{3}}{3}$ **3.** -1 **5.** 0.7869 **7.** -2.7475

9. 135° **11.** 36.9° **13.** 60.3° **15.** 121.0°

17. 78.7° **19.** 121.0° **21.** 0.1139, 1195 feet

23. (a) 18.4° (b) 15.8 meters **25.** 71.6°

27. 11.3° **29.** 81.9° **31.** 53.1° **33.** 46.9°

35. (2, 1), 42.3°; (4, 4), 78.7°; (6, 2), 59.0°

37. $(-4, -1)$, 11.9°; (3, 2), 21.8°; (1, 0), 146.3°

39. (a) 69° (b) 5970 feet, 2415 feet **41.** 0 **43.** 1.4

45. 7 **47.** $\dfrac{8\sqrt{37}}{37} \approx 1.3152$ **49.** (a) $\dfrac{7\sqrt{10}}{5}$ (b) 7

51. (a) $\dfrac{35\sqrt{37}}{74}$ (b) $\dfrac{35}{8}$ **53.** $2\sqrt{2}$

Section 6.2 *(page 431)*

Warm Up *(page 431)*

1. $x^2 - 10x + 5$	**2.** $x^2 + 6x + 8$	
3. $-x^2 - 8x - 6$	**4.** $-x^2 + 4x$	**5.** $(x + 3)^2 - 1$
6. $(x - 5)^2 - 4$	**7.** $2 - (x - 1)^2$	**8.** $-2(x - 1)^2$
9. $2x + 3y - 20 = 0$	**10.** $3x - 4y - 17 = 0$	

1. e **3.** d **5.** a

7. Vertex: (0, 0)
Focus: $\left(0, \frac{1}{2}\right)$
Directrix: $y = -\frac{1}{2}$

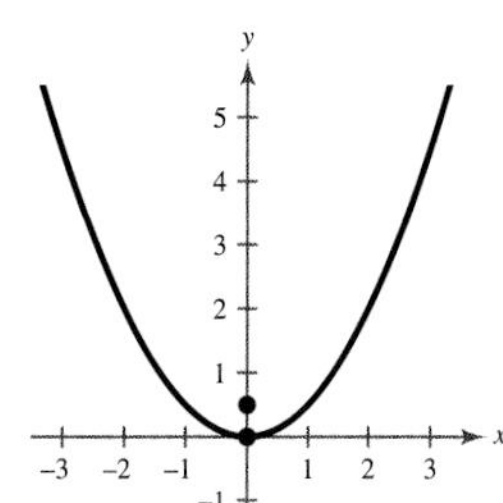

9. Vertex: (0, 0)
Focus: $\left(-\frac{3}{2}, 0\right)$
Directrix: $x = \frac{3}{2}$

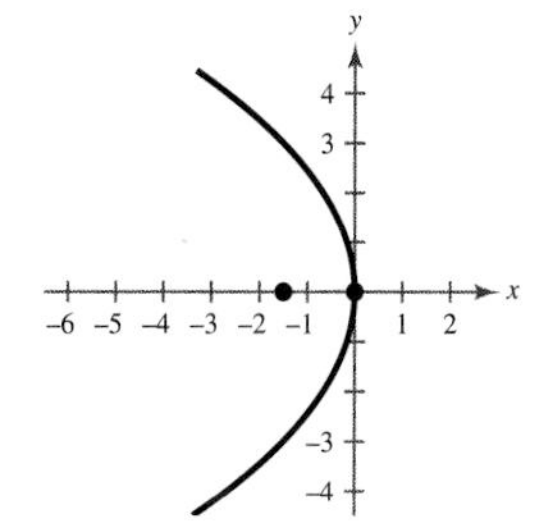

11. Vertex: $(0, 0)$
Focus: $(0, -2)$
Directrix: $y = 2$

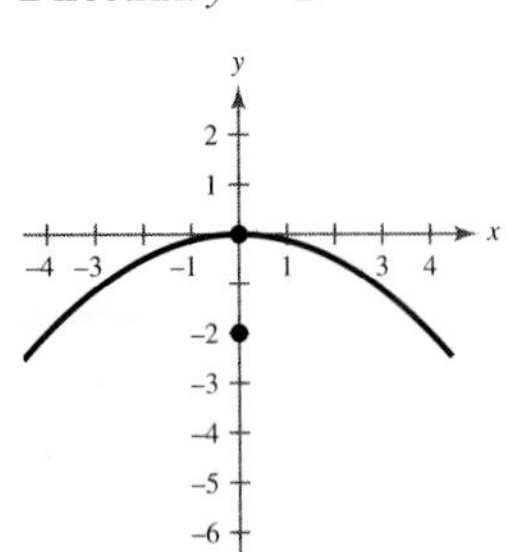

13. Vertex: $(1, -2)$
Focus: $(1, -4)$
Directrix: $y = 0$

15. Vertex: $\left(5, -\frac{1}{2}\right)$
Focus: $\left(\frac{11}{2}, -\frac{1}{2}\right)$
Directrix: $x = \frac{9}{2}$

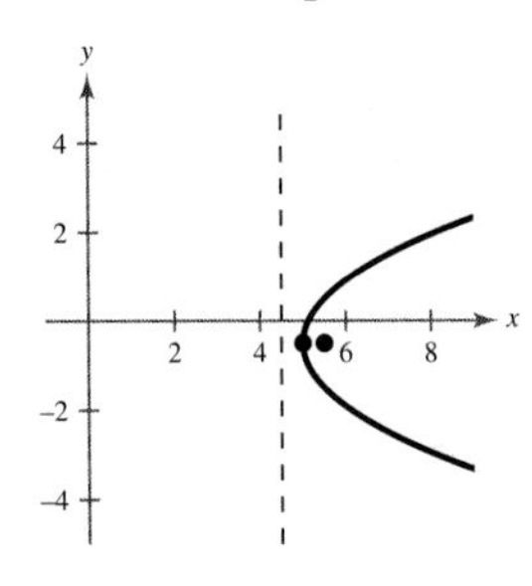

17. Vertex: $(1, 1)$
Focus: $(1, 2)$
Directrix: $y = 0$

19. Vertex: $(-2, -3)$
Focus: $(-4, -3)$
Directrix: $x = 0$

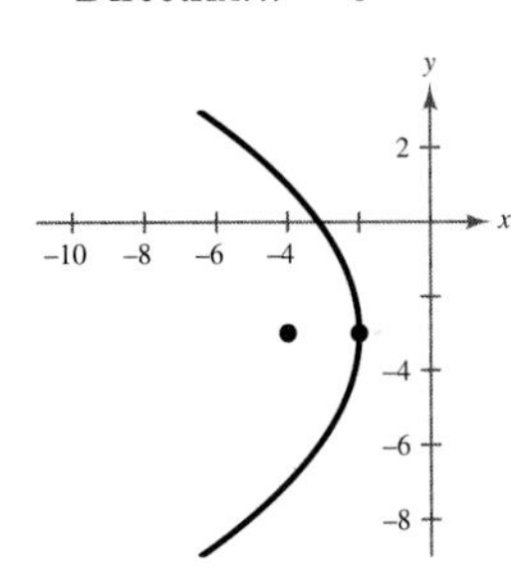

21. Vertex: $(-2, 1)$
Focus: $\left(-2, -\frac{1}{2}\right)$
Directrix: $y = \frac{5}{2}$

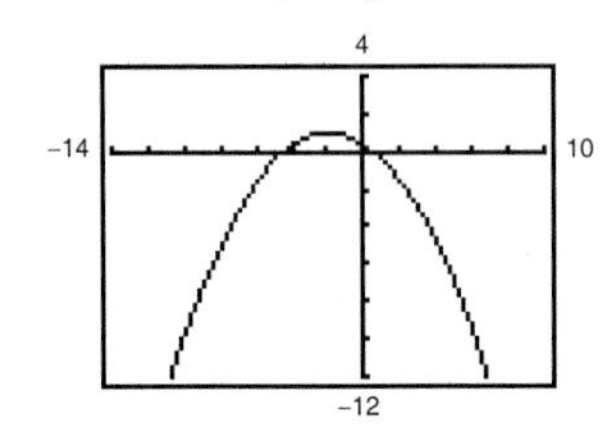

23. Vertex: $\left(\frac{1}{4}, -\frac{1}{2}\right)$
Focus: $\left(0, -\frac{1}{2}\right)$
Directrix: $x = \frac{1}{2}$

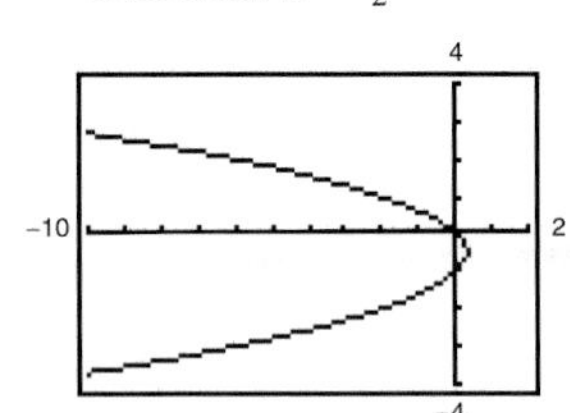

25. $(2, 4)$

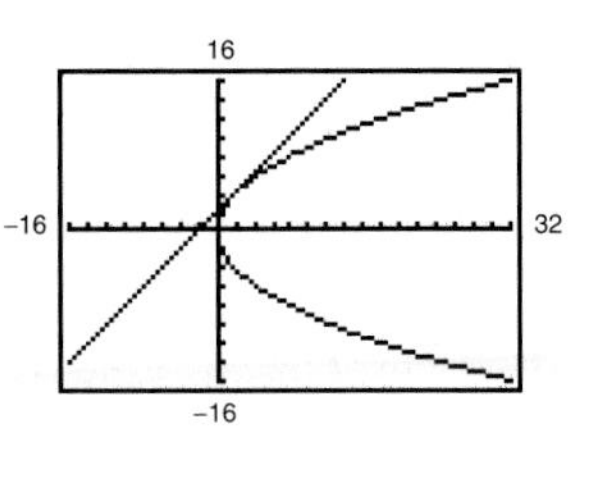

27. (a)

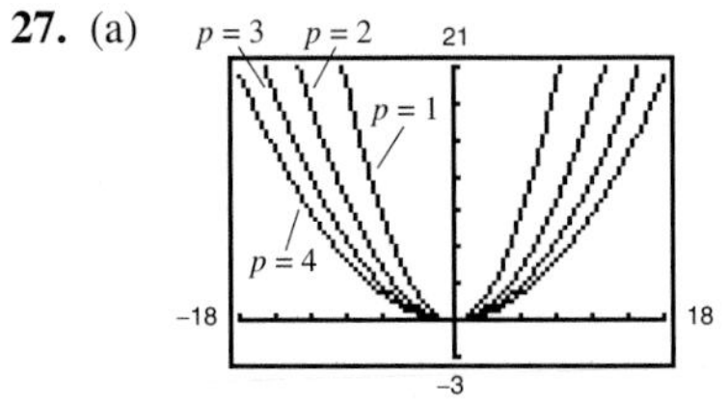

As p increases the graph becomes wider.

(b) $(0, 1)$, $(0, 2)$, $(0, 3)$, $(0, 4)$

(c) 4, 8, 12, 16; $4p$

(d) Easy way to determine two additional points on the graph.

29. $y = \frac{2}{3}x^2$ **31.** $x^2 = -6y$ **33.** $y^2 = -8x$

35. $x^2 = 4y$ **37.** $x^2 = -8y$ **39.** $y^2 = 9x$

41. $(x - 3)^2 = -(y - 1)$ **43.** $y^2 = 2(x + 2)$

45. $(y - 2)^2 = -8(x - 3)$ **47.** $x^2 = 8(y - 4)$

49. $(y - 2)^2 = 8x$ **51.** $y = \sqrt{6(x + 1)} + 3$

53. $y = \dfrac{1}{14}x^2$ **55.** (a) $y = \dfrac{3x^2}{640{,}000}$ (b) 462 centimeters

57.

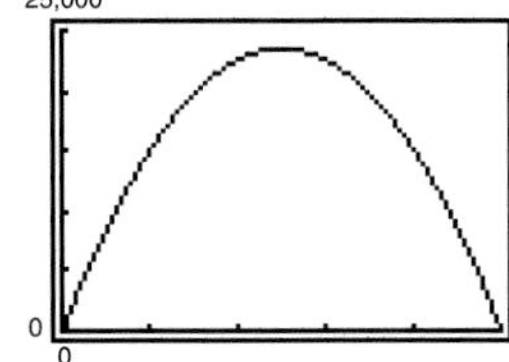

$x = 125$

59. (a)

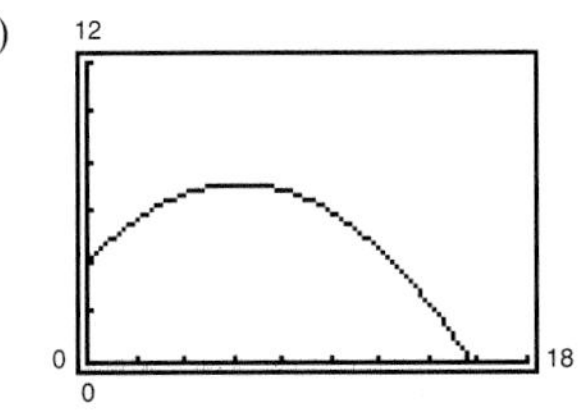

(b) Highest point: (6.25, 7.125)

Range: 15.69 feet

61. $\dfrac{x_1}{2p}$ **63.** $4x - y - 8 = 0;\ (2, 0)$

65. $4x - y + 2 = 0;\ \left(-\frac{1}{2}, 0\right)$

67. $y = x^3 - 7x^2 + 17x - 15$ **69.** $\pm 2,\ \frac{1}{2},\ -\frac{5}{3}$

Section 6.3 *(page 441)*

Warm Up *(page 441)*

1.

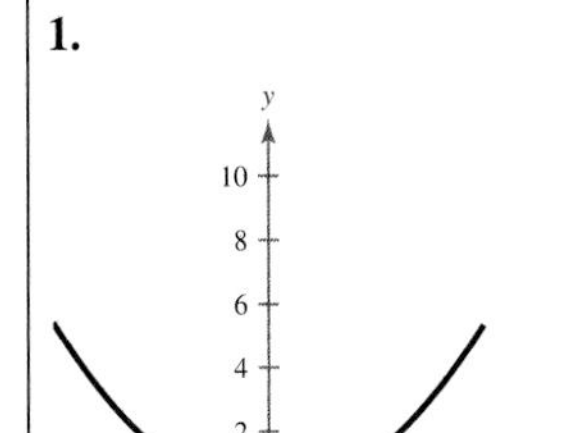

2.

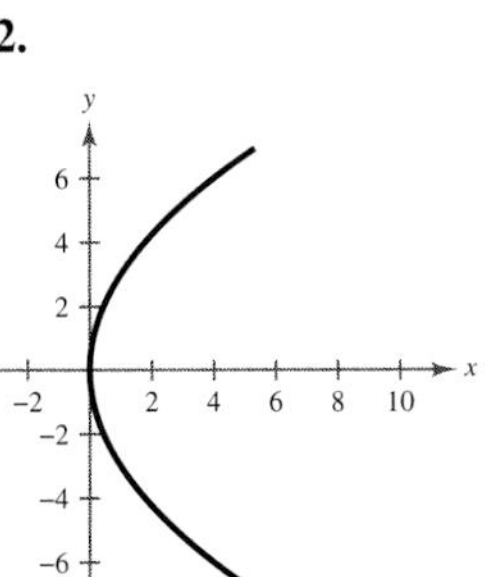

3.

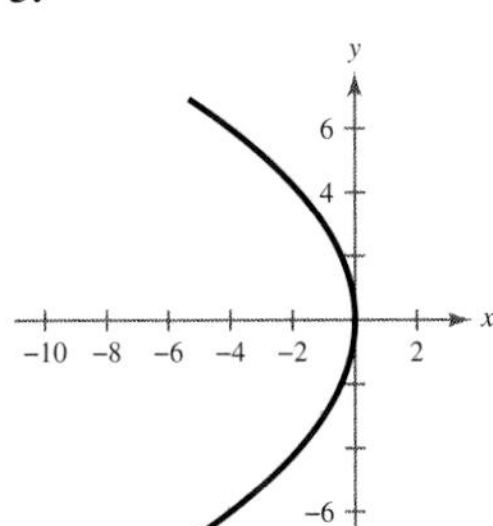

4.

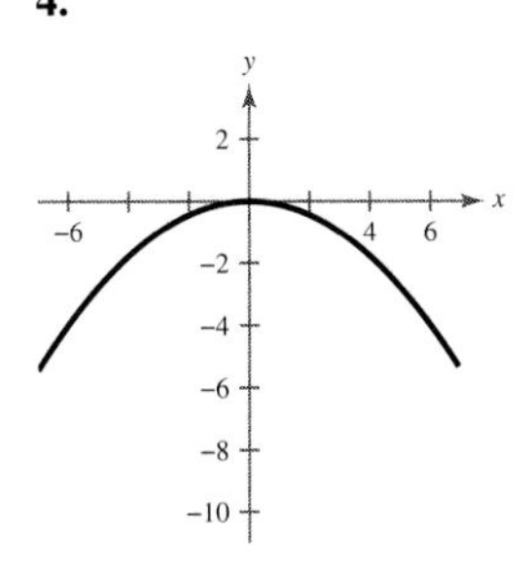

5. $c = 12$ **6.** $b = 1$ **7.** $a = 10$ **8.** $c = 2\sqrt{6}$

9. $4x^2 + 3y^2$ **10.** $\dfrac{9(x - 1)^2}{4} + 9(y + 2)^2$

1. b **3.** d **5.** a

7. Center: (0, 0)

Vertices: $(\pm 5, 0)$

Foci: $(\pm 3, 0)$

Eccentricity: $\frac{3}{5}$

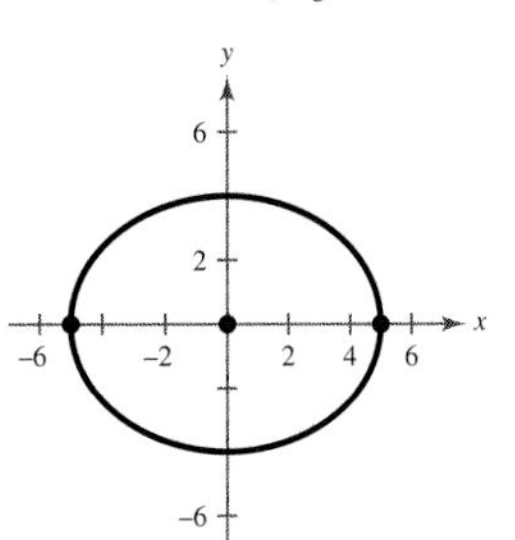

9. Center: (0, 0)

Vertices: $(0, \pm 5)$

Foci: $(0, \pm 3)$

Eccentricity: $\frac{3}{5}$

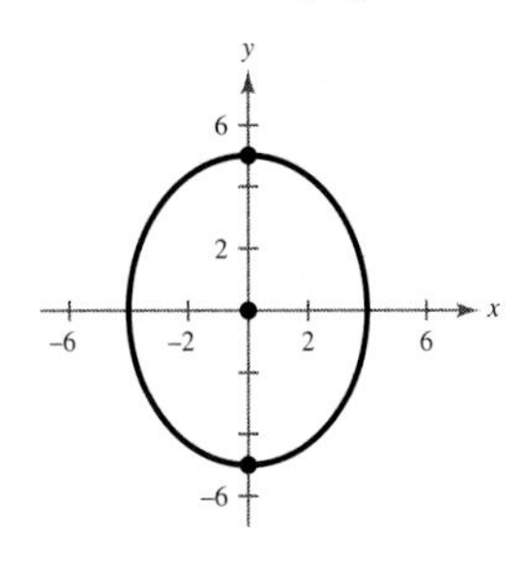

11. Center: (0, 0)

Vertices: $(\pm 3, 0)$

Foci: $(\pm 2, 0)$

Eccentricity: $\frac{2}{3}$

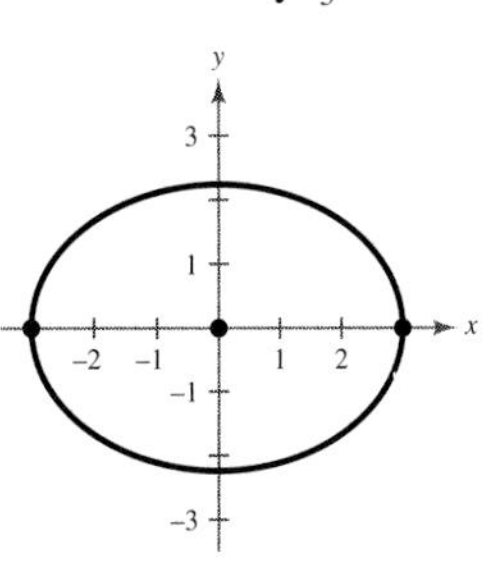

13. Center: (1, 5)

Vertices: (1, 10), (1, 0)

Foci: (1, 9), (1, 1)

Eccentricity: $\frac{4}{5}$

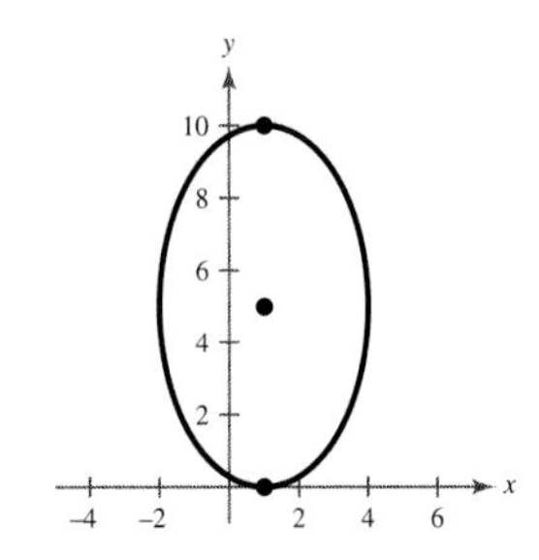

15. Center: $(-2, 3)$

Vertices: $(-2, 6),\ (-2, 0)$

Foci: $\left(-2, 3 \pm \sqrt{5}\right)$

Eccentricity: $\dfrac{\sqrt{5}}{3}$

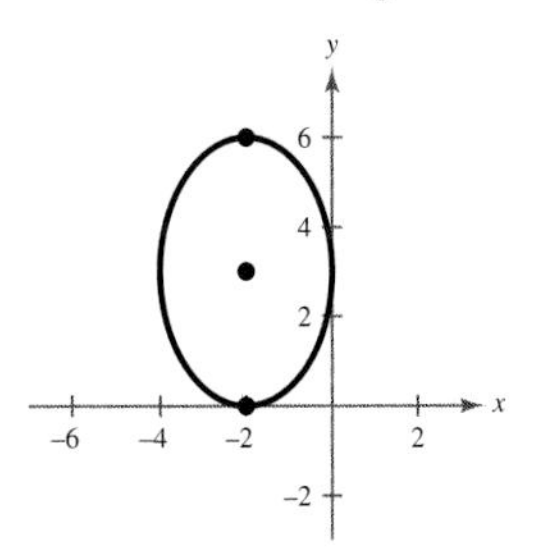

17. Center: $(1, -1)$
Vertices: $\left(\frac{9}{4}, -1\right), \left(-\frac{1}{4}, -1\right)$
Foci: $\left(\frac{7}{4}, -1\right), \left(\frac{1}{4}, -1\right)$
Eccentricity: $\frac{3}{5}$

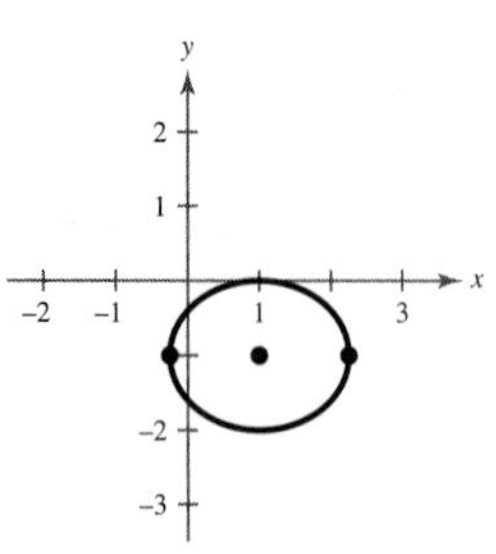

19. Center: $(0, 0)$
Vertices: $(0, \pm\sqrt{5})$
Foci: $(0, \pm\sqrt{2})$

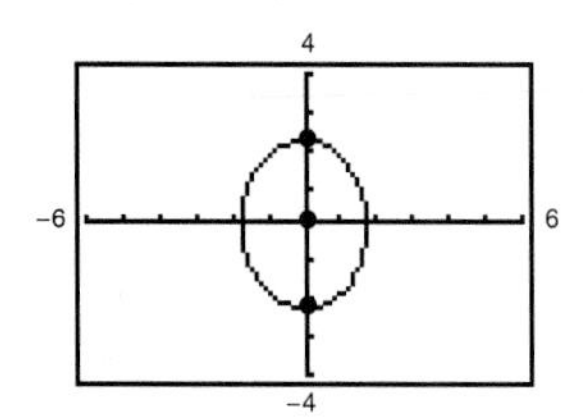

21. Center: $\left(\frac{1}{2}, -1\right)$
Vertices: $\left(\frac{1}{2} \pm \sqrt{5}, -1\right)$
Foci: $\left(\frac{1}{2} \pm \sqrt{2}, -1\right)$

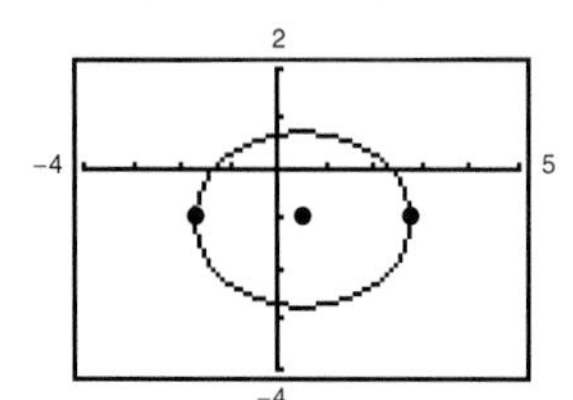

23. $x = \dfrac{3}{2}\left(2 + \sqrt{4 - y^2}\right)$ **25.** $\dfrac{x^2}{1} + \dfrac{y^2}{4} = 1$

27. $\dfrac{x^2}{25} + \dfrac{y^2}{21} = 1$ **29.** $\dfrac{x^2}{36} + \dfrac{y^2}{11} = 1$

31. $\dfrac{21x^2}{400} + \dfrac{y^2}{25} = 1$ **33.** $\dfrac{(x-2)^2}{1} + \dfrac{(y-3)^2}{9} = 1$

35. $\dfrac{(x-2)^2}{9} + \dfrac{(y-2)^2}{4} = 1$

37. $\dfrac{(x-2)^2}{4} + \dfrac{(y-2)^2}{1} = 1$

39. $\dfrac{x^2}{48} + \dfrac{(y-4)^2}{64} = 1$ **41.** $\dfrac{(x-3)^2}{9} + \dfrac{(y-5)^2}{16} = 1$

43. $\dfrac{x^2}{16} + \dfrac{(y-4)^2}{12} = 1$

45. (a) $2a$ (b) The sum of the distances from the two fixed points is constant.

47.

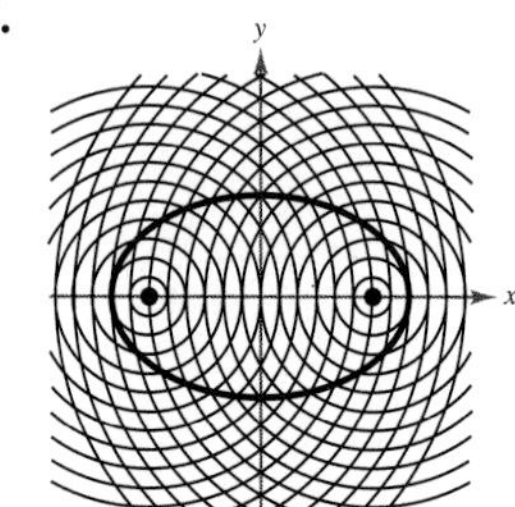

49. (a) $A = \pi a(20 - a)$ (b) $\dfrac{x^2}{196} + \dfrac{y^2}{36} = 1$

(c)

a	8	9	10	11	12	13
A	301.6	311.0	314.2	311.0	301.6	285.9

$a = 10$, circle

(d)

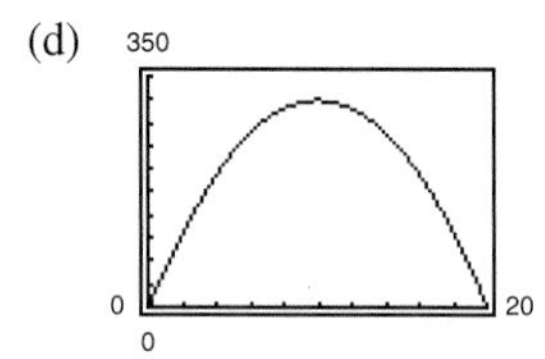

$a = 10$

51. $\dfrac{x^2}{25} + \dfrac{y^2}{16} = 1$ **53.** $\dfrac{x^2}{327.25} + \dfrac{y^2}{19.34} = 1$

55. $e = \dfrac{c}{a} \approx 0.052$

57. False. The equation of an ellipse is second degree in x and y.

59.

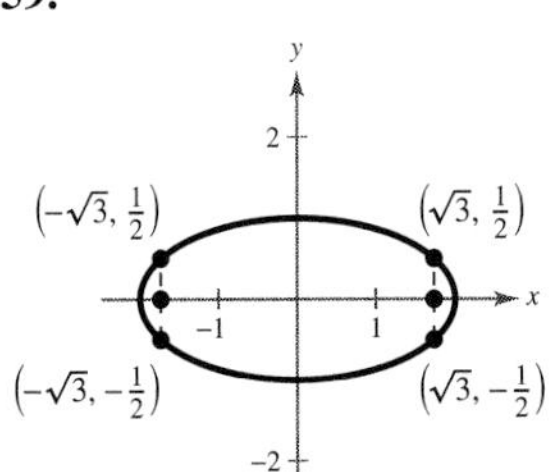

61.

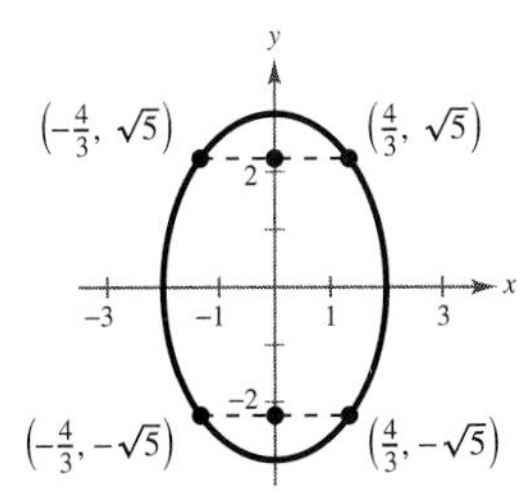

Section 6.4 *(page 452)*

Warm Up *(page 452)*

1. $\sqrt{61}$ **2.** $\sqrt{65}$

3.

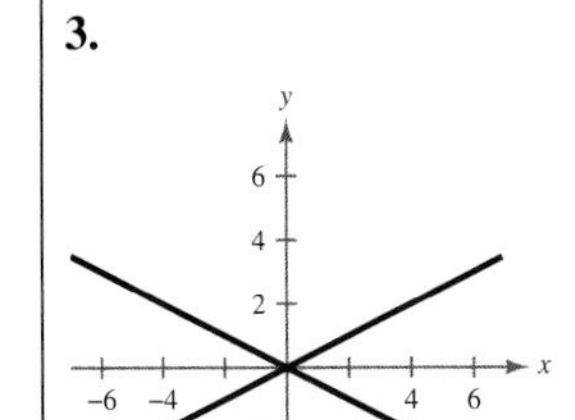

4.

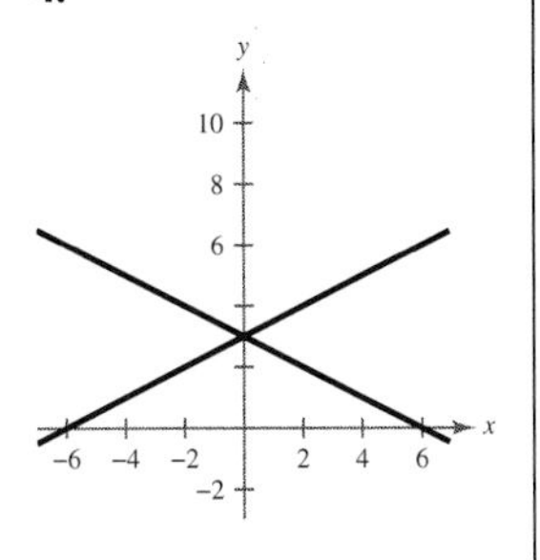

5.

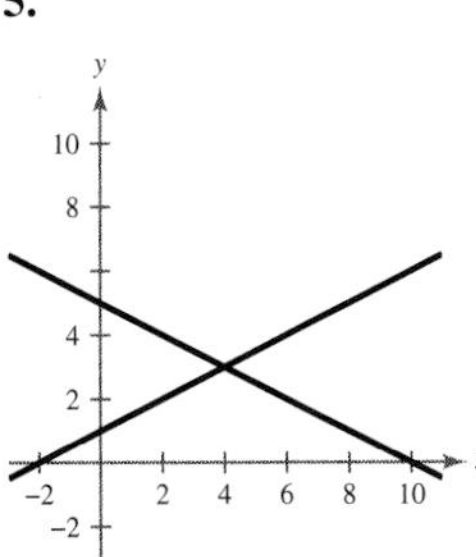

6.

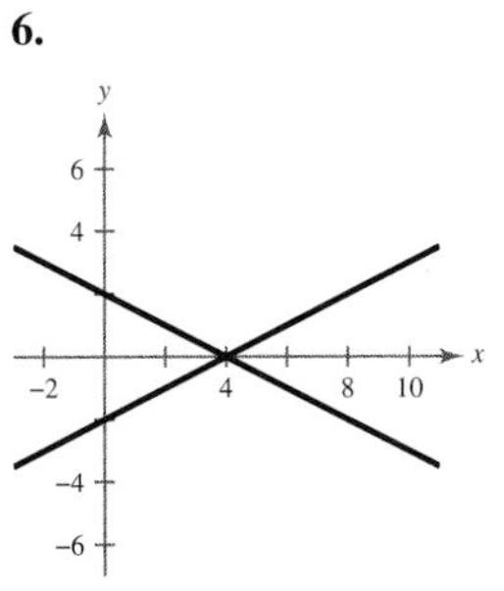

7. Parabola **8.** Ellipse **9.** Circle **10.** Parabola

1. b **3.** e **5.** a

7. Center: $(0, 0)$
Vertices: $(\pm 1, 0)$
Foci: $(\pm\sqrt{2}, 0)$

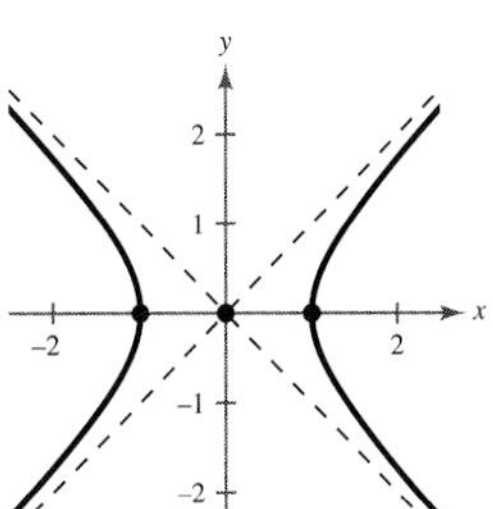

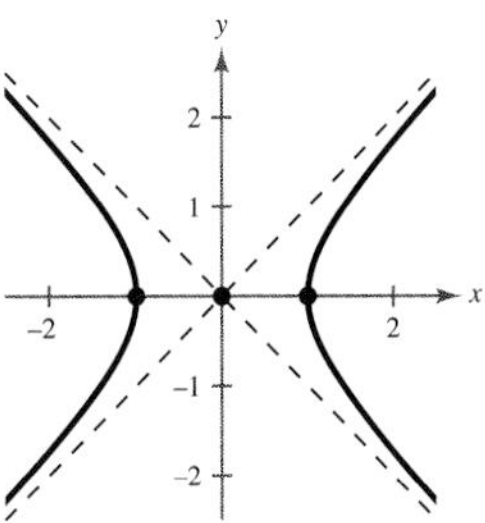

9. Center: $(0, 0)$
Vertices: $(0, \pm 1)$
Foci: $(0, \pm\sqrt{5})$

11. Center: $(0, 0)$
Vertices: $(0, \pm 5)$
Foci: $(0, \pm 13)$

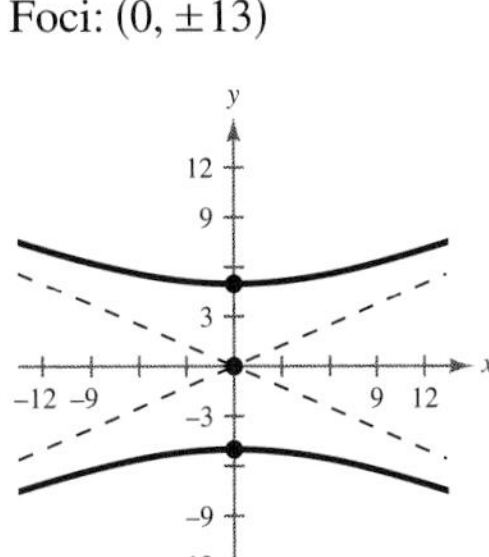

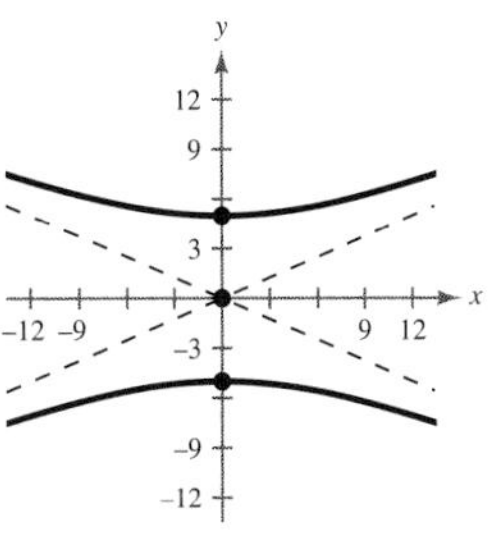

13. Center: $(1, -2)$
Vertices: $(3, -2), (-1, -2)$
Foci: $(1 \pm \sqrt{5}, -2)$

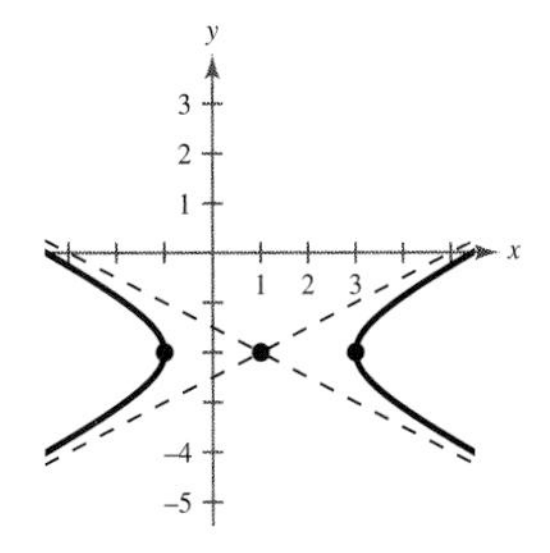

15. Center: $(2, -6)$
Vertices: $(2, -5), (2, -7)$
Foci: $(2, -6 \pm \sqrt{2})$

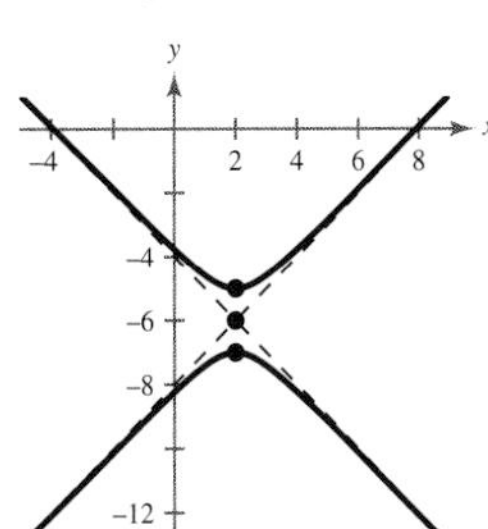

17. Center: $(2, -3)$
Vertices: $(3, -3), (1, -3)$
Foci: $(2 \pm \sqrt{10}, -3)$

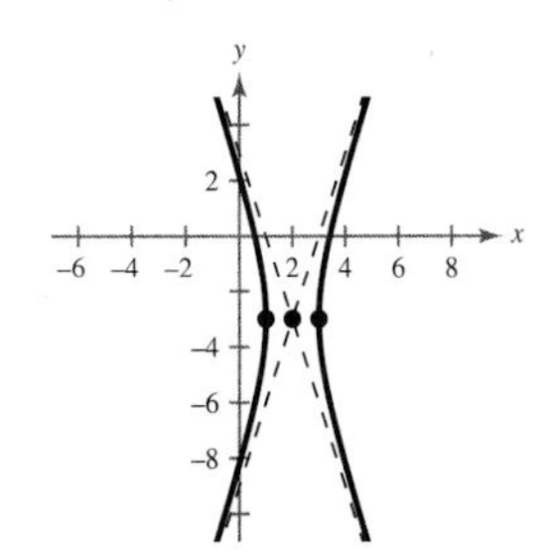

19. The graph of this equation is two lines intersecting at $(-1, -3)$.

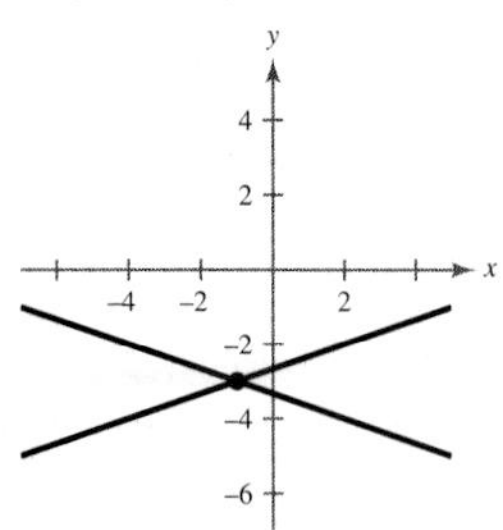

21. Center: $(0, 0)$
Vertices: $(\pm\sqrt{3}, 0)$
Foci: $(\pm\sqrt{5}, 0)$

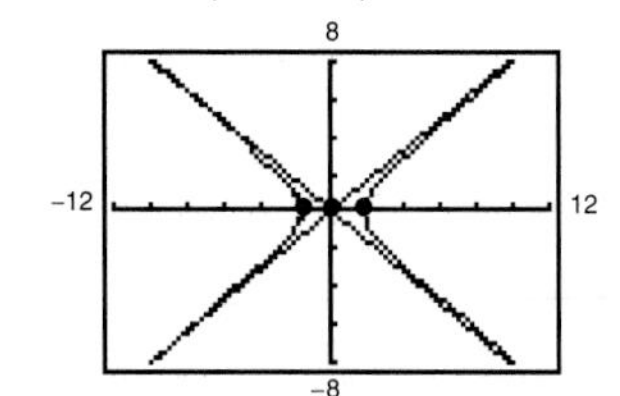

23. Center: $(1, -3)$
Vertices: $(1, -3 \pm \sqrt{2})$
Foci: $(1, -3 \pm 2\sqrt{5})$

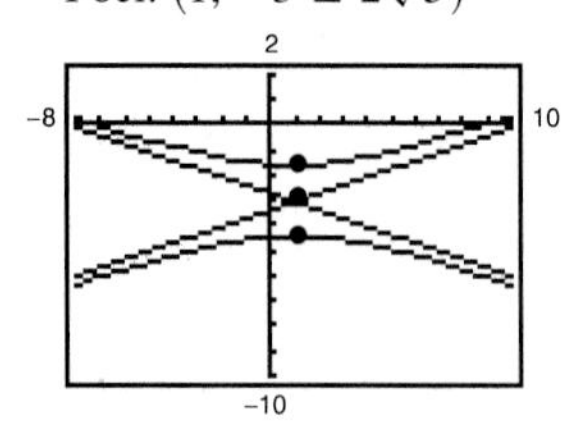

25. $\dfrac{y^2}{9} - \dfrac{x^2}{9/4} = 1$ **27.** $\dfrac{y^2}{4} - \dfrac{x^2}{12} = 1$

29. $\dfrac{x^2}{1} - \dfrac{y^2}{9} = 1$ **31.** $\dfrac{17y^2}{1024} - \dfrac{17x^2}{64} = 1$

33. $(y-1)^2 - x^2 = 1$ **35.** $\dfrac{(x-3)^2}{4} - \dfrac{(y-2)^2}{16/5} = 1$

37. $\dfrac{(x-4)^2}{4} - \dfrac{y^2}{12} = 1$ **39.** $\dfrac{(y-5)^2}{16} - \dfrac{(x-4)^2}{9} = 1$

41. $\dfrac{y^2}{9} - \dfrac{4(x-2)^2}{9} = 1$ **43.** $\dfrac{(x-3)^2}{9} - \dfrac{(y-2)^2}{4} = 1$

45. Left half **47.** $(4400, -4290)$

49. $(12(\sqrt{5} - 1), 0) \approx (14.83, 0)$ **51.** Circle

53. Hyperbola **55.** Ellipse **57.** Parabola

59. $x^3 + x^2 + 2x - 6$ **61.** $x^2 - 2x + 1 + \dfrac{2}{x+2}$

Section 6.5 *(page 461)*

Warm Up *(page 461)*

1.

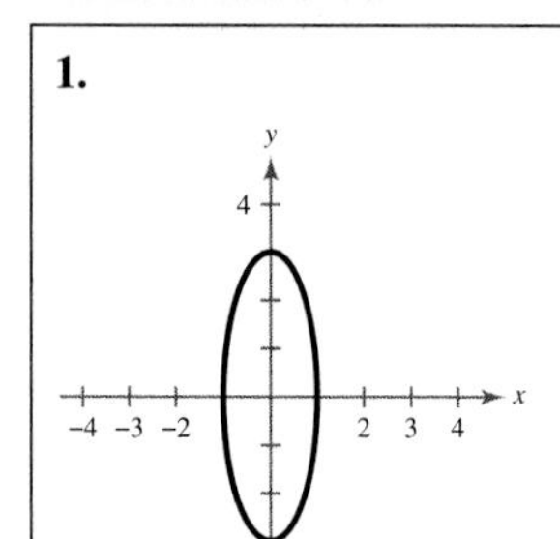

2.

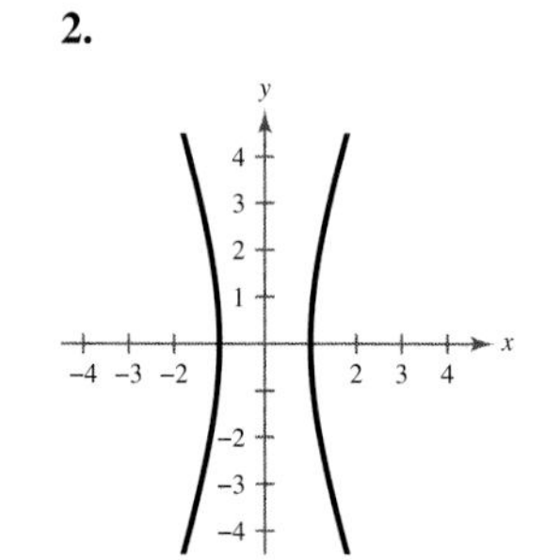

3.

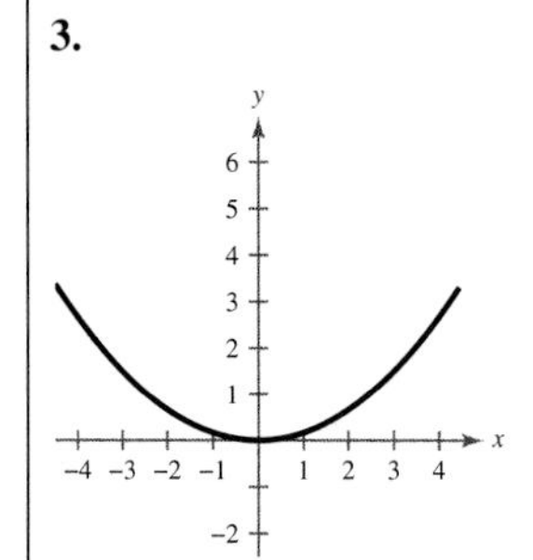

4.

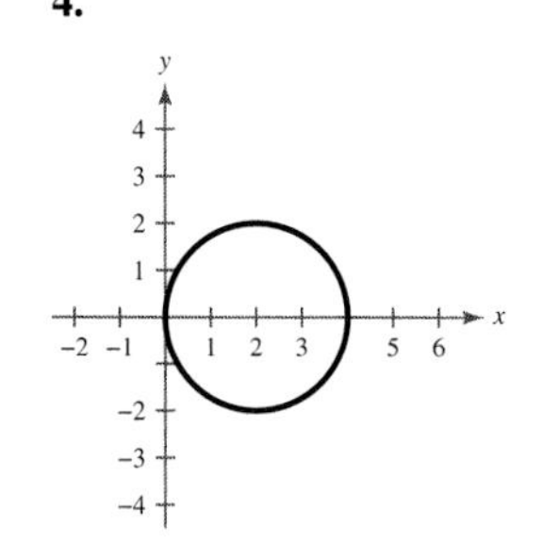

5.

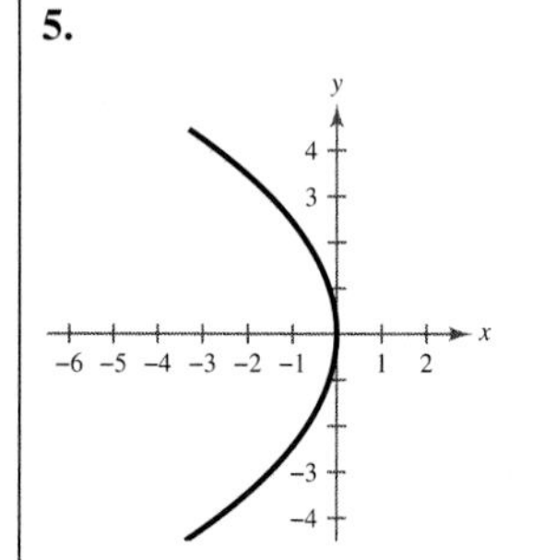

6.

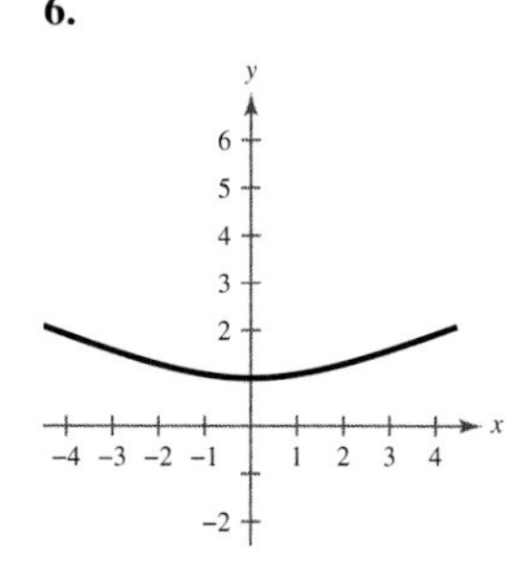

7. $\dfrac{1}{2}x - \dfrac{\sqrt{3}}{2}y$ **8.** $-\dfrac{1}{2}x + \dfrac{\sqrt{3}}{2}y$

9. $\dfrac{4x^2 - 12xy + 9y^2}{13}$ **10.** $\dfrac{x^2 - 2\sqrt{2}xy + 2y^2}{3}$

1. $(-3, 0)$ **3.** $\left(\frac{1}{2}(\sqrt{3} - 4), \frac{1}{2}(1 + 4\sqrt{3})\right)$

5. $\frac{(y')^2}{2} - \frac{(x')^2}{2} = 1$

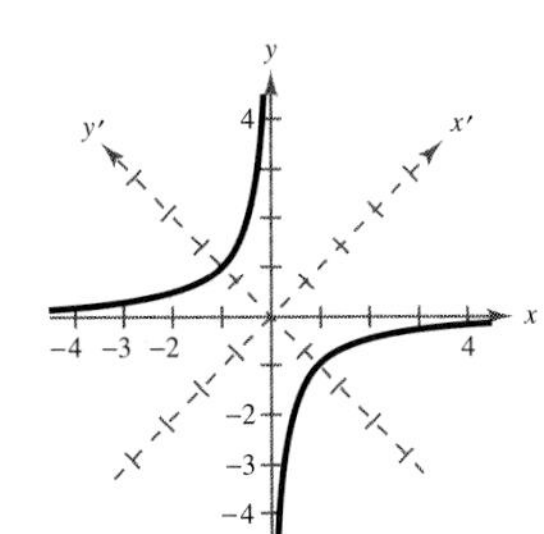

7. $\frac{(x')^2}{1/4} - \frac{(y')^2}{1/6} = 1$

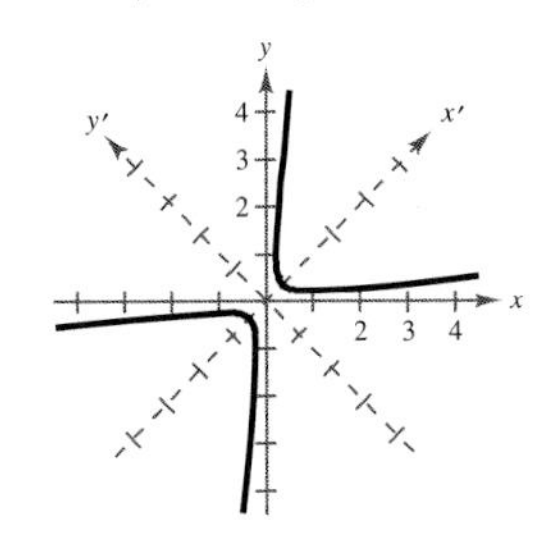

9. $\frac{(x' - 3\sqrt{2})^2}{16} - \frac{(y' - \sqrt{2})^2}{16} = 1$

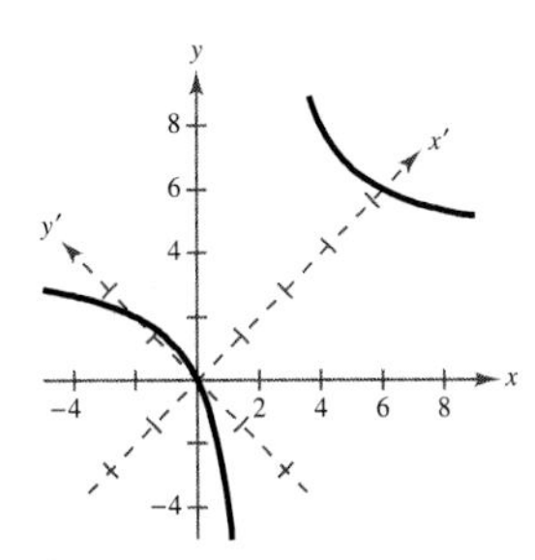

11. $\frac{(x')^2}{3} + \frac{(y')^2}{2} = 1$

13. $x' = -(y')^2$

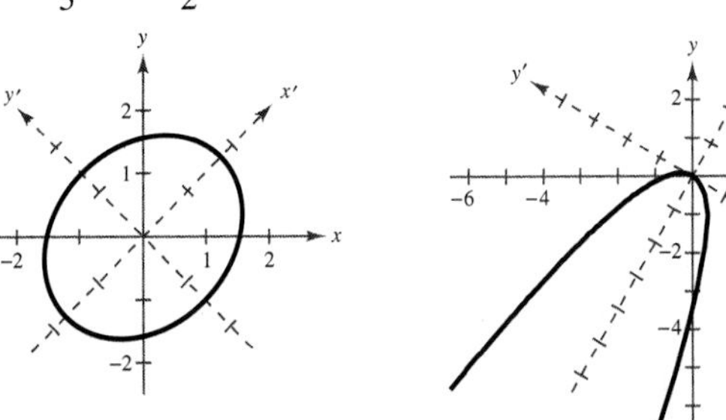

15. $y' = \frac{1}{6}(x')^2 - \frac{1}{3}x'$

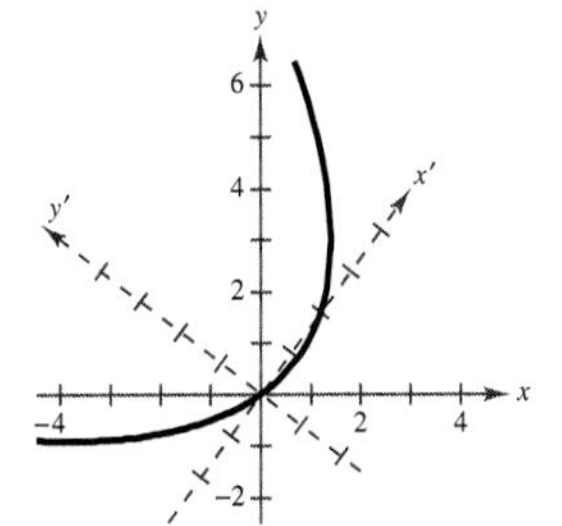

17. $\theta = 45°$

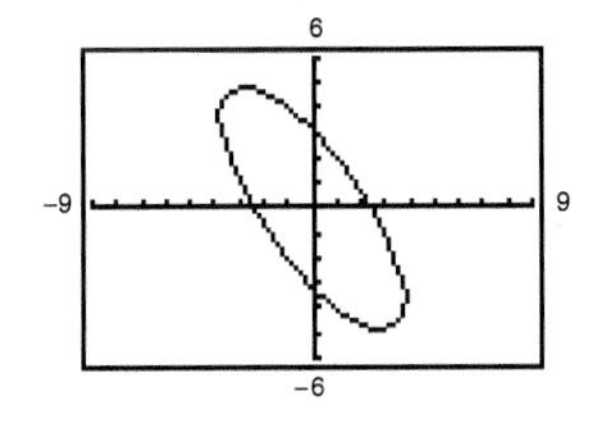

19. $\theta = 26.57°$

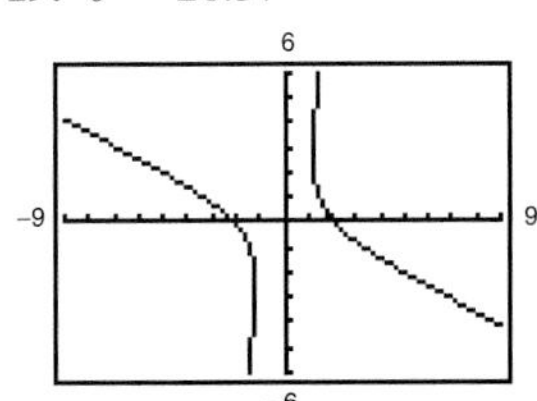

21. $\theta = 31.72°$

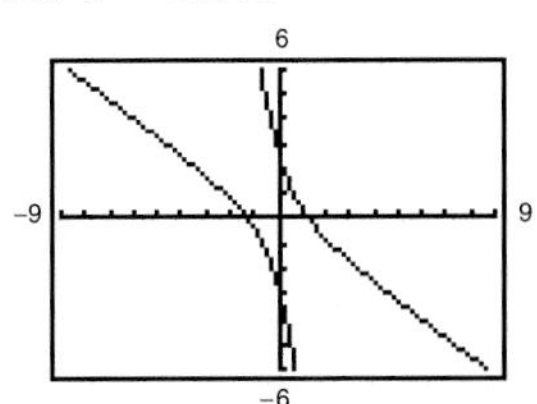

23. e **25.** b **27.** d **29.** Parabola **31.** Ellipse

33. Hyperbola **35.** Parabola

37.

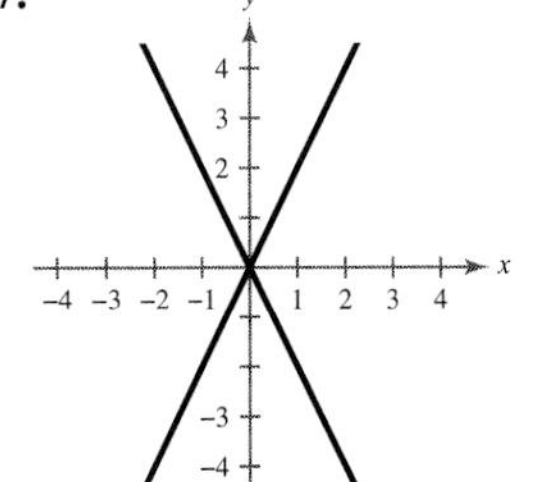

39.

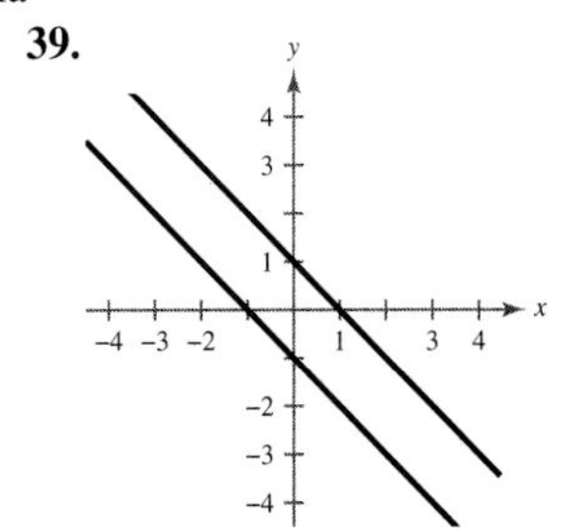

41.

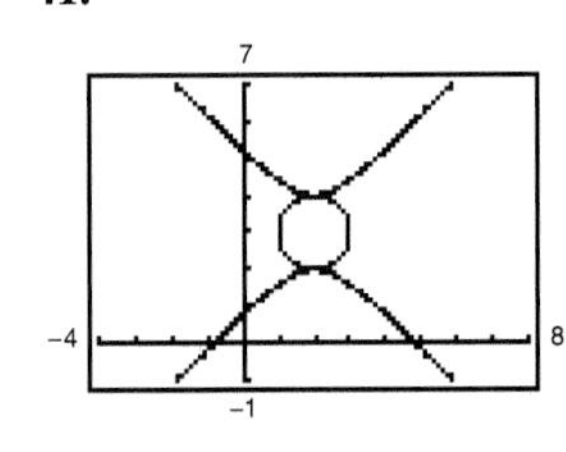

(2, 2), (2, 4)

43.

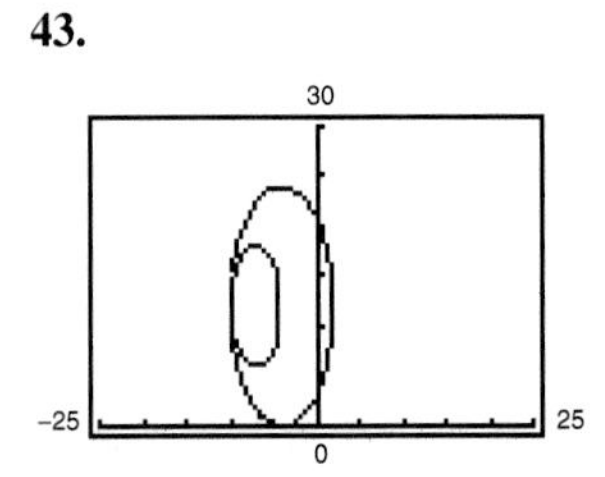

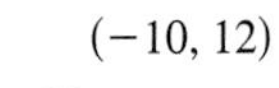
(−10, 12)

45.

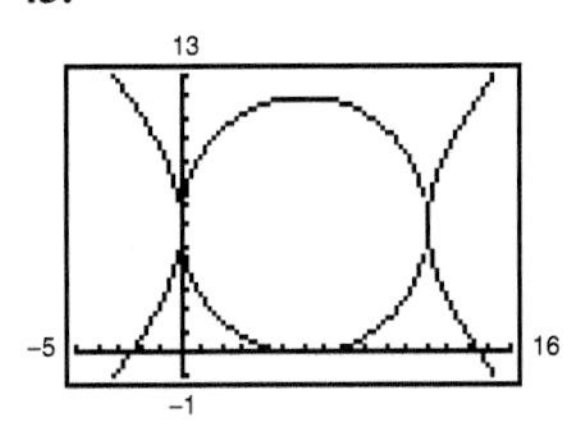

(0, 6), (12, 6)

47.

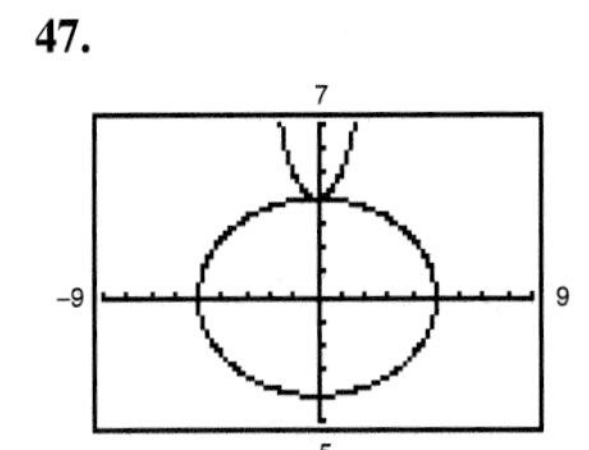

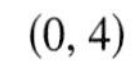
(0, 4)

49.

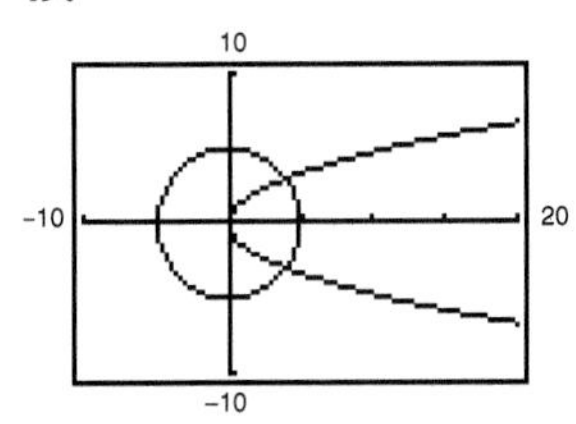

(4, ±3)

51.

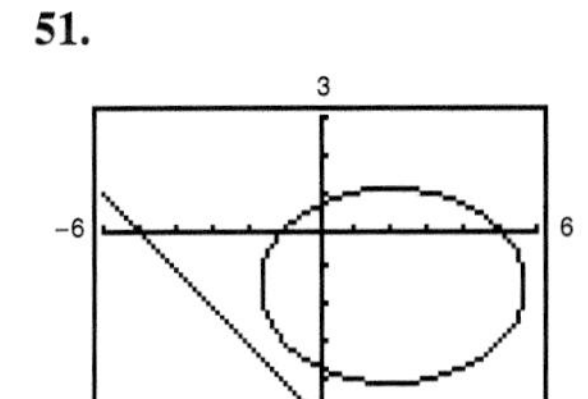

No solution

53.

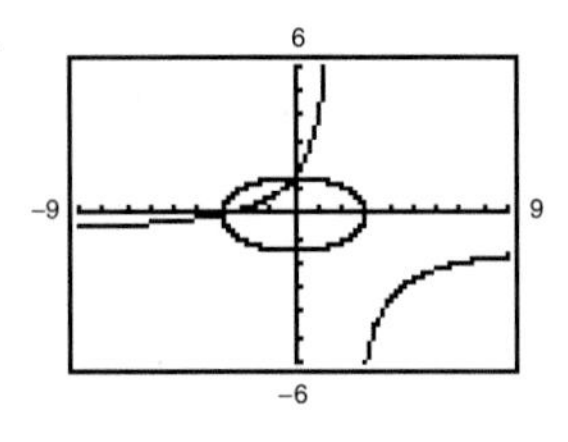

$(-3, 0), \left(0, \frac{3}{2}\right)$

55. Answers will vary.

57.

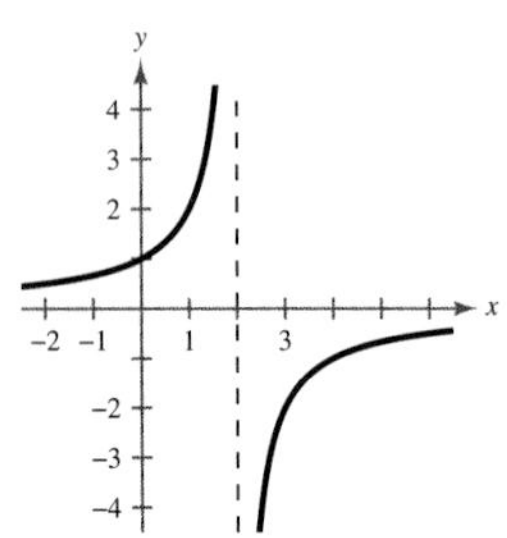

59.

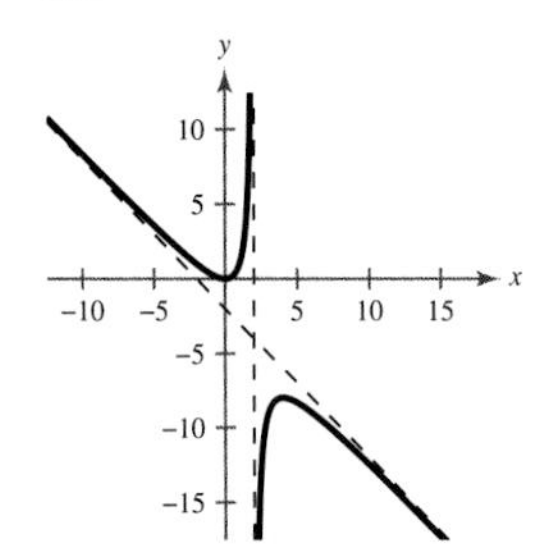

Section 6.6 *(page 469)*

Warm Up *(page 469)*

1.

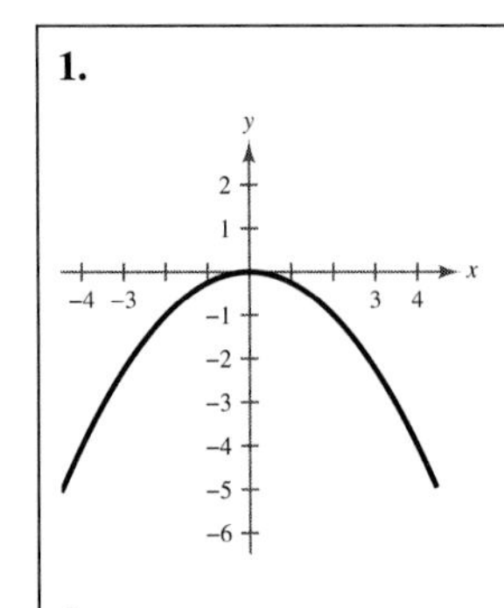

2.

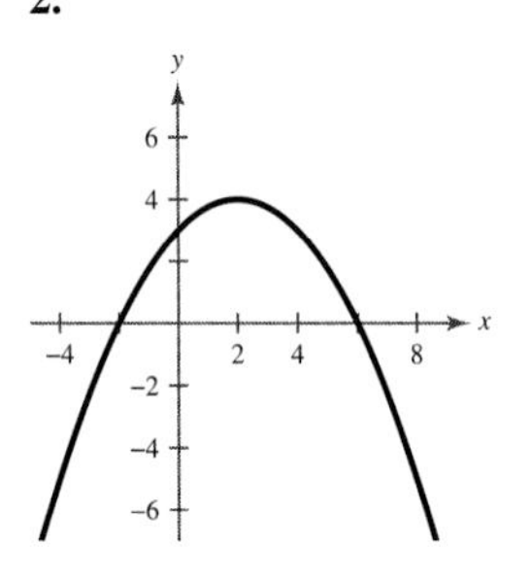

3.

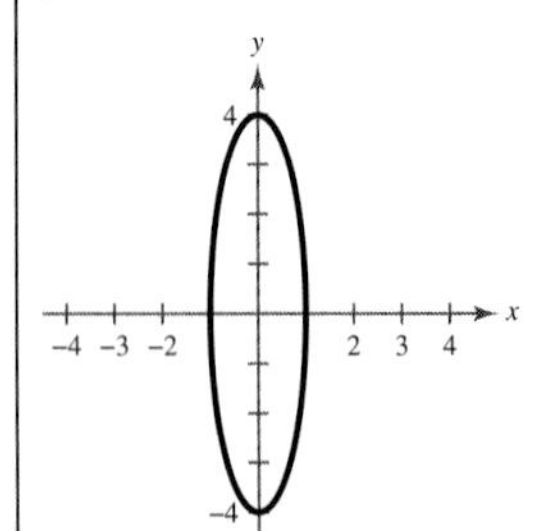

4.

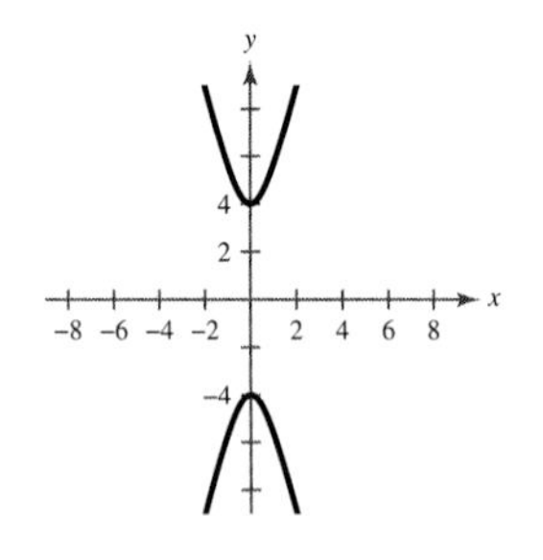

5.

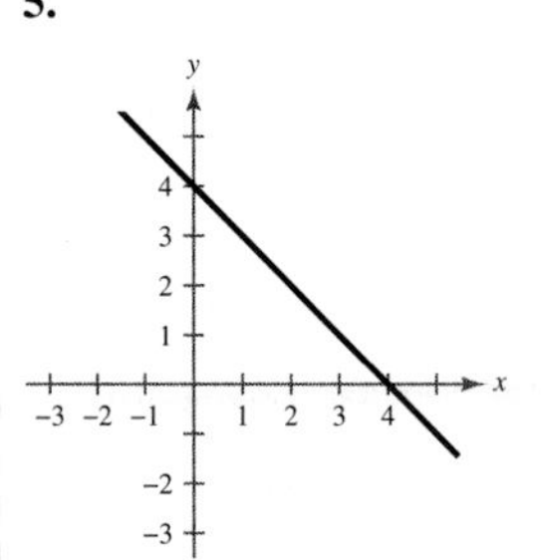

6.

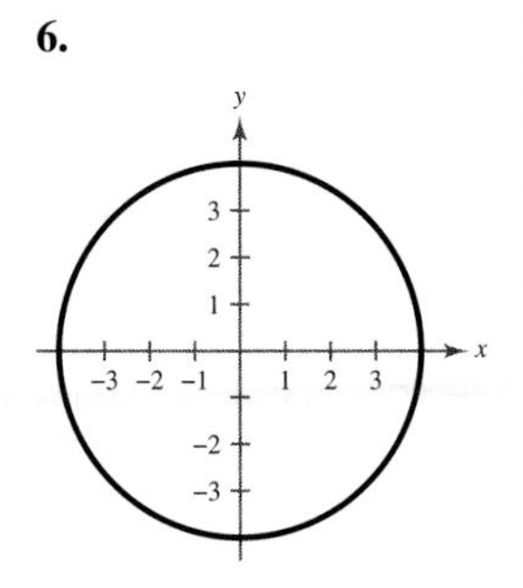

7. 10 **8.** $5 \tan^2 \theta$ **9.** $\sec^2 x + \tan^2 x$ **10.** $\frac{1}{2} \sin \theta$

1. (a)

t	0	1	2	3	4
x	0	1	$\sqrt{2}$	$\sqrt{3}$	2
y	1	0	-1	-2	-3

(b)

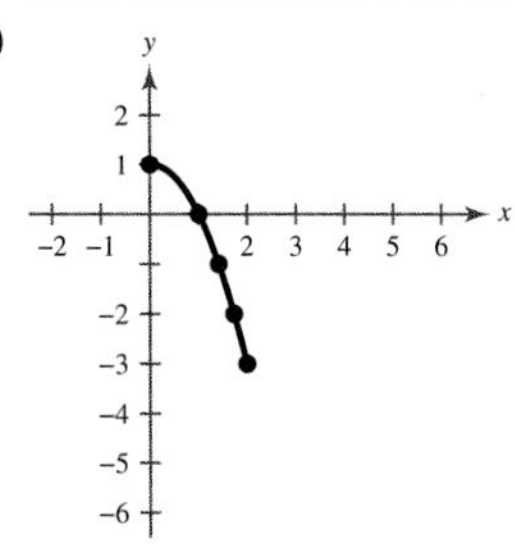

(c) $y = 1 - x^2$

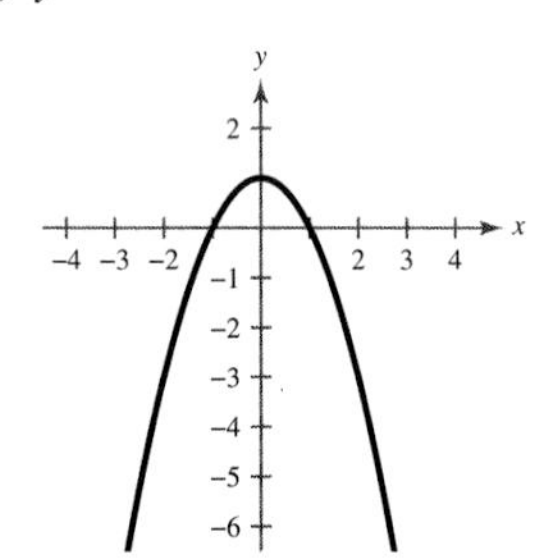

Entire parabola rather than just the right half

3.

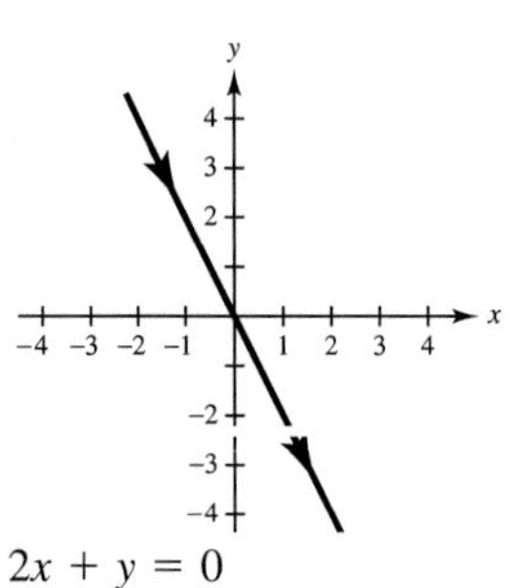

$2x + y = 0$

5.

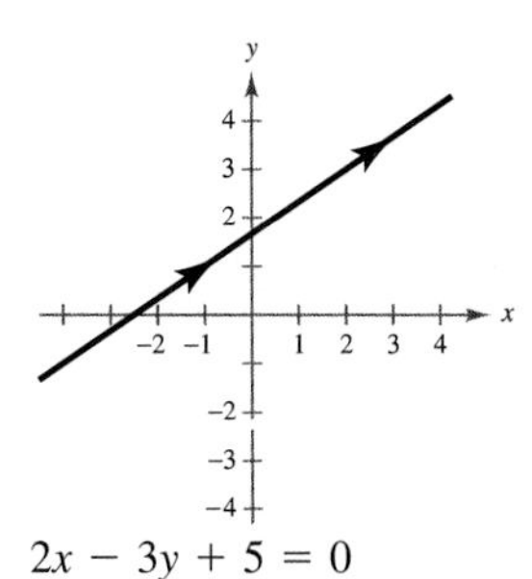

$2x - 3y + 5 = 0$

7.

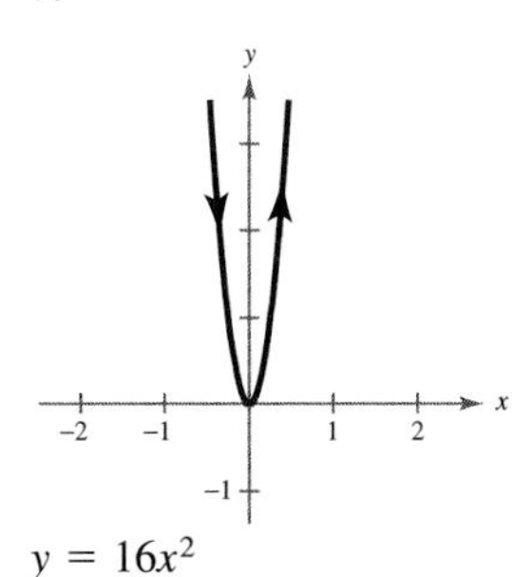

$y = 16x^2$

9.

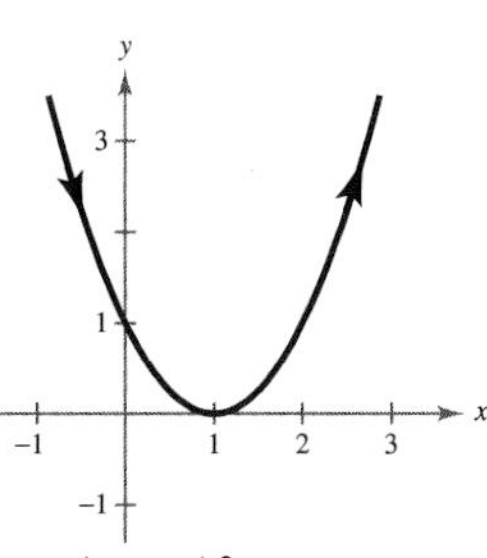

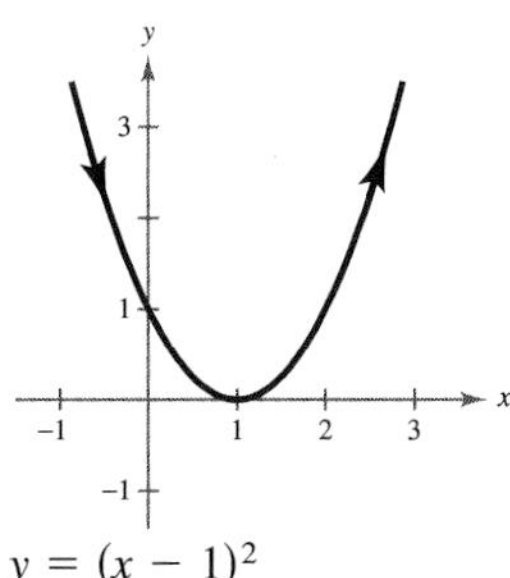

$y = (x - 1)^2$

11.

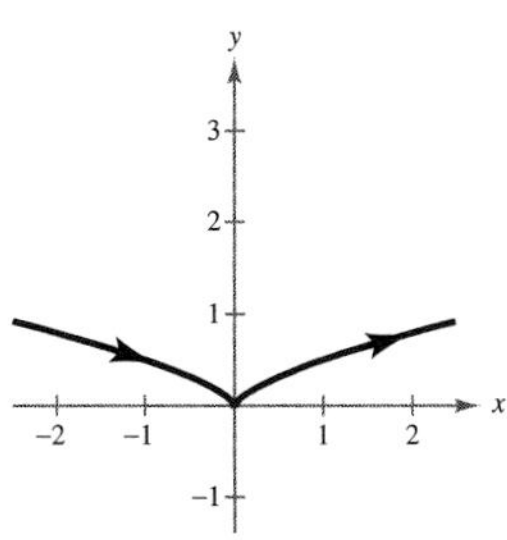

$y = \frac{1}{2}(\sqrt[3]{x})^2$

13.

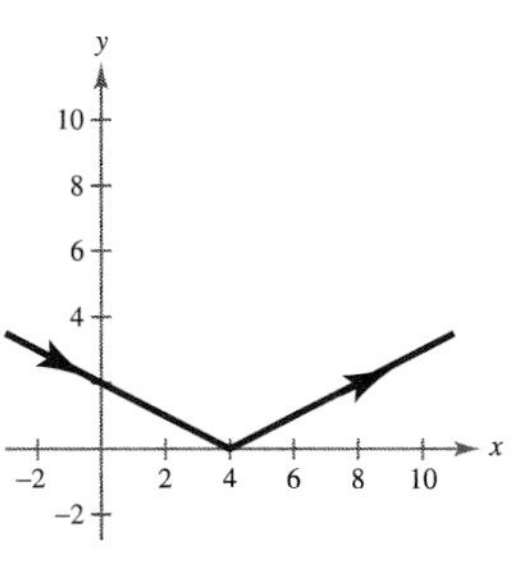

$y = \frac{1}{2}|x - 4|$

15.

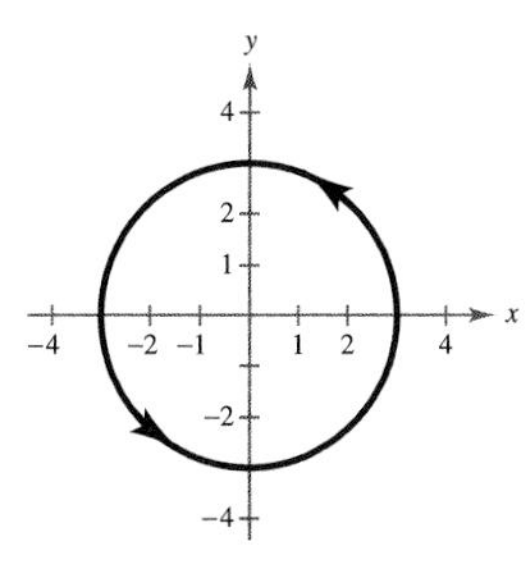

$x^2 + y^2 = 9$

17.

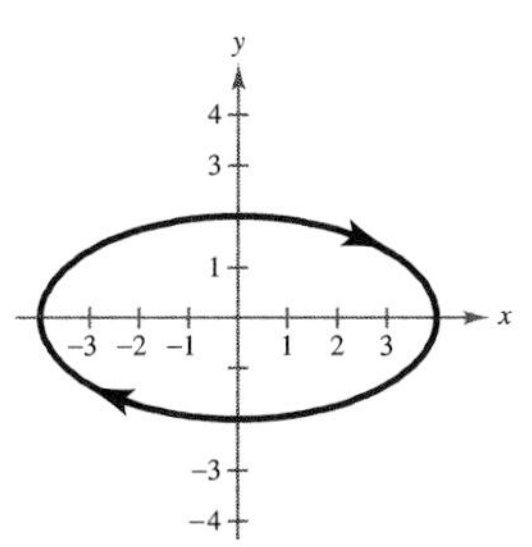

$\dfrac{x^2}{16} + \dfrac{y^2}{4} = 1$

19.

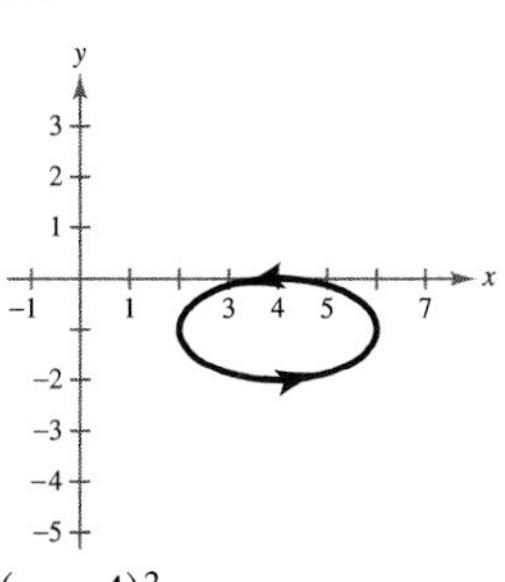

$\dfrac{(x - 4)^2}{4} + (y + 1)^2 = 1$

21.

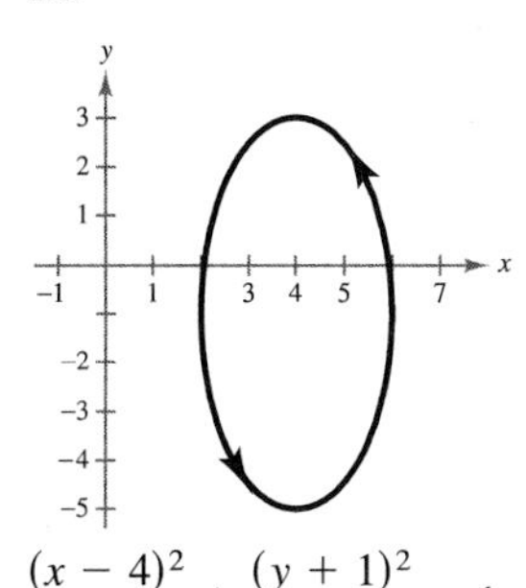

$\dfrac{(x - 4)^2}{4} + \dfrac{(y + 1)^2}{16} = 1$

23.

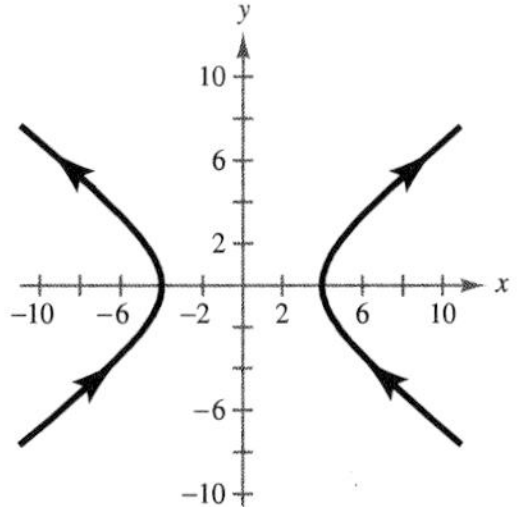

$\dfrac{x^2}{16} - \dfrac{y^2}{9} = 1$

25.

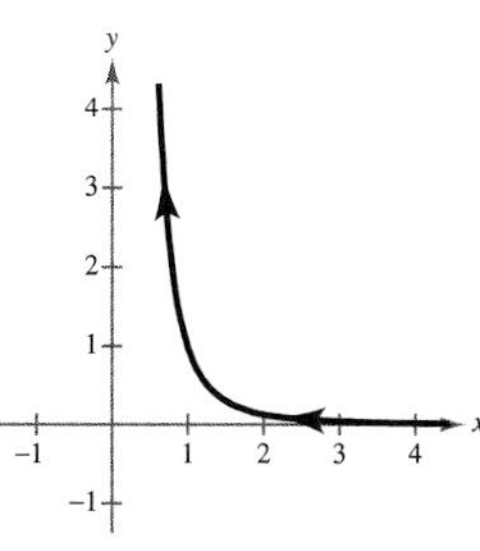

$y = \dfrac{1}{x^3},\ x > 0,\ y > 0$

27.

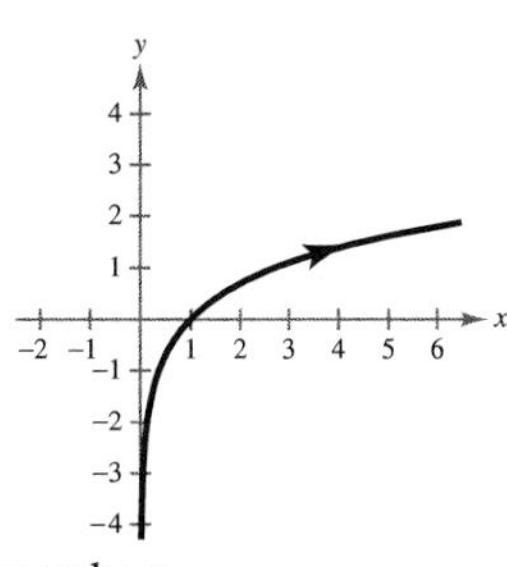

$y = \ln x$

29. Each curve represents a portion of the line $y = 2x + 1$.

	Domain	*Orientation*
(a)	$(-\infty, \infty)$	Left to right
(b)	$[-1, 1]$	Depends on θ
(c)	$(0, \infty)$	Right to left
(d)	$(0, \infty)$	Left to right

31. Parts (a), (b), and (d) represent the circle $x^2 + y^2 = 1$. The orientation is counterclockwise in part (a) and clockwise in parts (b) and (d). Part (c) represents the line $x + y = 1$ for $0 \le x \le 1$ and $0 \le y \le 1$.

33. $y - y_1 = m(x - x_1)$ **35.** $\dfrac{(x - h)^2}{a^2} + \dfrac{(y - k)^2}{b^2} = 1$

37. $x = 5t$, $y = -2t$ **39.** $x = 2 + 4\cos\theta$, $y = 1 + 4\sin\theta$ **41.** $x = 5\cos\theta$, $y = 3\sin\theta$

43. $x = 4\sec\theta$, $y = 3\tan\theta$ **45.** $x = t,\ y = 3t - 2$; $x = 2t,\ y = 6t - 2$

47. $x = t,\ y = t^3$; $x = \sqrt[3]{t},\ y = t$

49.

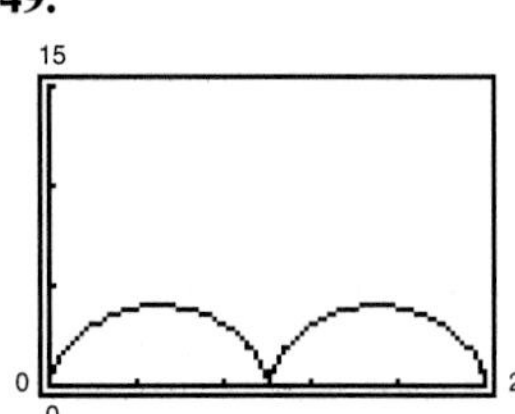

51.

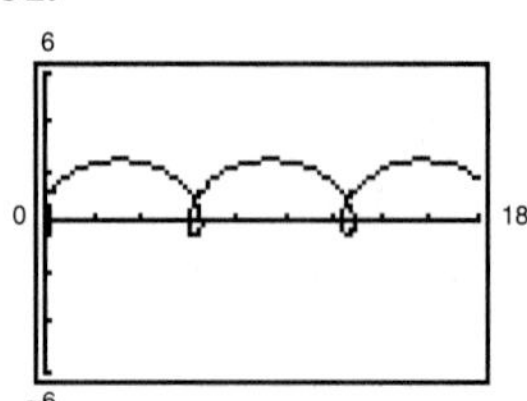

53.

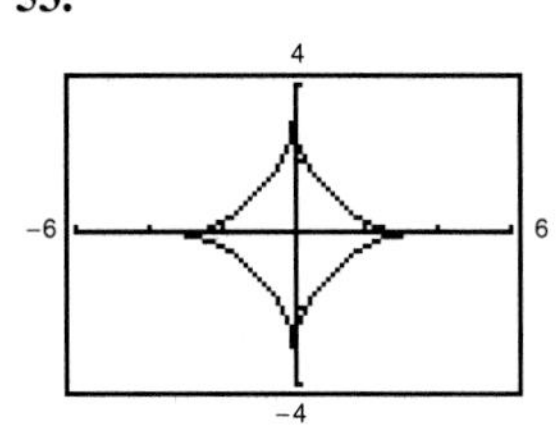

55.

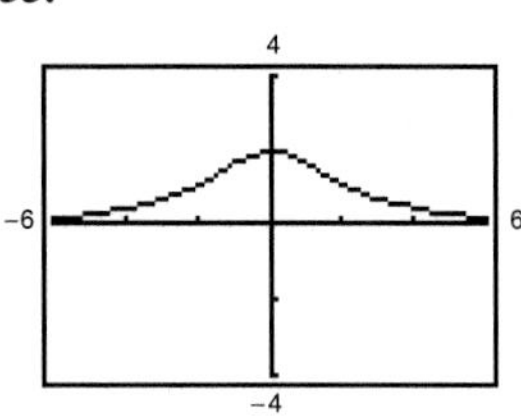

57. b **59.** d

61. (a)

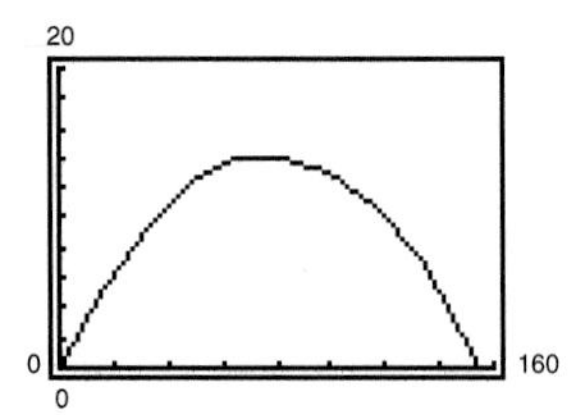

Maximum height: 14.2 feet

Range: 155.6 feet

(b)

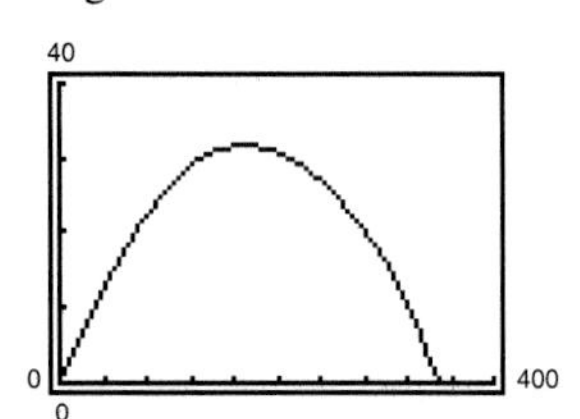

Maximum height: 31.8 feet

Range: 350.0 feet

(c)

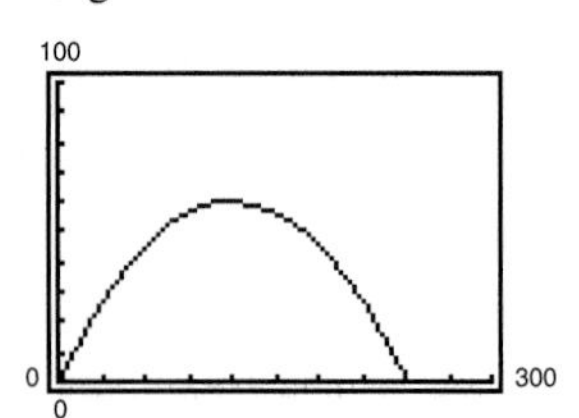

Maximum height: 60.5 feet

Range: 242.0 feet

(d)

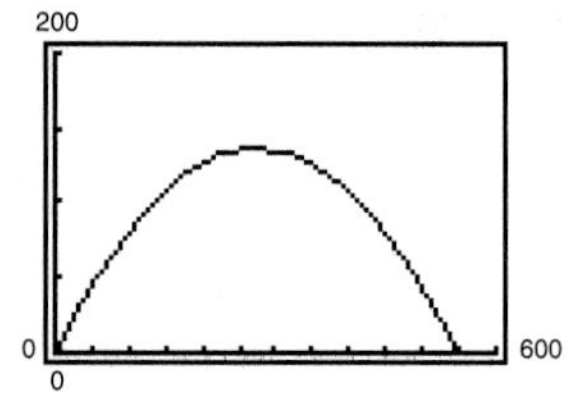

Maximum height: 136.1 feet

Range: 544.5 feet

63. (a) $x = (146.67 \cos \theta)t$

$y = 3 + (146.67 \sin \theta)t - 16t^2$

(b) $x = 141.7t$

$y = 3 + 38.0t - 16t^2$

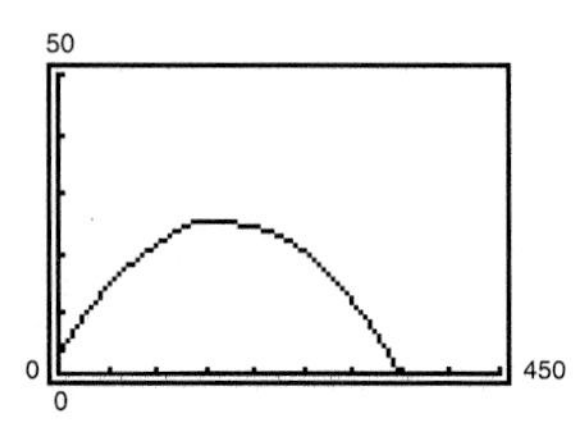

No

(c) $x = 135.0t$

$y = 3 + 57.3t - 16t^2$

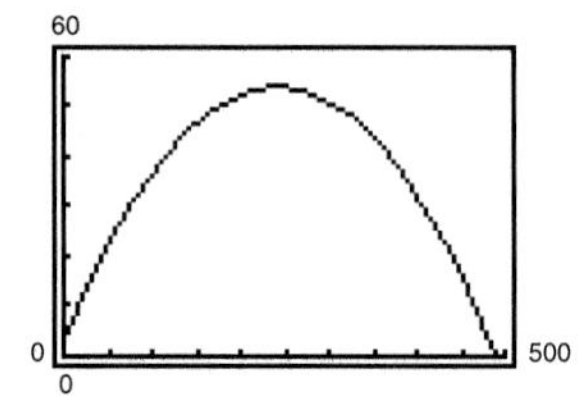

Yes

(d) 19.4°

65. Answers will vary.

67. $x = a\theta - b \sin \theta$

$y = a - b \cos \theta$

69. True **71.** (5, 2)

73. (1, −2, 1)

Section 6.7 *(page 479)*

Warm Up *(page 479)*

1. $\frac{3\pi}{4}$ **2.** $\frac{7\pi}{6}$ **3.** $\sin\theta = \frac{\sqrt{5}}{5}$; $\cos\theta = \frac{2\sqrt{5}}{5}$
4. $\sin\theta = -\frac{3}{5}$; $\cos\theta = \frac{4}{5}$ **5.** $\frac{3\pi}{4}$ **6.** 0.5880
7. $-\frac{\sqrt{3}}{2}$ **8.** $-\frac{\sqrt{2}}{2}$ **9.** -0.3090 **10.** 0.9735

1. $(0, 4)$ **3.** $\left(\frac{\sqrt{2}}{2}, \frac{\sqrt{2}}{2}\right)$

5.

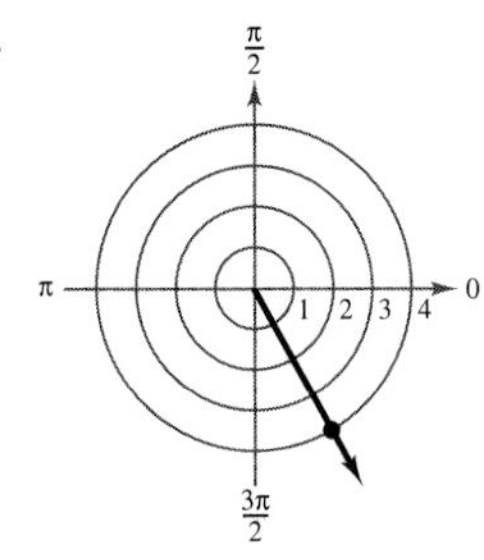

$(2, -2\sqrt{3})$

7.

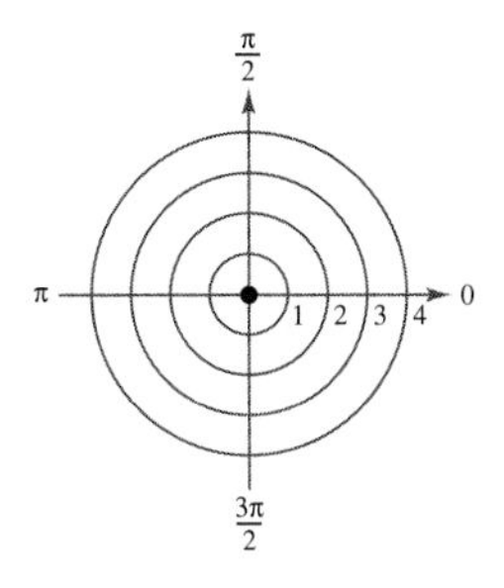

$(0, 0)$

9.

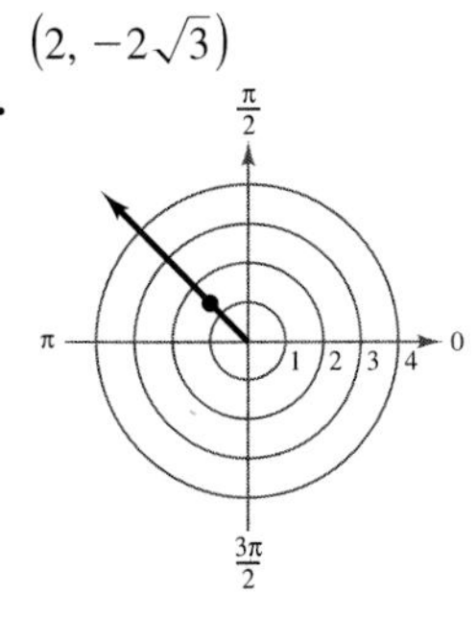

$(-1.004, 0.996)$

11. $(-\sqrt{2}, \sqrt{2})$ **13.** $(-1.204, -4.336)$

15.

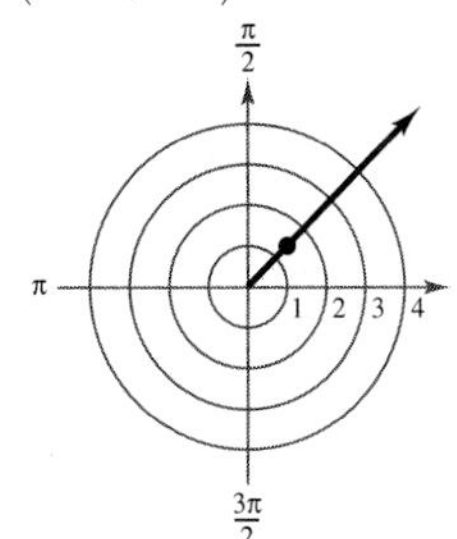

$\left(\sqrt{2}, \frac{\pi}{4}\right), \left(-\sqrt{2}, \frac{5\pi}{4}\right)$

17.

$(6, \pi), (-6, 0)$

19.

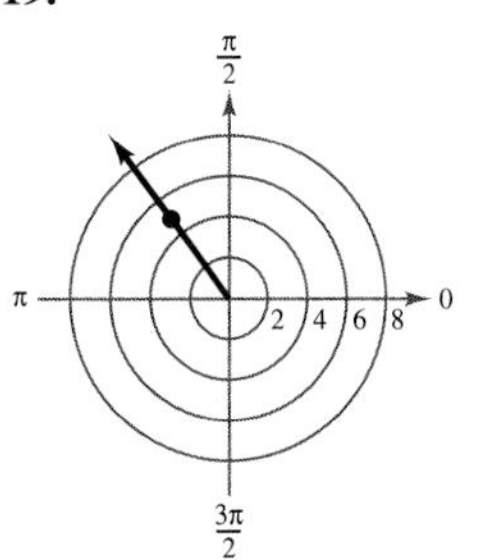

$(5, 2.214), (-5, 5.356)$

21.

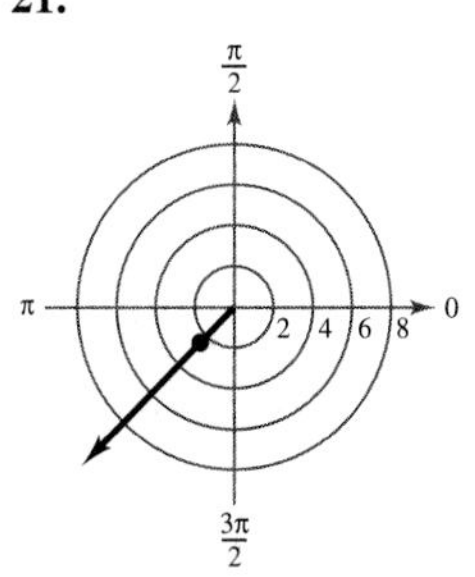

$\left(\sqrt{6}, \frac{5\pi}{4}\right), \left(-\sqrt{6}, \frac{\pi}{4}\right)$

23.

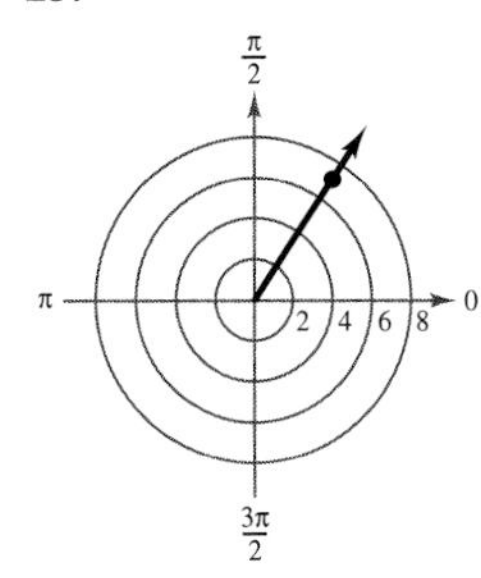

$(2\sqrt{13}, 0.983), (-2\sqrt{13}, 4.124)$

25. $(\sqrt{13}, -0.588)$ **27.** $(\sqrt{7}, 0.857)$ **29.** $\left(\frac{17}{6}, 0.490\right)$
31. True **33.** $r = 3$ **35.** $r = 2a\cos\theta$
37. $r = 4\csc\theta$ **39.** $r = 10\sec\theta$
41. $r = \dfrac{-2}{3\cos\theta - \sin\theta}$
43. $r^2 = 4\sec\theta\csc\theta = 8\csc 2\theta$ **45.** $r^2 = 9\cos 2\theta$
47. $x^2 + y^2 - 4y = 0$ **49.** $\sqrt{3}x - 3y = 0$
51. $y = 2$ **53.** $(x^2 + y^2)^2 = 6x^2y - 2y^3$
55. $4x^2 - 5y^2 - 36y - 36 = 0$
57. $x^2 + y^2 = 9$ **59.** $x - y = 0$

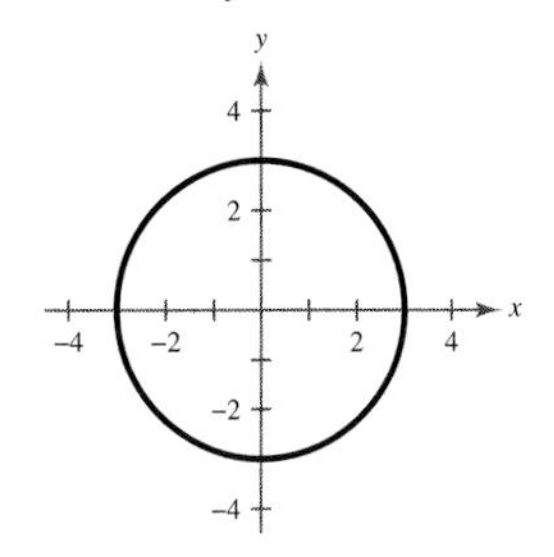

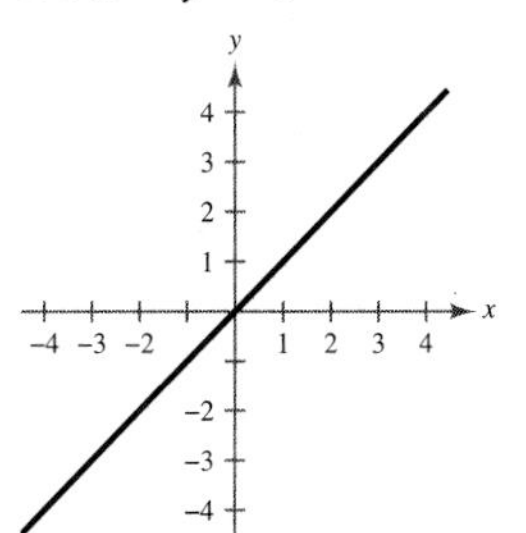

61. $x - 3 = 0$

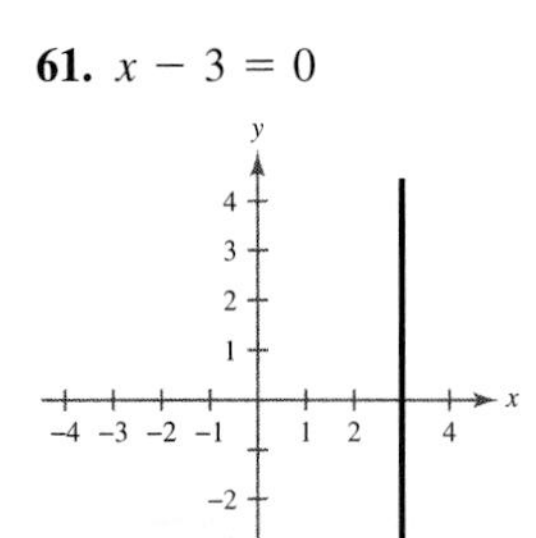

63. $(x - h)^2 + (y - k)^2 = h^2 + k^2$

Center: (h, k)

Radius: $\sqrt{h^2 + k^2}$

65. (a) Answers will vary.

(b) $d = \sqrt{r_1^2 + r_2^2 - 2r_1r_2} = |r_1 - r_2|$

The distance between two points on the line $\theta = \theta_1 = \theta_2$.

(c) $d = \sqrt{r_1^2 + r_2^2}$

Pythagorean Theorem

(d) Points: $(3, \pi/6)$, $(4, \pi/3)$

Distance: 2.053

Points: $(-3, 7\pi/6)$, $(-4, 4\pi/3)$

Distance: 2.053

67. $(2, 3)$ **69.** $\left(\frac{8}{7}, \frac{88}{35}, \frac{8}{5}\right)$ **71.** $(1, -4, 1, 2)$

Section 6.8 *(page 488)*

Warm Up *(page 488)*

1. Amplitude: 5

Period: $\dfrac{\pi}{2}$

2. Amplitude: 3

Period: 1

3. Amplitude: 5

Period: $\frac{4}{5}$

4. Amplitude: $\frac{1}{2}$

Period: 4π

5.

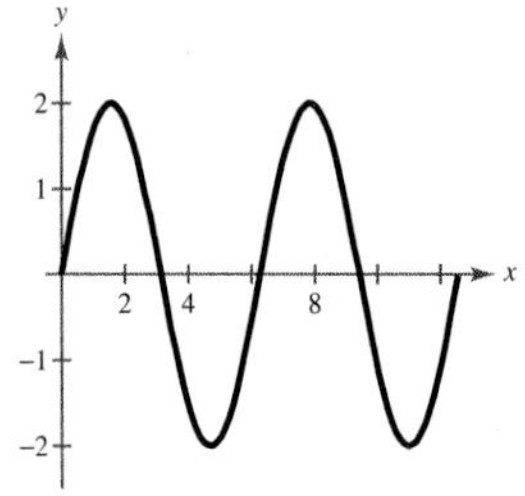

6.

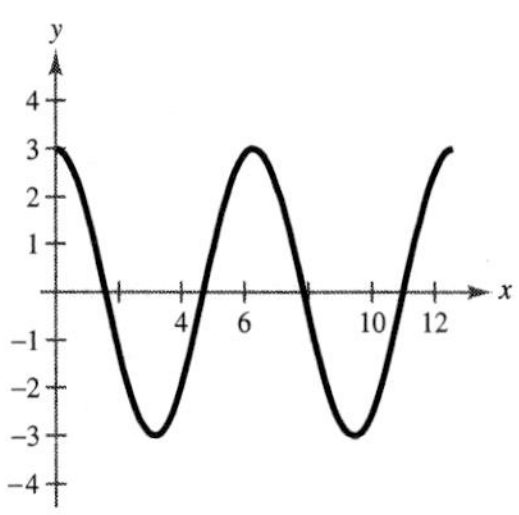

7.

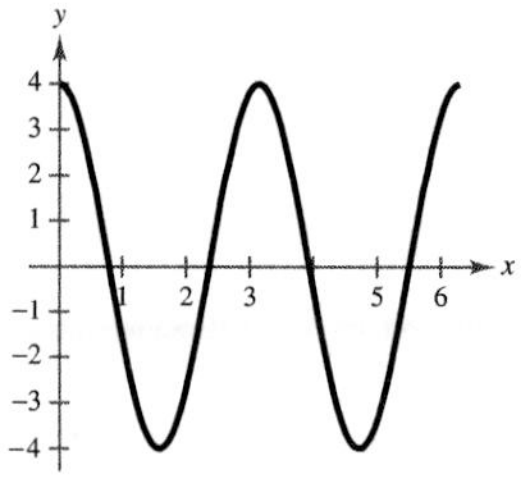

8.

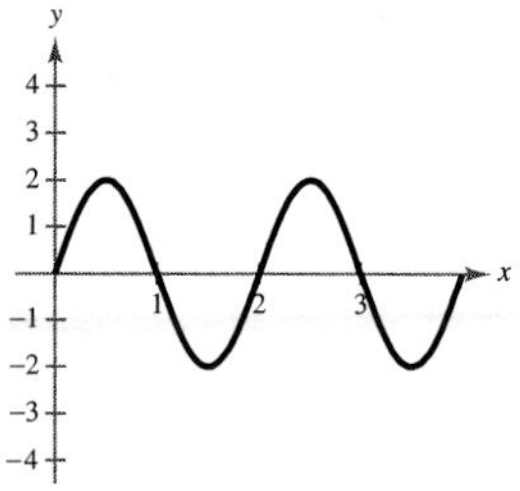

9. $\dfrac{1}{2}\left(\sqrt{3}\sin x - \cos x\right)$ **10.** $\dfrac{\sqrt{2}}{2}(\cos x + \sin x)$

1. Rose curve **3.** Limaçon **5.** Rose curve

7. Polar axis **9.** $\theta = \dfrac{\pi}{2}$ **11.** $\theta = \dfrac{\pi}{2}$, polar axis, pole

13. Maximum: $|r| = 20$ when $\theta = \dfrac{3\pi}{2}$

Zero: $r = 0$ when $\theta = \dfrac{\pi}{2}$

15. Maximum: $|r| = 4$ when $\theta = 0, \dfrac{\pi}{3}, \dfrac{2\pi}{3}$

Zero: $r = 0$ when $\theta = \dfrac{\pi}{6}, \dfrac{\pi}{2}, \dfrac{5\pi}{6}$

17. **19.**

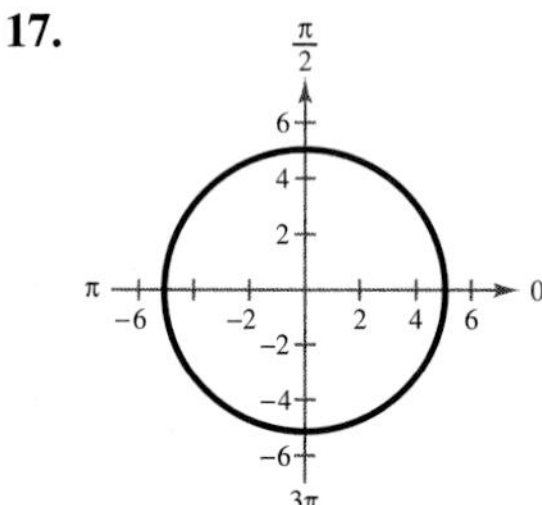

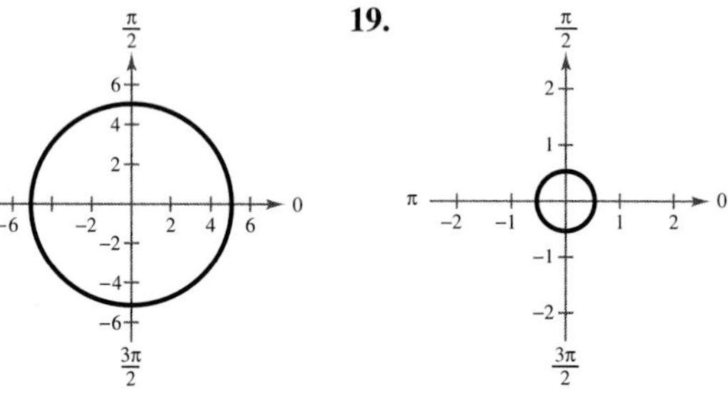

21. **23.**

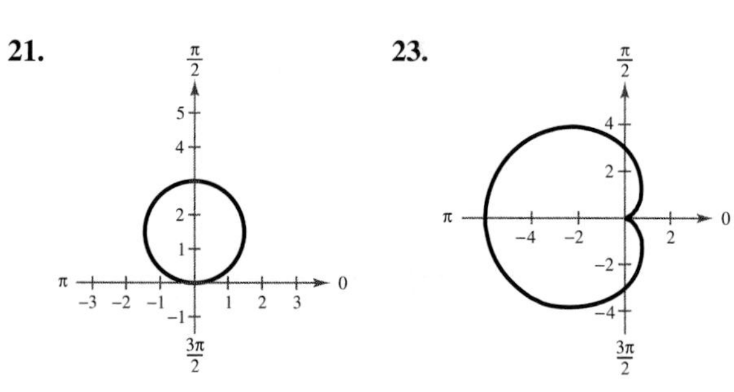

25.

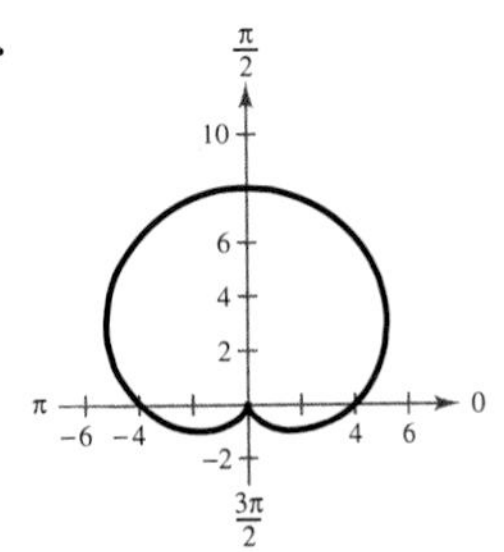

27.

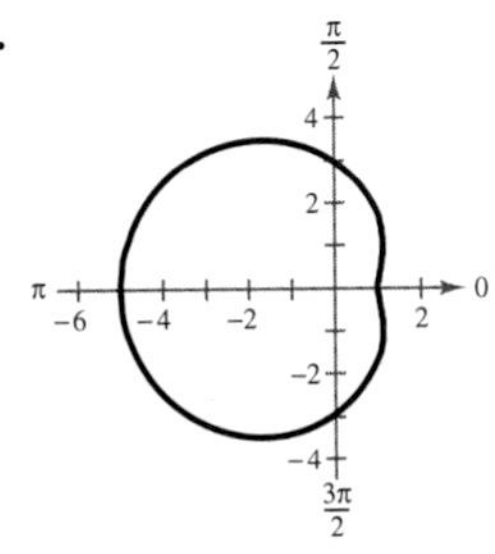

29.

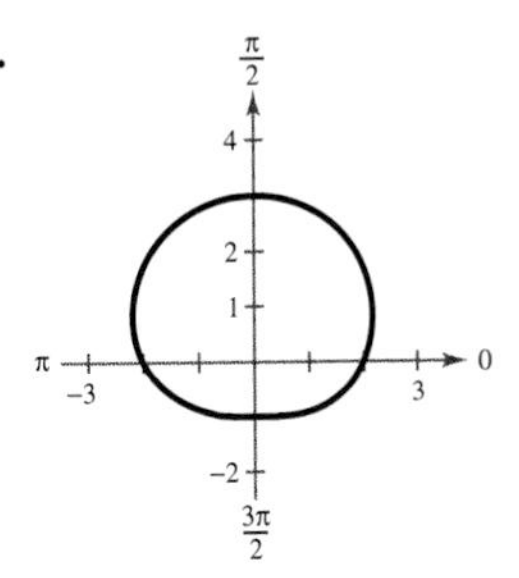

31.

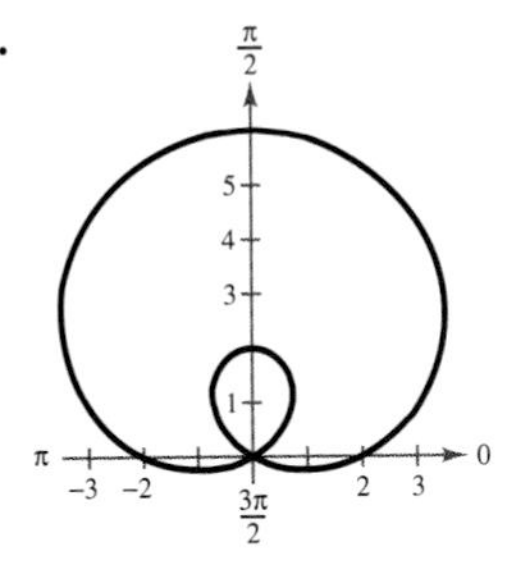

33.

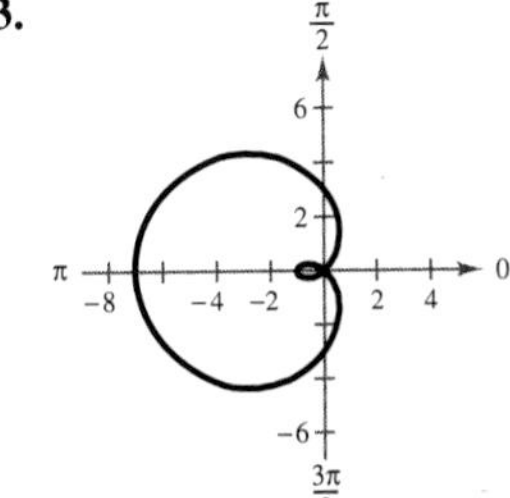

35.

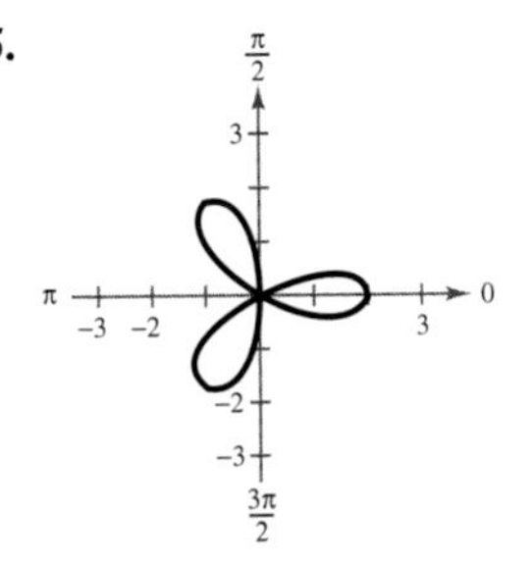

37.

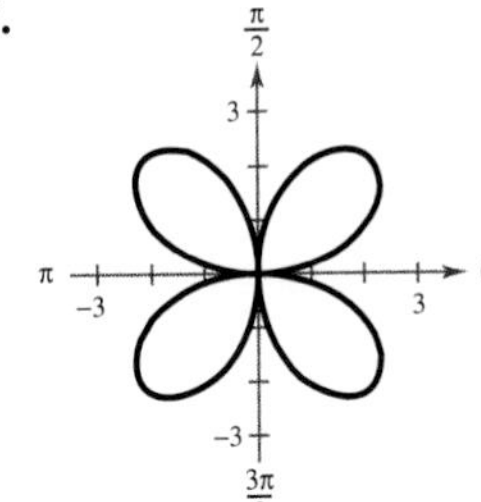

39.

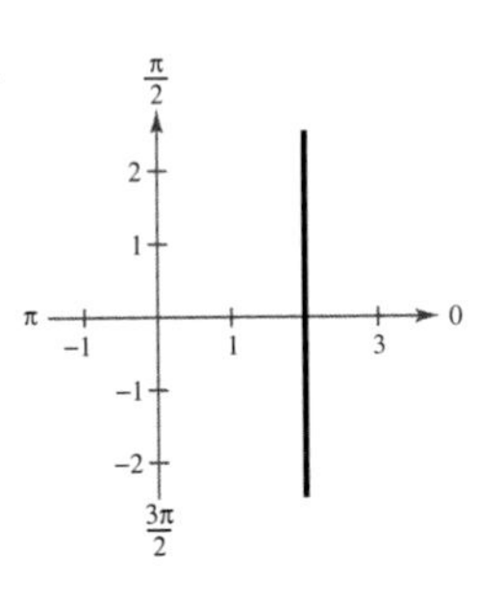

41.

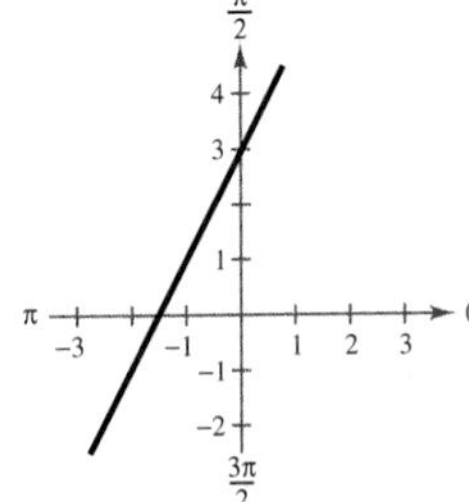

43.

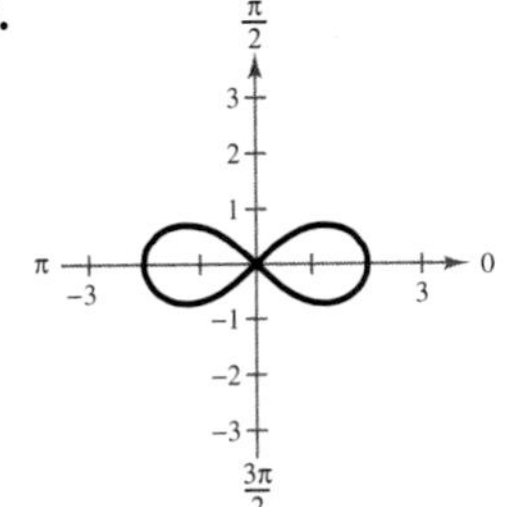

45.

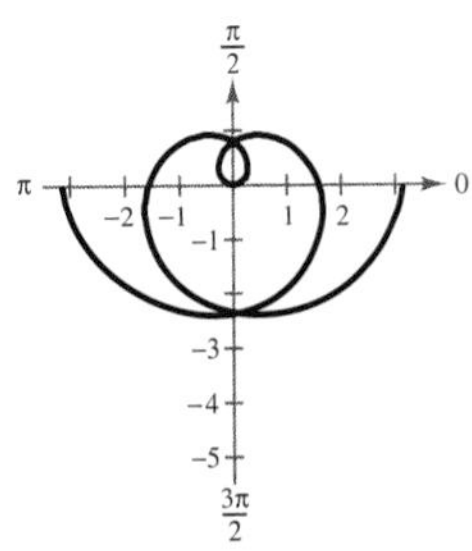

$-2\pi \leq \theta \leq 2\pi$

47.

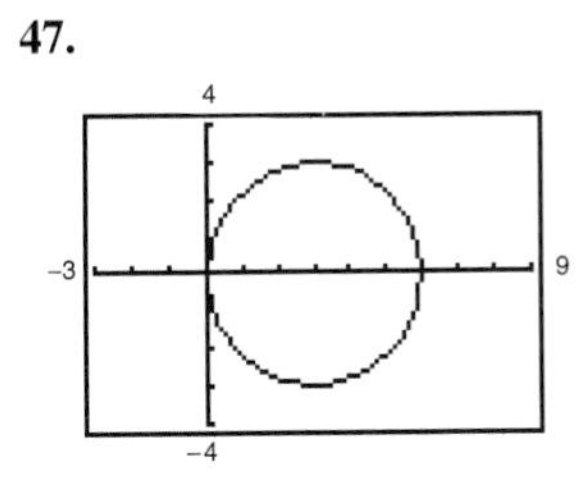

49.

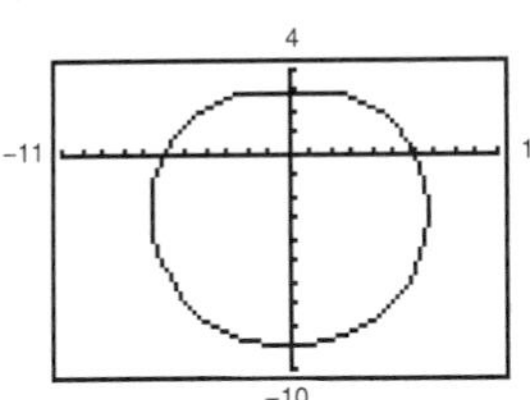

51.

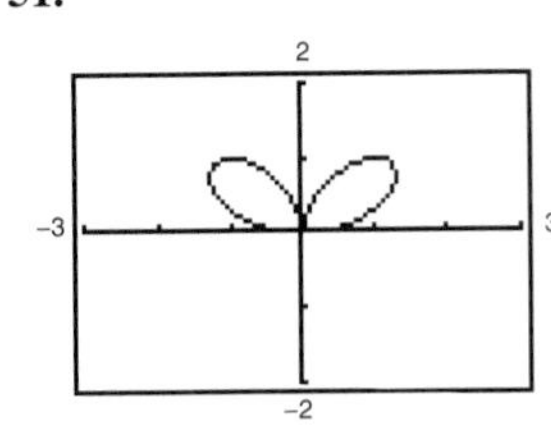

53.

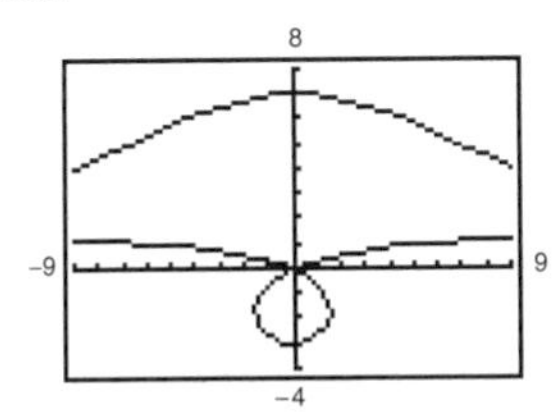

55.

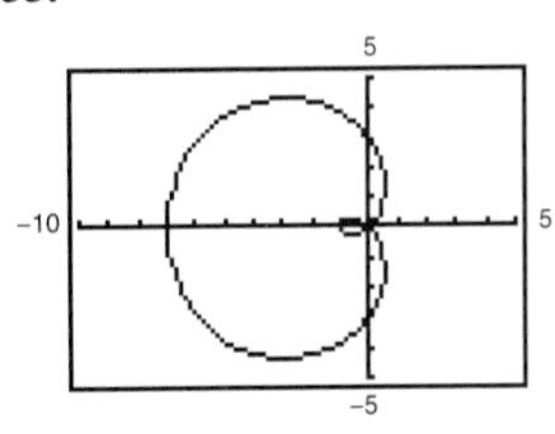

57.

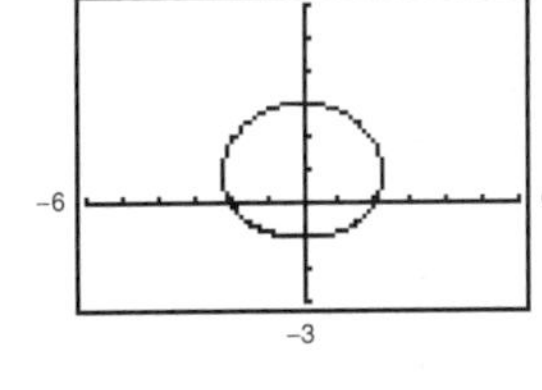

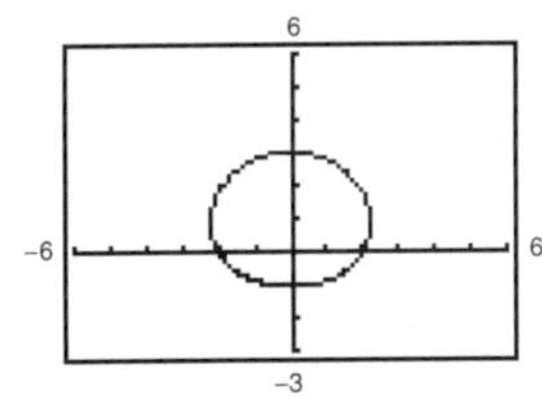

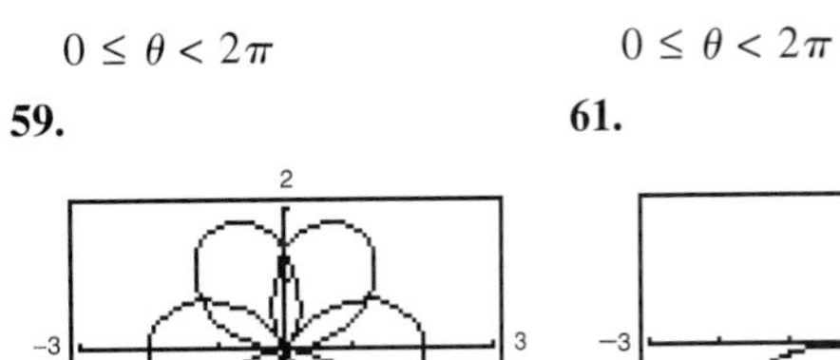

$0 \leq \theta < 2\pi$

$0 \leq \theta < 2\pi$

59.

$0 \leq \theta < 4\pi$

61.

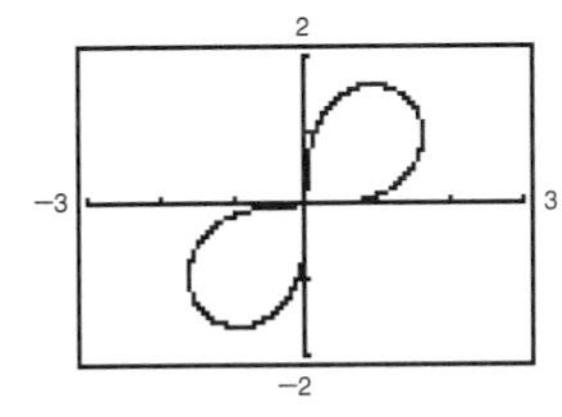

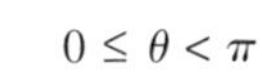

$0 \leq \theta < \pi$

63.

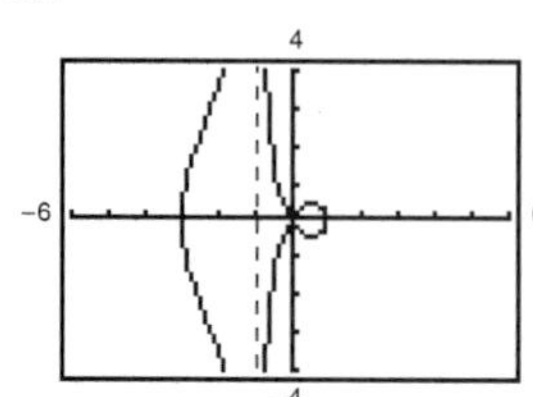

65.

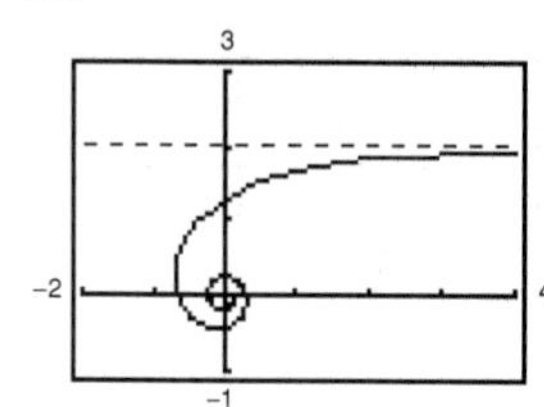

75.

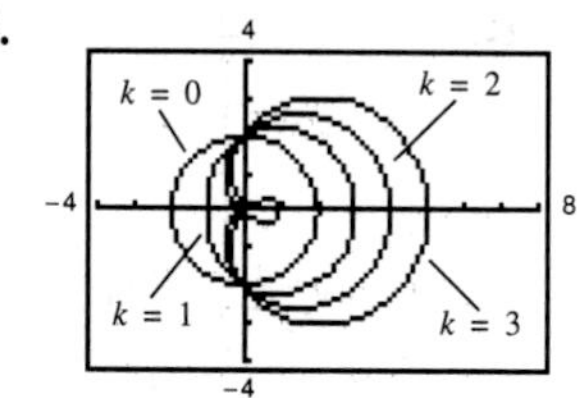

67. (a)

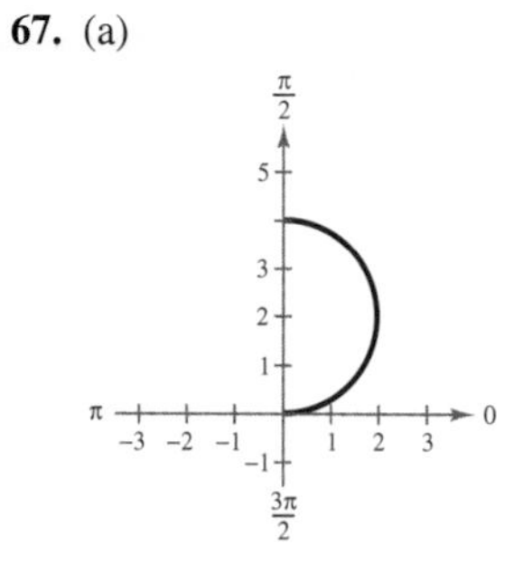

(b)

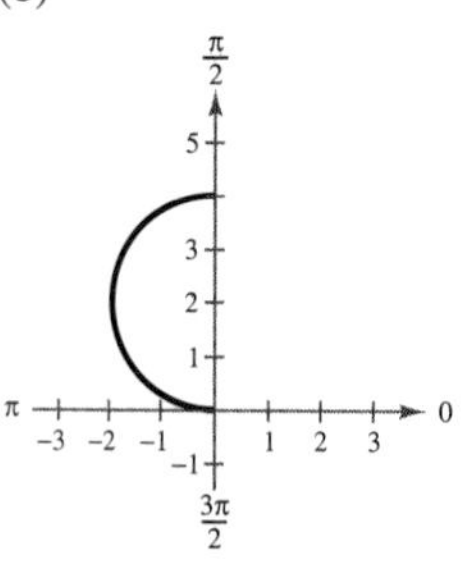

(c)

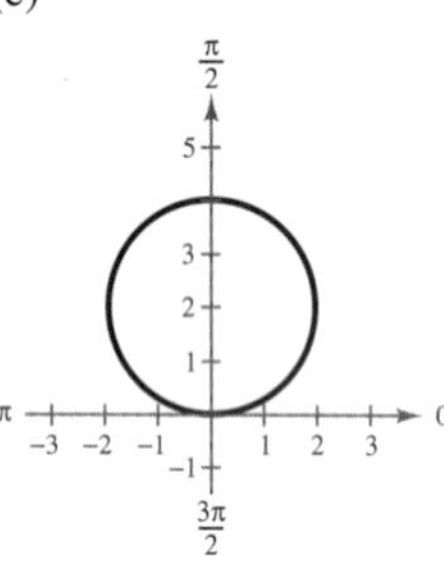

(d)

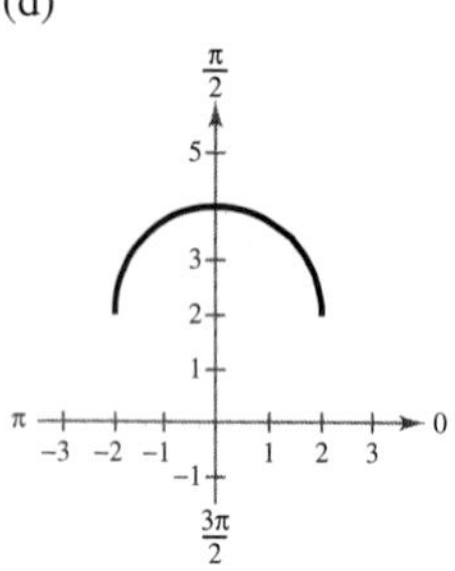

69. Answers will vary.

71. (a) $r = 2 - \sin\left(\theta - \dfrac{\pi}{4}\right)$

(b) $r = 2 + \cos\theta$

(c) $r = 2 + \sin\theta$

(d) $r = 2 - \cos\theta$

73. (a)

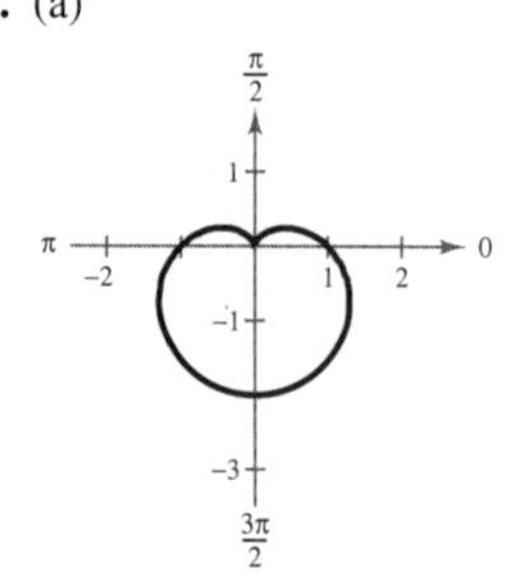

(b)

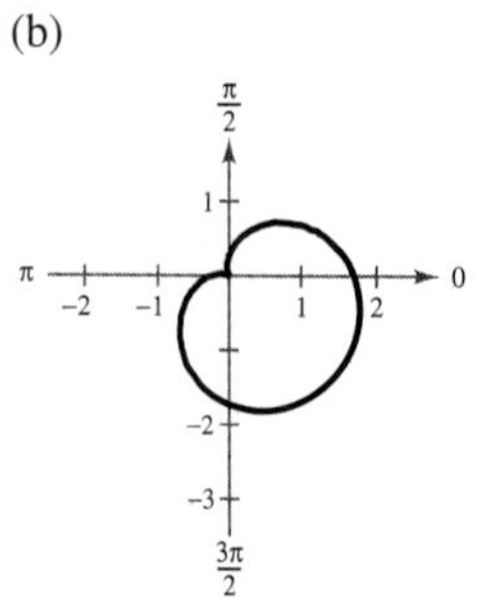

Section 6.9 *(page 497)*

Warm Up *(page 497)*

1. $\left(\dfrac{3\sqrt{2}}{2}, -\dfrac{3\sqrt{2}}{2}\right)$ **2.** $\left(-2, -2\sqrt{3}\right)$

3. $\left(3, \dfrac{3\pi}{2}\right), \left(-3, \dfrac{\pi}{2}\right)$ **4.** (13, 1.9656), (−13, 5.1072)

5. $r = 5$ **6.** $r^3 = 4\sec^2\theta\csc\theta$ **7.** $y = -4$

8. $x^2 + y^2 - 4x = 0$

9. Cardioid **10.** Limaçon with inner loop

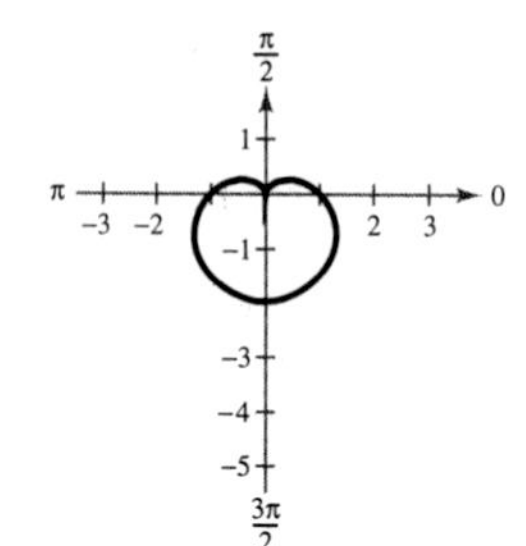

1.

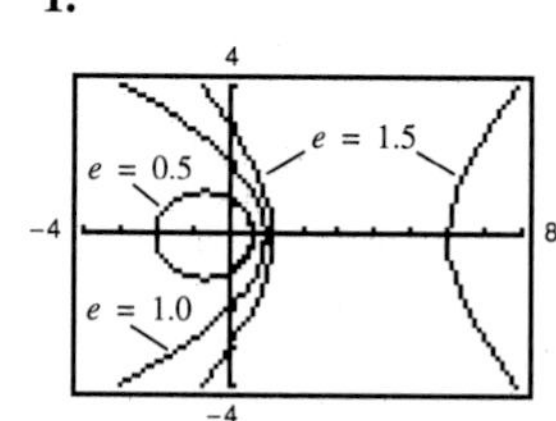

3.

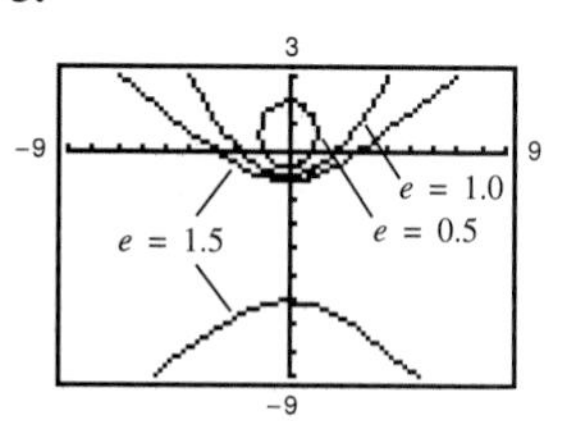

5. b **7.** d

9.

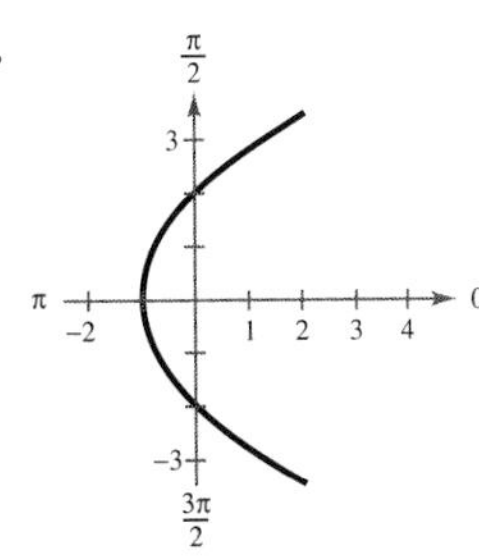

11.

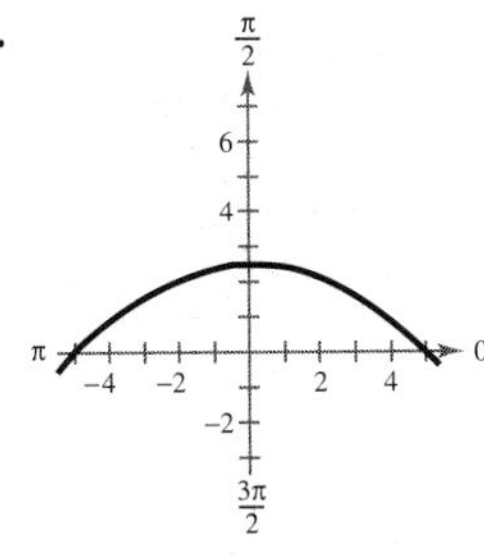

13.

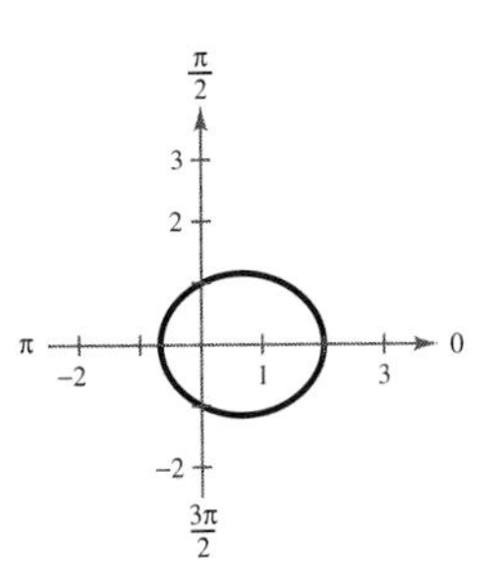

15.

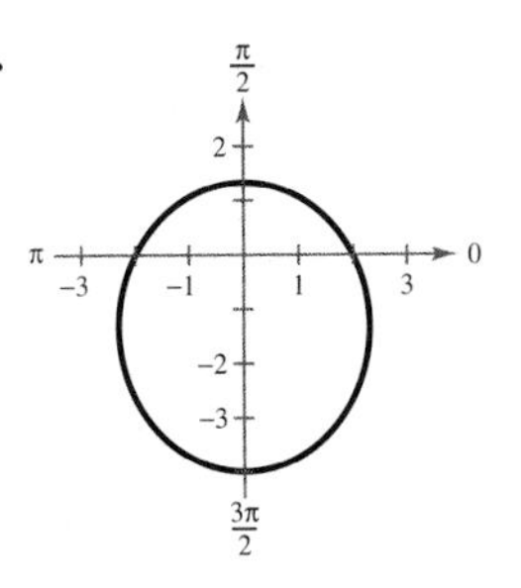

17.

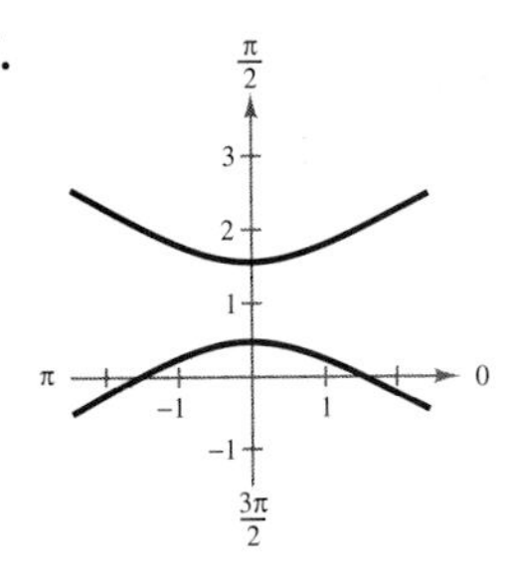

19.

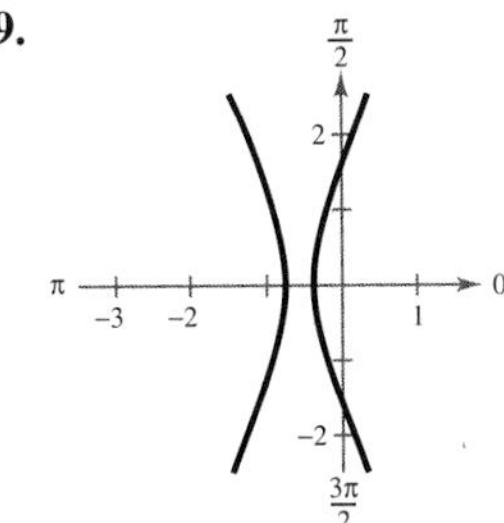

21.

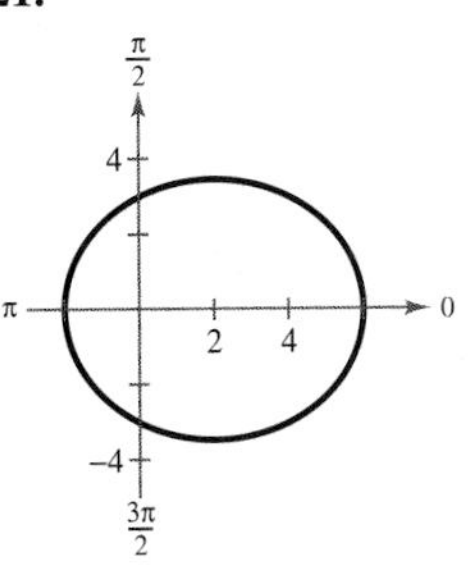

23.

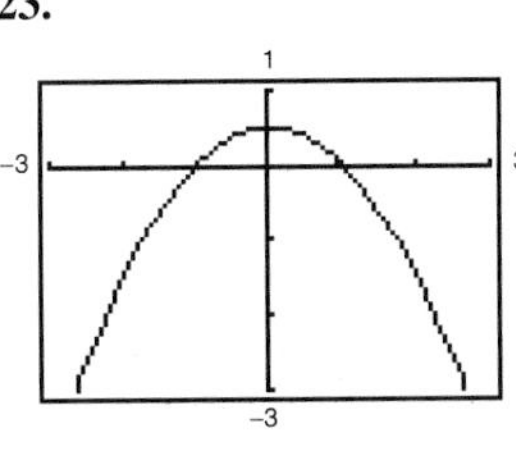

25.

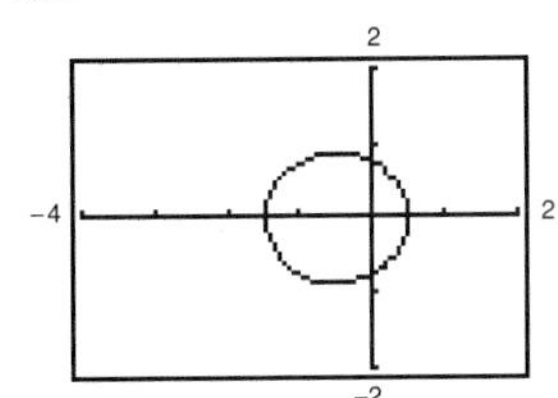

27.

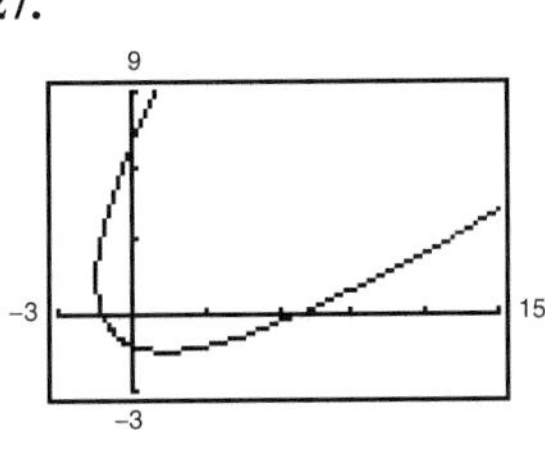

29.

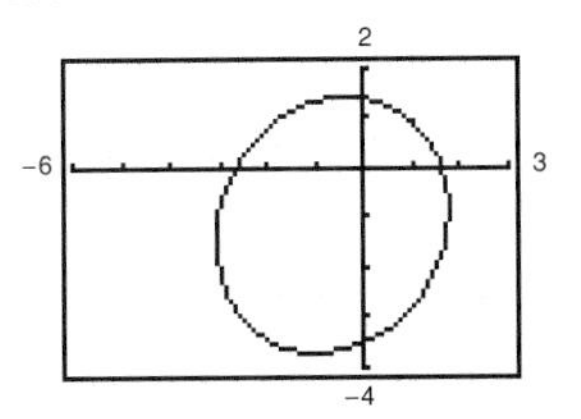

31. $r = \dfrac{1}{1 - \cos\theta}$ **33.** $r = \dfrac{1}{2 + \sin\theta}$

35. $r = \dfrac{2}{1 + 2\cos\theta}$ **37.** $r = \dfrac{2}{1 - \sin\theta}$

39. $r = \dfrac{10}{1 - \cos\theta}$ **41.** $r = \dfrac{16}{5 + 3\cos\theta}$

43. $r = \dfrac{20}{3 - 2\cos\theta}$ **45.** $r = \dfrac{9}{4 - 5\sin\theta}$

47. Answers will vary.

49. $r^2 = \dfrac{24{,}336}{169 - 25\cos^2\theta}$ **51.** $r^2 = \dfrac{144}{25\cos^2\theta - 9}$

53. $r^2 = \dfrac{144}{25\sin^2\theta - 16}$

55. Answers will vary.

57. $r = \dfrac{9.2931 \times 10^7}{1 - 0.0167\cos\theta}$

9.1405×10^7

9.4509×10^7

59. $r = \dfrac{5.5368 \times 10^9}{1 - 0.2481\cos\theta}$

4.4362×10^9

7.3638×10^9

61. $r = \dfrac{8200}{1 + \sin\theta}$

1467 miles

63.

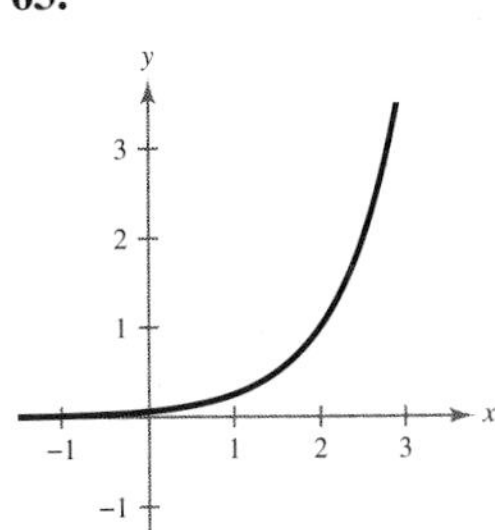

65.

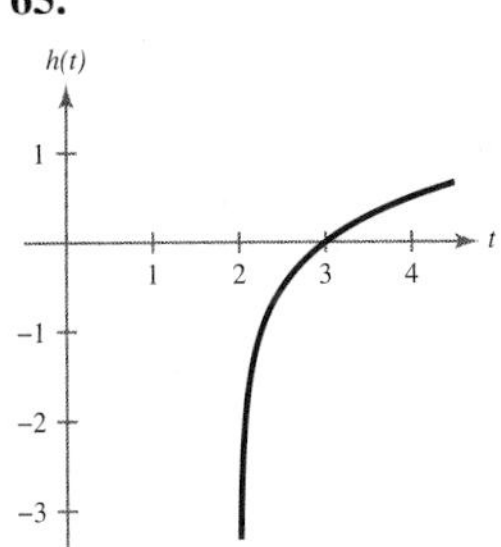

Focus on Concepts *(page 500)*

1. (a) Vertical translation
(b) Horizontal translation
(c) Reflection in the y-axis
(d) Parabola opens more slowly

2. (a) Major axis horizontal
(b) Circle
(c) Ellipse is flatter
(d) Horizontal translation

3. The extended diagonals of the central rectangle are asymptotes of the hyperbola.

4. 5. The ellipse becomes more circular and approaches a circle of radius 5.

5. The orientation would be reversed.

6. (a) The speed would double.
(b) The elliptical orbit would be flatter. The length of the major axis is greater.

7. False. The following are two sets of parametric equations for the line.
$x = t,\ y = 3 - 2t$
$x = 3t,\ y = 3 - 6t$

8. False. $(2, \pi/4)$, $(-2, 5\pi/4)$, and $(2, 9\pi/4)$ all represent the same point.

9. (a) Symmetric to the pole
(b) Symmetric to the polar axis
(c) Symmetric to $\pi/2$

10. (a) Same graphs
(b) Same graphs

Review Exercises *(page 501)*

1. $-\sqrt{3}$ **3.** $135°$ **5.** $2\sqrt{2}$ **7.** g **9.** f
11. d **13.** b **15.** h

17. Parabola

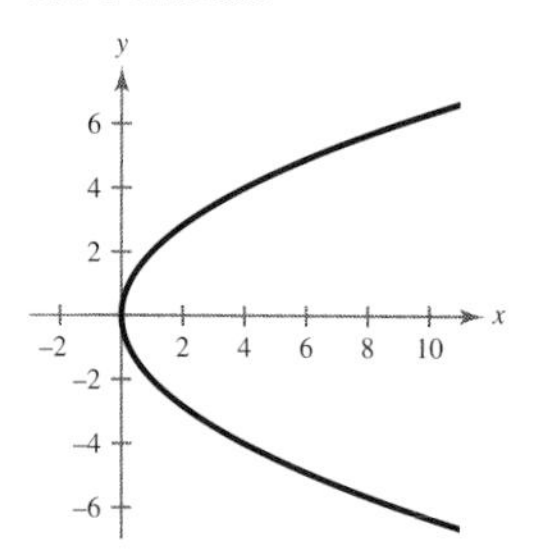

19. Parabola

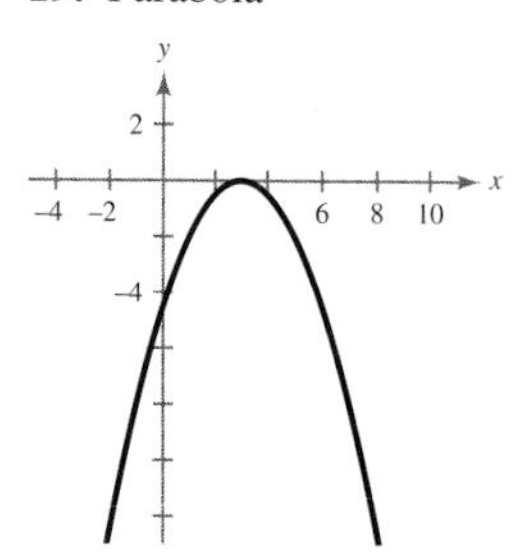

21. Degenerate circle (a point)

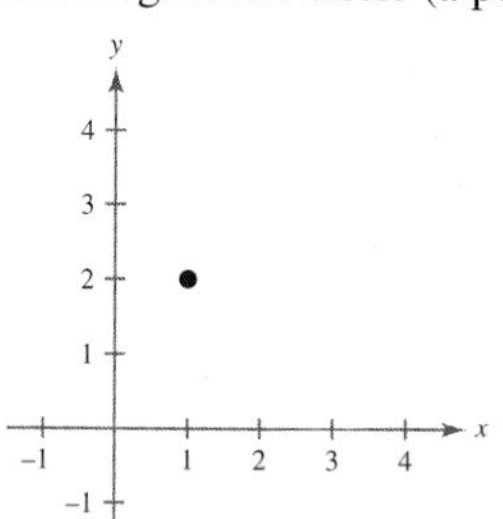

23. Ellipse

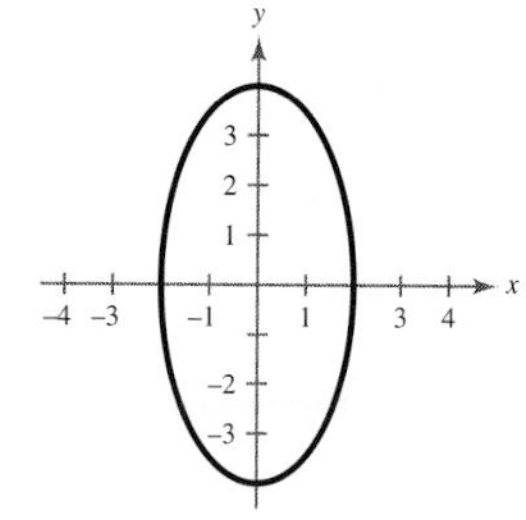

25. Ellipse

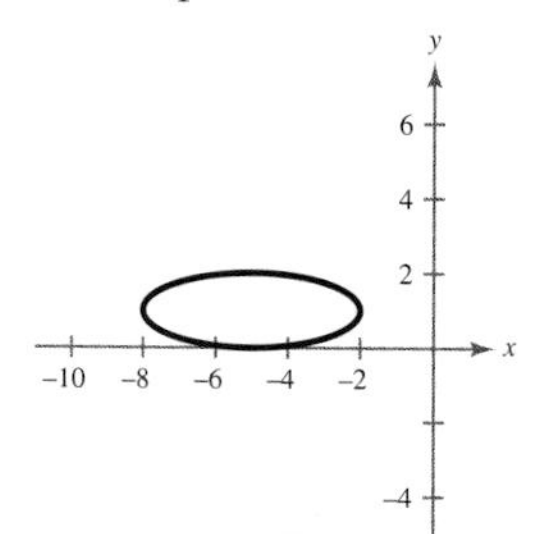

27. Hyperbola

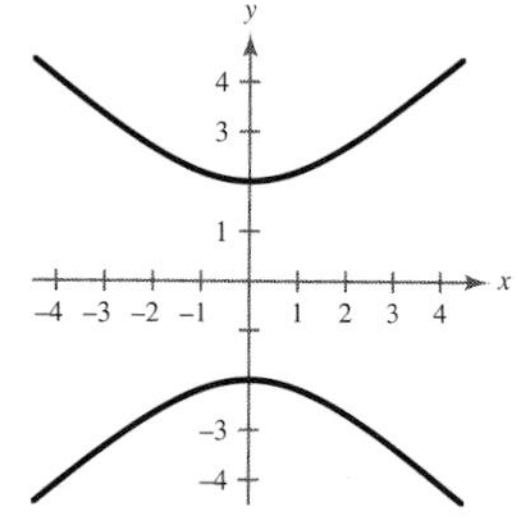

29. Parabola

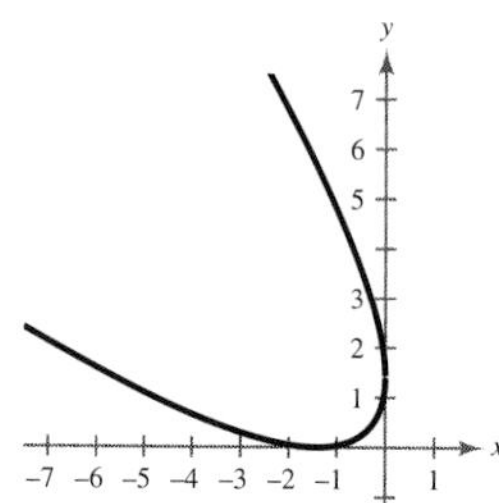

31. Ellipse

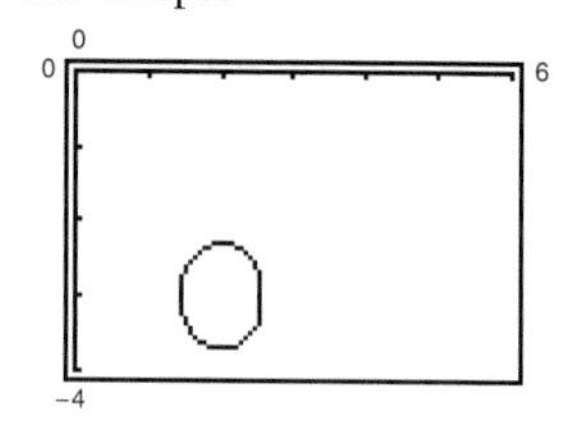

33. Hyperbola

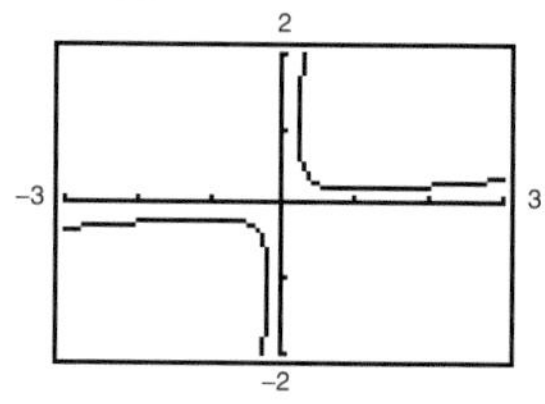

35. $(x - 4)^2 = -8(y - 2)$ **37.** $(y - 2)^2 = 12x$

39. $\dfrac{(x - 2)^2}{25} + \dfrac{y^2}{21} = 1$ **41.** $\dfrac{2x^2}{9} + \dfrac{y^2}{36} = 1$

43. $y^2 - \dfrac{x^2}{8} = 1$ **45.** $\dfrac{5(x - 4)^2}{16} - \dfrac{5y^2}{64} = 1$

47. $8\sqrt{6}$ meters

49. $\dfrac{576x^2}{25} - \dfrac{576y^2}{2279} = 1,\ \dfrac{64(x - 1)^2}{25} - \dfrac{64y^2}{39} = 1$

51.

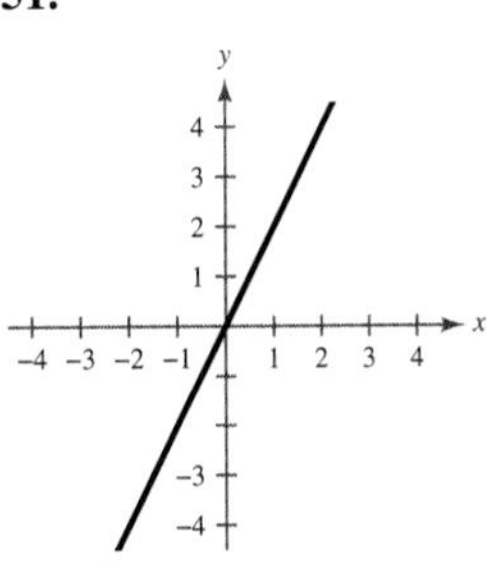

$y = 2x$

53.

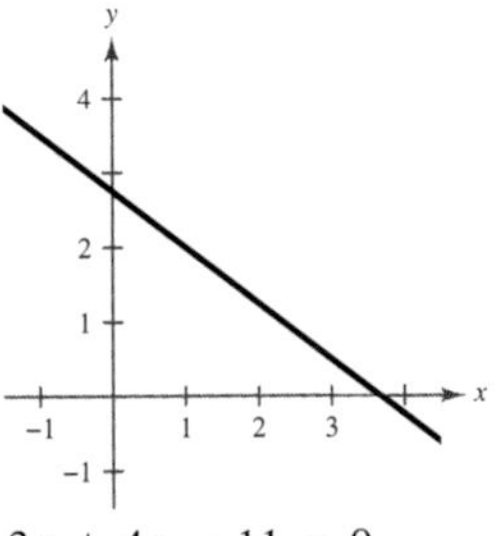

$3x + 4y - 11 = 0$

55.

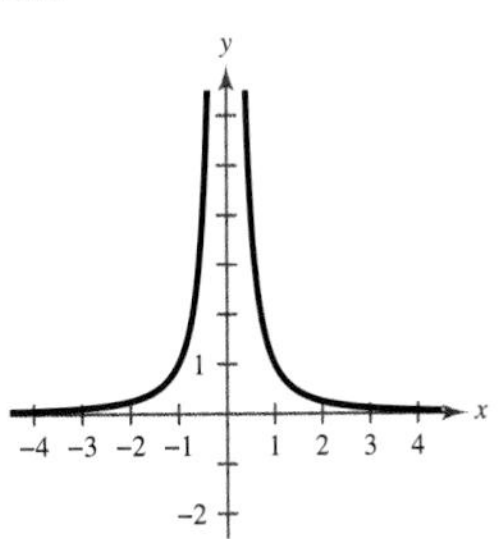

$y = \dfrac{1}{x^2}$

57.

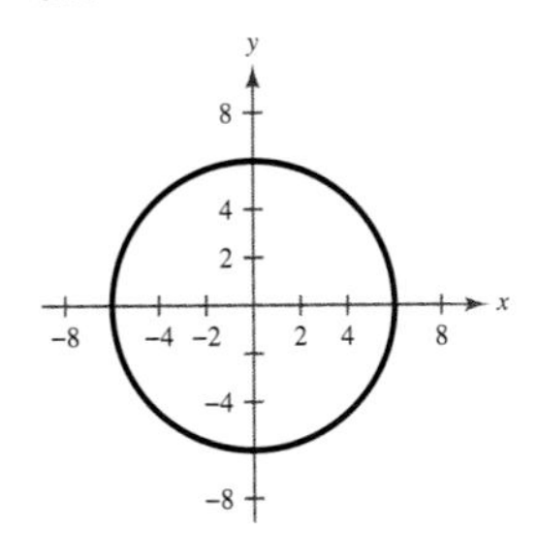

$x^2 + y^2 = 36$

59. $x = -3 + 4\cos\theta$

$y = 4 + 3\sin\theta$

61.

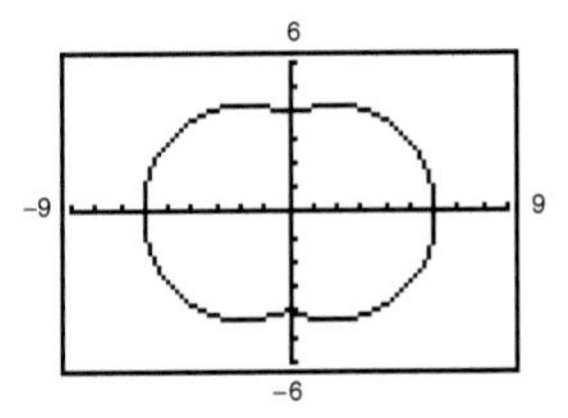

63. $x^2 + y^2 = 3x$ **65.** $(x^2 + y^2)^2 - x^2 + y^2 = 0$

67. $r = a\cos^2\theta\sin\theta$

69. Circle

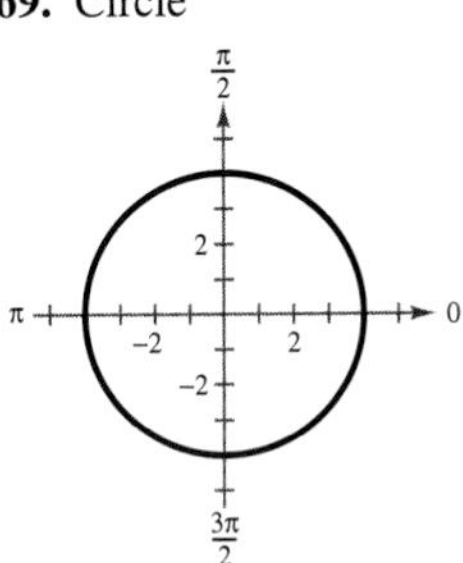

71. Rose curve

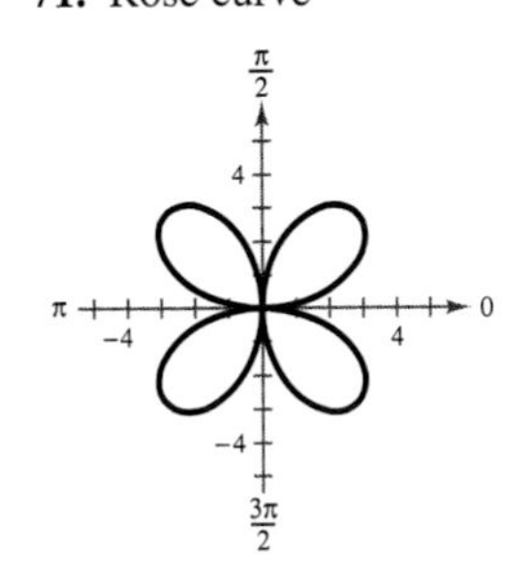

73. Cardioid

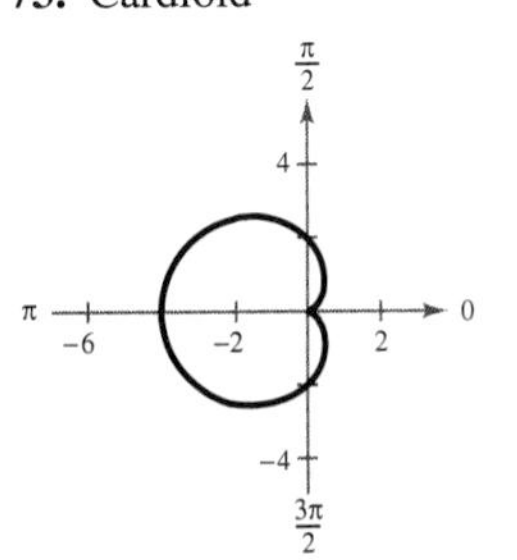

75. Limaçon

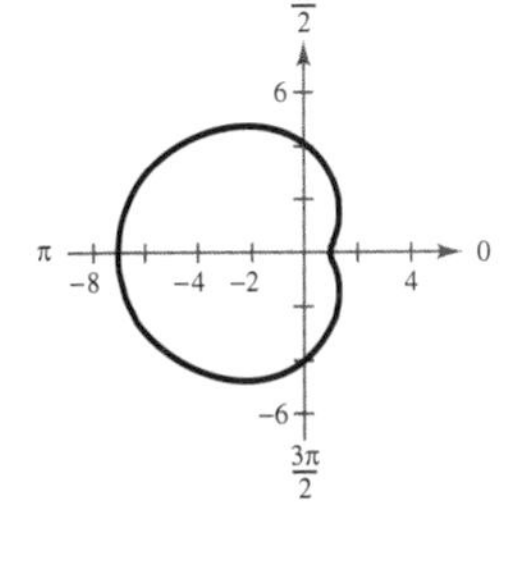

77. Rose curve

79. Parabola

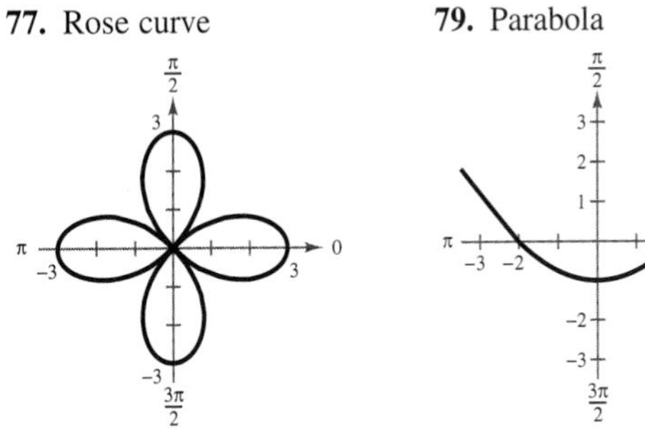

81.

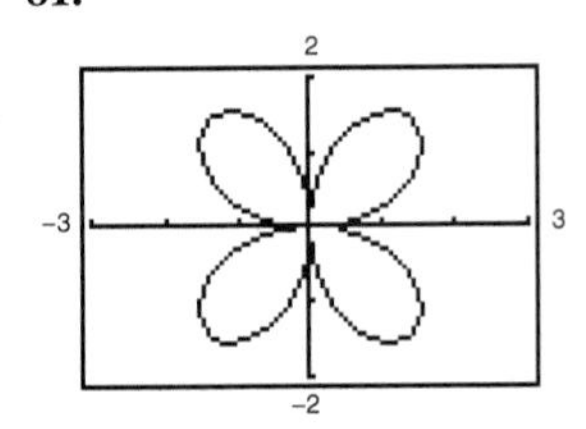

83.

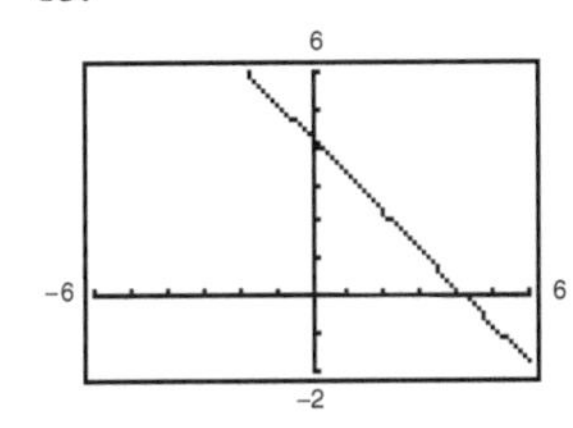

85. $r = 10\sin\theta$ **87.** $r = \dfrac{4}{1 - \cos\theta}$

89. $r = \dfrac{5}{3 - 2\cos\theta}$

Cumulative Test for Chapters 4–6

(page 506)

1. $3 - 5i$ **2.** $-\frac{8}{25} + \frac{6}{25}i$ **3.** $0, -2 \pm i$

4. $13(\cos 2.7468 + i\sin 2.7468)$ **5.** $-\dfrac{81}{2} + \dfrac{81\sqrt{3}}{2}i$

6. (a)

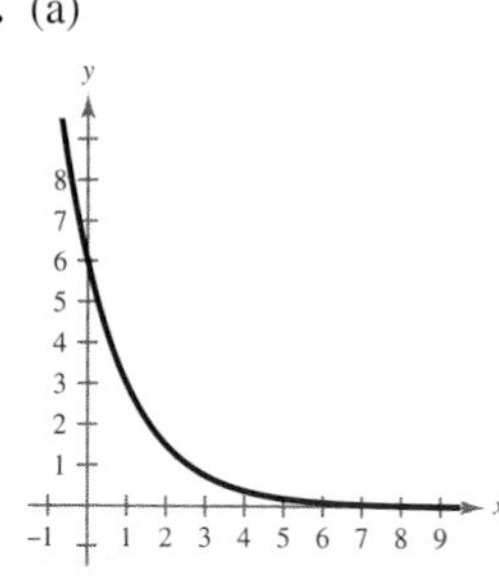

(b)

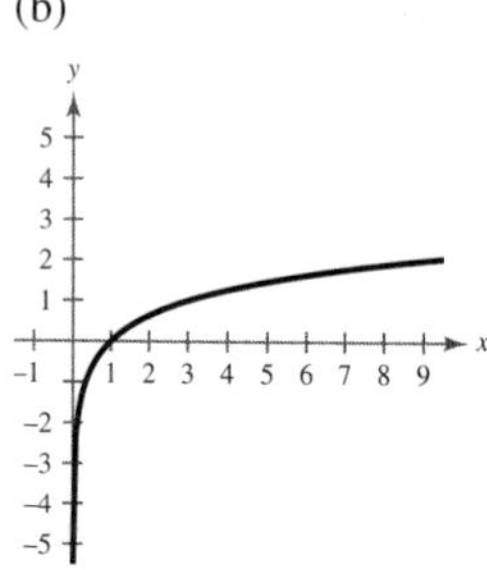

7. (a) $\frac{\ln 12}{2} \approx 1.2425$ (b) $\frac{64}{5}$

8.

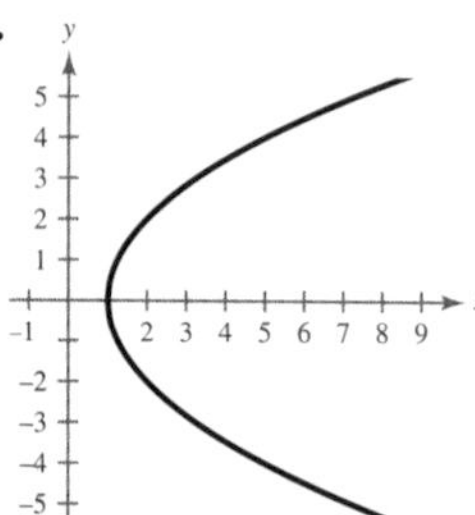

Vertex: (1,0)

Focus: (2, 0)

9.

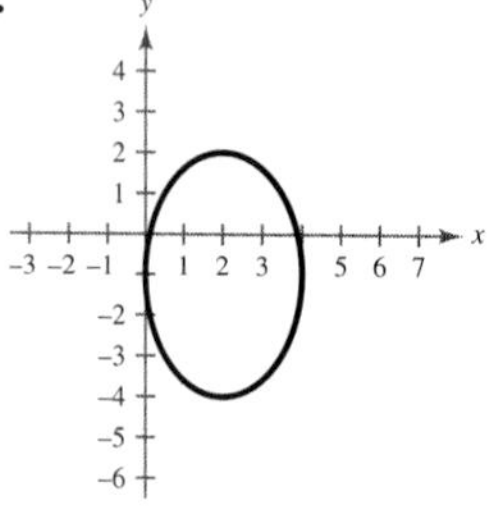

Vertex: (2, 2), (2, −4)

Foci: $\left(2, -1 \pm \sqrt{5}\right)$

10.

Vertex: $(\pm 1, 0)$

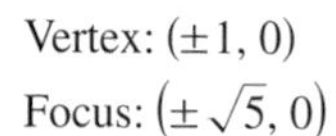

Focus: $\left(\pm \sqrt{5}, 0\right)$

11.

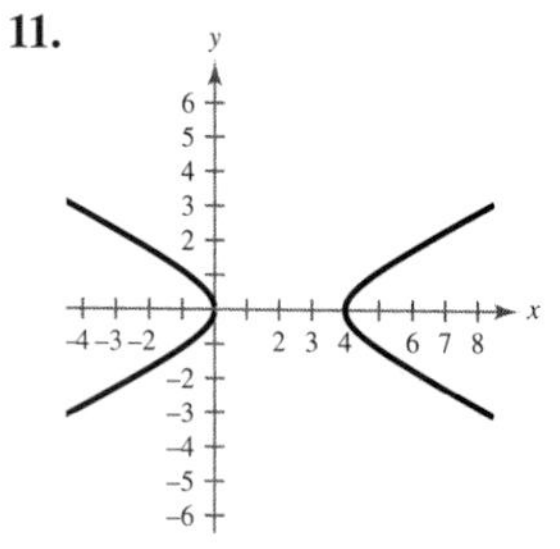

Vertex: (0, 0), (4, 0)

Foci: $\left(2 \pm \sqrt{5}, 0\right)$

12. $2x^2 - 12x - 3y + 12 = 0$

13. $5x^2 - 20y^2 + 80y - 64 = 0$

14. (a) 45° (b)

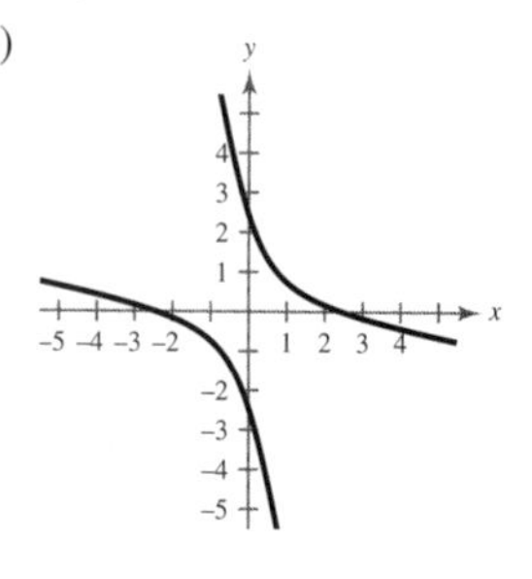

15.

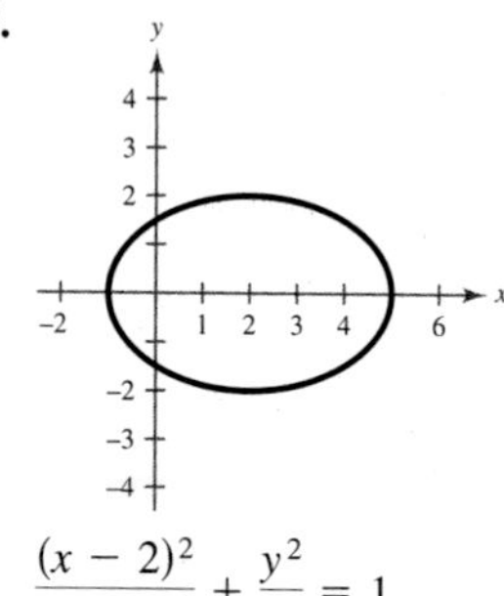

$$\frac{(x-2)^2}{9} + \frac{y^2}{4} = 1$$

16. $x = 2 + 4t$

$y = -3 + 7t$

17. $r = 6 \sin \theta$

18. (a)

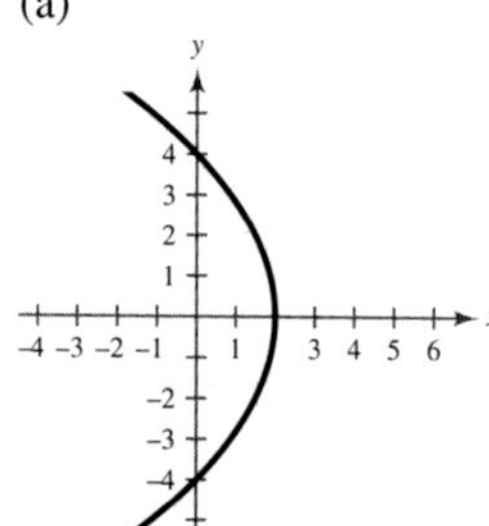

(b)

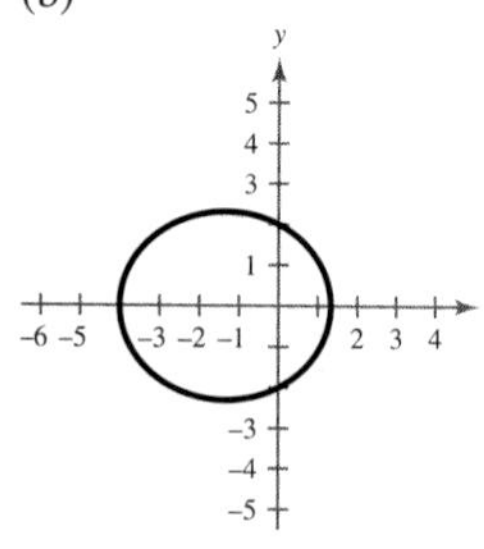

19. (a) (iii) (b) (i) (c) (ii)

APPENDIX A

Section A.1 *(page A10)*

1.

Stems	*Leaves*
7	0 5 5 5 7 7 8 8 8
8	1 1 1 1 2 3 4 5 5 5 5 7 8 9 9 9
9	0 2 8
10	0 0

3.

Stems	*Leaves*
6	18 68 71 94
7	16 19 25 25 42 57 58 76 84 88
8	13 28 35 41 46 59 61 62 66 81 83 89 91 92 92
9	05 06 15 18 25 25 26 28 41 44 83 90 92
10	04 10 62 95
11	51 78 86
12	23
13	
14	
15	
16	26

5.

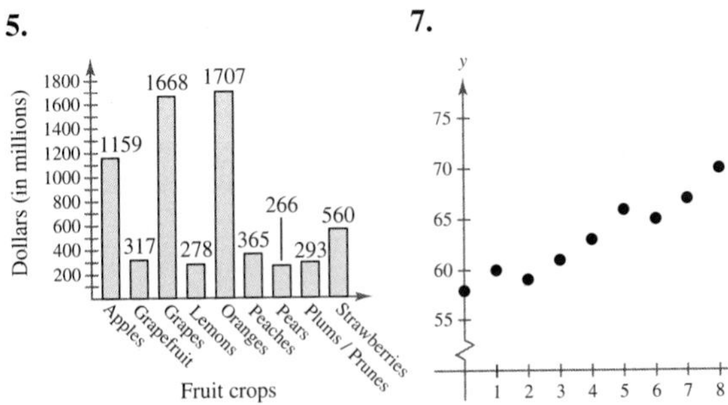

7.

9. $y = 57.49 + 1.43x$; 71.8

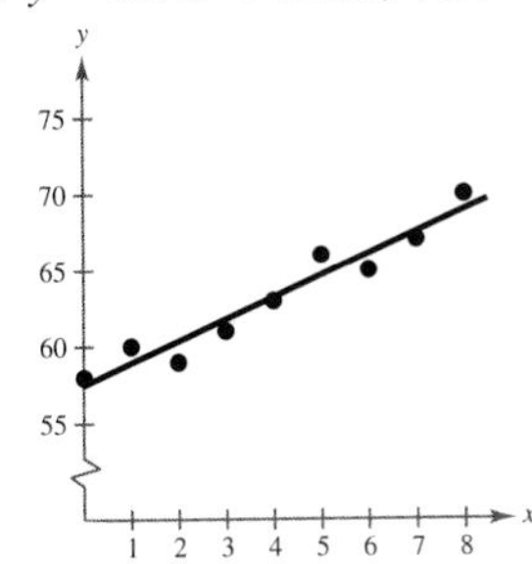

11.

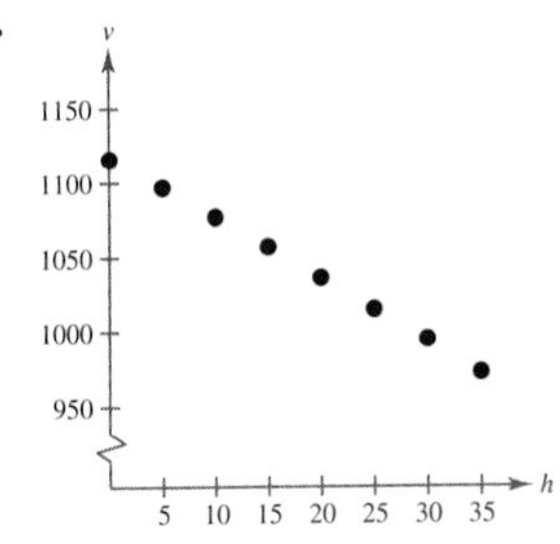

13. $v = 1117.3 - 4.1h$; 1006.6

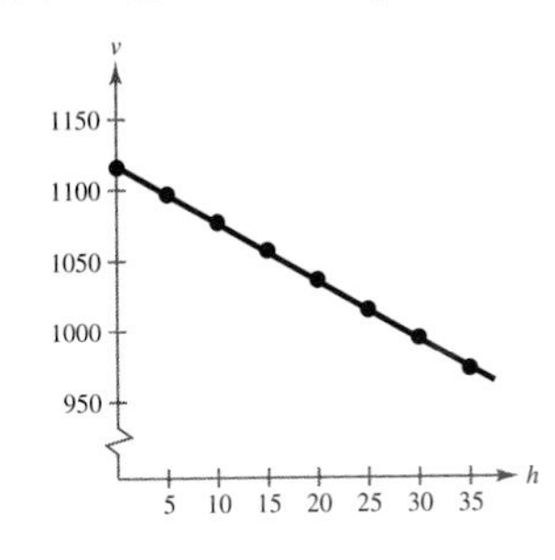

15. (a)

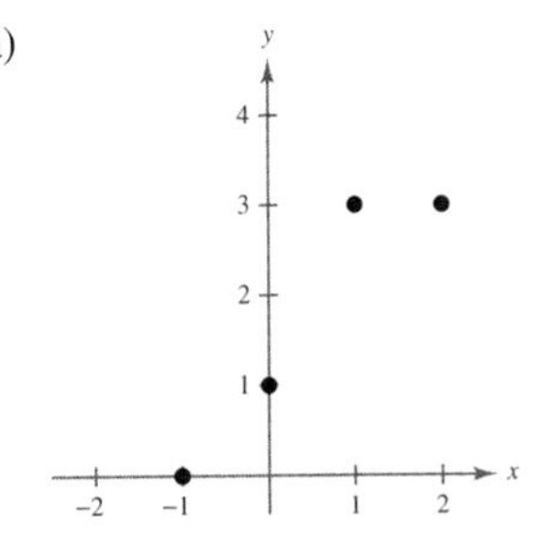

(b) $y = 1.1x + 1.2$; 0.7

(c) $y = \frac{11}{10}x + \frac{6}{5}$; 0.7

17. (a) and (c)

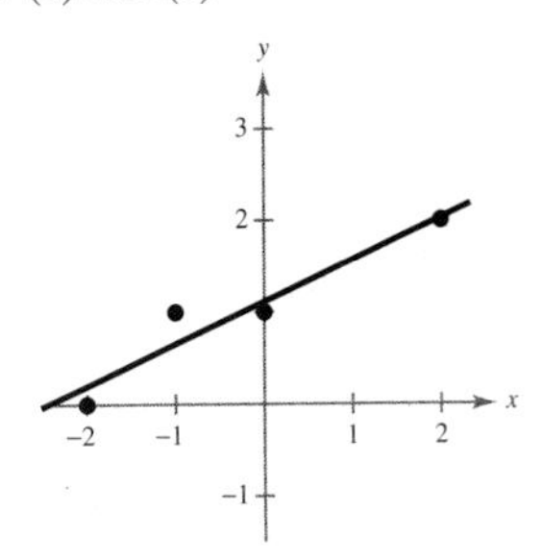

(b) $y = \frac{16}{35}x + \frac{39}{35}$

19. (a) and (c)

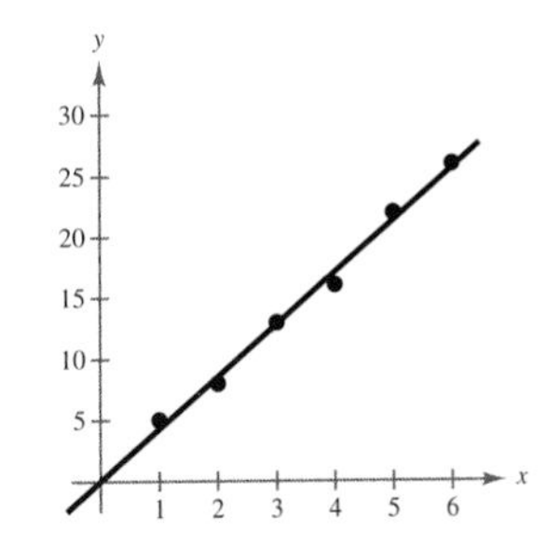

(b) $y = \frac{30}{7}x$

21. $y = -2.179x + 22.964$

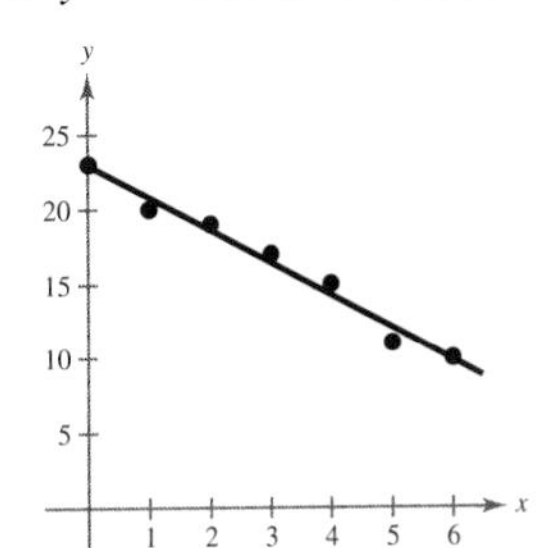

23. $y = 2.378x + 23.546$

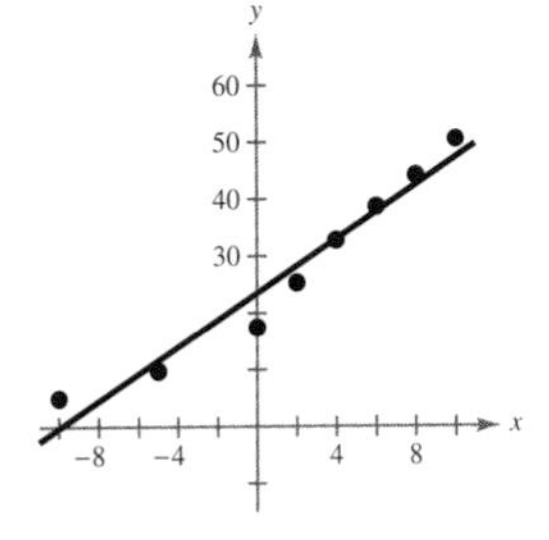

25. (a) $S = 384.1 + 21.2x$; \$95,784.10

(b)

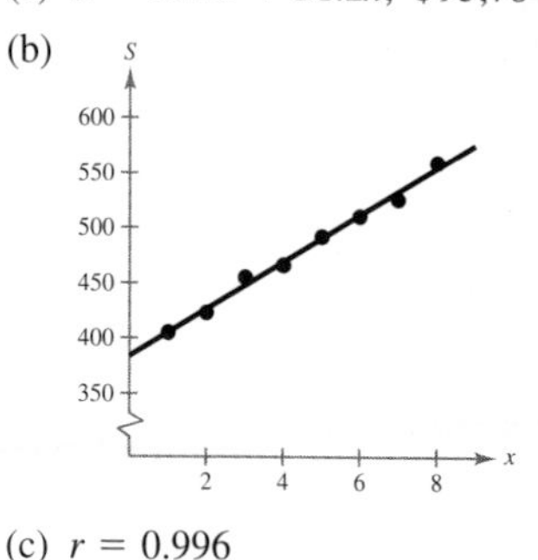

(c) $r = 0.996$

Section A.2 *(page A19)*

1. Mean: 8.86; median: 8; mode: 7

3. Mean: 10.29; median: 8; mode: 7

5. Mean: 9; median: 8; mode: 7

7. The mean is sensitive to extreme values.

9. Mean: \$67.14; median: \$65.35

11. Mean: 3.065; median: 3; mode: 3

13. One possibility: $\{4, 4, 10\}$

15. Mean: 76.6; median: 82; mode: 42
The median gives the most representative description.

17. $\bar{x} = 6, v = 10, \sigma = 3.16$ **19.** $\bar{x} = 2, v = \frac{4}{3}, \sigma = 1.15$

21. $\bar{x} = 4, v = 4, \sigma = 2$ **23.** $\bar{x} = 47, v = 226, \sigma = 15.03$

25. 3.42 **27.** 101.55 **29.** 1.65

31. $\bar{x} = 12$ and $|x_i - 12| = 8$ for all x_i

33. It will increase the mean by 5, but the standard deviation will not change.

35. With $\bar{x} = 235$ and $\sigma = 28$:

At least 75% of the scores in $[179, 291]$

At least 88.9% of the scores in $[151, 319]$

With $\bar{x} = 235$ and $\sigma = 16$:

At least 75% of the scores in $[203, 267]$

At least 88.9% of the scores in $[187, 283]$

INDEX OF APPLICATIONS

Biology and Life Sciences

Business

Chemistry and Physics

Construction

Consumer

Geometry

Interest Rate

Miscellaneous

Time and Distance

U.S. Demographics

INDEX

G

H

I

K

L

M

N

O

P

Q

R

S

T

U

V

W

X

Y

Z